Handbook of Mathematics and Computational Science

Springer
New York
Berlin
Heidelberg
Hong Kong
London
Milan
Paris
Tokyo

John W. Harris   Horst Stocker

Handbook of Mathematics and Computational Science

With 545 Illustrations

Springer

John W. Harris
Physics Department
Yale University
272 Whitney Avenue
New Haven, CT 06520
USA

Horst Stocker
Institut für Theoretische Physik
Johann Wolfgang Goethe Universität
Robert-Mayer Strasse 8-10
Frankfurt am Main, D60054
Germany

Translated from the German edition *Taschenbuch mathematischer Formeln und moderner Verfahren*, by Horst Stöcker, published by Verlag Harri Deutsch. A list of contributors for this volume is located at www.springer-ny.com/math/sampsup.html.

Library of Congress Cataloging-in-Publication Data
Harris, John W.
Handbook of mathematics and computational science / John W. Harris, Horst Stocker.
p. cm.
Includes bibliographical references and index.
ISBN 0-387-94746-9 (acid-free paper)
1. Mathematics-Handbooks, manuals, etc. 2. Computer science—Handbooks, manuals, etc. I. Stocker, Horst. II. Title.
QA40.S76 1998
510–dc21 98-20290

ISBN-13: 978-1-4612-5319-8 e-ISBN-13: 978-1-4612-5317-4
DOI:10.1007/978-1-4612-5317-4

Printed on acid-free paper.

9 8 7 6 5 4 3 SPIN 10909089

www.springer-ny.com

Springer-Verlag New York Berlin Heidelberg
A member of BertelsmannSpringer Science + Business Media GmbH

Introduction

The Harris and Stocker *Handbook of Mathematics and Computational Science* is a complete reference guide for working scientists, engineers and students. It serves as a Math Toolbox for rapid access to a wealth of mathematics information for everyday use, problem-solving, homework, examinations, etc. This Handbook has been compiled by professional scientists, engineers and lecturers who are experts in the day-to-day use of mathematics. In today's world, the application of mathematics goes hand-in-hand with the use of computers. In education and in practice today, methods of analytical mathematics are increasingly supplemented by the use of numerical or algebraic methods and computer software algorithms.

This Handbook represents only one part of a unique and Herculean effort to make all of the sciences accessible to the interested student, practicing engineer and scientist. This Handbook is internationally known for its clarity, completeness and its ability to integrate mathematics into one subject for the everyday use of mathematicians, scientists and engineers. The value of this compendium has been greatly enhanced by including hundreds of tables of commonly used functions, formulae, transformations, and series. There are also many applications to problems of daily interest of considerable value to the professional scientist and engineer, and others working daily with scientific formulae, and mathematical, or numerical, methods.

The Handbook contains:

- definitions of important words, formulae, facts, and rules commonly used in mathematics
- useful examples and practical applications
- suggestions regarding frequent sources of error
- important tips and crosslinks
- direct comparisons of analytical and numerical problem-solving methods
- sample programming sequences (computer pseudocode)
- elementary mathematics, financing, and other applications
- tables of important mathematical functions

The Handbook addresses a broad spectrum of everyday usage of mathematics. It provides:

- the basics of mathematics for high school students and beginning college/university students
- higher level mathematics for advanced students in science and engineering
- background material in mathematics for the working scientist and engineer making extensive use of:
  - numerical techniques
  - basic algorithms
  - computer codes
  - tables

The Handbook also covers basic elements of computer science, such as:

- boolean algebra
- programming languages: Pascal, FORTRAN, C, C++
- operating systems
- computer graphics
- symbolic computational software (Mathematica®, Maple®, etc.)
- graphs and trees
- fuzzy logic
- neural nets
- wavelets

and much more.

User access is direct and swift through the user-friendly layout, structured table of contents, and extensive index.

The authors would appreciate receiving comments and suggestions from readers for implementation in future editions at the following electronic mail addresses: John.Harris@Yale.edu or Stocker@Uni-Frankfurt.de.

New Haven, Connecticut — John W. Harris
Frankfurt, Germany — Horst Stocker

Contents

1

Numerical computation (arithmetics and numerics)

1.1 Sets

Set: Collection of definite, well-definable objects (**elements of the set**) to form a whole. Sets are denoted by capital letters ($X, Y, \ldots$), elements of the set are denoted by lower-case letters ($p, x, \ldots$).

For example, we write $p \in X$, if p is an **element set** X, and $p \notin X$, if p is not an element of X.

1.1.1 Representation of sets

Enumeration of the elements

□ $X = \{1, 2, 3, 4\}$

Characterization by a defining property E of the elements p, we write $X = \{p \mid p$ satisfies the property $E\}$ (read as "the set of p *such that* p satisfies the property E").

□ The set of all persons living in the USA = {person | person lives in the USA} is the set of all persons living in the USA.

$\mathbb{R} = \{z \mid z$ is real and $z > 0\}$ is the set of all positive real numbers.

Venn diagram: A diagram in which one or more sets are represented by sets of points, bordered by closed curves.

Venn diagrams

Empty set: $\emptyset = \{\}$.

▷ The empty set contains no elements, not even a zero.

$$\{0\} \neq \emptyset\,, \qquad \{0\} \neq \{\}$$

Equality of sets: Two sets X and Y are said to be **equal**, when every element of X is also an element of Y, **and** every element of Y is **also** an element of X. This is denoted by $X = Y$.

Cardinality or power of sets: Two sets X and Y are called **equivalent** if there is a one-to-one mapping of the elements of X onto the elements of Y, that is, if every element of X can be associated with **one and only one** element of Y (see also under the notion of "mapping").

▷ Finite sets are equivalent if they contain the same number of elements.

□ All sets of 20 elements are equivalent.
The set of all natural numbers and the set of all rational numbers are equivalent.

Subset: A set Y is contained in another set X (Y is a **proper** subset of X, notation: $Y \subset X$ or $X \supset Y$), if every element $y \in Y$ is also an element of X, and X and Y are not identical.

□ $\{1, 4\} \subset \{1, 2, 3, 4\}$
The set of all dwellings is a subset of all buildings.

▷ $X \subset Y$ and $Y \subset X \implies X = Y$
X is an **improper subset of** Y (and vice versa).

▷ According to DIN 5473, $\subseteq$ is the sign for **inclusion**, and $\subsetneq$ is the sign for **strict inclusion**.

Including set: If X is a subset of Y then, conversely, Y is the including set of X.

Reflexivity: $X \subset X$.

Transitivity: $X \subset Y$ and $Y \subset Z \implies X \subset Z$.

● Always: $\emptyset \subset X$.

1.1.2 Operations on sets

● Two sets, X and Y, are equal if and only if they contain exactly the same elements.

□ $\{1, 4\} = \{4, 1\}$. The elements can be arranged arbitrarily.
$\{1, 4\} = \{4, 1, 1\}$. 1 and 1 are not distinct, representing the same element.

Union: $Z = X \cup Y$ consists of all elements belonging to the set X or set Y.

□ $\{1, 2\} \cup \{1, 3, 4\} = \{1, 2, 3, 4\}$

● Always: $X \cup \emptyset = \emptyset \cup X = X$.

Intersection: $Z = X \cap Y$ contains all elements belonging to set X or to set Y.

□ $\{1, 2, 4\} \cap \{2, 3, 5\} = \{2\}$

● Always: $X \cap \emptyset = \emptyset \cap X = \emptyset$.

□ Intersecting set or average, $Z = X \cap Y$ consists of all elements contained in X as well as in Y.

● Always: $X \cap \emptyset = \emptyset \cap X = \emptyset$.

Disjointed sets: X and Y are said to be **disjointed** if they have no common element. This is equivalent to the statement $X \cap Y = \emptyset$.

□ $\{1, 3, 4\}$ and $\{2, 5, 7\}$ are disjointed sets.

Complementary set or **complement**: If we look at subsets of a given set X, the complement (with respect to set X) is A^C of a subset ($A \subset X$), defined as the set which contains all elements of X that are not elements of A.

□ $X = \{1, 2, 3, 4, 5, 6\}$,
$A = \{1, 2, 3\}$,
$A^C = \{4, 5, 6\}$, complement of A with respect to X.

- If X is the basic set, always:

$$X^C = \emptyset\,, \qquad \emptyset^C = X\,,$$
$$(A^C)^C = A\,,$$
$$A^C \cap A = \emptyset\,, \qquad A^C \cup A = X.$$

▷ Notations: A^C or $X - A$ or $\bar{A}$.

Difference: $X \setminus Y$, read X without Y, contains those elements of X that do not belong to Y:

$$X \setminus Y = \{x \mid x \in X \text{ and } x \notin Y\}$$

□ $\{1, 2, 3, 4, 5\} \setminus \{1, 2, 4, 6, 7\} = \{3, 5\}$

▷ The difference $X \setminus Y$ has to be distinguished clearly from the complement Y^C, because in forming the difference $X \setminus Y$ the set Y does not have to be a subset of X.

Symmetric difference or discrepancy: $X \Delta Y$, all elements contained in X or Y, but **not** in X **and** Y.

$$\begin{aligned} X \Delta Y &= (X \setminus Y) \cup (Y \setminus X) \\ &= (X \cup Y) \setminus (X \cap Y) = (X \cup Y) \setminus (Y \cap X) \end{aligned}$$

□ $\{1, 2, 3, 4, 5\} \Delta \{3, 4, 5, 6, 7\} = \{1, 2, 6, 7\}$

Product: $X \times Y$, the set of all ordered pairs (x, y) where $x \in X$ and $y \in Y$. The product is also called the **Cartesian product** of sets X and Y.

□ $X = \{1, 2\}, Y = \{5, 6\}, \qquad X \times Y = \{(1, 5), (1, 6), (2, 5), (2, 6)\}$

▷ In the case of ordered pairs, the sequence is important; in general $(x, y) \neq (y, x)$.

Power set: $\mathrm{P}(X)$, the set of all subsets of X.

□ $\mathrm{P}(X)\{1, 2\} = \big\{\{\}, \{1\}, \{2\}, \{1, 2\}\big\}$

▷ The empty set and the set itself belong to the power set!

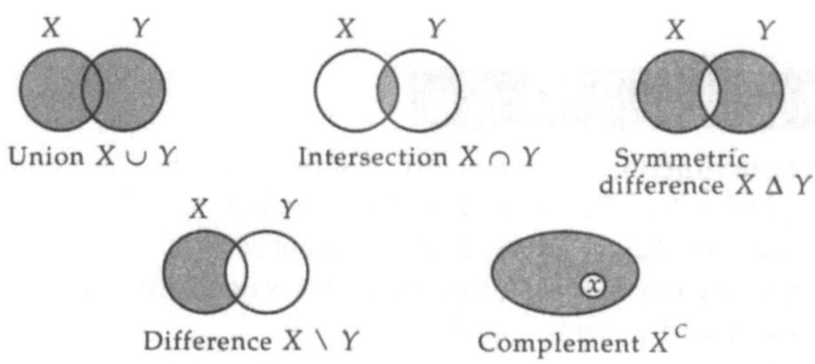

Set operations are applied in computer-aided design (CAD) systems. In three-dimensional CAD systems, complex composition bodies are created by the composition of basic bodies through the composition of union, intersection, and difference.

□ Composition of three bodies: $VV = (V_1 \cup V_2) \setminus V_3$.

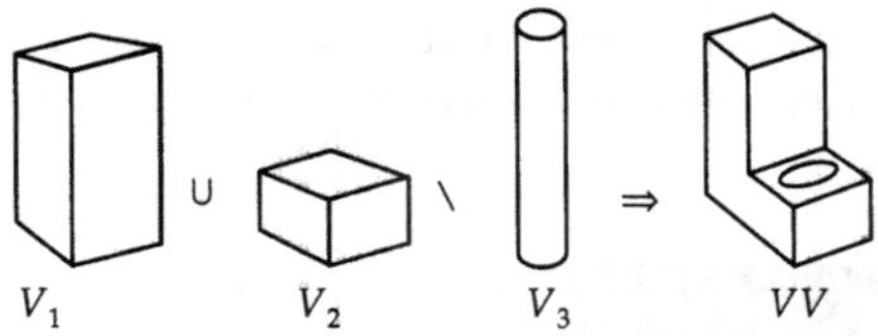

▷ In general, switching the order of the compositions leads to different results. In the above example, $(V_1 \setminus V_3) \cup V_2 = V_1 \cup V_2$ is the shape **without** the hole.

In connection with CAD, these compositions are also called Boolean operations.

1.1.3 Laws of the algebra of sets

Commutative laws:

$$X \cap Y = Y \cap X\,, \qquad X \cup Y = Y \cup X\,.$$

Associative laws:

$$X \cap Y \cap Z = (X \cap Y) \cap Z = X \cap (Y \cap Z)\,,$$
$$X \cup Y \cup Z = (X \cup Y) \cup Z = X \cup (Y \cup Z)\,.$$

● Union and intersection are commutative and associative.

▷ For the formation of the difference, the commutative, associative and distributive laws do **not** apply. For example, the following applies to the commutation:

□ $\{1, 2, 3\} \setminus \{1, 2\} = \{3\}$, but $\{1, 2\} \setminus \{1, 2, 3\} = \{\} \neq \{3\}$.

Distributive laws:

$$X \cap (Y \cup Z) = (X \cap Y) \cup (X \cap Z)\,,$$
$$X \cup (Y \cap Z) = (X \cup Y) \cap (X \cup Z)\,.$$

Idempotence laws:

$$X \cap X = X\,, \qquad X \cup X = X\,.$$

Absorption laws:

$$X \cup (X \cap Y) = X\,, \qquad X \cap (X \cup Y) = X\,.$$

De Morgan's laws:

$$(X \cap Y)^C = X^C \cup Y^C\,, \qquad (X \cup Y)^C = X^C \cap Y^C.$$

1.1.4 Mapping and function

Mapping (the term **function** is a synonym): A subset f of the Cartesian product $X \times Y$ is called a mapping or function from X to Y (in symbols: $f : X \to Y$) if from (x, y_1), $(x, y_2) \in f$ it follows that $y_1 = y_2$. Every element $x \in X$ is associated by f with a uniquely determined element $y \in Y$; this element y is denoted by $f(x)$, or $y = f(x)$.

▷ In particular, X and Y can be equal.

Inverse image: Set mapped to a second set; in this case set X.

Image: Set to which another set is mapped, in this case set Y.

Injective mapping: If $y_1 = f(x_1) = f(x_2) = y_2$, then $x_1 = x_2$ **always.**

Surjective mapping: For every element y of the image set Y there is at least one element x of the inverse image set that is mapped to y.

Bijective mapping: The mapping is injective and surjective.

□ Let $c \in Y$. Surjectivity of f implies that $f(x) = c$ has at least one solution for every element c of Y. Injectivity implies that every solution is unique.

1.2 Number systems

Number system (positional system): A natural number B and a set of B symbols, where B is the **base** and the symbols are the **digits** of the positional system.

- Representation of a number z in the number system with base B.

$$z_{(B)} = \sum_{i=-m}^{n} z_i B^i \ , \quad B \in \mathbb{N} \ , \ B \geq 2$$

1.2.1 Decimal number system

Decimal system: Base $B = 10$ and digits 0, 1, 2, 3, 4, 5, 6, 7, 8, 9.

- Every natural number can be represented as a combination of these symbols.

□ $1456 = 1 \cdot 1000 + 4 \cdot 100 + 5 \cdot 10 + 6 \cdot 1$
$= 1 \cdot 10^3 + 4 \cdot 10^2 + 5 \cdot 10^1 + 6 \cdot 10^0$
$= 1 \cdot B^3 + 4 \cdot B^2 + 5 \cdot B^1 + 6 \cdot B^0$, i. e.

in each case ten units are combined into one larger unit:

10 units (10^0)	=	1 ten (10^1)
10 tens (10^1)	=	1 hundred (10^2)
10 hundreds (10^2)	=	1 thousand (10^3)
10 thousands (10^3)	=	1 ten thousand (10^4)
1000 thousand (10^3)	=	1 million (10^6)
1000 million (10^6)	=	1 billion (10^9)
1000 billion (10^9)	=	1 trillion (10^{12})
1000 trillion (10^{12})	=	1 quadrillion (10^{15})
1000 quadrillion (10^{15})	=	1 quintillion (10^{18})

▷ Note: In North American usage, a billion is not the same as in European usage. The North American billion is 10^9, while the European billion is 10^{12}.

The unit "one" can be further subdivided as follows:

1 unit (10^0)	=	10 tenths ($10^{-1} = \frac{1}{10}$)
1 tenth (10^{-1})	=	10 hundredths ($10^{-2} = \frac{1}{100}$)
1 hundredth (10^{-2})	=	10 thousandths ($10^{-3} = \frac{1}{1000}$) etc.

- All real numbers can be represented within this system, sometimes only as infinite sums.

□ $12.94 = 1 \cdot 10 + 2 \cdot 1 + 9 \cdot \frac{1}{10} + 4 \cdot \frac{1}{100} = 1 \cdot B^1 + 2 \cdot B^0 + 9 \cdot B^{-1} + 4 \cdot B^{-2}$

The left-hand side notation is called **decimal fraction**, with the digits referring to negative basal exponents being placed to the right of the decimal point.

In connection with units of measurement, some **powers of ten** have special names:

value	name	abbreviation	value	name	abbreviation
10^1	Deca	da	10^{-1}	Deci	d
10^2	Hecto	h	10^{-2}	Centi	c
10^3	Kilo	k	10^{-3}	Milli	m
10^6	Mega	M	10^{-6}	Micro	μ
10^9	Giga	G	10^{-9}	Nano	n
10^{12}	Tera	T	10^{-12}	Pico	p
10^{15}	Penta	P	10^{-15}	Femto	f
10^{18}	Exa	E	10^{-18}	Atto	a
10^{21}	Zetta	Z	10^{-21}	Zepto	z
10^{24}	Yotta	Y	10^{-24}	Yocto	y

□ 1 mm is one millimeter, that is, $\frac{1}{1000}$ meter, or 1 hl = 1 hectoliter = 100 l = 100 liters, 1 MW is one Megawatt, i.e., a million Watts.

1.2.2 Other number systems

● One important system is the number system (also called **binary system**). It forms the basis for computer memory where only two states are possible. 2 is the base, while 0 and 1 (sometimes H for High and L for Low) are used as symbols.

decimal	binary	octal	hexadec.	decimal	binary	octal	hexadec.
0	0	0	0	10	1010	12	A
1	1	1	1	11	1011	13	B
2	10	2	2	12	1100	14	C
3	11	3	3	13	1101	15	D
4	100	4	4	14	1110	16	E
5	101	5	5	15	1111	17	F
6	110	6	6	16	10000	20	10
7	111	7	7	17	10001	21	11
8	1000	10	8	18	10010	22	12
9	1001	11	9	19	10011	23	13

□ $19_{(10)} = \mathbf{1} \cdot 2^4 + \mathbf{0} \cdot 2^3 + \mathbf{0} \cdot 2^2 + \mathbf{1} \cdot 2^1 + \mathbf{1} \cdot 2^0 \equiv 10011_{(2)}$ in this notation.

▷ In computer arithmetic, 2 is inconvenient as a natural base to represent larger numbers. As far as the **hexadecimal number system** is concerned, it combines 4 positions of the dual system and forms 16 as the base. Thus, 16 symbols are necessary to represent the digits: 0, 1, 2, 3, 4, 5, 6, 7, 8, 9, A, B, C, D, E, F.

□ $5AE_{(16)} = 5 \cdot 16^2 + 10 \cdot 16^1 + 14 \cdot 16^0 = 5 \cdot 256 + 10 \cdot 16 + 14 \cdot 1 = 1454_{(10)}$.

▷ To differentiate decimal from hexadecimal numbers, a dollar sign is often used as a prefix, e.g. $12, which is equivalent of 18 in the decimal system.

Octal number system: only 3 positions of the dual system are combined into one position. The base is 8, and the digits are 0, 1, 2, 3, 4, 5, 6, 7.

1.2.3 Computer representation

Bit: A binary digit in the computers or calculators; recognizes only two states, 0 and 1 (in technical terms: level of the applied voltage).

Byte: Grouping of 8 bits into a larger unit. Usually, the bits are numbered as 0–7, not 1–8!

Binary coded decimal (BCD) standard: Each decimal digit of a number is coded **separately** as a 4-bit binary number (**tetrad**). In order to code the numbers 0 to 9, only 10 tetrads are necessary; 6 **pseudo-tetrads** (1010, 1011, 1100, 1101, 1110, 1111) are not required.

□ The BCD code of the number 179: 0001 0111 1001.

IEEE Standard (IEEE = Institute of Elecrical and Electronics Engineers): The standard for representng numbers in the computer, used with most programming languages and in most computer systems.

□ Various types of data in C:
byte: representation of integers −128 . . . +127
int = 2 bytes, representation of integers −32768 . . . +32767

long = 4 bytes, representation of integers −2147483648 ... +2147483647
float = 4 bytes, approximation of real numbers x as decimal fractions with a maximum of 8 digits + 2 digits for the exponent in powers of ten: $|x| < 1.701411 \cdot 10^{38}$
double = 8 bytes, approximation of real numbers x as decimal fractions with a maximum of 16 digits + 3 digits for the exponent in powers of ten: $|x| < 1.701411 \cdot 10^{306}$.
There are also limits for float and double, $|x| > 10^{-38}$ for float and $|x| > 10^{-306}$ for double.

▷ Float and double are called **floating point numbers**. They have 8 or 16 significant digits.

▷ As a rule, pocket calculators have a key [EE] or [EXP] for the input of powers of ten. The sequence [1] [EXP] [5] is for the input of 10^5; the sequence [1] [0] [EXP] [5] is for the input of $10^6 = 10 \cdot 10^5$.

1.2.4 Horner's scheme for the representation of numbers

Horner's scheme for the representation of numbers:

$$z_{(B)} = \sum_{i=-m}^{n} z_i B^i$$
$$= (z_n + (z_{n-1} + \cdots + (z_{-m+1} + z_{-m} \cdot B^{-1}) \cdot B^{-1} \cdots) \cdot B^{-1}) \cdot B^n$$

▷ The symbols z_i are the digits of the number system with base B, i.e., 0, 1 in the dual system or 0, 1, 2, 3, 4, 5, 6, 7, 8, 9, A, B, C, D, E, F in the hexadecimal system.

□ Representation of 234.57 on the decimal number system:

$$234.57 = (2 + (3 + (4 + (5 + 7 \cdot 10^{-1}) \cdot 10^{-1}) \cdot 10^{-1}) \cdot 10^{-1}) \cdot 10^2$$
$$= 2 \cdot 10^2 + 3 \cdot 10^1 + 4 \cdot 10^0 + 5 \cdot 10^{-1} + 7 \cdot 10^{-2} .$$

1.3 Natural numbers

Set of natural numbers: $\mathbb{N} = \{0, 1, 2, 3, 4, \ldots\}$.

▷ DIN 5473 also defines zero as a natural number.

The set of natural numbers without zero is

$$\mathbb{N}^* = \mathbb{N} \setminus \{0\} = \{1, 2, 3, 4, \ldots\} .$$

Natural numbers include every kind of counting (**cardinal numbers**) and ordering (**ordinal numbers**, first, second, third, ...). Furthermore, natural numbers can be used for labeling ($a_1, a_2, \ldots$). The natural numbers also serve for indexing (a_1, a_2, etc.). Symbolically, natural numbers are often denoted by i, j, k, l, m, n.

▷ In FORTRAN, the letters `i-n` are reserved as a standard for `INTEGERS`.

Storing natural numbers: On a pocket calculator, large natural numbers cannot be represented exactly. For example, the budget of the USA for 1996, about $420,000,000,000, cannot be stored precisely, and a computer with eight **significant digits** cannot be distinguished from 420,000,000,348.

Similar restrictions apply to PCs and mainframe computers.

▷ In the computer, there is **always** a larger natural number which has **no** successor.

● Every natural number has a successor, e.g., 13 is the successor of 12.

- One and only one natural number, the zero, is not a successor of another natural number, i.e., it has no predecessor.

▷ In `PASCAL` predecessor of n: `pred(n)`
successor of n: `succ(n)`

- Different natural numbers have different successors.
- There are infinitely many natural numbers: The sequence of natural numbers has no upper limit. Natural numbers can be represented as isolated, equidistant points on a number ray (number half-line).

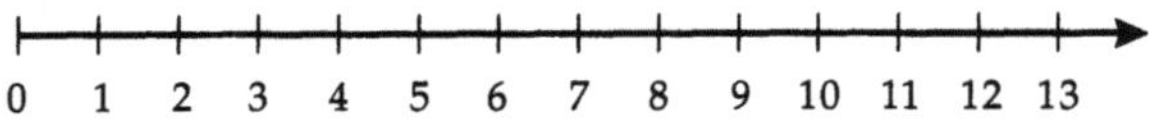

Number ray

1.3.1 Mathematical induction

- Always:
1. the natural number 0 (or 1 or any other natural number n_0) has a certain property E,
2. the natural number n has the property E, this is also correct for the successor $n' = n + 1$,
then all numbers m of $\mathbb{N}$ (or $\mathbb{N}^*$ or $m \in \mathbb{N}$, $m \geq n_0$) have this property.

□ **Bernoulli's inequality**: $(1 + x)^n \geq 1 + nx$ for $x \in \mathbb{R}$, $x > -1$ and $n \in \mathbb{N}$.
Proof: The inequality holds for $n = 2$: $(1 + x)^2 = 1 + 2x + x^2 > 1 + 2x$. If it is valid for n, then it is also valid for $n + 1$ because

$$\begin{aligned}(1 + x)^{n+1} &= (1 + x)^n (1 + x) > (1 + nx)(1 + x)\\ &= 1 + (n + 1)x + x^2 > 1 + (n + 1)x.\end{aligned}$$

Consequently, the relation holds for all $n \geq 1$.

1.3.2 Vectors and fields, indexing

***n*-dimensional vector**: n-tuple of numbers a_i, labeled from a_1 to a_n.
The numbers a_i are called the components of vector $a = (a_1, a_2, \ldots, a_n)$. Frequently, the vectors are also written as columns.

▷ The n-tuple is only an example of a vector; for the general definition of a vector, see Chapter 6 on vectors.

$n \times m$-dimensional field: $n \times m$-**matrix** or **array**, rectangular scheme of numbers a_{ik}, labeled from a_{11} to a_{nm}.
n is the number of **rows of the matrix**. m is the number of **columns of the matrix**.
The numbers a_{ik} are called the field **elements of the array (array elements)**.

$$\begin{matrix} a_{11} & a_{12} & \cdots & a_{1m} \\ a_{21} & a_{22} & \cdots & a_{2m} \\ \vdots & \vdots & \ddots & \vdots \\ a_{n1} & a_{n2} & \cdots & a_{nm} \end{matrix}$$

i is the **row index**, k is the **column index**.

1.3.3 Calculating with natural numbers

▷ The arithmetical operations used here are defined under "calculation with real numbers."

● Addition, multiplication, and exponentiation of natural numbers n and m each result in a natural number again:

$$m + n = m' , \quad m \cdot n = n' , \quad m^n = m'' , \qquad n' , m' , m'' \in \mathbb{N}.$$

● On the other hand, the subtraction $m - n$ $(m, n \in \mathbb{N})$ (the inverse to addition) may lead us out of $\mathbb{N}$.

□ $4 - 6 = -2 \notin \mathbb{N}$

● The division $m \div n$ $(m, n \in \mathbb{N})$ (the inverse to multiplication) may also lead us out of $\mathbb{N}$.

□ $3 \div 2 = \dfrac{3}{2} = 1.5 \notin \mathbb{N}$

If a natural number m is divisible by another natural number n **without remainder**, then n is a divisor of m, and m is divisible by n.
At the same time, m is a **multiple** of n.

□ 21 has divisors 1, 3, 7, 21.

● One is a divisor of every natural number,
every natural number has itself as a divisor, and
every natural number is a divisor of zero.
If m is a divisor of k and n is a divisor of m, then n is also a divisor of k.

□ 3 is a divisor of 6, and 6 is a divisor of 24; therefore, 3 is also a divisor of 24.

▷ In PASCAL special INTEGER divisions are possible:
`n div m`, $n, m \in \mathbb{N}$ gives the greatest natural number l, with $n/m \geq l$,
`n mod m` gives the remainder which is left in the division n/m.
In FORTRAN, the remainder of a division of integers is canceled automatically; that is, n/m in FORTRAN corresponds to `n div m` in PASCAL.

□ 7 `div` 3 = 2 and
7 `mod` 3 = 1.

Prime numbers $\mathbb{P}$**:** Numbers with two and only two divisors, namely 1 and itself. These divisors are called **improper divisors**. Nonprime numbers on the other hand may have many divisors. For example, 3 is a **proper divisor** of 15.

▷ One is not a prime number.

List of prime numbers < 100

2	3	5	7	11	13	17	19	23
29	31	37	41	43	47	53	59	61
67	71	73	79	83	89	97		

Composite numbers: Numbers with at least one proper divisor.

□ $15 = 3 \cdot 5$ is a composite number.

● Composite numbers are **not** prime numbers.

Prime factorization: Unique factorization of a composite number into a product of prime numbers.

□ Prime factorization of 44772:

$$\begin{aligned} 44772 &= 2 \cdot 22386 \\ &= 2 \cdot 2 \cdot 11193 \\ &= 2 \cdot 2 \cdot 3 \cdot 3731 \end{aligned}$$

$$
\begin{aligned}
&= 2 \cdot 2 \cdot 3 \cdot 7 \cdot 533 \\
&= 2 \cdot 2 \cdot 3 \cdot 7 \cdot 13 \cdot 41\,, \\
44772 &= 2^2 \cdot 3 \cdot 7 \cdot 13 \cdot 41\,.
\end{aligned}
$$

Relatively prime numbers: Prime numbers with no common real divisor (no common prime divisor).

□ $45 = 3 \cdot 3 \cdot 5$ and $26 = 2 \cdot 13$ are relatively prime.

Divisibility rules:

A number is **divisible by**... if and only if

- **2:** its last digit is divisible by 2.

□ 38394 is divisible by 2, because 4 is divisible by 2

- **3:** its transverse sum (the sum of all digits) is divisible by 3.

□ 435 is divisible by 3, because $4 + 3 + 5 = 12$ is divisible by 3.

- **4:** the number formed by the two last digits is divisible by 4.

□ 456724 is divisible by 4 because 24 is divisible by 4.

- **5:** its last digit is either 5 or 0.

□ 435 and 3400 are divisible by 5.

- **6:** it is divisible by 2 and 3.

□ 438 is divisible by 2 because 8 is divisible by 2, also $4 + 3 + 8 = 15$ is divisible by 3, thus 438 is divisible by 6.

- **7:** the rule is rather complicated, and it is better to try the division by 7 explicitly.
- **8:** the number formed by the last three digits is divisible by 8.

□ 342416 is divisible by 8 because 416 is divisible by 8.

- **9:** its transverse sum is divisible by 9.

□ 9414 is divisible by 9, because $9 + 4 + 1 + 4 = 18$ is divisible by 9.

- **10:** its last digit is a 0.

□ 4000 and 23412390 are divisible by 10.

Greatest common divisor (g.c.d.): The g.c.d. of several natural numbers is the product of the highest powers of prime factors that are common to all of these numbers.

□

$$
\begin{aligned}
660 &= 2^2 \cdot 3 \cdot 5 \quad \cdot 11 \\
420 &= 2^2 \cdot 3 \cdot 5 \cdot 7 \\
144 &= 2^4 \cdot 3^2
\end{aligned}
$$

Therefore, the g.c.d. of 660, 420, and 144 is $2^2 \cdot 3^1 = 12$.

▷ Program sequence for calculating the g.c.d. of two numbers a and b, $a \geq b$, according to Euclid's algorithm.

```
BEGIN Euclid (Let be a ≥ b; if b > a then interchange a and b.)
INPUT a,b
c := 1
WHILE (c≠0) DO
    c := a MOD b
    a := b
    b := c
ENDDO
OUTPUT a
END Euclid
```

Least common multiple (l.c.m.): The l.c.m. of several natural numbers is the product of all the highest powers of prime factors occurring in at least one of these numbers.

□

$$
\begin{aligned}
588 &= 2^2 \cdot 3 \cdot 7^2 \\
56 &= 2^3 \quad \cdot 7 \\
364 &= 2^2 \quad \cdot 7 \cdot 13
\end{aligned}
$$

Thus, the l.c.m. of 588, 56, and 364 is $2^3 \cdot 3^1 \cdot 7^2 \cdot 13^1 = 15288$.

● Always: g.c.d. $(a, b) \cdot$ l.c.m. $(a, b) = ab$.

▷ This relation can be employed for calculating the l.c.m. of two numbers according to Euclid's algorithm.

1.4 Integers

Extending $\mathbb{N}$ by the set $\{n \mid -n \in \mathbb{N}\}$ we obtain the set of integers $\mathbb{Z} = \{\ldots, -3, -2, -1, 0, 1, 2, 3, \ldots\}$.

● Within this set not only addition and multiplication but also subtraction can be performed without limitations: for all $n, m \in \mathbb{Z}$, $n - m = n' \in \mathbb{Z}$ applies as well.

▷ $\mathbb{N}$ is a subset of $\mathbb{Z}$: $\mathbb{N} \subset \mathbb{Z}$.

Division (the inverse to multiplication) cannot always be performed within the set of integers; that is, the result is generally no longer in the set of integers.

□ $1 \div (-2) = -\frac{1}{2} \notin \mathbb{Z}$.

▷ In programming languages, integers are called INTEGER **numbers**. Computers have a maximum INTEGER number which depends on the hardware and software used.

1.5 Rational numbers (fractional numbers)

If one constructs the new set $\mathbb{Q} = \left\{\frac{k}{m} \mid k, m \in \mathbb{Z}, m \neq 0\right\}$, the division can also be performed without limitation.

▷ **Exception**: Division by zero is not defined.

● $\mathbb{Q}$ contains the integers as a subset $\mathbb{Z} \subset \mathbb{Q}$. Each integer can also be written as a fraction (quotient), $m = \frac{m}{1} \in \mathbb{Q}$.

▷ The representation of a number $r \in \mathbb{Q}$ as a quotient $\frac{p}{q}$ is **not unique**.

□
$$\frac{1}{3} = \frac{2}{6} = \frac{709}{2127}$$

▷ In computers, rational numbers can generally be represented only approximately.

1.5.1 Decimal fractions

Terminating decimal fractions: If there are only zeros to the right of a digit, the sequence of digits terminates. They can be omitted if they occur after the decimal point.

□ $1.500 = 1.5$ is a terminating decimal fraction.
$1200.00 = 1200$ (the two zeros before the decimal point must not be omitted).

▷ However, 1.500 and 1.5 mean different accuracies, namely
$1.45 \leq 1.5 < 1.55$, but $1.4995 \leq 1.500 < 1.5005$.

Purely periodical decimal fractions: The sequence of digits does not terminate but repeats itself constantly.

□ Periodical decimal fractions

$$\frac{1}{11} = 0.09090909\ldots = 0.\overline{09}$$

$$\frac{4}{3} = 1.3333\ldots = 1.\overline{3}$$

$$\frac{3}{7} = 0.\overline{428571}$$

Mixed periodical decimal fractions: An irregular sequence of digits followed by a sequence of repetitive digits.

□ $$\frac{19}{15} = 1.26666... = 1.2\overline{6}$$

● All rational numbers can be represented as terminating, purely periodical, or mixed periodical decimal fractions.

Conversion:

Purely periodical decimal fraction:

$$0.\overline{\alpha_1\alpha_2\ldots\alpha_m} = \frac{\alpha_1\alpha_2\ldots\alpha_m}{\underbrace{99\ldots9}_{m \text{ times } 9}} .$$

Mixed periodical decimal fractions:

$$0.\beta_1\beta_2\ldots\beta_n\overline{\alpha_1\alpha_2\ldots\alpha_m} = \frac{\beta_1\beta_2\ldots\beta_n\alpha_1\alpha_2\ldots\alpha_m - \beta_1\beta_2\ldots\beta_n}{\underbrace{99\ldots9}_{m \text{ times } 9}\underbrace{00\ldots0}_{m \text{ times } 0}} .$$

Notation: $\gamma_1\gamma_2\ldots\gamma_n$ means the number with the sequence of digits γ_1 to γ_n and **not** the product of the numbers γ_1 to γ_n.

□
$$\begin{aligned} 1.43131\ldots &= 1.4\overline{31} \\ &= \frac{14}{10} + \frac{1}{10}\cdot\frac{31}{99} \\ &= \frac{14}{10} + \frac{31}{990} \\ &= \frac{14\cdot 99 + 31}{990} \\ &= \frac{1417}{990} \end{aligned}$$

▷ $0.\overline{9} = 1$

▷ In calculators, only terminating decimal fractions can be processed. All other types of numbers are approximated by means of terminating decimal fractions, which results in errors due to rounding and truncation.

□ Pocket calculator with 8 digits:
$1 \div 6 = 0.1666666 \neq 1/6$ **Truncation error**,
$1 \div 6 = 0.1666667 \neq 1/6$ **Rounding error**.

1.5.2 Fractions

Proper fractions:

$$\left|\frac{a}{b}\right| < 1,\ a, b \in \mathbb{Z},\ b \neq 0\,.$$

Improper fractions:

$$\left|\frac{a}{b}\right| \geq 1,\ a, b \in \mathbb{Z},\ b \neq 0\,.$$

Mixed number: Sum of an integer and a proper fraction.

□ $$4 + \frac{2}{3} = 4\frac{2}{3}$$

▷ **IMPORTANT**:

$$4\frac{2}{3} \neq 4 \cdot \frac{2}{3}\,, \qquad \text{but} \qquad 4\frac{2}{3} = 4 + \frac{2}{3}\,.$$

● The absolute value of a fraction is smaller than 1 if and only if the absolute value of the numerator is smaller than the absolute value of the denominator.
$|(-2/3)| < 1$, because $|-2| = 2 < 3 = |3|$.

Reciprocal of a fraction: Numerator and denominator are interchanged.

$$\frac{b}{a} \text{ is the reciprocal of } \frac{a}{b}\,, \qquad \frac{1}{a} \text{ is the reciprocal of } a\,.$$

Extension of fractions: Numerator **and** denominator are multiplied by the **same** number, or more generally, by the **same** mathematical expression.

$$\frac{a}{b} = \frac{a \cdot c}{b \cdot c}$$

□ $\frac{2}{3} = \frac{2 \cdot 5}{3 \cdot 5} = \frac{10}{15}$

▷ Extension by zero is **not** permitted!

Cancellation of fractions (the inverse to extension): Numerator **and** denominator are divided by the **same** number or, more generally, by the **same** mathematical expression.

$$\frac{a}{b} = \frac{a/c}{b/c}$$

□ $\frac{-8}{12} = \frac{-8/4}{12/4} = \frac{-2}{3}$

$$\frac{(a+b)c}{d(a+b)^2} = \frac{c}{d(a+b)}$$

● Extension or cancellation does not alter the value of the fraction.

▷ In numerical computations this rule can be violated sometimes due to rounding and truncation errors.

1.5.3 Calculating with fractions

Addition and subtraction: Fractions are added or subtracted by extending or reducing them to a common denominator the (**least common denominator**) and then adding or subtracting the numerators while keeping the denominator.

□ $$\frac{4}{15}+\frac{1}{9}=\frac{12}{45}+\frac{5}{45}=\frac{17}{45}$$

Multiplication: Fractions are multiplied by multiplying the numerator of one by the numerator of the other and the denominator of one by the denominator of the other.

$$\frac{a}{b}\cdot\frac{c}{d}=\frac{a\cdot c}{b\cdot d}=\frac{ac}{bd}$$

□ $\frac{2}{3}\cdot\frac{-1}{5}=\frac{-2}{15}=-\frac{2}{15}$

Division: Fractions are divided by each other by multiplying with the reciprocal of the divisor.

$$\frac{a}{b}\div\frac{c}{d}=\frac{a}{b}\cdot\frac{d}{c}=\frac{ad}{bc}$$

□ $\frac{1}{6}\div\frac{2}{3}=\frac{1}{6}\cdot\frac{3}{2}=\frac{1\cdot 3}{6\cdot 2}=\frac{1\cdot 1}{2\cdot 2}=\frac{1}{4}$

Double fractions: Can be regarded as a division of two fractions.

$$\frac{a/b}{c/d}=\frac{a}{b}\div\frac{c}{d}=(a\div b)\div(c\div d)=\frac{a}{b}\cdot\frac{d}{c}=\frac{ad}{bc}$$

Double fractions can be removed by dividing the product of the outer terms (a and d) by the product of the inner terms (b and c).

□ $\frac{1/2}{2/3}=\frac{1\cdot 3}{2\cdot 2}=\frac{3}{4}$

1.6 Calculating with quotients

1.6.1 Proportion

Proportion: The values a_i, b_i of paired values (a_i, b_i) are said to be proportional if $a_i : b_i = k = \text{const.}$

Proportionality factor, proportionality constant: The constant k in proportional relations.

□ 1 liter of milk costs $1.50, 2 liters of milk cost $3. The price of the milk is proportional to the quantity of milk purchased. The proportionality factor is $1.50/liter.
The circumference of a circle is proportional to its radius; the proportionality constant is 2π.

● Given a proportionality $a : b = c : d$, then the following statements hold with any constant $k \neq 0$:

1. Proportionality constant

$$\frac{a}{b}=\frac{c}{d}=k \quad \text{proportionality constant.}$$

2. Inversion

$$b : a = d : c \qquad \text{The proportionality constant is } 1/k\,.$$

3. Reducing and extending

$$ak : b = ck : d \qquad a : bk = c : dk$$
$$ak : bk = c : d \qquad a : b = ck : dk$$
$$a : c = b : d \qquad c : a = d : b$$

4. Corresponding addition and subtraction

$$(a \pm b) : b = (c \pm d) : d \qquad (a \pm b) : a = (c \pm d) : c$$

- **Geometric mean** s: If $a : s = s : b$, then $s = \sqrt{ab}$ is the geometric mean of a and b.
- **Harmonic mean** h: If $(a - h) : (h - b) = a : b$, then $h = 2ab/(a + b)$ is the harmonic mean of a and b.

$$h = \frac{2ab}{a+b} \quad \Longrightarrow \quad \frac{1}{h} = \frac{1}{2}\left(\frac{1}{a} + \frac{1}{b}\right)$$

1.6.2 Rule of three

Rule of three (also called calculation of the fourth proportional): Used for calculating of unknown quantities in proportional relations.

$$x : b = c : d \quad \Longrightarrow \quad x = \frac{b \cdot c}{d}$$
$$a : x = c : d \quad \Longrightarrow \quad x = \frac{a \cdot d}{c}$$
$$a : b = x : d \quad \Longrightarrow \quad x = \frac{a \cdot d}{b}$$
$$a : b = c : x \quad \Longrightarrow \quad x = \frac{b \cdot c}{a}$$

□ 1.5 m of rope cost \$1.20. How much of the rope can one get for \$5.00? The rule-of-three formula has to be used: a =1.5 m, b =\$1.20, and d =\$5. Then $x = (1.5\ \text{m}/\$1.20) \cdot \$5 = 6.25\ \text{m}$.

1.7 Mathematics of finance

Mathematical finance (better: mathematical methods in finance): the mathematical principles for

- **calculations,**
- **calculation of authorization,**
- **calculus of annuities,**
- **depreciation.**

Apart from the percentage calculation of percentage, essential mathematical elements are elementary equations and inequalities, in particular, arithmetic and geometric number sequences and series.

1.7.1 Calculations of percentage

Percent: From the Italian *per cento*, i.e., per 100 · p percent of M is $p \cdot M/100$. The symbol % is used for percent. $p\% \triangleq p/100$.

□ 5% of \$600 is \$30.
125% of 4 m is 5 m.

Markup of $p\%$ to K:

$$K' = K\left(1 + \frac{p}{100}\right).$$

Then K (referred to K') contains $p' = \dfrac{p \cdot 100}{100 + p}$ percent as an addition.

□ Merchandise valued at \$400 plus a 10% dealer markup would result in a price of \$440 to be paid by the customer. Thus, the final consumer price would include a markup of 9.1%.

Reduction or discount of $p\%$ to K:

$$K' = K\left(1 - \frac{p}{100}\right).$$

Then K (referred to K') contains $p' = \dfrac{p \cdot 100}{100 - p}$ percent reduction.

□ Merchandise valued at \$400 minus a 10% dealer discount would result in a price of \$360 to be paid by the customer. Thus, (in relation to the final price), the consumer has received a discount of 11.1%.

1.7.2 Interest and compound interest

Interest for a certain period: If an amount earns interest only during part of the year, then interest Z is only the corresponding fraction of the annual interest.

● Interest for parts of the year

$$Z = K \cdot \frac{p}{100} \cdot T\,, \quad \text{where}$$

$$T = \frac{d}{365}\,, \quad d \text{ denotes the number of days, or}$$

$$= \frac{m}{12}\,, \quad m \text{ denotes the number of months.}$$

□ At an interest rate of 5% for a principal of \$10,000, the interest after three months would be

$$Z = \$10{,}000 \cdot \frac{5}{100} \cdot \frac{3}{12} = \$125 \text{ in interest.}$$

Compound interest: Calculation of interest for longer periods, assuming that at the end of each year, the interest is added to the principal and yield interest itself. Initial capital K_0 at p percent yields $Z = K_0 \cdot p/100$ interest in the first year, and thus the new capital $K_1 = \ldots$ where $q = 1 + p/100$. In the second year, the yield is $K_2 = \ldots$ where $q = 1 + p/100$. After n years finally, the capital is

$$K_n = q^n \cdot K_0 = K_0 \cdot \left(1 + \frac{p}{100}\right)^n.$$

The quantity q is called the **accumulation factor**, by which the interest-bearing principal must be multiplied annually. n is the number of years in which the principal as well as the interest are untouched.

1.7.3 Amortization

Amortization amount T: Amount by which a remaining debt S is reduced by repayment.
Annuity A: The sum of amortization amount T and the interest Z accumulated at a certain due date:

$$A = T + Z.$$

Annuity due: Payment at the beginning of a period.
Ordinary annuity: Payment at the end of a period.
Repayment by instalments at diminishing amounts of interest: Repayment of an **initial debt** S_0 at a repayment rate (**amortization rate**) T that is equal for all interest periods n, for a term (**amortization period**) of N interest periods at a constant interest rate p per interest period:

$$T = \frac{S_0}{N} = \text{const.}$$

The interest arising for the n-th interest period diminishes as time goes on.

- Interest Z_n in the n-th interest period:

$$Z_n = S_0\left(1 - \frac{n-1}{N}\right) \cdot \frac{p}{100}, \qquad n = 1, \ldots, N.$$

- Annuity A_n in the n-th interest period:

$$A_n = T + Z_n = \frac{S_0}{N}\left[1 + (N - n + 1)\frac{p}{100}\right], \qquad n = 1, \ldots, N.$$

- **Remaining debt** S_n after the n-th interest period:

$$S_n = S_0\left(1 - \frac{n}{N}\right), \qquad (S_N = 0),\ n = 1, \ldots, N.$$

- **Total interest** Z:

$$Z = \sum_{n=1}^{N} Z_n = S_0 \frac{1+N}{2}\frac{p}{100}.$$

Amortization of annuity: Amortization of an initial debt S_0 with an (ordinary) annuity A, that is equal for all interest periods n, at a constant rate p per interest period (accumulation factor: $q = 1 + \frac{p}{100}$):

- Interest Z_n at the n-th interest period:

$$Z_n = A - \left(A - S_0\frac{p}{100}\right) q^{n-1}, \qquad n = 1, \ldots, N.$$

- Amortization money T_n for the n-th interest period:

$$T_n = \left(A - S_0\frac{p}{100}\right) q^{n-1} = T_1 q^{n-1}, \qquad n = 1, \ldots, N.$$

- **Remaining debt** S_n after n interest periods:

$$S_n = S_0 q^n - A\frac{q^{n-1}}{q-1}, \qquad n = 1, \ldots, N.$$

- **Amortization period** N (in interest periods):

$$N = \frac{-\lg\left(1 - \frac{(q-1)S_0}{A}\right)}{\lg(q)} .$$

□ At an interest rate of 6% per year, a debt of \$40,000 is to be repaid within 10 years at a constant amortization rate.
Amortization: \$4,000
Interest in the fifth year:

$$Z_5 = \$40{,}000 \left(1 - \frac{5-1}{10}\right) 0.06 = \$1{,}440 .$$

Annuity in the fifth year: $A_5 = \$5{,}440$.
Remaining debt after the fifth year:

$$S_5 = \$40{,}000 \left(1 - \frac{5}{10}\right) = \$20{,}000 .$$

Total interest:

$$Z = \$40{,}000 \frac{11}{2} 0.06 = \$13{,}200 .$$

1.7.4 Annuities

Annuity r: The payment of an amount r recurring at regular time intervals (in advance or when due), either as deposits to an account invested with compound interest or as payments from a capital invested with compound interest.
Final value of annuity R_n: The amount resulting from n annuity payments after the n-th interest period.
Cash value of annuity R_{0n}: Cash or present value of all annuity payments up to the n-th interest due date.
Final value of annuity factor s_n: Final value of n annuity payments in the amount of $r = 1$.
Ordinary final value of annuity factor: $s_n = \ldots$
Due final value of annuity factor: s'_n: ...
Cash value of annuity a_n: Cash or present value of n annuity payments in the amount of $r = 1$.
Ordinary cash value of annuity factor: $a_n = \ldots$
Due cash value of annuity factor: $a'_n =$
Final value of annuity R_n after n interest periods:

$$R_n = r\frac{q^n - 1}{q - 1} \qquad \text{(ordinary)} ,$$

$$R'_n = r \cdot q\frac{q^n - 1}{q - 1} \qquad \text{(due)} .$$

Cash value of annuity factor R_{0n}, after n interest periods:

$$R_{0n} = r\frac{1}{q^n}\frac{q^n - 1}{q} \qquad \text{(ordinary)}\,,$$

$$R'_{0n} = r\frac{1}{q^{n-1}}\frac{q^n - 1}{q} \qquad \text{(due)}\,.$$

Annuity payment from an initial capital K_0 invested with compound interest at $p\%$ interest per interest period:
Balance of account K_n after n ordinary annuity payments of amount r:

$$K_n = K_0 q^n - r\frac{q^{n-1}}{q-1} = K_0 q^n - R_n,$$

with $q = 1 + \frac{p}{100}$ as accumulation factor and R_n as the final value of annuity.
Perpetual annuity r_e: The accrued interest at least balances out the annuity payment, so that account K_0 is not reduced in time. (The capital grows in time if the interest exceeds the annuity payment.)

Condition for capital K_0 for a given (maximum) perpetual annuity r_e at an interest rate of accumulation p (factor $q = 1 + \frac{p}{100}$):

$$r_e = K_0(q-1) \qquad \text{(ordinary)},$$

$$r'_e = K_0\frac{q-1}{q} = \frac{1}{q}r_e \qquad \text{(due)}.$$

□ A capital of \$100,000 is invested at an interest of 0.5% per month. The maximum monthly (ordinary) annuity is

$$r_e = 100{,}000 \cdot 0.005 = \$500\,.$$

1.7.5 Depreciation

Depreciation: Annual reduction in value of an asset by wear and through aging.
Straight-line depreciation: Depreciation from acquisition value A_0 to scrap value A_N after N years of use at equal annual rates.
Scrap value A_n after n years:

$$A_n = A_0 - n\frac{A_0 - A_N}{N}, \qquad n = 1, \ldots, N.$$

Declining-balance method of depreciation: Depreciation of an asset with an acquisition value A_0 at the rate of $p\%$ annually of the previous year's scrap value.

Scrap value after n years:

$$A_n = A_0\left(1 - \frac{p}{100}\right)^n, \qquad n = 1, 2, 3, \ldots.$$

Depreciation period N for a given scrap value A_N and an annual depreciation rate p on the remaining scrap value:

$$N = \frac{\lg\left(\frac{A_N}{A_0}\right)}{\lg\left(1 - \frac{p}{100}\right)}.$$

Annual depreciation rate p in percent for a given depreciation period N and required scrap value A_N:

$$p = 100\left[1 - \left(\frac{A_N}{A_0}\right)^{1/N}\right].$$

□ A facility with an acquisition value of \$100,000 is depreciated annually at 6% of the scrap value.
Scrap value after 5 years:

$$A_5 = \$100{,}000\left(1 - \frac{6}{100}\right)^5 = \$73{,}390.40\,.$$

1.8 Irrational numbers

Numbers which cannot be represented as fractions are called **irrational numbers** $\mathbb{I}$. For example, no number $r \in \mathbb{Q}$ can be found for which $r \cdot r = r^2 = 2$. The solution $r = \sqrt{2} \notin \mathbb{Q}$.

□ $\sqrt{2}, \pi, \ln 7, \mathrm{e}, \ldots \in \mathbb{I}$

▷ Irrational numbers **cannot** be represented in the computer; they are approximated by finite decimal fractions.

□ While it is possible to calculate $\sqrt{2}$ with any accuracy, the results of each calculation are always **finite** (terminating) decimal fractions.

1.9 Real numbers

The set $\mathbb{R}$ of all real numbers is the union of rational and irrational numbers $\mathbb{I} \cup \mathbb{Q}$.
All numbers $x \in \mathbb{R}$ can be represented as (sometimes infinite) decimal fractions.

● Addition, multiplication, and subtraction as well as division can be performed in $\mathbb{R}$ without limitations (except for a division by zero).

▷ Even in $\mathbb{R}$, the equation $x^2 + 1 = 0$ has no solution. Thus, the domain of real numbers has to be extended again to become the range of complex numbers.

● Together with the operations $+$ and $-$ the set of real numbers $\mathbb{R}$ forms the **field of real numbers**.

▷ Numerically, there is **no** set of real numbers, only approximations of limited accuracy by finite decimal fractions. Nevertheless, such data are frequently called `REAL`. Notations include: 1.34, 0.134E1, 0.134D1.

1.10 Complex numbers

To find a solution to the equation $x^2 + 1 = 0$, the **imaginary unit** is defined:

$$\mathrm{j}^2 = -1$$

▷ In mathematics, the abbreviation for the imaginary unit is usually i, but in technical applications, this could be confused with the current intensity i. For that reason, j, is used here consistently.

A complex number $z \in \mathbb{C}$ can be written as follows:
Algebraically as a sum of real and imaginary parts $z = a + \mathrm{j}b$, where a and b are real numbers.
Exponentially, as exponential function $z = r\mathrm{e}^{\mathrm{j}\phi}$ with a real radius r and a real phase $\phi \in (\pi, \pi]$ $(-\pi < \phi \leq +\pi)$.
Trigonometrically (polar form), as $z = r\cos\phi + \mathrm{j}r\sin\phi$. or
Graphically, as a coordinate in the **Gaussian plane**.

- **Conversion:**

$$r = \sqrt{a^2 + b^2}\,, \quad \phi = \begin{cases} \arctan\dfrac{b}{a}, & \text{if } a > 0,\ b \geq 0 \\ \dfrac{\pi}{2}, & \text{if } a = 0,\ b > 0 \\ \pi + \arctan\dfrac{b}{a}, & \text{if } a < 0 \\ \dfrac{3\pi}{2}, & \text{if } a = 0,\ b < 0 \\ \arctan\dfrac{b}{a} + 2\pi, & \text{if } a > 0,\ b < 0 \end{cases}$$

$$a = r\cos\phi\,, \quad b = r\sin\phi\,.$$

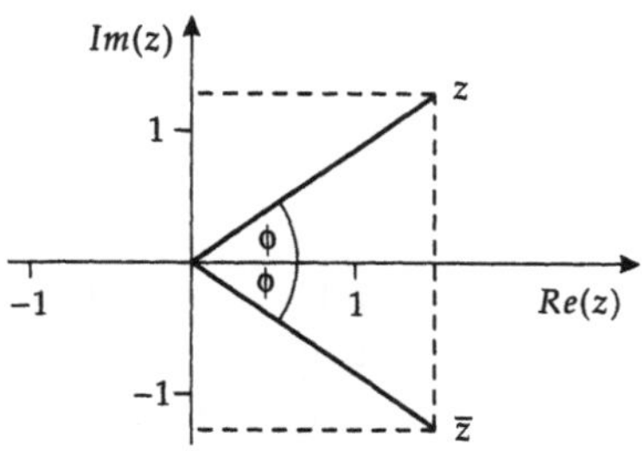

Gaussian plane

The **absolute value** of a complex number $z = a + \mathrm{j}b$ is defined by

$$|z| = \sqrt{a^2 + b^2} = r = \sqrt{z\bar{z}}\,.$$

▷ Always $|z| \geq 0$. Only the absolute values of complex numbers can be compared. For complex numbers the notation $z_1 > z_2$ is not defined.

Conjugate complex number: Reversing the sign of the imaginary part.

$$\bar{z} = a - \mathrm{j}b = \overline{a + \mathrm{j}b} = r\mathrm{e}^{-\mathrm{j}\phi}$$

▷ Other notation: $\bar{z} = z^*$.

▷ Programming languages do not always have a special data type for complex numbers.

1.10.1 Field of complex numbers

The set $\mathbb{C} = \{(a, b) \mid a, b \in \mathbb{R}\}$ together with the addition $(a, b) + (c, d) = (a + c, b + d)$ and the multiplication $(a, b) \cdot (c, d) = (ac - bd, ad + bc)$ form the **field of complex numbers**.

▷ Note the validity of customary rules when identifying:

$$(a, b) \equiv a + \mathrm{j}b \quad \text{with} \quad \mathrm{j} \cdot \mathrm{j} = -1.$$

In particular, $1 = (1, 0), j = (0, 1)$.

The inclusions $\mathbb{C} \supset \mathbb{R} \supset \mathbb{Q} \supset \mathbb{Z} \supset \mathbb{N}$ apply.

1.11 Calculating with real numbers

1.11.1 Sign and absolute value

Negative numbers are prefixed by the **sign** (−). In pocket calculators this sign corresponds to the key [+/−] and not to the key [−], which stands for subtraction.

Positive signs are generally omitted. Always: $-(-r) = +r = r$.

Absolute value, abs(r), or $|r|$, is defined as:

$$\text{abs}(r) = \text{abs}\, r = |r| = \begin{cases} -r\,, & \text{if } r < 0 \quad \text{and} \\ r\,, & \text{if } r \geq 0\,. \end{cases}$$

□ $\text{abs}(-3) = 3$ or $|4.43| = 4.43$.

▷ In programming languages `abs[x]` is often implemented as a function call.

Laws for calculations with absolute values:

$$\text{abs}(ab) = |ab| = |a| \cdot |b|\,,$$

$$\text{abs}(a/b) = \left|\frac{a}{b}\right| = \frac{|a|}{|b|}\,, \qquad b \neq 0\,.$$

● **Triangle inequality**

$$|a| - |b| \leq |a \pm b| \leq |a| + |b|$$

$$|a + b| \geq \big||a| - |b|\big|$$

$$|a + b + c + \cdots| \leq |a| + |b| + |c| + \cdots$$

The name comes from the geometry of plane triangles, to the sides of which the following applies:

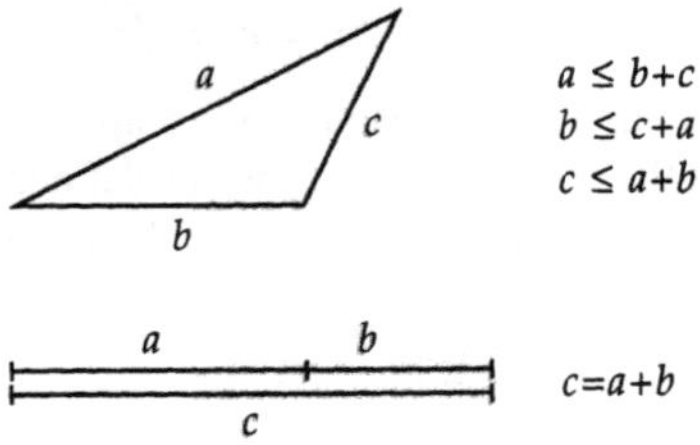

Triangle inequality

▷ **Numerical application of the absolute value function:** Program sequence for calculating the maximum and minimum of n numbers.

```
BEGIN MiniMax
INPUT n
INPUT x[i], i = 1...n
max := x[1]
min := x[1]
FOR i = 2 TO n DO
```

```
    max := (max + x[i] + abs(max - x[i]))/2
    min := (min + x[i] - abs(min - x[i]))/2
ENDDO
OUTPUT max, min
END MiniMax
```

1.11.2 Ordering relations

The set of real numbers can also be represented as a number line. Every real number corresponds to a point on the number line.

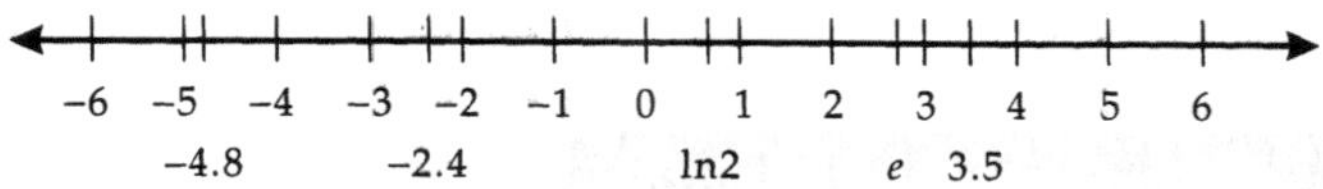

Number line

Zero point: Divides the number line in positive ($\mathbb{R}^+$) and negative real numbers ($\mathbb{R}^-$). Positive numbers lie to the right, negative numbers to the left of the zero point. Notation: $x > 0$ if x is positive, $x < 0$ if x is negative. Zero is neither positive nor negative, but is contained in $\mathbb{R}_0^+$ as well as in $\mathbb{R}_0^-$.

A real number r is **smaller than** a real number s (notation $r < s$), if, on the number line, r lies to the left of s. The relations **greater than** ($>$), **smaller than or equal to** ($\leq$), and **greater than or equal to** ($\geq$) are defined in an analogous manner.

Relation	Notation	Meaning
greater than	$r > s$	$r - s$ is positive
greater than or equal to	$r \geq s$	$r - s$ is positive or $= 0$
smaller than	$r < s$	$r - s$ is negative
smaller than or equal to	$r \leq s$	$r - s$ is negative or $= 0$

▷ Complex numbers **cannot** be ordered.

1.11.3 Intervals

Subsets of $\mathbb{R}$ bounded by two end-points a and b, $a < b$ on the number line are called **intervals**. One has to distinguish between
finite intervals

$I =$	$x \in I$, if	Name of the interval
$[a, b]$	$a \leq x \leq b$	closed interval
$[a, b)$	$a \leq x < b$	right-hand half-open interval
$(a, b]$	$a < x \leq b$	left-hand half-open interval
(a, b)	$a < x < b$	open interval

and **infinite intervals**

$I =$	$x \in I$, if	Special case
$[a, \infty)$	$a \leq x$	for $a = 0$, $I = \mathbb{R}_0^+$
(a, ∞)	$a < x$	for $a = 0$, $I = \mathbb{R}^+$
$(-\infty, b]$	$x \leq b$	for $b = 0$, $I = \mathbb{R}_0^-$
$(-\infty, b)$	$x < b$	for $b = 0$, $I = \mathbb{R}^-$
$(-\infty, \infty)$	$\|x\| < \infty$	$\mathbb{R}$

▷ Sometimes, open intervals are denoted by $]a, b[$ instead of (a, b).

1.11.4 Rounding and truncating

Scientific and technical representation of a real number: $x.xxx\mathrm{E}yyy$ (for double-precision in FORTRAN also $x.xxx\mathrm{D}yyy$).

□ 14357.34 = 1.435734E4 0.0003 = 3.0E−4

Mantissa: Decimal number represented by $x.xxx$ in the scientific/technical representation.

Exponent: In this case, the integer represented by yyy in scientific/technical representations.

▷ The length of the mantissa and the valid range of exponents depend on the system and the programming language.

Significant digits: Those digits of a decimal number which are known exactly in scientific/technical representations.

□ For 0.000001 = 1.E−6, only the 1 is significant.

□ Measurement with an ordinary ruler yields 12.300 cm. Only 12.3 is significant; a normal ruler does not allow a more accurate determination of length.

Often, calculations with computers yield results with more **significant digits** than the method or the input data allow.

▷ Often, the accuracy of the results is **not** improved by a large number of digits after the decimal point.

□ The lengths of the sides of a rectangle are known only up to two digits after the decimal point: $a = 3.12$ and $b = 1.53$. It is meaningless to write area $A = 3.12 \cdot 1.53 = 4.7736$. The accuracy of a and b does not allow this.

It is sufficient to write $A = a \cdot b \approx 4.77$.

Numerical approximations:

Truncation: Only a certain number of decimal places are taken into account; all following places are ignored.

□ $4.7456 \longrightarrow 4.74$ (truncation after two decimal places)

Rounding to a certain number (n) of decimal places: If the digit at place $n + 1$ is smaller than < 5, then the first n places after the decimal point are noted; otherwise (digit ≥ 5) the n-th digit is increased by one.

□ $4.7456 \longrightarrow 4.7$ (rounding to one place after the decimal point),
$4.7456 \longrightarrow 4.75$ (rounding to two places after the decimal point),
$4.99953 \longrightarrow 5.000$ (rounding to three places after the decimal point).

▷ In PASCAL there are functions `round(x)` and `trunc(x)` for rounding and truncating decimal fractions. The result is always an integer.

1.11.5 Calculating with intervals

Numerical calculations contain approximate numbers with a finite number of decimal places. Sometimes it is meaningful to give an upper and lower bound (i.e., an interval) within which the exact value can be found with certainty.

Interval $\tilde{a} = [\underline{a}, \overline{a}] = \{a \mid a \in \mathbb{R},\ \underline{a} \le a \le \overline{a}\}$,

$\underline{a}$ lower limit,

$\overline{a}$ upper limit.

Addition:

$$\tilde{a} + \tilde{b} = \tilde{c} = [\underline{a} + \underline{b}, \overline{a} + \overline{b}] .$$

Subtraction:

$$\tilde{a} - \tilde{b} = \tilde{c} = [\underline{a} - \overline{b}, \overline{a} - \underline{b}] .$$

Multiplication:

$$\tilde{a} \cdot \tilde{b} = \tilde{c} = [\underline{c}, \overline{c}] \quad \underline{c} = \min\{\underline{a}\underline{b}, \underline{a}\overline{b}, \overline{a}\underline{b}, \overline{a}\overline{b}\} \quad \overline{c} = \max\{\underline{a}\underline{b}, \underline{a}\overline{b}, \overline{a}\underline{b}, \overline{a}\overline{b}\} .$$

Division:

$$\frac{\tilde{a}}{\tilde{b}} = \tilde{c} = [\underline{c}, \overline{c}] \quad \underline{c} = \min\left\{\frac{\underline{a}}{\underline{b}}, \frac{\underline{a}}{\overline{b}}, \frac{\overline{a}}{\underline{b}}, \frac{\overline{a}}{\overline{b}}\right\} \quad \overline{c} = \max\left\{\frac{\underline{a}}{\underline{b}}, \frac{\underline{a}}{\overline{b}}, \frac{\overline{a}}{\underline{b}}, \frac{\overline{a}}{\overline{b}}\right\} .$$

□ Let quantity a be in the interval $\tilde{a} = [2, 4]$ with certainty. Let a second quantity b be a quantity in interval $\tilde{b} = [-1, 2]$ with certainty. Then, the following statements hold with certainty for $a \pm b$, $a \cdot b$ and a/b

$$\begin{aligned} a + b &\in [2 - 1, 4 + 2] = [1, 6] , \\ a - b &\in [2 - 2, 4 - (-1)] = [0, 5] , \\ a \cdot b &\in [4 \cdot (-1), 4 \cdot 2] = [-4, 8] , \\ a/b &\in [4/(-1), 4/2] = [-4, 2] . \end{aligned}$$

▷ For calculating with "uncertain" quantities, see also Chapter 22 under fuzzy logic/arithmetic.

1.11.6 Brackets

Brackets: Mathematical symbols occurring *always in pairs*. Operations enclosed by brackets nust be carried out first. Nesting of brackets is possible.

Symbols for **left-hand** brackets (in customary order): (, [, {,

Symbols for **right-hand** brackets (in customary order): },],).

Removal of brackets

Minus sign before a bracket:

$$-(a - b) = -a + b .$$

Removal of multiplication of brackets:

$$\begin{aligned} (a + b) \cdot (c + d) &= ac + bc + ad + bd , \\ (-a + b) \cdot (c - d) &= -ac + bc + ad - bd . \end{aligned}$$

● Brackets have to be used when an arithmetical operation of lower kind should be carried out before an arithmetical operation of higher kind.

□ Square of a sum. The brackets are necessary because the squaring has to be carried out before taking the addition. Generally:

$$(a+b)^2 \neq a^2 + b^2$$

▷ In numerical calculations, brackets also have to be put in places where they do not occur in analytical expressions.

□ Square root of a sum: $\sqrt{1+2}$,
Key sequence in pocket calculators: [(] [1] [+] [2] [)] [√‾] [=].

▷ In programming languages, arguments of functions, procedures, subroutines, and subscripts of fields are also enclosed by brackets.

□ $\sin(x)$ means $\sin x$,
$a(1,2)$ or $a[1,2]$ means the array or field element $a_{1,2}$.

1.11.7 Addition and subtraction

● Addition and subtraction are arithmetical operations of the first kind.

		Addition		
summand	plus	summand	equals	sum
a	$+$	b	$=$	c
4.5	$+$	3.2	$=$	7.7
		Subtraction		
minuend	minus	subtrahend	equals	difference
a	$-$	b	$=$	c
1.4	$-$	2.6	$=$	-1.2

▷ In pocket calculators the corresponding keys for the arithmetical operations are [+] and [−]. In all other programming languages $+$ and $-$ are also used as symbols for addition and subtraction.

● Subtraction can always be viewed as the addition of the negative subtrahend (subtrahend times -1):

$$1-2 = 1+(-2) = -1$$

▷ Caution is advisable when bracketed expressions are removed:

$$a-(b-c) = a-b+c$$

The **neutral element of addition** is zero.

$$0+a=a\,, \qquad a+0=a$$

● Addition is commutative and associative.

Commutative law:

$$a+b=b+a\,.$$

Associative law:

$$a+(b+c) = (a+b)+c = a+b+c\,.$$

▷ In general, these laws do not apply in calculators due to rounding and memory errors, because:

□ Pocket calculators with 10 places:
$(10^{12}+1)+(-10^{12}) = 0$, **but** $(10^{12}+(-10^{12}))+1 = 1$.

● Subtraction is neither commutative nor associative.

□ $2-3 \neq 3-2$

After converting the subtraction into an addition of negatives, the commutative as well as the associative laws apply again.

□ $2-3=2+(-3)=(-3)+2=-3+2$

1.11.8 Summation sign

Summation sign: Abbreviated notation for

$$\sum_{i=m}^{N} a_i = \begin{cases} a_m + a_{m+1} + a_{m+2} + a_{m+3} + \ldots + a_N, & \text{if } N \geq m \\ 0, & \text{if } N < m \,. \end{cases}$$

Computational rules for sums:

Sum of equal summands:

$$\sum_{i=m}^{N} a = (N-m+1)a\,, \quad m \leq N\,.$$

The index can be labeled arbitrarily, but must be preserved:

$$\sum_{i=m}^{N} a_i = \sum_{k=m}^{N} a_k\,.$$

Splitting of sums into partial sums:

$$\sum_{i=m}^{N} a_i = \sum_{i=m}^{l} a_i + \sum_{i=l+1}^{N} a_i\,, \quad m \leq l < N\,.$$

Factoring out constant factors C:

$$\sum_{i=m}^{N} Ca_i = C\sum_{i=m}^{N} a_i\,.$$

Sums of sums can be rewritten according to the commutative and associative laws:

$$\sum_{i=m}^{N}(a_i + b_i + c_i + \cdots) = \sum_{i=m}^{N} a_i + \sum_{i=m}^{N} b_i + \sum_{i=m}^{N} c_i + \cdots\,.$$

Indices can also be fixed by arithmetical operations:

$$\sum_{i=m}^{N} a_i = \sum_{i=l}^{N-m+l} a_{i+m-l}\,.$$

Double sums can be interchanged:

$$\sum_{i=1}^{N}\sum_{k=1}^{L} b_{ik} = \sum_{k=1}^{L}\sum_{i=1}^{N} b_{ik}\,.$$

▷ Sums are programmed as **loops**, double sums are coded as double loops. The structure of loops is:
Do... until... or
Do... from lower limit... to upper limit...

□ REPEAT ... UNTIL, WHILE ... DO and FOR ... TO ... DO the instruction in PASCAL; or DO ... ENDDO or CONTINUE in FORTRAN.

▷ Caution is advisable when computing longer sums. Due to truncating and rounding errors, the associative and commutative laws do **not** apply.

□ Numerical summation of the first 1000 terms in the **Leibniz series**

$$\frac{\pi}{4} = 1 - \frac{1}{3} + \frac{1}{5} - \frac{1}{7} + \frac{1}{9} - + \cdots .$$

First possibility: summation from left to right.
Second possibility: summation from right to left.
Third possibility: summation from left to right, but positive and negative terms separately with subsequent formation of difference.
Fourth possibility: summation from right to left, but positive and negative terms separately with subsequent formation of difference.

	Mainframe	PC
1	0.785121322	0.78514814
2	0.785148203	0.78514814
3	0.785149547	0.78514719
4	0.785145760	0.78514790

▷ Note that
1. The result depends on the order of summation.
2. The result depends on the computer used.
3. Computers often put out **more** places than their accuracy allows.
The exact result rounded to 8 places is 0.78514816.

1.11.9 Multiplication and division

Multiplication and division are arithmetical operations of the second kind:

		Multiplication		
multiplicand	times	multiplicator	equals	product
a	$\cdot$	b	$=$	c
2.0	$\cdot$	-7.2	$=$	-14.4
		Division		
dividend	divided by	divisor	equals	quotient
a	$\div$	b	$=$	c
5.1	$\div$	1.7	$=$	3.0

▷ In pocket calculators, the [×] key must be used for multiplication, **not** the key for the decimal point [.]. For division the key [÷] is used.
For programming in PASCAL, C, FORTRAN, ... the symbol '*' is used for multiplication, the symbol '/' is used for division.

Division by $b \neq 0$ can always be regarded as multiplication by $1/b$.

Neutral or identity element of multiplication is the unit 1.

$$1 \cdot a = a$$

Multiplication is commutative and associative.

Commutative law:

$$a \cdot b = b \cdot a \,.$$

Associative law:

$$a \cdot (b \cdot c) = (a \cdot b) \cdot c = a \cdot b \cdot c \,.$$

▷ However, these laws apply **only** analytically. They can be helpful in numerical calculations: $(10^{45} \cdot 10^{60}) \cdot 10^{-60}$ can generally not be calculated in pocket calculators because the intermediate result 10^{105} cannot be represented. However, $10^{45} \cdot (10^{60} \cdot 10^{-60})$ yields the correct result 10^{45}.

● A product is zero if and only if at least one factor is zero.

● Division is neither commutative nor associative.

□ $\frac{3}{2} \neq \frac{2}{3}$

After converting the division into a multiplication with the reciprocal of the divisor, the commutative and the associative laws apply again.

□
$$\frac{3}{2} = 3 \cdot \frac{1}{2} = \frac{1}{2} \cdot 3$$

1.11.10 Product sign

Product sign: Abbreviated notation for:

$$\prod_{i=m}^{N} a_i = \begin{cases} a_m \cdot a_{m+1} \cdot a_{m+2} \cdot a_{m+3} \cdot \cdots \cdot a_N, & \text{if } N \geq m \\ 1, & \text{if } N < m \,. \end{cases}$$

Computational laws for products:

Product of equal factors:

$$\prod_{i=m}^{N} a = a^{N-m+1} \,, \qquad m < N \,.$$

The index can be labeled arbitrarily, but must be preserved:

$$\prod_{i=1}^{N} a_i = \prod_{k=1}^{N} a_k \,.$$

Splitting of products into partial products:

$$\prod_{i=m}^{N} a_i = \prod_{i=m}^{l} a_i \cdot \prod_{i=l+1}^{N} a_i \,, \qquad m < l < N \,.$$

Constant factors can be separated out:

$$\prod_{i=m}^{N} C a_i = C^{N-m+1} \prod_{i=m}^{N} a_i \,.$$

Products of products can be rewritten according to the commutative law and the associative law:

$$\prod_{i=m}^{N} a_i b_i c_i \cdots = \prod_{i=m}^{N} a_i \cdot \prod_{i=m}^{N} b_i \cdot \prod_{i=m}^{N} c_i \cdots \,.$$

Indices can also be fixed by arithmetical operations:

$$\prod_{i=m}^{N} a_i = \prod_{i=l}^{N-m+l} a_{i+m-l} .$$

Double products can be interchanged:

$$\prod_{i=1}^{N}\prod_{k=1}^{L} b_{ik} = \prod_{k=1}^{L}\prod_{i=1}^{N} b_{ik} .$$

1.11.11 Powers and roots

Raising to a power and extracting a root are arithmetical operations of the third kind.
Powers with integral exponents: Abbreviated notation for a product of equal factors.

$$\underbrace{a \cdot a \cdot \cdots \cdot a}_{n \text{ factors}} = a^n$$

a is called the **base**, n is called the **exponent**.
Square numbers ($n = 2$): 1, 2, 4, 9, 16, 25, ...
Powers with a negative exponent: 1/power with a positive exponent,

$$\underbrace{\frac{1}{a} \cdot \frac{1}{a} \cdot \cdots \cdot \frac{1}{a}}_{n \text{ factors}} = \left(\frac{1}{a}\right)^n = a^{-n} .$$

- ● **Extraction of the root** is the inverse of raising to a power. Definition of the root: $\sqrt[n]{a} = x \Longleftrightarrow x^n = a; a \geq 0, n \in \mathbb{N}, x \geq 0$.
- □ Given a cube with a volume of 125 cm^3, then the edge length is $a = \sqrt[3]{125}$ cm = 5 cm.
- ● The n-th root of a number a is defined as the non-negative, uniquely determined number which n times multiplied by itself yields a. $-2 \neq \sqrt{4}$ although $(-2)^2 = 4$.

a is called the **radicand**, n is called the **index** of the root.

▷ If the index of the root is not given, then the root always means **square root** ($n = 2$).

Roots: Powers with fractional exponents,

$$\sqrt[n]{a} = a^{1/n} , \qquad \frac{1}{\sqrt[n]{a}} = a^{-1/n} , \qquad \text{in particular: } \sqrt[2]{a} = \sqrt{a} .$$

▷ Roots can always be represented as powers, which greatly simplifies the calculation with roots.

Computational laws for any powers

$$\begin{aligned} & 0^0 \text{ is not defined,} \\ a^0 &= 1, \text{ if } a \neq 0 , \\ 0^n &= 0, \text{ if } n \neq 0 , \\ a^1 &= a , \\ a^m \cdot a^n &= a^{m+n} , \\ \frac{a^m}{a^n} &= a^{m-n} = \frac{1}{a^{n-m}} , \\ a^n \cdot b^n &= (ab)^n , \end{aligned}$$

$$\frac{a^n}{b^n} = \left(\frac{a}{b}\right)^n = \left(\frac{b}{a}\right)^{-n},$$

$$(a^n)^m = (a^m)^n = a^{nm} \qquad \textit{but} \quad a^{n^m} = a^{(n^m)} \neq a^{nm}.$$

● These laws also apply to real exponents $n, m \in \mathbb{R}$!
Therefore, $a^{\sqrt{2}}$ is a well-defined expression.

▷ Powers in `FORTRAN`: `a**x`,
Powers in `PASCAL`: none. Hint: $a^x = \exp(x \cdot \ln a)$.

Computational rules for roots

$$\sqrt[n]{1} = 1, \qquad \sqrt[n]{0} = 0,$$

$$a^{m/n} = \sqrt[n]{a^m} = \left(\sqrt[n]{a}\right)^m, \qquad a^{-m/n} = \frac{1}{\sqrt[n]{a^m}} = \frac{1}{\left(\sqrt[n]{a}\right)^m},$$

$$\sqrt[n]{a} \cdot \sqrt[n]{b} = \sqrt[n]{ab},$$

$$\frac{\sqrt[n]{a}}{\sqrt[n]{b}} = \sqrt[n]{\frac{a}{b}},$$

$$\sqrt[n]{\sqrt[m]{a}} = \sqrt[m]{\sqrt[n]{a}} = \sqrt[nm]{a}.$$

▷ Square root in the computers: `sqrt(a)`.
Higher roots in `FORTRAN`: `a**(1.0/n)`,
Higher roots in `PASCAL`: `exp(1/n*ln(a))`.

Connection between root and absolute value:

$$|a| = \sqrt{a^2}, \text{ generally: } |a| = \sqrt[2n]{a^{2n}}.$$

Rationalizing the denominator: If the result of a calculation is a fraction with a root, the fraction is extended until no root occurs in the denominator.

Basic rules for rationalizing the denominator:

$$\frac{1}{\sqrt{a}} = \frac{\sqrt{a}}{a},$$

$$\frac{1}{\sqrt[n]{a}} = \frac{\sqrt[n]{a^{n-1}}}{a},$$

$$\frac{1}{b \pm \sqrt{a}} = \frac{b \mp \sqrt{a}}{b^2 - a}.$$

□
$$\frac{4ac}{2bc - \sqrt{ab}} = \frac{4ac \cdot (2bc + \sqrt{ab})}{4b^2c^2 - ab}$$

▷ Square roots (here of a) can be computed easily with an **iterative procedure** by applying the following rule: given the initial approximation $x_0 \neq 0$, then

$$x_{n+1} = \frac{1}{2}\left(x_n + \frac{a}{x_n}\right), \quad n = 0, 1, 2, \ldots.$$

▷ Subroutine for the numerical calculation of square roots.

```
BEGIN root
INPUT a
INPUT eps (absolute accuracy)
x := a
```

```
WHILE (abs(x*x - a) ≥ eps) DO
   x := (x + a/x)/2
ENDDO
OUTPUT x
END root
```

□ Calculation of the square root of 100, with an absolute accuracy of 10^{-6} using the starting value 1. The procedure converges very fast, in spite of the poor starting value.

iteration step	approximate value
0	1.0000000
1	50.5000000
2	26.2400990
3	15.0255301
4	10.8404347
5	10.0325785
6	10.0000529
7	10.0000000

1.11.12 Exponentiation and logarithms

Exponentiation and **taking the logarithm** are arithmetical operations of the third kind.

▷ Exponential expressions must be strictly distinguished from power expressions. In power expressions, e.g., a^3, the exponent is fixed; in exponential expressions, e.g., 3^a, the base is fixed.

Computational laws:

$$a^x \cdot a^y = a^{x+y}\,, \qquad \frac{a^x}{a^y} = a^{x-y}\,,$$
$$(a^x)^y = (a^y)^x = a^{xy}\,, \quad \textit{but } a^{x^y} = a^{(x^y)}\,.$$

Logarithm x of the **number** c to the **base** a ($c > 0$): A value for the exponent that satisfies the equation: $a^x = c$.

Notation: $x = \log_a c$.

▷ The logarithm is defined only for positive arguments; $\log(0)$ or $\log(-2)$ are not defined.

□ $10^x = 1000$
$\log_{10} 1000 = 3 \quad \Leftrightarrow \quad 10^3 = 1000.$

Computational laws

$$\log_a 1 = 0\,, \qquad \log_a a = 1\,, \qquad \log_a a^x = x\,,$$
$$\log_a (xy) = \log_a x + \log_a y\,, \qquad \log_a \frac{x}{y} = \log_a x - \log_a y\,,$$
$$\log_a x^n = n \cdot \log_a x\,, \qquad \log_a \sqrt[n]{x} = \frac{1}{n}\log_a x\,,$$
$$\log_a c = \frac{1}{\log_c a}\,, \qquad \log_a c \cdot \log_c a = 1\,,$$
$$\log_{1/a} x = -\log_a x\,.$$

Special logarithms:
Logarithm to base 10 also (**Brigg's logarithm**:), the base is 10, notation $\log_{10} x$, sometimes only $\lg x$ or $\log x$.
Logarithm to base 2, the base is 2, notation $\log_2 x$, $\mathrm{lb}\, x$, rarely $\mathrm{ld}\, x$.
Natural logarithm, the base is **Euler's number** e, notation $\log_e x$ or simply $\ln x$.

$$\begin{aligned} \mathrm{e} &= \lim_{n\to\infty}\left(1+\frac{1}{n}\right)^n \\ &= \sum_{n=0}^{\infty}\frac{1}{n!} \\ &= 2.71828\ldots \end{aligned}$$

▷ In many programming languages only the natural logarithm is used; the notation is `log(x)` in `FORTRAN` and `ln(x)` in `PASCAL`. Other logarithms can be computed easily by changing the base.

Changing to another base:

$$\log_b x = \frac{\log_a x}{\log_a b} .$$

□ Calculation of the base 10 logarithm for 23 in case the programming language used knows only the natural logarithm:

$$\log_{10} 23 = \ln(23)/\ln(10) \approx 3.135/2.303 \approx 1.362$$

Conversion factor from base ⇒ to base ⇓	2	e	10
2	1	1.443	3.322
e	0.693	1	2.303
10	0.301	0.434	1

□ The base 10 logarithm of 16 is about 1.204, then the base 2 logarithm for 16 equals $1.204 \cdot 3.322 \approx 4$.

1.12 Binomial theorem

1.12.1 *Binomial formulas*

Method for the removal of bracketed quadratic expressions:

$$\begin{aligned} (a \pm b)^2 &= a^2 \pm 2ab + b^2 , \\ (a+b)(a-b) &= a^2 - b^2 . \end{aligned}$$

Analogous treatment of higher powers:

$$\begin{aligned} (a+b)^3 &= a^3 + 3a^2b + 3ab^2 + b^3 , \\ (a-b)^3 &= a^3 - 3a^2b + 3ab^2 - b^3 , \\ (a+b)^4 &= a^4 + 4a^3b + 6a^2b^2 + 4ab^3 + b^4 , \\ (a-b)^4 &= a^4 - 4a^3b + 6a^2b^2 - 4ab^3 + b^4 , \\ (a+b)^2(a-b)^2 &= a^4 - 2a^2b^2 + b^4 . \end{aligned}$$

1.12.2 Binomial coefficients

Factorial: Symbol: !

$$\text{Definition:}\quad \begin{aligned} 0! &= 1\\ 1! &= 1\\ 2! &= 1\cdot 2\\ 3! &= 1\cdot 2\cdot 3 = 6\\ n! &= 1\cdot 2\cdot 3\cdot\ldots\cdot n\\ n! &= (n-1)!\cdot n \end{aligned}$$

Stirling's approximation formula (for large n):

$$n! \approx \sqrt{2\pi n}\left(\frac{n}{\mathrm{e}}\right)^n .$$

More accurate:

$$n! \approx \sqrt{2\pi n}\left(\frac{n}{\mathrm{e}}\right)^n \cdot \left(1 + \frac{1}{12n} + \frac{1}{288n^2}\right) .$$

Definition of **binomial coefficients** (read: n over k):

$$\binom{n}{k} = \frac{n}{1}\cdot\frac{n-1}{2}\cdot\frac{n-2}{3}\cdot\ldots\cdot\frac{n-(k-1)}{k} = \frac{n!}{(n-k)!\,k!} .$$

□ $$\binom{7}{3} = \frac{7\cdot 6\cdot 5}{1\cdot 2\cdot 3} = 35$$

1.12.3 Pascal's triangle

▷ **Pascal's triangle** or **binomial array** presents a simple scheme for calculating binomial coefficients.

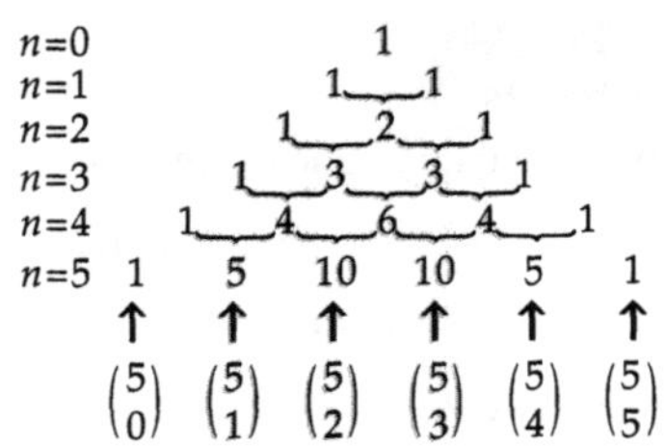

Pascal's triangle

● **Construction:** start with one; add one to the beginning and to the end of each row. All other elements of Pascal's triangle result from the sum of the two adjacent numbers in the row immediately above.

1.12.4 Properties of binomial coefficients

$$\binom{n}{k} = 0 \quad \text{if } k > n\,, \qquad \binom{n}{k+1} = \binom{n}{k}\frac{n-k}{k+1}$$

Symmetry of Pascal's triangle:

$$\binom{n}{k} = \binom{n}{n-k}\,.$$

The **first** and **last** elements in each row of Pascal's triangle are one.

$$\binom{n}{0} = \binom{n}{n} = 1$$

The **second** and the **next to last** elements of the n-th row have the value n.

$$\binom{n}{1} = \binom{n}{n-1} = n$$

Scheme for constructing Pascal's triangle:

$$\binom{n}{k-1} + \binom{n}{k} = \binom{n+1}{k}\,, \qquad \binom{n}{k} + \binom{n}{k+1} = \binom{n+1}{k+1}\,.$$

The **sum of all elements** in the n-th row of Pascal's triangle is 2^n.

$$\binom{n}{0} + \binom{n}{1} + \binom{n}{2} + \cdots + \binom{n}{n} = \sum_{k=0}^{n}\binom{n}{k} = 2^n$$

Further properties:

$$\binom{k}{k} + \binom{k+1}{k} + \binom{k+2}{k} + \cdots + \binom{n}{k} = \sum_{i=k}^{n}\binom{i}{k} = \binom{n+1}{k+1}\,,$$

$$\binom{n}{0} + \binom{n+1}{1} + \binom{n+2}{2} + \cdots + \binom{n+k}{k} = \sum_{i=0}^{k}\binom{n+i}{i} = \binom{n+k+1}{k}\,,$$

$$\binom{n}{0} + \binom{n}{2} + \binom{n}{4} + \cdots = 2^{n-1}\,, \qquad \binom{n}{1} + \binom{n}{3} + \binom{n}{5} + \cdots = 2^{n-1}\,.$$

▷ Program sequence for calculating the first n rows of Pascal's triangle.

```
BEGIN Pascal
INPUT n
u[i] := 0, i = 0...n+1
FOR k = 0 TO n DO
    s := 0
    t := 1
    FOR i = 0 TO k DO
       u[i] := s + t
       s := t
       t := u[i+1]
    ENDDO
    OUTPUT u[i], i = 0...k
ENDDO
END Pascal
```

1.12.5 Expansion of powers of sums

- **Binomial theorem**: The powers of the expression $(a \pm b)$ can be expanded as

$$(a+b)^n = a^n + \binom{n}{1} a^{n-1} b^1 + \binom{n}{2} a^{n-2} b^2 + \cdots + \binom{n}{n-1} a^1 b^{n-1} + b^n \,.$$

Taking into account

$$\binom{n}{0} = \binom{n}{n} = 1 \quad \text{, and} \quad a^0 = b^0 = 1$$

this formula can be generalized:

$$(a \pm b)^n = \sum_{k=0}^{n} \binom{n}{k} a^{n-k} (\pm b)^k \,.$$

- Further: $a^n - b^n = (a-b) \sum_{k=0}^{n-1} a^{n-k-1} b^k$.

2

Equations and inequalities (algebra)

Algebra: Solution of equations and inequalities containing **only** the basic operations addition, subtraction, multiplication, division, and raising to powers with natural exponents.
Algebraic equation of n-th (basic form):

$$a_n x^n + a_{n-1} x^{n-1} + \cdots + a_1 x + a_0 = 0, \qquad a_i \in \mathbb{R},\ a_n \neq 0 .$$

Compact notation:

$$\sum_{i=0}^{n} a_i x^i = 0 .$$

▷ Finding the zeros of polynomials leads to algebraic equations.
Diophantine equation: An algebraic equation with integral coefficients, for which only integers are wanted as solutions.
□ The Diophantine equation

$$ax + by = c \qquad a, b, c \in \mathbb{Z}$$

can be solved if and only if the greatest common divisor of a, b is a divisor of c.
▷ Diophantine equations occur, for example, when relations between numbers of pieces are described. Until now, general solutions for the Diophantine equations could be found only up to the second degree. For Diophantine equations of higher degree, only special solutions are available.

2.1 Fundamental algebraic laws

2.1.1 Nomenclature

Variable: Substitute for a particular nonspecified element.

□ x is a technical quantity standing for revolution per minute, voltage, time, etc. $2(x+1)-2=2x$. x can represent any real or complex number.

▷ Within the same task, the same symbols are always connected with the elements or numerical values.

Constant: Substitute for a certain element or a unique real or complex number.

□ π, e, -2, 3.22 are constants.

Coefficient: Real or complex number before a power of the variable.

□ $a_i x^i$; a_i is the coefficient of x^i.

Unknowns: Substitute for the element whose value is to be determined by solving an equation.

□ $3x = 9.3$. x is the unknown in this equation. $x = 3.1$ is the complex solution of the problem.

Most unknowns are represented by variables.

Term: Mathematical expression involving variables, constants, and arithmetic operations (+, −, etc.) in mathematically valid order.

□ $2\sqrt{4a+b}+a$, $1, 2$, a are terms,
$((3a$ is not a term, because part of the term is missing (brackets not closed).

Term is generic for numbers and complicated mathematical expressions which contain no equal signs or relational signs.

Equation: Two terms, S and T, linked by an equal sign = ($S = T$).

□ $7a+2=3$, $b=23$ are equations.

Inequality: Two terms, S and T, connected by one of the **relational signs** $<, >, \leq, \geq$ or by the **inequality sign** $\neq$.

□ $4>1$, $a<4b+2$, $5>1$ are inequalities.

Propositional form: Equation or inequality containing at least one variable.

□ $4x = 12$ is a propositional form.

Proposition: Equation or inequality containing no variables or unknowns. A proposition can be either true or false.

□ $2 \cdot 4 = 8$ is a true proposition,
$2 \cdot 4 = 7$ is a false proposition.

▷ In customary programming languages, the truth value of a proposition is represented by a **logical variable**, which can assume only the values `TRUE` or `FALSE`.

Defining equation: Assigns a value or a mathematical expression to a variable.

□ $x = 12y + 3$: Assigns 12 times the value of variable y plus 3, to variable x.

▷ In computer programs, defining equations are possible which make no mathematical sense, such as, `n=n+1` increases by the value stored under the variable name n. Hence, in `PASCAL` defining equations such as `n:=n+1` are distinguished from propositions in the form of an equation, as `n=3`, that is, if n has the value of 3, the proposition is true; otherwise the proposition is false.

Uniqueness: In general, the solutions of algebraic equations are not always unique, that is, various possible values may occur for the unknown.

□ $x^2 = 4$ has the solutions $x = +2$ and -2.

Fundamental set $\mathbb{G}$: Set of values allowed for the unknown.

▷ Frequently, **domain of definition** $\mathbb{D}$ is used synonymously with fundamental set, strictly speaking, although this notation applies only to functions.

Solution set $\mathbb{L}$: Set of values for the unknowns that make the equation into a true proposition. The solution set is always a subset of the fundamental set.

□ $2x = -4$ has no solution in $\mathbb{N}$ as a fundamental set, but solution -2 in fundamental set $\mathbb{Z}$.
$2x = 4$ becomes a true proposition with $x = -2$; -2 belongs to $\mathbb{L}$, but $x = 3$ makes $2x = -4$ into a wrong proposition and therefore does not belong to solution set $\mathbb{L}$.

2.1.2 Group

Group: Pair $(M, \circ)$ of a set M and an operation $\circ$, where the following applies:
Associative law:

$$(a \circ b) \circ c = a \circ (b \circ c) , \qquad a, b, c \in M .$$

Existence of a neutral element $e \in M$:

$$a \circ e = e \circ a = a , \qquad a \in M .$$

Existence of an inverse element $a^{-1} \in M$:

$$a \circ a^{-1} = a^{-1} \circ a = e , \qquad a \in M .$$

A group is called commutative or **Abelian**, if the following also applies:
Commutative law:

$$a \circ b = b \circ a , \qquad a, b \in M .$$

- **Existence of a composition**: For all $a, b \in M$ there exist $s, t \in M$ with $a \circ s = b$ and $t \circ a = b$.

□ The set of integers forms an Abelian group under ordinary addition $(\mathbb{Z}, +)$.
The set of all rotations in three-dimensional space forms a non-commutative group under the successive application as a composition of elements.

2.1.3 Ring

Ring: Triple $(M, \oplus, \otimes)$ of a set and two compositions $\oplus$ "addition" and $\otimes$ "multiplication," where the following holds:
Group property:
$(M, \oplus)$ is an Abelian group.
Associative law with respect to $\otimes$:

$$(a \otimes b) \otimes c = a \otimes (b \otimes c) , \qquad a, b, c \in M .$$

Distributive law:

$$\left.\begin{array}{r} a \otimes (b \oplus c) = (a \otimes b) \oplus (a \otimes c) \\ (a \oplus b) \otimes c = (a \otimes c) \oplus (b \otimes c) \end{array}\right\} , \qquad a, b, c \in M .$$

- **Existence of subtraction**: For $a, b \in M$ there is one and only one $s \in M$ with $s \oplus a = b$ and $a \oplus s = b$.

A ring is said to be commutative if the following applies in addition:
Commutative law with respect to $\otimes$:

$$a \otimes b = b \otimes a , \qquad a, b \in M .$$

□ The set of all integers with ordinary addition and multiplication $(\mathbb{Z}, +, \cdot)$ is a commutative ring.
The set of polynomials with ordinary addition and multiplication is a commutative ring.

2.1.4 Field

Field: Triple $(M, \oplus, \otimes)$ of a set and two compositions $\oplus$ and $\otimes$ where the following applies:

Group property $(M, \oplus)$:
$(M, \oplus)$ is an Abelian group with the neutral element zero (0; zero element).
Group property $(M\backslash\{0\}, \otimes)$:
$(M\backslash\{0\}, \otimes)$ is an Abelian group with the neutral element one (1; unit element).
Distributive law:

$$a \otimes (b \oplus c) = (a \otimes b) \oplus (a \otimes c) , \qquad a, b, c \in M .$$

- **Existence of division**: For $a, b \in M\backslash\{0\}$ there is one and only one $s \in M$ with $s \otimes a = b$ and $a \otimes s = b$.

□ The set of all real numbers with ordinary addition and multiplication is a field.

- $(\mathbb{R}, +, \cdot)$ is a field.
  $(\mathbb{C}, +, \cdot)$ is a field.

2.1.5 Vector space

Vector space: A set M of so-called **vectors** $\vec{x} \in M$ over a field $(\mathbb{K}, +, \cdot)$ with the following compositions defined for all $a, b \in \mathbb{K}$ and $\vec{x}, \vec{y} \in M$:
1. **Vector addition** $\oplus$:

$$\vec{x}, \vec{y} \in M \quad \Longrightarrow \quad \vec{x} \oplus \vec{y} \in M .$$

$(M, \oplus)$ must be an Abelian group.
2. **Scalar multiplication** $\odot$:

$$\vec{x} \in M, \ a \in \mathbb{K} \quad \Longrightarrow \quad a \odot \vec{x} \in M ;$$

The following laws apply:
Associative law

$$a \odot (b \odot \vec{x}) = (a \cdot b) \odot \vec{x}$$

1. First distributive law

$$(a \oplus b) \odot \vec{x} = (a \odot \vec{x}) \oplus (b \odot \vec{x})$$

2. Second distributive law

$$a \odot (\vec{x} \oplus \vec{y}) = (a \odot \vec{x}) \oplus (a \odot \vec{y}) .$$

▷ **Real vector space**: Vector space over the field $(\mathbb{R}, +, \cdot)$.
Complex vector space: Vector space over the field $(\mathbb{C}, +, \cdot)$.

□ Vector space of the three-dimensional vectors.

Re notation: Vectors are shown either, as in this case, in the form of $\vec{a}$, $\vec{x}$, or in the form of **a**, **x**.

For further explanations, see Chapter 6 on vector calculus.

2.1.6 Algebra

Algebra: Mathematical structure satisfying the properties of a vector space and a ring.

□ The set of all $n \times n$ matrices with ordinary matrix addition, and multiplication is an algebra.

2.2 Equations with one unknown

Equivalence of equations: Two equations are equivalent if they have the same fundamental set **and** the same solution set.

● **Isolation of the variable (unknown)**: Method for solving equations based on finding equivalent equations (**equivalence transformations**) that can be solved easily.

▷ At the end of each chain of equivalence transformations is an equation in which the variable is isolated on the left-hand side of the equation.

2.2.1 *Elementary equivalence transformations*

1. Interchange of the sides of an equation.

● By interchanging the sides of an equation, the new equation is equivalent to the old.

□ $x = 3$ and $3 = x$ are equivalent equations.

▷ This does not apply to inequalities.
$x > 3$ and $3 > x$ are **not** equivalent, and they are even contradictory.

2. Addition and subtraction of terms

● If the same term is added or subtracted on both sides of an equation, the new equation is equivalent to the old one.

□ $x = 3$ and $x + 4 = 3 + 4$ (or $x + 4 = 7$) are equivalent.
$x = 3$ and $x - x = 3 - x$ (or $0 = 3 - x$) are equivalent.

3. Multiplication and division of terms

● If both sides of an equation are multiplied or divided by the same term, which must not be zero, the new equation is equivalent to the old.

□ $x = 3$ and $x \cdot 4 = 3 \cdot 4$ (or $4x = 12$) are equivalent.
$x = 3$ and $x/3 = 3/3$ (or $\frac{1}{3}x = 1$) are equivalent.

▷ Caution is advisable when multiplying and dividing by terms which contain the variable itself.

□ $x = 3$ and $x(x + 1) = 3(x + 1)$ (or $x^2 + x = 3x + 3$) are equivalent only if for the second equation $x = -1$ is explicitly excluded as a possible solution.
$x^2 = 3x$ and $x^2/x = 3x/x$ (or $x = 3$). The division by x is permitted only if $x = 0$ is excluded, i.e., if $x = 0$ as a solution of $x^2 = 3x$ is known **before** the division. Only then are $x^2 = 3x$ and $x = 3$ equivalent.

4. Substitution

● If a term is substituted by a new variable, the solution of the original equation is obtained by solving the new equation and substituting back.

□ $x^2 - 6 = 3\,,$ substitution $y = x^2$
new equation: $y - 6 = 3\,,$ solution: $y = 9\,,$
back substitution: $x^2 = y = 9 \implies x_{1,2} = \pm\sqrt{y} = \pm\sqrt{9} = \pm 3.$

2.2.2 *Overview of the different kinds of equations*

Integral rational term: Sum or difference of integral powers of the variable. General term: $P(x) = a_n x^n + a_{n-1}x^{n-1} + \cdots + a_1 x + a_0\,, \quad n \in \mathbb{N}\,.$

□ $3x^2 + 4x - 12\,,\ 3x^{20}\,,\ 4$ are integral rational terms.

▷ If P is regarded as a function of the variable x, then P is also said to be a **polynomial**.

● Sums, differences, and products of integral rational terms are integral rational terms.

Integral rational equation: Equation expressing the equality of two integral rational terms:

$$P_1(x) = P_2(x) .$$

□ $2x^2 - 3 = 4x^3 - 12x + 1$ is an integral rational equation.

Fractional rational equation: Equation containing quotients of two integral rational terms:

$$\frac{P_1(x)}{P_2(x)} = \frac{P_3(x)}{P_4(x)} .$$

□ $$\frac{4x^2 - 3}{-3x^3 + 2} = 12x - 1 \text{ is a fractional rational equation}$$

Irrational equation: Equation containing terms with roots and/or rational powers of the variable.

□ $\sqrt{3x^2 - 2x} = x - 12$, $3x^2 + x - 3 = x^{2/3}$ are irrational equations.

Algebraic equations: All equations that can be converted to the basic form
$a_n x^n + a_{n-1}x^{n-1} + \cdots + a_1 x + a_0 = 0$
by an equivalence transformation, with the exception of possible restrictions in the domain of definition.

● Integral rational, fractional, and irrational equations are algebraic equations. They can be converted to the normal form by multiplication, exponentiation, and extraction of a root.

Transcendental equations: Equations containing expressions other than roots or powers, such as, $\lg x$, $\sin x$, e^x, or 3^x.

▷ A transcendental equation is not an algebraic equation.

2.3 Linear equations

Linear equations: The variable is restricted to the powers 0 (constant) or 1 (linear term).

2.3.1 *Ordinary linear equations*

Normal form of linear equations:

$$ax + b = 0,\ a \neq 0 \quad \text{with the solution} \quad x = -\frac{b}{a} .$$

● Linear equations can always be converted to the basic form.

□ $x + 2 = 2x - 1$ addition of 1,
$x + 3 = 2x$ subtraction of $2x$,
$-x + 3 = 0$ with the solution $x = 3$.

▷ Geometrical interpretation: abscissa of the point of intersection of two straight lines in the plane, or point of intersection of a straight line with the x-axis.

2.3.2 *Linear equations in fractional form*

Linear equations may be given in fractional form. In solving fractional equations, care must be taken not to divide by zero.

$$1 = \frac{1}{1 - x},\ x \neq 1$$

is equivalent to the linear equation

$$1 - x = 1.$$

2.3.3 Linear equations in irrational form

Linear equations may also contain roots. In solving irrational linear equations one has to take care that no negative radicands occur.

$$\sqrt{4-x} = 3, \; x \le 4$$

is equivalent to the linear equation
$4 - x = 9.$

▷ Taking the square of both sides of an equation is **not** an equivalence transformation because the basic set of permitted values for the variable may increase.

2.4 Quadratic equations

Quadratic equations contain not only a constant and/or linear term, but also the variable x at most to the second power.

$$ax^2 + bx + c = 0, \quad a \neq 0$$

Basic form of quadratic equations:

$$x^2 + px + q = 0 \quad \text{with} \quad p = \frac{b}{a}, \; q = \frac{c}{a}.$$

p-q-formula for the solution of quadratic equations:

$$x_{1,2} = -\frac{p}{2} \pm \sqrt{\left(\frac{p}{2}\right)^2 - q}, \qquad D = \left(\frac{p}{2}\right)^2 - q = \frac{p^2}{4} - q.$$

Discriminant D: Provides information about the type of solution:
1. $D > 0$ two distinct real solutions, $x_{1,2} = -p/2 \pm \sqrt{D}$;
2. $D = 0$ one real double solution, $x_1 = x_2 = -p/2$;
3. $D < 0$ two conjugate complex solutions, $x_{1,2} = -p/2 \pm \mathrm{j}\sqrt{-D}$.

A product representation of a quadratic equation

$$(x - x_1) \cdot (x - x_2) = 0$$

is always possible; the solutions x_1, x_2 are easy to read.

● **Vieta's root theorem**
Given the solutions x_1 and x_2 of the quadratic equation $x^2 + px + q = 0$, the sum of both solutions is equal to $-p$, and the product of both solutions is equal to q:
$x_1 + x_2 = -p$ and $x_1 x_2 = q$.

□ A quadratic equation can be taken to the basic form.
$3x^2 - 6x - 9 = 0$, dividing by 3 gives $x^2 - 2x - 3 = 0$, i.e., $p = -2$ and $q = -3$.
$D = 4 \Longrightarrow$ Case 1.
$x_{1,2} = 1 \pm \sqrt{4}$, so $x_1 = 3$ and $x_2 = -1$.
Vieta: $3 + (-1) = 2 = -p$ and $3 \cdot (-1) = -3 = q$.

▷ Geometrical interpretation: abscissas of the points of intersection of a parabola with a straight line, or points of intersection of a parabola with the x-axis, if $D \ge 0$.

2.4.1 Quadratic equations in fractional form

● A quadratic equation in fractional form can be converted to basic form by **term multiplication.**

□
$$\frac{3-x}{x+1} = 2x+3$$

1. Exclusion of $x = -1$ and multiplication by $x+1$:

$$3 - x = (2x+3)\cdot(x+1) = 2x^2 + 5x + 3\,.$$

2. Converting to normal form and solving:

$$0 = 2x^2 + 6x \implies 0 = x^2 + 3x\,,$$

with the solutions $x_1 = 0$ and $x_2 = -3$.

2.4.2 Quadratic equations in irrational form

● A quadratic equation containing radicals can be converted to normal form by **raising to the second power.**

▷ **Caution:** This may lead to **only spurious** solutions.

□ $\sqrt{-2x-3} = x+1$

1. Excluding $x \geq -3/2$, taking the square of both sides:

$$-2x - 3 = (x+1)^2 = x^2 + 2x + 1\,.$$

2. Converting to basic form and solving:

$$0 = x^2 + 4x + 4 = (x+2)^2$$

with the double solution $x_{1,2} = -2$.

A test by substituting into the initial equation shows that -2 is not a solution of the original equation.

2.5 Cubic equations

Cubic equation in basic form:

$$x^3 + ax^2 + bx + c = 0\,.$$

Substitution $y = x + a/3$ leads to the reduced form:

$$y^3 + py + q = 0 \quad \text{, with}$$
$$p = \frac{3b - a^2}{3} \quad \text{and} \quad q = c + \frac{2a^3}{27} - \frac{ab}{3}\,.$$

Discriminant:

$$D = \left(\frac{p}{3}\right)^3 + \left(\frac{q}{2}\right)^2$$

Cardano's formulas:

$$x_1 = -\frac{a}{3} + u + v\,,$$
$$x_{2,3} = -\frac{a}{3} - \frac{u+v}{2} \pm \sqrt{3}\,\frac{u-v}{2}\mathrm{j}\,,$$

where

$$u = \sqrt[3]{-\frac{q}{2} + \sqrt{D}} \quad \text{and} \quad v = \sqrt[3]{-\frac{q}{2} - \sqrt{D}}\,,$$

and

$$D = \left(\frac{p}{3}\right)^3 + \left(\frac{q}{2}\right)^2 \quad \text{is the discriminant.}$$

- **Vieta's root theorem for cubic equations**
  If x_1, x_2, x_3 are solutions of $x^3 + ax^2 + bx + c = 0$, then

$$a = -(x_1 + x_2 + x_3), \quad b = x_1x_2 + x_2x_3 + x_3x_1, \quad c = -x_1x_2x_3\,.$$

Classification of the solutions:
1. $D > 0$: one real and two complex conjugate solutions,
2. $D = 0$: three real solutions including a double solution;
3. $D < 0$: **irreducible case**, three distinct real solutions.

▷ For the irreducible case, complex expressions can be avoided with the **trigonometric form of solution**:

$$x_1 = -\frac{a}{3} + 2\sqrt{\frac{|p|}{3}}\cos\frac{\varphi}{3}\,,$$
$$x_2 = -\frac{a}{3} - 2\sqrt{\frac{|p|}{3}}\cos\frac{\varphi - \pi}{3}\,,$$
$$x_3 = -\frac{a}{3} - 2\sqrt{\frac{|p|}{3}}\cos\frac{\varphi + \pi}{3}\,,$$
$$\text{where} \quad \cos\varphi = -\frac{q}{2\sqrt{(|p|/3)^3}}\,.$$

□ Solving the equation $x^3 + 2x^2 - 5x - 6 = 0$

$$p = \frac{3\cdot(-5) - 2^2}{3} = -\frac{19}{3} \quad \text{and} \quad q = -6 + \frac{2\cdot 2^3}{27} - \frac{2\cdot(-5)}{3} = -\frac{56}{27}\,,$$
$$D = \left(-\frac{19}{9}\right)^3 + \left(-\frac{28}{27}\right)^2 \approx -8.333 < 0\,,$$
$$\cos\varphi = \frac{28}{27\sqrt{\left(\frac{19}{9}\right)^3}} \approx 0.338 \implies \varphi \approx 1.226\ \text{rad}\,,$$

$$x_1 = -\frac{2}{3} + 2\sqrt{\frac{19}{9}}\cos\phi_1 = 2, \qquad \phi_1 = \frac{\varphi}{3} \approx 0.409\ \text{rad}$$
$$x_2 = -\frac{2}{3} - 2\sqrt{\frac{19}{9}}\cos\phi_2 = -3, \qquad \phi_2 = \frac{\varphi - \pi}{3} \approx -0.639\ \text{rad}$$
$$x_3 = -\frac{2}{3} - 2\sqrt{\frac{19}{9}}\cos\phi_3 = -1, \qquad \phi_3 = \frac{\varphi + \pi}{3} \approx 1.456\ \text{rad}\,.$$

▷ Usually, cubic equations can be solved more quickly with numerical methods than with general analytical procedures.

2.6 Quartic equations

- There is also a general solution scheme for quartic equations. For equations of higher degree there are no general formulas by which the roots are expressed in terms of the coefficients only by means of algebraic operations (i.e., the four basic operations and taking the root).

2.6.1 General quartic equations

▷ The solution scheme is so complex that it is not widely used in modern computers, but it is described here nevertheless.

Solutions of the equation in **normal form** $x^4 + ax^3 + bx^2 + cx + d = 0$ agree with the solutions of

$$x^2 + \frac{a+D}{2} + \left(y + \frac{ay-c}{D}\right) = 0 .$$

where

$$D = \pm\sqrt{8y + a^2 - 4b} ,$$

and y is any **real** solution of the cubic equation

$$8y^3 - 4by^2 + (2ac - 8d)y + (4bd - a^2 d - c^2) = 0.$$

2.6.2 Biquadratic equations

Biquadratic equation: Special case of a quartic equation which is easy to solve.

$$x^4 + ax^2 + b = 0$$

▷ Solution method: substitution of $y = x^2$ and twofold solution of a quadratic equation.

Substitution: $y = x^2 \implies y^2 + ay + b = 0.$

Solution:

$$y_{1,2} = -\frac{a}{2} \pm \sqrt{\frac{a^2}{4} - b} .$$

Substituting back: $x = \pm\sqrt{y}$.
with the final result:

$$x_{1,2,3,4} = \pm\sqrt{y_{1,2}} = \pm\sqrt{-\frac{a}{2} \pm \sqrt{\frac{a^2}{4} - b}} .$$

▷ In general, the solutions are complex; i.e., negative radicands may occur.

2.6.3 Symmetric quartic equations

Symmetric quartic equation: Solvable special case of a quartic equation:

$$ax^4 + bx^3 + cx^2 + bx + a = 0 .$$

▷ Solution method: division by x^2 (possible if $a \neq 0$, otherwise the equation is not quartic at any rate), combination of terms, and substitution
$y = x + 1/x$.

Rewrite:

$$
\begin{aligned}
0 &= a\left(x^2 + \frac{1}{x^2}\right) + b\left(x + \frac{1}{x}\right) + c \\
&= a(y^2 - 2) + by + c \\
&= ay^2 + by + (c - 2a)\ .
\end{aligned}
$$

The final quadratic equation has the two solutions y_1 and y_2. In substituting back, the following quadratic equation must be solved in each case:

$$y_{1,2} = x + \frac{1}{x} \quad \Longleftrightarrow \quad x^2 - y_{1,2} \cdot x + 1 = 0\ .$$

▷ $y_{1,2}$ as well as the back substituted solutions of the original equation may be complex.

Solutions:

$$x_{1,2,3,4} = \frac{1}{2}\sqrt{y_{1,2} \pm \sqrt{y_{1,2}^2}} \quad \text{where} \quad y_{1,2} = \frac{1}{2a}\left(-b \pm \sqrt{b^2 - 4ac + 8a^2}\right)\ .$$

2.7 Equations of arbitrary degree

● **Fundamental theorem of algebra**
Every equation of degree n, $a_n x^n + a_{n-1}x^{n-1} + \cdots + a_1 x + a_0 = 0$, has exactly n solutions $x_1, x_2, \ldots, x_n$ in $\mathbb{C}$ if multiplicity is taken into account; i.e., multiple solutions may occur.

□ The equation $(x-2)^2 = 0$ has the solution $x_1 = x_2 = 2$, which occurs twice.

▷ When n is even, the equation does **not** need a real solution; when n is odd, the equation has **at least one** real solution.

For general algebraic equations (rational equations of degree n) there are **no general** solution methods.

2.7.1 Polynomial division

Factorization:

$$a_n x^n + a_{n-1}x^{n-1} + \cdots + a_1 x + a_0 = a_n(x - x_1)(x - x_2) \cdot \cdots \cdot (x - x_n)\,,$$

x_i, $i = 1, \ldots n$ are the solutions of the algebraic equations.

● Given a solution of an algebraic equation, a **linear factor** can be factored out and divided by the following, in analogy to the written division of decimal numbers.

□ $x = 4$ is solution of $x^4 - 4x^3 - 3x^2 + 10x + 8 = 0$. Factoring out $(x-4)$:

$$
\begin{array}{rrrrrl}
(x^4 & -4x^3 & -3x^2 & +10x & +8) & \div(x-4) = x^3 - 3x - 2 \\
-x^4 & +4x^3 & & & & \\ \cline{1-2}
 & 0x^3 & -3x^2 & +10x & & \\
 & & +3x^2 & -12x & & \\ \cline{2-4}
 & & & -2x & +8 & \\
 & & & +2x & -8 & \\ \cline{4-5}
 & & & & 0 &
\end{array}
$$

Thus:

$$x^4 - 4x^3 - 3x^2 + 10x + 8 = (x^3 - 3x - 2) \cdot (x - 4) .$$

- After $n - 1$ steps, successive factoring of linear factors leads to the factorized form. But $n - 1$ solutions must be known.

□ For the algebraic expression in the last example, the following applies:

$$\begin{aligned} x^4 - 4x^3 - 3x^2 + 10x + 8 &= (x - 4) \cdot (x + 1) \cdot (x + 1) \cdot (x - 2) \\ &= (x - 4) \cdot (x + 1)^2 \cdot (x - 2) . \end{aligned}$$

2.8 Fractional rational equations

General solution method:

1. Exclude all values of the variable values for which one denominator becomes zero.
2. Multiply the equation by the terms occurring in the denominator.
3. Convert the equation to basic form.
4. Solve the equation.

▷ Step 1 already implies the solution of an algebraic equation.

2.9 Irrational equations

General solution method:

1. Exclude all variable values for which a radicand can become negative.
2. Eliminate the roots and the powers with fractured exponents by **exponentiation** of the equation.
3. Convert the equation to basic form.
4. Solve the equation and check.

▷ The first step must be performed if **only real** solutions are wanted for the equation. For complex solutions, negative radicands are also permitted.

▷ In rewriting irrational equations new solutions may occur; therefore, a check should **always** be conducted.

2.9.1 Radical equations

Radical equations with one root

Solution method: Isolate the root and take the square (raise to a power).

□ Solving $\sqrt{x+7} = x + 1$:
Domain of definition: $\mathbb{D} = \{x \in \mathbb{R} \mid x \geq -7\}$.
Taking the square: $x + 7 = x^2 + 2x + 1 \mid -x - 7$.
Solving $0 = x^2 + x - 6$.
Solution: $x_1 = 2, \quad x_2 = -3$.
Check: $\sqrt{2+7} = \sqrt{9} = 3 = 2 + 1$ and $\sqrt{-3+7} = \sqrt{4} = 2 \neq -3 + 1 = -2$.
Set of solutions of $\sqrt{x+7} = x + 1$: $\mathbb{L} = \{2\}$.

▷ Taking the square/raising to a power can lead to non-equivalent equations. Further **apparent** solutions may occur. A check should **always** be conducted!

Radical equations with two roots

Solution method:

1. Isolate one root.
2. Take the square of the equation.
3. Isolate the remaining roots.
4. Take the square again.

□ Solving $3+\sqrt{x+3}=\sqrt{3x+6}+2$:
Domain of definition: $\mathbb{D}=\{x\in\mathbb{R}\mid x\geq -2\}$.

$$\begin{aligned}3+\sqrt{x+3}&=\sqrt{3x+6}+2 \mid \quad -2\\ 1+\sqrt{x+3}&=\sqrt{3x+6} \mid \quad \text{taking the square}\\ 1+2\sqrt{x+3}+(x+3)&=3x+6 \mid \quad -4-x\\ 2\sqrt{x+3}&=2x+2 \mid \quad \div 2\\ \sqrt{x+3}&=x+1 \mid \quad \text{taking the square}\\ x+3&=x^2+2x+1 \mid \quad -x-3\\ 0&=x^2+x-2\\ x_1=-2,&\quad x_2=1\end{aligned}$$

$$\text{Check:}\quad 3+\sqrt{-2+3}=4\neq\sqrt{3\cdot(-2)+6}+2=2\,,$$
$$3+\sqrt{1+3}=5=\sqrt{3\cdot 1+6}+2=5\,.$$

Set of solutions: $\mathbb{L}=\{1\}$.

2.9.2 *Power equations*

● If an equation contains powers with rational exponents, it must be converted to the basic form of an algebraic equation by raising to an appropriate power.

□ Rewriting $x^{2/3}-1=x$.

$$\begin{aligned}x^{2/3}-1&=x \mid \quad \text{isolating the power } (+1)\\ x^{2/3}&=x+1 \mid \quad \text{raising the equation to the third power } ()^3\\ x^2&=(x+1)^3 \mid \quad \text{multiplying and rewriting}\\ 0&=x^3+2x^2+3x+1\end{aligned}$$

2.10 Transcendental equations

Transcendental equation: Contains expressions that are more complicated than powers.

□ $\lg x=0$ and $\sin^2 x-x=4$ are transcendental equations.

▷ In general, there are no systematic methods of solving transcendental equations. Transcendental equations can often be solved only approximately with numerical methods.

2.10.1 *Exponential equations*

Equation containing one exponential expression

Structure: $a^T=S$, $\quad T, S$ are terms.

□ $3^x=x+5$ is an exponential equation.

Solvable special case: $S = b$ is constant and T is an integral rational term.
Solution method: Take the logarithm of the equation.

□ Solving the equation $2^{x^2-1} = 8$:

$$\begin{aligned} 2^{x^2-1} &= 8 \mid \quad \text{taking the logarithm} \\ (x^2-1)\lg 2 &= \lg 8 \mid = 3 \div \lg 2 \text{ (note: } \lg 8 = \lg 2^3 = 3 \cdot \lg 2) \mid \quad \text{divide by } \lg 2 \\ x^2 - 1 &= \frac{3 \cdot \lg 2}{\lg 2} \mid \quad +1 \\ x^2 &= 4 \mid \quad \text{taking the root} \\ x_{1,2} &= \pm 2 \,. \end{aligned}$$

▷ In taking the logarithm, ensure that both sides of the equation are > 0.

Equation with two exponential expressions
Structure: $a^T = b^S$, S, T are terms.
Solving the equation $2^{3x+1} = 4^{x-1}$:

$$\begin{aligned} 2^{3x+1} &= 4^{x-1} \mid \quad \text{taking the logarithm} \\ (3x+1)\lg 2 &= (x-1)\lg 4 \mid \quad \div \lg 2 \\ 3x+1 &= 2(x-1) \mid \quad -2x-1 \\ x &= -3 \,. \end{aligned}$$

2.10.2 *Logarithmic equations*

Equation containing the logarithm
Structure: $\log_a T = S$, S, T are terms.
Solvable special case: $S = b$ is constant and T is an integral rational term.
Solution method: Exponentiation of the equation.

□ Solving the equation $\log_2(x^2 - 1) = +3$:
Domain of definition: $\mathbb{D} = \{x \in \mathbb{R} \mid \text{abs}(x) > 1\}$.

$$\begin{aligned} \log_2(x^2-1) &= 3 \mid \quad \text{exponentiation } (2^{\cdots}) \\ x^2 - 1 &= 2^3 \mid \quad +1 \\ x^2 &= 9 \mid \quad \text{taking the root} \\ x_{1,2} &= \pm 3 \end{aligned}$$

Equation containing two logarithms
Structure: $\log_a T = \log_b S$, S, T are terms.
Solvable special case: $a = b$ and S, T are integral rational terms.

□ Solving the equation $\log_a(x+1) = \log_a(2x+4)$:
Domain of definition: $\mathbb{D} = \{x \in \mathbb{R} \mid x > -1\}$.

$$\begin{aligned} \log_a(x+1) &= \log_a(2x+4) \mid \quad \text{exponentiation } (a^{\cdots}) \\ x + 1 &= 2x + 4 \mid \quad -4 - x \\ -3 &= x \qquad \Longrightarrow \mathbb{L} = \{\ \}, \text{ because } -3 \notin \mathbb{D} \end{aligned}$$

▷ In case of logarithmic equations observe the exact domain of definition, since it is enlarged by exponentiation.

2.10.3 Trigonometric (goniometric) equations

Trigonometric (or goniometric) equation: Contains angle functions of the variable.

- ● Due to the periodicity of angle functions, trigonometric equations are generally of infinite ambiguity, i.e., there is there are an infinite number of solutions.
- □ Equation $\sin x = 0$ has the solutions $x = z \cdot \pi$ with $z \in \mathbb{Z}$.
- ● A simplification of trigonometric equations can be achieved by using the relations between trigonometric functions listed in Chapters 3 and 5, under "Angle Functions."

2.11 Equations with absolute values

- ● Usually, **equations with absolute values** are ambiguous because for all terms T: $|-T| = |+T|$.

Case distinction: Solution method for equations, assuming once a **positive** and once a **negative** argument within the absolute value signs.

2.11.1 Equations with one absolute value

Basic form: $S = |T|$, S, T are terms.
Solution method: Case distinction once.
1. Consider $S = +T$, for $T \geq 0$.
2. Consider $S = -T$, for $T < 0$.

□ Solving $2x + 3 = |x - 2|$:
Case 1: $2x + 3 = +(x - 2)$, for $x \geq 2$,
with the **apparent** solution $x_1 = -5$, because $(x_1 \not\geq 2)$;
Case 2: $2x + 3 = -(x - 2) = -x + 2$, for $x < 2$,
with **correct** solution $x_2 = -1/3$, because $(x < 2)$.

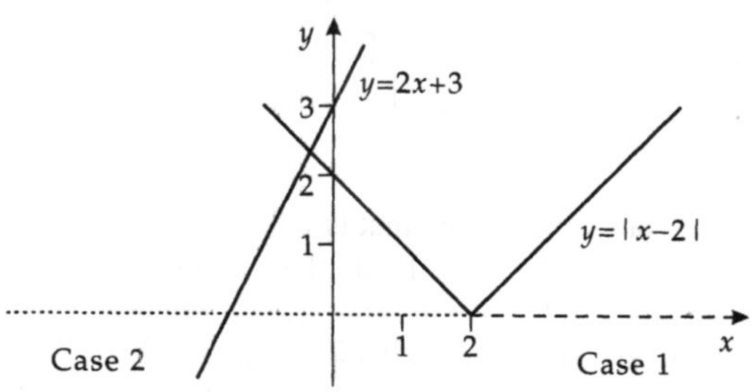

▷ The solution of an equation with absolute values by case distinction may lead to **apparent solutions**. Therefore, one has to ensure that the solution found lies in the interval considered, or one has to perform a check.

The task becomes more complicated if the absolute values of higher powers must be calculated.

□ Solving $1 = |x^2 - 2|$:
Case 1: $1 = +(x^2 - 2)$, for $x^2 \geq 2$,
with the solutions $x_{1,2} = \pm\sqrt{3}$,
check: $1 = |(\pm\sqrt{3})^2 - 2| = |3 - 2|$;
Case 2: $1 = -(x^2 - 2)$, for $x^2 < 2$,
with the solutions $x_{3,4} = \pm 1$,
check: $1 = |(\pm 1)^2 - 2|$.
All four solutions are proper solutions.

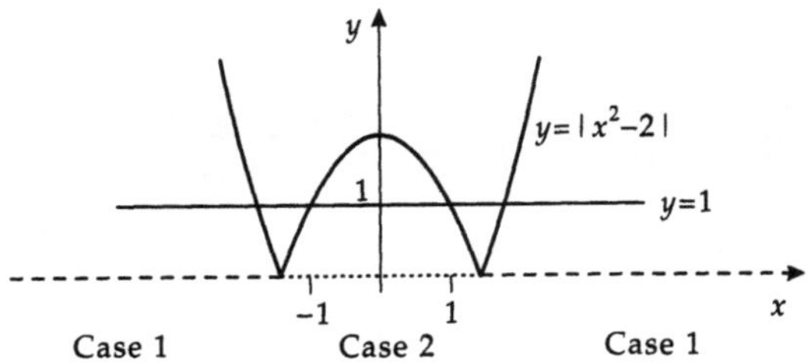

2.11.2 Equations with several absolute values

Solution method: By means of multiple case distinction, the domain of real numbers $\mathbb{R}$ is subdivided into several, sometimes improper intervals.

□ Solving $|x + 1| - 2 = |2x - 4|$:

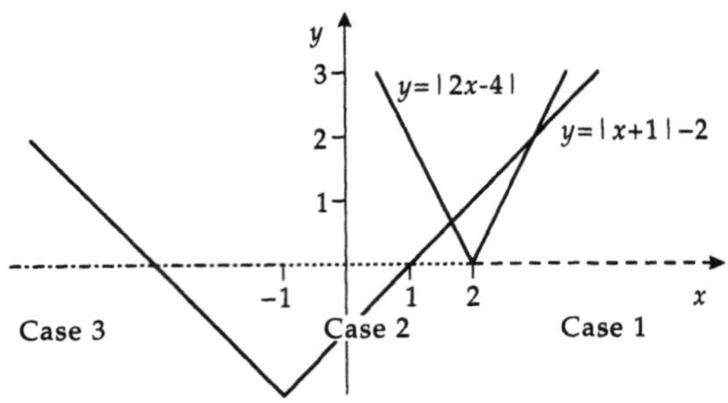

Case 1: $x + 1 \geq 0$ and $2x - 4 \geq 0$, that is $x \geq 2$,
solving: $+(x + 1) - 2 = +(2x - 4)$ with the solution $x_1 = 3$,
check: $|3 + 1| - 2 = 2 = |2 \cdot 3 - 4| \implies x_1$ is a solution;
Case 2: $x + 1 \geq 0$ and $2x - 4 < 0$, that is $-1 \leq x < 2$,
solving: $+(x + 1) - 2 = -(2x - 4)$ with the solution $x_2 = 5/3$,
check: $|5/3 + 1| - 2 = 2/3 = |2 \cdot 5/3 - 4| \implies x_2$ is a solution;
Case 3: $x + 1 < 0$ and $2x - 4 < 0$, that is $x < -1$,
solving: $-(x + 1) - 2 = -(2x - 4)$ with the solution $x_3 = 7$,
hence, $x_3 = 7 \not< -1$, cannot be a solution:
check: $|7 + 1| - 2 = 6 \neq 10 = |2 \cdot 7 - 4| \implies x_3$ is not a solution;
Case 4: $x + 1 < 0 \implies x < -1, 2x - 4 \geq 0 \implies x \geq 0, \implies$ there is no x-region containing a solution.

2.12 Inequalities

Inequality: Two terms S and T linked by one of the **relational symbols** $<$, $>$, $\leq$, $\geq$, or by the inequality sign $\neq$.

In general, the solutions of inequalities are unions of intervals.

● Inequalities can be converted similarly to, but not exactly as equations.

2.12.1 *Equivalence transformations for inequalities*

1. Interchanging the sides of inequalities

● When both sides of an inequality are interchanged, the new inequality becomes equivalent to the old if the relational sign is reversed.

□ $1 < x$ and $x > 1$ are equivalent inequalities.

▷ Too often, inequalities such as $1 < x$ and $x < 1$ are regarded as equivalent. This is **wrong**.

2. Addition and subtraction of terms

● When the same term is added or subtracted on both sides of an inequality, the new inequality is equivalent to the old without having to reverse the relational sign.

□ $x - 2 < 4$ and $x - 2 + 2 < 4 + 2$, or $x < 6$, are equivalent inequalities.

3. Multiplication and division by terms

● When both sides of an inequality are multiplied or divided by the same **always positive** term, the new inequality and the old are equivalent.
If the term is **always negative**, the old inequality and the new are equivalent if and only if the relational sign is reversed.

▷ As in the case of equations, multiplication or division by zero is **not permitted**!

● A **case distinction** must be carried out if the term by which the inequality is multiplied or divided can assume **positive as well as negative** values:
If $a < b$ must be multiplied by a term T, the following must be considered:
Case 1: $a \cdot T < b \cdot T$, for $T > 0$.
Case 2: $a \cdot T > b \cdot T$, for $T < 0$.

□ Solving the inequality $x - 1 < (x - 1)^{-1}$:
1. $x = 1$ must be excluded as a possible solution;
2. **Case 1**: $x - 1 > 0$, thus $x > 1$. Multiplication by $x - 1$, the relational sign does not have to be reversed;
$(x - 1)^2 < 1$ has the solution $1 < x < 2$ in the interval considered;
3. **Case 2**: $x - 1 < 0$, thus $x < 1$. Multiplication by $x - 1$, the relational sign must be reversed;
$(x - 1)^2 > 1$ has the solution $x < 0$ in the interval considered;
4. The inequality has all real numbers x with $x < 0$ or $1 < x < 2$ as a solution, $\mathbb{L} = \{x \mid x < 0;\ 1 < x < 2\}$.

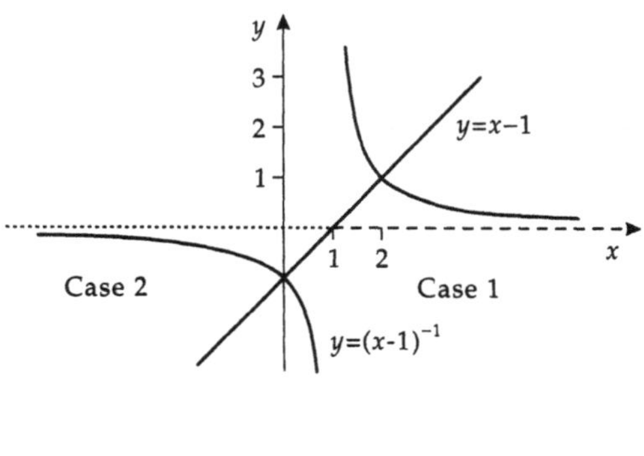

2.13 Numerical solution of equations

Determination of zeros: Solution method for equations of the form $T = 0$, where T is an algebraic or transcendental term.

- ● Every equation can be reduced to finding zeros because $T_1 = T_2$ is always equivalent to $T_1 - T_2 = 0$.

□ $e^x = x$ and $e^x - x = 0$ are equivalent.

2.13.1 *Graphical solution*

Graphical solution method: Both sides of an equation are viewed separately as a function of the variable and the corresponding graphs are drawn in a **common coordinate system**.

- ● Solutions are just the **values of the abscissas of the points of intersection** of both graphs.

▷ Although not accurate, graphical solutions provide a reasonable indication as to where the solutions can be found and where case distinctions may have to be carried out.

2.13.2 *Nesting of intervals*

Nesting of intervals: Method for successive restriction of a zero of a continuous function or of a solution of an equation.

- ● If the end-points of an interval are known in which **one and only one** zero of a continuous function lies, the location of the zero can be defined by successive bisection of the interval.

▷ If the continuous function has different signs at the end-points of an interval I, at least one zero occurs with $f' \neq 0$ in I.

Solving the equation $T(x) = 0$.

Sequence of the procedure:

Step 1: Fix two end-points, x_1 and x_2 so that at least one zero lies in the interval $[x_1, x_2]$; i.e., $T(x_1) < 0$ and $T(x_2) > 0$ or $T(x_1) > 0$ and $T(x_2) < 0$, thus certainly $T(x_1) \cdot T(x_2) < 0$.

Step 2: New intervals with $x_3 = (x_2 + x_1)/2$, check whether $T(x_3) \cdot T(x_1) < 0$ or $T(x_3) \cdot T(x_2) < 0$. In the first case choose $[x_1, x_3]$ otherwise $[x_3, x_2]$ as the new interval and iterate until the interval is sufficiently small.

▷ Program sequence for finding the zero of a function f in the interval $[x_u, x_o]$ by nesting of intervals.

```
BEGIN nesting of intervals
INPUT xu, xo ( let be f(xu)*f(xo)<0, xu<xo )
INPUT eps (limit of error)
error := (xo - xu)/2
WHILE (error > eps) DO
    xnew := (xo + xu)/2
    IF (f(xnew)*f(xu) < 0) THEN
       xo := xnew
    ELSE
       xu := xnew
    ENDIF
    error := 0.5*error
```

```
ENDDO
OUTPUT (xo+xu)/2
END nesting of intervals
```

2.13.3 Secant methods and method of false position

Secant method and **method of false position (regula falsi):** Iteration method for the approximate determination of the zeros of continuous functions f.
Idea of the methods: The new approximation is the zero of the secant of both preceding approximations.
Method of false position: A special secant method for which it is additionally required that the wanted zero lies always between two actual approximations, $f(x_{n-1})\cdot f(x_n) < 0$.

▷ Once the inclusion of the zero is achieved, this can be maintained in the next step: As a rule, the secant method becomes the method of false position.

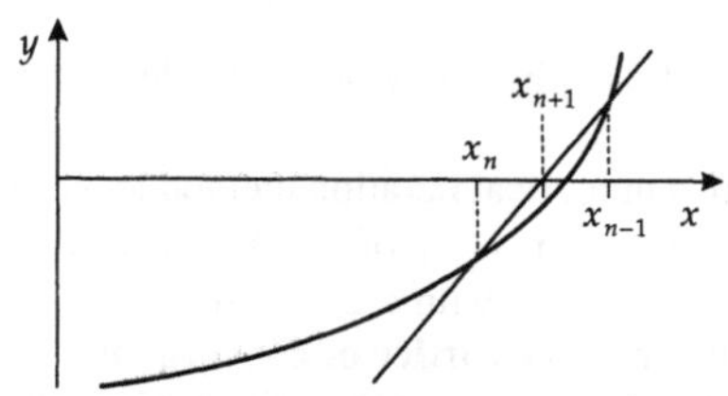

Principle of the method of false position

- **Iteration procedure for both methods**:

$$x_{n+1} = x_n - f(x_n)\frac{x_n - x_{n-1}}{f(x_n) - f(x_{n-1})} .$$

▷ Program sequence for the determination of the zero of a function f by the method of false position with the required accuracy ϵ.

```
  BEGIN Method of false position
INPUT xu, xo ( let be f(xu)*f(xo)<0, xu<xo )
INPUT eps (limit of error)
error = xo - xu WHILE (error > eps) DO
    xnew := xo - f(xo)*(xo - xu)/(f(xo) - f(xu))
    IF (f(xnew)*f(xu) < 0) THEN
      xo := xnew
    ELSE
      xu := xnew
    ENDIF
    error := abs(xo-xu)
ENDDO
OUTPUT xnew
END Method of false position
```

▷ In general, the method of false position and the related Newton's method converge often faster than the method of successive approximation.

2.13.4 Newton's method

Newton's method: Iteration method for the approximate determination of the zeros of a differentiable function f. The zero of the tangent to the curve at the old approximate value results in the new approximate value.

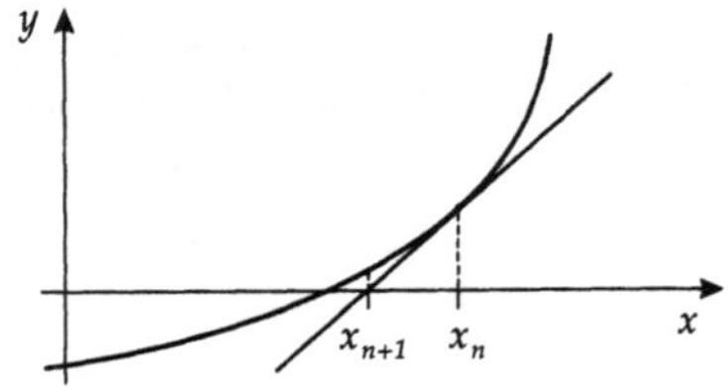

Principle of Newton's method

- **Iteration procedure**:

$$x_{n+1} = x_n - \frac{f(x_n)}{f'(x_n)} .$$

▷ Newton's method can be applied only when the derivative f' of the function f is known.

Order of convergence: In a numerical iteration method let ϵ_n and ϵ_{n+1} be the deviations from the required exact solution $x = x_l$ of $f(x) = 0$ in the n-th or $(n+1)$-th step. $x_n = x_l + \epsilon_n$, $x_{n+1} = x_l + \epsilon_{n+1}$. with $\epsilon_{n+1} = A\epsilon^p$.
The quantity p is called the method's **order of convergence**.

- Sufficient condition for the convergence of Newton's method:

$$\left| \frac{f(x) f(''(x)}{[f(x)]^2} \right| \leq q , \quad q < 1 , \qquad x \in U$$

for all x values in the neighborhood U of the zero that contains all points x_n. In some cases the method converges slowly or not converges at all. Then, the starting value x_0 is too far from the zero, or the zero is a multiple zero of f.

▷ If the zero x_l of f is simple, i.e., if $f(x_l) = 0$ and $f'(x_l) \neq 0$, then Newton's method converges locally quadratic,

$$\epsilon_{n+1} = \epsilon_n^2 \frac{f''(x_l)}{2f'(x_l)} .$$

In the case of multiple zeros Newton's method converges only linearly,

$$\epsilon_{n+1} = A\epsilon_n , \quad |A| < 1.$$

▷ Program sequence for finding the zeros of a function f by Newton's method with the required accuracy ϵ. The derivative of f is denoted f'.

```
BEGIN Newton
INPUT x0 (first approximation to the zero)
INPUT eps (limit of error)
x := x0
error := eps + 1
WHILE (error > eps) DO
    y := x - f(x)/f'(x)
    error := abs(y-x)
    x := y
```

```
ENDDO
OUTPUT x
END Newton
```

2.13.5 Successive approximation

Fixed point of a mathematical expression $T(x)$: The value x_0 to which $T(x_0) = x_0$ applies.
Successive approximation: Iteration method for the finding **fixed points** $x = T(x)$.

$$x_{n+1} = T(x_n), \quad n = 0, 1, 2, \ldots$$

□ Solving $x - \cos x = 0$: consider $\cos x = x$. Starting value: 1
1.$\cos 1.000 = 0.540$
2. $\cos 0.540 = 0.858$
3. $\cos 0.858 = 0.654$
4. $\cos 0.654 = 0.793$
⋮ ⋮
17. $\cos 0.738 = 0.740$
18. $\cos 0.740 = 0.739$
19. $\cos 0.739 = 0.739$ (fixed point accurate to 3 decimals)

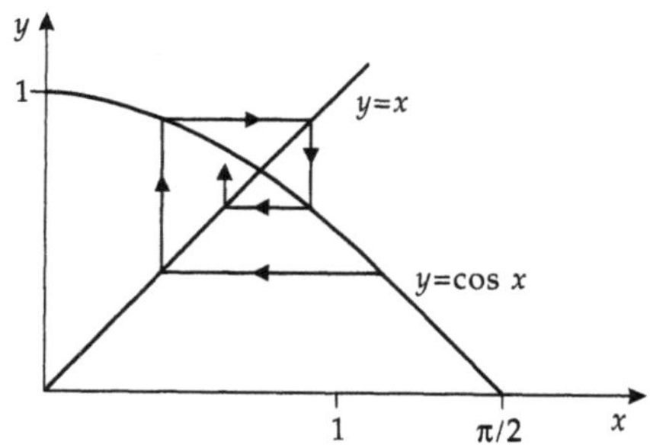

Example of a successive approximation

- **Banach's fixed point theorem**: If I is a closed interval and T a mapping, such that for all $x \in I$ one has $T(x) \in I$ and if T satisfies a **Lipschitz condition** or **contraction condition**

$$|T(x_2) - T(x_1)| \leq L \cdot |x_2 - x_1| \quad \text{with} \quad L < 1 \text{ for all } x_1, x_2 \in I\ ,$$

then a uniquely determined fixed point $x \in I$, $x = T(x)$ exists.
- Under the conditions of Banach's fixed point theorem, the method of successive approximation always converges to the (unique) fixed point of T.

▷ If I is a closed and bounded interval, and if T is differentiable, then $L = \max\{|T'(x)|,\ x \in I\}$.

▷ Program sequence for solving $\cos x = x$ by successive approximation.

```
BEGIN Successive Approximation
INPUT eps (required absolute accuracy)
x0 := 1
error := 1
WHILE (error > eps) DO
    x1 := cos(x0)
    error := abs(x1 - x0)
```

```
    x0 := x1
ENDDO
OUTPUT x0
END Successive Approximation
```

- Fixed point of $S(x)$ is the solution of the equation. Therefore, if the solution of the equation $T(x) = 0$ is wanted, an alternative to the direct solution can be to find the fixed point of $S(x) = T(x) + x$ by means of successive approximation.

▷ Often, equations cannot be solved by means of successive approximation since the corresponding $S(x)$ does **not** satisfy a Lipschitz condition.

Survey of numerical methods for the determining zeros:

Method	Number of starting values	Convergence	Stability	Applicability	Remarks
nesting of intervals	2	slow	always convergent	universal	
method of false position	2	moderate	always convergent	universal	
Newton's method	1	fast	not always convergent	limited if $f' = 0$	f' required
secant method	2	moderate to fast	not always convergent	universal	initial values need not include the zero

3

Geometry and trigonometry in the plane

While the basic elements of arithmetic are numbers, the basic elements of geometry are the point, the straight line, the plane, and the angle. **Point** (A, B, ...): Nondimensional, without any size, the intersection of two lines.

Line: One-dimensional point set: the translation of a point, or the intersection of two planes.

□ **Curved lines** (curves, circular lines, parabolas, hyperbolas, curves of higher order) or **straight lines**.

Straight line (g): A line unbounded on both sides.

Ray (s): A line bounded on one side.

Segment ($\overline{AB}$): A straight line bounded on both sides, distance between the points A and B.

Vector ($\vec{\mathbf{a}}$, $\vec{\mathbf{b}}$, ...): A directed segment. Represented by its initial point and end-point and capped with an arrow, or by bold (lowercase or capital) letters ($\overrightarrow{AB}$, $\vec{c}$, **a**, $\vec{\mathbf{b}}$).

Plane: Two-dimensional point set which arises from shifting a straight line along a second, nonparallel straight line, or as the bounds of certain bodies, the polyhedrons (wedge, cube, etc.).

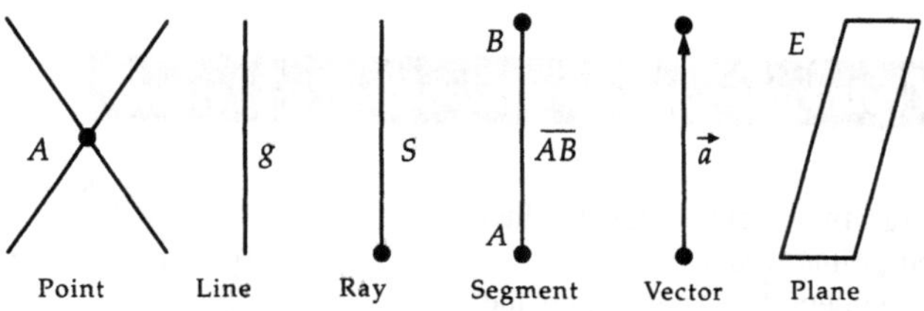

Angle (α, β, γ, ...): Results from turning (**rotating**) a ray about a point called the **vertex**; the angle measures the directional difference between two rays. Also, ratio between the arc rotation and the radius (arc measure).

Direction of rotation: Mathematically **positive angle**: counterclockwise direction of rotation; mathematically **negative angle**: clockwise direction of rotation.

3.1 Point curves

Point curves or conditional lines:
Those lines on which all points meet a certain condition.
Examples of point curves are:

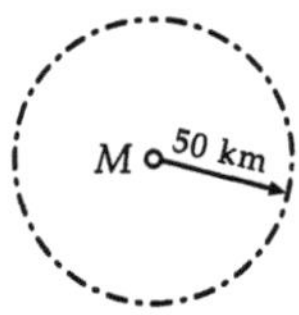

1. **Circle**: All points whose distance r from a fixed position, the **center** M is the same.

□ A radio station has a range of 50 km. All places where the station can be received lie within a circle of 50 km. radius

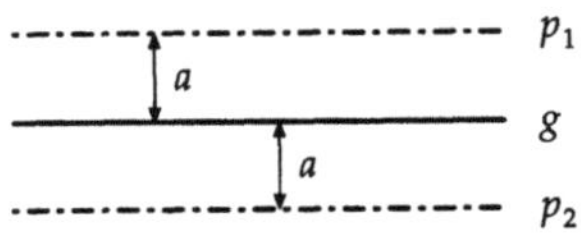

2. **Parallels** p_1, p_2: All points equidistant from a straight line g. Distance: a.

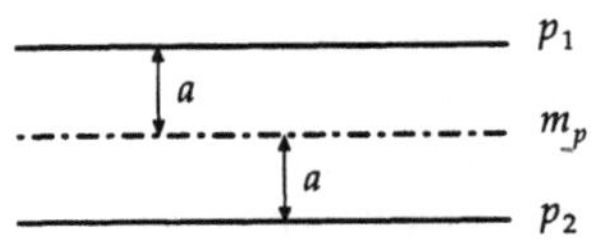

3. **Mid-parallel** m_p: All points equidistant from both parallels p_1 and p_2.
4. **Angle-bisector** w: All points equidistant from the straight lines g_1 and g_2 (the arms of an angle).

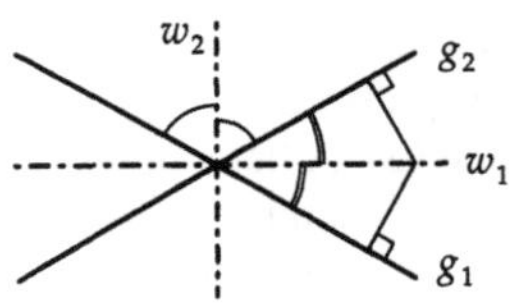

▷ Two intersecting geometrical point curves are required to fix a point (their point of intersection) in a plane.

3.2 Basic constructions

The basic constructions described here can be performed by the successive use of a compass and a ruler.

3.2.1 Construction of the midpoint of a segment

Compass: Draw two circles of equal radius about the end-points A and B.
Ruler: The segment $\overline{SS'}$, connecting the two points of intersection S, S', intersects $\overline{AB}$ in midpoint C.

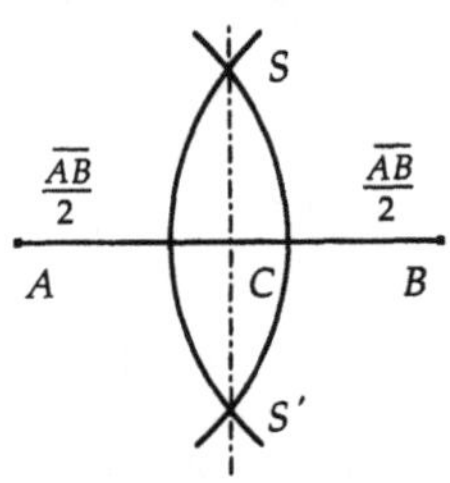

3.2.2 Construction of the bisector of an angle

Compass: A circular arc drawn about the vertex S intersects the arms in points A, B. Two further circular arcs of equal radius about the points A and B intersect in point C.
Ruler: Segment $\overline{SC}$ bisects the angle α.

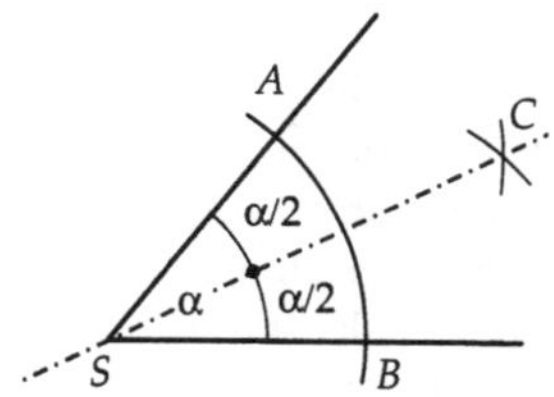

□ Bisection of a right angle yields an angle of 45°. Continued bisection yields 22.5°, etc. Bisection of an angle in an equilateral triangle ($\alpha = \beta = \gamma = 60°$) yields 30°. Continued bisection yields 15°, etc.

3.2.3 Construction of perpendiculars

Compass: Construct a semicircle about a point P on a straight line. The semicircle intersects straight line g in points A and B. Two further semicircles of equal radius about points A and B intersect in point C.
Ruler: The straight line through points P and C is perpendicular to the straight line g in point P.

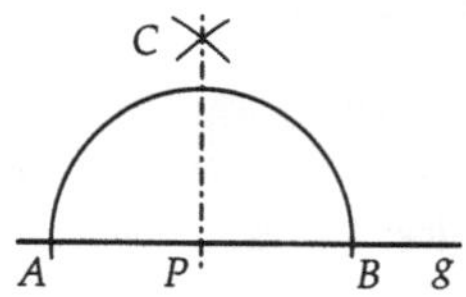

3.2.4 To drop a perpendicular

Compass: A semicircle about a point P not on the straight line g intersects g in the A and B. Two further semicircles of equal radius about points A and B intersect in point P'.
Ruler: The straight line through points P and P' is perpendicular to straight line g in point P.

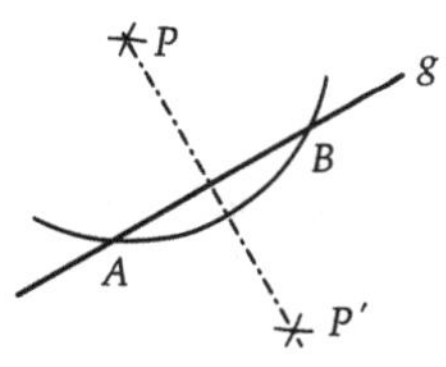

3.2.5 Construction of parallels at a given distance

Compass, ruler: Construct the perpendiculars to straight line g in two arbitrary points A and B ($A \neq B$) on g.
Ruler: On both of these perpendiculars mark an equal distance a ($|\overline{AD}| = |\overline{BC}| = a$). Segment $\overline{CD}$ is one of the parallels to a straight line g at a given distance a (construction of the rectangle with vertices $ABCD$).

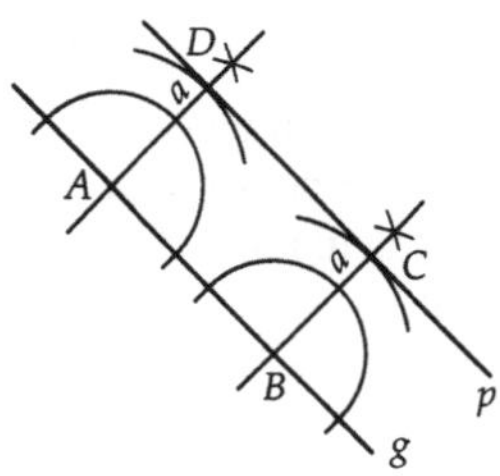

3.2.6 Parallels through a given point

Compass: A circular arc about point P not on the straight line g and with radius r intersects the straight line g in point C. A further circular arc of equal radius r about point C intersects straight line g in point B. A third circular arc of equal radius r about point B intersects the first circular arc about point P in point A.
Ruler: The extension of segment $\overline{PA}$ is the parallel to the straight line g through point P (construction of a rhombus with the vertices $PCBA$).

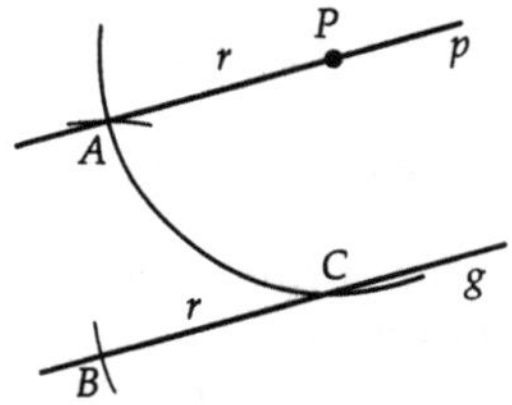

3.3 Angles

3.3.1 Specification of angles

Plane angles: The ratio between the length of the described circular arc b and the radius r. Plane angles are given in **degree measure** or in **arc measure**. A full revolution amounts to 360° in degree measure and to 2π in arc measure.

- **Degree measure**, φ in degrees, is used most often in geometry.

$$1\ \textbf{degree} = 1^\circ = 1/360\,\text{full rotation} \mathrel{\hat{=}} 0.01745\ \text{rad}$$

- **Arc measure**, $x = \dfrac{\partial b}{\partial r} = \text{arcus}\ \varphi = \text{arc}\ \varphi$, used mostly for trigonometric functions (sin, cos, etc.). For $b = r$ it follows that

$$1\ \textbf{radian} = 1\ \text{rad} = \frac{r}{r} = 1 \mathrel{\hat{=}} 180^\circ/\pi = 57.29578^\circ$$

▷ 1 rad corresponds to the angle at which the circular arc of the rotation equals the radius of the circle.

The angle in arc measure is a quantity of dimension 1. The supplement "rad" has been introduced only to distinguish it from other quantities which also have dimension 1.
The degree measure is subdivided as in time into **minutes** and **seconds**

$$1^\circ = 60' \quad \text{(60 minutes)}$$
$$1' = 60'' \quad \text{(60 seconds)}$$
$$1^\circ = 3600''$$

▷ In computers, fractions of an angle are given mostly as decimals.
When using pocket calculators, it is easy to forget switching from rad to degrees. **Before** calculating, check which unit should be used.

Conversions:
From arc measure x to degree measure φ:

$$\varphi = \frac{180^\circ}{\pi} \cdot x\,.$$

From degree measure φ to arc measure x:

$$x = \frac{\pi}{180^\circ} \cdot \varphi \,.$$

Conversion table: degree measure to arc measure:

one twelfth of a full revolution	$=$	30°	$\hat{=}$	$\pi/6$
one eighth of a full revolution	$=$	45°	$\hat{=}$	$\pi/4$
one sixth of a full revolution	$=$	60°	$\hat{=}$	$\pi/3$
one quarter of a full revolution	$=$	90°	$\hat{=}$	$\pi/2$
one-half of a full revolution	$=$	180°	$\hat{=}$	π
three-quarters of a full revolution	$=$	270°	$\hat{=}$	$3\pi/2$
one full revolution	$=$	360°	$\hat{=}$	2π
two full revolutions	$=$	720°	$\hat{=}$	4π
three full revolutions	$=$	1080°	$\hat{=}$	6π
n full revolutions	$=$	$n \cdot 360^\circ$	$\hat{=}$	$n \cdot 2\pi$

▷ Due to rounding errors in pocket calculators, the input in degrees is often more precise than the input in radians.

▷ In surveying, angles are given in **gon** (**grades**). A full angle $\varphi = 400$ grades;

$$1 \text{ gon} = 1 \text{ grade} = \frac{1}{400} \quad \text{of a full angle in surveying.}$$

3.3.2 *Types of angles*

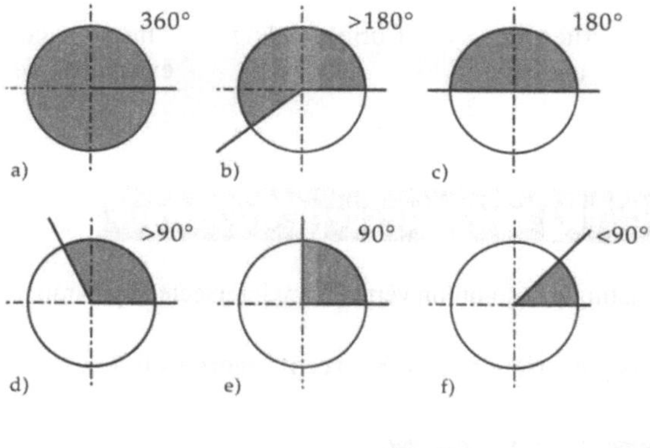

a) **full** angle	$\alpha = 360^\circ \hat{=} 2\pi$	d) **obtuse** angle	$90^\circ < \alpha < 180^\circ$
b) **reflex** angle	$180^\circ < \alpha < 360^\circ$	e) **right** angle	$\alpha = 90^\circ \hat{=} \frac{\pi}{2}$
c) **straight** angle	$\alpha = 180^\circ \hat{=} \pi$	f) **acute** angle	$0^\circ < \alpha < 90^\circ$

3.3.3 Angles between two parallels

- **Adjacent angles**: Neighboring angles between two intersecting straight lines, supplementing to 180° :

$$\alpha + \alpha' = 180^\circ \ .$$

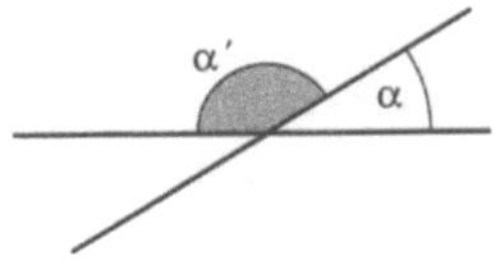

Adjacent angles

- **Vertically opposite angles**: Opposite angles between two intersecting straight lines are equal.
- **Supplementary angles**: Two angles whose sum is 180°: $\alpha + \beta = 180^\circ$.

□ The supplementary angle of $\alpha = 80^\circ$ is $180^\circ - 80^\circ = 100^\circ$.

- **Complementary angles**: Two angles whose sum is 90°.

□ The complementary angle of $\alpha = 30^\circ$ is $90^\circ - 30^\circ = 60^\circ$.

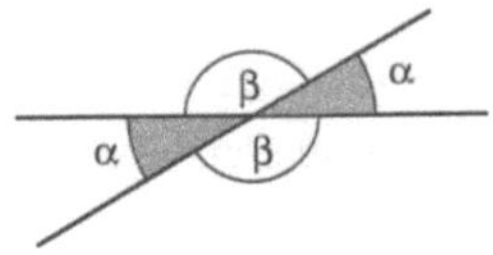

Opposite angles and supplementary angles

- **Alternate angles**: Opposite angles between two parallels and a straight line.
- **Corresponding angles** between two parallels and a straight are equal.
- **Interior** and **exterior angles** are supplementary.

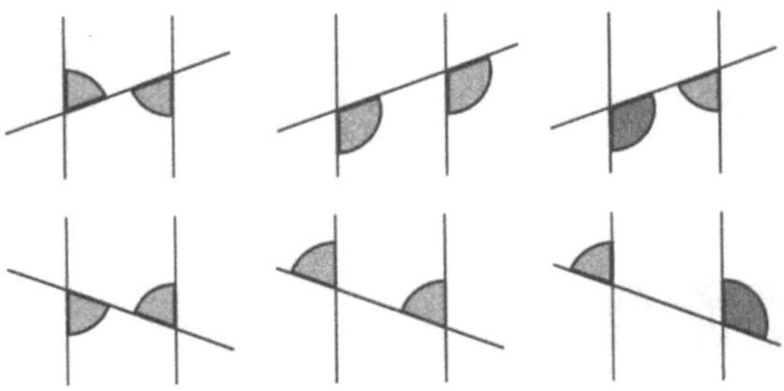

Alternate angles   Corresponding angles   Interior and exterior angles

3.4 Similarity and intercept theorems

If two rays originating in a common vertex S are intersected by parallel lines, then similar triangles result.
From the **similarity of triangles**, the **intercept theorems** follow.

3.4.1 Intercept theorems

- **First intercept theorem**: If two rays originating in a common vertex are intersected by parallel lines, the intercepts $a, a + b$ on one ray are in the same ratio as the intercepts $c, c + d$ on the other ray:

$$(a + b)/a = (c + d)/c \ .$$

- The first intercept theorem is also valid when the vertex lies between the parallel lines.

▷ To find the intercept on one ray we can use the ratio to the corresponding intercept on the other ray.

Application of the first intercept theorem: Four distinct proportions can be set up:

$$\frac{a}{a+b} = \frac{c}{c+d}, \qquad \frac{b}{a} = \frac{d}{c},$$
$$=$$
$$\frac{a+b}{a} = \frac{c+d}{c}, \qquad \frac{a}{b} = \frac{c}{d}.$$

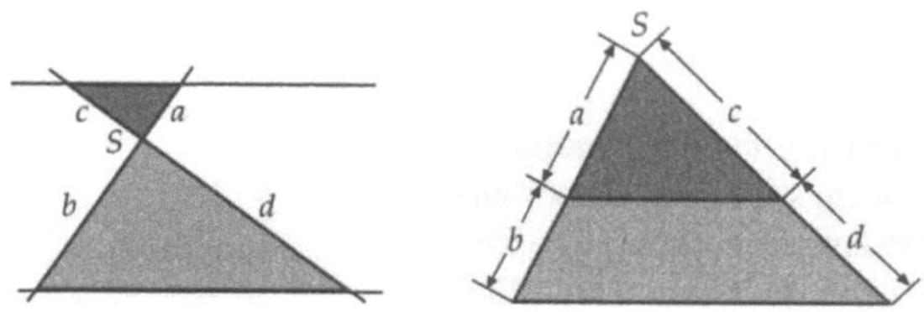

First intercept theorem

- **Second intercept theorem**: If two rays originating in a common vertex are intersected by parallel lines, the intercepts on the parallels are in the same ratio as the length of the corresponding ray sectors:

$$a/b = c/d .$$

- The second intercept theorem is also valid when the vertex lies between the two parallels.

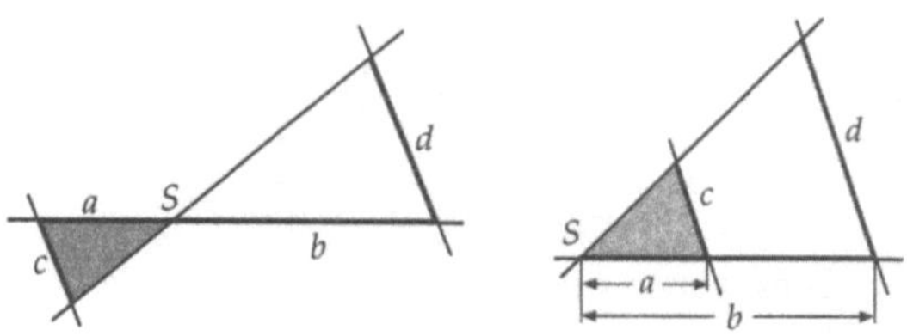

Second intercept theorem

3.4.2 *Division of a segment*

With the intercept theorems it is possible to divide any given segment $\overline{AB}$ in the ratio $m : n$. We differentiate:

Internal division: The dividing point T lies between the end-points of segment $\overline{AB}$ such that

$$\overline{AT} : \overline{TB} = m : n.$$

External division: The dividing point T is outside segment $\overline{AB}$ such that

$$\overline{AB} : \overline{BT} = m : n.$$

The division is performed by drawing two parallel lines through A and B on which the segments m and n are marked on opposite sides (internal division) or on the same side (external division). The resulting end-points are connected until the straight line through A and B is intersected in point T (second intercept theorem).

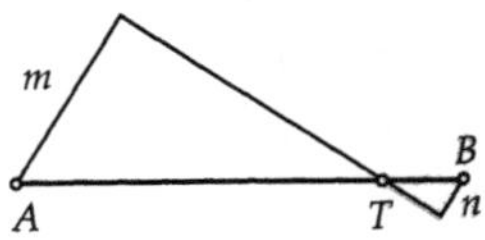

Internal division

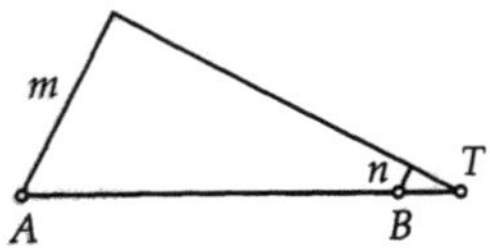

External division

Harmonic division: Segment $\overline{AB}$ is divided internally and externally in the same ratio $m : n$.

▷ The origin of the term **harmonic** is that three strings with the same tension, whose lengths are subject to harmonic division, produce a so-called harmonic sound. Given a harmonic division of 5:1, the lengths of the strings are as 10:12:15, the frequencies are as 1/10 : 1/12 : 1/15 = 6:5:4.

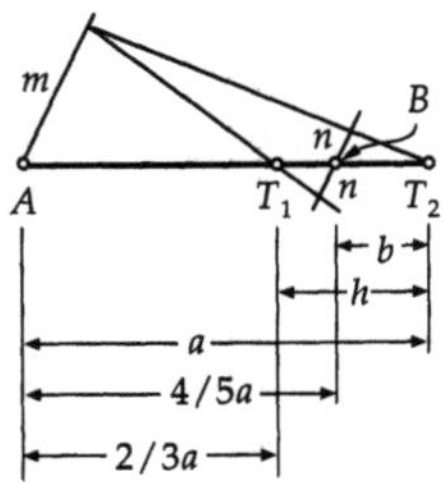

Harmonic division

3.4.3 Mean values

Harmonic mean h: According to harmonic division, the following applies according to second intercept theorem:

$$\frac{a}{b} = \frac{m}{n} = \frac{a-h}{h-b} \,.$$

Isolating the harmonic mean h of segments a and b:

$$\frac{a}{b} = \frac{a-h}{h-b} \quad \Rightarrow \quad h = \frac{2ab}{a+b} \,.$$

Geometric mean g: Construction with the help of Thales's circle on the segment $(a+b)$ by constructing the perpendicular in the end-point of a. According to the altitude theorem:

$$g = \sqrt{a \cdot b} \,.$$

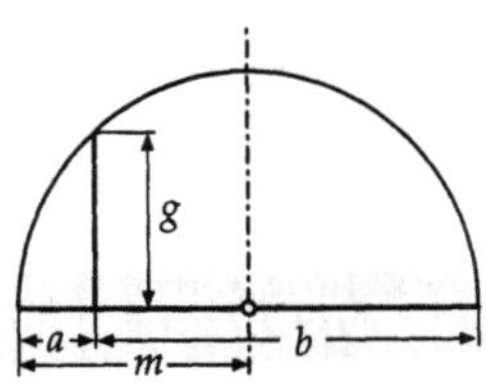

Arithmetic mean m, half of segment a and b:

$$m = \frac{a+b}{2} \,.$$

3.4.4 Golden Section

Golden Section: Division of a segment a in such a way that the ratio of the smaller part $c(= a - b)$ to the larger part b is equal to the ratio of the larger part b to the whole segment a. The affine ratio of the golden section is:

$$a : b = b : c = b : (a - b) \,, \quad \text{or}$$
$$\frac{a}{b} = \frac{b}{c} \frac{b}{(a-b)} \,, \quad \text{or}$$
$$b^2 = a \cdot c = a^2 - ab \,, \quad \text{or}$$
$$b^2 + ab - a^2 = 0$$

with the solution:

$$b_{1,2} = -\frac{a}{2} \pm \sqrt{\frac{a^2}{4} + a^2} \,,$$
$$b = \frac{a}{2}(\sqrt{5} - 1) \approx 0.618a \,,$$

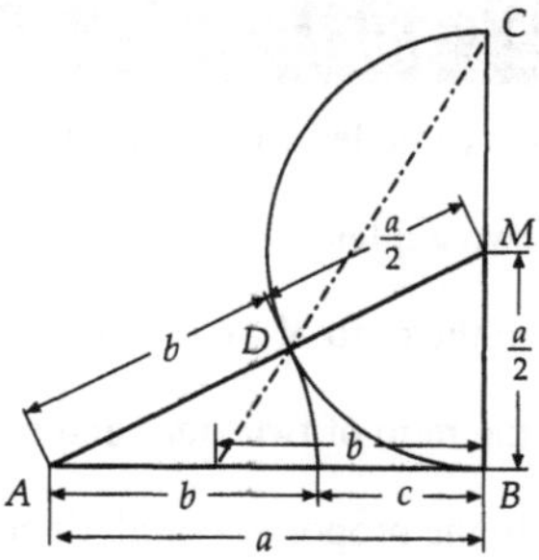

(only the positive solutions are of interest here).

Mean proportional: The larger segment $b = \sqrt{ac}$ of a. At the same time, segment b is the **geometric mean** of a and c.

▷ The Golden Section plays a special role in art and architecture; numerous temples and statues were designed according to the Golden Section.

3.5 Triangles

Triangle: Simplest geometric figure with

- three vertices (A, B, C) ,
- three sides (a, b, c) ,
- three angles (α, β, γ) .

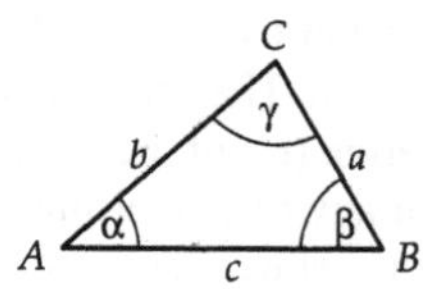

3.5.1 Congruence theorems

Congruent triangles can be completely superposed by rotation, parallel displacement (translation), inversion (reflection), or any combination of these.

Symbol of congruence: $\simeq$
$(\Delta ABC \simeq \Delta A'B'C')$

● Two triangles are congruent if they are equal in

a. three sides (SSS),

b. two sides and the included angle (SAS),

c. two sides and the angle opposite the longer side (SSA),

d. one side and two angles (ASA or SAA).

▷ Because the sum of the angles of a triangle is always $\pi \triangleq 180°$, triangles with three equal angles (AAA) are similar, but **not** necessarily congruent.

3.5.2 Similarity of triangles

● Two triangles are similar if they are equal in

a. two angles,

b. in the ratio of two sides and the included angle,

c. the ratio of two sides and the angle opposite the longer side,

d. in the proportion of all three sides.

3.5.3 Construction of triangles

Basic constructions of triangles: The conditions under which unique triangles can be constructed are determined by the congruence theorems. Three independent quantities must be known: the sides and the angles of the triangle are combined according to

SSS, SAS, SSA, ASA, SAA .

SSS:
Three sides (a, b, c) are *given* .
The end-points A, B of one side (c) are the centers of the circles with radii b about A and a about B. The intersection point is the third vertex (C) of the desired triangle. The sum of any two sides must be greater than the third side, otherwise the circular arcs do not intersect (triangle inequality).
Conditions:
$a < b + c,\ b < c + a,\ c < a + b.$

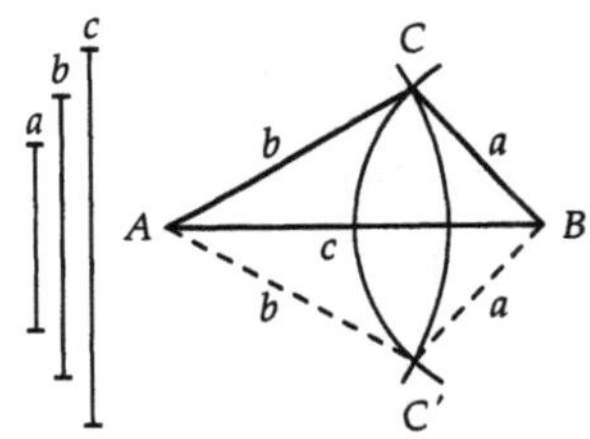

SAS:
Two sides (b, c) and the included angle (α) are *given*.
The segments b and c are marked on the arms of the angle (α). The points thus obtained are the missing vertices of the triangle.
Condition: $\alpha < 180°$.

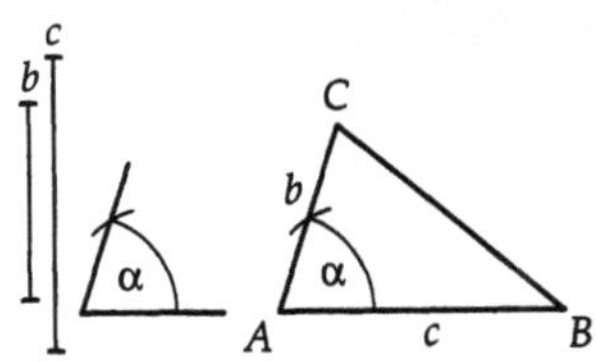

ASA / SAA:

1. ASA: One side (c) and two adjacent angles (α, β) are *given*.
   The angles (α, β) are marked off at the end-points (A, B) of side (c). The intersection point of their free arms is the third vertex (C) of the triangle.
   Condition: $\alpha + \beta < 180°$.

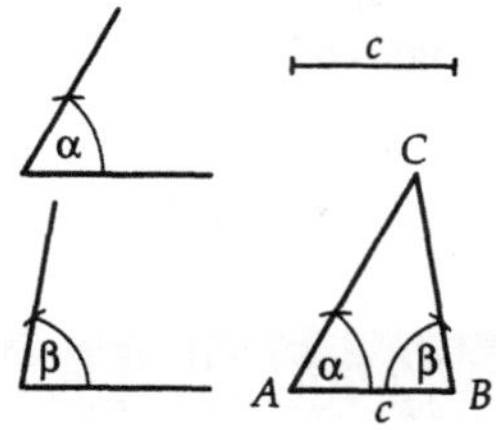

2. SAA: One side (c) and two angles (α, γ) are given.

   (a) Construct β as the supplementary angle of $\alpha + \gamma$. Subsequent construction as in ASA (1.).

   (b) Construct $c = |\overline{AB}|$ and mark off the angle α at point A. The angle γ is marked off at an arbitrary point C' of the free arm. The parallel to the free arm of the latter angle through point B intersects the free arm of the angle α at point C.

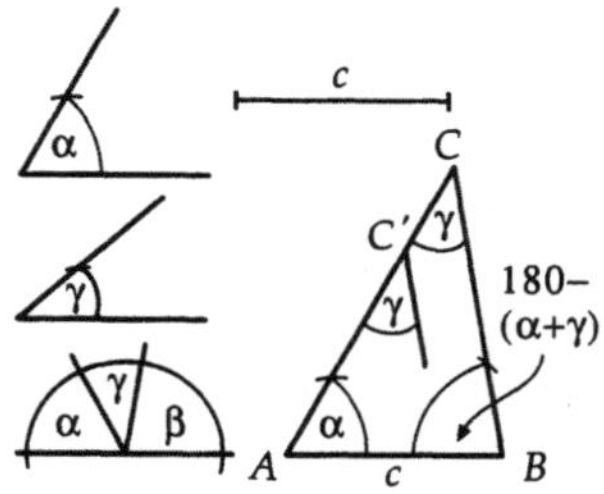

   Condition: $\alpha + \gamma < 180°$.

SSA:
Two sides (b, c) and one angle $\beta < 90°$ are *given*.
Construct $c = |\overline{AB}|$ and mark off β at point B. Draw a circular arc with radius b about point A .

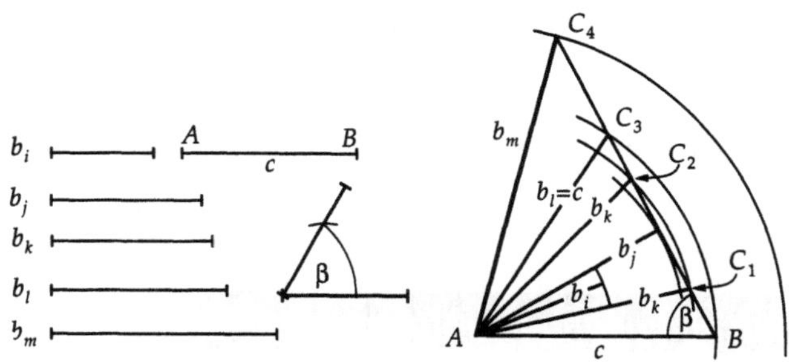

Case distinction according to length b:

i) The circle does not intersect the free arm of the given angle β. No solution results.

j) The circle touches the free arm. A right triangle ($\gamma = 90°$) results as the solution.

k) The circle intersects the free arm twice ($b = b_k < c$). Two triangles of different shapes (ABC_1 and ABC_2) result as solutions.

l) The circle intersects the free arm once in C_3 and traverses the vertex (B) of the given angle (β). An isosceles triangle $(b = b_l = c)$ results as a solution.

m) The circle intersects the free arm once in C_4 and encircles the vertex (B). A triangle with $b = b_m > c$ results as a solution.

3.5.4 Calculation of a right triangle

Let γ be a right angle and c the opposite side. Then, if two parameters (marked •) are known, the following applies to the remaining three parameters and the area A:

a	b	c	α	β	A
•	•	$\sqrt{a^2+b^2}$	$\arctan\left(\frac{a}{b}\right)$	$\arctan\left(\frac{b}{a}\right)$	$\frac{ab}{2}$
•	$\sqrt{c^2-a^2}$	•	$\arcsin\left(\frac{a}{c}\right)$	$\arccos\left(\frac{a}{c}\right)$	$\frac{a}{2}\sqrt{c^2-a^2}$
•	$a\cot(\alpha)$	$a\csc(\alpha)$	•	$\frac{\pi}{2}-\alpha$	$\frac{a^2}{2}\cot(\alpha)$
•	$a\tan(\beta)$	$a\sec(\beta)$	$\frac{\pi}{2}-\beta$	•	$\frac{a^2}{2}\tan(\beta)$
$\sqrt{c^2-b^2}$	•	•	$\arccos\left(\frac{b}{c}\right)$	$\arcsin\left(\frac{b}{c}\right)$	$\frac{b}{2}\sqrt{c^2-b^2}$
$b\tan(\alpha)$	•	$b\sec(\alpha)$	•	$\frac{\pi}{2}-\alpha$	$\frac{b^2}{2}\tan(\alpha)$
$b\cot(\beta)$	•	$b\csc(\beta)$	$\frac{\pi}{2}-\beta$	•	$\frac{b^2}{2}\cot(\beta)$
$c\sin(\alpha)$	$c\cos(\alpha)$	•	•	$\frac{\pi}{2}-\alpha$	$\frac{c^2}{4}\sin(2\alpha)$
$c\cos(\beta)$	$c\sin(\beta)$	•	$\frac{\pi}{2}-\beta$	•	$\frac{c^2}{4}\sin(2\beta)$

3.5.5 Calculation of an arbitrary triangle

For an arbitrary triangle the following hold:

1. Three sides a, b, c are known (**SSS**):

$$\alpha = 2\arctan\left(\sqrt{\frac{a^2-(b-c)^2}{(b+c)^2-a^2}}\right),$$

$$\beta = 2\arctan\left(\sqrt{\frac{b^2-(c-a)^2}{(c+a)^2-b^2}}\right),$$

$$\gamma = 2 \arctan\left(\sqrt{\frac{c^2 - (a-b)^2}{(a+b)^2 - c^2}}\right) ,$$

$$\text{area } A = \frac{\sqrt{(b+c)^2 - a^2}\sqrt{a^2 - (b-c)^2}}{4} .$$

2. Two sides a, b and the included angle γ are known (**SAS**):

$$c = \sqrt{a^2 + b^2 - 2ab\cos(\gamma)} ,$$

$$\alpha = \arcsin\left(\frac{a\sin(\gamma)}{\sqrt{a^2 + b^2 - 2ab\cos(\gamma)}}\right) ,$$

$$\beta = \arcsin\left(\frac{b\sin(\gamma)}{\sqrt{a^2 + b^2 - 2ab\cos(\gamma)}}\right) ,$$

$$\text{area } A = \frac{ab}{2}\sin(\gamma) .$$

3. Two angles α, β and side a opposite one of the angles are known (**SAA**):

$$b = a\csc(\alpha)\sin(\beta) ,$$
$$c = a\left[\cot(\alpha)\sin(\beta) + \cos(\beta)\right] ,$$
$$\gamma = \pi - \alpha - \beta ,$$

$$\text{area } A = \frac{a^2}{2}\sin^2(\beta)\left[\cot(\alpha) + \cot(\beta)\right] .$$

4. Two angles α, β and the enclosed side c are known (**ASA**):

$$a = \frac{c\csc(\beta)}{\cot(\alpha) + \cot(\beta)} ,$$
$$b = \frac{c\csc(\alpha)}{\cot(\alpha) + \cot(\beta)} ,$$
$$\gamma = \pi - \alpha - \beta ,$$

$$\text{area } A = \frac{a^2}{2\left(\cot(\alpha) + \cot(\beta)\right)} .$$

5. Two sides a, b, $(a \geq b)$ and angle α opposite the longer side are known (**SSA**):

$$c = \sqrt{a^2 + b^2\cos(2\alpha) + 2b\cos(\alpha)\sqrt{a^2 - b^2\sin^2(\alpha)}} ,$$

$$\beta = \arcsin\left(\frac{b\sin(\alpha)}{a}\right) ,$$

$$\gamma = \arccos\left(\frac{b}{a}\sin^2(\alpha) - \cos(\alpha)\sqrt{1 - \frac{b^2}{a^2}\sin^2(\alpha)}\right) ,$$

$$\text{area } A = \frac{b^2\sin(\alpha)}{2}\left(-\cos(\alpha) + \sqrt{\frac{a^2}{b^2} - \sin^2(\alpha)}\right) .$$

6. Two sides a, b, $(a \geq b)$ and angle β opposite the shorter side are known (**SSA**): There are two possibilities, distinguished by the $\pm$ sign:

$$c = \sqrt{b^2 + a^2\cos(2\beta) \pm 2a\cos(\beta)\sqrt{b^2 - a^2\sin^2(\beta)}}\,,$$

$$\alpha = \frac{\pi}{2} \pm \arccos\left(\frac{a\sin(\beta)}{b}\right),$$

$$\gamma = \arccos\left(\frac{a}{b}\sin^2(\beta) \mp \cos(\beta)\sqrt{1 - \frac{a^2}{b^2}\sin^2(\beta)}\right),$$

$$\text{area } A = \frac{a^2\sin(\beta)}{2}\left(\cos(\beta) \mp \sqrt{\frac{b^2}{a^2} - \sin^2(\beta)}\right).$$

3.5.6 Relations between angles and sides of a triangle

- **Triangle inequality**: The sum of the lengths of any two sides is greater than the length of the third side:

$$a + b > c\,, \quad b + c > a\,, \quad c + a > b\,.$$

- The difference between the lengths of any two sides of a triangle is smaller than the length of the third side:

$$a - b < c\,, \quad b - c < a\,, \quad c - a < b\,.$$

- **Sum of the angles** in a triangle is $180° \mathrel{\hat{=}} \pi$:

$$\alpha + \beta + \gamma = 180°\,.$$

▷ Due to measuring and rounding errors, these theorems are valid only approximately in practice.

Interior angles (α, β, γ): The angles inside the triangle. The **sum of the interior angles** is $180° \mathrel{\hat{=}} \pi$.

Exterior angles $(\alpha', \beta', \gamma')$: Any exterior angle is the supplement of its interior angle. The exterior and interior angle together become a straight angle ($180° \mathrel{\hat{=}} \pi$).

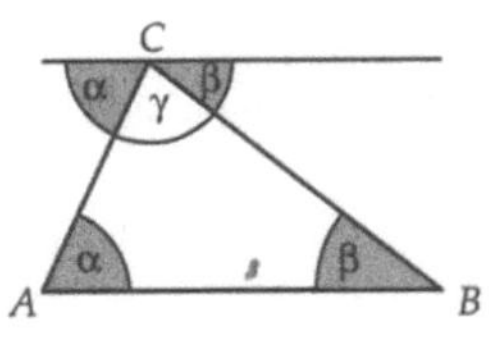

Interior angles

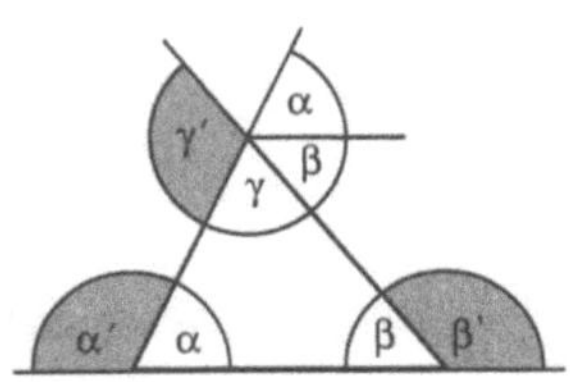

Exterior angles

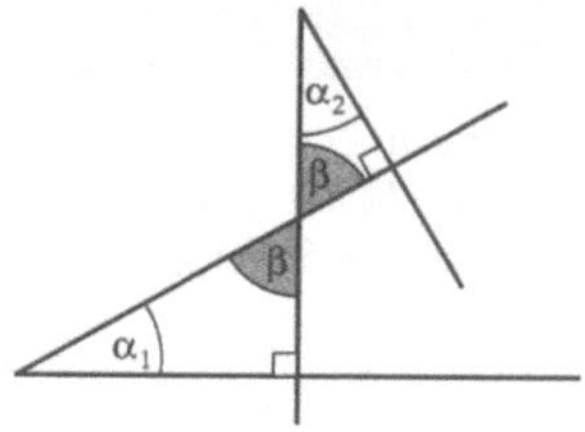

- The exterior angles of a triangle are equal to the sum of the two opposite angles:

$$\alpha' = \beta + \gamma \,,$$
$$\beta' = \alpha + \gamma \,,$$
$$\gamma' = \alpha + \beta \,.$$

- The sum of the exterior angles is 360°.
- Two angles are equal if their arms are pairwise perpendicular to each other, and the vertex of an angle lies outside the arms of the other:

$$\alpha_1 = \alpha_2 \,.$$

- The side opposite the greater (smaller) of two angles is longer (shorter) than that opposite the other:

$$a > b \quad \Rightarrow \quad \alpha > \beta \quad \text{etc.}$$

- Angles opposite equal sides are equal.

3.5.7 Altitude

Altitudes (h_a, h_b, h_c): Lines drawn from one **vertex** (A, B, C) perpendicular to the opposite **side of triangle** (a, b, c). Indices: sides of the triangle. **Altitude** h, perpendicular to the side of a triangle.

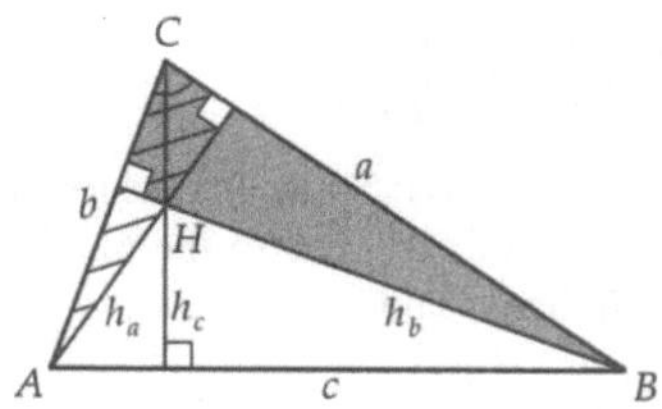

- All three sides are intersecting in the **orthocenter** H.
- The orthocenter H lies
  a. inside the triangle:
  Acute triangle,
  b. at a vertex of the triangle:
  Right triangle,
  c. outside the triangle:
  Obtuse triangle.

▷ Each of the altitudes subdivides a triangle into two right triangles, which can be used to construct a desired triangle (Thales's circle).

- The lengths of the altitudes and the corresponding sides are inversely proportional:

$$h_a : h_b = b : a; \quad h_b : h_c = c : b; \quad h_c : h_a = a : c \quad \text{and}$$
$$h_a : h_b : h_c = \frac{1}{a} : \frac{1}{b} : \frac{1}{c} \,.$$

From the definition follows:

$$h_a = b \cdot \sin\gamma = c \cdot \sin\beta \,,$$
$$h_b = c \cdot \sin\alpha = a \cdot \sin\gamma \,,$$
$$h_c = a \cdot \sin\beta = b \cdot \sin\alpha \,.$$

3.5.8 Angle-bisectors

Angle-bisectors ($w_\alpha, w_\beta, w_\gamma$): Straight lines bisecting the interior angles of a triangle. Indices: The bisected angles.

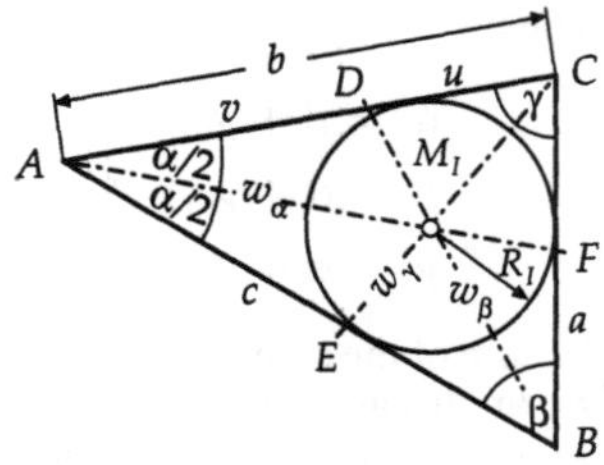

$$w_\alpha = \overline{AF} ,$$
$$w_\beta = \overline{BD} ,$$
$$w_\gamma = \overline{CE} ,$$

Lengths of the angle-bisectors. Indices: the angles to be bisected.

$$w_\alpha = \frac{1}{b+c}\sqrt{bc[(b+c)^2-a^2]} = \frac{2bc\cos(\alpha/2)}{b+c} ,$$
$$w_\beta = \frac{1}{c+a}\sqrt{ca[(c+a)^2-b^2]} = \frac{2ca\cos(\beta/2)}{c+a} ,$$
$$w_\gamma = \frac{1}{a+b}\sqrt{ab[(a+b)^2-c^2]} = \frac{2ab\cos(\gamma/2)}{a+b} .$$

- The angle-bisector divides the opposite side of the triangle in the ratio of the adjacent sides of the triangle.

$$\overline{CD} : \overline{AD} = \overline{BC} : \overline{AB}$$
$$u : v = a : c$$

Incenter (M_I): The point of intersection of three angle-bisectors.

3.5.9 Medians

Medians (s_a, s_b, s_c): Straight lines joining the midpoints of the sides to the opposite vertices. Indices: The bisected sides.

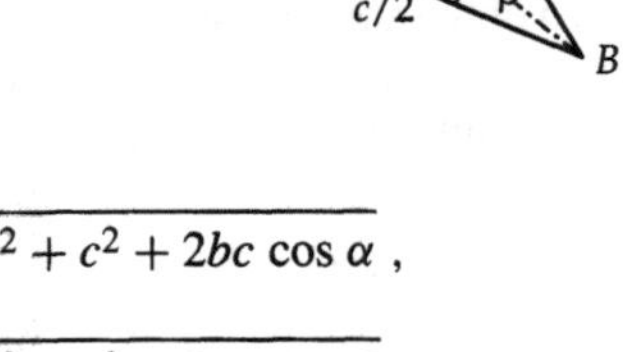

s_a goes from A to the midpoint of a ,

s_b goes from B to the midpoint of b ,

s_c goes from C to the midpoint of c ,

Lengths of the medians. Indices: The side to be bisected.

$$s_a = \frac{1}{2}\sqrt{2(b^2+c^2)-a^2} = \frac{1}{2}\sqrt{b^2+c^2+2bc\cos\alpha} ,$$
$$s_b = \frac{1}{2}\sqrt{2(a^2+c^2)-b^2} = \frac{1}{2}\sqrt{a^2+c^2+2ac\cos\beta} ,$$
$$s_c = \frac{1}{2}\sqrt{2(a^2+b^2)-c^2} = \frac{1}{2}\sqrt{a^2+b^2+2ab\cos\gamma} .$$

Center of gravity (S): The point of intersection of the three medians.

- The center of gravity S divides the median in the ratio of 2:1.

▷ If the medians for the construction of triangles are known, note: A median divides the triangle into two partial triangles, two sides of which are of equal length.

3.5.10 Mid-perpendiculars, incircle, circumcircle, excircle

Mid-perpendiculars (m_a, m_b, m_c): Set of all points with the same distance from two vertices of the triangle.

$$m_a^2 = R_C^2 - \frac{a^2}{4}, \quad m_b^2 = R_C^2 - \frac{b^2}{4}, \quad m_c^2 = R_C^2 - \frac{c^2}{4}$$

$$m_a^2 + m_b^2 + m_c^2 = 3R_C^2 - \frac{1}{4}(a^2 + b^2 + c^2) .$$

Incircle: Circle lying inside a triangle, where the three sides of the triangle are tangents.
Incenter M_I: Point of intersection of the angle-bisectors.
Inradius R_I (A is the area of the triangle):

$$R_I = \frac{2A}{a + b + c} .$$

Circumcircle: Circle through all vertices encircling the triangle.
Circumcenter M_C: Point of intersection of the mid-perpendiculars.
Circumradius: R_C:

$$R_C = \frac{abc}{4A} = \frac{bc}{2h_a} .$$

Excircle: Circle touching one side of the triangle and the extensions of its two other sides.

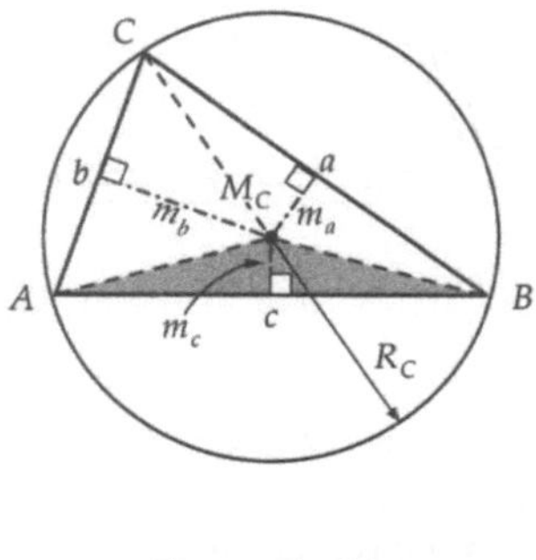

Circumcircle

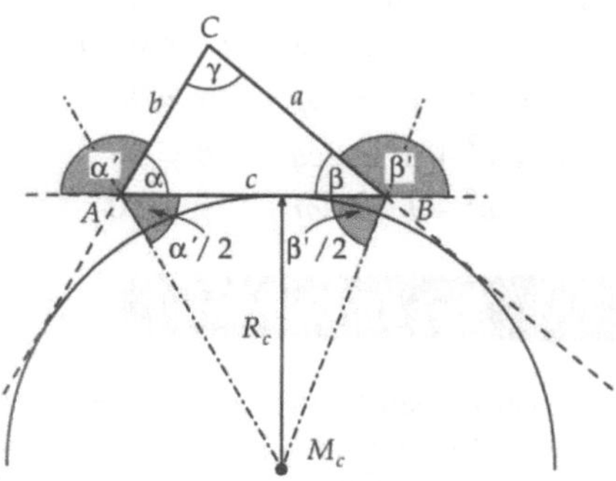

Excircle

Excenters M_a, M_b, M_c: Point of intersection of two angle-bisectors of the exterior angles, center of the described circle.

- If the circumcenter lies:
  a. inside the triangle: **Acute triangle**,
  b. on the hypotenuse c: **Right triangle**,
  c. outside the triangle: **Obtuse triangle**.

Exradii R_a, R_b, R_c:

$$R_a = \frac{2A}{b + c - a}, \quad R_b = \frac{2A}{a + c - b}, \quad R_c = \frac{2A}{a + b - c}$$

3.5.11 Area of a triangle

Area A: Half the product of the altitude and the base side:

$$A = \frac{a \cdot h_a}{2} = \frac{b \cdot h_b}{2} = \frac{c \cdot h_c}{2}$$
$$= \frac{ab}{2}\sin\gamma = \frac{bc}{2}\sin\alpha = \frac{ac}{2}\sin\beta$$
$$= \frac{a^2}{2}\frac{\sin\beta\sin\gamma}{\sin\alpha} = \frac{b^2}{2}\frac{\sin\alpha\sin\gamma}{\sin\beta} = \frac{c^2}{2}\frac{\sin\alpha\sin\beta}{\sin\gamma}$$

with $k = \frac{a+b+c}{2}$ (half the perimeter).

$$A = R_I \cdot k = 2R_C^2 \sin\alpha\sin\beta\sin\gamma = \frac{abc}{4R_C}$$

$$A = \sqrt{k(k-a)(k-b)(k-c)}$$ **Heron's formula** .

3.5.12 Generalized Pythagorean theorem

▷ In contrast to the specific **Pythagorean theorem** applicable only to right triangles, the following relations can be applied to oblique triangles generally.

p : projection c into b
q : projection a into c
r : projection b into a

$a^2 = b^2 + c^2 \pm 2bp \qquad \alpha \neq 90°$
$b^2 = c^2 + a^2 \pm 2cq \qquad \beta \neq 90°$
$c^2 = a^2 + b^2 \pm 2ar \qquad \gamma \neq 90°$

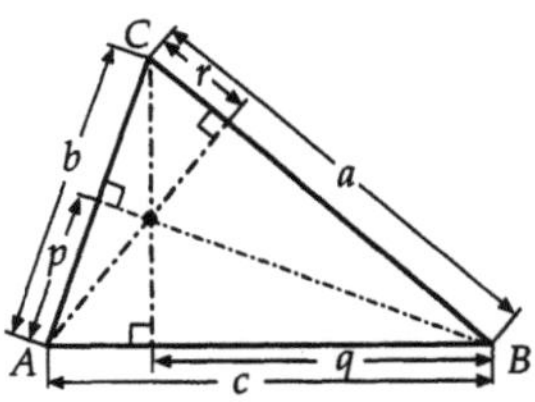

3.5.13 Angular relations

$$\begin{aligned}
\sin\alpha + \sin\beta + \sin\gamma &= 4\cos(\alpha/2)\cos(\beta/2)\cos(\gamma/2)\\
\sin 2\alpha + \sin 2\beta + \sin 2\gamma &= 4\sin\alpha\sin\beta\sin\gamma\\
\sin^2\alpha + \sin^2\beta + \sin^2\gamma &= 2(1 + \cos\alpha\cos\beta\cos\gamma)\\
\cos\alpha + \cos\beta + \cos\gamma &= 1 + 4\sin(\alpha/2)\sin(\beta/2)\sin(\gamma/2)\\
\cos 2\alpha + \cos 2\beta + \cos 2\gamma &= -(4\cos\alpha\cos\beta\cos\gamma + 1)\\
\cos^2\alpha + \cos^2\beta + \cos^2\gamma &= 1 - 2\cos\alpha\cos\beta\cos\gamma\\
\tan\alpha + \tan\beta + \tan\gamma &= \tan\alpha\tan\beta\tan\gamma
\end{aligned}$$

$$\cot\alpha\cot\beta + \cot\alpha\cot\gamma + \cot\beta\cot\gamma = 1$$

$$\cot\left(\frac{\alpha}{2}\right) + \cot\left(\frac{\beta}{2}\right) + \cot\left(\frac{\gamma}{2}\right) = \cot\left(\frac{\alpha}{2}\right)\cot\left(\frac{\beta}{2}\right)\cot\left(\frac{\gamma}{2}\right)$$

3.5.14 Sine theorem

The sides of a triangle are proportional to the sines of the opposite angles.

$$a : b : c = \sin\alpha : \sin\beta : \sin\gamma$$
$$\frac{a}{\sin\alpha} = \frac{b}{\sin\beta} = \frac{c}{\sin\gamma}$$

3.5.15 Cosine theorem

$$a^2 = b^2 + c^2 - 2bc\cos\alpha$$
$$b^2 = c^2 + a^2 - 2ca\cos\beta$$
$$c^2 = a^2 + b^2 - 2ab\cos\gamma$$

In the case of a right triangle (α, β, or $\gamma = 90°$), the Pythagorean theorem follows (because $\cos 90° = 0$).

3.5.16 Tangent theorem

$$\frac{a+b}{a-b} = \frac{\tan[(\alpha+\beta)/2]}{\tan[(\alpha-\beta)/2]} = \frac{\cot[\gamma/2]}{\tan[(\alpha-\beta)/2]}$$
$$\frac{b+c}{b-c} = \frac{\tan[(\beta+\gamma)/2]}{\tan[(\beta-\gamma)/2]} = \frac{\cot[\alpha/2]}{\tan[(\beta-\gamma)/2]}$$
$$\frac{a+c}{a-c} = \frac{\tan[(\alpha+\gamma)/2]}{\tan[(\alpha-\gamma)/2]} = \frac{\cot[\beta/2]}{\tan[(\alpha-\gamma)/2]}$$

3.5.17 Half-angle theorems

Half the perimeter: $k = \frac{a+b+c}{2}$

$$\sin\frac{\alpha}{2} = \sqrt{\frac{(k-b)(k-c)}{bc}}, \qquad \cos\frac{\alpha}{2} = \sqrt{\frac{k(k-a)}{bc}}$$
$$\tan\frac{\alpha}{2} = \sqrt{\frac{(k-b)(k-c)}{k(k-a)}}$$
$$\sin\frac{\beta}{2} = \sqrt{\frac{(k-a)(k-c)}{ac}}, \qquad \cos\frac{\beta}{2} = \sqrt{\frac{k(k-b)}{ac}}$$
$$\tan\frac{\beta}{2} = \sqrt{\frac{(k-a)(k-c)}{k(k-b)}}$$
$$\sin\frac{\gamma}{2} = \sqrt{\frac{(k-a)(k-b)}{ab}}, \qquad \cos\frac{\gamma}{2} = \sqrt{\frac{k(k-c)}{ab}}$$
$$\tan\frac{\gamma}{2} = \sqrt{\frac{(k-a)(k-b)}{k(k-c)}}$$

3.5.18 Mollweide's formulas

$$\frac{b+c}{a} = \frac{\cos[(\beta-\gamma)/2]}{\sin[\alpha/2]}, \qquad \frac{b-c}{a} = \frac{\sin[(\beta-\gamma)/2]}{\cos[\alpha/2]}$$
$$\frac{c+a}{b} = \frac{\cos[(\gamma-\alpha)/2]}{\sin[\beta/2]}, \qquad \frac{c-a}{b} = \frac{\sin[(\gamma-\alpha)/2]}{\cos[\beta/2]}$$
$$\frac{a+b}{c} = \frac{\cos[(\alpha-\beta)/2]}{\sin[\gamma/2]}, \qquad \frac{a-b}{c} = \frac{\sin[(\alpha-\beta)/2]}{\cos[\gamma/2]}$$

3.5.19 Theorems of sides

$$|a-b| < c < a+b$$
$$|b-c| < a < b+c$$
$$|a-c| < b < a+c$$

3.5.20 Isosceles triangle

The isosceles triangle has:

- two sides (a, b) of equal length, and
- two equal base angles (α, β).

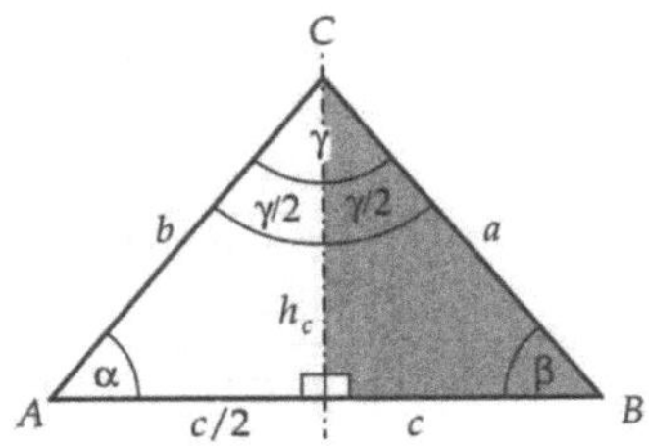

Axis (symmetry line): Bisects the base, bisects angle γ at the apex, and is perpendicular to the base.
Altitude:

$$h_a = h_b = \frac{2A}{a} = \frac{c}{a} \cdot h_c \,,$$
$$h_c = w_\gamma = \sqrt{a^2 - \left(\frac{c}{2}\right)^2} \,.$$

Median:

$$s_c = h_c \,,$$
$$s_a = s_b = \sqrt{\left(\frac{b}{2}\right)^2 + c^2 - bc\cos\alpha} \,.$$

Angle-bisector:

$$w_\gamma = h_c \,.$$

Mid-perpendicular:

$$m_c = \sqrt{a^2 - \left(\frac{c}{2}\right)^2} \,.$$

Inradius:

$$R_I = \frac{c}{4h_c}(2a - c) \,.$$

Circumradius:

$$R_C = \frac{a^2}{2h_c} \,.$$

Perimeter:

$$P = 2a + c \,.$$

Area:

$$A = \frac{c}{4}\sqrt{4a^2 - c^2} = \frac{c \cdot h_c}{2} \,.$$

Isosceles right triangle:

$$\gamma = 90^\circ .$$

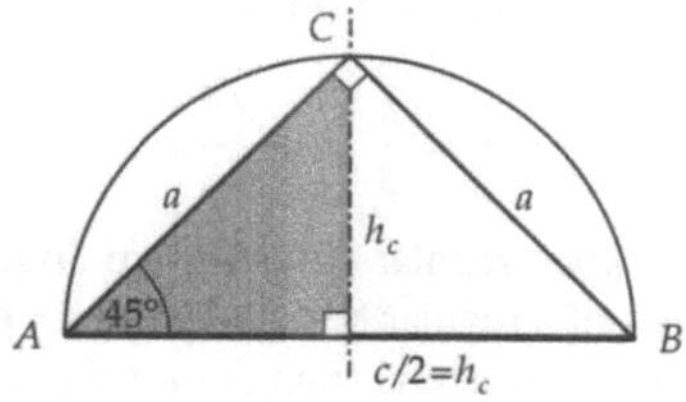

Thus, the base angles (α, β) are

$$\alpha = 90^\circ - \frac{\gamma}{2} = 45^\circ , \quad \alpha = \beta = 45^\circ .$$

3.5.21 Equilateral triangle

The equilateral triangle has:

- three sides of equal length

$$a = b = c ,$$

- three **axes of symmetry**: the altitudes, the angle-bisectors, and the medians, which all are equal in this case:

$$h_a = w_\alpha = s_a = \frac{a\sqrt{3}}{2} , \qquad h_b = w_\beta = s_b = \frac{a\sqrt{3}}{2} ,$$
$$h_c = w_\gamma = s_c = \frac{a\sqrt{3}}{2} .$$

● In an equilateral triangle, all interior angles are equal:

$$\alpha = \beta = \gamma = 60^\circ .$$

▷ This triangle is **central-symmetric** because all axes of symmetry intersect in a single point, the **orthocenter**. Each axis of symmetry subdivides the triangle into two congruent right triangles.

Altitude:

$$h = \frac{a}{2}\sqrt{3} .$$

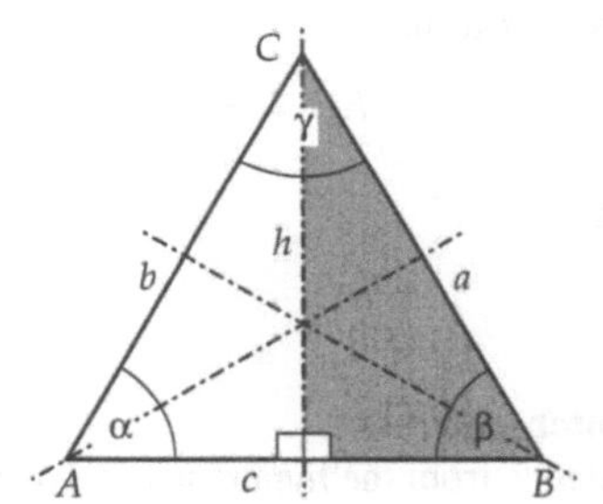

Inradius:

$$R_I = \frac{a}{6}\sqrt{3} .$$

Circumradius:

$$R_C = \frac{a}{2}\sqrt{3} - \frac{a}{2\sqrt{3}} = h - R_I .$$

Perimeter:

$$P = 3a .$$

Area:

$$A = \frac{a^2}{4}\sqrt{3}\,.$$

Center of gravity: The distance from each side is

$$x = \frac{h}{3} = a\frac{\sqrt{3}}{6}\,.$$

□ Special triangles result when regular polygons are divided. For example, the lines joining opposite vertices of a regular hexagon yield six **equilateral** triangles.

▷ Application in trusswork: Equilateral triangles are the faces of the tetrahedron.

3.5.22 *Right triangle*

Right angle, convention:

$$\gamma = 90^\circ \mathrel{\hat{=}} \pi/2 \quad (\text{angle at } C)\,,$$
$$\alpha + \beta = 90^\circ \mathrel{\hat{=}} \pi/2\,.$$

The **hypotenuse** c, and the opposite side a and the adjacent side b are related to the angle α and to each other through **trigonometric functions**:

$$\sin\alpha = \frac{\text{opposite side}}{\text{hypotenuse}} = \frac{a}{c}\,,$$
$$\cos\alpha = \frac{\text{adjacent side}}{\text{hypotenuse}} = \frac{b}{c}\,,$$
$$\tan\alpha = \frac{\text{opposite side}}{\text{adjacent side}} = \frac{a}{b}\,,$$
$$\cot\alpha = \frac{\text{adjacent side}}{\text{opposite side}} = \frac{b}{a} = \frac{1}{\tan\alpha}\,.$$

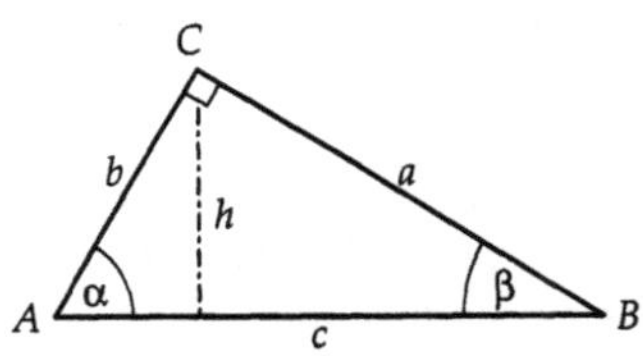

Altitude:

$$h = \frac{ab}{c}\,.$$

Inradius:

$$R_I = \frac{a+b-c}{2}\,.$$

Circumradius:

$$R_C = \frac{c}{2}\,.$$

Area:

$$A = \frac{1}{2}ab\,.$$

Center of gravity:
Distance from the hypotenuse c : $h/3$.
Distance from side a : $b/3$.
Distance from side b : $a/3$.

3.5.23 Theorem of Thales

- **Theorem of Thales**: In a right triangle, the right angle lies on the semicircle or circumcircle over the hypotenuse.

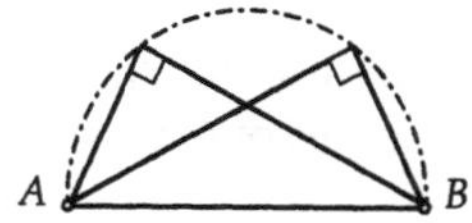

3.5.24 Pythagorean theorem

- **Pythagorean theorem**: In a right triangle the square of the hypotenuse c is equal to the sum of the squares on the other sides a and b.

$$c^2 = a^2 + b^2$$

Thus:

$$\sin^2\alpha + \cos^2\alpha = 1 .$$

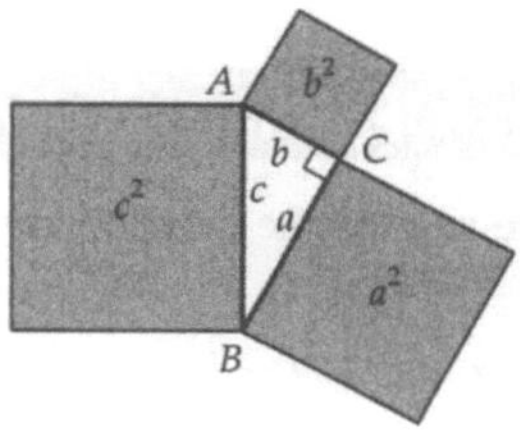

- The Pythagorean theorem is invertible: If the sum of the squares on two sides is equal to the square on the third side, then the triangle is a right triangle.

▷ Frequent mistake: The Pythagorean theorem does **not** hold for oblique triangles, to which the more general **cosine theorem** and the **generalized Pythagorean theorem** apply.

3.5.25 Theorem of Euclid

- **Hypotenuse intercept**: p is generalized by the orthogonal projection of sides a and b onto the hypotenuse.
- **Theorem of Euclid**: In a right triangle, the square on a shorter side is equal to the rectangle formed by the hypotenuse and the adjacent hypotenuse intercept.

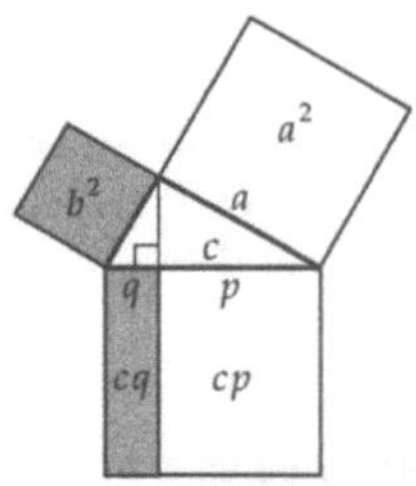

3.5.26 Altitude theorem

- **Altitude theorem**: In a right triangle, the square on the altitude is equal to the rectangle the hypotenuse intercepts.

$$h^2 = p \cdot q = \frac{a^2b^2}{c^2}$$

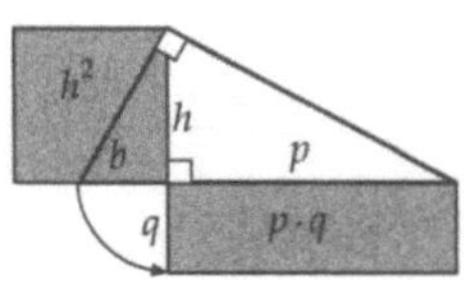

3.6 Quadrilaterals

Quadrilateral: Can be divided into two triangles by means of **diagonals** It has:

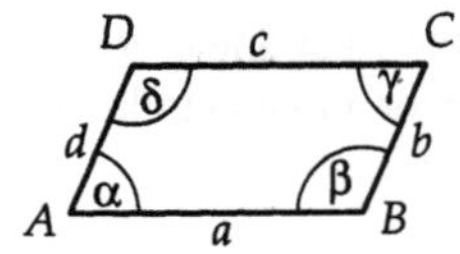

- four vertices (A, B, C, D) ,
- four sides (a, b, c, d) ,
- four angles $(\alpha, \beta, \gamma, \delta)$.

3.6.1 General quadrilateral

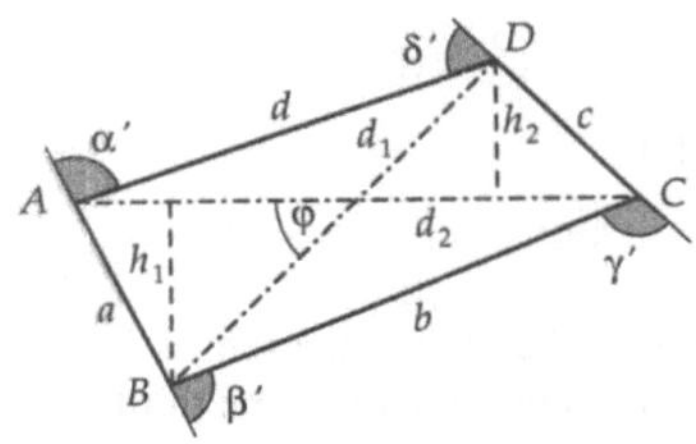

- **The sum of the interior angles** of a quadrilateral is:

$$\alpha + \beta + \gamma + \delta = 360^\circ \; .$$

- The sum of the exterior angles of a quadrilateral is also:

$$\alpha' + \beta' + \gamma' + \delta' = 360^\circ \; .$$

Perimeter for all quadrilaterals:

$$P = a + b + c + d \; .$$

h_1, h_2 : altitudes ,
d_1, d_2 : diagonals .

Area:

$$A = \frac{h_1 + h_2}{2} d_2 = \frac{1}{2} d_2 d_1 \sin\varphi \; ,$$

$$\text{with } k = \frac{P}{2} \text{ and } \varphi = \frac{\alpha + \gamma}{2} = \frac{\beta + \delta}{2} \; :$$

$$A = \sqrt{(k-a)(k-b)(k-c)(k-d) - abcd \cdot \cos^2\varphi} \; .$$

3.6.2 Trapezoid

Trapezoid: Quadrilateral with one pair of opposite parallel sides $(a||c)$.
Perimeter:

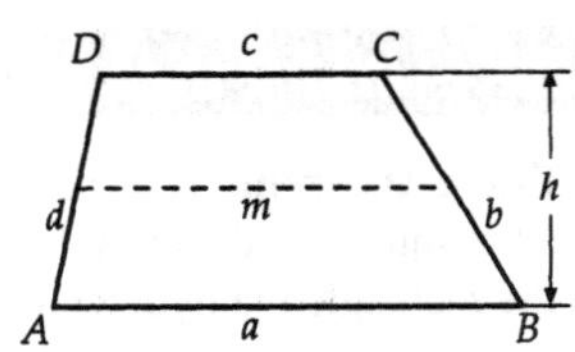

$$P = a + b + c + d \; .$$

Mid-line:

$$m = \frac{a + c}{2} \; .$$

Area:

$$A = \frac{a + c}{2} \cdot h = m \cdot h \; .$$

Center of gravity: Lies on the line joining the midpoint of the parallel baselines at distance

$$x = \frac{h}{3} \cdot \frac{a+2c}{a+c}$$

from the baseline (a).
Isosceles trapezoid: The base angles are equal and the mid-perpendiculars of the arms intersect at the circumcenter.

3.6.3 Parallelogram

A **parallelogram** has two pairs of equal opposite parallels: $a = c, b = d$ and $a||c, b||d$. The opposite interior angles are equal:

$$\alpha = \gamma \ , \ \beta = \delta \ .$$

The sum of two adjacent angles is 180°:

$$\alpha + \beta = 180^\circ \ , \qquad \beta + \gamma = 180^\circ \ ,$$
$$\gamma + \delta = 180^\circ \ , \qquad \delta + \alpha = 180^\circ \ .$$

Altitude:

$$h_a = b \cdot \sin \alpha \ .$$

Perimeter:

$$P = 2a + 2b \ .$$

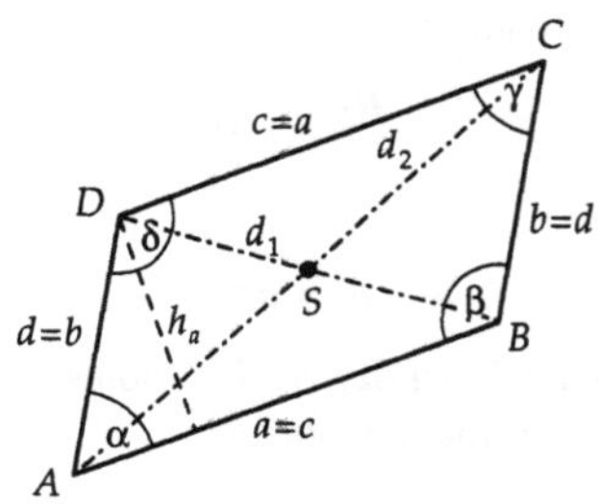

Diagonals:

$$d_{1,2} = \sqrt{a^2 + b^2 \pm 2a\sqrt{b^2 - h_a^2}} \ .$$

The diagonals bisect each other.
Area:

$$A = a \cdot h_a = ab \sin \alpha = b \cdot h_b \ .$$

Center of gravity S: Lies in the point of intersection of the diagonals.

3.6.4 Rhombus

Rhombus: Parallelogram with four equal sides a of equal length.
Altitude:

$$h_a = b \sin \alpha \ .$$

Perimeter:

$$P = 4a \ .$$

Diagonals:
The diagonals are bisected by the center of gravity and are perpendicular to each other:

$$d_1 = 2a\cos(\alpha/2)\ ,$$
$$d_2 = 2a\sin(\alpha/2)\ .$$

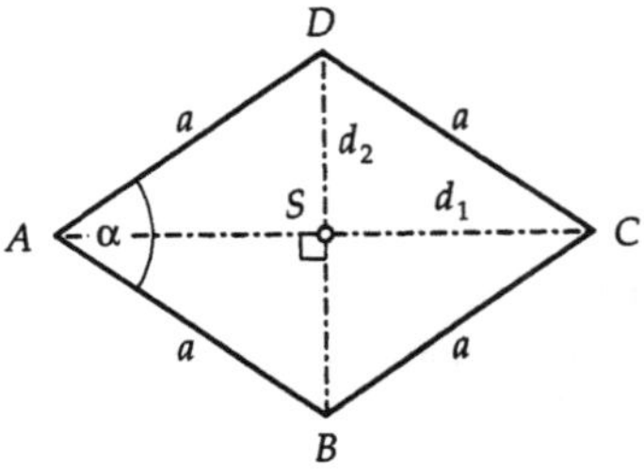

Area:

$$A = a^2 \sin\alpha = \frac{d_1 d_2}{2}\ .$$

Center of gravity: The point of intersection of the diagonals.

3.6.5 Rectangle

Rectangle: Parallelogram with two pairs of opposite, parallel sides and four right angles.
Altitude:

$$h_a = b\ ,\ h_b = a\ .$$

Perimeter:

$$P = 2a + 2b\ .$$

Diagonals:
The diagonals are bisected by the center of gravity (circumcenter).

$$d_1 = d_2 = \sqrt{a^2 + b^2}$$

Circumradius:

$$R_C = \frac{d}{2}\ .$$

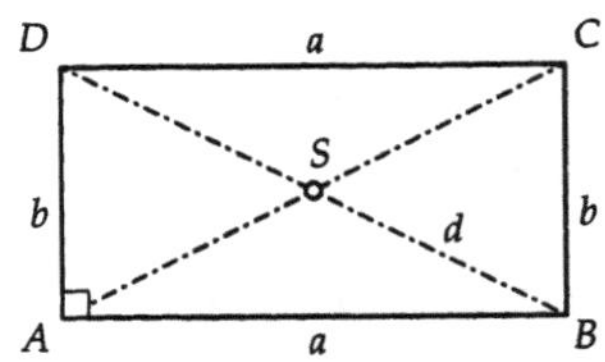

Area:

$$A = ab\ .$$

Center of gravity S**:** The point of intersection of the diagonals.

3.6.6 Square

Square: Rectangle with four equal sides.
Altitude:

$$h_a = a\ .$$

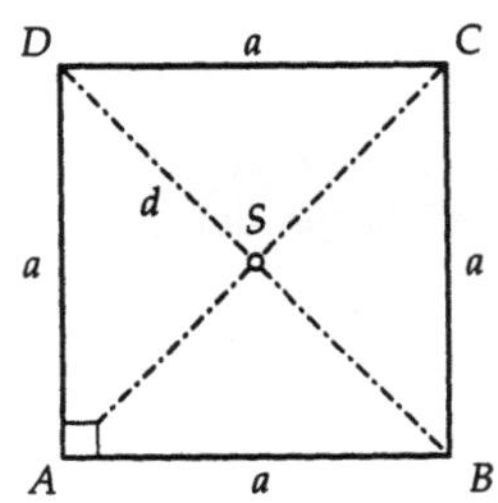

Perimeter:

$$P = 4a\ .$$

Diagonals:
The diagnonals are perpendicular to each other and bisected by the center of gravity (incenter and circumcenter).

$$d_1 = d_2 = \sqrt{2}a$$

Inradius:

$$R_I = \frac{a}{2} .$$

Circumradius:

$$R_C = \frac{\sqrt{2}a}{2} .$$

Area:

$$A = a^2 .$$

Center of gravity: S is the point of intersection of the diagonals.
Inradius and circumradius are given by the point of intersection of the diagonals.

3.6.7 Quadrilateral of chords

All vertices lie on the circumcircle. The sum of two opposite angles is 180°:

$$\alpha + \gamma = 180° , \ \beta + \delta = 180° .$$

▷ The converse also holds: All vertices lie on a circle if the sum of two opposite angles is 180°.

Perimeter:

$$P = a + b + c + d .$$

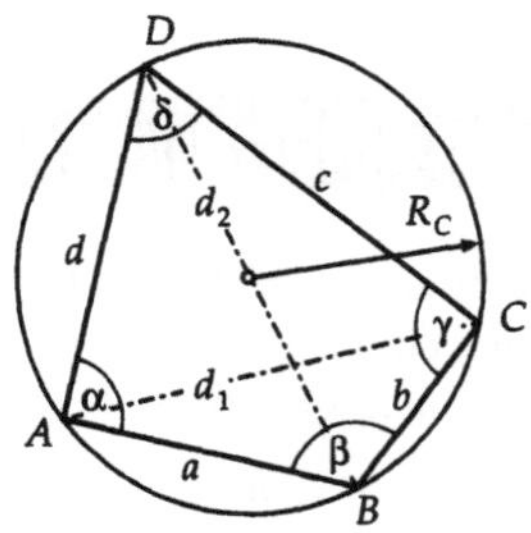

Circumradius:

$$R_C = \frac{1}{4A}\sqrt{(ab+cd)(ac+bd)(ad+bc)} .$$

Diagonals:

$$d_1 = \sqrt{\frac{(ac+bd)(bc+ad)}{ab+cd}} ,$$

$$d_2 = \sqrt{\frac{(ac+bd)(ab+cd)}{bc+ad}} .$$

Area:

$$A = \frac{d_1 d_2}{2} = \sqrt{(k-a)(k-b)(k-c)(k-d)} ,$$

with

$$k = \frac{1}{2}(a+b+c+d) .$$

Ptolemy's theorem: The product of the diagonals is equal to the sum of the products of the opposite sides:

$$d_1 d_2 = ac + bd .$$

3.6.8 Quadrilateral of tangents

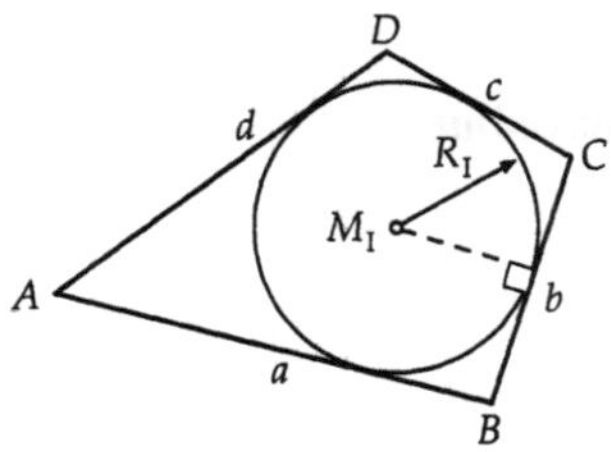

All four sides are tangents to the incircle. The sums of the lengths of two opposite sides are equal and equal to half the perimeter:

$$\frac{P}{2} = a + c = b + d \, .$$

Area:

$$A = \frac{P}{2} \cdot R_I = (a+c)R_I = (b+d)R_I \, .$$

▷ The converse also holds: A quadrilateral for which the sums of the lengths of two opposite sides are equal has an incircle. The sides are the tangents.

3.6.9 Kite

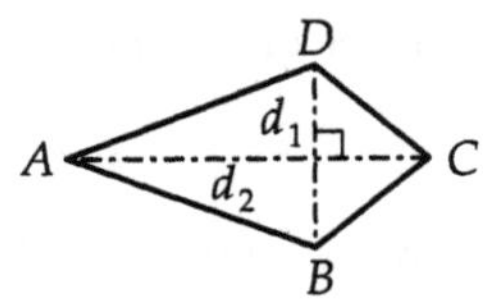

Area:

$$A = \frac{d_1 d_2}{2} \, .$$

▷ The kite is a special case of the quadrilateral of tangents.

3.7 Regular n-gons (polygons)

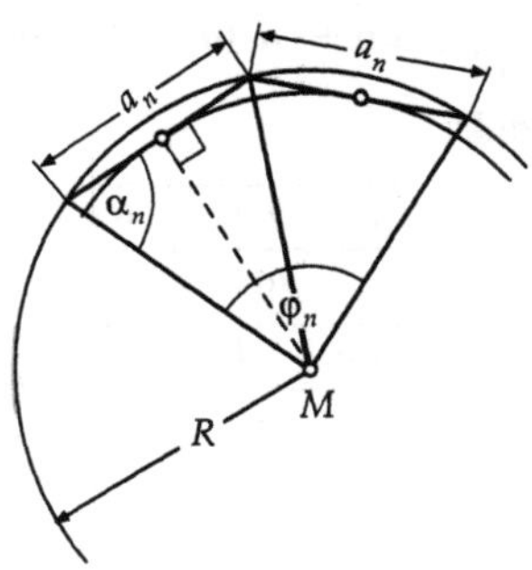

The vertices are equidistant on the circumference of a circle.
A **regular n-gon** of perimeter

$$P = n \cdot a_n$$

can be subdivided into n **isosceles** triangles with equal **base** a_n and equal **central angle**

$$\varphi_n = \frac{360°}{n} \mathrel{\hat{=}} \frac{2\pi}{n} \, .$$

The number of possible diagonals of an n-gon follows from

$$\frac{n(n-3)}{2} \, .$$

- **Euler's formula** for the plane: If V_p is the number of vertices and E_p is the number of edges, then:

$$V_p = E_p \, .$$

▷ All closed plane figures with straight boundary lines can be composed of triangles or subdivided into triangles.

□ Important application: In the **method of finite elements** for solving partial differential equations and for computer graphics.

3.7.1 General regular n-gons

There are n axes of symmetry. The n sides and n angles are equal.
Base angle:

$$\alpha_n = \left(1 - \frac{2}{n}\right) \cdot 90^\circ \,.$$

Exterior angle:

$$\alpha'_n = \frac{360^\circ}{n} \mathrel{\hat{=}} \frac{2\pi}{n} \,.$$

Sum of the angles:

$$(n-2)\pi \mathrel{\hat{=}} (n-2) \cdot 180^\circ \,.$$

Inradius:

$$R_I = \frac{a_n}{2} \cot \frac{180^\circ/n}{2} = R_C \cos\left(\frac{180^\circ}{n}\right) .$$

Circumradius:

$$R_C = \frac{a_n}{2\sin(90^\circ/n)} \,, \qquad R_C^2 = R_I^2 + \frac{1}{4} a_n^2 \,.$$

Perimeter:

$$P_n = n \cdot a_n \,.$$

Area:

$$A_n = n \frac{a_n^2}{4} \cot \frac{180^\circ/n}{2} = \frac{n}{2} a_n \cdot R_I \,.$$

Relations between the sides (a_n) and the areas (A_n) of an n-gon and a $2n$-gon:

$$a_{2n} = R_C \sqrt{2 - 2\sqrt{1 - \left(\frac{a_n}{2R_C}\right)^2}} \,, \qquad a_n = a_{2n} \sqrt{4 - \frac{a_{2n}^2}{R_C^2}} \,,$$

$$A_{2n} = \frac{n R_C^2}{\sqrt{2}} \sqrt{1 - \sqrt{1 - \frac{4A_n^2}{n^2 R_C^4}}} \,, \qquad A_n = A_{2n} \sqrt{1 - \frac{A_{2n}^2}{n^2 R_C^4}} \,.$$

3.7.2 Particular regular n-gons (polygons)

Regular triangle: See **equilateral triangle**.
Regular rectangle: See **square**.

Regular pentagon:
Length of side:

$$a_5 = \frac{R_C}{2}\sqrt{10 - 2\sqrt{5}} = 2R_I\sqrt{5 - 2\sqrt{5}}\,.$$

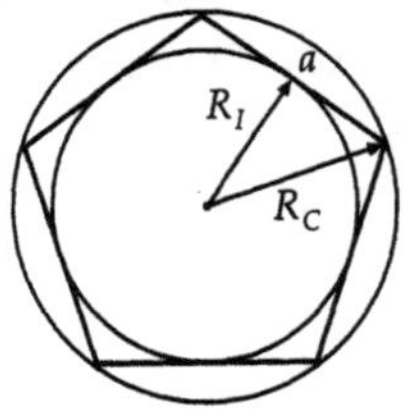

Circumradius:

$$R_C = \frac{a_5}{10}\sqrt{50 + 10\sqrt{5}} = R_I(\sqrt{5} - 1)\,.$$

Inradius:

$$R_I = \frac{a_5}{10}\sqrt{25 + 10\sqrt{5}} = \frac{R_C}{4}(\sqrt{5} + 1)\,.$$

Area:

$$A_5 = \frac{a_5^2}{4}\sqrt{25 + 10\sqrt{5}} = \frac{5R_C^2}{8}\sqrt{10 + 2\sqrt{5}} = 5R_I^2\sqrt{5 - 2\sqrt{5}}\,.$$

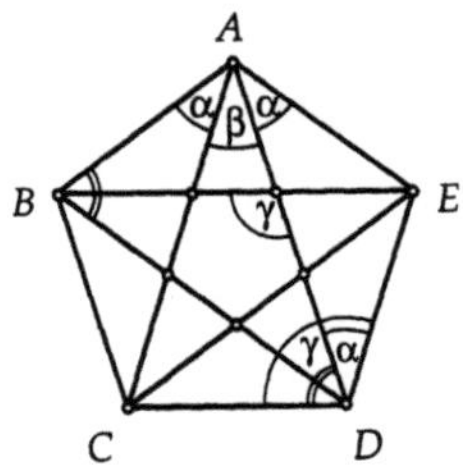

- In a regular pentagon, the diagonals form a star, the **pentagram**. Its interior is again a regular pentagon.
- The sides of the pentagram are divided according to the **golden section**.

Regular hexagon:
Length of side:

$$a_6 = \frac{2}{3}R_I\sqrt{3}\,.$$

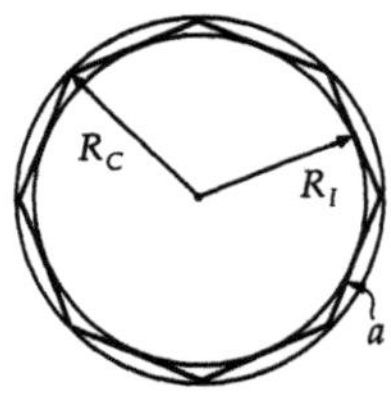

Circumradius:

$$R_C = \frac{2}{3}R_I\sqrt{3}\,.$$

Inradius:

$$R_I = \frac{R_C}{2}\sqrt{3}\,.$$

Area:

$$A_6 = \frac{3a_6^2}{2}\sqrt{3} = \frac{3R_C^2}{2}\sqrt{3} = 2R_I^2\sqrt{3}\,.$$

□ Exactly twenty regular hexagons and twelve regular pentagons can be used to construct stable polyhedrons (honeycombs, etc.).

Regular octagon:
Length of side:

$$a_8 = R_C\sqrt{2 - \sqrt{2}} = 2R_I(\sqrt{2} - 1)\,.$$

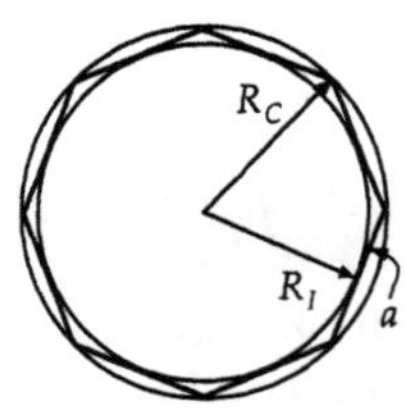

Circumradius:

$$R_C = \frac{a_8}{2}\sqrt{4 + 2\sqrt{2}} = R_I\sqrt{4 - 2\sqrt{2}}\,.$$

Inradius:

$$R_I = \frac{a_8}{2}(\sqrt{2}+1) = \frac{R_C}{2}\sqrt{2+\sqrt{2}}\,.$$

Area:

$$A_8 = 2a_8^2(\sqrt{2}+1) = 2R_C^2\sqrt{2} = 8R_I^2(\sqrt{2}-1)\,.$$

Regular decagon:
Length of side:

$$a_{10} = \frac{R_C}{2}(\sqrt{5}-1) = \frac{2R_I}{5}\sqrt{25-10\sqrt{5}}\,.$$

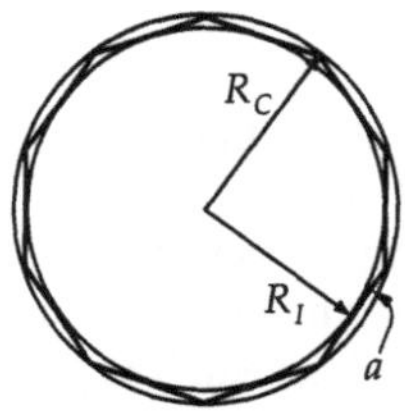

Circumradius:

$$R_C = \frac{a_{10}}{2}(\sqrt{5}+1) = \frac{R_I}{5}\sqrt{50-10\sqrt{5}}\,.$$

Inradius:

$$R_I = \frac{a_{10}}{2}\sqrt{5+2\sqrt{5}} = \frac{R_C}{4}\sqrt{10+2\sqrt{5}}\,.$$

Area:

$$A_{10} = \frac{5a_{10}^2}{2}\sqrt{5+2\sqrt{5}} = \frac{5R_C^2}{4}\sqrt{10-2\sqrt{5}} = 2R_I^2\sqrt{25-10\sqrt{5}}\,.$$

□ Complex numbers: In the complex plane, the vertices of a regular n-gon are the solution of the equation

$$z^n = a_0 e^{j\alpha}$$

with the circumradius $\sqrt[n]{a_0}$. The figure at the right indicates the solution of the equation

$$z^3 = a_0 \cdot e^{j\alpha}\,.$$

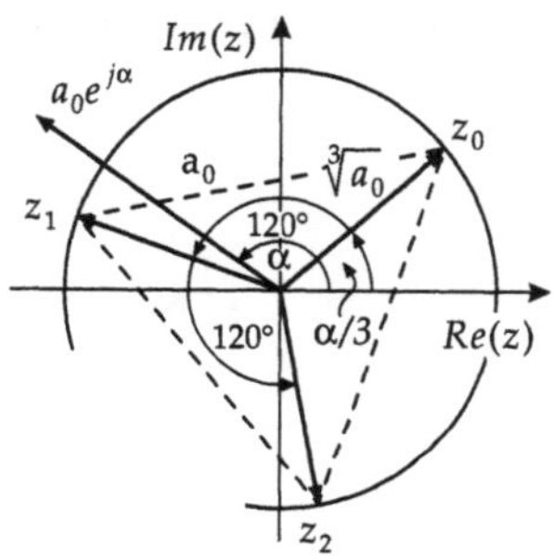

The sum for the n-th vertex is $(2n-4)\cdot 90°$ for the interior angles and 360° for the exterior angles.

3.8 Circular objects

3.8.1 Circle

Notation for circles:

Center :	M	Secant :	g
Radius :	r	Tangent :	t
Diameter :	d	Centerline :	z
Chord :	s		

Transcendental number: $\pi \approx 3.141592653$

Area: $A = \pi r^2 = \pi \frac{d^2}{4}$.

Circumference: $C = 2\pi r = \pi d$.

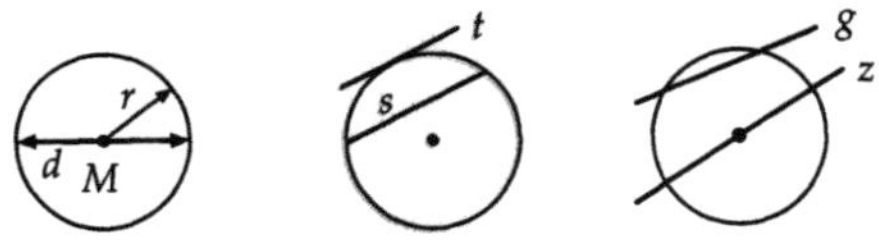

- The central **angle** ϵ is twice as large as the peripheral angle γ over the same arc ($\epsilon = 2\gamma$).
- The **angle** τ **between chord and intersept tangent** is equal to the peripheral **angle** γ belonging to the chord in the opposite τ.
- **Chord theorem**: The product of the intercepts of two intersecting chords is equal:

$$\overline{AS} \cdot \overline{SC} = \overline{BS} \cdot \overline{SD} = r^2 - s^2 .$$

- **Secant theorem**: The product of the lengths of two secants and their outer intercepts is equal:

$$\overline{PK} \cdot \overline{PL} = \overline{PE} \cdot \overline{PF} .$$

- **Secant-tangent theorem**: The product of the lengths of one secant and its outer intercept is equal to the square of the length of the tangent:

$$\overline{PE} \cdot \overline{PF} = t^2 .$$

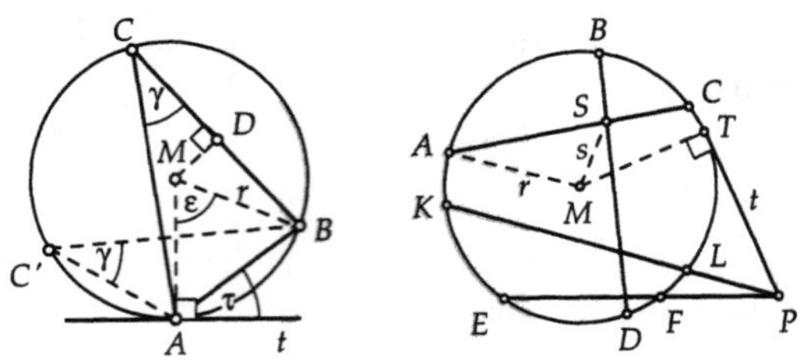

Chord, Secant and Secant-tangent theorem

3.8.2 Circular areas

Notation for circular areas:

Area :	A	Mean radius:	r_m
Circumference :	C	Outer diameter:	D
Arc length:	b	Inner diameter:	d
Center of gravity:	S	Width of the annulus:	d_M
Outer radius:	R	Altitude of the arc:	h_B
Inner radius:	r	Chord length:	s

3.8.3 Annulus, circular ring

Annulus or **circular ring:** Area between two circles with the same center (projection of doughnut shape).

Area:

$$A = \pi(R^2 - r^2) = \frac{\pi}{4}(D^2 - d^2)$$
$$= 2\pi d_M r_m \ .$$

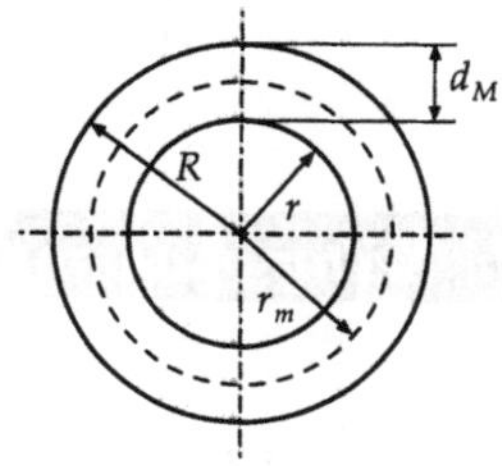

Width of the annulus:

$$d_M = R - r \ .$$

Mean radius:

$$r_m = \frac{R + r}{2} \ .$$

3.8.4 Sector of a circle

- **Sector of a circle:** Ratio of the area of the sector to the area of the circle is equal to the ratio of the angle at center φ to 360°:

$$A : A_{\text{circle}} = \varphi : 360° \ .$$

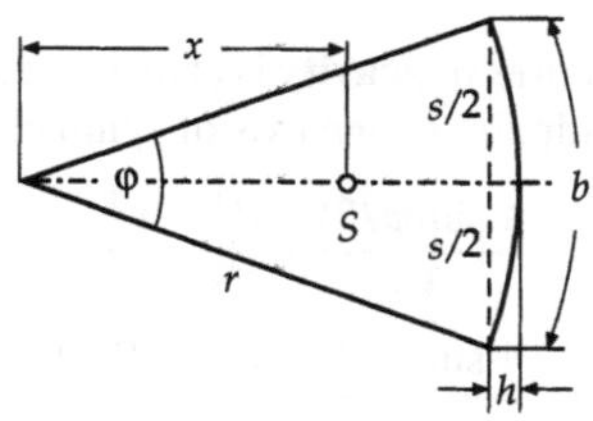

Area of the sector:

$$A = \frac{\varphi \cdot \pi r^2}{360°} \ .$$

For the central angle, analogous to the arc measure $\varphi = b/r$.

Analogous measures of area are the basis for the hyperbolic functions:

$$\sinh(x), \quad \tanh(x), \quad \coth(x)$$

and their inverse functions the area functions.

- The ratio of the **arc length of the sector** to the angle φ at the center (in degrees) is equal to the ratio of the circumference of the circle to the full angle:

$$\frac{b}{\varphi} = \frac{2\pi r}{360°} \quad \Rightarrow \quad b = \frac{\varphi}{360°} \cdot 2\pi r \ .$$

From the conversion radius to degrees follows the area of the sector of a circle:

$$A = \frac{br}{2} = \frac{r^2\varphi}{2} \quad (\varphi \text{ in radians})$$
$$= \frac{r^2}{2} \cdot \frac{\pi \cdot \varphi}{180°} \quad (\varphi \text{ in degrees}) \ .$$

Arc length:

$$b = r \cdot \varphi \quad (\varphi \text{ in radians})$$
$$= r \cdot \frac{\pi \cdot \varphi}{180°} \quad (\varphi \text{ in degrees}) \ .$$

The **center of gravity** is at distance x from the midpoint on the axis of symmetry,

$$x = \frac{4r\sin(\varphi/2)}{3\varphi} = \frac{2s}{3\varphi} = \frac{2rs}{3b} \quad (\varphi \text{ in radians})$$
$$= \frac{240^\circ}{\pi \cdot \varphi} \cdot r \cdot \sin\frac{\varphi}{2} = \frac{120^\circ}{\pi \cdot \varphi} s \quad (\varphi \text{ in degrees}) .$$

3.8.5 Sector of an annulus

Area:

$$A = \frac{\varphi}{2}(R^2 - r^2) = b_m \cdot d_M \qquad (\varphi \text{ in radians}).$$

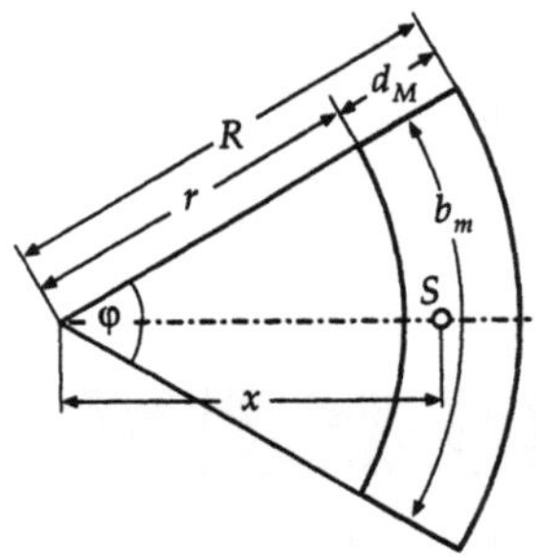

Mean arc length:

$$b_m = \frac{R+r}{2} \cdot \varphi .$$

Width of the annulus

$$d_M = R - r .$$

The **center of gravity** is at distance x from the midpoint on the axis of symmetry

$$x = \frac{4\sin(\varphi/2)}{3\varphi} \cdot \frac{R^3 - r^3}{R^2 - r^2}$$
$$= \frac{4\sin(\varphi/2)}{3\varphi} \cdot \frac{R^2 + Rr + r^2}{R+r} .$$

3.8.6 Segment of a circle

Area:

$$A = \frac{r^2}{2}(\varphi - \sin\varphi) = \frac{1}{2}[r(b-s) + sh], \quad (\varphi \text{ in radians}) ,$$
$$= \frac{r^2}{2}\left(\frac{\pi \cdot \varphi}{180^\circ} - \sin\varphi\right) \quad (\varphi \text{ in degrees}) .$$

Approximate formulas for the area:

- $A \approx \frac{2}{3} s \cdot h$ with

error $< 0.8\%$ for $0^\circ < \varphi \leq 45^\circ$;
error $< 3.3\%$ for $45^\circ < \varphi \leq 90^\circ$

- $A \approx \frac{2}{3} s \cdot h + \frac{h^3}{2s}$ with

error $< 0.1\%$ for $0^\circ < \varphi \leq 150^\circ$;
error $< 0.8\%$ for $150^\circ \leq \varphi \leq 180^\circ$.

Radius:

$$r = \frac{(s/2)^2 + h^2}{2h} \ .$$

Chord length:

$$s = 2r \sin\frac{\varphi}{2} \ .$$

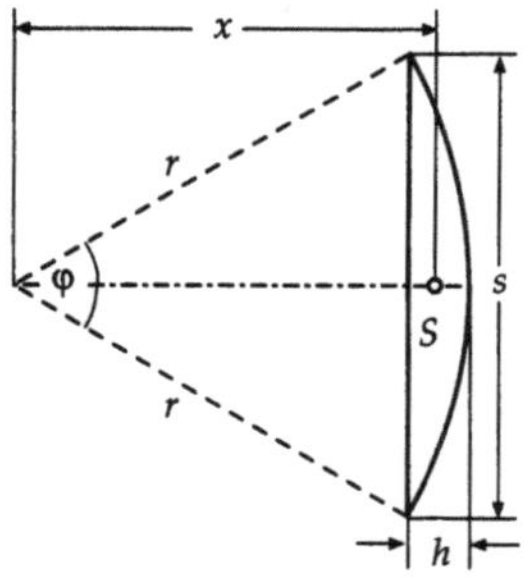

Altitude of the arc:

$$h_B = r\left(1 - \cos\frac{\varphi}{2}\right) = \frac{s}{2}\tan\frac{\varphi}{4} = 2r \sin^2\frac{\varphi}{4} \ .$$

The **center of gravity** is at distance x from the midpoint on the axis of symmetry,

$$x = \frac{s^3}{12 \cdot A} \ .$$

Arc length:

$$\begin{aligned} b &= r \cdot \varphi \quad (\varphi \text{ in radians}) \ , \\ &= r\frac{\pi \cdot \varphi}{180^\circ} \quad (\varphi \text{ in degrees}) \ , \\ &\approx \sqrt{s^2 + \frac{16}{3}h^2} \quad (\text{with error} < 0.3\% \text{ for } 0^\circ < \varphi \leq 90^\circ). \end{aligned}$$

3.8.7 *Ellipse*

Notation for ellipse:

Major semi-axis: a

Minor semi-axis: b

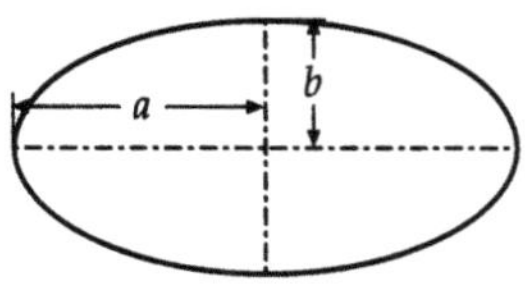

Area:

$$A = \pi ab \ .$$

Circumference:

$$C \approx \pi\left[1.5(a+b) - \sqrt{ab}\right] \ .$$

▷ The calculation of the circumference of the ellipse leads to the so-called **elliptic integral** of the second kind, which can be given in terms of an expansion in a series (see 15.6).

Center of gravity: The point of intersection of the major and minor axis.

4

Solid geometry

Notation for solid geometry:

Edges:	a, b, c
Slant height:	s
Base:	A_B
Side face:	A_S
Surface:	A_O
Top base:	A_T
Lateral surface:	A_L
Mean area:	A_m
Circumference (of the base):	C_B
Height of the solid:	h
Height of the side face:	h_S
Volume:	V
Radius of the circle or the sphere:	R
Diameter of the circle or the sphere:	d
Face diagonal:	d_F
Body diagonal:	d_B
Radius of the inscribed sphere, inradius:	R_I
Radius of the circumscribed sphere, circumradius:	R_C

4.1 General theorems

4.1.1 Cavalieri's theorem

- All solids with cross sections of equal areas at the same height have the same volume.

4.1.2 Simpson's rule

Simpson's rule can be used to perform approximate calculations for irregular solids. The rule also includes the calculation of the volume of a barrel and of the frustums of a pyramid or a cone.

Rule of thumb:

$$V \approx h \cdot A_B \,.$$

- Simpson's rule:

$$V \approx h\left(\frac{2}{3}A_m + \frac{1}{6}A_B + \frac{1}{6}A_T\right) \approx \frac{h}{6}(A_B + 4A_m + A_T)\,.$$

▷ In many cases this rule leads to inaccurate results! For more precise methods, integral calculus is required.

4.1.3 Guldin's rules

- **First rule:** The **area** of the lateral surface **of a solid of rotation** is equal to the product of the length of the generating curve on one side of the axis of rotation and the length of the path described by the center of gravity of the generating curve under a full revolution.
- **Second rule:** The **volume of a solid of rotation** is equal to the product of the area of the generating plane on one side of the rotation axis and the length of the path described by the center of gravity of the area under a full revolution about the axis.

□ **Torus:** Ring with a circular cross section,

$$A = 2\pi r \cdot 2\pi R\,.$$

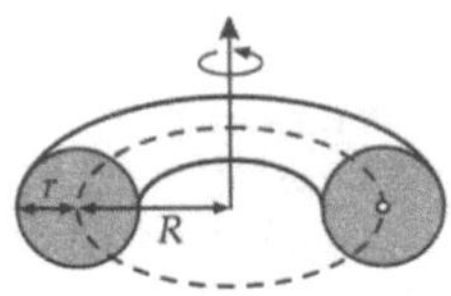

4.2 Prism

Prisms are bounded by two **congruent n-gons**, lying in parallel planes, and n **parallelograms**.

4.2.1 Oblique prism

Volume:

$$V = A_B \cdot h\,.$$

Surface:

$$A_O = A_L + 2A_B\,.$$

Center of gravity: Midpoint of the line joining the center of gravity of the two congruent n-gons.

□ **Parallelepiped**: An oblique prism with a parallelogram as its base.

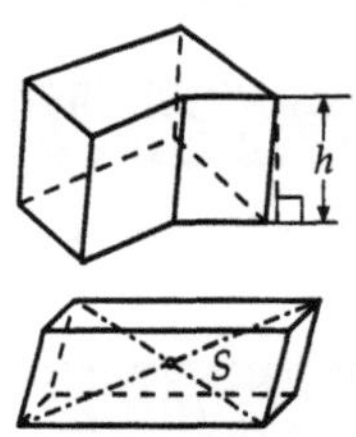

4.2.2 Right prism

Right prism: Edges are perpendicular to the parallel planes.
Volume:

$$V = A_B \cdot h \ .$$

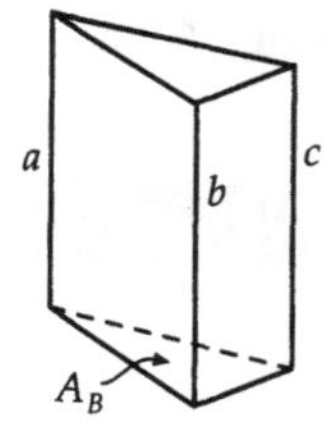

Lateral surface:

$$A_L = C_B \cdot h \ .$$

Surface:

$$A_O = C_B \cdot h + 2A_B \ .$$

Regular prisms: Right prisms whose bases are regular n-angles.
Cuboids and **cubes** are special cases of right prisms.

4.2.3 Cuboid

Cuboid: A right prism whose base is a rectangle.
Edge lengths: a, b, c .
Volume:

$$V = a \cdot b \cdot c \ .$$

Surface:

$$A_O = 2(ab + ac + bc) \ .$$

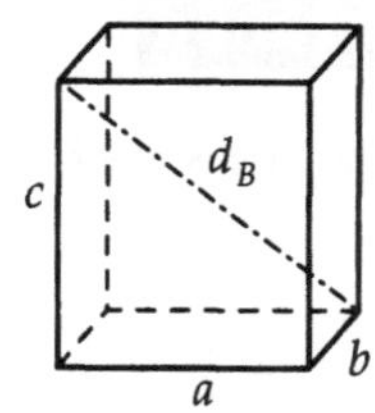

Body diagonal:

$$d_B = \sqrt{a^2 + b^2 + c^2} \ .$$

Center of gravity: The point of intersection of the body diagonals.

4.2.4 Cube

Cube: A cuboid with equal edges.
Edge lengths: a .
Volume:

$$V = a^3 \ .$$

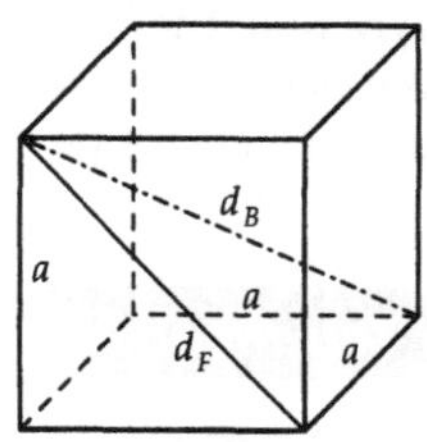

Surface:

$$A_O = 6a^2 \ .$$

Face diagonal:

$$d_F = a\sqrt{2} \ .$$

Body diagonal:

$$d_B = a\sqrt{3} \ .$$

Inradius:

$$R_I = \frac{a}{2} \ .$$

Circumradius:

$$R_C = \frac{d_B}{2} .$$

Center of gravity: Point of intersection of the body diagonals.

4.2.5 Obliquely truncated n-sided prism

Volume:

$$V = A_Q \cdot s_{DG} .$$

s_{DG}: line joining the centers of gravity of the base and the top base.
A_Q: area of a cross section perpendicular to s_{DG}.

□ Obliquely truncated 3-sided prism.
Volume:

$$V = A_B \frac{a+b+c}{3} \quad \text{(right)} ,$$

$$V = A_Q \frac{a+b+c}{3} \quad \text{(oblique)} .$$

4.3 Pyramid

Pyramid: The base is an arbitrary plane n-gon, the side-faces are triangles with a common vertex.

Volume:

$$V = \frac{1}{3} A_B \cdot h .$$

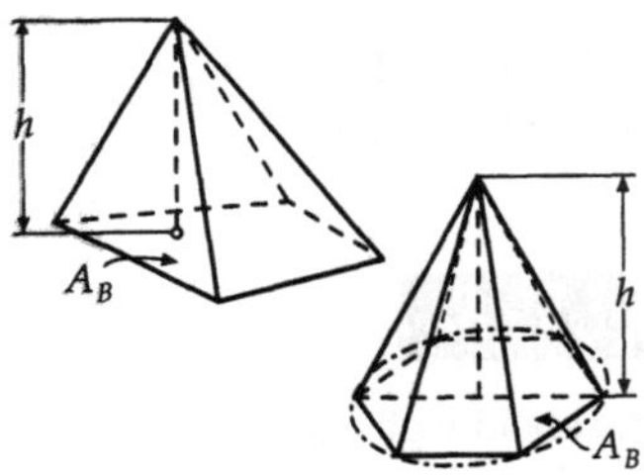

Surface:

$$A_O = A_B + A_L .$$

Center of gravity: At the distance $x = h/4$ from the base, on the line joining the vertex to the center of gravity of the base.
Regular pyramid: The base is a regular n-gon whose midpoint is the foot of the perpendicular through the vertex.

4.3.1 Tetrahedron

Tetrahedron: A pyramid whose base is a triangle.

Volume:

$$V = \frac{1}{3} A_B \cdot h .$$

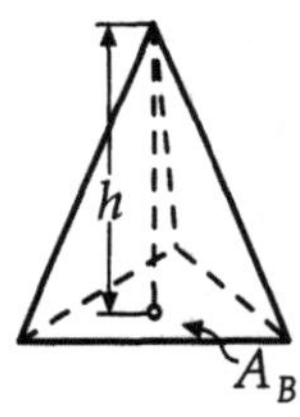

Regular tetrahedron: All surfaces are equilateral triangles.

4.3.2 Frustum of a pyramid

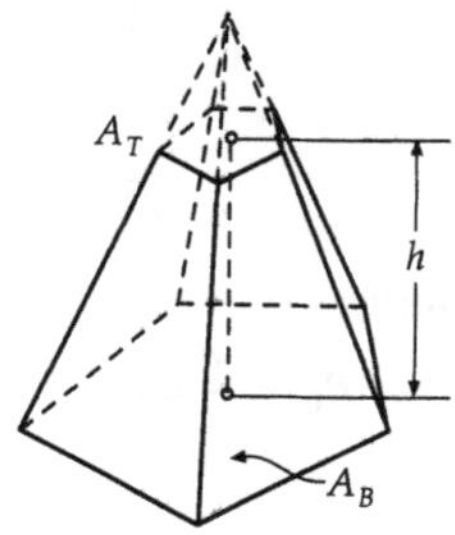

Frustum of a pyramid: Similar bases in parallel planes. Faces are trapezoids. The extensions of the side-edges traverse a single point.

Volume:

$$V = \frac{h}{3}(A_B + \sqrt{A_B A_T} + A_T) .$$

Surface:

$$A_O = A_T + A_B + A_L .$$

Center of gravity: At the distance x from the base on the line joining the center of gravity of the base to that of the top:

$$x = \frac{h}{4} \frac{A_B + 2\sqrt{A_T A_B} + 3A_T}{A_B + \sqrt{A_T A_B} + A_T} .$$

▷ Approximate formula for the volume (for $A_T \approx A_B$):

$$V \approx h \cdot \frac{A_T + A_B}{2} .$$

4.4 Regular polyhedron

Polyhedron: Three-dimensional solid bounded by planes.
Regular polyhedron: Bounded by regular congruent n-gons. The same number of faces meet at each vertex. A regular polyhedron can be circumscribed in a sphere.
Convex polyhedron: Polyhedron without reentrant vertices.
Edges: Cut lines of the planes bounding the solid.
Vertices: Points of intersection of edges.

4.4.1 Euler's theorem for polyhedrons

- If V_p is the number of vertices, F_p is the number of faces, and E_p is the number of edges, then for **convex polyhedrons**:

$$V_p + F_p = E_p + 2 .$$

▷ From Euler's theorem for polyhedrons it follows that there are only five regular polyhedrons: tetrahedron, hexahedron, octahedron, dodecahedrons, and icosahedron.

4.4.2 Tetrahedron

Tetrahedron: Bounded by four equilateral triangles with side a.
($V_p = 4$, $E_p = 6$, $F_p = 4$)

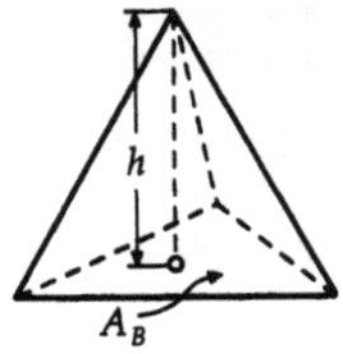

Volume:

$$V = \frac{a^3}{12}\sqrt{2}\,.$$

Surface:

$$A_O = a^2\sqrt{3}\,.$$

Circumradius:

$$R_C = \frac{a}{4}\sqrt{6}\,.$$

Inradius:

$$R_I = \frac{a}{12}\sqrt{6}\,.$$

4.4.3 Cube (hexahedron)

Cube: Bounded by six squares with side a.
($V_p = 8$, $E_p = 12$, $F_p = 6$)

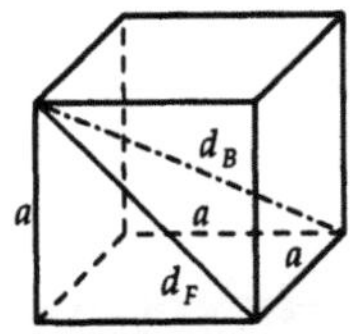

Volume:

$$V = a^3\,.$$

Surface:

$$A_O = 6a^2\,.$$

Circumradius:

$$R_C = \frac{a}{2}\sqrt{3}\,.$$

Inradius:

$$R_I = \frac{a}{2}\,.$$

4.4.4 Octahedron

Octahedron: Bounded by eight equilateral triangles with the side a.
($V_p = 6$, $E_p = 12$, $F_p = 8$)

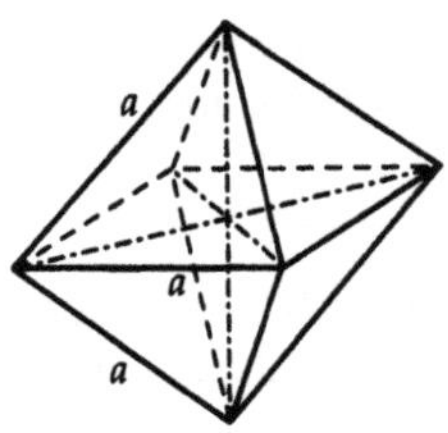

Volume:

$$V = \frac{a^3}{3}\sqrt{2}\,.$$

Surface:

$$A_O = 2a^2\sqrt{3}\,.$$

Circumradius:

$$R_C = \frac{a}{2}\sqrt{2}\,.$$

Inradius:

$$R_I = \frac{a}{6}\sqrt{6}\,.$$

4.4.5 Dodecahedron

Dodecahedron: Bounded by twelve regular pentagons with side a.
$(V_p = 20, E_p = 30, F_p = 12)$
Volume:

$$V = \frac{a^3}{4}(15 + 7\sqrt{5}) \,.$$

Surface:

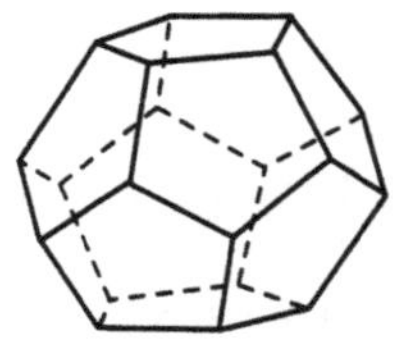

$$A_O = 3a^2\sqrt{5(5 + 2\sqrt{5})} \,.$$

Circumradius:

$$R_C = \frac{a}{4}\sqrt{3}(1 + \sqrt{5}) \,.$$

Inradius:

$$R_I = \frac{a}{4}\sqrt{10 + 22\sqrt{0.2}} \,.$$

4.4.6 Icosahedron

Icosahedron: Bounded by twenty isosceles triangles with side a.
$(V_p = 12, E_p = 30, F_p = 20)$
Volume:

$$V = \frac{5}{12}a^3(3 + \sqrt{5}) \,.$$

Surface:

$$A_O = 5a^2\sqrt{3} \,.$$

Circumradius:

$$R_C = \frac{a}{4}\sqrt{2(5 + \sqrt{5})} \,.$$

Inradius:

$$R_I = \frac{a}{12}\sqrt{3}(5 + \sqrt{5}) \,.$$

- ☐ Examples of **Archimedian solids**:
  (a) bounded by eight squares and eight isosceles triangles,
  (b) bounded by four isosceles triangles and four regular hexagons.
- ☐ C_{60}**-fullerenes** (**"Bucky balls"**), consisting of 12 pentagons and 20 hexagons. On all vertices of the truncated icosahedron there are carbon atoms. This is the recently discovered structure of pure carbon (besides diamonds [tetrahedron] and graphite), with unusual properties in terms of superconductivity and lubricity.

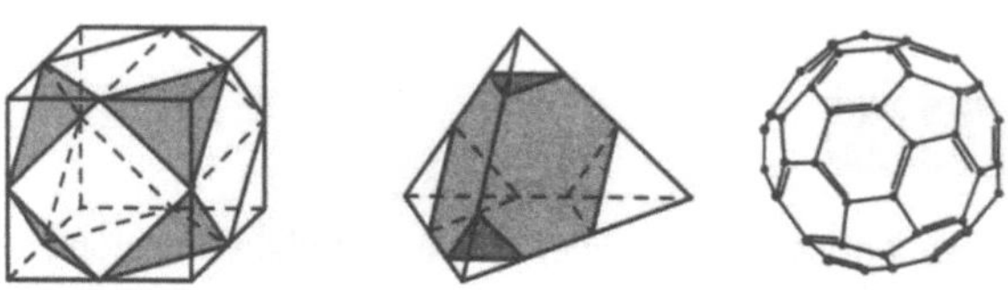

4.5 Other solids

4.5.1 Prismoid, prismatoid

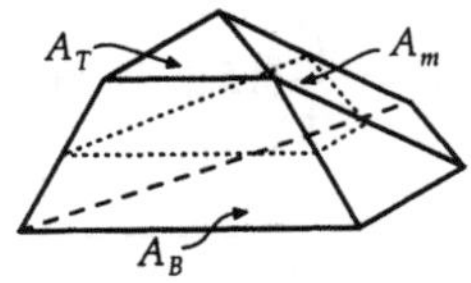

Prismoid (prismatoid): Solid with straight edges and plane, partly curved faces. Vertices or bases are lying on two parallel planes.
Volume:

$$V = \frac{h}{6}(A_T + 4A_m + A_B) .$$

A_m: mean area of cross section.
▷ The frustum of a cone, the wedge ($A_T = 0$), and the obelisk are special cases of prismoids.

4.5.2 Wedge

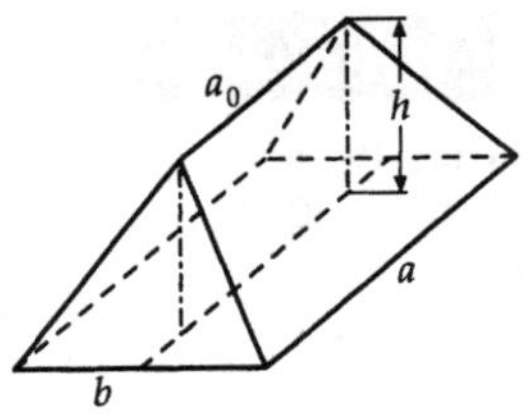

Wedge: The base is a rectangle, the side faces are isosceles triangles or trapezoids. The faces meet at the top in one edge a_0.
Volume:

$$V = \frac{bh}{6}(2a + a_0) .$$

The distance of the **center of gravity** from the base is:

$$x = \frac{h}{2}\frac{a + a_0}{2a + a_0} .$$

4.5.3 Obelisk

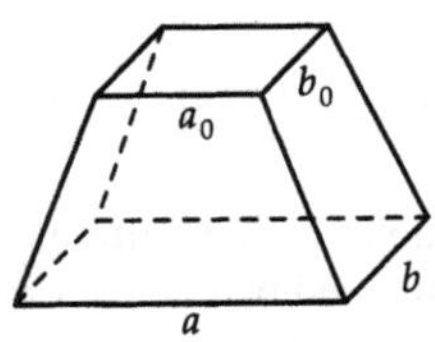

Obelisk: The base and the top base are non-similar parallel rectangles, the faces are trapezoids.

$$\begin{aligned} V &= \frac{h}{6}[(2a + a_0)b + (2a_0 + a)b_0] \\ &= \frac{h}{6}[ab + (a + a_0)(b + b_0) + a_0b_0] . \end{aligned}$$

The distance of the **center of gravity** from the base is

$$x = \frac{h}{2}\frac{ab + ab_0 + a_0b + 3a_0b_0}{2ab + ab_0 + a_0b + 2a_0b_0} .$$

4.6 Cylinder

Cylinder surfaces: Formed by a parallel shift of a straight line along a closed curve.
Cylinder: Formed if the cylindrical surface is intersected by two parallel planes.

4.6.1 General cylinder

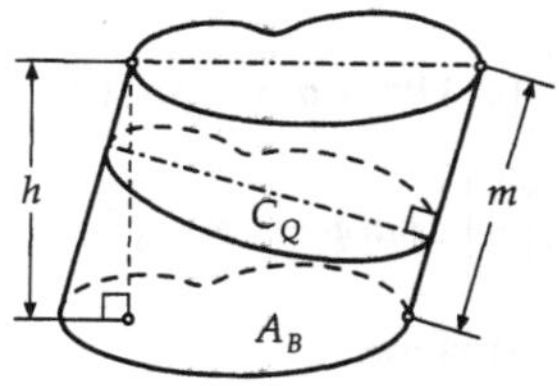

Volume:

$$V = A_B \cdot h \, .$$

Lateral surface:

$$A_L = m \cdot C_Q \, .$$

C_Q: Perimeter of the cross section normal to the axis.

Surface:

$$A_O = 2A_B + A_L \, .$$

4.6.2 Right circular cylinder

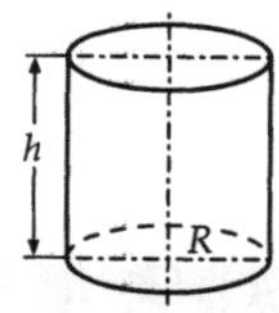

Volume:

$$V = \pi R^2 h \, .$$

Lateral surface:

$$A_L = 2\pi Rh \, .$$

Surface:

$$A_O = 2\pi R(R + h) \, .$$

Center of gravity: On the axis of symmetry at a distance $x = h/2$ from the base.

4.6.3 Obliquely cut circular cylinder

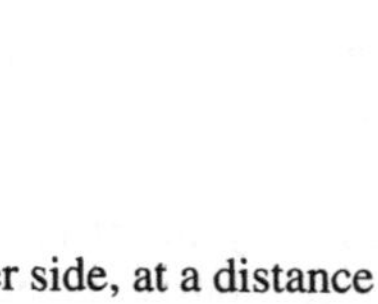

Volume:

$$V = \frac{\pi R^2}{2}(s_a + s_b) \, .$$

Lateral surface:

$$A_L = \pi R(s_a + s_b) \, .$$

Surface:

$$A_O = \pi R \left[s_a + s_b + R + \sqrt{R^2 + \left(\frac{s_a - s_b}{2}\right)^2} \right] \, .$$

Center of gravity: On the axis, shifted by x_s to the higher side, at a distance y_s from the base

$$x_s = \frac{R^2 \tan\alpha}{4s_b} \, ,$$

$$y_s = \frac{4s_b^2 + r^2 \tan^2\alpha}{8s_b} s_a + s_b \, .$$

Here α is the inclination of the top base with respect to the base.

4.6.4 Segment of a cylinder

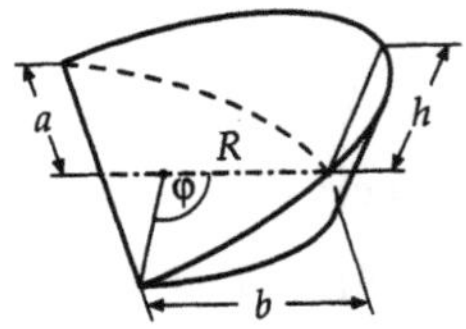

Volume:

$$V = \frac{h}{3b}[a(3R^2 - a^2) + 3R^2(b - R)\varphi]$$
$$= \frac{hR^3}{3b}\left[2\sin\varphi - \cos\varphi\left(3\varphi - \frac{1}{2}\sin 2\varphi\right)\right] .$$

Lateral surface:

$$A_L = \frac{2Rh}{b}[(b - R)\varphi + a]$$
$$= \frac{2Rh}{1 - \cos\varphi}(\sin\varphi - \varphi\cos\varphi) .$$

Special case: For $a = b = R$ (the base is half the area of a circle),

$$V = \frac{2}{3}hR^2 ,$$
$$A_L = 2Rh ,$$
$$A_O = A_L + \frac{\pi}{2}R^2 + \frac{\pi}{2}R\sqrt{R^2 + h^2} .$$

4.6.5 Hollow cylinder (tube)

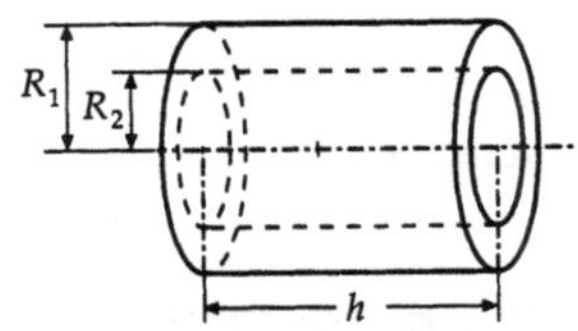

Volume:

$$V = \pi h(R_1^2 - R_2^2) .$$

Lateral surface:

$$A_L = 2\pi h(R_1 + R_2) .$$

Surface:

$$A_O = 2\pi(R_1 + R_2)(h + R_1 - R_2) .$$

Center of gravity:
On the axis of symmetry at a distance $x = h/2$ from the base.

4.7 Cone

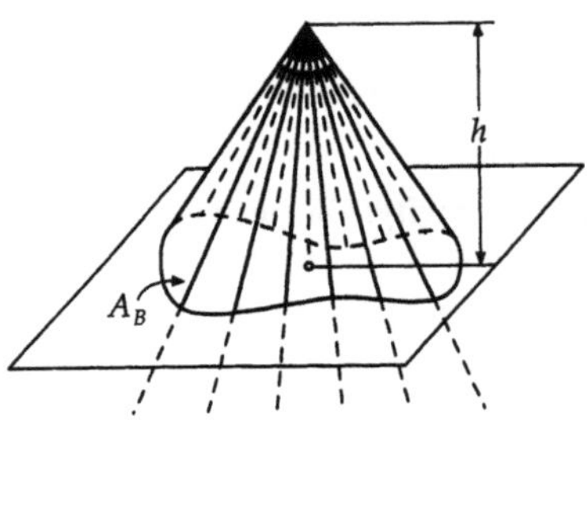

Conical surface: Described by a straight line that traverses a fixed point (the **vertex**) along a curve.
Cone: Formed by intersecting a conical surface with an arbitrary plane.
Volume:

$$V = \frac{1}{3}A_B \cdot h .$$

Surface:

$$A_O = A_B + A_L .$$

h is the perpendicular distance between the vertex and the base.

4.7.1 Right circular cone

Right circular cone: Circle whose base is radius R. The vertex at height h lies perpendicular to the center of the circle.

⊳ For conics (circles, ellipses, parabolas, and hyperbolas), see Chapter 5 on functions.

Volume:

$$V = \frac{\pi R^2 h}{3} .$$

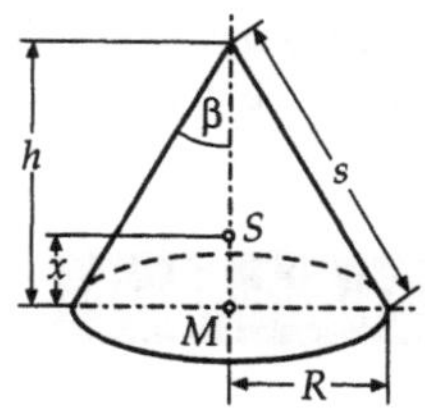

Lateral surface:

$$A_L = \pi \cdot R \cdot s .$$

Surface:

$$A_O = \pi R(R + s) .$$

Slant height:

$$s = \sqrt{R^2 + h^2} .$$

Center of gravity: On the axis of symmetry at a distance $x = h/4$ from the base.
The developed **surface of a right circular cone** is the sector of a circle with α as the central angle. Then

$$\alpha = 2\pi \frac{R}{s} = 2\pi \sin \beta ,$$

where β is half the vertex angle.

4.7.2 Frustum of a right circular cone

Frustum of a right circular cone: The base and the top are parallel circles.
Volume:

$$V = \frac{1}{3}\pi h(R_1^2 + R_1 R_2 + R_2^2) .$$

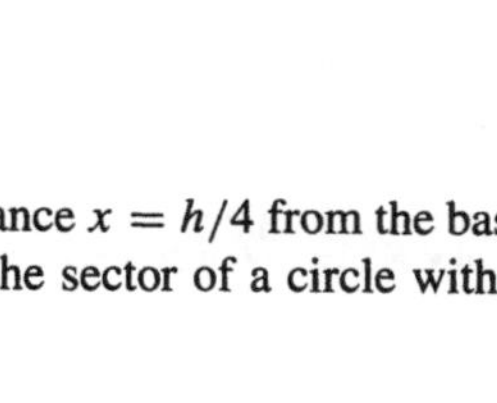

Lateral surface:

$$A_L = \pi s(R_1 + R_2) .$$

Surface:

$$A_O = \pi[R_1^2 + R_2^2 + s(R_1 + R_2)] .$$

Slant height:

$$s = \sqrt{h^2 + (R_1 - R_2)^2} .$$

Center of gravity: On the axis of symmetry, at a distance

$$x = \frac{h}{4} \frac{R_1^2 + 2R_1 R_2 + 3R_2^2}{R_1^2 + R_1 R_2 + R_2^2}$$

from the base.
Approximate formula for the volume of a frustum of a cone with $R_1 \approx R_2$:

$$V \approx \frac{\pi}{2} h(R_1^2 + R_2^2) \approx \frac{\pi}{4} h(R_1 + R_2)^2 .$$

4.8 Sphere

Solid angle: Can be measured by the ratio of area A that is cut off by a sphere about its vertex to the square of the radius of the sphere.
Unit of the solid angle: One **steradian** (1 sr), the angle for which the ratio of the spherical surface to the square of the radius of the sphere equals one.
Full solid angle ($4\pi = A_O$): The top of the unit sphere (analogous to the circumference 2π of the unit circle in two dimensions).

4.8.1 Solid sphere

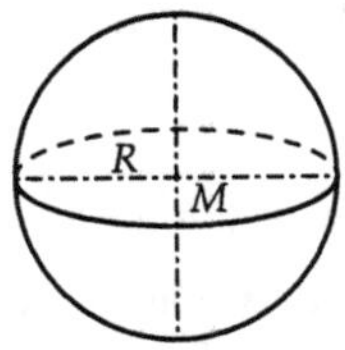

Volume:

$$V = \frac{4}{3}\pi R^3 = \frac{1}{6}\sqrt{\frac{A_O^3}{\pi}} = \frac{\pi}{6}d^3 .$$

Surface:

$$A_O = 4\pi R^2 = \sqrt[3]{36\pi V^2} = \pi d^2 .$$

Radius:

$$R = \frac{1}{2}\sqrt{\frac{A_O}{\pi}} = \sqrt[3]{\frac{3V}{4\pi}} .$$

Diameter:

$$d = \sqrt{\frac{A_O}{\pi}} = 2\sqrt[3]{\frac{3V}{4\pi}} .$$

Center of gravity: Lies in the center of the sphere.

4.8.2 Hollow sphere

Volume:

$$V = \frac{4}{3}\pi(r_2^3 - r_1^3) = \frac{1}{6}\pi(d_2^3 - d_1^3) .$$

Index 1 refers to the inner, index 2 to the outer radius or diameter.

4.8.3 Spherical sector

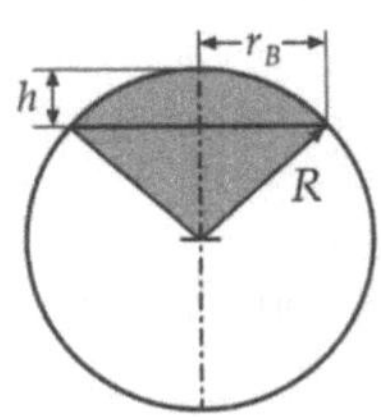

Volume:

$$V = \frac{2\pi R^2 h}{3} .$$

Surface:

$$A_O = \pi R(2h + r_B) ,$$

where r_B is the radius of the intersecting circle.

Center of gravity: On the axis of symmetry of the sector at a distance

$$x = \frac{3}{8}(2R - h)$$

from the center of the sphere.

4.8.4 Spherical segment (spherical cap)

Volume:

$$V = \frac{1}{6}\pi h(3r_B^2 + h^2) = \frac{1}{3}\pi h^2(3R - h)$$
$$= \frac{1}{6}\pi h^2(3d - 2h) .$$

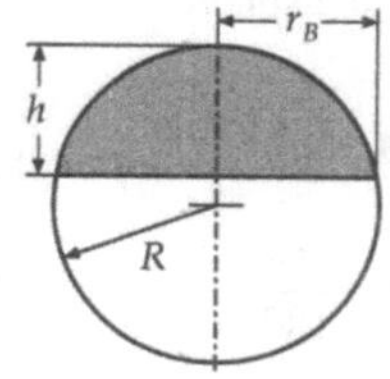

Surface:

$$A_O = \pi(2Rh + r_B^2) = \pi(h^2 + 2r_B^2)$$
$$= \pi h(4R - h) .$$

Lateral surface:

$$A_L = 2\pi Rh = \pi(r_B^2 + h^2) .$$

Radius of base circle:

$$r_B = \sqrt{h(2R - h)} .$$

Center of gravity: On the axis of symmetry at a distance

$$x = \frac{3}{4}\frac{(2R - h)^2}{(3R - h)}$$

from the center of the sphere.

4.8.5 Spherical zone (spherical layer)

Spherical zone: Difference between the volumes of spherical segments.
Volume:

$$V = \frac{\pi h}{6}(3r_1^2 + 3r_2^2 + h^2) .$$

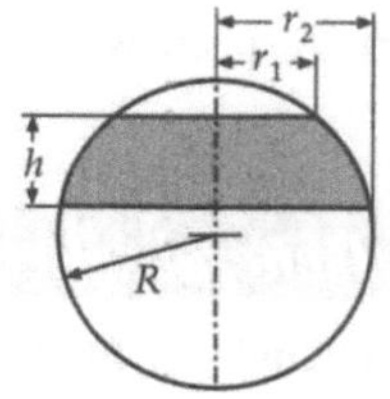

Surface:

$$A_O = \pi(2Rh + r_1^2 + r_2^2)$$
$$= \pi(dh + r_1^2 + r_2^2) .$$

Lateral surface:

$$A_L = 2\pi Rh .$$

Index 1 refers to the upper, index 2 to the lower radius of the limiting circle. If the center of the sphere is not located inside the spherical zone, the following also applies to the **altitude**

$$h = \sqrt{R^2 - r_1^2} - \sqrt{R^2 - r_2^2} \quad (r_1 < r_2) .$$

Radius of the sphere:

$$R^2 = r_2^2 + \left(\frac{r_2^2 - r_1^2 - h^2}{2h}\right)^2 \quad (r_1 < r_2)\,.$$

4.8.6 Spherical wedge

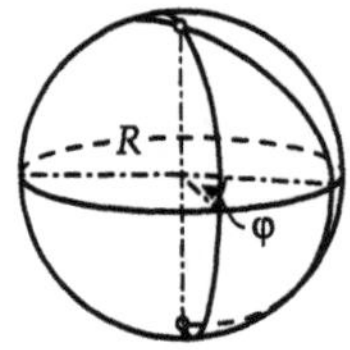

Spherical wedge: Bounded by a spherical surface and two planes intersecting in a diameter of the sphere.

Volume:

$$V = \frac{2}{3}R^3\varphi \qquad (\varphi \text{ in radians})$$

$$= \frac{2}{3}R^3 \cdot \frac{\pi\varphi}{180^\circ} \qquad (\varphi \text{ in degrees})\,.$$

Lateral surface:

$$A_L = 2R^2\varphi \qquad (\varphi \text{ in radians})$$

$$= 2R^2 \cdot \frac{\pi\varphi}{180^\circ} \qquad (\varphi \text{ in degrees})\,.$$

4.9 Spherical geometry

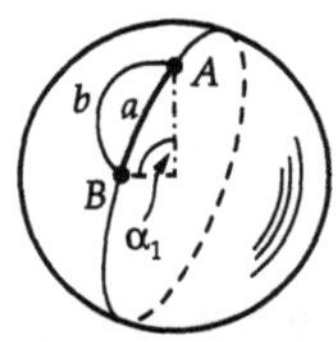

Small circle: Formed when a sphere is intersected by a plane that does not pass through the center.

Great circle: Arises when we intersect a sphere by a plane passing through the center. The radius of the great circle is the radius of the sphere.

Geodesic line: Shortest distance between two points on a spherical surface; lies on a great circle.

Unit sphere: A sphere with radius $R = 1$.

4.9.1 General spherical triangle (Euler's triangle)

Spherical triangle, or **Euler's triangle**: Determined by three **great circular arcs** not intersecting in the end-points of a spherical diameter or by three arbitrary points on the **spherical surface**, no two of which are diametrically opposite. The greater of two sides is opposite the greater angle.

- In a spherical triangle, by definition:

$$a, b, c < \pi \quad \text{and} \quad \alpha, \beta, \gamma < \pi R$$

where R is the radius of the sphere.

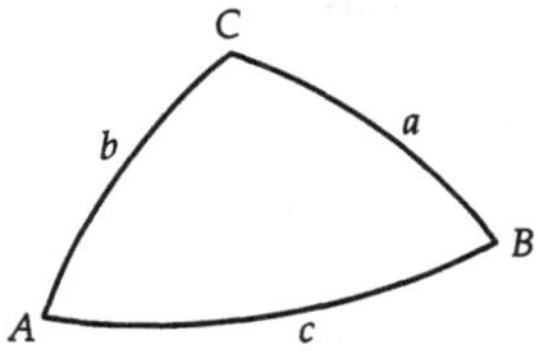

Sum of the sides:

$$S = a + b + c \,, \quad 0 < S < 2\pi .$$

Sum of the angles:

$$W = \alpha + \beta + \gamma \,, \quad \pi < W < 3\pi .$$

Spherical defect:

$$D = 2\pi - S \,.$$

Spherical excess:

$$E = W - \pi \,.$$

Perimeter:

$$P = R \cdot S \,.$$

Area of the surface:

$$A = R^2 \cdot E \,.$$

4.9.2 Right-angled spherical triangle

Right-angled spherical triangle: A **right** angle (γ) appears, since the sum of the angles is greater than π, α and/or β can be a right angle or an obtuse angle.

a, b: other sides,

c: hypotenuse,

α, β: angles opposite sides a, b.

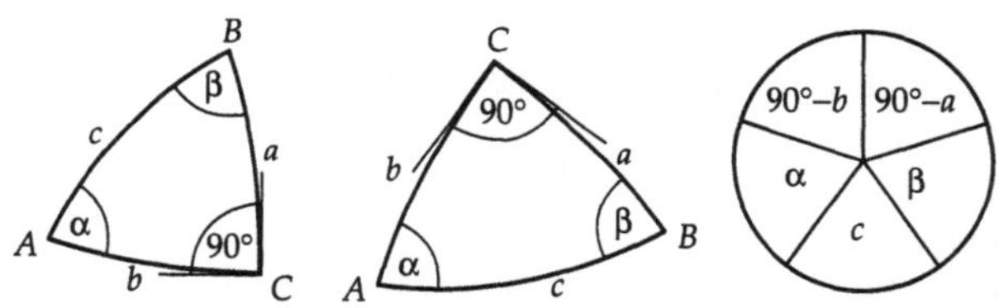

- **Napier's rules**: If one arranges the five parts of a right-angled spherical triangle (without the right angle) on a circle in the order in which they are arranged in the triangle, and if we replace the other sides by their complementary angles, then:

1. The cosine of every part is equal to the product of the cotangents of its adjacent parts.

2. The cosine of every part is equal to the product of the sines of the non-adjacent parts.

□ $\cos\alpha = \cot(90° - b)\cot c$, $\cos(90° - a) = \sin c \sin\alpha$

Main relations:

1. $\sin a = \sin c \sin\alpha$

2. $\sin b = \sin c \sin\beta$

3. $\tan a = \sin b \tan \alpha$

4. $\tan b = \sin a \tan \beta$

5. $\cos c = \cos a \cos b$

6. $\tan a = \tan c \cos \beta$

7. $\tan b = \tan c \cos \alpha$

8. $\cos \beta = \cos b \sin \alpha$

9. $\cos \alpha = \cos a \sin \beta$

10. $\cos c = \cot \alpha \cot \beta$.

When two parts of a right-angled spherical triangle are known, use these formulas to determine the remaining parts: *Given*: hypotenuse (c), angle (α).
Numbers of the formulas for determining the remaining parts:

$$a\ (1.) \quad b\ (7.) \quad \beta\ (10.)\ .$$

Given: side (a) and opposite angle (α).
Numbers of the formulas for determining the remaining parts:

$$b\ (3.) \quad c\ (1.) \quad \beta\ (9.)\ .$$

Given: side (a) and adjacent angle (β).
Numbers of the formulas for determining the remaining parts:

$$b\ (4.) \quad c\ (6.) \quad \alpha\ (9.)\ .$$

Given: side (a) and side (b).
Numbers of the formulas for determining the remaining parts:

$$c\ (5.) \quad \alpha\ (3.) \quad \beta\ (4.)\ .$$

Given: angle (α) and angle (β).
Numbers of the formulas for determining the remaining parts:

$$a\ (9.) \quad b\ (8.) \quad c\ (10.)\ .$$

4.9.3 Oblique spherical triangle

α, β, γ : angles,
a, b, c : sides opposite the angles.
Main relations:

1. $\dfrac{\sin a}{\sin \alpha} = \dfrac{\sin b}{\sin \beta} = \dfrac{\sin c}{\sin \gamma}$ **sine theorem**

2. $\cos a = \cos b \cos c + \sin b \sin c \cos \alpha$ **side cosine theorem**

3. $\cos \alpha = -\cos \beta \cos \gamma + \sin \beta \sin \gamma \cos a$ **angle cosine theorem**

4. $\sin a \cot b = \cot \beta \sin \gamma + \cos a \cos \gamma$

5. $\sin \alpha \cot \beta = \cot b \sin c - \cos \alpha \cos c$.

Given: side (a), side (b), side (c).
Numbers of the formulas for determining the remaining parts:

$$\alpha \ (2.) \quad \beta \quad \text{and} \quad \gamma \ (1.) \ .$$

Given: angle (α), angle (β), angle (γ).
Numbers of the formulas for determining the remaining parts:

$$a \ (3.) \quad b \quad \text{and} \quad c \ (1.) \ .$$

Given: side (a), side (b), included angle (γ).
Numbers of the formulas for determining the remaining parts:

$$\beta \ (4.) \quad \alpha \quad \text{and} \quad c \ (1.) \ .$$

Given: angle (α), angle (β), included side (c).
Numbers of the formulas for determining the remaining parts:

$$b \ (5.) \quad a \quad \text{and} \quad \gamma \ (1.) \ .$$

Given: side (a), side (b), one opposite angle (β).
Numbers of the formulas for determining the remaining parts:

$$\alpha \ (1.) \quad c \ (5.) \quad \gamma \ (1.) \ .$$

Given: angle (α), angle (β), one opposite side (b).
Numbers of the formulas for determining the remaining parts:

$$a \ (1.) \quad \gamma \ (4.) \quad c \ (1.) \ .$$

4.10 Solids of rotation

Solid of rotation: Formed by rotating a plane about an axis of symmetry.

▷ Rotations about different rotational axes lead to different solids of rotation.

4.10.1 Ellipsoid

Volume:

$$V = \frac{4}{3}\pi abc \ .$$

Ellipsoid of revolution:
Rotational axis $2a$:

$$V = \frac{4}{3}\pi ab^2 \ .$$

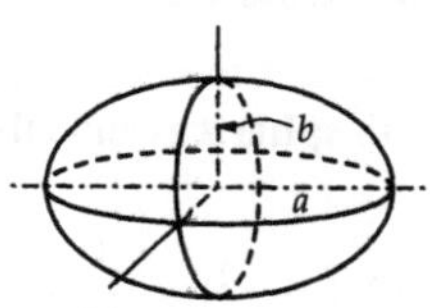

Rotational axis $2b$:

$$V = \frac{4}{3}\pi a^2 b \ .$$

Center of gravity: Point of intersection of the axes of symmetry.

4.10.2 Paraboloid of revolution

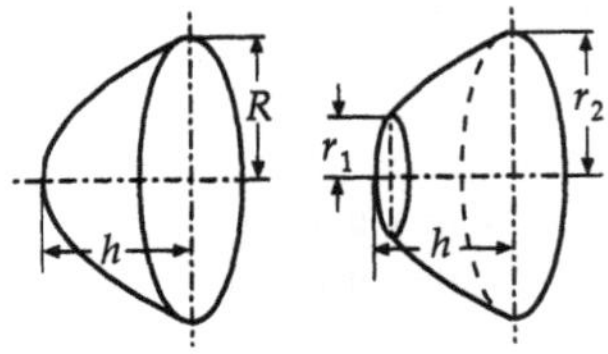

Volume:

$$V = \frac{1}{2}\pi R^2 h .$$

Center of gravity: On the axis at distance $x = (2/3) \cdot h$ from the vertex.
Truncated paraboloid of revolution: Base and top are parallel circles.
Volume:

$$V = \frac{1}{2}\pi h(r_1^2 + r_2^2) ,$$

where 1 labels the radius of the upper circle, and 2 labels the radius of the lower circle.

4.10.3 Hyperboloid of revolution

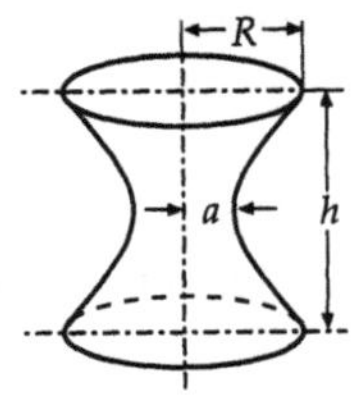

Hyperboloid of revolution: Base and top are parallel circles.
Linear eccentricity e: $e^2 = a^2 + b^2$.
Volume:
Hyperboloid of one sheet: rotation about the y-axis:

$$V = \pi h a^2 \left(1 + \frac{h^2}{12b^2}\right) = \frac{\pi h}{3}(2a^2 + R^2) ,$$

$$R^2 = a^2 \left(1 + \frac{h^2}{4b^2}\right) .$$

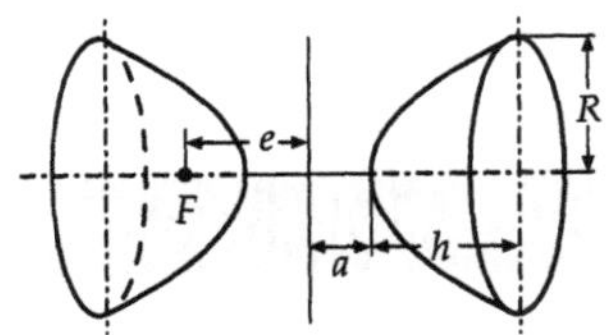

Hyperboloid of two sheets: rotation about the x-axis:

$$V = 2\pi h^2 \frac{b^2}{a^2}\left(a + \frac{h}{3}\right) = \pi h \left(R^2 - \frac{h^2 b^2}{3a^2}\right) ,$$

$$R^2 = \frac{hb^2}{a^2}(2a + h) .$$

4.10.4 Barrel

Barrel: Base and top are parallel circles.
Volume with **spherical** and **elliptic** curvature:

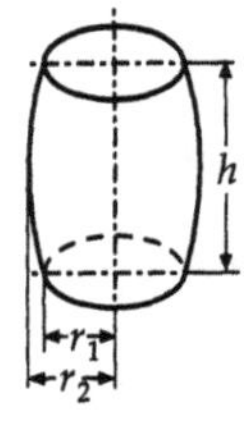

$$V = \frac{1}{3}\pi h(2r_2^2 + r_1^2) = \frac{1}{12}\pi h(2d_2^2 + r_1^2) .$$

Volume with **parabolic** curvature:

$$V = \frac{1}{15}\pi h(8r_2 + 4r_2 r_1 + 3r_1^2) .$$

4.10.5 Torus

Torus: Ring with a circular cross section.
Volume:

$$V = 2\pi^2 R^2 R_G \ .$$

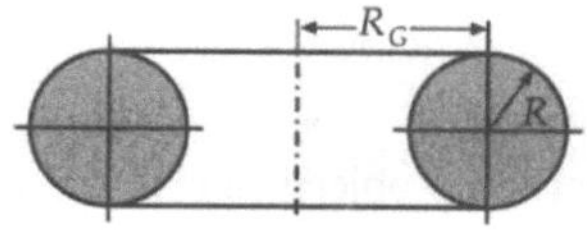

Surface:

$$A_O = 4\pi^2 R R_G \ .$$

4.11 Fractal geometry

Euclidian geometry deals with geometric structures and orderly dynamic systems with **integral dimensions**. But the term "dimension" can also be extended to include **fractional dimensions**, as for example the dimensions of a highly structured coastline. Fractured dimensions also play a role in **chaotic systems**.

4.11.1 Scaling invariance and self-similarity

Scaling invariance, or **self-similarity**: Phenomenon characteristic of many natural objects (clouds, plants, mountains, etc.). The same basic structures are always repeated on different scales.

□ **Self-similar fern leaf**: The pinnation consists of many leaflets, which (except for the reduction factor or scaling) are identical to the main leaf. The same applies to the pinnation of the leaflets.

Many chaotic systems exhibit self-similar behavior. This self-similarity is often the only approach to analyzing complex structures or dynamic systems.

4.11.2 Construction of self-similar objects

Start with a (simple) basic figure. This basic figure is replaced by a new figure constructed of N copies of the basic figure linearly scaled by a factor s. If in further steps we proceed in the same way with **all** scaled basic figures, we can continue this procedure ad infinitum. We obtain a **self-similar structure**, or **fractal**.

□ Koch's curves, Sierpinski gasket.

▷ In the case of natural fractals such as coastlines or blood vessels, the scaled objects usually do not agree **exactly** with the basic form, but they are very similar to the latter. They are called **statistically self-similar**.

4.11.3 Hausdorff dimension

● For self-similar structures, a power law, characterized by the parameters N and s, applies:

$$N = s^D \ .$$

Hausdorff dimension: The exponent D in this scaling law is given by:

$$D = \frac{\log N}{\log s}$$

▷ For many objects in classical geometry (such as segments, squares, and cubes), the Euclidean dimension agrees with the Hausdorff dimension.

□ **Segment**: $N = 3, s = 3$
Power law: $3 = 3^1$.
Hausdorff dimension: $D = 1$.

□ **Square**: $N = 9, s = 3$
Power law: $9 = 3^2$.
Hausdorff dimension: $D = 2$.

□ **Cube**: $N = 27, s = 3$
Power law: $27 = 3^3$.
Hausdorff dimension: $D = 3$.

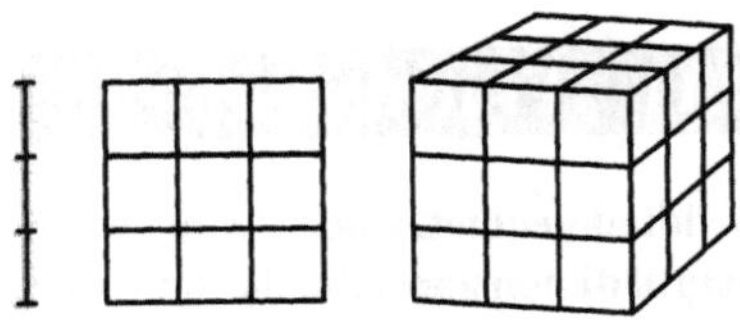

▷ The Hausdorff dimensions of segment, square, cube, triangle, etc., are **independent** of N or s. For example, if the square is subdivided into $N = 100$ smaller squares, scaled down by the factor $s = 10$, we also obtain the Hausdorff dimension 2.

▷ In general, complex self-similar objects have a **fractional** Hausdorff dimension. Therefore, they are also called **fractals**. Such objects include the Cantor set, Koch's curve, and the Sierpinski gasket.

4.11.4 Cantor set

Cantor set: Formed by removing the middle third of a segment. From **each** remaining third, the middle third is eliminated again, etc.

Identical objects	:	$N = 2$,
Scaling factor	:	$s = 3$,
Hausdorff dimension	:	$D = \ln 2/\ln 3 \approx 0.631$.

4.11.5 Koch's curve

Koch's curve: Construction principle: starting with a segment, remove the middle third of every segment, replace it by two segments which form an angle of 60°.

Identical objects	:	$N = 4$,
Scaling factor	:	$s = 3$,
Hausdorff dimension	:	$D = \ln 4/\ln 3 \approx 1.262$.

▷ The curve shown above, the original form of Koch's curve, is only one in a whole class of curves based on analogous construction principles.

4.11.6 Koch's snowflake

Koch's snowflake: Formed from an isosceles triangle by constructing three Koch's curves. In each iteration step, the perimeter increases by a factor of $4/3$, while the area remains finite.

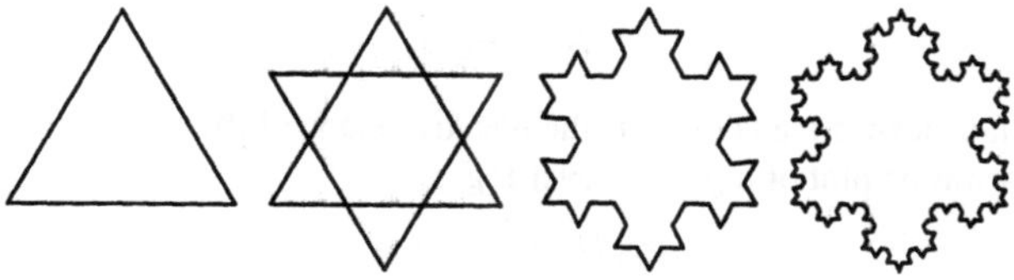

The first four steps in constructing Koch's snowflake

4.11.7 Sierpinski gasket

Sierpinski gasket: Obtained from an isosceles triangle by successive elimination of those triangles scaled down by a factor of 2, whose vertices are the midpoints of the sides of the triangles from the preceding iteration step. In each iteration step, the area is reduced by a factor of $3/4$; therefore, the Sierpinski gasket has an area of zero.

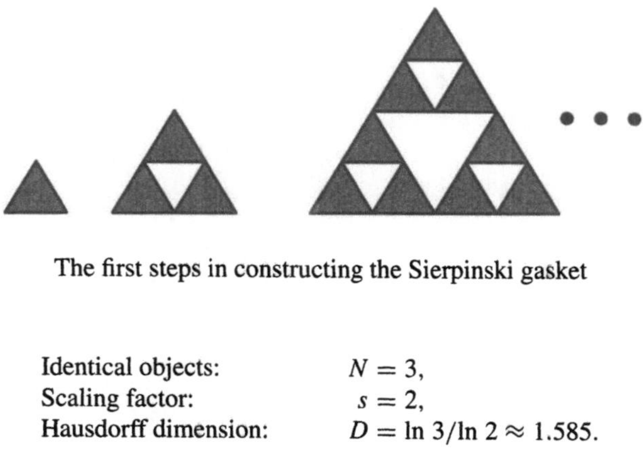

The first steps in constructing the Sierpinski gasket

Identical objects:	$N = 3$,
Scaling factor:	$s = 2$,
Hausdorff dimension:	$D = \ln 3/\ln 2 \approx 1.585$.

Advanced iteration of
the Sierpinski gasket

4.11.8 *Box-counting algorithm*

Box-counting algorithm: A procedure for the empirical determination of the Hausdorff dimension by means of a grid of a certain mesh size ϵ. For a given ϵ, the number $N(\epsilon)$ of mesh points is determined that involves a part of the fractal object. Then, if the power law

$$N \sim \frac{1}{\epsilon^D},$$

applies, the Hausdorff dimension D is the negative slope of the straight line obtained by a double-logarithmic plot of $\log N(\epsilon)$ over $\log \epsilon$.

$$\log N \sim -D \cdot \log \epsilon$$

- ☐ For example, count the squares that contain a part of the coastline. After decreasing the mesh size, count again. For different mesh sizes, different numbers of squares N are thus obtained.
- ☐ By means of this procedure, the fractal dimension of tissue folds, blood vessel branches, etc., has been determined as $D = 2.25$.

5

Functions

5.1 Sequences, series, and functions

5.1.1 Sequences and series

Sequence: A unique correspondence of a set of natural numbers to a given set A.

□ For example, in a table of contents, each chapter is assigned a page number.

Number sequence: If image set A is a subset of all real numbers, then the sequence is called a real number set.

Domain of definition of a sequence: The set of natural numbers for which a unique mapping rule exists.

Range of values of a sequence: The set of elements of A onto which the domain is mapped.

Term of a sequence: An element of the range of values of a sequence.

Finite number sequence: The domain of definition of the sequence that contains only a finite number of elements.

Infinite number sequence: The domain of definition of the sequence that is the set of natural numbers $\mathbb{N}$.

Notation of a number set:

$$(a_n) : n \mapsto a_n, \qquad n \text{ is mapped onto term } a_n.$$

▷ In general, term a_1 is taken as the first term of the sequence. But frequently one starts with the term a_0.

Kinds of representation:

- Explanation in words, description of the functional relation.

□ Every number is assigned to twice its value.

- Tabular representation:

n	1	2	3	4	5	$\cdots$
a_n	2	4	6	8	10	$\cdots$

- Functional relation:

$$(a_n): \ a_n = f(n) \qquad \text{e.g.,} \qquad a_n = 2n \ .$$

- Recursive representation:

$$(a_n): \ a_n = f(a_{n-1}) \qquad \text{e.g.,} \quad a_1 = 2, \quad a_n = a_{n-1} + 2 \ .$$

▷ One term must be given explicitly in a recursive representation.

Series: A recursively defined sequence that satisfies the following rule:

$$s_1 = f(1), \qquad s_n = s_{n-1} + f(n) \ .$$

By defining a sequence (a_n) by $a_n = f(n)$, one can write

$$s_n = s_{n-1} + a_n = \sum_{k=1}^{n} a_k \ .$$

▷ The term s_n is also said to be the n-th **partial sum** of the sequence (a_n).
For infinite series see also Chapter 12.
Examples of sequences:
Square numbers: 1, 4, 9, 16, 25, 36, 49, 64, ...

$$a_n = n^2 \ .$$

Prime numbers: 2, 3, 5, 7, 11, 13, 17, 19, 23, 29, ...
Fibonacci's numbers: 0, 1, 1, 2, 3, 5, 8, 13, 21, ...

$$a_1 = 0, \quad a_2 = 1, \quad a_n = a_{n-1} + a_{n-2} \ .$$

Factorial:1, 2, 6, 24, 120, 720, 5040, 40320, ...

$$a_n = n! \ , \qquad a_1 = 1, \quad a_n = n \cdot a_{n-1} \ .$$

Double factorial: 1, 2, 3, 8, 15, 48, 105, 384, ...

$$a_n = n!! \ , \qquad a_1 = 1, \quad a_2 = 2, \quad a_n = n \cdot a_{n-2} \ .$$

k**-fold factorial**: Analogous to the double factorial,

$$a_n = n \underbrace{! \cdots !}_{k} \ , \qquad a_m = m, \ m = 1, \cdots, k, \quad a_n = n \cdot a_{n-k} \ .$$

Difference sequence: If (a_n) is a sequence, then the difference sequence is given by

$$d_n = a_{n+1} - a_n \ .$$

Quotient sequence: If (a_n) is a sequence, then the quotient sequence is given by

$$q_n = \frac{a_{n+1}}{a_n} \ .$$

Constant sequence: A sequence (c_n), whose terms all have the same value c.

$$c_n = c \ .$$

Arithmetic sequence: A sequence that grows constantly; i.e., it has a constant difference sequence,

$$a_n = a_{n-1} + d \ .$$

Geometric sequence: A sequence that changes by a constant factor; i.e., it has a constant quotient sequence,

$$a_n = q \cdot a_{n-1} \ .$$

5.1.2 Properties of sequences, limits

Properties of sequences:

Positive (negative) definite sequences: All values are greater (smaller) than zero,

$$a_n > 0 \quad (a_n < 0) \qquad \text{for all } n \ .$$

▷ For $a_n \geq 0$ ($a_n \leq 0$), the sequence is positive (negative) semi-definite.

Bounded above (below): All values are smaller (greater) than or equal to a given bound S,

$$a_n \leq S \quad (a_n \geq S) \qquad \text{for all } n \ .$$

Supremum: The least upper bound of a sequence.

Infinum: The greatest lower bound of a sequence.

Maximum: The greatest value of a sequence.

Minimum: The least value of a sequence.

Alternating sequence: The sequence changes signs from term to term.

$$a_n \cdot a_{n-1} < 0 \ .$$

Monotonicity of a sequence: A sequence is called monotonically increasing (decreasing) if no term is smaller (greater) than its predecessor.

$$a_n \geq a_{n-1} \quad (a_n \leq a_{n-1}) \qquad \text{for all } n \ .$$

A sequence is called strictly monotonically increasing (decreasing) if each term is greater (smaller) than its predecessor,

$$a_n > a_{n-1} \quad (a_n < a_{n-1}) \qquad \text{for all } n \ .$$

Convergence: A sequence (a_n) is called convergent if there is a value a to which the following applies:

For an arbitrary value $\varepsilon > 0$, there exists an index $N(\varepsilon)$, so that all terms have a distance $n \geq N(\varepsilon)$ from a that is smaller than ε,

$$n \geq N(\varepsilon): \quad |a_n - a| < \varepsilon \ .$$

● a is called the **limit** of the sequence.

Representation:

$$a = \lim_{n\to\infty} a_n \ .$$

□ $a_n = \dfrac{n+1}{n}$ is a convergent sequence with the limit $\lim_{n\to\infty} a_n = 1$.

▷ A sequence is called divergent if it does not converge.

Null sequence: A sequence with the limit zero.

□ $a_n = \dfrac{1}{n}$ is a null sequence.

Cauchy's sequence:

$$n, m \geq N(\varepsilon): \quad |a_n - a_m| < \varepsilon \ .$$

▷ Every Cauchy sequence is convergent in the real (and complex) domain.

● Every monotonically increasing (decreasing) sequence bounded above (below) is convergent. The limit is the supremum (infinum).

Theorems for limits: If $\lim_{n\to\infty} a_n = a$ and $\lim_{n\to\infty} b_n = b$, then:

The limit of the multiple of a sequence is the multiple of the limit.

$$\lim_{n\to\infty} (c \cdot a_n) = c \cdot \lim_{n\to\infty} a_n = c \cdot a$$

The limit of the sum (difference) of two sequences is the sum (difference) of the limits.

$$\lim_{n\to\infty}(a_n \pm b_n) = \lim_{n\to\infty} a_n \pm \lim_{n\to\infty} b_n = a \pm b$$

The limit of the product of two sequences is the product of the limits.

$$\lim_{n\to\infty}(a_n \cdot b_n) = \lim_{n\to\infty} a_n \cdot \lim_{n\to\infty} b_n = a \cdot b$$

The limit of the quotient of two sequences is the quotient of the limits ($b \neq 0$).

$$\lim_{n\to\infty}\left(\frac{a_n}{b_n}\right) = \frac{\lim_{n\to\infty} a_n}{\lim_{n\to\infty} b_n} = \frac{a}{b}$$

- If beyond a certain index n_0 a sequence (c_n) lies between two sequences (a_n) and (b_n) converging to the same limit, then sequence (c_n) also converges to this limit.

$$n > n_0 \quad a_n \geq c_n \geq b_n\ , \qquad \lim_{n\to\infty}(a_n) = \lim_{n\to\infty}(b_n) = a\ , \qquad \lim_{n\to\infty}(c_n) = a$$

5.1.3 Functions

- **Function**: A unique mapping from a set X into a set Y.

X is called **inverse-image set** or **domain of definition** D_f.
Y is called **image set** or **range of values** W_f.
Every element $x \in X$ corresponds uniquely to an element $f(x) = y \in Y$ (read: f of x). Then, f is the set of **ordered pairs** (x, y) which has the characteristic that to each $x \in X$ there corresponds one and only one y. Notation:

$$f : X \to Y$$
$$x \mapsto f(x)\ .$$

▷ The statement pertaining to the domain of definition is part of the definition of a function.
▷ If the domain of definition $\mathbb{R}$, it is usually not included.
▷ Sequences and series can be regarded as functions with the domain of definition $\mathbb{N}$.

Representations of $f:\ x \mapsto 4x + 6$:
Ordered pairs:

$$f = \{(x, y) | y = 4x + 6\}\ .$$

Graph of a function diagram:
To every pair (x, y) of real numbers a pair is assigned in the plane.

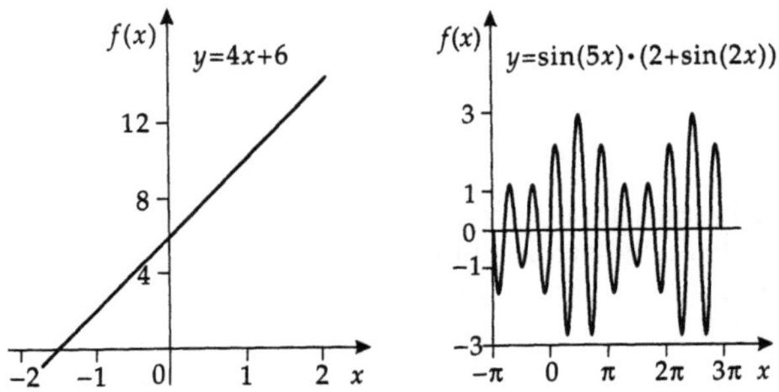

Graphs of functions

Functional relation:

$$f(x) = 4x + 6 \quad \text{or} \quad y = 4x + 6 \,.$$

Explicit representation of a function:

$$y = f(x) \,, \quad y = 4x + 6 \,.$$

Implicit representation of a function:

$$f(x, y) = 0 \,, \quad y - 4x - 6 = 0 \,.$$

● Not all functions given in an implicit representation can be represented explicitly.

□ The equation $y^5 + y - x^3 = 0$, y cannot be isolated, explicitly according to y.

Parametric representation:

$$x = g(t), \; y = h(t) \,, \quad x = t, \; y = 4t + 6 \,.$$

● Not all functions given in a parametric representation have an explicit or implicit representation.

□ $y = t^5 - t^2, x = t^4 + t$ $(t \geq 0)$ is such a function.

Other kinds of representation of $f : x \mapsto 2x$: $\{1, 2, 3\} \to \{2, 4, 6\}$

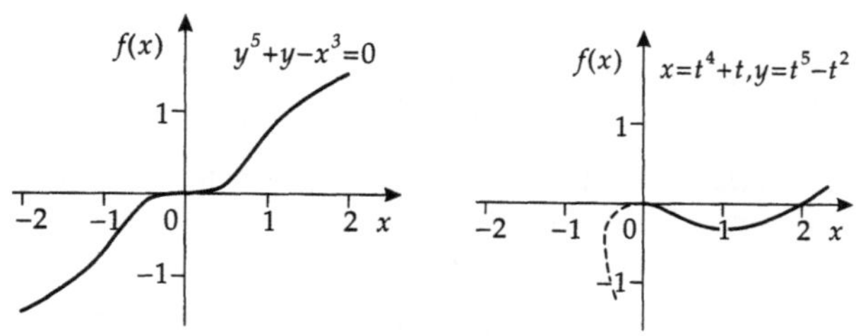

Graphs for $y^5 + y - x^3 = 0$ and $x = t^4 + t, y = t, y = t^2, t \geq 0$

Table of values:

x	1	2	3	all values x from the domain of definition
y	2	4	6	all values y from the range

Arrow diagram: The arrows indicate where the functions belong.

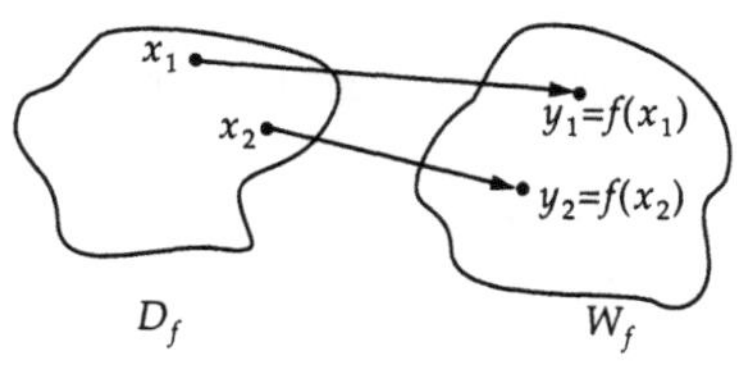

Arrow diagram of functions

▷ If not all pairs of values are given in the table of values or in the arrow diagram, then the function is not uniquely determined.

□ For functions defined on the entire set $\mathbb{R}$, the table of values given above can describe, for example, the function $f(x) = 2x$, the integer part function $f(x) = 2[x]$, or the function $f(x) = 2x + \sin(\pi x)$.

5.1.4 Classification of functions

Functions can be divided into various classes by their properties.

real functions

algebraic | transcendental

integral rational | fractional rational | roots | log. | exp. | hyperb. | trig.

rational functions | irrational functions

Integral rational functions or **polynomials:** Functions that can be represented by a finite number of additions, subtractions, and multiplications with the independent variable x.
General form:

$$f(x) = a_0 + a_1 x + a_2 x^2 + a_3 x^3 + \ldots + a_n x^n \qquad n \in \mathbb{N}\,.$$

□ Parabolas and straight lines are described by integral rational functions $f(x) = ax^2$ and $g(x) = mx + b$.

Rational functions: Functions that can be represented by a finite number of additions, subtractions, multiplications and divisions with the independent variable x.
Fractional rational functions: Functions that are rational but not integral rational.
General representation:

$$f(x) = \frac{a_0 + a_1 x + a_2 x^2 + a_3 x^3 + \ldots + a_n x^n}{b_0 + b_1 x + b_2 x^2 + b_3 x^3 + \ldots + b_m x^m} \qquad n, m \in \mathbb{N}\,.$$

Proper fractional rational function: The highest exponent in the denominator is greater than the highest exponent in the numerator: $m > n$.
Improper fractional rational function: $m \leq n$.

□ The hyperbola $f(x) = \dfrac{1}{x}$ is a proper fractional rational function.
$f(x) = \dfrac{x^2+1}{x-1}$ is an improper fractional rational function.

● Improper fractional rational functions can be divided into the sum of an integral rational function and a proper fractional rational function by means of polynomial division.

Algebraic functions: Obey an algebraic equation of any degree, the coefficients of which are integral rational functions (polynomials) in the independent variable x.
General expression:

$$\sum_{k=0}^{n} g_k(x) y^k = 0 \qquad g_k\text{: polynomials in } x.$$

▷ This equation also contains powers of y which is not the case for rational functions.

□ The square-root function $f(x) = \sqrt{x}$, $y^2 - x = 0$, is an algebraic function but not a rational function.

● All rational functions are algebraic functions.

Transcendental functions: Functions that are not algebraic.

▷ They can often be represented by infinite power series.

□ The exponential function $f(x) = \mathrm{e}^x$, the logarithmic function $f(x) = \ln(x)$, and all hyperbolic and trigonometric functions are transcendental.

Irrational functions: Functions that are not rational.

▷ All transcendental functions are irrational functions, but not all irrational functions are transcendental functions.

□ Power functions with fractional exponents are irrational. However, they are not transcendental, but algebraic functions.

5.1.5 *Limit and continuity*

Limit of a function: For every number $\varepsilon > 0$ there exists a number $\delta > 0$ such that $|f(x) - g| < \varepsilon$ for all x with $|x - x_0| < \delta$. g is called the limit of f at point x_0.

$$\lim_{x \to x_0} f(x) = g$$

□ The function $f(x) = \mathrm{e}^{-(1/x^2)}$ has the limit $\lim_{x \to 0} f(x) = 0$ at the point $x = 0$.

Right-hand and left-hand limit: Considering only values $x > x_0$ ($x < x_0$), we approach limit x_0 from the right (left).
For the limit, we often write

$$\lim_{x \to x_0 -} f(x) = \text{ left-hand limit;} \quad \lim_{x \to x_0 +} f(x) = \text{ right-hand limit.}$$

□ At point $x = 0$, the function $f(x) = \mathrm{e}^{-(1/x)}$ has the right-hand limit zero and is left-hand divergent.

If the right-hand limit equals the left-hand limit, then the limit of the function at point x_0 is

$$\lim_{x \to x_0 -} f(x) = \lim_{x \to x_0 +} f(x) = \lim_{x \to x_0} f(x) .$$

Continuity: A function is called continuous if every function value $f(a)$ is the limit of the function at point a.

$$\lim_{x \to a} f(x) = f(a) \quad \text{for all } a .$$

- **Epsilon-delta criterion**: A function f is continuous at x_0 if for every number $\varepsilon > 0$ there exists a number $\delta > 0$ such that $|f(x) - f(x_0)| < \varepsilon$ for all x with $|x - x_0| < \delta$.

□ The absolute value function $f(x) = |x|$ is continuous for all x.

The signum function $f(x) = \operatorname{sgn}(x) = \begin{cases} -1 & x < 0 \\ 0 & x = 0 \\ 1 & x > 0 \end{cases}$

is discontinuous at point $x = 0$.

Right-hand or left-hand continuity: If the function is discontinuous at x_0, but the function value $f(x_0)$ is a right-hand (left-hand) limit at point x_0, then the function is continuous from the right (left).

□ At point $x = 0$, the function $f(x) = \mathrm{e}^{-(1/x)}$ is continuous from the right.

Intermediate value theorem: If for the real function $f(x)$, defined and continuous in a closed interval $[a, b]$, the number c lies between the function values at a and b, then there exists a point x_0 in $[a, b]$ at which $f(x_0) = c$.

Bolzano's theorem: If a function $f(x)$, defined and continuous in a closed interval $[a, b]$, assumes function values with opposite signs at the end-points of the interval, then there is at least one point x_0 between a and b at which $f(x)$ becomes zero.

Calculating with limits: All rules for calculating with limits in the case of sequences apply here as well.

$$\lim_{x \to a} (f(x) \pm g(x)) = \lim_{x \to a} f(x) \pm \lim_{x \to a} g(x)$$

$$\lim_{x \to a} (f(x) \cdot g(x)) = \lim_{x \to a} f(x) \cdot \lim_{x \to a} g(x)$$

$$\lim_{x\to a}\frac{f(x)}{g(x)}=\frac{\lim\limits_{x\to a}f(x)}{\lim\limits_{x\to a}g(x)}\qquad \text{if } \lim_{x\to a}g(x)\neq 0$$

Further, if $g(x)$ is a continuous function, then

$$\lim_{x\to a}g(f(x))=g\left(\lim_{x\to a}f(x)\right).$$

▷ The following limit theorems are valid only if the limits exist.

□ The following applies:

$$\lim_{x\to a}(g(x))^n=\left(\lim_{x\to a}g(x)\right)^n\qquad n\in\mathbb{N},$$

$$\lim_{x\to a}\sqrt{g(x)}=\sqrt{\lim_{x\to a}g(x)},$$

$$\lim_{x\to a}\ln(g(x))=\ln\left(\lim_{x\to a}g(x)\right),\quad \text{if } \lim_{x\to a}g(x)>0.$$

De l'Hôpital's rule:
Always:
If $\lim\limits_{x\to a}f(x)=\lim\limits_{x\to a}g(x)=0$, or $\lim\limits_{x\to a}|f(x)|=\lim\limits_{x\to a}|g(x)|=\infty$,
then the limit of the quotient of the functions (if it exists) can be written as the limit of the quotient of the first derivatives.

$$\lim_{x\to a}\frac{f(x)}{g(x)}=\lim_{x\to a}\frac{f'(x)}{g'(x)}$$

□ Always:

$$\lim_{x\to 0}\frac{\sin x}{x}=\lim_{x\to 0}\frac{\cos x}{1}=1.$$

▷ If also for the first derivatives

$$\lim_{x\to a}f'(x)=\lim_{x\to a}g'(x)=0$$

or

$$\lim_{x\to a}|f'(x)|=\lim_{x\to a}|g'(x)|=\infty,$$

then the second derivative must be formed, etc.

5.2 Discussion of curves

5.2.1 *Domain of definition*

First, we determine the **domain** D of values x for which function fx is defined.

□ 1. For **integral rational functions**: $D=\mathbb{R}$.
2. For **fractional rational functions**, the zeros of the denominator must be excluded.
3. For **root functions**, the domains with negative radicands must be excluded.

□ $y=1/\sqrt{x-1},\quad D=\mathbb{R}\backslash\{x|x\leq 1\}$

5.2.2 *Symmetry*

Even function: Symmetric about the y-axis: $f(x)=f(-x)$.
Odd function: Point-symmetric about the origin: $f(x)=-f(-x)$.

The shape of the curve for negative values results from a reflection about the y-axis or the origin.

□ **Even function**: $x^2, x^4, \ldots, \cos x, \cosh x$.
Odd function: $x, x^3, \ldots, \sin x, \tan x, \cot x, \sinh x, \tanh x, \coth x$.

Determination of the symmetry of composite functions (e: even function; o: odd function):

$f(x)$	$g(x)$	$f(x) \pm g(x)$	$f(x) \cdot g(x)$	$f(x)/g(x)$	$f(x)^{g(x)}$
e	e	e	e	e	e
e	o	–	o	o	–
o	e	–	o	o	o
o	o	o	e	e	–

▷ Function that is neither even nor odd can be composed of an even and an odd part.

□ $e^x = \sinh x + \cosh x = \dfrac{e^x - e^{-x}}{2} + \dfrac{e^x + e^{-x}}{2}$, where $\sinh x$ is an odd function, and $\cosh x$ is an even function.

▷ Frequently, there are also symmetries with regard to marked points of a function such as zeros or poles.

□ $y = (x-1)^2$ is symmetric with regard to the vertical axis at $x = 1$.

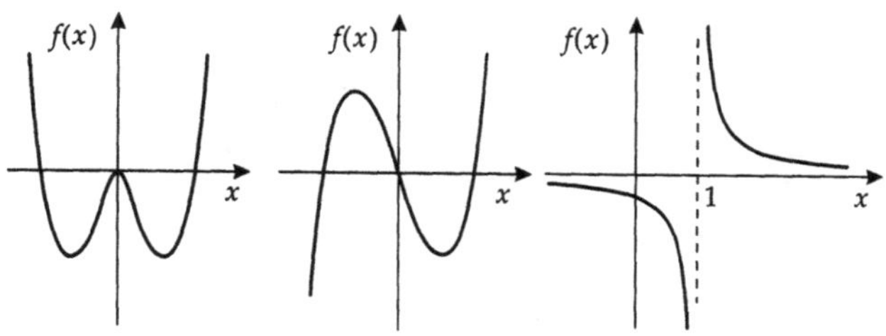

Even function Odd function Point symmetry at $x = 1$

5.2.3 Behavior at infinity

The limit of a function as $x \to +\infty$ and $x \to -\infty$ is investigated with the domain of definition being unbounded to one or both sides.

If necessary, **de l'Hôpital's rule** can be applied.

Integral rational function: The behavior at infinity is determined by the term with the highest exponent.

The sign of the limit at $x \to \infty$ corresponds to the sign of the highest term.

Even (odd) exponent: the limit at $x \to -\infty$ has the same (opposite) sign.

n	$\lim\limits_{x\to\infty} c \cdot x^n$	$\lim\limits_{x\to-\infty} c \cdot x^n$
even	$+\mathrm{sgn}(c) \cdot \infty$	$+\mathrm{sgn}(c) \cdot \infty$
odd	$+\mathrm{sgn}(c) \cdot \infty$	$-\mathrm{sgn}(c) \cdot \infty$

Fractional rational function: Quotient of **integral rational functions** with polynomials of degree m in the denominator and n in the numerator, respectively. The behavior at infinity is determined by the term with the highest exponent. If this term occurs in the numerator, the discussion is the same as that for integral rational functions. If the term occurs in the denominator, the function for large x-values goes toward zero.

In case of integral rational functions of the same degree, the total function goes toward a constant value (parallel to the x-axis), which is the quotient of the two previous factors of the highest terms.

▷ Split the fractional rational function into a proper fractional function and an integral rational function, called an asymptote. For large x-values, the curve approaches the function of the asymptote.

□ $y = \dfrac{x^3 - 2x^2 + 1}{x^2 - 4} = x - 2 + \dfrac{4x - 7}{x^2 - 4}$. The asymptote is given by the straight line $y = x - 2$.

degree of polynomials	asymptote
$n - m > 1$	parabola of $(n - m)$-th order
$n - m = 1$	straight line
$n = m$	parallel to x-axis
$n < m$	x-axis

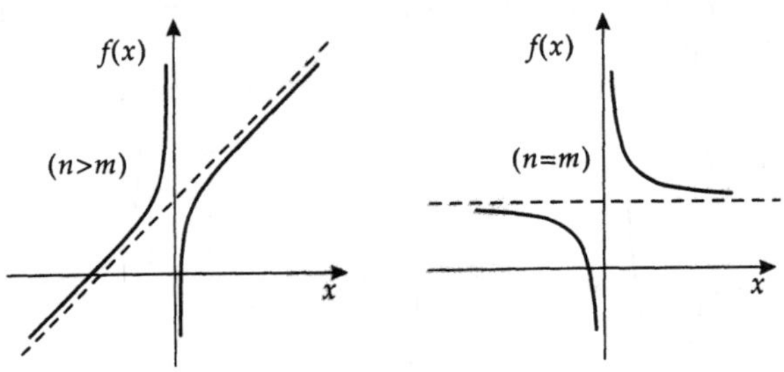

(a) Straight line (b) Constant

Asymptotes in case of fractional rational functions

5.2.4 Gaps of definition and points of discontinuity

Jumps occur in case of functions defined in sections and at individual points at which the function is not defined.

● The limits from the left and from the right are different in case of a function at the **jump discontinuity**.

$$\textbf{Jump discontinuity:} \quad \lim_{x\to c+} f(x) \neq \lim_{x\to c-} f(x) \,.$$

● **Pole**: One of the two limits is infinite.

$$\textbf{Pole:} \quad |\lim_{x\to c\pm} f(x)| = \infty \,.$$

□ $f(x) = \dfrac{\cos x}{x}$, $\lim\limits_{x\to\pm 0} f(x) = \pm\infty$ (pole at $x = 0$).

▷ Fractional rational functions have poles at the zeros of their denominator term if the numerator term at this point is other than zero points.

$$\frac{f(x)}{g(x)} \to \text{poles at } x_0 \text{ with } g(x_0) = 0\,, \text{ if } f(x_0) \neq 0\,.$$

● The gaps of definition can be eliminated if the limits from the left and from the right are equal and finite.

Removable gap: $\lim_{x \to c_+} f(x) = \lim_{x \to c_-} f(x)$.

□ $f(x) = \frac{\sin x}{x}$, $f(0) = \lim_{x \to 0\pm} f(x) = 1$ (see l'Hôpital's rule)

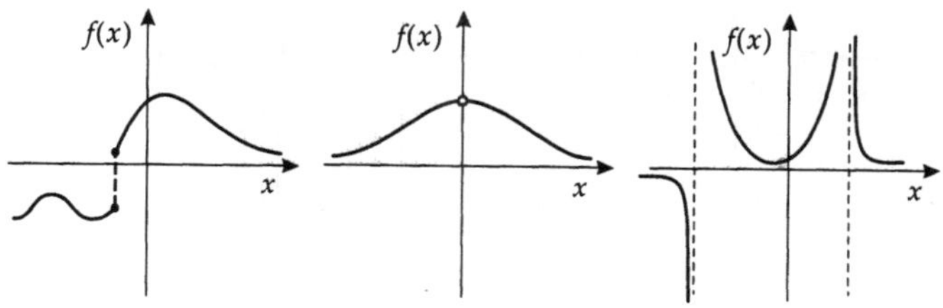

(a) Jump discontinuity (b) Removable gap (c) Poles

5.2.5 *Zeros*

- The zeros of a function f are the solution of the equation $f(x) = 0$.

▷ A function must not necessarily change its sign at a zero.

▷ Usually, numerical methods to determine zeros try to find a change in sign.

□ $y = (x-1)^2$ has a zero at $x = 1$, but the function values are never negative.

5.2.6 *Behavior of sign*

Determinating domains above and below the x-axis:

1. Zeros and poles subdivide the domain of definition into intervals.
2. Determine the sign for each subinterval.
3. Shade the domains.

   Point of intersection with the y-axis: $y_0 = f(0)$.

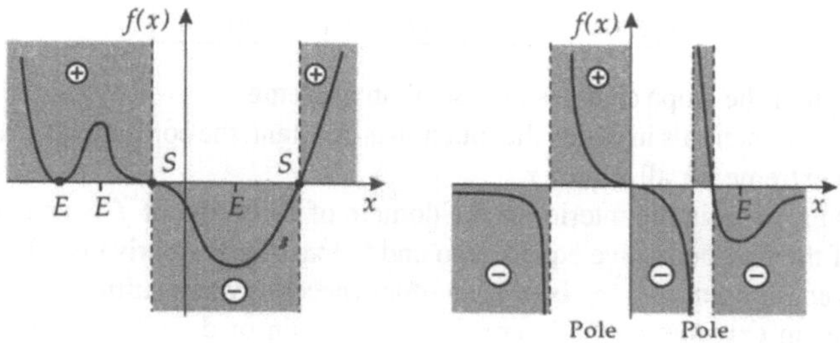

Behavior of sign

For a sum (difference) of functions, use graphic addition (subtraction).

□ $y = \frac{1}{x^2} - \frac{1}{(1-x)^2}$

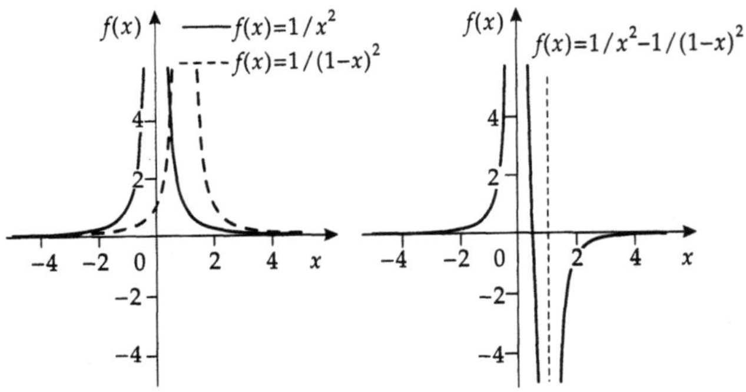

5.2.7 Behavior of slope, extremes

Monotonicity by section: Division into curve segments possessing only a positive or a negative slope (the function is monotonically increasing or monotonically decreasing by section).

- For a differentiable function:

$$f'(x) < 0 \implies \textbf{strictly monotonically decreasing},$$
$$f'(x) > 0 \implies \textbf{strictly monotonically increasing}.$$

- The function $f(x)$ has a relative extreme at point x_m if there exists a neighborhood U of x_m in which all function values are smaller or greater than $f(x_m)$:

$$\textbf{Relative maximum:} \quad f(x) \le f(x_m) \quad x \in U \,,$$
$$\textbf{Relative minimum:} \quad f(x) \ge f(x_m) \quad x \in U \,.$$

- In case of differentiable functions, a relative extreme lies between two intervals of distinct monotonicity.

$x < x_m$	$x > x_m$	extreme
$f'(x) > 0$	$f'(x) < 0$	maximum
$f'(x) < 0$	$f'(x) > 0$	minimum

- The sign of the slope changes in case of an **extreme**.

▷ Exception: Intervals in which the function is constant: the constant function $f(x) = c$ has an extreme for all values x.

- A point $x = x_0$ in the interior of the domain of definition of f is an extreme if and only if the first derivative equals zero and the next larger derivative, being different from zero at point $x = x_0$, is of even order (necessary and sufficient condition).
- If x_0 is an **extreme** in the interior of the domain of definition of a differentiable function f, then the following necessary condition holds:

$$f'(x_0) = 0.$$

At such an extreme, the tangent is parallel to the x-axis.

Theorem of Weierstraß: When a function $f(x)$, is defined and continuous in a closed interval $[a, b]$, $f(x)$ has an absolute maximum $M_{\max}$ and an absolute minimum $M_{\min}$ in the interval, i.e., there is in this interval at least one point x_0 and at least one point x_0' such that for all x with $a \le x \le b$

$$M_{\min} = f(x_0') \le f(x) \le f(x_0) = M_{\max} .$$

▷ Extremes can also be assumed at the boundary points of the domain of definition.

□ $f(x) = x^2$ in $[1, 2]$ assumes its maximum at $x = 2$ and its minimum at $x = 1$, although the derivative at these points does not vanish.

▷ If the first derivative equals zero, a saddle point may exist.

□ $f(x) = x^3$, $f'(x) = 3x^2$, $f'(0) = 0$; i.e., not an extreme but a saddle point at $x = 0$.

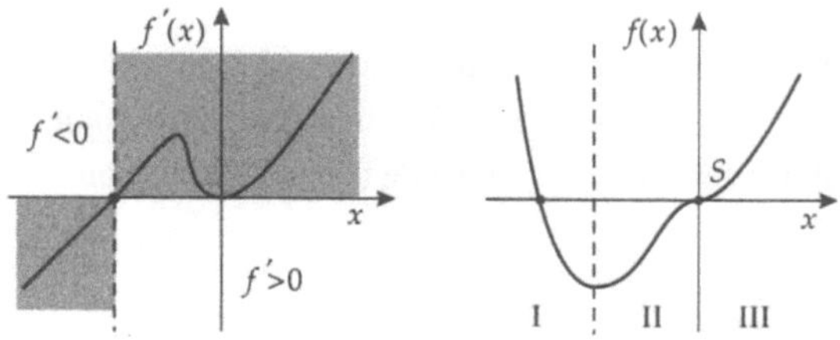

Derivative function

Original function I: monotonically decreasing; II: extreme; III: monotonically increasing. S: Saddle point

5.2.8 *Curvature*

Curvature of a curve: The change in slope. The curvature is obtained from the second derivative.

● A twice-differentiable function is
concave up (convex down) for $f''(x) > 0$,
convex up (concave down) for $f''(x) < 0$.

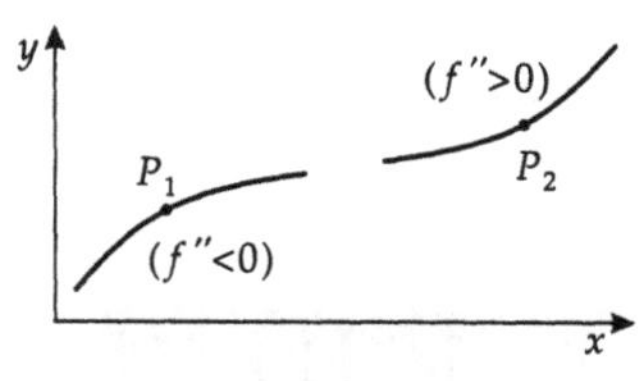

Curvature

▷ Memory aid: Think of a circle drawn within the segment of curvature of the function. If the circle lies *above* the function, the second derivative is **positive**. If the circle lies *below* the function, the second derivative is **negative**.

5.2.9 *Point of inflection*

● **Point of inflection**: Lies between two differently curved segments of a function that can be differentiated twice.

● The sign of the curvature changes at the **point of inflection**.

- The second derivative of $f(x)$ must be zero at the **point of inflection** x_I (necessary condition),

$$f''(x_I) = 0\,.$$

▷ The condition is not sufficient, i.e., a point of inflection does not necessarily exist.

□ $f(x) = x^4$, $f''(x) = 12x^2$, $f''(0) = 0$; i.e., the function has no point of inflection, but a minimum at $x = 0$.

- Sufficient condition for a point of inflection:

$$f''(x_I) = 0, \quad \text{and} \quad f^{(n)}(x_I) \neq 0,\ n \text{ odd, and } f^{(k)}(x_I) = 0,\ (2 < k < n).$$

- **Saddle point**: A particular point of inflection with a horizontal tangent:
  Saddle point: $f''(x_S) = 0$ then $f'(x_S) = 0$, and x_S is the point of inflection.
- An n-fold differentiable has the following behavior at the point x_0 with

$$f^{(n)}(x_0) \neq 0, \quad f^{(n-1)}(x_0) = f^{(n-2)}(x_0) = \cdots = f''(x_0) = f'(x_0) = 0\,,$$

Minimum: n even, $f^{(n)} > 0$,
Maximum: n even, $f^{(n)} < 0$,
Point of inflection: n odd, $f^{(n)} \neq 0$.

□ $f(x) = x^4$, $f'(x_0 = 0) = f''(x_0 = 0) = f'''(x_0 = 0) = 0$;
$f^{(4)}(x) = 4! = 24 > 0$;
i.e., there is a minimum at $x_0 = 0$.

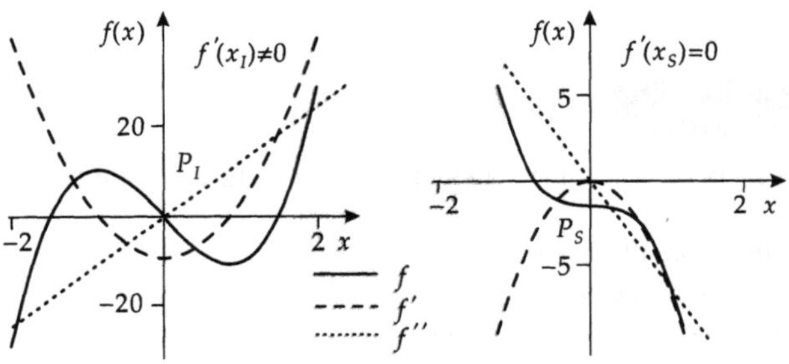

(a) Point of inflection (b) Saddle point

5.3 Basic properties of functions

The following sections describe the most important functions. The arrangement is based on the classification scheme for functions described above:

1. Simple functions: constant functions and some important functions that do not quite fit into this scheme, such as absolute value functions or rounding functions.
2. Integral rational functions: starting from the simplest rational functions, such as linear or quadratic functions, up to the general polynomial.
3. Fractional rational functions: from simple hyperbolas to quotients of polynomials.
4. Nonrational algebraic functions: from square roots to power functions with fractional exponents and roots of polynomials.

5. Particular transcendental functions: logarithmic, exponential and Gauss functions.
6. Hyperbolic functions: functions of the unit hyperbola.
7. Area functions: the inverse of hyperbolic functions.
8. Trigonometric functions: angle functions of the unit circle.
9. Arctrigonometric functions: the inverse of trigonometric functions.

First of all, functions will be represented by their general functional expressions, and then some of their applications will be explained.

(a) Definition

● Functions can be defined in explicit or implicit form.

Explicit form: Functions are represented by a direct correspondence of values as well as by combinations of other functions.

□ Definition of the signum function:

$$\operatorname{sgn}(x) = \begin{cases} -1 & x < 0 \\ 0 & x = 0 \\ 1 & x > 0 \, . \end{cases}$$

□ Definition of the absolute value function: $|x| = x \cdot \operatorname{sgn}(x)$.

Implicit form: A combination of x and $y = f(x)$ is described.

□ Cubic equation in x and y: $y^3 + y + x^3 - x = 0$.

▷ Implicit forms can often not be resolved explicitly.

▷ Caution is necessary because y can have several values.

□ The **equation of conics** allows two y-values for every x:

$$\sqrt{x^2 + y^2} + \varepsilon x = a \, .$$

Special case of an implicit definition: Description of a function as the inverse of other functions.

□ Square root: $y = \sqrt{x} \qquad x = y^2, \quad y \geq 0.$

Other possible definitions are representations as the **solutions of differential or integral equations.**

□ Equation of oscillations; differential equation for sine and cosine:

$$\frac{\mathrm{d}^2}{\mathrm{d}x^2} f(x) + f(x) = 0 \, , \qquad f(x) = a \sin(x) + b \cos(x) \, .$$

(b) Graphical representation

Graphical representations provide an important overview of the properties of a function.

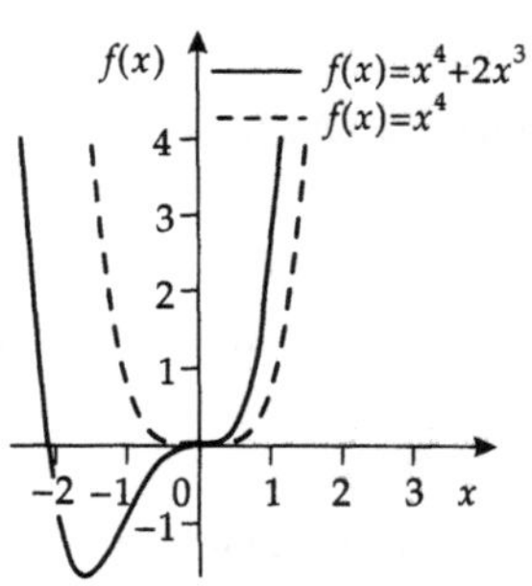

Polynomial of degree 4

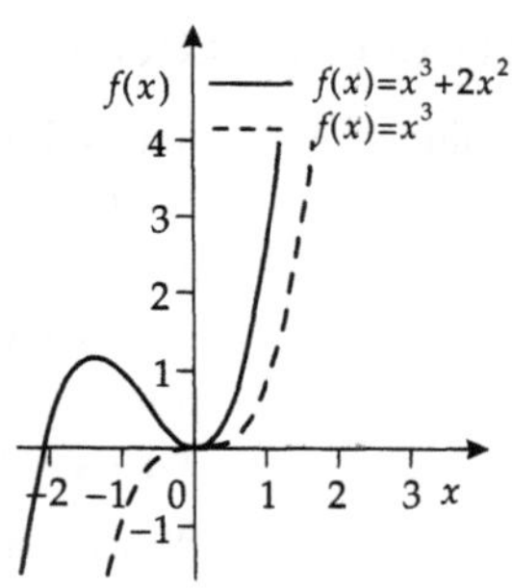

Polynomial of degree 3

▷ In most cases, two graphs are represented in one figure: one with a solid line and one with a dotted line. This allows a better representation of relationships between the functions on the one hand and the effect of parameters on the other.

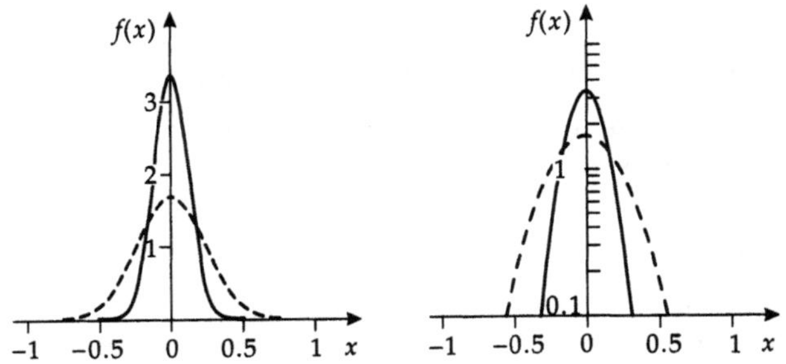

Gauss functions in linear and semilogarithmic representation

It is often practical to look at an axis in a logarithmic scale.

▷ One result is that this allows an overview over a large range of values.

Analogously, both axes may also be viewed in a logarithmic scale.

▷ This is of particular interest when looking at power functions, since a the double-logarithmic representation provides better information about prefactors and exponents.

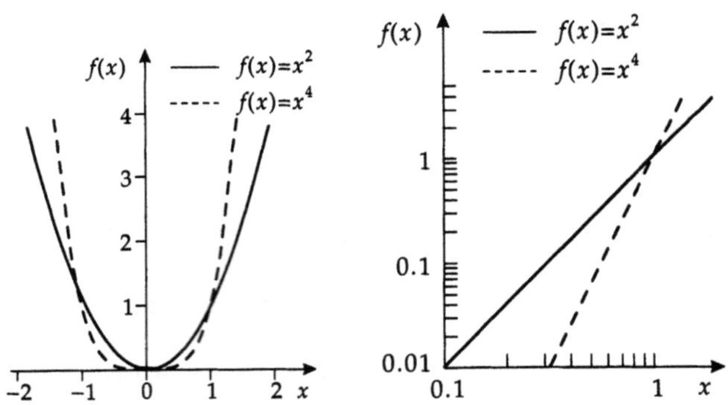

Power functions in linear and double-logarithmic representation

(c) Properties of functions

Domain of definition: Interval of x values for which the function has a defined value.
e.g., $\sqrt{x}$: $0 \leq x < \infty$.

Range of values: Interval of values $y = f(x)$ which the function may assume.
e.g., $\sqrt{x}$: $0 \leq f(x) < \infty$.

Quadrant: Where the graph of the function for $x > 0$ and $x < 0$
First quadrant: $x > 0$, $y > 0$, e.g., $y = \sqrt{x}$

second quadrant: $x < 0, y > 0$, e.g., $y = \sqrt{-x}$
third quadrant: $x < 0, y < 0$, e.g., $y = -\sqrt{-x}$
fourth quadrant: $x > 0, y < 0$, e.g., $y = -\sqrt{x}$

Periodicity: There is a point $x_0 \neq 0$ such that for all x:

$$f(x + x_0) = f(x) .$$

The lowest value $x_0 > 0$ is the period length or primitive period.

Monotonicity: For all x_1, x_2 with $x_1 < x_2$:

strictly monotonically increasing	$f(x_1) < f(x_2)$,
monotonically increasing	$f(x_1) \leq f(x_2)$,
strictly monotonically decreasing	$f(x_1) > f(x_2)$,
monotonically decreasing	$f(x_1) \geq f(x_2)$.

Symmetries: Mirror symmetry about the y-axis: $f(-x) = f(x)$, e.g., $f(x) = x^2$.
Point symmetry about the origin: $f(-x) = -f(x)$, e.g., at $f(x) = x^3$.

Asymptotes: Behavior of the function at infinity, e.g., $\sqrt{x^2+1} \to \pm x$ for $x \to \pm\infty$.

(d) Particular values

Zeros: Points at which $f(x) = 0$.

□ $f(x) = x^2 - 1$, has zeros at $x = 1, x = -1$.

Jump discontinuities: Points of discontinuity, at which the function performs a finite jump.

□ $f(x) = \mathrm{sgn}(x)$ at $x = 0$.

Poles: Points at which the function is divergent (its value tends to infinity).

□ $f(x) = \dfrac{1}{x^2}$ has a pole of second order at $x = 0$.

Extremes: Local maxima and minima of the function.

□ $f(x) = x^4 - 2x^2 + 1$
has a maximum at $x = 0$, minima at $x = \pm 1$.

Points of inflection: Points at which the curve changes the sign of curvature.

□ $f(x) = x^3$ has a saddle point at $x = 0$.

▷ Note that in this description, zeros and poles mean **real zeros** and **real poles**.

● Although a polynomial of degree n has **at most n zeros in the real domain**, the **fundamental theorem of algebra** states that such a polynomial has **exactly n zeros in the complex domain** if the multiplicity of zeros is taken into account.

(e) Reciprocal functions

Reciprocal of a function:

$$g(x) = \frac{1}{f(x)} .$$

▷ A reciprocal function can often be used to define new functions or to describe connections with the original function.

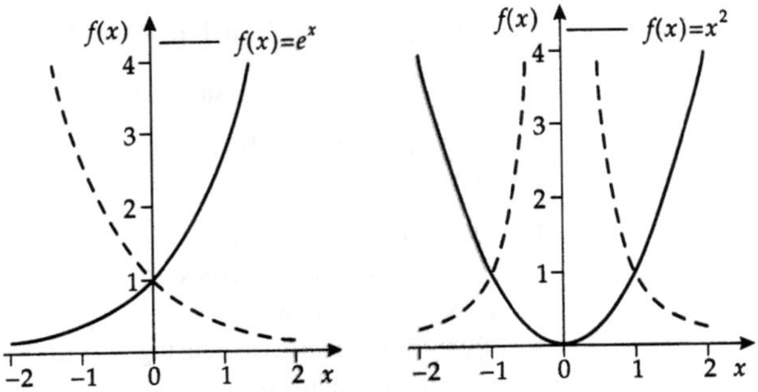

Reciprocal functions (- - - -)

(f) Inverse functions

The inverse function $f^{-1}(x)$ of $y = f(x)$ is determined by interchanging the variables x and y, $x = f(y)$, and then solving for y, $y = f^{-1}(x)$.
A function can be inverted if every image point $f(x) \in W_f$ occurs once, i.e., for $x_1 \neq x_2$ always: $f(x_1) \neq f(x_2)$. For periodic functions, the inverse function becomes multi-valued. By limiting it to the main value, it can be made single-valued.

□ $y = f(x) = ax + b$, $x = ay + b$, $y = f^{-1}(x) = \dfrac{1}{a}x - \dfrac{b}{a}$ $(a \neq 0)$;
$y = f(x) = \mathrm{e}^x$, $x = \mathrm{e}^y$, $y = f^{-1}(x) = \ln x$ $(x > 0)$

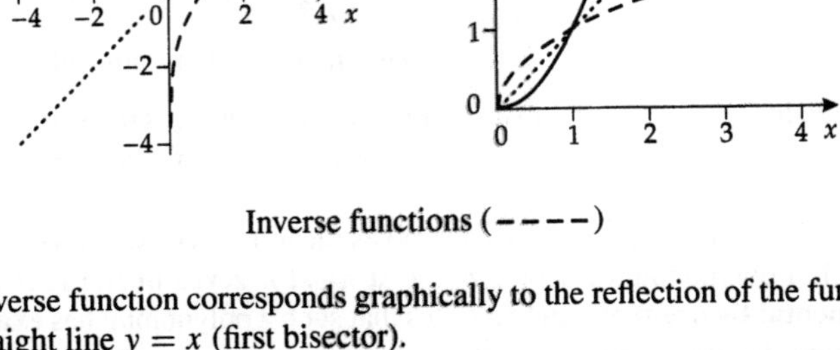

Inverse functions (- - - -)

● The inverse function corresponds graphically to the reflection of the function across the straight line $y = x$ (first bisector).

▷ However, it must be noted that the inversion is single-valued and has a limited domain of definitions and a limited range of values.

(g) Cognate functions

Many functions are related to other functions, e.g., the exponential function is related to logarithmic and hyperbolic functions.

(h) Conversion formulas

Conversion formulas are used for arithmetical operations functions,

□ quadratic function

$$f(x) = x^2 \qquad f(x) - f(y) = x^2 - y^2 = (x - y)(x + y) ,$$

or the argument of functions,

□ quadratic function

$$f(x) = x^2 \qquad f(x \pm y) = (x \pm y)^2 = x^2 \pm 2xy + y^2 .$$

(i) Approximation formulas (8 bits ≈ 0.4% of precision)

In certain areas, many complicated functions can be approximated by means of simple functions.

Below are formulas which—within the stated limits—are accurate to within 8 bits in computer applications, i.e., up to a relative error of 0.4%.

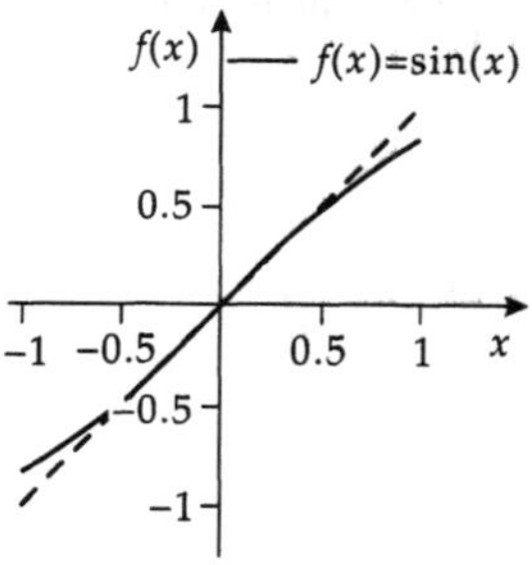

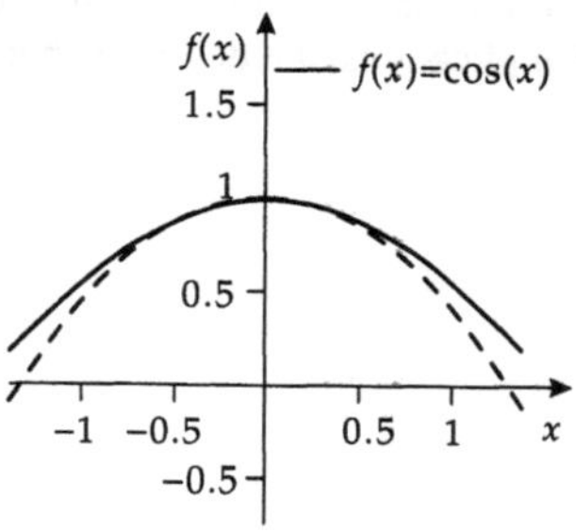

Simple approximation functions (– – – –)

● **Attention:** In calculating the formulas with the computer, potential rounding errors are not yet included in the limits of error.

(j) Expansion of series or products

Many functions can be represented as a finite or infinite series, as a finite or infinite product, or as a finite or infinite continued fraction.

□ Exponential function

$$\mathrm{e}^x = \sum_{k=0}^{\infty} \frac{x^k}{k!} = 1 + x + \frac{x^2}{2} + \frac{x^3}{6} + \frac{x^4}{24} + \frac{x^5}{120} + \ldots$$

(k) Derivative of functions

The first derivative is given here.

□ Power function:

$$\frac{\mathrm{d}}{\mathrm{d}x} ax^n = nax^{n-1}, \; n \neq 0.$$

Sometimes, a generalized form of higher derivatives can be given, e.g.,

$$\frac{\mathrm{d}^k}{\mathrm{d}x^k} ax^n = \frac{n!}{(n-k)!} ax^{n-k}, \qquad k \leq n .$$

(l) Primitive of the function

A primitive $F(x)$ of a given function $f(x)$ is a function defined in the same interval and whose derivative equals $f(x)$:

$$F(x) = \int f(x)\, \mathrm{d}x + c\,, \quad F'(x) = f(x)\,.$$

The integral of the function is given here as a particular primitive, e.g.,

$$\int_0^x \mathrm{e}^{at}\, \mathrm{d}t = \frac{1}{a}(\mathrm{e}^{ax} - 1)$$

Other important integrals are given, too.
For a detailed representation of integrals with the function, see the tables of integrals (Chapter 25) at the end of this book or (similar compilations of integrals).

(m) Particular extensions and applications

Some extensions of the function, e.g., the complex representation of the function as well as advanced applications, are described here.

Simple functions

In this context, simple functions are constant functions, jump functions, signum function, step function, break functions and absolute value functions.

▷ Some of the functions described below possess points of discontinuity or non-differentiable break points. However, all are integrable.

5.4 Constant function

$$f(x) = c$$

Constant functions are the simplest of all functions.

● Constants are invariants, i.e., invariable quantities. If not explicity written out, they are denoted by Roman letters (a, b, c) or Greek letters (α, β, γ).

(a) Definition

The same value c is assigned to all points x,

$$f(x) = c\ .$$

The function is a solution of the differential equation:

$$\frac{\mathrm{d}f}{\mathrm{d}x} = 0\ .$$

(b) Graphical representation

The graph of a constant function is a horizontal straight line that reads to infinity. Any values can be assumed on the y-axis.

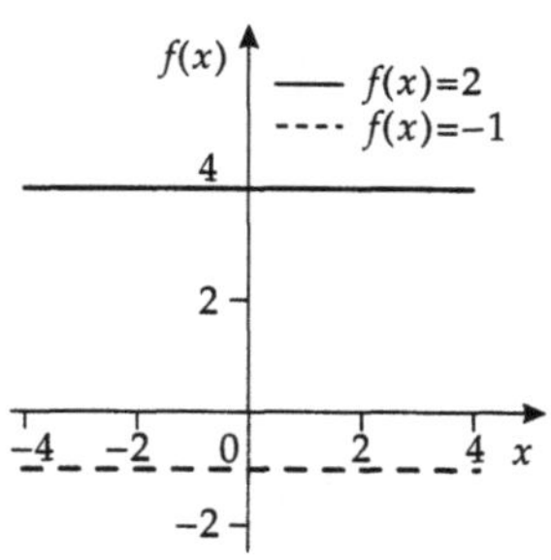

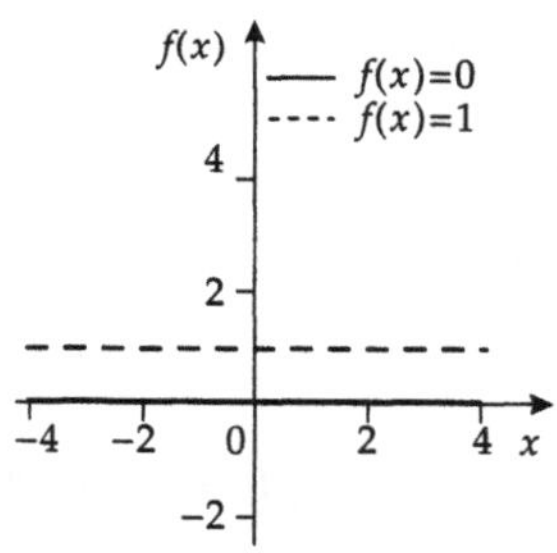

▷ Special cases are $f(x) = 0$ (**zero function**) and $f(x) = 1$ (**unity function**).

(c) Properties of the function

Domain of definition:	$-\infty < x < \infty$.
Range of values:	$f(x) = c$ (c can have any arbitrary value).
Quadrant:	$c > 0$ lies in the first and second quadrant, $c < 0$ lies in the third and fourth quadrant.
Periodicity:	Any periods can be defined.

Monotonicity: Monotonically increasing as well as monotonically decreasing, but never strictly monotonic.

Symmetries: Mirror-symmetric about the y-axis.
$f(x) = 0$ is also point-symmetric about the origin.

Asymptotes: The function $f(x) = c$ is its own asymptotic.

(d) Particular values

Zeros: $c \neq 0$ no zeros;
$c = 0$ all points are zeros.

Discontinuities: None; the function is continuous.

Poles: None; the function is defined for all x.

Extremes: Every $x \in \mathbb{R}$.

Points of inflection: None.

(e) Reciprocal function

The reciprocal function of $f(x) = c \neq 0$ is again a constant function.

$$\frac{1}{f(x)} = \frac{1}{c}$$

(f) Inverse function

The domain of definition of the inverse function is restricted to $x = c$; the range of values is $-\infty \leq y \leq +\infty$.
The graph of the inverse function is the parallel to the y-axis through point $x = c$.

(g) Cognate function

Step functions: functions possessing a certain constant value in a subdomain of the x-axis.

(h) Conversion formulas

The functions have the same value everywhere.

$$f(x+y) = f(x) = f(y) = f(ax) = c\,, \quad x, y, a, c \in \mathbb{R}\,.$$

(i) Approximation formulas for the function

The approximation formulas for constants apply, e.g., for Euler's number *e*.

$$\mathrm{e} \approx \left(\frac{131}{130}\right)^{130}$$

(j) Expansion of series or products of the function

The expansion of series and products of constants applies, e.g.,

$$\pi = 2 \cdot \frac{4}{3} \cdot \frac{16}{15} \cdot \frac{36}{35} \cdot \frac{64}{63} \cdot \ldots = 2 \prod_{k=1}^{\infty} \frac{4k^2}{4k^2 - 1}\,.$$

(k) Derivative of the function

The derivative of the constant function is the zero function.

$$\frac{\mathrm{d}}{\mathrm{d}x}c = 0$$

(l) Primitive of the function

The primitives of the constant function are the linear functions.

$$\int_0^x c\mathrm{d}t = c\,x$$

(m) Particular extensions and applications

▷ Constant functions are the "zero" approximation for interpolating in the immediate neighborhood of known function values.

Constant functions are of importance as asymptotes of other functions.

5.5 Step function

$$f(x) = H(x-a)$$

Step function $H(x-a)$, also called **Heaviside function**, jumps from 0 to 1 at $x-a=0$. Alternative notation: $E(x), \sigma(x), u(x)$, or $\Theta(x)$ instead of $H(x)$. The function $\Theta(x)$ must not be confused with the theta function.

(a) Definition

The function jumps from 0 to 1 at $x=a$.

$$H(x-a) = \begin{cases} 0 & x < a \\ \dfrac{1}{2} & x = a \\ 1 & x > a \end{cases}$$

The function can also be represented as an integral.

$$H(x) = \frac{1}{2} + \frac{1}{\pi}\int_0^{\infty} \frac{\sin(xt)}{t}\mathrm{d}t$$

(b) Graphical representation

The function is continuous, except for the jump discontinuity at $x=a$, and it is constant for $x>a$ and $x<a$.

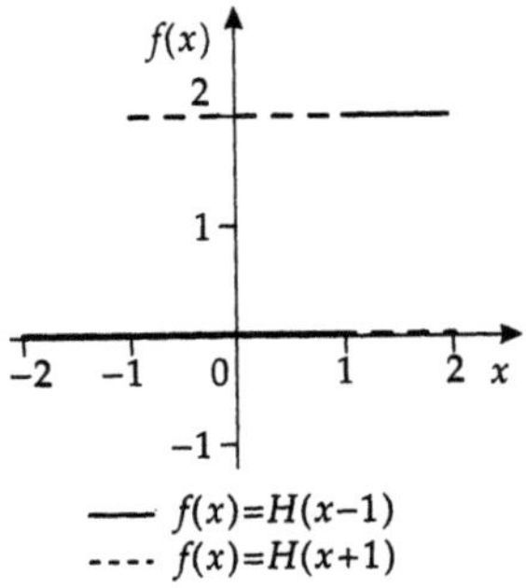

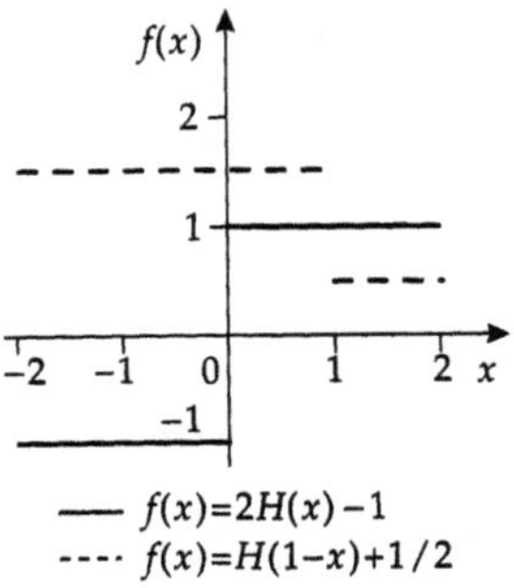

- Multiplication by a factor increases the function to the value of the factor for $x > a$. As a negative argument reflects, the function is reflected on the x-axis.

$$H(a - x) = 1 - H(x - a)$$

(c) Properties of the function

Domain of definition:	$-\infty < x < \infty$.
Range of values:	$f(x) = 0, \frac{1}{2}, 1$.
Quadrant:	Lies in the first or second quadrant (depending on α) or second quadrant.
Periodicity:	None.
Monotonicity:	Monotonically increasing (piecewise constant).
Symmetries:	Point symmetric about the point $\left(x = a, y = \frac{1}{2}\right)$.
Asymptotes:	$f(x) \to 1$ for $x \to +\infty$ $f(x) \to 0$ for $x \to -\infty$.

(d) Particular values

Zeros:	All $x < a$.
Discontinuities:	At $x = a$.
Poles:	None.
Extremes:	Maxima at all $x > a$, minima at all $x < a$.
Point of inflection:	None.

(e) Reciprocal function

For $x < a$, no reciprocal function is allowed.
For $x > a$, the value does not change.

(f) Inverse function

No inverse function exists.

(g) Cognate function

Constant function can be represented as a sum of two step functions.

$$cH(x-a)+cH(a-x)=c$$

Signum function (also called sign function) sgn(x) (also sign(x)): It describes the sign of x. See right-hand diagram under **b**.

$$\operatorname{sgn}(x)=2H(x)-1=\begin{cases}-1 & x<0\\ 0 & x=0\\ 1 & x>0\end{cases}$$

▷ The signum function can also be written as an integral:

$$\operatorname{sgn}(x)=\frac{2}{\pi}\int_0^\infty \frac{\sin(xt)}{t}\mathrm{d}t\ .$$

Break function: The primitive of the step function.
Alternative step function: Has the value 1 instead of $\frac{1}{2}$ at point $x=a$.
▷ In most cases, the difference is small.

$$\mathcal{H}(x-a)=\begin{cases}0 & x<a\\ 1 & x\geq a\end{cases}$$

Alternative sign function: Analogous to the signum function, formed with the alternative step function.

$$\operatorname{sgn}(x)=2\mathcal{H}(x)-1=\begin{cases}-1 & x<0\\ 1 & x\geq 0\end{cases}$$

(h) Conversion formulas

The product of two step functions is the function which jumps farther to the right!

$$H(x-a)\cdot H(x-b)=H(x-b),\qquad \text{if } b>a\ .$$

Negative argument and complementarity:

$$H(a-x)=1-H(x-a)\ ,\qquad H(x-a)+H(a-x)=1\ .$$

For the signum function, the following applies:

$$\operatorname{sgn}(-x)=-\operatorname{sgn}(x)\ ,\qquad \operatorname{sgn}(x)+\operatorname{sgn}(-x)=0\ .$$

(i) Approximation formulas for the function

For $f(x)=cH(x-a)$, the approximation formulas for constants apply.
Approximation to the step function by the hyperbolic tangent.

$$H(x-a)\approx\frac{1}{2}\left[1+\tanh\left(b(x-a)\right)\right]\ .$$

▷ The greater the value b, the better the approximation.

(j) Expansion in series or products of the function

If $f(x)$ can be represented as a series, then the same holds for $f(x)\cdot H(x)$.

$$f(x)\cdot H(x-b)=\sum_{k=0}^{\infty}H(x-b)\cdot a_kx^k\qquad \text{for } f(x)=\sum_{k=1}^{\infty}a_kx^k\ .$$

(k) Derivative of the function

The derivative of the step function is zero for all values x outside the discontinuity.

$$\frac{\mathrm{d}}{\mathrm{d}x} H(x-a) = 0 \,, \qquad x \neq a \,.$$

In a generalized sense, the derivative of the step function is the **delta function** described below.

$$\frac{\mathrm{d}}{\mathrm{d}x} H(x-a) = \delta(x-a) \,.$$

For the derivative of $f(x)H(x)$:

$$\frac{\mathrm{d}}{\mathrm{d}x}(f(x)H(x-a)) = \begin{cases} 0 & x < a \\ \dfrac{\mathrm{d}f}{\mathrm{d}x}(x) & x > a \,. \end{cases}$$

(l) Primitive of the function

The primitive of $H(x)$ is the **break function** (see diagram in the section on the absolute value function)

$$\int_{-\infty}^{x} H(t)\mathrm{d}t = x \cdot H(x) \,.$$

The integral of $f(x)H(x)$ is

$$\int_{b}^{x} f(t)H(t)\mathrm{d}t = \int_{a}^{x} f(t)\mathrm{d}t \,, \qquad b < a < x \,.$$

(m) Particular extensions and applications

Threshold function: The product $f(x)H(x-a)$ suppresses the value of $f(x)$ to 0 for $x < a$.

$$f(x)H(x-a) = \begin{cases} 0 & x < a \\ \dfrac{1}{2} f(a) & x = a \\ f(x) & x > a \end{cases}$$

Pulse function: Through the difference of two step functions, only one sector of the x-axis receives the value 1.

$$H(x-a) - H(x-b) = \begin{cases} 0 & x < a, b < x \\ \dfrac{1}{2} & x = a, x = b, \\ 1 & a < x < b \end{cases} \qquad a < b$$

Window function: Through the product of a function and a pulse function, the function is suppressed only in a certain domain.

$$f(x)[H(x-a)-H(x-b)] = \begin{cases} 0 & x<a, b<x \\ \frac{1}{2}f(a) & x=a \\ \frac{1}{2}f(b) & x=b \\ f(x) & a<x<b \end{cases} \qquad a<b$$

▷ Threshold functions, pulse functions and window functions can also be defined in analogy to the alternative step function.

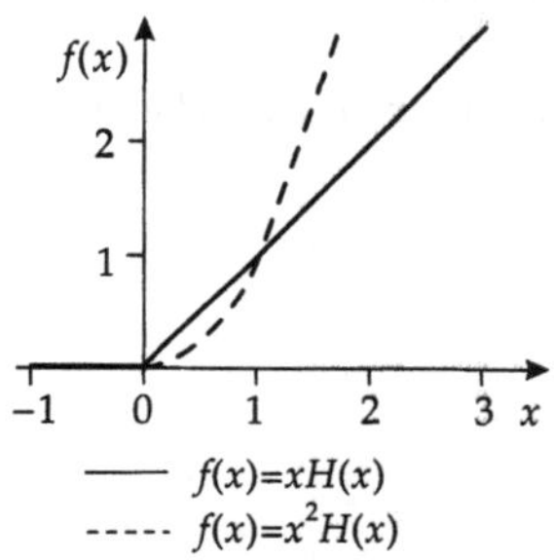

Threshold functions

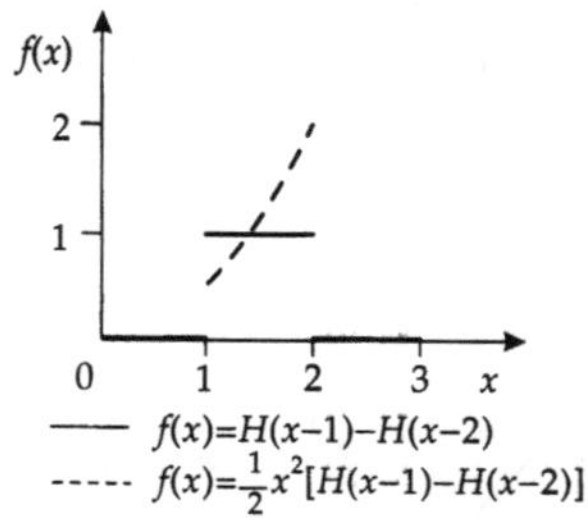

Pulse function (solid line), window function (dotted line)

Step functions: Usually defined in terms of alternative step functions.

$$f(x) = c_0 + \sum_{k=1}^{n}(c_k - c_{k-1})\mathcal{H}(x-a_k) \qquad a_1 < a_2 < \cdots < a_n ,$$

where c_0 is the value of the function for $x \to -\infty$, and c_k is the value of the function at the point a_k.

▷ Step functions and pulse functions find application in control technology. By means of these functions, the response behavior of controllers and controlled systems can be determined.

In systems theory, step functions create step responses.

5.6 Absolute value function

$$f(x) = |x-a|$$

In programming languages: `ABS(X-A)`.

Absolute value function $|x|$: Every value x is assigned its absolute value; i.e., the distance between points x and 0 on the number line.

● The function values are never negative.

Importance of the absolute value function: in particular for describing quantities which cannot be negatively defined.

□ Area between two curves; distance between two points.

(a) Definition

Another value is $+x$ for $x \geq 0$ and $-x$ for $x < 0$:

$$f(x) = |x| = (x) \cdot \operatorname{sgn}(x) = x \cdot H(x) - x \cdot H(-x) = \begin{cases} x & x \geq 0 \\ -x & x < 0 \end{cases}$$

Another definition uses the root of a square:

$$|x| = \sqrt{x^2} \qquad \text{for all } x \ .$$

(b) Graphical representation

For $x > a$, $|x - a|$ is a straight line that increases with slope 1; for $x < a$, it is a straight line that decreases with slope -1.

- A break point appears at $x = a$; at $x = a$, $f(x)$ is continuous but not differentiable.

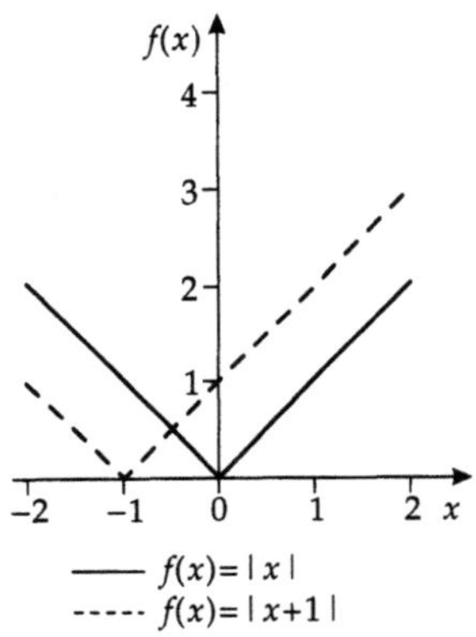

Absolute value functions

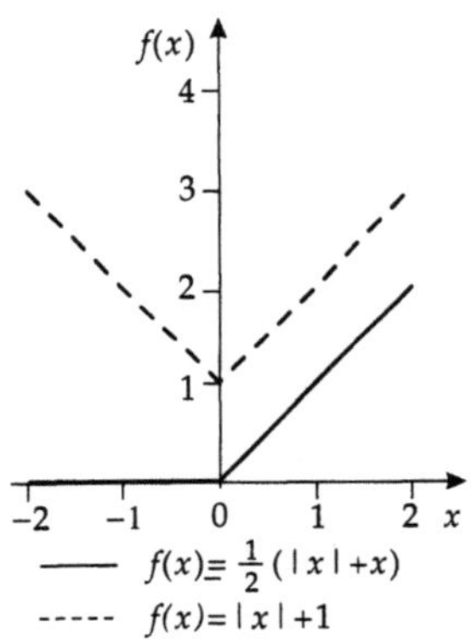

Break functions

An addition $|x + a|$ inside of the argument results in a shift to the left. An addition $|x| + a$ outside the argument results in a shift upwards.
Multiplication $c|x|$ by a factor of c changes the slope.

(c) Properties of the function

Domain of definition:	$-\infty < x < \infty$.
Range of values:	$0 \leq f(x) < \infty$.
Quadrant:	Lies in the first and second quadrants.
Periodicity:	None.
Monotonicity:	$x \geq a$ strictly monotonically increasing; $x \leq a$ strictly monotonically decreasing.
Symmetries:	Mirror-symmetric about axis $x = a$.
Asymptotes:	$f(x) \to \infty$ for $x \to \pm\infty$.

(d) Particular values

Zeros:	$x = a$.
Discontinuities:	None, the function is continuous.

Poles: None.

Extremes: Minimum at $x = a$, but not differentiable there.

Points of inflection: None.

(e) Reciprocal function

The reciprocal function (for $x \neq a$) is the absolute value of a hyperbola.

$$\frac{1}{|x-a|} = \left|\frac{1}{x-a}\right|$$

(f) Inverse function

The inverse function of $|x - a|$ can only be formed either for $x \geq a$ or for $x \leq a$.

$$\begin{array}{ll} x \geq a & f^{-1}(x) = x + a \text{ or} \\ x \leq a & f^{-1}(x) = -x + a \end{array}$$

(g) Cognate functions

Break function: The positive branch is remains; for $x < 0$ the function becomes zero. See the right-hand diagram of the graph.

$$f(x) = \frac{1}{2}(|x| + x) = |x| \cdot H(x) = \begin{cases} 0 & x \leq 0 \\ x & x > 0 \end{cases}$$

Triangle function: An isosceles triangle of altitude h and side $2a$ with the vertex above point x_0 with $a, h > 0$,

$$D(x; x_0, a, h) = h\left(1 - \frac{|x - x_0|}{a}\right) \cdot H(x - a) \cdot H(x + a)$$

$$= \begin{cases} 0 & x \leq x_0 - a, x \geq x_0 + a \\ \frac{h}{a}(x - x_0 + a) & x_0 - a < x \leq x_0 \\ \frac{h}{a}(x_0 - x + a) & x_0 < x < x_0 + a \end{cases}$$

Maximum function: A function of two or more arguments which assigns the argument with the greatest value to the arguments of the function. For two arguments, the maximum function can be written in terms of the absolute value function,

$$\max(x, y) = \frac{1}{2}(x + y + |x - y|) = \begin{cases} x & x \geq y \\ y & x < y \end{cases} .$$

For several arguments, it can be written recursively,

$$\max(x_1, x_2, \ldots, x_n) = \max(x_1, \max(x_2, \max(x_3, \ldots, \max(x_{n-1}, x_n)\ldots))) .$$

▷ Conversely, the absolute value function can be defined by means of the maximum function.

$$|x| = \max(x, -x)$$

Minimum function: A function of two or more arguments which assigns the argument with the smallest value to the arguments of the function. Analogous to the maximum function for two arguments, it can be written in terms of the absolute value function,

$$\min(x, y) = \frac{1}{2}(x + y - |x - y|) = \begin{cases} y & x \geq y \\ x & x < y \end{cases} .$$

For several arguments, it can be written recursively:

$$\min(x_1, x_2, \ldots, x_n) = \min(x_1, \min(x_2, \min(x_3, \ldots, \min(x_{n-1}, x_n) \ldots))) .$$

(h) Conversion formulas

The absolute value of the negative argument equals the absolute value of the positive argument.

$$|-x| = |x| \qquad |x - y| = |y - x|$$

● The absolute value of the product is the product of the absolute values.

$$|x \cdot y| = |x| \cdot |y|$$

● The absolute value of a sum (difference) does not have to be equal to the sum (difference) of the absolute values.

$$|x| - |y| \leq |(|x| - |y|)| \leq |x \pm y| \leq |x| + |y|$$

(i) Approximation formulas for the function

Normally not necessary.

(j) Expansion of series or products of the function

Contrary to step functions, the absolute value function of a power series is generally not the power series of the absolute values.

(k) Derivative of function

The derivative of $|x - a|$ equals the sign of $(x - a)$:

$$\frac{\mathrm{d}}{\mathrm{d}x}|x - a| = \begin{cases} -1 & x < a \\ 1 & x > a \end{cases} .$$

● The derivative for $x = a$ is not defined.

▷ In a generalized form of the derivative, the value of the derivative is equal to 0 at $x = a$, and the derivative function is the signum function.

$$\frac{\mathrm{d}}{\mathrm{d}x}|x - a| = \operatorname{sgn}(x - a) .$$

Derivative of a function of the absolute value function:

$$\frac{\mathrm{d}}{\mathrm{d}x} f(|x|) = \begin{cases} -\frac{\mathrm{d}f}{\mathrm{d}x}(-x) & x < a \\ \frac{\mathrm{d}f}{\mathrm{d}x}(x) & x > a . \end{cases}$$

Derivative of the absolute value of a function: The derivative of the function multiplied by the sign of the function value.

$$\frac{\mathrm{d}}{\mathrm{d}x}|f(x)| = \operatorname{sgn}[f(x)] \cdot \frac{\mathrm{d}f}{\mathrm{d}x} .$$

Again, the derivative is not defined at the zeros of the function.

(l) Primitive of the function

The primitive of $|x|$: The product of a parabola and the signum function.

$$\int_0^x |t|\mathrm{d}t = \frac{1}{2}x^2 \cdot \operatorname{sgn}(x) = \frac{1}{2}x \cdot |x| .$$

Integral of the absolute value of a function: Division into the intervals between the zeros $x_0^1, \ldots x_0^n$, where $a < x_0^1 < x_0^2 < \cdots < x_0^n < x$.

$$\int_a^x |f(t)|\mathrm{d}t = \left|\int_a^{x_0^1} f(t)\mathrm{d}t\right| + \left|\int_{x_0^1}^{x_0^2} f(t)\mathrm{d}t\right| + \cdots + \left|\int_{x_0^n}^{x} f(t)\mathrm{d}t\right| .$$

(m) Particular extensions and applications

The absolute value of a complex function: The root of the sum of the squares of real and imaginary parts

$$|x + \mathrm{j}y| = \sqrt{zz^*} = \sqrt{x^2 + y^2}$$

In electrical engineering, a rectifier converts electrical signals into the absolute values of the signals.

5.7 Delta function

$$f(x) = \delta(x - a)$$

The delta function, is also known as the **Dirac delta function**.

▷ The delta function must not be confused with Kronecker's symbols, δ_{ij}.
The delta function is not a function, but a **distribution**; i.e., a linearly continuous functional that affects per se, the elements of a function space.

● The delta function can be represented as a limit of functions.
The importance of the delta function lies in applications where it is multiplied by another function under an integral.

(a) Definition

Characterization of the delta function by the integral

$$\int_x^y \delta(t - a)\mathrm{d}t = \begin{cases} 1 & \text{if } x < a < y \\ 0 & \text{if } a < x \text{ or } y < a \end{cases} \qquad \text{for any } x < y .$$

Integral representation:

$$\delta(x - a) = \int_{-\infty}^{+\infty} \cos[2\pi(x - a)t]\mathrm{d}t .$$

Representation as the generalized derivative of the step function:

$$\delta(x-a) = \frac{\mathrm{d}}{\mathrm{d}x} H(x-a) \,.$$

Representation as a limit of functions becoming increasingly narrow, while keeping the area under the curve constant.
Limit of a triangle function of $D\left(x; a, \frac{1}{h}, h\right)$ altitude h and $\frac{1}{h}$ (defined in the section on absolute value functions):

$$\delta(x-a) = \lim_{h\to\infty} D\left(x; a, \frac{1}{h}, h\right) .$$

Limit of a pulse function of width $2b$:

$$\delta(x-a) = \lim_{b\to 0} \frac{1}{2b} \left(H(x-a+b) - H(x-a-b)\right) \,.$$

Limit of a Gauss function:

$$\delta(x-a) = \lim_{b\to\infty} \sqrt{\frac{b}{\pi}} \mathrm{e}^{-b(x-a)^2} \,.$$

Limit of the hyperbolic secant (reciprocal hyperbolic cosine):

$$\delta(x-a) = \lim_{b\to 0} \frac{1}{2b} \operatorname{sech}\left(\frac{x-a}{b}\right) .$$

(b) Graphical representation

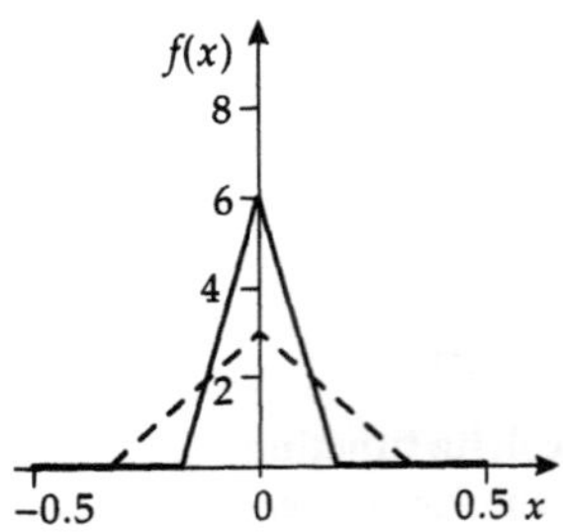

Triangle functions

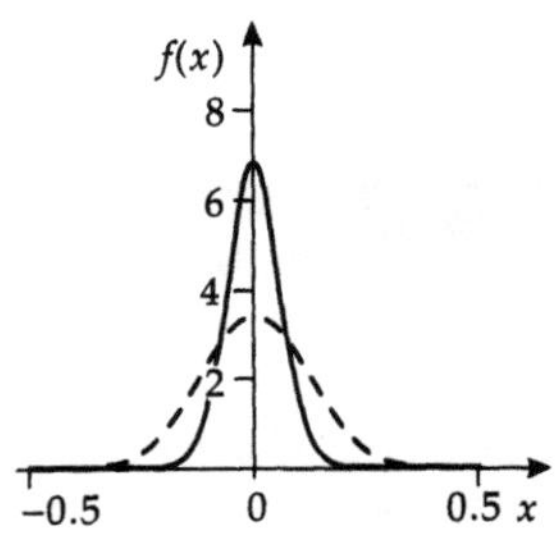

Gauss functions

A delta function cannot be represented graphically. It is zero everywhere except for $x = a$, where it is infinite.
However, it is possible to illustrate the limit formation which describes the delta function.

(c) Properties of the function

Domain of definition: $-\infty < x < \infty$.

Range of values: $f(x) = 0, \infty$.

Quadrant: Lies in the first and second quadrants.

Periodicity: None.

Monotonicity: Constant except for $x = a$.

Symmetries:	Mirror-symmetric about $x = a$.
Asymptotes:	$f(x) = 0$ for $x \to \pm\infty$.

(d) Particular values

Zeros:	All $x \neq a$.
Discontinuities:	$x = a$.
Poles:	$x = a$.
Extremes:	Not defineable.
Points of inflection:	None.

(e) Reciprocal function

Reciprocal functions do not exist.

(f) Inverse function

The function cannot be inverted.

(g) Cognate function

Related to delta functions via the limit formation are: Triangle functions, the pulse functions, Gauss functions and hyperbolic secants. Step function, primitive of the delta function. Constant function $f(x) = 1$ is the Fourier transform of the delta function.

(h) Conversion formulas

Mirror-symmetry of the delta function:

$$\delta(x-a) = \delta(a-x) = \delta(|x-a|) .$$

Factor in the argument:

$$\delta(cx) = \frac{1}{|c|}\delta(x) .$$

Function with simple zeros $x_0^1, \ldots, x_0^n$ in the argument:

$$\delta(f(x)) = \sum_{k=1}^{n} \left| \frac{1}{\frac{\mathrm{d}f}{\mathrm{d}x}(x_0^k)} \right| \delta(x - x_0^k) .$$

(i) Approximation formulas for the function

There are no approximation formulas for the delta function. If need be, the limit representations with finite parameters given under "definition" can be used as an approximation.

(j) Expansion of series or products of the function

There is no expansion of series or products of the function.

(k) Derivative of function

The derivative of the delta function is zero everywhere except at $x = a$, and it is not defined for $x = a$.

$$\frac{\mathrm{d}}{\mathrm{d}x}\delta(x-a) = 0 \qquad \text{for all } x \neq a\ .$$

Derivative in a generalized sense:

$$\frac{\mathrm{d}}{\mathrm{d}x}\delta(x-a) = -\delta(x-a)\frac{\mathrm{d}}{\mathrm{d}x}\ .$$

Application in an integral:

$$\int_x^y f(t)\frac{\mathrm{d}}{\mathrm{d}t}\delta(t-a)\mathrm{d}t = -\int_x^y \delta(t-a)\frac{\mathrm{d}}{\mathrm{d}t}f(t)\mathrm{d}t = -\frac{\mathrm{d}f}{\mathrm{d}t}(a)\ .$$

(l) Primitive of the function

The primitive of the delta function is the step function:

$$\int_{x_0}^x \delta(t-a)\mathrm{d}t = H(x-a)\ , \qquad x_0 \leq a\ .$$

Integral via the product of a function and the delta function:

$$\int_{x_0}^x f(t)\delta(t-a)\mathrm{d}t = f(a)H(x-a)\ , \qquad x_0 \leq a\ .$$

Integral over the entire number line:

$$\int_{-\infty}^{\infty} f(t)\delta(t-a)\mathrm{d}t = f(a)\ .$$

(m) Particular extensions and applications

A complex argument in the delta function corresponds to the product of a delta function for the real part and a delta function for the imaginary part.

$$\delta([x+\mathrm{j}y]-[a+\mathrm{j}b]) = \delta([x-a]+\mathrm{j}[y-b]) = \delta(x-a)\delta(y-b)$$

Kronecker's symbol: Analogous to the delta function for integers.

$$\delta_{ij} = \begin{cases} 0 & i \neq j \\ 1 & i = j \end{cases}$$

▷ The delta function is of importance in the treatment of partial linear differential equations.

The delta function (spike, signal pulse) is used to test controlled systems in automatic control technology.

5.8 Integer-part function, fractional-part function

$$f(x) = [x] = \mathrm{Int}(x)$$

Representation in programming languages: `INT(X)`, `TRUNC(X)`

▷ In FORTRAN, the digits after the decimal point are cut off by INT(X), corresponding to the alternative fractional-part function mentioned at the end of this section.

Integer-part function, $[x]$ or Int(x): Mapping of every real number x onto the nearest integer that is smaller than or equal to x.

The function is used for all kinds of rounding operations and for fitting continuous quantities into discrete lattices.

Fractional-part function, (x) or frac(x): The difference between x and $[x]$.

(a) Definition

The largest integer $n \in \mathbb{Z}$ that is smaller than or equal to x.

$$[x] = \text{Int}(x) = n, \qquad n \le x < n+1\,, \qquad n = 0, \pm 1, \pm 2, \ldots$$

Fractional-part function, the difference between x and $[x]$:

$$(x) = \text{Frac}(x) = x - [x]\,.$$

(b) Graphical representation

Integer-part function: Steps of constant value separated by one unit.
The steps have the same length of one unit.

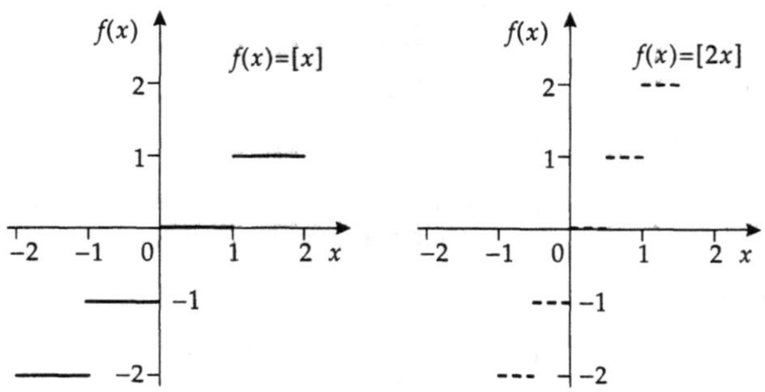

Integer-part functions

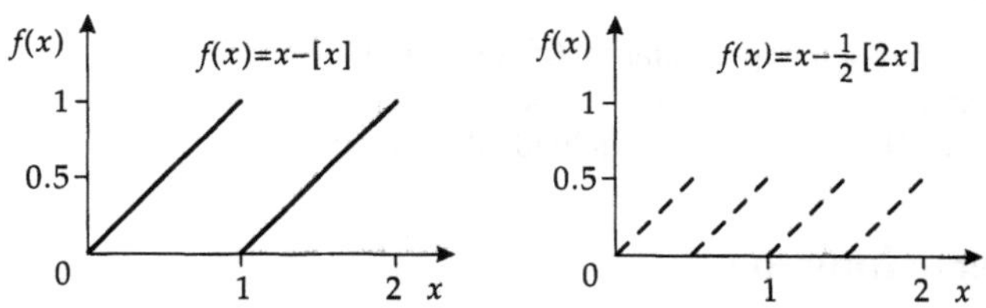

Fractional-part functions

- The function is piecewise continuous, but it also has a finite number of discontinuities.

A factor in the argument shortens the step length by that factor.

Fractional-part function: Segment of a straight line of slope 1 on an x-segment of one unit of length.

- The fractional-part function repeats after one unit of length.

(c) Properties of the function

Domain of definition:	$-\infty < x < \infty$.
Range of values:	$[x]$: $f(x) = 0, \pm 1, \pm 2, \ldots$ (x): $0 \leq f(x) < 1$.
Quadrant:	$[x]$: Lies in the first and third quadrants. (x): Lies in the first and second quadrants.
Periodicity:	$[x]$: None. (x): Length of the period: 1, segments $n \leq x < n + 1$.
Monotonicity:	$[x]$: Monotonically increasing. (x): Strictly monotonically also increasing in the interval $n \leq x < n + 1$.
Symmetries:	No mirror-symmetry, no point-symmetry (half-open intervals).
Asymptotes:	$[x]$: $f(x) \to \pm\infty$ for $x \to \pm\infty$. (x): No asymptotes.

(d) Particular values

Zeros:	$[x]$: $0 \leq x < 1$. (x): $x = 0, \pm 1, \pm 2, \ldots$
Discontinuities:	$x = 0, \pm 1, \pm 2, \ldots$
Poles:	None.
Extremes:	$[x]$: None. (x): Minima at $x = 0, \pm 1, \pm 2, \ldots$
Points of inflection:	None.
Values at $x = n$:	$[x]$: $[n] = n, \quad n = 0, \pm 1, \pm 2, \ldots$ (x): $(n) = 0, \quad n = 0, \pm 1, \pm 2, \ldots$

(e) Reciprocal function

No explicit conversion, representation ($f(x) \neq 0$) of
$[x]$: step functions vanishing as $x \to \pm\infty$.
(x): segments of hyperbolas in the individual intervals.

(f) Inverse function

There are no inverse functions.
For $0 \leq x \leq 1$, (x) is its own inverse function.

(g) Cognate function

Rounding function, ROUND(X), jumps at $\pm 0{,}5, \pm 1{,}5, \pm 2{,}5, \ldots$ instead of $0, \pm 1, \pm 2, \ldots$

$$f(x) = \left[x + \frac{1}{2}\right].$$

Modulo function $n \pmod m$: A function of two natural numbers; the modulo function is the remainder after the number m is subtracted from n so many times that the difference would change its sign if further subtractions were made.

$$n(\operatorname{mod} m) = m \cdot \operatorname{frac}\left(\frac{n}{m}\right)$$

(h) Conversion formulas

Complementarity:

$$[x] + (x) = x\ .$$

Addition of integers:

$$[x+n] = [x] + n \qquad (x+n) = (x)\ .$$

(i) Approximation formulas for the function

No approximation formulas are customary.

(j) Expansion of series or products of the function

Series expansion of $[f(x)]$ in terms of the inverse function F using the alternative step function $\mathcal{H}$:

When f is strictly monotonically increasing:

$$[f(x)] = \sum_{k=1}^{\infty} \mathcal{H}(x - F(k)) + \sum_{k=0}^{\infty} \{\mathcal{H}(x + F(-k)) - 1\}\ .$$

When f is strictly monotonically decreasing:

$$[f(x)] = -\sum_{k=1}^{\infty} \mathcal{H}(x + F(-k)) - \sum_{k=0}^{\infty} \{\mathcal{H}(x - F(k)) - 1\}\ .$$

(k) Derivative of function

Derivative of $[x]$ and (x), $x \neq \pm 1, \pm 2, \ldots$

$$[x]: \quad \frac{\mathrm{d}}{\mathrm{d}x}[x] = 0\ .$$

$$(x): \quad \frac{\mathrm{d}}{\mathrm{d}x}(x) = 1\ .$$

Function of an integer-part function:

$$\frac{\mathrm{d}}{\mathrm{d}x} f([x]) = 0\ , \qquad x \neq 0, \pm 1, \pm 2 \ldots$$

Function of a fractional-part function:

$$\frac{\mathrm{d}}{\mathrm{d}x} f(\operatorname{frac}(x)) = \frac{\mathrm{d}f}{\mathrm{d}x}(x - [x])\ , \qquad x \neq 0, \pm 1, \pm 2 \ldots$$

Integer-part function of a function (F is the inverse function):

$$\frac{\mathrm{d}}{\mathrm{d}x}[f(x)] = 0\ , \qquad x \neq F(0), F(\pm 1), F(\pm 2) \ldots$$

Fractional-part function of a function (F is the inverse function):

$$\frac{\mathrm{d}}{\mathrm{d}x}\operatorname{frac}(f(x)) = \frac{\mathrm{d}f}{\mathrm{d}x}(x-[x]), \qquad x \neq F(0), F(\pm 1), F(\pm 2)\ldots$$

(l) Primitive of the function

Function of an integer-part function:

$$\int_a^x f([t])\mathrm{d}t = ([a]+1-a)f([a]) + (x-[x])f([x]) + \sum_{k=[a]+1}^{[x]-1} f(k) .$$

Function of a fractional-part function:

$$\int_a^x f(\operatorname{frac}(t))\mathrm{d}t = \int_{a-[a]}^1 f(t)\mathrm{d}t + ([x]-[a]-2)\int_0^1 f(t)\mathrm{d}t + \int_0^{x-[x]} f(t)\mathrm{d}t .$$

Integer-part function of a strictly monotonically increasing function (with inverse function F):

$$\int_a^x [f(t)]\,\mathrm{d}t = [f(x)]\,x - [f(a)] - \sum_{k=1}^{[f(x)]-[f(a)]} F([f(a)]+k .$$

Integer-part function of a strictly monotonically decreasing function (with inverse function F):

$$\int_a^x [f(t)]\,\mathrm{d}t = [f(x)]\,x - [f(a)] + \sum_{k=1}^{[f(a)]-[f(x)]} F([f(x)]+k .$$

Fractional-part function of a function:

$$\int_a^x \operatorname{frac}(f(t))\mathrm{d}t = \int_a^x f(t)\mathrm{d}t - \int_a^x [f(t)]\,\mathrm{d}t .$$

(m) Particular extensions and applications

Alternative integer-part function: Describes the number before the decimal point in a decimal representation:

$$\operatorname{Ip}(x) = [|x|]\cdot \operatorname{sgn}(x)$$

Alternative fractional-part function: Complementary function of the alternative integer-part function.

$$\operatorname{Fp}(x) = x - \operatorname{Ip}(x)$$

▷ The difference lies in the treatment of negative numbers, e.g., for $x = -3.14$:

$$[-3.14] = -4, \qquad \operatorname{frac}(-3.14) = 0.86$$
$$\operatorname{Ip}(-3.14) = -3, \qquad \operatorname{Fp}(-3.14) = -0.14$$

▷ In `FORTRAN` this is accomplished by the function `INT(X)`.

Integral rational functions

Integral rational functions: Functions that can be written as a finite sum of power functions with integral, positive exponents.
Below, the simplest integral rational functions, the polynomials of degree 1, 2, and 3, are described first. This is followed by the power function of degree n and, finally, the polynomials of degree n.

5.9 Linear function—straight line

$$f(x) = ax + b$$

Polynomial of degree one.
The functional equation describes a straight line of slope a and with the y-intercept b.
Linear function: Mathematically not quite acccurate notation for the function defined by the equation of a straight line.
First approximation of a function near a known point.

▷ Determination of an unknown function value $f(x)$ with $x_1 < x < x_2$ form two known function values $f(x)$ and $f(x_1)$ is accomplished by drawing a straight line through points $(x_1, f(x_1))$ and $(x_2, f(x_2))$.

(a) Definition

Differential equation for a linear function:

$$\frac{\mathrm{d}}{\mathrm{d}x} f(x) = a \, .$$

Functional equation of a straight line:

$$f(x) = ax + b \qquad \text{with } a = f(x = 1) - f(x = 0), \quad b = f(x = 0) \, .$$

Slope of a straight line given by points x_1 and x_2:

$$a = \frac{f(x_2) - f(x_1)}{x_2 - x_1} = \frac{\Delta y}{\Delta x} \, , \qquad x_2 \neq x_1 \, .$$

▷ In geometry, the slope is frequently denoted by m.
Intercept of a straight line given by points $(x_1, f(x_1))$ and $(x_2, f(x_2))$:

$$b = \frac{x_2 f(x_1) - x_1 f(x_2)}{x_2 - x_1} \, , \qquad x_2 \neq x_1 \, .$$

(b) Graphical representation

The graph of the linear function is a straight line intersecting the y-axis at $x = 0$, $f(x) = b$, and the x-axis at $x = -\frac{b}{a}$, i.e., at $f(x) = 0$.
The straight line has slope a, and encloses an angle with the x-axis given by

$$\alpha = \arctan(a), \qquad a = \tan(\alpha).$$

- For $a > 0$ the straight line becomes steeper as a increases and flatter as a decreases.

▷ for $a = 0$, the linear function is the constant function.

A negative a means a negative slope; i.e., the function values $f(x)$ become smaller as x increases.

● The intercept b translates the function along the y-axis.

▷ For $b = 0$, the function goes through the origin.

For $b > 0$ shifts the function up, while $b < 0$ shifts it down.

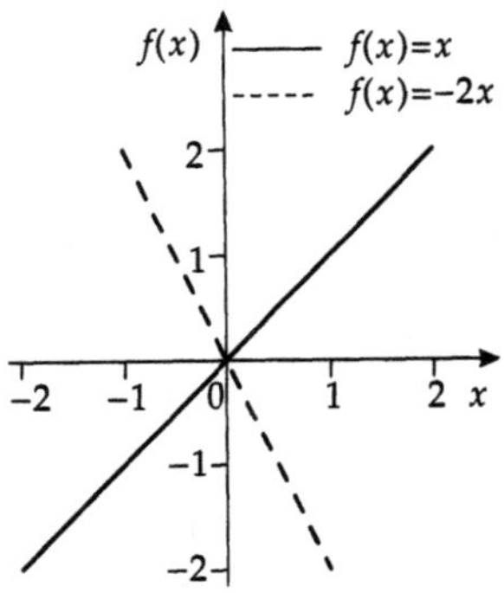

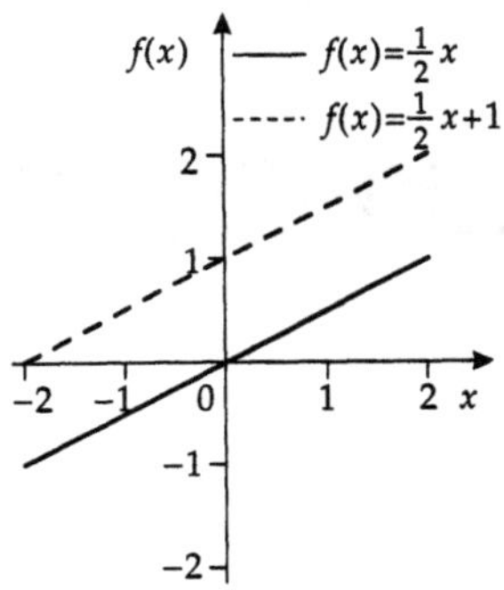

(c) Properties of the function

Domain of definition: $-\infty < x < \infty$.

Range of values: $-\infty < f(x) < \infty$.

Quadrant:

$a > 0, b = 0$	first and third quadrants
$a < 0, b = 0$	second and fourth quadrants
$a = 0, b > 0$	first and second quadrants
$a = 0, b < 0$	third and fourth quadrants
$a > 0, b > 0$	first, second, and third quadrants
$a > 0, b < 0$	first, third, and fourth quadrants
$a < 0, b > 0$	first, second, and fourth quadrants
$a < 0, b < 0$	second, third, and fourth quadrants

Periodicity: None.

Monotonicity:
$a > 0$: strictly monotonically increasing
$a < 0$: strictly monotonically decreasing
$a = 0$: constant

Symmetries: Point symmetries with respect to every point $(x, ax + b)$ of the straight line.

Asymptotes:
$a > 0$: $f(x) \to \pm\infty$ for $x \to \pm\infty$.
$a < 0$: $f(x) \to \mp\infty$ for $x \to \pm\infty$.
$a = 0$: $f(x) = b$ for $x \to \pm\infty$.

(d) Particular values

Zeros: $x_0 = -\dfrac{b}{a}$ for $a \neq 0$.

Discontinuities: None.

Poles:	None.
Extremes:	None.
Points of inflection:	None.
Value at $x = 0$:	$f(0) = b$.

(e) Reciprocal function

The reciprocal function is a hyperbola

$$\frac{1}{f(x)} = \frac{1}{ax+b} = (ax+b)^{-1} .$$

(f) Inverse function

Straight line of the slope $\frac{1}{a}$, for $a = 0$ the function has no inverse.

$$f^{-1}(x) = \frac{1}{a}x - \frac{b}{a} , \qquad a \neq 0 .$$

(g) Cognate functions

Constant functions, $a = 0$:
Absolute value function, $|x| = x\operatorname{sgn}(x)$:
Break function: $f(x) = xH(x)$.
Polynomials of higher degree: $ax + b$ is a special case.

(h) Conversion formulas

Reflection about the y-axis:

$$f(-x) = -f\left(x - \frac{2b}{a}\right)$$

The sum (difference) of two straight lines is itself a straight line.

$$(a_1x + b_1) \pm (a_2x + b_2) = (a_1 \pm a_2)x + (b_1 \pm b_2)$$

▷ In general, the product and the quotient are no longer linear functions.

(i) Approximation formulas for the function

Not necessary.

(j) Expansion of series or products of the function

Quotient of two linear functions:

$$\frac{a_1x+b_1}{a_2x+b_2} = \begin{cases} \dfrac{b_1}{b_2} + \dfrac{a_2b_1 - a_1b_2}{a_2b_2} \displaystyle\sum_{k=1}^{\infty} \left(\frac{-a_2x}{b_2}\right)^k & |x| < \left|\dfrac{b_2}{a_2}\right| \\ \dfrac{a_1}{a_2} - \dfrac{a_2b_1 - a_1b_2}{a_2b_2} \displaystyle\sum_{k=1}^{\infty} \left(\frac{-b_2}{a_2x}\right)^k & |x| > \left|\dfrac{b_2}{a_2}\right| \end{cases}$$

(k) Derivative of the function

The derivative is the slope of the straight line.

$$\frac{\mathrm{d}}{\mathrm{d}x}(ax+b) = a$$

▷ In geometry, the slope is often denoted by m.

(l) Primitive of the function

The primitive is the quadratic function.

$$\int_0^x (at+b)\mathrm{d}t = \frac{a}{2}x^2 + bx$$

(m) Particular extensions and applications

Linear interpolation: Approximating a function value $f(x)$ with $x_1 < x < x_2$ from the known function values $f(x_1)$ and $f(x_2)$.

$$f(x) = \frac{(x_2 - x)f(x_1) + (x - x_1)f(x_2)}{x_2 - x_1}$$

▷ When only the values at the interpolation nodes $x_1, x_2, \ldots, x_n$, a piecewise linear function can be defined by interpolation.

Linear regression: A straight line $ax + b$ is fitted to a plot of n data points $(x_1, f(x_1)), (x_2, f(x_2)), \ldots, (x_n, f(x_n))$ is the best approximation possible!

▷ The best possible fit is obtained when the sum of division squares $\sum_k (ax_k + b - f(x_k))^2$ reaches a minimum.
The values of a and b are:

$$a = \frac{n\left(\sum_{k=1}^n x_k f(x_k)\right) - \left(\sum_{k=1}^n x_k\right)\left(\sum_{k=1}^n f(x_k)\right)}{n\left(\sum_{k=1}^n x_k^2\right) - \left(\sum_{k=1}^n x_k\right)^2}$$

$$b = \frac{1}{n}\left[\left(\sum_{k=1}^n f(x_k)\right) - a\left(\sum_{k=1}^n x_k\right)\right].$$

See also Chapter 21 on statistics.

5.10 Quadratic function — parabola

$$f(x) = ax^2 + bx + c$$

Parabola: Curve of the quadratic function.
Next to the constant function and the linear function, this is the simplest polynomial.

▷ The space-time relationships $x(t)$ are quadratic in fields with constant forces.

The trajectory parabola, the free-fall solution curve with initial conditions is quadratic.

(a) Definition

Differential equation for the quadratic function:

$$\frac{\mathrm{d}^2}{\mathrm{d}x^2}f(x) = 2a$$

Functional equation:

$$f(x) = ax^2 + bx + c = a\left(x + \frac{b}{2a}\right)^2 + c - \frac{b^2}{4a}.$$

Discriminant of the quadratic function:

$$D = \frac{b^2}{4a^2} - \frac{c}{a}.$$

(b) Graphical representation

The graph of the function is a parabola shifted to the left by $\frac{b}{2a}$ and down by $\frac{b^2}{4a} - c = Da$.

- The curvature of the parabola is determined by a.

For $a > 0$, the parabola is open to the top; for $a < 0$, to the bottom. If $|a|$ is large, the parabola is narrow, if $|a|$ is small, the parabola becomes wide.

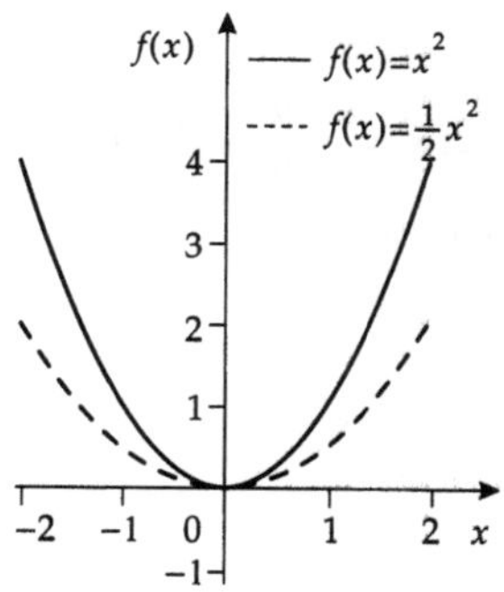

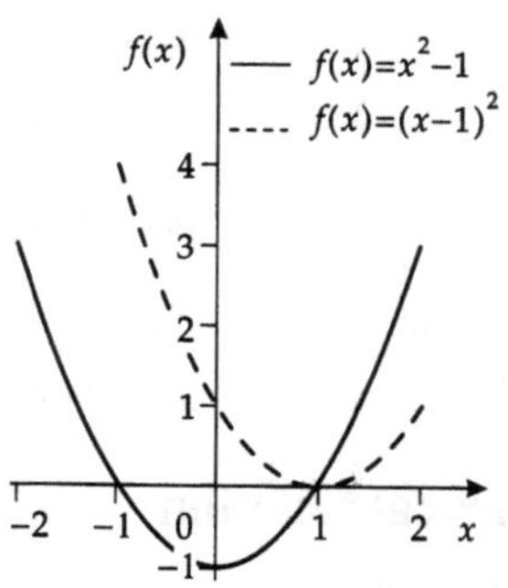

(c) Properties of the function

Domain of definition: $-\infty < x < \infty$.

Range of values: $a > 0$: $-Da \leq f(x) < \infty$.
$a < 0$: $-Da \geq f(x) > -\infty$.

Quadrant: $a > 0$: first, second, and possibly third and/or fourth quadrants.
$a < 0$: third, fourth, and possibly first and/or second quadrants.

Periodicity: None.

Monotonicity:
$a > 0,\ x > -\frac{b}{2a}$: strictly monotonically increasing.
$a > 0,\ x < -\frac{b}{2a}$: strictly monotonically decreasing.
$a < 0,\ x < -\frac{b}{2a}$: strictly monotonically increasing.
$a < 0,\ x > -\frac{b}{2a}$: strictly monotonically decreasing.

Symmetries: Mirror symmetry with respect to $x = -\frac{b}{2a}$

Asymptotes: $a > 0$: $f(x) \to \infty$ for $x \to \pm\infty$

$a < 0$: $f(x) \to -\infty$ for $x \to \pm\infty$

(d) Particular values

Zeros: 0 to 2 real zeros. Also see the section on particular extensions and applications.

$D < 0$: no real zeros

$D = 0$: $x_{1,2} = -\dfrac{b}{2a}$

$D > 0$: $x_{1,2} = -\dfrac{b}{2a} \pm \sqrt{D}$

▷ A zero of multiplicity 2 means that the curve touches the x-axis, i.e., that the curve has an extreme there as well.

Discontinuities: None.

Poles: None.

Extremes: $a > 0$: minimum at $x = -\dfrac{b}{2a}$ with $f(x) = -Da$.

P$a < 0$: maximum at $x = -\dfrac{b}{2a}$ with $f(x) = -Da$.

Points of inflection: None.

Value at $x = 0$: $f(0) = c$.

(e) Reciprocal function

The reciprocal function is a quadratic hyperbola,

$$\frac{1}{f(x)} = \frac{1}{ax^2 + bx + c}\,.$$

(f) Inverse function

The inverse function applies to $x \geq -\dfrac{b}{2a}$, with the plus sign and to $x \leq -\dfrac{b}{2a}$, with the minus sign.

$$f^{-1}(x) = -\frac{b}{2a} \pm \sqrt{\frac{x}{a} + \frac{b^2}{4a^2} - \frac{c}{a}}\,.$$

(g) Cognate functions

Root function: Inverse function of the quadratic function.
Quadratic hyperbola: Reciprocal function of the quadratic function.
Linear function: A special case of the quadratic function.

(h) Conversion formulas

The sum (difference) of two quadratic functions is a quadratic (or linear or constant) function.

$$\begin{aligned} f(x) &= (a_1x^2 + b_1x + c_1) \pm (a_2x^2 + b_2x + c_2) \\ &= (a_1 \pm a_2)x^2 + (b_1 \pm b_2)x + (c_1 \pm c_2) \end{aligned}$$

- Binomial formula:

$$(x+y)^2 = x^2 + 2xy + y^2$$

(i) Approximation formulas for the function

None.

(j) Expansion of series or products of the function

None.

(k) Derivative of the function

The derivative is a straight line:

$$\frac{\mathrm{d}}{\mathrm{d}x}(ax^2 + bx + c) = 2ax + b \, .$$

(l) Primitive of the function

The primitive function is a cubic function.

$$\int_0^x (at^2 + bt + c)\mathrm{d}t = \frac{a}{3}x^3 + \frac{b}{2}x^2 + cx \, .$$

(m) Particular extensions and applications

Quadratic function of a complex number:

$$a(x + \mathrm{j}y)^2 + b(x + \mathrm{j}y) + c = a(x^2 - y^2) + bx + c + \mathrm{j}y(2ax + b) \, .$$

Zeros of a quadratic function, $ax_0^2 + bx_0 + c = 0$.
The number of real zeros depends on the sign of the discriminant D.

$D = \dfrac{b^2}{4a^2} - \dfrac{c}{a} < 0$: No real zeros; two complex conjugate zeros.

$D = \dfrac{b^2}{4a^2} - \dfrac{c}{a} = 0$: One double zero at $x = -\dfrac{b}{2a}$.

$D = \dfrac{b^2}{4a^2} - \dfrac{c}{a} > 0$: Two zeros at $x = -\dfrac{b}{2a} \pm \sqrt{D}$.

p-q formula of the quadratic function, $x^2 + px + q = 0$.

$q > \dfrac{p^2}{4}$: No real zeros; two complex conjugate zeros.

$q = \dfrac{p^2}{4}$: One double zero at $x = -\dfrac{p}{2}$.

$q < \dfrac{p^2}{4}$: Two zeros at $x = -\dfrac{p}{2} \pm \sqrt{\dfrac{p^2}{4} - q}$.

- The quadratic function always has two zeros in the complex domain.

5.11 Cubic equation

$$f(x) = ax^3 + bx^2 + cx + d$$

Polynomial of degree three.
Primitive of the quadratic function.

(a) Definition

Solution of the differential equation:

$$\frac{d^3}{dx^3} f(x) = 6a \, .$$

General functional equation:

$$\begin{aligned} f(x) &= ax^3 + bx^2 + cx + d \\ &= a\left(x + \frac{b}{3a}\right)^3 + \left(c - \frac{b^2}{3a}\right)\left(x + \frac{b}{3a}\right) + d + \frac{2b^3}{27a^2} - \frac{cb}{3a} \, . \end{aligned}$$

Reduced equation:

$$f(x) = a\left(X^3 + pX + q\right), \qquad X = x + \frac{b}{3a},$$

$$p = \frac{c}{a} - \frac{1}{3}\left(\frac{b}{a}\right)^2, \quad q = \frac{d}{a} + \frac{2}{27}\left(\frac{b}{a}\right)^3 - \frac{1}{3}\frac{c}{a}\frac{b}{a}$$

Discriminant of the cubic function:

$$D = \left(\frac{p}{3}\right)^3 + \left(\frac{q}{2}\right)^2 .$$

(b) Graphical representation

For $a > 0$ ($a < 0$), the graph of the cubic function comes from minus (plus) infinity, intersects the x-axis at least at one and at most at three points, and then goes to plus (minus) infinity.

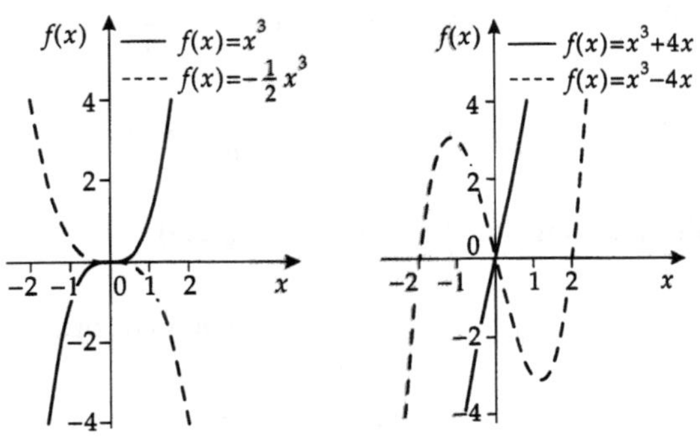

- ● The function has either no extreme or one maximum and one minimum.
- ▷ When the function has at least two distinct real zeros, it also has extremes.
- ● The function has one point of inflection.
- ▷ If the point of inflection is a saddle point, then the function has no extremes.

The functions in the left figure have one saddle point, no extremes, and one three-fold zero. The functions in the right figure have three simple zeros and two extremes (dotted line), or one simple zero and no extremes (solid line).

▷ Both functions in the right figure have no saddle point, but a different number of extremes.

(c) Properties of the function

Domain of definition:	$-\infty < x < \infty$.	
Range of values:	$-\infty < f(x) < \infty$.	
Quadrant:	$a > 0$:	first, third, and possibly second and/or fourth quadrants.
	$a < 0$:	second, fourth, and possibly first and/or third quadrants.
Periodicity:	None.	
Monotonicity:	$a > 0$:	strictly monotonically increasing for $x \to \pm\infty$ no extreme: strictly monotonically increasing for all x outside of the extremes: strictly monotonically increasing between the extremes: strictly monotonically increasing
	$a < 0$:	strictly monotonically decreasing for $x \to \pm\infty$ no extreme: strictly monotonically decreasing, for all x outside of the extremes: strictly monotonically increasing between the extremes: strictly monotonically increasing
Symmetries:	Point symmetry about the point of inflection.	
Asymptotes:	$a > 0$:	$f(x) \to \pm\infty$ for $x \to \pm\infty$
	$a < 0$:	$f(x) \to \mp\infty$ for $x \to \pm\infty$

(d) Particular values

Zeros:	At least one, at most three real zeros; see the section on particular extensions and applications: Cardano's formula.
Discontinuities:	None.
Poles:	None.
Extremes:	$p < 0$: two extremes at $x = -\dfrac{b}{3a} \pm p$ $p \geq 0$: no extreme
Points of inflection:	$p = 0$: saddle point at $x = -\dfrac{b}{3a}$ $p \neq 0$: point of inflection at $x = -\dfrac{b}{3a}$
Value at $x = 0$:	$f(0) = d$.

(e) Reciprocal function

The reciprocal function is a hyperbola of degree three:

$$\frac{1}{f(x)} = \frac{1}{ax^3 + bx^2 + cx + d} .$$

(f) Inverse function

$p \geq 0$: The function is invertible for all x.

$p < 0$: The function is invertible for $x \geq -\frac{b}{3a} - p$.

▷ For $p = 0$, the inverse function can be written as a cube root ($a \neq 0$).

$$a > 0: \quad f^{-1}(x) = -\frac{b}{3a} + \frac{1}{\sqrt[3]{a}} \sqrt[3]{|x - q|} \operatorname{sgn}(x - q)$$

$$a < 0: \quad f^{-1}(x) = -\frac{b}{3a} - \frac{1}{\sqrt[3]{-a}} \sqrt[3]{|x - q|} \operatorname{sgn}(x - q)$$

(g) Cognate functions

Quadratic and linear function: Special cases of the cubic function.
Hyperbola of degree three: reciprocal function.
Root function of degree three: inverse function.

(h) Conversion formulas

The sum and difference of two cubic functions is a cubic (or quadratic, linear or constant) function.

$$\begin{aligned}&(a_1x^3 + b_1x^2 + c_1x + d_1) \pm (a_2x^3 + b_2x^2 + c_2x + d_2)\\ &\quad = (a_1 \pm a_2)x^3 + (b_1 \pm b_2)x^2 + (c_1 \pm c_2)x + (d_1 \pm d_2)\end{aligned}$$

For a sum in the argument,

$$(x + y)^3 = x^3 + 3x^2y + 3xy^2 + y^3 .$$

(i) Approximation formulas for the function

Not necessary.

(j) Expansion of series or products of a function

A linear factor can be split off. If a zero x_0 of $ax^3 + bx^2 + cx + d$ is known (e.g., $x_0 = 0$ if $d = 0$), then the function can be factored into a product.

$$ax^3 + bx^2 + cx + d = (x - x_0)(a_0x^2 + b_0x + c_0)$$

Further possible zeros can be obtained by solving the quadratic equation $a_0x^2 + b_0x + c_0 = 0$.

(k) Derivative of the function

The derivative is a quadratic function:

$$\frac{\mathrm{d}}{\mathrm{d}x}(ax^3 + bx^2 + cx + d) = 3ax^2 + 2bx + c .$$

(l) Primitive of the function

The primitive is the quartic function:

$$\int_0^x (at^3 + bt^2 + ct + d)\mathrm{d}t = \frac{a}{4}x^4 + \frac{b}{3}x^3 + \frac{c}{2}x^2 + dx\,.$$

(m) Particular extensions and applications

Cubic splines: Piecewise cubic polynomials for the interpolation of data in given interpolation nodes. See Section 5.14.

Cardano's formula: Concept for finding the zeros of a cubic function.

The general cubic equation is rewritten as a reduced cubic equation.

$$ax^3 + bx^2 + cx + d = a\left(X^3 + pX + q\right) = 0,$$

$$X = x + \frac{b}{3a}, \quad p = \frac{c}{a} - \frac{1}{3}\left(\frac{b}{a}\right)^2, \quad q = \frac{d}{a} + \frac{2}{27}\left(\frac{b}{a}\right)^3 - \frac{1}{3}\frac{c}{a}\frac{b}{a}\,.$$

The discriminant of the cubic function is

$$D = \left(\frac{p}{3}\right)^3 + \left(\frac{q}{2}\right)^2\,.$$

Number of zeros:

$D > 0$: One real and two complex zeros.

$D = 0$: Three real zeros including one simple and one double zero, one triple zero at $x = -\dfrac{b}{3a}$ if $p = q = 0$.

$D < 0$: Three real simple zeros.

- The function always has three zeros in the complex domain, but the number of real zeros can be smaller.

Solution for $D = 0$, X_1 is the simple zero, X_2 is the double zero:

$$X_1 = 2\sqrt[3]{-\frac{q}{2}}\,, \qquad X_2 = -\sqrt[3]{-\frac{q}{2}}\,, \qquad x_{1/2} = X_{1/2} - \frac{b}{3a}\,.$$

Notation for the solutions with $D > 0$:

$$u = \sqrt[3]{-\frac{q}{2} + \sqrt{D}}\,, \qquad v = \sqrt[3]{-\frac{q}{2} - \sqrt{D}}\,.$$

Notation for $D > 0$, X_1 is the real solution, $X_{2/3}$ are complex:

$$X_1 = u + v\,, \qquad X_{2/3} = -\frac{u+v}{2} \pm \mathrm{j}\sqrt{3}\frac{u-v}{2}\,, \qquad x_{1/2/3} = X_{1/2/3} - \frac{b}{3a}\,.$$

Notation for the solution with $D < 0$:

$$\rho = \sqrt{\frac{-p^3}{27}}\,, \qquad \varphi = \arccos\left(\frac{-q}{2\rho}\right)\,.$$

Solution for $D < 0$, three real solutions:

$$X_{1/2/3} = 2\sqrt[3]{\rho}\cos\left(\frac{\varphi + k\pi}{3}\right)\,, \qquad k = 0, 2, 4\,, \qquad x_{1/2/3} = X_{1/2/3} - \frac{b}{3a}\,.$$

5.12 Power function of higher degree

$$f(x) = ax^n \qquad n = 1, 2, 3, \ldots$$

In programming languages: `A* X**N` or `A* X↑N`
Power function of degree n.
Description of curves which can grow faster than quadratic or cubic functions.

- □ Thermodynamics: The energy radiated from a blackbody is proportional to the fourth power of the temperature of the body $s = \sigma T^4$ (Stefan-Boltzmann law).

(a) Definition

Solution of the differential equation with constraints:

$$\frac{d^n}{dx^n} f(x) = n!a, \qquad f(0) = 0, \quad \frac{d^k f}{dx^k}(x = 0) = 0, \quad k = 1, \ldots, n-1 .$$

n-fold product of x:

$$f(x) = ax^n = a \prod_{k=1}^{n} x .$$

(b) Graphical representation

- Two important cases can be distinguished:
  n even: The function behaves similar to the quadratic function.
  n odd: The function behaves similar to the cubic function.

For $a > 0$ ($a < 0$), functions with even n come from the negative (positive) infinite; at $x = 0$, they have an n-fold zero which is a minimum (maximum), and they go into the positive (negative) infinite.
For $a > 0$ ($a < 0$), functions with odd n come from the negative (positive) infinite; at $x = 0$, they have an n-fold zero which is a saddle point, and they go into the positive (negative) infinite.

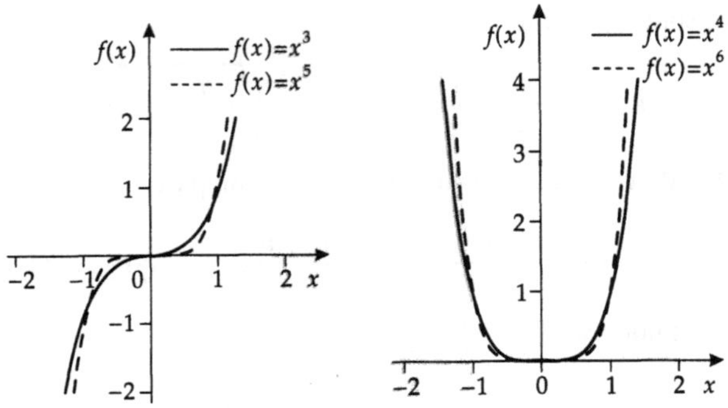

- At $x = 0$, all functions have the value of 0, and at $x = 1$, the value of a.

(c) Properties of the function

Domain of definition: $-\infty < x < \infty$.

Range of values:	n even	$a > 0$: $0 \le f(x) < \infty$
		$a < 0$: $-\infty < f(x) \le 0$
	n odd	$a \ne 0$: $-\infty < f(x) < \infty$
Quadrant:	n even	$a > 0$: first and second quadrants
		$a < 0$: third and fourth quadrants
	n odd	$a > 0$: first and third quadrants
		$a < 0$: second and fourth quadrants
Periodicity:	None.	
Monotonicity:	n even	$a > 0$: $x \ge 0$ strictly monotonically increasing
		$x \le 0$ strictly monotonically decreasing
		$a < 0$: $x \ge 0$ strictly monotonically decreasing
		$x \le 0$ strictly monotonically increasing
	n odd	$a > 0$: strictly monotonically increasing
		$a < 0$: strictly monotonically decreasing
Symmetries:	n even	mirror symmetry about $x = 0$
	n odd	point symmetry about $x = 0$, $y = 0$
Asymptotes:	n even	$a > 0$: $f(x) \to +\infty$ for $x \to \pm\infty$
		$a < 0$: $f(x) \to -\infty$ for $x \to \pm\infty$
	n odd	$a > 0$: $f(x) \to \pm\infty$ for $x \to \pm\infty$
		$a < 0$: $f(x) \to \mp\infty$ for $x \to \pm\infty$

(d) Particular values

Zeros:	For all functions: n-fold zero at $x_0 = 0$.	
Discontinuities:	None.	
Poles:	None.	
Extremes:	n even	$a > 0$: minimum at $x = 0$
		$a < 0$: maximum at $x = 0$
	n odd	no extreme
Points of inflection:	n even	no point of inflection
	n odd	saddle point at $x = 0$

(e) Reciprocal function

The reciprocal function is a hyperbola of power n.

$$\frac{1}{f(x)} = \frac{1}{ax^n} = \frac{1}{a}x^{-n}$$

(f) Inverse function

The inverse functions are the n-th roots ($a \ne 0$).

$$f^{-1}(x) = \sqrt[n]{\frac{x}{a}}$$

- If n is odd, then the function can be inverted for $-\infty < x < \infty$.
  If n is even, then the function can only be inverted for $x \ge 0$ (or with a negative sign for $x \le 0$).

(g) Cognate functions

Hyperbolas: The reciprocal functions of ax^n.
Roots: The inverse functions of ax^n.
General polynomials: Finite sums of $a_n x^n$.

(h) Conversion formulas

Reflection about the y-axis:

$$f(-x) = a(-x)^n = (-1)^n ax^n = \begin{cases} ax^n & n \text{ even} \\ -ax^n & n \text{ odd} \end{cases} .$$

Recursion relation:

$$ax^n = x \cdot ax^{n-1} .$$

The sum of two power functions of degree n is a power function of degree n.

$$ax^n \pm bx^n = (a \pm b)x^n$$

The product of a power function of degree n and a power function of degree m is a power function of degree $n + m$.

$$ax^n \cdot bx^m = (a \cdot b)x^{n+m}$$

The quotient of a power function of degree n and a power function of degree m is a power function of degree $n - m$.

$$\frac{ax^n}{bx^m} = \frac{a}{b}x^{n-m}$$

The m-th power of a power function of degree n is a power function of degree $n \cdot m$.

$$\left(ax^n\right)^m = a^m x^{n \cdot m}$$

(i) Approximation formulas for the function (8 bits ≈ 0.4% of precision)

For small x: the following approximation holds:

$$(a + x)^n \approx a^n \cdot \left(1 + \frac{n}{a}x\right) \qquad \text{for } |x| < \frac{|a|}{12\sqrt{n^2 - n}} .$$

(j) Expansion of series or products of the function

Binomial expansion of a power of a sum:

$$(x + y)^n = x^n + nx^{n-1}y + \frac{n(n-1)}{1 \cdot 2}x^{n-2}y^2 + \ldots + nxy^{n-1} + y^n$$

$$= \sum_{k=0}^{n} \binom{n}{k} x^{n-k}y^k , \qquad \binom{n}{k} = \frac{n!}{(n-k)!\,k!} .$$

$\binom{n}{k}$ are the binomial coefficients.

Finite expansion of the sum and difference of powers with odd exponents:

$$x^n \pm y^n = (x + y) \prod_{k=1}^{(n-1)/2} \left(x^2 \pm 2xy \cos\left(\frac{2\pi k}{n}\right) + y^2\right) .$$

Finite expansion of the sum of powers with even exponents:

$$x^n + y^n = \prod_{k=1}^{n/2} \left(x^2 + 2xy \cos\left(\frac{2\pi k - \pi}{n}\right) + y^2 \right) .$$

Finite expansion of the difference of powers with even exponents:

$$x^n - y^n = (x+y)(x-y) \prod_{k=1}^{(n-2)/2} \left(x^2 - 2xy \cos\left(\frac{2\pi k}{n}\right) + y^2 \right) .$$

Finite geometric series of powers:

$$\sum_{k=0}^{n} x^k = 1 + x + x^2 + \ldots + x^n = \frac{1 - x^{n+1}}{1 - x}, \quad \text{if} \quad x \neq 1 .$$

(k) Derivative of the function

The derivative of a power function of degree n is a power function of degree $n-1$:

$$\frac{\mathrm{d}}{\mathrm{d}x} a x^n = n a x^{n-1} .$$

The k-th derivative of a power function of degree n is

$$\frac{\mathrm{d}^k}{\mathrm{d}x^k} a x^n = \begin{cases} \dfrac{n!}{(n-k)!} a x^{n-k} & k \leq n \\ 0 & k > n \end{cases} .$$

(l) Primitive of the function

The primitive of a power function of degree n is a power function of degree $n+1$:

$$\int_0^x (a t^n) \mathrm{d}t = \frac{a}{n+1} x^{n+1} .$$

(m) Particular extensions and applications

Double-logarithmic representation: The power functions with $a > 0$ in the domain of $x > 0$ can be represented in a double-logarithmic manner; i.e., a logarithmic scale on the x-axis and the y-axis.

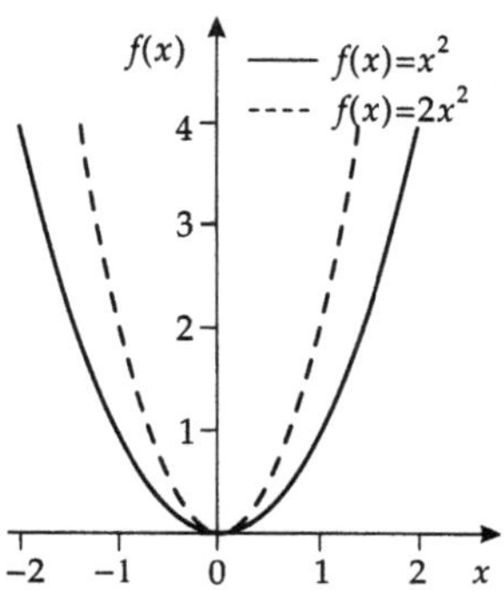

Linear representation

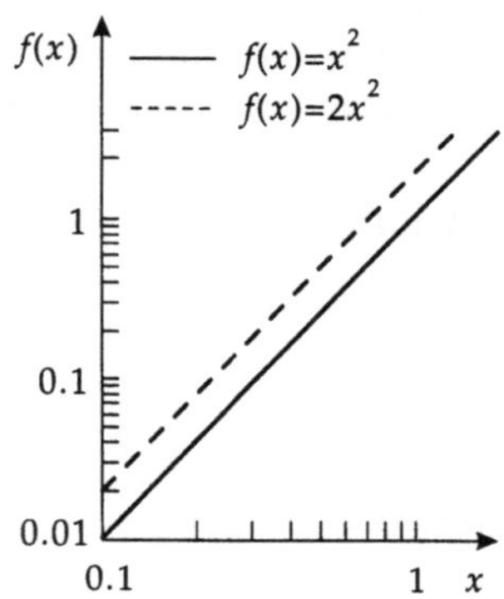

Double-logarithmic representation

- In a double-logarithmic representation, the graphs of $f(x) = ax^n$ are straight lines, whose slope is the exponent n of the power function.

The straight lines of equal power but different prefactors a have the same slope but are shifted with respect to each other in y-direction.
The straight lines of different powers have different slopes.

▷ If the pre-factors $b = \ln(c)$ are equal, the straight lines intersect each other in the point $(x = 1, y = a)$.

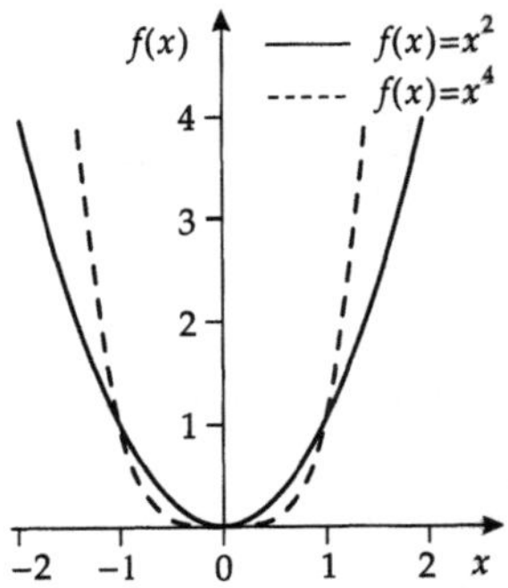

Linear representation

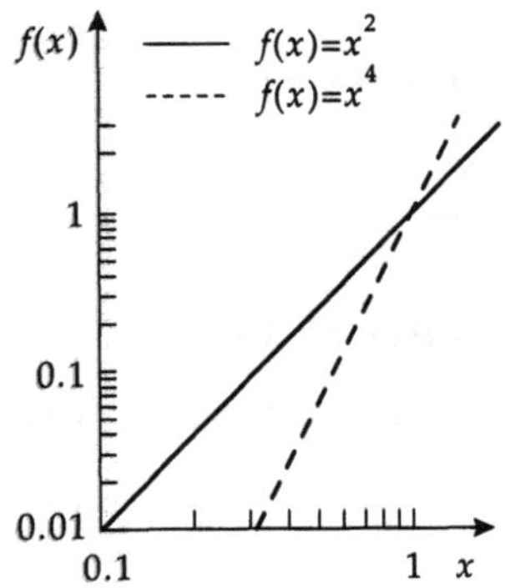

Double-logarithmic representation

In the case of equal prefactors, they intersect in point $(x = 1, y = a)$.
Double-logarithmic regression: A fit of a power function $f(x) = cx^a$ to data points $(x_k, f(x_k))$ can be achieved by means of a linear regression when the logarithms of points $(\ln(x_k), \ln(f(x_k)))$ are taken.

▷ Slope a obtained in a linear regression describes the exponent of the power function, and the intercept b on the y-axis describes the logarithm of prefactor $b = \ln(c)$.

5.13 Polynomials of higher degree

$$f(x) = a_n x^n + a_{n-1} x^{n-1} + \ldots + a_1 x + a_0$$

Polynomials of degree n.
Most general form of an integral rational function.
Sum of power functions of degree $m \leq n$.
Approximation of transcendental functions by polynomials of degree n.
Polynomials for fitting curves with an unknown functional representation.

(a) Definition

Finite sum of power functions:

$$f(x) = a_0 + \sum_{k=1}^{n} a_k x^k .$$

Solution of the differential equation

$$\frac{d^n}{dx^n} f(x) = n! a_n .$$

(b) Graphical representation

- The behavior of the function at infinity is determined by the power function $a_n x^n$ with the highest exponent n.

For large values of x, the polynomial of degree n essentially exhibits the properties of $a_n x^n$.

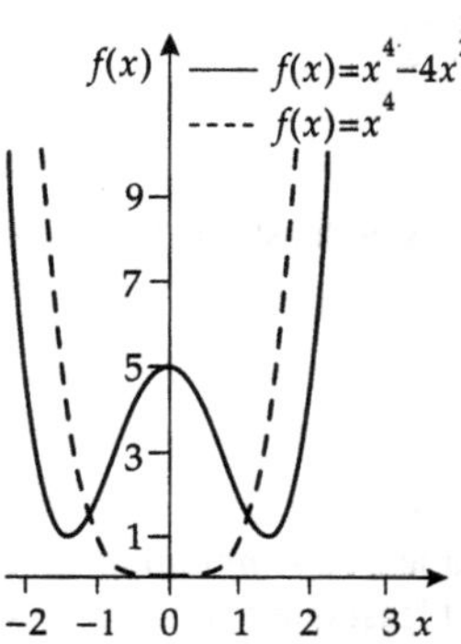

Linear representation

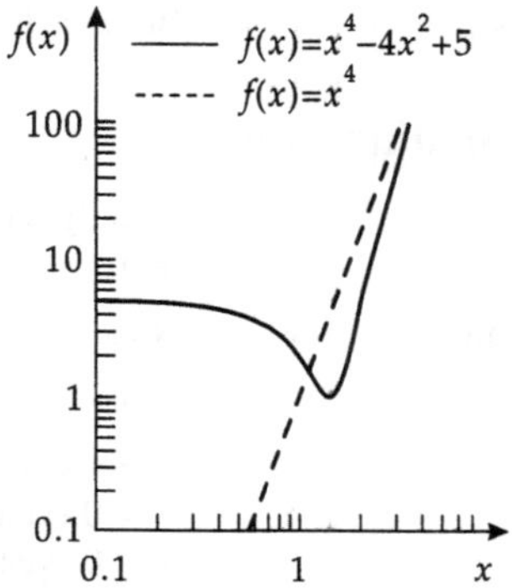

Double-logarithmic representation

▷ For x values close to zero, the term of the lowest power with $a_m \neq 0$ dominates.

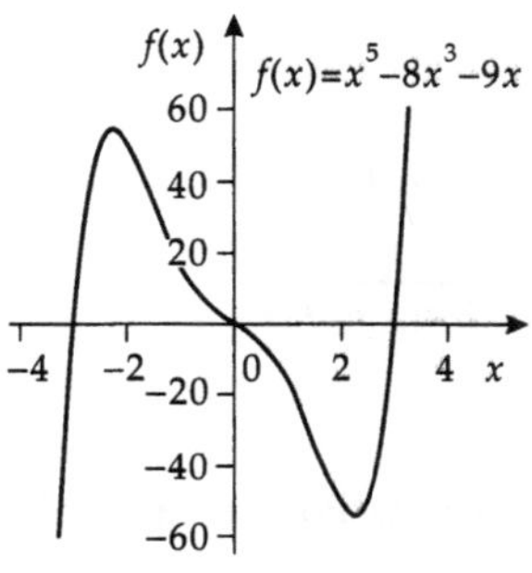

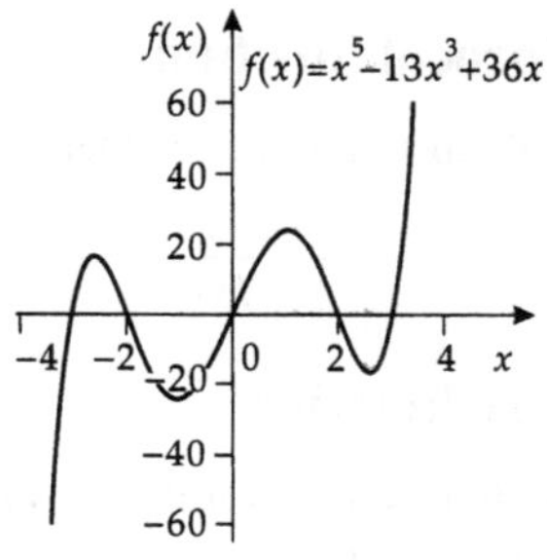

The function may have up to n real zeros, up to $n-1$ extremes, and up to $n-2$ points of inflection but it may also have less.

▷ When a twice-differentiable function possesses m real zeros, it has at least $m-1$ extremes, and, when it has k extremes, it has at least $k-1$ points of inflection.

(c) Properties of the function

Domain of definition:	$-\infty < x < \infty$.
Range of values:	Depends on the special case of polynomial. n odd: $-\infty < f(x) < \infty$
Quadrant:	Depends on the polynomial; but in any case, if n odd $a_n > 0$: first and third quadrants $a_n < 0$: second and fourth quadrants n even $a_n > 0$: first and second quadrants $a_n < 0$: third and fourth quadrants
Periodicity:	None.

Monotonicity:	Depends on the polynomial.
Symmetries:	In general, none. If the polynomial contains only even powers: mirror symmetry about the y-axis. If the polynomial contains only odd powers: point symmetry about the origin.
Asymptotes:	Behaves for $a_n x^n$ as $x \to \pm\infty$.

(d) Particular values

Zeros:	Exactly n complex zeros; at least one real zero when n is odd.
Discontinuities:	None.
Poles:	None.
Extremes:	At most, $n-1$ extremes. $n > 0$ even, $a_n > 0$: at least one minimum. $n > 0$ even, $a_n < 0$: at least one maximum.
Points of inflection:	At most, $n-2$ points of inflection. $n > 1$ odd: at least one point of inflection.
Value at $x = 0$:	$f(0) = a_0$.

(e) Reciprocal function

The reciprocal function is a fractional rational function:

$$\frac{1}{f(x)} = \frac{1}{a_n x^n + \ldots + a_1 x + a_0} .$$

(f) Inverse function

Generally, the function is invertible only in subdomains. For very large values, the inverse function approaches an n-th root:

$$f^{-1}(x) \approx \sqrt[n]{\frac{x}{a_n}} \qquad \text{for } x \to \infty .$$

(g) Cognate function

All integral rational functions discussed so far are special polynomials of degree n.
All power functions with a negative integral exponent are reciprocal functions special polynomials of degree n.
All k-th roots, $k \leq n$, are inverse functions of special polynomials of degree n.

(h) Conversion formulas

The sum (difference) of two polynomials of degrees n and m a polynomial itself, the degree of which is the greater value of n or m,

$$\sum_{i=0}^{n} a_i x^i \pm \sum_{j=0}^{m} b_j x^j = \sum_{k=0}^{n} (a_k \pm b_k) x^k \qquad n \geq m .$$

▷ For polynomials of lower degree, the missing coefficients are set equal to zero.

$$\text{For } m \neq n \text{ set } \begin{cases} m > n: & a_k = 0, \quad k = n+1, \ldots, m \\ m < n: & b_k = 0, \quad k = m+1, \ldots, n \end{cases} .$$

The product of two polynomials of degrees n and m is a polynomial of degree $m + n$:

$$\left(\sum_{i=0}^{n} a_i x^i\right) \cdot \left(\sum_{j=0}^{m} b_j x^j\right) = \sum_{k=0}^{m+n} \left(\sum_{l=0}^{k} a_l b_{k-l}\right) x^k ,$$

where the coefficients outside the range are set to zero,

$$a_i = 0 \text{ for } i < 0, i > n \quad \text{and} \quad b_j = 0 \text{ for } j < 0, j > m .$$

A sum in the argument can be calculated term-by-term:

$$f(x+y) = \sum_{m=0}^{n} a_m \sum_{k=0}^{m} \binom{m}{k} x^{m-k} y^k , \qquad \binom{n}{k} = \frac{n!}{(n-k)!\,k!} .$$

▷ This is an application of the binomial formula for each individual term $a_m(x+y)^m$.

(i) Approximation formulas for the function

For the individual terms, the approximation formulas given in Section 5.12 on power functions of higher degree can be used.

(j) Expansion of series or products of the function

For a product, the representation of a polynomial is given by

$$f(x) = a_n (x - x_1)^{n_1} (x - x_2)^{n_2} \cdot \ldots \cdot (x - x_k)^{n_k} ,$$

where x_i are the k zeros of multiplicity n_i of the polynomial.

(k) Derivative of the function

The derivative of a polynomial of degree n is a polynomial of degree $n - 1$.

$$\frac{\mathrm{d}}{\mathrm{d}x} f(x) = \sum_{k=1}^{n} k a_k x^{k-1} = n a_n x^{n-1} + (n-1) a_{n-1} x^{n-2} + \ldots + a_1$$

The m-th derivative is a polynomial of degree $n - m$.

$$\frac{\mathrm{d}^m}{\mathrm{d}x^m} f(x) = \sum_{k=m}^{n} \frac{k!}{(k-m)!} a_k x^{k-m}$$

(l) Primitive of the function

The primitive is a polynomial of degree $n + 1$.

$$\int_0^x \left(\sum_{k=0}^{n} a_k t^k\right) \mathrm{d}t = \sum_{k=0}^{n} \frac{a_k}{k+1} x^{k+1}$$

(m) Particular extensions and applications

Fundamental theorem of algebra:

● In the complex domain a polynomial of degree n has exactly n zeros when the multiplicity of the zeros is taken into account.

$$n_1 + n_2 + n_3 + \ldots + n_k = n\,, \qquad n_i = \text{multiplicities of the zero } x_i$$

▷ In the complex domain, a polynomial can be represented completely as a product of linear factors.

□ $x^4 - 2x^3 + 2x^2 - 2x + 1 = (x-1)^2(x^2+1) = (x-1)(x-1)(x+\mathrm{j})(x-\mathrm{j})$

5.14 Representation of polynomials and particular polynomials

5.14.1 Representation by sums and products

Representation by a sum: Representation of the polynomial ordered by powers:

$$f(x) = a_n x^n + a_{n-1}x^{n-1} + \cdots + a_1 x + a_0\,.$$

● This representation has the advantages that the calculations can be done clearly, that the asymptotic behavior is quickly recognized (important, for example, in the case of fractional rational functions), and that a comparison of coefficients is possible in a simple manner (e.g., in the decomposition of fractions into partial fractions).

Representation by product: Representation of a polynomial as a product of linear factors:

$$f(x) = a_n(x-x_1)^{n_1}(x-x_2)^{n_2}\cdots(x-x_k)^{n_k}\,.$$

x_i are the zeros of multiplicity n_i of the polynomial, k denotes the total number of distinct zeros, and n_k is their multiplicity.

● The advantage of this representation is that that all zeros x, with their multiplicities, are unknown.

□ Represented as products, the function $f(x) = x^5 - x^4 - x + 1$ is given by

$$\begin{aligned} f(x) &= (x-1)^2(x+1)(x^2+1) = (x-1)(x-1)(x+1)(x^2+1) \\ &= (x-1)^2(x+1)(x-\mathrm{j})(x+\mathrm{j})\,. \end{aligned}$$

The function has a double zero at $x = 1$, a simple zero at $x = -1$, and (in the complex domain) two simple zeros at $x = \mathrm{j}$ and $x = -\mathrm{j}$.

▷ For polynomials up to degree 4, the (real) zeros can be calculated analytically. For polynomials of higher degree, the determination of the zeros often has to be done numerically.

▷ Program sequence for calculating the zeros of a real polynomial

$$p(x) = a_0 x^n + a_1 x^{n-1} + \ldots + a_{n-1}x + a_n\,, \qquad n > 2\,.$$

Zeros found are removed, and the procedure is continued until a linear or quadratic equation remains. Important: Found zeros may be greatly distorted by rounding and truncation errors, etc., every zero should be reiterated again, for example, with the common Newton method.

```
BEGIN Bairstow
INPUT n (degree of the polynomial)
INPUT a[i], i=0,...,n (coefficients of the polynomial)
INPUT eps (truncation criterion)
INPUT maxit (maximum number of iterations)
deg:=n;
repeat
   iter := 0
   INPUT r,s (approximations for the coefficients of a
   square divisor [x^2+r*xts].)
   repeat
    iter := iter+1
    b[0]:=a[0]; b[1]:=a[1]+r*b[0];
    c[0]:=b[0]; c[1]:=b[1]+r*c[0];
    FOR i:=2 TO deg DO
      b[i]:=a[i]+r*b[i-1]+s*b[i-2];
      c[i]:=b[i]+r*c[i-1]+s*c[i-2];
    ENDDO;
    det:=sqr(c[deg-2])-c[deg-3]*c[deg-1];
    term:=(det=0) OR (iter>maxit);
    IF NOT term THEN
      dr:=(b[deg]*c[deg-3]-b[deg-1]*c[deg-2])/det;
      ds:=(b[deg-1]*c[deg-1]-b[deg]*c[deg-2])/det;
      acc:=sqrt(dr^2 + ds^2);
      r:=r+dr;s:=s+ds;
    ENDIF;
   UNTIL (acc<eps) OR term;
   IF NOT term THEN
    CALL LÖSE(x^2 + rx + s = 0)
OUTPUT x0,x1
deg:=deg-2;
FOR i:=1 TO deg DO a[i]:=b[i];
   ENDIF
until (deg<=2) or (iter>maxit) or term;
CALL LÖSE(quadratic or linear remaining function)
END BAIRSTOW
```

5.14.2 Taylor series

Representation of a polynomial in an expansion series about point x_0:

$$f(x) = f_0 + \frac{f_1}{1!}(x - x_0) + \frac{f_2}{2!}(x - x_0)^2 + \cdots + \frac{f_n}{n!}(x - x_0)^n$$
$$= \sum_{k=0}^{n} \frac{f_k}{k!}(x - x_0)^k .$$

The coefficients f_k are the k-th derivatives of $f(x)$ at $x = x_0$:

$$f_0 = f(x_0), \qquad f_1 = \frac{\mathrm{d}f}{\mathrm{d}x}(x_0), \qquad f_k = \frac{\mathrm{d}^k f}{\mathrm{d}x^k}(x_0) .$$

For this representation requires all derivative values of $f(x)$. It also allows an estimate of how the function goes close to x_0, since that is where the linear terms become important, then the quadratic terms, etc.

□ The function $f(x) = 3x^4 + 2x^3 - x^2 + 2x - 1$, expanded about point $x = -2$:

$$f(x) = \frac{72}{4!}(x+2)^4 - \frac{132}{3!}(x+2)^3 + \frac{118}{2!}(x+2)^2 - \frac{66}{1!}(x+2) + 23$$
$$= 3(x+2)^4 - 22(x+2)^3 + 59(x+2)^2 - 66(x+2) + 23\,.$$

□ The function $f(x) = 3x^4 + 2x^3 - x^2 + 2x - 1$, expanded about point $x = 0$:

$$f(x) = \frac{72}{4!}x^4 + \frac{12}{3!}x^3 - \frac{2}{2!}x^2 + \frac{2}{1!}x - 1\,.$$

5.14.3 *Horner's scheme*

Horner's scheme: A scheme for

1. simple calculation of the function values of a polynomial,
2. splitting off a linear factor from a polynomial,
3. determining the derivatives at an arbitrary point,
4. expanding a polynomial about an arbitrary point.

Representation of a polynomial of degree n,

$$f(x) = a_0 + x(a_1 + x(a_2 + x(\ldots + x\ldots(a_n))))\,.$$

Calculation of the function values

1. Write the coefficients $a_n, a_{n-1}, \ldots, a_0$ into the top row.
2. Write a 0 into the middle row (far left).
3. Add the numbers in the first and second rows (far left), and write the result in the third row.
4.

coefficient	a_n	a_{n-1}	a_{n-2}	$\ldots$	a_0
times x	0				
sum					

5. Multiply the result by x and write it in the middle row, one position to the right.
6. In this column, add the number from the top, add the result and write it in the bottom row.
7. Multiply this result by x again and write it in the middle row, then write one position farther to the right, and add it to the number immediately above, then write the sum below.
8. Repeat this procedure until reaching the far-right position in the bottom row.
9. The result on the bottom right is $f(x)$.

Schematically represented as:
$\downarrow$ means addition, $\nearrow$ means multiplication by x.

Coefficient.	a_n		a_{n-1}		a_{n-2}		$\dots$		a_0
times x	0		$a_n x$		$(a_{n-1}+a_n x)x$				
	$\downarrow$	$\nearrow$	$\downarrow$	$\nearrow$	$\downarrow$	$\nearrow$	$\downarrow$	$\nearrow$	$\downarrow$
sum	a_n		$a_{n-1}+a_n x$						$f(x)$

□ Calculate $f(x) = 2x^4 - x^3 - 2x^2 + 3x - 1$ for $x = 3$ using Horner's scheme:

coefficient	2	−1	−2	3	−1
times 3	0	6	15	39	126
sum	2	5	13	42	125

The result is $f(3) = 125$.

Splitting off a linear factor using Horner's scheme:
A zero x_1 of $f(x)$ is found by trial and error. A linear factor can be split off from $f(x)$

$$\begin{aligned} f(x) &= a_n x^n + a_{n-1}x^{n-1} + \cdots + a_1 x + a_0 \\ &= (x - x_1)\cdot(b_{n-1}x^{n-1} + \cdots + b_1 x + b_0) \end{aligned}$$

The coefficients of the reduced polynomial $b_n x_k$ can be obtained by means of Horner's scheme by applying it to the value x_1.

coefficient	a_n	a_{n-1}	a_{n-2}	$\dots$	a_0
times x_1	0				
sum	b_{n-1}	b_{n-2}	$\dots$	b_0	0

In the bottom row, the result is zero; next to it are the coefficients of the reduced polynomial.

□ $f(x) = 2x^4 - x^3 - 2x^2 + 3x - 2$ has the zero $x = 1$.

	a_4	a_3	a_2	a_1	a_0
coefficient	2	−1	−2	3	−2
times 1	0	2	1	−1	2
sum	2	1	−1	2	0
	b_3	b_2	b_1	b_0	$f(1)$

The decomposition is

$$f(x) = (x-1)\cdot(2x^3 + x^2 - x + 2)\,.$$

Simple polynomial division by Horner's scheme:
The polynomial $f(x)$ can be divided by the linear function $x - c$ by using the same method as for splitting off a linear factor. The function value obtained is the remainder term.

coefficient	a_n	a_{n-1}	a_{n-2}	$\dots$	a_0
times c	0				
sum	b_{n-1}	b_{n-2}	$\dots$	b_0	$f(c)$

$$\frac{a_n x^n + \cdots a_1 x + a_0}{x - c} = b_{n-1}x^{n-1} + \cdots + b_1 x + b_0 + \frac{f(c)}{x-c}$$

Complete Horner's scheme: By applying Horner's scheme to the value x_0, the value $f(x_0)$ and the reduced polynomial are obtained.
Using this polynomial for a further Horner's scheme, one obtains for x_0 the first derivative $f'(x_0)$. A continued application of this procedure yields the other coefficients of the Taylor series $\frac{1}{k!}\frac{\mathrm{d}^k f}{\mathrm{d}x^k}(x_0)$.

coefficient	a_n	a_{n-1}	a_{n-2}	$\dots$	a_0
times x_0	0				
sum	b_{n-1}	b_{n-2}	$\dots$	b_0	$f(x_0)$
times x_0	0				
sum	c_{n-2}	$\dots$	c_0	$f'(x_0)$	
times x_0	0				
sum	$\dots$	d_0	$f''(x_0)/2!$		
times x_0	0				
sum	$\dots$	$f^{(k)}(x_0)/k!$			
times x_0	0				
sum	$f^{(n)}(x_0)/n!$				

□ The derivatives of $f(x) = 2x^4 - x^3 - 2x^2 + 3x - 2$ at $x = -1$ are determined as follows:

coefficient	2	−1	−2	3	−2
times −1	0	−2	3	−1	−2
sum	2	−3	1	2	<u>−4</u>
times −1	0	−2	5	−6	
sum	2	−5	6	<u>−4</u>	
times −1	0	−2	7		
sum	2	−7	<u>13</u>		
times −1	0	−2			
sum	2	<u>−9</u>			
times −1	0				
sum	<u>2</u>				

The derivatives are
$f'(-1) = 1 \cdot (-4) = -4$, $f''(-1) = 2 \cdot 13 = 26$, $f'''(-1) = 6 \cdot (-9) = -54$,
$f^{(4)}(-1) = 24 \cdot 2 = 48$.
The expansion of $f(x)$ about $x = -1$ reads

$$f(x) = 2(x+1)^4 - 9(x+1)^3 + 13(x+1)^2 - 4(x+1) - 4$$

▷ Horner's scheme is also practical for converting a number into another number system (see Section 1.2). The value of the dual number $100\,101_2$ is obtained as follows: Considering the polynomial p expanded about 0,

$$p(x) = 1 \cdot x^5 + 0 \cdot x^4 + 0 \cdot x^3 + 1 \cdot x^2 + 0 \cdot x^1 + 1 \cdot x^0 .$$

The decimal value of the number $100\,101_2$ is equal to $p(2)$, which can be calculated easily with the help of Horner's scheme.

5.14.4 Newton's interpolation polynomial

Newton's interpolation scheme: A scheme for fitting a polynomial of degree n to $n+1$ data points, $x_0, \ldots, x_n$, with the values $y_0, \ldots, y_n$.
Representation of the polynomial

$$\begin{aligned} f(x) &= b_0 + b_1(x - x_0) + b_2(x - x_0)(x - x_1) \\ &\quad + b_3(x - x_0)(x - x_1)(x - x_2) + \cdots + b_n(x - x_1)\ldots(x - x_{n-1}) \end{aligned}$$

Divided differences:
Write down the values $x_0 \ldots x_n$ in a row, from left to right, and below that write the corresponding values $y_0 \ldots y_n$.
Calculate the differences between neighboring y-values and divide by the differences between the x-values. Write the new values between the subtracted values, subtracted from each other, one row below.

$$y_{0,1} = \frac{y_0 - y_1}{x_0 - x_1} \qquad y_{1,2} = \frac{y_1 - y_2}{x_1 - x_2} \quad \cdots \quad y_{n-1,n} = \frac{y_{n-1} - y_n}{x_{n-1} - x_n}\,.$$

Next, form the differences between the new values, and divide by the difference between the boundary values (at $y_{0,1}$ and $y_{1,2}$ the boundary values are x_0 and x_2).

$$y_{0,1,2} = \frac{y_{0,1} - y_{1,2}}{x_0 - x_2} \qquad y_{1,2,3} = \frac{y_{1,2} - y_{2,3}}{x_1 - x_3} \ldots$$

Once again, form the differences and divide by the boundary values.

$$y_{0,1,2,3} = \frac{y_{0,1,2} - y_{1,2,3}}{x_0 - x_3} \qquad y_{1,2,3,4} = \frac{y_{1,2,3} - y_{2,3,4}}{x_1 - x_4} \ldots$$

Continue this until only one difference remains.

$$y_{0,1,\ldots,n} = \frac{y_{0,1,2,\ldots,n-1} - y_{1,2,3,\ldots,n}}{x_0 - x_n}$$

The values with a zero as the first index, i.e., the terms at the far-left position of a row, can be identified with the parameters $b_0, \ldots, b_n$:

$$b_0 = y_0, \quad b_1 = y_{0,1}, \quad \ldots, \quad b_k = y_{0,1,\ldots,k}, \quad \ldots, \quad b_n = y_{0,1,\ldots,n}$$

Schematic representation:

$$\begin{array}{ccccccccccc}
x_0 & & x_1 & & x_2 & & x_3 & & & & x_n \\
y_0 & & y_1 & & y_2 & & y_3 & & \cdots & & y_n \\
& y_{0,1} & & y_{1,2} & & y_{2,3} & & \cdots & & y_{n-1,n} & \\
& & y_{0,1,2} & & y_{1,2,3} & & \cdots & & y_{n-2,n-1,n} & & \\
& & & \cdots & & \cdots & & \cdots & & & \\
& & & y_{0,1,\ldots,n-1} & & y_{1,2,\ldots,n} & & & & & \\
& & & & y_{0,1,\ldots,n} & & & & & &
\end{array}$$

□ A polynomial of degree 4 must be fitted to points $P(1, 30)$, $P(2, 27)$, $P(3, 25)$, $P(4, 24)$, and $P(5, 21)$.

$$\begin{array}{ccccccccc}
1 & & 2 & & 3 & & 4 & & 5 \\
30 & & 27 & & 25 & & 24 & & 21 \\
& -3 & & -2 & & -1 & & -3 & \\
& & \frac{1}{2} & & \frac{1}{2} & & -1 & & \\
& & & 0 & & -\frac{1}{2} & & & \\
& & & & -\frac{1}{8} & & & &
\end{array}$$

The polynomial has the form

$$30 - 3(x-1) + \frac{1}{2}(x-1)(x-2) - \frac{1}{8}(x-1)(x-2)(x-3)(x-4),$$

or written as a sum

$$p(x) = \frac{1}{8}x^4 + \frac{5}{4}x^3 - \frac{31}{8}x^2 + \frac{7}{4}x + 31.$$

Gregory-Newton method: Simplified difference scheme for equidistant interpolation nodes.
The x-values must be arranged by magnitude:

$$x_1 = x_0 + h, \quad x_2 = x_0 + 2h, \quad x_3 = x_0 + 3h, \ldots, \qquad h > 0.$$

Now only form the differences of the y-values. The division by the x-values is canceled.

$$y_{0,1} = y_1 - y_0, \qquad y_{1,2} = y_2 - y_1, \qquad \ldots, \qquad y_{n-1,n} = y_n - y_{n-1}$$

Form the other values in the same manner:

$$y_{0,1,2} = y_{1,2} - y_{0,1}, \qquad y_{1,2,3} = y_{2,3} - y_{1,2}, \qquad \ldots$$

Continue this until only one difference remains:

$$y_{0,1,\ldots,n} = y_{1,2,\ldots,n} - y_{0,1,2,\ldots,n-1}$$

Divide the values with a zero appearing as the start, i.e., the terms to the far-left of a row, by $k!h^k$ (for the k-th difference standing in the k-th row). The results can be identified with the parameters $b_0, \ldots, b_n$.:

$$b_0 = y_0, \qquad b_1 = \frac{y_{0,1}}{1!h}, \quad \ldots, \quad b_k = \frac{y_{0,1,\ldots,k}}{k!h^k}, \quad \ldots, \quad b_n = \frac{y_{0,1,\ldots,n}}{n!h^n}.$$

5.14.5 Lagrange polynomials

The Lagrange polynomial fits a polynomial of degree n to $n+1$ given points.
Let the points $P(x_0, y_0), P(x_1, y_1), \ldots, P(x_n, y_n)$ be the interpolation nodes. The polynomial can be represented by

$$f(x) = y_0 \cdot L_0(x) + y_1 \cdot L_1(x) + \ldots + y_n \cdot L_n(x).$$

The coefficients are given by the functions

$$L_k(x) = \prod_{\substack{i=0\\ i\neq k}}^{n} \frac{x - x_i}{x_k - x_i} = \frac{(x-x_0)(x-x_1)\cdots(x-x_{k-1})(x-x_{k+1})\cdots(x-x_n)}{(x_k-x_0)(x_k-x_1)\cdots(x_k-x_{k-1})(x_k-x_{k+1})\cdots(x_k-x_n)}.$$

The numerator contains all factors $(x - x_i)$ except $(x - x_k)$, and the denominator contains all factors $(x_k - x_i)$ except $(x_k - x_k)$.

□ A function through the points $P(0,0)$, $P(1,-1)$, and $P(3,3)$ should be fitted by a polynomial of degree 2.

The coefficients are:

$$L_0(x) = \frac{(x-1)(x-3)}{3} = \frac{1}{3}(x^2 - 4x + 3)$$
$$L_1(x) = \frac{x(x-3)}{-2} = -\frac{1}{2}(x^2 - 3x)$$
$$L_2(x) = \frac{x(x-1)}{6} = \frac{1}{6}(x^2 - x)\,.$$

The function is represented by

$$f(x) = 0 \cdot \frac{x^2 - 4x + 3}{3} - 1 \cdot \frac{x^2 - 3x}{-2} + 3 \cdot \frac{x^2 - x}{6} = x^2 - 2x = (x-1)^2 - 1\,.$$

▷ Program sequence for calculating the function $y = f(x)$ when the function is given at n interpolation nodes.

```
BEGIN Lagrange polynomial
y := 0
FOR i = 0 TO n DO
   product := f[i]
   FOR j = 0 TO n DO
      IF (i ≠ j) THEN
         product := product*(x - x[j])/(x[i] - x[j])
      ENDIF
   ENDDO
   y := y + product
ENDDO
END Lagrange polynomial
```

▷ In comparison with Newton's method, the method of Lagrange has the disadvantage that all coefficients have to be recalculated when a new pair of values is added. In the more efficient method of Newton only one additional row must be calculated.

● Caution is necessary when interpolating with a large number of interpolation nodes, since polynomials of high degree are formed. This causes instability of the function (outside the interpolation nodes) against small errors when the interpolation nodes are given, and it can lead to great function fluctuations in the area between the interpolation nodes.

5.14.6 Bezier polynomials and splines

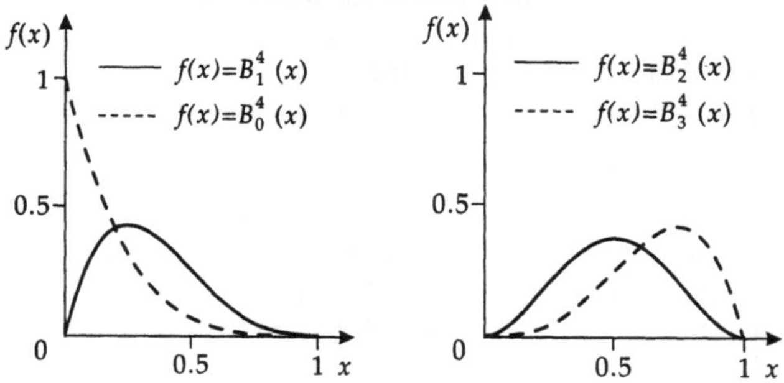

Bernstein polynomials

Bernstein polynomial: Polynomial of degree n,

$$B_k^n(x) = \binom{n}{k} x^k (1-x)^{n-k}, \qquad k = 0, 1, 2, \ldots, n .$$

These polynomials are used only in the domain $0 \leq x \leq 1$, where they have the following properties:

1. a k-fold zero at $x = 0$,
2. an $(n-k)$-fold zero at $x = 1$,
3. one and only one maximum at $x = \frac{k}{n}$,
4. all polynomials of the same degree form a **decomposition of the unit**, i.e.,
$$\sum_{k=0}^{n} B_k^n(x) = 1 ,$$
5. the polynomials B_k^n are linearly independent, and they form a basis for the polynomials of highest degree n; i.e., every polynomial of degree n can be described as a linear combination of Bernstein polynomials,
$$\sum_{k=0}^{n} a_k x^k = \sum_{k=0}^{n} b_k B_k^n(x) .$$

Bezier polynomial: Representation of a function in terms of Bernstein polynomials,

$$f(x) = \sum_{k=0}^{n} b_k B_k^n(x) .$$

Bezier points: Coordinates b_k of a representation of Bezier polynomials.
Bezier's polygon: Polygon with vertices $P(x = k/n, y = b_k)$ for $k = 0, \ldots, n$.
Interpolation of Bezier's polygon:
One has $n+1$ equidistant interpolation nodes, $(x_0, y_0), \ldots, (x_0 + n\Delta x, y_n)$, which are taken as the point $(x = k/n, y = y_k)$ of Bezier's polygon. Thus, for the first polynomial one obtains the Bezier polynomial $f(z) = \sum_{k=0}^{n} y_k B_k^n(z)$ with $z = \dfrac{x - x_0}{n\Delta x}$.

- The approximation of a curve in many points proceeds piecewise for segments of n points each (typically $n \approx 4$), which are fitted by curves of degree $n-1$.

At the boundaries, additional conditions must be met. This leads to additional conditions for the Bezier points, which are met by using auxiliary points.
Spline: Segmented curve in which every segment is fitted by a polynomial of degree n that is $n-1$ times continuously differentiable.
Subspline: Segmented curve with polynomials of degree n, which is differentiable less frequently, but at least once.

▷ Usually, cubic splines are used.

A cubic spline is a set of polynomials of degree three that go through two adjacent interpolation nodes x_j and $x_{j+1} = x_j + h$. Their first and second derivatives are continuously adjacent to each other at the end points of each interval.

- The interpolation nodes need not be equidistant, and $x_0 < x_1 < x_2 < \ldots < x_n$ must apply.

▷ A polynomial of degree three is completely determined by the function values and the first derivatives at the boundary points.

The spline function $s(x)$ is given by

$$s(x) = a_j(x)\ , x_j \le x \le x_{j+1} \quad \text{for all } j,$$

with

$$s_j(x) = a_0 + a_1(x - x_j) + a_2(x - x_j)^2 + a_3(x - x_j)^3$$

and

$$s_j'(x_j) = s_{j-1}'(x_j)\ , \qquad s_j''(x_j) = s_{j-1}''(x_j)\ .$$

The constants are fixed by the function values f_j and f_{j+1} and the derivatives f_j' and f_{j+1}' at the points x_j and x_{j+1}:

$$a_0 = f_j, \qquad a_1 = f_j', \qquad a_2 = \frac{3}{h^2}(f_{j+1} - f_j) - \frac{1}{h}(f_{j+1}' + 2f_j');$$
$$a_3 = \frac{2}{h^3}(f_j - f_{j+1}) + \frac{1}{h^2}(f_j' + f_{j+1}')\ .$$

The function values at all interpolation nodes $f_0, f_1, \ldots, f_n$ as well as the derivatives f_0' and f_n' at the end-points of the curve to be approximated are given.
The other derivatives are determined by the condition that the derivatives must be continuous at the interpolation nodes. Therefore,

$$f_{j-1}' + 4f_j' + f_{j+1}' = 3(f_{j+1} - f_{j-1})$$

▷ This equation defines a system of equations with a tridiagonal matrix of coefficients. The coefficients of the splines can be fixed by determining the values of the derivatives.

□ Approximating the function $f(x) = x^4$ by a cubic spline with the interpolation nodes $x_0 = -1, x_1 = 0, x_2 = 1$.

It holds that $f_0 = 1, f_1 = 0, f_2 = 1$ as well as $f_0' = -4, f_2' = 4$.
From $f_{1-1}' + 4f_1' + f_{1+1}' = -4 + 4f_1' + 4 = 3(f_{1+1} - f_{1-1}) = 3(1-1) = 0$ it follows that $f_1' = 0$.
In the interval $x_0 \le x < x_1$, substitution leads to $a_0 = 1, a_1 = -4, a_2 = 5, a_3 = 2$.

$$s_1(x) = 1 - 4(x+1) + 5(x+1)^2 - 2(x+1)^3 = -x^2 - 2x^3, \qquad -1 \le x < 0$$

In the interval $x_1 \le x \le x_2$, substitution leads to $a_0 = 0, a_1 = 0, a_2 = -1, a_3 = 2$.

$$s_2(x) = -x^2 + 2x^3, \qquad 0 \le x < 1$$

The function x^4 is approximated by $s(x) = -x^2 + 2|x|^3$.

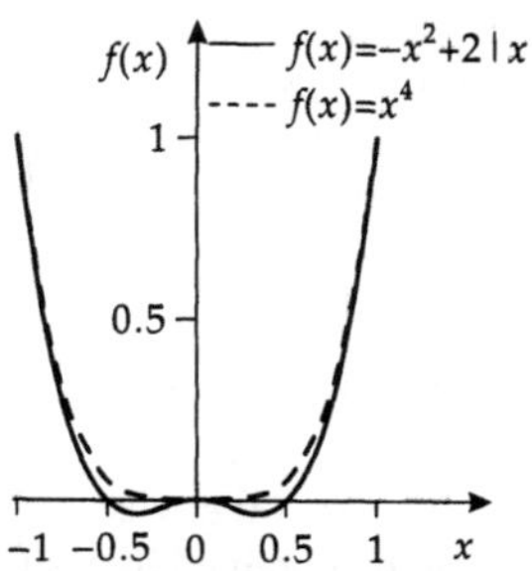

Function (dotted line) and approximating polynomial (solid line)

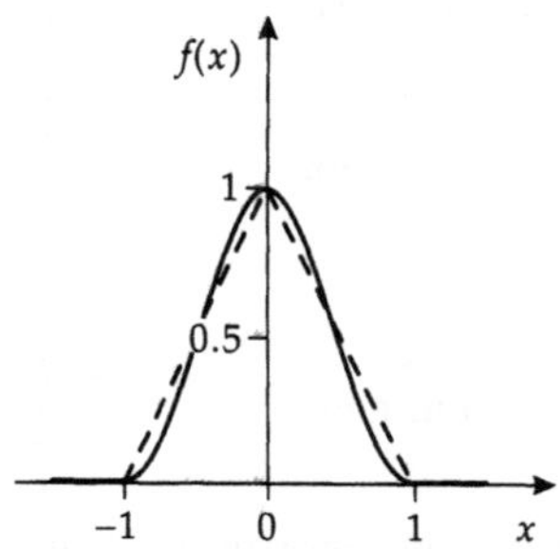

Triangle function (dotted line) and cubic spline function (solid line)

Basis spline or **B-spline**: Special spline function with low carrier: B-splines of degree k and defect 1 are uniquely fixed, up to a normalization constant, by the condition $B_{k,j}(t) = 0$ for $t \notin [t_j, t_{j+k+1}]$ $(j = -k, \ldots, n-1)$.
B-splines form a basis for the vector space of spline functions on the grid $t_0, \ldots, t_n$; therefore, any spline function may be written as a linear combination of B-splines:

$$s(t) = \sum_{j=-k}^{n-1} c_j B_{k,j}(t).$$

Really, for $t \in [t_i, t_{i+1}]$:

$$s(t) = \sum_{j=i-k}^{i} c_j B_{k,j}(t),$$

because $B_{k,j}(t) = 0$ for $t \notin [t_j, t_{j+k+1}]$.
The definition and calculation of the B-splines is done recursively:

$$B_{0,j} = \begin{cases} 1, & \text{for} \quad t_j \leq t < t_{j+1} \\ 0 & \text{otherwise} \end{cases} \qquad j = 0, 1, \ldots, n-1$$

(for $j = n-1$ replace $<$ by $\leq$)

$$B_{k,j}(t) = \omega_{k-1,j}(t) B_{k-1,j}(t) + [1 - \omega_{k-1,j+1}] B_{k-1,j+1}(t), \qquad j = -k, -k+1, \ldots, n-1.$$

Here,

$$\omega_{k-1,j}(t) = \begin{cases} \dfrac{t - t_j}{t_{j+k} - t_k}, & \text{for } t > t_j \\ 0, & \text{for } t \leq t_j. \end{cases}$$

The B-splines thus defined are even positive in the interval (t_j, t_{j+k+1}) and form a "decomposition of the unit" on $[t_0, t_n]$:

$$1 = \sum_{j=-k}^{n-1} B_{k,j}(t).$$

▷ B-splines are applied, e.g., as "test functions" in one-dimensional finite-range calculations in the framework of the finite-element method.

□ Simple B-splines:

$k = 1$ pulse function
$$B_{1,j} = \begin{cases} 1 & x_j \leq x < x_{j+1} \\ 0 & \text{otherwise} \end{cases}$$

$k = 2$ triangle function
$$B_{2,j} = \begin{cases} a(x - x_j) & x_j \leq x < x_{j+1} \\ a(x_{j+2} - x) & x_{j+1} \leq x < x_{j+2} \\ 0 & \text{otherwise} \end{cases}, \quad a = \frac{1}{h}$$

$k = 3$ parabolic area
$$B_{3,j} = \begin{cases} a(x - x_j)^2 & x_j \leq x < x_{j+1} \\ a(x_{j+2} - x)(x - x_j) \\ +a(x_{j+3} - x)(x - x_{j+1}) & x_{j+1} \leq x < x_{j+2} \\ a(x_{j+3} - x)^2 & x_{j+2} \leq x < x_{j+3} \\ 0 & \text{otherwise} \end{cases}, \quad a = \frac{1}{2h}$$

A curve $f(x)$ approximated to the points $f_0, \ldots, f_n$ can be obtained by a superposition of B-splines,

▷ Program sequence for calculating an interpolating cubic natural spline.

```
BEGIN spline function
INPUT n
INPUT (t[i],y[i]) i=0,...,n
h0:=t[1]-t[0];
h1:=t[2]-t[1];
a:=(y[1]-y[0])/SQR(h0);
v[1]:=-1/h0;
FOR i:=1 TO n-1 DO BEGIN
   h1:=t[i+1]-t[i];
   b:=(y[i+1]-y[i])/SQR(h1)
   u[i]=3*(a+b);
   hd[i]:=2*(1/ho+1/h1);
   nd[i]:=1/h1;
   a:=b; h0:=h1;
END;
w[n-1]:=-1/h1;
{forward elimination}
FOR i:=1 TO n-2 DO BEGIN
   q:=-nd[i]/hd[i];
   hd[i+1]:=hd[i+1]+nd[i]*q;
   u[i+1]:=u[i+1]q*u[i]; v[i+1]:=q*v[i];
END;
(backward elimination)
FOR i:=1 DOWNTO 2 DO BEGIN
   q:=hd[i];
   u[i]:=u[i]/q; v[i]:=v[i]/q; w[i]:=w[i]/q;
   q:=nd[i-1];
   u[i-1]:=u[i-1]-q*u[i]; v[i-1]:=v[i-1]-q*v[i];
   w[i-1]:=-q*w[i];
END;
   q:=hd[1];
   u[1]:=u[1]/q; v[1]:=v[1]/q; w[1]:=w[1]/q;
CALL boundary conditions (c,u,v,w,y,t,2,n);
{2 corresponds to natural boundary conditions}
END.
```

▷ Program sequence for fixing the spline coefficients for various boundary conditions:

1. derivatives at the boundary are given
2. natural boundary conditions
3. periodic boundary conditions
4. cubic polynomial as spline in $[t_0, t_2]$ and $[t_{n-2}, t_n]$, cubic splines.

The desired boundary condition can be chosen by giving `typ (1, 2, 3, or 4)`.

```
PROCEDURE boundary conditions(var c: ARRAY[0..n];
   u,v,w,y,t: ARRAY[0..n]; typ, n: INTEGER);
BEGIN
   h1:=t[1]-t[0]; hn:=t[n]-t[n-1];
   d1:=(y[1]-y[0])/h1; dn:=(y[n]-y[n-1])/hn;
```

```
    b1:=3*d1-u[1]; b2:=3*dn-u[n-1];
  CASE typ OF
  1: BEGIN
    {derivatives at the boundary are given}
    INPUT alpha, beta
    END;
  2: BEGIN
    {natural boundary conditions}
    a11:=2+v[1]; a12:=w[1]; a21:=v[n-1]; a22:=2+w[n-1];
    det:=a11*a22-a12*a21;
    alpha:=(b1*a22-a12*b2)/det;
    beta:=(a11*b2-b1*a21)/det;
    END;
  3: BEGIN
    {periodic boundary conditions}
    alpha:=(b1/h1+b2/hn)/((2+v[1]+w[1])/h1
        +(2+v[n-1]+w[n-1])/hn);
    beta:=alpha;
    END;
  4: BEGIN
    {three times differentiability in t[1] and t[n-1]}
    h2:=t[2]-t[1]; hm:=t[n-1]-t[n-2];
    d2:=(y[2]-y[1])/h2; dm:=(y[n-1]-y[n-2])/hm;
    a11:=((v[1]+1)/SQR(h1)-(v[1]+v[2])/SQR(h2));
    a12:=(w[1]/SQR(h1)-(w[1]+w[2])/SQR(h2));
    b1:=(u[1]+u[2]-2*d2)/SQR(h2)-(u[1]-2*d1)/SQR(h1);
    a21:=(v[n-1]+v[n-2])/SQR(hm)-v[n-1]/SQR(hn);
    a22:=(w[n-1]+w[n-2])/SQR(hm)-(1+w[n-1])/SQR(hn);
    det:=a11*a22-a12*a21;
    b2:=(2*dm-u[n-1]-u[n-2])/SQR(hm)+(u[n-1]-2*dn)/SQR(hn);
    alpha:=(b1*a22-a12*b2)/det;
    beta:=(a11*b2-b1*a21)/det;
    END:
  END;
  c[0]:=alpha;
  FOR i:=1 TO n-1 DO c[i]:=u[i]+alpha*v[i]+beta*w[i];
  c[n]:=beta;
  END; {boundary conditions}
```

▷ Program sequence for evaluating a cubic spline function

```
FUNCTION spline(x: REAL; t,y,c:ARRAY[0..n] OF REAL;
    n: INTEGER): REAL;
BEGIN
k:=n+1;
   REPEAT
   k:=k-1;
   UNTIL x>t[k];
   k:=k+1;
   h:=t[k]-t[k-1]; delta:=(y[k]-y[k-1])/h;
   a0:=y[k]; a1:=c[k]; a2:=(c[k]-delta)/h;
```

```
    a3:=(c[k]-2*delta+c[k-1])/SQR(h);
    spline:=((a3*(x-t[k-1])+a2)*(x-t[k])+a1)*(x-t[k])+a0;
END; {spline}
```

$$f(x) = \sum_{j=0}^{n} f_j B_{k,j}(x), \qquad f_j = f\left(x_j + \frac{k}{2}h\right).$$

5.14.7 Particular polynomials

Both Legendre and Chebyshev polynomials are of importance as fit polynomials and as bases of functions, because they have the characteristic that every polynomial of degree n can be represented as a linear combination of the first n Legendre or Chebyshev polynomials.

Legendre polynomials: Solutions of Legendre's differential equation

$$(1-x^2)y'' - 2xy' + n(n+1)y = 0 \qquad y = P_n(x) \quad n \in \mathbb{N}.$$

The polynomial can be represented by

$$P_n(x) = \frac{1}{2^n\, n!}\frac{\mathrm{d}^n}{\mathrm{d}x^n}((x^2-1)^n).$$

The first polynomials are

$$P_0(x) = 1, \qquad P_1(x) = x, \qquad P_2(x) = \frac{1}{2}(3x^2-1),$$
$$P_3(x) = \frac{1}{2}(5x^3-3x), \quad P_4(x) = \frac{1}{8}(35x^4-30x^2+3), \quad \ldots$$

- The Legendre polynomials obey the recurrence equation:

$$P_0(x) = 1, \quad P_1(x) = x,$$
$$(k+1)P_{k+1}(x) = (2k+1)xP_k(x) - kP_{k-1}(x), \; k = 1, 2, 3, \ldots$$

▷ The Legendre polynomials are orthogonal polynomials with respect to the weight function 1; that is, it holds that:

$$\int_{-1}^{1} P_n(x)P_m(x)\mathrm{d}x = \begin{cases} 0, & \text{if } m \neq n \\ \dfrac{2}{2m+1}, & \text{if } m = n \end{cases}.$$

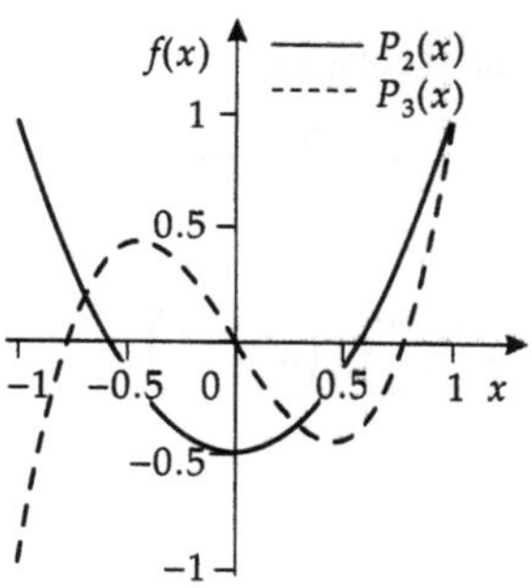

Legendre polynomials

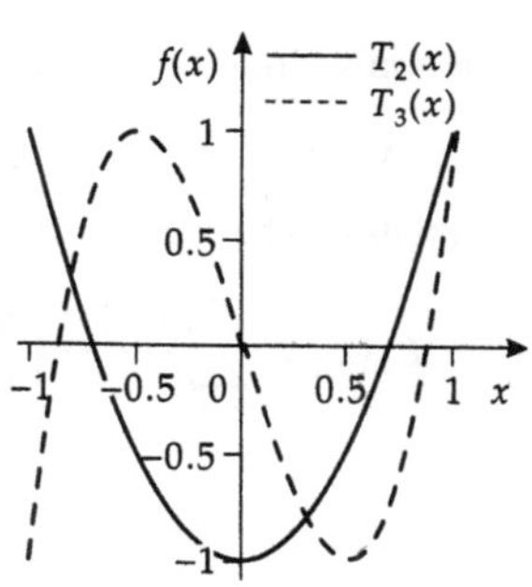

Chebyshev polynomials

Chebyshev polynomials: Polynomials which can be derived via the addition theorems of trigonometry.
Take $x = \cos(\varphi)$, $|x| \leq 1$, and postulate the polynomials

$$T_k(x) = \cos(k\varphi), \qquad \cos(\varphi) = x$$

Solution of the differential equation for $|x| \leq 1$:

$$(1 - x^2)y'' - xy' + n^2 y = 0 \,, \qquad y = T_n(x) \,, \quad n \geq 0 \text{ integer.}$$

We obtain the polynomials:

$$\begin{array}{lll} T_0(x) = 1 \,, & T_1(x) = x \,, & T_2(x) = 2x^2 - 1 \,, \\ T_3(x) = 4x^3 - 3x \,, & T_4(x) = 8x^4 - 8x^2 + 1 \,, & T_5 = 16x^5 - 20x^3 + 5x \quad \cdots \end{array}$$

- The Chebyshev polynomials satisfy the recurrence equation:

$$T_0(x) = 1, \quad T_1(x) = x, \quad T_{k+1}(x) = 2xT_k(x) - T_{k-1}(x), \; k = 1, 2, 3, \ldots$$

▷ Chebyshev polynomials are the orthogonal polynomials with respect to the weight function $\frac{1}{\sqrt{1-x^2}}$; i.e.,

$$\int_{-1}^{1} T_n(x) T_m(x) \frac{\mathrm{d}x}{\sqrt{1-x^2}} = \begin{cases} 0, & \text{if } m \neq n \\ \frac{\pi}{2}, & \text{if } m = n \neq 0 \\ \pi, & \text{if } m = n = 0 \end{cases} .$$

▷ T_n has the n zeros, $x_j = \cos\frac{(2k+1)\pi}{2n}$, $k = 0, 1, \ldots, n-1$, in the interval $[-1, 1]$; if these zeros are used as interpolation nodes, then the error in polynomial interpolation is particularly small.

Further polynomials which satisfy similar differential equations:
Hermite polynomials: Solutions of the differential equation

$$y'' - 2xy' + 2ny = 0, \qquad n = 0, 1, 2, \ldots$$

Hermite polynomials can be represented by

$$\begin{aligned} H_n(x) &= \sum_{k=0}^{m} \frac{(-1)^k}{k!} \frac{n!}{(n-2k)!} (2x)^{n-2k} \qquad m = \mathrm{Int}\left[\frac{n}{2}\right] \\ &= (-1)^n \mathrm{e}^{x^2} \frac{\mathrm{d}^n}{\mathrm{d}x^n} \mathrm{e}^{-x^2} . \end{aligned}$$

Laguerre polynomials: Solutions of the differential equation

$$xy'' + (1 - x)y' + ny = 0, \qquad n = 0, 1, 2, \ldots$$

Laguerre polynomials can be represented by

$$L_n(x) = \sum_{k=0}^{n} \frac{(-1)^k}{k!} \binom{n}{k} x^k = \frac{\mathrm{e}^x}{n!} \frac{\mathrm{d}^n}{\mathrm{d}x^n} \left(x^n \mathrm{e}^{-x}\right) .$$

Fractional rational functions

Fractional rational functions can be represented as a fraction of two polynomials. In the following, we describe first the hyperbola and reciprocal quadratic functions. This is followed by the description of the power function with a negative exponent, and finally the general quotient of two polynomials.

5.15 Hyperbola

$$f(x) = \frac{1}{ax+b}$$

Reciprocal linear function.

□ Spherically symmetric potential fields and cylindrically symmetric force fields usually have a $\frac{1}{r}$-dependence.

(a) Definition

Reciprocal linear function:

$$f(x) = \frac{1}{ax+b} = (ax+b)^{-1} .$$

Derivative of the logarithmic function:

$$f(x) = \frac{1}{a}\frac{\mathrm{d}}{\mathrm{d}x} \ln|ax+b| .$$

(b) Graphical representation

The function has a pole at $x = -\frac{b}{a}$.

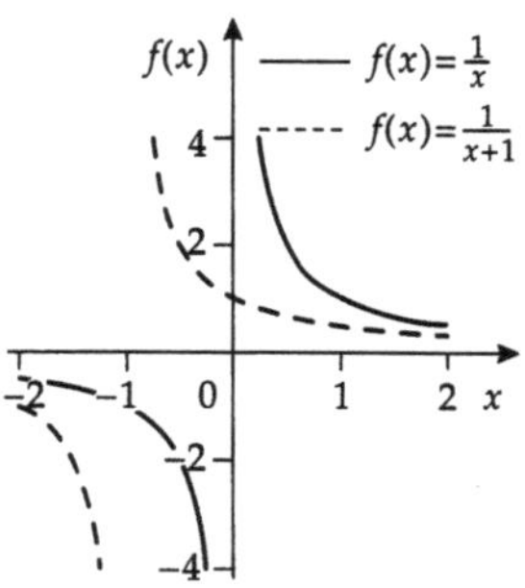

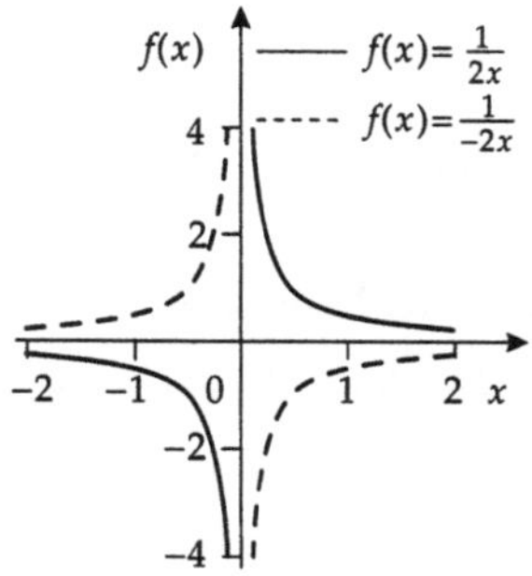

For $x > -\frac{b}{a}$ the function for $a > 0$ ($a < 0$) is positive (negative), and for $x < -\frac{b}{a}$ it is negative (positive).

● The function tends to zero as $x \to \pm\infty$.

▷ The function tends stronger to zero when the absolute value of a is increased.

(c) Properties of the function

Domain of definition: $-\infty < x < \infty, x \neq -\frac{b}{a}$.

Range of values: $-\infty < f(x) < \infty,\ f(x) \neq 0$.

Quadrant:
- $a > 0$
  - $b = 0$: first and third quadrants
  - $b > 0$: first, second, and third quadrants
  - $b < 0$: first, third, and fourth quadrants
- $a < 0$
  - $b = 0$: second and fourth quadrants
  - $b > 0$: second, third, and fourth quadrants
  - $b < 0$: first, second, and fourth quadrants

Periodicity: None.

Monotonicity:
- $a > 0$
  - $x > -\frac{b}{a}$: strictly monotonically decreasing
  - $x < -\frac{b}{a}$: strictly monotonically decreasing
- $a < 0$
  - $x > -\frac{b}{a}$: strictly monotonically increasing
  - $x < -\frac{b}{a}$: strictly monotonically increasing

Symmetries: Point symmetry about $P\left(x = -\frac{b}{a}, y = 0\right)$.

Asymptotes: $f(x) \to 0$ as $x \to \pm\infty$.

(d) Particular values

Zeros: None.

Discontinuities: None.

Poles: $x = -\frac{b}{a}$.

Extremes: None.

Points of inflection: None.

(e) Reciprocal function

The reciprocal function is the linear function

$$\frac{1}{f(x)} = \frac{1}{(ax+b)^{-1}} = ax + b\,.$$

(f) Inverse function

The inverse function of a hyperbola (shifted along the x-axis) is a hyperbola (shifted along the y-axis) ($a \neq 0$):

$$f^{-1}(x) = \frac{1}{ax} - \frac{b}{a}\,.$$

(g) Cognate functions

Linear function: the reciprocal function of $f(x)$.
Reciprocal quadratic function: the derivative of $f(x)$.
Logarithmic function: the primitive of $f(x)$.

(h) Conversion formulas

Reflection about the axis $x = -\frac{b}{a}$.

$$f(-x) = -f\left(x - 2\frac{b}{a}\right)$$

▷ The sum (difference) of two hyperbolas is generally no longer a hyperbola.

$$\frac{1}{x} \pm \frac{1}{y} = \frac{y \pm x}{xy}$$

(i) Approximation formulas for the function (8 bits ≈ 0.4% of precision)

For small x values, it holds that

$$\frac{1}{ax+b} \approx \frac{1}{b} - \frac{ax}{b^2} \qquad \text{for} \quad |x| < \frac{|b|}{16|a|} \,.$$

For large x values it holds that

$$\frac{1}{ax+b} \approx \frac{ax-b}{a^2x^2} \qquad \text{for} \quad |x| > \frac{16|a|}{|b|} \,.$$

(j) Expansion in series and products

Geometric series for small arguments:

$$\begin{aligned}\frac{1}{ax+b} &= \frac{1}{b} - \frac{ax}{b^2} + \frac{a^2x^2}{b^3} - \frac{a^3x^3}{b^4} + \ldots \\ &= \frac{1}{b}\sum_{k=0}^{\infty}\left(\frac{-ax}{b}\right)^k \qquad \text{for} \quad |x| < \left|\frac{b}{a}\right| \,.\end{aligned}$$

Geometric series for large arguments:

$$\begin{aligned}\frac{1}{ax+b} &= \frac{1}{ax} - \frac{b}{a^2x^2} + \frac{b^2}{a^3x^3} - \frac{b^3}{a^4x^4} + \ldots \\ &= \frac{1}{ax}\sum_{k=0}^{\infty}\left(\frac{-b}{ax}\right)^k \qquad \text{for} \quad |x| > \left|\frac{b}{a}\right| \,.\end{aligned}$$

Infinite product:

$$\frac{1}{ax+b} = \begin{cases} \dfrac{b-ax}{b^2}\displaystyle\prod_{k=1}^{\infty}\left[1 + \left(\frac{ax}{b}\right)^{2k}\right] & x < \left|\dfrac{b}{a}\right| \\ \dfrac{ax-b}{a^2x^2}\displaystyle\prod_{k=1}^{\infty}\left[1 + \left(\frac{b}{ax}\right)^{2k}\right] & |x| > \left|\dfrac{b}{a}\right| \end{cases} .$$

(k) Derivative of the function

The derivative is a reciprocal quadratic function:

$$\frac{\mathrm{d}}{\mathrm{d}x}\left(\frac{1}{ax+b}\right) = -\frac{a}{(ax+b)^2}\,.$$

(l) Primitive of the function

The primitive is a logarithmic function.

● The pole must be outside the domain of integration.

$$\int_0^x \frac{1}{at+b}\,\mathrm{d}t = \frac{1}{a}\ln\left|1+\frac{ax}{b}\right|$$

▷ When the pole is inside the domain of integration, then the integral can only be regarded as the Cauchy principal value (see Chapter 15 on integral calculus).

When $b = 0$, then

$$\int_1^x \frac{1}{at}\,\mathrm{d}t = \frac{1}{a}\ln|x|\,.$$

(m) Particular extensions and applications

Calculation of an integral by substitution: The properties of integrating $\frac{1}{x}$ allow for the calculation of special integrals.

$$\int_a^b \frac{\frac{\mathrm{d}}{\mathrm{d}t}f(t)}{f(t)}\,\mathrm{d}t = \int_{f(a)}^{f(b)} \frac{1}{u}\,\mathrm{d}u = \ln\left|\frac{f(b)}{f(a)}\right|\,.$$

Green's function: The function $f(\vec{\mathbf{r}}) = 1/r$ is the Green's function for the Poisson equation. See the Chapter 16 on vector analysis and Chapter 18 on partial differential equations.

5.16 Reciprocal quadratic function

$$f(x) = \frac{1}{ax^2+bx+c}.$$

Reciprocal function of a quadratic function.

□ Spherically symmetric force fields often have the form $f(\vec{\mathbf{r}}) = \frac{1}{r^2}$.

(a) Definition

Reciprocal quadratic function:

$$f(x) = \frac{1}{ax^2+bx+c} = (ax^2+bx+c)^{-1}\,.$$

Re-writing the function:

$$f(x) = \frac{1}{a\left(x+\frac{b}{2a}\right)^2 + c - \frac{b^2}{4a}} = \frac{1}{a\left[\left(x+\frac{b}{2a}\right)^2 - D\right]}\,.$$

Discriminant of the quadratic function:

$$D = \frac{b^2}{4a^2} - \frac{c}{a}\,.$$

(b) Graphical representation

- The properties of the function depend greatly on the discriminant D.
- If $D = 0$, the function is a quadratic hyperbola, $\frac{1}{x^2}$, shifted along the x-axis with a pole at $x = -\frac{b}{2a}$ it takes on positive values for $a > 0$ and negative values for $a < 0$.
- If $D < 0$, the function has no (real) poles, but an extreme at $x = -\frac{b}{2a}$, which is a maximum for $a > 0$ and a minimum for $a < 0$. The function is positive for $a > 0$ and negative for $a < 0$.
- If $D > 0$, then the function can be represented as a sum of two reciprocal linear functions. The function has two poles at the zeros of the quadratic function in the denominator $x = -\frac{b}{2a} \pm \sqrt{D}$, and an extreme at $x = -\frac{b}{2a}$ which is a maximum for $a > 0$ and a minimum for $a < 0$.

When $a > 0$ ($a < 0$), then the function is positive (negative) for $\left|x + \frac{b}{2a}\right| > \sqrt{D}$ and negative (positive) for $\left|x + \frac{b}{2a}\right| < \sqrt{D}$.

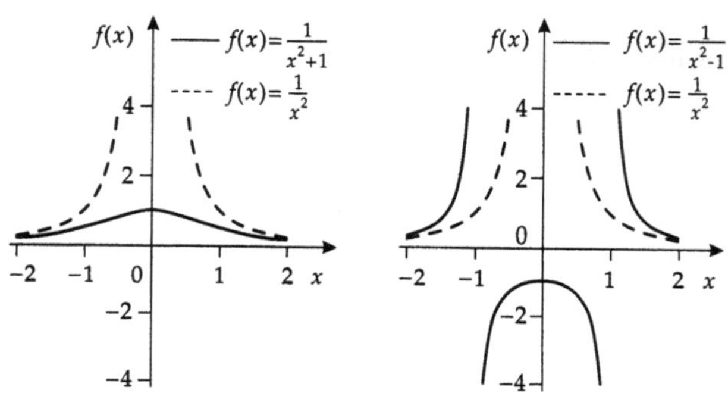

(c) Properties of the function

Domain of definition: $-\infty < x < \infty$.

$D < 0$: no restrictions

$D = 0$: $x \neq -\frac{b}{2a}$

$D > 0$: $x \neq -\frac{b}{2a} \pm \sqrt{D}$

Range of values: $a > 0$ $D < 0$: $0 < f(x) < -\frac{1}{Da}$

$D = 0$: $0 < f(x) < \infty$

$D > 0$: $0 < f(x) < \infty,\ -\infty < f(x) < -\frac{1}{Da}$

	$a < 0$ $D < 0$: $-\frac{1}{Da} < f(x) < 0$
	$D = 0$: $-\infty < f(x) < 0$
	$D > 0$: $-\infty < f(x) < 0, -\frac{1}{Da} < f(x) < \infty$
Quadrant:	$a > 0$ $D \leq 0$: first and second quadrants
	$D > 0$: first, second, third and/or fourth quadrants
	$a < 0$ $D \leq 0$: third and fourth quadrants
	$D > 0$: third, fourth, first and/or second quadrants
Periodicity:	None.
Monotonicity:	$a > 0$ $x > -\frac{b}{2a}$: strictly monotonically decreasing
	$x < -\frac{b}{2a}$: strictly monotonically increasing
	$a < 0$ $x > -\frac{b}{2a}$: strictly monotonically increasing
	$x < -\frac{b}{2a}$: strictly monotonically decreasing
Symmetries:	Mirror symmetry about the axis $x = -\frac{b}{2a}$.
Asymptotes:	$f(x) \to 0$ as $x \to \pm\infty$.

(d) Particular values

Zeros:	None.
Discontinuities:	None.
Poles:	$D < 0$: no real poles
	$D = 0$: $x = -\frac{b}{2a}$
	$D > 0$: $x = -\frac{b}{2a} \pm \sqrt{D}$
Extremes:	$a > 0$ $D = 0$: no extremes
	$D \neq 0$: maximum at $x = -\frac{b}{2a}$
	$a < 0$ $D = 0$: no extremes
	$D \neq 0$: minimum at $x = -\frac{b}{2a}$
Points of inflection:	None.
Value at $x = 0$:	$f(0) = \frac{1}{c}$.

(e) Reciprocal function

The reciprocal function is the quadratic function.

$$\frac{1}{f(x)} = \frac{1}{(ax^2 + bx + c)^{-1}} = ax^2 + bx + c\ .$$

(f) Inverse function

The function is invertible only for $x > -\dfrac{b}{2a}$ $\left(x < -\dfrac{b}{2a}, \text{ correspondingly}\right)$ only $(a \neq 0)$.
It is

$$f^{-1}(x) = \sqrt{\frac{1}{ax} + D} - \frac{b}{2a} .$$

(g) Cognate functions

Quadratic function and its cognate functions.
Root functions with a negative exponent.
Reciprocal linear and cubic functions.

(h) Conversion formulas

Decomposition for $c = 0$:

$$\frac{1}{ax^2 + bx} = \frac{1}{bx} - \frac{a}{abx + b^2} .$$

Decomposition for $D > 0$:

$$\frac{1}{ax^2 + bx + c} = \frac{1}{a} \cdot \frac{1}{x + \frac{b}{2a} + \sqrt{D}} \cdot \frac{1}{x + \frac{b}{2a} - \sqrt{D}} .$$

(i) Approximation formulas (8 bits ≈ 0.4% of precision)

For small values:

$$\frac{1}{ax^2 + bx + c} \approx \frac{c - bx}{c^2} \qquad \text{for} \quad |x| < \frac{|c|}{16\sqrt{|b^2 - ac - abx|}} .$$

For large x-values it holds that

$$\frac{1}{ax^2 + bx + c} \approx \frac{ax - b}{a^2 x^3} \qquad \text{for} \quad |x| > \frac{16\sqrt{\left|b^2 - ac + \frac{bc}{x}\right|}}{|a|} .$$

(j) Expansion in series and products

For small values:

$$\frac{1}{ax^2 + bx + c} = \frac{1}{c} - \frac{bx}{c^2} + \frac{(b^2 - ac)x^2}{c^3} - \frac{(b^3 - 2abc)x^3}{c^4} + \dots$$
$$= \frac{1}{c} \sum_{k=0}^{\infty} \left(\frac{-x(ax + b)}{c}\right)^k , \qquad \left|\frac{ax^2 + bx}{c}\right| < 1 .$$

For large x-values it holds that

$$\frac{1}{ax^2 + bx + c} = \frac{1}{ax^2} - \frac{b}{a^2 x^3} + \frac{b^2 - ac}{a^3 x^4} - \frac{b^3 - 2abc}{a^4 x^5} + \dots$$
$$= \frac{1}{ax^2} \sum_{k=0}^{\infty} \left(\frac{-bx - c}{ax^2}\right)^k , \qquad \left|\frac{bx + c}{ax^2}\right| < 1 .$$

(k) Derivative of the function

The derivative of the function is obtained by means of the chain rule:

$$\frac{\mathrm{d}}{\mathrm{d}x}\frac{1}{ax^2+bx+c} = -\frac{2ax+b}{(ax^2+bx+c)^2}$$

(l) Primitive of the function

Integral from the symmetry axis to x:

$$\int_{-b/2a}^{x}\frac{\mathrm{d}t}{at^2+bt+c} = \begin{cases} \dfrac{2}{\sqrt{\Delta}}\arctan\left(\dfrac{2ax+b}{\sqrt{\Delta}}\right) & \Delta = 4ac-b^2 > 0 \\ \dfrac{-2}{\sqrt{-\Delta}}\operatorname{artanh}\left(\dfrac{2ax+b}{\sqrt{-\Delta}}\right) & \Delta < 0, \quad x < \dfrac{\sqrt{-\Delta}-b}{2a}\,. \end{cases}$$

For $\Delta = 0$, the integral diverges, but the following integral can be given:

$$\int_{x}^{\infty}\frac{\mathrm{d}t}{at^2+bt+c} = \frac{2}{2ax+b}, \qquad \Delta = 0\,.$$

(m) Particular extensions and applications

Complex poles: In the complex domain the function $f(x)$ can always be decomposed into a product of two reciprocal linear functions:

$$\frac{1}{ax^2+bx+c} = \frac{1}{a}\cdot\frac{1}{x+\dfrac{b}{2a}+\sqrt{D}}\cdot\frac{1}{x+\dfrac{b}{2a}-\sqrt{D}}$$

- The denominator possesses either two different real zeros or one real double zero or two non-real zeros.

5.17 Power functions with a negative exponent

$$f(x) = ax^{-n} \qquad n = 1, 2, \ldots$$

Generalized hyperbolas.
Reciprocal functions of the integral-rational power functions.
In electrophysics, negative powers of higher order often occur in the interaction of charges, dipoles, quadrupoles, etc.

(a) Definition

Reciprocal function of x^n:

$$f(x) = ax^{-n} = \frac{a}{x^n} = a\left(\frac{1}{x}\right)^n = a\prod_{k=1}^{n}\frac{1}{x}\,.$$

(b) Graphical representation

An important distinguishing feature is the evenness of the exponent.

- For $a > 0$ ($a < 0$), the functions with even n have only positive (negative) values.
- If $a > 0$ ($a < 0$), the functions with odd n have positive (negative) values for positive x-values and negative (positive) values for negative x-values.

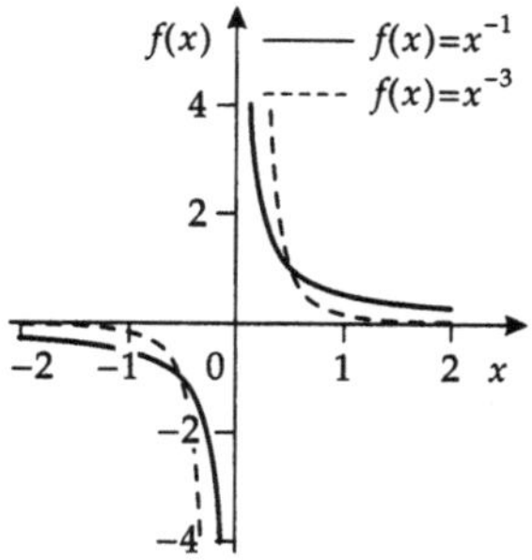

Odd functions

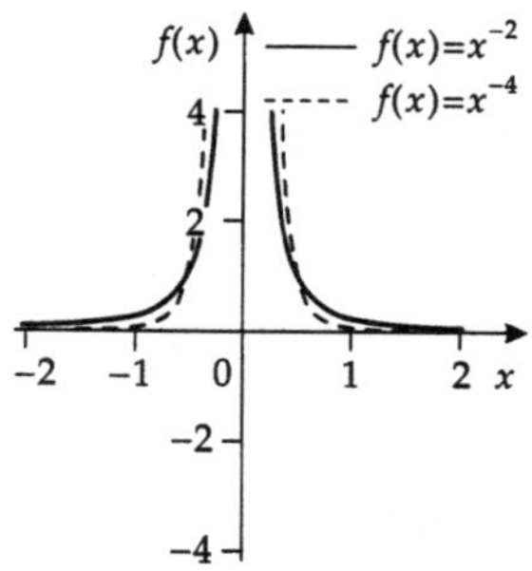

Even functions

- All functions have a pole of order n at $x = 0$.
- All functions go through point $x = 1$, $y = a$.

At $x \to \pm\infty$, functions with larger n approach the x-axis more rapidly than functions with smaller n.

(c) Properties of the function

Domain of definition:	$-\infty < x < \infty, x \neq 0.$	
Range of values:	n even	$a > 0$: $0 < f(x) < \infty$
		$a < 0$: $-\infty < f(x) < 0$
	n odd	$-\infty < f(x) < \infty$, $f(x) \neq 0$
Quadrant:	n even	$a > 0$: first and second quadrants
		$a < 0$: third and fourth quadrants
	n odd	$a > 0$: first and third quadrants
		$a < 0$: second and fourth quadrants
Periodicity:	None.	
Monotonicity for $x > 0$:	n even	$a > 0$: strictly monotonically decreasing
		$a < 0$: strictly monotonically increasing
	n odd	$a > 0$: strictly monotonically decreasing
		$a < 0$: strictly monotonically increasing
Monotonicity for $x < 0$:	n even	$a > 0$: strictly monotonically increasing
		$a < 0$: strictly monotonically decreasing
	n odd	$a > 0$: strictly monotonically decreasing
		$a < 0$: strictly monotonically increasing
Symmetries:	n even	mirror symmetry about the y-axis
	n odd	point symmetry about the origin
Behavior at infinity:	$f(x) \to 0$ as $x \to \pm\infty$.	

(d) Particular values

Zeros:	None.
Discontinuities:	None.

Poles: Pole of order n at $x = 0$.

Extrema: None.

Points of inflection: None.

(e) Reciprocal function

The reciprocal function is the integral-rational power function of degree n:

$$\frac{1}{f(x)} = \frac{1}{ax^{-n}} = \frac{1}{a}x^n .$$

(f) Inverse function

The inverse functions are the reciprocal n-th roots:

$$f^{-1}(x) = \sqrt[n]{\frac{a}{x}} .$$

▷ In the case of odd n the inverse function can be defined for all x, but in the case of even n, it can only be defined for $x > 0$ (or $x = 0$ with an additional minus sign).

(g) Cognate functions

Power functions of degree n, the reciprocal functions of $f(x)$.
n-th roots, the reciprocal inverse function.

(h) Conversion formulas

Reflection about the y-axis:

$$f(-x) = f(x) \quad \text{for } n \text{ even;} \qquad f(-x) = -f(x) \quad \text{for } n \text{ odd.}$$

▷ The same conversion formulas apply as for powers with positive exponents are valid.
The sum of two power functions of degree $-n$ is a third power function of degree $-n$:

$$ax^{-n} \pm bx^{-n} = (a \pm b)x^{-n} .$$

The product of a power function of degree $-n$ and a power function of degree $-m$ is a third power function of degree $-(n+m)$:

$$ax^{-n} \cdot bx^{-m} = (a \cdot b)x^{-(n+m)} .$$

The quotient of a power function of degree $-n$ and a power function of degree $-m$ is a third power function of degree $m - n$:

$$\frac{ax^{-n}}{bx^{-m}} = \frac{a}{b}x^{m-n} .$$

The m-th power of a power function of degree $-n$ is a third power function of degree $-n \cdot m$:

$$\left(ax^{-n}\right)^m = a^m x^{-n \cdot m} .$$

(i) Approximation formulas (8 bits ≈ 0.4% of precision)

For small x it holds that

$$(a+x)^{-n} \approx a^{-n} \cdot \left(1 - \frac{n}{a}x\right) \qquad \text{for } |x| < \frac{|a|}{12\sqrt{n^2+n}} .$$

(j) Expansion of series or products of the function

Finite geometric series of powers for $x \neq 0$:

$$\sum_{k=0}^{n} x^{-k} = 1 + x^{-1} + x^{-2} + \ldots + x^{-n} = \frac{x - x^{-n}}{x-1} .$$

(k) Derivative of the function

- The derivative of a power function of degree $-n$ is a power function of degree $-(n+1)$:

$$\frac{\mathrm{d}}{\mathrm{d}x} ax^{-n} = -nax^{-(n+1)} .$$

(l) Primitive of the function

The primitive of a power function of degree $-n$ is a power function of degree $-(n-1)$ $(n \neq 1)$:

$$\int_x^{\infty} \left(at^{-n}\right) \mathrm{d}t = \frac{a}{n-1} x^{1-n} .$$

For $n = 1$, the primitive is the logarithmic function:

$$\int_1^x \left(\frac{a}{t}\right) \mathrm{d}t = a \ln |x| .$$

(m) Particular extensions and applications

Double-logarithmic plot: As with integral-rational power functions, the reciprocal power functions can be plotted in a double-logarithmic scale. This results in straight lines with negative slopes.

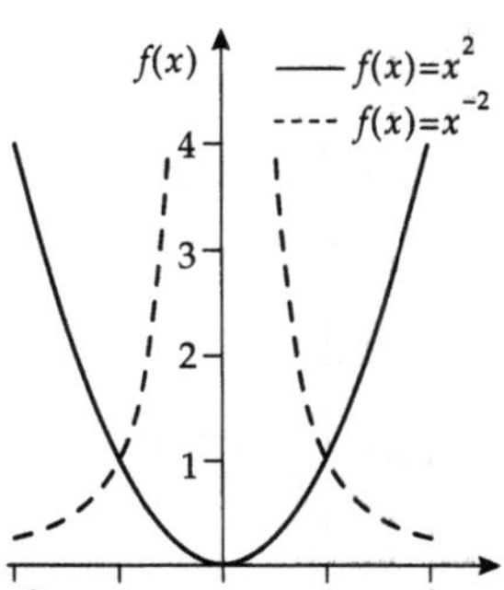

Linear representation

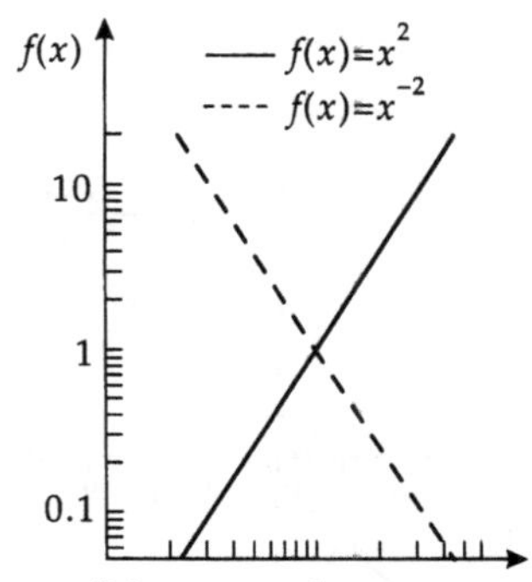

Double-logarithmic representation

5.18 Quotient of two polynomials

$$f(x) = \frac{a_n x^n + a_{n-1}x^{n-1} + \ldots + a_1 x + a_0}{b_m x^m + b_{m-1}x^{m-1} + \ldots + b_1 x + b_0}$$

This is the most general case of a fractional rational function.

(a) Definition

Representation of the function as a quotient of the polynomials $p(x)$ and $q(x)$:

$$f(x) = \frac{\sum_{k=0}^{n} a_k x^k}{\sum_{k=0}^{m} b_k x^k} = \frac{p(x)}{q(x)} .$$

Proper fractional rational function: $m > n$.
Improper fractional rational function: $m \leq n$.

(b) Graphical representation

- If in the representation of f, the following applies:
  Function $f(x)$ has the zeros of the numerator polynomial $p(x)$ as zeros, and the zeros of the denominator polynomial $q(x)$ as poles.
- If numerator and denominator have a zero of the same multiplicity at the same point x_0, then the definition gap can constantly be filled at that point.

▷ If the zero of the numerator is of higher multiplicity than the zero of the denominator, then the constant expansion is a zero of the function.

At infinity, the behavior of the function is determined essentially by the ratio of the terms with the greatest power in the numerator and denominator polynomials.

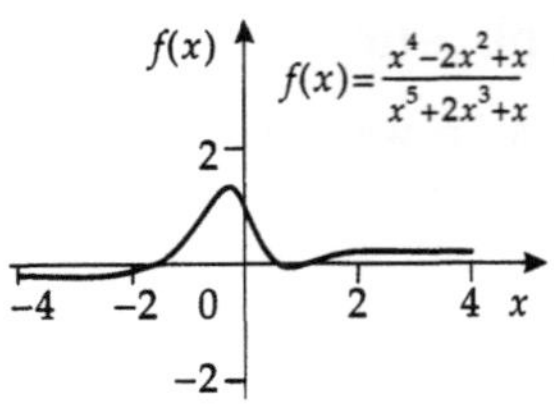

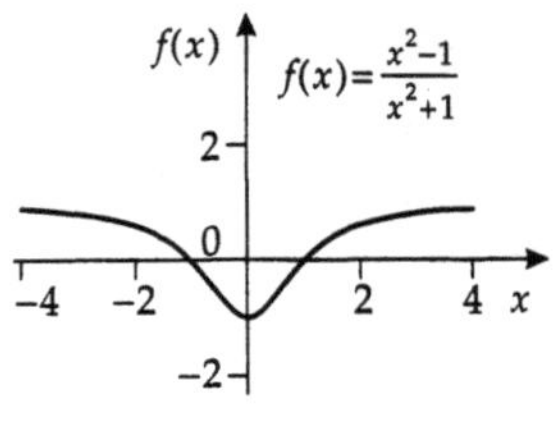

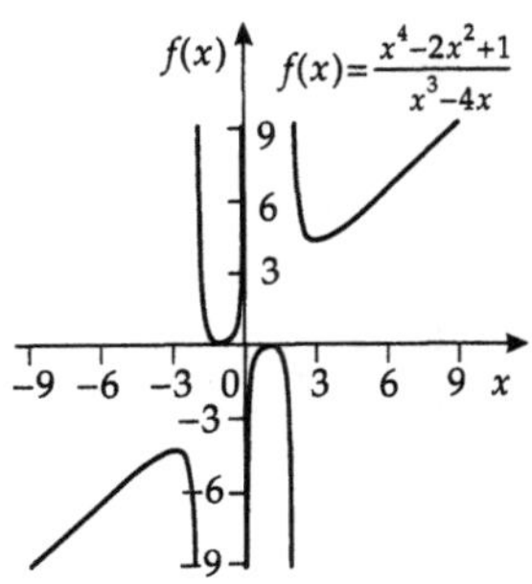

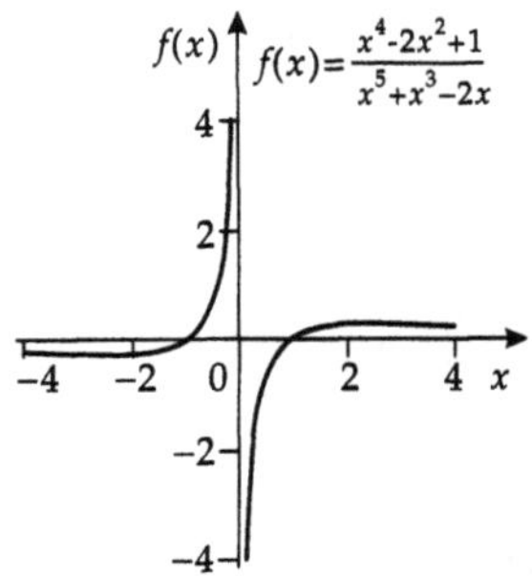

(c) Properties of the function

Domain of definition:	$-\infty < x < \infty$, except the zeros of $q(x)$.
Range of values:	Depends on the individual case.
Quadrant:	Often all four quadrants.
Periodicity:	None.
Monotonicity:	Depends on the individual case.
Symmetries:	In both the numerator and the denominator polynomials have only even or only odd powers, then $f(x)$ is mirror-symmetric about the y-axis. If one polynomial has only even powers and the other only odd powers, then $f(x)$ is point-symmetric about the origin.
Asymptotes:	For $x \to \pm\infty$, $f(x)$ tends to $\dfrac{a_n}{b_m} x^{n-m}$.

(d) Particular values

Zeros:	The zeros of p are the zeros of f, $q \neq 0$ at these points.
Discontinuities:	None.
Poles:	The zeros of $q(x)$ are the poles of $f(x)$.
Extremes:	Depends on the individual case. If $p(x)$ and $q(x)$ have an extreme at the same point x_E, then x_E is also an extreme of $f(x)$ if $\dfrac{\mathrm{d}^2 q}{\mathrm{d}x^2} p \neq \dfrac{\mathrm{d}^2 p}{\mathrm{d}x^2} q \neq 0$.
Points of inflection:	Depends on the individual case.
Value at $x = 0$:	$f(0) = \dfrac{a_0}{b_0}$.

(e) Reciprocal function

The reciprocal function of a fractional rational function is a fractional rational function itself.

$$\frac{1}{f(x)} = \frac{\sum_{k=0}^{m} b_k x^k}{\sum_{k=0}^{n} a_k x^k}$$

(f) Inverse function

Generally the function is only invertible in partial domains.

▷ For very large x, the inverse function tends to an $(n-m)$-th root ($a_n \neq 0$, $b_m \neq 0$)

$$x \to \infty: \qquad y = f(x) \to \frac{a_n}{b_m} x^{n-m} \qquad f^{-1}(x) \to \left(x \frac{b_m}{a_n}\right)^{1/n-m}$$

(g) Cognate functions

All polynomials are special cases of $f(x)$.
All reciprocal power functions are special cases of $f(x)$.
All roots are inverse functions of special cases of $f(x)$.

(h) Conversion formulas

Generally, no conversion formulas can be given.
In individual cases the conversion formulas for power functions can be used.

(i) Approximation formulas for the function

After a partial fraction decomposition, the approximations of the reciprocal linear functions can be used.

(j) Expansion of series or products of the function

Decomposition into a product of linear factors, poles, and remainder terms. The numerator polynomial $p(x)$ and the denominator polynomial $q(x)$ can be decomposed into linear factors. Thus, the total function can also be represented as a product:

$$f(x) = (x - p_1)^{n_1} \cdots (x - p_k)^{n_k} \frac{1}{(x - q_1)^{m_1}} \cdots \frac{1}{(x - q_l)^{m_l}} \cdots \frac{r_p(x)}{r_q(x)} .$$

$p_1 \dots p_k$ are the zeros of $p(x)$ (multiplicities $n_1 \dots n_k$); $q_1 \dots q_l$ are the zeros of $q(x)$ (multiplicities $m_1 \dots m_l$); $r_p(x)$ and $r_q(x)$ are the remainder terms of $p(x)$ and $q(x)$.
Partial fraction decomposition: A decomposition of $f(x)$ into a sum of fractions. See the following section.

(k) Derivative of the function

The derivative is obtained using the quotient rule:

$$\frac{\mathrm{d}}{\mathrm{d}x} f(x) = \frac{\dfrac{\mathrm{d}p(x)}{\mathrm{d}x} \cdot q(x) - p(x) \cdot \dfrac{\mathrm{d}q(x)}{\mathrm{d}x}}{(q(x))^2} .$$

▷ The derivatives of $p(x)$ and $q(x)$ are obtained as shown in Section 5.13 for polynomials:

$$\frac{\mathrm{d}}{\mathrm{d}x} \sum_{k=0}^{n} a_k x^k = \sum_{k=1}^{n} k a_k \, x^{k-1} .$$

(l) Primitive of the function

Generally, the integral of $f(x)$ is very complicated. We therefore refer to the tables of integrals at the end of this book (Chapter 25) and other tables of integrals (Gradstein, etc.).
Here, some simpler integrals (without limits of integration) are given:

$$\int \frac{ax + b}{cx + d} \mathrm{d}x = \frac{a}{c} x + \left(\frac{b}{c} - \frac{ad}{c^2} \right) \ln |cx + d| + \text{const.}$$

$$\int \frac{\mathrm{d}x}{(ax + b)(cx + d)} = \frac{1}{bc - ad} \ln \left| \frac{cx + d}{ax + b} \right| + \text{const.}$$

$$\int \frac{x \, \mathrm{d}x}{(ax + b)(cx + d)} = \frac{1}{bc - ad} \left(\frac{b}{a} \ln |ax + b| - \frac{\mathrm{d}}{c} \ln |cx + d| \right) + \text{const.}$$

More complicated integrals can also be solved if the partial fraction expansion of $f(x)$ is integrated.

(m) Particular extensions and applications

Polynomial division: Division of a polynomial $p(x)$ of degree n by a polynomial of degree $m < n$: See the following section.
Complex poles: Complex poles can be treated as described for the discussion of complex zeros of polynomials, since the poles of the function are given by the zeros of the denominator polynomials.
From the fundamental theorem of algebra:

- In the complex domain the quotient of a numerator polynomial of degree n and a denominator polynomial of degree m in reduced representation (no common linear factors in the numerator and denominator) has exactly n zeros and m poles if multiplicity is taken into account.

5.18.1 Polynomial division and partial fraction decomposition

Polynomial division and partial fraction decomposition are used to represent the quotient of two polynomials as a sum of power functions (shifted along the x-axis) with integral positive or negative exponents.
If the degree n of the numerator polynomial is greater than or equal to the degree m of the denominator polynomial, polynomial division is used to decompose the function into a polynomial of degree $n - m$ and a remainder function for which the degree of the numerator polynomial $r(x)$ is smaller than the degree of the denominator polynomial $q(x)$.

$$f(x) = \frac{p(x)}{q(x)} = \sum_{k=0}^{n-m} a_k x^k + \frac{r(x)}{q(x)}$$

By partial fraction decomposition, the remainder function can be written as a sum of simple fractions.

▷ Polynomial division and partial fraction decomposition are used, for example, in the integration of fractional rational functions.

Polynomial division: Division of a polynomial $p(x)$ of the degree n by a polynomial $q(x)$ of degree $m < n$. The division leads to a quotient polynomial $h(x)$ and a remainder term $r(x)/q(x)$.

$$f(x) = \frac{p(x)}{q(x)} = \frac{\sum_{i=0}^{n} a_i x^i}{\sum_{j=0}^{m} b_j x^j} = h(x) + \frac{r(x)}{q(x)}$$

The division is performed successively. Start with the highest terms of $p(x)$ and $q(x)$:

$$\frac{p(x)}{q(x)} = \frac{a_n}{b_m} x^{n-m} + \frac{p_1(x)}{q(x)} .$$

The remainder $p_1(x)$ is given by

$$p_1(x) = p(x) - \frac{a_n}{b_m} x^{n-m} q(x) = \sum_{i=0}^{k} c_i x^k .$$

If $p_1(x) = 0$, the division is ended; otherwise the division with the new polynomial $p_1(x)$. The result is the polynomial p_2:

$$\frac{p_1(x)}{q(x)} = \frac{c_k}{b_m}x^{k-m} + \frac{p_2(x)}{q(x)} .$$

Then, the solution for $p(x)$ is

$$\frac{p(x)}{q(x)} = \frac{a_n}{b_m}x^{n-m} + \frac{c_k}{b_m}x^{k-m} + \frac{p_2(x)}{q(x)} .$$

If $p_2(x) \neq 0$, then the division is continued until, finally, in the r-th step either $p_r(x) = 0$ (division without a remainder) or the degree of p_r becomes smaller than the degree of $q(x)$. In the latter case, p_r is the remainder polynomial $r(x)$. Then, the decomposition reads

$$f(x) = \frac{p(x)}{q(x)} = \frac{a_n}{b_m}x^{n-m} + \frac{c_k}{b_m}x^{k-m} + \frac{d_l}{b_m}x^{l-m} + \cdots + \frac{r(x)}{q(x)} .$$

□ Division of $p(x) = x^4 - x^2 + 3x + 1$ by $q(x) = x^2 + x - 1$:

$$\begin{aligned} x^4 - x^2 + 3x + 1 &= x^2 \cdot (x^2 + x - 1) - x^3 + 3x + 1 && x^2 \\ -x^3 + 3x + 1 &= -x \cdot (x^2 + x - 1) + x^2 + 2x + 1 && -x \\ x^2 + 2x + 1 &= 1 \cdot (x^2 + x - 1) + x + 2 && 1 \\ r(x) &= x + 2 \end{aligned}$$

$$\frac{x^4 - x^2 + 3x + 1}{x^2 + x - 1} = x^2 - x + 1 + \frac{x + 2}{x^2 + x - 1}$$

The expression $r(x)/q(x)$ can be decomposed further into partial fractions.

$$\frac{x^4 - x^2 + 3x + 1}{x^2 + x - 1} = x^2 - x + 1 + \frac{\sqrt{5} + 3}{\sqrt{5}(2x + 1) - 5} + \frac{\sqrt{5} - 3}{\sqrt{5}(2x + 1) + 5}$$

Partial fraction decomposition: Decomposition of $f(x)$ into partial fractions. No common polynomial factors should occur in the numerator $p(x)$ and the denominator $q(x)$, and n, the degree of $p(x)$, should be smaller than m, the degree of $q(x)$.
The product representation of $q(x)$ is

$$q(x) = (x - x_1)^{n_1}(x - x_2)^{n_2} \cdots (x - x_k)^{n_k} \cdot R(x) ,$$

where the values x_i are the k zeros of multiplicity n_i of the polynomial. Let the remainder term be represented as a product of irreducible (that is, in the domain of real numbers not factorable into linear terms) quadratic polynomials:

$$R(x) = r_1(x)^{l_1} \cdot \ldots \cdot r_s(x)^{l_s} , \qquad r_k = x^2 + p_k x + q_k , \qquad 4q_k > p_k^2 .$$

For every zero x_i of multiplicity n_i the following terms, up to the power n_i, are formed:

$$\frac{A_1^i}{x - x_i}, \quad \frac{A_2^i}{(x - x_i)^2}, \quad \ldots, \quad \frac{A_n^i}{(x - x_i)^{n_i}}, \qquad i = 1, \ldots, k .$$

Further, expressions for the remainder terms of multiplicity l_j are formed:

$$\frac{B_1^j x + C_1^j}{(r_j(x))}, \quad \frac{B_2^j x + C_2^j}{(r_j(x))^2}, \quad \ldots, \quad \frac{B_l^j x + C_l^j}{(r_j(x))^l}, \qquad j = 1, \ldots, s .$$

All these terms are added:

$$f(x) = \sum_{i=1}^{k} \left(\sum_{j=1}^{n_i} \frac{A_j^i}{(x - x_i)^j} \right) + \sum_{i=1}^{s} \left(\sum_{j=1}^{l_i} \frac{B_j^i x + C_j^i}{(r_i(x))^j} \right) .$$

▷ It is usually convenient to multiply the equation by the denominator polynomial first, to obtain a polynomial equation.

The coefficients of the partial fraction decomposition can be determined by using the method of equating coefficients.

▷ However, it is easier to eliminate all coefficients except one by substituting particular values (e.g., $x = 0$, the zeros of the denominator polynomials).

□ The function $f(x)$ is given by

$$f(x) = \frac{(x+1)^2}{(x-1)^3(x-2)} .$$

Sum of partial fractions:

$$f(x) = \frac{A_1}{x-1} + \frac{A_2}{(x-1)^2} + \frac{A_3}{(x-1)^3} + \frac{B}{x-2} .$$

Equating and multiplying by the denominator polynomial yields:

$$(x+1)^2 = A_1(x-1)^2(x-2) + A_2(x-1)(x-2) + A_3(x-2) + B(x-1)^3 .$$

Substitution method:
Substituting $x = 1$ and $x = 2$ yields $-A_3 = 4$ and $B = 9$.
Substituting $x = 0$ and $A_3 = -4$, $B = 9$ yields $2 = -2A_1 + 2A_2$, and therefore, $A_2 = A_1 + 1$.
Substituting $x = 3$ yields $-54 = 6A_1$, and therefore, $A_1 = -9$, $A_2 = -8$.
Comparison of coefficients:
Arranging the right-hand side by powers yields:

$$x^2 + 2x + 1 = (A_1 + B)x^3 + (-4A_1 + A_2 - 3B)x^2 + (5A_1 - 3A_2 + A_3 + 3B)x + (-2A_1 + 2A_2 - 2A_3 - B) .$$

Comparison of coefficients
at x^3: $A_1 + B = 0$, and thus $B = -A_1$,
at x^2: $-4A_1 + A_2 - 3B = 1$, and thus $A_2 = A_1 + 1$,
at x: $5A_1 - 3A_2 + A_3 + 3B = 2$, and thus $A_3 = A_1 + 5$,
and $-2A_1 + 2A_2 - 2A_3 - B = 1$, and thus $-A_1 = 9$.
Thus, the partial fraction decomposition reads

$$f(x) = -\frac{9}{x-1} - \frac{8}{(x-1)^2} - \frac{4}{(x-1)^3} + \frac{9}{x-2} .$$

5.18.2 Padé's approximation

Let f be a real function of a real argument x given by a power series:

$$f(x) = \sum_{i=0}^{\infty} a_i x^i .$$

Padé's approximation $f_{N,M}$ is the fractional rational function

$$f_{N,M}(x) = \frac{\sum_{n=0}^{N} b_n x^n}{\sum_{m=0}^{M} c_m x^m} ,$$

whose Taylor series about $x = 0$ agrees f to up and including the $(N + M)$-th power. No generally valid rules of convergence can be given for $N, M \to \infty$ (estimate the remainder).

▷ Without loss of generality, we can assume $c_0 = 1$; otherwise the numerator and denominator are divided by c_0.

▷ Dependending on the choice of N and M, different Padé approximants can be constructed which approximate the function to the power $(N + M)$.

Given N and M, Padé's approximations can be determined in the following way: The condition that $f_{N,M}(x)$ should approximate $f(x)$ up to the order of $N + M$ yields two equation systems to determine the coefficients $\{b_i\}$ and $\{c_i\}$:

$$\sum_{m=0}^{M} c_m a_{j-m} = 0 \quad \text{for} \quad j = N+1, N+2, \ldots, N+M$$

$$\sum_{m=0}^{M} c_m a_{j-m} = b_j \quad \text{for} \quad j = 0, \ldots, N .$$

Here, $a_i = 0$ for $i < 0$ should apply.
With $c_0 = 1$, the first equation reads as follows in matrix form:

$$\begin{pmatrix} a_N & a_{N-1} & \cdots & a_{N-M+1} \\ a_{N+1} & a_N & \cdots & a_{N-M+2} \\ \vdots & \vdots & & \vdots \\ a_{N+M-1} & a_{N+M-2} & \cdots & a_N \end{pmatrix} \begin{pmatrix} c_1 \\ c_2 \\ \vdots \\ c_M \end{pmatrix} = - \begin{pmatrix} a_{N+1} \\ a_{N+2} \\ \vdots \\ a_{N+M} \end{pmatrix}$$

From this the coefficients $\{c_i\}$ can be determined (e.g., by Gauss elimination). From the second equation we obtain the coefficients $\{b_i\}$:

$$\begin{aligned} b_0 &= c_0 a_0 \\ b_1 &= c_0 a_1 + c_1 a_0 \\ &\vdots \\ b_N &= \sum_{i=0}^{\min(N,M)} c_i a_{N-i} . \end{aligned}$$

The last two equations are called **Padé equations**. Removing the expansion coefficients $\{b_i\}$ and $\{c_i\}$, $f_{N,M}$ is determined completely.

□ Let

$$f(x) = \sqrt{\frac{1 + x/2}{1 + 2x}} = 1 - \frac{3}{4}x + \frac{39}{32}x^2 + 0(x^3)$$

be given by its power series. Then, the Padé approximations up to the second order read as follows.

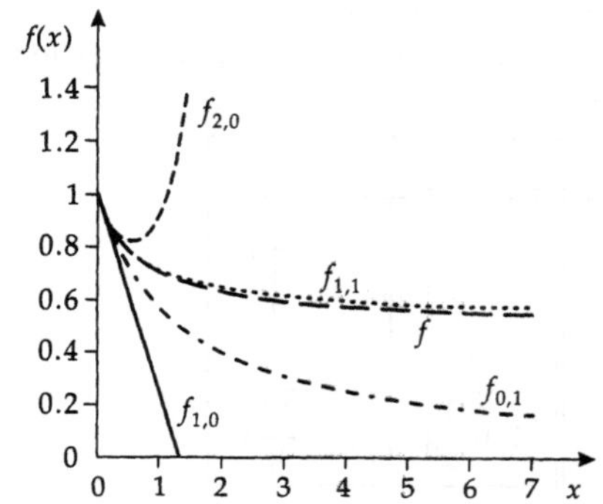

$$f_{0,0} = 1$$

$$f_{1,0} = 1 - \frac{3}{4}x$$

$$f_{0,1} = \frac{1}{1 + \frac{3}{4}x}$$

$$f_{2,0} = 1 - \frac{3}{4}x + \frac{39}{32}x^2$$

$$f_{1,1} = \frac{1 + \frac{7}{8}x}{1 + \frac{13}{8}x}$$

$$f_{0,2} = \frac{1}{1 + \frac{3}{4}x - \frac{21}{32}x^2}$$

$f_{2,0}(x)$ is identical to the power series expansion of f up to the second order.
An alternative possibility to determine the Padé approximations is to solve the equation systems for the coefficients $\{b_i\}$ or $\{c_i\}$ by applying Cramer's rule. With the definitions:

$$Q_{N,M}(x) = \begin{vmatrix} a_{N-M+1} & a_{N-M+2} & \cdots & a_{N+1} \\ a_{N-M+2} & a_{N-M+3} & \cdots & a_{N+2} \\ \vdots & \vdots & & \vdots \\ a_N & a_{N+1} & \cdots & a_{N+M} \\ x^M & x^{M-1} & \cdots & 1 \end{vmatrix}$$

$$P_{N,M}(x) = \begin{vmatrix} a_{N-M+1} & a_{N-M+2} & \cdots & a_{N+1} \\ a_{N-M+2} & a_{N-M+3} & \cdots & a_{N+2} \\ \vdots & \vdots & & \vdots \\ a_N & a_{N+1} & \cdots & a_{N+M} \\ \sum\limits_{i=0}^{N-M} a_i x^{i+M} & \sum\limits_{i=0}^{N-M+1} a_i x^{i+M-1} & \cdots & \sum\limits_{i=0}^{N} a_i x^i \end{vmatrix}$$

we obtain

$$\sum_{i=0}^{\infty} a_i x^i = \frac{P_{N,M}(x)}{Q_{N,M}(x)} + \sum_{i=N+M+1}^{\infty} x^i \begin{vmatrix} a_{N-M+1} & a_{N-M+2} & \cdots & a_{N+1} \\ a_{N-M+2} & a_{N-M+3} & \cdots & a_{N+2} \\ \vdots & \vdots & & \vdots \\ a_N & a_{N+1} & \cdots & a_{N+M} \\ a_{i-M} & a_{i-M+1} & \cdots & a_i \end{vmatrix} .$$

These relations are less suitable for computations than for theoretical considerations. Occasionally, they can be used for the estimation of errors.
The Padé approximations are often listed in a so-called **Padé table**:

$N =$	0	1	2	$\dots$
$M =$ 0	$f_{0,0}$	$f_{1,0}$	$f_{2,0}$	$\dots$
1	$f_{0,1}$	$f_{1,1}$	$f_{2,1}$	$\dots$
2	$f_{0,2}$	$f_{1,2}$	$f_{2,2}$	$\dots$
$\vdots$	$\vdots$	$\vdots$	$\vdots$	

□ The Padé table for the function in the last example reads as follows up to $N = 1$, $M = 1$:

$N =$	0	1
$M =$ 0	1	$1 - \frac{3}{4}x$
1	$\dfrac{1}{1 + \frac{3}{4}x}$	$\dfrac{1 + \frac{7}{8}x}{1 + \frac{13}{8}x}$

▷ Applications of the Padé approximation can be found in many areas of physics and chemistry. In particular, the Padé approximation is used when a functional relationship must be formed from the first terms of a perturbation expansion.

Irrational algebraic functions

Algebraic functions can be represented as a solution of an equation with coefficients which are the polynomials of the independent variables.

□ For $x \geq 0$, the equation $y \cdot y = x$ defines the root function $y = \sqrt{x}$.

An implicit algebraic function is defined, for example, by $y = a_0 + a_1 \cdot x + a_2 \cdot x \cdot x \cdots$. Below, we describe functions which are algebraic, but cannot be represented as a quotient or a (finite) sum of power functions. At first we describe square-root functions, then root functions, which are defined as the inverse functions of the power functions of higher degree, as well as power functions with fractional exponents, and finally, root functions whose radicands are rational functions. The relationships to conics are also pointed out.

5.19 Square-root function

$$f(x) = \sqrt{ax+b}$$

In **programming languages** `SQRT(A*X+B)`, sometimes also `SQR(A*X+B)`
Square root, sometimes called second root or, simply, root.
The inverse function of the quadratic function.
Of importance in geometry for calculating segments.
Of importance in vector analysis for forming the absolute values of vectors.
Of importance in statistics for adding the errors in independent measured variables.

(a) Definition

Inverse function of the quadratic function, on the positive branch:

$$y = f(x) = \frac{1}{a}x^2 - \frac{b}{a}, \qquad f^{-1}(x) = \sqrt{ax+b}\,.$$

Notation as a power function with fractional exponent:

$$f(x) = \sqrt{x} = x^{1/2}\,.$$

(b) Graphical representation

● For $a > 0$ $(a < 0)$ the square-root function is defined only for values $x \geq -\dfrac{b}{a}$ $\left(x \leq -\dfrac{b}{a}\right)$ only.

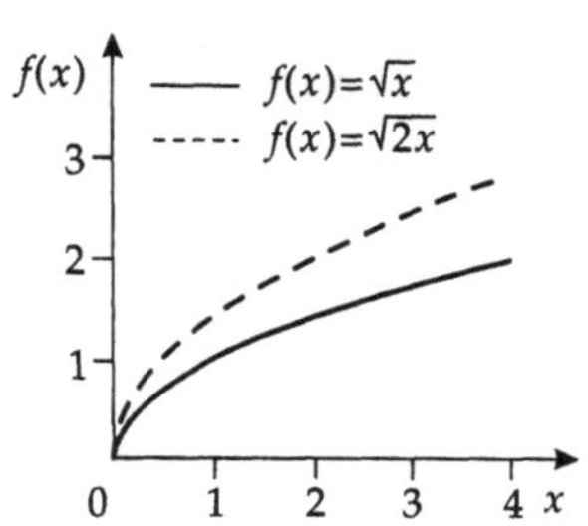

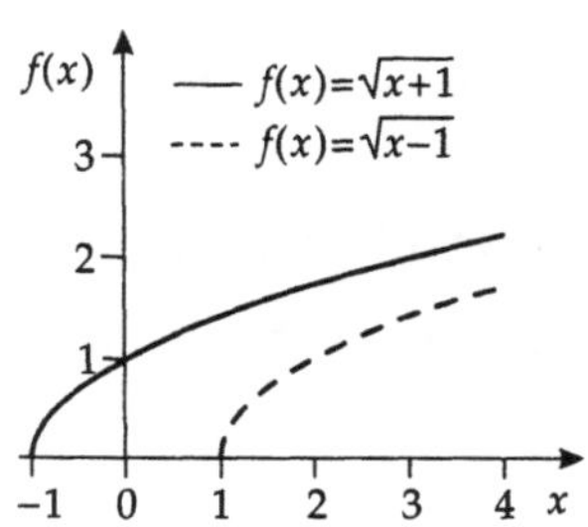

The function is zero at $x = -\frac{b}{a}$. At this point, the graph has a vertical tangent.

▷ The slope becomes infinite at $x = -\frac{b}{a}$.

For $a > 0$ ($a < 0$) the function is strictly monotonically increasing (and for $a < 0$ strictly monotonically decreasing). For large values of x, the function rises more slowly than any linear function $g(x) = ax,\ a > 0$.
For larger $a > 0$, the function rises faster.
At $x = 0$, the function has the value $\sqrt{b}$ (for $b > 0$).

(c) Properties of the function

Domain of definition:	$a > 0$: $-\frac{b}{a} \le x < \infty$ $a < 0$: $-\infty < x \le -\frac{b}{a}$
Range of values:	$0 \le f(x) < \infty$.
Quadrant:	$a > 0$ $b < 0$ first quadrant $b \ge 0$ first and second quadrants $a < 0$ $b \le 0$ second quadrant $b > 0$ first and second quadrants
Periodicity:	None.
Monotonicity:	$a > 0$: strictly monotonically increasing $a < 0$: strictly monotonically decreasing
Symmetries:	None.
Asymptotes:	$a > 0$: $f(x) \to \infty$ as $x \to \infty$ $a < 0$: $f(x) \to \infty$ as $x \to -\infty$

(d) Particular values

Zeros:	$x = -\frac{b}{a}$.
Discontinuities:	None, the function is continuous everywhere.
Poles:	None.
Extrema:	None, the function is strictly monotonic.
Points of inflection:	None, the function is concave everywhere.

(e) Reciprocal function

The reciprocal function is the root of the reciprocal:

$$\frac{1}{f(x)} = \frac{1}{\sqrt{x}} = \sqrt{\frac{1}{x}} = x^{-1/2} .$$

(f) Inverse function

The inverse function of the square-root function is the quadratic function:

$$f^{-1}(x) = \frac{1}{a}x^2 - \frac{1}{b}\,.$$

▷ For the inverse function only one branch of the parabola (generally the positive) is defined.

(g) Cognate functions

General root function, in analogy to the square root, the n-th root is the inverse function of the power function of degree n.
Quadratic function, the inverse function.
Powers of the root function, the derivative and primitive of the root function.

(h) Conversion formulas

The root of a (positive) product is the product of the roots of the absolute values:

$$\sqrt{x \cdot y} = \sqrt{|x|} \cdot \sqrt{|y|}\,.$$

The root of a (positive) quotient is the quotient of the roots of the absolute values:

$$\sqrt{\frac{x}{y}} = \frac{\sqrt{|x|}}{\sqrt{|y|}}\,.$$

▷ The absolute values are important, since both x and y can be negative.

(i) Approximation formulas (8 bits ≈ 0.4% of precision)

For small values of x it holds that

$$\sqrt{ax+b} \approx \sqrt{b} + \frac{ax}{2\sqrt{b}}, \qquad \text{for } |x| < \frac{b}{7|a|}\,.$$

For large values of x it holds that

$$\sqrt{ax+b} \approx \sqrt{ax} + \frac{b}{2\sqrt{ax}}, \qquad \text{for } ax > 7|b|\,.$$

(j) Expansion of series or products of the function

Using generalized binomial coefficients or double factorials, the following expansion applies to $b > 0$ and $|ax| < b$:

$$\begin{aligned}
\sqrt{ax+b} &= \sqrt{b}\left(1 + \frac{ax}{2b} - \frac{a^2x^2}{8b^2} + \frac{a^3x^3}{16b^3} - \ldots\right) \\
&= \sqrt{b}\sum_{k=0}^{\infty}\binom{\frac{1}{2}}{k}\left(\frac{ax}{b}\right)^k \\
&= \sqrt{b} - \sqrt{b}\sum_{k=1}^{\infty}\frac{(2k-3)!!}{(2k)!!}\left(-\frac{ax}{b}\right)^k, \qquad -b \le ax \le b\,.
\end{aligned}$$

For large values of x it holds for $ax > |b| \geq 0$ that

$$\sqrt{ax+b} = \sqrt{ax}\sum_{k=0}^{\infty}\binom{-\frac{1}{2}}{k}\left(\frac{b}{ax}\right)^k , \qquad -ax \leq c \leq ax .$$

Expansion of $\sqrt{x}$ with an exponential function that converges rapidly:

$$\sqrt{x} = \frac{\frac{1}{2} + \sum_{k=1}^{\infty} e^{-k^2\pi/x}}{\frac{1}{2} + \sum_{k=1}^{\infty} e^{-k^2\pi x}}$$

(k) Derivative of the function

The derivative is a reciprocal square root

$$\frac{d}{dx}\sqrt{ax+b} = \frac{a}{2\sqrt{ax+b}} .$$

(l) Primitive of the function

The primitive is the third power of the square root

$$\int_{-b/a}^{x} \sqrt{at+b}\,dt = \frac{2}{3a}\left(\sqrt{ax+b}\right)^3 .$$

(m) Particular extensions and applications

Roots of negative numbers: In the domain of real numbers, the roots of negative numbers are not defined. In the complex plane, the imaginary unit is defined by the roots of negative numbers.

$$j \stackrel{def}{=} \sqrt{-1}$$

▷ Thus all roots of negative numbers $a < 0$ can be represented:

$$\sqrt{a} = \sqrt{(-1)\cdot(-a)} = j \cdot \sqrt{-a} , \qquad a < 0 .$$

Complex argument:

$$\sqrt{x+jy} = \sqrt{\frac{1}{2}\left(\sqrt{x^2+y^2}+x\right)} + j \cdot \mathrm{sgn}(y)\sqrt{\frac{1}{2}\left(\sqrt{x^2+y^2}-x\right)} .$$

5.20 Root function

$$f(x) = \sqrt[n]{x}$$

General n-th root.

(a) Definition

Inverse function of the power function of degree n:

$$y = f(x) = x^n , \qquad n > 1 \quad \text{integer,} \qquad f^{-1}(x) = \sqrt[n]{x} .$$

Notation as a power:

$$f(x) = \sqrt[n]{x} = x^{1/n} .$$

Negative exponents:

$$f(x) = x^{-1/n} = \left(\frac{1}{x}\right)^{1/n} = \sqrt[n]{\frac{1}{x}} .$$

First, we define the function only for $x \geq 0$.

▷ For odd n it can be also represented for negative x values according to $f(x) = -f(-x)$:

$$f(x) = \begin{cases} \sqrt[n]{x} & x \geq 0 \\ -\sqrt[n]{-x} & x < 0 \end{cases} , \qquad n \text{ odd} .$$

(b) Graphical representation

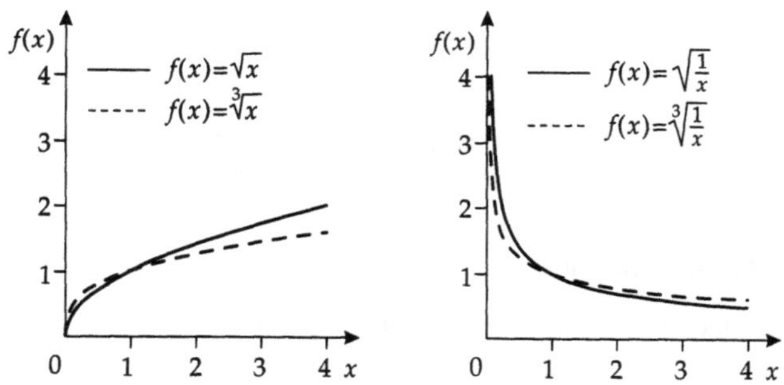

The functions with $n > 1$ are monotonically increasing; they pass the points $x = 1, f(x) = 1$ and $x = 0, f(x) = 0$.

▷ Graphs of $f(x) = \sqrt[n]{x}$ intersect the x-axis at the point $(x = 0, y = 0)$, perpendicular to the axis.

For greater n the curve rises faster between $x = 0$ and $x = 1$ and more slowly for $x > 1$ than for smaller n.

For $n < -1$ the graphs of the function look like a hyperbola decreasing from the pole at $x = 0$ and passing through $x = 1, f(x) = 1$.

(c) Properties of the function

Domain of definition:	n even	$n > 0$: $0 \leq x < \infty$
		$n < 0$: $0 < x < \infty$
	n odd	$n > 0$: $-\infty < x < \infty$
		$n < 0$: $-\infty < x < \infty, x \neq 0$
Range of values:	n even	$n > 0$: $0 \leq f(x) < \infty$
		$n < 0$: $0 < f(x) < \infty$
	n odd	$n > 0$: $-\infty < f(x) < \infty$
		$n < 0$: $-\infty < f(x) < \infty, x \neq 0$
Quadrant:	n even	first quadrant
	n odd	first and third quadrants
Periodicity:	None.	

Monotonicity:	$n > 0$:	strictly monotonically increasing
	$n < 0$:	strictly monotonically decreasing
Symmetries:	n even:	none
	n odd:	point symmetry about the origin
Asymptotes:	n even	$n > 0$: $f(x) \to \infty$ as $x \to \infty$
		$n < 0$: $f(x) \to 0$ as $x \to \infty$
	n odd	$n > 0$: $f(x) \to \pm\infty$ as $x \to \pm\infty$
		$n < 0$: $f(x) \to \pm 0$ as $x \to \pm\infty$

(d) Particular values

Zeros:	$n > 0$:	$x = 0$
	$n < 0$:	none
Discontinuities:	None.	
Poles:	$n > 0$:	none
	$n < 0$:	$x = 0$
Extremes:	None.	
Points of inflection:	$n > 0$ odd: a point of inflection of infinite slope at $x = 0$.	
	otherwise: no point of inflection	

(e) Reciprocal function

The reciprocal function changes the sign in the exponent:

$$\frac{1}{f(x)} = \frac{1}{x^{1/n}} = x^{-1/n} .$$

(f) Inverse function

The inverse function is a power function with the exponent n:

$$f^{-1}(x) = x^n .$$

(g) Cognate functions

Power functions with positive and negative exponents.
Power functions with a fractional exponent.

(h) Conversion formulas

Nesting of roots:

$$\sqrt[n]{\sqrt[m]{x}} = \sqrt[n\cdot m]{x} .$$

Multiplication of roots:

$$\sqrt[n]{x} \cdot \sqrt[m]{x} = \sqrt[n\cdot m]{x^{n+m}} .$$

Division of roots:

$$\frac{\sqrt[n]{x}}{\sqrt[m]{x}} = \sqrt[n\cdot m]{x^{m-n}} .$$

(i) Approximation formulas (8 bits ≈ 0.4% of precision)

For roots of higher order and x close to 1 it holds that

$$\sqrt[n]{x} = 1 + \frac{2(x-1)}{n(x+1)}\,, \qquad |n| \geq 5, \quad 0.7 \leq x \leq 1.4\,.$$

(j) Expansion of series or products of the function

Expansion of $\sqrt{x}$ with exponential functions which converges rapidly:

$$\sqrt{x} = \frac{\frac{1}{2} + \sum_{k=1}^{\infty} \mathrm{e}^{-k^2\pi/x}}{\frac{1}{2} + \sum_{k=1}^{\infty} \mathrm{e}^{-k^2\pi x}}\,.$$

(k) Derivative of the function

The derivative of a root function is the root function of a power:

$$\frac{\mathrm{d}}{\mathrm{d}x}\sqrt[n]{x} = \frac{1}{n}\sqrt[n]{x^{1-n}}\,.$$

Analogously, for the reciprocal function it holds that

$$\frac{\mathrm{d}}{\mathrm{d}x}\sqrt[n]{\frac{1}{x}} = -\frac{1}{n}\sqrt[n]{\left(\frac{1}{x}\right)^{n+1}}\,.$$

(l) Primitive of the function

The primitive of the root function is the root function of a power:

$$\int_0^x \sqrt[n]{t}\,\mathrm{d}t = \frac{n}{n+1}\sqrt[n]{x^{n+1}}\,.$$

- The integral from zero to x of the reciprocal function is finite, at $x = 0$ in spite of the pole.

$$\int_0^x \sqrt[n]{\frac{1}{t}}\,\mathrm{d}t = \frac{n}{n-1}\sqrt[n]{x^{n-1}}$$

(m) Particular extensions and applications

Root in the complex domain: In the domain of complex numbers, the real root of a number is only the principal value of the root.

- In the complex domain there are n complex values for an n-th root.

If $A = \sqrt[n]{|a|}$, $a \in \mathbb{R}$, be the real n-th root of $|a|$; in the complex domain it holds that

$$\sqrt[n]{a} = A \cdot \mathrm{e}^{\mathrm{j}\cdot k/n 2\pi}\,, \qquad k = 0, 1, \ldots, n-1\,.$$

Root of a complex number: Let a complex number z be represented by

$$z = x + \mathrm{j}y = \rho \cdot \mathrm{e}^{\mathrm{j}\cdot\varphi}, \qquad \rho = \sqrt{x^2 + y^2}, \quad \tan(\varphi) = \frac{y}{x}\,.$$

The n-th root is

$$\sqrt[n]{z} = \sqrt[n]{\rho} \cdot e^{j \cdot \varphi + 2\pi k/n}, \qquad k = 0, 1, \ldots, n-1 .$$

(See also Chapter 17 on complex numbers.)

5.21 Power functions with fractional exponents

$$f(x) = x^{m/n}$$

These are the most general algebraic power functions.
There are many applications in technology where powers of roots have to be calculated.

(a) Definition

Representation of the function with a reduced fraction as an exponent.
Interpretation as the n-th root of the m-th power:

$$f(x) = x^{m/n} = \sqrt[n]{x^m} .$$

- For this interpretation, we assume that $n > 0$, and that $m \neq 0$ may take positive and negative values (m and n integer).

If m and n are odd, the function can be defined for all x, and it can have positive and negative values.
If m is even and n is odd, the function can be defined for all x, but it does not have negative values.
If m is odd and n is even, the function can be defined only for $x \geq 0$ and it does not have negative values.

▷ The case for m and n even does not apply, since the fraction $\frac{m}{n}$ can still be reduced by two.

(b) Graphical representation

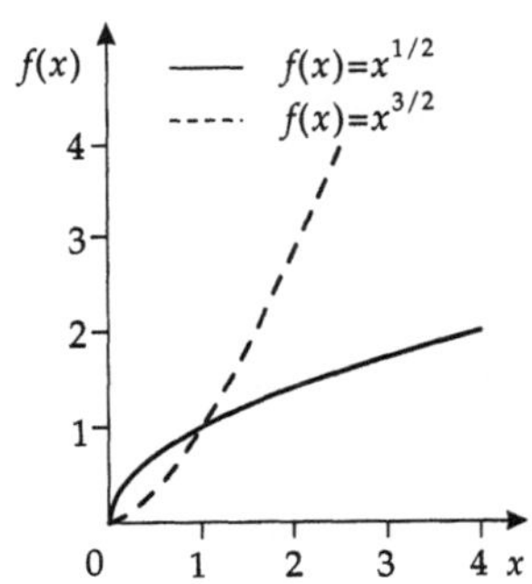

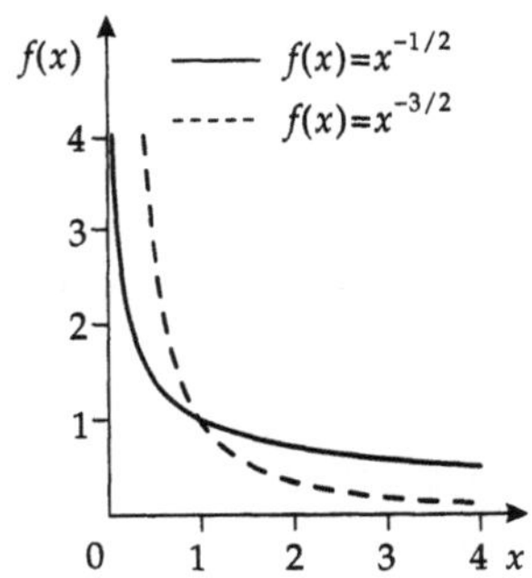

Functions (left) and corresponding reciprocal functions (right)

If $m > 0$, the function is strictly monotonically increasing for $x \geq 0$.
For $m > n$, the curve is convex, for $(0 < m < n)$, it is concave (for $x \geq 0$ in both cases).

For $m < 0$, the curve looks like a hyperbola. It is convex and strictly monotonically decreasing in the $x > 0$ domain.

The behavior for $x < 0$ follows from m and n:

- m and n odd: The function is point-symmetric about the origin. For $0 < m < n$, the graph intersects the x-axis perpendicularly; for $m > n$ it approaches horizontally.
- m even and n odd: The function is mirror-symmetric about the y-axis. If $0 < m < n$, a break appears at $x = 0$.
- m odd and n even: The function is not defined for $x < 0$.

□ Neil's parabola (semicubical parabola) with $m = 3$, $n = 2$ (see figure above).

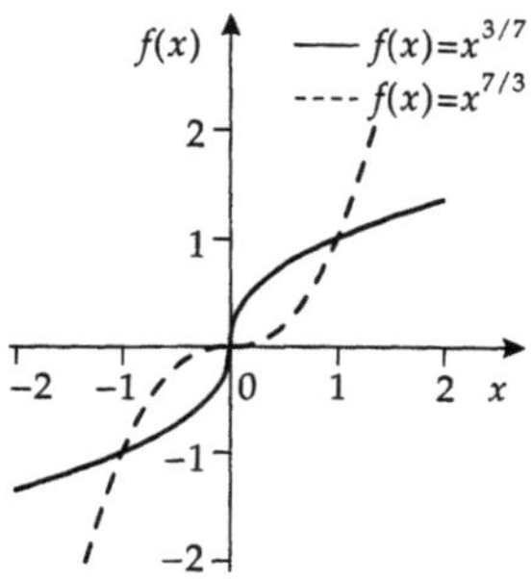

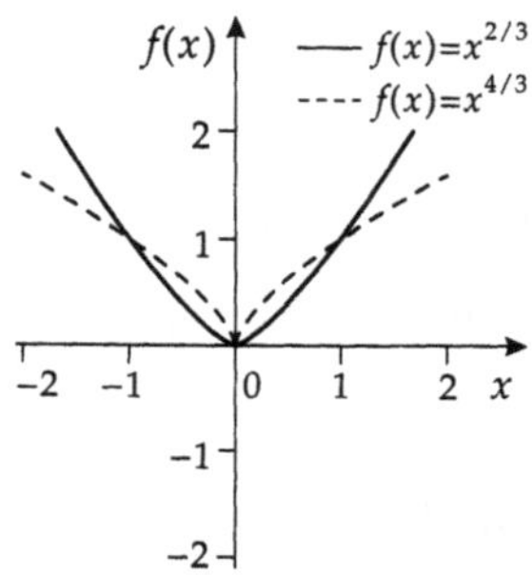

(c) Properties of the function

Domain of definition:	For $m < 0$, $x = 0$ has to be excluded. n odd: $-\infty < x < \infty$ n even: $0 \le x < \infty$
Range of values:	For $m < 0$, $f(x) = 0$ is excluded. m, n odd: $-\infty < f(x) < \infty$ m or n even: $0 < f(x) < \infty$
Quadrant:	m, n odd: first and third quadrants m even: first and second quadrants n even: first quadrant
Periodicity:	None.
Monotonicity:	$x > 0, m > 0$: strictly monotonically increasing $x > 0, m < 0$: strictly monotonically decreasing $x < 0$: according to the symmetries
Symmetries:	m, n odd: point symmetry about the origin m even: mirror symmetry about the y-axis n even: no symmetry
Asymptotes:	$m < 0$: $f(x) \to 0$ as $x \to \infty$ $m > 0$: $f(x) \to \infty$ as $x \to \infty$

(d) Particular values

Zeros:	$m > 0$: $x = 0$ $m < 0$: none

Discontinuities:	None.
Poles:	$m > 0$: none $m < 0$: $x = 0$
Extrema:	m even, $m > n$, minimum at $x = 0$.
Points of inflection:	m, n odd, $m > 2n$, saddle point at $x = 0$.

(e) Reciprocal function

The reciprocal function changes the sign in the exponent:

$$\frac{1}{f(x)} = \frac{1}{x^{m/n}} = x^{-m/n} .$$

(f) Inverse function

The inverse function has the reciprocal value in the exponent:

$$f^{-1}(x) = x^{n/m} .$$

(g) Cognate functions

All roots and power functions as well as reciprocal roots and reciprocal power functions can be regarded as special cases.

(h) Conversion formulas

Product of two functions:

$$x^a \cdot x^b = x^{a+b}, \qquad a = \frac{m_1}{n_1}, \quad b = \frac{m_2}{n_2} .$$

Quotient of two functions:

$$\frac{x^a}{x^b} = x^{a-b}, \qquad a = \frac{m_1}{n_1}, \quad b = \frac{m_2}{n_2} .$$

Power of a function:

$$\left(x^a\right)^b = x^{a \cdot b}, \qquad a = \frac{m_1}{n_1}, \quad b = \frac{m_2}{n_2} .$$

(i) Approximation formulas (8 bits ≈ 0.4% of precision)

For small $a = \dfrac{m}{n}$ and x close to 1:

$$x^a = 1 + \frac{2a(x-1)}{x+1}, \qquad |a| = \left|\frac{m}{n}\right| \le 0.2, \quad 0.7 \le x \le 1.4 .$$

(j) Expansion of series or products of the function

- Infinite binomial series with fractional exponent:

$$x^a = (1 + (x-1))^a , \qquad a = \frac{m}{n} , \qquad 0 < x < 2$$

$$= 1 + a(x-1) + \frac{a(a-1)}{2}(x-1)^2 + \frac{a(a-1)(a-2)}{3!}(x-1)^3 + \cdots$$

$$= \sum_{k=0}^{\infty} \binom{a}{k} (x-1)^k \,, \qquad \binom{a}{k} = \frac{a(a-1)\cdots(a-k+1)}{k(k-1)\cdots 1} \,.$$

▷ For integral exponents, the series terminates, since for $k > n$ each term contains a factor $k - k = 0$.

(k) Derivative of the function

Differentiation reduces the exponent by one:

$$\frac{\mathrm{d}}{\mathrm{d}x} x^a = a x^{a-1} \,, \qquad a = \frac{m}{n} \,.$$

(l) Primitive of the function

For $a > -1$ it holds that

$$\int_0^x t^a \mathrm{d}t = \frac{x^{a+1}}{a+1} \,, \qquad a = \frac{m}{n} \,.$$

For $a < -1$ it holds that

$$\int_x^{\infty} t^a \mathrm{d}t = \frac{-x^{a+1}}{a+1} \,.$$

▷ For $a = -1$ it holds that

$$\int_1^x t^{-1} \mathrm{d}t = \ln(x) \,.$$

(m) Particular extensions and applications

Extension to nonrational exponents: If a sequence of rational numbers a_n exists whose limit is an irrational number a,

$$a_n, \quad a_n \text{ rational, with } \lim_{n\to\infty} a_n = a \,,$$

the function x^a can be defined as the limit of the function sequences,

$$x^a = \lim_{n\to\infty} x^{a_n} \,.$$

▷ Then the function is no longer algebraic.
Alternative definition of the general power function by means of transcendental functions:

$$x^a = \mathrm{e}^{a \ln(x)} \,.$$

5.22 Roots of rational functions

$$f(x) = \sqrt{r(x)}, \qquad r(x) = \text{ rational function}$$

Square roots of polynomials.
Of importance for the representation of conics.

(a) Definition

General representation as a root of a rational function:

$$f(x) = \sqrt{r(x)} = \sqrt{\frac{p(x)}{q(x)}} = \sqrt{\frac{\sum_{k=0}^{n} a_k x^k}{\sum_{k=0}^{m} b_k x^k}} .$$

Special cases:

$$p(x) = (x^2 + a^2), \quad q(x) = 1 \qquad \text{and} \qquad p(x) = (a^2 - x^2), \quad q(x) = 1,$$

and their reciprocal functions

$$p(x) = 1 \quad q(x) = (x^2 + a^2) \qquad \text{and} \qquad p(x) = 1, \quad q(x) = (a^2 - x^2),$$

▷ The curve for $p(x) = a^6, q(x) = (x^2 + a^2)^2$ is called the **witch of Agnesi**.

(b) Graphical representation

The function $f(x)$ is determined essentially by the behavior of $p(x)$ and $q(x)$.
The zeros of $p(x)$ are the zeros of $f(x)$, and the zeros of $q(x)$ are the poles of $f(x)$.

▷ If the numerator and denominator polynomials have zeros at the same point x_0, the multiplicity of zeros has to be observed to decide whether it is a pole or a removable gap.

● The function is defined only when $p(x)$ and $q(x)$ have the same sign.

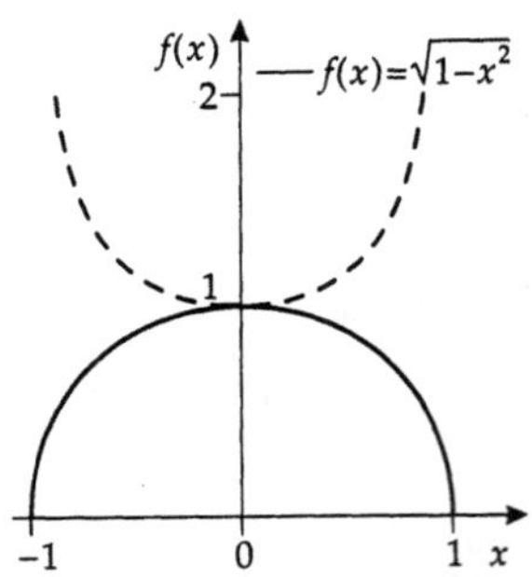

Reciprocal function (dotted line)

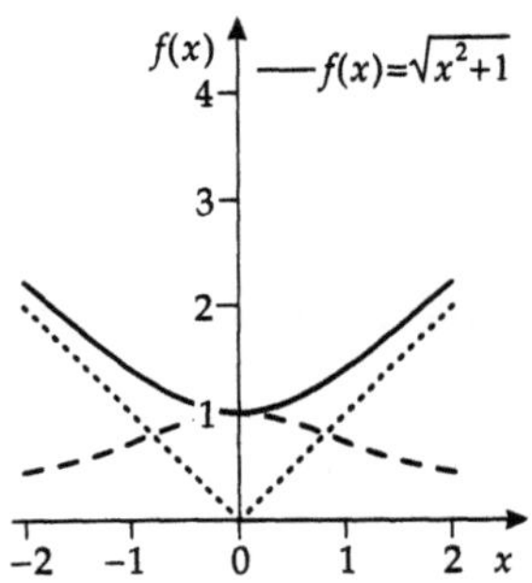

Reciprocal function (dotted line) and asymptote (pointed line)

The function $f(x) = \sqrt{a^2 - x^2}$ describes the upper half of a circle of radius $|a|$.
Its reciprocal function has a minimum at $x = 0$ and diverges at $x = \pm a$.
Function $f(x) = \sqrt{x^2 + a^2}$ describes the positive branches of a hyperbola with the asymptotes $y = \pm x$.
Its reciprocal function has a maximum at $x = 0$ and tends to zero as $x \to \pm\infty$.

(c) Properties of the function

Domain of definition: For $q(x) \neq 0$:

	$\sqrt{a^2 - x^2}$	$-a \le x \le a$
	$(\sqrt{a^2 - x^2})^{-1}$	$-a < x < a$
	$\sqrt{x^2 + a^2}$	$-\infty < x < \infty$
	$(\sqrt{x^2 + a^2})^{-1}$	$-\infty < x < \infty$

Range of values:	Range of positive values, depends on $p(x)$, $q(x)$:	
	$\sqrt{a^2 - x^2}$	$0 \leq f(x) \leq \lvert a\rvert$
	$(\sqrt{a^2 - x^2})^{-1}$	$\frac{1}{\lvert a\rvert} \leq f(x) < \infty$
	$\sqrt{x^2 + a^2}$	$\lvert a\rvert \leq f(x) < \infty$
	$(\sqrt{x^2 + a^2})^{-1}$	$0 < f(x) \leq \frac{1}{\lvert a\rvert}$

Quadrant: Generally the first and second quadrant; thus also for $(\sqrt{a^2 - x^2})^{\pm 1}$ and $(\sqrt{x^2 + a^2})^{\pm 1}$.

Periodicity: Generally periodicity; thus also for $(\sqrt{a^2 - x^2})^{\pm 1}$ and $(\sqrt{x^2 + a^2})^{\pm 1}$.

Monotonicity:	Depends on $p(x)$, $q(x)$, for $x > 0$:	
	$\sqrt{a^2 - x^2}$	strictly monotonically decreasing
	$(\sqrt{a^2 - x^2})^{-1}$	strictly monotonically increasing
	$\sqrt{x^2 + a^2}$	strictly monotonically increasing
	$(\sqrt{x^2 + a^2})^{-1}$	strictly monotonically decreasing

Symmetries: If $\frac{p(x)}{q(x)}$ has mirror symmetry, if $f(x)$ has it also. Mirror symmetry about the y-axis for $(\sqrt{a^2 - x^2})^{\pm 1}$ and $(\sqrt{x^2 + a^2})^{\pm 1}$.

Asymptotes:	$(\sqrt{a^2 - x^2})^{-1}$	$f(x) \to \infty$ as $x \to \pm a$
	$\sqrt{x^2 + a^2}$	$f(x) \to \lvert x\rvert$ as $x \to \pm\infty$
	$(\sqrt{x^2 + a^2})^{-1}$	$f(x) \to 0$ as $x \to \pm\infty$

(d) Particular values

Zeros:	Zeros of $p(x)$:	
	$\sqrt{a^2 - x^2}$	$x = \pm a$
	$(\sqrt{a^2 - x^2})^{-1}$	none
	$\sqrt{x^2 + a^2}$	no real zeros
	$(\sqrt{x^2 + a^2})^{-1}$	none

Discontinuities: Generally none, thus also for $(\sqrt{a^2 - x^2})^{\pm 1}$ and $(\sqrt{x^2 + a^2})^{\pm 1}$

Poles:	Zeros of $q(x)$;	
	$\sqrt{a^2 - x^2}$	none
	$(\sqrt{a^2 - x^2})^{-1}$	$x = \pm a$
	$\sqrt{x^2 + a^2}$	none
	$(\sqrt{x^2 + a^2})^{-1}$	no real poles.

Extremes: Extrema of $\frac{p(x)}{q(x)}$ with $q(x) \neq 0$;

	$\sqrt{a^2 - x^2}$	maximum at $x = 0$
	$(\sqrt{a^2 - x^2})^{-1}$	minimum at $x = 0$
	$\sqrt{x^2 + a^2}$	minimum at $x = 0$
	$(\sqrt{x^2 + a^2})^{-1}$	maximum at $x = 0$

Points of inflection: Depend on $p(x)$ and $q(x)$; $(\sqrt{a^2 - x^2})^{\pm 1}$ and $(\sqrt{x^2 + a^2})^{\pm 1}$ have no points of inflection.

Value at $x = 0$: $f(0) = \sqrt{\dfrac{a_0}{b_0}}$

$\sqrt{a^2 - x^2}$	$f(0) = \lvert a \rvert$
$(\sqrt{a^2 - x^2})^{-1}$	$f(0) = \dfrac{1}{\lvert a \rvert}$
$\sqrt{x^2 + a^2}$	$f(0) = \lvert a \rvert$
$(\sqrt{x^2 + a^2})^{-1}$	$f(0) = \dfrac{1}{\lvert a \rvert}$

(e) Reciprocal function

The reciprocal function is the root with numerator and denominator interchanged:

$$\frac{1}{f(x)} = \frac{1}{\sqrt{\frac{p(x)}{q(x)}}} = \sqrt{\frac{q(x)}{p(x)}} .$$

The exemplified functions are pairwise reciprocal.

(f) Inverse function

The inverse functions of the hyperbola and the circle are a hyperbola and circles. They are defined only for positive values, but may be extended also to negative values.

▷ Hyperbolae are the mirror images about the angle-bisector $y = x$. The new hyperbola intersects the x-axis at $x = |a|$.

$$y = f(x) = \sqrt{x^2 + a^2}\,, \qquad f^{-1}(x) = \sqrt{x^2 - a^2}\,.$$

The circle is its own inverse function:

$$y = f(x) = \sqrt{a^2 - x^2}\,, \qquad f^{-1}(x) = \sqrt{a^2 - x^2}\,.$$

(g) Cognate functions

Root function, special case $p(x) = x, q(x) = 1$.
Absolute value function, special case $p(x) = x^2, q(x) = 1$.

- The trigonometric functions and their inverse functions are closely related to the function $\sqrt{a^2 - x^2}$.
- The hyperbolic functions and their inverse functions are closely related to the function $\sqrt{x^2 + a^2}$.

▷ See Section 5.32 on trigonometric functions and Sections 5.15 and 5.26 on hyperbolic functions.

(h) Conversion formulas

For the exemplified function it holds that

$$f(-x) = f(x) .$$

For products in the argument it holds that

$$(\sqrt{a^2 - (cx)^2})^{\pm 1} = |c|^{\pm 1} \left(\sqrt{\left(\frac{a}{c}\right)^2 - x^2} \right)^{\pm 1} .$$

▷ In this way, for example, ellipses $y = \sqrt{a^2 - b^{-2}x^2}$ can be defined.
Analogously, it holds that

$$(\sqrt{(cx)^2 + a^2})^{\pm 1} = |c|^{\pm 1} \left(\sqrt{x^2 + \left(\frac{a}{c}\right)^2} \right)^{\pm 1} .$$

(i) Approximation formulas (8 bits ≈ 0.4% of precision)

For small x it holds that

$$\sqrt{a^2 - x^2} \approx \frac{2a^2 - x^2}{2|a|} , \qquad |x| < 0,15|a|$$

$$\frac{1}{\sqrt{a^2 - x^2}} \approx \frac{2a^2 + x^2}{2|a|^3} , \qquad |x| < 0,1|a|$$

$$\sqrt{x^2 + a^2} \approx \frac{2a^2 + x^2}{2|a|} , \qquad |x| \le 0,4|a|$$

$$\frac{1}{\sqrt{x^2 + a^2}} \approx \frac{2a^2 - x^2}{2|a|^3} , \qquad |x| \le 0,3|a| ,$$

and for large values it holds that

$$\sqrt{x^2 \pm a^2} \approx x \pm \frac{a^2}{2x} , \qquad |x| \ge 2,5|a|$$

$$\frac{1}{\sqrt{x^2 \pm a^2}} \approx \frac{2x^2 \mp a^2}{2|x|^3} , \qquad |x| \ge 3,3|a| .$$

(j) Expansion of series or products of the functions

Expansion of $\sqrt{a^2 - x^2}$ in a series:

$$\sqrt{a^2 - x^2} = |a| \left(1 - \frac{x^2}{2a^2} - \frac{x^4}{8a^4} - \frac{x^6}{16a^6} - \dots \right)$$
$$= |a| \sum_{k=0}^{\infty} \binom{\frac{1}{2}}{k} \left(\frac{x}{a}\right)^{2k} , \qquad -a \le x \le a .$$

Analogously, it holds for the reciprocal function that

$$\frac{1}{\sqrt{a^2 - x^2}} = \frac{1}{|a|} \sum_{k=0}^{\infty} \binom{-\frac{1}{2}}{k} \left(\frac{x}{a}\right)^{2k} , \qquad -a < x < a .$$

(k) Derivative of the function

▷ The derivative is taken by means of the chain rule:

$$\frac{\mathrm{d}}{\mathrm{d}x}\sqrt{r(x)} = \frac{1}{2\sqrt{r(x)}}\frac{\mathrm{d}}{\mathrm{d}x}r(x) .$$

Derivative of $\sqrt{a^2 - x^2}$:

$$\frac{\mathrm{d}}{\mathrm{d}x}\sqrt{a^2 - x^2} = -\frac{x}{\sqrt{a^2 - x^2}} .$$

Derivative of $(\sqrt{a^2 - x^2})^{-1}$:

$$\frac{\mathrm{d}}{\mathrm{d}x}\frac{1}{\sqrt{a^2 - x^2}} = \frac{x}{(\sqrt{a^2 - x^2})^3} .$$

Derivative of $\sqrt{x^2 + a^2}$:

$$\frac{\mathrm{d}}{\mathrm{d}x}\sqrt{x^2 + a^2} = \frac{x}{\sqrt{x^2 + a^2}} .$$

Derivative of $(\sqrt{x^2 + a^2})^{-1}$:

$$\frac{\mathrm{d}}{\mathrm{d}x}\frac{1}{\sqrt{x^2 + a^2}} = -\frac{x}{(\sqrt{x^2 + a^2})^3} .$$

(l) Primitive of the function

No general integration formula can be given. See the tables of integrals at the end of this book (Chapter 25) or other tables of integrals.
The primitives of the exemplified functions contain inverse trigonometric and the hyperbolic functions.
Primitive of $\sqrt{a^2 - x^2}$ for $-a \le x \le a$:

$$\int_0^x \sqrt{a^2 - t^2}\,\mathrm{d}t = \frac{x}{2}\sqrt{a^2 - x^2} + \frac{a^2}{2}\arcsin\left(\frac{x}{a}\right) .$$

Primitive of $(\sqrt{a^2 - x^2})^{-1}$ for $-a \le x \le a$:

$$\int_0^x \frac{\mathrm{d}t}{\sqrt{a^2 - t^2}} = \arcsin\left(\frac{x}{a}\right) .$$

Primitive of $\sqrt{x^2 + a^2}$:

$$\int_0^x \sqrt{t^2 + a^2}\,\mathrm{d}t = \frac{x}{2}\sqrt{x^2 + a^2} + \frac{a^2}{2}\sinh^{-1}\left(\frac{x}{a}\right) .$$

Primitive of $(\sqrt{x^2 + a^2})^{-1}$:

$$\int_0^x \frac{\mathrm{d}t}{\sqrt{t^2 + a^2}} = \sinh^{-1}\left(\frac{x}{a}\right) .$$

(m) Particular extensions and applications

Complex arguments: As described for the root functions, the root can also be defined for complex arguments.
Interesting special case: For a purely imaginary argument $z = \mathrm{j}x$ the two exemplified functions interchange their interpretation:

- The function $\sqrt{a^2 + z^2}$ represents the arc of a circle:

$$\sqrt{a^2 + (\mathrm{j}x)^2} = \sqrt{a^2 - x^2} \, .$$

- The function $\sqrt{a^2 - z^2}$ represents the arc of a hyperbola:

$$\sqrt{a^2 - (\mathrm{j}x)^2} = \sqrt{a^2 + x^2} \, .$$

▷ With this descriptive representation, the connections between trigonometric and hyperbolic functions in the complex domain can be better understood on the basis of the geometrical interpretation of these functions (see Sections 5.26 and 5.32).

(n) Conics

See also Chapter 8 on analytic geometry and the figures in Section 5.38.

Conics: Figures arising from the intersection of a (double) cone with a plane.
Equation of the double cone:

$$(x - x_0)^2 + (y - y_0)^2 = c^2(z - z_0)^2 \, ,$$

where c determines the apex angle and x_0, y_0, z_0 determines the vertex, respectively.
Equation of the plane:

$$z = m(x \cos(\varphi) + y \sin(\varphi)) + a \, ,$$

where m is the slope, a is the intercept on the z-axis, and φ is the angle between the x-axis and the axis of greatest slope.
General implicit representation of the intersection curve:

$$a_{11}x^2 + 2a_{12}xy + a_{22}y^2 + b_1x + b_2y + d = 0 \, .$$

The term $2a_{12}xy$ can be eliminated by rotating the coordinate system.

▷ For $a_{11} \neq 0 \neq a_{22}$, the terms b_1x and b_2y can be eliminated by a translation of the coordinate system:

$$a_{11}x^2 + a_{22}y^2 + d = 0 \, .$$

If in the equations of the cone and the plane it is assumed that $x_0 = y_0 = z_0 = 0$, $c = 1$, as well as $\varphi = 0$ and $m = \varepsilon \geq 0$, then the equation of intersection is simplified according to

$$x^2(1 - \varepsilon^2) + y^2 - 2a\varepsilon x = a^2 \, ,$$

where ε is the **numerical eccentricity** given by

$\varepsilon = 0$	**circle**
$0 < \varepsilon < 1$	**ellipse**
$\varepsilon = 1$	**parabola**
$\varepsilon > 1$	**hyperbola**

For $\varepsilon = 1$, the quadratic term in x disappears. The curve consists of two branches of a root:

$$y = \pm\sqrt{a^2 + 2ax} \, .$$

In the other cases, the linear term can be eliminated; we then obtain the **central representation.**

Ellipse:

$$\frac{x^2}{a^2} + \frac{y^2}{b^2} = 1 \qquad b^2 = a^2(1 - \varepsilon^2) ,$$

where a is the major semiaxis and b is the minor semiaxis.

▷ For $\varepsilon = 0$, the curve is a circle with $a^2 = b^2$.

The ellipse can be described by the following parameterization:

$$x = a\cos(t) , \qquad y = b\sin(t) , \qquad 0 \le t < 2\pi .$$

Hyperbola:

$$\frac{x^2}{a^2} - \frac{y^2}{b^2} = 1 , \qquad b^2 = a^2(\varepsilon^2 - 1) .$$

The branches of the hyperbola are described by the following parameterization (+ for the right and − for the left branch):

$$x = \pm a\cosh(t) , \qquad y = \pm b\sinh(t) , \qquad -\infty < t < \infty .$$

In central representation the foci of the figures have the coordinates $x = \pm\varepsilon a$, $y = 0$.
Foci of the figures in the center point representation lie at $x = \pm\epsilon a$, $y = 0$.

● In the ellipse, the sum of the distances from a point of a curve the two foci is a constant.

□ The orbits of earth satellites are circular or elliptical curves with the center of the earth in one focus.

● In the hyperbola, difference of distances between one point of a curve and the two foci is constant.

□ The variation of two repulsively charged particles has the trajectory curve of a hyperbola.

Origin of the coordinate system in the focus of the conic:

Representation in polar coordinates using the eccentricity ε:

$$r = \frac{p}{1 + \varepsilon\cos(\varphi)} .$$

▷ The pole lies in one of the foci of the conic, φ is the angle between the radius vector and the line between this focus and the **nearest** vertex. This equation yields only one branch of the hyperbola (see Section 5.38)

Implicit equation in Cartesian coordinates:

$$\sqrt{x^2 + y^2} - \varepsilon x = a .$$

Isolation of the positive y-values yields

$$y = \sqrt{(a + \varepsilon x)^2 - x^2} .$$

Distinction according to the eccentricity ε:

$\varepsilon = 0$	circle	$f(x) = \sqrt{a^2 - x^2}$
$0 < \varepsilon < 1$	ellipse	$f(x) = \sqrt{\frac{a^2}{1-\varepsilon^2} - \left(x\sqrt{1-\varepsilon^2} - \frac{\varepsilon}{\sqrt{1-\varepsilon^2}a}\right)^2}$
$\varepsilon = 1$	parabola	$f(x) = \sqrt{a^2 + 2ax}$
$\varepsilon > 1$	hyperbola	$f(x) = \sqrt{\left(x\sqrt{\varepsilon^2 - 1} + \frac{\varepsilon}{\sqrt{\varepsilon^2-1}}a\right)^2 - a^2}.$

The inverse functions lead to the equation of a parabola for $\varepsilon = 1$; to the function $\sqrt{a^2 - x^2}$ for $\varepsilon = 0$; and to a hyperbola shifted along the y-axis, $\sqrt{x^2 + a^2} - \sqrt{2}a$, for $\varepsilon = \sqrt{2}$.

▷ Analogous considerations are possible in spaces of three or more dimensions. The resulting solids arising in this way are called ellipsoids, hyperboloids, etc.

General equation in central representation:

$$a_{11}x^2 + a_{22}y^2 + a_{33}z^2 + 2a_{12}xy + 2a_{13}xz + 2a_{23}yz = d .$$

▷ The coefficients can be written as the elements of a symmetric matrix.

Through the transformation to principal axes, the matrix can be brought into diagonal form, and the solid can be represented in a coordinate system given by the principal axes.

□ The tensor of inertia is such a matrix. The elements outside the diagonal are called the deviation moments. They are responsible for unbalances. In diagonal form, the diagonal elements describe the moment of inertia during rotation about the corresponding axis.

Transcendental functions

Transcendental functions are functions that cannot be represented as a finite combination of algebraic terms.
Important examples of these functions are exponential and logarithmic functions, hyperbolic functions, the trigonometric functions, as well as their inverse functions.
In this section, the logarithmic functions, exponential functions, and exponential function of powers are described. The other transcendental functions are discussed in separate sections.

5.23 Logarithmic function

$$f(x) = \log_a(x)$$

In programming languages: `LOG(X)`, in `PASCAL` `LN(X)`
Simplest transcendental function.
Before computers were in general use, logarithmic functions were of special importance and used for many arithmetic operations.
Tabulation of logarithms in compilations of formulas simplifies the multiplication and exponentiation of complicated numbers.

□ **Slide rule**: Multiplication and division of complicated quantities through the graphic addition of logarithmically plotted values.

Logarithmic scales: Plotting of values ranging over many orders of magnitude in such a form that, for example, 1 and 10 have the same relative distance on the scale as 10 and 100 or 100 and 1000.

▷ Some physical and chemical quantities, such as decibels (dB) and the pH-value, are defined logarithmically.

General logarithm: Defined to different bases, which has to do with the importance of the logarithmic function as the inverse function of the exponential function. The base is written as an index of the functional expression.

$$\log_a x \ : \ \text{logarithm of } x \text{ to the base } a,\ 0 < a, a \neq 1$$

Natural logarithm, also called **Neper's logarithm** or **hyperbolic logarithm**: Logarithm to the base e (Euler's number):

$$\ln(x) = \log_e(x)$$

▷ Logarithms without stating the base generally refer to the natural logarithm.

● In mathematics, logarithm always means the natural logarithm.

Logarithms to different bases can be expressed by the natural logarithm.
Common logarithm, also called Brigg's logarithm or decimal logarithm: Logarithm to base 10; special case of the logarithm often used in the compilation of formulas.

$$\lg(x) = \log_{10}(x) \qquad \text{in FORTRAN sometimes LOG10(X)}$$

Dual logarithm, also called **binary logarithm**: Logarithm to base 2, used in computer applications.

$$\text{lb}(x) = \log_2(x)$$

● The logarithmic increases more slowly than any power of x.

(a) Definition

Inverse function of the exponential function:

$$e^{\ln(x)} = x \quad \text{for all } x > 0; \qquad \ln(e^x) = x \quad \text{for all } x.$$

Primitive of the hyperbola $\frac{1}{x}$:

$$\ln(x) = \int_1^x \frac{1}{t}\,dt\,.$$

Definition by a limiting procedure:

$$\ln(x) = \lim_{n\to\infty}\left[n\left(x^{1/n} - 1\right)\right]\,.$$

General logarithm, inverse function of general exponential function:

$$a^{\log_a(x)} = x \quad \text{for all } x > 0; \qquad \log_a(a^x) = x \quad \text{for all } x.$$

General logarithm, reduction to the natural logarithm:

$$\log_a(x) = \frac{\ln(x)}{\ln(a)} = \frac{\lg(x)}{\lg(a)}\,.$$

(b) Graphical representation

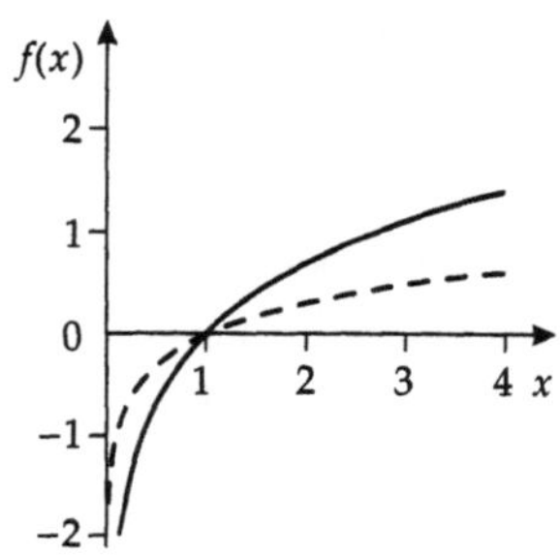

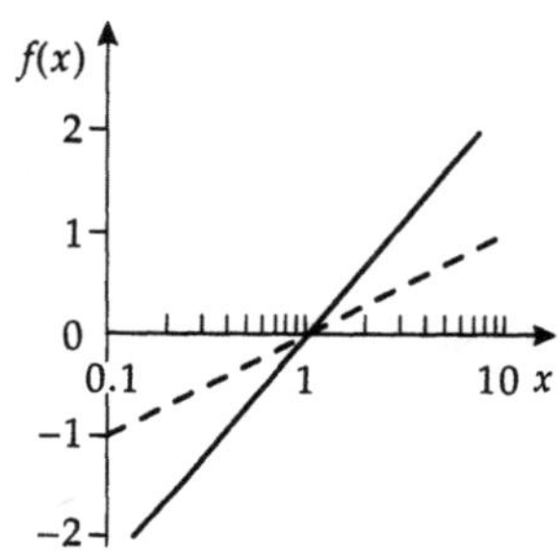

Natural logarithm (solid line) and common logarithm (dotted line).

In a semilogarithmic plot (logarithmic scale on the x-axis), the graphs of the logarithmic functions are straight lines.

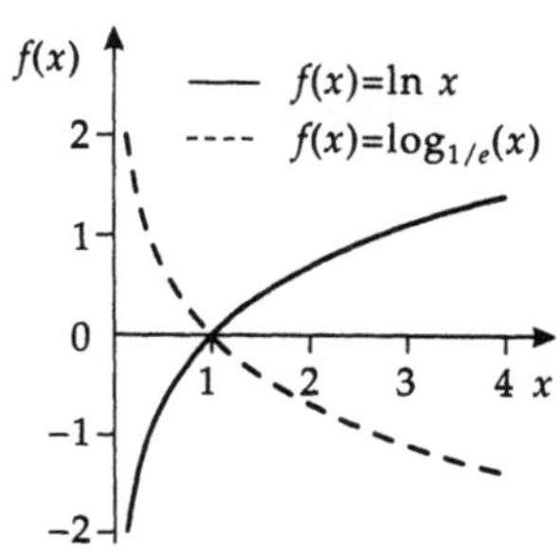

▷ $\log_{1/e}(x)$ and $\ln\frac{1}{x}$ are identical.

The natural logarithm increases faster than the common logarithm and slower than the binary logarithm.
Logarithms to numbers $a < 1$ are decreasing. They correspond to the logarithms of the reciprocal arguments to the reciprocal base (see conversions).

(c) Properties of the function

Domain of definition:	$0 < x < \infty$
Range of values:	$-\infty < f(x) < \infty$
Quadrant:	Lies in the first and fourth quadrants.
Monotonicity:	$a > 1$: strictly monotonically increasing $0 < a < 1$: strictly monotonically decreasing
Symmetries:	No point or mirror symmetries.
Asymptotes:	$a > 1$ $\quad \log_a(x) \to +\infty$ as $x \to \infty$ $0 < a < 1$ $\quad \log_a(x) \to -\infty$ as $x \to \infty$
Periodicity:	No periodicity.

(d) Particular values

Zeros:	$x = 1$, as $\log_a(1) = 0$ for all a.
Discontinuities:	None, continuous function.
Poles:	$x = 0$: $a > 1$ $\quad \log_a(x) \to -\infty$ $0 < a < 1$ $\quad \log_a(x) \to +\infty$
Extrema:	None, strictly monotonic function.
Points of inflection:	None, function everywhere is convex if $a > 1$ concave if $0 < a < 1$

(e) Reciprocal function

The reciprocal function interchanges the index and the argument for $x \neq 0$.

$$\frac{1}{\log_a(x)} = \log_x(a)$$

(f) Inverse function

The inverse function is the exponential function.

$$f^{-1}(x) = a^x \; ; \quad \text{i.e., it holds that}$$
$$a^{\log_a(x)} = x, \qquad e^{\ln(x)} = x \qquad \text{for all } x > 0 \, .$$

▷ In the complex domain, the restriction $x > 0$ does not apply.

(g) Cognate function

Close relationships with the exponential function as inverse function.
Area functions, the inverse functions of hyperbolic functions, can be represented by logarithm functions (see 5.29ff.).

Logarithmic integral function: Defined by the Cauchy principal value of the integral:

$$\mathrm{li}\,(x) = \int_0^x \frac{\mathrm{d}t}{\ln(t)} = x\int_0^1 \frac{\mathrm{d}t}{\ln(x)+\ln(t)}\,.$$

Logarithmic exponential function, $\mathrm{Ei}(x)$ (see section 5.24): Logarithmic function plus a power series plus Euler's constant C:

$$\mathrm{Ei}\,(x) = \int_{-\infty}^x \frac{\mathrm{e}^t}{t}\mathrm{d}t = C + \ln|x| + \sum_{k=1}^{\infty}\frac{x^k}{k!\cdot k}, \qquad x < 0\,,$$

also connected with the logarithmic integral logarithm.

$$\mathrm{li}\,(x) = \mathrm{Ei}\,[\ln(x)]\,.$$

(h) Conversion formulas

- All conversions are valid for $x, y > 0$. For the treatment of negative arguments see item **m.** extensions.

Conversion between different bases:

$$\log_a(x) = \frac{\ln(x)}{\ln(a)}\,.$$

Interchange of argument and base:

$$\log_a(b) = \frac{1}{\log_b(a)}\,.$$

Product or power in the argument:

$$\ln(x\cdot y) = \ln(x) + \ln(y)\,, \qquad \ln(x^y) = y\cdot\ln(x)\,.$$

For the common logarithm, the following applies in particular:

$$\lg(a\cdot 10^m) = m + \lg(a)\,.$$

▷ This is used in tables of logarithms containing only the logarithms of numbers from 1 to 10 (mantissas).

Reciprocal value and quotient:

$$\ln\left(\frac{1}{x}\right) = -\ln(x)\,, \qquad \ln\left(\frac{x}{y}\right) = \ln(x) - \ln(y)\,.$$

Finite product of functions, or infinite product of absolute convergence:

$$\ln\left(\prod_i f_i\right) = \sum_i \ln(f_i)\,.$$

(i) Approximation formulas (8 bits ≈ 0.4% of precision)

For $\frac{3}{4} \le x \le \frac{4}{3}$, it holds that $\ln(x) \approx \frac{x-1}{\sqrt{x}}$.

For $\frac{1}{2} \le x \le 2$, it holds that $\ln(x) \approx (x-1)\left(\frac{6}{1+5x}\right)$.

(j) Expansion of series or products of the function

Power series expansion for $-1 < x \leq 1$:

$$\ln(1+x) = x - \frac{x^2}{2} + \frac{x^3}{3} - \cdots = -\sum_{k=1}^{\infty} \frac{(-x)^k}{k} .$$

Expansion for $x \geq \frac{1}{2}$:

$$\ln(x) = \frac{x-1}{x} + \frac{(x-1)^2}{2x^2} + \frac{(x-1)^3}{3x^3} + \cdots = \sum_{k=1}^{\infty} \frac{(x-1)^k}{k \cdot x^k} .$$

For $x > 0$, it holds that

$$\ln(x) = 2\left[\frac{x-1}{x+1} + \frac{(x-1)^3}{3(x+1)^3} + \frac{(x-1)^5}{(x+1)^5} + \cdots\right] = 2\sum_{k=0}^{\infty} \frac{1}{2k+1}\left(\frac{x-1}{x+1}\right)^{2k+1} .$$

Expansion into a continued fraction:

$$\ln(1+x) = \cfrac{1}{1 + \cfrac{x}{1 - x + \cfrac{x}{2 - x + \cfrac{4x}{3 - 2x + \cfrac{9x}{4 - 3x + \frac{16x}{5-4x+\ldots}}}}}}$$

$$\frac{\ln(1+x)}{x} = \cfrac{1}{1 + \cfrac{x}{2 + \cfrac{x}{3 + \cfrac{4x}{4 + \cfrac{4x}{5 + \cfrac{9x}{6 + \cfrac{9x}{7 + \frac{16x}{8+\ldots}}}}}}}}$$

(k) Derivative of the function

The derivative of the function is the hyperbola $\frac{1}{x}$:

$$\frac{\mathrm{d}}{\mathrm{d}x} \ln(ax+b) = \frac{a}{ax+b} .$$

n-th derivative:

$$\frac{\mathrm{d}^n}{\mathrm{d}x^n} \ln(ax+b) = -(n-1)!\left(\frac{-a}{ax+b}\right)^n .$$

(l) Primitive of the function

$$\int_{(1-c)/b}^{x} \ln(bt+c)\mathrm{d}t = \left(x + \frac{c}{b}\right)[\ln(bx+c) - 1]$$

$$\int_1^x \ln^n(t)\mathrm{d}t = (-1)^n n! x \sum_{k=0}^{n} \frac{[-\ln(x)]^k}{k!}$$

(m) Particular extensions and applications

Complex logarithm: The principal value of the logarithm is (see also Chapter 17 on complex functions)

$$\ln(x + \mathrm{j}y) = \frac{1}{2}\ln(x^2 + y^2) + \mathrm{j} \cdot \mathrm{sgn}\,(y)\mathrm{arccot}\left(\frac{x}{|y|}\right).$$

Thus, the logarithm of negative numbers has the principal value

$$\ln(-x) = \ln(x) + \mathrm{j}\pi, \qquad x > 0\,.$$

Integrals with substitution, often a solution by substitution with the logarithm.

$$\int_a^x \frac{\frac{\mathrm{d}}{\mathrm{d}t} f}{f} \mathrm{d}t = \ln|f(x)| + \mathrm{const}$$

Linear regression of data from which the logarithm was taken. A power function $y = ax^b$ can be fitted to the data points, when a linear regression is performed with $(\ln(x), \ln(y))$ instead of with (x, y).

▷ Analogously, the function $y = a \cdot b^x$ can be fitted when the values $(x, \ln(y))$ are used instead of (x, y).

5.24 Expansion function

$$f(x) = \mathrm{e}^{ax+b}$$

In programming languages: `EXP(A*X+B)`

Description of many natural growth processes (e.g., bacterial reproduction), decay processes (radioactive decay of atoms), damping or resonance processes as well as economic processes in which a given set grows (or diminishes) in proportion to the number of its elements.

● The derivative of the function is proportional to the function itself.

▷ This leads to the description of exponential growth or exponential decay.

● The exponential function increases more rapidly than any polynomial in x.

Natural antilogarithm: Common term based on a possible definition as the inverse function of the natural logarithm.

▷ The general exponential function a^x is included in the general notation e^{ax+b} for the exponential function.

Self-exponential function, x^x: A special case of general exponential function, which is briefly mentioned under item **g**.

(a) Definition

Power of Euler's number e:

$$e^x: \qquad e = \lim_{n\to\infty}\left(\frac{n+1}{n}\right)^n = 2.7182818284\ldots$$

Inverse function of the natural logarithm:

$$e^y = x\,, \qquad y = \ln(x), \quad x > 0\,.$$

Limit of a polynomial product (slowly convergent):

$$e^x = \lim_{n\to\infty}\left(1+\frac{x}{n}\right)^n .$$

Power series (rapidly convergent):

$$e^x = \sum_{i=0}^{\infty}\frac{x^i}{i!} .$$

Solution of the differential equation

$$\frac{\mathrm{d}f(x)}{\mathrm{d}x} = a\cdot f(x); \qquad f(x) = \mathrm{e}^{ax+b}.$$

(b) Graphical representation

There are three important differences:

$a > 0$	The function very rapidly increases toward infinity.
$a < 0$	The function very rapidly decreases toward zero.
$a = 0$	Constant function.

$a > 1$ ($a < -1$) leads to a faster increase (faster decrease) of the function.
$0 < a < 1$ ($-1 < a < 0$) leads to a slower increase (slower decrease).

▷ At point ($x = 0$), the functions $f(x) = \mathrm{e}^{ax+b}$ have the value e^b; that is, they have the value 1 for $b = 0$.

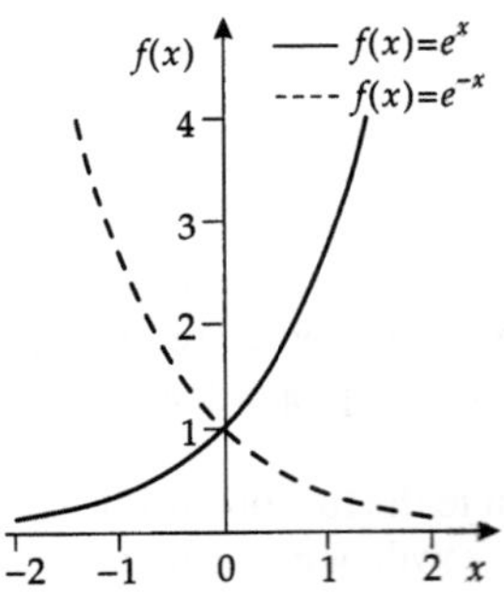

Linear representation

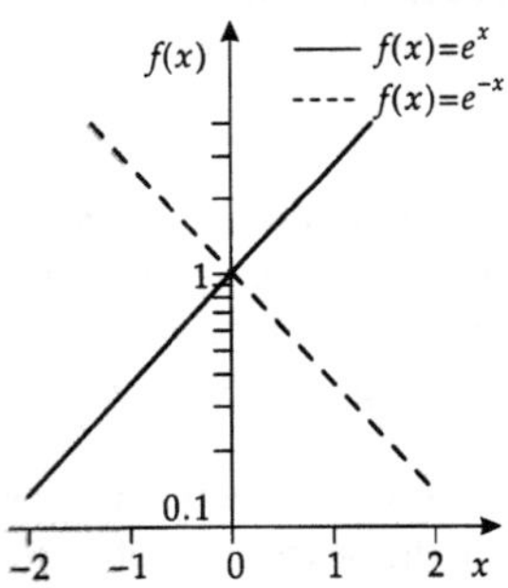

Semilogarithmic representation

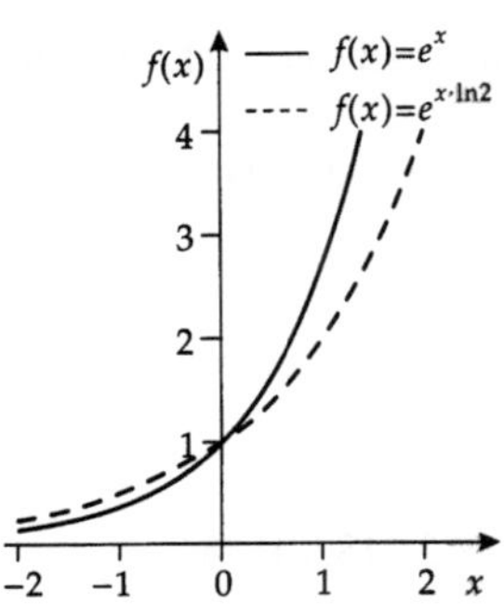

Linear representation

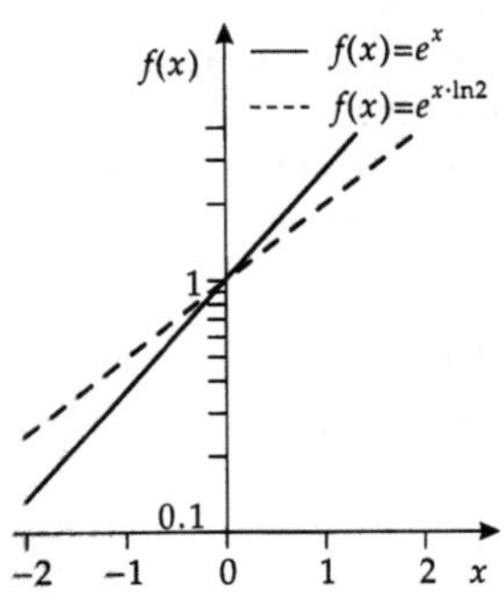

Semilogarithmic representation

Semilogarithmic representation (y-axis logarithmic):

- The graph of the exponential function is a straight line.

The slope of the straight line corresponds to the intensity of the rise of e^{ax+b}, and it is proportional to a. Negative a leads to negative slopes.

▷ For $b = 0$, all straight lines intersect in the point $(x = 0, y = 1)$. For $b \neq 0$, the straight lines are translated.

$b \neq 0$ corresponds to a prefactor of the exponential function $e^b \cdot e^{ax}$.

(c) Properties of the function

Domain of definition:	$-\infty < x < \infty$
Range of values:	$0 < f(x) < \infty$
Quadrant:	Lies in the first and second quadrants.
Periodicity:	No periodicity.
Monotonicity:	$a > 0$: strictly monotonically increasing $a < 0$: strictly monotonically decreasing $a = 0$: constant function
Symmetries:	No point symmetries or mirror symmetries.
Asymptotes:	$a > 0$: $e^{ax+b} \to 0$ as $x \to -\infty$ $a < 0$: $e^{ax+b} \to 0$ as $x \to +\infty$

(d) Particular values

Zeros:	No zeros; the function is always positive.
Discontinuities:	No jump discontinuities; the function is continuous everywhere.
Poles:	No poles.
Extremes:	No extremes; the function is strictly monotonic.
Points of inflection:	No points of inflection; the function is concave everywhere.
Value at $x = 0$:	$f(0) = e^b$, for $b = 0$: $f(0) = 1$.

Table of some values of the exponential function.
1.23-4 means $1.23 \cdot 10^{-4}$.
Further values can be obtained by multiplication of the tabular values: e.g.,
$e^{6.28} = e^x(6) \cdot e^x(0.2) \cdot e^x(0.08) = 403.41 \cdot 1.2214 \cdot 1.0833 \approx 533.8$;
$e^{-3.14} = e^{-x}(3) \cdot e^{-x}(0.1) \cdot e^{-x}(0.04) = 0.0498 \cdot 0.9048 \cdot 0.9608 \approx 0.0433$.

x	e^x	e^{-x}	x	e^x	e^{-x}	x	e^x	e^{-x}
1	2.7183	0.3679	0.1	1.1052	0.9048	0.01	1.0101	0.9901
2	7.3891	0.1353	0.2	1.2214	0.8187	0.02	1.0202	0.9802
3	20.086	0.0498	0.3	1.3499	0.7408	0.03	1.0305	0.9705
4	54.598	0.0183	0.4	1.4918	0.6703	0.04	1.0408	0.9608
5	148.21	6.74-3	0.5	1.6487	0.6065	0.05	1.0513	0.9523
6	403.41	2.48-3	0.6	1.8221	0.5488	0.06	1.0618	0.9418
7	1096.6	9.12-4	0.7	2.0138	0.4966	0.07	1.0725	0.9324
8	2981.0	3.35-4	0.8	2.2255	0.4493	0.08	1.0833	0.9231
9	8103.1	1.23-4	0.9	2.4596	0.4066	0.09	1.0942	0.9139

(e) Reciprocal function

The reciprocal function is an exponential function with a negative argument,

$$\frac{1}{e^{ax+b}} = e^{-ax-b} .$$

Thus, the exponential function has the remarkable property

$$f(x) \cdot f(-x) = 1 = e^x \cdot e^{-x} .$$

(f) Inverse function

The inverse function of $f(x)$ is the natural logarithm ($a \neq 0$),

$$f^{-1}(x) = -\frac{b}{a} + \frac{1}{a} \ln x . \quad \text{i.e., it holds that}$$

$$\ln e^x = x \quad \text{for all } x, \qquad e^{\ln(x)} = x \quad \text{for all } x > 0.$$

(g) Cognate function

General exponential functions a^x can be represented directly as (natural) exponential functions.

$$a^x = e^{x \cdot \ln a}$$

Self-exponential function, x^x: can be reduced analogously to the exponential function.

$$x^x = e^{x \ln x}$$

x^x is defined only for $x > 0$, but can be extended continuously to $x \geq 0$, with 0^0 det 1. The function has a minimum at $x = \frac{1}{e} = 0.36787$ with $x^x = 0.69220$; then it increases farther than the exponential function toward infinity.
$x^{\left(\frac{1}{x}\right)}$ can be rewritten analogously, with a maximum at $x = e$, approaching 1 for large values.
Hyperbolic functions: $\sinh(x)$, $\cosh(x)$, $\tanh(x)$, representable by sums, differences, and quotients of exponential functions.
Logarithmic integral function: $\mathrm{Ei}(x)$, defined by the integral

$$\mathrm{Ei}\,(x) = \int_{-\infty}^{x} \frac{e^t}{t} \mathrm{d}t = C + \ln|x| + \sum_{k=1}^{\infty} \frac{x^k}{k! \cdot k} \qquad x < 0 .$$

with Euler's constant $C = 0.577215665\ldots$
Theta functions: Infinite sums of exponential functions with the perfect squares $(1, 4, 9, \ldots)$ or the squares of odd numbers $(1, 9, 25, \ldots)$ as factors in the argument, e.g.:

$$\Theta_3\left(0; \frac{x}{\pi^2}\right) = 1 + 2 \cdot \sum_{k=1}^{\infty} e^{-k^2 x} .$$

Probability curve, or alternatively **Gauss functions**: Exponential functions with squared argument e^{-ax^2}.

(h) Conversion formulas

Conversions for addition and multiplication in the argument:

$$e^{x+y} = e^x \cdot e^y$$

$$e^{-x} = \frac{1}{e^x}, \qquad \text{and thus} \qquad e^{x-y} = \frac{e^x}{e^y}$$

$$e^{x \cdot y} = \left(e^x\right)^y = \left(e^y\right)^x$$

Geometric series expansions: Infinite sums of exponential functions with negative argument e^{-nx}, $x > 0$.

▷ They can be rewritten in terms of hyperbolic functions.

$$e^{-x} + e^{-2x} + e^{-3x} + e^{-4x} + \cdots = \frac{1}{e^x - 1} = \frac{1}{2}\coth\left(\frac{x}{2}\right) - \frac{1}{2}$$

$$e^{-x} - e^{-2x} + e^{-3x} - e^{-4x} + \cdots = \frac{1}{e^x + 1} = -\frac{1}{2}\tanh\left(\frac{x}{2}\right) + \frac{1}{2}$$

$$e^{-x} + e^{-3x} + e^{-5x} + e^{-7x} + \cdots = \frac{e^x}{e^{2x} - 1} = \frac{1}{2}\operatorname{cosech}(x)$$

$$e^{-x} - e^{-3x} + e^{-5x} - e^{-7x} + \cdots = \frac{e^x}{e^{2x} + 1} = \frac{1}{2}\operatorname{sech}(x)$$

(i) Approximation formulas (8 bits ≈ 0.4% of precision)

In the domain $-1 < x < 1$, it holds that

$$e^x \approx \left(1 + \frac{x}{130}\right)^{130}$$

(j) Expansion of series or products of the function

Power series expansion of the exponential function, valid for all arguments between $-\infty$ and ∞.

$$e^{ax+b} = 1 + \frac{ax+b}{1!} + \frac{(ax+b)^2}{2!} + \ldots = \sum_{k=0}^{\infty} \frac{(ax+b)^k}{k!}.$$

Continued fraction decomposition:

$$e^x = \cfrac{1}{1 - \cfrac{x}{1 + \cfrac{x}{2 - \cfrac{x}{3 + \cfrac{x}{2 - \cfrac{x}{5 + \ldots}}}}}}, \qquad e^x = 1 + \cfrac{x}{1 - \cfrac{x}{2 + \cfrac{x}{3 - \cfrac{x}{2 + \cfrac{x}{5 - \ldots}}}}}$$

(k) Derivative of the function

The derivative is proportional to the function itself:

$$\frac{d}{dx} e^{ax+b} = a \cdot e^{ax+b}.$$

n-th derivative:

$$\frac{d^n}{dx^n} e^{ax+b} = a^n \cdot e^{ax+b}.$$

Derivative of the self-exponential function x^x:

$$\frac{\mathrm{d}}{\mathrm{d}x}x^x = x^x(1+\ln(x)) \,.$$

(l) Primitive of the function

Integral function of the exponential function, proportional to the exponential function, divergent for $a < 0$:

$$\int_{-\infty}^{x} \mathrm{e}^{at+b}\,\mathrm{d}t = \frac{1}{a}\mathrm{e}^{ax+b} \,, \qquad a > 0 \,.$$

Generally valid representation for all a:

$$\int_{x}^{y} \mathrm{e}^{at+b}\,\mathrm{d}t = \frac{1}{a}\left(\mathrm{e}^{ay+b} - \mathrm{e}^{ax+b}\right) \,.$$

Error function (error integral), erf(x): As solution of the integral for $b < 0$:

$$\int_{0}^{x} \frac{\mathrm{e}^{at+b}}{\sqrt{t}}\mathrm{d}t = \sqrt{\frac{\pi}{-a}}\mathrm{e}^{b}\operatorname{erf}(\sqrt{-bx}) \,, \qquad b < 0 \,.$$

Other important integrals:

$$\int_{0}^{x} \frac{\mathrm{d}t}{a+\mathrm{e}^{bt+c}} = \frac{x}{a} - \frac{1}{ab}\ln\left(\frac{a+\mathrm{e}^{bx+c}}{a+\mathrm{e}^{c}}\right) .$$

$$\int_{-\infty}^{x} \frac{\mathrm{d}t}{\mathrm{e}^{bt}+a\mathrm{e}^{-bt}} = \begin{cases} \dfrac{1}{b\sqrt{a}}\arctan\left(\dfrac{\mathrm{e}^{bx}}{\sqrt{a}}\right) & \text{for } a > 0, \\ \dfrac{1}{b\sqrt{-a}}\arctan\left(\dfrac{\mathrm{e}^{bx}}{\sqrt{-a}}\right) & \text{for } a < 0. \end{cases}$$

Gamma function: Defined by the integral:

$$\int_{0}^{\infty} \mathrm{e}^{t}t^{x-1}\mathrm{d}t = \Gamma(x) \,.$$

- **Recursion property:** $\Gamma(x+1) = x\Gamma(x)$.

▷ For $x \in \mathbb{N}$, we obtain the factorial:
$\Gamma(n+1) = n!, \ n \in \mathbb{N}$.
For half-integral arguments, it holds that:

$$\Gamma\left(\frac{1}{2}\right) = \sqrt{\pi}, \quad \Gamma\left(n+\frac{1}{2}\right) = \frac{1\cdot3\cdot5\cdot7\cdot\cdots\cdot(2n-1)}{2n}\sqrt{\pi}$$

For further integrals, see Chapter 25 or other tables.

Laplace transformation (see Chapter 20 on Laplace transformations): Calculates integrals such as

$$\int_{0}^{\infty} f(t)\mathrm{e}^{at+b} = \mathrm{e}^{b}\hat{f}_L(-a)$$

by using the Laplace transform

$$\hat{f}_L(s) = \int_{0}^{\infty} f(t)\mathrm{e}^{-st}\mathrm{d}t \,.$$

(m) Particular extensions and applications

Euler's formula: Complex extension of the exponential function leading to trigonometric functions:

$$e^{jx} = \cos(x) + j\sin(x) .$$

▷ In particular, this representation is required to describe networks in alternating-current engineering (see Chapter 15).

See also Chapter 17 on complex functions.

Superposition of exponential functions and trigonometric functions: Application in electrical engineering and in the theory of elasticity, in order to represent damping ($e^{-ax} \cdot \ldots$) and resonance ($e^{+ax} \cdot \ldots$).

□ The charging and discharging curves of capacitors are exponential functions.

Gauss curve: Used in probability theory and in statistics as the weight function for convolution (in an integral) with other functions.

Bernoulli's numbers B_k: Definable through the expansion of

$$\frac{x}{e^x - 1} = B_0 + B_1\frac{x}{1!} + B_2\frac{x^2}{2!} + B_3\frac{x^3}{3!} + B_4\frac{x^4}{4!} + \cdots = \sum_{n=0}^{\infty} B_n\frac{x^n}{n!} \quad \text{for } |x| < 2\pi.$$

(See also Chapters 20 and 25.)

Table of the first twelve Bernoulli's numbers with integral index ($B_0 = 1$, $B_1 = -\frac{1}{2}$, $B_{2n+1} = 0$).

k	B_k	k	B_k	k	B_k	k	B_k
2	$\frac{1}{6}$	8	$-\frac{1}{30}$	14	$\frac{7}{6}$	20	$-\frac{174611}{330}$
4	$-\frac{1}{30}$	10	$\frac{5}{66}$	16	$-\frac{3617}{510}$	22	$\frac{854513}{138}$
6	$\frac{1}{42}$	12	$-\frac{691}{2730}$	18	$\frac{43867}{798}$	24	$-\frac{236364091}{2730}$

5.25 Exponential functions of powers

$$e^{-ax^n}$$

The exponential functions of powers are especially important for $a > 0$, particularly in the case of $n = 2$ and $n = -1$.

Gauss function, or **bell-shaped curve**: Exponential function with $a > 0, n = 2$, used mainly in statistics.

▷ Other notation with $2\sigma^2$ as denominator in the exponent, (see item **m** and Chapter 19). See also the remarks on statistics in the section on particular extensions (item **m**).

$$f(x) = e^{-ax^2} = e^{-x^2/2\sigma^2} .$$

Temperature-dependent functions in thermodynamics often are represented by exponential functions with $a > 0, n = -1$.

$$f(T) = e^{-E/kT}$$

□ Boltzmann factor: Weight function of statistical mechanics, important in describing distributions in thermal systems (exponential function with $n = 2$ with respect to the velocity v and an exponential function with $n = -1$ with respect to the temperature T):

$$f_B(v, T) = e^{-mv^2/2k_BT} .$$

(a) Definition

The definitions of the exponential function can be taken over.
(For the definition of the Gauss curve ($n = 2$) in terms of the parameter σ, see also item **m**.

$$n = 2: \quad e^{-ax^2} = e^{-x^2/2\sigma^2}$$
$$n = -1: \quad e^{-ax^{-1}} = e^{-a/x}$$

▷ In our further discussion, we assume that $a > 0$.

(b) Graphical representation

The Gauss curve e^{-ax^2} has a maximum at $x = 0$ and decreases rapidly toward 0. As a becomes larger (smaller σ), it decreases more rapidly toward 0. The area under the curve becomes smaller.

▷ If we wish to keep the area under the curve constant (see item **m**), the curve becomes narrower and higher in the middle.

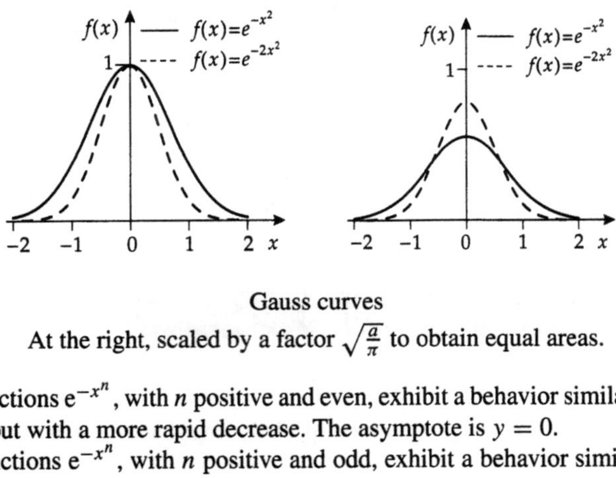

Gauss curves
At the right, scaled by a factor $\sqrt{\frac{a}{\pi}}$ to obtain equal areas.

- The functions e^{-x^n}, with n positive and even, exhibit a behavior similar to the Gauss curve, but with a more rapid decrease. The asymptote is $y = 0$.

▷ The functions e^{-x^n}, with n positive and odd, exhibit a behavior similar to e^{-x}, but with a more rapid decrease.

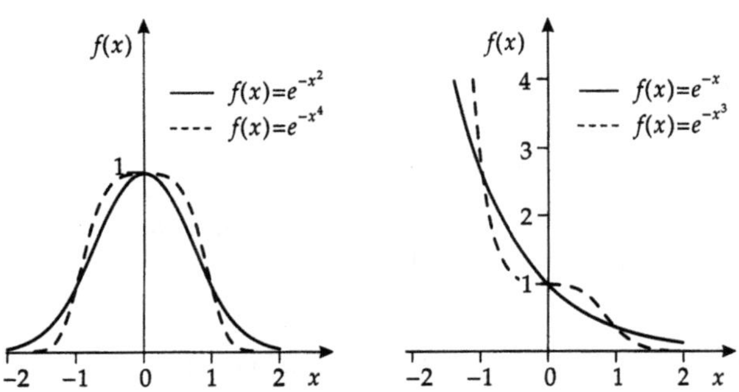

The function $e^{-1/x}$ has a gap of definition at $x = 0$, which can be joined semi-continuously from the right ($x \to 0^+$) through point ($x = 0$, $y = 0$). From the left, the function diverges toward $+\infty$. For very large positive $x \to +\infty$ and negative $x \to -\infty$, the function tends to 1. With increasing a, the function increases faster for negative x-values and slower for positive x-values.

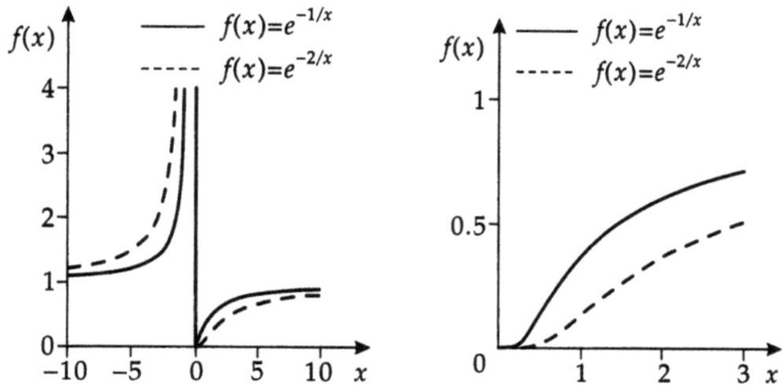

At the right, a detail for $x > 0$.

- The faster functions e^{-x^n} with n negative and odd exhibit a behavior similar to $e^{-1/x}$, but with a more rapid increase.

A remarkable characteristic is shown by function e^{-1/x^2} (and analogously, e^{-x^n} with n even and negative). The function can be extended continuously through point ($x = 0$, $f(x) = 0$), and then it is often differentiable everywhere x (also at point $x = 0$).

▷ For very large (and very small) values, the function approaches 1.

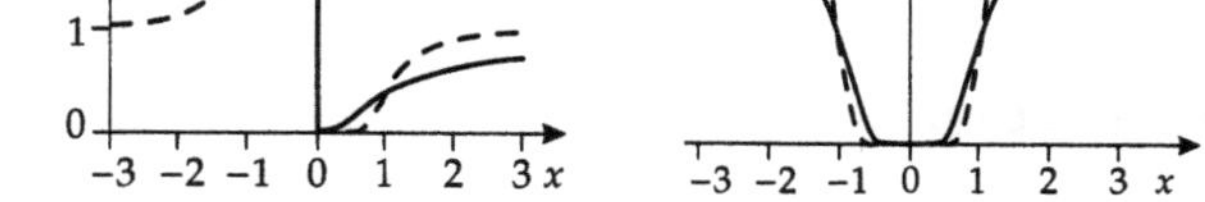

(c) Properties of the function

Domain of definition:	$n = 2$:	$-\infty < x < +\infty$
	$n = -1$:	$\infty < x < 0$ and $0 < x < +\infty$
		can be joined semi-continuously at $x = 0$
Range of values:	$n = 2$:	$0 < f(x) \leq 1$
	$n = -1$:	$0 < f(x) < 1$ and $1 < f(x) < +\infty$
		$0 \leq f(x) < 1$ if joined semi-continuously
Quadrant:	$n = 2$ lies in the first and second quadrants.	
	$n = -1$ lies in the second and fourth quadrants.	
Periodicity:	No periodicity.	

Monotonicity:	$n = 2$:	$x > 0$: strictly monotonically decreasing
		$x < 0$: strictly monotonically increasing
	$n = -1$:	$x > 0$: strictly monotonically increasing
		$x < 0$: strictly monotonically increasing
Symmetries:	$n = 2$:	mirror-symmetric about the y-axis
	$n = -1$:	no symmetries
Asymptotes:	$n = 2$:	$f(x) \to 0$ as $x \to \pm\infty$
	$n = -1$:	$f(x) \to 1$ as $x \to \pm\infty$

(d) Particular values

Zeros:	$n = 2$:	no zeros
	$n = -1$:	semi-continuous extension at $x = 0$
Discontinuities:	$n = 2$:	no jump discontinuities
	$n = -1$:	discontinuous at $x = 0$
Poles:	$n = 2$:	no poles
	$n = -1$:	$x = 0$ when approaching from the left
Extremes:	$n = 2$:	maximum at $x = 0$
	$n = -1$:	no extreme
Points of inflection:	$n = 2$:	$x = \pm\frac{1}{\sqrt{2a}}$
	$n = -1$:	$x = \frac{1}{2a}$

(e) Reciprocal function

The reciprocal functions no longer have a minus sign in the exponent.

$$n = 2: \quad \frac{1}{e^{-ax^2}} = e^{ax^2}$$

$$n = -1: \quad \frac{1}{e^{-ax^{-1}}} = e^{ax^{-1}}$$

(f) Inverse function

$n = 2$: The function is invertible only in the domain $x \geq 0$ (or $x \leq 0$).

$$f^{-1}(x) = \sqrt{-\frac{1}{a}\ln(x)} \quad \left(\text{or}\, f^{-1}(x) = -\sqrt{-\frac{1}{a}\ln(x)}\right)$$

$n = -1$: $\quad f^{-1}(x) = \dfrac{-a}{\ln(x)}$

(g) Cognate function

Cognate functions are the exponential functions e^x and their related functions.
Error function $\operatorname{erf}(x)$: Definite integral of the Gauss function,

$$\operatorname{erf}(x) = \frac{2}{\sqrt{\pi}}\int_0^x e^{-t^2}\,dt = \frac{\operatorname{sgn}(x)}{\sqrt{\pi}}\int_0^{x^2}\frac{e^{-t}}{\sqrt{t}}dt\,.$$

Conjugate function erfc(x): Integral of the residual area under a Gauss curve,

$$\operatorname{erfc}(x) = \frac{2}{\sqrt{\pi}} \int_x^{\infty} \mathrm{e}^{-t^2}\,\mathrm{d}t = 1 - \operatorname{erf}(x)\,.$$

Delta function $\delta(x)$: Not actually a function, but it can be regarded as a limit of the Gauss function,

$$\delta(x) = \lim_{a\to\infty} \sqrt{\frac{a}{\pi}}\,\mathrm{e}^{-ax^2}\,.$$

(h) Conversion formulas

All conversion formulas of exponential functions hold.

(i) Approximation formulas for the function

- Surprisingly, the simple triangle approximation of the Gauss function is accurate to 9% (or better).

$$\mathrm{e}^{-x^2} = \begin{cases} 1 - \dfrac{|x|}{\sqrt{\pi}} & |x| \le \sqrt{\pi} \\ 0 & |x| \ge \sqrt{\pi} \end{cases}$$

▷ The area under the triangle function equals exactly the integral of the Gauss function.

(j) Expansion of series or products of the function

The power series expansion of the exponential function is valid for all arguments between $-\infty$ and ∞.

$$\mathrm{e}^{-ax^n} = 1 + \frac{-ax^n}{1!} + \frac{(-ax^n)^2}{2!} + \cdots = \sum_{k=0}^{\infty} \frac{(-ax^n)^k}{k!}$$

(k) Derivative of the function

The functions can be differentiated by means of the chain rule, like exponential functions.

$$\frac{\mathrm{d}}{\mathrm{d}x}\mathrm{e}^{-ax^n} = -a\,n\,x^{n-1}\mathrm{e}^{-ax^n}$$

(l) Primitive of the function

The integral of the Gauss function is the **error function**:

$$\int_0^x \mathrm{e}^{-t^2}\mathrm{d}t = \frac{\sqrt{\pi}}{2}\operatorname{erf}(x)\,.$$

The integral over the whole x-axis is

$$\int_{-\infty}^{+\infty} \mathrm{e}^{-ax^2}\mathrm{d}x = \sqrt{\frac{\pi}{a}}\,.$$

The integral of $\mathrm{e}^{-x^{-1}}$ is ($x > 0$):

$$\int_0^x \mathrm{e}^{-1/t}\mathrm{d}t = x\mathrm{e}^{-1/x} + \operatorname{Ei}\left(-\frac{1}{x}\right),$$

where $\mathrm{Ei}(x)$ is the **integral exponential function** (see Section 5.24 on the exponential function).

(m) Particular extensions and applications

In statistics, the Gauss function is generally represented as follows:

$$f(x) = \frac{1}{\sqrt{2\pi}\,\sigma} \mathrm{e}^{-x^2/2\sigma^2} .$$

The area under the Gauss curve has the value 1. The curve becomes narrower as σ decreases (see also the figure under item **b** in "graphical representation"). The curve has points of inflection at $x = \pm\sigma$, where σ is the standard deviation of the distribution. See also Chapter 21.

▷ For $\sigma \to 0$, the function becomes a delta function.

Hyperbolic functions

Hyperbolic functions are transcendental functions. They are closely related to exponential functions.

They can be written as sums and quotients of exponential functions.

▷ In the complex plane, hyperbolic and trigonometric functions can be represented as functions of the same kind. They can be transformed into each other by using complex arguments.

Hyperbolic functions are closely connected with conic sections, as will be discussed below.

Consider a branch of the hyperbola on the positive x-axis,

$$y = \pm\sqrt{x^2 - 1}\,.$$

Let this branch be intersected by two straight lines, $g_1 = g$ and $g_2 = -g$.

$$g_1(x) = g(x) = T \cdot x, \qquad g_2(x) = -g(x) = -T \cdot x, \qquad -1 < T < 1\,.$$

The intersection point of g with the hyperbola has the components $x = C$, $y = S = T \cdot C$, where, according to the equation of the hyperbola, $|y| = \sqrt{x^2 - 1}$ is valid, and the sign is determined by the sign of T.

- The area enclosed by $g_1 = g$, $g_2 = -g$, and the hyperbola has the magnitude $|A|$.

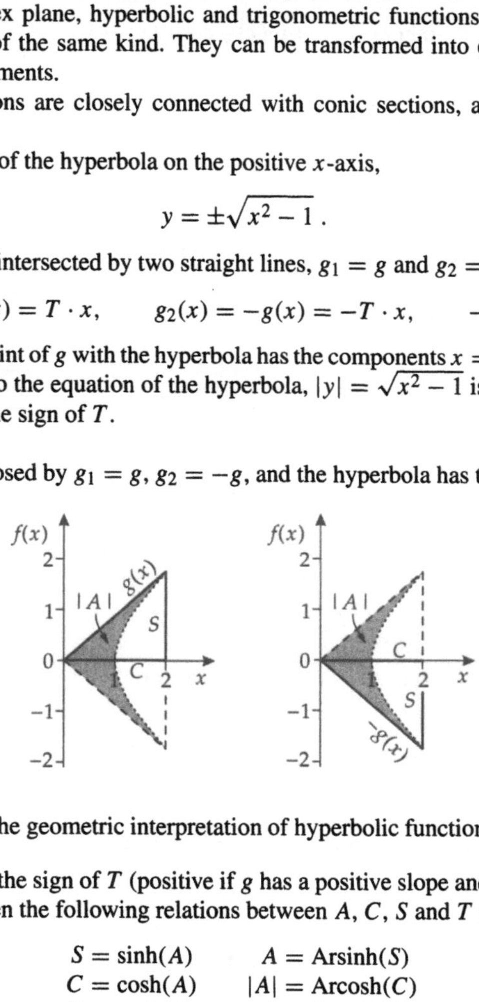

The geometric interpretation of hyperbolic functions.

If value A is given the sign of T (positive if g has a positive slope and negative if g has a negative slope), then the following relations between A, C, S and T can be established:

$$\begin{aligned} S &= \sinh(A) & A &= \operatorname{Arsinh}(S) \\ C &= \cosh(A) & |A| &= \operatorname{Arcosh}(C) \\ T &= \tanh(A) & A &= \operatorname{Artanh}(T) \end{aligned}$$

The functions **hyperbolic sine, hyperbolic cosine,** and **hyperbolic tangent** can be interpreted geometrically as the y-coordinate, the x-coordinate of the point of intersection, and as the slope of a straight line intersected by the "unit hyperbola."

▷ This interpretation corresponds to the interpretation of sine, cosine, and tangent in the intersection of a straight line with the unit circle (see Section 5.11 on trigonometric functions).

This also explains the names of the functions and the alternative notations of hyperbolic functions.

- The two most important conversion rules of hyperbolic functions can be inferred from the geometric interpretation,

$$\sinh^2(x) = \cosh^2(x) - 1 \qquad \tanh(x) = \frac{\sinh(x)}{\cosh(x)}\,.$$

- Tables of conversion formulas

function	sinh	cosh	tanh
$\sinh(x)$	$\sinh(x)$	$\operatorname{sgn}(x)\sqrt{\cosh^2(x)-1}$	$\frac{\tanh(x)}{\sqrt{1-\tanh^2(x)}}$
$\cosh(x)$	$\sqrt{1+\sinh^2(x)}$	$\cosh(x)$	$\frac{1}{\sqrt{1-\tanh^2(x)}}$
$\tanh(x)$	$\frac{\sinh(x)}{\sqrt{1+\sinh^2(x)}}$	$\operatorname{sgn}(x)\frac{\sqrt{\cosh^2(x)-1}}{\cosh(x)}$	$\tanh(x)$
$\operatorname{sech}(x)$	$\frac{1}{\sqrt{1+\sinh^2(x)}}$	$\frac{1}{\cosh(x)}$	$\sqrt{1-\tanh^2(x)}$
$\operatorname{csch}(x)$	$\frac{1}{\sinh(x)}$	$\frac{\operatorname{sgn}(x)}{\sqrt{\cosh^2(x)-1}}$	$\frac{\sqrt{1-\tanh^2(x)}}{\tanh(x)}$
$\coth(x)$	$\frac{\sqrt{1+\sinh^2(x)}}{\sinh(x)}$	$\frac{\operatorname{sgn}(x)\cosh(x)}{\sqrt{\cosh^2(x)-1}}$	$\frac{1}{\tanh(x)}$

function	sech	csch	coth
$\sinh(x)$	$\operatorname{sgn}(x)\frac{\sqrt{1-\operatorname{sech}^2(x)}}{\operatorname{sech}(x)}$	$\frac{1}{\operatorname{csch}(x)}$	$\frac{\operatorname{sgn}(x)}{\sqrt{\coth^2(x)-1}}$
$\cosh(x)$	$\frac{1}{\operatorname{sech}(x)}$	$\frac{\sqrt{1+\operatorname{csch}^2(x)}}{\lvert\operatorname{csch}(x)\rvert}$	$\frac{\lvert\coth(x)\rvert}{\sqrt{\coth^2(x)-1}}$
$\tanh(x)$	$\operatorname{sgn}(x)\sqrt{1-\operatorname{sech}^2(x)}$	$\frac{\operatorname{sgn}(x)}{\sqrt{1+\operatorname{csch}^2(x)}}$	$\frac{1}{\coth(x)}$
$\operatorname{sech}(x)$	$\operatorname{sech}(x)$	$\frac{\lvert\operatorname{csch}(x)\rvert}{\sqrt{1+\operatorname{csch}^2(x)}}$	$\frac{\sqrt{\coth^2(x)-1}}{\lvert\coth(x)\rvert}$
$\operatorname{csch}(x)$	$\frac{\operatorname{sgn}(x)\operatorname{sech}(x)}{\sqrt{1-\operatorname{sech}^2(x)}}$	$\operatorname{csch}(x)$	$\operatorname{sgn}(x)\sqrt{\coth^2(x)-1}$
$\coth(x)$	$\frac{\operatorname{sgn}(x)}{\sqrt{1-\operatorname{sech}^2(x)}}$	$\operatorname{sgn}(x)\sqrt{1+\operatorname{csch}^2(x)}$	$\coth(x)$

The functions **hyperbolic secant**, **hyperbolic cosecant** and **hyperbolic cotangent** are the reciprocal functions of the hyperbolic cosine, the hyperbolic sine, and the hyperbolic tangent, respectively.

The area functions are the inverse functions of the hyperbolic functions, they assign the values S, C, T to the enclosed area.

5.26 Hyperbolic sine and cosine functions

$$f(x) = \sinh(x)\ , \qquad g(x) = \cosh(x)$$

Other abbreviations used for the hyperbolic sine and the hyperbolic cosine are sh(x) or Sin(x) and ch(x) or Cos(x), respectively.

▷ The notations sh and ch are found mainly in the English literature, while the notations Sin and Cos are hardly ever used, to avoid confusion.

Catenary curve: A massive rope (chain), fixed at the end-points, that can be represented in the form of a hyperbolic cosine.

(a) Definition

Representation by means of the exponential function,

$$f(x) = \sinh(x) = \frac{e^x - e^{-x}}{2}\ ; \qquad g(x) = \cosh(x) = \frac{e^x + e^{-x}}{2}\ .$$

General solution of a second-order differential equation,

$$\frac{d^2}{dx^2} f(x) = a^2 f(x); \qquad f(x) = c_1 \sinh(ax) + c_2 \cosh(ax)\ .$$

(b) Graphical representation

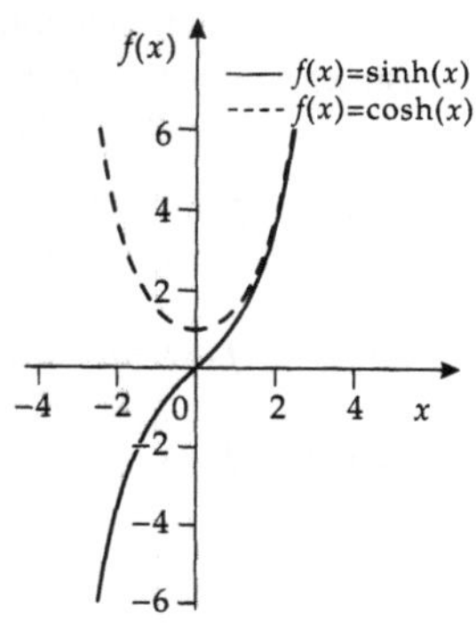

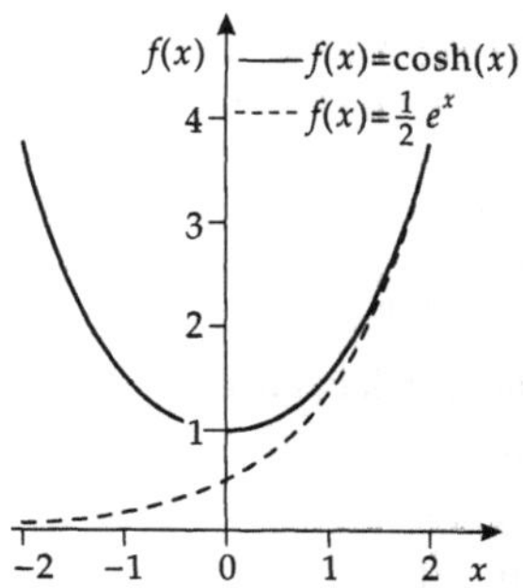

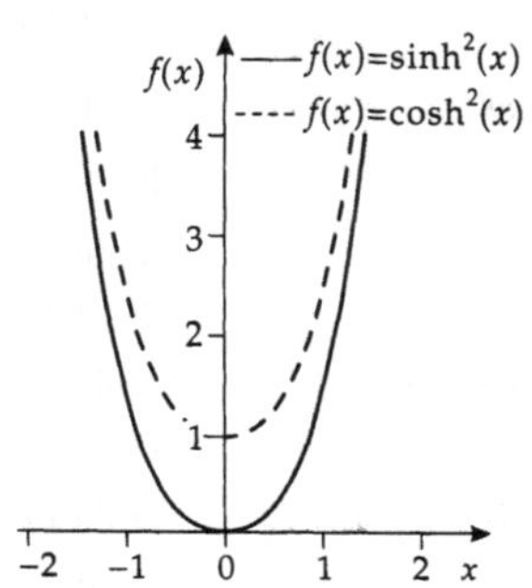

● For large x-values, both functions increase exponentially toward infinity.
The hyperbolic sine is point-symmetric about the origin, its function value is $f(x) = 0$ at $x = 0$, and it tends to $-\infty$ as $x \to -\infty$. The function can be approximated by a straight line $y = x$ in the neighborhood of the origin.
The hyperbolic cosine is mirror-symmetric about the y-axis and it has a minimum at $x = 0$, $y = 1$. For large negative values, the function tends to $+\infty$ as $x \to -\infty$.

▷ For large positive values, the hyperbolic sine and hyperbolic cosine always approach each other.

A factor $0 < a < 1$ in the argument broadens the curve and reduces the slope.
For $a > 1$, the slope becomes larger and the figure becomes narrower.
For $a < 0$, the hyperbolic sine becomes a decreasing function; the hyperbolic cosine does not change if the sign is changed.
For $x > 0$, the square of the function increases rapidly.

● The functions $\sinh^2(x)$ and $\cosh^2(x)$ differ only in an additive constant, $\cosh^2(x) - \sinh^2(x) = 1$.

(c) Properties of the function

Domain of definition:	$-\infty < x < \infty$	
Range of values:	$\sinh(x)$	$-\infty < f(x) < \infty$
	$\cosh(x)$	$1 \le f(x) < \infty$
Quadrant:	$\sinh(x)$	first and third quadrants
	$\cosh(x)$	first and second quadrants
Periodicity:	None.	
Monotonicity:	$\sinh(x)$	strictly monotonically increasing
	$\cosh(x)$	$x > 0$ strictly monotonically increasing
		$x < 0$ strictly monotonically decreasing
Symmetries:	$\sinh(x)$	point symmetry about the origin
	$\cosh(x)$	mirror symmetry about the y-axis
Asymptotes:	$\sinh(x)$	$f(x) \to \pm\frac{1}{2}e^{\pm x}$ as $x \to \pm\infty$.
	$\cosh(x)$	$f(x) \to +\frac{1}{2}e^{\pm x}$ as $x \to \pm\infty$.

(d) Particular values

Zeros:	$\sinh(x)$	$f(x) = 0$ at $x = 0$
	$\cosh(x)$	none
Discontinuities:	None.	
Poles:	None.	
Extremes:	$\sinh(x)$	none
	$\cosh(x)$	minimum at $x = 0$
Points of inflection:	$\sinh(x)$	point of inflection at $x = 0$
	$\cosh(x)$	none
Value at $x = 0$:	$\sinh(x)$	$f(0) = 0$
	$\cosh(x)$	$f(0) = 1$

Value of	$x \to -\infty$	$x = -1$	$x = 0$	$x = 1$	$x \to \infty$
$\sinh(x)$	$-\infty$	$\dfrac{1-e^2}{2e}$	0	$\dfrac{e^2-1}{2e}$	$+\infty$
$\cosh(x)$	$+\infty$	$\dfrac{1+e^2}{2e}$	1	$\dfrac{e^2+1}{2e}$	$+\infty$

(e) Reciprocal function

- The hyperbolic cosecant and hyperbolic secant are the reciprocal functions of the hyperbolic sine and hyperbolic cosine, respectively:

$$\frac{1}{\sinh(x)} = \operatorname{csch}(x)\,, \qquad \frac{1}{\cosh(x)} = \operatorname{sech}(x)\,.$$

(f) Inverse function

The inverse functions are the corresponding inverse hyperbolic functions:

$$x = \operatorname{Arsinh}(y), \qquad y = \sinh(x), \qquad -\infty < x < \infty\,.$$

The hyperbolic cosine has an inverse for $x \geq 0$ only (or for $x < 0$ with a minus sign).

$$x = \operatorname{Arcosh}(y), \qquad y = \cosh(x), \qquad 0 \leq x < \infty\,.$$

(g) Cognate function

Other hyperbolic functions are related to the hyperbolic sine and the hyperbolic cosine according to the conversion rules (see the beginning of this section).
The area functions are the corresponding inverse hyperbolic functions.
In the complex plane the trigonometric functions are closely correlated with the hyperbolic functions.
The hyperbolae $f(x) = \sqrt{x^2 \pm a^2}$ are closely related via the geometrical interpretation of the hyperbolic functions.

(h) Conversion formulas

Reflection in the x-axis:

$$\sinh(-x) = -\sinh(x)\,, \qquad \cosh(-x) = \cosh(x)\,.$$

The sum of the hyperbolic sine and hyperbolic cosine is

$$\cosh(x) \pm \sinh(x) = e^{\pm x}\,.$$

De Moivre's formula:

$$\cosh(nx) \pm \sinh(nx) = [\cosh(x) \pm \sinh(x)]^n = e^{\pm nx}$$

- **Addition theorems**

$$\sinh(x \pm y) = \sinh(x)\cosh(y) \pm \cosh(x)\sinh(y)$$
$$\cosh(x \pm y) = \cosh(x)\cosh(y) \pm \sinh(x)\sinh(y)$$

Integral multiples of x in the argument of $\sinh(x)$:

$$\sinh(2x) = 2\sinh(x)\cosh(x) = 2\sinh(x)\sqrt{1+\sinh^2(x)}$$
$$\sinh(3x) = 4\sinh^3(x) + 3\sinh(x) = \sinh(x)\left[4\sinh^2(x) - 1\right]$$

$$\sinh(4x) = 4\sinh(x)\cosh(x)\left[2\cosh^2(x) - 1\right]$$
$$\sinh(5x) = \sinh(x)\left[16\cosh^4(x) - 12\cosh^2(x) + 1\right]$$

Integral multiples of x in the argument of $\cosh(x)$:

$$\cosh(2x) = \sinh^2(x) + \cosh^2(x) = 2\cosh^2(x) - 1$$
$$\cosh(3x) = 4\cosh^3(x) - 3\cosh(x) = \cosh(x)\left[4\sinh^2(x) + 1\right]$$
$$\cosh(4x) = 8\cosh^4(x) - 8\cosh^2(x) + 1$$
$$\cosh(5x) = \cosh(x)\left[16\cosh^4(x) - 20\cosh^2(x) + 5\right]$$

n**-fold argument**:

$$\sinh(nx) = \binom{n}{1}\cosh^{n-1}(x)\sinh(x) + \binom{n}{3}\cosh^{n-3}(x)\sinh^3(x) + \binom{n}{5}\cosh^{n-5}(x)\sinh^5(x) + \binom{n}{7}\cosh^{n-7}(x)\sinh^7(x) + \dots$$

$$\cosh(nx) = \binom{n}{0}\cosh^{n}(x) + \binom{n}{2}\cosh^{n-2}(x)\sinh^2(x) + \binom{n}{4}\cosh^{n-4}(x)\sinh^4(x) + \binom{n}{6}\cosh^{n-6}(x)\sinh^6(x) + \dots$$

Half-argument:

$$\sinh\left(\frac{x}{2}\right) = \operatorname{sgn}(x)\sqrt{\frac{\cosh(x) - 1}{2}} = \frac{\sinh(x)}{\sqrt{2\cosh(x) + 2}}$$
$$\cosh\left(\frac{x}{2}\right) = \sqrt{\frac{\cosh(x) + 1}{2}} = \frac{|\sinh(x)|}{\sqrt{2\cosh(x) - 2}}$$

Addition of functions:

$$\sinh(x) \pm \sinh(y) = 2\sinh\left(\frac{x \pm y}{2}\right)\cosh\left(\frac{x \mp y}{2}\right)$$
$$\cosh(x) + \cosh(y) = 2\cosh\left(\frac{x + y}{2}\right)\cosh\left(\frac{x - y}{2}\right)$$
$$\cosh(x) - \cosh(y) = 2\sinh\left(\frac{x + y}{2}\right)\sinh\left(\frac{x - y}{2}\right)$$

Multiplication of functions:

$$\sinh(x)\sinh(y) = \frac{1}{2}\cosh(x + y) - \frac{1}{2}\cosh(x - y)$$
$$\sinh(x)\cosh(y) = \frac{1}{2}\sinh(x + y) + \frac{1}{2}\sinh(x - y)$$
$$\cosh(x)\cosh(y) = \frac{1}{2}\cosh(x + y) + \frac{1}{2}\cosh(x - y)$$

Powers of $\sinh(x)$:

$$\sinh^2(x) = \cosh^2(x) - 1 = \frac{1}{2}[\cosh(2x) - 1]$$
$$\sinh^3(x) = \frac{1}{4}[\sinh(3x) - 3\sinh(x)]$$
$$\sinh^4(x) = \frac{1}{8}[\cosh(4x) - 4\cosh(2x) + 3]$$

$$\sinh^5(x) = \frac{1}{16}[\sinh(5x) - 5\sinh(3x) + 10\sinh(x)]$$

$$\sinh^6(x) = \frac{1}{32}[\cosh(6x) - 6\cosh(4x) + 15\cosh(2x) - 10]$$

Powers of $\cosh(x)$:

$$\cosh^2(x) = \sinh^2(x) + 1 = \frac{1}{2}[\cosh(2x) + 1]$$

$$\cosh^3(x) = \frac{1}{4}[\cosh(3x) + 3\cosh(x)]$$

$$\cosh^4(x) = \frac{1}{8}[\cosh(4x) + 4\cosh(2x) + 3]$$

$$\cosh^5(x) = \frac{1}{16}[\cosh(5x) + 5\cosh(3x) + 10\cosh(x)]$$

$$\cosh^6(x) = \frac{1}{32}[\cosh(6x) + 6\cosh(4x) + 15\cosh(2x) + 10]$$

n-th **power**:

▷ Note that $\sinh(-x) = -\sinh(x)$ and $\cosh(-x) = \cosh(x)$.

$$\sinh^n(x) = \frac{1}{2^n}\sum_{k=0}^{n}(-1)^k\binom{n}{k}\sinh([n-2k]x) \qquad \text{for odd } n$$

$$= \frac{1}{2^n}\sum_{k=0}^{n}(-1)^k\binom{n}{k}\cosh([n-2k]x) \qquad \text{for even } n$$

$$\cosh^n(x) = \frac{1}{2^n}\sum_{k=0}^{n}\binom{n}{k}\cosh([n-2k]x) \qquad \text{for all } n$$

(i) Approximation formulas (8 bits ≈ 0.4% of accuracy)

For small values of x, it holds that

$$\sinh(x) \approx x + \frac{x^3}{6}, \qquad |x| < 0.84$$

$$\cosh(x) \approx \left(1 + \frac{x^2}{4}\right)^2, \qquad |x| < 0.70$$

For large values of x, it holds that

$$\sinh(x) \approx \operatorname{sgn}(x)\frac{e^{|x|}}{2}, \qquad |x| > 2.78$$

$$\cosh(x) \approx \sinh(|x|) \approx \frac{e^{|x|}}{2}, \qquad |x| > 2.78$$

(j) Expansion of series or products of the function

Power series expansion:

$$\sinh(x) = x + \frac{x^3}{3!} + \frac{x^5}{5!} + \frac{x^7}{7!} + \cdots = \sum_{k=0}^{\infty}\frac{x^{2k+1}}{(2k+1)!}$$

$$\cosh(x) = 1 + \frac{x^2}{2!} + \frac{x^4}{4!} + \frac{x^6}{6!} + \cdots = \sum_{k=0}^{\infty}\frac{x^{2k}}{(2k)!}$$

Product expansion:

$$\sinh(x) = x\left(1+\frac{x^2}{\pi^2}\right)\left(1+\frac{x^2}{4\pi^2}\right)\left(1+\frac{x^2}{9\pi^2}\right)\cdots$$
$$= x\prod_{k=1}^{\infty}\left(1+\frac{x^2}{k^2\pi^2}\right)$$
$$\cosh(x) = \left(1+\frac{4x^2}{\pi^2}\right)\left(1+\frac{4x^2}{9\pi^2}\right)\left(1+\frac{4x^2}{25\pi^2}\right)\cdots$$
$$= \prod_{k=1}^{\infty}\left(1+\frac{4x^2}{(2k-1)^2\pi^2}\right)$$

(k) Derivative of the function

Hyperbolic sine and hyperbolic cosine are derivatives of each other.

$$\frac{\mathrm{d}}{\mathrm{d}x}\sinh(ax) = a\cosh(ax) \qquad \frac{\mathrm{d}}{\mathrm{d}x}\cosh(ax) = a\sinh(ax)$$

Higher derivative:

$$\frac{\mathrm{d}^n}{\mathrm{d}x^n}\sinh(ax) = \begin{cases} a^n\cosh(ax) & n \text{ odd} \\ a^n\sinh(ax) & n \text{ even} \end{cases}$$
$$\frac{\mathrm{d}^n}{\mathrm{d}x^n}\cosh(ax) = \begin{cases} a^n\sinh(ax) & n \text{ odd} \\ a^n\cosh(ax) & n \text{ even} \end{cases}$$

(l) Primitive of the function

The primitive of the hyperbolic sine and hyperbolic cosine are

$$\int_0^x \sinh(at)\,\mathrm{d}t = \frac{1}{a}\left[\cosh(ax) - 1\right]$$
$$\int_0^x \cosh(at)\,\mathrm{d}t = \frac{1}{a}\sinh(ax)$$

(m) Particular extensions and applications

Complex argument:

$$\sinh(x+\mathrm{j}y) = \sinh(x)\cos(y) + \mathrm{j}\cosh(x)\sin(y)$$
$$\cosh(x+\mathrm{j}y) = \cosh(x)\cos(y) + \mathrm{j}\sinh(x)\sin(y)$$

Purely imaginary argument:

$$\sinh(\mathrm{j}x) = \mathrm{j}\sin(x) \qquad \cosh(\mathrm{j}x) = \cos(x)$$

▷ The direct connection of hyperbolic sine and hyperbolic cosine with sine and cosine is obvious.

5.27 Hyperbolic tangent and cotangent function

$$f(x) = \tanh(x)\,, \qquad g(x) = \coth(x)$$

Other abbreviations for the hyperbolic tangent are $\mathrm{th}(x)$ and $\mathrm{Tan}(x)$, and for the hyperbolic cotangent they are $\mathrm{cth}(x)$, $\mathrm{ctnh}(x)$, $\mathrm{ctgh}(x)$ and $\mathrm{Cot}(x)$.

▷ The notations th and cth can be found mainly in the English literature; the notations Tan and Cot are hardly ever used, to avoid confusion.

(a) Definition

Representation in terms of exponential functions:

$$f(x) = \tanh(x) = \frac{e^x - e^{-x}}{e^x + e^{-x}} = \frac{e^{2x} - 1}{e^{2x} + 1}$$
$$g(x) = \coth(x) = \frac{e^x + e^{-x}}{e^x - e^{-x}} = \frac{e^{2x} + 1}{e^{2x} - 1}$$

Representation in terms of hyperbolic functions:

$$\tanh(x) = \frac{\sinh(x)}{\cosh(x)} = \frac{\mathrm{sech}\,(x)}{\mathrm{csch}\,(x)}$$
$$\coth(x) = \frac{1}{\tanh(x)} = \frac{\cosh(x)}{\sinh(x)} = \frac{\mathrm{csch}\,(x)}{\mathrm{sech}\,(x)}$$

General solution of a differential equation of the first order:

$$\frac{d}{dx} f(x) = \frac{a}{c}(c^2 - f^2(x)); \qquad f(x) = c \tanh(ax) \text{ or } c \coth(ax)$$

(b) Graphical representation

● Both functions possess a restricted range of values.
$\tanh(x)$ has values between -1 and 1, $\coth(x)$ has values with absolute value greater than one.

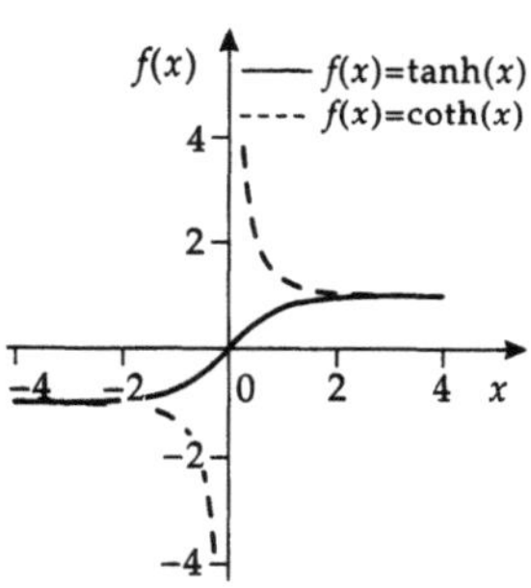

$f(x) = \tanh(x)$ (solid line),
$f(x) = \coth(x)$ (dashed line)

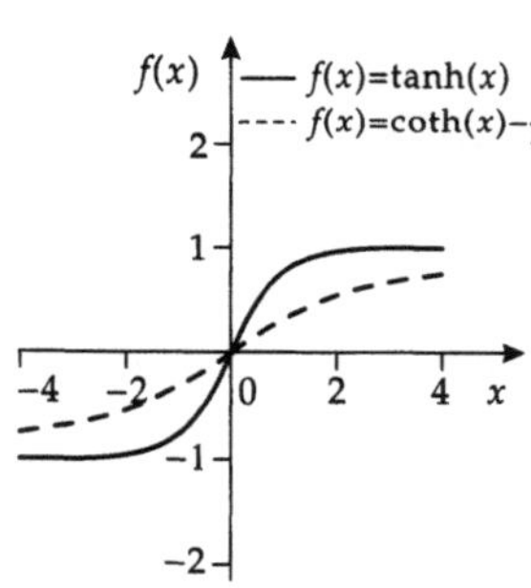

$f(x) = \coth(x) - \frac{1}{x}$
(Langevin's function, dashed line),
$f(x) = \tanh(x)$ (solid line)

● Both functions are point-symmetric about the origin.
$\tanh(x)$ is strictly monotonically increasing and has a zero at $x = 0$.
$\coth(x)$ has a pole at $x = 0$, and is strictly monotonically decreasing for $x > 0$ and $x = 0$.
$\tanh(x)$: For values near $x = 0$, the behavior is similar to that of $\sinh(x)$ and to the straight line $y = x$.

$\coth(x)$ For values near $x = 0$, the behavior is similar to $\frac{1}{x}$.

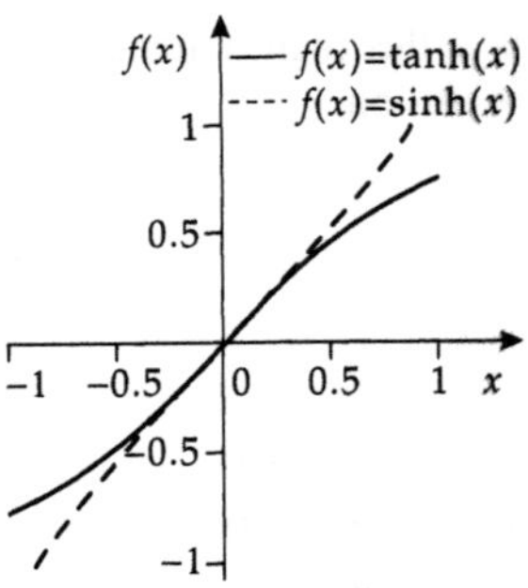

$f(x) = \tanh(x)$ (solid line) in comparison with $f(x) = \sinh(x)$ (dashed line)

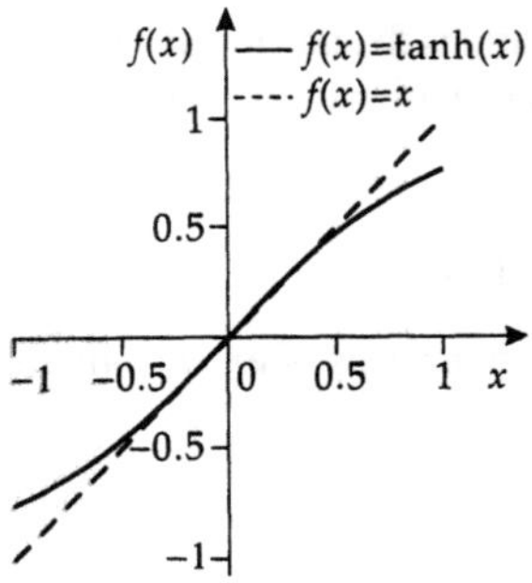

$f(x) = \tanh(x)$ (solid line) in comparison with $f(x) = x$ (dashed line)

(c) Properties of the function

Domain of definition:	$\tanh(x)$	$-\infty < x < \infty$
	$\coth(x)$	$-\infty < x < \infty, x \neq 0$
Range of values:	$\tanh(x)$	$-1 < f(x) < 1$
	$\coth(x)$	$-\infty < x < 1; \quad 1 < x < \infty$
Quadrant:	Lies in the first and third quadrants.	
Periodicity:	None.	
Monotonicity:	$\tanh(x)$	strictly monotonically increasing
	$\coth(x)$	strictly monotonically decreasing for $x < 0, x > 0$
Symmetries:	Point symmetry about the origin.	
Asymptotes:	$f(x) \to \pm 1$ as $x \to \pm\infty$	

(d) Particular values

Zeros:	$\tanh(x)$	$x = 0$
	$\coth(x)$	none
Discontinuities:	None.	
Poles:	$\tanh(x)$	none
	$\coth(x)$	$x = 0$
Extremes:	None.	
Points of inflection:	$\tanh(x)$	$x = 0$
	$\coth(x)$	none

Value of	$x \to -\infty$	$x = -1$	$x = -\frac{1}{2}$	$x = 0$	$x = \frac{1}{2}$	$x = 1$	$x \to \infty$
$\tanh(x)$	-1	$\frac{1-\mathrm{e}^2}{1+\mathrm{e}^2}$	$\frac{1-\mathrm{e}}{1+\mathrm{e}}$	0	$\frac{\mathrm{e}-1}{\mathrm{e}+1}$	$\frac{\mathrm{e}^2-1}{\mathrm{e}^2+1}$	$+1$
$\coth(x)$	-1	$\frac{1+\mathrm{e}^2}{1-\mathrm{e}^2}$	$\frac{1+\mathrm{e}}{1-\mathrm{e}}$	$\mp\infty$	$\frac{\mathrm{e}+1}{\mathrm{e}-1}$	$\frac{\mathrm{e}^2+1}{\mathrm{e}^2-1}$	$+1$

(e) Reciprocal function

The hyperbolic tangent and the hyperbolic cotangent are reciprocals of each other:

$$\frac{1}{\tanh(x)} = \coth(x)\ , \qquad \frac{1}{\coth(x)} = \tanh(x)\ .$$

(f) Inverse function

The inverse functions are the corresponding area functions.

$$x = \operatorname{artanh}(y), \qquad y = \tanh(x)$$
$$x = \operatorname{arcoth}(y), \qquad y = \coth(x)$$

- For the inverse function, the different domains of definition must be taken into consideration.

(g) Cognate functions

Langevin's function: Of importance in electrodynamics in the treatment of dielectrics, can be represented as (see the figure under **b**, "graphical representation")

$$f(x) = \coth(x) - \frac{1}{x}\ .$$

Other hyperbolic functions are related to the hyperbolic tangent and the hyperbolic cotangent according to the conversion rules.
In the complex plane, the trigonometric functions are closely correlated with the hyperbolic functions.
The hyperbolas $f(x) = \sqrt{x^2 \pm a^2}$ are closely related to the hyperbolic functions via the geometrical interpretation of the latter.

(h) Conversion formulas

Both functions are point-symmetric about the origin:

$$\tanh(-x) = -\tanh(x)\ , \qquad \coth(-x) = -\coth(x)\ .$$

Addition in the argument:

$$\tanh(x \pm y) = \frac{\tanh(x) \pm \tanh(y)}{1 \pm \tanh(x)\tanh(y)}$$
$$\coth(x \pm y) = \frac{1 \pm \coth(x)\coth(y)}{\coth(x) \pm \coth(y)}$$

Double argument:

$$\tanh(2x) = \frac{2\tanh(x)}{1 + \tanh^2(x)} = \frac{2}{\tanh(x) + \coth(x)}$$
$$\coth(2x) = \frac{1 + \coth^2(x)}{2\coth(x)} = \frac{\tanh(x) + \coth(x)}{2}$$

Half-argument:

$$\tanh\left(\frac{x}{2}\right) = \frac{\sinh(x)}{\cosh(x) + 1} = \frac{\cosh(x) - 1}{\sinh(x)}$$
$$= \sqrt{\frac{\cosh(x) - 1}{\cosh(x) + 1}}\operatorname{sgn}(x) = \frac{1 - \sqrt{1 - \tanh^2(x)}}{\tanh(x)}$$

$$= \coth(x) - \operatorname{sgn}(x)\sqrt{\coth^2(x) - 1} = \coth(x) - \operatorname{csch}(x)$$

$$\coth\left(\frac{x}{2}\right) = \frac{\sinh(x)}{\cosh(x) - 1} = \frac{\cosh(x) + 1}{\sinh(x)}$$

$$= \sqrt{\frac{\cosh(x) + 1}{\cosh(x) - 1}}\operatorname{sgn}(x) = \frac{1 + \sqrt{1 - \tanh^2(x)}}{\tanh(x)}$$

$$= \coth(x) + \operatorname{sgn}(x)\sqrt{\coth^2(x) - 1} = \coth(x) + \operatorname{csch}(x)$$

Addition of functions:

$$\tanh(x) \pm \tanh(y) = \frac{\sinh(x \pm y)}{\cosh(x)\cosh(y)}$$

$$\coth(x) \pm \tanh(y) = \frac{\cosh(x \pm y)}{\sinh(x)\cosh(y)}$$

$$\coth(x) \pm \coth(y) = \frac{\sinh(x \pm y)}{\sinh(x)\sinh(y)}$$

Product of functions:

$$\tanh(x)\tanh(y) = \frac{\tanh(x) + \tanh(y)}{\coth(x) + \coth(y)}$$

$$\coth(x)\coth(y) = \frac{\coth(x) + \coth(y)}{\tanh(x) + \tanh(y)}$$

(i) Approximation formulas (8 bits $\approx$ 0.4% of accuracy)

For small absolute values of x, it holds that

$$\tanh(x) \approx x\left(1 - \frac{x^2}{3}\right), \qquad |x| \le 0.41$$

$$\coth(x) \approx \frac{1}{x}\left(1 + \frac{x^2}{3}\right), \qquad |x| \le 0.65$$

For large absolute values of x, it holds that

$$\tanh(x) \approx \operatorname{sgn}(x)\left(1 - 2\mathrm{e}^{-2|x|}\right), \qquad |x| \ge 1.6$$

$$\coth(x) \approx \operatorname{sgn}(x)\left(1 + 2\mathrm{e}^{-2|x|}\right), \qquad |x| \ge 1.6$$

(j) Expansion of series or products of the function

Power series expansion in terms of Bernoulli's numbers B_k:

$$\tanh(x) = x - \frac{x^3}{3} + \frac{2x^5}{15} - \frac{17x^7}{315} + \dots$$

$$= \frac{1}{x}\sum_{k=1}^{\infty} \frac{(4^k - 1)B_k(4x^2)^k}{(2k)!}, \qquad -\frac{\pi}{2} < x < \frac{\pi}{2}$$

$$\coth(x) = \frac{1}{x} + \frac{x}{3} - \frac{x^3}{45} + \frac{2x^5}{945} - \dots$$

$$= \frac{1}{x}\sum_{k=0}^{\infty} \frac{B_k(4x^2)^k}{(2k)!}, \qquad -\pi < x < \pi$$

⊳ The Bernoulli's numbers are defined in Section 5.24.

Series expansion in terms of exponential functions:

$$\begin{aligned}\tanh(x) &= \operatorname{sgn}(x)\left[1 - 2e^{-2|x|} + 2e^{-4|x|} - 2e^{-6|x|} + \cdots\right]\\ &= \operatorname{sgn}(x)\left[1 + \sum_{k=1}^{\infty}(-1)^k 2e^{-2k|x|}\right]\\ &= \operatorname{sgn}(x)\sum_{k=-\infty}^{\infty}(-1)^k e^{-2|kx|}\\ \coth(x) &= \operatorname{sgn}(x)\left[1 + 2e^{-2|x|} + 2e^{-4|x|} + 2e^{-6|x|} + \cdots\right]\\ &= \operatorname{sgn}(x)\left[1 + \sum_{k=1}^{\infty} 2e^{-2k|x|}\right]\\ &= \operatorname{sgn}(x)\sum_{k=-\infty}^{\infty} e^{-2|kx|}, \qquad x \neq 0\end{aligned}$$

Partial fraction decomposition:

$$\begin{aligned}\tanh(x) &= \frac{8x}{\pi^2 + 4x^2} + \frac{8x}{9\pi^2 + 4x^2} + \frac{8x}{25\pi^2 + 4x^2} + \frac{8x}{49\pi^2 + 4x^2} + \cdots\\ &= \sum_{k=0}^{\infty}\frac{8x}{(2k+1)^2\pi^2 + 4x^2}\\ \coth(x) &= \frac{1}{x} + \frac{2x}{\pi^2 + x^2} + \frac{2x}{4\pi^2 + x^2} + \frac{2x}{9\pi^2 + x^2} + \frac{2x}{16\pi^2 + x^2} + \cdots\\ &= \frac{1}{x} + \sum_{k=1}^{\infty}\frac{2x}{k^2\pi^2 + x^2} = \sum_{k=-\infty}^{\infty}\frac{x}{k^2\pi^2 + x^2}\end{aligned}$$

Continued fraction decomposition:

$$\tanh(x) = \cfrac{x}{1 + \cfrac{x^2}{3 + \cfrac{x^2}{5 + \cfrac{x^2}{7 + \cdots}}}}$$

(k) Derivative of the function

The derivatives of the hyperbolic tangent and the hyperbolic cotangent are given by the squares of the functions:

$$\frac{d}{dx}\tanh(ax) = a\operatorname{sech}^2(ax) = a[1 - \tanh^2(ax)]$$

$$\frac{d}{dx}\coth(ax) = -a\operatorname{csch}^2(ax) = a[1 - \coth^2(ax)]$$

(l) Primitive of the function

The primitives are the logarithms of hyperbolic functions:

$$\int_0^x \tanh(at)\,dt = \frac{1}{a}\ln(\cosh(ax)), \quad a \neq 0$$

$$\int_{x_0}^{x} \coth(t)\, dt = \ln(\sinh(x)) \quad x > 0,\ x_0 = \ln(1+\sqrt{2}) = 0.88137\ldots$$

The primitives of squared functions contain the function itself:

$$\int_0^x \tanh^2(at)\, dt = x - \frac{\tanh(ax)}{a}, a \neq 0$$

$$\int_{x_0}^{x} \coth^2(t)\, dt = x - \coth(x)\,, \qquad x > 0, \quad x_0 = 1.19967864\,.$$

The integrals of the difference between the functions and the asymptote $y = 1$ are

$$\int_x^\infty [1 - \tanh(t)]dt = \ln\left[1 + e^{-2x}\right]$$

$$\int_x^\infty [\coth(t) - 1]dt = \ln\left[1 - e^{-2x}\right]\,, \qquad x > 0\,.$$

The integral of the difference between $\coth(x)$ and $\frac{1}{x}$ is

$$\int_0^x \left[\coth(t) - \frac{1}{t}\right] dt = \ln\left(\frac{\sinh(x)}{x}\right)\,.$$

(m) Particular extensions and applications

Complex argument:

$$\tanh(x + jy) = \frac{\sinh(x)\cos(y) + j\cosh(x)\sin(y)}{\cosh(x)\cos(y) + j\sinh(x)\sin(y)}$$

$$\coth(x + jy) = \frac{\cosh(x)\cos(y) + j\sinh(x)\sin(y)}{\sinh(x)\cos(y) + j\cosh(x)\sin(y)}$$

Purely imaginary argument:

$$\tanh(jy) = j\tan(y) \qquad \coth(jy) = -j\coth(y)$$

5.28 Hyperbolic secant and hyperbolic cosecant functions

$$f(x) = \text{sech}\,(x)\,, \qquad g(x) = \text{csch}\,(x)$$

Hyperbolic functions used less frequently:

▷ For the hyperbolic cosecant, the notations $\text{cosch}(x)$ and $\text{cosech}(x)$ are used.

(a) Definition

The hyperbolic secant is the reciprocal function of the hyperbolic cosine:

$$\text{sech}\,(x) = \frac{2}{e^x + e^{-x}} = \frac{1}{\cosh(x)}$$

The hyperbolic cosecant is the reciprocal function of the hyperbolic sine:

$$\text{csch}\,(x) = \frac{2}{e^x - e^{-x}} = \frac{1}{\sinh(x)}\,.$$

First-order differential equation:

$$\frac{\mathrm{d}f(x)}{\mathrm{d}x}+\frac{a}{c}f(x)\sqrt{a^2+f^2(x)}=0; \qquad f(x)=c\cdot \operatorname{csch}(ax)$$

$$\frac{\mathrm{d}f(x)}{\mathrm{d}x}+\frac{a}{c}f(x)\sqrt{a^2-f^2(x)}=0; \qquad f(x)=c\cdot \operatorname{sech}(ax)$$

(b) Graphical representation

- For large absolute values of x, both functions tend to zero.

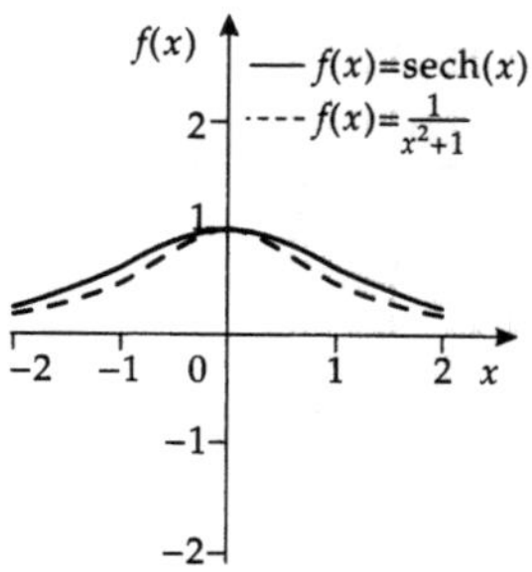

$f(x)=\operatorname{sech}(x)$ (solid line),
$f(x)=\dfrac{1}{x^2+1}$ (dashed line)

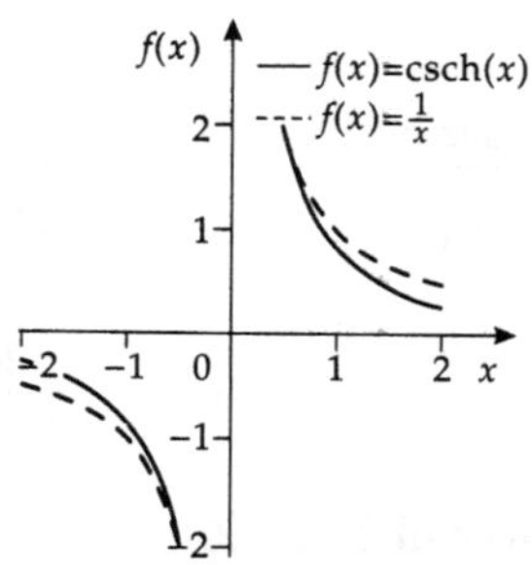

$f(x)=\operatorname{csch}(x)$ (solid line),
$f(x)=\frac{1}{x}$ (dashed line)

sech (x) is limited to values between 0 and 1, is mirror-symmetric about the y-axis and has a maximum at $x=0$, $y=1$.
csch (x) has positive as well as negative values, is point-symmetric about the origin and has a pole at $x=0$.

(c) Properties of the function

Domain of definition:	sech (x)	$-\infty<x<\infty$
	csch (x)	$-\infty<x<\infty$, but $x\neq 0$
Range of values:	sech (x)	$0<f(x)\leq 1$
	csch (x)	$-\infty<f(x)<\infty$, $f(x)\neq 0$
Quadrant:	sech (x)	first and second quadrants
	csch (x)	first and third quadrants
Periodicity:	None.	
Monotonicity:	sech (x)	$x>0$ strictly monotonically decreasing
		$x<0$ strictly monotonically increasing
	csch (x)	$x>0$ strictly monotonically decreasing
		$x<0$ strictly monotonically decreasing
Symmetries:	sech (x)	mirror symmetry about the y-axis
	csch (x)	point symmetry about the origin
Asymptotes:	$f(x)\to 0$ as $x\to\pm\infty$	

(d) Particular values

Zeros: None.

Discontinuities: None.

Poles: sech (x) none; csch (x) pole at $x = 0$

Extremes: sech (x) maximum at $x = 0$; csch (x) none

Points of inflection: sech (x) at $x = \pm\frac{1}{2}\ln\left(\frac{2+\sqrt{2}}{2-\sqrt{2}}\right) = \pm 0,8813\ldots$; csch (x) none

Value of	$x \to -\infty$	$x = -1$	$x = 0$	$x = 1$	$x \to \infty$
csch (x)	0	$\frac{2e}{1-e^2}$	$\mp\infty$	$\frac{2e}{e^2-1}$	0
sech (x)	0	$\frac{2e}{1+e^2}$	1	$\frac{2e}{e^2+1}$	0

(e) Reciprocal function

The hyperbolic cosine and hyperbolic sine are the reciprocal functions of the hyperbolic secant and hyperbolic cosecant, respectively:

$$\frac{1}{\operatorname{sech}(x)} = \cosh(x) , \qquad \frac{1}{\operatorname{csch}(x)} = \sinh(x) .$$

(f) Inverse function

The inverse functions are the corresponding inverse hyperbolic functions.

▷ sech (x) can be inverted only for $x \geq 0$ (with an additional minus sign for $x < 0$)

$$x = \operatorname{Arsech}(y) \qquad y = \operatorname{sech}(x)$$

$$x = \operatorname{Arcsch}(y) \qquad y = \operatorname{csch}(x)$$

The distinct domains of definition of the inverse functions have to be taken into account.

(g) Cognate functions

The other hyperbolic functions and their cognates.
The area functions.

(h) Conversion formulas

Reflection in the y-axis:

$$\operatorname{sech}(-x) = \operatorname{sech}(x) , \qquad \operatorname{csch}(-x) = -\operatorname{csch}(x)$$

Double argument:

$$\operatorname{sech}(2x) = \frac{\operatorname{sech}^2(x)}{2 - \operatorname{sech}^2(x)}$$

$$\operatorname{csch}(2x) = \frac{\operatorname{sech}(x)\operatorname{csch}(x)}{2}$$

Interrelation formula:

$$\left(1 - \operatorname{sech}^2(x)\right)\left(1 + \operatorname{csch}^2(x)\right) = 1$$

(i) Approximation formulas (8 bits $\approx$ 0.4% of accuracy)

For large absolute values of x, it holds that

$$\operatorname{sech}(x) = 2\mathrm{e}^{-|x|}, \qquad |x| \geq 2.8$$
$$\operatorname{csch}(x) = 2\operatorname{sgn}(x)\mathrm{e}^{-|x|}, \qquad |x| \geq 2.8$$

(j) Expansion of series or products

Partial fraction decomposition:

$$\operatorname{sech}(x) = \frac{4\pi}{\pi^2 + 4x^2} - \frac{12\pi}{9\pi^2 + 4x^2} + \frac{20\pi}{25\pi^2 + 4x^2} - \ldots$$
$$= \sum_{k=0}^{\infty} (-1)^k \frac{(8k+4)\pi}{(2k+1)^2\pi^2 + x^2}$$
$$\operatorname{csch}(x) = \frac{1}{x} - \frac{2x}{\pi^2 + x^2} + \frac{2x}{4\pi^2 + x^2} - \frac{2x}{9\pi^2 + x^2} + \ldots$$
$$= \sum_{k=-\infty}^{\infty} (-1)^k \frac{x}{k^2\pi^2 + x^2}$$

(k) Derivative of the function

For the derivatives, it holds that

$$\frac{\mathrm{d}}{\mathrm{d}x}\operatorname{sech}(x) = -\operatorname{sech}(x)\tanh(x) = -\operatorname{sech}(x)\sqrt{1 - \operatorname{sech}^2(x)}$$
$$\frac{\mathrm{d}}{\mathrm{d}x}\operatorname{csch}(x) = -\operatorname{csch}(x)\coth(x) = -\operatorname{csch}(x)\sqrt{1 + \operatorname{csch}^2(x)}$$

(l) Primitive of the function

The primitive of sech is the inverse tangent of the hyperbolic sine:

$$\int_0^x \operatorname{sech}(t)\,\mathrm{d}t = \arctan(\sinh(x)) \equiv \operatorname{gd}(x)\,.$$

▷ The primitive is also called the **Gudermannian function**, $\operatorname{gd}(x)$.
The primitive of $\operatorname{csch}(x)$ is the logarithm of the hyperbolic cotangent:

$$\int_x^\infty \operatorname{csch}(t)\,\mathrm{d}t = \ln\left(\coth\left(\frac{x}{2}\right)\right), \qquad x > 0\,.$$

The primitives of the squares are the hyperbolic tangent and the hyperbolic cotangent:

$$\int_0^x \operatorname{sech}^2(t)\,\mathrm{d}t = \tanh(x)$$
$$\int_x^\infty \operatorname{csch}^2(t)\,\mathrm{d}t = -\coth(x)\,, \qquad x > 0\,.$$

(m) Particular extensions and applications

Complex argument:

$$\operatorname{sech}(x+\mathrm{j}y) = \frac{1}{\cosh(x)\cos(y) + \mathrm{j}\sinh(x)\sin(y)}$$

$$\operatorname{csch}(x+\mathrm{j}y) = \frac{1}{\sinh(x)\cos(y) + \mathrm{j}\cosh(x)\sin(y)}$$

Purely imaginary argument:

$$\operatorname{sech}(\mathrm{j}y) = \sec(y) \qquad \operatorname{csch}(\mathrm{j}y) = -\mathrm{j}\csc(y)$$

Area hyperbolic functions

Area **hyperbolic functions** are the inverse functions of the hyperbolic functions.
Inverse hyperbolic functions: Another term for area-hyperbolic functions (see preceding section on hyperbolic functions). As with the hyperbolic functions, there are also conversion formulas; these are shown in the conversion tables below.

Function	$f(x) = \text{Arsinh}(x)$	$f(x) = \text{Arcosh}(x)$	$f(x) = \text{Artanh}(x)$
Arsinh (x)	$f(x)$	$\text{sgn}(x) f(\sqrt{1+x^2})$	$f\left(\frac{x}{\sqrt{1+x^2}}\right)$
Arcosh (x)	$f(\sqrt{x^2-1})$	$f(x)$	$f\left(\frac{\sqrt{x^2-1}}{x}\right)$
Artanh (x)	$f\left(\frac{x}{\sqrt{1-x^2}}\right)$	$\text{sgn}(x) f\left(\frac{1}{\sqrt{1-x^2}}\right)$	$f(x)$
Arsech (x)	$f\left(\frac{\sqrt{1-x^2}}{x}\right)$	$f\left(\frac{1}{x}\right)$	$f(\sqrt{1-x^2})$
Arcsch (x)	$f\left(\frac{1}{x}\right)$	$\text{sgn}(x) f\left(\sqrt{1+\frac{1}{x^2}}\right)$	$\text{sgn}(x) f\left(\frac{1}{\sqrt{1+x^2}}\right)$
Arcoth (x)	$\text{sgn}(x) f\left(\frac{1}{\sqrt{x^2-1}}\right)$	$\text{sgn}(x) f\left(\sqrt{\frac{x^2}{x^2-1}}\right)$	$f\left(\frac{1}{x}\right)$

Function	$f(x) = \text{Arsech}(x)$	$f(x) = \text{Arcsch}(x)$	$f(x) = \text{Arcoth}(x)$
Arsinh (x)	$\text{sgn}(x) f\left(\frac{1}{\sqrt{1+x^2}}\right)$	$f\left(\frac{1}{x}\right)$	$f\left(\frac{\sqrt{1+x^2}}{x}\right)$
Arcosh (x)	$f\left(\frac{1}{x}\right)$	$f\left(\frac{1}{\sqrt{x^2-1}}\right)$	$f\left(\frac{x}{\sqrt{x^2-1}}\right)$
Artanh (x)	$\text{sgn}(x) f(\sqrt{1-x^2})$	$f\left(\frac{\sqrt{1-x^2}}{x}\right)$	$f\left(\frac{1}{x}\right)$
Arsech (x)	$f(x)$	$f\left(\frac{x}{\sqrt{1-x2}}\right)$	$f\left(\frac{1}{\sqrt{1-x^2}}\right)$
Arcsch (x)	$\text{sgn}(x) f\left(\sqrt{\frac{x^2}{1+x^2}}\right)$	$f(x)$	$\text{sgn}(x) f(\sqrt{1+x^2})$
Arcoth (x)	$\text{sgn}(x) f\left(\sqrt{1-\frac{1}{x^2}}\right)$	$\text{sgn}(x) f(\sqrt{x^2-1})$	$f(x)$

Area-hyperbolic functions are characterized by the prefix Ar, such as, Arsinh (x). Notations with capital initial letters, Arsinh, as well as small letters, arsinh(x), are used.

▷ Here, notations with capital first letters are used to distinguish clearly between the area-hyperbolic functions and the inverse arc functions.

There are also notations with the prefix arc or arg, arcsinh(x), argsinh(x), but these can lead to confusion.
Notations with -1 as an exponent, such as $\sinh^{-1}(x)$, are used mainly in North America.

5.29 Area hyperbolic sine and hyperbolic cosine

$$f(x) = \text{Arsinh}\,(x)\,, \qquad g(x) = \text{Arcosh}\,(x)$$

Inverse functions of the hyperbolic sine and the hyperbolic cosine.
Alternative notation: arsh(x), arcsinh(x); arch(x), arccosh(x); and $\sinh^{-1}(x)$, $\cosh^{-1}(x)$.

(a) Definition

Inverse function of the hyperbolic function:

$$\begin{aligned} x &= \text{Arsinh}\,(y)\,, \quad y = \sinh(x) \\ x &= \text{Arcosh}\,(y)\,, \quad y = \cosh(x)\,, \qquad x > 0 \end{aligned}$$

Representation in terms of logarithms:

$$\begin{aligned} \text{Arsinh}\,(x) &= \ln\left(x + \sqrt{x^2+1}\right) \\ \text{Arcosh}\,(x) &= \ln\left(x \pm \sqrt{x^2-1}\right)\,, \qquad (+ : x \geq 1,\ - : x \leq -1) \end{aligned}$$

Integral representation:

$$\begin{aligned} \text{Arsinh}\,(x) &= \int_0^x \frac{\mathrm{d}t}{\sqrt{1+t^2}} \\ \text{Arcosh}\,(x) &= \int_0^x \frac{\mathrm{d}t}{\sqrt{t^2-1}}\,, \qquad x \geq 1 \end{aligned}$$

Extension of the domain of definition: sometimes Arcosh (x) is defined also for $x \leq -1$.

$$\text{Arcosh}\,(x) \equiv \text{Arcosh}\,(|x|)$$

(b) Graphical representation

Both functions are strictly monotonically increasing, approaching infinity as $x \to \infty$.

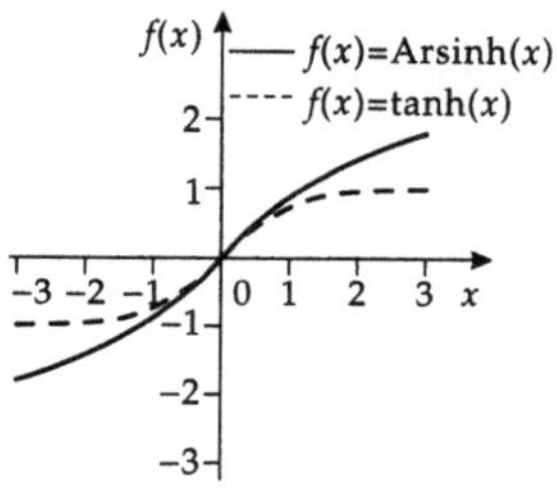

$f(x) = \text{Arsinh}(x)$ (solid line),
$f(x) = \tanh(x)$ (dashed line)

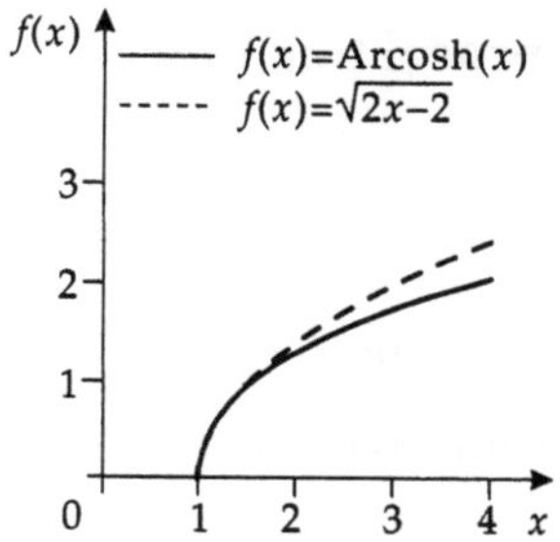

$f(x) = \text{Arcosh}(x)$ (solid line),
$f(x) = \sqrt{2x-2}$ (dashed line)

▷ While values of x are allowed for Arsinh (x), Arcosh (x) is defined only for values $x \geq 1$.

Arsinh (x) goes through the origin in whose vicinity it behaves like a straight line with slope 1.

Arcosh (x) goes into point $x = 1$, $y = 0$ with a vertical tangent.

(c) Properties of the function

Domain of definition:	Arsinh (x)	$-\infty < x < \infty$
	Arcosh (x)	$1 \leq x < \infty$
Range of values:	Arsinh (x)	$-\infty < f(x) < \infty$
	Arcosh (x)	$0 \leq f(x) < \infty$
Quadrant:	Arsinh (x)	first and third quadrants
	Arcosh (x)	first quadrant
Periodicity:	None.	
Monotonicity:	Strictly monotonically increasing.	
Symmetries:	Arsinh (x)	point symmetry about the origin
	Arcosh (x)	no symmetry
Asymptotes:	Arsinh (x)	$f(x) \to \pm \ln(2\|x\|)$ as $x \to \pm\infty$
	Arcosh (x)	$f(x) \to \ln(2x)$ as $x \to +\infty$

(d) Particular values

Zeros:	Arsinh (x)	$x = 0$
	Arcosh (x)	$x = 1$
Discontinuities:	None.	
Poles:	None.	
Extrema:	None.	
Points of inflection:	Arsinh (x)	$x = 0$
	Arcosh (x)	none

(e) Reciprocal function

The reciprocal functions cannot be written as other area-hyperbolic functions or hyperbolic functions.

(f) Inverse function

The inverse functions are the hyperbolic functions:

$$x = \sinh(y) , \qquad y = \operatorname{Arsinh}(x) .$$

- The inverse function of $\operatorname{Arcosh}(x)$ defines only the positive branch of $\cosh(x)$:

$$x = \cosh(y) , \qquad y = \operatorname{Arcosh}(x) , \qquad y \geq 0 .$$

(g) Cognate function

Hyperbolic functions, the inverse functions.
Other area-hyperbolic functions.
Logarithmic function, a representation of area-hyperbolic functions.

(h) Conversion formulas

A reflection about the y-axis is possible only for $\operatorname{Arsinh}(x)$ only:

$$\operatorname{Arsinh}(-x) = -\operatorname{Arsinh}(x) .$$

Addition of functions:

$$\operatorname{Arsinh}(x) \pm \operatorname{Arsinh}(y) = \operatorname{Arsinh}\left(x\sqrt{1+y^2} \pm y\sqrt{1+x^2}\right)$$

$$\operatorname{Arsinh}(x) \pm \operatorname{Arcosh}(y) = \operatorname{Arsinh}\left(xy \pm \sqrt{(x^2+1)(y^2-1)}\right)$$

$$\operatorname{Arcosh}(x) \pm \operatorname{Arcosh}(y) = \operatorname{Arcosh}\left(xy \pm \sqrt{(x^2-1)(y^2-1)}\right)$$

(i) Approximation formulas (8 bits ≈ 0.4% of accuracy)

For large values of x, it holds that

$$\operatorname{Arsinh}(x) \approx \operatorname{sgn}(x)\left(\ln(2|x|) + \frac{1}{4x^2}\right) , \qquad |x| \geq 2$$

$$\operatorname{Arcosh}(x) \approx \ln(2|x|) - \frac{1}{4x^2} , \qquad |x| \geq 2.2 .$$

(j) Expansion of series or products of the function

Series expansion of $\operatorname{Arsinh}(x)$ for $|x| < 1$:

$$\begin{aligned}\operatorname{Arsinh}(x) &= x - \frac{x^3}{6} + \frac{3x^5}{40} - \frac{5x^7}{112} + \cdots \\ &= x\sum_{k=0}^{\infty} \frac{(2k-1)!!}{(2k)!!}\,\frac{(-x^2)^k}{2k+1}\end{aligned}$$

Expansion of Arsinh (x) for $|x| > 1$:

$$\operatorname{Arsinh}(x) = \operatorname{sgn}(x) \cdot \left[\ln(2|x|) + \frac{1}{4x^2} - \frac{3}{32x^4} + \frac{5}{96x^6} - \cdots \right]$$

$$= \operatorname{sgn}(x) \cdot \left[\ln(2|x|) - \sum_{k=1}^{\infty} \frac{(2k-1)!!}{2k\,(2k)!!\,(-x^2)^k} \right]$$

(k) Derivative of the function

The derivatives of Arsinh (x) and Arcosh (x) are the roots of reciprocal quadratic functions.

$$\frac{\mathrm{d}}{\mathrm{d}x}\operatorname{Arsinh}(x) = \frac{1}{\sqrt{1+x^2}}$$

$$\frac{\mathrm{d}}{\mathrm{d}x}\operatorname{Arcosh}(x) = \frac{1}{\sqrt{x^2-1}}, \qquad x > 1$$

(l) Primitive of the function

Primitives of Arsinh (x) and Arcosh (x)

$$\int_0^x \operatorname{Arsinh}(t)\,\mathrm{d}t = x\operatorname{Arsinh}(x) - \sqrt{x^2+1} + 1$$

$$\int_1^x \operatorname{Arcosh}(t)\,\mathrm{d}t = x\operatorname{Arcosh}(x) - \sqrt{x^2-1}\,, \qquad x \geq 1$$

(m) Particular extensions and applications

Purely imaginary argument: we obtain the inverse trigonometric functions,

$$\operatorname{Arsinh}(\mathrm{j}x) = \mathrm{j}\,\arcsin(x)\,.$$

5.30 Area-hyperbolic tangent and hyperbolic cotangent

$$f(x) = \operatorname{Artanh}(x)\,, \qquad g(x) = \operatorname{Arcoth}(x)$$

The inverse functions of the hyperbolic tangent and hyperbolic cotangent.
Alternative notation: arth(x), arctanh(x); arth(x), arccoth(x); $\tanh^{-1}(x)$, $\coth^{-1}(x)$.

(a) Definition

Inverse function of the hyperbolic functions:

$$x = \operatorname{Artanh}(y) \qquad y = \tanh(x)$$

$$x = \operatorname{Arcoth}(y) \qquad y = \coth(x)$$

Representation in terms of logarithms:

$$\operatorname{Artanh}(x) = \ln\left(\sqrt{\frac{1+x}{1-x}}\right) = \frac{1}{2}\ln\left(\frac{1+x}{1-x}\right), \qquad |x| < 1$$

$$\text{Arcoth}(x) = \ln\left(\sqrt{\frac{x+1}{x-1}}\right) = \frac{1}{2}\ln\left(\frac{x+1}{x-1}\right), \qquad |x| > 1$$

Integral representation:

$$\text{Artanh}(x) = \int_0^x \frac{dt}{1-t^2}$$

$$\text{Arcoth}(x) = \int_x^\infty \frac{dt}{t^2-1} \qquad x > 1$$

$$= \int_{-\infty}^x \frac{dt}{t^2-1} \qquad -1 > x$$

(b) Graphical representation

Artanh (x) is strictly monotonically increasing; Arcoth (x) is strictly monotonically decreasing in its domain of definition.

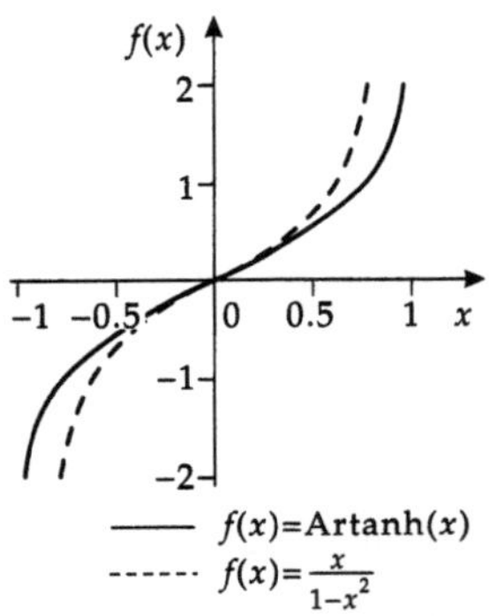

$f(x) = $ Artanh (x) (solid line), $f(x) = \dfrac{x}{1-x^2}$ (dashed line)

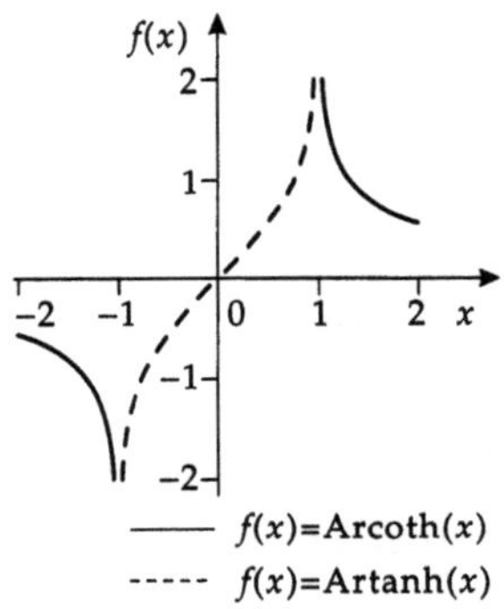

$f(x) = $ Arcoth (x) (solid line), $f(x) = $ Artanh (x) (dashed line)

- For Artanh (x), only values $|x| < 1$ are allowed; Arcoth (x) is defined only for values $|x| > 1$.

Artanh (x) goes through the origin and behaves like a straight line of slope 1 in the vicinity of the origin.

Arcoth (x) approaches 0 as $x \to \pm\infty$.

(c) Properties of the function

Domain of definition:	Artanh (x)	$-1 < x < 1$
	Arcoth (x)	$-\infty < x < -1$ and $1 < x < \infty$
Range of values:	Artanh (x)	$-\infty < f(x) < \infty$
	Arcoth (x)	$-\infty < f(x) < \infty$, $f(x) \neq 0$
Quadrant:	First and third quadrants.	
Periodicity:	None.	
Monotonicity:	Artanh (x)	strictly monotonically increasing
	Arcoth (x)	strictly monotonically decreasing

Symmetries:	Point symmetry about the origin.	
Asymptotes:	Artanh (x)	$f(x) \to \infty$ as $x \to 0^+$
	Arcoth (x)	$f(x) \to 0$ as $x \to \pm\infty$

(d) Particular values

Zeros:	Artanh (x)	$x = 0$
	Arcoth (x)	none
Discontinuities:	None.	
Poles:	Pole at $x = -1, x = +1$	
Extremes:	None.	
Points of inflection:	Artanh (x)	$x = 0$
	Arcoth (x)	none

(e) Reciprocal function

The reciprocal functions cannot be written in terms of area-hyperbolic or hyperbolic functions.

(f) Inverse function

The inverse functions are the hyperbolic functions:

$$x = \tanh(y) \qquad y = \operatorname{Artanh}(x)$$

$$x = \coth(y) \qquad y = \operatorname{Arcoth}(x)$$

(g) Cognate function

Hyperbolic functions, the inverse functions.
Other inverse hyperbolic functions.
Logarithmic functions.

(h) Conversion formulas

Reflection about the y-axis:

$$\operatorname{Artanh}(-x) = -\operatorname{Artanh}(x)\,, \qquad \operatorname{Arcoth}(-x) = -\operatorname{Arcoth}(x)$$

Addition of functions:

$$\operatorname{Artanh}(x) \pm \operatorname{Artanh}(y) = \operatorname{Artanh}\left(\frac{x \pm y}{1 \pm xy}\right)$$

$$\operatorname{Artanh}(x) \pm \operatorname{Arcoth}(y) = \operatorname{Artanh}\left(\frac{1 \pm xy}{x \pm y}\right)$$

$$\operatorname{Arcoth}(x) \pm \operatorname{Arcoth}(y) = \operatorname{Arcoth}\left(\frac{1 \pm xy}{x \pm y}\right)$$

(i) Approximation formulas (8 bits ≈ 0.4% of accuracy)

For small values of x, it holds that

$$\operatorname{Artanh}(x) \approx \frac{x}{\sqrt{1 - \dfrac{2x^3}{3}}}, \qquad |x| \leq \frac{1}{2}$$

(j) Expansion of series or products in the function

Series expansion of Artanh (x) for $|x| < 1$:

$$\begin{aligned}\operatorname{Artanh}(x) &= x + \frac{x^3}{3} + \frac{x^5}{5} + \frac{x^7}{7} + \frac{x^9}{9} + \cdots \\ &= \sum_{k=0}^{\infty} \frac{x^{2k+1}}{2k+1}\end{aligned}$$

Continual fraction expansion of Artanh (x):

$$\operatorname{Artanh}(x) = x \cfrac{1}{1 - \cfrac{x^2}{3 - \cfrac{(2x)^2}{5 - \cfrac{(3x)^2}{7 - \cfrac{(4x)^2}{9 - \cdots}}}}}$$

(k) Derivative of the function

The derivatives of Artanh (x) and Arcoth (x) are the reciprocal quadratic functions:

$$\frac{\mathrm{d}}{\mathrm{d}x}\operatorname{Artanh}(x) = \frac{1}{1 - x^2}, \qquad |x| < 1$$

$$\frac{\mathrm{d}}{\mathrm{d}x}\operatorname{Arcoth}(x) = \frac{1}{1 - x^2}, \qquad |x| > 1$$

(l) Primitive of the function

Primitives of Artanh (x) and Arcoth (x):

$$\int_0^x \operatorname{Artanh}(t)\,\mathrm{d}t = x\operatorname{Artanh}(x) + \ln\left(\sqrt{1 - x^2}\right), \qquad |x| < 1$$

$$\int_2^x \operatorname{Arcoth}(t)\,\mathrm{d}t = x\operatorname{Arcoth}(x) + \ln\left(\sqrt{\frac{x^2 - 1}{27}}\right), \qquad x > 1$$

(m) Particular extensions and applications

Purely imaginary argument: we obtain the inverse trigonometric functions.

$$\operatorname{Artanh}(\mathrm{j}x) = \mathrm{j}\arctan(x) \qquad \operatorname{Arcoth}(\mathrm{j}x) = -\mathrm{j}\operatorname{arccot}(x)$$

5.31 Area-hyperbolic secant and hyperbolic cosecant

$$f(x) = \operatorname{Arsech}(x)\,, \qquad g(x) = \operatorname{Arcsch}(x)$$

The inverse functions of the hyperbolic secant and hyperbolic cosecant.
Alternative notation: arcsech(x), arccsch(x); $\operatorname{sech}^{-1}(x)$, $\operatorname{csch}^{-1}(x)$.

(a) Definition

Inverse function of the hyperbolic function:

$$x = \operatorname{Arsech}(y)\,, \qquad y = \operatorname{sech}(x)\,, \qquad x > 0$$
$$x = \operatorname{Arcsch}(y)\,, \qquad y = \operatorname{csch}(x)$$

Representation in terms of logarithms:

$$\operatorname{Arsech}(x) = \ln\left(\frac{1+\sqrt{1-x^2}}{x}\right)\,, \qquad 0 < x \le 1$$
$$\operatorname{Arcsch}(x) = \ln\left(\frac{1}{x} + \sqrt{1+\frac{1}{x^2}}\right)$$

Integral representation:

$$\operatorname{Arsech}(x) = \int_0^x \frac{\mathrm{d}t}{t\sqrt{1-t^2}}$$
$$\operatorname{Arcsch}(x) = \int_x^\infty \frac{\mathrm{d}t}{t\sqrt{t^2+1}}\,, \qquad x > 0$$
$$= \int_{-\infty}^x \frac{\mathrm{d}t}{t\sqrt{t^2+1}}\,, \qquad x < 0$$

Extension of the domain of definition; sometimes Arsech (x) is also defined for $x \le 0$.

$$\operatorname{Arsech}(x) \equiv \operatorname{Arsech}(|x|)\,, \qquad -1 < x < 1$$

(b) Graphical representation

Both functions are strictly monotonically decreasing in their domain of definition.

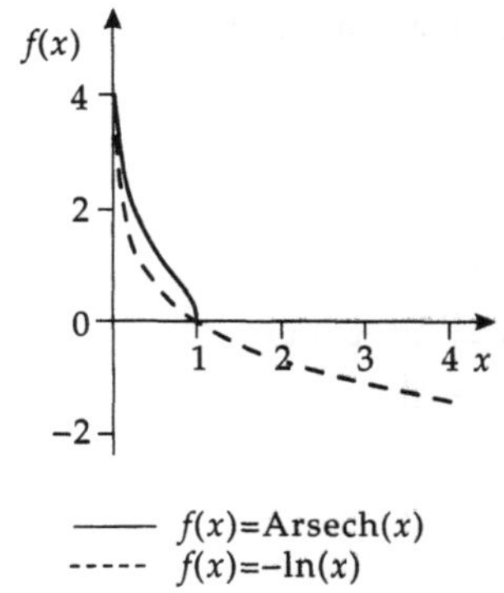

$f(x) = \operatorname{Arsech}(x)$ (solid line), $f(x) = -\ln(x)$ (dashed line)

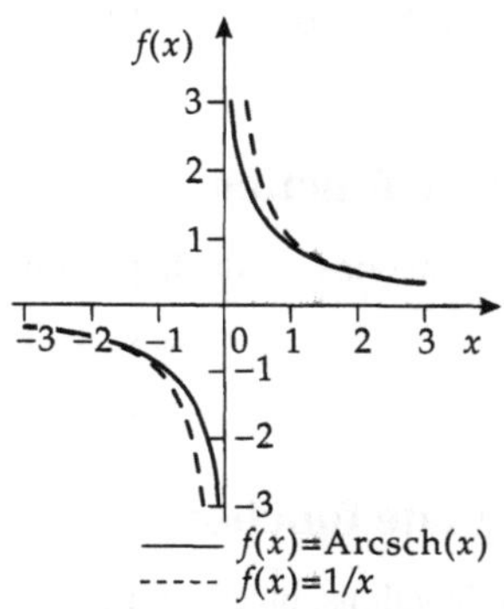

$f(x) = \operatorname{Arcsch}(x)$ (solid line), $f(x) = \frac{1}{x}$ (dashed line)

- While for Arsech (x), only the values $0 < x \leq 1$ are allowed; Arcsch (x) is defined for all values $x \neq 0$.

Arsech (x) diverges at $x = 0$ and intersects the x-axis perpendicularly ($x = 1, y = 0$).
Arcsch (x) is point-symmetric about the origin, diverges at $x = 0$, and approaches $y = 0$ for $x \to \pm\infty$.

(c) Properties of the function

Domain of definition:	Arsech (x)	$0 < x \leq 1$
	Arcsch (x)	$-\infty < x < \infty, x \neq 0$
Range of values:	Arsech (x)	$0 \leq f(x) < \infty$
	Arcsch (x)	$-\infty < f(x) < \infty,\ f(x) \neq 0$
Quadrant:	Arsech (x)	first quadrant
	Arcsch (x)	first and third quadrant
Periodicity:	None.	
Monotonicity:	Strictly monotonically decreasing.	
Symmetries:	Arsech (x)	no symmetry
	Arcsch (x)	point symmetry about the origin
Asymptotes:	Arsech (x)	$f(x) \to 0$ for $x \to +1$
	Arcsch (x)	$f(x) \to 0$ for $x \to \pm\infty$

(d) Particular values

Zeros:	Arsech (x)	$x = 1$
	Arcsch (x)	no
Discontinuities:	None.	
Poles:	Pole at $x = 0$.	
Extrema:	None.	
Points of inflection:	Arsech (x)	$x = \sqrt{\frac{1}{2}}$
	Arcsch (x)	none

(e) Reciprocal function

The reciprocal functions cannot be written in terms of area-hyperbolic or hyperbolic functions.

(f) Inverse function

The inverse functions are the hyperbolic functions:

$$x = \operatorname{sech}(y) \qquad y = \operatorname{Arsech}(x)$$

$$x = \operatorname{csch}(y) \qquad y = \operatorname{Arcsch}(x)$$

(g) Cognate function

Hyperbolic functions, inverse functions.
Other area-hyperbolic functions.
Logarithmic functions, representation of the area-hyperbolic functions.

(h) Conversion formulas

The reflection in the y-axis is possible only for Arcsch (x):

$$\text{Arcsch}\,(-x) = -\text{Arcsch}\,(x)\ .$$

(i) Approximation formulas for the function

For small values of x, it holds that

$$\text{Arsech}\,(x) \approx -\ln\left(\frac{x}{2}\right) - \frac{x^2}{4}\,, \qquad 0 < x \le 0.45$$

$$\text{Arcsch}\,(x) \approx \text{sgn}(x) \cdot \left[-\ln\left(\frac{x}{2}\right) + \frac{x^2}{4}\right]\,, \qquad |x| \le \frac{1}{2}$$

(j) Expansion of series or products of the function

Series expansion of Arsech (x) for $|x| < 1$:

$$\begin{aligned}\text{Arsech}\,(x) &= \ln\left(\frac{2}{x}\right) - \frac{x^2}{4} - \frac{3x^4}{32} - \frac{5x^6}{96} - \cdots \\ &= \ln\left(\frac{2}{x}\right) - \sum_{k=1}^{\infty} \frac{(2k-1)!!}{(2k)!!}\,\frac{x^{2k}}{2k}\,, \qquad 0 < x < 1\end{aligned}$$

(k) Derivative of the function

Derivatives of Arsech (x) and Arcsch (x)

$$\frac{\mathrm{d}}{\mathrm{d}x}\text{Arsech}\,(x) = -\frac{1}{x\sqrt{1-x^2}}\,, \qquad 0 < x < 1$$

$$\frac{\mathrm{d}}{\mathrm{d}x}\text{Arcsch}\,(x) = -\frac{1}{|x|\sqrt{x^2+1}}$$

(l) Primitive of the function

Primitives of Arsech (x) and Arcsch (x):

$$\int_0^x \text{Arsech}\,(t)\,\mathrm{d}t = x\,\text{Arsech}\,(x) + \arcsin(x)\,, \qquad 0 < x < 1$$

$$\int_0^x \text{Arcsch}\,(t)\,\mathrm{d}t = x\,\text{Arcsch}\,(x) + |\text{Arsinh}\,(x)|$$

(m) Particular extensions and applications

Purely imaginary argument: We obtain the inverse trigonometric functions:

$$\text{Arcsch}\,(\mathrm{j}x) = -\mathrm{j}\,\text{arccsc}(x)$$

Trigonometric functions

As well as the hyperbolic functions, **trigonometric functions** belong to the transcendental functions. They are closely related to the exponential functions.

⊳ In the complex plane, the hyperbolic and trigonometric functions can be represented as functions of the same kind, which can be transformed into each other by using complex arguments.

Like the hyperbolic functions, the trigonometric functions are closely connected with conic sections. But instead of intersections with branches of the hyperbola $y^2 - x^2 = 1$, in this case, they are based on intersections with the arcs of the circle $y^2 + x^2 = 1$.

Consider the unit circle about the origin:

$$y = \pm\sqrt{1 - x^2}\,.$$

This circle is intersected by a straight line, and at first only the first and second quadrants, $y \geq 0$, are of interest.

$$g(x) = T \cdot x, \qquad y \geq 0\,.$$

The intersection point of g with the upper circular arc has the coordinates $x = c$, $y = s = t \cdot c$.

⊳ c and s are interrelated through the equation $s = \sqrt{1 - c^2}$.

The length of the circular arc between the points $(x = 1, y = 0)$ and $(x = c, y = s)$ has the value a.

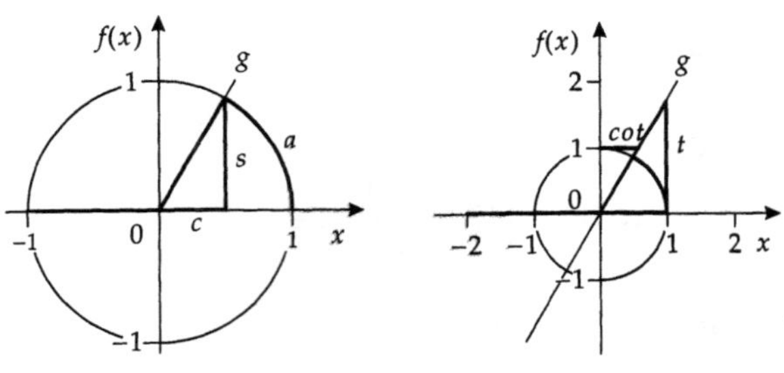

Geometric interpretation of the trigonometric functions

The values of s, c, and t can be related to a:

$$S = \sin(A) \qquad C = \cos(A) \qquad T = \tan(A)$$

We can treat the lower circular arc analogously. To guarantee the distinctiveness of the function $f(a)$ we determine that the segments, plotted clockwise, have negative values, and the segments, plotted counterclockwise, have positive values.

⊳ If we consider only counterclockwise rotations and not only multiple revolutions, i.e., $0 \leq a \leq 2\pi$, then a can also be interpreted as the area enclosed by the segments $p(x = 0, y = 0)$, $p(x = 1, y = 0)$, and the circular arc.
This interpretation is analogous to the interpretation of the hyperbolic functions (see the section on hyperbolic functions).

Thus, one point $P(x, y)$ can possess several values a (actually an infinite number).

$P(x = 0, y = -1)$ can be described, for example by the arcs $A = -\dfrac{\pi}{2}$, $A = \dfrac{3\pi}{2}$.

In particular, any number of circular revolutions can be performed, thus enlarging the arc length a by an integral multiple n of the circumference 2π of the circle.

$P(x = 0, y = -1)$ can be described by the arcs $A = \frac{3\pi}{2} \pm 2n\pi$.

▷ In computers and pocket calculators, functions with larger arguments (i.e., after the addition of many revolutions) are calculated with less accuracy than functions with smaller arguments.

Thus, to every value a a value s, c and t can be assigned uniquely.

The assigning functions are called the **sine function**, **cosine function**, and **tangent function** (to emphasize the function character, x is used instead of a).

$$S = \sin(A)\,, \qquad C = \cos(A)\,, \qquad T = \tan(A)$$

Further trigonometric functions are the **secant function**, $\sec(x)$, the **cosecant function**, $\csc(x)$, and the **cotangent function**, $\cot(x)$, which are the reciprocal functions of cosine, sine and tangent, respectively.

$$\sec(x) = \frac{1}{\cos(x)} \qquad \csc(x) = \frac{1}{\sin(x)} \qquad \cot(x) = \frac{1}{\tan(x)}$$

- The backward assignment of s, c, and t to a is no longer unique.

▷ To avoid this, we define the so-called **principal value**. For details, see the section on inverse trigonometric functions.

- Two important relations of the trigonometric functions have already been discussed above:

$$\sin^2(x) = 1 - \cos^2(x) \qquad \sin(x) = \tan(x) \cdot \cos(x)$$

The geometric interpretation of the trigonometric functions (see Figures on preceding page).

In the figure on the left, the straight line g intersects the circular arc at point $P(x = C, y = S)$.

In the figure on the right, the same straight line g intersects the x-axis in the origin $O = P(x = 0, y = 0)$, and the straight line at point $K = P(x = k, y = 1)$.

The straight line $x = 1$ intersects the x-axis in point $E = P(x = 1, y = 0)$ and the straight line g in point $T = P(x = 1, y = t)$.

Furthermore, the segments c, s, and the segment joining the origin $P(x = 0, y = 0)$ to the intersection point $P(x = C, y = S)$ of the straight line with the circle can be combined to form a **right triangle**, where the segment between the origin and the intersection point is the hypotenuse, and the segments c and s are the other sides.

The function values of the various trigonometric functions can be obtained as follows:

$\sin(x)$: y-value of the intersection point $(x = c, y = s)$ of the straight line and the circle (figure at left). Ratio of the side s to the hypotenuse.

$\cos(x)$: x-value of the intersection point $(x = c, y = s)$ of the straight line and the circle (figure at left). Ratio of side c to the hypotenuse.

$\tan(x)$: y-value of the intersection point t of straight line g and the vertical tangent $x = 1$; length of the segment $\overline{ET}$ (figure at right).
Ratio of side s to side c.

$\cot(x)$: y-value of the intersection points K of the straight line g and the horizontal tangent $y = 1$; length of the segment from $(x = 0, y = 1)$ to K (figure at right). Ratio of the side c to the side s.

$\sec(x)$: length of the segment $\overline{OT}$ (figure at right); the sign is determined by the sign of c. Ratio of hypotenuse to side s.

csc(x): length of the segment $\overline{OK}$ (figure at right); the sign is determined by the sign of s. Ratio of hypotenuse to the side s.

Conversion of trigonometric functions

In the conversion of trigonometric functions, one has to note the sign of the functions. The following table shows the **sign factors** s_2, s_3, s_4 (which depend on the value of x) often must be taken into account.

Domain:	$0 \to \frac{\pi}{2}$	$\frac{\pi}{2} \to \pi$	$\pi \to \frac{3\pi}{2}$	$\frac{3\pi}{2} \to 2\pi$
s_2	$+1$	$+1$	-1	-1
s_3	$+1$	-1	$+1$	-1
s_4	$+1$	-1	-1	$+1$

Function	sin	cos	tan
$\sin(x)$	$\sin(x)$	$s_2\sqrt{1-\cos^2(x)}$	$\frac{s_4\tan(x)}{\sqrt{1+\tan^2(x)}}$
$\cos(x)$	$s_4\sqrt{1-\sin^2(x)}$	$\cos(x)$	$\frac{s_4}{\sqrt{1+\tan^2(x)}}$
$\tan(x)$	$\frac{s_4\sin(x)}{\sqrt{1-\sin^2(x)}}$	$s_2\frac{\sqrt{1-\cos^2(x)}}{\cos(x)}$	$\tan(x)$
$\cot(x)$	$\frac{s_4\sqrt{1-\sin^2(x)}}{\sin(x)}$	$\frac{s_2\cos(x)}{\sqrt{1-\cos^2(x)}}$	$\frac{1}{\tan(x)}$
$\sec(x)$	$\frac{s_4}{\sqrt{1-\sin^2(x)}}$	$\frac{1}{\cos(x)}$	$s_4\sqrt{1+\tan^2(x)}$
$\csc(x)$	$\frac{1}{\sin(x)}$	$\frac{s_2}{\sqrt{1-\cos^2(x)}}$	$\frac{s_4\sqrt{1+\tan^2(x)}}{\tan(x)}$

Function	cot	sec	csc
$\sin(x)$	$\frac{s_2}{\sqrt{\cot^2(x)+1}}$	$\frac{s_3\sqrt{\sec^2(x)-1}}{\sec(x)}$	$\frac{1}{\csc(x)}$
$\cos(x)$	$\frac{s_2\cot(x)}{\sqrt{\cot^2(x)+1}}$	$\frac{1}{\sec(x)}$	$\frac{s_2\sqrt{\csc^2(x)-1}}{\csc(x)}$
$\tan(x)$	$\frac{1}{\cot(x)}$	$s_3\sqrt{\sec^2(x)-1}$	$\frac{s_3}{\sqrt{\csc^2(x)-1}}$
$\cot(x)$	$\cot(x)$	$\frac{s_3}{\sqrt{\sec^2(x)-1}}$	$s_3\sqrt{\csc^2(x)-1}$
$\sec(x)$	$\frac{s_2\sqrt{\cot^2(x)+1}}{\cot(x)}$	$\sec(x)$	$\frac{s_3\csc(x)}{\sqrt{\csc^2(x)-1}}$
$\csc(x)$	$s_2\sqrt{\cot^2(x)+1}$	$\frac{s_3\sec(x)}{\sqrt{\sec^2(x)-1}}$	$\csc(x)$

Further interrelations can be formed with $\tau = \tan\left(\frac{x}{2}\right)$:

$$\tau = \tan\left(\frac{x}{2}\right) = (-1)^m \sqrt{\frac{1-\cos(x)}{1+\cos(x)}}, \qquad m = \operatorname{int}\left[\frac{2x}{\pi}\right]$$

Conversions with $\tau = \tan\left(\frac{x}{2}\right)$

$$\sin(x) = \frac{2\tau}{1+\tau^2} \qquad \csc(x) = \frac{1+\tau^2}{2\tau}$$

$$\cos(x) = \frac{1-\tau^2}{1+\tau^2} \qquad \sec(x) = \frac{1+\tau^2}{1-\tau^2}$$

$$\tan(x) = \frac{1}{1-\tau} - \frac{1}{1+\tau} \qquad \cot(x) = \frac{1-\tau^2}{2\tau}$$

In mathematics, the argument of trigonometric functions is usually stated in the differential of arc A.

This dimensionless unit is called the **radian**, abbreviated as rad.

▷ In programming languages, the argument of trigonometric functions is usually stated in radians.

In geometry and in technical applications, the angle α belonging to the arc is often used as an argument.

Usually, the angle is given in **degrees** (in pocket calculators deg = "degrees"), e.g., 30° = 30 degrees.

▷ Note which mode (deg/rad) is used in pocket calculators.

Conversion formula:

$$\alpha(\text{deg}) = \frac{180^\circ}{\pi} x(\text{rad}) = 57.2958 x(\text{rad})$$

$$x(\text{rad}) = \frac{\pi}{180^\circ} \alpha(\text{deg}) = 0.01745 \alpha(\text{deg})$$

▷ In pocket calculators with degree and radian modes, the calculation of functions with large arguments is more accurate in degree mode than in radian mode.

Short table of important sine, cosine, and tangent values.

Other values can be found under the description of the function.

Argument x (deg)	(rad)	$\sin(x)$ $\frac{1}{\csc(x)}$	$\cos(x)$ $\frac{1}{\sec(x)}$	$\tan(x)$ $\frac{1}{\cot(x)}$
$0°$	0	0	1	0
$15°$	$\frac{\pi}{12}$	$\frac{1}{\sqrt{6}+\sqrt{2}}$	$\frac{\sqrt{3}+1}{\sqrt{8}}$	$2-\sqrt{3}$
$30°$	$\frac{\pi}{6}$	$\frac{1}{2}$	$\frac{\sqrt{3}}{2}$	$\frac{1}{\sqrt{3}}$
$45°$	$\frac{\pi}{4}$	$\frac{1}{\sqrt{2}}$	$\frac{1}{\sqrt{2}}$	1
$60°$	$\frac{\pi}{3}$	$\frac{\sqrt{3}}{2}$	$\frac{1}{2}$	$\sqrt{3}$
$75°$	$\frac{5\pi}{12}$	$\frac{1}{\sqrt{6}-\sqrt{2}}$	$\frac{\sqrt{3}-1}{\sqrt{8}}$	$2+\sqrt{3}$
$90°$	$\frac{\pi}{2}$	1	0	∞

5.32 Sine and cosine functions

$$f(x) = \sin(x) , \qquad g(x) = \cos(x)$$

Representation in programming languages: `F=SIN(X)`, `G=COS(X)`
Most important trigonometric functions.
Important for describing oscillating systems such as springs, pendulums, oscillating electronic circuits, etc.
Important for describing periodic functions.

(a) Definition

Basic geometric definition according to the notation used above:

$$\sin(a) = s , \qquad \cos(a) = c .$$

Interpretation in a right triangle:

$$\sin(\alpha) = \frac{\text{side } s}{\text{hypotenuse}} , \qquad \cos(\alpha) = \frac{\text{side } c}{\text{hypotenuse}} .$$

Hyperbolic function with purely imaginary argument:

$$\sin(x) = -\mathrm{j} \sinh(\mathrm{j}x) , \qquad \cos(x) = \cosh(\mathrm{j}x) .$$

Euler's formula: Exponential function with imaginary argument:

$$\mathrm{e}^{\mathrm{j}x} = \cos(x) + \mathrm{j}\sin(x) .$$

Solutions of the "oscillation equation," a second-order differential equation (written here with the argument t)

$$\frac{\mathrm{d}^2}{\mathrm{d}t^2} f(t) + \omega^2 f(t) = 0\,, \quad f(t) = c_1 \sin(\omega t) + c_2 \cos(\omega t), \quad c_1, c_2 \text{ any coefficients}$$

Other representations, see series expansion and complex representation.

(b) Graphical representation

Sine and cosine functions have a period of 2π; i.e., for all x, it holds that

$$\sin(x + 2\pi) = \sin(x - 2\pi) = \sin(x) \qquad \cos(x + 2\pi) = \cos(x - 2\pi) = \cos(x)$$

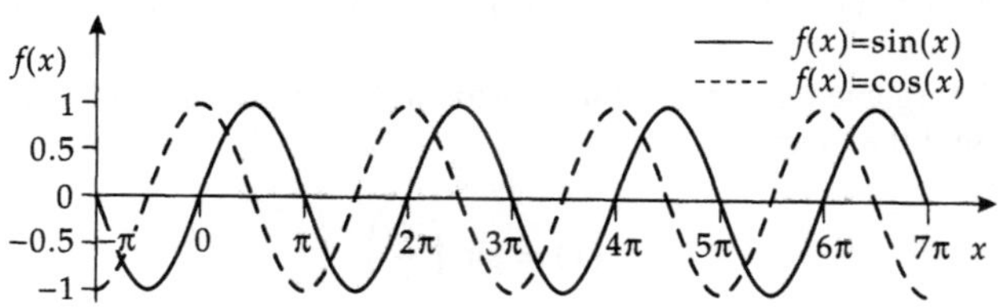

Sine and cosine are shifted with respect to one another by the value $\frac{\pi}{2}$:

$$\sin\left(x + \frac{\pi}{2}\right) = \cos(x)\,, \qquad \cos\left(x - \frac{\pi}{2}\right) = \sin(x)\,.$$

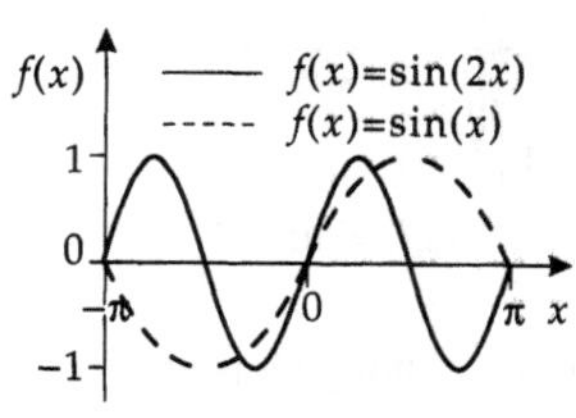

$f(x) = \sin(2x)$ (solid line),
$f(x) = \sin(x)$ (dashed line)

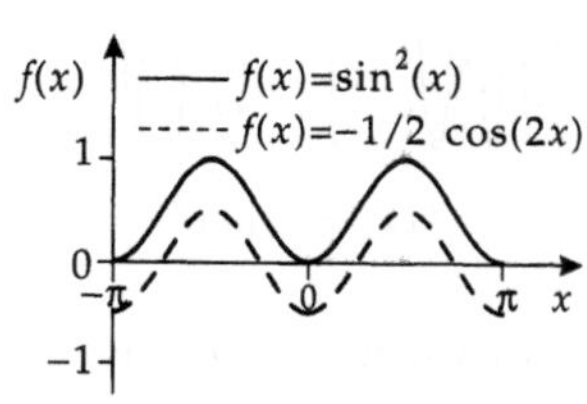

$f(x) = \sin^2(x)$ (solid line),
$f(x) = \frac{1}{2}\cos(2x)$ (dashed line)

The sine function is point-symmetric about the origin; the cosine function is mirror symmetric about the y-axis.

- ● Both functions have values between -1 and 1.
- ▷ Near $x = 0$, the sine function behaves as a straight line of slope 1.
  Near $x = 0$, the cosine function behaves as a parabola opening downward.

A factor $a > 1$ in the argument diminishes the period by factor a^{-1}. A factor $a < 1$ enlarges the period.

- ▷ The square of the functions is positive and has the same shape as a sine (cosine) of half the period.

A pre-factor b before the function changes the maxim value of the function. Negative factor $b < 0$ acts as a shift of the function by value π: $b\sin(x) = -b\sin(x + \pi)$.

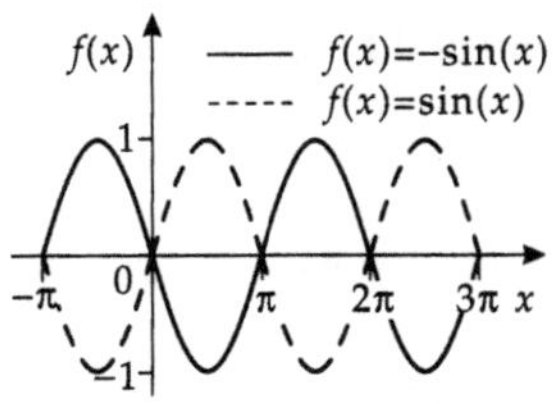

$f(x) = -\sin(x)$ (solid line),
$f(x) = \sin(x)$ (dashed line)

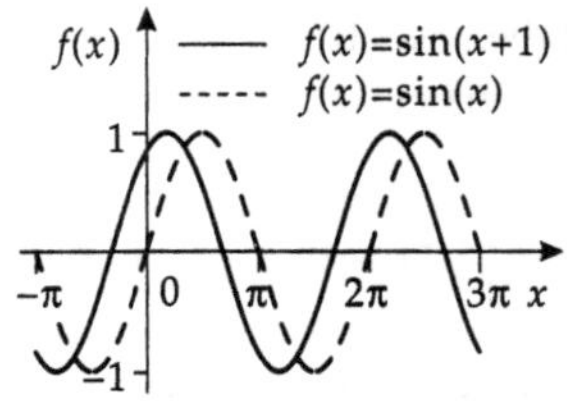

$f(x) = \sin(x + 1)$ (solid line),
$f(x) = \sin(x)$ (dashed line)

A positive summand φ in the argument $\sin(x + \varphi)$ shifts the function by φ to the left. $\varphi = \pi$ acts as a negative sign for the unshifted function.

- For $\varphi = \dfrac{\pi}{2}$, the sine transforms to a cosine, and the cosine transforms to a sine with a negative sign.

A nonlinear function of x in the argument changes the period of the function depending on x.
A square in the argument causes the function to oscillate more rapidly for larger x-values.
A **special case** if the function

$$f(x) = \sin\left(\frac{1}{x}\right)$$

For $x \to 0$, the function oscillates more and more rapidly. At the point $x = 0$, the function is not defined and cannot be extended continuously.

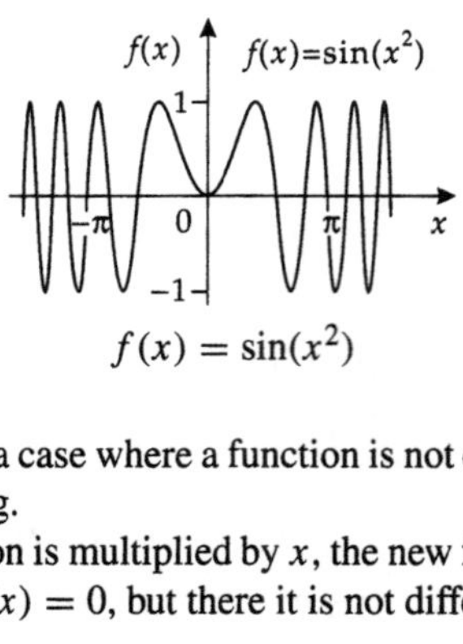

$f(x) = \sin(x^2)$

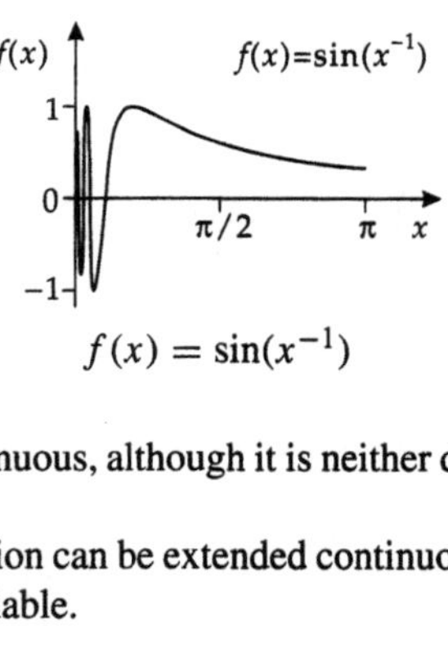

$f(x) = \sin(x^{-1})$

▷ This is a case where a function is not continuous, although it is neither divergent nor jumping.

If the function is multiplied by x, the new function can be extended continuously at point $x = 0$ by $f(x) = 0$, but there it is not differentiable.

(c) Properties of the functions

Domain of definition:	$-\infty < x < \infty$
Range of values:	$-1 \leq f(x) \leq 1$
Quadrant:	Lies in all four quadrants.
Periodicity:	Period 2π.
Monotonicity:	Strictly monotonically increasing and strictly monotonically decreasing intervals.

Symmetries:

$\sin(x)$ point symmetry about the origin
point symmetry about $x = n\pi$, $y = 0$, n integer,
mirror symmetry about $x = \left(n + \frac{1}{2}\right)\pi$, n integer,

$\cos(x)$ mirror symmetry about the y-axis
mirror symmetry about $x = n\pi$, n integer,
point symmetry about $x = \left(n + \frac{1}{2}\right)\pi$, $y = 0$.

Asymptotes:

No fixed behavior as $x \to \pm\infty$.
The functions oscillate between -1 and 1.

(d) Particular values

deg	rad	sin(x)	cos(x)	rad	deg	sin(x)	cos(x)
0	0	0	1	0.05	2.87	0.0500	0.9988
5	0.087	0.0872	0.9962	0.1	5.73	0.0998	0.9950
10	0.175	0.1736	0.9848	0.15	8.59	0.1494	0.9888
15	0.261	0.2588	0.9659	0.2	11.46	0.1987	0.9801
20	0.349	0.3420	0.9397	0.25	14.32	0.2474	0.9689
25	0.436	0.4226	0.9063	0.3	17.19	0.2955	0.9553
30	0.524	0.5	0.8660	0.35	20.05	0.3429	0.9394
35	0.611	0.5736	0.8192	0.4	22.92	0.3894	0.9211
40	0.698	0.6428	0.7660	0.5	28.65	0.4794	0.8776
45	0.785	0.7071	0.7071	0.6	34.38	0.5646	0.8253
50	0.873	0.7660	0.6428	0.7	40.12	0.6442	0.7648
55	0.960	0.8192	0.5736	0.8	45.84	0.7174	0.6967
60	1.047	0.8660	0.5	0.9	51.57	0.7833	0.6216
65	1.135	0.9063	0.4226	1.0	57.30	0.8415	0.5403
70	1.222	0.9397	0.3420	1.1	63.03	0.8912	0.4536
75	1.309	0.9659	0.2588	1.2	68.75	0.9320	0.3624
80	1.396	0.9848	0.1736	1.3	74.48	0.9636	0.2675
85	1.484	0.9962	0.0872	1.4	80.21	0.9855	0.1700
90	1.571	1	0	1.5	85.94	0.9975	0.0707

Zeros:

$\sin(x)$ $x = n\pi$, n integer

$\cos(x)$ $x = \left(n + \frac{1}{2}\right)\pi$, n integer

Discontinuities: None.

Poles: None.

Extremes:

$\sin(x)$ Maxima $x = \left(2n + \frac{1}{2}\right)\pi$.
Minima $x = \left(2n - \frac{1}{2}\right)\pi$.

$\cos(x)$ Maxima $x = 2n\pi$.
Minima $x = (2n - 1)\pi$.

Points of inflection: $\sin(x)$ $x = n\pi$, n integer

$\cos(x)$ $x = \left(n + \frac{1}{2}\right)\pi$, n integer

(e) Reciprocal functions

The reciprocal functions of the cosine and sine are the secant and cosecant, respectively:

$$\frac{1}{\sin(x)} = \csc(x)\,, \qquad \frac{1}{\cos(x)} = \sec(x)$$

(f) Inverse functions

- Sine and cosine can be inverted only on half a period (π).

The sine function is usually inverted in the domain $-\frac{\pi}{2} \le x \le \frac{\pi}{2}$:

$$x = \arcsin(y) \qquad y = \sin(x) \qquad -\frac{\pi}{2} \le x \le \frac{\pi}{2}$$

The cosine function is usually inverted in the domain $0 \le x \le \pi$:

$$x = \arccos(y) \qquad y = \cos(x) \qquad 0 \le x \le \pi$$

(g) Cognate function

Tangent and cotangent quotients of sine and cosine.
Secant and cosecant reciprocal functions of cosine and sine.
Arc functions inverse functions of the trigonometric functions.
Hyperbolic functions, related to the trigonometric functions.
Sine integral: Defined by the integral

$$\begin{aligned}\mathrm{Si}(x) &= \int_0^x \frac{\sin(t)}{t}\mathrm{d}t = \frac{\pi}{2} - \int_x^\infty \frac{\sin(t)}{t}\mathrm{d}t \\ &= x - \frac{x^3}{3\cdot 3!} + \frac{x^5}{5\cdot 5!} - \frac{x^7}{7\cdot 7!} + \cdots \\ &= \sum_{k=0}^{\infty} \frac{(-1)^k}{2k+1}\frac{x^{2k+1}}{(2k+1)!}\end{aligned}$$

Integral cosine $\mathrm{Ci}(x)$: Defined by the integral

$$\mathrm{Ci}\,(x) = -\int_x^\infty \frac{\cos(t)}{t}\mathrm{d}t = C + \ln(x) + \sum_{k=1}^{\infty}\frac{(-1)^k}{2k}\frac{x^{2k}}{(2k)!}$$

with Euler's constant $C = 0.577215665\ldots$.

(h) Conversion formulas

- **Pythagorean theorem in the unit circle**:

$$\sin^2(x) + \cos^2(x) = 1$$

Reflection in the y-axis:

$$\sin(-x) = -\sin(x)\,, \qquad \cos(-x) = \cos(x).$$

Translation along the x-axis:

$$\sin\left(x \pm \frac{n\pi}{2}\right) = \begin{cases} \sin(x) & n = 0, 4, 8, \ldots \\ \pm\cos(x) & n = 1, 5, 9, \ldots \\ -\sin(x) & n = 2, 6, 10, \ldots \\ \mp\cos(x) & n = 3, 7, 11, \ldots \end{cases}$$

$$\cos\left(x \pm \frac{n\pi}{2}\right) = \begin{cases} \cos(x) & n = 0, 4, 8, \ldots \\ \mp\sin(x) & n = 1, 5, 9, \ldots \\ -\cos(x) & n = 2, 6, 10, \ldots \\ \pm\sin(x) & n = 3, 7, 11, \ldots \end{cases}$$

The application of the translation and reflection properties yields some interrelations important for geometry.

$$\sin(x) = \cos\left(\frac{\pi}{2} - x\right), \qquad \cos(x) = \sin\left(\frac{\pi}{2} - x\right).$$

- **Addition theorems:**

$$\sin(x \pm y) = \sin(x)\cos(y) \pm \cos(x)\sin(y)$$
$$\cos(x \pm y) = \cos(x)\cos(y) \mp \sin(x)\sin(y)$$

Integral factor in the argument of the sine:

$$\begin{aligned} \sin(2x) &= 2\sin(x)\cos(x) \\ \sin(3x) &= 3\sin(x) - 4\sin^3(x) = \sin(x)[4\cos^2(x) - 1] \\ \sin(4x) &= 8\sin(x)\cos^3(x) - 4\sin(x)\cos(x) \\ \sin(5x) &= \sin(x) - 12\sin(x)\cos^2(x) + 16\sin(x)\cos^4(x) \\ \sin(nx) &= \binom{n}{1}\sin(x)\cos^{n-1}(x) - \binom{n}{3}\sin^3(x)\cos^{n-3}(x) \\ &\quad + \binom{n}{5}\sin(x)^5\cos^{n-5}(x) - \binom{n}{7}\sin^7(x)\cos^{n-7}(x) + \cdots \end{aligned}$$

Integral factor in the argument of the cosine:

$$\begin{aligned} \cos(2x) &= 2\cos^2(x) - 1 = 1 - 2\sin^2(x) \\ \cos(3x) &= 4\cos^3(x) - 3\cos(x) \\ \cos(4x) &= 1 - 8\cos^2(x) + 8\cos^4(x) \\ \cos(5x) &= 5\cos(x) - 20\cos^3(x) + 16\cos^5(x) \\ \cos(nx) &= \binom{n}{0}\cos^n(x) - \binom{n}{2}\sin^2(x)\cos^{n-2}(x) \\ &\quad + \binom{n}{4}\sin(x)^4\cos^{n-4}(x) - \binom{n}{6}\sin^6(x)\cos^{n-6}(x) + \cdots \end{aligned}$$

Half arguments:

$$\sin\left(\frac{x}{2}\right) = (-1)^m\sqrt{\frac{1 - \cos(x)}{2}} \qquad m = \operatorname{int}\left[\frac{\pi + |x|}{2\pi}\right]$$

$$\cos\left(\frac{x}{2}\right) = (-1)^m\sqrt{\frac{1 + \cos(x)}{2}} \qquad m = \operatorname{int}\left[\frac{|x|}{2\pi}\right]$$

Nonintegral factor $a \neq 0, \pm 1, \pm 2, \ldots$, for $-\pi < x < \pi$:

$$\sin(ax) = -\frac{2}{\pi}\sin(a\pi)\sum_{k=1}^{\infty}(-1)^k \frac{k}{k^2 - a^2}\sin(kx)$$

$$\cos(ax) = -\frac{2}{\pi}\sin(a\pi)\left[\frac{1}{2a^2} - \sum_{k=1}^{\infty}(-1)^k \frac{k}{k^2 - a^2}\cos(kx)\right]$$

Sum of functions:

$$\sin(x) \pm \sin(y) = 2\sin\left(\frac{x \pm y}{2}\right)\cos\left(\frac{x \mp y}{2}\right)$$

$$\cos(x) + \cos(y) = 2\cos\left(\frac{x + y}{2}\right)\cos\left(\frac{x - y}{2}\right)$$

$$\cos(x) - \cos(y) = -2\sin\left(\frac{x + y}{2}\right)\sin\left(\frac{x - y}{2}\right)$$

Sum of functions with equal arguments:

$$\cos(x) \pm \sin(x) = \sqrt{2}\sin\left(x \pm \frac{\pi}{4}\right) = \sqrt{2}\cos\left(x \mp \frac{\pi}{4}\right)$$

Product of functions:

$$\sin(x)\sin(y) = \frac{1}{2}\cos(x - y) - \frac{1}{2}\cos(x + y)$$

$$\cos(x)\sin(y) = \frac{1}{2}\sin(x + y) - \frac{1}{2}\sin(x - y)$$

$$\cos(x)\cos(y) = \frac{1}{2}\cos(x + y) + \frac{1}{2}\cos(x - y)$$

Triple products:

$$\sin(x)\sin(y)\sin(z) = -\frac{1}{4}\sin(x + y + z) + \frac{1}{4}\sin(x + y - z) + \frac{1}{4}\sin(x - y + z) + \frac{1}{4}\sin(-x + y + z)$$

$$\sin(x)\sin(y)\cos(z) = -\frac{1}{4}\cos(x + y + z) - \frac{1}{4}\cos(x + y - z) + \frac{1}{4}\cos(x - y + z) + \frac{1}{4}\cos(-x + y + z)$$

$$\sin(x)\cos(y)\cos(z) = \frac{1}{4}\sin(x + y + z) + \frac{1}{4}\sin(x + y - z) + \frac{1}{4}\sin(x - y + z) - \frac{1}{4}\sin(-x + y + z)$$

$$\cos(x)\cos(y)\cos(z) = \frac{1}{4}\cos(x + y + z) + \frac{1}{4}\cos(x + y - z) + \frac{1}{4}\cos(x - y + z) + \frac{1}{4}\cos(-x + y + z)$$

Powers of the sine function:

$$\sin^2(x) = \frac{1 - \cos(2x)}{2}$$

$$\sin^3(x) = \frac{3\sin(x) - \sin(3x)}{4}$$

$$\sin^4(x) = \frac{3 - 4\cos(2x) + \cos(4x)}{8}$$

$$\sin^5(x) = \frac{10\sin(x) - 5\sin(3x) + \sin(5x)}{16}$$

$$\sin^6(x) = \frac{10 - 15\cos(2x) + 6\cos(4x) - \cos(6x)}{32}$$

$$\sin^n(x) = \frac{(-1)^{n/2}}{2^n}\sum_{k=0}^{\infty}(-1)^k\binom{n}{k}\cos((n-2k)x)\,, \qquad n \text{ even}$$

$$= \frac{(-1)^{(n-1)/2}}{2^n}\sum_{k=0}^{\infty}(-1)^k\binom{n}{k}\sin((n-2k)x)\,, \qquad n \text{ odd}$$

Powers of the cosine function:

$$\cos^2(x) = \frac{1 + \cos(2x)}{2}$$

$$\cos^3(x) = \frac{3\cos(x) + \cos(3x)}{4}$$

$$\cos^4(x) = \frac{3 + 4\cos(2x) + \cos(4x)}{8}$$

$$\cos^5(x) = \frac{10\cos(x) + 5\cos(3x) + \cos(5x)}{16}$$

$$\cos^6(x) = \frac{10 + 15\cos(2x) + 6\cos(4x) + \cos(6x)}{32}$$

$$\cos^n(x) = \frac{1}{2^n}\sum_{k=0}^{\infty}\binom{n}{k}\cos((n-2k)x)$$

(i) Approximation formulas (8 bits $\approx$ 0.4% of accuracy)

For small values of x, it holds that

$$\sin(x) \approx x\left(1 - \frac{x^2}{2\pi}\right)\,, \qquad -1.1 \le x \le 1.1$$

$$\cos(x) \approx \sqrt{\left(1 - \frac{x^2}{3}\right)}\,, \qquad -0.9 \le x \le 0.9$$

▷ Often, simple approximation formulas are sufficient:

$$\sin(x) \approx x\,, \quad \cos(x) \approx 1 - \frac{x^2}{2}\,,\ |x| \le 0.1\,.$$

(j) Expansion of series or products

Power series expansion of the sine and cosine:

$$\sin(x) = x - \frac{x^3}{3!} + \frac{x^5}{5!} - \frac{x^7}{7!} + \cdots$$

$$= x\sum_{k=0}^{\infty}\frac{(-x^2)^k}{(2k+1)!}$$

$$\cos(x) = 1 - \frac{x^2}{2!} + \frac{x^4}{4!} - \frac{x^6}{6!} + \cdots$$
$$= \sum_{k=0}^{\infty} \frac{(-x^2)^k}{(2k)!}$$

Product expansion of sine and cosine:

$$\sin(x) = x\left(1 - \frac{x^2}{\pi^2}\right)\left(1 - \frac{x^2}{4\pi^2}\right)\left(1 - \frac{x^2}{9\pi^2}\right)\cdots = x\prod_{k=1}^{\infty}\left(1 - \frac{x^2}{k^2\pi^2}\right)$$

$$\cos(x) = \left(1 - \frac{4x^2}{\pi^2}\right)\left(1 - \frac{4x^2}{9\pi^2}\right)\left(1 - \frac{4x^2}{25\pi^2}\right)\cdots = \prod_{k=1}^{\infty}\left(1 - \frac{4x^2}{(2k-1)^2\pi^2}\right)$$

Product expansion of sine with cosine as factors:

$$\sin(x) = x\cos\left(\frac{x}{2}\right)\cos\left(\frac{x}{4}\right)\cos\left(\frac{x}{8}\right)\cdots = x\prod_{k=1}^{\infty}\cos\left(\frac{x}{2^k}\right)$$

(k) Derivative of the functions

Sine and cosine are derivatives of each other (except for a constant and sign):

$$\frac{\mathrm{d}}{\mathrm{d}x}\sin(ax) = a\cos(ax) = a\sin\left(ax + \frac{\pi}{2}\right)$$
$$\frac{\mathrm{d}}{\mathrm{d}x}\cos(ax) = -a\sin(ax) = a\cos\left(ax + \frac{\pi}{2}\right)$$

As a generalization, the n-th derivative can be represented as a shift by $n\frac{\pi}{2}$ along the x-axis:

$$\frac{\mathrm{d}^n}{\mathrm{d}x^n}\sin(ax) = a^n\sin\left(ax + \frac{n\pi}{2}\right)$$
$$\frac{\mathrm{d}^n}{\mathrm{d}x^n}\cos(ax) = a^n\cos\left(ax + \frac{n\pi}{2}\right)$$

(l) Primitive of the function

Sine and cosine are primitives to each other (except for the sign)

$$\int_0^x \sin(at)\mathrm{d}t = \frac{1-\cos(ax)}{a}, \qquad \int_0^x \cos(at)\mathrm{d}t = \frac{\sin(ax)}{a}$$

Orthogonality of sine and cosine: In the function space, sine and cosine are orthogonal to each other; i.e., they satisfy the following conditions (n, m are integers):

$$\int_0^{2\pi} \cos(nt)\sin(mt)\,\mathrm{d}t = 0$$

$$\int_0^{2\pi} \cos(nt)\cos(mt)\,\mathrm{d}t = \begin{cases} 0 & n \neq m \\ 2\pi & m = n = 0 \\ \pi & m = n \neq 0 \end{cases}$$

$$\int_0^{2\pi} \sin(nt)\sin(mt)\,\mathrm{d}t = \begin{cases} 0 & n \neq m \\ 0 & m = n = 0 \\ \pi & m = n \neq 0 \end{cases}$$

▷ $-\pi$ and π can also be used for the end-points of the interval. This property is of importance for Fourier series.

(m) Particular extensions and applications

Complex representation, representation by means of the exponential function:

$$\sin(x) = \frac{e^{jx} - e^{-jx}}{2j} \qquad \cos(x) = \frac{e^{jx} + e^{-jx}}{2}$$

Complex argument:

$$\sin(x + jy) = \sin(x)\cosh(y) + j\cos(x)\sinh(y)$$
$$\cos(x + jy) = \cos(x)\cosh(y) - j\sin(x)\sinh(y)$$

Purely imaginary argument:

$$\sin(jy) = j\sinh(y) \qquad \cos(jy) = \cosh(y)$$

Fourier transformation: The transformation of functions by convolution with an exponential function with imaginary argument:

$$F(s) = \int_{-\infty}^{\infty} f(t)e^{-jst}\,dt$$

See also Chapter 19 on the Fourier transformation.
Fourier series: The representation of a periodic function in terms of sine and cosine functions. See also the following section and the chapter on Fourier series.

5.32.1 *Superpositions of oscillations*

In this section, some applications of the sine function will be explained.
The amplitude of an oscillation is the positive maximum magnitude of the oscillation during one period.

$$f(x) = A\sin(x)\,, \quad \text{where} \quad |A| \text{ is the amplitude}$$

The angular frequency ω of an oscillation is the pre-factor in the argument. Often, the factor 2π is removed from the circular frequency, so that the **frequency** is the reciprocal value of the period

$$f(t) = A\sin(\omega t) = A\sin(2\pi f t)\,, \quad \text{where} \quad f \text{ is the frequency}$$

▷ Smaller frequencies imply longer periods.
The phase of an oscillation is a summand in the argument, which shifts the function along the t-axis.

$$f(t) = A\sin(\omega t + \varphi)\,, \quad \text{where} \quad \varphi \text{ is the phase}$$

Product of the sine function with other functions

$$f(x) = g(x) \cdot \sin(x)$$

The functions $g(x)$ and $-g(x)$ form the envelopes of the oscillation.

□ **The resonance** of excited oscillations often proceeds with amplitudes growing linearly.

$$f(x) = x \cdot \sin(x)$$

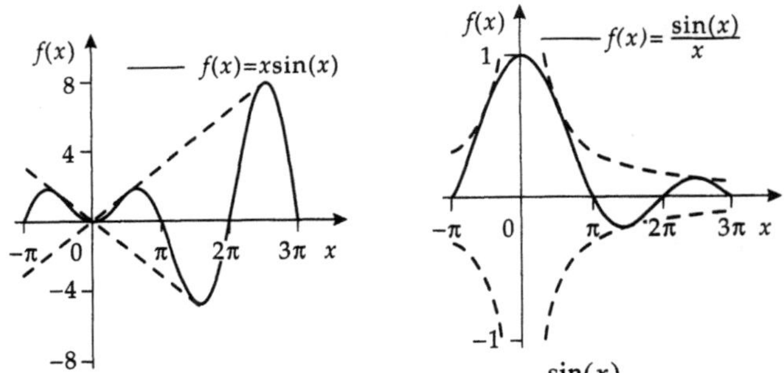

$f(x) = x\sin(x)$ (solid line) and envelope (dashed line)

$f(x) = \dfrac{\sin(x)}{x}$ (solid line) and envelope (dashed line)

- The nodes of the oscillation originate at the zeros of $g(x)$.

The curve must not necessarily diverge at the poles of $g(x)$ if $g(x)$ has a pole of first order at x_P and if the sine function has a zero at the same point.

□ In spectroscopy, the following function is important:

$$f(x) = \frac{\sin(x)}{x} .$$

This function does not diverge at $x = 0$, but takes the value 1 there.

In general, the **damping of an oscillation** occurs exponentially with a damping factor $\gamma > 0$. The function can be written as

$$f(t) = \mathrm{e}^{-\gamma t} \sin(\omega t) .$$

Wave pockets have an exponential increase and an exponential decrease.

$$f(x) = \mathrm{e}^{-ax^2} \sin(bx)$$

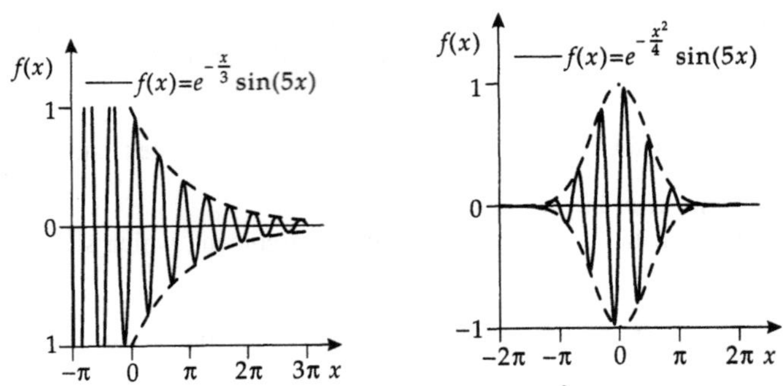

$f(x) = \mathrm{e}^{-\frac{x}{3}} \sin(5x)$ (solid line) and envelope (dashed line)

$f(x) = \mathrm{e}^{\frac{-x^2}{4}} \sin(5x)$ (solid line) and envelope (dashed line)

Modulation: The product of two oscillations with different frequencies is of importance. The result is an oscillation with "oscillating amplitudes."

$$f(t) = A\sin(\Omega t)\sin(\omega t)$$

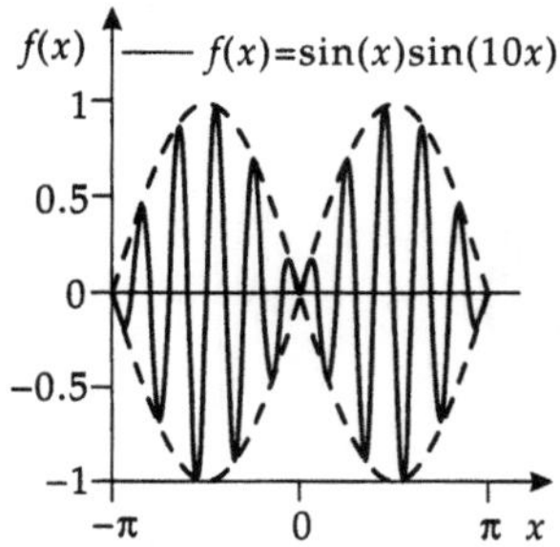

$f(x) = \sin(x)\sin(10x)$ (solid line) and envelope (dashed line)

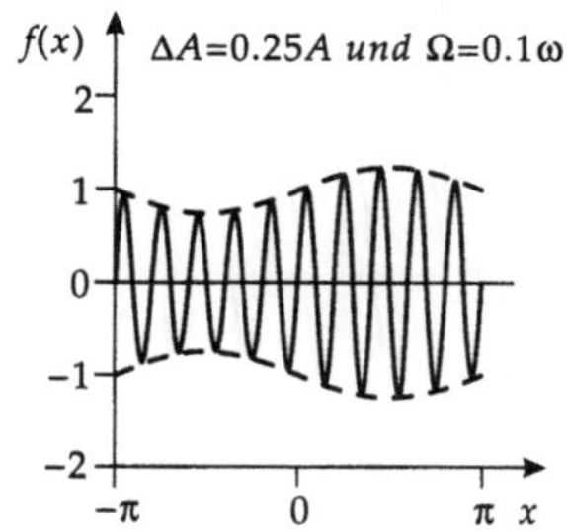

Amplitude modulation with $\Delta A = 0,25A$ and $\Omega = 0,1\omega$ (solid line) and envelope (dashed line)

Carrier frequency $\omega, \omega \gg \Omega$: Determines the bandwidth of a transmitter in broadcasting.
Modulation frequency Ω: Lower frequency, transmits the signal in radio engineering.
Amplitude modulation (AM): The amplitude A is modified by a part ΔA, depending on the modulation frequency Ω.

$$f(t) = (A + \Delta A \sin(\Omega t))\sin(\omega t)$$

▷ In broadcasting the amplitude modulation is used in the middle and short wave ranges.

$$f(x) = A\sin([\omega + \Delta\omega \sin(\Omega t)]t)$$

Frequency modulation (FM): Used mainly in the very high frequency range. In this case the oscillation frequency is modulated with a certain frequency.

▷ Transmission with frequency modulation is less susceptible to atmospheric disturbances, but it requires a greater bandwidth.

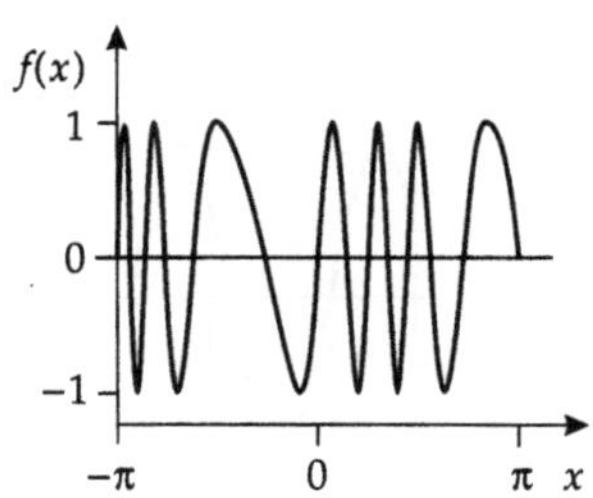

$f(x) = \sin(6x)$

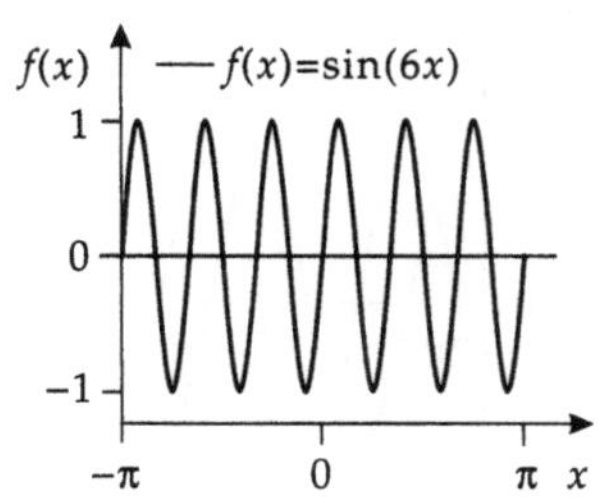

Frequency modulation with $\Delta\omega = 0.5\omega$ and $\Omega = 0.17\omega$

Beat: The superposition of oscillations with similar (slightly different) frequencies results in an oscillation with a new frequency that lies between both original frequencies, and with an amplitude that oscillates with half the difference of the frequencies.

$$\begin{aligned} f(x) &= \sin(\omega t) + \sin([\omega + \Delta\omega]t) \\ &= 2\sin\left(\left[\omega + \frac{\Delta\omega}{2}\right]t\right)\cos\left(\frac{\Delta\omega}{2}t\right) \end{aligned}$$

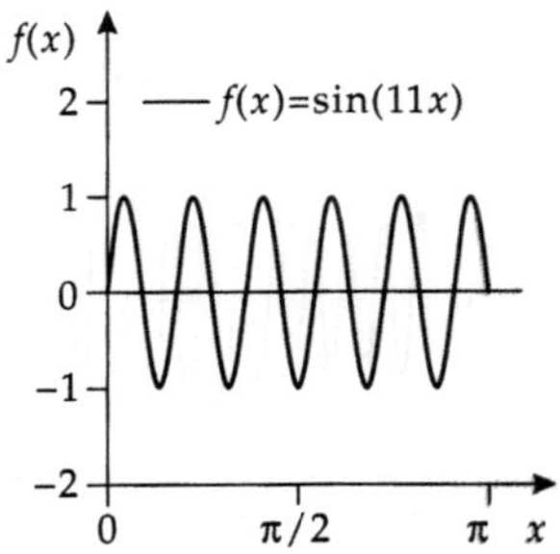

$f(x) = \sin(11x)$

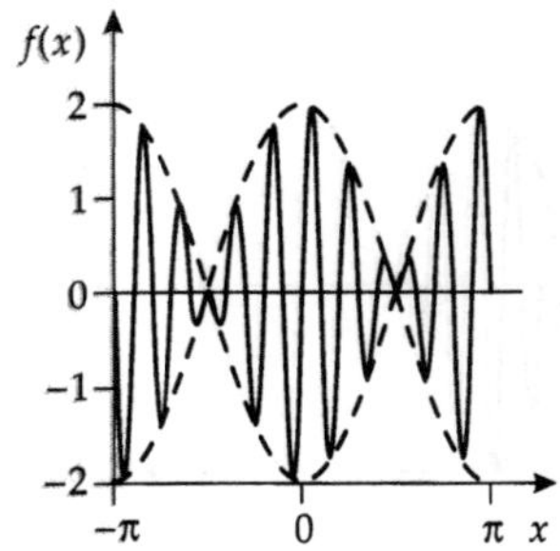

Beat of the oscillations $\sin(11x)$ and $\sin(9x)$ (solid line) and envelope (dashed line)

Superposition of oscillations of equal frequency: If two oscillations of different amplitude and phase but of equal frequency are superposed, the result is another oscillation of the same frequency.

$$\begin{aligned} f(t) &= A_1 \sin(\omega t + \varphi_1) + A_2 \sin(\omega t + \varphi_2) \\ &= A \sin(\omega t + \varphi)\,. \end{aligned}$$

Then the amplitude and frequency are:

$$A = \sqrt{A_1^2 + A_2^2 + 2A_1A_2 \cos(\varphi_2 - \varphi_1)}$$

$$\tan\varphi = \frac{A_1 \sin(\varphi_1) + A_2 \sin(\varphi_2)}{A_1 \cos(\varphi_1) + A_2 \cos(\varphi_2)} = \frac{B_0}{A_0}\,.$$

The quadrant in which the wanted angle φ lies results from the combination of the signs of A_0 and B_0.

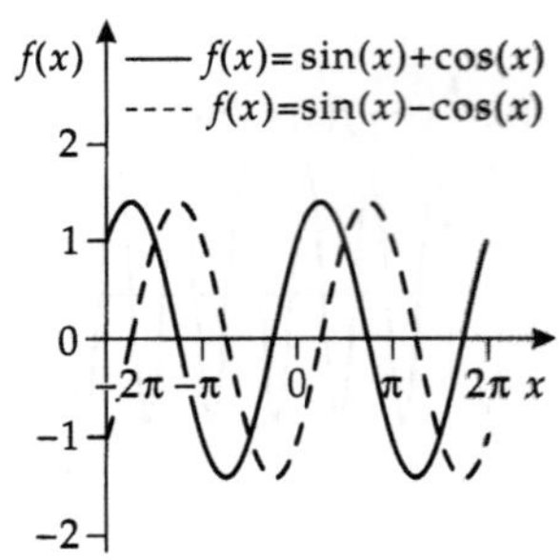

$f(x) = \sin(x) + \cos(x)$ (solid line),
$f(x) = \sin(x) - \cos(x)$ (dashed line)

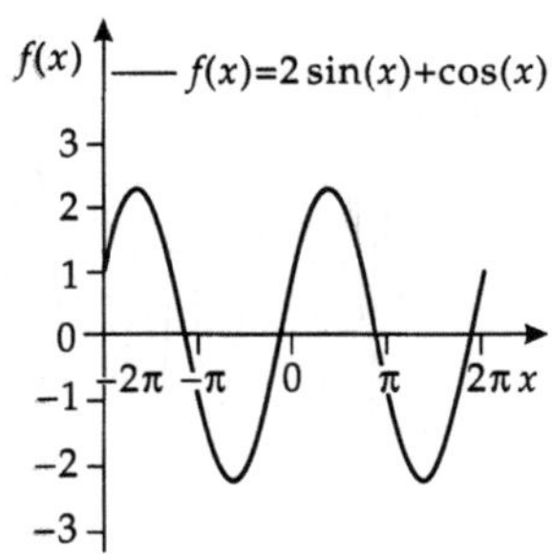

$f(x) = 2\sin(x) + \cos(x)$

Lissajous figures: Curves of second order; that is, y must not necessarily depend on x, but y and x can be represented as functions of a parameter, such as t.

$$y = y(t)\,, \qquad x = x(t)$$

Lissajous figures can be represented by

$$y = A_y \sin(\omega_y t + \varphi_y)\,, \qquad x = A_x \sin(\omega_x t + \varphi_x)\,.$$

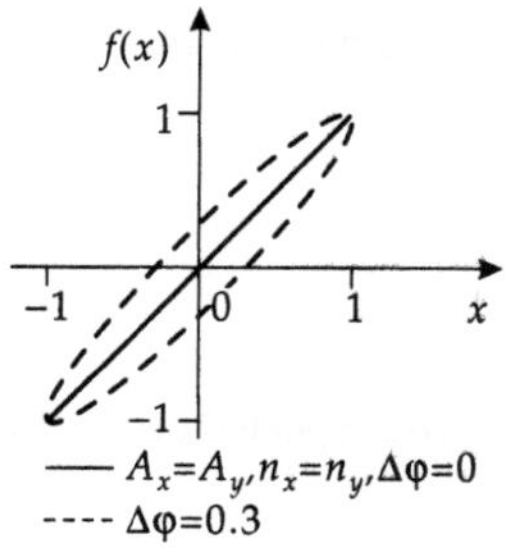

$A_x = A_y$, $n_x = n_y$ and $\Delta\varphi = 0$ (solid line), $\Delta\varphi = 0.3$ (dashed line)

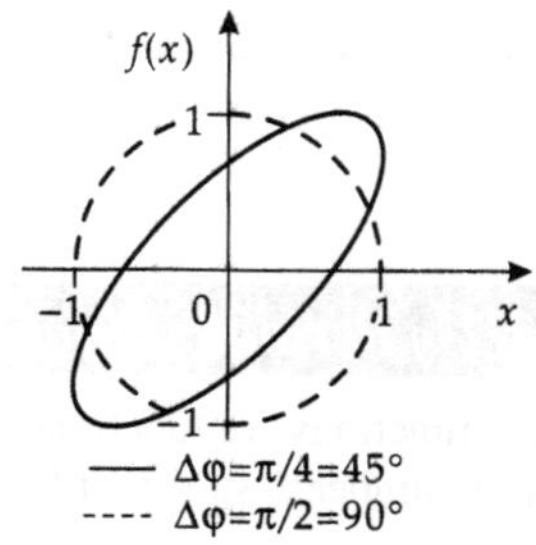

$\Delta\varphi = \frac{\pi}{4} = 45°$ (solid line) and $\Delta\varphi = \frac{\pi}{2} = 90°$ (dashed line)

Lissajous figures

For the curves to be closed, i.e., $y(t+T) = y(t), x(t+T) = x(t)$ for $T > 0$, it is required that

$$\omega_y = n_y \omega_0 \,, \qquad \omega_x = n_x \omega_0 \,, \qquad n, m \text{ integers}$$

▷ Lissajous figures can be used to analyze the interrelation of two oscillations graphically (e.g., on an oscillograph).

For $n_x = n_y$, $\Delta\varphi = \varphi_x - \varphi_y = \dfrac{\pi}{2}$ the curve is an ellipse with the semi-axes A_x and A_y.

For $n_x = n_y$, $\Delta\varphi = 0$ the curve is a straight line segment with length $2\sqrt{A_x^2 + A_y^2}$ and slope $m = \dfrac{A_y}{A_x}$.

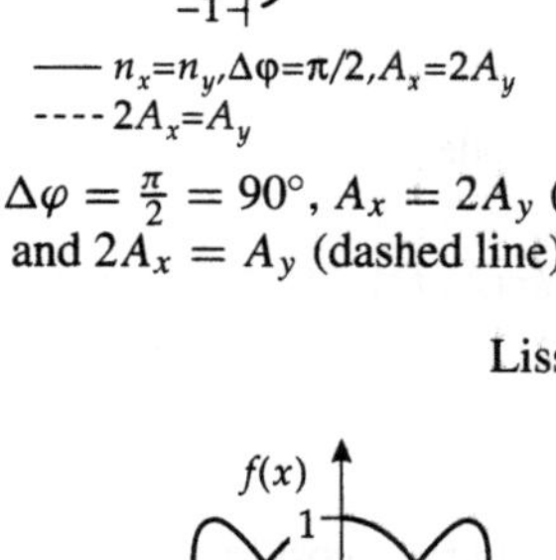

$n_x = n_y$, $\Delta\varphi = \frac{\pi}{2} = 90°$, $A_x = 2A_y$ (solid line) and $2A_x = A_y$ (dashed line)

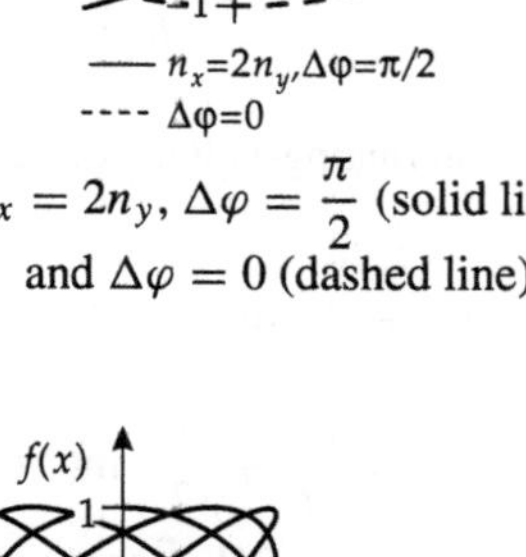

$n_x = 2n_y$, $\Delta\varphi = \dfrac{\pi}{2}$ (solid line) and $\Delta\varphi = 0$ (dashed line)

Lissajou figures

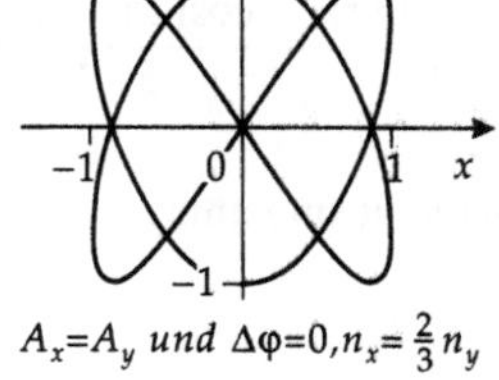

$A_x = A_y$ and $\Delta\varphi = 0$, $n_x = \frac{2}{3} n_y$

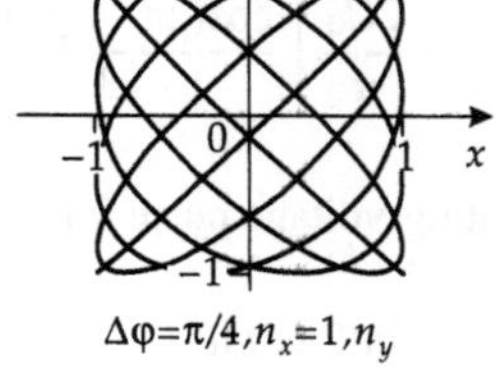

$\Delta\varphi = \dfrac{\pi}{4}$, $n_x = 1$, $1n_y$

Lissajous figures

□ For $n_x = 2n_y$, $\Delta\varphi = 0$, the curve has the form of a figure eight.

5.32.2 *Periodic functions*

Periodic functions are functions on $\mathbb{R}$ that are piecewise repeated cyclically: there is a number L such that for all $x \in \mathbb{R}$:

$$f(x + 2L) = f(x) = f(x - 2L).$$

Period: The smallest value $2L$ which the above equation applies.

□ $\sin(ax)$, $a \neq 0$, has the period $2\pi/a$.

The functions considered must be piecewise continuous and not divergent.

● Every function thus defined can be represented by (infinite) sums of sine and cosine functions.

Fourier series: Representation of a function of period $2L$ by means of sine and cosine functions.

$$f(x) = \frac{a_0}{2} + \sum_{k=1}^{\infty} \left[a_k \cos\left(\frac{k\pi x}{L}\right) + b_k \sin\left(\frac{k\pi x}{L}\right) \right]$$

The series converges toward all continuous points of $f(x)$; at jump discontinuities it converges toward the mean value of the right and left boundary points.

Fourier coefficients: The coefficients a_k, b_k of the Fourier series. They are calculated as follows:

$$a_k = \frac{1}{L} \int_{-L}^{L} f(x) \cos\left(\frac{k\pi x}{L}\right) \mathrm{d}x\,, \qquad k = 0, 1, 2, \ldots$$

$$b_k = \frac{1}{L} \int_{-L}^{L} f(x) \sin\left(\frac{k\pi x}{L}\right) \mathrm{d}x\,, \qquad k = 1, 2, \ldots$$

Alternating-voltage pulse at the output of a diode:

$$f(x) = U_0 \sin(2\pi f t) \cdot H(\sin(2\pi f t)) = \begin{cases} \sin(2\pi f t) & 0 \le t \le (\pi/f) \\ 0 & (\pi/f) \le t \le (2\pi/f) \end{cases}$$

$$= \frac{U_0}{\pi} + \frac{U_0}{2} \sin(2\pi \cdot f t)$$

$$- \frac{2U_0}{\pi} \left[\frac{\cos(2\pi \cdot 2ft)}{1 \cdot 3} + \frac{\cos(2\pi \cdot 4ft)}{3 \cdot 5} + \frac{\cos(2\pi \cdot 6ft)}{5 \cdot 7} + \cdots \right].$$

Alternating-voltage pulse for a rectifier, without smoothing

$$f(x) = U_0 |\sin(2\pi f t)|$$

$$= \frac{2U_0}{\pi} - \frac{4U_0}{\pi} \left[\frac{\cos(2\pi \cdot 2ft)}{1 \cdot 3} + \frac{\cos(2\pi \cdot 4ft)}{3 \cdot 5} + \frac{\cos(2\pi \cdot 6ft)}{5 \cdot 7} + \cdots \right].$$

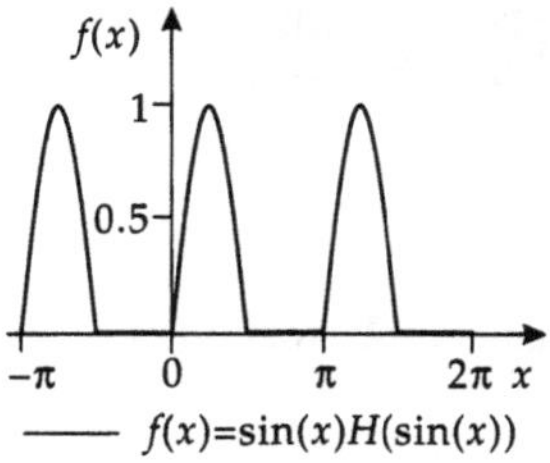

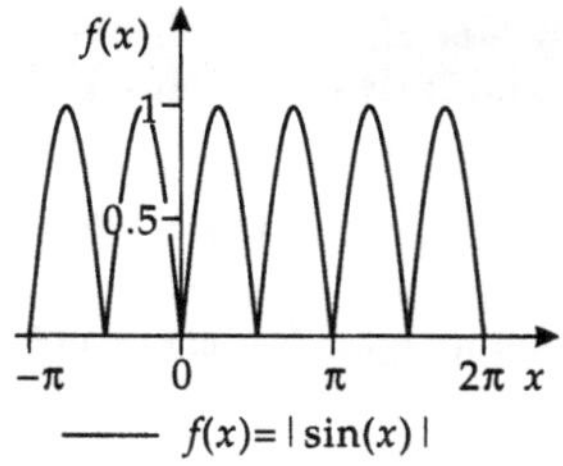

$f(x) = \sin(x)H(\sin(x))$ $\qquad$ $f(x) = |\sin(x)|$

Sawtooth pulse: Idealized relaxation oscillation:

$$f(x) = \frac{U_0}{2L}\left(x - 2L\,\mathrm{int}\left[\frac{x}{2L}\right]\right) = \frac{U_0}{2L}x\,, \qquad 0 \le x \le 2L$$
$$= \frac{U_0}{2} - \frac{U_0}{\pi}\left[\sin\left(\frac{\pi x}{L}\right) + \frac{1}{2}\sin\left(\frac{2\pi x}{L}\right) + \frac{1}{3}\sin\left(\frac{3\pi x}{L}\right) + \cdots\right].$$

Triangle pulse:

$$f(x) = \frac{U_0}{L}\left|x - 2L\mathrm{int}\left[\frac{x}{2L}\right] - \frac{1}{2}\right| = U_0 - \frac{U_0}{L}|x|\,, \qquad -L \le x \le L$$
$$= \frac{U_0}{2} + \frac{4U_0}{\pi^2}\left[\frac{1}{1^1}\cos\left(\frac{\pi x}{L}\right) + \frac{1}{3^2}\cos\left(\frac{3\pi x}{L}\right) + \frac{1}{5^2}\cos\left(\frac{5\pi x}{L}\right) + \cdots\right].$$

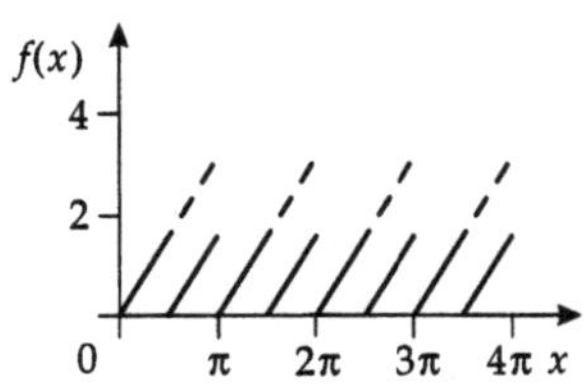

Sawtooth function

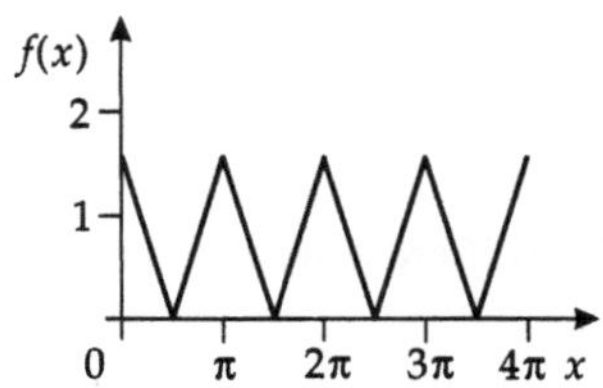

Triangle function

Square-wave pulse:

$$f(x) = U_0 H\left(\sin\left(\frac{\pi x}{L}\right)\right) = \begin{cases} 0 & -L < x < 0 \\ U_0 & 0 \le x \le L \end{cases}$$
$$= \frac{U_0}{2} + \frac{2U_0}{\pi}\left[\sin\left(\frac{\pi x}{L}\right) + \frac{1}{3}\sin\left(\frac{3\pi x}{L}\right) + \frac{1}{5}\sin\left(\frac{5\pi x}{L}\right) + \ldots\right].$$

Narrow square-wave pulses can be calculated using the integral relations for a_k, b_k. They contain additional factors in their arguments, taking into account the pulse width.

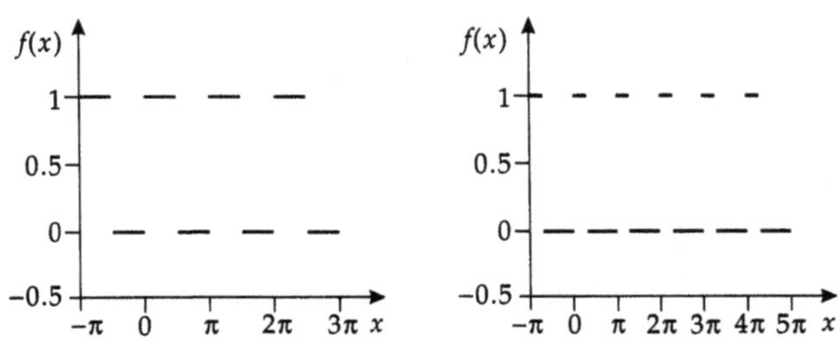

Square-wave pulses

5.33 Tangent and cotangent functions

$$f(x) = \tan(x) , \qquad g(x) = \cot(x)$$

Other notations are $\operatorname{tg}(x)$ for $\tan(x)$, $\operatorname{cotan}(X)$ and $\operatorname{ctg}(X)$ for $\cot(x)$.

(a) Definition

Geometric definition: See explanations at the beginning of this section.
Interpretation within the right triangle:

$$\tan(\alpha) = \frac{\text{side } S}{\text{side } C} , \qquad \cot(\alpha) = \frac{\text{side } C}{\text{side } S}$$
$$\tan(A) = T , \qquad \cot(A) = \frac{1}{T} .$$

Definition in terms of sine and cosine functions:

$$\tan(x) = \frac{\sin(x)}{\cos(x)} , \qquad \cot(x) = \frac{\cos(x)}{\sin(x)} .$$

Exponential function with imaginary argument:

$$\tan(x) = \frac{2\mathrm{j}}{\mathrm{e}^{2\mathrm{j}x} + 1} - \mathrm{j} , \qquad \cot(x) = \frac{2\mathrm{j}}{\mathrm{e}^{2\mathrm{j}x} - 1} + \mathrm{j} .$$

Solution of the differential equation:

$$\frac{\mathrm{d}f}{\mathrm{d}x} = af^2 + bf + c , \qquad b^2 < 4ac ,$$

by means of the functions:

$$f_1(x) = \frac{\sqrt{4ac - b^2}}{2a} \tan\left(\frac{x\sqrt{4ac - b^2}}{2}\right) - \frac{b}{2a}$$
$$f_2(x) = -\frac{\sqrt{4ac - b^2}}{2a} \cot\left(\frac{x\sqrt{4ac - b^2}}{2}\right) - \frac{b}{2a}$$

Indefinite integral of the secant and cosecant functions:

$$\tan(x) = \int_0^x \sec^2(t)\,\mathrm{d}t , \qquad \cot(x) = \int_x^{\frac{\pi}{2}} \csc^2(t)\,\mathrm{d}t .$$

Integral of a power function:

$$\tan(x) = \frac{2}{\pi} \int_0^\infty \frac{t^{2x/\pi} - 1}{t^2 - 1}\mathrm{d}t , \qquad x < \frac{\pi}{2}$$
$$\cot(x) = \frac{2}{\pi} \int_0^\infty \frac{t^{2x/\pi}}{t - t^3}\mathrm{d}t , \qquad x < \pi .$$

(b) Graphical representation

- Tangent and cotangent functions are periodic functions with period π.

▷ This is in contrast to the other trigonometric functions, which have period 2π.

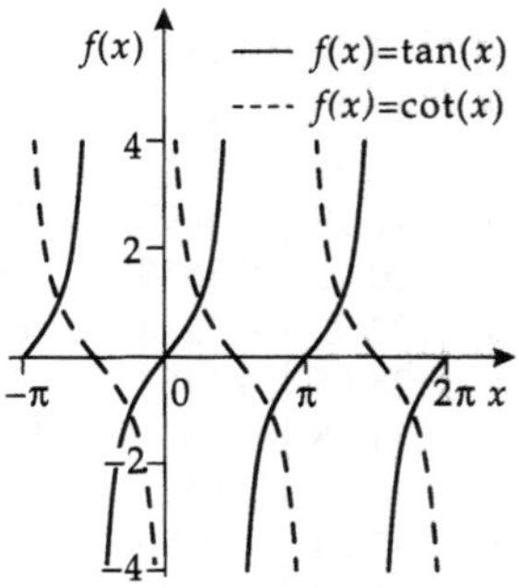

$f(x) = \tan(x)$ (solid line) and $f(x) = \cot(x)$ (dashed line)

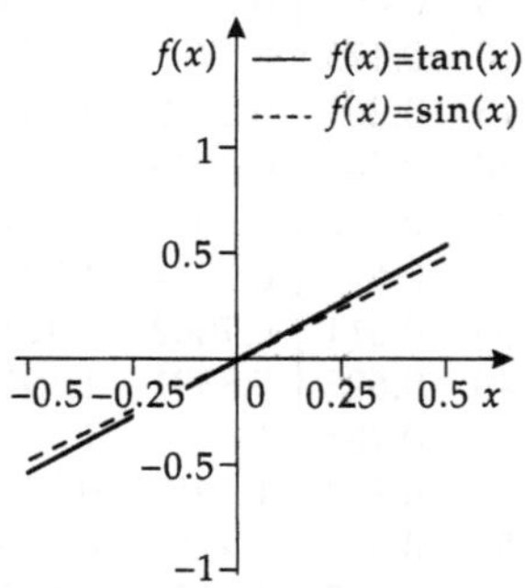

$f(x) = \tan(x)$ (solid line) and $f(x) = \sin(x)$ (dashed line)

The tangent function has zeros $0, \pm\pi, \pm 2\pi, \ldots$ and poles $\pm\frac{\pi}{2}, \pm\frac{3\pi}{2}, \ldots$.
The cotangent function has zeros $\pm\frac{\pi}{2}, \pm\frac{3\pi}{2}, \ldots$ and poles $0, \pm\pi, \pm 2\pi, \ldots$.

▷ In the vicinity of the zeros the tangent function behaves like a straight line of slope 1, and the cotangent function behaves like a straight line of slope -1.

The tangent (cotangent) function is strictly monotonically increasing (decreasing) in the interval between two zeros.

(c) Properties of the functions

Domain of definition:	$-\infty < x < \infty$
	$\tan(x)$: $x \neq \pm\dfrac{\pi}{2}, \pm\dfrac{3\pi}{2} \ldots$
	$\cot(x)$: $x \neq 0, \pm\pi, \pm 2\pi \ldots$
Range of values:	$-\infty < f(x) < \infty$
Quadrant:	All four quadrants.
Periodicity:	Period π.
Monotonicity:	In the interval between the poles
	$\tan(x)$: strictly monotonically increasing
	$\cot(x)$: strictly monotonically decreasing
Symmetries:	Point symmetry about the origin.
Asymptotes:	None for $x \to \pm\infty$.

(d) Particular values

deg	rad	$\tan(x)$	$\cot(x)$	rad	deg	$\tan(x)$	$\cot(x)$
0	0	0	∞	0.05	2.87	0.0500	19.983
5	0.087	0.0875	11.430	0.1	5.73	0.1003	9.9666
10	0.175	0.1763	5.6713	0.15	8.59	0.1511	6.6166
15	0.261	0.2679	3.7321	0.2	11.46	0.2027	4.9332
20	0.349	0.3640	2.7475	0.25	14.32	0.2553	3.9163
25	0.436	0.4663	2.1445	0.3	17.19	0.3093	3.2327
30	0.524	0.5774	1.7321	0.35	20.05	0.3650	2.7395
35	0.611	0.7002	1.4281	0.4	22.92	0.4228	2.3652
40	0.698	0.8391	1.1918	0.5	28.65	0.5463	1.8305
45	0.785	1	1	0.6	34.38	0.6841	1.4617
50	0.873	1.1918	0.8391	0.7	40.12	0.8423	1.1872
55	0.960	1.4281	0.7002	0.8	45.84	1.0296	0.9721
60	1.047	1.7321	0.5774	0.9	51.57	1.2602	0.7936
65	1.135	2.1445	0.4663	1.0	57.30	1.5574	0.6421
70	1.222	2.7475	0.3640	1.1	63.03	1.9684	0.5090
75	1.309	3.7321	0.2679	1.2	68.75	2.5722	0.3888
80	1.396	5.6713	0.1763	1.3	74.48	3.6021	0.2776
85	1.484	11.430	0.0875	1.4	80.21	5.7979	0.1725
90	1.571	∞	0	1.5	85.94	14.101	0.0709

Zeros:
$\tan(x)$: $x = 0, \pm\pi, \pm 2\pi \ldots$
$\cot(x)$: $x = \pm\frac{\pi}{2}, \pm\frac{3\pi}{2} \ldots$

Discontinuities: None.

Poles:
$\tan(x)$: $x = \pm\frac{\pi}{2}, \pm\frac{3\pi}{2} \ldots$
$\cot(x)$: $x = 0, \pm\pi, \pm 2\pi \ldots$

Extremes: None.

Points of inflection:
$\tan(x)$: $x = 0, \pm\pi, \pm 2\pi \ldots$
$\cot(x)$: $x = \pm\frac{\pi}{2}, \pm\frac{3\pi}{2} \ldots$

(e) Reciprocal functions

- Tangent and cotangent functions are reciprocals of each other:

$$\frac{1}{\tan(x)} = \cot(x) , \qquad \frac{1}{\cot(x)} = \tan(x) .$$

(f) Inverse functions

The inverse functions are the corresponding functions.

- The function can only be inverted on one length of the period.

$$x = \arctan(y) , \qquad y = \tan(x)$$
$$x = \operatorname{arccot}(y) , \qquad y = \cot(x)$$

(g) Cognate functions

Sine and cosine functions: Their quotients yield the tangent and cotangent functions.
Secant and cosecant functions: The reciprocal functions.
Inverse trigonometric functions.
Hyperbolic functions: Related to the trigonometric functions.

(h) Conversion formulas

Reflection about the y-axis:

$$\tan(-x) = -\tan(x) , \qquad \cot(-x) = -\cot(x)$$

Shift along the x-axis:

$$\tan\left(x \pm \frac{n\pi}{4}\right) = \begin{cases} \tan(x) & n = 0, 4, 8, \ldots \\ \dfrac{\tan(x) \pm 1}{1 \mp \tan(x)} & n = 1, 5, 9, \ldots \\ -\cot(x) & n = 2, 6, 10, \ldots \\ \dfrac{\tan(x) \mp 1}{1 \pm \tan(x)} & n = 3, 7, 11, \ldots \end{cases}$$

$$\cot\left(x \pm \frac{n\pi}{4}\right) = \begin{cases} \cot(x) & n = 0, 4, 8, \ldots \\ \dfrac{\cot(x) \mp 1}{1 \pm \cot(x)} & n = 1, 5, 9, \ldots \\ -\tan(x) & n = 2, 6, 10, \ldots \\ \dfrac{\cot(x) \pm 1}{1 \mp \cot(x)} & n = 3, 7, 11, \ldots \end{cases}$$

Together with the reflection properties, the result is the relation of complementary angles, which is important in geometry.

$$\tan(x) = \cot\left(\frac{\pi}{2} - x\right) , \qquad \cot(x) = \tan\left(\frac{\pi}{2} - x\right)$$

Addition in the argument:

$$\tan(x \pm y) = \frac{\tan(x) \pm \tan(y)}{1 \mp \tan(x)\tan(y)} , \qquad \cot(x \pm y) = \frac{\cot(x)\cot(y) \mp 1}{\cot(y) \pm \cot(x)}$$

Integral multiples in the argument of the tangent function:

$$\tan(2x) = \frac{2\tan(x)}{1 - \tan^2(x)} = \frac{2\cot(x)}{\cot^2(x) - 1}$$

$$\tan(3x) = \frac{3\tan(x) - \tan^3(x)}{1 - 3\tan^2(x)}$$

$$\tan(4x) = \frac{4\tan(x) - 4\tan^3(x)}{1 - 6\tan^2(x) + \tan^4(x)}$$

$$\tan(5x) = \frac{5\tan(x) - 10\tan^3(x) + \tan^5(x)}{1 - 10\tan^2(x) + 5\tan^4(x)}$$

Half argument:

$$\tan\left(\frac{x}{2}\right) = (-1)^m \sqrt{\frac{1 - \cos(x)}{1 + \cos(x)}} = \tau , \qquad m = \operatorname{int}\left[\frac{2x}{\pi}\right]$$

Conversions with $\tau = \tan\left(\frac{x}{2}\right)$:

$$\sin(x) = \frac{2\tau}{1+\tau^2} \qquad \csc(x) = \frac{1+\tau^2}{2}$$
$$\cos(x) = \frac{1-\tau^2}{1+\tau^2} \qquad \sec(x) = \frac{1+\tau^2}{1-\tau^2}$$
$$\tan(x) = \frac{1}{1-\tau} - \frac{1}{1+\tau} \qquad \cot(x) = \frac{1-\tau^2}{2\tau}$$
$$\cos(x)+1 = \frac{2}{1+\tau^2} \qquad \sec(x)+1 = \frac{1}{1-\tau} + \frac{1}{1+\tau}$$
$$\sec(x)+\tan(x) = \frac{1+\tau}{1-\tau} \qquad \sec(x)-\tan(x) = \frac{1-\tau}{1+\tau}$$
$$\csc(x)+\cot(x) = \frac{1}{\tau} \qquad \csc(x)-\cot(x) = \tau$$

For multiples in the argument of the cotangent function, check the corresponding formulas for the tangent function using

$$\cot(ax) = \frac{1}{\tan(ax)}\,.$$

Addition of functions:

$$\tan(x) \pm \tan(y) = \sin(x \pm y)\sec(x)\sec(y)$$
$$\cot(x) \pm \cot(y) = \sin(x \pm y)\csc(x)\csc(y)$$
$$\cot(x) + \tan(y) = 2\csc(2x)$$
$$\cot(x) - \tan(y) = 2\cot(2x)$$

Product of functions:

$$\tan(x)\tan(y) = \frac{\cos(x-y)-\cos(x+y)}{\cos(x-y)+\cos(x+y)}$$
$$\tan(x)\cot(y) = \frac{\sin(x+y)+\sin(x-y)}{\sin(x+y)-\sin(x-y)}$$
$$\cot(x)\cot(y) = \frac{\cos(x-y)+\cos(x+y)}{\cos(x-y)-\cos(x+y)}$$

(i) Approximation formulas (8 bits ≈ 0.4% of accuracy)

For small values of x:

$$\tan(x) \approx \frac{x}{1-\frac{x^2}{3}}\,, \qquad -0.6 \le x \le 0.6$$
$$\cot(x) \approx \frac{\left(1-\frac{x^2}{6}\right)^2}{x}\,, \qquad -0.5 \le x \le 0.5\,.$$

In the vicinity of poles x_P, it holds that

$$\tan(x) \approx \frac{1}{x_P - x}, \qquad x_P = \pm\frac{\pi}{2}, \pm\frac{3\pi}{2}, \ldots, \qquad -0.1 \leq x_P - x \leq 0.1$$

$$\cot(x) \approx \frac{1}{x - x_P}, \qquad x_P = 0, \pm\pi, \pm 2\pi, \ldots, \qquad -0.1 \leq x_P - x \leq 0.1$$

(j) Expansion of series or products

Series decomposition with Bernoulli's numbers B_k:
For the tangent $-\frac{\pi}{2} < x < \frac{\pi}{2}$, it holds that

$$\tan(x) = x + \frac{x^3}{3} + \frac{2x^5}{15} + \frac{17x^7}{315} + \cdots = \sum_{k=1}^{\infty} \frac{4^k(4^k - 1)|B_k|}{(2k)!} x^{2k-1} .$$

For the cotangent $-\pi < x < \pi$, it holds that

$$\cot(x) = \frac{1}{x} - \frac{x}{3} - \frac{x^3}{45} - \frac{x^5}{945} - \cdots = \frac{1}{x} - \sum_{k=1}^{\infty} \frac{4^k |B_{2k}|}{(2k)!} x^{2k-1} .$$

Partial fraction decomposition for the tangent and cotangent functions:

$$\tan(x) = \frac{8x}{\pi^2 - 4x^2} + \frac{8x}{9\pi^2 - 4x^2} + \frac{8x}{25\pi^2 - 4x^2} + \cdots + \frac{8x}{(2k-1)^2\pi^2 - 4x^2} + \cdots$$

$$\cot(x) = \frac{1}{x} - \frac{2x}{\pi^2 - x^2} - \frac{2x}{4\pi^2 - x^2} - \frac{2x}{9\pi^2 - x^2} - \cdots - \frac{2x}{k^2\pi^2 - x^2} - \cdots$$

Continued fraction decomposition of the tangent (outside the poles):

$$\tan(x) = \cfrac{x}{1 - \cfrac{x^2}{3 - \cfrac{x^2}{5 - \cfrac{x^2}{7 - \cfrac{x^2}{9 - \cdots}}}}}$$

(k) Derivative of the functions

The derivatives of the tangent and cotangent functions result in quadratic terms.

$$\frac{\mathrm{d}}{\mathrm{d}x}\tan(x) = \sec^2(x) = 1 + \tan^2(x)$$

$$\frac{\mathrm{d}}{\mathrm{d}x}\cot(x) = -\csc^2(x) = -1 - \cot^2(x)$$

(l) Primitive of the functions

The primitive of the tangent and cotangent functions are logarithms of secant and cosecant functions:

$$\int_0^x \tan(t)\mathrm{d}t = \ln|\sec(x)| = \frac{1}{2}\ln(1 + \tan^2(x))$$

$$\int_x^{\pi/2} \cot(t)\mathrm{d}t = \ln|\csc(x)| = \frac{1}{2}\ln(1 + \cot^2(x)) .$$

For any a, it holds that

$$\int_0^{\pi/4} \tan^a(t)\mathrm{d}t = \int_{\pi/4}^{\pi/2} \cot^a(t)\mathrm{d}t \ .$$

For $-1 < a < 1$, it further holds that

$$\int_0^{\pi/2} \tan^a(t)\mathrm{d}t = \int_0^{\pi/2} \cot^a(t)\mathrm{d}t = \frac{\pi}{2}\sec\left(\frac{a\pi}{2}\right) .$$

(m) Particular extensions and applications

Complex argument:

$$\tan(x+\mathrm{j}y) = \frac{\sin(2x)+\mathrm{j}\sinh(2y)}{\cos(2x)+\cosh(2y)}$$
$$\cot(x+\mathrm{j}y) = \frac{\sin(2x)-\mathrm{j}\sinh(2y)}{\cosh(2x)-\cos(2y)}$$

Purely imaginary argument:

$$\tan(\mathrm{j}y) = \mathrm{j}\tanh(y) \qquad \cot(\mathrm{j}y) = -\mathrm{j}\coth(y)$$

5.34 Secant and cosecant

$$f(x) = \sec(x) \ , \qquad g(x) = \csc(x)$$

The notations $\csc(x)$ and $\mathrm{cosec}(x)$ are used for the cosecant function. Rarely used functions.

(a) Definition

Geometric interpretation: See the beginning of this chapter.

$$\sec(x) = \text{ segment } \overline{OT} \ , \qquad \csc(x) = \text{ segment } \overline{OK}$$

Interpretation within the right triangle:

$$\sec(\alpha) = \frac{\text{hypotenuse}}{\text{side } C} \ , \qquad \csc(\alpha) = \frac{\text{hypotenuse}}{\text{side } S} \ .$$

Reciprocal functions of sine and cosine functions:

$$\sec(x) = \frac{1}{\cos(x)} \ , \qquad \csc(x) = \frac{1}{\sin(x)} \ .$$

Exponential functions with negative argument:

$$\sec(x) = \frac{2\mathrm{e}^{\mathrm{j}x}}{\mathrm{e}^{2\mathrm{j}x}+1} \ , \qquad \csc(x) = \frac{2\mathrm{j}\mathrm{e}^{\mathrm{j}x}}{\mathrm{e}^{2\mathrm{j}x}+1} \ .$$

Integral representation:

$$\sec(x) = \frac{2}{\pi}\int_0^{\infty} \frac{t^{2x/\pi}}{t^2+1}\mathrm{d}t \ , \qquad -\frac{\pi}{2} < x < \frac{\pi}{2}$$
$$\csc(x) = \frac{1}{\pi}\int_0^{\infty} \frac{t^{x/\pi}}{t^2+1}\mathrm{d}t \ , \qquad 0 < x < \pi \ .$$

(b) Graphical representation

The functions can be described as arcs, alternately running in the positive and negative y-domain. The functions have a period of 2π, adopting all values except those between -1 and 1.

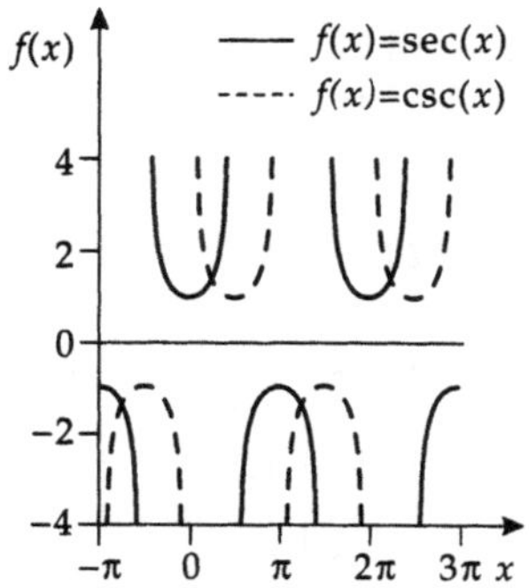

$f(x) = \sec(x)$ (solid line), $f(x) = \csc(x)$ (dashed line)

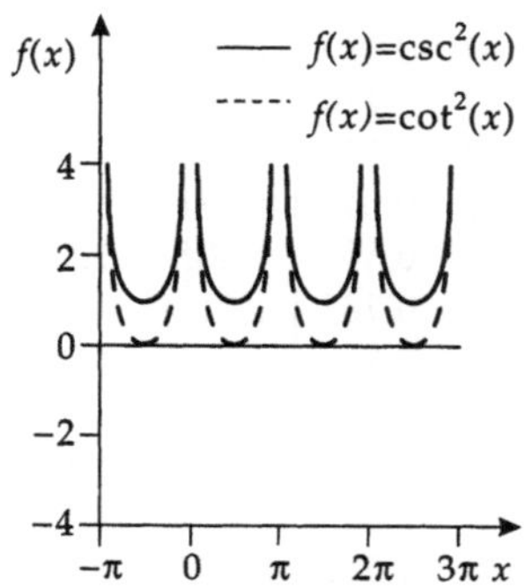

$f(x) = \csc^2(x)$ (solid line), $f(x) = \cot^2(x)$ (dashed line)

(c) Properties of the functions

Domain of definition: $-\infty < x < \infty$, except

sec(x) $x = n\pi$, n integer

csc(x) $x = \left(n + \frac{1}{2}\right)\pi$, n integer

Range of values: $-\infty < f(x) \le -1$ and $1 \le f(x) < \infty$

Quadrant: All four quadrants.

Periodicity: Period 2π.

Monotonicity: Strictly monotonically increasing and strictly monotonically decreasing pieces.

Symmetries:

sec(x) point symmetry about the origin
point symmetry about $x = n\pi$, $y = 0$, n integer.
mirror symmetry about $x = \left(n + \frac{1}{2}\right)\pi$, n integer.

csc(x) mirror symmetry about the y-axis
mirror symmetry about $x = n\pi$, n integer.
point symmetry about $x = \left(n + \frac{1}{2}\right)\pi$, $y = 0$.

Asymptotes: No definite behavior as $x \to \pm\infty$.

(d) Particular values

deg	rad	sec(x)	csc(x)	rad	deg	sec(x)	csc(x)
0	0	1	∞	0.05	2.87	1.0013	20.008
5	0.087	1.004	11.47	0.1	5.73	1.0050	10.017
10	0.175	1.015	5.759	0.15	8.59	1.0114	6.6917
15	0.261	1.035	3.864	0.2	11.46	1.0203	5.0335
20	0.349	1.064	2.924	0.25	14.32	1.0321	4.0420
25	0.436	1.103	2.366	0.3	17.19	1.0468	3.3839
30	0.524	1.155	2	0.35	20.05	1.0645	2.9163
35	0.611	1.221	1.743	0.4	22.92	1.0857	2.5679
40	0.698	1.305	1.550	0.5	28.65	1.1395	2.0858
45	0.785	1.414	1.414	0.6	34.38	1.2116	1.7710
50	0.873	1.556	1.305	0.7	40.12	1.3075	1.5523
55	0.960	1.743	1.221	0.8	45.84	1.4353	1.3940
60	1.047	2	1.155	0.9	51.57	1.6087	1.2766
65	1.135	2.366	1.103	1.0	57.30	1.8508	1.1884
70	1.222	2.924	1.064	1.1	63.03	2.2046	1.1221
75	1.309	3.864	1.035	1.2	68.75	2.7597	1.0729
80	1.396	5.759	1.015	1.3	74.48	3.7383	1.0378
85	1.484	11.47	1.004	1.4	80.21	5.8835	1.0148
90	1.571	∞	1	1.5	85.94	14.137	1.0025

Zeros: None.

Discontinuities: None.

Poles:
sec(x) $x = n\pi$, n integer
csc(x) $x = \left(n + \frac{1}{2}\right)\pi$, n integer

Extremes:
sec(x) minima $x = \left(2n + \frac{1}{2}\right)\pi$
maxima $x = \left(2n - \frac{1}{2}\right)\pi$
csc(x) minima $x = 2n\pi$
maxima $x = (2n - 1)\pi$

Points of inflection: None.

(e) Reciprocal functions

The reciprocal functions are the cosine and sine functions:

$$\frac{1}{\sec(x)} = \cos(x)\,, \qquad \frac{1}{\csc(x)} = \sin(x)\,.$$

(f) Inverse functions

The inverse functions are the inverse trigonometric functions.

- The functions are invertible only on half the length of a period.

$$x = \operatorname{arcsec}(y)\,, \qquad y = \sec(x)$$
$$x = \operatorname{arccsc}(y)\,, \qquad y = \csc(x)$$

(g) Cognate functions

Sine and cosine functions, the reciprocal functions.
Tangent and cotangent functions, the quotient of secant and cosecant functions.
Inverse trigonometric functions.
Hyperbolic functions, related to trigonometric functions.

(h) Conversion formulas

Relationship between secant and cosecant functions:

$$\sec(x) = \csc(x)\tan(x) = \frac{\csc(x)}{\cot(x)}$$

Shift along the x-axis:

$$\sec\left(x \pm \frac{n\pi}{2}\right) = \begin{cases} \sec(x) & n = 0, 4, 8, \ldots \\ \mp\csc(x) & n = 1, 5, 9, \ldots \\ -\sec(x) & n = 2, 6, 10, \ldots \\ \pm\csc(x) & n = 3, 7, 11, \ldots \end{cases}$$

$$\csc\left(x \pm \frac{n\pi}{2}\right) = \begin{cases} \csc(x) & n = 0, 4, 8, \ldots \\ \pm\sec(x) & n = 1, 5, 9, \ldots \\ -\csc(x) & n = 2, 6, 10, \ldots \\ \mp\sec(x) & n = 3, 7, 11, \ldots \end{cases}$$

Double argument:

$$\sec(2x) = \frac{\sec^2(x)}{2 - \sec^2(x)}$$

$$\csc(2x) = \frac{\sec(x)\csc(x)}{2} = \frac{\csc^2(x)}{2\sqrt{\csc^2(x) - 1}}$$

Half argument:

$$\sec\left(\frac{x}{2}\right) = (-1)^m \sqrt{\frac{2\sec(x)}{1 + \sec(x)}} \qquad m = \mathrm{int}\left[\frac{\pi + |x|}{2\pi}\right]$$

$$\csc\left(\frac{x}{2}\right) = (-1)^m \sqrt{\frac{\sec(x)}{\sec(x) - 1}} \qquad m = \mathrm{int}\left[\frac{|x|}{2\pi}\right]$$

(i) Approximation formulas (8 bits ≈ 0.4% of accuracy)

For small values of x, it holds that

$$\sec(x) \approx \left(1 - \frac{x^2}{3}\right)^{-3/2}, \qquad -0.9 \le x \le 0.9$$

$$\csc(x) \approx \frac{1}{x} + \frac{x}{6}, \qquad -\frac{2}{3} \le x \le \frac{2}{3}$$

(j) Expansion of series or products

Series expansion of the secant for $-\pi < 2x < \pi$:

$$\sec(x) = 1 + \frac{x^2}{2} + \frac{5x^4}{24} + \frac{61x^6}{720} + \cdots$$

Series expansion of the cosecant with Bernoulli's numbers B_k:

$$\csc(x) = \frac{1}{x} + \frac{x}{6} + \frac{7x^3}{360} + \frac{31x^5}{15120} + \cdots$$
$$= \sum_{k=0}^{\infty} \frac{|(4^k - 2)B_k|x^{2k-1}}{(2k)!}, \qquad -\pi < x < \pi .$$

Partial fraction decomposition:

$$\sec(x) = \frac{4\pi}{\pi^2 - 4x^2} - \frac{12\pi}{9\pi^2 - 4x^2} + \frac{20\pi}{25\pi^2 - 4x^2} - \cdots$$
$$= \pi \sum_{k=0}^{\infty} \frac{(-1)^k(8k+4)}{(2k+1)^2\pi^2 - 4x^2}$$
$$\csc(x) = \frac{1}{x} + \frac{2x}{\pi^2 - x^2} - \frac{2x}{4\pi^2 - x^2} + \frac{2x}{9\pi^2 - x^2} - \cdots$$
$$= \frac{1}{x} - 2x \sum_{k=0}^{\infty} \frac{(-1)^k}{k^2\pi^2 - x^2} .$$

Partial fraction decomposition of the squares:

$$\sec^2(x) = \frac{4}{(\pi - 2x)^2} + \frac{4}{(\pi + 2x)^2} + \frac{4}{(3\pi - 2x)^2} + \frac{4}{(3\pi + 2x)^2} + \cdots$$
$$= \sum_{k=-\infty}^{\infty} \frac{1}{\left[x + \left(k + \frac{1}{2}\right)\pi\right]^2}$$
$$\csc^2(x) = \frac{1}{x^2} + \frac{1}{(\pi - x)^2} + \frac{1}{(\pi + x)^2} + \frac{1}{(2\pi - x)^2} + \frac{1}{(2\pi + x)^2} + \cdots$$
$$= \sum_{k=-\infty}^{\infty} \frac{1}{[x + k\pi]^2} .$$

(k) Derivative of the functions

Differentiation via the chain rule:

$$\frac{\mathrm{d}}{\mathrm{d}x}\sec(x) = \sec(x)\tan(x) = \frac{\sec^2(x)}{\csc(x)}$$
$$\frac{\mathrm{d}}{\mathrm{d}x}\csc(x) = -\csc(x)\cot(x) = -\frac{\csc^2(x)}{\sec(x)} .$$

(l) Primitive of the functions

The primitives are logarithms of trigonometric functions:

$$\int_0^x \sec(t)\mathrm{d}t = \ln\left|\tan\left(\frac{x}{2} + \frac{\pi}{4}\right)\right| = \ln[\sec(x) + \cot(x)]$$
$$\int_{\pi/2}^x \csc(t)\mathrm{d}t = \ln\left|\tan\left(\frac{x}{2}\right)\right| = \ln[\csc(x) - \cot(x)]$$

The primitives of the squares are tangent and cotangent functions:

$$\int_0^x \sec^2(t)\mathrm{d}t = \tan(x) , \qquad \int_x^{\pi/2} \csc^2(t)\mathrm{d}t = \cot(x)$$

(m) Particular extensions and applications

Complex argument:

$$\sec(x + \mathrm{j}y) = \frac{\coth(y) + \mathrm{j}\tan(x)}{\sec(x)\sinh(y) + \cos(x)\operatorname{csch}(y)}$$

$$\csc(x + \mathrm{j}y) = \frac{\coth(y) - \mathrm{j}\cot(x)}{\sin(x)\operatorname{csch}(y) + \csc(x)\sinh(y)}$$

Purely imaginary argument:

$$\sec(\mathrm{j}y) = \operatorname{sech}(y) \qquad \csc(\mathrm{j}y) = -\mathrm{j}\operatorname{csch}(y)$$

Euler's numbers can be defined by the series expansion of the secant:

$$\sec(x) = E_0 + E_2\frac{x^2}{2!} + E_4\frac{x^4}{4!} + E_6\frac{x^6}{6!} + \cdots$$

Table of the first twelve Euler's numbers with integral index ($E_0 = 1$, $E_{2n+1} = 0$)

k	E_k	k	E_k	k	E_k
2	−1	10	−50, 521	18	−2, 404, 879, 675, 441
4	5	12	2, 702, 765	20	370, 371, 188, 237, 525
6	−61	14	−199, 360, 981	22	−69, 348, 874, 393, 137, 901
8	1, 385	16	19, 391, 512, 145	24	15, 514, 34, 163, 557, 086, 905

Inverse trigonometric functions

Inverse trigonometric functions are the inverse of trigonometric functions. They assign the length of the corresponding arc to the values defined by the trigonometric functions.

- Because this assignment is not unique, a **principal value** is assigned to an inverse trigonometric function.

▷ For $\arcsin(x)$, $\arctan(x)$, and $\operatorname{arccsc}(x)$ the principal value lies between $-\frac{\pi}{2}$ and $\frac{\pi}{2}$; for $\arccos(x)$, $\operatorname{arccot}(x)$, and $\operatorname{arcsec}(x)$ it lies between 0 and π.

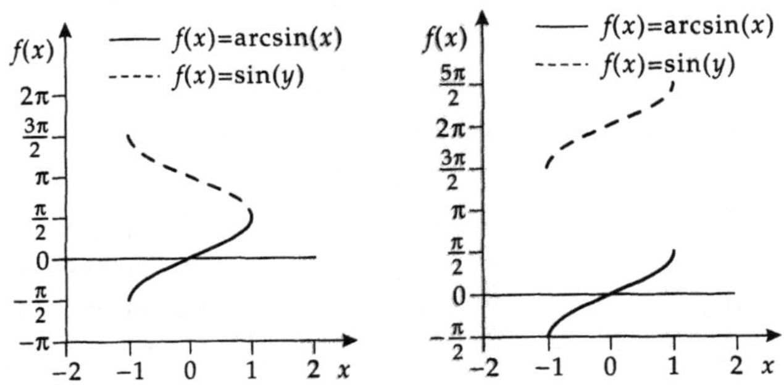

Principal value of $f(x) = \arcsin(x)$ (——)(solid line) and further possibilities with $x = \sin(y)$ (- - -)(dashed line)

Similar to the trigonometric functions, there are also conversion formulas for the inverse trigonometric functions.

$$(x) = \mathrm{sgn}(x),$$

$$p(x) = \pi \cdot H(-x) = \begin{cases} \pi & x < 0 \\ 0 & x < 0 \end{cases} .$$

Function	$f(x) = \arcsin(x)$	$f(x) = \arccos(x)$	$f(x) = \arctan(x)$
$\arcsin(x)$	$f(x)$	$\frac{\pi}{2} - f(x)$	$f\left(\frac{x}{\sqrt{1-x^2}}\right)$
$\arccos(x)$	$\frac{\pi}{2} - f(x)$	$f(x)$	$f\left(\frac{\sqrt{1-x^2}}{x}\right) + p(x)$
$\arctan(x)$	$f\left(\frac{x}{\sqrt{1+x^2}}\right)$	$s(x) f\left(\frac{1}{\sqrt{1+x^2}}\right)$	$f(x)$
$\operatorname{arccot}(x)$	$f\left(\frac{s(x)}{\sqrt{1+x^2}}\right) + p(x)$	$f\left(\frac{x}{\sqrt{1+x^2}}\right)$	$\frac{\pi}{2} - f(x)$
$\operatorname{arcsec}(x)$	$f\left(\frac{\sqrt{x^2-1}}{x}\right) + p(x)$	$f\left(\frac{1}{x}\right)$	$s(x) f\left(\sqrt{x^2-1}\right) + p(x)$
$\operatorname{arccsc}(x)$	$f\left(\frac{1}{x}\right)$	$f\left(\frac{\sqrt{x^2-1}}{x}\right) - p(x)$	$f\left(\frac{s(x)}{\sqrt{x^2-1}}\right)$

- An important conversion characteristic is the constant sum of two complementary inverse trigonometric functions.

$$\arcsin(x) + \arccos(x) = \arctan(x) + \operatorname{arccot}(x) = \operatorname{arcsec}(x) + \operatorname{arccsc}(x) = \frac{\pi}{2}$$

▷ Some program libraries include only the inverse tangent function as a standard function. Other inverse trigonometric functions must be derived via the conversion functions.

Function	$f(x) = \operatorname{arccot}(x)$	$f(x) = \operatorname{arcsec}(x)$	$f(x) = \operatorname{arccsc}(x)$
$\arcsin(x)$	$f\left(\frac{\sqrt{1-x^2}}{x}\right) - p(x)$	$s(x) f\left(\frac{1}{\sqrt{1-x^2}}\right)$	$f\left(\frac{1}{x}\right)$
$\arccos(x)$	$f\left(\frac{x}{\sqrt{1-x^2}}\right)$	$f\left(\frac{1}{x}\right)$	$f\left(\frac{s(x)}{\sqrt{1-x^2}}\right)$
$\arctan(x)$	$\frac{\pi}{2} - f(x)$	$s(x) f\left(\sqrt{x^2+1}\right)$	$f\left(\frac{\sqrt{x^2+1}}{x}\right)$
$\operatorname{arccot}(x)$	$f(x)$	$f\left(\frac{\sqrt{x^2+1}}{x}\right)$	$s(x) f\left(\sqrt{1+x^2}\right) + p(x)$
$\operatorname{arcsec}(x)$	$f\left(\frac{s(x)}{\sqrt{x^2-1}}\right)$	$f(x)$	$\frac{\pi}{2} - f(x)$
$\operatorname{arccsc}(x)$	$f\left(\sqrt{x^2-1}\right) - p(x)$	$\frac{\pi}{2} - f(x)$	$f(x)$

▷ In order to distinguish clearly between the inverse hyperbolic functions and the inverse trigonometric functions, notation with capital letters is used for the inverse hyperbolic functions.

Notation with -1 as exponent, $\arcsin(x) = \sin^{-1}(x)$, is used mainly in North America.

▷ This abbreviated form can also be found on many pocket calculators.

5.35 Inverse sine and cosine functions

$$f(x) = \arcsin(x)\,, \qquad g(x) = \arccos(x)$$

The inverse functions of the sine and cosine functions.

(a) Definition

Inverse functions of sine and cosine functions:

$$y = \arcsin(x), \qquad x = \sin(y), \qquad -\tfrac{\pi}{2} \le y \le \tfrac{\pi}{2}$$
$$y = \arccos(x), \qquad x = \cos(y), \qquad 0 \le y \le \pi.$$

Representation by means of definite integrals

$$\arcsin(x) = \int_0^x \frac{dt}{\sqrt{1-t^2}}, \qquad |x| < 1$$

$$\arccos(x) = \int_x^1 \frac{dt}{\sqrt{1-t^2}}, \qquad |x| < 1.$$

(b) Graphical representation

- The functions are defined only in the domain $-1 \leq x \leq 1$.
  They are the mirror images of arcs of the sine and cosine functions about the angle-bisector $y = x$.

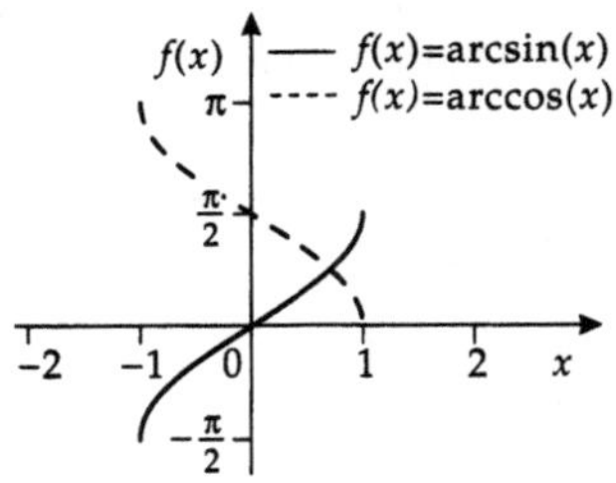

arcsin(x) (solid line),
arccos(x) (dashed line)

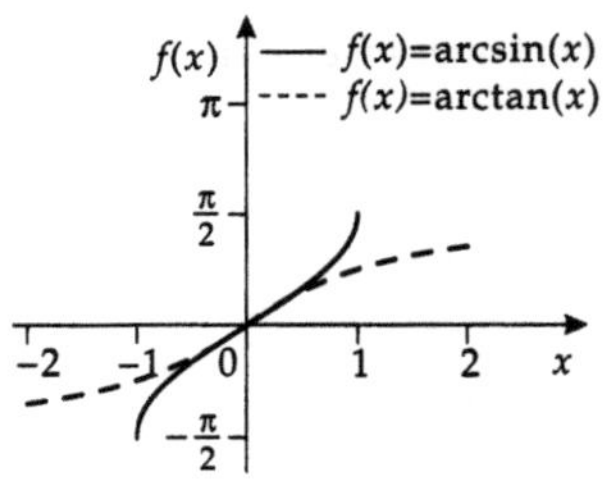

arcsin(x) (solid line),
arctan(x) (dashed line)

At the boundaries of the domain of definition the functions have a vertical tangent, i.e., the first derivative is divergent.
arcsin(x) is strictly monotonically increasing while arccos(x) is strictly monotonically decreasing.

▷ Close to $x = 0$, the functions behave like straight lines whose slopes have the absolute value 1.

(c) Properties of the functions

Domain of definition:	$-1 \leq x \leq 1$	
Range of values:	arcsin(x)	$-\frac{\pi}{2} \leq f(x) \leq \frac{\pi}{2}$
	arccos(x)	$0 \leq f(x) \leq \pi$
Quadrant:	arcsin(x)	first and third quadrants
	arccos(x)	first and second quadrants
Periodicity:	None.	
Monotonicity:	arcsin(x)	strictly monotonically increasing
	arccos(x)	strictly monotonically decreasing
Symmetries:	arcsin(x)	point symmetry about the origin $x = 0, y = 0$
	arccos(x)	point symmetry about point $x = 0$, $y = \frac{\pi}{2}$
Other:	arcsin(x)	$f(x) \to \pm\frac{\pi}{2}$ as $x \to \pm 1$
	arccos(x)	$f(x) \to \pi \mp \frac{\pi}{2}$ as $x \to \pm 1$

(d) Particular values

Zeros:	arcsin(x)	$x = 0$
	arccos(x)	$x = 1$
Discontinuities:	None.	
Poles:	None.	

Extremes: None.

Points of inflection: At $x = 0$.

Value of	$x=-1$	$x=-\frac{1}{\sqrt{2}}$	$x=0$	$x=\frac{1}{\sqrt{2}}$	$x=1$
$\arcsin(x)$	$-\frac{\pi}{2}$	$-\frac{\pi}{4}$	0	$\frac{\pi}{4}$	$\frac{\pi}{2}$
$\arccos(x)$	π	$\frac{3\pi}{4}$	$\frac{\pi}{2}$	$\frac{\pi}{4}$	0

(e) Reciprocal functions

The reciprocal functions cannot be written in terms of inverse trigonometric or trigonometric functions.

(f) Inverse functions

The inverse functions are the sine and cosine functions:

$$\begin{aligned} x &= \sin(y), & y &= \arcsin(x) \\ x &= \cos(y), & y &= \arccos(x) . \end{aligned}$$

(g) Cognate functions

Trigonometric functions: Inverse of the inverse hyperbolic functions.
The roots of reciprocal quadratic functions are related to inverse sine and cosine functions via differentiation and integration.

(h) Conversion formulas

Reflection in the y-axis:

$$\begin{aligned} \arcsin(-x) &= -\arcsin(x) \\ \arccos(-x) &= \pi - \arccos(x) \end{aligned}$$

Addition of complementary functions:

$$\arcsin(x) + \arccos(x) = \frac{\pi}{2}$$

Conversion of functions:

$$\arcsin(x) = \frac{\operatorname{sgn}(x)}{2}\arccos(1-2x^2) = \operatorname{sgn}(x)\arccos(\sqrt{1-x^2})$$

$$\arccos(x) = 2\arcsin\left(\sqrt{\frac{1-x}{2}}\right) = \begin{cases} \arcsin\left(\sqrt{1-x^2}\right), & x>0 \\ \pi - \arcsin\left(\sqrt{1-x^2}\right), & x>0 . \end{cases}$$

Conversion in the argument:

$$\arcsin(x) = \frac{1}{2}\arcsin\left(2x\sqrt{1-x^2}\right), \qquad -\frac{1}{\sqrt{2}} \le x \le \frac{1}{\sqrt{2}}$$

$$\arccos(x) = 2\arccos\left(\sqrt{\frac{1+x}{2}}\right), \qquad -1 \le x \le 1 .$$

Addition of functions:

$$\arcsin(x) \pm \arcsin(y) = k\pi + \arcsin\left(x\sqrt{1-y^2} \pm y\sqrt{1-x^2}\right)$$

$$k = \mathrm{Int}\left[\frac{1}{2} + \frac{\arcsin(x) + \arcsin(y)}{\pi}\right]$$

$$\arccos(x) \pm \arccos(y) = \left(\frac{1}{2} - k\right)\pi + \arccos\left(xy \mp \sqrt{1-x^2}\sqrt{1-y^2}\right)$$

$$k = \mathrm{Int}\left[\frac{\arccos(x) + \arccos(y)}{\pi}\right].$$

(i) Approximation formulas (8 bits ≈ 0.4% of accuracy)

For small values of x, it holds that

$$\arcsin(x) \approx x\sqrt{\frac{3}{3-x^2}}, \qquad -0.5 \le x \le 0.5$$

(j) Expansion in series and products

Series expansion of the inverse sine function:

$$\arcsin(x) = x + \frac{1}{2}\frac{x^3}{3} + \frac{1\cdot 3}{2\cdot 4}\frac{x^5}{5} + \frac{1\cdot 3\cdot 5}{2\cdot 4\cdot 6}\frac{x^7}{7} + \cdots$$
$$= \sum_{k=0}^{\infty} \frac{(2k-1)!!}{(2k)!!}\frac{x^{2k+1}}{2k+1}.$$

Continued fraction representation:

$$\arcsin(x) = \cfrac{x\sqrt{1-x^2}}{1 - \cfrac{(1\cdot 2)x}{3 - \cfrac{(1\cdot 2)x}{5 - \cfrac{(3\cdot 4)x}{7 - \cfrac{(3\cdot 4)x}{9 - \cfrac{(5\cdot 6)x}{11 - \cdots}}}}}}$$

(k) Derivative of the functions

$\arcsin(x)$ and $\arccos(x)$ have equal derivatives except for the sign.

$$\frac{\mathrm{d}}{\mathrm{d}x}\arcsin(ax+b) = -\frac{\mathrm{d}}{\mathrm{d}x}\arccos(ax+b) = \frac{a}{\sqrt{1-(ax+b)^2}}$$

(l) Primitive of the functions

The primitives are $(-c-1 \le x \le 1-c)$:

$$\int_{-c/b}^{x} \arcsin(bx+c) = \left(x+\frac{c}{b}\right)\arcsin(bx+c) - \frac{1}{b} + \sqrt{\frac{1}{b^2} - \left(x+\frac{c}{b}\right)^2}$$

$$\int_{-c/b}^{x} \arccos(bx+c) = \left(x+\frac{c}{b}\right)\arccos(bx+c) + \frac{1}{b} - \sqrt{\frac{1}{b^2} - \left(x+\frac{c}{b}\right)^2} .$$

(m) Particular extensions and applications

Link to the inverse hyperbolic functions:

$$\arcsin(x) = -\mathrm{j}\,\mathrm{Arsinh}\,(\mathrm{j}x) , \qquad \arccos(x) = \pm\mathrm{j}\,\mathrm{Arcosh}\,(\mathrm{j}x) .$$

5.36 Inverse tangent and cotangent functions

$$f(x) = \arctan(x) , \qquad g(x) = \mathrm{arccot}\,(x)$$

The inverse functions of the tangent and cotangent functions.

(a) Definition

Inverse functions of the tangent and cotangent functions are

$$\begin{array}{lll} y = \arctan(x), & x = \tan(y), & -\frac{\pi}{2} \le y \le \frac{\pi}{2} \\ y = \mathrm{arccot}\,(x), & x = \cot(y), & 0 \le y \le \pi . \end{array}$$

Representation in terms of indefinite integrals:

$$\arctan(x) = \int_0^x \frac{\mathrm{d}t}{1+t^2}$$

$$\mathrm{arccot}\,(x) = \int_x^\infty \frac{\mathrm{d}t}{1+t^2} .$$

(b) Graphical representation

- The functions are defined for all x. They are the mirror images of branches of the tangent and cotangent functions about the angle-bisector $y = x$.

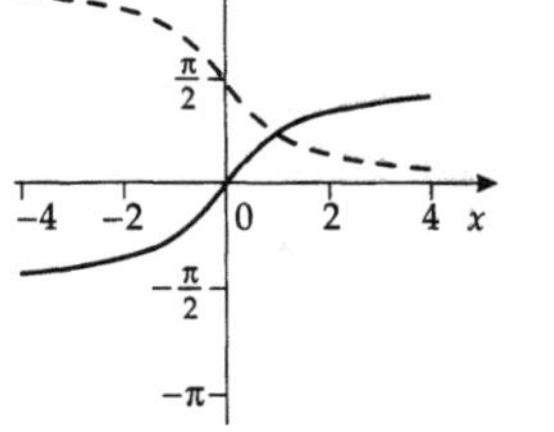

arctan(x) (solid line), arccot (x) (dashed line)

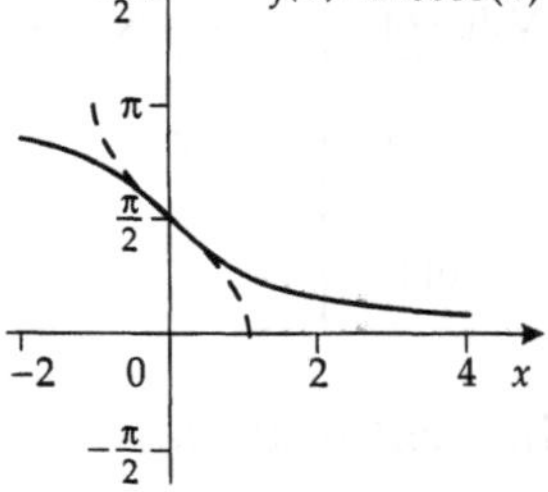

arccot (x) (solid line), arccos(x) (dashed line)

The functions have horizontal asymptotes, i.e., the first derivative approaches 0 as $x \to \pm\infty$.
arctan(x) is strictly monotonically increasing, whereas arccot (x) is strictly monotonically decreasing.

▷ In the vicinity of $x = 0$, the functions behave like straight lines whose slope has absolute value 1.

(c) Properties of the functions

Domain of definition:	$-\infty < x < \infty$	
Range of values:	arctan(x)	$-\frac{\pi}{2} \le f(x) \le \frac{\pi}{2}$
	arccot (x)	$0 \le f(x) \le \pi$
Quadrant:	arctan(x)	first and third quadrants
	arccot (x)	first and second quadrants
Periodicity:	None.	
Monotonicity:	arctan(x)	strictly monotonically increasing
	arccot (x)	strictly monotonically decreasing
Symmetries:	arctan(x)	point symmetry about the origin $x = 0, y = 0$
	arccot (x)	point symmetry about $x = 0$, $y = \frac{\pi}{2}$
Asymptotes:	arctan(x)	$f(x) \to \pm\frac{\pi}{2}$ as $x \to \pm\infty$
	arccot (x)	$f(x) \to \pi \mp \frac{\pi}{2}$ as $x \to \pm\infty$

(d) Particular values

Zeros:	arctan(x)	$x = 0$
	arccot (x)	none
Discontinuities:	None.	
Poles:	None.	
Extrema:	None.	
Points of inflection:	At $x = 0$.	

Value of	$x \to -\infty$	$x = -1$	$x = 0$	$x = 1$	$x =\to \infty$
arctan(x)	$-\frac{\pi}{2}$	$-\frac{\pi}{4}$	0	$\frac{\pi}{4}$	$\frac{\pi}{2}$
arccot (x)	π	$\frac{3\pi}{4}$	$\frac{\pi}{2}$	$\frac{\pi}{4}$	0

(e) Reciprocal functions

The reciprocal functions cannot be written in terms of inverse trigonometric or trigonometric functions.

(f) Inverse functions

The inverse functions are the tangent and cotangent functions:

$$x = \tan(y), \qquad y = \arctan(x)$$
$$x = \cot(y), \qquad y = \operatorname{arccot}(x) .$$

(g) Cognate functions

Trigonometric functions.
Reciprocal quadratic functions are linked to the inverse tangent or cotangent functions via differentiation and integration.

(h) Conversion formulas

Reflection about the y-axis:

$$\arctan(-x) = -\arctan(x)$$
$$\operatorname{arccot}(-x) = \pi - \operatorname{arccot}(x) .$$

Addition of complementary functions:

$$\arctan(x) + \operatorname{arccot}(x) = \frac{\pi}{2} .$$

Conversion of functions:

$$\arctan(x) = \operatorname{arccot}\left(\frac{1}{x}\right) - \begin{cases} 0 & x > 0 \\ \pi & x < 0 \end{cases}$$
$$\operatorname{arccot}(x) = \arctan\left(\frac{1}{x}\right) + \begin{cases} 0 & x > 0 \\ \pi & x < 0 . \end{cases}$$

Conversion in the argument:

$$\arctan\left(\frac{1}{x}\right) = -\arctan(x) + \frac{\pi \operatorname{sgn}(x)}{2} , \qquad x \neq 0$$
$$\operatorname{arccot}\left(\frac{1}{x}\right) = -\operatorname{arccot}(x) + \pi - \frac{\pi \operatorname{sgn}(x)}{2} , \qquad x \neq 0 .$$

Addition of functions:

$$\arctan(x) \pm \arctan(y) = k\pi + \arctan\left(\frac{x \pm y}{1 \mp xy}\right) ,$$
$$k = \operatorname{Int}\left[\frac{1}{2} + \frac{\arctan(x) + \arctan(y)}{\pi}\right]$$
$$\operatorname{arccot}(x) \pm \operatorname{arccot}(y) = \left(\frac{1}{2} - k\right)\pi + \operatorname{arccot}\left(\frac{x \pm y}{1 \mp xy}\right) ,$$
$$k = \operatorname{Int}\left[\frac{\operatorname{arccot}(x) + \operatorname{arccot}(y)}{\pi}\right] .$$

(i) Approximation formulas (8 bits $\approx$ 0.4% of accuracy)

For small values of x, it holds that

$$\arctan(x) \approx \frac{3x}{3+x^2}\,, \qquad -0.45 \le x \le 0.45\,.$$

For $x \le 1.8$, it holds that

$$\operatorname{arccot}(x) \approx \frac{3x}{3x^2-1}\,, \qquad x \le 1.8\,.$$

(j) Expansion of series or products

Series expansion of the inverse tangent function for $|x| \le 1$:

$$\begin{aligned}\arctan(x) &= x - \frac{x^3}{3} + \frac{x^5}{5} - \frac{x^7}{7} + \cdots \\ &= x \sum_{k=0}^{\infty} \frac{(-x^2)^k}{2k+1}\,.\end{aligned}$$

Series expansion of the inverse tangent functions for $|x| \ge 1$:

$$\begin{aligned}\arctan(x) &= \frac{\pi\,\operatorname{sgn}(x)}{2} - \frac{1}{x} + \frac{1}{3x^3} - \frac{1}{5x^5} + \frac{1}{7x^7} - \cdots \\ &= \frac{\pi\,\operatorname{sgn}(x)}{2} - \frac{1}{x} \sum_{k=0}^{\infty} \frac{(-x^2)^{-k}}{2k+1}\,.\end{aligned}$$

Continued fraction representation:

$$\arctan(x) = \cfrac{x}{1 + \cfrac{x^2}{3 + \cfrac{4x^2}{5 + \cfrac{9x^2}{7 + \cfrac{16x^2}{9 + \cfrac{25x^2}{11 + \cdots}}}}}}$$

(k) Derivative of the functions

$\arctan(x)$ and $\operatorname{arccot}(x)$ have equal derivatives except for the sign:

$$\frac{\mathrm{d}}{\mathrm{d}x} \arctan(ax+b) = -\frac{\mathrm{d}}{\mathrm{d}x} \operatorname{arccot}(ax+b) = \frac{a}{1-(ax+b)^2}\,.$$

(l) Primitive of the functions

The primitives are

$$\int_{-c/b}^{x} \arctan(bx+c) = \left(x + \frac{c}{b}\right) \arctan(bx+c) - \frac{1}{b} \ln \sqrt{1+(bx+c)^2}$$

$$\int_{-c/b}^{x} \operatorname{arccot}(bx+c) = \left(x + \frac{c}{b}\right) \operatorname{arccot}(bx+c) + \frac{1}{b} \ln \sqrt{1+(bx+c)^2}\,.$$

(m) Particular extensions and applications

Link to the inverse hyperbolic functions:

$$\arctan(x) = -\mathrm{j}\,\mathrm{Artanh}\,(\mathrm{j}x) \qquad \operatorname{arccot}(x) = \mathrm{j}\,\mathrm{Arcoth}\,(\mathrm{j}x)$$

5.37 Inverse secant and cosecant functions

$$f(x) = \operatorname{arcsec}(x)\,, \qquad g(x) = \operatorname{arccsc}(x)$$

The inverse functions of the secant and cosecant functions.

(a) Definition

Inverse functions of the secant and cosecant functions.:

$$y = \operatorname{arccsc}(x), \qquad x = \csc(y), \qquad -\frac{\pi}{2} \leq y \leq \frac{\pi}{2}$$

$$y = \operatorname{arcsec}(x), \qquad x = \sec(y), \qquad 0 \leq y \leq \pi\,.$$

Representation in terms of definite integrals:

$$\operatorname{arcsec}(x) = \int_1^x \frac{\mathrm{d}t}{t\sqrt{t^2-1}}$$

$$\operatorname{arccsc}(x) = \begin{cases} \displaystyle\int_{-\infty}^{x} \frac{\mathrm{d}t}{t\sqrt{t^2-1}} & -\infty < x \leq -1 \\ \displaystyle\int_{x}^{\infty} \frac{\mathrm{d}t}{t\sqrt{t^2-1}} & 1 \leq x < \infty\,. \end{cases}$$

(b) Graphical representation

- The functions are defined for all x with $|x| \geq 1$. They are mirror images of branches of the secant and cosecant functions about the angle-bisector $y = x$.

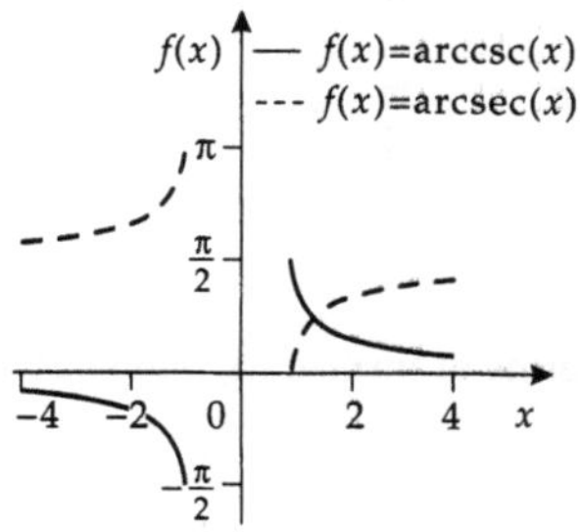

arccsc (x) (solid line), arcsec (x) (dashed line)

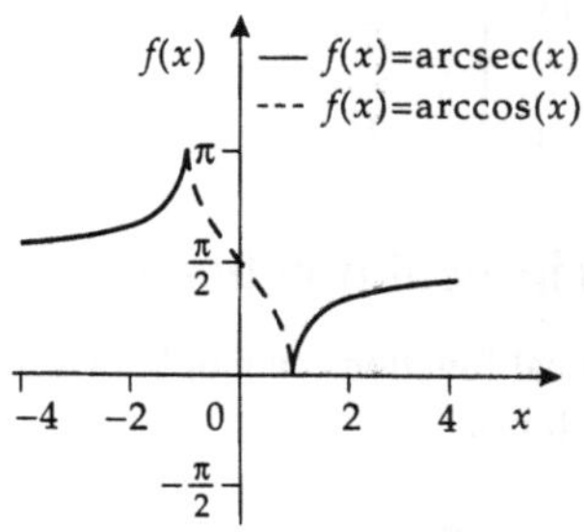

arcsec (x) (solid line), arccos(x) (dashed line)

For large values of x, the functions have a horizontal tangent; i.e., the first derivative approaches zero.

In both domains, arcsec (x) is strictly monotonically increasing, whereas arccsc (x) is strictly monotonically decreasing.

(c) Properties of the functions

Domain of definition:	$-\infty < x \leq -1, 1 \leq x < \infty$	
Range of values:	arcsec (x)	$0 \leq f(x) \leq \pi$
	arccsc (x)	$-\frac{\pi}{2} \leq f(x) \leq \frac{\pi}{2}$
Quadrant:	arcsec (x)	first and second quadrants
	arccsc (x)	first and third quadrants
Periodicity:	None.	
Monotonicity:	For $x \leq -1$ and $x \geq 1$,	
	arcsec (x)	strictly monotonically increasing
	arccsc (x)	strictly monotonically decreasing
Symmetries:	arcsec (x)	point symmetry about $x = 0$, $y = \frac{\pi}{2}$
	arccsc (x)	point symmetry about the origin $x = 0, y = 0$
Asymptotes:	arcsec (x)	$f(x) \to \pi \mp \frac{\pi}{2}$ as $x \to \pm\infty$
	arccsc (x)	$f(x) \to \pm\frac{\pi}{2}$ as $x \to \pm\infty$

(d) Particular values

Zeros:	None.
Discontinuities:	None.
Poles:	None.
Extremes:	None.
Points of inflection:	None.

Value of	$x \to -\infty$	$x = -\sqrt{2}$	$x = -1$	$x = 1$	$x = \sqrt{2}$	$x =\to \infty$
arcsec (x)	$\frac{\pi}{2}$	$\frac{3\pi}{4}$	π	0	$\frac{\pi}{4}$	$\frac{\pi}{2}$
arccsc (x)	0	$-\frac{\pi}{4}$	$-\frac{\pi}{2}$	$\frac{\pi}{2}$	$\frac{\pi}{4}$	0

(e) Reciprocal functions

The reciprocal functions cannot be written in terms of inverse trigonometric or trigonometric functions.

(f) Inverse functions

The inverse functions are the secant and cosecant functions.

$$x = \sec(y), \qquad y = \operatorname{arcsec}(x)$$
$$x = \csc(y), \qquad y = \operatorname{arccsc}(x)\,.$$

(g) Cognate functions

Trigonometric functions.
The roots of the reciprocal functions of second degree are linked to the inverse secant and cosecant functions via differentiation and integration.

(h) Conversion formulas

Reflection about the y-axis:

$$\begin{aligned}\operatorname{arcsec}(-x) &= \pi - \operatorname{arcsec}(x)\\ \operatorname{arccsc}(-x) &= -\operatorname{arccsc}(x)\,.\end{aligned}$$

Addition of complementary functions:

$$\operatorname{arcsec}(x) + \operatorname{arccsc}(x) = \frac{\pi}{2}\,.$$

Conversion of functions:

$$\begin{aligned}\operatorname{arcsec}(x) &= \operatorname{arccsc}\left(\frac{x}{\sqrt{x^2-1}}\right) + \begin{cases}0 & x>0\\ \pi & x<0\end{cases}\\ \operatorname{arccsc}(x) &= \operatorname{arcsec}\left(\frac{x}{\sqrt{x^2-1}}\right) - \begin{cases}0 & x>0\\ \pi & x<0\,.\end{cases}\end{aligned}$$

Addition of functions:

$$\begin{aligned}\operatorname{arcsec}(x) \pm \operatorname{arcsec}(y) &= \left(\frac{1}{2} - k\right)\pi + \operatorname{arcsec}\left(\frac{xy}{1 \mp \sqrt{x^2-1}\sqrt{y^2-1}}\right)\\ k &= \operatorname{Int}\left[\frac{\operatorname{arccot}(x) + \operatorname{arccot}(y)}{\pi}\right]\\ \operatorname{arccsc}(x) \pm \operatorname{arccsc}(y) &= k\pi + \operatorname{arccsc}\left(\frac{xy}{\sqrt{y^2-1} \mp \sqrt{x^2-1}}\right)\\ k &= \operatorname{Int}\left[\frac{1}{2} + \frac{\arctan(x) + \arctan(y)}{\pi}\right]\,.\end{aligned}$$

(i) Approximation formulas (8 bits $\approx$ 0.4% of accuracy)

For large values of x, it holds that

$$\operatorname{arccsc}(x) \approx \frac{1}{x}\sqrt{\frac{3x^2}{3x^2-1}}\,, \qquad |x| \geq 2\,.$$

(j) Expansion of series or products

Series expansion of the inverse cosecant function for $|x| > 1$:

$$\begin{aligned}\operatorname{arccsc}(x) &= \frac{1}{x} + \frac{1}{2}\frac{1}{3x^3} + \frac{1\cdot 3}{2\cdot 4}\frac{1}{5x^5} + \frac{1\cdot 3\cdot 5}{2\cdot 4\cdot 6}\frac{1}{7x^7} + \ldots\\ &= \sum_{k=0}^{\infty}\frac{(2k-1)!!}{(2k)!!}\frac{x^{-(2k+1)}}{2k+1}\,.\end{aligned}$$

(k) Derivative of the functions

arctan(x) and arccot (x) have equal derivatives except for the sign:

$$\frac{\mathrm{d}}{\mathrm{d}x}\operatorname{arcsec}(ax+b) = -\frac{\mathrm{d}}{\mathrm{d}x}\operatorname{arccsc}(ax+b) = \frac{a}{|ax+b|\sqrt{(ax+b)^2-1}} \, .$$

(l) Primitive of the function

The primitives are ($bx > 1 - c$):

$$\int_{(1-c)/b}^{x} \operatorname{arcsec}(bx+c) = \left(x + \frac{c}{b}\right)\operatorname{arcsec}(bx+c) + \frac{1}{b}\operatorname{Arcosh}(bx+c)$$

$$\int_{(1-c)/b}^{x} \operatorname{arccsc}(bx+c) = \left(x + \frac{c}{b}\right)\operatorname{arccsc}(bx+c) - \frac{1}{b}\operatorname{Arcosh}(bx+c)$$

(m) Particular extensions and applications

Link to the inverse hyperbolic secant and hyperbolic cosecant functions:

$$\operatorname{arcsec}(x) = \pm \mathrm{j}\operatorname{Arsech}(\mathrm{j}x) \, , \qquad \operatorname{arccsc}(x) = \mathrm{j}\operatorname{Arcsch}(\mathrm{j}x) \, .$$

Plane curves

Some important special curves in the plane will be discussed below.
In contrast to functions, several y-values can be defined for a fixed x-value.
Curves may be given in various representations.
Functional representation in terms of Cartesian coordinates: $f(x, y) = 0$.

□ Ellipse, center at the origin: $\frac{x^2}{a^2} + \frac{y^2}{b^2} = 1$.

Functional representation in terms of polar coordinates: $f(r, \varphi) = 0$.

□ Ellipse, central representation: $r - \frac{b}{\sqrt{1 - \varepsilon^2 \cos^2(\varphi)}} = 0, b > 0, 0 \leq \varepsilon < 1$.

Parametric representation: $x = x(t)$, $y = y(t)$.

□ Ellipse, center at the origin: $x = a\sin(t), \quad y = b\sin(t), \quad 0 \leq t < 2\pi$.

5.38 Algebraic curves of the n-th order

Algebraic curves of the n-th order can be described by means of the following general equation:

$$\sum_{k=0}^{n} \sum_{m=0}^{n-k} a_{km} x^k y^m = 0 .$$

i.e., n is the highest power appearing in the sum.
In the following, curves of the second, third, and fourth orders are introduced.

5.38.1 Curves of the second order

Conics: General equation:

$$ax^2 + by^2 + cxy + dx + ey + f = 0 .$$

For a discussion of conics, see the sections in Chapters 3 and 8 and Section 5.22.

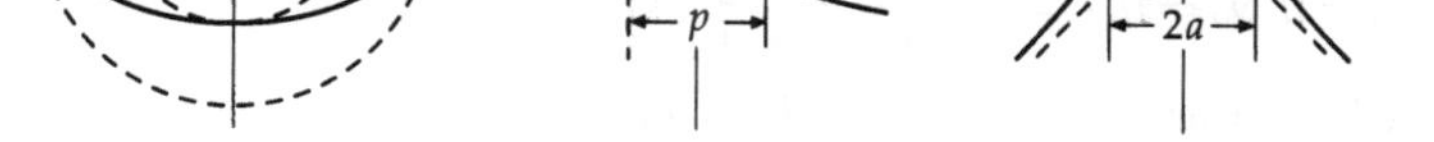

Ellipse Parabola Hyperbola

Ellipse:

General representation: $\frac{(x - x_0)^2}{a^2} + \frac{(y - y_0)^2}{b^2} = 1$.

Parametric representation: $x = x_0 + a\cos(t)$, $y = y_0 + b\sin(t)$.

Assumption: $a \geq b$, a major semi-axis, b minor semi-axis.
Numerical eccentricity: $\varepsilon = \dfrac{\sqrt{a^2 - b^2}}{a} < 1$.
Central representation: $x_0 = 0$, $y_0 = 0$.
Representation in terms of polar coordinates: $r = \dfrac{b}{\sqrt{1 - \varepsilon^2 \cos^2(\varphi)}}$.
Position of the right-hand (+) and left-hand (−) focus:

$$x = x_0 \pm \sqrt{a^2 - b^2} = x_0 \pm \varepsilon a, \quad y = y_0 \,.$$

Equation in polar coordinates: Polar axis is the z axis, pole in left-hand focus: (−), pole in right-hand focus (+):

$$r = \frac{p}{1 \pm \varepsilon \cos(\varphi)} \qquad p = \frac{b^2}{a} \,.$$

Circle:

Special form of the ellipse with $p = a = b$, $\varepsilon = 0$.
General representation : $(x - x_0)^2 + (y - y_0)^2 = a^2, r = a$.
Parametric representation $x = x_0 + a\cos(t)$, $y = y_0 + a\sin(t)$.
Central representation: $x_0 = 0$, $y = 0$, $r = |a|$.
The two foci coincide in the center.

Parabola:

General representation $(y - y_0)^2 = 2(x - x_0)p$.
Parametric representation: $x = x_0 + \frac{1}{2p}t^2$, $y = y_0 + t$.
Numerical eccentricity: $\varepsilon = 1$.
Vertex representation: $x_0 = 0$, $y_0 = 0$, $r = 2p\cos(\varphi(1 + \cot^2(\varphi)))$.
Position of the focus: $x = x_0 + \dfrac{p}{2}$, $\quad y = y_0$.
Equation in polar coordinates, pole in focus, polar axis is the z axis. $r = \dfrac{p}{1 - \cos(\varphi)}$.

Hyperbola:

General representation: $\dfrac{(x - x_0)^2}{a^2} - \dfrac{(y - y_0)^2}{b^2} = 1$.
Parametric representation: $x = x_0 + a\cosh(t)$, $y = y_0 + b\sinh(t)$.
Numerical eccentricity: $\varepsilon = \dfrac{\sqrt{a^2 + b^2}}{a} > 1$.
Asymptotes: $y = y_0 \pm \dfrac{b}{a}(x - x_0)$.
Center: $x_0 = 0$, $y_0 = 0$, $r = \dfrac{b}{\varepsilon^2 \cos^2(\varphi) - 1}$.
Position of the focus: $x = x_0 \pm \sqrt{a^2 + b^2} = x_0 \pm \varepsilon a$, $y = y_0$.
Equation in polar coordinate: Pole in left-hand focus, polar axis is the z axis, left-hand branch: +, right-hand branch: −, left-hand branch:

$$r = \frac{p}{1 \pm \varepsilon \cos(\varphi)} \,, \qquad p = \frac{b^2}{a} \,.$$

Pole in right-hand focus: Replace p by $-p$.
Pole in center; polar axis is the x axis.

5.38.2 Curves of the third order

- All cubic functions $y = ax^3 + bx^2 + cx + d$ and all reciprocal quadratic functions as well as their inverse functions are curves of the third order.

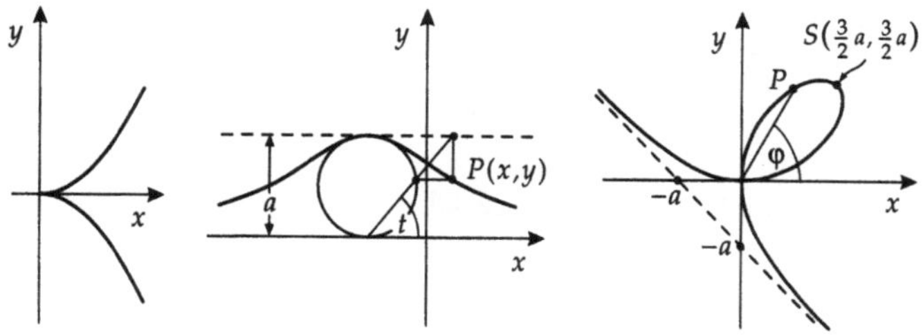

Neil's parabola Witch of Agnesi Folium of Descartes

Neil's parabola:

$ax^3 - y^2 = 0 \quad a > 0.$
Parametric representation: $x = t^2,\ y = at^3,\ -\infty < t < \infty.$
Length of the arc from $x = 0,\ y = 0$ to x, y: $l = \dfrac{(4 + 9a^2x)^{3/2}) - 8)}{27a^2}.$
Curvature at point x, y: $k = \dfrac{6a}{\sqrt{x}(4 + 9a^2x)^{3/2}}.$

Witch of Agnesi:

$(x^2 + a^2)y - a^3 = 0, \quad a > 0.$
Vertex: $x = 0,\ y = a.$
Asymptote: $y = 0.$
Area between the curve and the asymptote: $A = \pi a^2.$
Radius of curvature at the vertex: $r = \dfrac{a}{2}.$
Parametric representation: $x = a\cot(t),\ y = a\sin^2(t).$

Folium of Descartes:
$x^3 + y^3 - 3axy = 0\,, \quad a > 0.$
Parametric representation: $x = \dfrac{3at}{1 + t^3},\ y = \dfrac{3at^2}{1 + t^3},\ t \neq -1.$
Interpretation: $t = \tan(\varphi)$, φ: angle between the positive x axis and the line joining $x = 0,\ y = 0$ to $x = x(t),\ y = y(t).$
Vertex: $x = \dfrac{3}{2}a,\ y = \dfrac{3}{2}a.$
Double point: $x = 0,\ y = 0.$
Tangents at the double point: x axis ($y = 0$) and y axis ($x = 0$).
Radius of curvature at the double point: $r = \dfrac{3}{2}a$ for both branches of the curve.
Asymptote: $y = -x - a.$
Area inside the loop: $A = \dfrac{3}{2}a^2.$
Area between the asymptote and the curve (without the loop): $A = \dfrac{3}{2}a^2.$

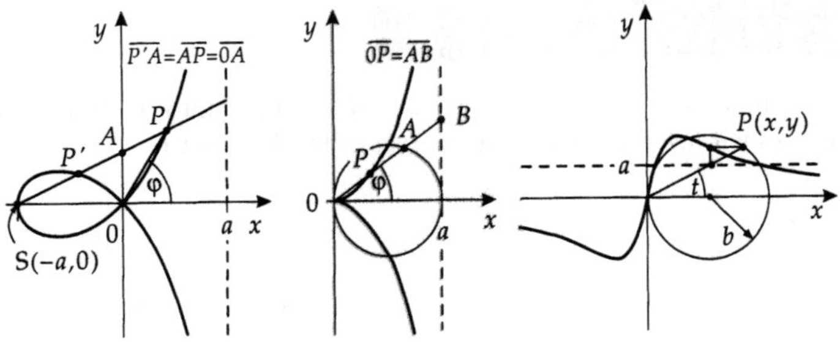

Strophoid Cissoid Serpentine

Strophoid:

$(x+a)x^2 - (x-a)y^2 = 0 \qquad a > 0.$

Equation in polar coordinates: $r = a\dfrac{\cos(2\varphi)}{\cos(\varphi)}$.

Parametric representation: $x = \dfrac{a(t^2-1)}{t^2+1}$, $y = \dfrac{at(t^2-1)}{t^2+1}$, $-\infty < t < \infty$.

Interpretation: $t = \tan(\varphi)$, φ: angle between the positive x axis and the line joining $x=0$, $y=0$ to $x = x(t)$, $y = y(t)$.

Vertex: $x = -a$, $y = 0$.

Double point: $x = 0$, $y = 0$.

Tangents at the double point: $(y = x)$ and $(y = -x)$.

Asymptote: $x = a$.

Area inside the loop: $A = 2a^2 - \dfrac{\pi a^2}{2}$.

Area between curve and asymptote (without the loop): $A = 2a^2 + \dfrac{\pi a^2}{2}$.

Cissoid:

$y^2(x-a) + x^3 = 0\,, \quad a > 0.$

Equation in polar coordinates: $r = a\sin(\varphi)\tan(\varphi)$.

Parametric representation: $x = \dfrac{at^2}{1+t^2}$, $y = \dfrac{at^3}{1+t^2}$, $-\infty < t < \infty$.

Interpretation: $t = \tan(\varphi)$, φ: angle between the positive x axis and the line joining $x=0$, $y=0$ to $x = x(t)$, $y = y(t)$.

Cusp of the curve: $x = 0$, $y = 0$.

Asymptote: $x = a$.

Area between curve and asymptote: $A = \dfrac{3}{4}\pi a^2$.

Serpentine:

$(a^2 + x^2)y = abx.$

Parametric representation: $x = a\cot(t)$, $y = b\sin(t)\cos(t)$, $0 \le t \le \pi$.

Mirror symmetry about the origin.

Let a circle with radius b be drawn about the point $x = b$, $y = 0$.

A ray, starting at $x = 0$, $y = 0$, and enclosing the angle t with the x axis, intersects the circle at point $y = y(t)$ and the straight line $y = a$ at point $x = x(t)$.

5.38.3 Curves of the fourth and higher order

Conchoid of a curve K with respect to point x_0, y_0:
A straight line is drawn through the point (x_0, y_0) and through a point P of the curve. The segment is s. Then, the segments $s + b$ and $s - b$ are marked on the straight line. These points, P,P; belong to the conchoid of K.

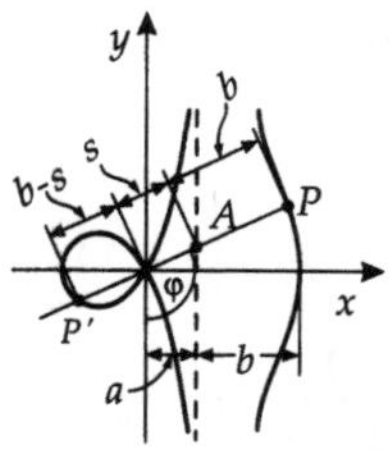

Conchoid

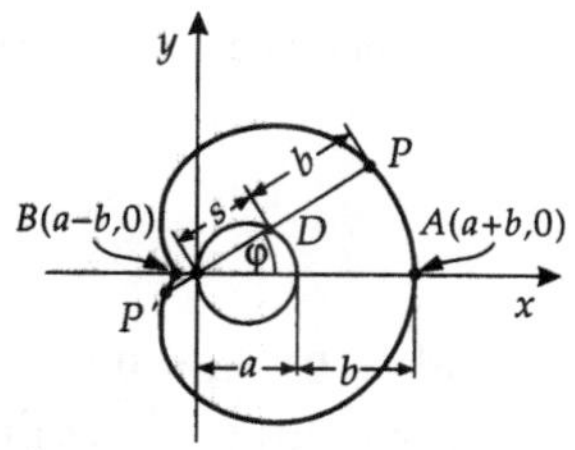

Pascal's limacon

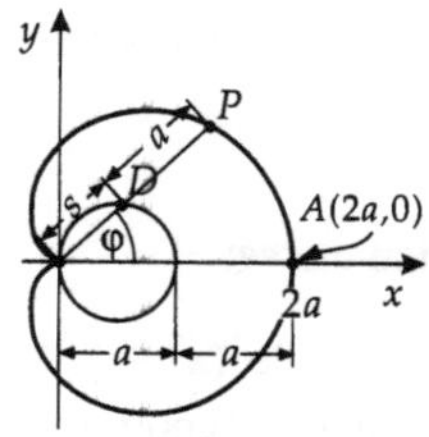

Cardioid

Conchoid of Nicomedes:

$(x - a)^2(x^2 + y^2) - b^2x^2 = 0\,, \quad a > 0\,, \quad b > 0.$
Conchoid of a straight line with respect to $x = 0$, $y = 0$.
Parametric representation: $x = a + b\cos(t)$, $y = a\tan(t) + b\sin(t)$.
Left-hand branch $-\frac{\pi}{2} < t < \frac{\pi}{2}$; right-hand branch $\frac{\pi}{2} < t < \frac{3\pi}{2}$.
Asymptote for both branches: $x = a$.
Vertices: $x = a \pm b$, $y = 0$, left-hand: $-$, right-hand: $+$.
The point $x = 0$, $y = 0$ is an isolated point for $b < a$, a cusp for $b = a$, and a double point for $b > a$.

Pascal's limacon:

$(x^2 + y^2 - ax)^2 - b^2(x^2 + y^2) = 0.$
Conchoid of a circle with radius $\frac{a}{2}$ with respect to $x = 0$, $y = 0$.
Parametric representation: $x = a\cos^2(t) + b\cos(t)$, $y = a\cos(t)\sin(t) + b\sin(t)$, $0 \le t < 2\pi$.
The point $x = 0$, $y = 0$ is an isolated point for $a < b$ and a double point for $a > b$.
For $a = b$, the curve is a cardioid.

Cardioid:

$(x^2 + y^2)(x^2 + y^2 - 2ax) - a^2y^2 = 0, \quad a > 0.$
Equation in polar coordinates: $r = a(1 + \cos(\varphi))$.
Special case of Pascal's limacon, and special case of an epicycloid.
Parametric representation: $x = a\cos[t(1 + \cos(t))]$, $y = a\sin[t(1 + \cos(t))]$,
$0 \le t < 2\pi$.
Vertex: $x = 2a$, $y = 0$.
Cusp: $x = 0$, $y = 0$.
Length of the curve: $8a$.
Area: $\frac{3}{2}\pi a^2$.

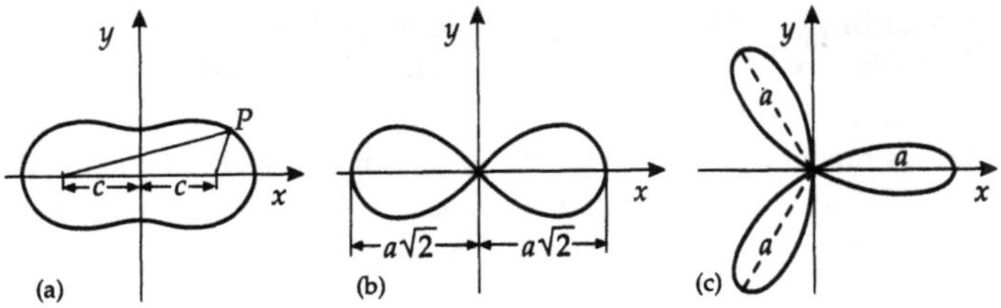

Cassinian curve Lemniscate Three-leaved rose

Cassinian curve:

$(x^2+y^2)^2 - 2c^2(x^2-y^2) - (a^4-c^4) = 0, \quad a > 0, \quad c > 0.$
Equation in polar coordinates: $r^2 = c^2\cos(2\varphi) \pm \sqrt{c^4\cos^2(2\varphi) + a^4 - c^4}$.
A class of curves, the product of the distances of the points from the curve to points $x = -c$, $y = 0$ and $x = c$, $y = 0$ is a given constant value a^2.

$\overline{PF} \cdot \overline{PF} = a^2.$

For $a < c$, the curve splits into two separate closed curves: $a = c$ describes the **lemniscate**; $a > c$ describes a closed curve that has a constriction in the neighborhood of $x = 0$.

Bernoulli's lemniscate:

$(x^2+y^2)^2 - 2a^2(x^2-y^2) = 0, \quad a > 0.$
Equation in polar coordinates: $r = a\sqrt{2\cos(2\varphi)}$.
Special case of a Cassinian curve, for $a = c$.
Double point and at the same time point of inflection at $x = 0$, $y = 0$.
Radius of curvature: $R = \dfrac{2a^2}{3r}$
Area of each loop: $A = a^2$.

Three-leaved rose:

$r = a\cos(3\varphi), \quad a > 0.$
General equation: $r = a\cos(n\varphi)$.
The curve encloses n leaves (here $n = 3$).
Area of a leaf: $A = \dfrac{1}{n}\dfrac{\pi a^2}{4}$; here $A = \dfrac{\pi a^2}{12}$.
Rose with $2n$ leaves: $r = |a\cos(n\varphi)|$.

5.39 Cycloidal curves

Cycloidal curves are formed when a circle rolls along a second curve without sliding. The coordinates of a point, fixed inside or outside the circular disk, are plotted.

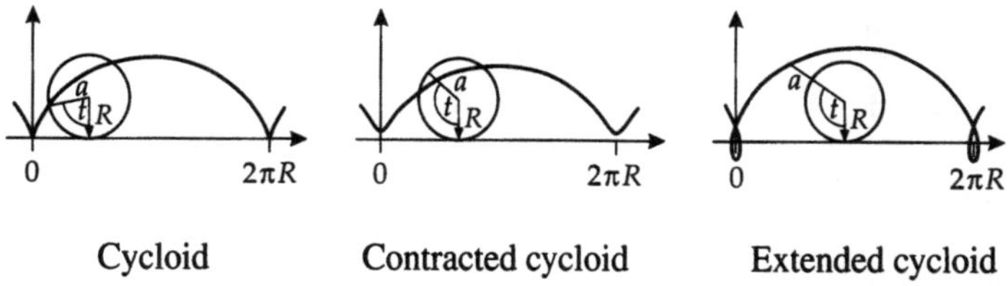

Cycloid:

$$x = R\arccos\left(\frac{R-y}{R}\right) - \sqrt{y(2R-y)}, \quad R > 0.$$

Cycloidal curve of point P of a circle rolling along a straight line.
Parametric representation: $x = R(t - \sin(t))$, $y = R(1 - \cos(t))$,
R: radius of the circle, t: rolling angle in rad.
Period of the curve: circumference $2\pi R$ of the circle.
Length of an arc: $l = 8R$.
Area under an arc: $A = 3\pi R^2$.

Trochoid:

Extended or contracted cycloid.
Parametric representation: $x = Rt - a\sin(t)$, $y = R - a\cos(t)$.
a: distance between the point and the center of the circular disk.

Contracted cycloid:

$a < R$, the point lies inside the circular disk.

Extended cycloid:

$a > R$, the point lies outside the circular disk.

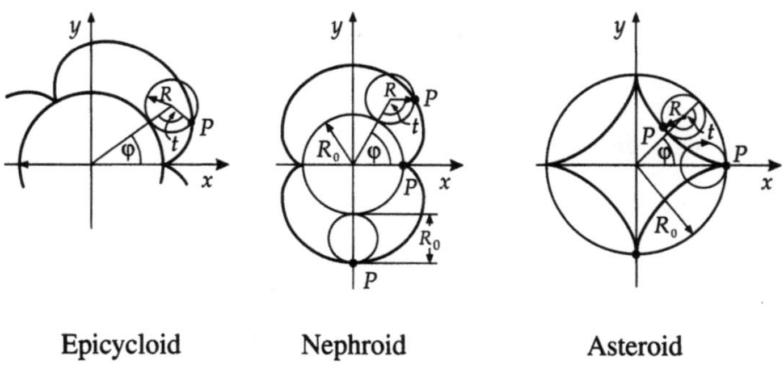

Epicycloid:

A circle with radius R rolls along the outside with a circle of radius R_0.

Parametric representation: $x = (R_0 + R)\cos(t) - R\cos\left(\frac{R_0 + R}{R}t\right)$.

$$y = (R_0 + R)\sin(t) - R\sin\left(\frac{R_0 + R}{R}t\right),$$

where φ is the rotation angle (rolling angle: $m \cdot \varphi$).

If the ratio $m = \dfrac{R_0}{R}$ is an integer, the curve consists of exactly m arcs; if m is rational, the curve is closed.

Cardioid: $R = R_0, \quad m = 1.$

Nephroid: $R_0 = 2R, \quad m = 2.$

Length of an arc: $\dfrac{8R(R_0 + R)}{R_0} = \dfrac{8(R_0 + R)}{m}.$

Area between arc and fixed circle: $A = \dfrac{\pi R^2(3R_0 + 2R)}{R_0} = \dfrac{\pi R(3R_0 + 2R)}{m}.$

Epitrochoid:

Extended ($a > R$) or contracted ($a < R$) epicycloid.

Parametric representation:
$$x = (R_0 + R)\cos(t) - a\cos\left(\frac{R_0 + R}{R}t\right)$$
$$y = (R_0 + R)\sin(t) - a\sin\left(\frac{R_0 + R}{R}t\right).$$

a: distance between the point and the center of the rolling circular disk.

Hypocycloid:

A circle with radius R rolling along the inside of a circle with radius R_0.

Parametric representation:
$$x = (R_0 - R)\cos(t) + R\cos\left(\frac{R_0 - R}{R}t\right)$$
$$y = (R_0 - R)\sin(t) - R\sin\left(\frac{R_0 - R}{R}t\right),$$

where φ is the rotation angle (rolling angle: $m \cdot \varphi$).

If the ratio $m = \dfrac{R_0}{R}$ is an integer, then the curve consists of exactly m arcs; if m is rational, then the curve is closed.

Deltoid: $R_0 = 3R, \quad m = 3.$

Asteroid: $R_0 = 4R, \quad m = 4.$

Length of an arc: $\dfrac{8R(R_0 - R)}{R_0} = \dfrac{8(R_0 - R)}{m}.$

Area between arc and fixed circle: $A = \dfrac{\pi R^2(3R_0 - 2R)}{R_0} = \dfrac{\pi R(3R_0 - 2R)}{m}.$

Hypotrochoid:

Extended ($a > R$) or contracted ($a < R$) hypocycloid.

Parametric representation:
$$x = (R_0 - R)\cos(t) + a\cos\left(\frac{R_0 - R}{R}t\right)$$
$$y = (R_0 - R)\sin(t) - a\sin\left(\frac{R_0 - R}{R}t\right),$$

where a is the distance between the point and the center of the rolling circular disk.

5.40 Spirals

Spirals: Can usually be represented by $r = f(\varphi)$, where f is a strictly monotonic function.
The parametric representation is $x = f(t)\cos(t)$, $y = f(t)\sin(t)$.
Let points $P_1 = P(r = r_1, \varphi = \varphi_1)$, and $P_2 = P(r = r_2, \varphi = \varphi_2)$ be any two points on the spiral, and $O = P(x = 0, y = 0)$ be the origin.
Arc $\overline{P_1 P_2}$: The segment between P_1 and P_2.
Sector $P_1 O P_2$: The area enclosed by segments OP_1, OP_2, with the arc. $P_1 P_2$.

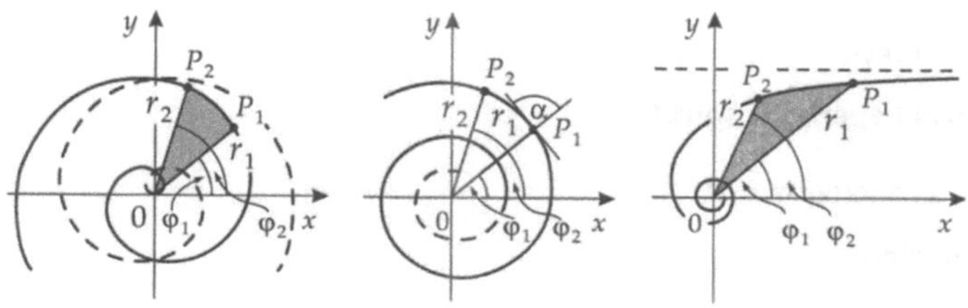

Spiral of Archimedes Logarithmic spiral Hyperbolic spiral

Spiral of Archimedes:

$r = a\varphi, \quad a = \dfrac{v}{\omega} > 0.$
Curve of a point which moves at a constant angular velocity ω with a constant radial velocity v.
Length of the arc $\overline{OP(r,\varphi)}$: $l = \dfrac{a}{2}(\varphi\sqrt{\varphi^2+1} + \operatorname{Arsinh}(\varphi))$.
Approximation for large values of φ: $s \approx \dfrac{a}{2}\varphi^2$.
Area of the sector $\overline{P_1 O P_2}$: $A = \dfrac{a^2}{6}(\varphi_2^3 - \varphi_1^3)$.

Logarithmic spiral:

$r = a e^{k\varphi}, \quad k > 0, \quad a > 0.$
All rays coming from the origin intersect the curve at the same angle:
α: $\cot(\alpha) = k$.
Length of the arc $\overline{P_1 P_2}$: $l = \dfrac{\sqrt{k^2+1}}{k}(r_2 - r_1) = \dfrac{r_2 - r_1}{\cos(\alpha)}$.
Limit $r_1 \to 0$: $s \to \dfrac{r_2}{\cos(\alpha)}$, enclosed area: $A \to \dfrac{r_2^2}{4k}$.
Radius of curvature $r\sqrt{1+k^2}$.

Hyperbolic spiral:

$r = \dfrac{a}{\varphi}$ and $r = \dfrac{a}{|\varphi - \pi|}$, $a > 0$.

Asymptote: $y = a$. Area of the sector $P_1 O P_2$: $A = \dfrac{a^2}{2}\left(\dfrac{1}{\varphi_1} - \dfrac{1}{\varphi_2}\right)$.

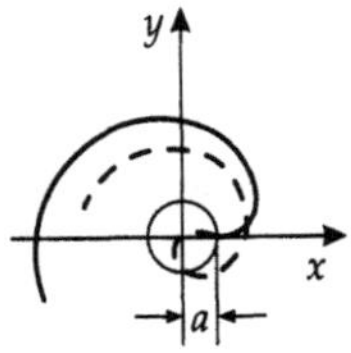

Parabolic spiral

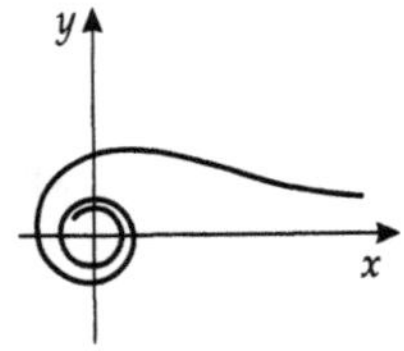

Quadratic hyperbolic spiral

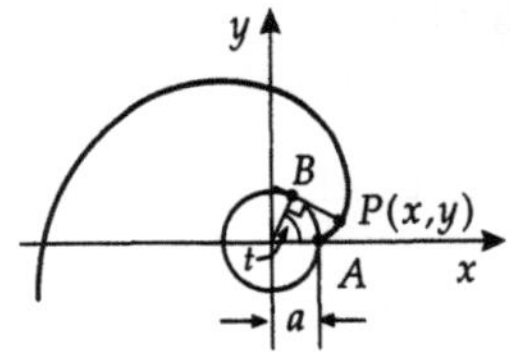

Involute of circle

Parabolic spiral:

$(r-a)^2 = 4ak\varphi$.

Quadratical hyperbolic spiral:

$r^2\varphi = a^2$, asymptote $y = 0$.

Involute of circle:

The evolute (locus of the centers of curvature) of this curve is a circle with radius a.
Parametric representation: $x = a\cos(t) + at\sin(t)$, $y = a\sin(t) - at\cos(t)$.

5.41 Other curves

Catenary:

$y = a\cosh\left(\frac{x}{a}\right), \qquad a > 0.$
Line formed by a flexible thread suspended at its two ends.

In the vicinity of the vertex, it holds that $y \approx \frac{x^2}{2a} + a$.

Length of the arc between $x = 0$ and $x = x_0$: $l = a\sinh\left(\frac{x_0}{a}\right)$.

Area under the arc: $A = a^2\sinh\left(\frac{x_0}{a}\right)$.

See also Section 5.26 on the hyperbolic sine.

Tractrix:

The evolute of the catenary.
Equation: $x = a\text{Arcosh}\left(\frac{a}{y}\right) \mp \sqrt{a^2 - y^2}$.
Asymptote: $y = 0$.
Curve that arises when a heavy object is suspended from a nonstretchable thread the end of which is moved along the x axis.

Lissajous figure:

Superposition of two oscillations along the x axis and y axis.
Parametric representation: $x = A_x\sin(n_x t + \varphi_x)$, $y = A_y\sin(n_y t + \varphi_y)$.

The figures depend on $\dfrac{A_x}{A_y}$, $\dfrac{n_x}{n_y}$ and $\varphi_x - \varphi_y$.
See also Section 5.32 on the superposition of oscillations.

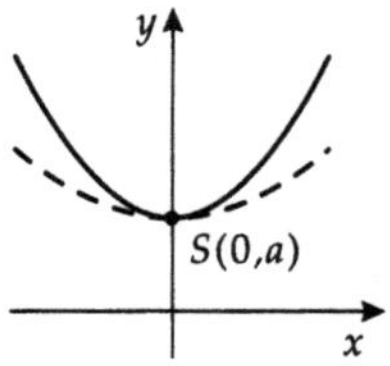

Catenary

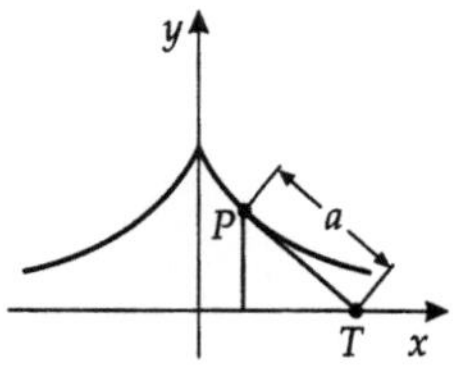

Tractrix

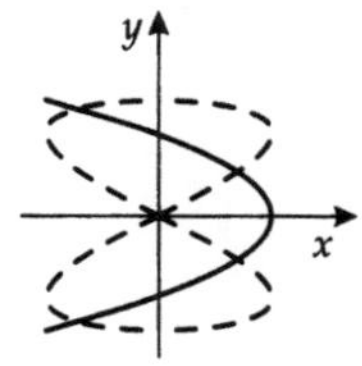

Lissajous figure

6

Vector analysis

6.1 Vector algebra

6.1.1 Vector and scalar

Scalar and **vector quantities**:

Scalar: A quantity that is determined completely by a numerical value (real measure) and the unit of measurement.

□ Length, time, temperature, mass, work, energy, potential, capacity.

Many quantities require not only measure and unit, but also direction.

Vector: A directed quantity (segment) represented by an arrow. The direction of the arrow determines the direction of the vector. The length of the vector corresponds to the measure.

Notation: Usually, a vector is denoted by boldfaced Roman letters, supplied with an arrow ($\vec{\mathbf{a}}$, $\vec{\mathbf{b}}$), but other notations ($\mathbf{a}$, $\underline{\mathbf{a}}$, etc.) can also be found.

A unique specification is also obtained by specifying the starting point P and the end-point Q: $\vec{\mathbf{a}} = \overrightarrow{PQ}$.

□ Velocity, acceleration, electric and magnetic field strength, gravitational force, position vector, torque.

▷ Many vectors are given in two or three dimensions (plane, space), but the vector concept can be extended (as most rules of arithmetic) directly to n dimensions. n-dimensional vectors (even with $n > 1000$) are of great practical importance in dealing with large linear equation systems.

Magnitude or **standard, length of a vector**: The distance between the initial point and the end-point determines the length of the vector:

$$a = |\vec{\mathbf{a}}| = \overline{PQ} \geq 0 .$$

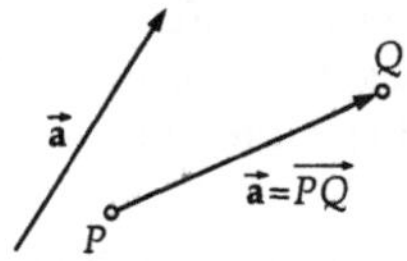

● A vector $\vec{\mathbf{a}}$ is uniquely determined by its **magnitude** and **direction**.

▷ *Only* magnitude and direction of a vector are of relevance for its specification, not its exact position in space. The "vector" symbolizes all arrows which can result from it through parallel translation.

6.1.2 Particular vectors

Null vector $\vec{\mathbf{0}}$: A vector of magnitude zero; its initial and end-point coincide. The null vector is of zero length and of infinite direction.
Unit vector $\vec{\mathbf{e}}$: A vector whose length is one unit:

$$|\vec{\mathbf{e}}| = 1 .$$

Position vector (bound vector) of the point P:

$$\vec{\mathbf{r}}(P) = \vec{OP} .$$

For each vector there is a representative, whose initial point (O) lies at the origin of the coordinate system, and whose head leads to the P.
Free vector: A vector that may be translated arbitrarily, but must not be reflected, rotated, or scaled (see Section 7.2 on coordinate transformation).
Sliding vector: A vector that can be translated arbitrarily along its line of action.

□ Forces at the rigid body are sliding vectors.

Equality of vectors: Two vectors are regarded as equal if they have the same magnitude and direction, $\vec{\mathbf{a}} = \vec{\mathbf{b}}$.

● Two vectors of the same length are equal if and only if they can be superposed by only one parallel displacement without rotation.

Parallel vectors, $\vec{\mathbf{a}}||\vec{\mathbf{b}}$: Can be brought to the same straight line by means of parallel displacement.
Vectors of equal direction, $\vec{\mathbf{a}} \uparrow\uparrow \vec{\mathbf{b}}$: Equidirectional vectors.
Opposite (antiparallel) vectors, $\vec{\mathbf{a}} \uparrow\downarrow \vec{\mathbf{b}}$: Vectors of opposite direction.
Inverse vector, $-\vec{\mathbf{a}}$: A vector the same magnitude as $\vec{\mathbf{a}}$, but of opposite direction. The inverse vector $-\vec{\mathbf{a}}$ is antiparallel to $\vec{\mathbf{a}}$, $\vec{\mathbf{a}} \uparrow\downarrow (-\vec{\mathbf{a}})$.

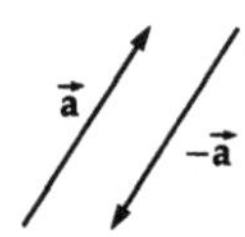

Vector field: The set of (field) vectors is assigned to the points in space; function of $\mathbb{R}^n$ into $\mathbb{R}^n$.

□ Gravitational field strength, magnetic field strength, electric field strength.

6.1.3 Multiplication of a vector by a scalar

The product of a vector $\vec{\mathbf{a}}$ and a scalar r results in another vector $\vec{\mathbf{b}} = r\vec{\mathbf{a}}$ whose magnitude is $|r|$ times that of vector $\vec{\mathbf{a}}$:

$$|r\vec{\mathbf{a}}| = |r| \cdot |\vec{\mathbf{a}}|$$

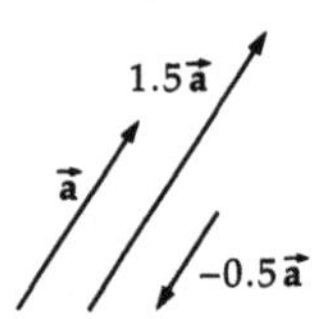

▷ Often, the multiplication sign is omitted to avoid confusion with the scalar product of two vectors $\vec{\mathbf{a}} \cdot \vec{\mathbf{b}}$.

$r > 0$: $r\vec{\mathbf{a}}$ and $\vec{\mathbf{a}}$ have the same direction ($\vec{\mathbf{a}} \uparrow\uparrow r\vec{\mathbf{a}}$).

□ Newton's force law of mechanics: Force equals mass times acceleration, $\vec{\mathbf{F}} = m\vec{\mathbf{a}}$.

$r < 0$: $r\vec{\mathbf{a}}$ and $\vec{\mathbf{a}}$ have the opposite direction ($\vec{\mathbf{a}} \uparrow\downarrow r\vec{\mathbf{a}}$).
$r = -1$: Inverse vector $r\vec{\mathbf{a}} = -\vec{\mathbf{a}}$ with the same magnitude as $\vec{\mathbf{a}}$ but of different direction.
▷ The two vectors $\vec{\mathbf{a}}$ and $r\vec{\mathbf{a}}$ ($r \neq 0$) are collinear; i.e., they are linearly dependent.

6.1.4 Vector addition

Addition of two vectors $\vec{\mathbf{a}}$ and $\vec{\mathbf{b}}$: The vector $\vec{\mathbf{b}}$ is translated until its initial point coincides with the end-point of vector $\vec{\mathbf{a}}$.
Sum vector $\vec{\mathbf{c}}$: The resulting vector which reaches from the initial point of $\vec{\mathbf{a}}$ to the end-point of the translated vector $\vec{\mathbf{b}}$,

$$\vec{\mathbf{c}} = \vec{\mathbf{a}} + \vec{\mathbf{b}} \, .$$

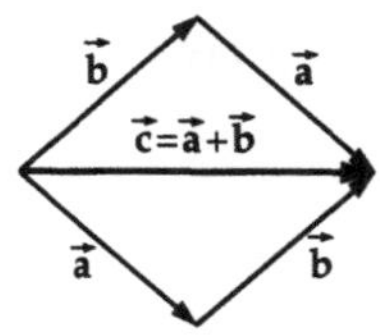

□ Parallelogram of forces in trusswork.

▷ The computational laws (algebra) of real numbers apply to calculations with scalar quantities, but in calculations with vectors, **vector algebra** is required for calculations with vectors: $|\vec{\mathbf{c}}| \neq |\vec{\mathbf{a}}| + |\vec{\mathbf{b}}|$.

● The addition of several vectors follows the **polygon rule** (vector polygon). The initial point of the vector to be added is placed at the end-point of the preceding vector. This leads to a polygon.

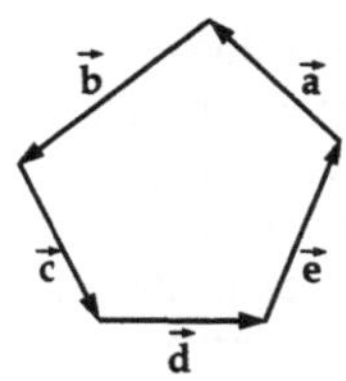

● If the vector polygon is closed, then the sum vector is the null vector.

□ When vector addition of forces leads to the null vector, then the forces are in equilibrium.

6.1.5 Vector subtraction

Subtraction of two vectors $\vec{\mathbf{a}}$ and $\vec{\mathbf{b}}$: Addition of vector $\vec{\mathbf{a}}$ and $-\vec{\mathbf{b}}$, (the inverse vector of $\vec{\mathbf{b}}$). The result is **difference vector** $\vec{\mathbf{d}}$:

$$\vec{\mathbf{d}} = \vec{\mathbf{a}} - \vec{\mathbf{b}} = \vec{\mathbf{a}} + (-\vec{\mathbf{b}}) \, .$$

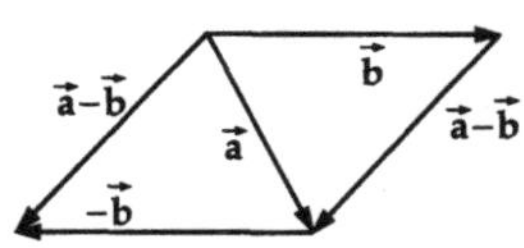

The difference vector $\vec{\mathbf{d}}$ points from the end-point of $\vec{\mathbf{b}}$ to the end-point of $\vec{\mathbf{a}}$ if $\vec{\mathbf{a}}$ and $\vec{\mathbf{b}}$ have a common initial point.

▷ The difference vector can also be represented in the form of a parallelogram.

□ Vector of the distance $\vec{\mathbf{r}}_2 - \vec{\mathbf{r}}_1$ between two points with the position vectors $\vec{\mathbf{r}}_1$ and $\vec{\mathbf{r}}_2$: $|\vec{\mathbf{r}}_2 - \vec{\mathbf{r}}_1|$.

6.1.6 Calculating laws

Axioms for vector spaces: Usually following laws are satisfied for calculations with vectors in n-dimensional space (**n-dimensional vector space** $\mathbb{R}^n$, $n = 2, 3, \ldots$).

● **Commutative law**:

$$\vec{\mathbf{a}} + \vec{\mathbf{b}} = \vec{\mathbf{b}} + \vec{\mathbf{a}}$$

- **Associative laws**:

$$\vec{\mathbf{a}} + (\vec{\mathbf{b}} + \vec{\mathbf{c}}) = (\vec{\mathbf{a}} + \vec{\mathbf{b}}) + \vec{\mathbf{c}}$$

$$r \cdot (s \cdot \vec{\mathbf{a}}) = r \cdot s \cdot \vec{\mathbf{a}} \quad r, s \in \mathbb{R}$$

- **Neutral element**:

$$\vec{\mathbf{0}} + \vec{\mathbf{a}} = \vec{\mathbf{a}} + \vec{\mathbf{0}} = \vec{\mathbf{a}}\,.$$

- **Inverse element**:

$$\vec{\mathbf{a}} + (-\vec{\mathbf{a}}) = (-\vec{\mathbf{a}}) + \vec{\mathbf{a}} = \vec{\mathbf{0}}\,.$$

- **Distributive laws**:

$$r \cdot (\vec{\mathbf{a}} + \vec{\mathbf{b}}) = r \cdot \vec{\mathbf{a}} + r \cdot \vec{\mathbf{b}}$$

$$(r + s) \cdot \vec{\mathbf{a}} = r \cdot \vec{\mathbf{a}} + s \cdot \vec{\mathbf{a}}$$

$$0 \cdot \vec{\mathbf{a}} = \vec{\mathbf{a}} \cdot 0 = \vec{\mathbf{0}}$$

$$1 \cdot \vec{\mathbf{a}} = \vec{\mathbf{a}} \cdot 1 = \vec{\mathbf{a}}\,.$$

- **Triangle inequality**:

$$||\vec{\mathbf{a}}| - |\vec{\mathbf{b}}|| \le |\vec{\mathbf{a}} \pm \vec{\mathbf{b}}| \le |\vec{\mathbf{a}}| + |\vec{\mathbf{b}}|$$

□ Two dimensions ($n = 2$): The plane, also called the "plane of solution vectors," of homogeneous equation systems with two unknowns.

□ Three dimensions ($n = 3$): The three-dimensional space, also called the space of the solution vectors of a system of three equations with three unknowns.

▷ Contrast to calculations with real numbers; the magnitude of the sum of two vectors can be smaller than the magnitude of their difference, namely, if the angle enclosed by the vectors is larger than 90°.

6.1.7 Linear dependence/independence of vectors

Linear combination $\vec{\mathbf{b}}$ of vectors $(\vec{\mathbf{a}}_1, \vec{\mathbf{a}}_2, \ldots, \vec{\mathbf{a}}_n)$: Sum of n vectors $\vec{\mathbf{a}}_i = \{\vec{\mathbf{a}}_1, \ldots, \vec{\mathbf{a}}_n\}$ with scalar coefficients $r_i = \{r_1, \ldots, r_n\}$:

$$\vec{\mathbf{b}} = r_1\vec{\mathbf{a}}_1 + r_2\vec{\mathbf{a}}_2 + \cdots + r_i\vec{\mathbf{a}}_i + \ldots + r_n\vec{\mathbf{a}}_n, \quad r_i \in \mathbb{R}\,,\ \vec{\mathbf{a}}_i \in \mathbb{R}^m\,,\ m \neq n\,.$$

□ $n = 3$:

$$\vec{\mathbf{b}} = r_1\vec{\mathbf{a}}_1 + r_2\vec{\mathbf{a}}_2 + r_3\vec{\mathbf{a}}_3\,.$$

Linearly dependent vectors: Vectors $\vec{\mathbf{a}}_1,\ \vec{\mathbf{a}}_2,\ \ldots, \vec{\mathbf{a}}_n$ ($\vec{\mathbf{a}}_i \neq \vec{\mathbf{0}},\ i = 1, \ldots n$): There exists a linear combination of the n vectors which vanishes:

$$r_1\vec{\mathbf{a}}_1 + r_2\vec{\mathbf{a}}_2 + \cdots + r_n\vec{\mathbf{a}}_n = \vec{\mathbf{0}}\,,$$

although not all coefficients $r_1, r_2, \ldots, r_n$ are zero at the same time.

- In the case of linearly dependent vectors, at least one of the vectors can be represented as a linear combination of the other vectors, i.e., the sum of multiples of the other vectors.

Collinear vectors: Parallel or antiparallel vectors $\vec{\mathbf{a}}$ and $\vec{\mathbf{b}} \neq \vec{0}$, where one of the two vectors is a multiple of the other vector.

$$\vec{\mathbf{a}} = r\vec{\mathbf{b}}\,, \quad r \in \mathbb{R}\,,\ \vec{\mathbf{a}}, \vec{\mathbf{b}} \in \mathbb{R}^2$$

- Collinear vectors are linearly dependent: every vector can be represented by a multiple of a parallel vector.

Coplanar vectors: Two (or more vectors), $\vec{\mathbf{a}}$, $\vec{\mathbf{b}}$, and $\vec{\mathbf{c}} \neq \vec{0}$, lying in one plane.

- Three or more vectors in $\mathbb{R}^2$ are linearly dependent: Every vector in a plane can be represented by a sum of multiples of two linearly independent vectors in the plane.

$$r\vec{\mathbf{a}} + s\vec{\mathbf{b}} + t\vec{\mathbf{c}} = \vec{0}$$

- Three vectors of $\mathbb{R}^3$ lie in one plane when the determinant of their components vanishes:

$$\begin{vmatrix} a_x & b_x & c_x \\ a_y & b_y & c_y \\ a_z & b_z & c_z \end{vmatrix} = 0$$

Linearly independent vectors, $\vec{\mathbf{a}}_1, \vec{\mathbf{a}}_2, \ldots, \vec{\mathbf{a}}_i, \ldots, \vec{\mathbf{a}}_n$: All linear combinations vanish if and only if **all** r_i are zero, $r_i \equiv 0$:

$$r_1\vec{\mathbf{a}}_1 + r_2\vec{\mathbf{a}}_2 + \cdots + r_i\vec{\mathbf{a}}_i + \ldots + r_n\vec{\mathbf{a}}_n = \vec{\mathbf{0}} \quad \Longrightarrow \quad r_i = 0 \quad \forall i \; .$$

- Two noncollinear vectors or three non-coplanar vectors are linearly independent.
- If the number n of vectors is larger than the dimension m of the vector space, the vectors are always linearly dependent.

□ Three vectors in one plane and four vectors in a three-dimensional space are always linearly dependent; i.e., one vector can be expressed as a sum of multiples of the other vectors.

6.1.8 Basis

Basis of an n-dimensional vector space $\mathbb{R}^n$: A system of n linearly independent vectors in $\mathbb{R}^n$.

- Every vector of a vector space can be described by a linear combination of its base vectors.

Orthogonal basis: The base vectors are perpendicular to each other.
Orthonormal basis: The base vectors are perpendicular to each other and are **normalized unit vectors**:

$$|\vec{\mathbf{e}}_1| = |\vec{\mathbf{e}}_2| = \cdots = |\vec{\mathbf{e}}_n| = 1 \; .$$

Cartesian basis (unit vectors): Rectilinear orthonormal basis which allows a simple geometric interpretation of the problems to be described.

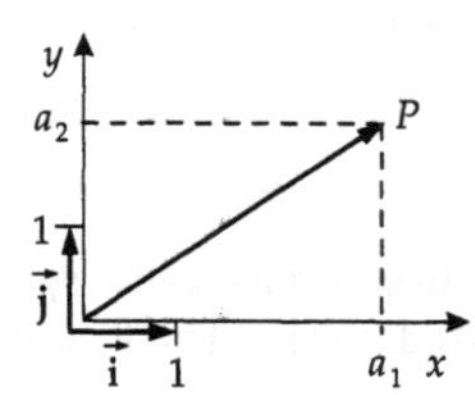

Normalized basis of the two-dimensional vector space $\mathbb{R}^2$ (e.g., x, y of the plane): Two linearly independent (i.e., nonparallel) unit vectors $\vec{\mathbf{e}}_1, \vec{\mathbf{e}}_2$ (not necessarily perpendicular to each other).

Representation by components

Cartesian coordinate system in the n-dimensional vector space $\mathbb{R}^n$ allows for the representation of n-dimensional position vectors (whose initial points lie at the origin) by specifying the coordinates $(a_1, a_2, \ldots, a_n)$ of point P at which the head of the vector ends.

The vectors $\mathbf{a}_i\vec{\mathbf{e}}_i$, $i = l, \ldots, n$ are called the components of vector $\vec{\mathbf{a}}$.

□ In the representation of a vector by means of a row matrix or column matrix, the numbers $\mathbf{a}_i$ themselves are often called the components of the vector.

Representation of a vector by component in n dimensions:

a) Notation as a column:

$$\vec{\mathbf{a}} = \begin{pmatrix} a_1 \\ a_2 \\ \vdots \\ a_n \end{pmatrix} .$$

b) Notation as a row (transposed column vector):

$$\vec{\mathbf{a}}^{\mathsf{T}} = (a_1, a_2, \dots, a_n) .$$

▷ Calculation by components is discussed here only for Cartesian coordinates; the formulas do not apply to polar, spherical, or cylindrical coordinates!

□ Vectors are added or subtracted by adding or subtracting their components.

Representation by components:

$$\begin{pmatrix} a_1 \\ a_2 \\ a_3 \end{pmatrix} \pm \begin{pmatrix} b_1 \\ b_2 \\ b_3 \end{pmatrix} = \begin{pmatrix} a_1 & \pm & b_1 \\ a_2 & \pm & b_2 \\ a_3 & \pm & b_3 \end{pmatrix} .$$

□ Total force as the result of individual forces:

$$F_x = F_{x_1} + F_{x_2} + \cdots + F_{x_n} = \sum_{\alpha=1}^{n} F_{x\alpha}$$

$$F_y = F_{y_1} + F_{y_2} + \cdots + F_{y_n} = \sum_{\alpha=1}^{n} F_{y\alpha}$$

$$F_z = F_{z_1} + F_{z_2} + \cdots + F_{z_n} = \sum_{\alpha=1}^{n} F_{z\alpha} .$$

□ Difference of position vectors in representation by components:

$$\begin{aligned} \vec{\mathbf{a}} - \vec{\mathbf{b}} &= \vec{\mathbf{a}}_x + \vec{\mathbf{a}}_y + \vec{\mathbf{a}}_z - \vec{\mathbf{b}}_x - \vec{\mathbf{b}}_y - \vec{\mathbf{b}}_z \\ &= (a_x - b_x)\vec{\mathbf{i}} + (a_y - b_y)\vec{\mathbf{j}} + (a_z - b_z)\vec{\mathbf{k}} . \end{aligned}$$

□ Addition of row vectors:

$$\vec{\mathbf{a}} = (3, -5, 7) \qquad \vec{\mathbf{b}} = (5, 7, -8)$$

$$\vec{\mathbf{a}} + \vec{\mathbf{b}} = [3 + 5, (-5) + 7, 7 + (-8)] = (8, 2, -1)$$

Basis in three-dimensional space:

Normalized basis: Three arbitrary, linearly independent (that is, non-coplanar) unit vectors $\vec{\mathbf{e}}_1, \vec{\mathbf{e}}_2, \vec{\mathbf{e}}_3$,

$$\vec{\mathbf{a}} = a_1 \cdot \vec{\mathbf{e}}_1 + a_2 \cdot \vec{\mathbf{e}}_2 + a_3 \cdot \vec{\mathbf{e}}_3 .$$

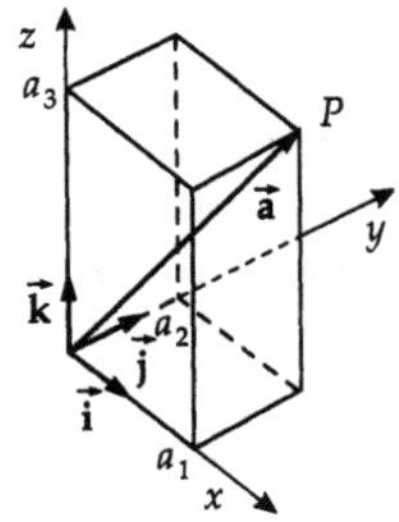

Orthonormal basis in Cartesian coordinate systems: Three unit vectors $\vec{\mathbf{i}}, \vec{\mathbf{j}}, \vec{\mathbf{k}}$, perpendicular to each other, with

$$\vec{\mathbf{i}} = \begin{pmatrix} 1 \\ 0 \\ 0 \end{pmatrix} , \ \vec{\mathbf{j}} = \begin{pmatrix} 0 \\ 1 \\ 0 \end{pmatrix} , \ \vec{\mathbf{k}} = \begin{pmatrix} 0 \\ 0 \\ 1 \end{pmatrix} .$$

i.e., $\vec{\mathbf{i}}, \vec{\mathbf{j}}$, and $\vec{\mathbf{k}}$ are unit vectors along the x, y, and z axis, respectively (lying on the coordinate axis).

Right-hand rule: The extended thumb ($\vec{\mathbf{e}}_1$), the index finger ($\vec{\mathbf{e}}_2$), and the middle finger ($\vec{\mathbf{e}}_3$) of the right hand, flexed by 90°, form a right-hand system.

The vectors in the three-dimensional space are given by their components a_1, a_2, a_3 with respect to a rectangular Cartesian coordinate system. By omitting the third component, a two-dimensional vector in the plane is generated, to which the same calculating operations apply.

$$\text{3D}: \begin{pmatrix} a_1 \\ a_2 \\ a_3 \end{pmatrix} \qquad \text{2D}: \begin{pmatrix} a_1 \\ a_2 \end{pmatrix}.$$

- Every vector $\vec{\mathbf{a}}$ in the plane can be represented as a **linear combination** (sum of multiples of $\vec{\mathbf{e}}_1$ and $\vec{\mathbf{e}}_2$) of the two **base vectors** ($\vec{\mathbf{e}}_1, \vec{\mathbf{e}}_2$):

$$\vec{\mathbf{a}} = a_1 \cdot \vec{\mathbf{e}}_1 + a_2 \cdot \vec{\mathbf{e}}_2 \,; \quad a_1, a_2 \in \mathbb{R}$$

Affine coordinates of $\vec{\mathbf{a}}$ with respect to the basis $\vec{\mathbf{e}}_1, \vec{\mathbf{e}}_2$ are a_1, a_2.

□ **Cartesian coordinate system** in two dimensions: The base vectors $\vec{\mathbf{i}}, \vec{\mathbf{j}}$ are **unit vectors** perpendicular to each other,

$$\vec{\mathbf{i}} = \begin{pmatrix} 1 \\ 0 \end{pmatrix}, \vec{\mathbf{j}} = \begin{pmatrix} 0 \\ 1 \end{pmatrix}.$$

The components (coordinates a_1, a_2) determine the vector completely,

$$\vec{\mathbf{a}} = a_1 \cdot \vec{\mathbf{i}} + a_2 \cdot \vec{\mathbf{j}}$$

Column vector:

$$\vec{\mathbf{a}} = \begin{pmatrix} a_1 \\ a_2 \end{pmatrix}.$$

Row vector:

$$\vec{\mathbf{a}}^{\mathsf{T}} = (a_1, a_2) .$$

- In the three-dimensional vector space, any vector $\vec{\mathbf{a}}$ can be described by means of three Cartesian components (a_x, a_y, a_z), which are the pre-factors of the Cartesian unit vectors $\vec{\mathbf{i}}, \vec{\mathbf{j}}, \vec{\mathbf{k}}$:

$$\vec{\mathbf{a}} = (a_x, a_y, a_z) = a_x\vec{\mathbf{i}} + a_y\vec{\mathbf{j}} + a_z\vec{\mathbf{k}}, \text{ with}$$
$$\vec{\mathbf{i}} = \begin{pmatrix} 1 \\ 0 \\ 0 \end{pmatrix}, \quad \vec{\mathbf{j}} = \begin{pmatrix} 0 \\ 1 \\ 0 \end{pmatrix}, \quad \vec{\mathbf{k}} = \begin{pmatrix} 0 \\ 0 \\ 1 \end{pmatrix}.$$

- **Linear independence** in the plane: Two vectors in the plane,

$$\vec{\mathbf{a}} = \begin{pmatrix} a_1 \\ a_2 \end{pmatrix} \quad \vec{\mathbf{b}} = \begin{pmatrix} b_1 \\ b_2 \end{pmatrix}$$

are **linearly independent** if the vector equation $r\vec{\mathbf{a}} + s\vec{\mathbf{b}} = \vec{0}$ has **only** the trivial solution $r = s = 0$. This vector equation is equivalent to the system of linear equations in r and s:

$$ra_1 + sb_1 = 0$$
$$ra_2 + sb_2 = 0\,.$$

● The vectors $\vec{\mathbf{a}}$ and $\vec{\mathbf{b}}$ are **linearly dependent (collinear)** if the determinant vanishes; i.e., the system of linear equations has no unique solution!

$$\begin{vmatrix} a_1 & a_2 \\ b_1 & b_2 \end{vmatrix} = a_1 \cdot b_2 - a_2 \cdot b_1 = 0$$

▷ **Linear dependence** means that the two straight lines on which $\vec{\mathbf{a}}$ and $\vec{\mathbf{b}}$ lie are parallel (do not intersect each other) or are superposed (have an infinite number of intersection points). In that case, the system of linear equations given above has no solution. The determinant is zero!

□ $\vec{\mathbf{a}} = \begin{pmatrix} -2 \\ 7 \end{pmatrix}$; $\vec{\mathbf{b}} = \begin{pmatrix} 1 \\ 3 \end{pmatrix}$; $\vec{\mathbf{c}} = \begin{pmatrix} 8 \\ -15 \end{pmatrix}$.

$\begin{vmatrix} -2 & 1 \\ 7 & 3 \end{vmatrix} = -6 - 7 = -13 \neq 0$

$\rightarrow$ $\vec{\mathbf{a}}$ and $\vec{\mathbf{b}}$ are linearly independent and can be used as a basis.

$\vec{\mathbf{c}} = r\vec{\mathbf{a}} + s\vec{\mathbf{b}}$

$-2r + s = 8$ and $7r + 3s = -15$.

Solving the system of equations yields:

$\vec{\mathbf{c}} = -3\vec{\mathbf{a}} + 2\vec{\mathbf{b}}$. Here, $\vec{\mathbf{c}}$ can be represented as an integral multiple of $\vec{\mathbf{a}}$ and $\vec{\mathbf{b}}$.

6.2 Scalar product or inner product

Scalar product: The multiplication of two vectors $\vec{\mathbf{a}}, \vec{\mathbf{b}} \in \mathbb{R}^n$ with n components such that the result is a **scalar**; i.e., it is a **real number**:

$$\vec{\mathbf{a}} \cdot \vec{\mathbf{b}} = c\,,\; c \in \mathbb{R}\,.$$

● The two vectors must have an equal number of components!

If φ is the angle between $\vec{\mathbf{a}}$ and $\vec{\mathbf{b}}$, then:

$$\vec{\mathbf{a}} \cdot \vec{\mathbf{b}} = |\vec{\mathbf{a}}| \cdot |\vec{\mathbf{b}}| \cdot \cos\varphi \qquad (0^\circ \leq \varphi \leq 180^\circ)\,.$$

▷ Significance of the scalar product:
Projection of $\vec{\mathbf{b}}$ onto $\vec{\mathbf{a}}$,
multiplied by $|\vec{\mathbf{a}}|$, or
projection of $\vec{\mathbf{a}}$ onto $\vec{\mathbf{b}}$,
multiplied by $|\vec{\mathbf{b}}|$.

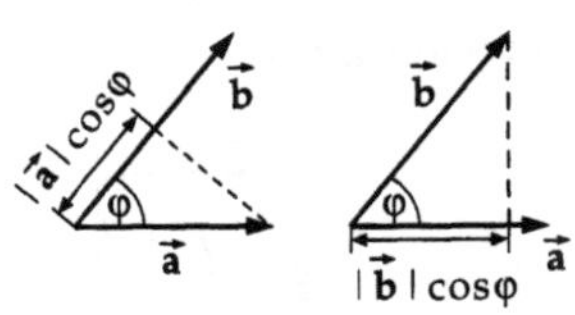

Representation by components, "row times column":

$$\vec{\mathbf{a}}^{\mathsf{T}} \cdot \vec{\mathbf{b}} = (a_1, a_2, \ldots, a_n) \cdot \begin{pmatrix} b_1 \\ b_2 \\ \vdots \\ b_n \end{pmatrix} = a_1b_1 + a_2b_2 + \cdots + a_nb_n\,.$$

▷ In matrix notation, the order of the vectors is extremely important! $\vec{\mathbf{a}}^{\mathsf{T}} \cdot \vec{\mathbf{b}}$ yields a scalar (row times column), but $\vec{\mathbf{b}} \cdot \vec{\mathbf{a}}^{\mathsf{T}}$ yields an $n \times n$ matrix (**the dyadic product**).

6.2.1 *Calculating laws*

● **Commutative law**:

$$\vec{\mathbf{a}} \cdot \vec{\mathbf{b}} = \vec{\mathbf{b}} \cdot \vec{\mathbf{a}}$$

● **Distributive laws**:

$$(r\vec{\mathbf{a}}) \cdot \vec{\mathbf{b}} = \vec{\mathbf{a}} \cdot (r\vec{\mathbf{b}}) = r(\vec{\mathbf{a}} \cdot \vec{\mathbf{b}})$$
$$(rs)\vec{\mathbf{a}} = r(s\vec{\mathbf{a}}) = rs\vec{\mathbf{a}} \quad \text{for} \quad r, s \in \mathbb{R}\,,\ \vec{\mathbf{a}}, \vec{\mathbf{b}} \in \mathbb{R}^n$$

● **Associative law** for multiplication with real number **r**:

▷ Because $\vec{\mathbf{a}} \cdot \vec{\mathbf{b}} = c$ results in a scalar ($c \in \mathbb{R}$), the scalar product of three vectors cannot be formed: $\vec{\mathbf{a}} \cdot \vec{\mathbf{b}} \cdot \vec{\mathbf{c}}$ never exists! The product $(\vec{\mathbf{a}} \cdot \vec{\mathbf{b}})\vec{\mathbf{c}}$ does exist: It is the product of *number* $\vec{\mathbf{a}} \cdot \vec{\mathbf{b}} = |\vec{\mathbf{a}}||\vec{\mathbf{b}}| \cos\varphi$ and *vector* $\vec{\mathbf{c}}$.

● The scalar product of two vectors $\vec{\mathbf{a}}, \vec{\mathbf{b}}$ can be formed only when $\vec{\mathbf{a}}$ and $\vec{\mathbf{b}}$ belong to the same vector space.

□ $\vec{\mathbf{a}} = (a_1, a_2)\ ,\ \vec{\mathbf{b}} = (b_1, b_2, b_3) \Rightarrow \vec{\mathbf{a}} \cdot \vec{\mathbf{b}}$ does not exist!

6.2.2 *Properties and applications of the scalar product*

● **Calculation of length (magnitude)**:

$$\sqrt{\vec{\mathbf{a}} \cdot \vec{\mathbf{a}}} = |\vec{\mathbf{a}}| \geq 0\,.$$

□ $\vec{\mathbf{a}} = (1, 1, 1) \Rightarrow |\vec{\mathbf{a}}| = \sqrt{1^2 + 1^2 + 1^2} = \sqrt{3}$

□ Magnitude, normalization and equality of vectors in $\mathbb{R}^3$:
Magnitude or **length, norm** of a vector:

$$|\vec{\mathbf{a}}| = \sqrt{\vec{\mathbf{a}} \cdot \vec{\mathbf{a}}} = \sqrt{a_1^2 + a_2^2 + a_3^2}\,.$$

Normalization of a vector, unit vector:

$$\frac{\vec{\mathbf{a}}}{|\vec{\mathbf{a}}|} = \frac{1}{|\vec{\mathbf{a}}|} \cdot \begin{pmatrix} a_1 \\ a_2 \\ a_3 \end{pmatrix} = \frac{1}{\sqrt{a_1^2 + a_2^2 + a_3^2}} \cdot \begin{pmatrix} a_1 \\ a_2 \\ a_3 \end{pmatrix}.$$

Magnitude:

$$\left| \frac{\vec{\mathbf{a}}}{|\vec{\mathbf{a}}|} \right| = 1\,.$$

□ Normalizing the magnitude of the vector $\vec{\mathbf{a}} = 3\vec{\mathbf{i}} - \vec{\mathbf{j}} + 2\vec{\mathbf{k}}$

$$|\vec{\mathbf{a}}| = \sqrt{3^2 + (-1)^2 + 2^2} = \sqrt{14} \approx 3.742\,.$$

Equality of vectors $\vec{\mathbf{a}}$ and $\vec{\mathbf{b}}$:

● Vectors are equal if and only if all their components are equal:

$$\vec{\mathbf{a}} = \vec{\mathbf{b}} \Longleftrightarrow a_1 = b_1,\ a_2 = b_2,\ a_3 = b_3\ ,$$

or

$$\begin{pmatrix} a_1 \\ a_2 \\ a_3 \end{pmatrix} = \begin{pmatrix} b_1 \\ b_2 \\ b_3 \end{pmatrix} .$$

- **Calculation of angles**:

$$\cos\varphi = \frac{\vec{\mathbf{a}} \cdot \vec{\mathbf{b}}}{|\vec{\mathbf{a}}| \cdot |\vec{\mathbf{b}}|} = \frac{a_1 b_1 + a_2 b_2 + \cdots + a_n b_n}{\sqrt{a_1^2 + a_2^2 + \cdots + a_n^2} \cdot \sqrt{b_1^2 + b_2^2 + \cdots + b_n^2}} .$$

□ Wanted: The angle φ enclosed between the two vectors

$$\vec{\mathbf{a}} = (0, 1, 1) \quad \vec{\mathbf{b}} = (1, 1, 0) :$$

$$\cos\varphi = \frac{\vec{\mathbf{a}} \cdot \vec{\mathbf{b}}}{|\vec{\mathbf{a}}| \cdot |\vec{\mathbf{b}}|} = \frac{1}{\sqrt{2} \cdot \sqrt{2}} = \frac{1}{2} \quad \to \quad \varphi = 60^\circ .$$

▷ Application in 3-D computer graphics:
Hidden surface elimination of convex bodies. If the scalar product of a vector normal to the surface and a vector normal to the screen is negative, i.e., the surface points away from the screen, then the surface is not shown on the screen.

- **Triangle inequality**:

$$|\vec{\mathbf{a}} + \vec{\mathbf{b}}| \leq |\vec{\mathbf{a}}| + |\vec{\mathbf{b}}|$$

$$\sqrt{(a_1 + b_1)^2 + \cdots + (a_n + b_n)^2} \leq \sqrt{a_1^2 + \cdots + a_n^2} + \sqrt{b_1^2 + \cdots + b_n^2} .$$

- **Cauchy-Schwarz inequality**:

$$|\vec{\mathbf{a}} \cdot \vec{\mathbf{b}}| \leq |\vec{\mathbf{a}}| \cdot |\vec{\mathbf{b}}| .$$

- Collinear vectors as a special case:
The scalar product of two parallel vectors $\vec{\mathbf{a}}$ and $\vec{\mathbf{b}}$ is equal to the product of the magnitude of the two vectors ($\varphi = 0, \cos\varphi = 1$),

$$\vec{\mathbf{a}} \cdot \vec{\mathbf{b}} = |\vec{\mathbf{a}}| \cdot |\vec{\mathbf{b}}| , \qquad \vec{\mathbf{a}} \uparrow\uparrow \vec{\mathbf{b}} .$$

The scalar product of opposite vectors is equal to the negative product of the magnitudes of $\vec{\mathbf{a}}$ and $\vec{\mathbf{b}}$ ($\varphi = 180^\circ, \cos\varphi = -1$),

$$\vec{\mathbf{a}} \cdot \vec{\mathbf{b}} = -|\vec{\mathbf{a}}| \cdot |\vec{\mathbf{b}}| , \qquad \vec{\mathbf{a}} \uparrow\downarrow \vec{\mathbf{b}} .$$

- Orthogonal vectors as a special case:
The scalar product of orthogonal vectors vanishes
($\varphi = 90^\circ, \cos\varphi = 0$),

$$\vec{\mathbf{a}} \cdot \vec{\mathbf{b}} = 0 \quad \Longleftrightarrow \quad \vec{\mathbf{a}} \perp \vec{\mathbf{b}} .$$

□ Construction of orthogonal vectors; determining one component (b_3) of $\vec{\mathbf{b}}$ such that $\vec{\mathbf{a}} \perp \vec{\mathbf{b}}$:

$$\vec{\mathbf{a}} = (3, 2, -2) \quad \vec{\mathbf{b}} = (2, 1, b_3)$$
$$\vec{\mathbf{a}} \cdot \vec{\mathbf{b}} = 6 + 2 - 2b_3 = 0 \quad \to \quad b_3 = 4 .$$

- **Law of cosines**:

$$(\vec{\mathbf{a}} \pm \vec{\mathbf{b}})^2 = \vec{\mathbf{a}}^2 \pm 2\vec{\mathbf{a}} \cdot \vec{\mathbf{b}} + \vec{\mathbf{b}}^2 = |\vec{\mathbf{a}}|^2 + |\vec{\mathbf{b}}|^2 \pm 2|\vec{\mathbf{a}}||\vec{\mathbf{b}}| \cos(\vec{\mathbf{a}}, \vec{\mathbf{b}})$$

$$(\vec{\mathbf{a}} + \vec{\mathbf{b}})^2 - (\vec{\mathbf{a}} - \vec{\mathbf{b}})^2 = 4\vec{\mathbf{a}} \cdot \vec{\mathbf{b}} = 4|\vec{\mathbf{a}}||\vec{\mathbf{b}}| \cos(\vec{\mathbf{a}}, \vec{\mathbf{b}})$$

$$|\vec{\mathbf{a}} \pm \vec{\mathbf{b}}| = \sqrt{\vec{\mathbf{a}}^2 \pm 2\vec{\mathbf{a}} \cdot \vec{\mathbf{b}} + \vec{\mathbf{b}}^2}$$

- The Cartesian unit vectors $\vec{\mathbf{i}}, \vec{\mathbf{j}}, \vec{\mathbf{k}}$ in the space $\mathbb{R}^3$ are mutually perpendicular:

$$\vec{\mathbf{i}}\cdot\vec{\mathbf{i}}=\vec{\mathbf{j}}\cdot\vec{\mathbf{j}}=\vec{\mathbf{k}}\cdot\vec{\mathbf{k}}=1 \quad \Longleftrightarrow \quad (\text{since } \varphi = 0^\circ, \text{also} \cos\varphi = 1)$$
$$\vec{\mathbf{i}}\cdot\vec{\mathbf{j}}=\vec{\mathbf{i}}\cdot\vec{\mathbf{k}}=\vec{\mathbf{j}}\cdot\vec{\mathbf{k}}=0 \quad \Longleftrightarrow \quad (\text{since } \varphi = 90^\circ, \text{also} \cos\varphi = 0)$$

- **Kronecker's symbol summarizes orthonormality**:

$$\delta_{lm} = \vec{\mathbf{e}}_l \cdot \vec{\mathbf{e}}_m = \begin{cases} 0 & (l \neq m) \\ 1 & (l = m) \end{cases} \quad \text{with } \vec{\mathbf{e}}_1 = \vec{\mathbf{i}}, \vec{\mathbf{e}}_2 = \vec{\mathbf{j}}, \vec{\mathbf{e}}_3 = \vec{\mathbf{k}} \text{ and } l, m = \{1, 2, 3\}\,.$$

□ The multiplication of a vector by a scalar is carried out by components:

$$r\vec{\mathbf{a}} = r\begin{pmatrix} a_1 \\ a_2 \\ a_3 \end{pmatrix} = \begin{pmatrix} ra_1 \\ ra_2 \\ ra_3 \end{pmatrix}$$

$$(r \cdot s)\vec{\mathbf{a}} = r(s\vec{\mathbf{a}})\,.$$

$$r(\vec{\mathbf{a}} + \vec{\mathbf{b}}) = r\vec{\mathbf{a}} + r\vec{\mathbf{b}} = \begin{pmatrix} ra_1 \\ ra_2 \\ ra_3 \end{pmatrix} + \begin{pmatrix} rb_1 \\ rb_2 \\ rb_3 \end{pmatrix}$$

$$(r + s)\vec{\mathbf{a}} = r\vec{\mathbf{a}} + s\vec{\mathbf{a}}\quad.$$

Null vector:

$$\vec{\mathbf{0}} = \begin{pmatrix} 0 \\ 0 \\ 0 \end{pmatrix}.$$

Unit vector:

$$|\vec{\mathbf{e}}| = 1\,.$$

□ Unit vectors along the x, y, z-axis:

$$\vec{\mathbf{i}} = \vec{\mathbf{e}}_1 = \begin{pmatrix} 1 \\ 0 \\ 0 \end{pmatrix} \quad \vec{\mathbf{j}} = \vec{\mathbf{e}}_2 = \begin{pmatrix} 0 \\ 1 \\ 0 \end{pmatrix} \quad \vec{\mathbf{k}} = \vec{\mathbf{e}}_3 = \begin{pmatrix} 0 \\ 0 \\ 1 \end{pmatrix}.$$

6.2.3 Schmidt's orthonormalization method

- **Schmidt's orthonormalization method**: Systematic method to construct a set of orthonormalized vectors $\vec{y}_i$, $i = 1, \ldots, n$, from a set of linearly independent vectors $\vec{x}_i$, $i = 1, \ldots, n$.

$$\textbf{1.}\quad \vec{z}_i = \vec{x}_i - \sum_{k=1}^{i-1} (\vec{x}_i \cdot \vec{y}_k)\vec{y}_k$$

$$\textbf{2.}\quad \vec{y}_i = \frac{\vec{z}_i}{\vec{z}_i \cdot \vec{z}_i}$$

The method reiterates 1. and 2. until i has taken the values from 1 to n.

6.2.4 Direction cosine

Direction cosine: The **projections** of $\vec{\mathbf{a}}$ onto the x, y, z axis yield the cosines of the angles α, β, γ between vector $\vec{\mathbf{a}}$ and the positive x, y, z axis, respectively.

$$\cos\alpha = \frac{a_1}{|\vec{\mathbf{a}}|}$$
$$\cos\beta = \frac{a_2}{|\vec{\mathbf{a}}|}$$
$$\cos\gamma = \frac{a_3}{|\vec{\mathbf{a}}|}$$

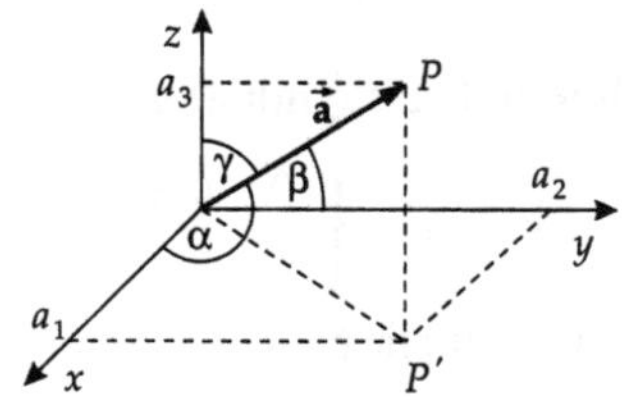

x coordinate : $a_1 = |\vec{\mathbf{a}}|\cos\alpha$
y coordinate : $a_2 = |\vec{\mathbf{a}}|\cos\beta$
z coordinate : $a_3 = |\vec{\mathbf{a}}|\cos\gamma$

- For the direction cosines, it holds that $\cos^2\alpha + \cos^2\beta + \cos^2\gamma = 1$.

□ Direction cosines for $\vec{\mathbf{a}} = 3\vec{\mathbf{i}} - \vec{\mathbf{j}} + 2\vec{\mathbf{k}}$:

$$|\vec{\mathbf{a}}| = \sqrt{3^2 + (-1)^2 + 2^2} = \sqrt{14} = 3.742$$
$$\cos\alpha \approx \frac{3}{3.742} \approx 0.802\,, \quad \alpha \sim 36.7^\circ$$
$$\cos\beta \approx \frac{-1}{3.742} \approx -0.267\,, \quad \beta \sim 105.5^\circ$$
$$\cos\gamma \approx \frac{2}{3.742} \approx 0.534\,, \quad \gamma \sim 57.7^\circ$$

6.2.5 Application hypercubes of vector analysis

Hypercube: A cube in the Euclidean space of arbitrary dimension.

□ $n = 2 \implies$ square,
$n = 3 \implies$ cube,
$n = 4 \implies$ hypercube in 4 dimensions.

▷ Hypercubes are used in the design of high-performance computers if several processors of the same kind have to be connected at low cost but high efficiency (few connections but short signal propagation delay).

- The hypercube in n dimensions has 2^n vertices (= processors) and $n \cdot 2^{n-1}$ edges (= connections between the processors).
  The longest signal path to be covered is between two processors lying on an n-dimensional space diagonal; it passes n connections.

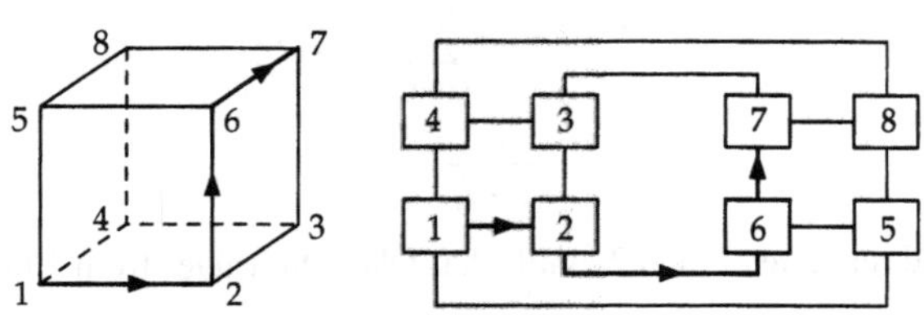

Three-dimensional "hypercube"

▷ The connections between the nodes are of equal length; only the scheme of connections is shown here.

□ $n = 10$. $2^{10} = 1024$ processors can be connected with only $10 \cdot 2^9 = 5120$ connections, i.e., by 5 connections per processor (with 10 points of connection for each processor), in such a way that the longest signal propagation delay between any two processors corresponds merely to the signal propagation delay associated with 10 connection paths.

6.3 Vector product of two vectors

▷ The vector product is defined only in the three-dimensional space $\mathbb{R}^3$!

The vector product (also called **cross product, outer product**) of two vectors $\vec{\mathbf{a}}$ and $\vec{\mathbf{b}}$

$$\vec{\mathbf{a}} \times \vec{\mathbf{b}} = \vec{\mathbf{c}}$$

is a vector, $\vec{\mathbf{c}}$.

The magnitude of $\vec{\mathbf{c}} = \vec{\mathbf{a}} \times \vec{\mathbf{b}}$ is the area of the parallelogram spanned by vectors $\vec{\mathbf{a}}$ and $\vec{\mathbf{b}}$, where φ is the angle between $\vec{\mathbf{a}}$ and $\vec{\mathbf{b}}$,

$$|\vec{\mathbf{a}} \times \vec{\mathbf{b}}| = |\vec{\mathbf{a}}| \cdot |\vec{\mathbf{b}}| \cdot |\sin \varphi| \ .$$

The direction of $\vec{\mathbf{c}} = \vec{\mathbf{a}} \times \vec{\mathbf{b}}$ is perpendicular to the plane spanned by the vectors $\vec{\mathbf{a}}$ and $\vec{\mathbf{b}}$. $\vec{\mathbf{a}}, \vec{\mathbf{b}}, \vec{\mathbf{c}} = \vec{\mathbf{a}} \times \vec{\mathbf{b}}$ form a **right-handed system** in this order.

In contrast to the scalar product ("$\vec{\mathbf{a}}$ *times* $\vec{\mathbf{b}}$") the vector product is read ("$\vec{\mathbf{a}}$ *cross* $\vec{\mathbf{b}}$").

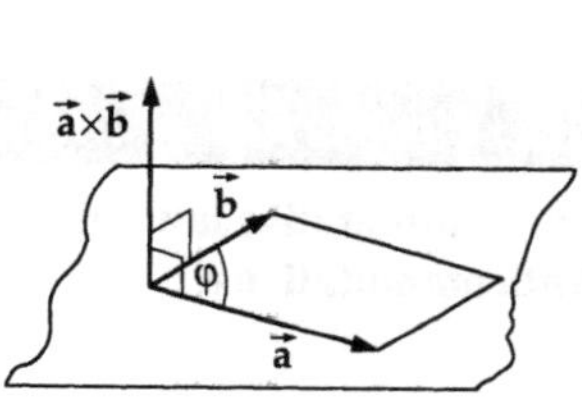

Geometrical interpretation: The vector $\vec{\mathbf{a}} \times \vec{\mathbf{b}} = \vec{\mathbf{c}}$ is **perpendicular** to the vectors $\vec{\mathbf{a}}, \vec{\mathbf{b}}$. The magnitude of this vector ($|\vec{\mathbf{c}}|$) equals the area of the parallelogram formed by $\vec{\mathbf{a}}, \vec{\mathbf{b}}$.

▷ Notice the difference between vector product and scalar product: the vector product is a vector, not a scalar, and due to the $\sin\varphi$-factor (instead of $\cos\varphi$), the vector product becomes a maximum for vectors $\vec{\mathbf{a}}, \vec{\mathbf{b}}$ which are perpendicular to each other.

Representation by components:

$$\begin{aligned}\vec{\mathbf{a}} \times \vec{\mathbf{b}} &= (a_1\vec{\mathbf{i}} + a_2\vec{\mathbf{j}} + a_3\vec{\mathbf{k}}) \times (b_1\vec{\mathbf{i}} + b_2\vec{\mathbf{j}} + b_3\vec{\mathbf{k}}) \\ &= (a_2b_3 - a_3b_2) \cdot \vec{\mathbf{i}} + (a_3b_1 - a_1b_3) \cdot \vec{\mathbf{j}} + (a_1b_2 - a_2b_1) \cdot \vec{\mathbf{k}}\end{aligned}$$

or:

$$\vec{\mathbf{a}} \times \vec{\mathbf{b}} = \begin{pmatrix} a_1 \\ a_2 \\ a_3 \end{pmatrix} \times \begin{pmatrix} b_1 \\ b_2 \\ b_3 \end{pmatrix} = \begin{pmatrix} a_2b_3 & - & a_3b_2 \\ a_3b_1 & - & a_1b_3 \\ a_1b_2 & - & a_2b_1 \end{pmatrix}$$

▷ Formally, the cross product can be written as a "**determinant**," with the base vectors appearing in the first column.

$$\vec{\mathbf{a}} \times \vec{\mathbf{b}} = \begin{vmatrix} \vec{\mathbf{i}} & a_1 & b_1 \\ \vec{\mathbf{j}} & a_2 & b_2 \\ \vec{\mathbf{k}} & a_3 & b_3 \end{vmatrix}$$

$$= \vec{\mathbf{i}} \cdot \begin{vmatrix} a_2 & b_2 \\ a_3 & b_3 \end{vmatrix} - \vec{\mathbf{j}} \cdot \begin{vmatrix} a_1 & b_1 \\ a_3 & b_3 \end{vmatrix} + \vec{\mathbf{k}} \cdot \begin{vmatrix} a_1 & b_1 \\ a_2 & b_2 \end{vmatrix}$$

▷ Note: The vector product $\vec{\mathbf{c}} = \vec{\mathbf{a}} \times \vec{\mathbf{b}}$ is not a proper determinant, i.e., it is not a scalar.

□ $\vec{\mathbf{a}} = \begin{pmatrix} 2 \\ -1 \\ -2 \end{pmatrix}$; $\vec{\mathbf{b}} = \begin{pmatrix} -1 \\ 2 \\ 3 \end{pmatrix}$

$$\vec{\mathbf{c}} = \vec{\mathbf{a}} \times \vec{\mathbf{b}} = \begin{vmatrix} \vec{\mathbf{i}} & 2 & -1 \\ \vec{\mathbf{j}} & -1 & 2 \\ \vec{\mathbf{k}} & -2 & 3 \end{vmatrix}$$

$$= \vec{\mathbf{i}} \cdot \begin{vmatrix} -1 & 2 \\ -2 & 3 \end{vmatrix} - \vec{\mathbf{j}} \cdot \begin{vmatrix} 2 & -1 \\ -2 & 3 \end{vmatrix} + \vec{\mathbf{k}} \cdot \begin{vmatrix} 2 & -1 \\ -1 & 2 \end{vmatrix}$$

$$= \vec{\mathbf{i}} \cdot (-3+4) - \vec{\mathbf{j}} \cdot (6-2) + \vec{\mathbf{k}} \cdot (4-1)$$

$$= \begin{pmatrix} 1 \\ -4 \\ 3 \end{pmatrix}$$

Cyclic symmetry of the vector product of the orthonormalized base vectors $\vec{\mathbf{i}}, \vec{\mathbf{j}}, \vec{\mathbf{k}}$:

$$\begin{array}{lclclcl} \vec{\mathbf{i}} \times \vec{\mathbf{i}} & = & \vec{\mathbf{j}} \times \vec{\mathbf{j}} & = & \vec{\mathbf{k}} \times \vec{\mathbf{k}} & = & \vec{0} \\ \vec{\mathbf{i}} \times \vec{\mathbf{j}} & = & \vec{\mathbf{k}}\,; \quad \vec{\mathbf{j}} \times \vec{\mathbf{k}} & = & \vec{\mathbf{i}}\,; \quad \vec{\mathbf{k}} \times \vec{\mathbf{i}} & = & \vec{\mathbf{j}} \\ \vec{\mathbf{j}} \times \vec{\mathbf{i}} & = & -\vec{\mathbf{k}}\,; \quad \vec{\mathbf{k}} \times \vec{\mathbf{j}} & = & -\vec{\mathbf{i}}\,; \quad \vec{\mathbf{i}} \times \vec{\mathbf{k}} & = & -\vec{\mathbf{j}} \end{array} .$$

6.3.1 Properties of the vector product

- **The commutative law** does not apply! Instead:
- **Anticommutativity**:

$$\vec{\mathbf{b}} \times \vec{\mathbf{a}} = -\vec{\mathbf{a}} \times \vec{\mathbf{b}}$$

- **The associative law** does not apply:

$$\vec{\mathbf{a}} \times (\vec{\mathbf{b}} \times \vec{\mathbf{c}}) \neq (\vec{\mathbf{a}} \times \vec{\mathbf{b}}) \times \vec{\mathbf{c}}$$

- **Distributive laws**:

$$\vec{\mathbf{a}} \times (\vec{\mathbf{b}} + \vec{\mathbf{c}}) = \vec{\mathbf{a}} \times \vec{\mathbf{b}} + \vec{\mathbf{a}} \times \vec{\mathbf{c}}$$
$$(r\vec{\mathbf{a}}) \times \vec{\mathbf{b}} = \vec{\mathbf{a}} \times (r\vec{\mathbf{b}}) = r(\vec{\mathbf{a}} \times \vec{\mathbf{b}})$$

Angle φ between $\vec{\mathbf{a}}$ and $\vec{\mathbf{b}}$:

$$\sin\varphi = \sin(\vec{\mathbf{a}}, \vec{\mathbf{b}}) = \frac{|\vec{\mathbf{a}} \times \vec{\mathbf{b}}|}{|\vec{\mathbf{a}}| \cdot |\vec{\mathbf{b}}|}$$

- The vector product of two parallel vectors is zero.
  $\vec{\mathbf{a}} \times \vec{\mathbf{b}} = 0 \quad (\vec{\mathbf{a}}, \vec{\mathbf{b}} \neq \vec{\mathbf{0}}) \iff \vec{\mathbf{a}} \| \vec{\mathbf{b}}$
  $\vec{\mathbf{a}}$, $\vec{\mathbf{b}}$ are linearly dependent.
- The vector product of two orthogonal vectors $\vec{\mathbf{a}}, \vec{\mathbf{b}}, \vec{\mathbf{a}} \perp \vec{\mathbf{b}}$, is a vector of magnitude

$$|\vec{\mathbf{a}} \times \vec{\mathbf{b}}| = |\vec{\mathbf{a}}||\vec{\mathbf{b}}| .$$

- $(\vec{\mathbf{a}} \times \vec{\mathbf{b}})^2 = \vec{\mathbf{a}}^2\vec{\mathbf{b}}^2 - (\vec{\mathbf{a}} \cdot \vec{\mathbf{b}})^2$

6.4 Multiple products of vectors

▷ Multiple products are defined only in **three** dimensions $\mathbb{R}^3$!

6.4.1 Scalar triple product

▷ The scalar triple product is defined only in **three** dimensions!

Scalar triple product, the **scalar product** of a vector $\vec{\mathbf{a}}$ with the **vector product** of the two vectors $\vec{\mathbf{b}}$ and $\vec{\mathbf{c}}$:

$$\vec{\mathbf{a}} \cdot (\vec{\mathbf{b}} \times \vec{\mathbf{c}})$$

- The scalar triple product is a **scalar, not** a vector!

Equivalent notations:

$$(\vec{\mathbf{a}}, \vec{\mathbf{b}}, \vec{\mathbf{c}}) =< \vec{\mathbf{a}}, \vec{\mathbf{b}}, \vec{\mathbf{c}} >= [\vec{\mathbf{a}}, \vec{\mathbf{b}}, \vec{\mathbf{c}}] = \det(\vec{\mathbf{a}}, \vec{\mathbf{b}}, \vec{\mathbf{c}}) \ .$$

Notation by component:

$$\begin{aligned} \vec{\mathbf{a}} \cdot (\vec{\mathbf{b}} \times \vec{\mathbf{c}}) &= (a_1, a_2, a_3) \cdot [(b_1, b_2, b_3) \times (c_1, c_3, c_3)] \ , \\ &= (a_1, a_2, a_3) \cdot (b_2c_3 - b_3c_2, -b_1c_3 + b_3c_1, b_1c_2 - b_2c_1) \\ &= a_1(b_2c_3 - b_3c_2) - a_2(b_1c_3 - b_3c_1) + a_3(b_1c_2 - b_2c_1) \ . \\ (\vec{\mathbf{a}} \times \vec{\mathbf{b}}) \cdot \vec{\mathbf{c}} &= \begin{vmatrix} a_1 & a_2 & a_3 \\ b_1 & b_2 & b_3 \\ c_1 & c_2 & c_3 \end{vmatrix} = \begin{pmatrix} a_2b_3 & - & a_3b_2 \\ a_3b_1 & - & a_1b_3 \\ a_1b_2 & - & b_1a_2 \end{pmatrix} \cdot \begin{pmatrix} c_1 \\ c_2 \\ c_3 \end{pmatrix} \\ &= c_1 \cdot (a_2b_3 - a_3b_2) - c_2 \cdot (a_1b_3 - a_3b_1) + c_3 \cdot (a_1b_2 - b_1a_2) \ . \end{aligned}$$

Geometric interpretation: The three **non-coplanar** i.e., (**linearly independent**) vectors $\vec{\mathbf{a}}$, $\vec{\mathbf{b}}$, $\vec{\mathbf{c}}$ form the edges of a **parallelepiped**.

- **Area** of the parallelogram of the base:

$$A_G = |\vec{\mathbf{a}} \times \vec{\mathbf{b}}| = |\vec{\mathbf{a}}| \cdot |\vec{\mathbf{b}}| \cdot |\sin\varphi| \ .$$

- **Height**, or length of the projection of vector $\vec{\mathbf{c}}$ onto $\vec{\mathbf{a}} \times \vec{\mathbf{b}}$:

$$h = \frac{|(\vec{\mathbf{a}} \times \vec{\mathbf{b}}) \cdot \vec{\mathbf{c}}|}{|\vec{\mathbf{a}} \times \vec{\mathbf{b}}|} \ .$$

- **Volume** of the parallelepiped (base times height):

$$V = |(\vec{\mathbf{a}} \times \vec{\mathbf{b}}) \cdot \vec{\mathbf{c}}| \ .$$

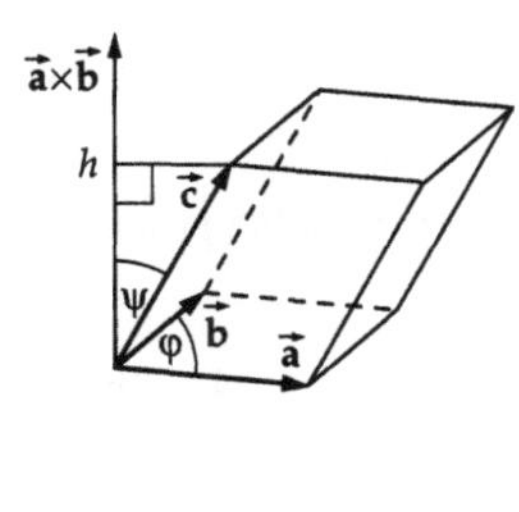

- The scalar triple product (the volume of the parallelepiped spanned by the three vectors) of three linearly dependent vectors $\vec{\mathbf{a}}$, $\vec{\mathbf{b}}$, $\vec{\mathbf{c}}$ (coplanar vectors) equals zero.
- Volume of a tetrahedron:

$$V_T = \left| \frac{1}{6} (\vec{\mathbf{a}} \times \vec{\mathbf{b}}) \cdot \mathbf{c} \right|$$

An (oriented) volume of the parallelepiped $<\vec{\mathbf{a}}, \vec{\mathbf{b}}, \vec{\mathbf{c}}>$ with a sign can be defined if the absolute-value signs of the sine are left out, $<\vec{\mathbf{a}}, \vec{\mathbf{b}}, \vec{\mathbf{c}}> < 0 \Longleftrightarrow \vec{\mathbf{a}}, \vec{\mathbf{b}}, \vec{\mathbf{c}}$ form a left-handed system.

Properties of the scalar triple product

- Three vectors $\vec{\mathbf{a}}, \vec{\mathbf{b}}, \vec{\mathbf{c}}$ are **coplanar** if and only if their **scalar triple product** vanishes, i.e., if $(\vec{\mathbf{a}}, \vec{\mathbf{b}}, \vec{\mathbf{c}}) = 0$.
- The scalar triple product changes signs when two vectors are interchanged:

$$(\vec{\mathbf{b}}, \vec{\mathbf{a}}, \vec{\mathbf{c}}) = -(\vec{\mathbf{a}}, \vec{\mathbf{b}}, \vec{\mathbf{c}})$$
$$(\vec{\mathbf{c}}, \vec{\mathbf{b}}, \vec{\mathbf{a}}) = -(\vec{\mathbf{a}}, \vec{\mathbf{b}}, \vec{\mathbf{c}})$$
$$(\vec{\mathbf{a}}, \vec{\mathbf{c}}, \vec{\mathbf{b}}) = -(\vec{\mathbf{a}}, \vec{\mathbf{b}}, \vec{\mathbf{c}})$$

- **A cyclic permutation** of the vectors does not change the scalar triple product:

$$(\vec{\mathbf{a}}, \vec{\mathbf{b}}, \vec{\mathbf{c}}) = (\vec{\mathbf{b}}, \vec{\mathbf{c}}, \vec{\mathbf{a}}) = (\vec{\mathbf{c}}, \vec{\mathbf{a}}, \vec{\mathbf{b}}) .$$

Decomposition into components of vector $\vec{\mathbf{a}}$ with respect to a general basis $\vec{\mathbf{a}}, \vec{\mathbf{b}}, \vec{\mathbf{c}}$:

$$\vec{\mathbf{d}} = \frac{1}{(\vec{\mathbf{a}}, \vec{\mathbf{b}}, \vec{\mathbf{c}})} \cdot [(\vec{\mathbf{b}}, \vec{\mathbf{c}}, \vec{\mathbf{d}}) \cdot \vec{\mathbf{a}} + (\vec{\mathbf{c}}, \vec{\mathbf{a}}, \vec{\mathbf{d}}) \cdot \vec{\mathbf{b}} + (\vec{\mathbf{a}}, \vec{\mathbf{b}}, \vec{\mathbf{d}}) \cdot \vec{\mathbf{c}}] .$$

□ The volume of a prism spanned by the following vectors

$$\vec{\mathbf{a}} = 3\vec{\mathbf{i}} + 2\vec{\mathbf{j}} - 1\vec{\mathbf{k}}$$
$$\vec{\mathbf{b}} = 5\vec{\mathbf{i}} + 3\vec{\mathbf{j}} + 2\vec{\mathbf{k}}$$
$$\vec{\mathbf{c}} = 4\vec{\mathbf{i}} + 1\vec{\mathbf{j}} + 3\vec{\mathbf{k}}$$

is obtained by

$$(\vec{\mathbf{a}}, \vec{\mathbf{b}}, \vec{\mathbf{c}}) = [\vec{\mathbf{a}}\vec{\mathbf{b}}\vec{\mathbf{c}}] = \begin{vmatrix} 3 & 2 & -1 \\ 5 & 3 & 2 \\ 4 & 1 & 3 \end{vmatrix} = 14 .$$

□ **Double vector product**, the vector product of three vectors $\vec{\mathbf{a}} \times \vec{\mathbf{b}} \times \vec{\mathbf{c}}$:

$$\vec{\mathbf{a}} = \begin{pmatrix} a_1 \\ a_2 \\ a_3 \end{pmatrix} , \quad \vec{\mathbf{b}} = \begin{pmatrix} b_1 \\ b_2 \\ b_3 \end{pmatrix} , \quad \vec{\mathbf{c}} = \begin{pmatrix} c_1 \\ c_2 \\ c_3 \end{pmatrix}$$

Multiplying the right-hand partial product yields:

$$\vec{\mathbf{b}} \times \vec{\mathbf{c}} = \begin{pmatrix} b_2c_3 & - & b_3c_2 \\ -b_1c_3 & + & b_3c_1 \\ b_1c_2 & - & b_2c_1 \end{pmatrix}$$

Multiplying the double product yields:

$$\vec{\mathbf{a}} \times (\vec{\mathbf{b}} \times \vec{\mathbf{c}}) = \begin{vmatrix} \vec{\mathbf{i}} & a_1 & b_2c_3 - b_3c_2 \\ \vec{\mathbf{j}} & a_2 & -b_1c_3 + b_3c_1 \\ \vec{\mathbf{k}} & a_3 & b_1c_2 - b_2c_1 \end{vmatrix}$$

▷ Is calculated like a determinant; however, it is not a scalar but a **vector**.

- **Expansion theorems of the vector product:**

$$\vec{\mathbf{a}} \times (\vec{\mathbf{b}} \times \vec{\mathbf{c}}) = (\vec{\mathbf{a}} \cdot \vec{\mathbf{c}})\vec{\mathbf{b}} - (\vec{\mathbf{a}} \cdot \vec{\mathbf{b}})\vec{\mathbf{c}} .$$

The vector $\vec{\mathbf{a}} \times (\vec{\mathbf{b}} \times \vec{\mathbf{c}})$ lies in the plane spanned by $\vec{\mathbf{b}}$ and $\vec{\mathbf{c}}$:

$$(\vec{\mathbf{a}} \times \vec{\mathbf{b}}) \times \vec{\mathbf{c}} = (\vec{\mathbf{a}} \cdot \vec{\mathbf{c}})\vec{\mathbf{b}} - (\vec{\mathbf{b}} \cdot \vec{\mathbf{c}})\vec{\mathbf{a}} .$$

- $\vec{\mathbf{a}}$, $\vec{\mathbf{b}}$, and $\vec{\mathbf{c}}$ are **linearly dependent** vectors.

Scalar product of two vector products (Lagrange's identity):

$$\begin{aligned}(\vec{\mathbf{a}} \times \vec{\mathbf{b}}) \cdot (\vec{\mathbf{c}} \times \vec{\mathbf{d}}) &= (\vec{\mathbf{a}} \cdot \vec{\mathbf{c}})(\vec{\mathbf{b}} \cdot \vec{\mathbf{d}}) - (\vec{\mathbf{a}} \cdot \vec{\mathbf{d}}) \cdot (\vec{\mathbf{b}} \cdot \vec{\mathbf{c}}) \\ &= (\vec{\mathbf{a}}, \vec{\mathbf{b}}, \vec{\mathbf{d}}) \cdot \vec{\mathbf{c}} - (\vec{\mathbf{a}}, \vec{\mathbf{b}}, \vec{\mathbf{c}}) \cdot \vec{\mathbf{d}} \\ &= (\vec{\mathbf{a}}, \vec{\mathbf{c}}, \vec{\mathbf{d}}) \cdot \vec{\mathbf{b}} - (\vec{\mathbf{b}}, \vec{\mathbf{c}}, \vec{\mathbf{d}}) \cdot \vec{\mathbf{a}} , \\ &= \begin{vmatrix} \vec{\mathbf{a}}\vec{\mathbf{c}} & \vec{\mathbf{a}}\vec{\mathbf{d}} \\ \vec{\mathbf{b}}\vec{\mathbf{c}} & \vec{\mathbf{b}}\vec{\mathbf{d}} \end{vmatrix} .\end{aligned}$$

with $(\vec{\mathbf{a}}, \vec{\mathbf{b}}, \vec{\mathbf{c}}) = [\vec{\mathbf{a}}\vec{\mathbf{b}}\vec{\mathbf{c}}] = (\vec{\mathbf{a}} \times \vec{\mathbf{b}}) \cdot \vec{\mathbf{c}}$.

Vector product of two vector products:

$$\begin{aligned}(\vec{\mathbf{a}} \times \vec{\mathbf{b}}) \times (\vec{\mathbf{c}} \times \vec{\mathbf{d}}) &= [\vec{\mathbf{a}}\vec{\mathbf{c}}\vec{\mathbf{d}}]\vec{\mathbf{b}} - [\vec{\mathbf{b}}\vec{\mathbf{c}}\vec{\mathbf{d}}]\vec{\mathbf{a}} \\ &= [\vec{\mathbf{a}}\vec{\mathbf{b}}\vec{\mathbf{d}}]\vec{\mathbf{c}} - [\vec{\mathbf{a}}\vec{\mathbf{b}}\vec{\mathbf{c}}]\vec{\mathbf{d}}\end{aligned}$$

- Four vectors in $\mathbb{R}^3$ are always linearly dependent.

$$\vec{\mathbf{a}}[\vec{\mathbf{b}}\vec{\mathbf{c}}\vec{\mathbf{d}}] - \vec{\mathbf{b}}[\vec{\mathbf{a}}\vec{\mathbf{c}}\vec{\mathbf{d}}] + \vec{\mathbf{c}}[\vec{\mathbf{a}}\vec{\mathbf{b}}\vec{\mathbf{d}}] - \vec{\mathbf{d}}[\vec{\mathbf{a}}\vec{\mathbf{b}}\vec{\mathbf{c}}] = 0$$

Notation in terms of a determinant:

$$\begin{vmatrix} \vec{\mathbf{a}} & a_x & a_y & a_z \\ \vec{\mathbf{b}} & b_x & b_y & b_z \\ \vec{\mathbf{c}} & c_x & c_y & c_z \\ \vec{\mathbf{d}} & d_x & d_y & d_z \end{vmatrix} = 0 .$$

Decomposition of the vector $\vec{\mathbf{d}}$ according to $\vec{\mathbf{a}}\vec{\mathbf{b}}$ and $\vec{\mathbf{c}}$:

$$\vec{\mathbf{d}} = \frac{\vec{\mathbf{a}}[\vec{\mathbf{d}}\vec{\mathbf{b}}\vec{\mathbf{c}}] + \vec{\mathbf{b}}[\vec{\mathbf{a}}\vec{\mathbf{d}}\vec{\mathbf{c}}] + \vec{\mathbf{c}}[\vec{\mathbf{a}}\vec{\mathbf{b}}\vec{\mathbf{d}}]}{[\vec{\mathbf{a}}\vec{\mathbf{b}}\vec{\mathbf{c}}]}$$

7

Coordinate systems

7.1 Coordinate systems in two dimensions

Two oriented **axes** subdivide the plane into four **quadrants**.
▷ This applies also to oblique coordinate systems.
Origin, **direction** (+/−), and **scale** must be fixed.

7.1.1 Cartesian coordinates

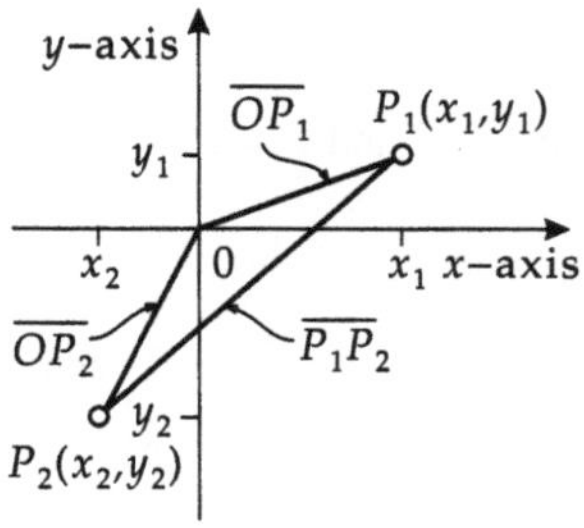

Coordinate axes: Usually two **orthogonal axes** (perpendicular to each other).
Origin, pole, or **zero point**: Point of intersection O of the two axes.
Coordinates x and y of P.
Abscissa: x-axis, the horizontal axis.
Ordinate: y-axis, the vertical axis.
▷ The point P with the coordinates x and y is written as $P(x, y)$. The value of the abscissa is always given first.

Distance between point $P(x, y)$ and the origin (Pythagorean theorem):

$$\overline{OP} = \sqrt{x^2 + y^2} .$$

Distance between two points $P_1(x_1, y_1)$ and $P_2(x_2, y_2)$:

$$d = \overline{P_1 P_2} = \sqrt{(x_1 - x_2)^2 + (y_1 - y_2)^2} .$$

7.1.2 Polar coordinates

Polar coordinates specify the position of a point $P(r, \varphi)$.

r: **Distance coordinate, distance**, or **radius**: Length of the radius vector from O to P.

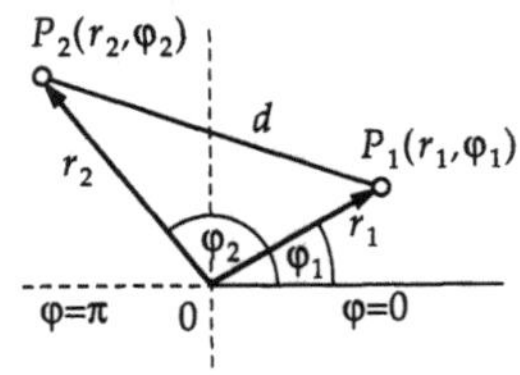

φ: **Angular coordinate, directional angle, polar angle** or **argument**: Angle between the polar axis (abscissa) and the segment $\overline{PO}$.

□ x-axis (abscissa).

Polar axis: A ray starting from point O (**pole**).

Angle φ: Positive for counterclockwise rotations; negative for clockwise rotations.

▷ Because in plane geometry, an angle (**in arc measure**) is defined only up to multiples of 2π, only the **principal value** lying in the interval $0 \leq \varphi < 2\pi$ is usually given (in degrees $0° \leq \varphi < 360°$).

Distance d between two points $P_1(r_1, \varphi_1)$, $P_2(r_2, \varphi_2)$ in polar coordinates:

$$d = \sqrt{r_1^2 + r_2^2 - 2r_1r_2\cos(\varphi_2 - \varphi_1)}\ .$$

7.1.3 Conversions between two-dimensional coordinate systems

From polar coordinates (r, φ) to Cartesian coordinates (x, y):

$$x = r \cdot \cos\varphi\ , \qquad y = r \cdot \sin\varphi\ .$$

From Cartesian coordinates (x, y) to polar coordinates (r, φ):

$$r = \sqrt{x^2 + y^2}\ , \qquad \varphi = \begin{cases} \arctan\dfrac{b}{a}, & \text{if } a > 0 \\ \dfrac{\pi}{2}, & \text{if } a = 0,\ b > 0 \\ \pi + \arctan\dfrac{b}{a}, & \text{if } a < 0 \\ \dfrac{3\pi}{2}, & \text{if } a = 0,\ b < 0 \end{cases}$$

□ For $r = 2$ and $\varphi = 30°$, the Cartesian coordinates can be determined as follows:

$$\sin 30° = \frac{1}{2}\ , \quad \cos 30° = \frac{\sqrt{3}}{2}$$

$$\to x = 2 \cdot \cos 30° = \sqrt{3}\ , \quad y = 2 \cdot \sin 30° = 1\ .$$

□ For $x = 3$, $y = 4$, the polar coordinates are

$$r = \sqrt{3^2 + 4^2} = \sqrt{25} = 5$$

$$\varphi = \arctan 4/3 = 53°\ .$$

7.2 Two-dimensional coordinate transformation

Coordinate transformation: The transition from one coordinate system to another in the same representation. We differentiate:

- **Parallel displacement (translation)**
- **Rotation**
- **Reflection**
- **Scaling**

▷ Important applications in CAD/CAM, computer graphics, rotational motions, oscillations.

One distinguishes between **active** and **passive** transformations.

$$\textbf{Active coordinate transformation:}\quad P \to P', \quad K = K'$$
$$\textbf{Passive coordinate transformation:}\quad P = P', \quad K \to K',$$

Where K, P are the points (or coordinate systems) before the coordinate transformation, and K', P' are those after.

7.2.1 Parallel displacement (translation)

Parallel displacement can be represented best in Cartesian coordinates.
Active parallel displacement of the vector $\vec{\mathbf{r}} = (x, y)$ by the **displacement vector** $\vec{\mathbf{T}} = (a, b)$ creates the new vector $\vec{\mathbf{r}}' = (x', y')$ with

$$x' = x + a, \qquad y' = y + b$$

in the **same** coordinate system.
Passive parallel displacement: The displacement vector (a, b) moves the origin O of the coordinate system from K to $\vec{\mathbf{T}} = (a, b)$, creating the new coordinate system K'. Since $P(x, y) = P'(x', y')$, the coordinates are transformed according to:

$$x' = x - a, \qquad y' = y - b$$

Passive parallel displacement by the displacement vector $\vec{\mathbf{T}} = (a, b) = (2, 4)$.
The coordinate system (x, y) is shifted and point $P = (8, 8)$ is represented in the new coordinate system (x', y') by the position vector $\vec{\mathbf{p}}$.

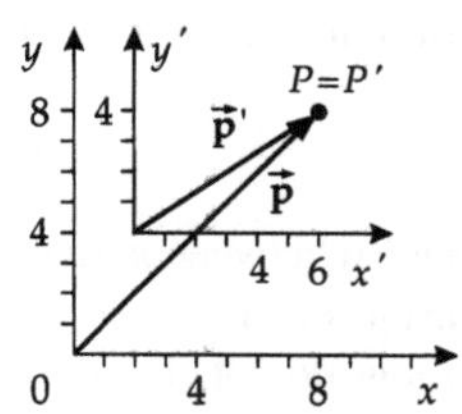

$$\vec{\mathbf{p}} = \begin{pmatrix} 8 \\ 8 \end{pmatrix}_{\{x,y\}}$$

$$\vec{\mathbf{p}}' = \begin{pmatrix} 8-2 \\ 8-4 \end{pmatrix}_{\{x',y'\}} = \begin{pmatrix} 6 \\ 4 \end{pmatrix}_{\{x',y'\}}$$

The coordinate system is shifted by $a = 2, b = 4$.
Point coordinates change by $-2, -4$ in the new coordinate system.
Active parallel displacement:
Point P (position vector $\vec{\mathbf{p}}$) is shifted, but the coordinate system of point remains unchanged.

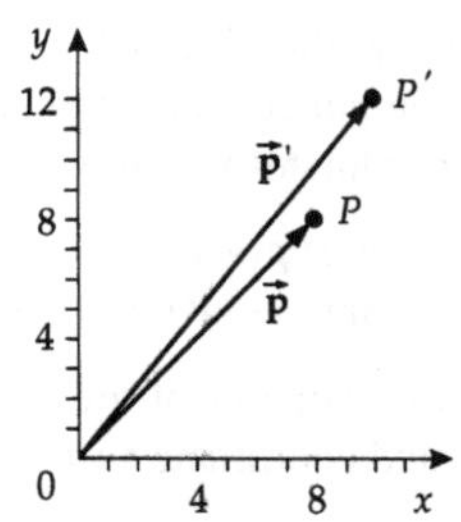

Translation of P by $a = 2, b = 4$, $P \to P'$ in the same system.

$$\vec{\mathbf{p}} = \begin{pmatrix} 8 \\ 8 \end{pmatrix}$$

$$\vec{\mathbf{p}}' = \begin{pmatrix} 8+2 \\ 8+4 \end{pmatrix} = \begin{pmatrix} 10 \\ 12 \end{pmatrix}$$

Case 1: The coordinate system is translated by $\vec{\mathbf{T}} = (2, 4)$, and the straight line is represented in the new system (x', y'):

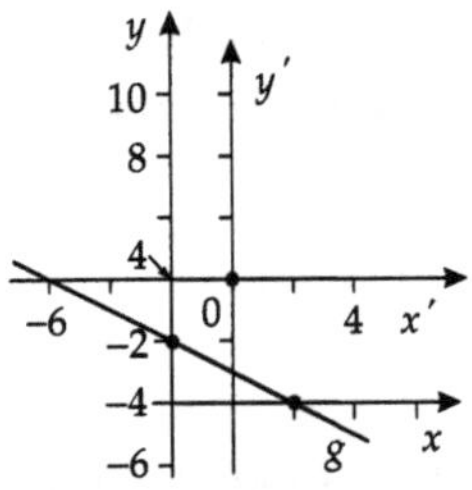

$$g_{\{x,y\}} : y = -\frac{1}{2}x + 2$$

$$g_{\{x',y'\}} : y' = -\frac{1}{2}x' - 3$$

The straight line remains, and the coordinate system has changed.

Case 2: The straight line is shifted by $\vec{\mathbf{T}} = (2, 4)$:

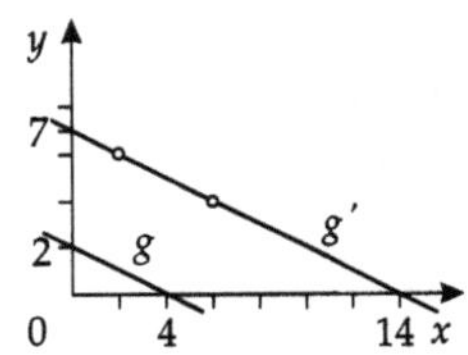

$$g : y = -\frac{1}{2}x + 2$$

$$g' : y' = -\frac{1}{2}x' + 7$$

The straight line has been shifted, and the coordinate system remains.

7.2.2 Rotation

The **rotation** of a vector $\vec{\mathbf{r}}$ can be represented best in polar coordinates:

$$x = r \cdot \cos\varphi\,, \qquad y = r \cdot \sin\varphi\,.$$

We distinguish between **active** and **passive** rotations (transformations), as in the case of parallel displacements.

▷ Rotations do not alter the length of vectors!

In an **active rotation**, the vector $\vec{\mathbf{r}}$ is rotated clockwise about the rotation angle β. The angle of direction of $\vec{\mathbf{r}}$ is $\varphi' = \varphi - \beta$.

A passive rotation about angle β corresponds to a counterclockwise rotation of the coordinate system K into K'.

Active: Components of the rotated vector $\vec{\mathbf{r}}$ in the coordinate system K:

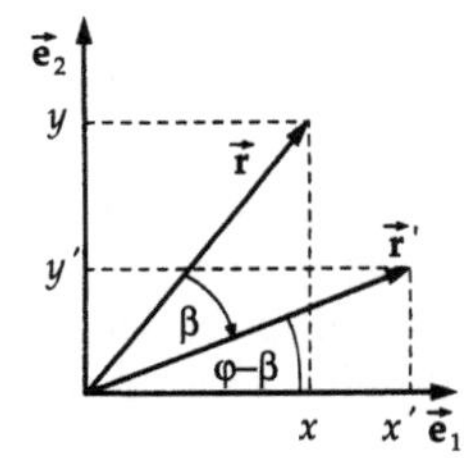

$$x' = r \cdot \cos(\varphi - \beta) = r \cdot \cos\varphi'\,,$$
$$y' = r \cdot \sin(\varphi - \beta) = r \cdot \sin\varphi'\,.$$

Passive: Components of the vector $\vec{\mathbf{r}}' = \vec{\mathbf{r}}$ in the rotated coordinate system K':

$$x' = x \cdot \cos\beta + y \cdot \sin\beta$$

$$y' = -x \cdot \sin\beta + y \cdot \cos\beta$$

or the converse:

$$x = x' \cdot \cos\beta - y' \cdot \sin\beta$$
$$y = x' \cdot \sin\beta + y' \cdot \cos\beta$$

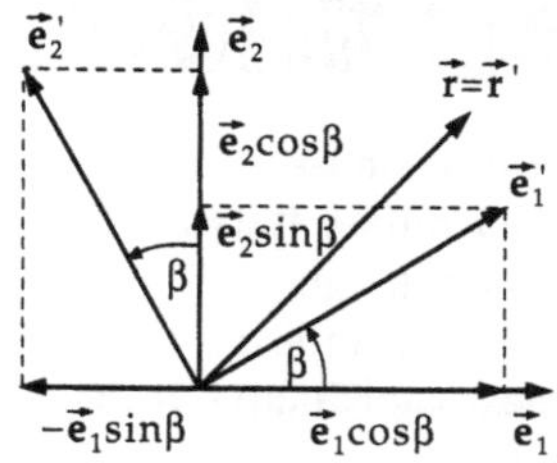

□ Rotation of the vector $\vec{r}(r, \varphi) = (1, 0°)$ by $\beta = 45°$:
For an **active** rotation we obtain, with $\vec{r}(r, \varphi) = (1, 0°)$:

$$x' = 1 \cdot \cos(-45°) = 0.707$$
$$y' = 1 \cdot \sin(-45°) = -0.707$$

□ **Passive** rotation:

$$x' = 1 \cdot 0.707 + 0 = 0.707$$
$$y' = -1 \cdot 0.707 + 0 = -0.707$$

▷ For matrix and vector notation, see also Section 7.5, application in computer graphics.

7.2.3 Reflection

- **Reflection** on the x-axis:
The x-components remain, but the signs of the y-components are reversed. Transformation equations:

$$x' = x \quad \text{and} \quad y' = -y .$$

- **Reflection** on the y-axis:
The sign of the x-components are reversed, but the y-components remain.
Transformation equation:

$$x' = -x \quad \text{and} \quad y' = y .$$

- **Reflection** at the angle-bisector of the first quadrant:
The x- and y-components are interchanged. Transformation equations:

$$x' = y \quad \text{and} \quad y' = x .$$

- **Reflection on the origin**:
The x- and y-components change their signs. Transformation equations:

$$x' = -x \quad \text{and} \quad y' = -y .$$

7.2.4 Scaling

Every individual component of the position vector is stretched or compounded by a scaling factor.

□ In Cartesian coordinates, the vector $\vec{r} = (x, y)$ is scaled to $\vec{r}' = (x', y')$ by the factors S_x, S_y through:

$$\vec{r}' = \begin{pmatrix} x' \\ y' \end{pmatrix} = \begin{pmatrix} S_x \cdot x \\ S_y \cdot y \end{pmatrix} = \begin{pmatrix} S_x & 0 \\ 0 & S_y \end{pmatrix} \cdot \begin{pmatrix} x \\ y \end{pmatrix}$$

▷ Here the successive application of different transformations leads to complex calculation steps. In computer graphics, this problem is considerably simplified by means of homogeneous coordinates.

7.3 Coordinate systems in three dimensions

Origin, **direction** (+/−), and **scale** must be fixed.
Linear coordinate systems: In three-dimensional space determined by three arbitrary, non-coplanar straight lines.

□ Cartesian coordinates: The x-, y-, z-axes are pairwise mutually perpendicular.

Curvilinear coordinates: In three-dimensional space determined by three one-parameter pencils of surfaces.

□ Cylindrical coordinates, spherical coordinates.

7.3.1 *Cartesian coordinates*

Three pairwise **orthogonal** (perpendicular) **coordinate axes**: **abscissa**, x-axis; **ordinate**, y-axis; and **applicate**, z-axis. The three axes form a **right-handed system** in that order.
Distance between point $P(x, y, z)$ and origin O:

$$\overline{OP} = \sqrt{x^2 + y^2 + z^2}\,.$$

Shortest **distance** between two points $P_1(x_1, y_1, z_1)$ and $P_2(x_2, y_2, z_2)$:

$$d = \overline{P_1P_2} = \sqrt{(x_1 - x_2)^2 + (y_1 - y_2)^2 + (z_1 - z_2)^2}\,.$$

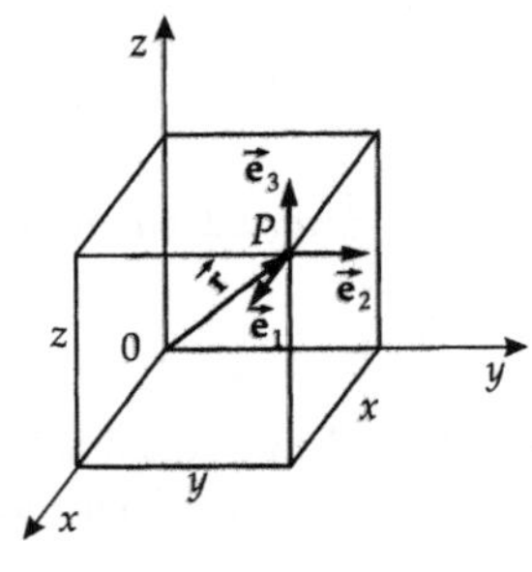

7.3.2 *Cylindrical coordinates*

- **Cylindrical coordinates** (ρ, ϕ, z): Describe the position of a point by the **radius of the cylinder** $\rho \geq 0$, distance between the point and the z-axis.

▷ ρ is *not* the distance from the origin!

ϕ ; $0 \leq \phi < 2\pi$:
Azimuth: Angle between the positive x-axis and projection of the segment ρ, onto the (x, y)-plane.

▷ For $z = 0$, ρ and ϕ are the polar coordinates in the x, y-plane.

z ; $-\infty < z < \infty$:
Applicate: Oriented distance from the point to the $(x$-$y)$ plane.

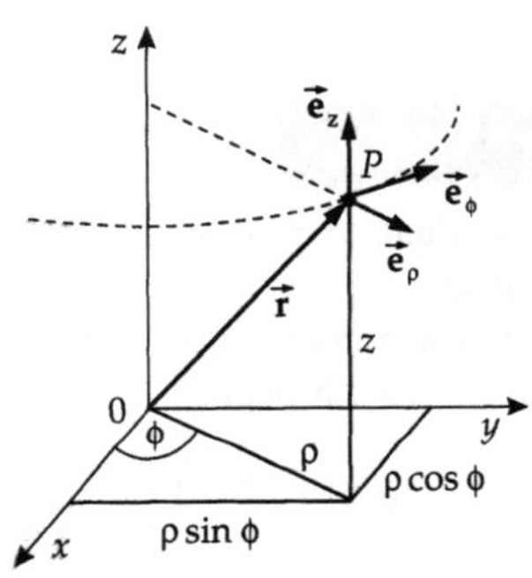

▷ Cylindrical coordinates are of importance for three-dimensional problems with **cylindrical symmetry**, i.e., with rotational symmetry about the z-axis.

● **Coordinate surfaces**: Surfaces defined in space are determined by fixing one coordinate.

$\rho =$ const.: A cylindrical surface is formed with the z-axis as the axis of symmetry.
$\phi =$ const.: A half-plane is formed, with the z- axis as the bounding straight line.
$z =$ const.: A plane parallel to the $(x$-$y)$ plane is formed.

7.3.3 *Spherical coordinates*

● **Spherical coordinates** (r, ϑ, φ): Describe the position of a point $P(r, \vartheta, \varphi)$ by means of
$r = \overline{OP}$; $r \geq 0$:
Distance between the point and the origin.

▷ Note that there is a difference between the ρ in cylindrical coordinates and the r in spherical coordinates.

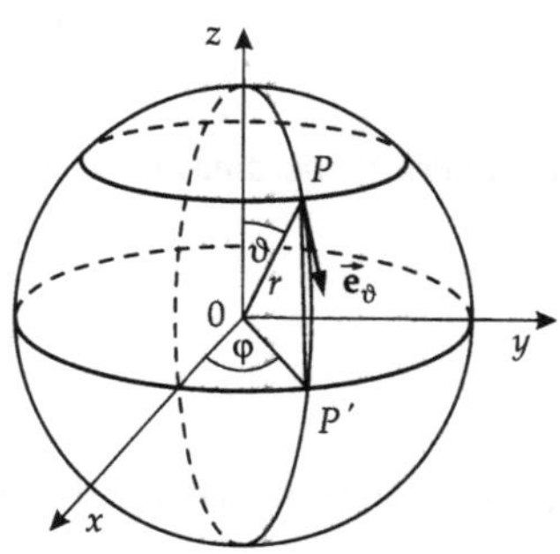

φ ; $0 \leq \varphi < 2\pi$:
Angle between the positive x-axis and the projection $\overline{OP'}$ onto the (x, y) plane (also called **geographical longitude**).

▷ Note the difference from the following latitude coordinate described below!

ϑ ; $0 \leq \vartheta \leq \pi$:
Angle between the polar axis (positive z-axis) and $\overline{OP}$ (**latitude coordinate**, $\vartheta =$ 90°: **equatorial plane**). This latitude coordinate must **not** be confused with the latitude coordinate in geography, which is called φ, where $\varphi = 0°$ is the equator!

● **Coordinate surfaces**: By fixing one coordinate, surfaces in space are determined.

$r =$ const.: a sphere about coordinate origin O formed.
$\varphi =$ const.: a half-plane bounded by the z-axis is created.
$\vartheta =$ const.: a cone is formed with the z-axis as the axis of symmetry.

7.3.4 *Conversions between three-dimensional coordinate systems*

From cylindrical coordinates (ρ, ϕ, z') to Cartesian coordinates (x, y, z):

$$x = \rho \cdot \cos\phi \quad , \quad y = \rho \cdot \sin\phi \quad , \quad z = z' \, .$$

From Cartesian coordinates (x, y, z) to cylindrical coordinates (ρ, ϕ, z'):

$$\rho = \sqrt{x^2 + y^2} \quad , \quad \phi = \arccos\frac{x}{\sqrt{x^2 + y^2}} = \arctan\frac{y}{x} \quad , \quad z' = z \, . \qquad (7.1)$$

▷ Here z' and z are distinguished for clarity.

From spherical coordinates (r, ϑ, φ) to Cartesian coordinates (x, y, z):

$$\begin{aligned} x &= r \cdot \sin\vartheta \cdot \cos\varphi \qquad & 0 \leq \varphi < 2\pi \\ y &= r \cdot \sin\vartheta \cdot \sin\varphi \qquad & 0 \leq \vartheta \leq \pi \\ z &= r \cdot \cos\vartheta \, . \end{aligned}$$

From Cartesian coordinates (x, y, z) to spherical coordinates (r, ϑ, φ):

$$r = \sqrt{x^2 + y^2 + z^2}$$

$$\cos\vartheta = \frac{z}{r} = \frac{z}{\sqrt{x^2 + y^2 + z^2}}$$

$$\cos\varphi = \frac{x}{\sqrt{x^2 + y^2}} \quad ; \quad \sin\varphi = \frac{y}{\sqrt{x^2 + y^2}} \quad ; \quad \tan\varphi = \frac{y}{x} \,.$$

From spherical coordinates (r, ϑ, φ) to cylindrical coordinates (ρ, ϕ, z') :

$$\rho = r \sin\vartheta \,,$$

$$\phi = \varphi \qquad (0 \le \varphi < 2\pi) \,,$$

$$z' = r\cos\vartheta \qquad (0 \le \vartheta \le \pi) \,.$$

From cylindrical coordinates (ρ, ϕ, z') to spherical coordinates (r, ϑ, φ):

$$r = \sqrt{\rho^2 + z'^2}$$

$$\vartheta = \arctan\frac{\rho}{z'}$$

$$\varphi = \phi \,.$$

▷ Here φ and ϕ are distinguished for clarity.

7.4 Coordinate transformation in three dimensions

7.4.1 Parallel displacement (translation)

An **active parallel displacement** of the Cartesian vector $\vec{\mathbf{r}} = (x, y, z)$ to $\vec{\mathbf{r}}' = (x', y', z')$ is achieved by a component-wise addition of the **displacement vector** $\vec{\mathbf{T}} = (a, b, c)$:

$$x' = x + a \,, \qquad y' = y + b \,, \qquad z' = z + c \,.$$

Passive parallel displacement is achieved by the opposite displacement of the coordinate system K to K'. The translated coordinate system with the axes $\vec{\mathbf{e}}'_x, \vec{\mathbf{e}}'_y, \vec{\mathbf{e}}'_z$, parallel to the axes $\vec{\mathbf{e}}_x, \vec{\mathbf{e}}_y, \vec{\mathbf{e}}_z$ of K is obtained by a component-wise subtraction,

$$x' = x - a \,, \qquad y' = y - b \,, \qquad z' = z - c \,.$$

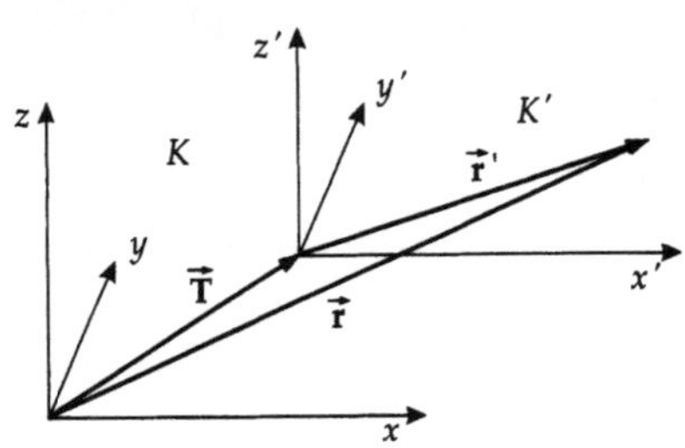

Three-dimensional parallel displacement

▷ All angles and distances are preserved.

7.4.2 Rotation

- In three dimensions the result of the rotations depends on the order of the partial rotations about the x-, y-, z-axis (see Section 7.5, application in computer graphics).

Passive rotation transforms the old coordinates (x, y, z) into the new coordinates (x', y', z'):

$$\begin{aligned}\alpha_x, \beta_x, \gamma_x&: \text{angle between the } x' \text{ axis and the } x, y, z \text{ axes.}\\ \alpha_y, \beta_y, \gamma_y&: \text{angle between the } y' \text{ axis and the } x, y, z \text{ axes.}\\ \alpha_z, \beta_z, \gamma_z&: \text{angle between the } z' \text{ axis and the } x, y, z \text{ axes.}\end{aligned}$$

Transformation equations apply only to the given order of rotations about the various axes!

$$\begin{aligned}x &= x' \cos\alpha_x + y' \cos\alpha_y + z' \cos\alpha_z & x' &= x \cos\alpha_x + y \cos\beta_x + z \cos\gamma_x\\ y &= x' \cos\beta_x + y' \cos\beta_y + z' \cos\beta_z \quad \text{or} & y' &= x \cos\alpha_y + y \cos\beta_y + z \cos\gamma_y\\ z &= x' \cos\gamma_x + y' \cos\gamma_y + z' \cos\gamma_z & z' &= x \cos\alpha_z + y \cos\beta_z + z \cos\gamma_z\,.\end{aligned}$$

Relations between the **direction cosines** of the new axes:

$$\begin{aligned}\cos^2\alpha_x + \cos^2\beta_x + \cos^2\gamma_x &= 1\\ \cos^2\alpha_x + \cos^2\alpha_y + \cos^2\alpha_z &= 1\end{aligned}$$

or

$$\begin{aligned}\cos\alpha_x\cos\alpha_y + \cos\beta_x\cos\beta_y + \cos\gamma_x\cos\gamma_y &= 0\\ \cos\alpha_x\cos\beta_x + \cos\alpha_y\cos\beta_y + \cos\alpha_z\cos\beta_z &= 0\end{aligned}$$

or

$$D = \begin{vmatrix} \cos\alpha_x & \cos\beta_x & \cos\gamma_x\\ \cos\alpha_y & \cos\beta_y & \cos\gamma_y\\ \cos\alpha_z & \cos\beta_z & \cos\gamma_z \end{vmatrix} = 1$$

▷ In practice, the transformation equations are often given by means of matrix representation.

7.5 Application in computer graphics

Computer graphics: Methods to generate, process and analyze images.
Definition (ISO = International Standard Organization): Computer graphics consists of methods for the conversion of data between computers and graphic input and output devices.

- Main applications of computer graphics:
  - **Computer Aided Design (CAD)**,
  - **presentation graphics and technical documentation**,
  - **cartography**,
  - **animation** and **photo-realistic representations**, **virtual reality**.

7.6 Transformations

Two- and **three-dimensional transformations** are important mathematical foundations for interactive graphic programs.

7.6.1 Object representation and object description

Transformation pipeline: Successively applied transformations to represent graphic objects on an **output device** (screen, plotter, printer).

- **Transformation of objects**: Graphic objects are moved (translation), enlarged or reduced (**scaling**), turned (**rotation**), or mirrored (**reflection**) in a world coordinate system that depends on the problem.
- **Transformation of coordinate systems**: Transformation from the x, y, z world coordinate system to a x_d, y_d, z_d **viewing coordinate system**. Representation of three-dimensional scenarios according to various angles of view and sites.

□ **Side views, front views, spatial views.**

- **Projection transformation**: Mapping of three-dimensional objects onto a two-dimensional **projection plane** corresponding to the screen or the drawing plane.

▷ To simplify projection transformation, the coordinate system is generally displayed in such a way that the z_d axis is perpendicular to the view plane, which is the x_p, y_p-projection plane.

□ – Producing perspective views,
– producing scaled drawings,
– producing orthogonal and oblique parallel projections.

- **Workstation transformation**: Conversion from device-independent x_p, y_p image coordinates to device coordinates.

□ For raster displays, these are the addresses of **pixels**.

- **Window-viewport transformation**: Mapping of a usually rectangular or cuboidal section of the world coordinate space (viewport) onto a limited area (window) in the device's coordinate system.

This is necessary if the full screen is not available for the picture, or if several views must be represented in parts at the same time.

Graphics standard GKS-3-D (GKS = Graphic Kernel System) is the three-dimensional extension of GKS-: At the beginning of the viewing pipeline, the world coordinates are converted into normalized device coordinates. In this way, a cuboid in the world coordinate system is mapped onto the unit cube (box-to-box mapping). Thus, every coordinate value lies in the interval [0, 1].

Both in **GKS** and in **PHIGS** (**P**rogrammer's **H**ierarchical **I**nteractive **G**raphics **S**tandard), the projected coordinates are normalized (normalized projection coordinates). Mapping on a two-dimensioned image plane is unnecessary, when high-performance graphic workstations are available that can process a third coordinate (3-D graphic workstations).

If only parts of the world (windows) have to be represented, or if the representation is restricted to a portion (viewport), then **clipping** takes place in the viewing coordinate system or in the image plane.

Clipping: Suppression of objects and cutting off those parts of the object that do not appear in the window, or cutting off those parts of the image that do not appear in the viewport.

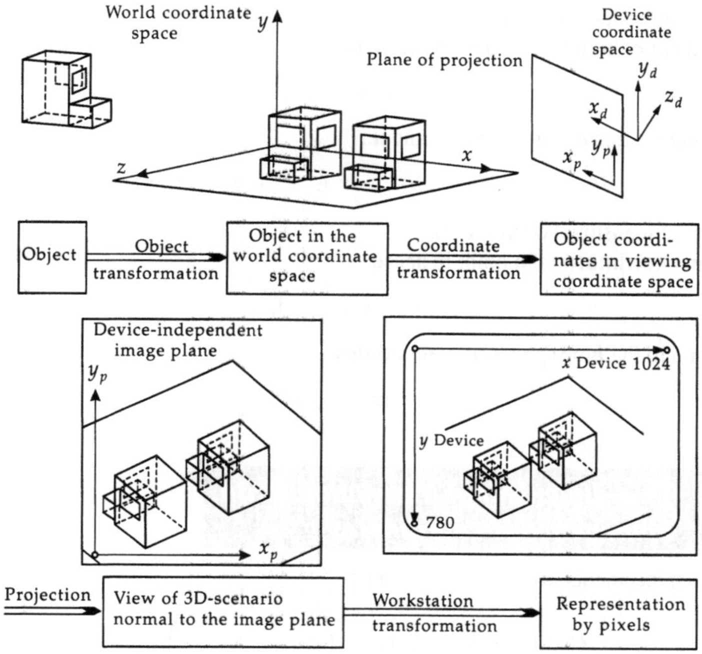

7.6.2 Homogeneous coordinates

Homogeneous coordinates: Means for the standardized description of **geometrical transformations** by **matrix multiplication**. Geometric transformations and total scaling are combined in a 4 × 4 matrix.

Rotation Scaling	Translation
Perspective Transformation	Total Scaling

Advantages:

Standardized treatment of all transformations.

Complex transformations are composed from simple transformations by means of matrix multiplication.

Instead of successively carrying out several transformations to all points of a three-dimensional scenario, a total transformation matrix can be calculated only once, which is multiplied by the homogeneous coordinates of the points.

Simple inversion of the transformation by matrix inversion.

In a hierarchical order of objects, the positions of subsidiary units with respect to the primary units can be stored by the corresponding transformation matrices (position dependence of subunits and single components, kinematic dependence of arm and gripper motions of industrial robots).

Support of operations with 4 × 4 matrices by graphics standards (PHIGS).

Extremely fast, hardware-supported matrix operations in high-power graphic workstations.

● Two-dimensional coordinate system:
Cartesian coordinates of a point in the plane:

$$P = (x, y) \in \mathbb{R}^2 .$$

Homogeneous coordinates are not unique:

$$P_H = (h \cdot x, h \cdot y, h) , \quad h \in \mathbb{R},\ h \neq 0 \ \text{arbitrary}.$$

● Three-dimensional coordinates:
Cartesian coordinates of a point in space:

$$P = (x, y, z) \in \mathbb{R}^3.$$

Correspond to the **homogeneous coordinates**:

$$P_H = (h \cdot x, h \cdot y, h \cdot z, h).$$

7.6.3 *Two-dimensional translations with homogeneous coordinates*

T_x, T_y: Displacements in x and y directions:

$$\text{Cartesian coordinates:} \quad \begin{pmatrix} x & + & T_x \\ y & + & T_y \end{pmatrix}$$

$$\text{homogeneous coordinates:} \quad \begin{pmatrix} 1 & 0 & T_x \\ 0 & 1 & T_y \\ 0 & 0 & 1 \end{pmatrix} \cdot \begin{pmatrix} x \\ y \\ 1 \end{pmatrix}$$

7.6.4 *Two-dimensional scaling with homogeneous coordinates*

S_x, S_y: Components of the scaling.

$$\text{Cartesian coordinates:} \quad \begin{pmatrix} S_x & \cdot & x \\ S_y & \cdot & y \end{pmatrix}$$

$$\text{Homogeneous coordinates:} \quad \begin{pmatrix} S_x & 0 & 0 \\ 0 & S_y & 0 \\ 0 & 0 & 1 \end{pmatrix} \cdot \begin{pmatrix} x \\ y \\ 1 \end{pmatrix}$$

□ Scaling of an object, consisting of two points, with respect to the x-axis by factor of $S_x = 2$, and with respect to the y-axis by the factor of $S_y = 0.5$.

$$P_1 : \begin{pmatrix} 1 \\ 1 \end{pmatrix} \to P_1' : \begin{pmatrix} 2 \cdot 1 \\ 0.5 \cdot 1 \end{pmatrix} = \begin{pmatrix} 2 \\ 0.5 \end{pmatrix} ;$$

$$P_2 : \begin{pmatrix} 3 \\ 1 \end{pmatrix} \to P_2' : \begin{pmatrix} 2 \cdot 3 \\ 0.5 \cdot 1 \end{pmatrix} = \begin{pmatrix} 6 \\ 0.5 \end{pmatrix} .$$

▷ Algorithm for the scaling of a square:

```
BEGIN
   scaling factor Sx = 2 ;
   scaling factor Sy = 0.5 ;
   set up the scaling matrix
```

```
S = ( 2   0    0 )
    ( 0   0.5  0 )  ;
    ( 0   0    1 )
FOR i = 1 TO 4 DO
BEGIN
  determine the (x, y)-coordinates of the i-th vertex;
  perform the scaling: multiplication of S by the
  vector of the homogeneous coordinates (x, y, 1) ;
  store the scaled coordinates as the vertex
  coordinates;
END ;
plot the scaled object with vertices P'_i;
END .
```

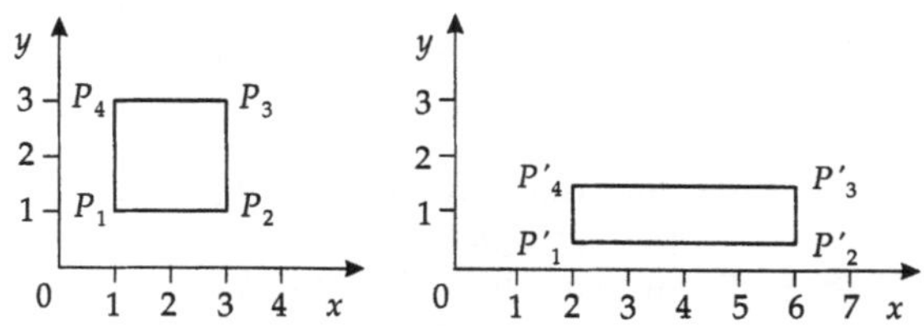

The scaling of an object changes its position.

7.6.5 Three-dimensional translation with homogeneous coordinates

Matrix representation of a translation by the vector (T_x, T_y, T_z):

$$\begin{pmatrix} 1 & 0 & 0 & T_x \\ 0 & 1 & 0 & T_y \\ 0 & 0 & 1 & T_z \\ 0 & 0 & 0 & 1 \end{pmatrix} \cdot \begin{pmatrix} x \\ y \\ z \\ 1 \end{pmatrix} \rightarrow \begin{pmatrix} x + T_x \\ y + T_y \\ z + T_z \\ 1 \end{pmatrix} .$$

▷ The **inverse transformation** (displacement in the opposite direction) is accomplished by changing the sign of the matrix elements T_x, T_y, and T_z.

7.6.6 Three-dimensional scaling with homogeneous coordinates

Matrix representation of points in a **scaling** with factors S_x, S_y, S_z:

$$\begin{pmatrix} S_x & 0 & 0 & 0 \\ 0 & S_y & 0 & 0 \\ 0 & 0 & S_z & 0 \\ 0 & 0 & 0 & 1 \end{pmatrix} \cdot \begin{pmatrix} x \\ y \\ z \\ 1 \end{pmatrix} \rightarrow \begin{pmatrix} S_x \cdot x \\ S_y \cdot y \\ S_z \cdot z \\ 1 \end{pmatrix}$$

▷ If all three axes must be scaled by the same factor, then this **total scaling** can also be described by the following matrix multiplication:

$$\begin{pmatrix} 1 & 0 & 0 & 0 \\ 0 & 1 & 0 & 0 \\ 0 & 0 & 1 & 0 \\ 0 & 0 & 0 & 1/S_g \end{pmatrix} \cdot \begin{pmatrix} x \\ y \\ z \\ 1 \end{pmatrix} \rightarrow \begin{pmatrix} x \\ y \\ z \\ 1/S_g \end{pmatrix}$$

Division by the fourth homogeneous coordinate $1/S_g$ yields the Cartesian coordinates.

▷ **Inverse transformation** is achieved by the reciprocal values of the diagonal elements.

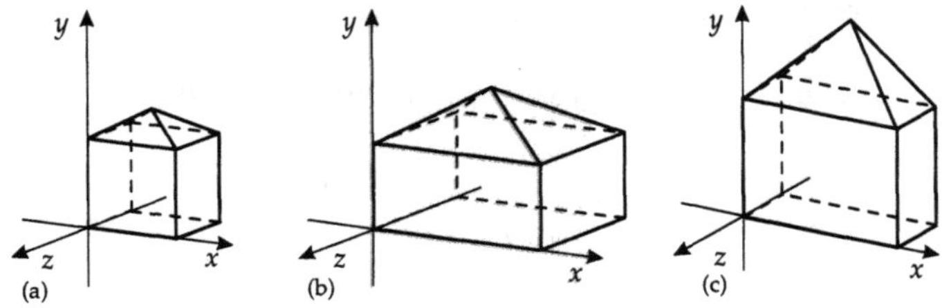

Three-dimensional scaling of an object:
(a) Original object.
(b) Scaling with respect to the x-axis by $S_x = 2$.
(c) Scaling of (b) with respect to the y-axis by $S_y = 1.5$, and with respect to the z-axis by $S_z = 0.5$.

7.6.7 Three-dimensional rotation of points with homogeneous coordinates

The **angles of rotation** about the x-, y-, z-axis are denoted by $\alpha_x, \alpha_y, \alpha_z$.

▷ In a successive application of rotations, note that matrix multiplication is **not** commutative.

● Rotation of a point about the x-axis:

$$\begin{pmatrix} 1 & 0 & 0 & 0 \\ 0 & \cos(\alpha_x) & -\sin(\alpha_x) & 0 \\ 0 & \sin(\alpha_x) & \cos(\alpha_x) & 0 \\ 0 & 0 & 0 & 1 \end{pmatrix} \cdot \begin{pmatrix} x \\ y \\ z \\ 1 \end{pmatrix} = \begin{pmatrix} x \\ y \cdot \cos(\alpha_x) - z \cdot \sin(\alpha_x) \\ y \cdot \sin(\alpha_x) + z \cdot \cos(\alpha_x) \\ 1 \end{pmatrix}$$

● Rotation of a point about the y-axis:

$$\begin{pmatrix} \cos(\alpha_y) & 0 & \sin(\alpha_y) & 0 \\ 0 & 1 & 0 & 0 \\ -\sin(\alpha_y) & 0 & \cos(\alpha_y) & 0 \\ 0 & 0 & 0 & 1 \end{pmatrix} \cdot \begin{pmatrix} x \\ y \\ z \\ 1 \end{pmatrix} = \begin{pmatrix} x \cdot \cos(\alpha_y) + z \cdot \sin(\alpha_y) \\ y \\ -x \cdot \sin(\alpha_y) + z \cdot \cos(\alpha_y) \\ 1 \end{pmatrix}$$

● Rotation of a point about the z-axis:

$$\begin{pmatrix} \cos(\alpha_z) & -\sin(\alpha_z) & 0 & 0 \\ \sin(\alpha_z) & \cos(\alpha_z) & 0 & 0 \\ 0 & 0 & 1 & 0 \\ 0 & 0 & 0 & 1 \end{pmatrix} \cdot \begin{pmatrix} x \\ y \\ z \\ 1 \end{pmatrix} = \begin{pmatrix} x \cdot \cos(\alpha_z) - y \cdot \sin(\alpha_z) \\ y \cdot \cos(\alpha_z) + x \cdot \sin(\alpha_z) \\ z \\ 1 \end{pmatrix}$$

▷ The **inverse** of an **orthogonal transformation** is equal to the **transposed transformation matrix**.

□ **Inverse rotation**: The angles are replaced by their inverse values:

$$\mathbf{R}^{-1}(\alpha) = \mathbf{R}(-\alpha)$$

▷ This generally applies only to rotations about one principal axis. For an inverse rotation the order of the rotations usually must be reversed:

$$(\mathbf{R}_1\mathbf{R}_2\mathbf{R}_3)^{-1} = \mathbf{R}_3^{-1}\mathbf{R}_2^{-1}\mathbf{R}_1^{-1}$$

7.6.8 Positioning of an object in space

▷ An object can be moved from an original position to a required position along an infinite number of paths.

● Approach in computer graphics: The object is first rotated about the principal axes, and then translated. The order of principal axes can be chosen to facilitate different ways of moving the object.

□ Positioning or moving an object in space implies that every vertex is subject to transformation. With a detailed modeling of three-dimensional objects for example, a machine with tools and a workpiece, there could be thousands of vectors.

▷ If only the final result matters, i.e., the oriented position in space, but not the intermediate steps, then the total transformation is calculated from the individual transformations by matrix multiplication, and only then is the transformation applied to the vectors.

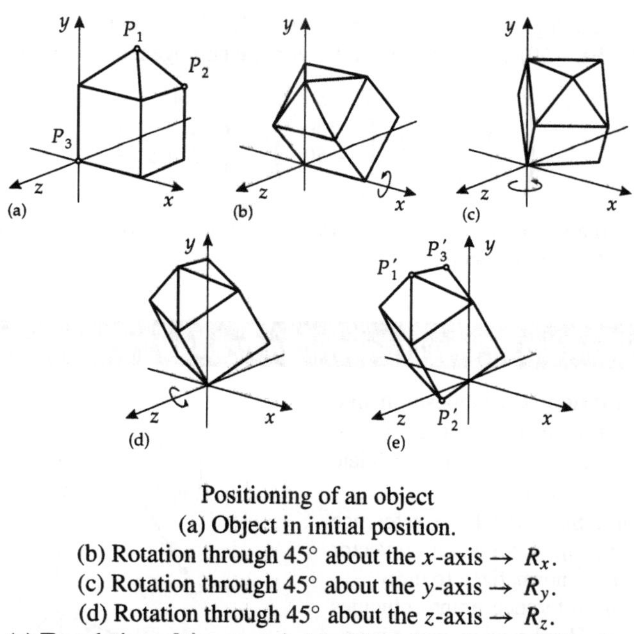

Positioning of an object
(a) Object in initial position.
(b) Rotation through 45° about the x-axis → R_x.
(c) Rotation through 45° about the y-axis → R_y.
(d) Rotation through 45° about the z-axis → R_z.
(e) Translation of the rotated object in x-direction by 100 units, in y-direction by 50 units, and in z-direction by 200 units.

- Calculation of the transformation matrix from rotations and translation:

$$R = \left(\begin{array}{c|c} \text{Rotation} & \text{Translation} \\ \hline & \\ 0\ 0\ 0 & 1 \end{array}\right) = T \cdot R_z \cdot R_y \cdot R_x$$

$$= \begin{pmatrix} 1 & 0 & 0 & T_x \\ 0 & 1 & 0 & T_y \\ 0 & 0 & 1 & T_z \\ 0 & 0 & 0 & 1 \end{pmatrix} \cdot \begin{pmatrix} \cos(\alpha_z) & -\sin(\alpha_z) & 0 & 0 \\ \sin(\alpha_z) & \cos(\alpha_z) & 0 & 0 \\ 0 & 0 & 1 & 0 \\ 0 & 0 & 0 & 1 \end{pmatrix} \cdot$$

$$\cdot \begin{pmatrix} \cos(\alpha_y) & 0 & \sin(\alpha_y) & 0 \\ 0 & 1 & 0 & 0 \\ -\sin(\alpha_y) & 0 & \cos(\alpha_y) & 0 \\ 0 & 0 & 0 & 1 \end{pmatrix} \cdot \begin{pmatrix} 1 & 0 & 0 & 0 \\ 0 & \cos(\alpha_x) & -\sin(\alpha_x) & 0 \\ 0 & \sin(\alpha_x) & \cos(\alpha_x) & 0 \\ 0 & 0 & 0 & 1 \end{pmatrix}$$

$$R_x = \begin{pmatrix} 1 & 0 & 0 & 0 \\ 0 & \sqrt{2}/2 & -\sqrt{2}/2 & 0 \\ 0 & \sqrt{2}/2 & \sqrt{2}/2 & 0 \\ 0 & 0 & 0 & 1 \end{pmatrix}, \quad R_y = \begin{pmatrix} \sqrt{2}/2 & 0 & \sqrt{2}/2 & 0 \\ 0 & 1 & 0 & 0 \\ -\sqrt{2}/2 & 0 & \sqrt{2}/2 & 0 \\ 0 & 0 & 0 & 1 \end{pmatrix},$$

$$R_z = \begin{pmatrix} \sqrt{2}/2 & -\sqrt{2}/2 & 0 & 0 \\ \sqrt{2}/2 & \sqrt{2}/2 & 0 & 0 \\ 0 & 0 & 1 & 0 \\ 0 & 0 & 0 & 1 \end{pmatrix}, \quad T = \begin{pmatrix} 1 & 0 & 0 & 100 \\ 0 & 1 & 0 & 50 \\ 0 & 0 & 1 & 200 \\ 0 & 0 & 0 & 1 \end{pmatrix},$$

$$R = T \cdot R_z \cdot R_y \cdot R_x = \begin{pmatrix} 0.5 & -0.15 & 0.85 & 100 \\ 0.5 & 0.85 & -0.15 & 50 \\ -0.71 & 0.5 & 0.5 & 200 \\ 0 & 0 & 0 & 1 \end{pmatrix}.$$

□ After positioning, therefore, the top vertex of the object, point $[P_1(50, 150, -50)]$, goes into point P_1' with the homogeneous coordinates:

$$\begin{pmatrix} x_1' \\ y_1' \\ z_1' \\ 1 \end{pmatrix} = R \cdot \begin{pmatrix} 50 \\ 150 \\ -50 \\ 1 \end{pmatrix} = \begin{pmatrix} 60 \\ 210 \\ 214.5 \\ 1 \end{pmatrix},$$

□ $P_2(0, 0, 0)$ and $P_3(100, 100, -100)$ are mapped onto $P_2'(100, 50, 200)$ and $P_3'(50, 200, 129)$, respectively.

7.6.9 Rotation of objects about an arbitrary axis in space

Arbitrary rotation: Decomposition into rotations about the **principal axes** (x-, y-, z-axis) of the three-dimensional coordinate system and translation.

Axis rotation: Specified by a reference point $P_0(x_0, y_0, z_0)$ and a direction vector $\vec{v} = (v_x, v_y, v_z)^T$ normalized to unity.

Approach: The reference point coincides with the origin. The direction vector is oriented along the positive z-axis. The rotation

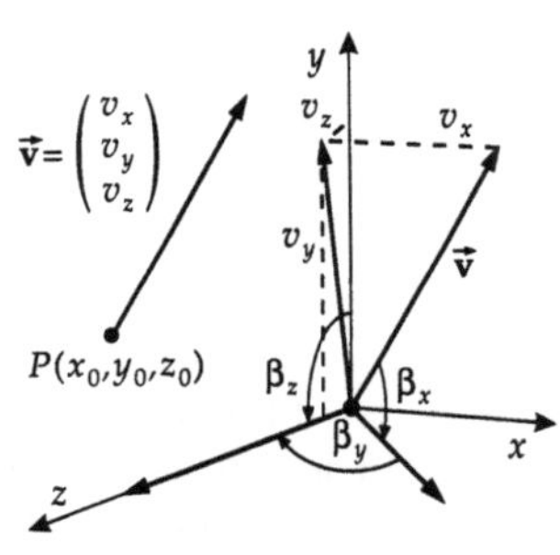

axis and the z-axis coincide. Rotation of the object is through the angle β_v about the z-axis as rotation axis.
Transformations have moved the axis of rotation to the z-axis can be reversed by multiplication with the inverse matrices in reverse order.

▷ In programming the objects are not subjected to all individual transformations. The decomposition into individual transformations only simplifies the calculation of the total transformation matrix. This matrix can then be used to transform the object.

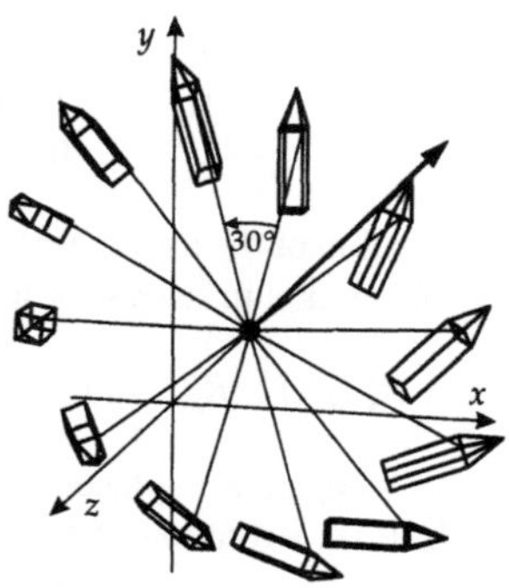

Rotating of an object about an axis of rotation in 30° steps.

Determination of the transformation matrix in the following steps:

- T: Translation of the axis vector such that the reference point $P(x_0, y_0, z_0)$ coincides with the origin:

$$T = \begin{pmatrix} 1 & 0 & 0 & -x_0 \\ 0 & 1 & 0 & -y_0 \\ 0 & 0 & 1 & -z_0 \\ 0 & 0 & 0 & 1 \end{pmatrix} .$$

- R_x: Rotation of the axis vector through an angle β_x so that $\vec{v}$, with $|\vec{v}| = 1$, lies in the z, x plane. For β_x, it must hold that:

$$\sin(\beta_x) = \frac{v_y}{\sqrt{v_y^2 + v_z^2}} , \quad \cos(\beta_x) = \frac{v_z}{\sqrt{v_y^2 + v_z^2}} ,$$

$$R_x = \begin{pmatrix} 1 & 0 & 0 & 0 \\ 0 & \cos(\beta_x) & -\sin(\beta_x) & 0 \\ 0 & \sin(\beta_x) & \cos(\beta_x) & 0 \\ 0 & 0 & 0 & 1 \end{pmatrix} .$$

- R_y: Rotation about the y-axis such that the (rotated) vector of the rotation axis coincides with the z-axis. This requires rotation through the angle $\beta_y = -\gamma_y$ with

$$\sin(\gamma_y) = \frac{v_x}{\sqrt{v_x^2 + v_y^2 + v_z^2}} = v_x , \quad \cos(\gamma_y) = \frac{\sqrt{v_y^2 + v_z^2}}{\sqrt{v_x^2 + v_y^2 + v_z^2}} = \sqrt{v_y^2 + v_z^2} ,$$

$$\sin(\beta_y) = -\sin(\gamma_y) = -v_x , \quad \cos(\beta_y) = \cos(\gamma_y) = \sqrt{v_y^2 + v_z^2} ,$$

$$R_y = \begin{pmatrix} \cos(\beta_y) & 0 & \sin(\beta_y) & 0 \\ 0 & 1 & 0 & 0 \\ -\sin(\beta_y) & 0 & \cos(\beta_y) & 0 \\ 0 & 0 & 0 & 1 \end{pmatrix} .$$

- R_z: Rotation about the given rotation angle β_v about the rotation axis now proceeds about the z-axis:

$$R_z = \begin{pmatrix} \cos(\beta_v) & -\sin(\beta_v) & 0 & 0 \\ \sin(\beta_v) & \cos(\beta_v) & 0 & 0 \\ 0 & 0 & 1 & 0 \\ 0 & = & 0 & 1 \end{pmatrix} .$$

- **Inversion**: Stepwise transformation of the axis of rotation into the z-axis is reversed by applying the inverse transformation in reverse order, where

$$R_y^{-1}(\beta_y) = R_y(-\beta_y) = R_y^T(\beta_y) , \; R_x^{-1}(\beta_x) = R_x(-\beta_x) = R_x^T(\beta_x)$$

and T^{-1} results by inserting the negative displacement vector.

- The **general rotation matrix** R for a rotation of objects about an arbitrary rotation axis in space is achieved by multiplication with these seven matrices:

$$R = T^{-1} R_x^{-1} R_y^{-1} R R_z R_y R_x T .$$

7.6.10 Animation

Animation can be achieved by repeated transformations and representations of graphic objects.

□ **Graphic investigation of collisions** (for example in the **simulation of an NC-program (NC = numerical control)** for the automatic control of machine tools, for monitoring the collision of a tool with a workpiece, for demonstrating the function of a machine or product).

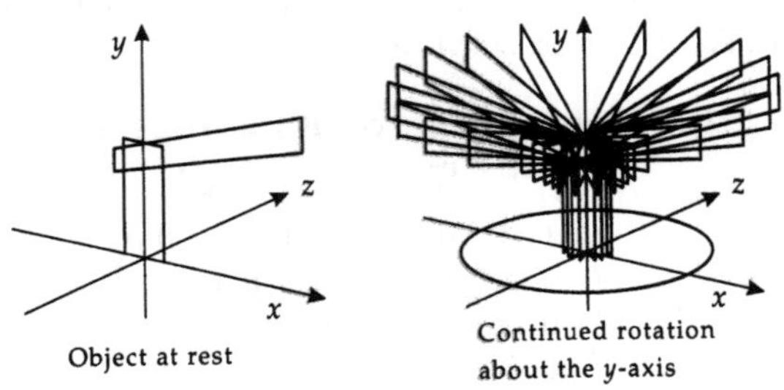

Animation

7.6.11 Reflections

Reflections: Special cases of **scaling**.

- The absolute value of scaling factors is one.

- Reflection at the origin of the coordinate system:

$$\begin{pmatrix} -1 & 0 & 0 & 0 \\ 0 & -1 & 0 & 0 \\ 0 & 0 & -1 & 0 \\ 0 & 0 & 0 & 1 \end{pmatrix} \cdot \begin{pmatrix} x \\ y \\ z \\ 1 \end{pmatrix} \rightarrow \begin{pmatrix} -x \\ -y \\ -z \\ 1 \end{pmatrix} .$$

- Reflection in the plane $x = 0$ (y, z plane):

$$\begin{pmatrix} -1 & 0 & 0 & 0 \\ 0 & 1 & 0 & 0 \\ 0 & 0 & 1 & 0 \\ 0 & 0 & 0 & 1 \end{pmatrix} \cdot \begin{pmatrix} x \\ y \\ z \\ 1 \end{pmatrix} = \begin{pmatrix} -x \\ y \\ z \\ 1 \end{pmatrix} .$$

Successive application of transformations: Multiplication of individual transformation matrices.

▷ The **order of the transformations** must be maintained. Only in special cases can transformations be commuted.

7.6.12 Transformation of coordinate systems

Passive transformation: **Change of the coordinate system** to determine the position of an object in another coordinate system.

□ The point with coordinates $P(x, y, z)$ is given in the coordinate system K, while the coordinates of the same point $P(x', y', z')$ are required in the coordinate system K'.

□ In computer graphics, the objects are arranged in a problem-related world coordinate system. The object coordinates relate to the world coordinate system. To create different spatial views, the corresponding viewing coordinate systems are changed.

7.6.13 Translation of a coordinate system

▷ The **translation** of a coordinate system by the vector (T_x, T_y, T_z) corresponds to a **displacement** of the geometric object by the vector $(-T_x, -T_y, -T_z)$.

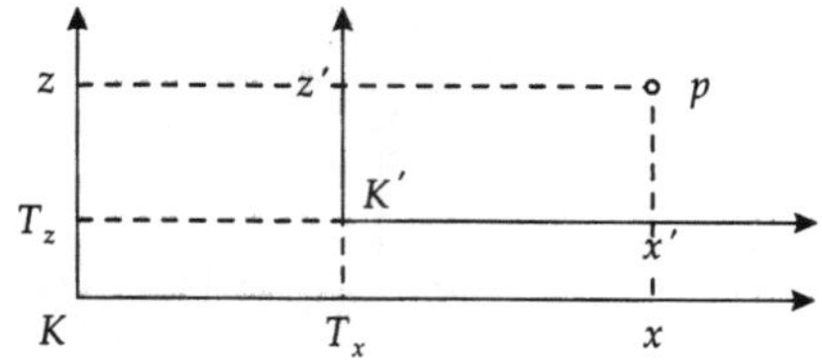

Translation of a coordinate system
(without y-coordinate)

- Matrix representation of **translation** in homogeneous coordinates:

$$\begin{pmatrix} 1 & 0 & 0 & -T_x \\ 0 & 1 & 0 & -T_y \\ 0 & 0 & 1 & -T_z \\ 0 & 0 & 0 & 1 \end{pmatrix} \cdot \begin{pmatrix} x \\ y \\ z \\ 1 \end{pmatrix} \rightarrow \begin{pmatrix} x' \\ y' \\ z' \\ 1 \end{pmatrix} .$$

7.6.14 Rotation of a coordinate system about a principal axis

The coordinate system K' has been rotated through an angle $\alpha_x, \alpha_y, \alpha_z$ in relation to the system K. x', y', z' are the coordinates in the rotated coordinate system.

Rotations of the coordinate system about the principal axes:

- Rotation about the x-axis:

$$\begin{pmatrix} 1 & 0 & 0 & 0 \\ 0 & \cos(\alpha_x) & \sin(\alpha_x) & 0 \\ 0 & -\sin(\alpha_x) & \cos(\alpha_x) & 0 \\ 0 & 0 & 0 & 1 \end{pmatrix} \cdot \begin{pmatrix} x \\ y \\ z \\ 1 \end{pmatrix} \to \begin{pmatrix} x' \\ y' \\ z' \\ 1 \end{pmatrix}.$$

- Rotation about the y-axis:

$$\begin{pmatrix} \cos(\alpha_y) & 0 & -\sin(\alpha_y) & 0 \\ 0 & 1 & 0 & 0 \\ \sin(\alpha_y) & 0 & \cos(\alpha_y) & 0 \\ 0 & 0 & 0 & 1 \end{pmatrix} \cdot \begin{pmatrix} x \\ y \\ z \\ 1 \end{pmatrix} \to \begin{pmatrix} x' \\ y' \\ z' \\ 1 \end{pmatrix}.$$

- Rotation about the z-axis:

$$\begin{pmatrix} \cos(\alpha_z) & \sin(\alpha_z) & 0 & 0 \\ -\sin(\alpha_z) & \cos(\alpha_z) & 0 & 0 \\ 0 & 0 & 1 & 0 \\ 0 & 0 & 0 & 1 \end{pmatrix} \cdot \begin{pmatrix} x \\ y \\ z \\ 1 \end{pmatrix} \to \begin{pmatrix} x' \\ y' \\ z' \\ 1 \end{pmatrix}.$$

▷ The rotation of a coordinate system through an angle α corresponds to the rotation of a geometric object through angle $-\alpha$.

□ If the processing of a workpiece on a milling machine has to be graphically simulated by means of an NC program to control a machine tool, note the following:

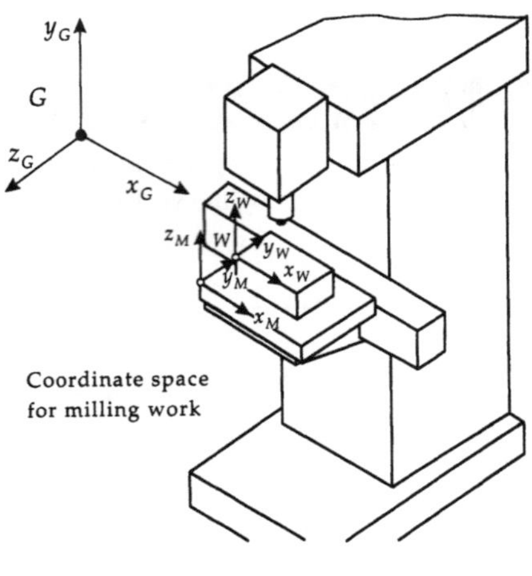

M : Machine coordinate system
W : Workpiece coordinate system
P : Tool position

In an NC program, the tool information relates to the tool coordinate system. The information must be transferred to the machine coordinate system to which the control commands for the machine axes, and, thus for the position of the tool refer. The workpiece, the tool and the machine are in turn positioned in the graphics system in accordance with a word coordinate system.

Starting with the NC-program, coordinate system transformations from the workpiece coordinate system W to the machine coordinate system M and to world coordinate system G must be carried out, to determine for example, the next tool position P in the world coordinate system G.

□ Example for the transformation of a coordinate system: In milling work, the tool should work so that the center of the cutter P assumes coordinates (80,80,120) in the workpiece coordinate system. The machine coordinate system is translated by $T_{W_x} = -160.5$, $T_{W_y} = -120.5$, $T_{W_z} = -150$ in relation to the tool coordinate system. The world coordinate system is given by the machine coordinate system by the translation with $T_{M_x} = -2105$, $T_{M_y} = 1300$, $T_{M_z} = -1400$, and rotation through 90° about the x_M axis. The position vector $\vec{\mathbf{p}}' = (x', y', z')^T$ of P in the world coordinate system is obtained from $\vec{\mathbf{p}} = (80, 80, 120)^T$ in the workpiece coordinate system by successive application of two translations and one rotation.

$$\begin{pmatrix} x' \\ y' \\ z' \\ 1 \end{pmatrix} = \begin{pmatrix} 1 & 0 & 0 & 0 \\ 0 & \cos(\pi/2) & \sin(\pi/2) & 0 \\ 0 & -\sin(\pi/2) & \cos(\pi/2) & 0 \\ 0 & 0 & 0 & 1 \end{pmatrix} \cdot \begin{pmatrix} 1 & 0 & 0 & -T_{M_x} \\ 0 & 1 & 0 & -T_{M_y} \\ 0 & 0 & 1 & -T_{M_z} \\ 0 & 0 & 0 & 1 \end{pmatrix}$$

$$\cdot \begin{pmatrix} 1 & 0 & 0 & -T_{W_x} \\ 0 & 1 & 0 & -T_{W_y} \\ 0 & 0 & 1 & -T_{W_z} \\ 0 & 0 & 0 & 1 \end{pmatrix} \cdot \begin{pmatrix} 80 \\ 80 \\ 120 \\ 1 \end{pmatrix}$$

$$= \begin{pmatrix} 1 & 0 & 0 & 0 \\ 0 & 0 & 1 & 0 \\ 0 & -1 & 0 & 0 \\ 0 & 0 & 0 & 1 \end{pmatrix} \cdot \begin{pmatrix} 1 & 0 & 0 & 2105 \\ 0 & 1 & 0 & -1300 \\ 0 & 0 & 1 & 1400 \\ 0 & 0 & 0 & 1 \end{pmatrix}$$

$$\cdot \begin{pmatrix} 1 & 0 & 0 & 160.5 \\ 0 & 1 & 0 & 120.5 \\ 0 & 0 & 1 & 150 \\ 0 & 0 & 0 & 1 \end{pmatrix} \begin{pmatrix} 80 \\ 80 \\ 120 \\ 1 \end{pmatrix} = \begin{pmatrix} 2345.5 \\ 1670 \\ 1099.5 \\ 1 \end{pmatrix} .$$

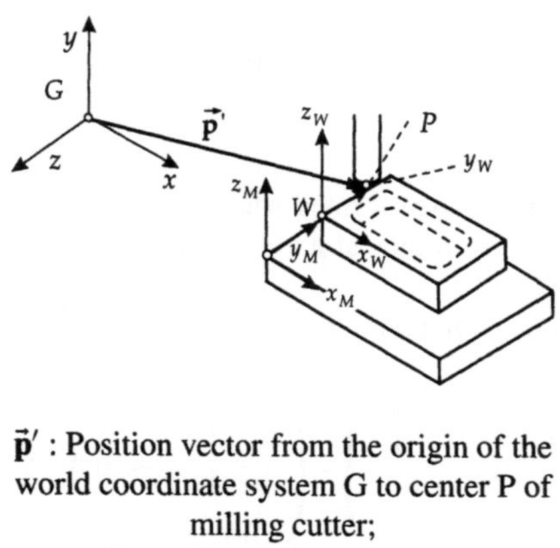

$\vec{\mathbf{p}}'$: Position vector from the origin of the world coordinate system G to center P of milling cutter;
G: world coordinate system;
M: machine coordinate system;
W: workpiece coordinate system
P: center of the cutter

7.7 Projections

7.7.1 Fundamental principles

Projection: Mathematical method of mapping the geometry of three-dimensional objects onto a two-dimensional drawing plane.
Projection ray: A ray from a projection center to a point of the object.
Image point: Intersection point of the projection ray with the projection plane.
Parallel projection: Represents the real size and shape of the object. Projection rays are parallel when they impinge on the projection plane.

□ Providing different views for scaled drawings.

Central or **perspective projection**: Object is represented as it appears to the human eye or a camera.
Properties of perspective projection: **Perspective contraction** and **vanishing points**.

7.7.2 Parallel projection

Two groups, divided into **projectional direction** and **normal vector of the projection plane**:
Orthogonal projection: Projecting rays impinge perpendicularly on the projection plane.
Oblique parallel projection: **Direction of projection** and **direction of the normal vector** do not coincide; the projection rays impinge obliquely on the projection plane.

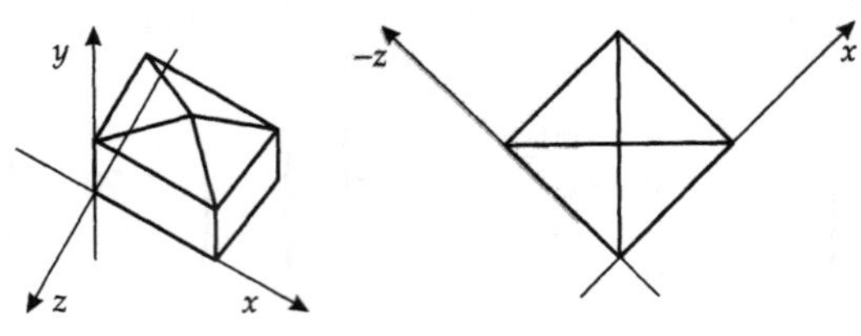

Orthogonal parallel projection

▷ **Multiplane projections**: Are formed when the projectional direction is chosen to be parallel to one of the principal axes. Elevations, plan views and side views in technical drawings.

● **Orthogonal parallel projection** on the (x, y)-plane:

$$\begin{pmatrix} 1 & 0 & 0 & 0 \\ 0 & 1 & 0 & 0 \\ 0 & 0 & 0 & 0 \\ 0 & 0 & 0 & 1 \end{pmatrix} \cdot \begin{pmatrix} x \\ y \\ z \\ 1 \end{pmatrix} \rightarrow \begin{pmatrix} x \\ y \\ 0 \\ 1 \end{pmatrix} .$$

Location of the projection plane in an **arbitrary** position in space: Unique definition by a point in the plane and the normal vector.
System of representation coordinates (x_d, y_d, z_d) fixed by the **base vectors** $\vec{\mathbf{e}}_x, \vec{\mathbf{e}}_y, \vec{\mathbf{n}}$.

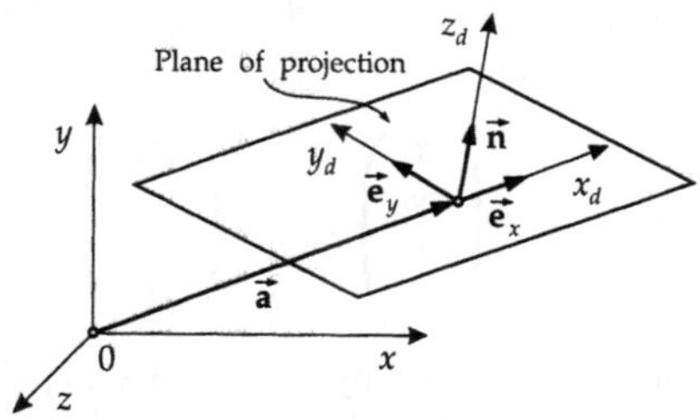

Projection plane in an arbitrary position in space

- **Orthogonal parallel projection** in vector notation:

$$\vec{\mathbf{x}}' = \vec{\mathbf{x}} + (d - \vec{\mathbf{n}} \cdot \vec{\mathbf{x}})\cdot$$

The coordinates of vector $\vec{\mathbf{x}}'$ refer to the **world coordinate system**.

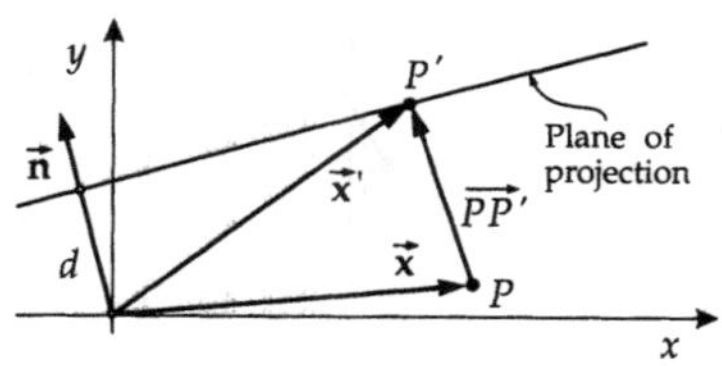

Orthogonal parallel projection in an arbitrary plane

Representation with respect to the **plane coordinate system** (x_d, y_d) by projecting the difference vector $\vec{\mathbf{x}}' - \vec{\mathbf{a}}$, lying in the plane, onto the axes of the (x_d, y_d) plane:

$$x_p = \vec{\mathbf{e}}_x \cdot (\vec{\mathbf{x}}' - \vec{\mathbf{a}})$$
$$y_p = \vec{\mathbf{e}}_y \cdot (\vec{\mathbf{x}}' - \vec{\mathbf{a}})$$

Representation of the projection as a special transformation by (4×4) matrices:
The transition from the world coordinate system to the viewing coordinate system corresponds to a transformation of the coordinate system.
Orientation of the axes can be combined from rotations about the principal axes.
The positioning of the origin takes place by a **translation.**
Projection in the (x_d, y_d) **plane** by the matrix multiplication given above.
R: **combined rotation matrix,**
T: **translation matrix,**
P: **projection matrix**:

$$R = \begin{pmatrix} c_{11} & c_{12} & c_{13} & 0 \\ c_{21} & c_{22} & c_{23} & 0 \\ c_{31} & c_{32} & c_{33} & 0 \\ 0 & 0 & 0 & 1 \end{pmatrix}, \quad T = \begin{pmatrix} 1 & 0 & 0 & -a_x \\ 0 & 1 & 0 & -a_y \\ 0 & 0 & 1 & -a_z \\ 0 & 0 & 0 & 1 \end{pmatrix}, \quad P = \begin{pmatrix} 1 & 0 & 0 & 0 \\ 0 & 1 & 0 & 0 \\ 0 & 0 & 0 & 0 \\ 0 & 0 & 0 & 1 \end{pmatrix}.$$

- Orthogonal parallel projection of a three-dimensional vector onto an arbitrary projection plane in matrix notation:

$$P_{\text{orth}} \cdot \begin{pmatrix} x \\ y \\ z \\ 1 \end{pmatrix} = P \cdot T \cdot R \cdot \begin{pmatrix} x \\ y \\ z \\ 1 \end{pmatrix} = P \cdot \begin{pmatrix} x_d \\ y_d \\ z_d \\ 1 \end{pmatrix} \rightarrow \begin{pmatrix} x_p \\ y_p \\ 0 \\ 1 \end{pmatrix}.$$

□ The projection ray can be determined by the user of a graphic system by specifying the coordinates of observer's position. The projectional direction corresponds to the line of sight to the origin of the coordinate system of the object.

Oblique projection P' of a point P in space onto the (x, y)-plane: Point of intersection between the (x, y)-plane and the **projection ray** $\vec{\mathbf{v}}$ direction coming from P. Coordinates:

$$x_p = x - z \cdot \cot\alpha,$$
$$y_p = y - z \cdot \cot\beta.$$
$$z_p = z = 0$$

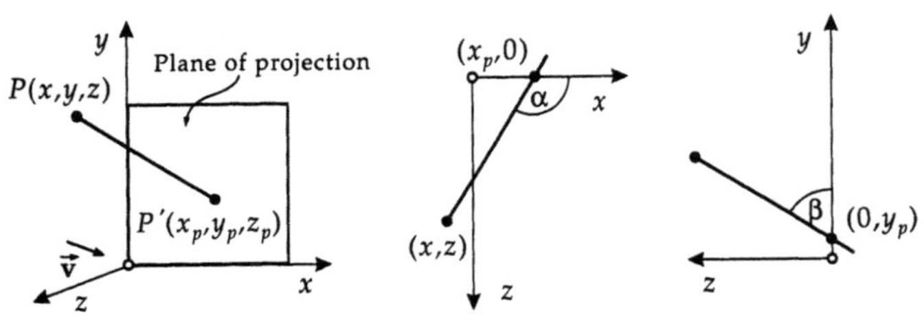

Oblique parallel projection onto the (x, y)-plane

Oblique parallel projection onto the (x, y)-plane, fixed by directional vector $\vec{\mathbf{v}}$ of the projection direction, not equal to the **normal vector** of the (x, y)-plane. Oblique parallel projection preserves the angles and lengths of the planes which lie parallel to the projection plane. Segments perpendicular to the projection plane are contracted in dependence on $\vec{\mathbf{v}}$.

- **Oblique parallel projection** onto the (x, y)-plane in matrix representation:

$$\begin{pmatrix} 1 & 0 & -\cot\alpha & 0 \\ 0 & 1 & -\cot\beta & 0 \\ 0 & 0 & 0 & 0 \\ 0 & 0 & 0 & 1 \end{pmatrix} \cdot \begin{pmatrix} x \\ y \\ z \\ 1 \end{pmatrix} \rightarrow \begin{pmatrix} x_p \\ y_p \\ 0 \\ 1 \end{pmatrix}$$

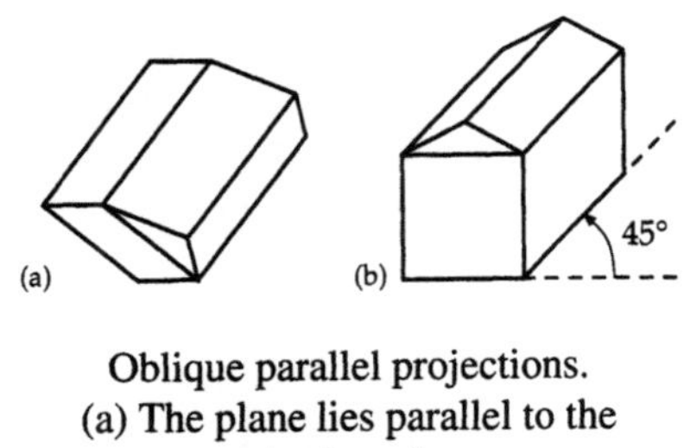

Oblique parallel projections.
(a) The plane lies parallel to the projection plane.
(b) The elevation lies parallel to the projection plane (cavalier perspective).

7.7.3 Central projection

Projection of a three-dimensional object from point P_0 (at infinity) the **center of projection** or **eye position**, by means of **projection rays** onto the image plane.
The resulting image is the **central view** or **perspective view** of the object.

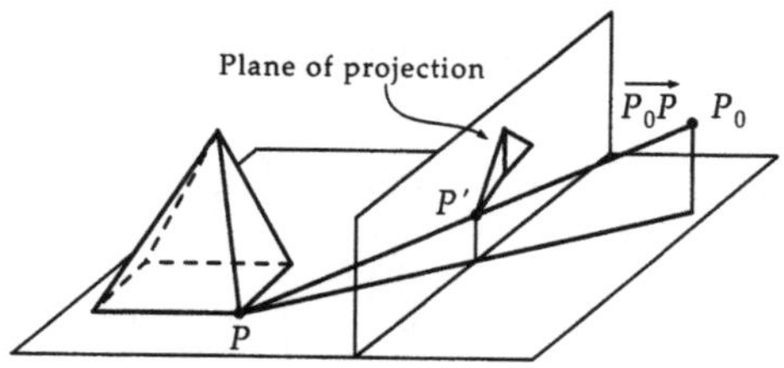

Central projection

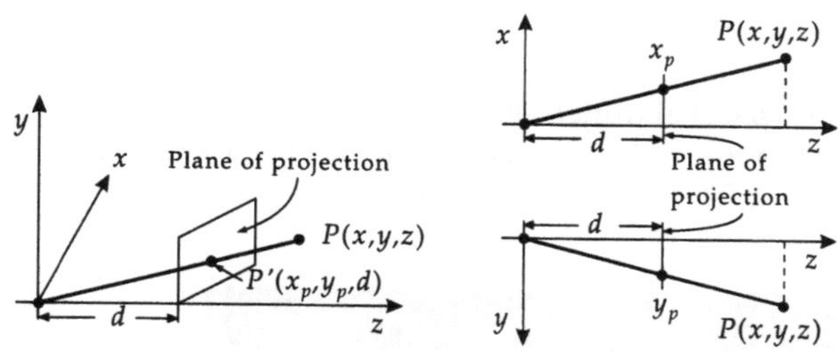

Central projection with projection center at the origin

- **Central projection** onto the parallel to the (x, y)-plane at a distance d with the projection center in the origin of the coordinates.

$$\begin{pmatrix} 1 & 0 & 0 & 0 \\ 0 & 1 & 0 & 0 \\ 0 & 0 & 0 & 0 \\ 0 & 0 & 1/d & 0 \end{pmatrix} \cdot \begin{pmatrix} x \\ y \\ z \\ 1 \end{pmatrix} \rightarrow \begin{pmatrix} x \\ y \\ 0 \\ (z/d) \end{pmatrix} .$$

To calculate the Cartesian projected coordinates we must divide by the fourth homogeneous coordinate:

$$x_p = \frac{x \cdot d}{z} \quad , \quad y_p = \frac{y \cdot d}{z} \; .$$

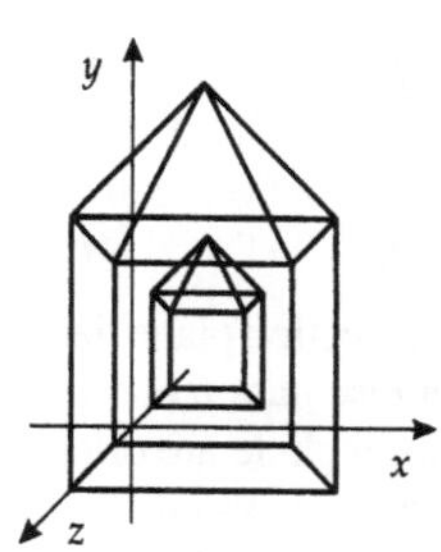

▷ Distance d has the effect of a **scaling factor** by which we must multiply x and y.
Division by z has the result that objects farther away are represented smaller than near objects. This corresponds to the **optical impression** a viewer has.

▷ With this projection, all points before or behind the projection center can be mapped, but not the points in the (x, y)-plane because this would lead to a division by $z = 0$.

- **Central projection** onto the (x, y)-plane with the projection center at $z = -d$:

$$\begin{pmatrix} 1 & 0 & 0 & 0 \\ 0 & 1 & 0 & 0 \\ 0 & 0 & 0 & 0 \\ 0 & 0 & 1/d & 1 \end{pmatrix} \cdot \begin{pmatrix} x \\ y \\ z \\ 1 \end{pmatrix} \rightarrow \begin{pmatrix} x \\ y \\ 0 \\ z/d+1 \end{pmatrix} .$$

The Cartesian coordinates are formed through division by the fourth homogeneous coordinate:

$$x_P = \frac{x}{z/d+1} , \quad y_P = \frac{x}{z/d+1} = \frac{d \cdot x}{z+d} .$$

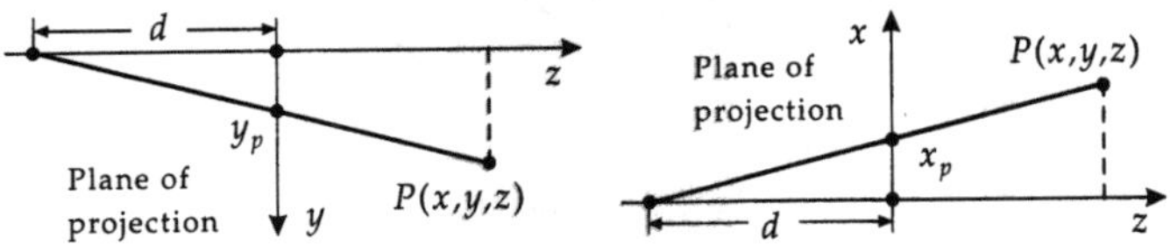

Central projection onto the (x,y)-plane

▷ In this form, **parallel projection** is contained in the central projection as a special case when the **projection center** is infinitely far away from the origin of the coordinate, i.e., for $d \rightarrow \infty$ as $1/d \rightarrow 0$.

7.7.4 General formulation of projections

A general formulation of projections contains both parallel and central projections. The starting point is projectional P' with the coordinates $[x_P, y_P, z_P]$ of an arbitrary point P onto a projection plane. The projection plane is perpendicular to the z-axis at a distance z_P from the origin of the coordinator. Let g be the distance between the projection center P_0 and point $(0, 0, z_P)$ in the projection plane, and let (d_x, d_y, d_z) be the direction vector from that point to the projection center.

- **Projection** onto a projection plane perpendicular to the z-axis **with an arbitrary projection center**:

$$P_{\text{proj}} \cdot \begin{pmatrix} x \\ y \\ z \\ 1 \end{pmatrix} = \begin{pmatrix} 1 & 0 & -d_x/d_z & z_P \cdot d_x/d_z \\ 0 & 1 & -d_y/d_z & z_P \cdot d_y/d_z \\ 0 & 0 & -(z_P)/(q \cdot d_z) & (z_P^2)/(q \cdot d_z) + z_P \\ 0 & 0 & -(1)/(q \cdot d_z) & (z_P)/(q \cdot d_z) + 1 \end{pmatrix} \cdot \begin{pmatrix} x \\ y \\ z \\ 1 \end{pmatrix}$$

$$\rightarrow \begin{pmatrix} x - z \cdot d_x/d_z + z_P \cdot d_x/d_z \\ y - z \cdot d_y/d_z + z_P \cdot d_y/d_z \\ -z \cdot (z_P)/(q \cdot d_z) + (z_P^2 + z_P q d_z)/(q \cdot d_z) \\ (z_P - z)/(q \cdot d_z) + 1 \end{pmatrix} .$$

▷ If the projection plane lies arbitrarily in the world coordinate system, then we proceed to a representation coordinate system where the z_d axis and the normal vector of the projection plane are oriented in the same direction (see Section 7.6, transformation of coordinate systems).

□ **Special cases** of P_{proj}:

Orthogonal parallel projection onto the (x, y)-plane:

$z_P = 0, \quad q = \infty, \quad (d_x, d_y, d_z) = (0, 0, -1).$

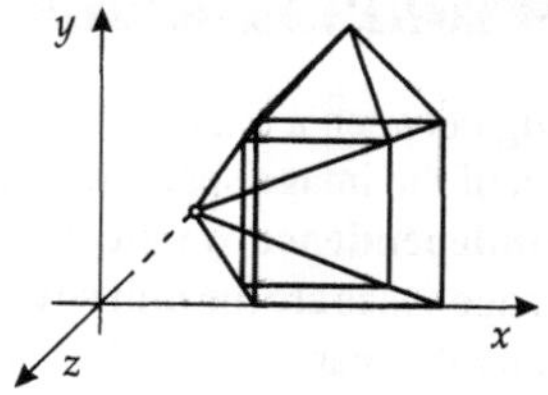

Central projection onto a plane parallel to the (x, y)-plane, projection center at the origin:
$z_P = d$, $q = d$, $(d_x, d_y, d_z) = (0, 0, -1)$.
Central projection of the (x, y)-plane, projection center on the negative z-axis:
$z_P = 0$, $q = d$, $(d_x, d_y, d_z) = (0, 0, -1)$.
Provided $q \neq \infty$, the central projection defines a so-called

One-point perspective: Parallel straight lines, not parallel to the projection plane, converge during projection in one point to a single point in the projection. The point where the parallels intersect with the z-axis is called the **vanishing point**.

▷ In three-dimensional space, parallels intersect only at infinity. Therefore, the vanishing point can be regarded as the projection of a point at infinity on the z-axis, with homogeneous coordinates $(0, 0, 1, 0)^T$.
Multiplication by P_{proj} and subsequent division by the homogeneous coordinate w yields the vanishing point coordinates:

$$x = q \cdot d_x \quad , \quad y = q \cdot d_y \quad , \quad z = z_P .$$

● If a required vanishing point (x, y) is given, and its distance g from the projection center is known, then, due to $\sqrt{d_x^2 + d_y^2 + d_z^2} = 1$, the direction vector (d_x, d_y, d_z), and thus also the projection matrix, are uniquely defined.

If the projection plane is not perpendicular to the z-axis but lies arbitrarily in space, then there are up to three vanishing points. Accordingly, the projections are called one-point, two-point or three-point perspectives:

Two-point perspective: The projection plane intersects two principal axes. The parallels to these principal axes intersect in two vanishing points.

Three-point perspective: The projection plane intersects three principal axes. Three vanishing points result.

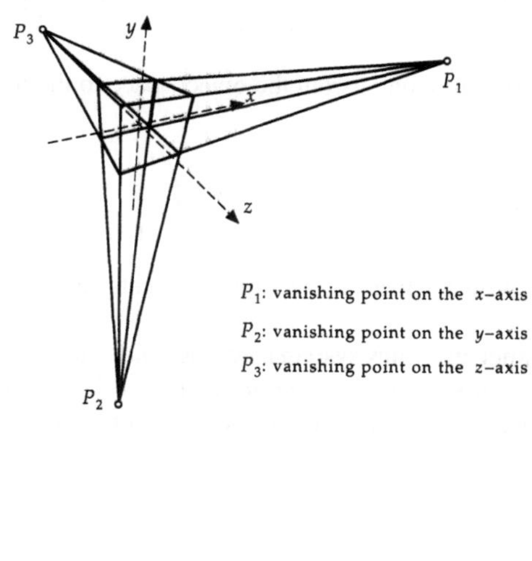

7.8 Window/viewport transformation

Mapping between a two- or three-dimensional sector of the object space (**window**) and a portion of the image space, the **viewport** of a graphics output device.
Device independence by introducing standardized device or image coordinates: Assigned to the image space are the **unit square** $[0,1]\times[0,1]$ or the **unit cube** $[0,1]\times[0,1]\times[0,1]$. Then, only the mappings of the standardized coordinates onto the special device coordinates are dependent on the device.

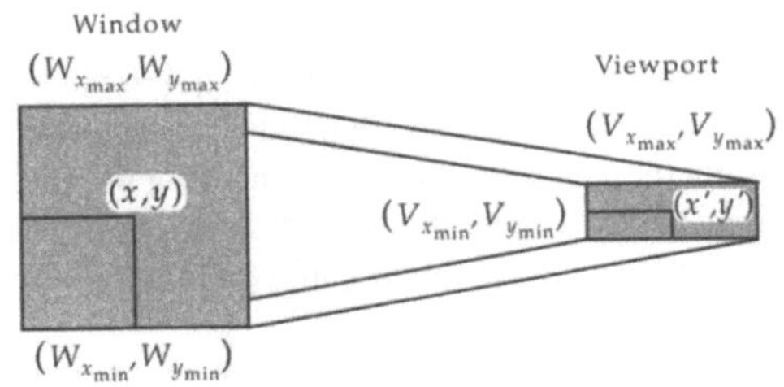

Window/viewport transformation

Window/viewport transformation, performed in three partial steps:
1. **Translation** of the window in the object space (onto the **origin**) → T_1,
2. **Scaling** between object space and image space → S,
3. **Translation** of the viewport onto the required position → T_2.

Condition: **The object space must be two-dimensional.**
Scaling factors:

$$S_x = \frac{V_{xmax} - V_{xmin}}{W_{xmax} - W_{xmin}} \qquad S_y = \frac{V_{ymax} - V_{ymin}}{W_{ymax} - W_{ymin}}$$

Transformation matrices:

$$T_1 = \begin{pmatrix} 1 & 0 & -W_{xmin} \\ 0 & 1 & -W_{ymin} \\ 0 & 0 & 1 \end{pmatrix}, \quad S = \begin{pmatrix} S_x & 0 & 0 \\ 0 & S_y & 0 \\ 0 & 0 & 1 \end{pmatrix}, \quad T_2 = \begin{pmatrix} 1 & 0 & -V_{xmin} \\ 0 & 1 & -V_{ymin} \\ 0 & 0 & 1 \end{pmatrix}.$$

Overall transformation of a point from the object space to the image space by matrix multiplication:

$$\begin{pmatrix} x' \\ y' \\ 1 \end{pmatrix} = T_2 \cdot S \cdot T_1 \cdot \begin{pmatrix} x \\ y \\ 1 \end{pmatrix}.$$

Three-dimensional object space: Projecting the points in space onto the plane and mapping.

▷ **High-performance graphics systems** are also able to store a z value (z buffer) for each image point, in addition to other information, such as color. Here, no projection of the three-dimensional object is needed because a **quasi-three-dimensional image space** is available.

8

Analytic geometry

8.1 Elements of the plane

8.1.1 Distance between two points

Cartesian coordinate system: Distance between $P_1(x_1, y_1)$ and $P_2(x_2, y_2)$:

$$d = \overline{P_1P_2} = \sqrt{(x_2 - x_1)^2 + (y_2 - y_1)^2}\ .$$

Polar coordinates: Distance between $P_1(r_1, \varphi_1)$ and $P_2(r_2, \varphi_2)$:

$$d = \overline{P_1P_2} = \sqrt{r_1^2 + r_2^2 - 2r_1 \cdot r_2 \cdot \cos(\varphi_1 - \varphi_2)}\ .$$

8.1.2 Division of a segment

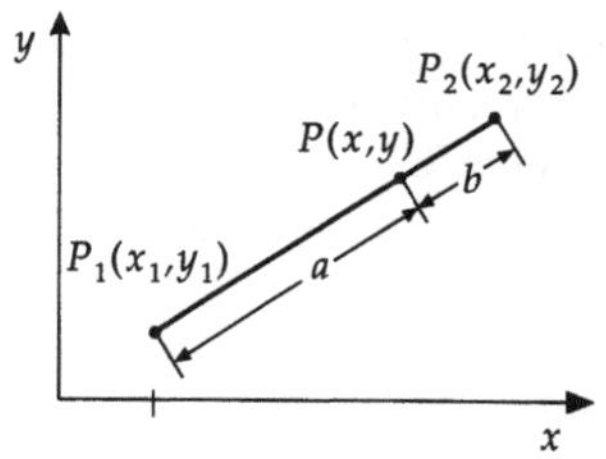

Segment

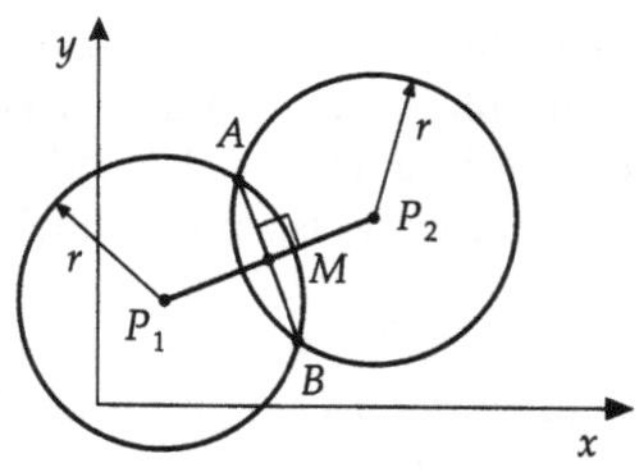

Right bisector

Ratio of division: Division of a segment in a given ratio, coordinates of point P with $\dfrac{\overline{P_1P}}{\overline{PP_2}} = \dfrac{a}{b} = \lambda$:

$$x = \frac{bx_1 + ax_2}{b+a} = \frac{x_1 + \lambda x_2}{1+\lambda}, \quad y = \frac{by_1 + ay_2}{b+a} = \frac{y_1 + \lambda y_2}{1+\lambda}.$$

Midpoint of the segment $\overline{P_1 P_2}$:

$$x = \frac{x_1 + x_2}{2}, \qquad y = \frac{y_1 + y_2}{2}.$$

□ **Center of gravity** of a system of material points $M_i(x_i, y_i)$ with the masses m_i $(i = 1, 2, \ldots, n)$:

$$x = \frac{\sum m_i x_i}{\sum m_i}, \qquad y = \frac{\sum m_i y_i}{\sum m_i}.$$

Right bisector: The straight line that joins the points of intersection (A, B) of two circles of equal radius r $(r > AB)$ to each other whose centers are the end-points of the line segment that is bisected: M: midpoint of $\overline{P_1 P_2}$ and $\overline{AB}$ (**bisection of segment**).

8.1.3 *Area of a triangle*

Area of a triangle with vertices $P_1(x_1, y_1)$, $P_2(x_2, y_2)$, and $P_3(x_3, y_3)$:

$$A = \frac{1}{2} \begin{vmatrix} x_1 & y_1 & 1 \\ x_2 & y_2 & 1 \\ x_3 & y_3 & 1 \end{vmatrix} = \frac{1}{2}\Big(x_1(y_2 - y_3) + x_2(y_3 - y_1) + x_3(y_1 - y_2)\Big).$$

Three points lie in a straight line if

$$\begin{vmatrix} x_1 & y_1 & 1 \\ x_2 & y_2 & 1 \\ x_3 & y_3 & 1 \end{vmatrix} = 0.$$

Area of a polygon:

$$A = \frac{1}{2}\Big((x_1 - x_2)(y_1 + y_2) + (x_2 - x_3)(y_2 + y_3) + \ldots + (x_n - x_1)(y_n + y_1)\Big).$$

8.1.4 *Equation of a curve*

Functional equation: An equation with two or more variables.
Independent variable: Can be substituted by arbitrarily chosen values.
Dependent variable: Quantity whose value must be calculated from an equation.
Case distinction for the functional equation $F(x, y) = 0$:
Algebraic curve: $F(x, y)$ is a polynomial; the degree of the polynomial is the **order of the curve.**
□ $F(x, y) = x^2 + y^2 - 4 = 0$ (circle of radius $R = 2$ about the origin).
Transcendental curve: $F(x, y)$ is not a polynomial.
□ $F(x, y) = y^2 + 6 \ln x = 0$.

8.2 Straight line

- If a point lies on a straight line (or on a curve in the general sense), then its coordinates satisfy the equation of a straight line (curve).
- If a pair of values (x, y) obeys the equation of a line (curve), then the corresponding point lies on this line (curve).

8.2.1 Forms of straight-line equations

General form of the straight-line equation:

$$Ax + By + C = 0\,.$$

Point-direction equation or **point-gradient equation**: A straight line is determined by a point and by its own direction.

$$y - y_1 = m(x - x_1)\,, \quad \text{or} \quad m = \tan\alpha = \frac{y - y_1}{x - x_1}\,,$$

$m = \tan\alpha$: slope of the straight line (angular coefficient),
α: angle of slope.

▷ Application: Determination of the line when one point and the angular coefficient of the straight line are known.

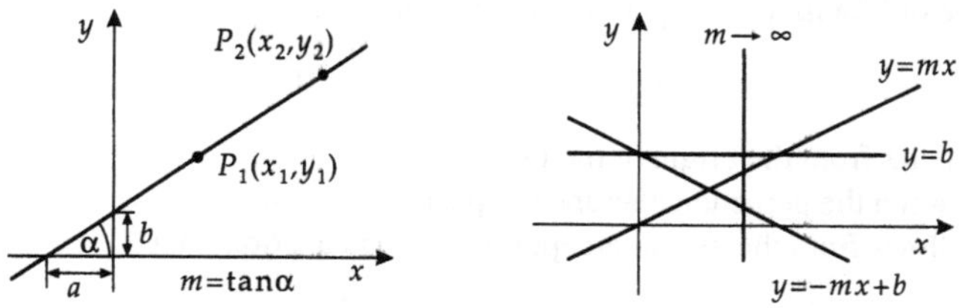

Types of straight lines

Normal form of the equation of a straight line:

$$y = mx + b \quad \text{or} \quad y - b = m(x - 0)$$

m: slope of the line (angular coefficient).
b: intercept on the ordinate (absolute term).

□ **Line through the origin:** $b = 0 \quad \rightarrow y = mx.$
Line parallel to the x-axis: $m = 0 \quad \rightarrow y = b.$
Line parallel to the y-axis: $x = a$, for all y.
Equation of the x-axis: $m = b = 0 \quad \rightarrow y = 0$
Equation of the y-axis: $x = 0$, for all y.
Line of a 45° angle with the $+x$ axis: $m = 1 \quad \rightarrow y = x + b.$
Line of a $90° + 45° = 135°$ angle with the $+x$ axis:
$m = -1 \quad \rightarrow y = -x + b.$

Two-point equation of a straight line:

$$\frac{y - y_1}{x - x_1} = \frac{y_2 - y_1}{x_2 - x_1}\,.$$

▷ Application: Only when two points of the line are known, every point of the line can be calculated.

Axis intercept form of the equation of a straight line:

$$\frac{x}{a} + \frac{y}{b} = 1\,,$$

a: Point of intersection with the x-axis,
b: Point of intersection with the y-axis.

▷ Application: Every point of the straight line can be calculated if the points of intersection with the axis (a, b) are given.

▷ Not applicable to a line through the origin!

Equation of the straight line in polar coordinates:

$$r = \frac{d}{\cos(\varphi - \alpha)},$$

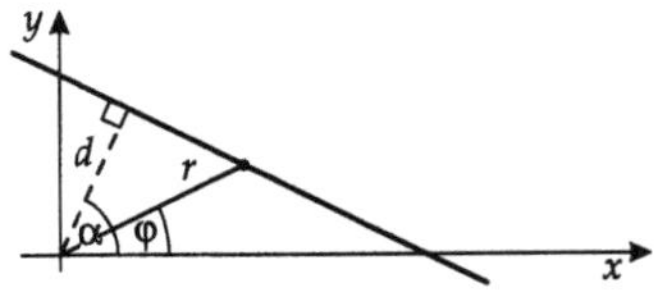

d: distance from the pole to the line,
φ: angle between the polar axis and the perpendicular from the pole to the line.

8.2.2 Hessian normal form

Hessian normal form of the equation of a straight line:

$$x \cos\varphi + y \sin\varphi - p = 0,$$

p: perpendicular from the origin to the line,
φ: angle between the perpendicular and the positive x-axis.

▷ This follows from the axis intercept form of the equation with

$$a = \frac{p}{\cos\varphi}, \quad b = \frac{p}{\sin\varphi}.$$

□ Application: Calculation of the shortest distance d between a point $P_1(x_1, y_1)$ and a line.

$$d = x_1 \cos\varphi + y_1 \sin\varphi - p$$

▷ φ is not unique if the line goes through the origin.

● From the general equation of a straight line

$$Ax + By + C = 0$$

it also follows for the **Hessian normal form**:

$$\frac{Ax + By + C}{\mp\sqrt{A^2 + B^2}} = 0.$$

Distance from a point P_1 to a line:

$$d = \frac{Ax_1 + By_1 + C}{\mp\sqrt{A^2 + B^2}} \quad \text{for} \quad \frac{C}{\mp\sqrt{A^2 + B^2}} < 0.$$

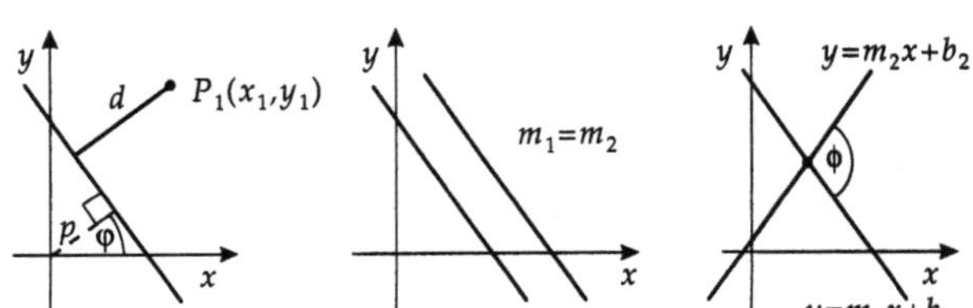

Hessian normal form Point of intersection of lines

8.2.3 Point of intersection of straight lines

Point of intersection of straight lines: By solving both equations of the straight line

$$A_1x + B_1y + C_1 = 0 \quad \text{and} \quad A_2x + B_2y + C_2 = 0.$$

We obtain the coordinates (x_0, y_0) of the point of intersection

$$x_0 = \frac{B_1C_2 - B_2C_1}{A_1B_2 - A_2B_1} \quad \text{and} \quad y_0 = \frac{C_1A_2 - C_2A_1}{A_1B_2 - A_2B_1}.$$

Parallel lines for

$$A_1B_2 - A_2B_1 = 0.$$

Equal lines for

$$\frac{A_1}{A_2} = \frac{B_1}{B_2} = \frac{C_1}{C_2}.$$

Point of intersection of three lines exists for

$$\begin{vmatrix} A_1 & B_1 & C_1 \\ A_2 & B_2 & C_2 \\ A_3 & B_3 & C_3 \end{vmatrix} = 0.$$

Equation of a pencil of lines: Equation of all lines that pass through the point of intersection of two given lines:

$$(A_1x + B_1y + C_1) + k \cdot (A_2x + B_2y + C_2) = 0 \quad (k \in \mathbb{R}).$$

▷ All lines of the pencil are obtained when k runs from $-\infty$ to ∞.

8.2.4 Angle between straight lines

Angle between two lines: For equations of the straight line in general form $Ax + By + C = 0$:

$$\tan\phi = \frac{A_1B_2 - A_2B_1}{A_1A_2 + B_1B_2}$$

$$\cos\phi = \frac{A_1A_2 + B_1B_2}{\sqrt{A_1^2 + B_1^2} \cdot \sqrt{A_2^2 + B_2^2}}$$

$$\sin\phi = \frac{A_1B_2 - A_2B_1}{\sqrt{A_1^2 + B_1^2} \cdot \sqrt{A_2^2 + B_2^2}}$$

Angle between two straight lines, for equations of the straight line in normal form: $y = mx + b$:

$$\tan\phi = \frac{m_1 - m_2}{1 + m_1m_2}$$

$$\cos\phi = \frac{1 + m_1m_2}{\sqrt{1 + m_1^2} \cdot \sqrt{1 + m_2^2}}$$

$$\sin\phi = \frac{m_1 - m_2}{\sqrt{1 + m_1^2} \cdot \sqrt{1 + m_2^2}}$$

▷ The angle ϕ is measured counterclockwise between the two lines.

8.2.5 Parallel and perpendicular straight lines

For **parallel lines,**

$$\frac{A_1}{A_2} = \frac{B_1}{B_2}, \quad \text{or} \quad m_1 = m_2.$$

For **perpendicular straight lines,**

$$A_1 A_2 + B_1 B_2 = 0, \quad \text{or} \quad m_1 = -\frac{1}{m_2}.$$

8.3 Circle

8.3.1 Equations of a circle

Equation of a circle with radius R and with center at the origin, in **Cartesian coordinates**:

$$x^2 + y^2 = R^2.$$

Equation of a circle with radius R and center (x_0, y_0):

$$(x - x_0)^2 + (y - y_0)^2 = R^2.$$

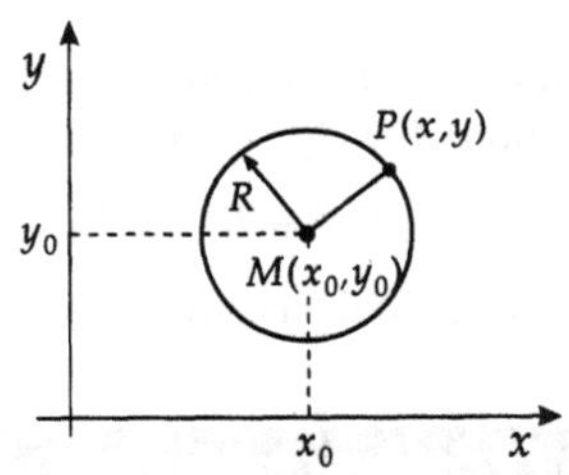

The general equation of the second degree

$$ax^2 + 2bxy + cy^2 + 2dx + 2ey + f = 0$$

describes a circle **only** for $a = c$ and $b = 0$. General form of the circle:

$$x^2 + y^2 + 2dx + 2ey + f = 0,$$

with radius $R = \sqrt{d^2 + e^2 - f}$ and center $(x_0, y_0) = (-d, -e)$.
Case distinction:
$f < d^2 + e^2$: real curve,
$f = d^2 + e^2$: only the center $(x_0, y_0) = (-d, -e)$ is a solution of the equation,
$f > d^2 + e^2$: no real curve.
Parametric representation of the circle:

$$x = x_0 + R\cos t, \quad y = y_0 + R\sin t,$$

t: angle between the variable radius and the x-axis.
Equation of a circle in polar coordinates, general equation:

$$r^2 - 2rr_0\cos(\phi - \phi_0) + r_0^2 = R^2,$$

with center (r_0, ϕ_0).

If the center lies on the polar axis and the circle goes through the origin, then the equation of the circle is given by:

$$r = 2R\cos\phi .$$

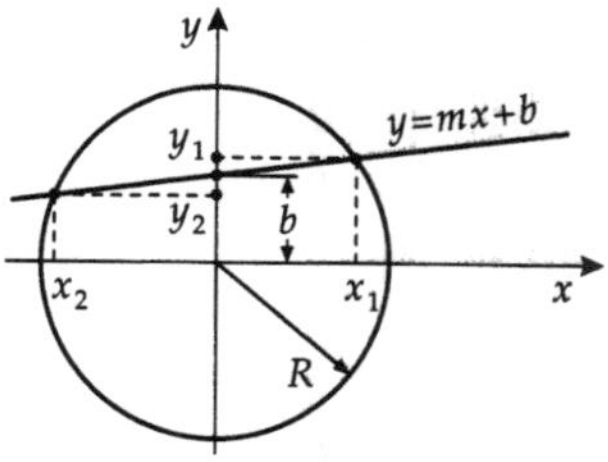

Circle and straight line

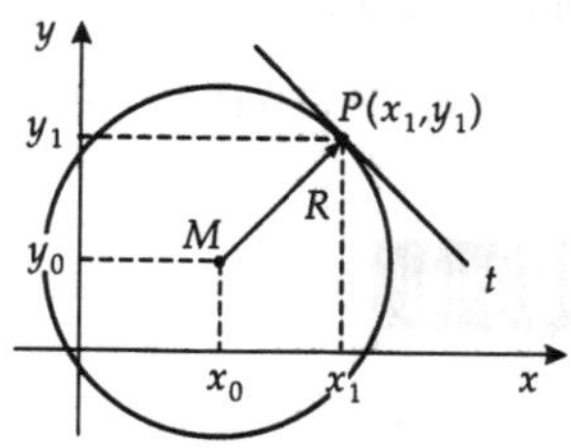

Tangent to a circle

8.3.2 Circle and straight line

Points of intersection of a circle $x^2 + y^2 = r^2$ with a line $y = mx + b$ at

$$x_{1,2} = \frac{-mb \pm \sqrt{\Delta}}{1+m^2}, \quad y_{1,2} = \frac{b \pm m\sqrt{\Delta}}{1+m^2},$$

with the discriminant

$$\Delta = r^2(1+m^2) - b^2 .$$

Case distinction:
Secant for $\Delta > 0$ (two points of intersections),
Tangent for $\Delta = 0$ (one point of intersection),
no point of intersection for $\Delta < 0$.

8.3.3 Intersection of two circles

Two circles whose centers M_1 and M_2 have a distance from each other that is smaller than the sum of the radii, $\overline{M_1M_2} < (R_1 + R_2)$, intersect each other in two points $P_1(x_1, y_1)$ and $P_2(x_2, y_2)$.
Radical axis of the two circles, the straight line through the two intersection points P_1 and P_2 (common secant of both circles).

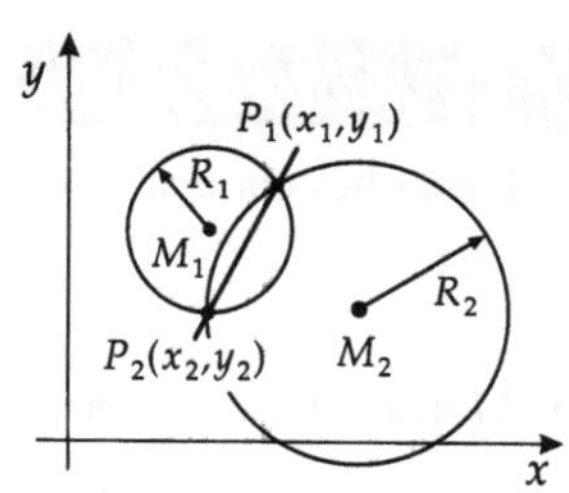

For the circles

$$x^2 + y^2 + 2d_1x + 2e_1y + f_1 = 0 , \quad M_1(-d_1, e_1)$$
$$x^2 + y^2 + 2d_2x + 2e_2y + f_2 = 0 , \quad M_2(d_2, e_2)$$

the equation of the radical axis is

$$(d_1 - d_2)x + (e_1 - e_2)y + \frac{1}{2}(f_1 - f_2) = 0 .$$

8.3.4 Equation of the tangent to a circle

Equation of the tangent at point $P(x_1, y_1)$ to a circle:

$$xx_1 + yy_1 = R^2.$$

With a general location of the circle, with center $M(x_0, y_0)$:

$$(x - x_0)(x_1 - x_0) + (y - y_0)(y_1 - y_0) = r^2.$$

8.4 Ellipse

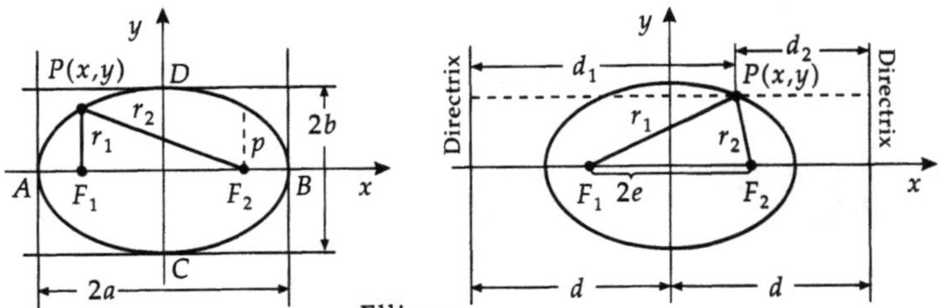

Ellipse

Elements of the ellipse:
Vertices of the ellipse: Points A, B, C, D,
Major axis: Length $AB = 2a$,
Minor axis: Length $CD = 2b$,
Foci (points on the major axis with the distance $e = \sqrt{a^2 - b^2}$ from the center): Points F_1, F_2,
Linear eccentricity: e,
Numerical eccentricity of the **ellipse**: $\varepsilon = \dfrac{e}{a} < 1$,
Semifocal chord (half the length of the chord through a focus parallel to the minor axis):

$$p = \frac{b^2}{a}.$$

8.4.1 Equations of the ellipse

Normal form (the coordinate axes coincide with the axes of the ellipse):

$$\frac{x^2}{a^2} + \frac{y^2}{b^2} = 1,$$

Principal form (ellipse in general position with center (x_0, y_0)):

$$\frac{(x - x_0)^2}{a^2} + \frac{(y - y_0)^2}{b^2} = 1,$$

Parametric form (center (x_0, y_0)):

$$x = x_0 + a\cos t\,, \qquad y = y_0 + b\sin t\,, \qquad 0 \le t \le 2\pi\,.$$

Polar coordinates (pole at the center, the polar axis is the x-axis):

$$r = \frac{b}{\sqrt{1 - \varepsilon^2 \cos^2 \varphi}} \qquad (\varepsilon < 1),$$

Polar coordinates (pole at the left focus, the polar axis is the x-axis):

$$r = \frac{p}{1 - \varepsilon \cos \varphi} \quad (\varepsilon < 1), \quad p = \frac{b^2}{a}.$$

Polar coordinates (pole in the right focus, the polar axis is the x-axis):

$$r = \frac{p}{1 + \varepsilon \cos \varphi}.$$

▷ The equation of the ellipse in polar coordinates, in this last form, applies to all curves of the second order (see Section 5.38).

8.4.2 Focal properties of the ellipse

Focal properties of the ellipse: The ellipse is the geometric locus of all points for which the sum of distances between two given fixed points (foci) is a constant equal to $2a$. Distance between an arbitrary point $P(x, y)$ on the ellipse and the foci:

$$r_1 = a - \varepsilon x, \quad r_2 = a + \varepsilon x, \quad r_1 + r_2 = 2a.$$

Directrices: Line parallel to the minor axis at a distance $d = \dfrac{a}{\varepsilon}$.

Properties of the directrices of the ellipse: For an arbitrary point $P(x, y)$ on the ellipse, it holds that:

$$\frac{r_1}{d_1} = \frac{r_2}{d_2} = \varepsilon.$$

8.4.3 Diameters of the ellipse

The diameters of the ellipse: Chords that pass through the center of the ellipse and are bisected by the center. **Conjugate diameter**: Midpoints of all chords parallel to a diameter of the ellipse. For the slopes of the diameter and the conjugate diameter, it holds that:

$$m_1 \cdot m_2 = -\frac{b^2}{a^2}.$$

- **Theorem of Apollonius**: If $2a_1$ and $2b_1$ are the lengths, and α, β ($m_1 = -\tan\alpha$, $m_2 = \tan\beta$) are the slope angles of the two conjugate diameters, it holds that:

$$a_1 b_1 \sin(\alpha + \beta) = ab \quad \text{and} \quad a_1^2 + b_1^2 = a^2 + b^2.$$

8.4.4 Tangent and normal to the ellipse

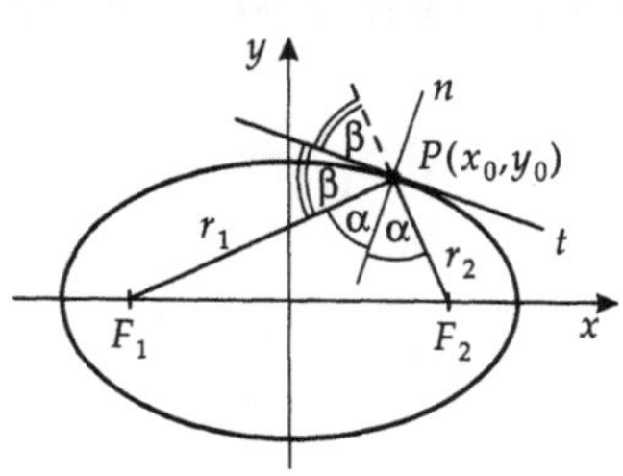

Ellipse: tangent and normal

Equation of the tangent at the point $P(x_0, y_0)$:

$$\frac{xx_0}{a^2} + \frac{yy_0}{b^2} = 1.$$

▷ The normal and tangent of the ellipse are the angle-bisectors of the interior and exterior angles between the radius vectors from the foci to the point of contact.

A straight line of the general form $Ax + By + C = 0$ is a tangent to the ellipse if

$$A^2a^2 + B^2b^2 - C^2 = 0.$$

Equation of the normal at the point $P(x_0, y_0)$:

$$y - y_0 = \frac{a^2}{b^2}\frac{y_0}{x_0}(x - x_0) .$$

8.4.5 Curvature of the ellipse

Radius of curvature of an ellipse at point $P(x_0, y_0)$:

$$R = a^2b^2\left(\frac{x_0^2}{a^4} + \frac{y_0^2}{b^4}\right)^{3/2} = \frac{(r_1r_2)^{3/2}}{ab} = \frac{p}{\sin^3\phi},$$

ϕ: angle between tangent and radius vector from one of the foci to the point of contact.
Minimum curvature for the vertices C and D: $R = a^2/b$,
Maximum curvature for the vertices A and B: $R = b^2/a = p$.

8.4.6 Areas and circumference of the ellipse

Area of an ellipse:

$$A = \pi ab.$$

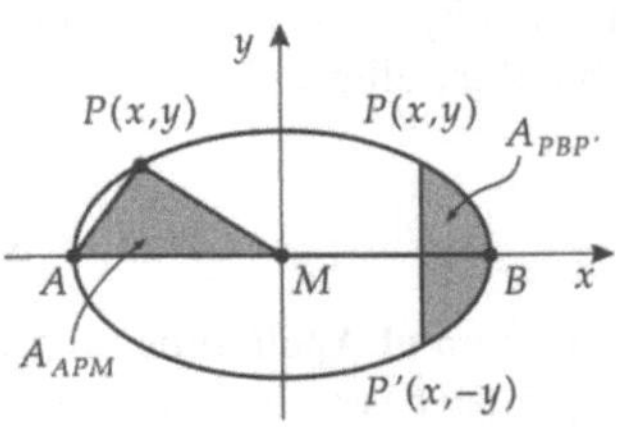

Area of the sector APM with the point $P(x, y)$ on the ellipse:

$$A_{APM} = \frac{ab}{2}\arccos\frac{x}{a}.$$

Area of the segment PBP' with the two points $P(x, y)$ and $P'(-x, y)$ on the ellipse:

$$A_{PBP'} = ab \cdot \arccos\frac{x}{a} - xy.$$

The circumference of the ellipse cannot be represented analytically! Has to be calculated numerically by the elliptic integral of the second kind.

$$U = 4aE(\varepsilon, 2\pi) = a\int_0^{2\pi}\sqrt{1 - \varepsilon^2\sin^2 t}\,\mathrm{d}t$$

$$= 2\pi a\left(1 - \frac{1}{4}\varepsilon^2 - \frac{3}{64}\varepsilon^4 - \frac{5}{256}\varepsilon^6 - \frac{175}{16384}\varepsilon^8 - \ldots\right).$$

Other approximation formula:

$$U \approx \pi(1.5(a + b) - \sqrt{ab}), \quad U \approx \pi(a + b)\frac{64 - 3l^4}{64 - 16l^2},$$

with

$$l = \frac{a-b}{a+b}.$$

8.5 Parabola

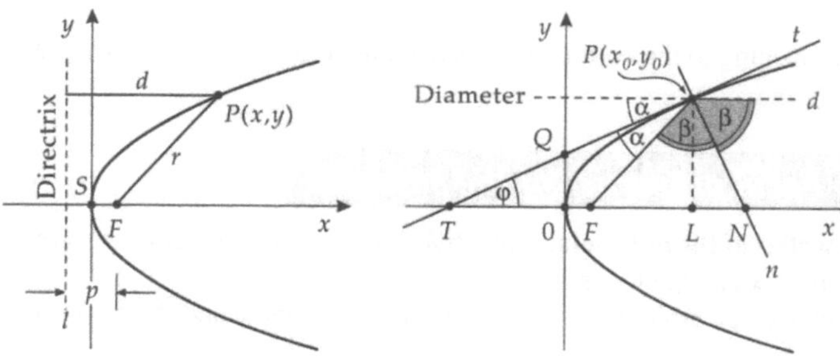

Parabola

Elements of the parabola:
Vertex of the parabola: Point S,
Axis of the parabola: x-axis
Focus (point on the x-axis at distance $p/2$ from the vertex): Point F,
Numerical eccentricity of the parabola: $\varepsilon = 1$,
Semifocal chord (half the length of a chord through the focus perpendicular to the axis of the parabola): p,
Directrix (line perpendicular to the axis of the parabola) l: The directrix intersects the axis at a distance $p/2$ from the vertex).

8.5.1 *Equations of the parabola*

Normal form (the origin corresponds to the vertex of the parabola pointing to the left):

$$y^2 = 2px,$$

Principal form (vertex (x_0, y_0)):

$$(y - y_0)^2 = 2p(x - x_0).$$

Parametric representation (vertex (x_0, y_0)):

$$x = x_0 + \frac{1}{2p}t^2, \quad y = y_0 + t, \quad 0 \le t \le \infty.$$

Polar coordinates (pole at the vertex, the polar axis is the x-axis):

$$r = 2p\cos\varphi(1 + \cot^2\varphi).$$

Polar coordinates (pole at the focus, the polar axis is the x-axis):

$$r = \frac{p}{1 - \cos\varphi}.$$

▷ $p > 0$: opening to the right; $p < 0$: opening to the left.

Vertical axis (with the parameter $p = \frac{1}{2|a|}$):

$$y = ax^2 + bx + c.$$

Coordinates of the vertex $S(x_0, y_0)$ in this representation:

$$x_0 = -\frac{b}{2a}, \quad y_0 = \frac{4ac - b^2}{4a}.$$

▷ $a > 0$: opening to the bottom; $a < 0$: opening to the top; $a = 0$: straight line.

8.5.2 *Focal properties of the parabola*

Focal properties of the parabola: A parabola is the locus of all points that are equidistant from the directrix and the focus.
Distance between an arbitrary point $P(x, y)$ on the parabola and the focus or directrix:

$$FP = PL = d = x + \frac{p}{2}.$$

8.5.3 *Diameters of the parabola*

Diameters of the parabola: Lines parallel to the axis of the parabola. A diameter bisects the chords with slope m, parallel to a tangent.
Equation of the diameter of the parabola:

$$y = \frac{p}{m}.$$

8.5.4 *Tangent and normal of the parabola*

Equation of the tangent at the point $P(x_0, y_0)$:

$$yy_0 = p(x + x_0).$$

▷ The normal and tangent of the parabola are angle-bisectors of the interior and exterior angle between the radius vector and the diameter through the point of contact.

The segment of the tangent of the parabola between contact point P and the point of intersection T with the axis of the parabola (x-axis) is bisected by the tangent in the vertex (y-axis):

$$\overline{TQ} = \overline{QP}. \quad \text{Furthermore,} \quad \overline{TO} = \overline{OL}.$$

A line $y = mx + b$ is tangent to the parabola for

$$p = 2bm.$$

Equation of the normal at the point $P(x_0, y_0)$:

$$y = -\frac{y_0}{p}(x - x_0) + y_0 .$$

8.5.5 Curvature of a parabola

Radius of curvature of a parabola at the point $P(x_0, y_0)$:

$$R = \frac{(p + 2x_0)^{3/2}}{\sqrt{p}} = \frac{p}{\sin^3 \varphi} = \frac{N^3}{p^2}.$$

φ: angle between the tangent and the x-axis,
N: length of the normal segment PN.
Maximum curvature at the vertex S: $R = p$.

8.5.6 Areas and arc lengths of the parabola

Area of the segment PSQ of the parabola corresponds to 2/3 of the area of the parallelogram $QRTP$. Area of the segment PSP' of the parabola with the points $P(x, y)$ and $P'(x, -y)$:

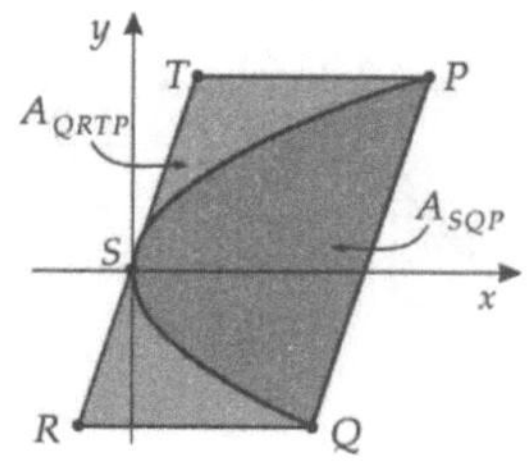

$$A_{PSP'} = \frac{4}{3}xy.$$

Length of the arc of the parabola between vertex S and point $P(x, y)$:

$$L = \frac{p}{2}\left[\sqrt{\frac{2x}{p}\left(1 + \frac{2x}{p}\right)} + \ln\left(\sqrt{\frac{2x}{p}} + \sqrt{1 + \frac{2x}{p}}\right)\right]$$

$$= -\sqrt{x\left(x + \frac{p}{2}\right)} + \frac{p}{2}\operatorname{Arsinh}\sqrt{\frac{2x}{p}}.$$

Approximation formula for small values of x/y:

$$L \approx y\left[1 + \frac{2}{3}\left(\frac{x}{y}\right)^2 - \frac{2}{5}\left(\frac{x}{y}\right)^4\right].$$

8.5.7 Parabola and straight line

Coordinates of the point of intersection of a straight line $y = mx + b$ with a parabola $y^2 = 2px$:

$$x_{1,2} = \frac{p - mb \pm \sqrt{\Delta}}{m^2}, \quad y_{1,2} = \frac{p \pm \sqrt{\Delta}}{m},$$

with the discriminant

$$\Delta = p(p - 2mb).$$

Case distinction:

- two points of intersection for $\Delta > 0$,
- one point of intersection for $\Delta = 0$,
- no point of intersection for $\Delta < 0$.

- Every straight line parallel to the x-axis intersects the parabola in only one point.
- A line parallel to the y-axis with $x = a \neq 0$ (a and p have the same size) intersects the parabola in two points.

8.6 Hyperbola

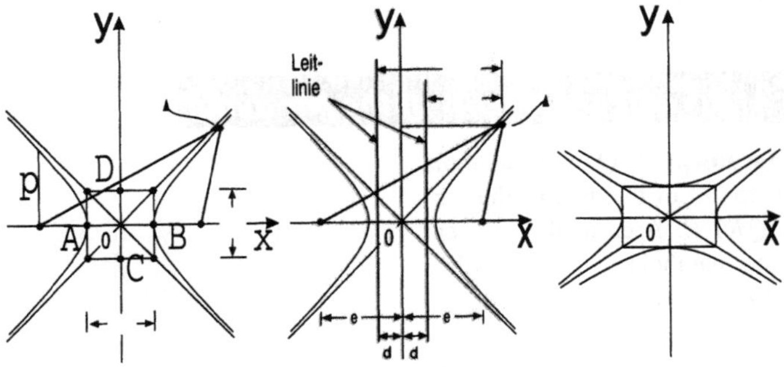

Hyperbola

Elements of the hyperbola:
Vertices of the hyperbola: Points A, B,
Center of the hyperbola: Point M,
Real axis: $AB = 2a$,
Imaginary axis: $CD = 2b$,
Foci (points on the real axis at distance $e = \sqrt{b^2 + a^2} > a$ from the center): Points F_1, F_2,
Numerical eccentricity of the hyperbola: $\varepsilon = \dfrac{e}{a} > 1$,
Semifocal chord (half the length of the chord through a focus parallel to the imaginary axis): $p = \dfrac{b^2}{a}$.

8.6.1 Equations of the hyperbola

Normal form (the coordinate axes correspond to the axes of the hyperbola):

$$\frac{x^2}{a^2} - \frac{y^2}{b^2} = 1,$$

Principal form (center (x_0, y_0)):

$$\frac{(x - x_0)^2}{a^2} - \frac{(y - y_0)^2}{b^2} = 1,$$

Parametric form (upper sign: right branch; lower sign: left branch):

$$x = x_0 \pm a \cosh t, \qquad y = y_0 + b \sinh t.$$

Parameter t: The area enclosed by the two straight lines from the center of the unit hyperbola to points $P(x, y)$, $P'(x, -y)$ and the segment 0 from P to P' (see also the geometric interpretation of the hyperbolic functions, p. . .).

Polar coordinates (pole at the center, the polar axis is the x-axis):

$$r = \frac{b}{\sqrt{\varepsilon^2 \cos^2 \varphi - 1}} \qquad (\varepsilon > 1)\,.$$

Polar coordinates (pole at the left focus, the polar axis is the x-axis) only left branch:

$$r = \frac{p}{1 + \varepsilon \cos \varphi}\,.$$

▷ The equation of the hyperbola in polar coordinates in the last form applies to all curves of the second order.

Equilateral hyperbola has axes of equal length, $a = b$; equation of equilateral hyperbole:

$$x^2 - y^2 = a^2.$$

The asymptotes are mutually perpendicular:

$$y = \pm x.$$

Equation of the equilateral hyperbola if the asymptotes are chosen as coordinate axes (rotation through 45°):

$$xy = \frac{a^2}{2}\,, \quad \text{or} \quad y = \frac{a^2}{2x}.$$

8.6.2 *Focal properties of the hyperbola*

Focal properties of the hyperbola: A hyperbola is the geometric locus of all points for which the difference of the distances between two given fixed points (foci) is a constant equal to $2a$.

Distance between an arbitrary point $P(x, y)$ on the left branch of the hyperbola and the foci:

$$r_1 = a - \varepsilon x, \quad r_2 = -a - \varepsilon x, \quad r_1 - r_2 = 2a,$$

and for points on the right branch:

$$r_1 = \varepsilon x - a, \quad r_2 = \varepsilon x + a, \quad r_2 - r_1 = 2a.$$

Directrices: Parallel to the real axis and at a distance $d = \dfrac{a}{\varepsilon}$ from the center.

Directrix property of the hyperbola: For an arbitrary point $P(x, y)$ of the hyperbola it holds that:

$$\frac{r_1}{d_1} = \frac{r_2}{d_2} = \varepsilon.$$

Asymptotes of the hyperbola: Straight lines approached without limit by the branches of the hyperbola at infinity. The equation of the two asymptotes is:

$$y = \pm \frac{b}{a} x.$$

The segments between the points of contact P of the tangent to the hyperbola and the points of intersection with asymptotes T and T' are equal on both sides (the point of contact P bisects the tangent segments between the asymptotes):

$$\overline{PT} = \overline{PT'}.$$

8.6.3 Tangent and normal to the hyperbola

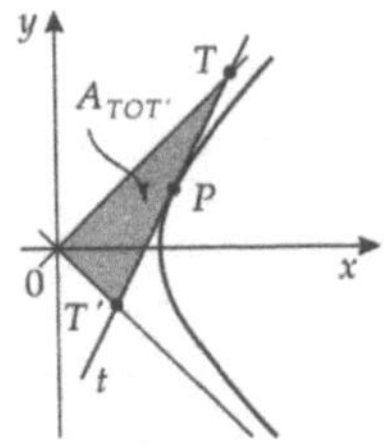

Equation of the tangent at the point $P(x_0, y_0)$:

$$\frac{xx_0}{a^2} - \frac{yy_0}{b^2} = 1.$$

▷ The normal and tangent to the hyperbola are angle-bisectors of the interior and exterior angles between the radius vectors of the foci and point of contact.

A straight line given in the general form $Ax + By + C = 0$ is a tangent to the hyperbola for:

$$A^2a^2 - B^2b^2 - C^2 = 0.$$

8.6.4 Conjugate hyperbolas and diameter

Conjugate hyperbolas described by

$$\frac{x^2}{a^2} - \frac{y^2}{b^2} = 1 \quad \text{and} \quad \frac{y^2}{b^2} - \frac{x^2}{a^2} = 1$$

have common asymptotes. Real and imaginary axes are interchanged.
Diameters of the hyperbola: Chords that pass through the center of the hyperbola; and are bisected it.
Conjugate diameters: The midpoints of all chords parallel to a diameter of the hyperbola.
For the gradients of the diameters and conjugate diameters, it holds that:

$$m_1 \cdot m_2 = \frac{b^2}{a^2}.$$

- If $2a_1$ and $2b_1$ are the lengths, and if α, β $(m_1 = -\tan\alpha, m_2 = \tan\beta)$ are the slope angles of two conjugate diameters, it holds that:

$$a_1b_1\sin(\alpha - \beta) = ab \quad \text{and} \quad a_1^2 - b_1^2 = a^2 - b^2.$$

8.6.5 Curvature of a hyperbola

Radius of curvature of a hyperbola at the point $P(x, y)$:

$$R = a^2b^2\left(\frac{x_0^2}{a^4} + \frac{y_0^2}{b^4}\right)^{3/2} = \frac{(r_1r_2)^{3/2}}{ab} = \frac{p}{\sin^3\phi}.$$

ϕ: angle between the tangent and radius vector of the point of contact.
Maximum curvature at the vertices A and B: $R = b^2/a = p$.

8.6.6 Areas of hyperbola

Area of the triangle between the tangents and both asymptotes:

$$A_{TOT'} = ab.$$

Area of the parallelogram with sides parallel to the asymptotes and with a point of the hyperbola as vertex:

$$A_{QORM} = \frac{a^2 + b^2}{4} = \frac{e^2}{4}.$$

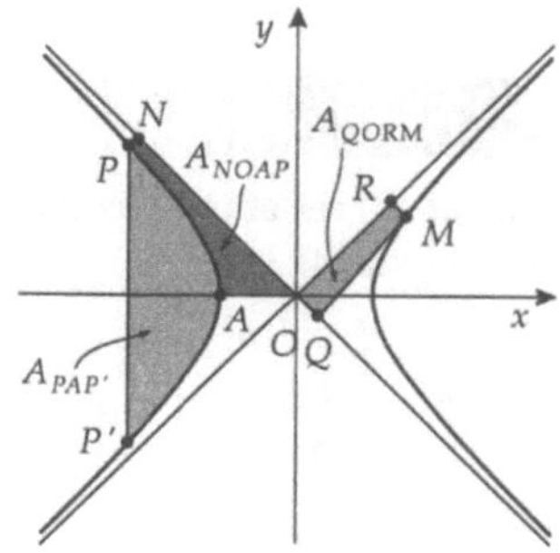

Area of the sector $NOAP$ with the point $P(x, y)$ of the hyperbola and the parallel PN to the asymptote:

$$A_{NOAP} = \frac{ab}{4} + \frac{ab}{2} \ln\left(\frac{2PP'}{e}\right).$$

Area of the segment PAP' of the hyperbola with the points $P(x, y)$ and $P'(-x, y)$ of the hyperbola:

$$A_{PAP'} = xy - ab \ln\left(\frac{x}{a} + \frac{y}{b}\right).$$

8.6.7 Hyperbola and straight line

Coordinates of the point of intersection of the hyperbola in the form of

$$\frac{x^2}{a^2} - \frac{y^2}{b^2} = 1\,,$$

with a straight line given in the form $y = mx + n$:

$$x_{1,2} = \frac{-a^2 mn \pm ab\sqrt{\Delta}}{a^2m^2 - b^2}, \qquad y_{1,2} = \frac{-b^2 n \pm abm\sqrt{\Delta}}{a^2m^2 - b^2}$$

with the discriminant:

$$\Delta = b^2 + n^2 - a^2m^2.$$

Case distinction:
two points of intersection for $\Delta > 0$,
one point of intersection for $\Delta = 0$,
no point of intersection for $\Delta < 0$.

8.7 General equation of conics

General equation of curves of the second order (conics):

$$ax^2 + 2bxy + cy^2 + 2dx + 2ey + f = 0\,.$$

Defines a hyperbola, a parabola, an ellipse, a circle, or a pair of lines (**singular curve of the second order**).

8.7.1 Form of conics

Invariants of curves of the second order (do not change when the coordinate system is translated or rotated):

$$A = \begin{vmatrix} a & b & d \\ b & c & e \\ d & e & f \end{vmatrix}, \quad B = \begin{vmatrix} a & b \\ b & c \end{vmatrix} = ac - b^2, \quad C = a + c.$$

Case	Case	Shape of the curve
	$B < 0$	Hyperbola
$A \neq 0$	$B = 0$	Parabola
	$B > 0$	Ellipse
	$B < 0$	Pair of intersecting lines
$A = 0$	$B = 0$	Pair of parallel real straight lines
	$B > 0$	Pair of imaginary straight lines with a real point of intersection

▷ Here, imaginary means that the equation cannot be satisfied for any real values.

8.7.2 Transformation to principal axes

Transformation of the general equation to principal axes (the symmetry axes of the curve correspond to the coordinate axes) by
1. translation of the origin (elimination of the linear terms) and
2. rotation of the coordinate axes (elimination of the mixed quadratic term) by the angle φ,

$$\operatorname{tg}(2\varphi) = \frac{2b}{a - c}.$$

Normal form in the case of **ellipse** and **hyperbola**:

$$a'x'^2 + c'y'^2 + \frac{A}{B} = 0,$$

with

$$a' = \frac{a + c + \sqrt{(a-c)^2 + 4b^2}}{2}, \quad c' = \frac{a + c - \sqrt{(a-c)^2 + 4b^2}}{2}.$$

Normal form in the case of **parabola**:

$$y'^2 = 2px', \quad \text{where} \quad p = \frac{ae - bd}{(a+c)\sqrt{a^2 + b^2}}.$$

8.7.3 Geometric construction (conic section)

General properties of the curves of the second order:
The curves of the second order are plane sections through a right circular cone (**conic sections**).

Singular conic: The plane section passes through the vertex of the cone (**point**) or touches the cone in a generator (**straight line**).

Non-singular conics: The plane of section

- intersects all generators: **ellipse** (E) or,
- is parallel to one generator: **parabola** (P) or,
- is parallel to two generators, intersects both halves of the double cone: **hyperbola** (H)

▷ The **circle** (C) is a special case of the ellipse: The plane section is orthogonal to the axis of the cone. **A pair of straight lines** intersecting each other is formed when the plane section runs parallel to two generators through the vertex.

▷ Generators of the cone: Two lines on the surface of the cone that arise by a plane section through the vertex parallel to the z-axis.

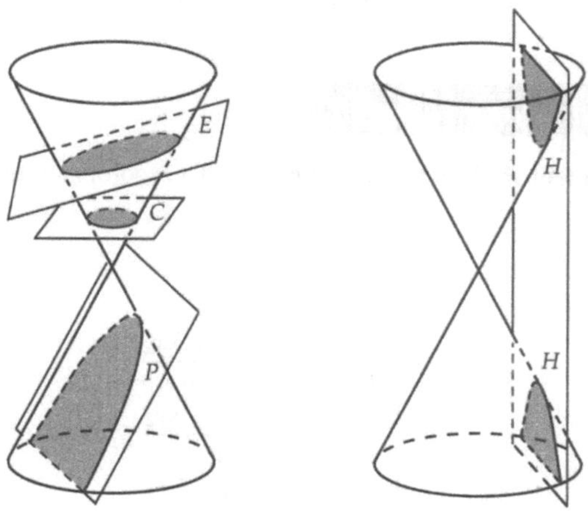

Conic sections:
E = ellipse; C = circle;
P = parabola; H = hyperbole

8.7.4 Directrix property

Directrix property: The geometric locus of all points with constant ratio ε (**excentricity**) of the distances to a given line (the directrix) and a given point (the focus) is a curve of the second order. For the eccentricity, it holds that

$\varepsilon > 1$: **Hyperbola,**

$\varepsilon = 1$: **Parabola,**

$\varepsilon < 1$: **Ellipse,**

$\varepsilon = 0$: **Circle.**

8.7.5 Polar equation

Polar equation: In polar coordinates, the equation for a curve of the second order reads:

$$r = \frac{p}{1 + \varepsilon \cos\varphi},$$

p: semifocal chord of the curve,
ε: eccentricity of the curve.
The pole lies in the focus, the polar axis is directed toward the nearest vertex.
▷ In this form, only one branch of the curve is obtained for the hyperbola.

8.8 Elements in space

8.8.1 Distance between two points

Cartesian coordinate system: Distance between $P_1(x_1, y_1, z_1)$ and $P_2(x_2, y_2, z_2)$:

$$d = \sqrt{(x_2 - x_1)^2 + (y_2 - y_1)^2 + (z_2 - z_1)^2}\,.$$

8.8.2 Division of a segment

Division of a segment $\overline{P_1P_2}$ in a given ratio $a : b$: The coordinates of point P with:

$$\frac{\overline{P_1P}}{\overline{PP_2}} = \frac{a}{b} = \lambda :$$

$$x = \frac{bx_1 + ax_2}{b + a} = \frac{x_1 + \lambda x_2}{1 + \lambda},$$
$$y = \frac{by_1 + ay_2}{b + a} = \frac{y_1 + \lambda y_2}{1 + \lambda},$$
$$z = \frac{bz_1 + az_2}{b + a} = \frac{z_1 + \lambda z_2}{1 + \lambda}.$$

Midpoint of segment $\overline{P_1P_2}$:

$$x = \frac{x_1 + x_2}{2}, \quad y = \frac{y_1 + y_2}{2}, \quad z = \frac{z_1 + z_2}{2}.$$

□ **Center of gravity** of a system of material points $M_i(x_i, y_i, z_i)$ with masses m_i ($i = 1,2,..,n$):

$$x = \frac{\sum m_i x_i}{\sum m_i}, \quad y = \frac{\sum m_i y_i}{\sum m_i}, \quad z = \frac{\sum m_i z_i}{\sum m_i}, .$$

8.8.3 Volume of a tetrahedron

Volume of a tetrahedron with vertices $P_1(x_1, y_1, z_1)$, $P_2(x_2, y_2, z_2)$, $P_3(x_3, y_3, z_3)$, and $P_4(x_4, y_4, z_4)$:

$$V = \frac{1}{6}\begin{vmatrix} x_1 & y_1 & z_1 & 1 \\ x_2 & y_2 & z_2 & 1 \\ x_3 & y_3 & z_3 & 1 \\ x_4 & y_4 & z_4 & 1 \end{vmatrix} = \frac{1}{6}\begin{vmatrix} x_1 - x_2 & y_1 - y_2 & z_1 - z_2 \\ x_1 - x_3 & y_1 - y_3 & z_1 - z_3 \\ x_1 - x_4 & y_1 - y_4 & z_1 - z_4 \end{vmatrix}.$$

Four points lie in one plane if

$$\begin{vmatrix} x_1 - x_2 & y_1 - y_2 & z_1 - z_2 \\ x_1 - x_3 & y_1 - y_3 & z_1 - z_3 \\ x_1 - x_4 & y_1 - y_4 & z_1 - z_4 \end{vmatrix} = 0.$$

8.9 Straight lines in space

8.9.1 Parametric representation of a straight line

Parametric representation of a straight line: Passing through the end-point of $\vec{\mathbf{a}}$ in direction of $\vec{\mathbf{b}} \neq \vec{\mathbf{0}}$:

$$g : \vec{\mathbf{x}} = \vec{\mathbf{a}} + t \cdot \vec{\mathbf{b}}, \quad t \,\varepsilon\, \mathbb{R}.$$

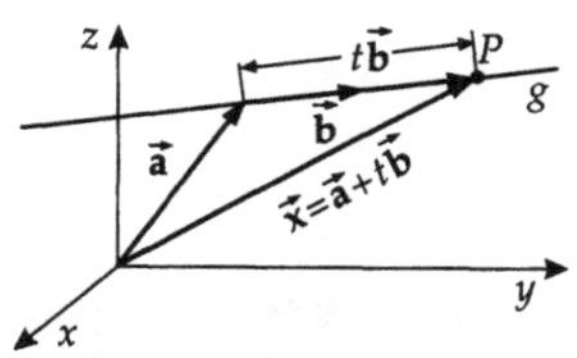

Two straight lines g_1, g_2 are parallel if and only if their **direction vectors** $\vec{\mathbf{b}}_1, \vec{\mathbf{b}}_2$ are linearly dependent.
Symbolic notation:

$$g_1 || g_2 \iff \exists s \neq 0 \quad \text{with} \quad \vec{\mathbf{b}}_1 = s\vec{\mathbf{b}}_2 .$$

Parametric representation of a line passing through two points P_1, P_2:

$$g : \vec{\mathbf{x}} = \vec{\mathbf{p}}_1 + t \cdot (\vec{\mathbf{p}}_2 - \vec{\mathbf{p}}_1) \quad \text{with} \quad t \,\varepsilon\, \mathbb{R}.$$

8.9.2 Point of intersection of two straight lines

Two straight lines intersect if and only if there is a vector $\vec{\mathbf{x}}_0$ whose end-point lies on line (g_1) as well as on line (g_2).
Equations of straight lines:

$$g_1 : \vec{\mathbf{x}} = \vec{\mathbf{a}}_1 + r\vec{\mathbf{b}}_1 ,$$
$$g_2 : \vec{\mathbf{x}} = \vec{\mathbf{a}}_2 + s\vec{\mathbf{b}}_2 .$$

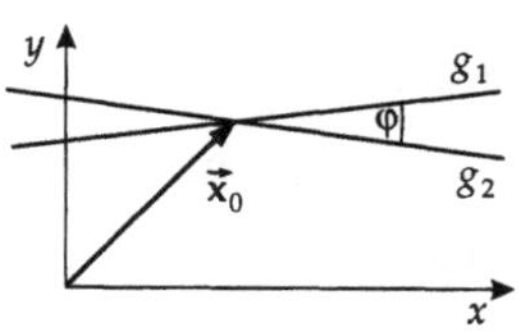

System of equations for r, s:

$$\vec{\mathbf{a}}_1 + r \cdot \vec{\mathbf{b}}_1 = \vec{\mathbf{a}}_2 + s \cdot \vec{\mathbf{b}}_2 , \quad \text{or}$$
$$r \cdot \vec{\mathbf{b}}_1 - s \cdot \vec{\mathbf{b}}_2 = \vec{\mathbf{a}}_2 - \vec{\mathbf{a}}_1 .$$

Case distinction in solving the system of equations:

1. The system of equations has *no* solution:
   The lines do not intersect.
   The cut set is empty (∅).
2. The system of equations has *exactly one* solution (r_0, s_0):
   The lines intersect in the end-point of

$$\vec{\mathbf{x}}_0 = \vec{\mathbf{a}}_1 + r_0 \cdot \vec{\mathbf{b}}_1 = \vec{\mathbf{a}}_2 + s_0 \cdot \vec{\mathbf{b}}_2$$

The cut set is $\{\vec{\mathbf{x}}_0\}$.

3. The system of equations has an *infinite number* of solutions.
   The lines are identical.
   The cut set is $g_1 = g_2$.

8.9.3 Angle of intersection between two intersecting straight lines

The angle φ between two intersecting lines

$$g_1 : \ \vec{\mathbf{x}} = \vec{\mathbf{a}}_1 + t \cdot \vec{\mathbf{b}}_1$$
$$g_2 : \ \vec{\mathbf{x}} = \vec{\mathbf{a}}_2 + t \cdot \vec{\mathbf{b}}_2$$

is the angle between the direction vectors

$$\cos\varphi = \frac{\vec{\mathbf{b}}_1 \cdot \vec{\mathbf{b}}_2}{|\vec{\mathbf{b}}_1| \cdot |\vec{\mathbf{b}}_2|} .$$

8.9.4 Foot of a perpendicular (perpendicular line)

The **foot** $\vec{\mathbf{x}}_0$ of the perpendicular from P to g is determined by two statements:

1. The end-point of $\vec{\mathbf{x}}_0$ lies on g; hence, thus is a parameter value t_0 such that

$$\vec{\mathbf{x}}_0 = \vec{\mathbf{a}} + t_0 \vec{\mathbf{b}} .$$

2. The perpendicular $\vec{\mathbf{x}}_0 - \vec{\mathbf{p}}$ is perpendicular to $\vec{\mathbf{b}}$, i.e.:

$$(\vec{\mathbf{x}}_0 - \vec{\mathbf{p}}) \cdot \vec{\mathbf{b}} = 0 \quad \text{or} \quad \vec{\mathbf{x}}_0 \cdot \vec{\mathbf{b}} = \vec{\mathbf{p}} \cdot \vec{\mathbf{b}} .$$

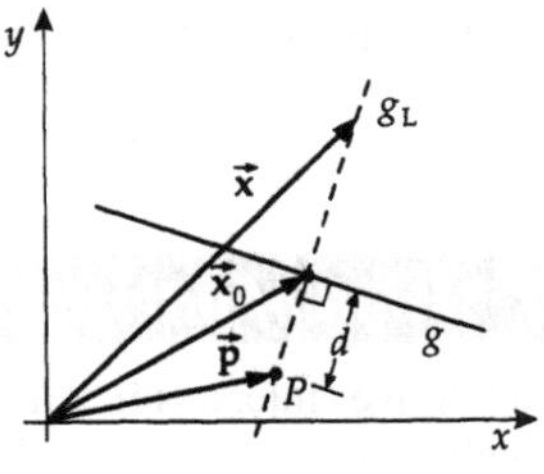

Perpendicular line

Foot:

$$\vec{\mathbf{x}}_0 = \vec{\mathbf{a}} + t_0 \cdot \vec{\mathbf{b}} \quad \text{with} \quad t_0 = \frac{(\vec{\mathbf{p}} - \vec{\mathbf{a}}) \cdot \vec{\mathbf{b}}}{\vec{\mathbf{b}}^2} .$$

Perpendicular:

$$\vec{\mathbf{x}}_0 - \vec{\mathbf{p}} .$$

Perpendicular line:

$$g_L : \ \vec{\mathbf{x}} = \vec{\mathbf{p}} + s \cdot (\vec{\mathbf{x}}_0 - \vec{\mathbf{p}}) .$$

8.9.5 Distance between a point and a straight line

For point P and line $g : \ \vec{\mathbf{x}} = \vec{\mathbf{a}} + t \cdot \vec{\mathbf{b}}$:
the **distance** is:

$$d = \frac{|\vec{\mathbf{b}} \times (\vec{\mathbf{p}} - \vec{\mathbf{a}})|}{|\vec{\mathbf{b}}|}$$

8.9.6 Distance between two lines

Parallel lines: The distance is obtained by determining the distance between an arbitrary point of one line and the other line.
Non-parallel lines:

$$g_1 : \vec{\mathbf{x}} = \vec{\mathbf{a}}_1 + t \cdot \vec{\mathbf{b}}_1$$
$$g_2 : \vec{\mathbf{x}} = \vec{\mathbf{a}}_2 + t \cdot \vec{\mathbf{b}}_2$$

Distance: Height of the parallelepiped spanned by the vectors $\vec{\mathbf{a}}_2 - \vec{\mathbf{a}}_1, \vec{\mathbf{b}}_1, \vec{\mathbf{b}}_2$:

$$d(g_1, g_2) = \frac{|<\vec{\mathbf{a}}_2 - \vec{\mathbf{a}}_1, \vec{\mathbf{b}}_1, \vec{\mathbf{b}}_2>|}{|\vec{\mathbf{b}}_1 \times \vec{\mathbf{b}}_2|} = \frac{\text{Volume of the parallelepiped}}{\text{Base of the parallelepiped}}$$

If g_1 and g_2 intersect, then $\vec{\mathbf{a}}_2 - \vec{\mathbf{a}}_1, \vec{\mathbf{b}}_1, \vec{\mathbf{b}}_2$ lie in one plane; the scalar triple product and thus the distance between g_1 and g_2 is zero.
Skew lines: Three non-parallel lines that do not intersect each other.
Symbolic notation:

$$g_1, g_2 \ \textbf{skew} \iff <\vec{\mathbf{a}}_2 - \vec{\mathbf{a}}_1, \vec{\mathbf{b}}_1, \vec{\mathbf{b}}_2> \neq 0 .$$

8.10 Planes in space

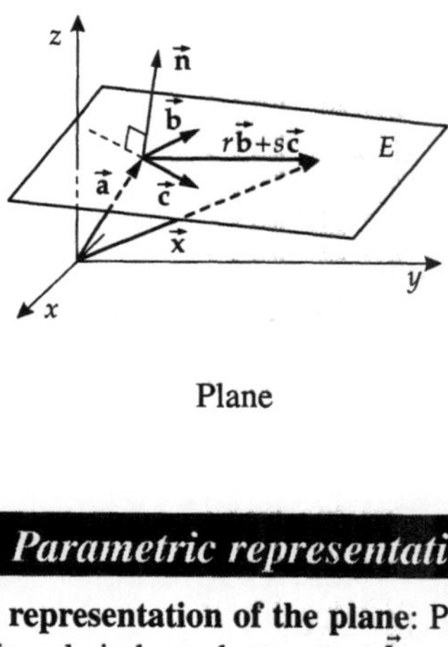

Plane

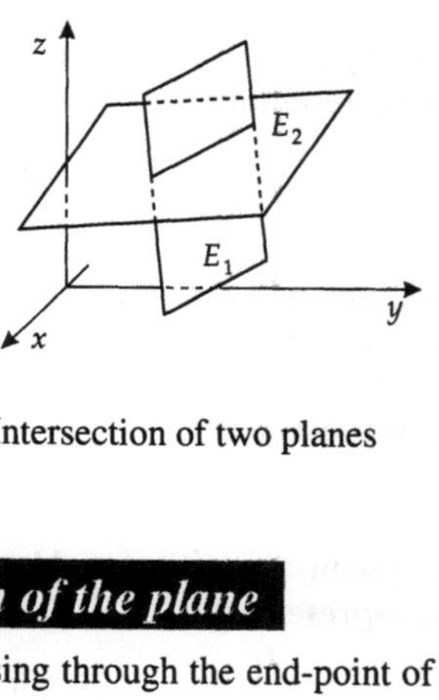

Intersection of two planes

8.10.1 Parametric representation of the plane

Parametric representation of the plane: Passing through the end-point of $\vec{\mathbf{a}}$, spanned by the two linearly independent vectors $\vec{\mathbf{b}}$ and $\vec{\mathbf{c}}$:

$$E : \vec{\mathbf{x}} = \vec{\mathbf{a}} + r \cdot \vec{\mathbf{b}} + s \cdot \vec{\mathbf{c}} \text{ , with } r, s \in \mathbb{R} .$$

Parametric representation of the plane: Through three **non-collinear** points, P_1, P_2, P_3:

$$E : \vec{\mathbf{x}} = \vec{\mathbf{p}}_1 + r \cdot (\vec{\mathbf{p}}_2 - \vec{\mathbf{p}}_1) + s \cdot (\vec{\mathbf{p}}_3 - \vec{\mathbf{p}}_1) \text{ with } r, s \in \mathbb{R} .$$

8.10.2 Coordinate representation of the plane

Normal vector of the plane:

$$\vec{\mathbf{n}} = (a, b, c) \neq (0, 0, 0) .$$

Coordinate representation of the plane,

$$E:\ d = \vec{\mathbf{n}} \cdot \vec{\mathbf{x}} = ax + by + cz\ .$$

- Two planes E_1 and E_2 are parallel if and only if their normal vectors $\vec{\mathbf{n}}_1$ and $\vec{\mathbf{n}}_2$ are linearly dependent.

Symbolic notation:

$$E_1 || E_2 \quad \Longleftrightarrow \quad \exists r \neq 0 \text{ with } \vec{\mathbf{n}}_1 = r\vec{\mathbf{n}}_2\ .$$

8.10.3 Hessian normal form of the plane

$$E:\ \vec{\mathbf{n}} \cdot \vec{\mathbf{x}} = d, \quad \text{with} \quad (|\vec{\mathbf{n}}| = 1, d \geq 0)\ ,$$

d = distance between plane and origin,
$\vec{\mathbf{n}}$ points from the origin to the plane.

8.10.4 Conversions

From parametric to coordinate representation:
Multiplication of the parametric representation by a vector $\vec{\mathbf{n}}$ perpendicular to the direction vectors $\vec{\mathbf{b}}$ and $\vec{\mathbf{c}}$ (**normal vector**):
Parametric representation

$$\vec{\mathbf{x}} = \vec{\mathbf{a}} + r \cdot \vec{\mathbf{b}} + s \cdot \vec{\mathbf{c}}\ .$$

Multiplication by $\vec{\mathbf{n}} = \vec{\mathbf{b}} \times \vec{\mathbf{c}}$ yields the
Coordinate representation

$$\vec{\mathbf{n}} \cdot \vec{\mathbf{x}} = \vec{\mathbf{n}} \cdot \vec{\mathbf{a}}$$
$$ax + by + cz = d\ .$$

From coordinate to parametric representation:
Solving the equation system of $ax + by + cz = d$ by isolating $\vec{\mathbf{x}}$. The result is written as a vector ($\vec{\mathbf{x}} = (x, y, z) = \ldots$):
Coordinate representation

$$ax + by + cz = d\ .$$

Solving the system of equations yields
Parametric representation

$$\vec{\mathbf{x}} = \vec{\mathbf{a}} + r \cdot \vec{\mathbf{b}} + s \cdot \vec{\mathbf{c}}\ .$$

From coordinate representation to Hessian normal form:
Division of the coordinate representation

$$ax + by + cz = d$$

by the magnitude

$$\sqrt{a^2 + b^2 + c^2} \quad \text{of} \quad \vec{\mathbf{n}} = (a, b, c)\ .$$

From parametric representation to Hessian normal form:
1. Convert parametric representation to coordinate representation.
2. Convert coordinate representation to Hessian normal form.

8.10.5 Distance between a point and a plane

Representation of plane E and point P: $E : \vec{\mathbf{n}} \cdot \vec{\mathbf{x}} = d$, P: end-point of the vector $\overline{p}$.
Distance between point P and plane E:

$$A = |\vec{\mathbf{n}} \cdot \vec{\mathbf{p}} - d| .$$

Case distinction:
$\vec{\mathbf{n}} \cdot \vec{\mathbf{p}} > d$: P lies on the one side of E, the origin on the other.
$\vec{\mathbf{n}} \cdot \vec{\mathbf{p}} = d$: P lies on plane E,
$\vec{\mathbf{n}} \cdot \vec{\mathbf{p}} < d$: P lies on the same side of E as the origin.

8.10.6 Point of intersection of a line and a plane

Line:

$$g : \vec{\mathbf{x}} = \vec{\mathbf{a}} + t \cdot \vec{\mathbf{b}} .$$

Plane

$$E : \vec{\mathbf{n}} \cdot \vec{\mathbf{x}} = d .$$

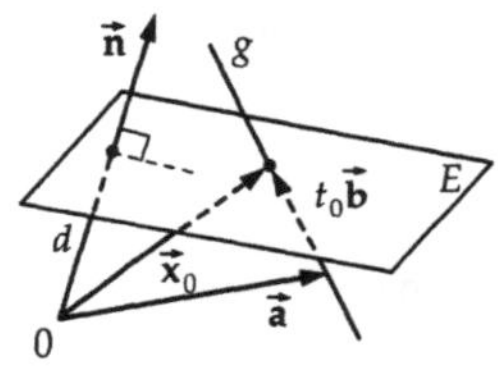

Point of intersection

$$\vec{\mathbf{x}}_0 = \vec{\mathbf{a}} + t_0\vec{\mathbf{b}} , \quad \text{with} \quad t_0 = \frac{d - \vec{\mathbf{n}} \cdot \vec{\mathbf{a}}}{\vec{\mathbf{n}} \cdot \vec{\mathbf{b}}},$$

if g and E are not parallel; thus it holds that:

$$G \not\parallel E , \quad \text{or } \vec{\mathbf{n}} \cdot \vec{\mathbf{b}} \neq 0 .$$

8.10.7 Angle of intersection between two intersecting planes

Planes:

$$E_1 : \vec{\mathbf{n}}_1 \cdot \vec{\mathbf{x}} = d_1$$
$$E_2 : \vec{\mathbf{n}}_2 \cdot \vec{\mathbf{x}} = d_2$$

The angle φ between two intersecting planes is the angle between their normal vectors:

$$\cos\varphi = \frac{\vec{\mathbf{n}}_1 \cdot \vec{\mathbf{n}}_2}{|\vec{\mathbf{n}}_1| \cdot |\vec{\mathbf{n}}_2|}.$$

8.10.8 Foot of the perpendicular (perpendicular line)

The foot $\vec{\mathbf{x}}_0$ of the perpendicular of P on E is determined by two statements:
1. The end-point of $\vec{\mathbf{x}}_0$ lies on E. Thus, $\vec{\mathbf{n}} \cdot \vec{\mathbf{x}}_0 = d$.
2. The end-point of $\vec{\mathbf{x}}_0$ lies on the perpendicular line

$$\vec{\mathbf{x}} = \vec{\mathbf{p}} + t \cdot \vec{\mathbf{n}}.$$

This results in a parameter value t_0 such that

$$\vec{\mathbf{x}}_0 = \vec{\mathbf{p}} + t_0 \cdot \vec{\mathbf{n}}.$$

Foot:

$$\vec{\mathbf{x}}_0 = \vec{\mathbf{p}} + t_0 \cdot \vec{\mathbf{n}}\,, \quad \text{with} \quad t_0 = \frac{d - \vec{\mathbf{n}} \cdot \vec{\mathbf{p}}}{\vec{\mathbf{n}}^2}\,.$$

Perpendicular:

$$\vec{\mathbf{x}}_0 - \vec{\mathbf{p}}\,.$$

Perpendicular line:

$$g_L : \vec{\mathbf{x}} = \vec{\mathbf{p}} + t \cdot \vec{\mathbf{n}}\,.$$

8.10.9 Reflection

Reflection of the point P in **Line**

$$g : \vec{\mathbf{x}} = \vec{\mathbf{a}} + t \cdot \vec{\mathbf{b}}$$

or in the **plane**

$$E : \vec{\mathbf{n}} \cdot \vec{\mathbf{x}} = d$$

Reflection point P':

$$\vec{\mathbf{p}}' = 2\vec{\mathbf{x}}_0 - \vec{\mathbf{p}}\,.$$

$\vec{\mathbf{x}}_0$: foot of the perpendicular from P to g or E.
Reflection of line g in plane E by reflecting two points of g in E and determining line g' through these two reflection points.

8.10.10 Distance between two parallel planes

Representation of two planes:

$$E_1 : \vec{\mathbf{n}}_1 \cdot (\vec{\mathbf{r}} - \vec{\mathbf{r}}_1) = 0, \quad E_2 : \vec{\mathbf{n}}_2 \cdot (\vec{\mathbf{r}} - \vec{\mathbf{r}}_2) = 0.$$

Parallel planes: If their normal vectors $\vec{\mathbf{n}}_1$ and $\vec{\mathbf{n}}_2$ are collinear, i.e.

$$\vec{\mathbf{n}}_1 \times \vec{\mathbf{n}}_2 = 0.$$

Distance between two parallel planes:

$$d = \frac{|\vec{\mathbf{n}}_1 \cdot (\vec{\mathbf{r}}_1 - \vec{\mathbf{r}}_2)|}{|\vec{\mathbf{n}}_1|} = \frac{|\vec{\mathbf{n}}_2 \cdot (\vec{\mathbf{r}}_1 - \vec{\mathbf{r}}_2)|}{|\vec{\mathbf{n}}_2|}.$$

8.10.11 Cut set of two planes

The cut set of two planes is obtained by solving a system of equations. If the planes are parallel, the cut set is a straight line.

▷ Converting the equations of the plane to coordinate the representation simplifies the calculation of the cut set.

8.11 Plane of the second order in normal form

Normal form of a plane of the second order: The **center** (the point that bisects the chords passing through it) coincides with the origin, and the coordinate axes lie in with the symmetry axes of the plane.

8.11.1 Ellipsoid

Ellipsoid (a, b, c: lengths of the semi-axes):

$$\frac{x^2}{a^2} + \frac{y^2}{b^2} + \frac{z^2}{c^2} = 1,$$

Special cases:
1. $a = b > c$ Oblate **ellipsoid of rotation**, results from the rotation of the ellipse in the x, z-plane about the minor axis,

$$\frac{x^2}{a^2} + \frac{z^2}{c^2} = 1.$$

2. $a = b < c$: Prolate **ellipsoid of rotation**, results from the rotation of the ellipse in the x, z-plane about the major axis,

$$\frac{x^2}{a^2} + \frac{z^2}{c^2} = 1.$$

3. $a = b = c$:
Sphere of radius $R = a$,

$$x^2 + y^2 + z^2 = a^2 .$$

An arbitrary plane intersects the ellipsoid in an ellipse (in a special case, a circle appears as a special case).
Volume of the ellipsoid:

$$V = \frac{4}{3}\pi abc.$$

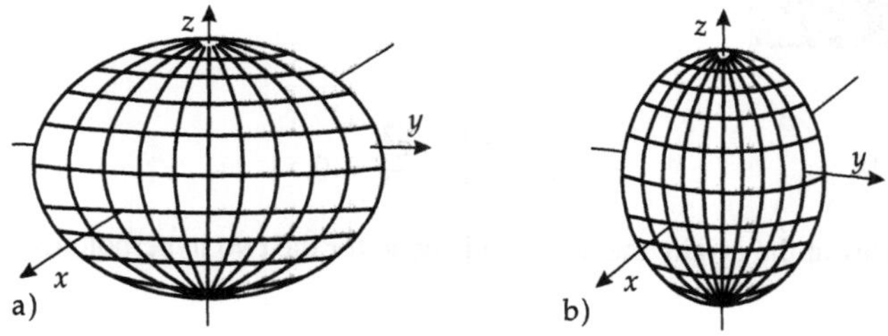

Ellipsoids

8.11.2 Hyperboloid

Hyperboloid of one sheet:

$$\frac{x^2}{a^2} + \frac{y^2}{b^2} - \frac{z^2}{c^2} = 1,$$

a, b: lengths of the real semi-axes; c: length of the imaginary semi-axis.
Linear generator of a plane: Line that is entirely in the plane.
The two linear generators of a hyperboloid of one sheet:

$$\frac{x}{a}+\frac{z}{c}=u\left(1+\frac{y}{b}\right), \quad u\left(\frac{x}{a}-\frac{z}{c}\right)=1-\frac{y}{b},$$

$$\frac{x}{a}+\frac{z}{c}=v\left(1-\frac{y}{b}\right), \quad v\left(\frac{x}{a}-\frac{z}{c}\right)=1+\frac{y}{b},$$

where u and v are arbitrary quantities. Every point of the plane is passed by two linear generators.
Hyperboloid of two sheets,

$$\frac{x^2}{a^2}+\frac{y^2}{b^2}-\frac{z^2}{c^2}=-1,$$

c: length of the real semi-axis; a, b: length of the imaginary semi-axes.
Sections parallel to the z-axis, in the case of the two hyperboloids, are again hyperboles (for the hyperboloid of one sheet, two intersecting straight lines are also possible).
Sections parallel to the x, y plane are ellipses.
Hyperboloid of rotation: In the case of $a = b$, rotation causes a hyperbola with the semi-axes a and c about the axis $2c$.

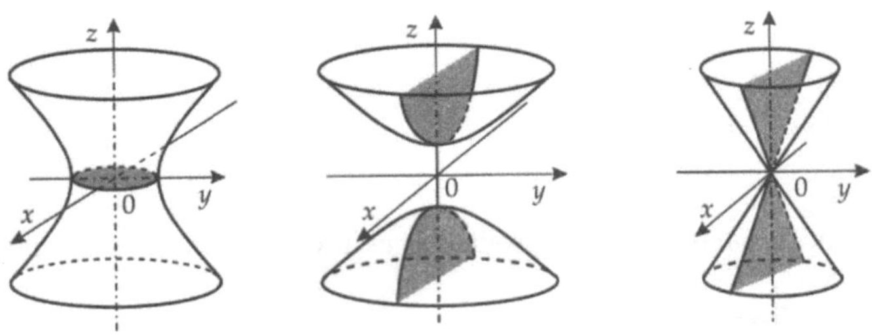

8.11.3 Cone

Cone:

$$\frac{x^2}{a^2}+\frac{y^2}{b^2}-\frac{z^2}{c^2}=0,$$

with the vertex at the origin, the asymptotic cone for the two hyperboloids

$$\frac{x^2}{a^2}+\frac{y^2}{b^2}-\frac{z^2}{c^2}=\pm 1\,.$$

Directing curve: Ellipse with the semi-axes a and b parallel to the x, y plane at distance c. **Right circular cone**: In the case of $a = b$.

8.11.4 Paraboloid

Paraboloids have no center. The vertex of the paraboloid lies in the origin; and the z-axis is the axis of symmetry.

Elliptic paraboloid:

$$z = \frac{x^2}{a^2} + \frac{y^2}{b^2}.$$

Sections parallel to the z-axis are parabolas; cuts perpendicular to the z-axis are ellipses.
Paraboloid of rotation: In the case of $a = b$, obtained by rotation of the parabola

$$z = \frac{x^2}{a^2},$$

in the x, z-plane about its symmetry axis.
Volume of the section of the paraboloid up to a plane at height h that is perpendicular to the z-axis:

$$V = \frac{1}{2}\pi abh.$$

Hyperbolic paraboloid:

$$z = \frac{x^2}{a^2} - \frac{y^2}{b^2}.$$

Sections perpendicular to the x- or y-axis are congruent parabolas, the sections perpendicular to the z-axis are hyperbolas (and an intersecting pair of straight lines for $z = 0$).

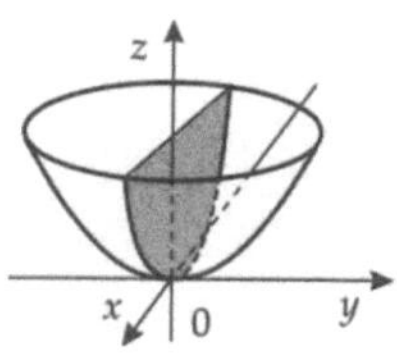

Elliptic Paraboloid

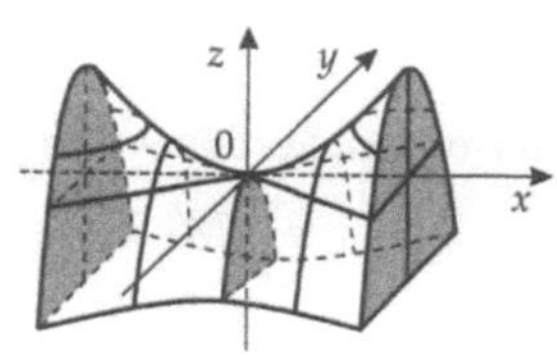

Hyperbolic Paraboloid

Linear generator of a plane: Straight line that lies entirely in the plane.
The families of linear generators of the hyperbolic paraboloid:

$$\frac{x}{a} + \frac{y}{b} = u, \quad u\left(\frac{x}{a} - \frac{y}{b}\right) = 2z,$$

$$\frac{x}{a} - \frac{y}{b} = v, \quad v\left(\frac{x}{a} + \frac{y}{b}\right) = 2z,$$

where u and v are arbitrary quantities. Each point of the plane is traversed by two linear generators.

8.11.5 Cylinder

Elliptic cylinder: $\frac{x^2}{a^2} + \frac{y^2}{b^2} = 1.$

Hyperbolic cylinder: $\frac{x^2}{a^2} - \frac{y^2}{b^2} = 1.$

Parabolic cylinder: $y^2 = 2px.$

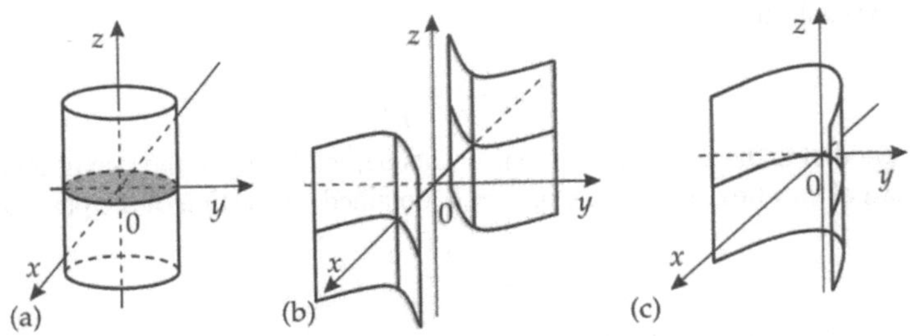

(a) Elliptic Cylinder (b) Hyperbolic Cylinder (c) Parabolic Cylinder

8.12 General plane of the second order

8.12.1 General equation

General equation of a plane of the second order:

$$a_{11}x^2 + a_{22}y^2 + a_{33}z^2 + 2a_{12}xy + 2a_{23}yz + 2a_{31}zx$$
$$+2a_{14}x + 2a_{24}y + 2a_{34}z + a_{44} = 0.$$

Matrix notation (see also Section 9.1 on matrices):

$$\vec{\mathbf{x}}^{\mathrm{T}} \cdot \mathbf{M} \cdot \vec{\mathbf{x}} + 2 \cdot \vec{\mathbf{a}}^{\mathrm{T}} \cdot \vec{\mathbf{x}} + a_{44} = 0,$$

with

$$\mathbf{M} = \begin{pmatrix} a_{11} & a_{12} & a_{13} \\ a_{21} & a_{22} & a_{23} \\ a_{31} & a_{32} & a_{33} \end{pmatrix}, \quad \vec{\mathbf{x}} = \begin{pmatrix} x \\ y \\ z \end{pmatrix}, \quad \vec{\mathbf{a}} = \begin{pmatrix} a_{14} \\ a_{24} \\ a_{34} \end{pmatrix}.$$

with $a_{ik} = a_{ki}$.

8.12.2 Transformation to principal axes

Transformation to principal axes: Rotation (elimination of the mixed quadratic terms) and translation (elimination of the linear terms) of the coordinate system.
Geometric interpretation: Transformation of the coordinate axes into the symmetry axes of the planes.
Elimination of the mixed quadratic terms by demanding that M be diagonal (see Sections 9.1 and 9.2 on matrices):

$$\mathbf{M} \cdot \vec{\mathbf{x}}_i = \lambda_i \cdot \vec{\mathbf{x}}_i,$$

with eigenvalues λ_i and eigenvectors $\vec{\mathbf{x}}_i$.
Calculation of the eigenvalues λ_i from:

$$\begin{vmatrix} a_{11} - \lambda & a_{12} & a_{13} \\ a_{21} & a_{22} - \lambda & a_{23} \\ a_{31} & a_{32} & a_{33} - \lambda \end{vmatrix} = 0.$$

▷ For double (triple) eigenvalues with multiplicity two (three), two (three) [−] arbitrary axes [mutually perpendicular] can be chosen as the principal axes.

Rotation of the coordinate axes into the direction of the eigenvectors of **M** yields the following form of the equation:

$$\vec{\mathbf{x}}'^{\mathsf{T}} \cdot \mathbf{M}' \cdot \vec{\mathbf{x}}' + 2 \cdot \vec{\mathbf{a}}'^{\mathsf{T}} \cdot \vec{\mathbf{x}}' + a'_{44} = 0,$$

with

$$\mathbf{M}' = \begin{pmatrix} \lambda_1 & 0 & 0 \\ 0 & \lambda_2 & 0 \\ 0 & 0 & \lambda_3 \end{pmatrix} .$$

Elimination of the linear terms by the following translation:

$$x'' = x' + \frac{a'_{14}}{a'_{11}} \quad (a'_{11} \neq 0),$$

$$y'' = y' + \frac{a'_{24}}{a'_{22}} \quad (a'_{22} \neq 0),$$

$$z'' = z' + \frac{a'_{34}}{a'_{33}} \quad (a'_{33} \neq 0).$$

Final form after rotation and translation:

$$\lambda_1 (x'')^2 + \lambda_2 (y'')^2 + \lambda_3 (z'')^2 + d = 0,$$

with the eigenvalues λ_i of the matrix **M**.

8.12.3 Shape of a surface of the second order

Classification according to the **invariants of a surface of the second order** (invariants do not change through translation and rotation of the coordinate systems):

$$A = \begin{vmatrix} a_{11} & a_{12} & a_{13} & a_{14} \\ a_{21} & a_{22} & a_{23} & a_{24} \\ a_{31} & a_{32} & a_{33} & a_{34} \\ a_{41} & a_{42} & a_{43} & a_{44} \end{vmatrix}, \quad B = \begin{vmatrix} a_{11} & a_{12} & a_{13} \\ a_{21} & a_{22} & a_{23} \\ a_{31} & a_{32} & a_{33} \end{vmatrix}, \quad C = a_{11} + a_{22} + a_{33},$$

$$D = a_{22}a_{33} + a_{33}a_{11} + a_{11}a_{22} - a_{23}^2 + a_{31}^2 + a_{12}^2, \quad \text{with } a_{ik} = a_{ki}.$$

1. $B \neq 0$ (**central surfaces**):

Case	$BC > 0,\ D > 0$	$BC < 0$ and/or $D < 0$
$A < 0$	Ellipsoid $\frac{x^2}{a^2} + \frac{y^2}{b^2} + \frac{z^2}{c^2} = 1$	Hyperboloid of two sheets $\frac{x^2}{a^2} + \frac{y^2}{b^2} - \frac{z^2}{c^2} = -1$
$A > 0$	Imaginary ellipsoid $\frac{x^2}{a^2} + \frac{y^2}{b^2} + \frac{z^2}{c^2} = -1$	Hyperboloid of one sheet $\frac{x^2}{a^2} + \frac{y^2}{b^2} - \frac{z^2}{c^2} = 1$
$A = 0$	Imaginary cone (with real vertex) $\frac{x^2}{a^2} + \frac{y^2}{b^2} + \frac{z^2}{c^2} = 0$	Cone $\frac{x^2}{a^2} + \frac{y^2}{b^2} - \frac{z^2}{c^2} = 0$

2. $B = 0$ (**Paraboloids, cylinders** and **pairs of planes**):

Case	$A < 0\,(D > 0)$	$A > 0\,(D < 0)$
$A \neq 0$	Elliptic paraboloid $\frac{x^2}{a^2} + \frac{y^2}{b^2} = \pm z$	Hyperbolic paraboloid $\frac{x^2}{a^2} - \frac{y^2}{b^2} = \pm z$
$A = 0$	$D > 0:$ real or imaginary elliptic cylinder $D = 0:$ parabolic cylinder $D < 0:$ hyperbolic cylinder	

▷ The last row (case $A = B = 0$) is valid only if the surface divided into two (real, imaginary or singular planes). Condition equation for the decomposition of the surface of the second order.

$$\begin{vmatrix} a_{11} & a_{12} & a_{14} \\ a_{21} & a_{22} & a_{24} \\ a_{41} & a_{42} & a_{44} \end{vmatrix} + \begin{vmatrix} a_{11} & a_{13} & a_{14} \\ a_{31} & a_{33} & a_{34} \\ a_{41} & a_{43} & a_{44} \end{vmatrix} + \begin{vmatrix} a_{22} & a_{23} & a_{24} \\ a_{32} & a_{33} & a_{34} \\ a_{42} & a_{43} & a_{44} \end{vmatrix} = 0.$$

9

Matrices, determinants, and systems of linear equations

9.1 Matrices

Matrix: A rectangular array of $m \cdot n$ numbers indexed according to their position in the array. For computing operations see Section 9.3.

System of linear equations: A system of several linear equations for several unknowns where the unknowns must be determined in such a way that the equations are satisfied simultaneously.

Matrices are used for the treatment of systems of linear equations (linear maps, transformations, discretization of differential equations, finite elements, etc.). Large systems of equations with more than 100 equations often occur in calculations for electrical networks and structural analysis of machine components (shafts, motors). See also applications in the chapters on differential equations, vector calculus, and computer graphics.

□ System of linear equations with three unknowns x_1, x_2, x_3:

$$\begin{aligned} a_{11}x_1 + a_{12}x_2 + a_{13}x_3 &= c_1 \\ a_{21}x_1 + a_{22}x_2 + a_{23}x_3 &= c_2 \\ a_{31}x_1 + a_{32}x_2 + a_{33}x_3 &= c_3 \end{aligned}$$

Coefficient matrix A: The matrix formed in a system of linear equations with the known coefficients a_{ij} before the unknowns x_j:

$$\mathbf{A} = \begin{pmatrix} a_{11} & a_{12} & a_{13} \\ a_{21} & a_{22} & a_{23} \\ a_{31} & a_{32} & a_{33} \end{pmatrix} .$$

Abbreviated symbolic matrix notation:

$$\mathbf{A}\mathbf{x} = \mathbf{c} .$$

Sometimes also:

$$[\mathbf{A}]\{\mathbf{x}\} = \{\mathbf{c}\} .$$

Solution vector x, also {**x**}: The vector that has to be determined.

Vector of constants c, also {**c**}. (See discussion below on vectors.)

▷ In vector analysis, due to their geometrical interpretation (arrow in plane or space), a vector is specified by an arrow ($\vec{\mathbf{a}}$) or by underlining ($\underline{\mathbf{a}}$). In matrix calculus this arrow is usually omitted because here, a vector is regarded as a row or column matrix.

Matrix $\mathbf{A}_{(m,n)}$, rectangular arrangement of $m \cdot n$ **matrix elements** a_{ij} in m rows (horizontal) and n columns (vertical).

Type of a matrix $\mathbf{A}_{(m,n)}$, $m \times n$ matrix (read "m by n-matrix")**A**: Classification of matrices according to the number of rows ($= m$) and the number of the columns ($= n$). All matrices with the same number of rows and columns belong to the same type.

$$\mathbf{A} = \begin{pmatrix} a_{11} & \cdots & a_{1n} \\ \vdots & & \vdots \\ a_{m1} & \cdots & a_{mn} \end{pmatrix} \qquad m, n \in \mathbb{N}$$

Abbreviated notation: $\mathbf{A} = (a_{ij})_{i=1,\ldots,m, j=1,\ldots,n} = (A_{ij}) = (a_{ij})$. **Matrix elements** a_{ij}, also A_{ij}, are indexed (natural numbers): Index $i \in \mathbb{N}$: number of the row ($i = 1, \ldots, m$), index $j \in \mathbb{N}$: number of the column ($j = 1, \ldots, n$).

□ Matrix element a_{32} stands at the intersection of the 3rd row and the 2nd column.

▷ Matrix elements may be real or complex numbers, or they may consist of other mathematical objects, such as polynomials, differentials, vectors or operators (e.g., functional or differential operators).

Real matrix: Matrix with real matrix elements.

Complex matrix: Matrix with complex matrix elements.

▷ In computers, matrices are often represented as indexed quantities (fields or arrays). Initialization of a real matrix with m rows and n columns:
in FORTRAN: `DIMENSION A(m,n)`,
in PASCAL: `TYPE A=ARRAY[1,..,m,1,..,n] OF REAL;`.
The matrix element a_{ij} is represented
in FORTRAN by `A(i,j)` and
in PASCAL and C by `A[i,j]` or `A[i][j]`.

▷ In computers, the representation of a complex matrix is often possible only by the separate treatment of its real and imaginary parts.
Initialization of a complex matrix with m rows and n columns:
in FORTRAN: `COMPLEX A(m,n)`,
in PASCAL:

```
VAR A: ARRAY[1..m,1..n] OF
         RECORD
           Re: REAL;
           Im: REAL
         END;
```

In FORTRAN, operations with COMPLEX-type variables are defined, but in PASCAL and C, operations with the real and imaginary parts must be declared separately.
In FORTRAN, there is access to the matrix element $_{ij}$ just as with real matrices. In PASCAL, there is access to the real and imaginary parts separately:

$$\text{real part of } a_{ij} = \texttt{A[i, j].Re}$$
$$\text{imaginary part of } a_{ij} = \texttt{A[i, j].Im}$$

Main diagonal elements: Matrix elements lying on the diagonal from top left to bottom right:

$$a_{11},\ a_{22},\ a_{33}, \ldots, \text{ i.e., all } a_{ij} \text{ with } i = j:$$
$$a_{ii} = (a_{11},\ a_{22}, \ldots, a_{nn}).$$

Secondary diagonal elements: The matrix elements lying on the diagonal from top right to bottom left:

$$a_{1,n},\ a_{2,n-1},\ a_{3,n-2}, \ldots, \text{ i.e., all } a_{ij} \text{ with } j = n-(i-1),\ i, j \in \{1, \ldots, n\}:$$

$$a_{i,n-(i-1)} = (a_{1,n-1},\ a_{2,n-2}, \ldots, a_{n,1}).$$

Trace of a matrix: Sum of the diagonal elements of an $n \times n$ matrix:

$$\mathrm{Sp}\mathbf{A} = a_{11} + a_{22} + \cdots + a_{nn} = \sum_{i=1}^{n} a_{ii}$$

□ Trace of the 3×3 matrix **A**:

$$\mathrm{Sp}(\mathbf{A}) = \mathrm{Sp}\begin{pmatrix} 1 & 3 & 4 \\ 2 & 4 & 2 \\ 3 & 4 & 7 \end{pmatrix} = 1 + 4 + 7 = 12$$

9.1.1 Row and column vectors

Row vector: A matrix with only one horizontal row (**row matrix**) or one single row of a matrix.

Column vector: Matrix with only one vertical column (**column matrix**) or one single column of a matrix.

$$\text{Column vector:}\quad \mathbf{a} = \begin{pmatrix} a_1 \\ \vdots \\ a_n \end{pmatrix} \qquad \text{Row vector:}\quad \mathbf{a}^{\mathrm{T}} = (a_1, \cdots, a_n)$$

$\mathbf{a}^{\mathrm{T}}$ denotes the vector to be transposed to **a** (see Section 9.2 on transposed matrices)

□ If the elements of matrix **A** with n rows and m columns are denoted by a_{ij}, and k is a natural number between 1 and m, or l is a natural number between 1 and n, then

$$\begin{pmatrix} a_{1k} \\ \vdots \\ a_{nk} \end{pmatrix}$$

is called the k-th column vector or

$$(a_{l1}, \cdots, a_{lm})$$

the l-th row vector of matrix **A**.

▷ Every **single-row matrix** can be regarded as a vector. Conversely, every vector can be regarded as a $1 \times n$ matrix. **Scalars** can be regarded as a 1×1 matrix.

▷ Initialization of a vector with n elements:
in FORTRAN: `DIMENSION A(n)`
in PASCAL: `TYPE A=ARRAY[1..n] OF REAL;`

The vector element a_i is represented:
in `FORTRAN` by `A(i)` and
in `PASCAL` by `A[i]`.

▷ **Index overflow**: If the index i does not lie in the permitted range (between 1 and n inclusive), `PASCAL` (but not `FORTRAN`) cancels with an error message (range check error), if the check has not been shut off. Therefore, index overflows must be excluded explicitly in `FORTRAN` programs.

9.2 Special matrices

9.2.1 Transposed, conjugate, and adjoint matrices

Transposed matrix $\mathbf{A}^\mathsf{T}$, arises from the matrix **A** by interchanging the column vectors and the row vectors (for square matrices, a "reflection in the main diagonal").

● The transposed matrix is obtained by interchanging the indices:

$$a_{ji}^\mathsf{T} = a_{ij} \ .$$

● If **A** is a $m \times n$ matrix, then $\mathbf{A}^\mathsf{T}$ is a $n \times m$ matrix.

□ The transpose of a 2×3 matrix is a 3×2 matrix :

$$\begin{pmatrix} 1 & 3 & 9 \\ 2 & 4 & 6 \end{pmatrix}^\mathsf{T} = \begin{pmatrix} 1 & 2 \\ 3 & 4 \\ 9 & 6 \end{pmatrix} .$$

● The transpose of a transposed matrix $(\mathbf{A}^\mathsf{T})^\mathsf{T}$ is the original matrix **A**:

$$(\mathbf{A}^\mathsf{T})^\mathsf{T} = \mathbf{A} \ .$$

Complex conjugate matrix $\mathbf{A}^*$ is obtained from matrix **A** through a complex conjugation (see Section **??** on complex numbers) of the individual matrix elements:

$$a_{ij}^* = \mathrm{Re}(a_{ij}) - \mathrm{j} \cdot \mathrm{Im}(a_{ij}) \ .$$

Adjoint matrix $\overline{\mathbf{A}}$ is obtained by complex conjugating and transposing the matrix **A**:

$$\overline{\mathbf{A}} = (\mathbf{A}^*)^\mathsf{T} \ .$$

9.2.2 Square matrices

Square matrices are of fundamental importance for linear mapping and transformations (see Section 24.9 on computer graphics).

The square matrix (or $n \times n$ **matrix**) has exactly as many rows as it has columns:

$$\mathbf{A} = \begin{pmatrix} a_{11} & \cdots & a_{1n} \\ \vdots & & \vdots \\ a_{n1} & \cdots & a_{nn} \end{pmatrix} \qquad n \in \mathbb{N}$$

▷ Every nonsquare matrix can be made into a square matrix by adding zeros. Note: this does not change its rank.

▷ For every square matrix there exists a complex or real number det **A**, also written as $|\mathbf{A}|$, the determinant of **A**.

Symmetric matrix: Square matrix that is identical with its transpose:

$$\mathbf{A} = \mathbf{A}^{\mathsf{T}} .$$

● Symmetric matrices are mirror-symmetric about this main diagonal; it holds that:

$$a_{ij} = a_{ji} \quad \text{for all } i,\ j .$$

□ Symmetric matrix:

$$\begin{pmatrix} 3 & -4 & 0 \\ -4 & 1 & 2 \\ 0 & 2 & 5 \end{pmatrix}$$

Antisymmetric or **skew-symmetric matrix**: A square matrix equal to its negative transpose:

$$\mathbf{A} = -\mathbf{A}^{\mathsf{T}}$$

● For antisymmetric matrices, it holds that $a_{ij} = -a_{ji}$, and all diagonal elements equal zero ($a_{ii} = 0$). Thus even, the trace vanishes: $\mathrm{Sp}(\mathbf{A}) = \mathbf{0}$.

□ Antisymmetric matrix:

$$\begin{pmatrix} 0 & -3 & 2 \\ 3 & 0 & -5 \\ -2 & 5 & 0 \end{pmatrix}$$

● Every square matrix can be separated into a sum of a symmetric matrix $\mathbf{A}_s$ and an antisymmetric matrix $\mathbf{A}_{as}$, $\mathbf{A} = \mathbf{A}_s + \mathbf{A}_{as}$, with:

$$\mathbf{A}_s = \frac{1}{2}(\mathbf{A} + \mathbf{A}^{\mathsf{T}}) ,$$

$$\mathbf{A}_{as} = \frac{1}{2}(\mathbf{A} - \mathbf{A}^{\mathsf{T}}) .$$

□ Separation of a quadratic matrix into the sum of a symmetric and antisymmetric matrix:

$$\begin{pmatrix} 2 & -5 & 8 \\ -3 & 1 & -9 \\ 2 & 5 & 3 \end{pmatrix} = \begin{pmatrix} 2 & -4 & 5 \\ -4 & 1 & -2 \\ 5 & -2 & 3 \end{pmatrix} + \begin{pmatrix} 0 & -1 & 3 \\ 1 & 0 & -7 \\ -3 & 7 & 0 \end{pmatrix}$$

Hermitian or self-adjoint matrix: Square matrix that is equal to its adjoint matrix $\overline{\mathbf{A}}$ (notation also A^H):

$$\mathbf{A} = \overline{\mathbf{A}} = (\mathbf{A}^*)^{\mathsf{T}}$$

● For Hermitian matrices, it holds that $a_{ij} = a_{ji}^*$.

Anti-Hermitian or skew-Hermitian matrix: Square matrix that is equal to the negative of its adjoint:

$$\mathbf{A} = -\overline{\mathbf{A}}$$

● For skew-Hermitian matrices, it holds that $a_{ij} = -a_{ji}^*,\ a_{ii} = 0,\ \mathrm{Sp}(\mathbf{A}) = \mathbf{0}$.

- Every square matrix can be divided into a sum of Hermitian matrix $\mathbf{A}_h$ and an anti-Hermitian matrix $\mathbf{A}_{ah}$, $\mathbf{A} = \mathbf{A}_h + \mathbf{A}_{ah}$, with:

$$\mathbf{A}_h = \frac{1}{2}(\mathbf{A} + \overline{\mathbf{A}})$$

$$\mathbf{A}_{ah} = \frac{1}{2}(\mathbf{A} - \overline{\mathbf{A}})$$

▷ For real square matrices, the Hermitian matrices are reduced to the symmetric matrices, and the skew-Hermitian matrices are reduced to the skew-symmetric matrices.

Unitary matrix A: It is unitary when the multiplication of **A** by its adjunct $\overline{\mathbf{A}}$ results in the unit matrix.

9.2.3 *Triangular matrices*

Upper (or right) triangular matrix U (U = upper): Matrix in which all elements below (or to the left of) the main diagonal are equal to zero:

$$\mathbf{u}_{ij} = 0 \text{ for all } i > j\ .$$

Lower (or left) triangular matrix (L = lower): Square matrix for which all elements above (or to the right of) the main diagonal are equal to zero:

$$\mathbf{l}_{ij} = 0 \text{ for all } i < j\ .$$

$$\mathbf{L} = \begin{pmatrix} a_{11} & 0 & \cdots & 0 \\ a_{21} & a_{22} & \ddots & \vdots \\ \vdots & & \ddots & 0 \\ a_{n1} & \cdots & \cdots & a_{nn} \end{pmatrix}, \qquad \mathbf{U} = \begin{pmatrix} a_{11} & a_{12} & \cdots & a_{1n} \\ 0 & a_{22} & & \vdots \\ \vdots & \ddots & \ddots & \vdots \\ 0 & \cdots & 0 & a_{nn} \end{pmatrix}.$$

□ Lower or upper triangular matrix:

$$\mathbf{L} = \begin{pmatrix} 1 & 0 & 0 \\ 3 & 4 & 0 \\ 3 & 2 & 1 \end{pmatrix} \quad \text{or} \quad \mathbf{U} = \begin{pmatrix} 1 & 7 & 2 \\ 0 & 4 & 2 \\ 0 & 0 & 1 \end{pmatrix}.$$

- The transpose of an upper triangular matrix is a lower triangular matrix and vice versa:

$$\mathbf{U}^{\mathsf{T}} = \mathbf{L}$$

$$\mathbf{L}^{\mathsf{T}} = \mathbf{U}\,.$$

□ Transpose of a triangular matrix:

$$\mathbf{L}^{\mathsf{T}} = \begin{pmatrix} 5 & 0 & 0 \\ 3 & 4 & 0 \\ 1 & 2 & 4 \end{pmatrix}^{\mathsf{T}} = \begin{pmatrix} 5 & 3 & 1 \\ 0 & 4 & 2 \\ 0 & 0 & 4 \end{pmatrix} = \mathbf{U}\,.$$

▷ **LU-decomposition**: Every matrix can be converted by interchanging the rows in such a way that it can be decomposed as the product of a left/lower and right/upper triangular matrix. This is used in the numerical solution of systems of linear equations (compare the Gaussian algorithm, Section 9.6).

The determinant of a triangular matrix is the **product** of the diagonal elements (not the sum, as for the trace):

$$\det \mathbf{L} = a_{11} \cdot a_{22} \cdots a_{nn} = \prod_{i=1}^{n} a_{ii} \; ; \quad \text{holds analogously also for} \quad \det \mathbf{U} \; !$$

Step matrix: A matrix with steps in the rows and columns:

$$\begin{pmatrix} * & \cdot & \cdot & \cdot & \cdot \\ 0 & * & * & \cdot & \cdot \\ 0 & 0 & 0 & * & \cdot \\ \cdot & \cdot & 0 & 0 & * \\ \cdot & \cdot & \cdot & 0 & 0 \end{pmatrix}$$

The boundaries of the steps are marked by stars (∗). The stars replace numbers $a_{ij} \neq 0$. Below the stars, only zeros appear.
Trapezoidal matrix: $m \times n$ matrix in trapezoidal form, with $m - k$ null rows:

$$\left(\begin{array}{cccc|cccc} a_{11} & a_{12} & \cdots & a_{1k} & a_{1k+1} & \cdots & \cdots & a_{1n} \\ 0 & a_{22} & & \vdots & \vdots & & & \vdots \\ \vdots & \ddots & \ddots & \vdots & \vdots & & & \vdots \\ 0 & \cdots & 0 & a_{kk} & a_{kk+1} & \cdots & \cdots & a_{kn} \\ \hline 0 & \cdots & \cdots & 0 & 0 & \cdots & \cdots & 0 \\ \vdots & \ddots & & \vdots & \vdots & \ddots & & \vdots \\ \vdots & & \ddots & \vdots & \vdots & & \ddots & \vdots \\ 0 & \cdots & \cdots & 0 & 0 & \cdots & \cdots & 0 \end{array}\right)$$

▷ The trapezoidal matrix is used often as the extended matrix of coefficients in solving systems of linear equations.

▷ Step matrices, trapezoidal matrices, and triangular matrices are of great practical importance in solving systems in row echelon; see Section 9.6 on the numerical solution of systems of linear equations.

9.2.4 Diagonal matrices

Diagonal matrix: Square matrix in which all elements outside the main diagonal are zero.

$$\mathbf{D} = \begin{pmatrix} a_{11} & 0 & \cdots & 0 \\ 0 & a_{22} & \ddots & \vdots \\ \vdots & \ddots & \ddots & 0 \\ 0 & \cdots & 0 & a_{nn} \end{pmatrix}, \quad a_{ij} = 0 \quad \text{for all} \quad i \neq j \; .$$

- Every diagonal matrix is at the same time a right and left triangular matrix.
- Every diagonal matrix is symmetric.
- The transpose of a diagonal matrix is the same matrix:

$$\mathbf{D}^{\mathsf{T}} = \mathbf{D}$$

□

$$\mathbf{D}^\mathsf{T} = \begin{pmatrix} 5 & 0 & 0 \\ 0 & 9 & 0 \\ 0 & 0 & 4 \end{pmatrix}^\mathsf{T} = \begin{pmatrix} 5 & 0 & 0 \\ 0 & 9 & 0 \\ 0 & 0 & 4 \end{pmatrix} = \mathbf{D}$$

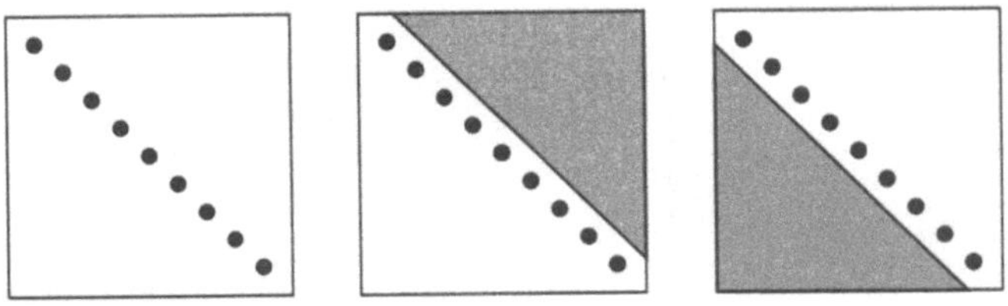

Diagonal matrix, right triangular matrix, left triangular matrix.
• Main diagonal element

Tridiagonal matrix: A square matrix that has nonzero matrix elements only on the main diagonal $a_{ij},\ i = j$ and on the subdiagonal immediately above or below the main diagonal $a_{ij},\ i = j \pm 1$:

$$a_{ij} = 0 \text{ for } |i - j| > 1 \ .$$

Band matrix: A square matrix that has nonzero elements on the main diagonal and on the diagonal lying k above, $a_{ij},\ j \leq i + k$, or on the diagonal lying l below, $a_{ij},\ i \leq j + l$, the main diagonal. The nonzero matrix elements are arranged along a diagonal "band":

$$a_{ij} = 0 \text{ for } i - j > l \ , \ j - i > k, \ \ l < n \ .$$

Upper (right) Hessenberg matrix: A square matrix that has vanishing matrix elements below the first diagonal to the left, $a_{ij},\ i = j + 1$:

$$a_{ij} = 0 \text{ for } j - i > 1 \ .$$

Lower (left) Hessenberg matrix: A square matrix that has vanishing matrix elements above the first diagonal to the right, $a_{ij},\ i = j - 1$:

$$a_{ij} = 0 \text{ for } j - i > 1 \ .$$

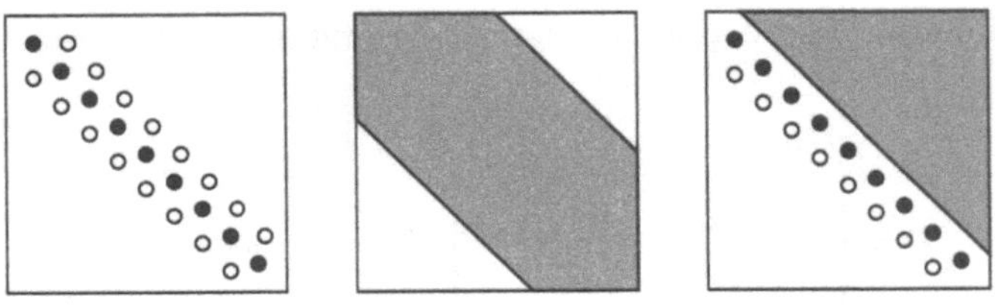

Tridiagonal matrix, band matrix, right Hessenberg matrix.
• Element on the main diagonal, ∘ Element on the subdiagonal

Unit matrix I (denoted also by **E**, see identity map in Section 9.3): A square matrix whose main diagonal elements equal 1 and all other matrix elements equal zero:

$$\mathbf{I} = \begin{pmatrix} 1 & 0 & \cdots & 0 \\ 0 & 1 & \ddots & \vdots \\ \vdots & \ddots & \ddots & 0 \\ 0 & \cdots & 0 & 1 \end{pmatrix} .$$

□ The 3×3 unit matrix:

$$\mathbf{I}_{(3,3)} = \begin{pmatrix} 1 & 0 & 0 \\ 0 & 1 & 0 \\ 0 & 0 & 1 \end{pmatrix} .$$

Scalar matrix: A diagonal matrix whose main diagonal elements are all equal to c and all other matrix elements equal zero:

$$\mathbf{S} = \begin{pmatrix} c & 0 & \cdots & 0 \\ 0 & c & \ddots & \vdots \\ \vdots & \ddots & \ddots & 0 \\ 0 & \cdots & 0 & c \end{pmatrix} = c \begin{pmatrix} 1 & 0 & \cdots & 0 \\ 0 & 1 & \ddots & \vdots \\ \vdots & \ddots & \ddots & 0 \\ 0 & \cdots & 0 & 1 \end{pmatrix} = c\mathbf{I} .$$

□ 2×2 scalar matrix:

$$\mathbf{S} = \begin{pmatrix} 2 & 0 \\ 0 & 2 \end{pmatrix} .$$

Null matrix: All matrix elements equal zero:

$$\mathbf{0} = \begin{pmatrix} 0 & 0 & \cdots & 0 \\ 0 & 0 & \ddots & \vdots \\ \vdots & \ddots & \ddots & 0 \\ 0 & \cdots & 0 & 0 \end{pmatrix} .$$

□ A 2×4 null matrix:

$$\mathbf{0} = \begin{pmatrix} 0 & 0 & 0 & 0 \\ 0 & 0 & 0 & 0 \end{pmatrix}$$

A row null matrix (corresponds to the null vector):

$$\mathbf{0} = \begin{pmatrix} 0 & 0 & 0 & 0 & 0 \end{pmatrix}$$

▷ For any matrices that can be linked arbitrarily (see Section 9.3 for addition and multiplication of matrices), it holds that:

$$\mathbf{A} + \mathbf{0} = \mathbf{0} + \mathbf{A} = \mathbf{A} \quad \text{and} \qquad \mathbf{A0} = \mathbf{0A} = \mathbf{0} .$$

- **Equality of matrices**: Two matrices **A** and **B** are said to be **equal**, $\mathbf{A} = \mathbf{B}$, if and only if they are of the same **type** (i.e., they have the same number of rows and columns, and if all corresponding matrix elements are equal:

$$\mathbf{A} = \mathbf{B} \Leftrightarrow a_{ij} = b_{ij} \text{ for all } i = 1, \ldots, n,\ j = 1, \ldots, m$$

▷ From $\det \mathbf{A} = \det \mathbf{B}$ it **does not** follow that $\mathbf{A} = \mathbf{B}$ (see Section 9.4 on determinants)!

9.3 Operations with matrices

▷ Matrices can be added, subtracted, multiplied, raised to a power, differentiated, and integrated only under certain suitable conditions!

9.3.1 *Addition and subtraction of matrices*

Matrix **B** can be added to or subtracted from matrix **A** if their matrix elements are of the same type (e.g., real numbers), and if they have the same number of rows and columns.

● Matrices are added or subtracted **by element** by adding or subtracting their matrix elements.

$$\mathbf{A} \pm \mathbf{B} = \mathbf{C} \qquad \Leftrightarrow \qquad a_{ij} \pm b_{ij} = c_{ij}$$

▷ Program sequence for the addition of matrices:

```
BEGIN Addition of matrices
INPUT A[1...m,1...n],B[1...m,1...n]
 FOR i:= 1 TO m DO
    FOR j:= 1 TO n DO
       C[i,j]:= A[i,j] + B[i,j]
    ENDDO
 ENDDO
  OUTPUT C[1,...,m,1,...,n]
END Addition of matrices
```

□ Addition of two matrices:

$$\begin{pmatrix} 1 & 3 \\ 2 & 4 \end{pmatrix} + \begin{pmatrix} 2 & 3 \\ 0 & 4 \end{pmatrix} = \begin{pmatrix} 3 & 6 \\ 2 & 8 \end{pmatrix}$$

Calculating rules for $m \times n$ matrices:

● The sum matrix $\mathbf{A} + \mathbf{B}$ and the difference matrix $\mathbf{A} - \mathbf{B}$ are themselves $m \times n$ matrices.

● Commutative law of matrix addition.

$$\mathbf{A} + \mathbf{B} = \mathbf{B} + \mathbf{A}, \quad \text{in components: } a_{ij} + b_{ij} = b_{ij} + a_{ij} \; .$$

▷ Contrary to addition, the multiplication of two matrices **is not commutative** (see the section below on multiplication of matrices).

● Associative law of matrix addition:

$$(\mathbf{A} + \mathbf{B}) + \mathbf{C} = \mathbf{A} + (\mathbf{B} + \mathbf{C}), \qquad (a_{ij} + b_{ij}) + c_{ij} = a_{ij} + (b_{ij} + c_{ij}) \; .$$

● **Null element**: $\mathbf{A} + \mathbf{0} = \mathbf{0} + \mathbf{A} = \mathbf{A}$.

● **Negative matrix**: $\mathbf{A} + (-\mathbf{A}) = \mathbf{0}$.

● **Transpose of a sum**: $(\mathbf{A} + \mathbf{B})^{\mathsf{T}} = \mathbf{A}^{\mathsf{T}} + \mathbf{B}^{\mathsf{T}}$.

● The set of all $m \times n$ matrices whose elements are real numbers form an Abelian group under matrix addition.

9.3.2 *Multiplication of a matrix by a scalar factor c*

A matrix can be multiplied by a real or comples number c (scalar factor).

● A matrix is multiplied by a factor c by multiplying **every** individual element a_{ij} by c.

$$\mathbf{C} = c\,\mathbf{A} = c\begin{pmatrix} a_{11} & \cdots & a_{1n} \\ \vdots & \ddots & \vdots \\ a_{m1} & \cdots & a_{mn} \end{pmatrix} = \begin{pmatrix} ca_{11} & \cdots & ca_{1n} \\ \vdots & \ddots & \vdots \\ ca_{m1} & \cdots & ca_{mn} \end{pmatrix}$$

▷ Multiplication of a matrix by a scalar can also be regarded as a multiplication by a scalar matrix:

$$c\mathbf{A} = (c\mathbf{I})\mathbf{A} .$$

▷ **Division** by a scalar factor $d \neq 0$ can be regarded as multiplication by $1/d$.

● A factor c common to all matrix elements can be placed before the matrix. Division by zero is not permissible.

□ Scalar matrix:

$$\begin{pmatrix} c & 0 & \cdots & 0 \\ 0 & c & \ddots & \vdots \\ \vdots & \ddots & \ddots & 0 \\ 0 & \cdots & 0 & c \end{pmatrix} = c\,\mathbf{I} = c\begin{pmatrix} 1 & 0 & \cdots & 0 \\ 0 & 1 & \ddots & \vdots \\ \vdots & \ddots & \ddots & 0 \\ 0 & \cdots & 0 & 1 \end{pmatrix}$$

Calculating rules for matrix addition and **multiplication** by a **scalar**:

● Multiplication of a matrix by factors c and d (real or complex numbers) is commutative, associative and distributive.

● **Commutative law**:

$$c\,\mathbf{A} = \mathbf{A}\,c$$

● **Associative law**:

$$c(d\,\mathbf{A}) = (cd)\mathbf{A}$$

● **Distributive laws**:

$$(c \pm d)\mathbf{A} = c\,\mathbf{A} \pm d\,\mathbf{A}\,,$$

$$c(\mathbf{A} \pm \mathbf{B}) = c\mathbf{A} \pm c\mathbf{B}$$

● Multiplication by the unit element, unit matrix **I**, known as an identity map:

$$1\mathbf{A} = \mathbf{I}\mathbf{A} = \mathbf{A}$$

● The matrices (linear maps) form a **vector space** under addition and multiplication by a scalar.

9.3.3 Multiplication of vectors, scalar product

The multiplication of two vectors is defined in three ways: Depending on whether forming the **scalar product**, the **vector product** or the **dyadic product** is formed, the result is a scalar, a vector or a matrix.

▷ Scalar product and the vector product of two vectors **a** and **b** are defined only if **a** and **b** have the same number of elements (components). For the dyadic product, **a** and **b** may have different numbers of elements.

The scalar product of the vectors **a** and **b** with n elements **each** denoted by a point, is a **scalar** (= real or complex number!). It is the sum of the product of corresponding elements (see Chapter 6 on vector calculus for applications in geometry, and see Section 24.9 on computer graphics):

$$\mathbf{a}\cdot\mathbf{b} = a_1b_1 + a_2b_2 + \cdots + a_nb_n = \sum_{i=1}^{n} a_ib_i \ .$$

Sometimes, the product of two matrices is also called a scalar product.

▷ The scalar product of two vectors (denoted by a point) can also be regarded as a scalar product (not a point) of a row matrix and a column matrix; but then the **order** of $\mathbf{a}^\mathsf{T}$ and $\mathbf{b}$ is **of importance**:

$$\mathbf{a}\cdot\mathbf{b} = \mathbf{a}^\mathsf{T}\mathbf{b} \neq \mathbf{b}\mathbf{a}^\mathsf{T}$$

The scalar product $\mathbf{a}^\mathsf{T}\mathbf{b}$ is a **number**, to be distinguished clearly from the dyadic product $\mathbf{b}\mathbf{a}^\mathsf{T}$, which yields a **matrix**.

□ Scalar product of a row vector (row matrix) and a column vector (column matrix):

$$\mathbf{a}^\mathsf{T}\mathbf{b} = (a_1 \cdots a_n)\begin{pmatrix} b_1 \\ \vdots \\ b_n \end{pmatrix} = a_1b_1 + a_2b_2 + a_3b_3 + \cdots + a_nb_n = \sum_{i=1}^{n} a_ib_i \ .$$

The vector product or **cross product** $\mathbf{a}\times\mathbf{b}$ of two vectors $\mathbf{a}$ and $\mathbf{b}$ with three components, denoted by a cross between the vectors (see Chapter 6 on vector calculus), yields a vector.

Dyadic product or **tensor product**: A product of two vectors $\mathbf{a}$ with m elements, and $\mathbf{b}$, with n elements, that yields an $m \times n$ matrix. The elements c_{ij} of the dyad $\mathbf{C}$ of the vectors $\mathbf{a}$ and $\mathbf{b}$ are given by

$$c_{ij} = a_ib_j \ .$$

● The dyadic product can be defined as a matrix product (see section below on multiplication of matrices)

$$\mathbf{C} = \mathbf{a}\mathbf{b}^\mathsf{T} \ ,$$

$\mathbf{a}$: matrix with one column (= column vector). $\mathbf{b}^\mathsf{T}$: matrix with one row (= row vector).

□ Dyadic product:

$$\mathbf{a} = \begin{pmatrix} 1 \\ 3 \\ 2 \end{pmatrix} \ , \quad \mathbf{b}^\mathsf{T} = (2,\ 1,\ 4) \ , \quad \mathbf{C} = \mathbf{a}\mathbf{b}^\mathsf{T}$$

$$\mathbf{a}\mathbf{b}^\mathsf{T} = \begin{pmatrix} 1 \\ 3 \\ 2 \end{pmatrix}(2,\ 1,\ 4) = \begin{pmatrix} 2 & 1 & 4 \\ 6 & 3 & 12 \\ 4 & 2 & 8 \end{pmatrix} = \mathbf{C}$$

□ The dyadic product of a vector $\mathbf{a}$ with a $a_i = i,\ i = 1, \ldots, 10$ by the vector itself (= matrix product of $\mathbf{a}$ and $\mathbf{a}^\mathsf{T}$) yields the symmetric square 10×10 matrix:

$$\mathbf{C} = \mathbf{a}\mathbf{a}^\mathsf{T} \text{ with } c_{ij} = i \cdot j,$$

whose matrix elements are all possible products of the numbers 1 to 10; i.e., it yields the multiplication table up to ten:

$$\mathbf{C}=\begin{pmatrix} 1 & 2 & 3 & 4 & 5 & 6 & 7 & 8 & 9 & 10 \\ 2 & 4 & 6 & 8 & 10 & 12 & 14 & 16 & 18 & 20 \\ 3 & 6 & 9 & 12 & 15 & 18 & 21 & 24 & 27 & 30 \\ 4 & 8 & 12 & 16 & 20 & 24 & 28 & 32 & 36 & 40 \\ 5 & 10 & 15 & 20 & 25 & 30 & 35 & 40 & 45 & 50 \\ 6 & 12 & 18 & 24 & 30 & 36 & 42 & 48 & 54 & 60 \\ 7 & 14 & 21 & 28 & 35 & 42 & 49 & 56 & 63 & 70 \\ 8 & 16 & 24 & 32 & 40 & 48 & 56 & 64 & 72 & 80 \\ 9 & 18 & 27 & 36 & 45 & 54 & 63 & 72 & 81 & 90 \\ 10 & 20 & 30 & 40 & 50 & 60 & 70 & 80 & 90 & 100 \end{pmatrix}$$

9.3.4 Multiplication of a matrix by a vector

The multiplication of an $m \times n$ matrix A by an n-dimensional column vector x yields a column vector **b** with m rows (components):

$$\mathbf{Ax}=\mathbf{b}$$

$$\begin{pmatrix} a_{11} & \cdots & a_{1n} \\ \vdots & & \vdots \\ a_{m1} & \cdots & a_{mn} \end{pmatrix}\begin{pmatrix} x_1 \\ \vdots \\ x_n \end{pmatrix}=\begin{pmatrix} a_{11}x_1+a_{12}x_2+\cdots+a_{1n}x_n \\ \vdots \\ a_{m1}x_1+a_{m2}x_2+\cdots+a_{mn}x_n \end{pmatrix}=\begin{pmatrix} b_1 \\ \vdots \\ b_m \end{pmatrix}.$$

▷ This is the abbreviated notation for systems of linear equations, **A** is called the system matrix, **b** the vector of constants, and **x** the solution vector or system vector.

The components b_i are obtained as a scalar product of the row vectors $\mathbf{a}_i$ of matrix **A** and the vector **x**:

$$b_i=\mathbf{a}_i\cdot\mathbf{x}=a_{i1}x_1+a_{i2}x_2+\cdots+a_{in}x_n=\sum_{j=1}^{n}a_{ij}x_j\ ,\ i=1,\ldots,m\ .$$

□ Multiplication of a matrix by a vector:

$$\begin{pmatrix} 1 & 3 & 2 \\ 2 & 4 & 0 \\ 3 & 2 & 1 \end{pmatrix}\begin{pmatrix} 2 \\ 1 \\ 4 \end{pmatrix}=\begin{pmatrix} 13 \\ 8 \\ 12 \end{pmatrix}$$

▷ The product of a matrix **A** and the vector **x** is defined only when **A** has exactly as many columns as **x** has rows (components).

9.3.5 Multiplication of matrices

▷ The product **AB** of matrix **A** and matrix **B** is defined only when **A** has exactly as many colmuns as **B** has rows.

□ $\mathbf{A}_{(3,3)}\mathbf{B}_{(2,3)}$ is not defined!
$\mathbf{A}_{(3,3)}$ has three columns, but $\mathbf{B}_{(2,3)}$ has only two rows!

The matrix product, or **scalar product**, of matrix **A** and matrix **B** yields matrix $\mathbf{C}=\mathbf{AB}$, whose elements are the scalar products of the row vectors of **A** and the column vectors of the matrix **B**. If $\mathbf{a}_i$ denotes the i-th row vector of **A** and $\mathbf{b}_j$ the j-th column vector of

B, then the elements c_{ij} of $\mathbf{C} = \mathbf{AB}$ are given by the number (scalar product!):

$$c_{ij} = \mathbf{a}_i \cdot \mathbf{b}_j = \sum_{l=1}^{n} a_{il} b_{lj} \ .$$

Here, the scalar product of the row vectors $\mathbf{a}_i$ $i = 1, \ldots, n$ and the column vectors $\mathbf{b}_j$ $j = 1, \ldots, m$ is denoted by a point.

□ Matrix product:

$$\begin{pmatrix} 1 & 3 \\ 2 & 4 \end{pmatrix} \begin{pmatrix} 2 & 3 \\ 0 & 4 \end{pmatrix} = \begin{pmatrix} 1\cdot 2 + 3\cdot 0 & 1\cdot 3 + 3\cdot 4 \\ 2\cdot 2 + 4\cdot 0 & 2\cdot 3 + 4\cdot 4 \end{pmatrix} = \begin{pmatrix} 2 & 15 \\ 4 & 22 \end{pmatrix}$$

● **Type-rule**: If a matrix $\mathbf{A}_{(m,l)}$ with m rows and l columns is multiplied by a matrix $\mathbf{B}_{(l,n)}$ with l rows and n columns, then the result is a matrix with m rows and n columns:

$$\mathbf{C}_{(m,n)} = \mathbf{A}_{(m,l)} \mathbf{B}_{(l,n)} \ .$$

Number of columns = number of elements in a **row**.
Number of rows = number of elements in a **column**.

● The number of columns l of **A** must be equal to the number of rows l of **B**:

$$\mathbf{A}_{(m,l)} \mathbf{B}_{(l,n)} = \mathbf{C}_{(m,n)} \ .$$

▷ Multiplication is only possible if the inner dimensions l are equal.

▷ The outer dimensions m and n correspond to the dimensions $m \times n$ of the resulting matrix.

▷ The order of matrix multiplications is important: $\mathbf{AB} \neq \mathbf{BA}$

$$\mathbf{A}_{(m,l)} \mathbf{B}_{(l,m)} \neq \mathbf{B}_{(l,m)} \mathbf{A}_{(m,l)} \ .$$

▷ Program sequence for the multiplication of an $m \times n$ matrix **A** by an $n \times l$ matrix **B**. The result is stored in an $m \times l$ matrix **C**.

```
BEGIN matrix multiplication
INPUT m, n, l
INPUT a[i,k], i=1,...,m, k=1,...,n
INPUT b[i,k], i=1,...,n, k=1,...,l
FOR i = 1 TO m DO
    FOR j = 1 TO l DO
       sum := 0
       FOR k = 1 TO n DO
          sum := sum + a[i,k]*b[k,j]
       ENDDO
       c[i,j] := sum
    ENDDO
ENDDO
OUTPUT c[i,k], i=1,...,m, k=1,...,l
END matrix multiplication
```

9.3.6 Calculating rules of matrix multiplication

● Matrix multiplication generates a **linear map** or a **linear transformation**, $x \to x' = \mathbf{A}x$, if

$$\mathbf{A}(\mathbf{x}+\mathbf{y}) = \mathbf{A}\mathbf{x}+\mathbf{A}\mathbf{y}$$
$$\mathbf{A}(c\mathbf{x}) = c\mathbf{A}\mathbf{x}$$

▷ Application for rotation, translation, inversion, and scaling in computer graphics.

- Sums and products which occur must be defined!
- $c\mathbf{AB} = \mathbf{A}(c\mathbf{B}) = c(\mathbf{AB})$
- **Associative law:** $(\mathbf{AB})\mathbf{C} = \mathbf{A}(\mathbf{BC}) = \mathbf{ABC}$
- Matrix multiplication is usually **not** commutative!

$$\mathbf{AB} \neq \mathbf{BA}\,.$$

▷ The order of matrices must not be interchanged:

$$\mathbf{A}_{(2,3)}\mathbf{B}_{(3,3)} = \mathbf{C}_{(2,3)}$$

$$\mathbf{B}_{(3,3)}\mathbf{A}_{(2,3)} \quad \text{is not defined!}$$

□ Commuted product (see above):

$$\begin{pmatrix} 2 & 3 \\ 0 & 4 \end{pmatrix}\begin{pmatrix} 1 & 3 \\ 2 & 4 \end{pmatrix} = \begin{pmatrix} 2\cdot 1+3\cdot 2 & 2\cdot 3+3\cdot 4 \\ 0\cdot 1+4\cdot 2 & 0\cdot 3+4\cdot 4 \end{pmatrix} = \begin{pmatrix} 8 & 18 \\ 8 & 16 \end{pmatrix}$$

▷ Important difference:
scalar product $\mathbf{a}^\mathsf{T}\mathbf{b}$: the result is a **scalar**,
dyadic product $\mathbf{b}\mathbf{a}^\mathsf{T}$: the result is a **matrix**!

$$\mathbf{a}^\mathsf{T}\mathbf{b} \neq \mathbf{b}\mathbf{a}^\mathsf{T}$$

▷ Scalar products and dyadic products are special cases of the general matrix product.

Commutative matrices: Two square matrices for which it holds that:

$$\mathbf{AB} = \mathbf{BA}$$

▷ The commuted product $\mathbf{B}_{(l,n)}\mathbf{A}_{(m,l)}$ is defined only for $n = m$, but the matrix multiplication is not commutative.

- **Distributive laws:**

$$\mathbf{A}(\mathbf{B}+\mathbf{C}) = \mathbf{AB}+\mathbf{AC}$$

$$(\mathbf{B}+\mathbf{C})\mathbf{A} = \mathbf{BA}+\mathbf{CA}$$

- **Unit element: unit matrix** $\mathbf{AI} = \mathbf{A}$
- **Null element: null matrix** $\mathbf{A0} = \mathbf{0}$
- **Zero divisor:** $\mathbf{AB} = \mathbf{0} \not\Rightarrow \mathbf{A} = \mathbf{0}$ or $\mathbf{B} = \mathbf{0}$
- $\mathbf{AB} = \mathbf{AC} \not\Rightarrow \mathbf{B} = \mathbf{C}$
- The transposed matrix of a product of two matrices is equal to the product of the two transposed matrices in reverse order:

$$(\mathbf{AB})^\mathsf{T} = \mathbf{B}^\mathsf{T}\mathbf{A}^\mathsf{T}\,.$$

▷ If $\mathbf{y}$ satisfies the system of equations $\mathbf{Ay} = \mathbf{b}$, and $\mathbf{x}$ satisfies the system of equations $\mathbf{Bx} = \mathbf{y}$, then $(\mathbf{AB})\mathbf{x} = \mathbf{b}$.

9.3.7 Multiplication by a diagonal matrix

- A matrix **A** is multiplied left or right by a diagonal matrix **D** by multiplying every element of a row or column of **A** by the diagonal element of the corresponding row or column of **D**.

$$(\mathbf{DA})_{ij} = d_{ii}a_{ij}\ , \quad \text{or} \quad (\mathbf{AD})_{ij} = a_{ij}d_{jj}$$

- **Identity map** of matrix **A**; i.e., multiplication of matrix **A** by the unity matrix **I**.
- Multiplications by the scalar matrix $\mathbf{S} = c\mathbf{I}$ or the unity matrix **I** are commutative:

$$\mathbf{AS} = \mathbf{SA} = c\mathbf{A}$$

$$\mathbf{AI} = \mathbf{IA} = \mathbf{A}$$

- The multiplication of a matrix **A** by the null matrix **0** yields the null matrix.

$$\mathbf{A0} = \mathbf{0A} = \mathbf{0}$$

▷ However, the converse usually does not hold.

- From $\mathbf{AB} = \mathbf{BA} = \mathbf{0}$ it does not necessarily follow that **A** or **B** equals zero!

Zero divisor A and **B**: Neither **A** nor **B** are null matrices, but **AB** is the null matrix.

$$\mathbf{AB} = \mathbf{BA} = \mathbf{0} \qquad \text{but} \qquad \mathbf{A} \neq \mathbf{0} \quad \text{and} \quad \mathbf{B} \neq \mathbf{0}\,.$$

□ Two matrices with nonzero matrix elements, but whose product yields zero:

$$\begin{pmatrix} 2 & 0 & 1 & 0 \\ 0 & -3 & 0 & 1 \\ 4 & 0 & 2 & 0 \\ 0 & 9 & 0 & -3 \end{pmatrix} \begin{pmatrix} -1 & 0 & 1 & 0 \\ 0 & 1 & 0 & 2 \\ 2 & 0 & -2 & 0 \\ 0 & 3 & 0 & 6 \end{pmatrix} = \begin{pmatrix} 0 & 0 & 0 & 0 \\ 0 & 0 & 0 & 0 \\ 0 & 0 & 0 & 0 \\ 0 & 0 & 0 & 0 \end{pmatrix} = \mathbf{0}\,.$$

9.3.8 Matrix multiplication according to Falk's scheme

Falk's scheme offers better clarity in the multiplication of two matrices "by hand."

□ **Matrix notation** of Falk's scheme for the product of matrices **A** and **B** :

$$\begin{array}{c|c} & \mathbf{B} \\ \hline \mathbf{A} & \mathbf{C} \end{array}\,, \quad \text{where} \quad \mathbf{C} = \mathbf{AB}\,.$$

To calculate the product $\mathbf{C} = \mathbf{AB}$, matrix **A** is written into the bottom left rectangle, and matrix **B** is written into the top right rectangle. Every matrix element c_{ij} of **C** thus results from the scalar product of the left-hand row vector $\mathbf{a}_i$ of **A** and the column vector $\mathbf{b}_j$ of **B** above at the node of row i of **A** and column j of **B**.

Vector notation according to Falk's scheme:

$$\begin{array}{c|ccc|c|ccc} & \mathbf{b}_1 & & \cdots & \mathbf{b}_j & \cdots & & \mathbf{b}_n \\ \hline \mathbf{a}_1 & \mathbf{a}_1\cdot\mathbf{b}_1 & & \cdots & & \cdots & & \mathbf{a}_1\cdot\mathbf{b}_n \\ \vdots & \vdots & & & \vdots & & & \vdots \\ \hline \mathbf{a}_i & & & \cdots & c_{ij} & \cdots & & \\ \hline \vdots & \vdots & & & \vdots & & & \vdots \\ \mathbf{a}_m & \mathbf{a}_m\cdot\mathbf{b}_1 & & \cdots & & \cdots & & \mathbf{a}_m\cdot\mathbf{b}_n \end{array} \qquad c_{ij} = \mathbf{a}_i\mathbf{b}_j = \sum_{k=1}^{l} a_{ik}b_{kj}$$

Matrix notation of Falk's scheme:
The matrix element c_{ij} of matrix $\mathbf{C} = \mathbf{AB}$ stands at the node of the i-th row of matrix $\mathbf{A}$ and the j-th row of matrix $\mathbf{B}$:

$$\begin{array}{c|cc|c|cc}
 & b_{11} & \cdots & b_{1j} & \cdots & b_{1n} \\
 & \vdots & & \vdots & & \vdots \\
 & b_{l1} & \cdots & b_{lj} & \cdots & b_{ln} \\ \hline
a_{11}\cdots a_{1l} & \sum_{k=1}^{l} a_{1k}b_{k1} & \cdots & & \cdots & \sum_{k=1}^{l} a_{1k}b_{km} \\
\vdots & \vdots & & \vdots & & \vdots \\ \hline
a_{i1}\cdots a_{il} & & \cdots & c_{ij} & \cdots & \\ \hline
\vdots & \vdots & & \vdots & & \vdots \\
a_{m1}\cdots a_{ml} & \sum_{k=1}^{l} a_{mk}b_{k1} & \cdots & & \cdots & \sum_{k=1}^{l} a_{mk}b_{kn}
\end{array}$$

$$\text{with } c_{ij} = a_{i1}b_{1j} + a_{i2}b_{2j} + \cdots a_{il}b_{lj} = \sum_{k=1}^{l} a_{ik}b_{kj}$$

□ Multiplication of two 2×2 matrices by means of Falk's scheme:

$$\begin{array}{cc|cc}
 & & 2 & 3 \\
 & & 0 & 4 \\ \hline
1 & 3 & 1\cdot 2+3\cdot 0 & 1\cdot 3+3\cdot 4 \\
2 & 4 & 2\cdot 2+4\cdot 0 & 2\cdot 3+4\cdot 4
\end{array}
\Leftrightarrow
\begin{array}{c|c}
 & \mathbf{B} \\ \hline
\mathbf{A} & \mathbf{C}
\end{array}
\text{ with: } \mathbf{C} = \mathbf{AB} = \begin{pmatrix} 2 & 15 \\ 4 & 22 \end{pmatrix}$$

□ Multiplication of a 2×3 matrix $\mathbf{A}$ by a 3×4 matrix $\mathbf{B}$ by means of Falk's scheme:

$$\begin{array}{ccc|cccc}
 & & & 1 & 4 & 3 & 0 \\
 & & & 1 & 1 & -1 & 3 \\
 & & & 0 & -2 & -3 & 2 \\ \hline
1 & 0 & 3 & 1 & -2 & -6 & 6 \\
2 & 1 & -4 & 3 & 17 & 17 & -5
\end{array}
\Rightarrow \mathbf{C} = \begin{pmatrix} 1 & -2 & -6 & 6 \\ 3 & 17 & 17 & -5 \end{pmatrix}$$

The resulting matrix $\mathbf{C}$ is a 2×4 matrix.

9.3.9 Checking of row and column sums

The check of the row sum or **the column sum** can be used to control matrix multiplication.
Row-sum vector a of matrix $\mathbf{A} = (a_{ij})$: The column vector whose matrix elements a_i are the sum of the matrix elements of the i-th row:

$$a_i = a_{i1} + a_{i2} + \cdots + a_{in} .$$

Column-sum vector b of matrix $\mathbf{B} = (b_{ij})$: The row vector whose matrix elements b_j are the sum of the matrix elements of the j-th column:

$$b_j = b_{1j} + b_{2j} + \cdots + b_{nj} .$$

Check of the row sum of the product

$$\mathbf{AB} = \mathbf{C}$$

- The product of matrix **A** with the row-sum vector **b** of matrix **B** has to be equal to the row-sum vector **c** of matrix **C**!

$$\mathbf{Ab} = \mathbf{c}$$

▷ From $\mathbf{Ab} \neq \mathbf{c}$ it follows that a computational error is present. The converse conclusion is not necessarily valid, but the possibility of a computational error is lower. In particular, for large matrices the check is relatively easy.

Checking the row sum in five steps:

1. Calculate the product $\mathbf{C} = \mathbf{AB}$;
2. Calculate the row-sum vector **c** of **C**;
3. Calculate the row-sum vector **b** of **B**;
4. Multiply **b** by the matrix **A**from the left;
5. Comparison: The result **Ab** must equal **c**.

□ Checking the row sum:

1. $\mathbf{AB} = \mathbf{C}$

$$\begin{pmatrix} 1 & 3 \\ 2 & 4 \end{pmatrix} \begin{pmatrix} 2 & 3 \\ 0 & 4 \end{pmatrix} = \begin{pmatrix} 2 & 15 \\ 4 & 22 \end{pmatrix}$$

2. Calculate the row-sum vector **c** of matrix **C**:

$$\mathbf{c} = \begin{pmatrix} 2+15 \\ 4+22 \end{pmatrix} = \begin{pmatrix} 17 \\ 26 \end{pmatrix}$$

3. Calculate the row-sum vector **b** of matrix **B**:

$$\mathbf{b} = \begin{pmatrix} 2+3 \\ 0+4 \end{pmatrix} = \begin{pmatrix} 5 \\ 4 \end{pmatrix}$$

4. Calculate the product of matrix **A** and row-sum vector **b**:

$$\mathbf{Ab} = \begin{pmatrix} 1 & 3 \\ 2 & 4 \end{pmatrix} \begin{pmatrix} 5 \\ 4 \end{pmatrix} = \begin{pmatrix} 5+12 \\ 10+16 \end{pmatrix} = \begin{pmatrix} 17 \\ 26 \end{pmatrix}$$

5. Comparison of the row sums:

$$\mathbf{Ab} = \mathbf{c}$$

$$\begin{pmatrix} 17 \\ 26 \end{pmatrix} = \begin{pmatrix} 17 \\ 26 \end{pmatrix}$$

- **Checking the column sum**: The product of the column-sum vector **a** of matrix **A** and matrix **B** has to be equal to the column-sum vector **c** of matrix **C**.

9.4 Determinants

Determinants: One purpose of determinants is to investigate the **solvability** of systems of linear equations (with an arbitrary number of unknowns).

Determinants also permit the solution of a system of equations by Cramer's rule, but this procedure is extermely inefficient!

Determinant det **A** ($|\mathbf{A}|$, $|a_{ik}|$) of a matrix **A**: A real or complex number, the determinant det **A**, can be uniquely assigned to every $n \times n$ square matrix **A**. The determinant is equal to the sum of all possible products of n matrix elements so that every summand contains exactly one element of each row and each column of the matrix, and each summand is weighted by $+1$ or -1, depending on whether the corresponding permutation $\pi(i)$ of the numbers $1, 2, \ldots, n$ is generated by an even (+) or odd (−) number $j(\pi)$ of permutations of two neighboring indices.

Permutations:

$$(1, 2, 3, 4, 5, \ldots),\ (2, 1, 3, 4, 5, \ldots),\ (2, 3, 1, 5, 4, \ldots)\ , \text{ etc.}$$

Notation:

$$\det \mathbf{A} = \det \begin{pmatrix} a_{11} & \cdots & a_{1n} \\ \vdots & \ddots & \vdots \\ a_{n1} & \cdots & a_{nn} \end{pmatrix} = \begin{vmatrix} a_{11} & \cdots & a_{1n} \\ \vdots & \ddots & \vdots \\ a_{n1} & \cdots & a_{nn} \end{vmatrix} \stackrel{def}{=} \sum_{\pi}^{n!} (-1)^{j(\pi)} \prod_{i=1}^{n} a_{i,\pi(i)} \,.$$

▷ For every set of indices $(1, 2, 3, \ldots, n)$ there are exactly $n!$ permutations that have to be summed up. Therefore, numerical methods involving the determinant of a matrix require an effort of $O(n!)$ operations generally and are therefore very time consuming. In practice, determinants are generally calculated after the matrix is converted to diagonal form.

▷ It is clearly less of an effort to calculate determinants numerically with $n > 3$ by means of the Gaussian algorithm.

9.4.1 Two-row determinants

The determinant of a two-row matrix A , also called a **two-row** determinant or determinant of the **second** order, is the number:

$$\det \mathbf{A} = |\mathbf{A}| = \det \begin{pmatrix} a_{11} & a_{12} \\ a_{21} & a_{22} \end{pmatrix} = \begin{vmatrix} a_{11} & a_{12} \\ a_{21} & a_{22} \end{vmatrix} = a_{11}a_{22} - a_{12}a_{21} \,.$$

The number $a_{11}a_{22} - a_{12}a_{21}$ is equal to the **product** of the elements on the **main diagonal minus** the product of the elements on the **secondary diagonal**.

● Criterion for the solvability of systems of equations:
A system of linear equations with two equations and two unknowns has **one** solution if and only if a **number**, the **determinant** det **A** of coefficient matrix **A**, is **nonzero**, det $\mathbf{A} \neq 0$.

9.4.2 General computational rules for determinants

The following computational rules are valid for all determinants, i.e., two-row determinants. In that case they are particularly easy to calculate.

▷ Different matrices can have the same determinant.

● $\det \mathbf{A} = \det \mathbf{B}$ does **not** imply $\mathbf{A} = \mathbf{B}$.

□ $\det \begin{pmatrix} 1 & 0 \\ 0 & -1 \end{pmatrix} = -1 = \det \begin{pmatrix} 4 & 0 \\ 0 & -1/4 \end{pmatrix} = -1\,,$
although the two matrices are different, $\mathbf{A} \neq \mathbf{B}$!

- **Transposition of a matrix**: The value of a determinant remains unchanged even if the main diagonal is transposed:

$$\det \mathbf{A} = \begin{vmatrix} a_{11} & a_{12} \\ a_{21} & a_{22} \end{vmatrix} = \begin{vmatrix} a_{11} & a_{21} \\ a_{12} & a_{22} \end{vmatrix} = \det \mathbf{A}^{\mathsf{T}}.$$

- **Exchange theorem**: If two rows or columns are interchanged, the determinant changes its sign:

$$\begin{vmatrix} a_{11} & a_{12} \\ a_{21} & a_{22} \end{vmatrix} = (a_{11}a_{22} - a_{12}a_{21}) = -(a_{12}a_{21} - a_{11}a_{22}) = -\begin{vmatrix} a_{21} & a_{22} \\ a_{11} & a_{12} \end{vmatrix}$$

- **Factor rule**: If the elements of an arbitrary row or column of a determinant are multiplied by a real scalar c, then the determinant is multiplied by c:

$$\begin{vmatrix} ca_{11} & ca_{12} \\ a_{21} & a_{22} \end{vmatrix} = (ca_{11}a_{22} - ca_{12}a_{21}) = c(a_{11}a_{22} - a_{12}a_{21}) = c\begin{vmatrix} a_{11} & a_{12} \\ a_{21} & a_{22} \end{vmatrix}$$

- Conversely, it holds that: A determinant is multiplied by a real scalar c, when all elements of **one and only one** arbitrary row or columns are multiplied by c.
- If the matrix elements of a row or a column of the matrix have a common factor, then this factor can be factored out of the determinant.

▷ A matrix is multiplied by a scalar by multiplying each individual element by the factor.
For the **determinant**, **one and only one** row or column of the determinant may be multiplied by the scalar!

General factor rule:

$$\begin{aligned} c\det\mathbf{A} &= c\det\begin{pmatrix} a_{11} & \cdots & a_{1n} \\ \vdots & & \vdots \\ a_{n1} & \cdots & a_{nn} \end{pmatrix} = \det\begin{pmatrix} ca_{11} & \cdots & ca_{1n} \\ \vdots & & \vdots \\ a_{n1} & \cdots & a_{nn} \end{pmatrix} \\ &= \det\begin{pmatrix} ca_{11} & \cdots & a_{1n} \\ \vdots & & \vdots \\ ca_{n1} & \cdots & a_{nn} \end{pmatrix} = \det\begin{pmatrix} a_{11} & \cdots & ca_{1n} \\ \vdots & & \vdots \\ a_{n1} & \cdots & ca_{nn} \end{pmatrix} \\ &= \det\begin{pmatrix} a_{11} & \cdots & a_{1n} \\ \vdots & & \vdots \\ ca_{n1} & \cdots & ca_{nn} \end{pmatrix} \end{aligned}$$

□ Multiplication by a scalar factor c:

$$4\det\begin{pmatrix} 1 & 2 & 3 \\ 2 & 1 & 4 \\ 3 & 4 & 1 \end{pmatrix} = \det\begin{pmatrix} 4 & 8 & 12 \\ 2 & 1 & 4 \\ 3 & 4 & 1 \end{pmatrix}$$

$$= \det\begin{pmatrix} 1 & 2 & 3 \\ 8 & 4 & 16 \\ 3 & 4 & 1 \end{pmatrix} = \det\begin{pmatrix} 4 & 2 & 3 \\ 8 & 1 & 4 \\ 12 & 4 & 1 \end{pmatrix}$$

□ The determinants of the following matrices vanish:

$$\mathbf{A} = \begin{pmatrix} 2 & 6 \\ 0 & 0 \end{pmatrix}, \quad \mathbf{B} = \begin{pmatrix} 2 & 4 \\ 3 & 6 \end{pmatrix}, \quad \mathbf{C} = \begin{pmatrix} 3 & 3 \\ 6 & 6 \end{pmatrix}, \quad \mathbf{D} = \begin{pmatrix} 0 & 3 \\ 0 & 6 \end{pmatrix}$$

Reason:
$\det \mathbf{A} = 0$: The matrix elements of complete row are zero.
$\det \mathbf{B} = 0$: Two rows are proportional.
$\det \mathbf{C} = 0$: Two columns are equal.
$\det \mathbf{D} = 0$: The matrix elements of complete column are zero.

- **Rule for linear combinations**: The value of a determinant does not change if an arbitrary multiple of elements of a row or a column is added to (or subtracted from) by elements of another row or column.

$$\begin{vmatrix} a_{11} & a_{12} \\ a_{21} & a_{22} \end{vmatrix} = \begin{vmatrix} a_{11} \pm c a_{12} & a_{12} \\ a_{21} \pm c a_{22} & a_{22} \end{vmatrix}$$

- **Multiplication theorem for determinants**: The determinant of the product of two matrices $\mathbf{A}$ and $\mathbf{B}$ is equal to the product of the determinants of the individual matrices:

$$\det(\mathbf{A} \cdot \mathbf{B}) = \det(\mathbf{A}) \det(\mathbf{B}) = |\mathbf{A}|\ |\mathbf{B}|$$

▷ The multiplication theorem provides a possibility of calculating the matrix product directly from the determinants of the individual matrices. Thus, it may not be necessary to calculate the matrix product.

- **Decomposition theorem**: If a row (or column) consists of a sum of elements, then the determinant can be decomposed as follows into a sum of two determinants:

$$\begin{vmatrix} (a_{11} + b_1) & a_{12} \\ (a_{21} + b_2) & a_{22} \end{vmatrix} = \begin{vmatrix} a_{11} & a_{12} \\ a_{21} & a_{22} \end{vmatrix} + \begin{vmatrix} b_1 & a_{12} \\ b_2 & a_{22} \end{vmatrix}$$

- **Determinant of triangular matrices**: The determinant of a triangular matrix equals the product of the elements of the main diagonal and has the value

$$\det \mathbf{A} = \begin{vmatrix} a_{11} & a_{12} \\ 0 & a_{22} \end{vmatrix} = a_{11} a_{22} = \Pi_i a_{ii} \ .$$

- **Determinant of the diagonal matrix D** (also the upper and lower triangular matrix):

$$\det \mathbf{D} = d_{11} d_{22} = \prod_i d_{ii} \ .$$

□ For the unit matrix it holds analogously that: $\det \mathbf{I} = 1 \cdot 1$.
□ For the null matrix it holds that: $\det \mathbf{0} = 0$.

9.4.3 Zero value of the determinant

An n-row determinant is zero if and only if one or more of the following conditions are met:

- All matrix elements of a row (or column) equal zero.
- Two rows (or columns) are identical.
  Multiplication of a row (or column) by -1 and addition yields a row (or column) of all zeros.
- Two rows (or columns) are proportional.
  If the proportionality factor is c, then the c-fold of a row (or column) can be subtracted from the other row (column). Because the value of the determinant does not change, but a row (or column) with all zeros appears, the determinant equals zero.
- One row (or column) can be represented as a linear combination of the other rows (or columns): A linear combination of the other row or column vectors can be subtracted from this row (or column). This yields a row (or column) with all zeros.

9.4.4 Three-row determinants

Three-row determinants are calculated in order to determine the solvability of systems of linear equations of the third order, i.e., of three equations with three unknowns.

Determinant of the third degree or **determinant of the third order**: The number derived from a three-row matrix:

$$\begin{aligned}\det \mathbf{A} &= \begin{vmatrix} a_{11} & a_{12} & a_{13} \\ a_{21} & a_{22} & a_{23} \\ a_{31} & a_{32} & a_{33} \end{vmatrix} \\ &= a_{11}a_{22}a_{33} + a_{12}a_{23}a_{31} + a_{13}a_{21}a_{32} - a_{13}a_{22}a_{31} - a_{11}a_{23}a_{32} - a_{12}a_{21}a_{33} .\end{aligned}$$

□ The triple scalar product of three vectors can be written as the determinant of a three-row matrix whose columns are the components of the vectors:

$$\begin{aligned}\vec{\mathbf{a}} \cdot (\vec{\mathbf{b}} \times \vec{\mathbf{c}}) &= \begin{vmatrix} a_x & b_x & c_x \\ a_y & b_y & c_y \\ a_z & b_z & c_z \end{vmatrix} \\ &= a_x b_y c_z + a_y b_z c_x + a_z b_x c_y - a_z b_y c_x - a_x b_z c_y - a_y b_x c_z .\end{aligned}$$

● Solvability of systems of linear equations of the third order:

For a system of equations

$$a_{11}x_1 + a_{12}x_2 + a_{13}x_3 = b_1$$

$$a_{21}x_1 + a_{22}x_2 + a_{23}x_3 = b_2$$

$$a_{31}x_1 + a_{32}x_2 + a_{33}x_3 = b_3$$

to be solved uniquely for the unknowns x_1, x_2, x_3 the determinant of the coefficient matrix of the system of equations must be nonzero, i.e.,

$$\det \mathbf{A} = \begin{vmatrix} a_{11} & a_{12} & a_{13} \\ a_{21} & a_{22} & a_{23} \\ a_{31} & a_{32} & a_{33} \end{vmatrix} \neq 0 .$$

● **Sarrus's rule**: To calculate the determinant of a three-row matrix, the first two columns of the matrix are repeated on the right of the determinant.

Value of the determinant: The products of the elements along the **diagonals** sloping from left top to right bottom are **added**, and the products of the elements on the **diagonals** sloping from right top to left bottom are **subtracted.**

$$\det \mathbf{A} = \left| \begin{matrix} a_{11} & a_{12} & a_{13} \\ a_{21} & a_{22} & a_{23} \\ a_{31} & a_{32} & a_{33} \end{matrix} \right| \begin{matrix} a_{11} & a_{12} \\ a_{21} & a_{22} \\ a_{31} & a_{32} \end{matrix}$$

$= a_{11}a_{22}a_{33} + a_{12}a_{23}a_{31} + a_{13}a_{21}a_{32}$ products from the main diagonals

$-a_{13}a_{22}a_{31} - a_{11}a_{23}a_{32} - a_{12}a_{21}a_{33}$ products from the secondary diagonals

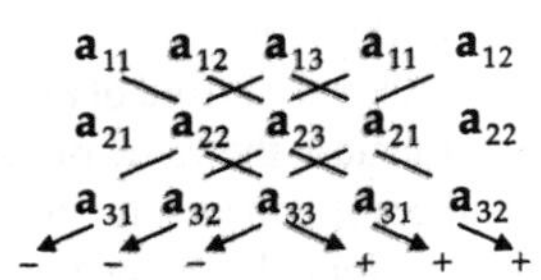

Illustration of Sarrus's rule

▷ Sarrus's rule holds **only** for **three-row** determinants, but **not** (!) for determinants of higher order, which are calculated by means of Laplace's expansion theorem.

□ Calculation of the determinant of a three-row matrix by means of Saruss's rule:

$$\det \mathbf{A} = \left| \begin{array}{ccc|cc} 5 & -1 & 4 & 5 & -1 \\ 1 & -2 & 7 & 1 & -2 \\ 0 & 3 & 2 & 0 & 3 \end{array} \right|$$

$$= 5 \cdot (-2) \cdot 2 + (-1) \cdot 7 \cdot 0 + 4 \cdot 1 \cdot 3 + \quad \text{products from the main diagonals}$$

$$-4 \cdot (-2) \cdot 0 - 5 \cdot 7 \cdot 3 - (-1) \cdot 1 \cdot 2 \quad \text{product from the secondary diagonals}$$

$$= -111 \,.$$

- Determinants of higher order can be calculated by the determinants of lower order, the so-called **minor determinants**.
- The calculation of determinants of the third order can be reduced to the calculation of determinants of the second order. Proceed analogously, with determinants of higher order.

▷ **Minor determinant of the second order**, $\det \mathbf{U}_{ik}$: Arises from deleting the i-th row and the k-th column of a three-row determinant: The remaining elements form a two-row determinant.

▷ **Cofactor** $\mathcal{A}_{ik}$ of the element a_{ik}: The minor determinant $\det \mathbf{U}_{ik}$ multiplied by the sign factor $(-1)^{i+k}$

$$\mathcal{A}_{ik} = (-1)^{i+k} \cdot \det \mathbf{U}_{ik}$$

□ Three-row determinants can be represented by factoring out the elements of the first row with minor determinants of the second order.

$$\begin{aligned} \det \mathbf{A} &= a_{11} \begin{vmatrix} a_{22} & a_{23} \\ a_{32} & a_{33} \end{vmatrix} + a_{12} \begin{vmatrix} a_{23} & a_{21} \\ a_{33} & a_{31} \end{vmatrix} + a_{13} \begin{vmatrix} a_{21} & a_{22} \\ a_{31} & a_{32} \end{vmatrix} \\ &= a_{11}a_{22}a_{33} + a_{12}a_{23}a_{31} + a_{13}a_{21}a_{32} \\ &\quad -a_{13}a_{22}a_{31} - a_{11}a_{23}a_{32} - a_{12}a_{21}a_{33} \\ &= a_{11}(a_{22}a_{33} - a_{23}a_{32}) - a_{12}(a_{21}a_{33} - a_{23}a_{31}) + a_{13}(a_{21}a_{32} - a_{22}a_{31}) \\ &= a_{11} \det \mathbf{U}_{11} - a_{12} \det \mathbf{U}_{12} + a_{13} \det \mathbf{U}_{13} \,. \end{aligned}$$

Or

$$\det \mathbf{A} = a_{11}\mathcal{A}_{11} + a_{12}\mathcal{A}_{12} + a_{13}\mathcal{A}_{13} = \sum_{k=1}^{3} a_{1k}\mathcal{A}_{1k}$$

$$\mathcal{A}_{11} = +\det \mathbf{U}_{11} \,, \quad \mathcal{A}_{12} = -\det \mathbf{U}_{12} \,, \quad \mathcal{A}_{13} = +\det \mathbf{U}_{13} \,,$$

Expansion of the determinant in terms of the elements of **the first row**: The elements of the first row of matrix **A** are factored out. The determinant is an expression in which the elements of the first row and their cofactors occur.

Corresponding expansion formulas exist for any row or column:

- **Laplace's expansion**:

  A three-row determinant can be **expanded** according to any of the rows or columns as follows:

  Expansion according to the i-th row:

$$\det \mathbf{A} = \sum_{k=1}^{3} a_{ik}\mathcal{A}_{ik} \,, \quad i = 1, \ldots, 3$$

Expansion according to the k-th column:

$$\det \mathbf{A} = \sum_{i=1}^{3} a_{ik}\mathcal{A}_{ik} \ , \quad k = 1, \ldots, 3$$

Here, $\mathcal{A}_{ik} = (-1)^{i+k} \det \mathbf{U}_{ik}$ denotes the cofactor of the element a_{ik} of $\mathbf{A}$, and $\det \mathbf{U}_{ik}$ denotes the minor determinant formed by deleting the i-th row and the k-th column.

□ The minor of a 3×3 matrix $\mathbf{A}$ is a 2×2 matrix $\mathbf{U}_{ik}$:

$$\mathbf{U}_{32} = \left(\begin{array}{c|c|c} a_{11} & a_{12} & a_{13} \\ a_{21} & a_{22} & a_{23} \\ \hline a_{31} & a_{32} & a_{33} \end{array}\right) \Leftarrow \text{delete row 3} = \begin{pmatrix} a_{11} & a_{13} \\ a_{21} & a_{23} \end{pmatrix}$$

⇑ delete column 2

Minor determinant:

$$\det \mathbf{U}_{32} = \begin{vmatrix} a_{11} & a_{13} \\ a_{21} & a_{23} \end{vmatrix} = a_{11}a_{23} - a_{13}a_{21}$$

- Three-row determinants satisfy the same calculating rules as two-row determinants.

9.4.5 Determinants of higher (n-th) order

The term "determinants" can be generalized for any square matrix of n-th order:
The determinant of n-th order, or n-row **determinant**, assigns **one** number, $\det \mathbf{A}$, to an n-row square matrix:
Determinant of n-th order:

$$\det \mathbf{A} = \begin{vmatrix} a_{11} & \cdots & a_{1n} \\ \vdots & \ddots & \vdots \\ a_{n1} & \cdots & a_{nn} \end{vmatrix} .$$

- **Criterion for the solvability of systems of equations** of n-th order: A system of linear equations of n-th order (i.e., a system of n equations and n unknowns) has a **unique** solution **if and only if** the determinant of the matrix of coefficients is **nonzero**, $\det \mathbf{A} \neq 0$!

Minor determinant $\det \mathbf{U}_{ik}$ **of order** $n-1$ of the $n \times n$ matrix $\mathbf{A}$: The i-th row and the k-th column of $\mathbf{A}$ are deleted and a determinant is formed from the resulting $(n-1) \times (n-1)$ matrix. The indices i or j are used to denote the row or column deleted.
Algebraic complement, adjuncts, minor determinant, cofactor of the element a_{ij}: The $(n-1)$-row minor determinant $\det \mathbf{U}_{ij}$ resulting from the matrix $\mathbf{A}$ by deleting the i-th row and j-th column and forming the determinant of the resulting matrix $\mathbf{U}_{ij}$, multiplied by the sign factor $(-1)^{i+j}$:

$$\mathcal{A}_{ij} = (-1)^{i+j} \det \mathbf{U}_{ij} \ .$$

Analogously to two-row and three-row determinants, the n-row determinants can also be defined. They can be used to set up criteria for the solvability of systems of linear equations with n unknowns.

- **Laplace's expansion**: An n-row determinant can be expanded according to the matrix elements of any row or column.

□ Expansion in terms of the i-th row:

$$\det \mathbf{A} = \sum_{k=1}^{n} a_{ik} \mathcal{A}_{ik} \quad (i = 1, \ldots, n) .$$

□ Expansion in terms of the k-th column:

$$\det \mathbf{A} = \sum_{i=1}^{n} a_{ik} \mathcal{A}_{ik} \quad (k = 1, \ldots, n) ,$$

$\mathcal{A}_{ik} = (-1)^{i+k} \det \mathbf{U}_{ik}$: algebraic complement of a_{ik} in $\det A$.
$\det \mathbf{U}_{ik}$: $(n-1)$-row minor determinant.

- For three-row determinants, Sarrus's rule results.
- The value of an n-row determinant is independent of the row or column according to which the expansion is done.

▷ The calculation work can be considerably reduced by expanding according to the row or column that contains most of the zeros.

▷ The numerical calculation of determinants with $n > 3$ is most efficient with elementary conversions. The matrix is made triangular. Then the determinant is the product of the diagonal elements.

▷ The sign factor in the cofactor can be determined by means of the chessboard rule.

□ $n \times n$ matrix:

+	−	+	−	...
−	+	−	+	
+	−	+	−	
−	+	−	+	
⋮				

- **The computational rules for determinants of order** n are analogous to those for two-row and three-row determinants.

▷ In multiplying a determinant by a factor c, one has to multiply **merely one** arbitrary row or column of the determinant by c.
This is in contrast to the multiplication of a matrix by a scalar c, in which **every** element (a_{ij}) has to be multiplied by c: $c \det \mathbf{A} \neq \det(c\,\mathbf{A})$; instead one would have $c^n \det \mathbf{A} = \det(c\,\mathbf{A})$;

9.4.6 Calculation of n-row determinants

- Laplace expansion in terms of a row or a column with as many zeros as possible.
- The value of the determinants remains unchanged by elementary conversions.
- **Generate** zeros in rows by addition (subtraction) of multiples of rows.
- **Generate** zeros in columns by addition (subtraction) of multiples of columns.

▷ **Practical calculation of n-row determinants by elementary conversions**: The **calculation** of determinants of higher order is performed by **numerical methods**. The matrix is transformed into triangular form by elementary conversions, which do not alter the value of the determinant. Then the determinant can be calculated easily.

- The determinant of a triangular matrix is the product of the **elements on the main diagonal**:

$$\det \mathbf{A} = a_{11} \cdot a_{22} \cdot a_{33} \cdot \cdots \cdot a_{nn}$$

for

$$\mathbf{A} = \begin{pmatrix} a_{11} & a_{12} & \cdots & a_{1n} \\ 0 & a_{22} & \cdots & \vdots \\ \vdots & \vdots & \ddots & \vdots \\ 0 & \cdots & 0 & a_{nn} \end{pmatrix} \quad \text{or} \quad \mathbf{A} = \begin{pmatrix} a_{11} & 0 & \cdots & 0 \\ a_{21} & a_{22} & \ddots & \vdots \\ \vdots & \vdots & \ddots & \vdots \\ a_{n1} & a_{n2} & \cdots & a_{nn} \end{pmatrix}.$$

□ Expansion in terms of the first column:

$$\mathbf{A} = \begin{pmatrix} 1 & 4 & 5 \\ 0 & 2 & 6 \\ 0 & 0 & 3 \end{pmatrix} = 1 \cdot (2 \cdot 3 - 0 \cdot 6) - 0 \cdot (\cdots) + 0 \cdot (\cdots) = 6\,.$$

▷ Such methods for matrix manipulation are utilized in solving systems of linear equations.

9.4.7 Regular and inverse matrix

Regular matrix: An n-row square matrix **A** whose determinant $\det \mathbf{A}$ **does not** vanish:

$$\det \mathbf{A} \neq 0 \;\Leftrightarrow\; \mathbf{A} \text{ is regular} \;\Leftrightarrow\; \mathbf{A} \text{ is not singular}$$

● Systems of linear equations can only be solved uniquely if the matrix of coefficients is regular, i.e., if the determinant of the matrix of coefficients is nonzero, $\det \mathbf{A} \neq 0$.

Singular matrix: A matrix **A** whose determinant $\det \mathbf{A}$ is equal to zero:

$$\det \mathbf{A} = 0 \;\Leftrightarrow\; \mathbf{A} \text{ is singular} \;\Leftrightarrow\; \mathbf{A} \text{ is not regular.}$$

● Systems of linear equations $\mathbf{Ax} = \mathbf{b}$ (**b**: arbitrary vector of constants) whose matrix of coefficients **A** is singular, $\det \mathbf{A} = 0$, **cannot be solved uniquely**.

Inverse matrix $\mathbf{A}^{-1}$ or **reciprocal matrix**: The square matrix $\mathbf{A}^{-1}$ that satisfies:

$$\mathbf{A}\mathbf{A}^{-1} = \mathbf{A}^{-1}\mathbf{A} = \mathbf{I}$$

● If matrix **A** is multiplied by its inverse matrix $\mathbf{A}^{-1}$, then we obtain the unit matrix.

● The matrix product of **A** and $\mathbf{A}^{-1}$ is commutative.

● Not every matrix has an inverse.

● A square matrix has at most one inverse.

Invertibility of a matrix: If for an n-row matrix **A** an n-row square matrix $\mathbf{A}^{-1}$ exists with:

$$\mathbf{A}\mathbf{A}^{-1} = \mathbf{I}\,,$$

then **A** is called an **invertible** matrix.

● Only regular matrices are invertible.

● Singular matrices **are not** invertible.

● Only those systems of linear equations with an invertible matrix of coefficients can be solved **uniquely**!

● Systems of linear equations can be solved directly by means of the inverse matrix:

$$\mathbf{Ax} = \mathbf{c} \Leftrightarrow \mathbf{A}^{-1}\mathbf{Ax} = \mathbf{A}^{-1}\mathbf{c} \Rightarrow \mathbf{x} = \mathbf{A}^{-1}\mathbf{c}$$

▷ Thus, systems of linear equations with a fixed matrix of coefficients **A**, but different vectors of constants **c**, can be solved for many different elements of disturbance **c** by calculating the inverse $\mathbf{A}^{-1}$ only once (see the Gauss-Jordan method in Section 9.6).

- **Similarity of matrices**: Two matrices **A** and **B** are **similar** if and only if there is at least one invertible (regular) matrix **U** for which it holds that:

$$\mathbf{A} = \mathbf{U}^{-1}\mathbf{B}\mathbf{U}\,.$$

▷ The **similarity transformation U** can be a rotation, translation, elongation, compression or reflection. See also transformations and computer graphics.

9.4.8 *Calculation of the inverse matrix in terms of determinants*

- For every regular n-row square matrix **A** there is exactly one inverse matrix $\mathbf{A}^{-1}$ with

$$\mathbf{A}^{-1} = \frac{1}{\det \mathbf{A}}\mathcal{A}^{\mathsf{T}} = \frac{1}{\det \mathbf{A}}\begin{pmatrix} \mathcal{A}_{11} & \cdots & \mathcal{A}_{n1} \\ \vdots & & \vdots \\ \mathcal{A}_{1n} & \cdots & \mathcal{A}_{nn} \end{pmatrix}$$

The matrix elements of $\mathcal{A}^{\mathsf{T}}$ are obtained by transposing the matrix of the **cofactors** $(\mathcal{A}_{ik})_{i,k=1,\ldots,n}$ of a_{ik}. **Cofactors** in turn can be expressed by subdeterminants:

$$\mathcal{A}_{ik} = (-1)^{i+k} \cdot \det \mathbf{U}_{ik}$$

$\det \mathbf{U}_{ik}$ is the $n-1$-row **subdeterminant** of **A**; it is calculated from the **square submatrix** $\mathbf{U}_{ik}$ of shape $(n-1) \times (n-1)$, which is obtained from **A** by deleting the i-th row and the k-th column.

▷ Exactly those matrix elements of **A** whose row or column indices are given in the submatrix $\mathbf{U}_{ik}$ are absent in matrix **A**.

□ The submatrix of a 3×3 matrix **A** is a 2×2 matrix $\mathbf{U}_{ik}$:

$$\mathbf{U}_{32} = \left(\begin{array}{c|c|c} a_{11} & a_{12} & a_{13} \\ a_{21} & a_{22} & a_{23} \\ \hline a_{31} & a_{32} & a_{33} \end{array}\right) \begin{array}{l} \\ \\ \Leftarrow \text{delete row 3} \end{array} = \begin{pmatrix} a_{11} & a_{13} \\ a_{21} & a_{23} \end{pmatrix}.$$
$$\Uparrow \text{ delete column 2}$$

Subdeterminant:

$$\det \mathbf{U}_{32} = \begin{vmatrix} a_{11} & a_{13} \\ a_{21} & a_{23} \end{vmatrix} = a_{11}a_{23} - a_{13}a_{21}\,.$$

Cofactor:

$$\mathcal{A}_{32} = (-1)^{2+3} \det \mathbf{U}_{32}\,.$$

▷ Note that the inverse matrix $\mathbf{A}^{-1}$ involves the **transpose** $\mathcal{A}^{\mathsf{T}}$ of the square matrix of the algebraic complement $\mathcal{A}$ (see transposed matrices in Section 9.2), which is obtained by reflection in the main diagonal or by interchanging the indices.

▷ In practice, the inverse—especially in case of large matrices—is not calculated of a determinant (very inefficient!) but by means of elementary conversions. For the numerical calculation of the inverse matrix, see the Gauss-Jordan method (Section 9.6).

□ Rotation of a position vector **x** through the angle α about the z-axis:

$$\mathbf{A} = \begin{pmatrix} \cos\alpha & -\sin\alpha & 0 \\ \sin\alpha & \cos\alpha & 0 \\ 0 & 0 & 1 \end{pmatrix} \qquad \mathbf{y} = \mathbf{A}\mathbf{x}$$

$$\mathbf{A}^{-1} = \frac{1}{\det\mathbf{A}} \begin{pmatrix} \cos\alpha & \sin\alpha & 0 \\ -\sin\alpha & \cos\alpha & 0 \\ 0 & 0 & 1 \end{pmatrix}$$

$$\det\mathbf{A} = \cos^2\alpha + \sin^2\alpha = 1$$

Characteristic for operators (transformations) with $|\mathbf{x}| = |\mathbf{y}|$; the standard remains.

$$\Longrightarrow \mathbf{A}^{-1} = \begin{pmatrix} \cos\alpha & \sin\alpha & 0 \\ -\sin\alpha & \cos\alpha & 0 \\ 0 & 0 & 1 \end{pmatrix} = \begin{pmatrix} \cos(-\alpha) & -\sin(-\alpha) & 0 \\ \sin(-\alpha) & \cos(-\alpha) & 0 \\ 0 & 0 & 1 \end{pmatrix}.$$

Note the sine terms change signs!

- **Properties of invertible matrices:**

$$\left(\mathbf{A}^{-1}\right)^{-1} = \mathbf{A}$$
$$(\mathbf{A}\cdot\mathbf{B})^{-1} = \mathbf{B}^{-1}\mathbf{A}^{-1}$$
$$\left(\mathbf{A}^{-1}\right)^{T} = \left(\mathbf{A}^{T}\right)^{-1}$$

9.4.9 Rank of a matrix

Subdeterminant of order p: The determinant $\det\mathbf{U}$ obtained from a square $p \times p$ submatrix. The hypermatrix **A** can be also a nonsquare $m \times n$ matrix ($m, n \geq p$).
Rank of a matrix, rank(**A**): The highest order of all subdeterminants of **A** that is.

- For an $m \times n$ matrix **A** of rank(**A**) $= r$, it holds that: **A** has at least one subdeterminant $\det\mathbf{U}$ of order r which is nonzero. Subdeterminants of an order larger than r vanish.
- **Criterion for the solvability of systems of linear equations**: A system of linear equations $\mathbf{A}\mathbf{x} = \mathbf{c}$ can be solved uniquely if and only if the rank of the $n \times n$ matrix of coefficients **A** is equal to n:

$$\text{rank}(\mathbf{A}_{(n,n)}) = n \ \Leftrightarrow\ \mathbf{A}\mathbf{x} = \mathbf{c} \text{ can be solved uniquely.}$$

- The rank of an $m \times n$ matrix **A** with rank $(\mathbf{A}_{(m,n)}) = r$ is at most equal to the smaller of the numbers m and n:

$$r \leq \begin{cases} m & \text{for} \quad m \leq n \\ n & \text{for} \quad n \leq m\,. \end{cases}$$

- Rank of a regular square matrix **A** with n rows and n columns:

$$\text{rank}(\mathbf{A}) = n \ \Leftrightarrow\ \det\mathbf{A} \neq 0\,.$$

- Rank of a singular square matrix **A** with n rows and n columns:

$$\text{rank}(\mathbf{A}) < n \ \Leftrightarrow\ \det\mathbf{A} = 0\,.$$

- Rank of the null matrix **0**:

$$\text{rank}(\mathbf{0}) = 0\,.$$

Column rank: Maximum number of linearly independent column vectors.
Row rank: Maximum number of linearly independent row vectors.

- If the row rank of a matrix is equal to the column rank, then this is also the rank (**A**) of matrix **A**.

$$\text{Row rank}(\mathbf{A}) = \text{column rank}(\mathbf{A}) \Rightarrow \text{rank}(\mathbf{A}) = \text{row rank}(\mathbf{A}) \,.$$

- The rank of a trapezoidal matrix is equal to the number of nonvanishing rows.

Elementary conversions of matrices:

- Two rows (or columns) are interchanged.
- A row (or column) is multiplied by a nonzero number.
- A row (column) or its multiple is added to a row (or column).
- The rank of a matrix does not change when subjected to elementary conversions!

▷ The elementary conversions are also used in the Gaussian method for solving systems of linear equations (Section 9.5).

9.4.10 Determination of the rank by means of minor determinants

The rank r of an $m \times n$ matrix **A** can be determined in three steps, which may have to be repeated if necessary:

- At first, let $m \leq n$ (otherwise, m and n are interchanged in the following procedure).
- Check whether all subdeterminants of order m are vanishing.
- If this is not the case, then $r = m$.
- If all subdeterminants of order m are vanishing, repeat this procedure for $m - 1$.

□ Determination of the rank of a 2×3 matrix:

$$\mathbf{A} = \begin{pmatrix} 7 & 4 & 9 \\ 7 & 3 & 0 \end{pmatrix}, \quad m = 2,\ n = 3$$

$$\Rightarrow \text{rank}(\mathbf{A}) < 3, \ \text{because } m = 2.$$

Does $\text{rank}(\mathbf{A}) = 2$? Find a two-row subdeterminant with $\det \mathbf{U} \neq 0$.

$$\det \begin{pmatrix} 4 & 9 \\ 3 & 0 \end{pmatrix} = 4 \cdot 0 - 9 \cdot 3 = -27 \neq 0$$

$$\Rightarrow \text{rank}(\mathbf{A}) = 2 \,.$$

9.5 Systems of linear equations

Systems of linear equations occur in electrical engineering (calculating the stability of machine components and load-breaking structures), in mechanical engineering, as well as in many other fields, also in the numerical solution of discretized systems of differential equations.

System of linear equations with m equations, m constants c_i, n unknowns x_i, and $m \times n$ coefficients a_{ij}:

$$\begin{aligned} a_{11}x_1 + a_{12}x_2 + \cdots + a_{1n}x_n &= c_1 \\ a_{21}x_1 + a_{22}x_2 + \cdots + a_{2n}x_n &= c_2 \\ \vdots \qquad\qquad & \quad \vdots \\ a_{m1}x_1 + a_{m2}x_2 + \cdots + a_{mn}x_n &= c_m \,. \end{aligned}$$

Matrix notation:

$$\begin{pmatrix} a_{11} & \cdots & a_{1n} \\ \vdots & \ddots & \vdots \\ a_{m1} & \cdots & a_{mn} \end{pmatrix} \begin{pmatrix} x_1 \\ \vdots \\ x_n \end{pmatrix} = \begin{pmatrix} c_1 \\ \vdots \\ c_m \end{pmatrix} .$$

Symbolic notation:

$$\mathbf{Ax} = \mathbf{c},$$

sometimes also:

$$[\mathbf{A}]\{\mathbf{x}\} = \{\mathbf{c}\} \,.$$

Coefficient or system matrix A: Matrix of the coefficients a_{ij}.
Solution or system vector x: Vector of the unknowns which must be determined.
Vector of constants or **right-hand side c**: The vector contains the constants c_i describing the conditions (such as, external forces) to which the system is subjected.

□ Electrical network:

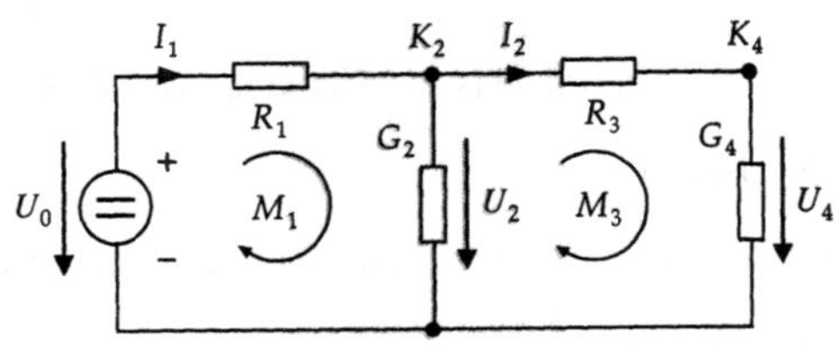

Electrical network

$$\begin{array}{lccccl} \text{mesh equation for } M_1: & R_1 I_1 & +U_2 & & & = U_0 \\ \text{nodal equation for } K_2: & -I_1 & +G_2 U_2 & +I_3 & & = 0 \\ \text{mesh equation for } M_3: & & -U_2 & +R_3 I_3 & -U_4 & = 0 \\ \text{outset equation for } K_4: & & & -I_3 & +G_4 U_4 & = 0 \end{array}$$

Matrix notation of the system of equations:

$$\underbrace{\begin{pmatrix} R_1 & 1 & 0 & 0 \\ -1 & G_2 & 1 & 0 \\ 0 & -1 & R_3 & 1 \\ 0 & 0 & -1 & G_4 \end{pmatrix}}_{\mathbf{A}} \underbrace{\begin{pmatrix} I_1 \\ U_2 \\ I_3 \\ U_4 \end{pmatrix}}_{\mathbf{x}} = \underbrace{\begin{pmatrix} U_0 \\ 0 \\ 0 \\ 0 \end{pmatrix}}_{\mathbf{c}} .$$

▷ A system of linear equations can be solved only under certain conditions. Numerous different methods are available (e.g., graphical method, Gauss-Jordan method).

Homogeneous system of equations: The right side of the equation is zero.

$$\mathbf{c} = \mathbf{0}$$

$$c_1 = \cdots = c_m = 0.$$

□ Homogeneous system of equations with two unknowns:

$$a_{11}x_1 + a_{12}x_2 = 0$$

$$a_{21}x_1 + a_{22}x_2 = 0$$

Inhomogeneous system of equations:

$$\mathbf{c} \neq \mathbf{0}\,,$$

$$c_i \neq 0 \text{ for at least one } i = 1, \ldots, m.$$

□ Inhomogeneous system of equations with two unknowns:

$$a_{11}x_1 + a_{12}x_2 = 3$$

$$a_{21}x_1 + a_{22}x_2 = 6$$

Overdetermined system of linear equations: Necessary (but not in sufficient) condition: $n < m$.
Underdeterminate system of linear equations: Necessary (but in sufficient) condition: $n > m$.

9.5.1 Systems of two equations with two unknowns

Graphical solution: The determination of the point of intersection of two straight lines corresponds to the solution of a system of linear equations with two equations and two unknowns:

$$a_{11}x_1 + a_{12}x_2 = c_1$$

$$a_{21}x_1 + a_{22}x_2 = c_2\,.$$

Three types of solution can occur:

1. Two straight lines intersect in **one** point.
   ⇒ The coordinates of the point of intersection are unique solutions of the system of equations!
2. The straight lines do **not** intersect. They are parallel.
   ⇒ No solution of the system of equations exists!
3. The straight lines coincide. **Every point** of one straight line lies on the other and is therefore a point of intersection (solution).
   ⇒ There are an **infinite number of solutions** of the system of equations!

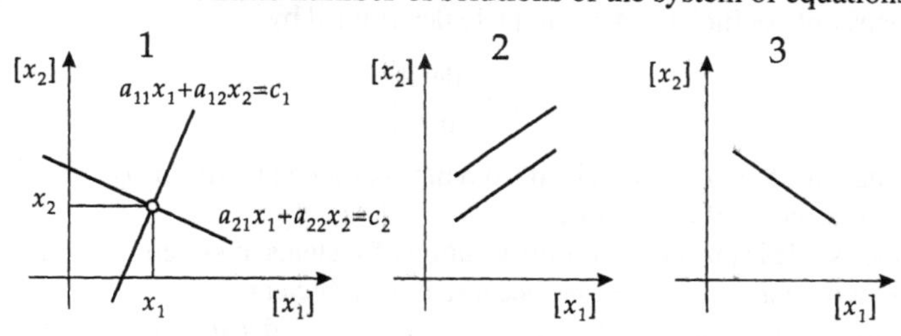

Determination of the point of intersection of the straight lines

Analytic solution according to the substitution method:
Matrix notation of the system of equations:

$$\mathbf{A}\mathbf{x} = \mathbf{c}$$

or in components:

$$\begin{pmatrix} a_{11} & a_{12} \\ a_{21} & a_{22} \end{pmatrix} \begin{pmatrix} x_1 \\ x_2 \end{pmatrix} = \begin{pmatrix} c_1 \\ c_2 \end{pmatrix}.$$

1. Multiply the first equation of the system by a_{22}.
2. Multiply the second equation by $-a_{12}$.
3. Add the resulting equations $(a_{11}a_{22} - a_{12}a_{21}) \cdot x_1 = c_1a_{22} - a_{12}c_2$.
4. Isolate according to x_1:

$$x_1 = \frac{c_1a_{22} - a_{12}c_2}{a_{11}a_{22} - a_{12}a_{21}}.$$

Unique solution for x_1!
▷ Division is possible only if $\det \mathbf{A} = a_{11}a_{22} - a_{12}a_{21} \neq 0$.

5. Multiply the first equation of the system by $-a_{21}$.
6. Multiply the second equation by a_{11}.
7. Add the resulting equation:

$$(a_{11}a_{22} - a_{12}a_{21}) \cdot x_2 = c_2a_{11} - a_{21}c_1 .$$

8. Isolate according to x_2:

$$x_2 = \frac{c_2a_{11} - a_{21}c_1}{a_{11}a_{22} - a_{12}a_{21}}.$$

Unique solution for x_2!
▷ Division is possible only if $\det \mathbf{A} = a_{11}a_{22} - a_{12}a_{21} \neq 0$!
Thus, the value of $\det \mathbf{A} = a_{11}a_{22} - a_{12}a_{21}$ determines whether the system of equations can be solved **uniquely**.
Cramer's rule, the solution vector $\mathbf{x}$ of a system of linear equations $\mathbf{A}\mathbf{x} = \mathbf{c}$ with the regular matrix of coefficients $\mathbf{A}$ is uniquely determined by:

$$x_i = \frac{\det \mathbf{A}_i}{\det \mathbf{A}},$$

where $\mathbf{A}_i$ denotes the matrix obtained from the coefficient matrix $\mathbf{A}$ by replacing the i-th column by the vector of constants $\mathbf{c}$.

▷ Cramer's rule is unsuitable for the solution of systems of systems of linear equations with more than three equations because it is inefficient.

□ For two linear equations with $(a_{11}a_{22} - a_{12}a_{21}) = \det(a_{ij}) \neq 0$ the equations can be solved according to x_1 and x_2, using Cramer's rule, and the system of equations has the unique solution:

$$x_1 = \frac{a_{22}c_1 - a_{12}c_2}{\det(a_{ij})} = \frac{1}{\det(a_{ij})}\begin{vmatrix} c_1 & a_{12} \\ c_2 & a_{22} \end{vmatrix}$$

$$x_2 = \frac{a_{11}c_2 - a_{21}c_1}{\det(a_{ij})} = \frac{1}{\det(a_{ij})}\begin{vmatrix} a_{11} & c_1 \\ a_{21} & c_2 \end{vmatrix}$$

9.6 Numerical solution methods

Classification of numerical methods for the solution of systems of linear equations: **Direct and iterative methods.**
Direct methods: Extensions of the substitution method. Standard methods for the solution of systems of linear equations: **Gaussian elimination method, LU-decomposition, Gauss-Jordan method**.
Iterative methods: An approximate solution of the system of equations is improved until the desired accuracy is achieved (**Gauss-Seidel method**).

- **Unique solvability**: A system of linear equations with n equations and n unknowns has a unique solution if and only if the determinant of the coefficient is nonzero, i.e., if the rank of the matrix is equal to n:

$$\det \mathbf{A} \neq 0 \Leftrightarrow \operatorname{rank}(\mathbf{A}_{(n,n)}) = n \, .$$

9.6.1 *Gaussian algorithm for systems of linear equations*

Solving a system of linear equations $\mathbf{Ax} = \mathbf{c}$ by transformation in two stages.

9.6.2 *Forward elimination*

By elementary transformations the system of equations is transformed into the row echelon form (top right means triangular matrix).
Elementary transformations lead to an equivalent system of equations. The equivalent system of equations has the same solutions for the unknowns as the original system of equations.

- Multiplying an equation by a nonzero factor.
- Adding or subtracting multiples of equations.
- Interchanging the order of equations.

For elementary matrix operations, this is equivalent to:

- Multiplying a row of the matrix by a scalar factor.
- Adding or subtracting multiples of rows of a matrix.
- Interchanging two rows of a matrix.

Transforming the original system of equations into an equivalent system of equations in triangular form:

$$\begin{array}{ccccccccccc} a'_{11}x_1 & + & a'_{12}x_2 & + & a'_{13}x_3 & + & \cdots & + & a'_{1n}x_n & = & c'_1 \\ & & a'_{22}x_2 & + & a'_{23}x_3 & + & \cdots & + & a'_{2n}x_n & = & c'_2 \\ & & & & & & \cdots & & \cdots & = & \cdots \\ & & & & & & a'_{n-1,n-1}x_{n-1} & + & a'_{n-1,n}x_n & = & c'_{n-1} \\ & & & & & & & & a'_{nn}x_n & = & c'_n\,. \end{array}$$

Process of forward elimination:

- First elimination step:

1. Use the first row to eliminate the first coefficient a_{i1} $(i > 1)$ of the other rows: If $a_{11} \neq 0$, subtract the a_{21}/a_{11}-fold of the first row from the second row, subtract the a_{31}/a_{11}-fold of the first row from the third row, etc.

▷ The first elimination step generates zeros in the first column of the matrix, all $a'_{i1} = 0$ (except a_{11})!

- Second elimination step:

2. Use the second row to eliminate the coefficients a_{i2} $(i > 2)$: If $a_{22} \neq 0$,

then subtract the a_{32}/a_{22}-fold of the second row from the third row, subtract the a_{42}/a_{22}-fold of the second row from the fourth row, etc.

Continue these elimination steps to the n-th row.

- **Transforming into a triangular form** in $n-1$ steps of eliminations: Newly resulting coefficients of the k-th step, $k = 1, \ldots, n-1$, are calculated by the following recursive formulas:

$$\begin{array}{ll} a'_{ij} := 0; & i = k+1, \ldots, n; \; j = k \\ a'_{ij} := a'_{ij} - a'_{kj} \cdot (a'_{ik}/a'_{kk}); & i = k+1, \ldots, n; \; j = k+1, \ldots, n \\ c'_i := c'_i - c'_k \cdot (a'_{ik}/a'_{kk}); & i = k+1, \ldots, n. \end{array}$$

▷ A program sequence for Gaussian elimination is given after the following section, together with the sequence for backsubstitution.

□ Gaussian method for a system of three equations with three unknowns:

$$\begin{array}{rrrl} -x_1 & +x_2 & +2x_3 & = 2 \\ 3x_1 & -x_2 & +x_3 & = 6 \\ -x_1 & +3x_2 & +4x_3 & = 4 \end{array} \qquad \begin{pmatrix} -1 & 1 & 2 & 2 \\ 3 & -1 & 1 & 6 \\ -1 & 3 & 4 & 4 \end{pmatrix}$$

1. First step: Elimination of x_1 in the last two equations:

$$\begin{array}{rrrl} -x_1 & +x_2 & +2x_3 & = 2 \\ & 2x_2 & +7x_3 & = 12 \\ & 2x_2 & +2x_3 & = 2 \end{array} \qquad \begin{pmatrix} -1 & 1 & 2 & 2 \\ 0 & 2 & 7 & 12 \\ 0 & 2 & 2 & 2 \end{pmatrix}$$

2. Second step: Elimination of x_2 in the third equation:

$$\begin{array}{rrrl} -x_1 & +x_2 & +2x_3 & = 2 \\ & 2x_2 & +7x_3 & = 12 \\ & & -5x_3 & = -10 \end{array} \qquad \begin{pmatrix} -1 & 1 & 2 & 2 \\ 0 & 2 & 7 & 12 \\ 0 & 0 & -5 & -10 \end{pmatrix}$$

9.6.3 Pivoting

▷ The Gaussian method fails if the diagonal element or the pivot element a'_{kk} of an elimination step is equal to zero, $a'_{kk} = 0$ (termination of the method by means of division by zero).

Pivoting: If a diagonal element is zero, $a'_{kk} = 0$, then, before performing the k-th elimination step, interchange **pivot row** k is with the row $m > k$ that has the maximum absolute value for the coefficient for x_k:

New **pivot row** m, new **pivot element** a'_{mk}.

An even more far-reaching pivoting strategy (**full pivoting**) that assumes a far-reaching equilibration of the matrix seeks the largest absolute value of all matrix elements. To transfer this element to the position (k, k), interchanges of rows and columns are required; the interchange of columns in a system of linear equations corresponds to a renumbering of the unknowns, which must be done additively.

▷ For reasons of numerical stability, pivoting is often meaningful for $a'_{kk} \neq 0$ because for "poorly conditioned" systems of equations (in two dimensions: two lines with almost the same slope), a strong propagation error can occur due to division by small diagonal elements.

For numerical calculations with the Gaussian algorithm, row pivoting is always practical, and full pivoting is recommended. If the matrix **A** is symmetric and positive definite, then pivoting is not necessary.

▷ Program sequence for pivoting matrix **A** and vector **c** of the system of equations $\mathbf{Ax} = \mathbf{c}$.

```
BEGIN Pivoting
INPUT n
INPUT a[i,k], i=1...n, k=1...n
INPUT c[i], i=1...n FOR k=1 TO n
pivot := k
maxa := abs(a[k,k])
FOR i = k+1 TO n DO
   dummy := abs(a[i,k])
   IF (dummy > maxa) THEN
      maxa := dummy
      pivot := i
   ENDIF
ENDDO
IF (pivot ≠ k) THEN
   FOR j = k TO n DO
      dummy := a[pivot,j]
      a[pivot,j] := a[k,j]
      a[k,j] := dummy
   ENDDO
   dummy := c[pivot]
   c[pivot] := c[k]
   c[k] := dummy
ENDIF
OUTPUT a[i,k], i=1,...,n, k=1,...,n
OUTPUT c[i], i=1,...,n
END Pivoting
```

▷ This part of the program can be applied to the Gaussian as well as the Gauss-Jordan method.

9.6.4 Backsubstitution

The solutions x_i are obtained from the triangular form by stepwise backsubstitution: First, x_n is isolated in the lowest row. The other elements x_i , $i = n-1, \ldots, 1$, of the solution vector $\mathbf{x}$ are determined recursively using the equation

$$x_n = \frac{c'_n}{a'_{nn}}, \qquad x_i = \frac{c'_i}{a'_{ii}} - \sum_{k=i+1}^{n} x_k \frac{a'_{ik}}{a'_{ii}}, \quad i = n-1, \ldots, 1$$

□ Let $\mathbf{A}$ be a 4×4 matrix. Then the system of equations $\mathbf{A}\mathbf{x} = \mathbf{c}$ reads, with the unknown vector $\mathbf{x} = (x_1, x_2, \ldots, x_n)^{\mathsf{T}}$:

$$\begin{aligned} a_{11}x_1 + a_{12}x_2 + a_{13}x_3 + a_{14}x_4 &= c_1 \\ a_{22}x_2 + a_{23}x_3 + a_{24}x_4 &= c_2 \\ a_{33}x_3 + a_{34}x_4 &= c_3 \\ a_{44}x_4 &= c_4 \end{aligned}$$

1. Solving by substitution starts with the last equation:

$$x_4 = \frac{c_4}{a_{44}}$$

2. The value derived for x_4 is substituted into the penultimate row, which is solved according to x_3:

$$x_3 = \frac{c_3 - a_{34}x_4}{a_{33}}$$

3. This procedure is continued until the first row is solved according to x_1. Thus, vector $\mathbf{x}$ is known, and the system of equations has been solved.

▷ Program sequence for solving the system $\mathbf{Ax} = \mathbf{c}$ with an $n \times n$ matrix $\mathbf{A}$ and a vector $\mathbf{c}$ by Gaussian elimination without pivoting.

```
BEGIN Gaussian elimination
INPUT n
INPUT a[i,k], i=1...n, k=1...n
INPUT c[i], i=1...n
BEGIN forward elimination
FOR k = 1 TO n-1 DO
   FOR i = k+1 TO n DO
      CALL pivoting (* if necessary *)
      factor := a[i,k]/a[k,k]
      FOR j = k+1 TO n DO
         a[i,j] := a[i,j] - factor*a[k,j]
      ENDDO
   c[i] := c[i] - factor*c[k]
   ENDDO
ENDDO
END forward elimination
BEGIN backsubstitution
x[n] := c[n]/a[n,n]
FOR i = n-1 TO 1 STEP -1 DO
```

```
      sum := 0
      FOR j = i+1 TO n DO
         sum := sum + a[i,j]*x[j]
      ENDDO
      x[i] := (c[i] - sum)/a[i,i]
   ENDDO
   END backsubstitution
   OUTPUT x[i], i=1,...,n
   END Gaussian elimination
```

▷ In the Gaussian algorithm the forward elimination requires about $n^3/3$ operations; backsubstitution requires about $n^2/2$ operations.

9.6.5 LU-decomposition

LU-decomposition (lower-upper-decomposition): The decomposition of a matrix **A** into a product of a right-triangular matrix **U** and a left-triangular matrix **L**, based on the Gaussian method.

● Every matrix **A** can be written as a product of a lower and an upper triangular matrix, provided that no pivoting is required in performing the Gaussian algorithm:

$$\mathbf{A} = \mathbf{L}\mathbf{U}$$

$$\mathbf{L} = \begin{pmatrix} l_{11} & 0 & \cdots & 0 \\ l_{21} & l_{22} & \ddots & \vdots \\ \vdots & & \ddots & 0 \\ l_{n1} & \cdots & \cdots & l_{nn} \end{pmatrix} \quad \text{or} \quad \mathbf{U} = \begin{pmatrix} u_{11} & u_{12} & \cdots & u_{1n} \\ 0 & u_{22} & & \vdots \\ \vdots & \ddots & \ddots & \vdots \\ 0 & \cdots & 0 & u_{nn} \end{pmatrix}.$$

▷ Not every matrix has an LU-decomposition.

□ $\begin{pmatrix} 0 & 1 \\ 1 & 0 \end{pmatrix}$ has no LU-decomposition.

Algorithm for solving the system of equations $\mathbf{Ax} = \mathbf{c}$

Step 1: LU-decomposition of the coefficient matrix **A**:

$$\mathbf{Ax} = \mathbf{LUx} = \mathbf{c}\,.$$

Step 2: Introducing an (unknown) auxiliary vector **y**, and solving the system of equations $\mathbf{Ly} = \mathbf{c}$:

$$\Rightarrow \quad \mathbf{y} = \mathbf{L}^{-1}\mathbf{c}$$

Step 3: Solving the system of equations $\mathbf{Ux} = \mathbf{y}$:

$$\Rightarrow \quad \mathbf{x} = \mathbf{U}^{-1}\mathbf{y}$$

▷ LU-decomposition is recommended if several systems of linear equations with the **same** coefficient matrix **A** but **different** inhomogeneities **c** must be solved; then, Step 1 is needed only once.

The decomposition of the coefficient matrix **A** into the lower triangular matrix **L** and the upper triangular matrix **U** is not unique. Based on the Gaussian elimination, the most important decompositions are the Doolittle, Crout and Cholesky decompositions.

Doolittle's decomposition: The diagonal elements of the **lower** triangular matrix **L** are equal to one , $l_{jj} = 1$, $\det \mathbf{L} = 1$, i.e., **L** has the form

$$\mathbf{L} = \begin{pmatrix} 1 & 0 & \cdots & 0 \\ l_{21} & 1 & \ddots & \vdots \\ \vdots & \ddots & \ddots & 0 \\ l_{n1} & \cdots & l_{n,n-1} & 1 \end{pmatrix}.$$

- In Doolittle's decomposition, **U** is the matrix of the coefficients that are formed in the Gaussian elimination:

$$\mathbf{U} = \begin{pmatrix} u_{11} & u_{12} & \cdots & u_{1n} \\ 0 & u_{22} & & \vdots \\ \vdots & \ddots & \ddots & \vdots \\ 0 & \cdots & 0 & u_{nn} \end{pmatrix}$$

Calculation of the coefficients r_{jk}:

$$\bullet \quad u_{1k} = a_{1k}, \quad k = 1, \ldots, n\,,$$

$$\bullet \quad u_{jk} = a_{jk} - \sum_{s=1}^{j-1} l_{js} u_{sk}, \quad k = j, \ldots, n; \; j \geq 2\,.$$

Calculation of the coefficients l_{jk}, the constants of elimination:

$$\bullet \quad l_{j1} = \frac{a_{j1}}{u_{11}}, \quad j = 2, \ldots, n\,,$$

$$\bullet \quad l_{jk} = \frac{1}{u_{kk}} \left(a_{jk} - \sum_{s=1}^{k-1} l_{js} u_{sk} \right), \quad j = k+1, \ldots, n; \; k \geq 2\,.$$

Crout's decomposition: The diagonal elements of the **upper** triangular matrix **U** are equal to one, $r_{jj} = 1$, $\det \mathbf{U} = 1$, that is, **U** has the form

$$\mathbf{U} = \begin{pmatrix} 1 & r_{12} & \cdots & r_{1n} \\ 0 & 1 & \ddots & \vdots \\ \vdots & \ddots & \ddots & r_{n-1,n} \\ 0 & \cdots & 0 & 1 \end{pmatrix}.$$

1. Calculation of the coefficients l_{jk}:

$$\bullet \quad l_{j1} = a_{j1}, \quad j = 1, \ldots, n\,,$$

$$\bullet \quad l_{jk} = a_{jk} - \sum_{s=1}^{k-1} l_{js} u_{sk}, \quad j = k, \ldots, n; \; k \geq 2\,.$$

2. Calculation of the coefficients u_{jk}:

   - $u_{1k} = \dfrac{a_{1k}}{l_{11}}, \quad k = 2, \ldots, n$,
   - $u_{jk} = \dfrac{1}{l_{jj}} \left(a_{jk} - \sum_{s=1}^{j-1} l_{js} u_{sk} \right), \quad k = j+1, \ldots, n; \; j \geq 2$.

▷ Program sequence for the LU-decomposition of an $n \times n$ matrix **A**, according to Crout's algorithm.

```
BEGIN Crout's decomposition
INPUT n
INPUT a[i,k], i=1...n, k=1...n
FOR j = 2 TO n DO
   a[1,j] := a[1,j]/a[1,1]
ENDDO
FOR j = 2 TO n-1 DO
   FOR i = j TO n DO
      sum := 0
      FOR k = 1 TO j-1 DO
         sum := sum + a[i,k]*a[k,j]
      ENDDO
      a[i,j] := a[i,j] - sum
   ENDDO
   FOR k = j + 1 TO n DO
      sum := 0
      FOR i = 1 TO j-1 DO
         sum := sum + a[j,i]*a[i,k]
      ENDDO
      a[j,k] := (a[j,k] - sum)/a[j,j]
   ENDDO
ENDDO
sum := 0
FOR k = 1 TO n-1 DO
   sum := sum + a[n,k]*a[k,n]
ENDDO
a[n,n] := a[n,n] - sum
OUTPUT a[i,k], i=1,...,n, k=1,...,n
END Crout's decomposition
```

Cholesky's decomposition: LU-decomposition for a symmetric, positively definite matrix $\mathbf{A}$, $\mathbf{A} = \mathbf{A}^{\mathsf{T}}$ and $\mathbf{x}^{\mathsf{T}}\mathbf{A}\mathbf{x} > 0$ for all $\mathbf{x} \neq 0$; $Q = \mathbf{x}^{\mathsf{T}}\mathbf{A}\mathbf{x}$ ia a **quadratic form**.

● The upper triangular matrix **U** can be chosen according to

$$\mathbf{U} = \mathbf{L}^{\mathsf{T}} ,$$

$$\mathbf{A} = \mathbf{L}\mathbf{L}^{\mathsf{T}} \text{ with } r_{jk} = l_{kj} .$$

Calculation of the coefficients l_{jk}:

$$l_{11} = \sqrt{a_{11}}\,,$$

$$l_{jj} = \sqrt{a_{jj} - \sum_{s=1}^{j-1} l_{js}^2}\,, \quad j = 2, \ldots, n\,,$$

$$l_{j1} = \frac{a_{j1}}{u_{11}}, \quad j = 2, \ldots, n\,,$$

$$l_{jk} = \frac{1}{l_{kk}} \left(a_{jk} - \sum_{s=1}^{k-1} l_{js} l_{ks} \right), \quad j = k+1, \ldots, n; \; k \geq 2\,.$$

□ Decomposition of a symmetric 3 × 3 matrix **A** into the product $\mathbf{LL}^{\mathsf{T}}$

$$\mathbf{A} = \begin{pmatrix} 4 & 2 & 14 \\ 2 & 17 & -5 \\ 14 & -5 & 83 \end{pmatrix} = \begin{pmatrix} 2 & 0 & 0 \\ 1 & 4 & 0 \\ 7 & -3 & 5 \end{pmatrix} \begin{pmatrix} 2 & 1 & 7 \\ 0 & 4 & -3 \\ 0 & 0 & 5 \end{pmatrix} = \mathbf{LL}^{\mathsf{T}}$$

▷ Program sequence for the LU-decomposition of a symmetric positive definite matrix **A** following Cholesky's algorithm.

```
BEGIN Cholesky's LU-decomposition
INPUT n
INPUT a[i,k], i=1...n, k=1...n
FOR k = 1 TO n DO
   FOR i = 1 TO k-1 DO
      sum := 0
      FOR j = 1 TO i-1 DO
         sum := sum + a[i,j]*a[k,j]
      ENDDO
      a[k,i] := (a[k,i] - sum)/a[i,i]
   ENDDO
   sum := 0
   FOR j = 1 TO k-1 DO
      sum := sum + a[k,j]*a[k,j]
   ENDDO
   a[k,k] := sqrt(a[k,k] - sum)
ENDDO
OUTPUT a[i,k], i=1...n, k=1...n
END Cholesky's LU-decomposition
```

9.6.6 Solvability of (m × n) systems of equations

The Gaussian method is applicable not only for quadratic systems of equations (equal number of equations and unknowns):

Rectangular systems of equations with n unknowns and m equations:

$$\begin{aligned} a_{11}x_1 + a_{12}x_2 + a_{13}x_3 + \cdots + a_{1n}x_n &= c_1 \\ a_{21}x_1 + a_{22}x_2 + a_{23}x_3 + \cdots + a_{2n}x_n &= c_2 \\ &\vdots \\ a_{m1}x_1 + a_{m2}x_2 + a_{m3}x_3 + \cdots + a_{mn}x_n &= c_m \end{aligned}$$

The final form of the system of equations in row-echelon form can be provided by elementary transformations:

$$\begin{aligned}
a'_{11}x_1 + a'_{12}x_2 + \cdots + a'_{1r}x_r + a'_{1,r+1}x_{r+1} + \cdots + a'_{1n}x_n &= c'_1 \\
a'_{22}x_2 + \cdots + a'_{2r}x_r + a'_{2,r+1}x_{r+1} + \cdots + a'_{2n}x_n &= c'_2 \\
\vdots \qquad \vdots \qquad &= \vdots \\
a'_{rr}x_r + a'_{r,r+1}x_{r+1} + \cdots + a'_{rn}x_n &= c'_r \\
0 &= c'_{r+1} \\
0 &= c'_{r+2} \\
&\vdots \\
0 &= c'_m
\end{aligned}$$

- **Conditions for the solvability of** $(m \times n)$ **systems of equations:**
  There is no solution of the system of equations if one or more $c'_{r+1}, c'_{r+2}, \ldots, c'_m$ are nonzero: not all equations are satisfied.
  Unique solution for $r = n$.
  There is more than one solution for $r < n$ if all $c'_{r+1}, c'_{r+2}, \ldots, c'_m$ are zero. Assume abritrary numbers for the excess $n-r$ unknowns and calculate the remaining r unknowns.
  General solution of the system of equations, $n - r$ excess unknowns

  $$x_{r+1}, x_{r+2}, \ldots, x_n$$

  remain free parameters (no values are substituted).
  A specific solution is obtained by substituting concrete number values for the parameters: **an infinite** number of specific solutions is possible.
- The condition for the solvability of the system of linear equations is always satisfied if the right side is the null vector. All $c'_{r+1}, \ldots, c'_m$ are also equal to zero under elementary transformations.

Homogeneous system of linear equations:

$$\mathbf{Ax} = \mathbf{0},$$

Here, $\mathbf{c}$ is the null vector $\mathbf{c} = \mathbf{0}$.

- **Homogeneous systems of equations** $\mathbf{Ax} = \mathbf{0}$ have nontrivial solutions $\mathbf{x} \neq 0$ only if $\mathbf{A}$ is singular,

  $$\det \mathbf{A} = 0.$$

- **Inhomogeneous system of linear equations**

  $$\mathbf{Ax} = \mathbf{c} \neq \mathbf{0},$$

  at least one component of $\mathbf{c}$ is nonzero, $\mathbf{c} \neq \mathbf{0}$. In the case of square matrix $\mathbf{A}$, a unique solution is possible only if $\mathbf{A}$ is regular, $\det \mathbf{A} \neq 0$.
- Every homogeneous system is solvable. **The trivial solution** is $\mathbf{x} = 0$.
- **The rank** r of the system of equations is the number r from the trapezoidal form of the system of linear equations.
  r is the maximum number of equations that are not interdependent.
  At the same time, r is the rank of the coefficient matrix $\mathbf{A}$, i.e., the maximum number of linearly independent row vectors of $\mathbf{A}$.
- If the rank of the matrix $\mathbf{A}$ and the rank of the matrix extended by the vector $\mathbf{c}$ of the right side $(\mathbf{A}|\mathbf{c})$ coincide, then the system of linear equations $\mathbf{Ax} = \mathbf{c}$ can be solved.

If the systems of equations can be solved, then the rank of the augmented matrix $\mathbf{A}|\mathbf{c}$ is equal to the rank of the coefficient matrix $\mathbf{A}$.

$$\operatorname{rank}(\mathbf{A}) = \operatorname{rank}(\mathbf{A}|\mathbf{b})$$

□ Augmented 3×3 matrix:

$$(\mathbf{A}|\mathbf{b}) = \left(\begin{array}{ccc|c} a_{11} & a_{12} & a_{13} & c_1 \\ a_{21} & a_{22} & a_{23} & c_2 \\ a_{31} & a_{32} & a_{33} & c_3 \end{array}\right)$$

- For quadratic $(n \times n)$ systems of equations, the $(n \times n)$ matrix of coefficients $\mathbf{A}$ is regular if the rank r is equal to the number n of rows. This is the case for $\det \mathbf{A} \neq 0$.

▷ In the Gaussian method, $\mathbf{A}$ is transformed into an equivalent triangular matrix by means of elementary conversions and possibly by interchanging rows or columns.

- The determinants of $\mathbf{A}$ and $\mathbf{A}'$ differ at most in sign. Their magnitude is the product of the elements on the main diagonal of $\mathbf{A}'$:

$$\det \mathbf{A} = (-1)^m \det \mathbf{A}', \qquad \det \mathbf{A}' = a'_{11} \cdot a'_{22} \cdot \ldots \cdot a'_{nn} = \prod_{i=1}^{n} a'_{ii} .$$

Here, m is the number of interchanges of rows/columns.

9.6.7 Gauss-Jordan method for matrix inversion

Gauss-Jordan method: By elementary conversions, the system of linear equations

$$\mathbf{A}\mathbf{x} = \mathbf{c}$$

is transformed stepwise into the diagonal form of the equivalent system of equations

$$\mathbf{x} = \mathbf{A}^{-1}\mathbf{c} = \mathbf{c}''$$

with the same solutions as the original system of equations!

$$\begin{array}{lllllll} a'_{11}x_1 & & & & & = & c'_1 \\ & a'_{22}x_2 & & & & = & c'_2 \\ & & \cdots & & & & \cdots \\ & & & a'_{n-1,n-1}x_{n-1} & & = & c'_{n-1} \\ & & & & a'_{nn}x_n & = & c'_n . \end{array}$$

▷ The unknowns in the equations are reduced in such a way that each equation contains only **one** unknown.

Transformation of the system of equations into diagonal form:
The new coefficients of the k-th elimination step are calculated recursively in n elimination steps, $k = 1, \ldots, n$:

$$\begin{array}{ll} a'_{ij} := 0; & i = 1, \ldots, n;\ \ i \neq k;\ \ j = k \\ a'_{ij} := a'_{ij} - a'_{kj} \cdot (a'_{ik}/a'_{kk}); & i = 1, \ldots, n;\ \ i \neq k;\ \ j = 1, \ldots, n \\ c'_i := c'_i - c'_k \cdot (a'_{ik}/a'_{kk}); & i = 1, \ldots, n;\ \ i \neq k \end{array}$$

Solutions x_i in diagonal form:

$$x_i = c''_i = \frac{c'_i}{a'_{ii}}; \qquad i = 1, \ldots, n$$

▷ The solution of the x_i can be read off directly if the k-th row is normalized in the k-th elimination step.

$$a'_{ki} = a'_{ki}/a'_{kk}\,, \quad i = k, \ldots, n\,; \qquad \text{diagonal elements } a'_{kk} = 1.$$

▷ In contrast to the Gaussian method, the Gauss-Jordan method eliminates the unknowns during the elimination step also from the rows above.

Normalization of the equations by means of division by the diagonal elements (the diagonal matrix becomes the unit matrix).

Representation of the transformed system of equations $\mathbf{x} = \mathbf{A}^{-1}\mathbf{c} = \mathbf{c}''$ in matrix notation:

$$\begin{pmatrix} 1 & 0 & \cdots & 0 \\ 0 & 1 & \ddots & \vdots \\ \vdots & \ddots & \ddots & 0 \\ 0 & \cdots & 0 & 1 \end{pmatrix} \begin{pmatrix} x_1 \\ x_2 \\ \vdots \\ x_n \end{pmatrix} = \begin{pmatrix} c''_1 \\ c''_2 \\ \vdots \\ c''_n \end{pmatrix}$$

● Direct reading off the solution vectors $\mathbf{x} = \mathbf{c}''$ of the system of equations:

$$x_i = b''_i\,, \; i = 1, \ldots, n.$$

□ Gauss-Jordan method:

1	$x_1 + x_2 - x_3 = 1$	$a_{11}x_1 + a_{12}x_2 + a_{13}x_3 = c_1$
2	$2x_1 + x_2 + x_3 = 4$	$a_{21}x_1 + a_{22}x_2 + a_{23}x_3 = c_2$
3	$4x_1 - 4x_2 + 2x_3 = 2$	$a_{31}x_1 + a_{32}x_2 + a_{33}x_3 = c_3$

Step 1: Elimination of x_1, except in $\boxed{1}$

$$\begin{aligned}
\boxed{1} \times (-\,a_{21}/a_{11}) = \boxed{1} \times (-2) &= -2x_1 - 2x_2 + 2x_3 = -2 \\
+\boxed{2} &= 2x_1 + x_2 + x_3 = 4 \\
\Rightarrow \boxed{2}' &= \underline{0}x_1 - x_2 + 3x_3 = 2 \\
\boxed{1} \times (-\,a_{31}/a_{11}) = \boxed{1} \times (-4) &= -4x_1 - 4x_2 + 4x_3 = -4 \\
+\boxed{3} &= 4x_1 - 4x_2 + 2x_3 = 2 \\
\Rightarrow \boxed{3}' &= \underline{0}x_1 - 8x_2 + 6x_3 = -2
\end{aligned}$$

Results of Step 1:

1	$x_1 + x_2 - x_3 = 1$	$a_{11}x_1 + a_{12}x_2 + a_{13}x_3 = c_1$
2′	$\boxed{0x_1} - x_2 + 3x_3 = 2$	$0x_1 + a'_{22}x_2 + a'_{23}x_3 = c'_2$
3′	$\boxed{0x_1} - 8x_2 + 6x_3 = -2$	$0x_1 + a'_{32}x_2 + a'_{33}x_3 = c'_3$

Step 2: Elimination of x_2, except in $\boxed{2}'$

$$
\begin{array}{rcrcrcrcrcrcr}
\boxed{2}'\times\left(-a_{12}/a'_{22}\right) & = & \boxed{2}'\times 1 & = & 0x_1 & - & x_2 & + & 3x_3 & = & 2 \\
 & & +\boxed{1} & = & x_1 & + & x_2 & - & x_3 & = & 1 \\
 & & \Rightarrow \boxed{1}' & = & x_1 & + & \underline{0}x_2 & + & 2x_3 & = & 3 \\
\boxed{2}'\times\left(-a'_{32}/a'_{22}\right) & = & \boxed{2}'\times(-8) & = & 0x_1 & + & 8x_2 & - & 24x_3 & = & -16 \\
 & & +\boxed{3}' & = & 0x_1 & - & 8x_2 & + & 6x_3 & = & -2 \\
 & & \Rightarrow \boxed{3}'' & = & 0x_1 & + & \underline{0}x_2 & - & 18x_3 & = & -18
\end{array}
$$

Results of Step 2:

$\boxed{1}'$	x_1	$+$	$\boxed{0x_2}$	$+$	$2x_3$	$=$	3	$a'_{11}x_1 + 0a_{12}x_2 + a'_{13}x_3 = c'_1$
$\boxed{2}'$	$0x_1$	$-$	x_2	$+$	$3x_3$	$=$	2	$0x_1 + a'_{22}x_2 + a'_{23}x_3 = c'_2$
$\boxed{3}''$	$0x_1$	$+$	$\boxed{0x_2}$	$-$	$18x_3$	$=$	-18	$0x_1 + 0x_2 + a''_{33}x_3 = c''_3$

Step 3: Elimination of x_3, except in $\boxed{3}''$

$$
\begin{array}{rcrcrcrcrcrcr}
\boxed{3}''\times\left(-a'_{13}/a''_{33}\right) & = & \boxed{3}''\times 1/9 & = & 0x_1 & + & 0x_2 & - & 2x_3 & = & -2 \\
 & & +\boxed{1}' & = & x_1 & + & 0x_2 & + & 2x_3 & = & 3 \\
 & & \Rightarrow \boxed{1}'' & = & x_1 & + & 0x_2 & + & \underline{0}x_3 & = & 1 \\
\boxed{3}''\times\left(-a'_{23}/a'_{33}\right) & = & \boxed{3}''\times 1/6 & = & 0x_1 & + & 0x_2 & - & 3x_3 & = & -3 \\
 & & +\boxed{2}' & = & 0x_1 & - & x_2 & + & 3x_3 & = & 2 \\
 & & \Rightarrow \boxed{2}'' & = & 0x_1 & - & x_2 & + & \underline{0}x_3 & = & -1
\end{array}
$$

Results of Step 3:

$\boxed{1}''$	x_1	$+$	$0x_2$	$-$	$\boxed{0x_3}$	$=$	1	$\boxed{a''_{11}x_1} + 0x_2 + 0x_3 = c''_1$
$\boxed{2}''$	$0x_1$	$-$	x_2	$+$	$\boxed{0x_3}$	$=$	-1	$0x_1 + \boxed{a''_{22}x_2} + 0x_3 = c''_2$
$\boxed{3}''$	$0x_1$	$+$	$0x_2$	$-$	$18x_3$	$=$	-18	$0x_1 + 0x_2 + \boxed{a''_{33}x_3} = c''_3$

9.6.8 *Calculation of the inverse matrix* $\mathbf{A}^{-1}$

All conversions by which **A** is transformed into the unit matrix **I** are performed simultaneously with the unit matrix **I**:
The unit matrix **I** is transformed to the inverse matrix $\mathbf{A}^{-1}$.

▷ Thus, it is possible to solve the systems of equations efficiently for complex systems (with a fixed system or "coupling" matrix **A**) for distinct external conditions (different vectors **b**, **c**, **d**, ...).

□ Calculate the behavior of an automobile with springs and shock absorbers on various road surfaces.

Systems of equations with a constant, invertible matrix **A**, but distinct right-hand sides

$$\begin{aligned} \mathbf{A}\mathbf{x} &= \mathbf{b}, \\ \mathbf{A}\mathbf{y} &= \mathbf{c}, \\ \mathbf{A}\mathbf{z} &= \mathbf{d}, \qquad \text{etc.} \\ \mathbf{A} &= \text{const.} \end{aligned}$$

can be solved numerically by inverting the matrix **A**.

- The solution vectors **x**, **y**, **z**, etc., are the products of the inverted (fixed) matrix $\mathbf{A}^{-1}$ and the various vectors of constants on the right-hand side:

$$\begin{aligned} \mathbf{x} &= \mathbf{A}^{-1}\mathbf{b} \\ \mathbf{y} &= \mathbf{A}^{-1}\mathbf{c} \\ \mathbf{z} &= \mathbf{A}^{-1}\mathbf{d} \qquad \text{etc.} \end{aligned}$$

□ Determination of the inverse matrix:

$$\mathbf{A} = \begin{pmatrix} 1 & 1 & -1 \\ 2 & 1 & 1 \\ 4 & -4 & 2 \end{pmatrix}, \qquad \mathbf{I} = \begin{pmatrix} 1 & 0 & 0 \\ 0 & 1 & 0 \\ 0 & 0 & 1 \end{pmatrix}$$

$$\Downarrow \text{ elementary transformations } \Downarrow$$

$$\mathbf{I} = \begin{pmatrix} 1 & 0 & 0 \\ 0 & 1 & 0 \\ 0 & 0 & 1 \end{pmatrix}, \qquad \mathbf{A}^{-1} = \text{the inverse of } \mathbf{A}$$

$$\begin{array}{l} \boxed{1} \\ \boxed{2} \\ \boxed{3} \end{array} \quad \left(\begin{array}{rrr|rrr} 1 & 1 & -1 & 1 & 0 & 0 \\ 2 & 1 & 1 & 0 & 1 & 0 \\ 4 & -4 & 2 & 0 & 0 & 1 \end{array}\right)$$

$$\begin{array}{l} \boxed{1} \\ \boxed{2}' = \boxed{1} \times (-2) + \boxed{2} \\ \boxed{3}' = \boxed{1} \times (-4) + \boxed{3} \end{array} \quad \left(\begin{array}{rrr|rrr} 1 & 1 & -1 & 1 & 0 & 0 \\ 0 & -1 & 3 & -2 & 1 & 0 \\ 0 & -8 & 6 & -4 & 0 & 1 \end{array}\right)$$

$$\begin{array}{l} \boxed{1}' = \boxed{2}' \times 1 + \boxed{1} \\ \boxed{2}' \\ \boxed{3}'' = \boxed{2}' \times (-8) + \boxed{3}' \end{array} \quad \left(\begin{array}{rrr|rrr} 1 & 0 & 2 & -1 & 1 & 0 \\ 0 & -1 & 3 & -2 & 1 & 0 \\ 0 & 0 & -18 & 12 & -8 & 1 \end{array}\right)$$

$$\begin{array}{l} \boxed{1}'' = \boxed{3}'' \times (1/9) + \boxed{1}' \\ \boxed{2}'' = \boxed{3}'' \times (1/6) + \boxed{2}' \\ \boxed{3}'' \end{array} \quad \left(\begin{array}{rrr|rrr} 1 & 0 & 0 & 1/3 & 1/9 & 1/9 \\ 0 & -1 & 0 & 0 & -1/3 & 1/6 \\ 0 & 0 & -18 & 12 & -8 & 1 \end{array}\right)$$

Normalization:

$$\begin{array}{l} \boxed{1}'' \times 1 \\ \boxed{2}'' \times (-1) \\ \boxed{3}'' \times (-1/18) \end{array} \quad \left(\begin{array}{rrr|rrr} 1 & 0 & 0 & 1/3 & 1/9 & 1/9 \\ 0 & 1 & 0 & 0 & 1/3 & -1/6 \\ 0 & 0 & 1 & -2/3 & 4/9 & -1/18 \end{array}\right)$$

Result:

$$\mathbf{A}^{-1} = \begin{pmatrix} 1/3 & 1/9 & 1/9 \\ 0 & 1/3 & -1/6 \\ -2/3 & 4/9 & -1/18 \end{pmatrix}.$$

Check $\mathbf{A}^{-1} \cdot \mathbf{A} = \mathbf{I}$:

$$\begin{pmatrix} 1/3 & 1/9 & 1/9 \\ 0 & 1/3 & -1/6 \\ -2/3 & 4/9 & -1/18 \end{pmatrix} \cdot \begin{pmatrix} 1 & 1 & -1 \\ 2 & 1 & 1 \\ 4 & -4 & 2 \end{pmatrix} = \begin{pmatrix} 1 & 0 & 0 \\ 0 & 1 & 0 \\ 0 & 0 & 1 \end{pmatrix}$$

▷ Program sequence for the inversion of an $n \times n$ matrix **A** by the Gauss-Jordan method.

```
BEGIN Gauss-Jordan
INPUT n
INPUT a[i,k], i=1,...,n, k=1,...,n
FOR k = 1 TO n DO
   FOR j = 1 TO n DO
      b[k,j] := 0
   ENDDO
   b[k,k] := 1
ENDDO
FOR k = 1 TO n DO
   dummy := a[k,k]
   FOR j = 1 TO n DO
      a[k,j] := a[k,j]/dummy
      b[k,j] := b[k,j]/dummy
   ENDDO
   FOR i = 1 TO n DO
      IF (i ≠ k) THEN
         dummy := a[i,k]
         FOR j = 1 TO n+1 DO
            a[i,j] := a[i,j] - dummy*a[k,j]
            b[i,j] := b[i,j] - dummy*b[k,j]
         ENDDO
      ENDIF
   ENDDO
ENDDO
OUTPUT b[i,k], i=1,...,n, k=1,...,n
END Gauss-Jordan
```

▷ This program can be used in conjunction with the pivoting program of the Gaussian method.

9.7 Iterative solution of systems of linear equations

In calculating networks and electric power grids in electrical engineering and mechanical engineering, balances in mathematics of finance, in trusswork statics (e.g., high-voltage transmission towers), as well as working with finite elements, very large systems of equations (for example, 1000 equations are not a rarity).

▷ In practice, many coefficients of the vector of constants **c** and of the system matrix **A** are equal to zero in large systems of equations.

Sparse $n \times n$ **matrix**: The **sparsity**, i.e., the number of nonzero elements a_{ij}, is smaller than $\approx 8\%$.

□ For a high-voltage transmission tower one obtains the matrix equation:

$$\begin{aligned} \mathbf{Sx} = \mathbf{K} \quad & \text{result} \\ n \times n \text{ stiffness matrix}: \quad & \mathbf{S} \\ \text{load vector}: \quad & \mathbf{K} \\ \text{displacement vector}: \quad & \mathbf{x}. \end{aligned}$$

Bandwidth of the system matrix: The **band diagonal matrix** contains a strip of constant width with nonzero elements.

□ Electrical network

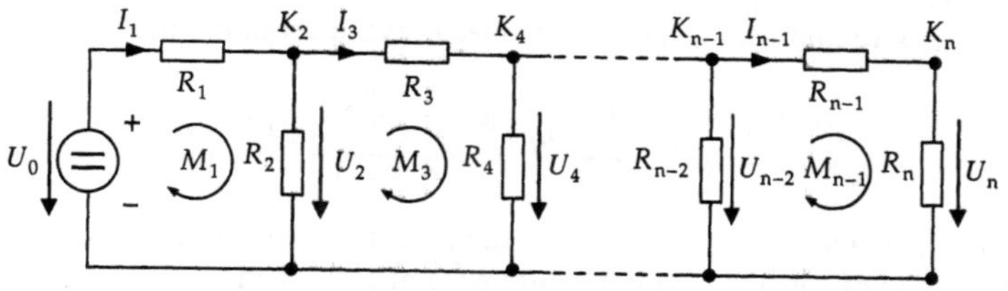

Electrical network

$$\begin{array}{llllllll} \text{mesh equation for } M_1: & R_1 I_1 & +U_2 & & & & & = U_0 \\ \text{nodal equation for } K_2: & -I_1 & +U_2/R_2 & +I_3 & & & & = 0 \\ \text{mesh equation for } M_3: & & -U_2 & +R_3 I_3 & +U_4 & & & = 0 \\ \vdots & \vdots & \vdots & \vdots & \vdots & & \vdots & \vdots \\ \text{cut set equation:} & & & & & & -I_{n-1} + U_n/R_n & = 0 \end{array}$$

Matrix form of the system of equations:

$$\begin{pmatrix} R_1 & 1 & 0 & 0 & 0 & \dots & 0 & 0 \\ -1 & 1/R_2 & 1 & 0 & 0 & \dots & 0 & 0 \\ 0 & -1 & R_3 & 1 & 0 & \dots & 0 & 0 \\ \vdots & \vdots & \vdots & \vdots & \vdots & \ddots & \vdots & \vdots \\ 0 & 0 & 0 & 0 & 0 & \dots & R_{n-1} & 1 \\ 0 & 0 & 0 & 0 & 0 & \dots & -1 & 1/R_n \end{pmatrix} \cdot \begin{pmatrix} I_1 \\ U_2 \\ I_3 \\ \vdots \\ I_{n-1} \\ U_n \end{pmatrix} = \begin{pmatrix} U_0 \\ 0 \\ 0 \\ \vdots \\ 0 \\ 0 \end{pmatrix}$$

For the system matrix we obtain a bandwidth of $m = 3$ where three coefficients lie in a row beside each other or in a column above each other.

Rounding error: The direct methods for solving systems of linear equations (the Gaussian algorithm and varieties thereof) may possibly yield a falsified result due to rounding errors and the propagation of errors; convergent iterative methods have proven to be less susceptible to rounding errors. If there are rounding errors in the input data, then iterative methods cannot help either.

For iterative methods we must distinguish between total-step and single-step methods.

9.7.1 Total-step methods (Jacobi)

▷ Condition for total-step methods:
All diagonal elements

$$a_{ii} \neq 0 \text{ for } i = 1, \ldots, n.$$

● If the system of equations can be solved uniquely, then the condition $a_{ii} \neq 0$ (for all $i = 1, \ldots, n$) can be met by an interchange of lines:

Step 1: Isolating x_i in the i-th row:

$$\begin{array}{lllllll}
x_1 = (c_1 & & -a_{12}x_2 & -a_{13}x_3 & \cdots & -a_{1n}x_n &)/a_{11} \\
x_2 = (c_2 & -a_{21}x_1 & & -a_{23}x_3 & \cdots & -a_{2n}x_n &)/a_{22} \\
\cdots & \cdots & \cdots & & \cdots & \cdots & \cdots \\
\cdots & \cdots & \cdots & \cdots & & \cdots & \cdots \\
x_n = (c_n & -a_{n1}x_1 & -a_{n2}x_2 & \cdots & -a_{n,n-1}x_{n-1} & &)/a_{nn} .
\end{array}$$

▷ Note that the empty places would correspond to the term $a_{ii}x_i$ in the matrix.

Step 2: Giving the starting vector $\mathbf{x}^{(0)}$.

Step 3: Successive improvement of $\mathbf{x}^{(0)}$ by means of the iteration rule:

$$x_i^{(k+1)} = \left[c_i - \sum_{j=1}^{i-1} a_{ij}\, x_j^{(k)} - \sum_{j=i+1}^{n} a_{ij}\, x_j^{(k)} \right] / a_{ii}$$

for $i = 1, \ldots, n$ and $k = 0, 1, 2, \ldots$, i.e., the determined vector $\mathbf{x}^{(k)}$ is substituted into the same system of equations until the difference of $\mathbf{x}^{(k+1)}$ and $\mathbf{x}^{(k)}$ is sufficiently small.

● For calculating $x_i^{(k+1)}$ the component $x_i^{(k)}$ is not needed on the right side of the equation.

▷ Frequently, total-step methods converge slowly.

9.7.2 Single-step methods (Gauss-Seidel)

▷ A faster convergence can usually be achieved by means of single-step methods!
Instead of the k-th approximation (as in total-step methods) the $(k+1)$-th approximations of the components x_1 to x_{i-1} calculated already are used to determine the solution vector.
Iteration prescription of the Gauss-Seidel method:

$$x_i^{(k+1)} = \left[b_i - \sum_{j=1}^{i-1} a_{ij}\, x_j^{(k+1)} - \sum_{j=i+1}^{n} a_{ij}\, x_j^{(k)} \right] / a_{ii} .$$

▷ Program sequence for solving the system of equations $\mathbf{Ax} = \mathbf{c}$ by the Gauss-Seidel iteration method with relaxation.

```
BEGIN Gauss-Seidel
INPUT n, eps ,maxit
INPUT a[i,k], i=1,...,n, k=1,...,n
INPUT c[i], i=1,...,n
INPUT lambda (relaxation parameter)
epsa := 1.1*eps
FOR i = 1 TO n DO
   dummy := a[i,i]
   FOR j = 1 TO n DO
      a[i,j] := a[i,j]/dummy
```

```
    ENDIF
    c[i] := c[i]/dummy
  ENDDO
  iter := 0
  WHILE (iter < maxit AND epsa > eps) DO
    iter := iter + 1
    FOR i = 1 TO n DO
      old := x[i]
      sum := c[i]
      FOR j = 1 TO n DO
        IF (i ≠ j) THEN
          sum := sum - a[i,j]*x[j]
        ENDDO
      ENDDO
      x[i] := lambda*sum + (1 - lambda)*old
      IF (x[i] ≠ 0) THEN
        epsa := abs((x[i] - old)/x[i])*100
      ENDIF
    ENDDO
  ENDDO
  OUTPUT x[i], i=1,...,n
  END Gauss-Seidel
```

- **Relaxation parameter**: Λ with $0 < \Lambda < 2$ may considerably improve the convergence of the Gauss-Seidel method essentially.

Underrelaxation, $\Lambda \le 1$: Used to obtain convergence for nonconvergent systems, as well as for damping oscillations to achieve faster convergence.
Overrelaxation, $1 < \Lambda \le 2$: Used to improve convergence for already convergent systems.

9.7.3 Criteria of convergence for iterative methods

Sufficient criteria of convergence:
Row-sum criterion:

$$\sum_{j=1, j\neq i}^{n} \left| \frac{a_{ij}}{a_{ii}} \right| < 1\,, \qquad \text{for } i = 1, \ldots, n.$$

Column-sum criterion:

$$\sum_{i=1, i\neq j}^{n} \left| \frac{a_{ij}}{a_{ii}} \right| < 1\,, \qquad \text{for } j = 1, \ldots, n.$$

Schmidt-von Mises criterion:

$$\sum_{j=1}^{n} \sum_{j=1, j\neq 1}^{n} \left(\frac{a_{ij}}{a_{ii}} \right)^2 < 1$$

- If the absolute value of all diagonal elements of matrix **A** is large compared with the sum of the remaining coefficients of the corresponding rows or columns, then the iteration methods are convergent.

▷ If the criteria of convergence are not fulfilled, then it cannot be stated whether there is convergence or not; i.e., the method may converge nevertheless.

Sometimes, it is possible by means of elementary transformations to produce an equivalent system of equations with large diagonal elements. If this is not possible, the solution must be determined by means of a direct method.

Thomas's algorithm for tridiagonal matrices: Application of the Gaussian algorithm for sparse matrices in which the computing operations for the zero elements become unnecessary.

▷ Program sequence for solving the system of equations $\mathbf{Ax} = \mathbf{c}$ with an $n \times n$ tridiagonal band matrix **A**. Only the main diagonal and the two subdiagonals must be stored. These are the vectors **e**, **f**, and **g**.

```
BEGIN Tridiagonal Thomas's algorithm
INPUT n
INPUT a[i,k], i=1,...,n, k=1,...,n
INPUT c[i], i=1,...,n
e[i] := a[i,i-1], i=2,...,n
f[i] := a[i,i], i=1,...,n
g[i] := a[i,i+1], i=1,...,n-1
BEGIN decomposition
FOR k = 2 TO n DO
   e[k] := e[k]/f[k-1]
   f[k] := f[k] - e[k]*g[k-1]
ENDDO
END decomposition
BEGIN forward elimination
FOR k = 2 TO n DO
   c[k] := c[k] - e[k]*c[k-1]
ENDDO
END forward elimination
BEGIN backsubstitution
x[n] := c[n]/f[n]
FOR k = n-1 TO 1 STEP -1 DO
   x[k] := (c[k] - g[k]*x[k+1])/f[k]
ENDDO
END backsubstitution
OUTPUT x[i], i=1,...,n
END tridiagonal Thomas's algorithm
```

9.7.4 Storage of the coefficient matrix

▷ In iterative method (total-step and single-step method) **no** new nonzero elements occur in the calculation. The coefficient matrix **A** is not changed in the course of the calculation. Here, a nonchained data structure can be used to advantage.

□ In conventional memory techniques, a coefficient matrix with $n = 2000$ and a sparsity of nonzero elements of 0.4% needs $n \cdot n = 4{,}000{,}000$ memory locations, although only 16,000 nonzero elements occur.

▷ Nonchained listings (ARRAYS) usually require less storage space than the direct storage of chained structures. This memory technique is used to avoid access to slow background memories.

▷ For sparse matrices, try to store only nonzero elements a_{ij} (i = row, j = column). For a matrix with n nonzero elements in PASCAL this is done often as a record:

```
VAR A: ARRAY[1..n]
         OF RECORD
              row: INTEGER;
              column: INTEGER;
              value: REAL
             END;
```

Access to the i-th nonzero element is by

```
row index := A[i].row
column index := A[i].column
matrix element:= A[i].value
```

This very practical solution does not exist in FORTRAN, where two ARRAYS have to be defined: One contains the nonzero elements, the other the indices.
Arrangement of the data structure:

row i	column j	value a_{ij}

□ **A** is a 5 × 5 matrix with seven nonzero elements

$$\mathbf{A} = \begin{pmatrix} 0 & 0 & a_{13} & 0 & 0 \\ a_{21} & 0 & 0 & a_{24} & 0 \\ 0 & 0 & a_{33} & 0 & 0 \\ a_{41} & 0 & 0 & 0 & a_{45} \\ 0 & a_{52} & 0 & 0 & 0 \end{pmatrix}$$

Then, the memory contains:

1	3	a_{13}	2	1	a_{21}	2	4	a_{24}	...	5	2	a_{52}	0	0	0.0

▷ Here, the end of memory is denoted by (0, 0, 0.0).
The memory requirement can be further reduced if every new column or row index is stored only once.

9.8 Table of solution methods

Categories of algorithms for solving systems of linear equations: Direct and iterative methods.
It is possible to solve systems of linear equations with large, sparse matrices by iterative methods.
Gaussian elimination method:

$$\left(\begin{array}{ccc|c} a_{11} & a_{12} & a_{13} & c_1 \\ a_{21} & a_{22} & a_{23} & c_2 \\ a_{31} & a_{32} & a_{33} & c_3 \end{array}\right) \Longrightarrow$$

$$\left(\begin{array}{ccc|c} a_{11} & a_{12} & a_{13} & c_1 \\ & a'_{22} & a'_{23} & c'_2 \\ & & a''_{33} & c''_3 \end{array}\right) \Longrightarrow \begin{array}{l} x_3 = c''_3/a''_3 \\ x_2 = (c'_2 - a'_{23}x_3)/a'_{22} \\ x_1 = (c_1 - a_{12}x_2 - a_{13}x_3)/a_{11} \end{array}$$

Gauss-Jordan (matrix inversion)

$$\begin{pmatrix} a_{11} & a_{12} & a_{13} & | & 1 & 0 & 0 \\ a_{21} & a_{22} & a_{23} & | & 0 & 1 & 0 \\ a_{31} & a_{32} & a_{33} & | & 0 & 0 & 1 \end{pmatrix} \Longrightarrow \begin{pmatrix} 1 & 0 & 0 & | & a_{11}^{-1} & a_{12}^{-1} & a_{13}^{-1} \\ 0 & 1 & 0 & | & a_{21}^{-1} & a_{22}^{-1} & a_{23}^{-1} \\ 0 & 0 & 1 & | & a_{31}^{-1} & a_{32}^{-1} & a_{33}^{-1} \end{pmatrix}$$

LU-decomposition

$$\begin{pmatrix} a_{11} & a_{12} & a_{13} \\ a_{21} & a_{22} & a_{23} \\ a_{31} & a_{32} & a_{33} \end{pmatrix} \overset{\text{LU}}{\Longrightarrow} \begin{pmatrix} l_{11} & 0 & 0 \\ l_{21} & l_{22} & 0 \\ l_{31} & l_{32} & l_{33} \end{pmatrix} \begin{pmatrix} d_1 \\ d_2 \\ d_3 \end{pmatrix} = \begin{pmatrix} c_1 \\ c_2 \\ c_3 \end{pmatrix} \overset{\text{LU, F.–Sub.}}{\Longrightarrow}$$

$$\begin{pmatrix} 1 & r_{12} & r_{13} \\ 0 & 1 & r_{23} \\ 0 & 0 & 1 \end{pmatrix} \begin{pmatrix} x_1 \\ x_2 \\ x_3 \end{pmatrix} = \begin{pmatrix} d_1 \\ d_2 \\ d_3 \end{pmatrix} \overset{\text{B.–Sub.}}{\Longrightarrow} \begin{pmatrix} x_1 \\ x_2 \\ x_3 \end{pmatrix},$$

where F.-Sub. and B.-Sub. denote the forward substitution to d_i and the backsubstitution to x_i, respectively, and LU means the LU-decomposition.
Gauss-Seidel method

$$\left.\begin{array}{l} x_1^{(j)} = (c_1 - a_{12}x_2^{(j-1)} - a_{13}x_3^{(j-1)})/a_{11} \\ x_2^{(j)} = (c_2 - a_{21}x_1^{(j)} - a_{23}x_3^{(j-1)})/a_{22} \\ x_3^{(j)} = (c_3 - a_{31}x_1^{(j)} - a_{32}x_2^{(j)})/a_{33} \end{array}\right\} \quad \text{until} \quad \left|\frac{x_i^{(j)} - x_i^{(j-1)}}{x_i^{(j)}}\right| \le \epsilon \quad \text{for all } i\,,$$

where j and ϵ denote the j-th iteration and ϵ a given accuracy limit, respectively.

Method	Accuracy, error sources	Use	Programming efforts	Remarks
graphical	poor	very limited		more effort than with numerical methods
Cramer's rule	rounding errors	limited		excessive calculation for $n > 3$
elimination of unknowns	rounding errors	limited		
Gaussian elimination	rounding errors	universal	moderate	
Gauss-Jordan	rounding errors	universal	moderate	also calculation of the inverse
LU-decomposition	rounding errors	universal	moderate	preferred elimination method
Gauss-Seidel	very good	diagonally dominant systems	easy	possibly no convergence*

*if not diagonal dominant

9.9 Eigenvalue equations

Eigenvalue equations occur, in technology, for example, for carrying out transformations to principal axes, to calculate the fundamental oscillations of coupled systems, etc.
The eigenvector x of a square matrix is the solution vector that satisfies the equation

$$\mathbf{Ax} = \lambda\mathbf{x}\,, \qquad \mathbf{x} \neq 0$$

called the **eigenvalue** equation.
Eigenvalue λ of **A**: The real or complex number giving the scaling factor λ by which the eigenvector **x** is elongated or contracted due to the **linear map Ax**.

▷ The **generalized eigenvalue problem**

$$\mathbf{Ax} = \lambda\mathbf{Bx}$$

is important for the method of finite elemente.

Characteristic equation of **A**: The system of equations that defines the eigenvalues and eigenvectors:

$$(\mathbf{A} - \lambda\mathbf{I})\mathbf{x} = \mathbf{0}, \quad \mathbf{x} \neq \mathbf{0}$$

$$\det(\mathbf{A} - \lambda\mathbf{I}) = \begin{vmatrix} a_{11} - \lambda & a_{12} & \dots & a_{1n} \\ a_{21} & a_{22} - \lambda & & \vdots \\ \vdots & & \ddots & \vdots \\ a_{n1} & \dots & \dots & a_{nn} - \lambda \end{vmatrix} = 0.$$

- Necessary condition for the existence of a solution $\mathbf{x} \neq \mathbf{0}$ of this system of equations: The **characteristic matrix** $(\mathbf{A} - \lambda\mathbf{I})$ must be singular:

$$\det(\mathbf{A} - \lambda\mathbf{I}) = 0.$$

Characteristic polynomial or **characteristic determinant** $P(\lambda)$ of **A**: Polynomial of degree n, where n is the number of rows of **A**, with

$$P(\lambda) = \det(\mathbf{A} - \lambda\mathbf{I})\,.$$

- Every real or complex root (zero) of the characteristic polynomial $P(\lambda)$ is a solution of the characteristic equation of **A**, i.e., an eigenvalue λ of **A**.

Spectrum of **A**: The set of all eigenvalues of **A** .

- The $n \times n$ matrix **A** has at least one and at most n (usually complex) numerically different eigenvalues.
- The eigenvectors $\mathbf{x}_1, \mathbf{x}_2, \dots, \mathbf{x}_n$ of a matrix **A** belonging to different eigenvalues $\lambda_1, \lambda_2, \dots, \lambda_n$ are linearly independent.
- The eigenvalues of symmetric (antisymmetric) matrices are real (purely imaginary or zero); the eigenvectors $\mathbf{x}_1, \mathbf{x}_2, \dots, \mathbf{x}_n$ belonging to different eigenvalues $\lambda_1, \lambda_2, \dots, \lambda_n$ are mutually orthogonal.
- The sum of the n eigenvalues λ_i is equal to the trace (sum of the elements on the main diagonal) of **A**:

$$\mathrm{Sp}(\mathbf{A}) = \sum_{i=1}^{n} a_{ii} = \sum_{i=1}^{n} \lambda_i\,.$$

- The determinant of **A** is equal to the product of the (possibly numerically equal) eigenvalues:

$$\det \mathbf{A} = \lambda_1 \cdot \lambda_2 \cdot \cdots \cdot \lambda_n = \Pi_{i=1}^{n} \lambda_i \ .$$

Product representation of the characteristic polynomial:

$$P(\lambda) = (-1)^n (\lambda - \lambda_1)(\lambda - \lambda_2) \ldots (\lambda - \lambda_n) \ .$$

Combination of equal factors (eigenvalues)

$$P(\lambda) = (-1)^n (\lambda - \lambda_1)^{m_1} (\lambda - \lambda_2)^{m_2} \ldots (\lambda - \lambda_r)^{m_r} \ .$$

with the numerically different eigenvalues $\lambda_1, \ldots, \lambda_r$,

$$r \le n \ , \quad m_1 + \cdots + m_r = n \ .$$

Algebraic multiplicity m_i of an eigenvalue.
Orthogonal matrices: Square matrices for which the inverse $\mathbf{P}^{-1}$ is equal to the transposed matrix $\mathbf{P}^{\mathsf{T}}$.

$$\mathbf{P} \cdot \mathbf{P}^{\mathsf{T}} = \mathbf{P}^{\mathsf{T}} \cdot \mathbf{P} = \mathbf{I}$$

- The eigenvalue of orthogonal matrices have the absolute value of one.
- The eigenvectors $\mathbf{x}$ of a symmetric matrix $\mathbf{S}$ can be regarded as the column vectors of an orthogonal matrix $\mathbf{P} = (\mathbf{x}_1, \mathbf{x}_2, \ldots, \mathbf{x}_n)$, with

$$\begin{aligned} \mathbf{SP} &= \mathbf{S}(\mathbf{x}_1, \mathbf{x}_2, \ldots, \mathbf{x}_n) \\ &= (\lambda_1 \mathbf{x}_1, \lambda_2 \mathbf{x}_2, \ldots, \lambda_n \mathbf{x}_n) \\ &= (\mathbf{x}_1, \mathbf{x}_2, \ldots, \mathbf{x}_n) \begin{pmatrix} \lambda_1 & 0 & \ldots & 0 \\ 0 & \lambda_2 & \ddots & \vdots \\ \vdots & \ddots & \ddots & 0 \\ 0 & \cdots & 0 & \lambda_n \end{pmatrix} \\ &= (\mathbf{x}_1, \mathbf{x}_2, \ldots, \mathbf{x}_n) \cdot \Lambda \ . \end{aligned}$$

with the diagonal matrix Λ, whose diagonal elements are the eigenvalues of $\mathbf{S}$:

$$\Lambda = \begin{pmatrix} \lambda_1 & 0 & \ldots & 0 \\ 0 & \lambda_2 & \ddots & \vdots \\ \vdots & \ddots & \ddots & 0 \\ 0 & \cdots & 0 & \lambda_n \end{pmatrix} .$$

- Diagonalization of $\mathbf{S}$:

$$\mathbf{S} = \mathbf{P} \Lambda \mathbf{P}^{\mathsf{T}} \quad , \quad \Lambda = \mathbf{P}^{-1} \mathbf{S} \mathbf{P} \ .$$

- The product of all eigenvalues of $\mathbf{S}$ is equal to the absolute element of the characteristic polynomial,

$$a_0 = \det \mathbf{S} = \det \Lambda \ .$$

Numerical calculation of the eigenvalues of a matrix A:
For the numerical solution of the eigenvalue problem, specific methods are available that are *not* based on the definition of eigenvalues as zeros of the characteristic polynomial. Some methods yield only individual eigenvalues, other methods yield all eigenvalues.

Inverse iteration: Method to calculate individual eigenvalues. The method is based on the following algorithm:
1. Choose the starting vector $\mathbf{v}_k$ of length 1. Choose the number μ in the vicinity of which an eigenvalue is expected.
2. Solve the system of linear equations

$$(\mathbf{A} - \mu\mathbf{w}_{k+1} = \mathbf{v}_k.$$

3. Set

$$\mathbf{v}_{k+1} := \mathbf{w}_{k+1}/|\mathbf{w}_{k+1}|, \quad k := k+1$$

and go to 2.
(Note that the matrix in Step 2 is the same in every iteration step; therefore, it is of advantage to use the LU-decomposition of this matrix.)
Assuming that for the eigenvalue λ_i of $\mathbf{A}$ it holds that:

$$|\lambda_1 - \mu| < |\lambda_i - \mu|, \quad i \neq 1,$$

the sequence

$$\left\{\mathbf{w}_{k+1}^{\mathsf{T}}\mathbf{e}_i / \mathbf{v}_k^{\mathsf{T}}\mathbf{e}_i\right\}$$

converges to the number $\dfrac{1}{\lambda_1 - \mu}$ from which λ_1 may be calculated, assuming that the i-th component of the eigenvector referring to λ_1 is different from zero: for sufficiently large k also $\mathbf{v}_k^{\mathsf{T}}\mathbf{e}_i \neq 0$.
Under the above conditions, the sequence of vectors $\{\mathbf{v}_k\}$ converges to an eigenvector belonging to λ_1.
QR method: Method to calculate all eigenvalues. The basic algorithm consists of the following steps:
1. Set $\mathbf{A}_0 := \mathbf{A}; \ k := 0$
2. Calculate the QR-decomposition of $\mathbf{A}_k = \mathbf{Q}_k\mathbf{R}_k$ ($\mathbf{Q}_k$ is an orthogonal matrix, $\mathbf{R}_k$ is a top right triangle matrix).
3. Set $\mathbf{A}_{k+1} := \mathbf{R}_k\mathbf{Q}_k, \ k := k+1$. Go to 2.
If $\mathbf{A}$ has only real eigenvalues differing in their absolute value, then the diagonal elements of $\mathbf{A}$, their absolute value, converge to the eigenvalues of $\mathbf{A}$. To improve the convergence properties a shift strategy (similar to that for the inverse iteration)is actually still used:
2′. Choose μ_k, calculate the QR-decomposition of $\mathbf{A}_k - \mu_k\mathbf{E} = \mathbf{Q}_k\mathbf{R}_k$.
3′. Set $\mathbf{A}_{k+1} := \mathbf{R}_k\mathbf{Q}_k + \mu_k\mathbf{E}, \ k := k+1$. Go to 2′.
As a shift parameter, choose the eigenvalue of the 2×2 matrix appearing in the bottom right corner of $\mathbf{A}_k$ which is closest to the diagonal element a_m^k.
In practice the effort of the QR-algorithm may be reduced further if matrix $\mathbf{A}$ is first transformed to a Hessenberg matrix by means of a similarity transformation.

▷ The practical numerical problems in calculating the of eigenvalues should not be underestimated; it is therefore strongly recommended to use known, proven codes.

9.10 Tensors

Tensor T of rank n: A quantity on an m-dimensional vector space whose components carrying n subscripts, $\mathbf{T}_{i_1 i_2 \ldots i_n}$, $i_k = 1, 2, \ldots, m$, in linear coordinate transformations $\vec{\mathbf{x}}' = \mathbf{A}\vec{\mathbf{x}}$ transform like products $x_{i_1}x_{i_2}, \ldots, x_{i_n}$ of n coordinates of the vector $\vec{\mathbf{x}} = (x_1, x_2, \ldots, x_m)$.

Linear coordinate transformation:

$$x_i' = \sum_{k=1}^{m} a_{i_k} x_k \,, \qquad i = 1, 2, \ldots, m \,.$$

Transformation of the tensor components:

$$T'_{i_1 i_2 \ldots i_n} = \sum_{k_1, k_2, \ldots, k_n = 1}^{m} a_{i_1 k_1} a_{i_2 k_2} \ldots a_{i_n k_n} T_{k_1 k_2 \ldots k_n} \,.$$

▷ A tensor of rank n on a three-dimensional vector space has 3^n components.

Tensor of rank 0: A quantity with only one component that has the same value in all coordinate systems. A tensor of rank 0 is an **invariant scalar**.

Tensor of rank 1: A quantity with m components that transform like a vector. A tensor of rank 1 can be represented as a **vector**:

$$\vec{\mathbf{T}} = (T_1, T_2, \ldots, T_m) \,.$$

Tensor of rank 2: A quantity with m^2 components that can be represented as an $m \times m$ **matrix**:

$$\mathbf{T} = \begin{pmatrix} T_{11} & T_{12} & \ldots & T_{1m} \\ T_{21} & T_{22} & \ldots & T_{2m} \\ \vdots & \vdots & & \vdots \\ T_{m1} & T_{m2} & \ldots & T_{mm} \end{pmatrix} .$$

▷ The representation of a vector $\vec{\mathbf{v}}$ (e.g., velocity vector) in $\mathbb{R}^3$ in different bases, $\vec{\mathbf{e}}_\mathbf{i}$, $\vec{\mathbf{e}}'_\mathbf{i}$, $i = 1, 2, 3$,

$$\vec{\mathbf{v}} = \sum_{i=1}^{3} v_i \vec{\mathbf{e}}_\mathbf{i} = \sum_{i=1}^{3} v_i' \vec{\mathbf{e}}'_\mathbf{i} \,,$$

does not change its actual value; $\vec{\mathbf{v}}$ is called an **invariant quantity**. The mathematical form of physical laws that can be written in terms of equations between tensors does not change in transitions to another coordinate system.

Symmetric tensor of rank 2: $T_{ik} = T_{ki}$ for all i, k.

Antisymmetric or **skew-symmetric** tensor of rank 2: $T_{ik} = -T_{ki}$ for all i, k.

▷ The diagonal elements of a skew-symmetric tensor of rank 2 are zero, $T_{ii} = 0$.

Trace of a tensor of rank 2: The sum of diagonal elements,

$$\mathrm{Tr}(\mathbf{T}) = \sum_{i=1}^{m} T_{ii} \,.$$

The trace is a **tensor invariant**. It is preserved in the case of linear coordinate transformations.

□ **Inertia tensor** in mechanics: A symmetric tensor of rank 2 by means of which the angular momentum of a rigid body with respect to an arbitrary rotational axis can be calculated,

$$\mathbf{J} = \begin{pmatrix} J_x & J_{xy} & J_{xz} \\ J_{xy} & J_y & J_{yz} \\ J_{xz} & J_{yz} & J_z \end{pmatrix} .$$

J_x, J_y, J_z: Moments of inertia with respect to the coordinate axes. J_{xy}, J_{xz}, J_{yz}: Deviation moments with respect to the coordinate axes.

▷ The deviation moments vanish in the system of principal axes.

□ **Pressure tensor** in hydromechanics, **stress tensor** in elastomechanics, **field tensor** in electrodynamics.

9.10.1 Algebraic operations with tensors

● **Addition** and **subtraction** of tensors of equal rank: Analogous to vectors and matrices, tensors of equal rank are added and subtracted by components:

$$\mathbf{T} = \mathbf{T}^{(1)} \pm \mathbf{T}^{(2)}\,, \qquad T_{i_1 i_2 \ldots i_n} = T^{(1)}_{i_1 i_2 \ldots i_n} \pm T^{(2)}_{i_1 i_2 \ldots i_n}\,.$$

● **Multiplication** of a tensor **by a scalar**: The multiplication of a tensor by a scalar c is also by components:

$$\mathbf{T}^{(2)} = c\mathbf{T}^{(1)}\,, \qquad T^{(2)}_{i_1 i_2 \ldots i_n} = cT^{(1)}_{i_1 i_2 \ldots i_n}\,.$$

● **Multiplication** of two tensors: The product $\mathbf{T} = \mathbf{T}^{(1)} \times \mathbf{T}^{(2)}$ of a tensor $\mathbf{T}^{(1)}$ of rank n and a tensor $\mathbf{T}^{(2)}$ of rank r is a tensor of rank $(n + r)$ with the components

$$T_{i_1 i_2 \ldots i_n k_1 k_2 \ldots k_r} = T^{(1)}_{i_1 i_2 \ldots i_n} T^{(2)}_{k_1 k_2 \ldots k_r}\,.$$

The following distributive and associative laws apply:

$$\mathbf{T}^{(1)} \times (\mathbf{T}^{(2)} + \mathbf{T}^{(3)}) = \mathbf{T}^{(1)} \times \mathbf{T}^{(2)} + \mathbf{T}^{(1)} \times \mathbf{T}^{(3)}\,,$$
$$\mathbf{T}^{(1)} \times (\mathbf{T}^{(2)} \times \mathbf{T}^{(3)}) = (\mathbf{T}^{(1)} \times \mathbf{T}^{(2)}) \times \mathbf{T}^{(3)}\,.$$

● **Contraction** of a tensor: The summation over two equal indices in a tensor of rank n $(n \geq 2)$ leads to a tensor of rank $n - 2$.

● **Inner tensor product**: Two tensors are multiplied with each other and subsequently contracted.

□ The contraction of the dyadic product (tensor of rank 2) of two vectors (tensors of rank 1) $\vec{\mathbf{x}}$ and $\vec{\mathbf{y}}$ (see Section 9.3) yields the scalar product $\vec{\mathbf{x}} \cdot \vec{\mathbf{y}}$, i.e., a tensor of rank 0:

$$\mathbf{z} = \vec{\mathbf{x}}\vec{\mathbf{y}}^T\,, \quad z_{ik} = x_i y_k\,, \quad i, k = 1, 2, \ldots, m$$

tensor of rank 2.
Contraction: $i = k = j$, summation over j:

$$\sum_{j=1}^{m} z_{jj} = \sum_{j=1}^{m} x_j y_j\,,$$

tensor of rank 0.

□ The inner product of two tensors of rank 2 (matrices) **A** and **B** corresponds to matrix multiplication **A**. **B**.:
Tensor multiplication:

$$\mathbf{A} \times \mathbf{B} = \mathbf{C}\,, \quad c_{ikrs} = a_{ik} b_{rs}\,, \quad i, k, r, s = 1, 2, \ldots, m$$

tensor of rank 4.
Contraction: $k = r = l$, summation over l.

$$\sum_l a_{il} b_{ls} = (\mathbf{AB})_{is}$$

tensor of rank 2.

10

Boolean algebra-application in switching algebra

10.1 Basic notions

Computers are constructed of a multitude of digital circuits. This chapter deals with the formal description of a subset of these circuits, the **digital combinational networks**. Digital networks are characterized by two essential features:

- They are constructed of logical components (gates).
- The circuit contains no feedbacks.

▷ Feedbacks may lead to storing, so that the initial values of a circuit depend not only on the actual input but also on the state of the memory. Circuits with feedbacks are called **sequential circuits**.

These circuits, here called **logical circuits**, may be described formally by means of **Boolean algebra**.

Boolean algebra is employed in the design, verification and documentation of logic circuits. An important aspect of circuit design is the **minimization of the circuit**. It can be performed by means of Boolean algebra.

▷ In most programming languages it is possible to use Boolean expressions.

10.1.1 Propositions and truth values

Proposition: A sentence in natural language to which a **truth value**; i.e., **true** or **false** can be added. A proposition is always either true or false (double-valuedness theorem) and cannot be true as well as false (excluded contradiction).

Notations for truth values:

false	true
false	true
0	1
O	L
L(ow)	H(igh)

Below, the truth values are denoted by 0 and 1.

▷ In `PASCAL` the truth values are denoted by the constants `false` and `true`; in `FORTRAN` by `.false.` and `.true.`

▷ Technical applications:

switch	closed	open	
current	flows	does not flow	
voltage	high	low	
hole	exists	does not exist	(punched cards)
lamp	lights up	does not light up	

10.1.2 Proposition variables

Proposition variable, Boolean variable: Used for propositions as long as the assignment of truth values to propositions is left open. Usually, the variables are denoted by capital letters A, B, C, etc.

▷ When proposition variables are agreed upon, Boolean data are used in PASCAL, and logical data in FORTRAN.

10.2 Boolean connectives

There are three basic types of Boolean operators, for which different notations are used connectives.

Notation for Boolean operators				
Operation	Notation			Remarks
Negation	NOT	$\neg$	$\overline{}$	The bar is above the proposition, $\neg$ is placed in front of the proposition.
Conjunction	AND	$\wedge$	$\cdot$	The conjunction sign is not necessary.
Disjunction	OR	$\vee$	$+$	

Below, the notation shown in the last column of notations will be used. As with products, the points for conjunctions are often left out. The notations for negation, conjunction and disjunction are read as "not," "and" and "or."

▷ The symbol $\vee$ is a stylized v based on Latin: vel (either... or). $\wedge$ is simply an upside down $\vee$ (for "and").

▷ In PASCAL, these operators are written as `NOT`, `AND`, `OR` while in FORTRAN they are written as `.not.`, `.and.`, `.or.`

Below, the **switching symbols** for the logical function are given. According to ISO-standards, every gate is represented by a rectangular box in which the function is given. In addition, we also show the obsolete symbol, which is still sometimes used.

The operators can be described by means of **truth tables** or **logic tables**. They list all possible assignments of Boolean variables together with the corresponding operators. For n variables, the truth table has 2^n rows.

10.2.1 Negation: not

A	$\overline{A}$
0	1
1	0

New and old symbols for a **not**-gate

In combination with other gates, denied inputs or outputs are marked simply by a small circle at the corresponding connection.

▷ $\overline{A}$ is true if and only if A is **not** true.

□ Technical application: Opening switch. The current flows if the switch is **not** activated.

10.2.2 Conjunction: and

A	B	A · B
0	0	0
0	1	0
1	0	0
1	1	1

New and old symbols for an **and**-gate

▷ $A + B$ is true if and only if A **as well as** B are true.

□ Technical application: Series connection of switches. The current flows only if all switches are activated.

□ An elevator (E) moves if the door (D) is closed and a floor button (S) has been pushed:

$$E = D \cdot S.$$

Taking into account that activation of the button (FG) for the (ground) floor on which we are at present must not make the elevator move, then it follows that

$$E = D \cdot F \cdot \overline{FG}.$$

10.2.3 Disjunction (inclusive): or

A	B	A+B
0	0	0
0	1	1
1	0	1
1	1	1

New and old symbols for an **or**-gate

▷ $A + B$ is true if and only if A **or** B (or both) are true.

▷ Sometimes, disjunction is called **alternative**.

□ Technical application: Parallel connection of switches. A current flows if at least one switch is activated.

▷ Boolean operators are related to set operation (where sets are denoted by A and B):

Boolean	Sets	
Conjunction	intersection	$A \cap B = \{x \mid x \in A \wedge x \in B\}$
Disjunction	union	$A \cup B = \{x \mid x \in A \vee x \in B\}$
Negation	complement	$\overline{A} = \{x \mid x \notin A\} = \{x \mid \neg(x \in A)\}$

10.2.4 Calculating rules

Within Boolean functions, the "point before a" rule always applies. The negation is always evaluated with priority. Deviations from these rules must be identified by brackets.

Connective	Priority
$\overline{}$	1
$\cdot$	2
$+$	3

▷ The same priorities also apply to PASCAL and FORTRAN.

● The following calculating rules hold:

$$\overline{0} = 1$$
$$\overline{1} = 0$$
$$\overline{\overline{A}} = A$$

$$A + 1 = 1$$
$$A \cdot 0 = 0$$

Identities:	$A + 0 = A$ $A \cdot 1 = A$
Idempotence laws:	$A + A = A$ $A \cdot A = A$
Tautology:	$A + \overline{A} = 1$
Contradiction:	$A \cdot \overline{A} = 0$
Commutative laws:	$A + B = B + A$ $A \cdot B = B \cdot A$
Associative laws:	$A + (B + C) = (A + B) + C$ $A \cdot (B \cdot C) = (A \cdot B) \cdot C$
Distributive laws:	$A \cdot (B + C) = A \cdot B + A \cdot C$ $A + (B \cdot C) = (A + B) \cdot (A + C)$
de Morgan's rules: (inversion laws)	$\overline{A \cdot B} = \overline{A} + \overline{B}$ $\overline{A + B} = \overline{A} \cdot \overline{B}$

Tautology: A proposition that is always true.
Contradiction: A proposition that is always false.

▷ Boolean notations should not be confused with arithmetical notations! When there is danger of confusion, it is better to use $\wedge$, $\vee$ and $\neg$.

□ The distributive law for disjunction must be tested. For that purpose, the following truth table is presented:

A	B	C	$(B \cdot C)$	$A + (B \cdot C)$	$A + B$	$A + C$	$(A + B) \cdot (A + C)$
0	0	0	0	0	0	0	0
0	0	1	0	0	0	1	0
0	1	0	0	0	1	0	0
0	1	1	1	1	1	1	1
1	0	0	0	1	1	1	1
1	0	1	0	1	1	1	1
1	1	0	0	1	1	1	1
1	1	1	1	1	1	1	1
				↑			↑

The two columns labeled by ↑ are identical for every assignment of the proposition variable; i.e., the propositions are equal, or the distributive law holds also for the disjunction.

▷ The application of the distributive law seems to be very obvious in the case of conjunction since it appears to be identical to the distributive law of multiplication. On the other hand, we must become accustomed to the distributive law of the disjunction (the "or" operation). Note that conjunction and disjunction have **nothing** to do with the multiplication and addition of numbers.

10.3 Boolean functions

Boolean function: Assigns one output value to several input variables. All Boolean functions can be represented by the basic Boolean operators. There are $2^{(2^n)}$ different functions with n input variables. If there is only one input variable, the four possible functions are: Identity ($f(A) = A$), negation ($f(A) = \overline{A}$), tautology ($f(A) = 1$) and contradiction ($f(A) = 0$).

In the case of two input variables, there are 16 different functions. The most important functions have their own names and symbols:

Name of the function	Sheffer or NAND	Peirce or NOR	Exclusive or XOR	Equivalence	Impli-cation
Alternative notations for function A B	$\overline{A \cdot B}$ NAND(A, B)	$\overline{A + B}$ NOR(A, B)	$\overline{A}B + A\overline{B}$ A XOR B, $A \not\equiv B$, $A \oplus B$	$\overline{A}\,\overline{B} + AB$ $A \equiv B$ $A \leftrightarrow B$	$\overline{A} + B$ $A \to B$
0 0	1	1	0	1	1
0 1	1	0	1	0	1
1 0	1	0	1	0	0
1 1	0	0	0	1	1
New symbol	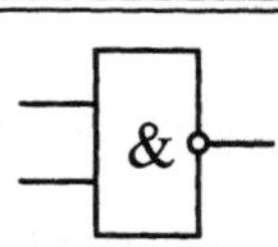	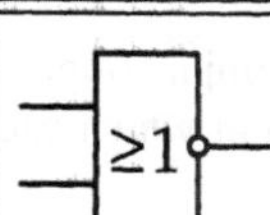	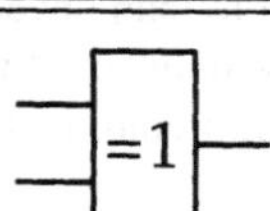	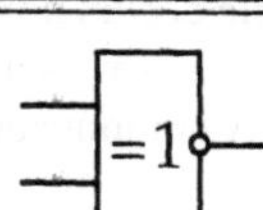	
Old symbol	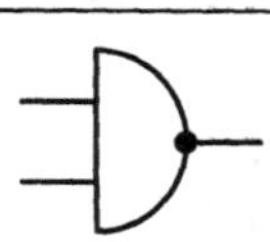	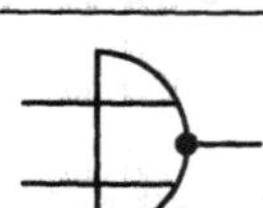	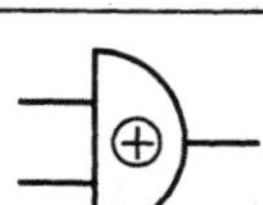	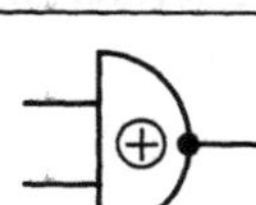	

Among the functions not mentioned above are the and (conjunction), the or (disjunction), the constant functions (tautology and contradiction), and functions depending on only one variable.

□ In data processing, a negative sign (sgn) of a number is denoted by the logical value 1, a positive sign is denoted by the value 0. For multiplication or division of two numbers, the following logical function for the determination of the result sign (res. sgn) is obtained:

Signs:

sgn1	sgn2	res. sgn.
+	+	+
+	−	−
−	+	−
−	−	+

Truth table:

sgn1	sgn2	res. sgn.
0	0	0
0	1	1
1	0	1
1	1	0

$$\text{res.sgn} = \overline{\text{sgn1}} \cdot \text{sgn2} + \text{sgn1} \cdot \overline{\text{sgn2}} = \text{sgn1} \oplus \text{sgn2}$$

The result sign is determined by means of an exclusive "or" function.

10.3.1 Operator basis

Operator basis: The set of operators by means of which any Boolean function can be represented.

□ The sets

$$\{\overline{\ },+,\cdot\},\ \{\overline{\ },+\},\ \{\overline{\ },\cdot\},\ \{\text{NAND}\}, \text{ and } \{\text{NOR}\}$$

are operator bases. In practice, this means that any Boolean function can be represented by using exclusively NAND-gates or exclusively NOR-gates. In most families of digital circuits, NAND- and NOR-gates are predominant.

10.4 Normal forms

Using the calculating rules for Boolean operations, different representations of the same function can be transformed into each other.

Normal forms are introduced to achieve a unified notation.

10.4.1 Disjunctive normal forms

Component of a conjunction k: The conjunction of primitive or denied variables, where each variable occurs no more than once.

□ $k = x_1 \cdot \overline{x_3} \cdot x_6$ is a component of a conjunction.

Minterm: A component of a conjunction in which each variable occurs once and only once.

Disjunctive normal form, DN: A disjunctive operator of the components of conjunctions $k_0, k_1, \ldots, k_j$.

$$\text{DN} = k_0 + k_1 + \ldots + k_j = \sum_{i=0}^{j} k_i \ .$$

Principal disjunctive normal form, PDN: A disjunctive operator of minterms.

▷ Every DN can be expanded to a PDN.

□ The expression

$$ABC + A\overline{B} + \overline{A}C$$

is a DN, but not a PDN, since the second component of the conjunction does not contain the variable C, and the third component does not contain the variable B. Thus, it is not a minterm. The expansion to a PDN is very simple, as follows:

$$\begin{aligned} ABC + A\overline{B} + \overline{A}C &= ABC + A\overline{B}(\overline{C} + C) + \overline{A}C(\overline{B} + B) \\ &= ABC + A\overline{B}\,\overline{C} + A\overline{B}C + \overline{A}C\overline{B} + \overline{A}CB\ . \end{aligned}$$

10.4.2 Conjunctive normal form

Component of a disjunction d: The disjunction of primitive or denied variables, where each variable occurs no more than once.

□ $d = \overline{x_1} + \overline{x_2} + x_5$ is the component of a disjunction.

Maxterm: The component of a disjunction in which each variable appears once and only once.

Conjunctive normal form, CN: A conjunctive connection of the components of disjunctions $d_0, d_1, \ldots, d_j$.

$$\text{CN} = d_0 \cdot d_1 \cdots d_j = \prod_{i=0}^{j} d_i\ .$$

Principal conjunctive normal form, PCN: A conjunctive connection of maxterms.

▷ Every CN can be extended to a PCN.

□ The expression

$$(A + B + C) \cdot (A + \overline{B})$$

is a CN, but not a PCN, since the second component of the disjunction does not contain the variable C. Thus, it is not a maxterm. The expansion to a PCN is very simple, as follows:

$$\begin{aligned} (A + B + C) \cdot (A + \overline{B}) &= (A + B + C) \cdot (A + \overline{B} + \overline{C}C) \\ &= (A + B + C) \cdot (A + \overline{B} + \overline{C}) \cdot (A + \overline{B} + C)\ . \end{aligned}$$

10.4.3 Representation of functions by normal forms

Every Boolean function can be represented in a (principal) disjunctive normal form as well as in a (principal) conjunctive normal form.

Algorithm for the construction of the disjunctive normal form

1. Set up the truth table for the function.

2. Delete all rows whose function value is equal to zero.

3. Translate each remaining row into a minterm. For that purpose, in each row deny every input variable that contains a 0 in the truth table but do not deny and take over into a minterm any input variable that contains a 1 in the truth table.

4. The disjunction of all minterms thus obtained yields the required principal disjunctive normal form.

□ Determine the disjunctive normal form of the exclusive or (XOR).

1. Truth table:

Row number	A	B	XOR(A, B)
1	0	0	0
2	0	1	1
3	1	0	1
4	1	1	0

2. Delete the rows containing a zero.

3. Translate each row into a minterm:

Row number	A	B	XOR(A, B)	$\longrightarrow$	minterm
2	0	1	1	$\longrightarrow$	$\overline{A}B$
3	1	0	1	$\longrightarrow$	$A\overline{B}$

4. Disjunction of all minterms:

$$\mathrm{XOR}(A, B) = \overline{A}B + A\overline{B} .$$

Algorithm for the construction of the conjunctive normal form

1. Set up the truth table for the function.

2. Delete all rows whose function value is equal to 1.

3. Translate each remaining row into a maxterm. For that purpose, in each row, deny every variable that contains a 1 in the truth table but do not deny and take over into a maxterm any input variable that contains a 0 in the truth table.

4. The conjunction of all maxterms obtained in this way yields the required principal conjunctive normal form.

□ Determine the conjunctive normal form of the Boolean function that is 1 and only 1 when an odd number of the three input variables is 1.

▷ This function is also called a **parity function**. It is sometimes used to recognize errors during the transmission of digital signals.

1. Truth table:

Row number	A	B	C	X
1	0	0	0	0
2	0	0	1	1
3	0	1	0	1
4	0	1	1	0
5	1	0	0	1
6	1	0	1	0
7	1	1	0	0
8	1	1	1	1

2. Delete the rows containing a 1.

3. Translate every row into a maxterm:

Row number	A	B	C	X	$\longrightarrow$	maxterm
1	0	0	0	0	$\longrightarrow$	$A+B+C$
4	0	1	1	0	$\longrightarrow$	$A+\overline{B}+\overline{C}$
6	1	0	1	0	$\longrightarrow$	$\overline{A}+B+\overline{C}$
7	1	1	0	0	$\longrightarrow$	$\overline{A}+\overline{B}+C$

4. Conjunction of all maxterms:

$$f(A, B, C) = \\ = (A+B+C)\cdot(A+\overline{B}+\overline{C})\cdot(\overline{A}+B+\overline{C})\cdot(\overline{A}+\overline{B}+C)\,.$$

▷ For brevity, functions with many units as function values are usually represented in conjunctive normal form.
Functions with many zeros as function values are usually represented in disjunctive normal form.

▷ Conjunctive and disjunctive normal forms can be transformed into each other by using the distributive laws and a subsequent combination of terms. But this procedure may be very time-consuming.

□ Determine the disjunctive normal form of the function

$$f(A, B, C) = \\ = (A+B+C)\cdot(A+\overline{B}+\overline{C})\cdot(\overline{A}+B+\overline{C})\cdot(\overline{A}+\overline{B}+C)$$

given in conjunctive normal form.
Application of the distributive law for the conjunctive ("multiplying out") yields a total of 81 terms:

$$\begin{aligned} f(A, B, C) = {} & A\cdot A\cdot\overline{A}\cdot\overline{A} + A\cdot A\cdot\overline{A}\cdot\overline{B} + \cdots \\ & \cdots + A\cdot\overline{B}\cdot\overline{C}\cdot\overline{B} + \cdots \\ & \cdots + C\cdot\overline{C}\cdot\overline{C}\cdot C \end{aligned}$$

Due to the fact that $X\cdot\overline{X}=0$, only eight terms remain,

$$\begin{aligned} f(A, B, C) = {} & A\overline{B}\,\overline{C}\,\overline{B} + A\overline{C}\,\overline{C}\,\overline{B} + BABC + B\overline{C}B\overline{A} \\ & + B\overline{C}\,\overline{C}\,\overline{A} + CABC + C\overline{B}\,\overline{A}\,\overline{A} + C\overline{B}\,\overline{A}\,\overline{B}, \end{aligned}$$

which due to the fact that $X\cdot X = X$ and commutativity can be further combined:

$$f(A, B, C) = A\overline{B}\,\overline{C} + \overline{A}B\overline{C} + \overline{A}\,\overline{B}C + ABC\,.$$

This is the required DN.

▷ The transformation is simpler if the truth table is set up first, and then the required normal form is constructed from the truth table.

10.5 Karnaugh-Veitch diagrams

Karnaugh-Veitch diagrams (KV-diagrams): A further possibility to represent functions, but in practice they are used more to minimize Boolean functions than to represent them.

Minimization of Boolean functions: Equivalence transformation of a Boolean function such that a minimal number of disjunctions and conjunctions are required for its representation.

10.5.1 Producing a KV-diagram

To produce a KV-diagram for a function $f(x_1, \ldots, x_n)$, starts with two adjacent fields denoted by x_1 and $\overline{x_1}$.

x_1	$\overline{x_1}$

In the next step, the diagram is reflected in the double line. The inverse image is denoted by x_2 and the image is denoted by $\overline{x_2}$. The double line is rotated from the lower boundary to the right boundary.

	x_1	$\overline{x_1}$
x_2		
$\overline{x_2}$		

The diagram is again reflected in the double line, and the inverse image and the image are again denoted by a variable and its negation. The double line is shifted from the right boundary to the lower.

	x_3		$\overline{x_3}$	
	x_1	$\overline{x_1}$	$\overline{x_1}$	x_1
x_2				
$\overline{x_2}$				

Next, the inverse image and image are again denoted by a variable and its negation, and we obtain:

		x_3		$\overline{x_3}$	
		x_1	$\overline{x_1}$	$\overline{x_1}$	x_1
x_4	x_2				
	$\overline{x_2}$				
$\overline{x_4}$	$\overline{x_2}$				
	x_2				

Proceed in this manner until all n variables are placed in the KV-diagram.

10.5.2 Entering a function in a KV-diagram

A Boolean function f can be represented by a KV-diagram. Each field of the KV-diagram represents the minterm obtained by a conjunctive connection of the notations of rows and columns.

□ The minterms assigned to the fields of a KV-diagram read:

		x_3		$\overline{x_3}$	
		x_1	$\overline{x_1}$	$\overline{x_1}$	x_1
x_4	x_2	$x_1\,x_2\,x_3\,x_4$	$\overline{x_1}\,x_2\,x_3\,x_4$	$\overline{x_1}\,x_2\,\overline{x_3}\,x_4$	$x_1\,x_2\,\overline{x_3}\,x_4$
	$\overline{x_2}$	$x_1\,\overline{x_2}\,x_3\,x_4$	$\overline{x_1}\,\overline{x_2}\,x_3\,x_4$	$\overline{x_1}\,\overline{x_2}\,\overline{x_3}\,x_4$	$x_1\,\overline{x_2}\,\overline{x_3}\,x_4$
$\overline{x_4}$	$\overline{x_2}$	$x_1\,\overline{x_2}\,x_3\,\overline{x_4}$	$\overline{x_1}\,\overline{x_2}\,x_3\,\overline{x_4}$	$\overline{x_1}\,\overline{x_2}\,\overline{x_3}\,\overline{x_4}$	$x_1\,\overline{x_2}\,\overline{x_3}\,\overline{x_4}$
	x_2	$x_1\,x_2\,x_3\,\overline{x_4}$	$\overline{x_1}\,x_2\,x_3\,\overline{x_4}$	$\overline{x_1}\,x_2\,\overline{x_3}\,\overline{x_4}$	$x_1\,x_2\,\overline{x_3}\,\overline{x_4}$

In each field enter a 1 if the corresponding minterm is a component of the principal disjunctive normal form of function f. Otherwise, enter a 0 or — to improve clarity — leave the field free.

10.5.3 Minimization with the help of KV-diagrams

Minimal function: The function $f_{\min}$ that has originated from function f by an equivalence transformation and requires a minimal number of conjunctions and disjunctions for its representation.

For KV-diagrams with four or fewer variables, the following algorithm determines the minimal function:

Algorithm for the determination of minimal functions

1. Produce the KV-diagram.

2. Enter the function in the diagram.

3. Draw rectangles with the side lengths 1, 2, or 4 into the diagram, enclosing as many fields as possible and containing only fields with units. The rectangles may extend beyond the boundary to the opposite side; i.e., the right and left edges as well as the upper and lower edges are connected with each other.

4. Find an unshaded field containing a 1 that is covered by as few rectangles as possible.

   Shade the largest of these rectangles (including the parts which may lie on the opposite side and still belong to the rectangle.)

   Repeat the last step until all fields containing a 1 are shaded.

5. Form a component of a conjunction for each shaded rectangle. In this term conjunctively combine those variables which uniquely denote one of the two edges. If one edge has maximal length, then the term consists only in the notation for the other edge. (If both edges have maximal length, then the function is always 1, a tautology.)

6. All components of the conjunction resulting from the preceding step are combined disjunctively. The result is the disjunctive normal form of the minimal function required.

□ Required is the Boolean function that compares the two numbers A and B in their binary representations $A = a_1 a_0$ and $B = b_1 b_0$, and that results in a logical 1 if $A < B$.

The truth table reads:

a_1	a_0	b_1	b_0	A	B	$A<B$	a_1	a_0	b_1	b_0	A	B	$A<B$
0	0	0	0	0	0	0	1	0	0	0	2	0	0
0	0	0	1	0	1	1	1	0	0	1	2	1	0
0	0	1	0	0	2	1	1	0	1	0	2	2	0
0	0	1	1	0	3	1	1	0	1	1	2	3	1
0	1	0	0	1	0	0	1	1	0	0	3	0	0
0	1	0	1	1	1	0	1	1	0	1	3	1	0
0	1	1	0	1	2	1	1	1	1	0	3	2	0
0	1	1	1	1	3	1	1	1	1	1	3	3	0

Thus, the principal disjunctive normal form reads:

$$f(a_1, a_0, b_1, b_0) = \overline{a_1}\,\overline{a_0}\,\overline{b_1}\,b_0 + \overline{a_1}\,\overline{a_0}\,b_1\,\overline{b_0} + \overline{a_1}\,\overline{a_0}\,b_1\,b_0 + \\ + \overline{a_1}\,a_0\,b_1\,\overline{b_0} + \overline{a_1}\,a_0\,b_1\,b_0 + a_1\,\overline{a_0}\,b_1\,b_0\,.$$

The figure below shows the KV-diagram with the function entered. Groups of two (crossing the edge) and four units have been formed by units already. The assignment of the groups to the components of conjunction is indicated by arrows. The disjunction of these components of the conjunction is the required minimized function.

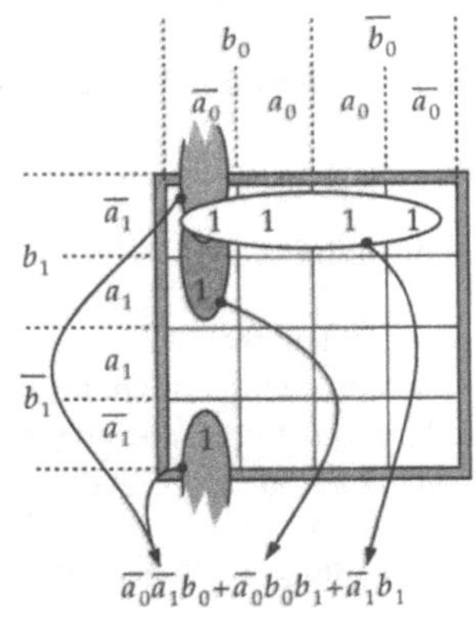

10.6 Minimization according to Quine and McCluskey

Another method to minimize Boolean functions has been proposed by Quine and McCluskey. This method can also be applied to functions with more than four variables.
Algorithm to determine the minimal function

1. Determinate the prime implicants
2. Determinate the minimal overlap.

Implicant of f: A component of conjunction k is called an implicant of f if it follows from $k = 1$ that also $f = 1$.
Contraction of a component of conjunction k: Results when one or more variables in k are left out.

▷ A component of conjunction k is always an implicant of every contraction of k.

Prime implicant: An implicant of f for which no contraction is still an implicant of f.

□ $f = abc + a\bar{b}c + bcd$.

abc is an implicant of f since $f = 1$ if $abc = 1$. But abc is not a prime implicant of f since the contraction ac is still an implicant. ac is a prime implicant since the further contractions a and c are no longer implicants.

Algorithm to determine the prime implicants

1. Transform the function into the disjunctive normal form.

2. Enter all components of conjunction into a column of a table.

3. For all components of conjunction, check whether there are other components of conjunction in that column distinguished from each other by one variable. One component of conjunction must contain this variable in a primitive form, the other one in a denied form.

   □ abc and $a\bar{b}c$ are such a pair.

   Below, such components of conjunction are called **similar**.

   If for one component of conjunction a similar component is found, then both are marked. In the adjacent column, a component of conjunction is entered which results when the variable is left out that is distinct in the two terms just considered.

   □

⋮		⋮
abc	*	ac
$a\bar{b}c$	*	
⋮		⋮

   abc and $a\bar{b}c$ are recognized as similar and marked by *. The two terms are combined:
   $abc + a\bar{b}c = ac(b + \bar{b}) = ac$
   The term ac is entered in the adjacent column.

   If for one component of conjunction all others have been checked for similarity, the next component of conjunction is considered.

   This procedure is repeated until all components of conjunction of the column have been considered.

   ▷ In comparing the components of conjunction, all must be considered, including those already marked by *. The multiple marking of components of conjunction is not necessary.

4. If no new terms have been entered in the new column, the algorithm is terminated.

   Otherwise, delete all multiple marks of a term in the last column, so that every term is contained once and only once in this column.

   Repeat the algorithm for this new column, starting with step 3.

The table produced by this algorithm is called the **prime-implicant table**. All terms of the prime-implicant table showing no mark are the prime implicants of the function to be minimized.

□ Find the prime implicants of the function

$$f(a_1, a_0, b_1, b_0) = \overline{a_1}\,\overline{a_0}\,\overline{b_1}\,b_0 + \overline{a_1}\,\overline{a_0}\,b_1\,\overline{b_0} + \overline{a_1}\,\overline{a_0}\,b_1\,b_0 + \\ + \overline{a_1}\,a_0\,b_1\,\overline{b_0} + \overline{a_1}\,a_0\,b_1\,b_0 + a_1\,\overline{a_0}\,b_1\,b_0$$

given in disjunctive normal form.
Producing the prime-implicant table:

(1)	$\overline{a_1}\,\overline{a_0}\,\overline{b_1}\,b_0*$	(13)	$\overline{a_1}\,\overline{a_0}\,b_0$	(23, 45)	$\overline{a_1}\,b_1$
(2)	$\overline{a_1}\,\overline{a_0}\,b_1\,\overline{b_0}*$	(23)	$\overline{a_1}\,\overline{a_0}\,b_1*$	(24, 35)	$\overline{a_1}\,b_1$
(3)	$\overline{a_1}\,\overline{a_0}\,b_1\,b_0*$	(24)	$\overline{a_1}\,b_1\,\overline{b_0}*$		
(4)	$\overline{a_1}\,a_0\,b_1\,\overline{b_0}*$	(35)	$\overline{a_1}\,b_1\,b_0*$		
(5)	$\overline{a_1}\,a_0\,b_1\,b_0*$	(36)	$\overline{a_0}\,b_1\,b_0$		
(6)	$a_1\,\overline{a_0}\,b_1\,b_0*$	(45)	$\overline{a_1}\,a_0\,b_1*$		

The numbers in parentheses show which terms of the preceding column gave rise to a new term. For example, the first component of conjunction of the second column originates from the combination of the terms in the first and third rows of the first column.
In the last column the same component of conjunction appears twice, and is therefore deleted once. The prime implicants are exactly the unmarked components of conjunction
$\overline{a_1}b_1$, $\overline{a_1}\,\overline{a_0}\,b_0$, and $\overline{a_0}b_1b_0$.

All prime implicants found are not always necessary for representing the function. The necessary prime implicants are determined by the
algorithm to determine the minimal overlap:

1. Produce a new table. Mark the columns with the components of conjunction. The rows are denoted by the prime implicants. The prime implicants are in order of ascending length.
2. Enter an × in the cross fields of all rows and columns if the corresponding component of conjunction is an implicant of the corresponding prime implicant.
3. Cancel all rows and columns in the table for which the column contains only one ×. The corresponding prime implicant is necessary and it is denoted. Repeat this step until none of the columns contains an individual ×.
4. Choose rows from the table in such a way that a new table results with the following properties:
   1. Every column contains at least one ×.
   2. The sum of the lengths of the prime implicants is as small as possible.
5. All prime implicants of the new table are denoted as necessary prime implicants.
6. The disjunctive operator of all necessary prime implicants represents the required minimal function.

□ The components of conjunction and the prime implicants of the preceding example form the following table:

	$\overline{a_1}\,\overline{a_0}\,\overline{b_1}\,b_0$	$\overline{a_1}\,\overline{a_0}\,b_1\,\overline{b_0}$	$\overline{a_1}\,\overline{a_0}\,b_1\,b_0$	$\overline{a_1}\,a_0\,b_1\,\overline{b_0}$	$\overline{a_1}\,a_0\,b_1\,b_0$	$a_1\,\overline{a_0}\,b_1\,b_0$
$\overline{a_1}\,b_1$		×	×	×	×	
$\overline{a_1}\,\overline{a_0}\,b_0$	×		×			
$\overline{a_0}\,b_1\,b_0$			×			×

Apart from the third column, all columns contain one and only one ×. Hence, all the rows are deleted, and all prime implicants are included in the list of necessary prime implicants.
The minimal function reads:

$$f(a_1, a_0, b_1, b_0) = \overline{a_1}\, b_1 + \overline{a_1}\, \overline{a_0}\, b_0 + \overline{a_0}\, b_1\, b_0$$

10.7 Multi-valued logic and fuzzy logic

In the informal language there are many propositions for which it is difficult to assign uniquely one of the truth values true or false. In such cases it may be reasonable to permit also values in between true and false, e.g., 'possibly'. If the truth value of a proposition is unknown, then the value 'unknown' may be introduced, too.

10.7.1 Multi-valued logic

Multi-valued logic, an extension of the Boolean algebra by additional truth values. The definition of the logic connectives has to be extended in such a way that the application to 'normal' truth values leads to the common Boolean algebra.

▷ Not all of the calculation rules given for the Boolean algebra can be maintained. Which of the calculating rules are no longer valid in an extension depends on the manner in which the Boolean connectives are defined for the new truth values.

□ In the language SQL for data bases there is the value ? standing for 'unknown'. The extended truth tables for negation, conjunction, and disjunction read

A	$\overline{A}$
0	1
?	?
1	0

A	B	$A \wedge B$	$A \vee B$
0	0	0	0
0	?	0	?
0	1	0	1
?	0	0	?
?	?	?	?
?	1	?	1
1	0	0	1
1	?	?	1
1	1	1	1

10.7.2 Fuzzy logic

Fuzzy logic, multi-valued logic with truth values from the closed interval [0, 1]. There are many fine graduations between true and false.

▷ 0 means 'completely false', 1 means 'completely true'.

□ Some properties of an object as bright or dark, big or small, warm and cold sometimes cannot be denoted uniquely to be true or false. One could call a temperature above 20 °Cto be 'warm', below 20 °Cto be 'not warm'. This leads to a discontinuity in the assignment for a temperature of exactly 20 °C. This discontinuity may be avoided by increasing the assignment of the temperatures to the truth value of the proposition 'is warm' continuous by and monotonically from 0 to 1 for temperature values of, e.g., between 15 °Cand 25 °C.

Fuzzy set A, a subset of a given fundamental set G whose elements are contained in A only to a certain degree.

Membership function m_A, a map of the fundamental set G onto the closed interval $[0, 1]$. $m_A(x)$ expresses 'how much' the element $x \in G$ is contained in A.
Thus, the truth value of the proposition $x \in A$ is $m_A(x)$.

▷ For common classical sets the membership function is

$$m_A(x) = \begin{cases} 0 & \text{falls } x \notin A \\ 1 & \text{falls } x \in A \end{cases}$$

the **characteristic function** of the set A.

Linguistic variable, takes truth values between 0 and 1. The assignment of a measured quantity to a truth value of a linguistic variable proceeds via a suitable membership function.

□ In order to describe the temperature T of the air, frequently, one uses the fuzzy notion cold, moderate, warm.

In principle, one can think of any membership function. In practice, one is restricted mostly to piecewise linear functions. Only the vertices are given; in between these points the function runs linearly.

□ $m_{\text{cold}}(T = 5\ {}^\circ\text{C}) = 1$ cold
$m_{\text{cold}}(T = 15\ {}^\circ\text{C}) = 0$ not cold

$m_{\text{moderate}}(T = 5\ {}^\circ\text{C}) = 0$ not moderate
$m_{\text{moderate}}(T = 15\ {}^\circ\text{C}) = 1$ moderate
$m_{\text{moderate}}(T = 25\ {}^\circ\text{C}) = 0$ not moderate

$m_{\text{warm}}(T = 15\ {}^\circ\text{C}) = 0$ not warm
$m_{\text{warm}}(T = 25\ {}^\circ\text{C}) = 1$ warm

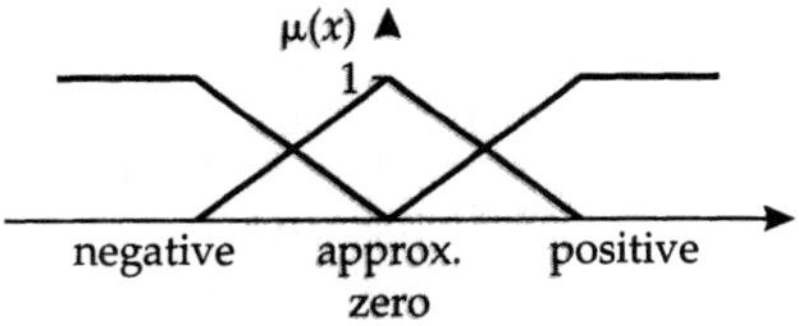

Membership function

Generalization of Boolean operators to fuzzy logic: Restricting to values of 0 and 1, the common Boolean algebra has to be reached as a limit. There are many possibilities to generalize the Boolean operators. Frequently, one uses

	Boole	Fuzzy
AND	$C = A \wedge B$	$C = \min(A, B)$
OR	$C = A \vee B$	$C = \max(A, B)$
NOT	$C = \neg A$	$C = 1 - A$

□ In Japan control systems based on fuzzy methods are in use already. Examples, are the start or brake of a subway without jolt for a minimum consumption of energy, the control of a refrigerator, or the avoidance to spoil a video picture by camera shakes.

11

Graphs and Algorithms

11.1 Graphs

11.1.1 Basic definitions

Graph $G(E, K)$: A combination of two sets E and K, where E means a (finite) set, the **vertices** or **nodes**, and $K \subseteq \binom{E}{2}$ is a set of pairs $\{u, v\}$, $u \neq v$, the **edges or arcs**.
$G(E, K)$ is called finite if E and K contain only a finite number of elements; otherwise graph $G(E, K)$ is called infinite.
Empty graph: E and K are empty sets.
Loop: An edge u whose end-points coincide.
Multiple edge: The same ordered or unordered pair of vertices is assigned to several edges.
Multigraph: Loops, $\{u, u\} \in K$, and multiple edges between vertices u, v are possible.
Complete graph $K_{(n)}$: A graph with $|E| = n$ number of vertices and $|K| = \binom{n}{2}$ number of edges.
Bipartite graph: E consists of two disjoint sets S and T, and every edge has one vertex in S and one in T.
Complete bipartite graph $K_{m,n}$, $|S| = m$, $|T| = n$: All kinds of edges between S and T are present;

$$|E| = m + n, \quad |K| = m \cdot n .$$

□

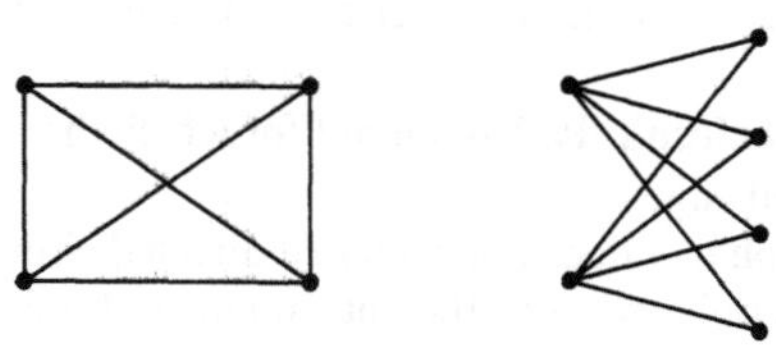

Complete graph K_4
Complete bipartite graph $K_{2,4}$

□ Problems in coordinating persons and jobs.

Q_n-cube: Set of vertices consisting of a sequence of the numbers 0 and 1 of length n with $|E| = 2^n$ vertices and $|K(Q_n)| = n \cdot 2^{n-1}$ edges.

□ Q_3-cube:
$|E| = 2^3 = 8$
$|K(Q_n)| = 12$

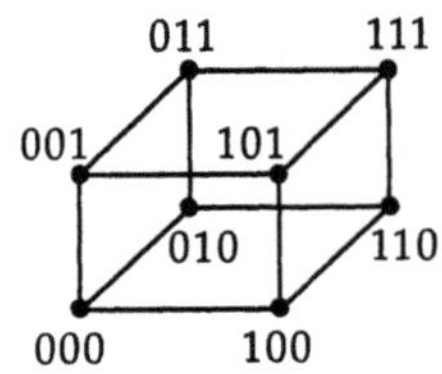

Path of length n: A sequence of $n + 1$ different vertices $u_1, u_2, \ldots, u_{n+1}$ and n edges $\{u_i, u_{i+1}\} \in K, i = 1, 2, \ldots, n$.

Circuit of length n: A sequence of n different vertices $u_1, u_2, \ldots, u_n$ and n different edges $u_i, u_{i+1} \in K, i = 1, 2, \ldots, n$.

Adjacent or **neighboring vertices** u, v: The end-points of one and the same element of K.

Incidence of vertex n **and edge** k: u is the terminal vertex of k.

Incidence of edge k **and edge** l: Edge k and edge l have one and the same terminal vertex.

Degree or **valence of vertex** u, $d(u) = |\mathrm{N}(u)|$: The number of the set of **neighbors** $N(u)$ of $u \in E$.

□ Isolated vertex: $d(u) = 0$.

Subgraph $H(E', K')$ **of** $G(E, K)$: If $E' \subseteq E$ and $K' \subseteq K$.

Induced subgraph: Contains all edges between the vertices in E' that are also contained in G.

▷ Important subgraphs are formed when vertices and edges are left out.

□ For given $G(E, K)$, the graph $G \setminus A$, $A \subseteq E$ results from G when A and all edges incident to this vertex are left out.

Components of $G(E, K)$: The subgraphs induced on the equivalence classes of K.

Bridge: Edge of a graph which when removed increases the components of a graph.

Distance $d(u, v)$ **of two vertices** u and v, $d(u, v)$: The length of the shortest path from u to v.

▷ If u and v belong to different components, then such a path does not exist; write: $d(u, v) = \infty$.

Diameter of $G(E, K)$: $D(G) = \max_{u,v \in E} d(u, v)$.

Directed graph, digraph, oriented graph $\vec{G}(E, K)$: A graph consisting of a set of vertices E and a set of edges $K \subseteq E^2$ of ordered pairs (directed, oriented edges); in $k = (u, v)$, u is the initial vertex (tail), and v the terminal vertex of k.

▷ Every directed edge occurs no more than once; $u \neq v$.

▷ In $\vec{G}$ both of the directed edges (u, v) and (v, u) may occur.

Connected graph: For two vertices differing from each other, at least one of the paths is between the two vertices.

Strongly connected graph: From every vertex u there is a directed path to every other vertex v.

Directed path, directed circuit: Path or circuit for which the length of the path is equal to the number of directed edges.

Acyclical graph: A graph that does not contain a directed circuit.

□ Transportation schedules for the efficient shipment of goods from producer to consumer.

- **Order n and size q of the graph** $G(E, K)$ are determined by

$$n = |E| \quad \text{and} \quad q = |K|$$

where $|E|$ is the cardinality of E, and $|K|$ the cardinality of K. (If E and K are finite sets, then n and q are the number of their elements.)

- **Connection of the degrees of vertices $d(u)$ and edges** K of a graph

$$G(E, K): \sum_{u \in E} d(u) = 2|K|.$$

▷ Every graph has an even number of edges with odd degrees.

11.1.2 Representation of graphs

Graphical representation: Pictorial representation by vertices and edges, but this is not suitable for computers; for this purpose, use matrix representation with consecutive numbering of vertices $u_1, u_2, \ldots, u_n$ and edges $k_1, k_2, \ldots, k_q$.
Adjacency matrix A: $n \times n$ matrix

$$\mathbf{A} = (a_{ij}), \quad a_{ij} = \begin{cases} 1 & \text{for} \quad \{u_i, u_j\} \in K \\ 0 & \text{otherwise} \end{cases}$$

▷ **A** is a symmetric matrix with $a_{ii} = 0$ and row sums and column sums that coincide with the degrees.

□

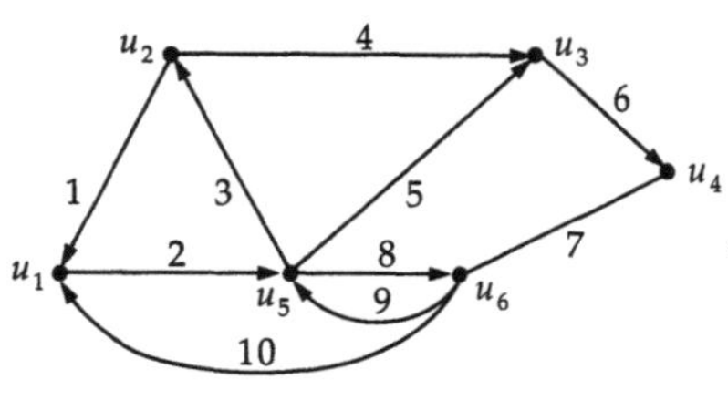

$A=$

$i \backslash j$	1	2	3	4	5	6
1	0	0	0	0	1	0
2	1	0	0	0	0	0
3	0	1	0	1	0	0
4	0	0	0	0	0	0
5	0	1	1	0	0	1
6	1	0	0	1	1	0

Graphs and its adjacency matrix

Incidence matrix B: $n \times q$ matrix $\mathbf{B} = (b_{ij})$,

$$b_{ij} = \begin{cases} 1 & \text{if edge } k_j \text{ leaves } u_i \\ -1 & \text{if edge } k_j \text{ terminates in } u_i \\ 0 & \text{otherwise}. \end{cases}$$

▷ The columns of **B** correspond to the edges, the rows of **B** correspond to the vertices of the graph; each column contains exactly one element $+1$, one element -1, and otherwise elements 0. For each row the number of elements $+1$ indicates how many edges lead away from the corresponding vertex. The number of elements -1 indicates how many edges terminate in the corresponding vertex.

11.1.3 Trees

Tree: Connected graph containing no circuits.

□ Branching-out networks: underline network, family tree, telephone network; suitable data structure for search and sort applications.

Spanning tree: Subgraph of graph $G(E, K)$ i.e., a tree of order $n = |E|$.

● A complete graph K_n has

$$t(n) = n^{n-2}$$

spanning trees.

□ G is a tree of order $n = 3$, $t(3) = 3$.

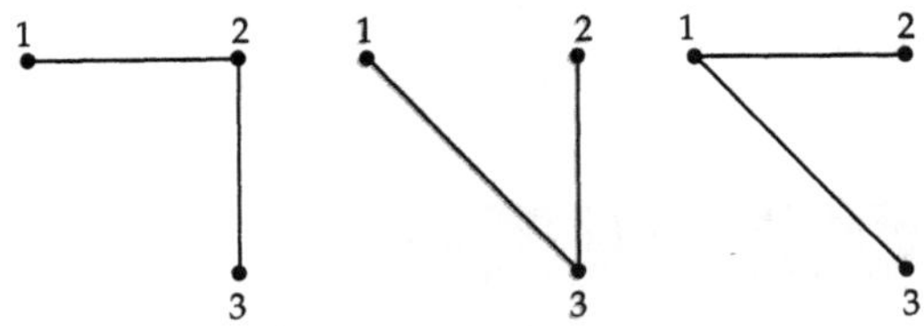

Breadth-First Search (BFS): Algorithm for the construction of a spanning tree for a graph given by its adjacency matrix by searching for the vertices (according to the breadth).

Depth-First Search (DFS): Algorithm for the construction of a spanning tree for a given graph by searching for the edges (according to the depth) .

Greedy algorithm: Algorithm for the construction of a spanning tree with minimum weight for a weighted graph (connected graph with weighting function).

□ Plan of routes with minimum total costs; circuit diagram with potential for all elements to communicate with each other, at minimum total costs.

Algorithm of Dijkstra: Algorithm for the construction of a spanning tree for a weighted graph with a unique minimum path from $u \in E$ to $v \in E$, where $d(u, v) = l(v)$ for all v holds for the lenght of the shortest path.

11.2 Matchings

Matching $M \subseteq K$ for a bipartite graph $G(S+T, K)$: Set of edges that is pairwise, not incident.

□ Job-coordination problems with an $S = (P_1, \ldots, P_n)$ set of persons, a $T(J_1, \ldots, J_n)$ set of jobs, and $P_i J_j \in K$ as the coordination of P_i to J_j. Aim: an optimum coordination of elements of S and elements of T, taking into account the weights on the edges.

Matching number $m(G)$ of G: The number of edges in a maximum size matching.

▷ $m(G)$ may be lower than $|S|$ or $|T|$.

Maximum matching M: If

$$|M| = m(G) \text{ holds,}$$

● for a bipartite graph $G(S + T, K)$, $m(G) = |S|$ if for a subset A of S $|A| \leq |N(A)|$ with $N(A) = \{v \in T | nv \in K, u \in A\}$. ($N(A)$ is the set of neighbors of A.)

● for a bipartite graph $G(S + T, K)$

$$m(G) = |S| - \max_{A \subseteq S}(|A| - |N(A)|).$$

□ $m(G) < |M|$

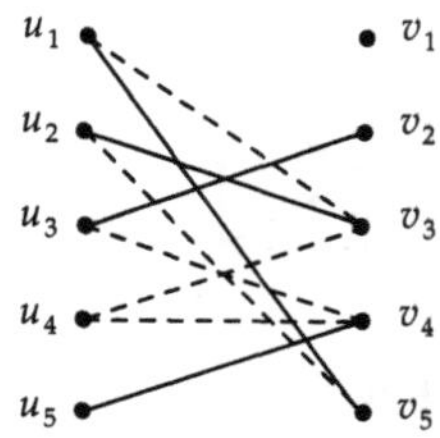

$m(G) = 4 < |M| = 5$, because three elements of $S(u_3, u_4, u_5)$ have only two neighbors (v_2, v_3) as elements of T.

Carrier of $G(S+T, K)$: The set of vertices $D \subseteq S+T$, where D hits (covers) every edge.

▷ For every carrier D and every matching M, D covers every edge of M, and $|D| \geq |M|$.

● Equilibrium theorem:
For a bipartite graph $G(S+Z, K)$

$$\max(|M|) = \min(|D|).$$

▷ There are methods for the general construction of a maximum of matchings and a minimum of carriers.

11.3 Networks

11.3.1 Flows in networks

Network from u to v: A **directed graph** $\vec{G}(E, K)$ with source u, sink v, $u \neq v$, $uv \in K$, **and** a **weighting function** c, the capacity of the graph that includes an evaluation of the individual edges ($c : K \to \mathbb{N}_0$).

□ $\vec{G}$ as a road system with $K = \{k_1, k_2, \ldots, k_m\}$, where a value $c(k_i)$ is assigned to any directed k_i such that $c(k_i)$ characterizes a certain property of k_i, such as permeability.

▷ Networks supplement the previous structural decription of graphs. They are maps of real systems by describing processes and states within systems.

Permissible flow from source u **to sink** v in network $\vec{G}(E, K)$: A function $f\colon\ K \to \mathbb{N}_0$ that does not exceed the capacity of the edges and for which the net flow, i.e., the difference between in-flow and out-flow for all vertices exept u and v, is zero.

Value of a permissible flow: The net flow into the sink.

▷ The main goal of an algorithm for **maximum flow** is to maximize the value of a permissible flow.

11.3.2 Eulerian line and Hamiltonian circuit

Eulerian line: An edge train $k_1, k_2, \ldots, k_n$ in a multigraph within which k_i is incident to k_{i+1}, every edge is contained once and only once, and which returns to the inital vertex. If initial vertex and terminal vertex are not identical, then the route is called an **open Eulerian line**.

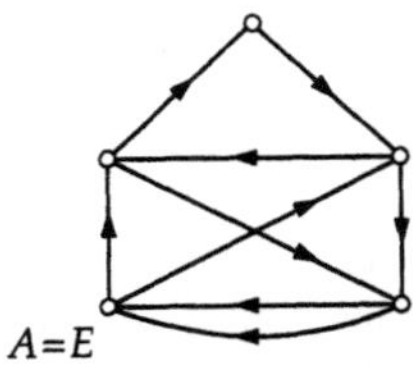

Eulerian line

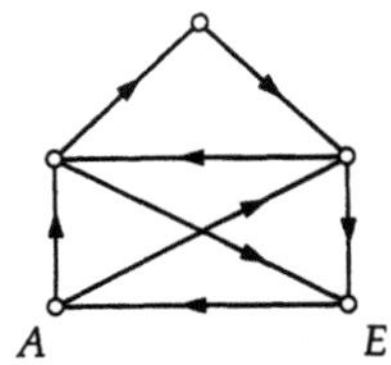

Open Eulerian line

Eulerian graph: A graph that has an Eulerian line; a connected multigraph $G(E, K)$ with an even degree of vertices, where loops must be counted twice.

- In a graph with an open Eulerian line, the initial terminal points are vertices of odd degree, all remaining vertices have an even degree.

□ Koenigsberg Bridge Problem:
Can seven bridges be crossed in such a way that each bridge is crossed only once?

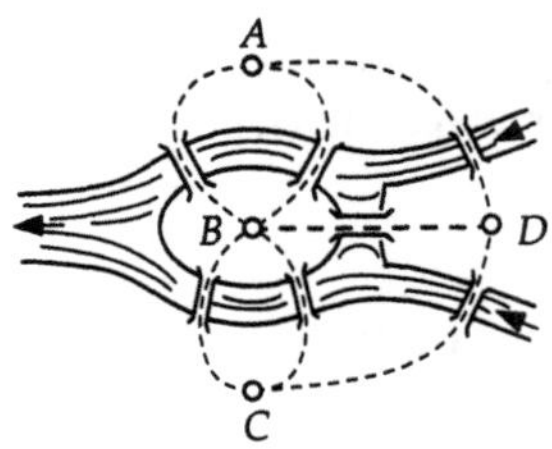

No, the graph of the Koenigsberg bridges can be drawn neither by an Eulerian line nor by an open Eulerian line.

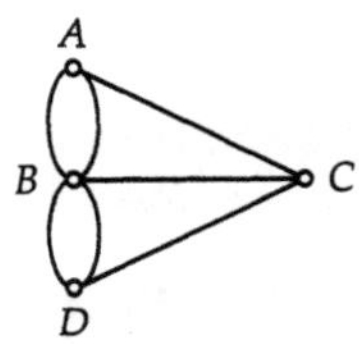

▷ Besides the edge-trains in the graphs considered so far, there are also edge-trains that do not cross each other, these are the **circular chains**.

Hamiltonian cycle: A closed edge-train in a graph that contains all vertices once and only once.

Hamiltonian graph: A graph that possesses a Hamiltonian cycle.

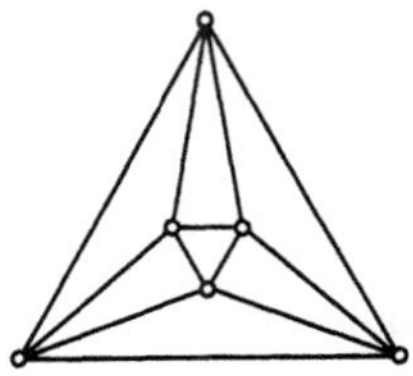

Graph with Hamiltonian cycle.

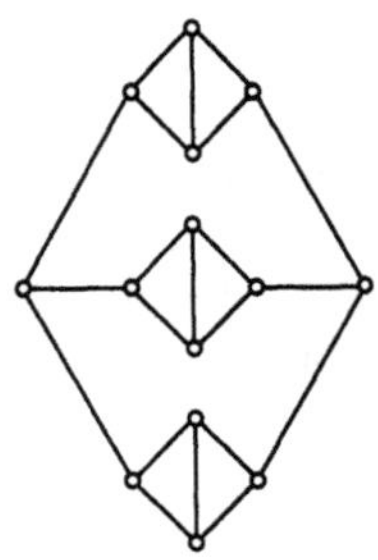

Graph without Hamiltonian cycle.

- **Theorem of Dirac** (sufficient condition):
If within a graph $G(E, K)$ all vertices have a degree greater than or equal to k, $k \in \mathbb{N} \setminus \{1\}$ and if $|E| \leq 2k$, then G contains a Hamiltonian cycle.

12

Differential calculus

12.1 Derivative of a function

The derivative of a function in point x gives the slope of the graph of the function at this point. It is defined as the slope of the tangent to the curve at the point x.

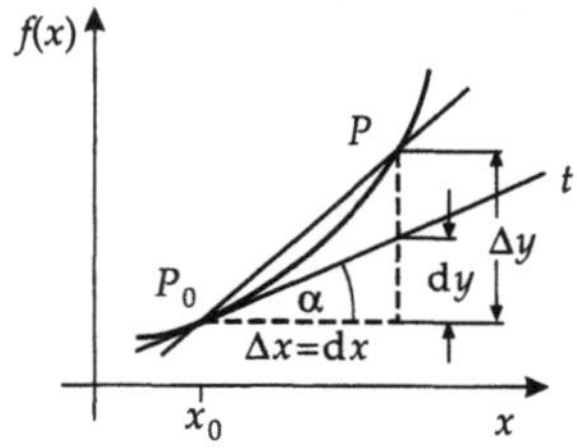

Secant

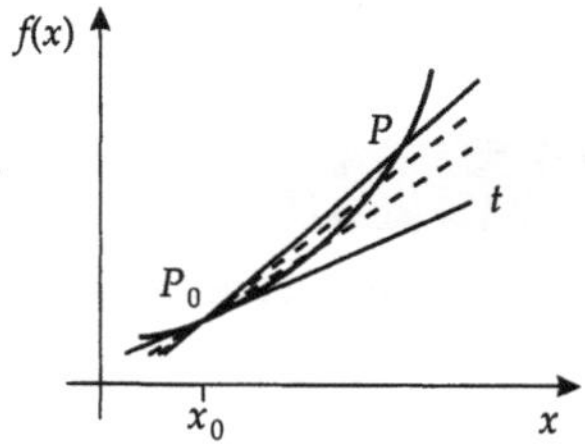

Tangent

Difference quotient:

$$\frac{\Delta y}{\Delta x} = \frac{\Delta f(x)}{\Delta x} = \frac{y - y_0}{x - x_0} = \frac{f(x) - f(x_0)}{x - x_0}$$

Describes the slope of the secant (in two-point form) through the points $P(x, y)$ and $P_0(x_0, y_0)$.

The **derivative** of the function $f(x)$ at the point x_0 equals the slope of the tangent at the point P_0 by the limit $P \to P_0$:

$$\lim_{x \to x_0} \frac{\Delta f(x)}{\Delta x} = \lim_{x \to x_0} \frac{f(x) - f(x_0)}{x - x_0} = \tan \alpha = f'(x_0),$$

if the limit exists and is unique.
The derivative of a function corresponds to the **slope of its graph** at the point x_0.
The **derivative** corresponds to the limit of the **difference quotient**:

$$f'(x) = \lim_{\Delta x \to 0} \frac{f(x + \Delta x) - f(x)}{\Delta x} = \lim_{\Delta x \to 0} \frac{\Delta y}{\Delta x} = \frac{\mathrm{d}y}{\mathrm{d}x} .$$

Importance: The slope of the tangent (derivative) corresponds to the limit of the slope of the secant (difference quotient).

□ If the function $f(x)$ is a straight line (a polynomial of the first degree), then the difference quotient is equal to the derivative.
The derivative $f'(x)$ is equal to the slope of the straight line $m = \tan \alpha$.

▷ The approximation of the tangent by the secant is the foundation of numerical differentiation and is exact for linear functions.

Notation for the **derivative** at the point x_0:

$$y'(x_0) = f'(x_0) = \left.\frac{\mathrm{d}y}{\mathrm{d}x}\right|_{x_0} = \left.\frac{\mathrm{d}}{\mathrm{d}x} f(x)\right|_{x_0} .$$

Derivatives with respect to time t are often denoted by a point instead of a prime.

$$\dot{x} = \frac{\mathrm{d}x}{\mathrm{d}t}$$

▷ The derivative is defined pointwise. The value of the derivative depends on the point at which it is taken. Thus, the derivative is itself a function.

□ Velocity as the derivative of path x with respect to time: $v = \dot{x}$, acceleration as the derivative of velocity with respect to time: $a = \dot{v}$, current as the derivative of charge q with respect to time: $i = \dot{q}$, gradient of a road, force as the derivative of the potential with respect to the location, minimizing and maximizing problems.

12.1.1 Differential

Differential of a function: Defined by:

$$\mathrm{d}y = f'(x_0)\mathrm{d}x .$$

Here, $\mathrm{d}x$ is the differential of the independent variable, $\mathrm{d}x = \Delta x$.
The differential $\mathrm{d}y$ is the change of the ordinate of the tangent to the curve at point P_0, between x_0 to $x_0 + \mathrm{d}x$.
Application: With a sufficiently small step size, the change of the function can be estimated by:

$$\Delta y \approx f'(x_0)\Delta x = \mathrm{d}y.$$

The estimation states how much the function value changes between x_0 and $x_0 + \Delta x$ if the tangent at x_0 is linearly continued (see also Taylor series).

▷ The basis of numerical integration of differential equations in the computer.

12.1.2 *Differentiability*

- A function $f(x)$ is said to be differentiable at x_0 if, for **all** sequences converging to x_0, the limit

$$y'(x_0) = f'(x_0) = \lim_{x \to x_0} \frac{f(x) - f(x_0)}{x - x_0}$$

exists, i.e., is finite.

- If the **left-hand derivative**

$$\lim_{x \to x_0 - 0} \frac{f(x) - f(x_0)}{x - x_0}$$

and the **right-hand derivative**

$$\lim_{x \to x_0 + 0} \frac{f(x) - f(x_0)}{x - x_0}$$

coincide, then the function is **differentiable** at x_0.

- A necessary (but not a sufficient) condition for the differentiability of a function is its continuity at that point.

▷ Differentiable functions are represented by "smooth" curves. A function is not **differentiable** at a point if it is **discontinuous**. If the curve has a **corner**, then only the left-hand derivative and the right-hand derivative exist; i.e., the slope of the tangent changes abruptly at the corner.

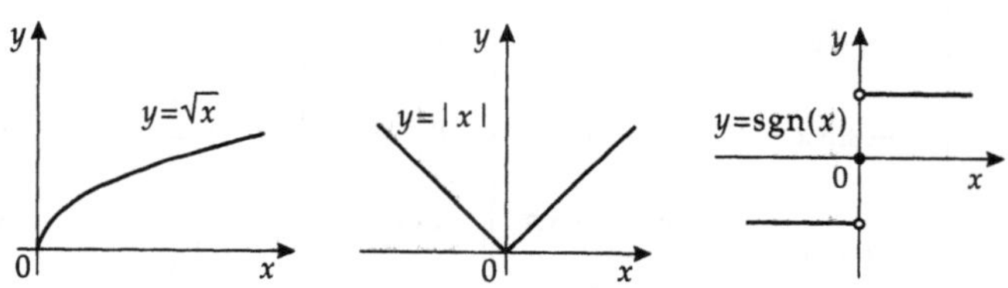

Non-differentiable functions at $x = 0$
Square root function Absolute value Signum function

□ The **absolute value function**, $y = |x|$, is not differentiable at point $x_0 = 0$ because the left-hand derivative and the right-hand derivative are different (-1 and $+1$, respectively). But the function is continuous at this point!
The **square-root function**, $y = \sqrt{x}$, is not differentiable at point $x_0 = 0$ because the slope is not finite (the tangent is perpendicular to the x-axis).

- If $f(x)$ is differentiable at any point of the domain of definition D, then $f(x)$ is called **differentiable**.

□ **Differentiable functions** are: ax^n, $\sin x$, $\cos x$, e^x, $\ln x$ and all algebraic compositions of these functions.

Function f' is called the **derived function** of $f(x)$. The domain of definition of f' is equal to that of $f(x)$, if $f(x)$ is differentiable in the entire domain D.

12.2 Differentiation rules

12.2.1 *Derivatives of elementary functions*

Function	Derivative	Function	Derivative	Function	Derivative
c	0	$\frac{1}{x}$	$-\frac{1}{x^2}$	$\sqrt{ax+b}$	$\frac{a}{2\sqrt{ax+b}}$
x	1	$\frac{1}{ax+b}$	$-\frac{a}{(ax+b)^2}$	$\sqrt{ax^2+b}$	$\frac{ax}{\sqrt{ax^2+b}}$
$ax+b$	a	$\frac{1}{x^2}$	$-\frac{2}{x^3}$	$\sqrt{r^2-x^2}$	$-\frac{x}{\sqrt{r^2-x^2}}$
x^2	$2x$	$\frac{1}{x^3}$	$-\frac{3}{x^4}$	e^x	e^x
x^3	$3x^2$	$\frac{1}{x^n}$	$-\frac{n}{x^{n+1}}$	a^x	$a^x \ln a$
x^n	nx^{n-1}	$\sqrt{x}$	$\frac{1}{2\sqrt{x}}$	$\ln x$	$\frac{1}{x}$
$(f(x))^2$	$2f(x)f'(x)$	$\sqrt[3]{x}$	$\frac{1}{3\cdot\sqrt[3]{x^2}}$	$\log_a x$	$\frac{1}{x\ln a}$
$(f(x))^n$	$nf(x)^{n-1}f'(x)$	$\sqrt[n]{x}$	$\frac{1}{n\cdot\sqrt[n]{x^{n-1}}}$	$\ln f(x)$	$\frac{f'(x)}{f(x)}$

12.2.2 *Derivatives of trigonometric functions*

Function	Derivative	Domain of definition	Function	Derivative	Domain of definition
$\sin x$	$\cos x$	$x \in \mathbb{R}$	$\arcsin x$	$\frac{1}{\sqrt{1-x^2}}$	$\lvert x\rvert < 1$
$\cos x$	$-\sin x$	$x \in \mathbb{R}$	$\arccos x$	$-\frac{1}{\sqrt{1-x^2}}$	$\lvert x\rvert < 1$
$\tan x$	$\frac{1}{\cos^2 x}$	$x \neq \frac{\pi}{2} + k\pi$	$\arctan x$	$\frac{1}{1+x^2}$	$x \in \mathbb{R}$
$\cot x$	$-\frac{1}{\sin^2 x}$	$x \neq k\pi$	$\operatorname{arccot} x$	$-\frac{1}{1+x^2}$	$x \in \mathbb{R}$

▷ Note the different domains of definition of the functions.

12.2.3 *Derivatives of hyperbolic functions*

Function	Derivative	Domain of definition	Function	Derivative	Domain of definition
$\sinh x$	$\cosh x$	$x \in \mathbb{R}$	$\text{Arsinh}x$	$\frac{1}{\sqrt{1+x^2}}$	$x \in \mathbb{R}$
$\cosh x$	$\sinh x$	$x \in \mathbb{R}$	$\text{Arcosh}x$	$\frac{1}{\sqrt{x^2-1}}$	$x > 1$
$\tanh x$	$\frac{1}{\cosh^2 x}$	$x \in \mathbb{R}$	$\text{Artanh}x$	$\frac{1}{1-x^2}$	$\lvert x\rvert < 1$
$\coth x$	$-\frac{1}{\sinh^2 x}$	$x \neq 0$	$\text{Arcoth}x$	$\frac{1}{1-x^2}$	$\lvert x\rvert > 1$

▷ One should note the different domains of definition of the functions.

▷ To avoid confusion, the inverse trigonometric functions are denoted by lowercase letters, whereas the inverse hyperbolic functions are denoted by capital letters.

12.2.4 Constant rule

● **Constant rule**: $c' = 0$.
The derivative of a constant is equal to zero.

12.2.5 Factor rule

● **Factor rule**: $(c \cdot f(x))' = c \cdot f'(x)$.
A constant factor remains during differentiation.

□ $(\sqrt{3} \cdot x^2)' = \sqrt{3}(x^2)' = \sqrt{3} \cdot 2x$

12.2.6 Power rule

● **Power rule**: In calculating the derivative of a power function, the exponent is diminished by one and the old exponent appears as a factor.

$$\frac{\mathrm{d}}{\mathrm{d}x}x^n = n \cdot x^{n-1}, \quad (n \in \mathbb{R}).$$

For any differentiable function it holds that:

$$\frac{\mathrm{d}}{\mathrm{d}x}(f(x))^n = n \cdot (f(x))^{n-1} \cdot f'(x), \quad (n \in \mathbb{R}).$$

□ $\left((x^3+2)^2\right)' = 2 \cdot (x^3+2) \cdot 3x^2$

□ Derivative of the square-root function: $(\sqrt{x})' = (x^{1/2})' = \frac{1}{2}x^{-1/2} = \frac{1}{2\sqrt{x}}$.

12.2.7 Sum rule

● **Sum rule**: $(f(x) \pm g(x))' = f'(x) \pm g'(x)$
The derivative of a sum (difference) is equal to the sum (difference) of the derivatives.

12.2.8 Product rule

● **Product rule**,
for two functions:

$$(f(x) \cdot g(x))' = f(x) \cdot g'(x) + f'(x) \cdot g(x),$$

for three functions:

$$(fgh)' = fgh' + fg'h + f'gh.$$

□ $(x^2 \sin x)' = 2x \sin x + x^2 \cos x$

12.2.9 Quotient rule

- **Quotient rule**: $\left(\frac{f}{g}\right)' = \frac{gf' - fg'}{g^2}, \quad \left(\frac{1}{g}\right)' = \frac{-g'}{g^2}$

▷ The quotient rule can also be written as a product rule with a negative exponent.

$$\left(\frac{1}{g}\right)' = (g^{-1})' = -g^{-2} \cdot g'$$

□ $f(x) = 2x + a, \; g(x) = 3x^2 + b$

$$\left(\frac{f(x)}{g(x)}\right)' = \frac{(3x^2+b)\cdot 2 - (2x+a)\cdot 6x}{(3x^2+b)^2}$$

▷ Fractional functions can be differentiated at x_0 only if $g(x_0) \neq 0$.

▷ Errors in sign due to interchanging the functions are avoided if the function $g(x)$ in the denominator is always written first into the numerator of the derivatives.

12.2.10 Chain rule

Used for calculating the derivative of computer functions

$$y = f(g(x)) = f(g)$$

with $y = f(g)$ as the outer function and $g = g(x)$ as the inner function.

- **Chain rule**:

$$(f(g(x))' = g'(x) \cdot f'(g(x)) \; ; \qquad \frac{\mathrm{d}f}{\mathrm{d}x} = \frac{\mathrm{d}g}{\mathrm{d}x} \cdot \frac{\mathrm{d}f}{\mathrm{d}g}$$

with $\frac{\mathrm{d}f}{\mathrm{d}g}$ as the outer derivative and $\frac{\mathrm{d}g}{\mathrm{d}x}$ as the inner derivative.

- **Chain rule** for a 3-fold nested function:

$$(f(g(h(x)))' = h'(x) \cdot g'(h(x)) \cdot f'(g(h(x))) \; ; \qquad \frac{\mathrm{d}f}{\mathrm{d}x} = \frac{\mathrm{d}h}{\mathrm{d}x} \cdot \frac{\mathrm{d}g}{\mathrm{d}h} \cdot \frac{\mathrm{d}f}{\mathrm{d}g} \, .$$

- **Chain rule** for the nesting of N functions:

$$\begin{aligned}&(f_1(f_2(\cdots(f_N(x)))))' \\ &\quad = f_N'(x) \cdot f_{N-1}'(f_N(x)) \cdot f_{N-2}'(f_{N-1}(f_N(x))) \cdot \cdots \cdot f_1'(f_2(\cdots(f_N(x)))) \, .\end{aligned}$$

▷ The rule also follows from the symbolic extension of the fraction (see differential). The derivative of a composite function corresponds to the product of the outer derivatives and the inner derivatives.

▷ The following approach is recommended: Differentiate from inside to outside, i.e., use the sequence in which the functions would be calculated **stepwise** with a pocket calculator.

□ $f(x) = \sin^2(2x) = (\sin(2x))^2 = f(g(h(x))) \quad \text{with}$

$$h(x) = 2x \quad g(h(x)) = \sin(2x) \quad f(x) = f(g(h(x))) = \sin^2(2x).$$

Calculation steps		Sequence of derivatives
First intermediate result	$u = 2x$	$u' = h'(x) = 2$
Second intermediate result	$v = \sin(u)$	$v' = g'(u) = \cos(u)$ $= g'(h(x)) = \cos(2x)$
Third intermediate result	$w = v^2$	$w' = f'(v) = 2v$ $= f'(g(x)) = 2\sin(u)$ $= f'(g(h(x))) = 2\sin(2x)$
Derivative: $f'(x) = u' \cdot v' \cdot w' = 2 \cdot \cos(2x) \cdot 2\sin(2x)$		

□ $f(x) = \exp\{\sin x^2\}, \qquad f'(x) = 2x \cos x^2 \exp\{\sin x^2\}$

12.2.11 Logarithmic differentiation of functions

- **Logarithmic derivative**: Differentiation of the logarithm $\ln y$ of a differentiable function y with $y(x) > 0$,

$$(\ln y)' = \frac{y'}{y}.$$

Functions of the type $y = h(x)^{g(x)}$ are differentiated according to the following rule: First, take the logarithm,

$$y = h(x)^{g(x)} \quad \to \quad \ln y = g(x) \cdot \ln h(x)$$

then differentiate, and then solve for y':

$$\frac{\mathrm{d} \ln y}{\mathrm{d}x} = \frac{y'}{y} = \left(g' \ln h + g\frac{h'}{h}\right) \quad \to \quad y' = y \cdot \left(g' \ln h + g\frac{h'}{h}\right)$$

□ $y = x^x, \quad \ln y = x \ln x, \quad \frac{y'}{y} = 1 \cdot \ln x + x \cdot \frac{1}{x}, \quad y' = x^x(\ln x + 1)$

□ $y = (\sin x)^{\cos x}, \sin x > 0 \quad \to \quad \ln y = \cos x \cdot \ln(\sin x),$

$$\frac{y'}{y} = -\sin x \cdot \ln(\sin x) + \cos x \cos x \cdot \frac{1}{\sin x},$$

$$y' = (\sin x)^{\cos x}\left(\frac{\cos^2 x}{\sin x} - \sin x \cdot \ln(\sin x)\right)$$

12.2.12 Differentiation of functions in parametric representation

- Parametric representation of a curve, with parameter t:

$$x = x(t), \qquad y = y(t).$$

- Parametric derivative with respect to parameter t:

$$\frac{\mathrm{d}x}{\mathrm{d}t} = \dot{x}, \qquad \frac{\mathrm{d}y}{\mathrm{d}t} = \dot{y}.$$

- Parametric representation of the first derivative with respect to x of a function:

$$y' = \frac{\mathrm{d}y}{\mathrm{d}x} = \frac{\mathrm{d}y/\mathrm{d}t}{\mathrm{d}x/\mathrm{d}t} = \frac{\dot{y}}{\dot{x}}.$$

- Second derivative of a function in parametric representation, with respect to x by means of the parametric derivative:

$$y'' = \frac{d^2 y}{dx^2} = \frac{d}{dx}\frac{\dot{y}}{\dot{x}} = \frac{d(\dot{y}/\dot{x})/dt}{dx/dt} = \frac{\dot{x}\ddot{y} - \ddot{x}\dot{y}}{\dot{x}^3}.$$

□ Trajectory of a parachutist jumping from a plane that flies at velocity v_0 at altitude h:

$$x(t) = v_0 t, \quad y(t) = h - \frac{1}{2}gt^2,$$
$$y' = \dot{y}/\dot{x} = -gt/v_0 = -gx/v_0^2, \quad y'' = -v_0 g/v_0^3 = -g/v_0^2.$$

12.2.13 Differentiation of functions in polar coordinates

Representation of a function in polar coordinates with angle φ:

$$r = r(\varphi).$$

Conversion to Cartesian coordinates according to

$$x = r(\varphi)\cos\varphi, \qquad y = r(\varphi)\sin\varphi.$$

Differentiation with respect to the angle φ:

$$\frac{dx}{d\varphi} = \frac{dr(\varphi)}{d\varphi}\cdot\cos\varphi - r(\varphi)\sin\varphi = \dot{x}\,,$$
$$\frac{dy}{d\varphi} = \frac{dr(\varphi)}{d\varphi}\cdot\sin\varphi + r(\varphi)\cos\varphi = \dot{y}\,.$$

- First derivative of a function in polar coordinates:

$$y' = \frac{\dot{y}}{\dot{x}} = \frac{\dot{r}\sin\varphi + r\cos\varphi}{\dot{r}\cos\varphi - r\sin\varphi}.$$

- Second derivative of a function in polar coordinates:

$$y'' = \frac{\dot{x}\ddot{y} - \ddot{x}\dot{y}}{\dot{x}^3} = \frac{r^2 + 2\dot{r}^2 - r\ddot{r}}{(\dot{r}\cos\varphi - r\sin\varphi)^3}.$$

□ Archimedean spiral: $r(\varphi) = \varphi$,

$$y' = \frac{\sin\varphi + \varphi\cos\varphi}{\cos\varphi - \varphi\sin\varphi}, \quad y'' = \frac{\varphi^2 + 2}{(\cos\varphi - \varphi\sin\varphi)^3}.$$

12.2.14 Differentiation of an implicit function

- **Differentiation of an implicit function** $F(x, y) = 0$:

$$\frac{d}{dx}F(x, y) = \frac{\partial F}{\partial x} + \frac{\partial F}{\partial y}\frac{dy}{dx} = F_x + F_y \cdot y' = 0\,; \quad y' = -\frac{F_x}{F_y}.$$

Approach: Differentiation by terms according to the chain rule and solving for y' (see also partial derivative).

□ Equation of a circle: $F(x, y) = x^2 + y^2 - r^2 = 0$,

$$\frac{d}{dx}F(x, y) = \frac{d}{dx}x^2 + \frac{d}{dx}y^2 = 2x + \frac{d}{dy}y^2\frac{dy}{dx} = 2x + 2yy' = 0, \rightarrow y' = -x/y$$

12.2.15 Differentiation of the inverse function

- **Differentiation of the inverse function** U of $f(x)$ for $y = f(x)$ and $x = U(y)$,

$$f'(x) = \frac{1}{U'(y)}, \qquad \frac{dy}{dx} = \frac{1}{dx/dy}, \qquad U' = \frac{dx}{dy} \neq 0 \quad .$$

□ $f(x) = \text{arcsin}x$, $U(y) = \sin y$

$$(\text{arcsin}x)' = \frac{1}{(\sin y)'} = \frac{1}{\cos y} = \frac{1}{\sqrt{1-(\sin y)^2}} = \frac{1}{\sqrt{1-x^2}}$$

□ $x - y + y^5 = 0$: This function cannot be solved simply for y, but can for x.

$$x = U(y), \; U'(y) = (y - y^5)' = 1 - 5y^4,$$
$$f'(x) = 1/(1 - 5y^4).$$

Thus, the derivative can be calculated for a given point (x, y) without recognizing the function $y = f(x)$ explicitly.

12.2.16 Table of differentiation rules

$u = u(x)$ and $v = v(x)$ are functions of x.

Constant rule	$c' = 0$
Factor rule	$(c \cdot u)' = c \cdot u'$
Power rule	$(u^n)' = n \cdot u^{n-1} \cdot u' \quad (n \in \mathbb{R},\ n \neq 0)$
Sum rule	$(u \pm v)' = u' \pm v'$
Product rule for two functions	$(uv)' = uv' + u'v$
Product rule for three functions	$(uvw)' = uvw' + uv'w + u'vw$
Product rule for *N* functions	$(u_1u_2 \ldots u_N)' = \sum_{i=1}^{N} u_1 \ldots u_i' \ldots u_N$
Quotient rule	$\left(\frac{u}{v}\right)' = \frac{vu' - uv'}{v^2}, \quad \left(\frac{1}{v}\right)' = \frac{-v'}{v^2}$
Logarithmic derivative	$(u^v)' = u^v \left(v' \ln u + v\frac{u'}{u}\right) \quad u(x), v(x) > 0$
Chain rule for two functions	$\frac{\mathrm{d}u}{\mathrm{d}x} = \frac{\mathrm{d}u}{\mathrm{d}v} \cdot \frac{\mathrm{d}v}{\mathrm{d}x}$
Chain rule for three functions	$\frac{\mathrm{d}u}{\mathrm{d}x} = \frac{\mathrm{d}u}{\mathrm{d}v} \cdot \frac{\mathrm{d}v}{\mathrm{d}w} \cdot \frac{\mathrm{d}w}{\mathrm{d}x}$
Differentiation in parametric representation	$u'(x(t)) = \frac{\mathrm{d}u}{\mathrm{d}x} = \frac{\dot{u}}{\dot{x}}$
Differentiation in polar coordinates	$u' = \frac{\mathrm{d}u}{\mathrm{d}x} = \frac{\dot{r}\sin\varphi + r\cos\varphi}{\dot{r}\cos\varphi - r\sin\varphi}$
Differentiation of an implicit function	$\frac{\mathrm{d}}{\mathrm{d}x}F(x, y) = F_x + F_y \cdot \frac{\mathrm{d}y}{\mathrm{d}x} = 0$
Differentiation of the inverse function	$u' = \frac{\mathrm{d}u}{\mathrm{d}x} = \frac{1}{\mathrm{d}x/\mathrm{d}u} = \frac{1}{(u^{-1})'}$

12.3 Mean value theorems

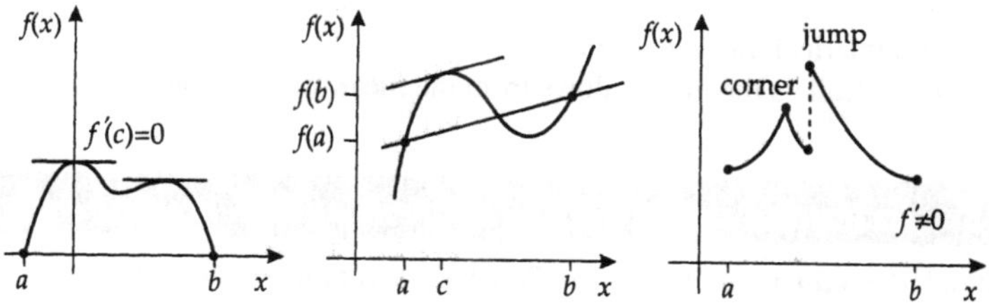

Rolle's theorem Mean value theorem Opposite example

12.3.1 Rolle's theorem

- **Rolle's theorem**: If a function $y = f(x)$ is
  1. differentiable in the open interval (a, b),
  2. continuous in the closed interval $[a, b]$, and if
  3. $f(a) = f(b)$,

  then there exists at least one c between a and b with

$$f'(c) = 0, \qquad c \in (a, b).$$

□ Between two zeros of a differentiable function there is at least one extremum, except for the constant function $y = 0$.

▷ Rolle's theorem does not apply to corners and jumps in (a, b), because the condition of differentiability has been violated here.

▷ The function must be continuous at the end-points a and b.

▷ The function may have further, absolute extremes at a and b.

12.3.2 Mean value theorem of differential calculus

- **Mean value theorem of differential calculus** (theorem of Lagrange): If a function $y = f(x)$ is

  1. differentiable in (a, b), and

  2. continuous in $[a, b]$,

  then there exists at least one c between a and b, such that

$$\frac{f(b) - f(a)}{b - a} = f'(c), \qquad c \in (a, b).$$

▷ Geometric interpretation: In the interval (a, b) there exists a point x_0 at which the **slope of the tangent** is equal to the slope of the **secant** through $f(a)$ and $f(b)$. This is the mean slope.

▷ The notation *Mean value theorem* is misleading. If would be better to use "intermediate value theorem," because c lies only **between** a and b and not necessarily in the middle of the interval.

□ A car moves from point A to point B at a mean velocity of 80 km/h. For in at least one point C between A and B, the instantaneous velocity coincides with the mean velocity.

▷ Basis of numerical differentiation.

If $f(a) = f(b)$, Rolle's theorem follows from the mean value theorem.

12.3.3 Extended mean value theorem of differential calculus

● **Extended mean value theorem** (Cauchy's theorem): If two functions $f(x)$ and $g(x)$ are

1. differentiable in (a, b),
2. continuous in $[a, b]$,
3. $g'(x) \neq 0$, for all $x \in (a, b)$, then there exists at least one c between a and b, such that

$$\frac{f(b) - f(a)}{g(b) - g(a)} = \frac{f'(c)}{g'(c)}, \qquad c \in (a, b).$$

Conclusions from the mean value theorem:

● If a function $f(x)$ is differentiable in an interval I, and if $f'(x) = 0$ for all points of the interval I, then the function is constant in the interval I:

$$f'(x) = 0 \quad \text{for all } x \in I \quad \rightarrow \quad f(x) = c.$$

These considerations are valid in an interval I.

● If two functions $f(x)$ and $g(x)$ are differentiable in the interval I and their derivatives coincide, then they differ at most by an additive constant:

$$f'(x) = g'(x) \quad \rightarrow \quad f(x) = g(x) + c, \quad x \in I.$$

12.4 Higher derivatives

Higher derivative $f^{(n)}$: Multiple application of differentiation.

Second derivative: Represents the change of the slope at a point, the sense of **curvature of a curve**.

□ Acceleration as the second derivative of the path with respect to time $\ddot{x} = F/m$.

● If the second derivative is negative, the function is convex up.
If the second derivative is positive, the function is concave up.

Notations for the second derivative:

$$f''(x) = (f'(x))' = \frac{\mathrm{d}}{\mathrm{d}x} f'(x) = \frac{\mathrm{d}^2}{\mathrm{d}x^2} f(x) = \frac{\mathrm{d}^2 y}{\mathrm{d}x^2}.$$

For time derivatives: $\ddot{x}$, $\ddot{q}$.

□ 1. $(\sin x)' = \cos x$, $(\sin x)'' = (\cos x)' = -\sin x$

2. $y = x^3$, $y' = 3x^2$, $y'' = 6x$

n-th derivative, notation:

$$f^{(n)}(x) = \frac{\mathrm{d}^n}{\mathrm{d}x^n} f(x) = \frac{\mathrm{d}^n y}{\mathrm{d}x^n}.$$

- n-th derivative of a product (**Leibniz's rule**):

$$(f(x) \cdot g(x))^{(n)} = \sum_{k=0}^{n} \binom{n}{k} f^{(n-k)}(x) \cdot g^{(k)}(x) .$$

Explicitly for $n = 2$:

$$(fg)'' = f''g + 2f'g' + fg'' .$$

□ $(x^2 \sin x)'' = 2\sin x + 2 \cdot 2x\cos x + x^2 \cdot (-\sin x)$
Explicitly for $n = 3$:

$$(fg)''' = f'''g + 3f''g' + 3f'g'' + fg''' .$$

Function	n-th derivative
x^m	$\begin{cases} m(m-1)\dots(m-n+1)x^{m-n} & m > n \\ n! & n = m \in \mathbb{N} \\ 0 & m < n \end{cases}$
$(a_n x^n + \dots + a_1 x + a_0)$	$a_n n!$
$\ln x$	$(-1)^{n+1}(n-1)! \cdot x^{-n}$
$\log_a x$	$(-1)^{n+1}\dfrac{(n-1)!}{\ln a}x^{-n}, \quad a, x > 0,\ a \neq 1$
e^{kx}	$k^n e^{kx}$
a^{kx}	$(\ln a)^n k^n a^{kx}$

Function	n-th derivative
$\sin kx$	$k^n \sin(kx + n\pi/2)$
$\cos kx$	$k^n \cos(kx + n\pi/2)$
$\cosh kx$	$\begin{cases} k^n \sinh kx & n \text{ odd} \\ k^n \cosh kx & n \text{ even} \end{cases}$
$\sinh kx$	$\begin{cases} k^n \cosh kx & n \text{ odd} \\ k^n \sinh kx & n \text{ even} \end{cases}$
$f(x) \cdot g(x)$	$\sum_{k=0}^{n} \binom{n}{k} f^{(n-k)}(x) \cdot g^{(k)}(x)$ Leibniz's product rule

Second differential:

$$d^2y = d(dy) = f''(x)dx^2 .$$

Third differential:

$$d^3y = d(d^2y) = f'''(x)dx^3 .$$

***n*-th differential**:

$$d^ny = d(d^{n-1}y) = f^{(n)}(x)dx^n .$$

12.4.1 Slope, extremes

Piecewise monotonicity: Decomposition of the curve into segments possessing only a positive or a negative slope (the function is piecewise monotonically increasing or monotonically decreasing).

- For a differentiable function it holds that:

strictly monotonically decreasing:	$f'(x) < 0$
strictly monotonically increasing:	$f'(x) > 0$

- The function $f(x)$ has a **relative extreme** at point x_m; all function values are smaller (maximum) or greater (minimum) than $f(x_m)$:

Relative maximum:	$f(x) \leq f(x_m)$	$x \in U$
Relative minimum:	$f(x) \geq f(x_m)$	$x \in U$

- For differentiable functions, a relative extremum lies between two segments of different monotonicity.
- The sign of the slope changes at an **extremum**:

$x < x_m$	$x > x_m$	Extreme
$f'(x) > 0$	$f'(x) < 0$	Maximum
$f'(x) < 0$	$f'(x) > 0$	Minimum

Sufficient condition for a maximum/minimum:

Relative maximum:	$f'(x_0) = 0$, and $f''(x_0) < 0$,
Relative minimum:	$f'(x_0) = 0$, and $f''(x_0) > 0$.

- There is an extreme at $x = x_0$ if and only if the first derivative is equal to zero and the next higher derivative which is nonzero at $x = x_0$, is of even order.
  Necessary and **sufficient** condition for a:

Relative maximum:
$$\begin{cases} f^{(k)}(x_0) = 0, & \text{if } 1 \leq k < n \text{ and} \\ f^{(n)}(x_0) < 0, & n \text{ even,} \end{cases}$$

Relative minimum:
$$\begin{cases} f^{(k)}(x_0) = 0, & \text{if } 1 \leq k < n \text{ and} \\ f^{(n)}(x_0) > 0, & n \text{ even,} \end{cases}$$

- If x_0 is an **extreme** of a differentiable function, then the following holds (necessary condition):

$$f'(x_0) = 0\,, \quad \text{or } f'(x_0) \text{ is not defined.}$$

At an extreme the tangent is either parallel to the x-axis or not defined.

▷ If the first derivative is equal to zero, then there may be also a saddle point.

□ $f(x) = x^3$, $f'(x) = 3x^2$, $f'(0) = 0$, there is no extreme but a saddle point at $x = 0$.

▷ The extremes as well as the behavior of the slope of a function can be determined from a sign sketch of the derivative.

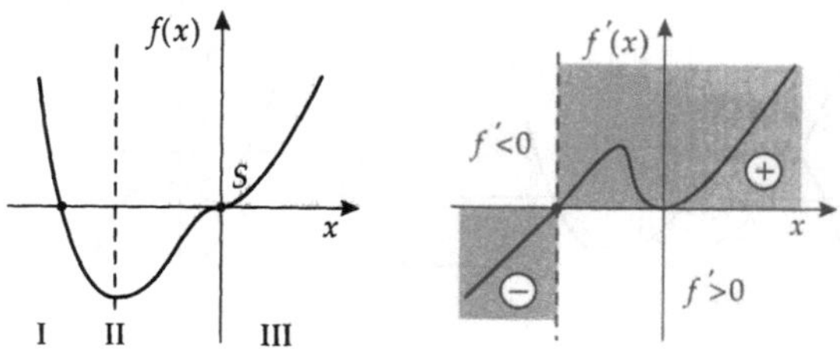

left original function
(I: monotonic decreasing
II: extremum
III: monotonic increasing)

right derivative

12.4.2 Curvature

Curvature of a curve: The change of slope, obtained from the second derivative.

- A twice-differentiable function is:

$$\textbf{concave up} \text{ for } \quad f''(x) > 0\,,$$
$$\textbf{convex up} \text{ for } \quad f''(x) < 0\,.$$

- Geometric interpretation of **convexity**: A function f is **convex up** in an interval if the graph of f in the whole interval lies below every tangent to the curve. According to this definition a curve **convex down** is concave up.

▷ Both definitions of **convexity** are used in the literature.

▷ Think of a circle drawn within the segment of curvature of the function. If the circle is positioned **above** the curve, the second derivative is **positive**; if the circle is positioned **below** the curve, the second derivative is **negative**.

12.4.3 Point of inflection

Point of inflection: Lies between two curve segments of different curvature of a two-fold differentiable function.

- The sign of the curvature changes at the **point of inflection**.
- The second derivative of $f(x)$ must be zero at the **point of inflection** x_I (necessary condition):

$$f''(x_I) = 0\,.$$

▷ The condition is not sufficient; i.e., a point of inflection does not necessarily exist.

□ $f(x) = x^4$, $f''(x) = 12x^2$, $f''(0) = 0$, but the function has no point of inflection, only a minimum at $x = 0$.

● Necessary and sufficient condition for a point of inflection:

$$f''(x_I) = 0 \text{ and } f^{(n)}(x_I) \neq 0,\ n \text{ odd, and } f^{(k)}(x_I) = 0,\ (2 < k < n).$$

▷ Points of inflection with a vertical tangent usually do not satisfy this condition (e.g., $\sqrt[3]{x}$).

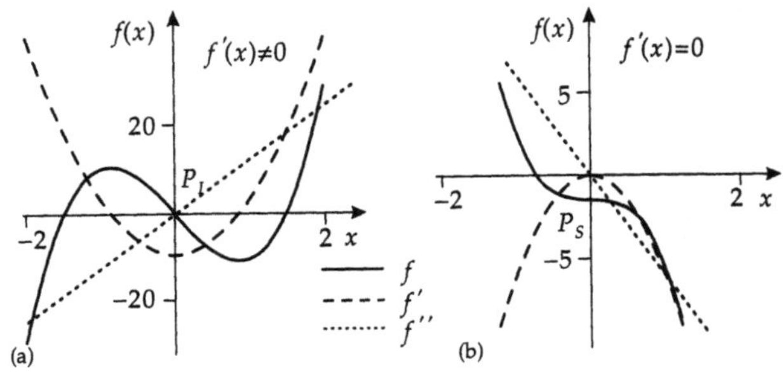

Point of inflection — Saddle point

● **Saddle point**: Special point of inflection with a horizontal tangent:

Saddle point: $f''(x_S) = 0$, $f'(x_S) = 0$, and x_S point of inflection.

● An n-fold differentiable function has the following behavior at a point x_0 with

$$f^{(n)}(x_0) \neq 0, \quad f^{(n-1)}(x_0) = f^{(n-2)}(x_0) = \cdots = f''(x_0) = f'(x_0) = 0\,,$$

Minimum:	n even,	$f^{(n)} > 0$
Maximum:	n even,	$f^{(n)} < 0$
Point of inflection:	n odd,	$f^{(n)} \neq 0$

□ $f(x) = x^4$, $f'(x_0 = 0) = f''(x_0 = 0) = f'''(x_0 = 0) = 0$,
$f^{(4)}(x) = 4! = 24 > 0$,
i.e., there is a minimum at $x_0 = 0$.

12.5 Approximation method of differentiation

12.5.1 *Graphical differentiation*

If a curve or series of measurements is known, but not the equation of the function, graphical differentiation is a possibilty.

1. Choose the point x_i of the curve (or the i-th measuring point of the series of measurements).

2. Draw the secant through two neighboring points, enclosing x_i, which are equally distant from x_i.

3. Repeat the construction of secants for closer points.

4. Draw approximately the tangent at x_i.

5. Make a parallel shift of the tangent to point $(x, y) = (-1, 0)$.

6. The approximate value $f'(x_i)$ corresponds to the y-coordinate of the intersection point with the y-axis.

7. Repeat the whole procedure for further points of the curve.

▷ For extremes or saddle points of a curve, the shifted tangent lies on the x-axis (the value of the derivative is equal to zero).

▷ The slope of the tangent is also equal to the quotient of the intercept on the y-axis and the intercept on the x-axis.

12.5.2 *Numerical differentiation*

Functions can be differentiated numerically which is practical when the analytic solution takes some effort or cannot be determined at all. Approximates the derivative by the difference quotient (difference formula of the first order):

$$f'(x) = \lim_{h\to 0} \frac{f(x+h) - f(x)}{h} = \frac{f(x+h) - f(x)}{h} + F(x, h)$$

$F(x, h)$ is the error term of the approximation; it is of order $O(h)$; i.e., the error is linear in h. The error term depends on step size h and on the point x considered.

Two-point difference formula (forward difference):

$$f'(x) = \frac{f(x+h) - f(x)}{h} + O(h)$$

Two-point difference formula (backward difference):

$$f'(x) = \frac{f(x) - f(x-h)}{h} + O(h)$$

For the above approximations, the error is reduced in proportion to the reduction of step size h.

▷ If step size h is too small, then the result can be affected by an error due to rounding. Rule of thumb for the ordinary difference quotient:

$$h_{\text{opt}} \approx 10^{-nmax/2},$$

where $nmax$ is the number of digits in the computer. Typically for computers, one has $nmax = 8$ (single precision), $nmax = 16$ (double precision).

3-point formula (improved difference formula):

$$f'(x) = \frac{f(x+h) - f(x-h)}{2h} + O(h^2)$$

Second derivative:

$$f''(x) = \frac{f(x+h) - 2f(x) + f(x-h)}{h^2} + O(h^2)$$

Here, the error changes quadratically with h.
Differentiation formulas of the fourth order:
First derivative:

$$f'(x) = \frac{-f(x+2h) + 8f(x+h) - 8f(x-h) + f(x-2h)}{12h} + O(h^4)$$

Second derivative:

$$f''(x) = \frac{-f(x+2h) + 16f(x+h) - 30f(x) + 16f(x-h) - f(x-2h)}{12h^2} + O(h^4)\,.$$

Procedure for tabulated data:

(a) Differentiation formula of the first order for fast estimations or

(b) numerical fit with subsequent differentiation.

12.6 Differentiation of functions with several variables

Functions with several variables occur often in practice (e.g., a surface in space, representation by contour lines as on maps, etc.).

12.6.1 Partial derivative

First partial derivative of a function $z = f(x, y)$ with respect to x and y; limits of the difference quotients are:

$$f_x(x, y) = \frac{\partial}{\partial x} f(x, y) = \lim_{h\to 0} \frac{f(x+h, y) - f(x, y)}{h},$$

$$f_y(x, y) = \frac{\partial}{\partial y} f(x, y) = \lim_{h\to 0} \frac{f(x, y+h) - f(x, y)}{h},$$

if they exist.
Geometric interpretation: Slope of the tangents to a surface at points (x, y) parallel to the x-axis (f_x) and the y-axis (f_y).

The partial derivatives are defined pointwise; they represent functions themselves.

● Partial derivatives are formed by differentiation with respect to one independent variable while regarding the other constant.

▷ Numerical approximation proceeds for each variable. Simple difference quotient:

$$f_x \approx \frac{f(x+h, y) - f(x, y)}{h}, \quad f_y \approx \frac{f(x, y+h) - f(x, y)}{h}.$$

Improved difference formulas are formed accordingly.

□ $z = f(x, y) = x^2 + y^2$: $f_x = 2x$, $f_y = 2y$

First partial derivatives of a function $f(x_1, x_2, \ldots, x_n)$ of n variables with respect to x_i, $i = 1, \ldots, n$:

$$\frac{\partial}{\partial x_i} f(x_1, \ldots, x_n) = \lim_{h \to 0} \frac{f(x_1, \ldots, x_i + h, \ldots, x_n) - f(x_1, \ldots, x_n)}{h} = f_{x_i}$$

□ $f(x, y, z) = x^3 + xy + y \ln z$:
$f_x = 3x^2 + y$, $f_y = x + \ln z$, $f_z = y/z$

Differentiation of an implicit function, $F(x, y) = 0$:

$$\frac{\mathrm{d}y}{\mathrm{d}x} = y' = -\frac{F_x(x, y)}{F_y(x, y)}, \quad \text{if} \quad F_y(x, y) \neq 0.$$

□ Equation of the ellipse:

$$F(x, y) = \frac{x^2}{a^2} + \frac{y^2}{b^2} - 1 = 0, \; y' = -\frac{2x/a^2}{2y/b^2} = -\frac{xb^2}{ya^2}, \; y'(0, b) = 0.$$

Higher partial derivatives of a function with two independent variables (see also under vector analysis):

$$f_{xx} = \frac{\partial^2}{\partial x^2} f(x, y) = \frac{\partial}{\partial x}\left(\frac{\partial}{\partial x} f(x, y)\right)$$
$$f_{xy} = \frac{\partial^2}{\partial x \partial y} f(x, y) = \frac{\partial}{\partial y}\left(\frac{\partial}{\partial x} f(x, y)\right)$$
$$f_{yx} = \frac{\partial^2}{\partial y \partial x} f(x, y) = \frac{\partial}{\partial x}\left(\frac{\partial}{\partial y} f(x, y)\right)$$
$$f_{yy} = \frac{\partial^2}{\partial y^2} f(x, y) = \frac{\partial}{\partial y}\left(\frac{\partial}{\partial y} f(x, y)\right)$$

● **Theorem of Schwarz**: If the function $f(x, y)$ and its partial derivatives are continuous, then the order of the partial differentiations is interchangable:

$$f_{xy} = f_{yx}$$
$$f_{xyy} = f_{yxy} = f_{yyx}$$
$$f_{yxx} = f_{xyx} = f_{xxy}, \quad \ldots$$

□ $f(x, y) = y \sin x + x^2 \cos y$:
$f_x = y \cos x + 2x \cos y$, $f_y = \sin x - x^2 \sin y$,
$f_{xy} = \cos x - 2x \sin y = f_{yx}$

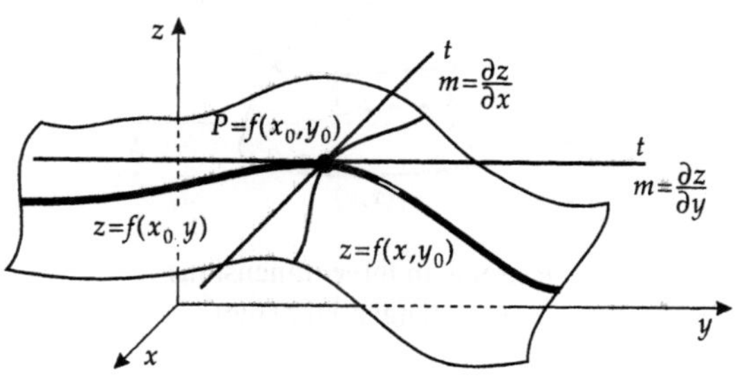

Partial derivative

12.6.2 Total differential

Total differential of functions in two independent variables:

$$df(x, y) = \frac{\partial}{\partial x} f(x, y)dx + \frac{\partial}{\partial y} f(x, y)dy = f_x dx + f_y dy,$$

of functions with n independent variables:

$$df(x_1, x_2, \ldots, x_n) = \sum_{i=1}^{n} \frac{\partial}{\partial x_i} f(x_1, x_2, \ldots, x_n)dx_i = \sum_{i=1}^{n} f_{x_i} dx_i$$

Application: Linear approximation of a function at a point $(x_0 + \Delta x, y_0 + \Delta y)$ in terms of the plane spanned by the two tangents:

$$f(x_0 + \Delta x, y_0 + \Delta y) \approx f(x_0, y_0) + f_x(x_0, y_0)\Delta x + f_y(x_0, y_0)\Delta y\,.$$

□ Paraboloid of rotation $z = f(x, y) = x^2 + y^2$.

Total differential: $df(x, y) = 2x\,dx + 2y\,dy$

Approximation for the point $(2, 2)$ in case of expansion about the point $(1, 1)$ $(\Delta x = \Delta y = 1)$:

$$f(2, 2) \approx f(1, 1) + 2 \cdot 1\Delta x + 2 \cdot 1\Delta y = 2 + 2 + 2 = 6.$$

12.6.3 Extremes of functions in two dimensions

- If there is an **extreme** at the point (x_m, y_m), then the following must hold (necessary condition):

$$f_x(x_m, y_m) = 0, \qquad f_y(x_m, y_m) = 0$$

- The sufficient condition for an extreme reads:

$$f_x(x_m, y_m) = 0, \qquad f_y(x_m, y_m) = 0\,, \quad \text{and}$$
$$D = f_{xx}(x_m, y_m) \cdot f_{yy}(x_m, y_m) - f_{xy}^2(x_m, y_m) > 0$$

For $f_{xx} < 0$ ($f_{xx} > 0$) we have a maximum (minimum).

Geometric interpretation: The function has a horizontal tangential plane at the extreme.

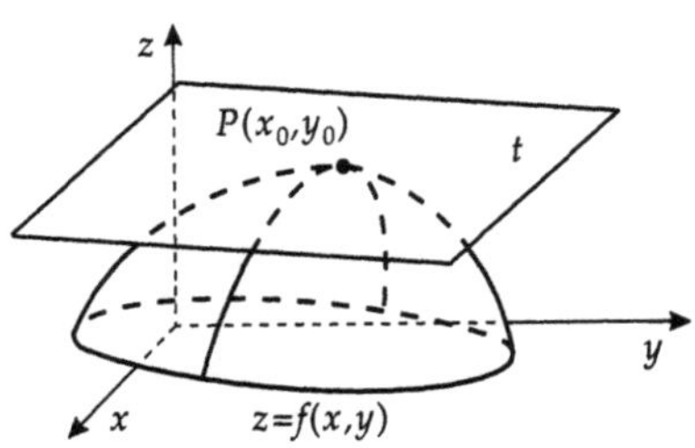

Extreme in three dimensions
(t = tangential plane)

▷ A point (x_0, y_0) with vanishing first-order partial derivatives is a saddle point if $D < 0$. The saddle point also has a horizontal tangential plane

□ $f(x, y) = x^3 - 3x - y^2 + 5$:
$f_x = 3x^2 - 3 = 0,\ f_y = -2y = 0$
Possible extremes at $(1, 0)$ and $(-1, 0)$
Check: $f_{xx} = 6x,\ f_{xy} = 0,\ f_{yy} = -2$,
$D = f_{xx} \cdot f_{yy} - f_{xy}^2 = -12x$,
At $(1, 0)$ there is no extreme, because $D(1, 0) = -12 < 0$.
At $(-1, 0)$ there is $D(-1, 0) = 12 > 0$, thus no extreme.
Type of the extreme at $(-1,0)$:
$f_{yy} = -2 < 0$, there is a maximum at this point $(-1,0)$.

12.6.4 Extremes with constraints

Search for the extremes of a function $f(x, y)$ subject to the constraint $g(x, y) = 0$. Thus, a two-dimensional function, a space curve, is given whose extremes are required.

□ Required is the maximum of altitude $h = f(x, y)$ for a hiking trip: $y = g(x)$.

● **Method of Lagrange multipliers**: To determine the extreme consider the function

$$F(x, y) = f(x, y) + L \cdot g(x, y)$$

with an (unknown) multiplier L and equates its partial derivatives to zero (necessary condition):

$$F_x = f_x + L \cdot g_x = 0, \quad F_y = f_y + L \cdot g_y = 0.$$

Thus, one must solve three conditional equations for three unknowns, x, y, and L.

□ Rectangle with maximum area at constant circumference U:
Area: $f(x, y) = xy$
Circumference U: $g(x, y) = 2x + 2y - U = 0$,
$F(x, y) = xy + L(2x + 2y - U)$,
$F_x = y + 2L = 0,\ F_y = x + 2L = 0 \rightarrow x_0 = y_0 = U/4$,
the square is the rectangle with maximum area for a given circumference.

▷ The multiplier L is only an auxiliary quantity whose value is not required. Therefore, it is best to eliminate it at the beginning of the calculation.

● Generalization: The function $f(x_1, x_2, \dots, x_n)$ and $m < n$ constraints of the type $g_i(x_1, x_2, \dots, x_n) = 0$ $(i = 1, \dots m)$ are given. Consider the function

$$F(x_1, x_2, \dots, x_n) = f(x_1, x_2, \dots, x_n) + \sum_{i=1}^{m} L_i \cdot g_i(x_1, x_2, \dots, x_n)$$

with m multipliers L_i, and determine the required extremes from

$$F_{x_i}(x_1, x_2, \dots, x_n) = 0 \qquad (i = 1, \dots, n).$$

Then, we have $n + m$ conditional equations for the m values L_i and the n values x_i, which then have to be solved for x_i.

▷ First eliminate the auxiliary quantities L_i.

12.7 Application of differential calculus

12.7.1 Calculation of indefinite expressions

Evaluation of limits of indefinite expressions of the type $\frac{0}{0}$ or $\frac{\infty}{\infty}$ by the Bernoulli-De l'Hôpital rule.

- **Bernoulli-De l'Hôpital rule**: If $\lim\limits_{x\to a} f(x) = \lim\limits_{x\to a} g(x) = 0$ or $\lim\limits_{x\to a} |f(x)| = \lim\limits_{x\to a} |g(x)| = \infty$, then (if it exists) the limit of the quotient of the functions can be written as the limit of the quotient of the first derivatives.

$$\lim_{x\to a} \frac{f(x)}{g(x)} = \lim_{x\to a} \frac{f'(x)}{g'(x)} .$$

▷ Denominator and numerator must be differentiated separately. Do not differentiate the fraction by means of the quotient rule!

The rule also applies to right and left limits if

$$\lim_{x\to 0\pm} f(x) = \lim_{x\to 0\pm} g(x) = 0.$$

If after applying the rule an indefinite form is obtained, then the rule can be applied several times, if the conditions for the derivatives are met once more.

□ $\lim\limits_{x\to\infty} \frac{x^n}{e^x} = \lim\limits_{x\to\infty} \frac{nx^{n-1}}{e^x} = \ldots = \lim\limits_{x\to\infty} \frac{n!}{e^x} = 0,$

the exponential function approaches infinity faster than any power function.

▷ Under certain circumstances, the limit cannot be determined by this rule, although a limit does exists.

For calculating with limits of sums, diffferences, products or quotients, theorems apply that are similar to those for number sequences.

	Definite form	Indefinite form
Sum	$c + \infty \to \infty, \quad \infty + \infty \to \infty$	$\infty - \infty$
Product	$\infty \cdot \infty \to \infty$	$0 \cdot \infty$
Quotient	$\frac{0}{\infty} \to 0, \quad \frac{\infty}{0} \to \infty$	$\frac{0}{0}, \quad \frac{\infty}{\infty}$
Power	$\infty^{\infty} \to \infty, \quad \infty^{-\infty} \to 0, \quad 0^{\infty} \to 0$	$1^{\infty}, \quad 0^0, \quad \infty^0$

▷ For improper quotients, the sign is sometimes not unique, because the right and left limits may differ.

□ $\lim\limits_{x\to 0+} \frac{\sin x}{x^2} = \infty, \quad \lim\limits_{x\to 0-} \frac{\sin x}{x^2} = -\infty$

The last row contains *indefinite forms*, which can be brought to the indefinite forms $\frac{0}{0}$ or $\frac{\infty}{\infty}$ as follows:

Indefinite form	Translation	Limit (if it exists)
$\infty - \infty$	$\frac{1}{f} - \frac{1}{g} = \frac{g-f}{f \cdot g}$	$\frac{g' - f'}{f \cdot g' + f' \cdot g}$
$0 \cdot \infty$	$f \cdot g = \frac{f}{1/g} = \frac{g}{1/f}$	$-\frac{f' \cdot g^2}{g'}, \quad -\frac{g' \cdot f^2}{f'}$
$1^\infty, \quad 0^0, \quad \infty^0$	$f^g = e^{g \cdot \ln f}$	$\exp\left\{-\frac{f' \cdot g^2}{f \cdot g'}\right\}$

The last column gives the limits according to the Bernoulli-De l'Hôpital rule (if the limit exists).

▷ The limit applies only if the numerator and the denominator of the conversion are differentiable.

□ $\lim\limits_{x \to 0+} x^x = \lim\limits_{x \to 0+} e^{x \ln x} = \exp\{\lim\limits_{x \to 0+} x \ln x\}$,

$$\lim_{x \to 0+} x \ln x = \lim_{x \to 0+} \frac{\ln x}{1/x} = \lim_{x \to 0+} \frac{1/x}{-1/x^2} = \lim_{x \to 0+} (-x) = 0,$$

$$\lim_{x \to 0+} x^x = e^0 = 1$$

▷ If an indefinite expression results again, the Bernoulli-De l'Hôpital rule must be applied several times.

□ $$\lim_{x \to 0} \left(\frac{1}{x^3} - \frac{1}{x^2}\right) = \lim_{x \to 0} \left(\frac{x^2 - x^3}{x^5}\right)$$

$$= \lim_{x \to 0} \left(\frac{2x - 3x^2}{5x^4}\right) = \lim_{x \to 0} \left(\frac{2 - 6x}{20x^3}\right) \to \infty$$

12.7.2 Discussion of curves

Required is an overview of the course of the path of a curve of a function curve.
Order in the discussion of curves (**not** obligatory):

1. **Domain of definition** (e.g., exclude zeros in the denominator or negative values in a radical).
2. **Symmetry** (even functions $f(-x) = f(x)$ or odd functions $f(-x) = -f(x)$): The path for negative x-values is obtained by reflection in the y-axis or the origin.
3. **Zeros**: The intersection points with the x-axis.
4. **Poles**: The function value approaches infinity.
5. **Behavior at infinity**: Asymptotes (if need be, apply the Bernoulli-De l'Hôpital rule).
6. **Behavior of the sign**: Sign of the function varies between zeros and poles.
7. **Monotonicity**: From the behavior of the sign of the derivative

   strictly monotonically decreasing: $f'(x) < 0$

   strictly monotonically increasing: $f'(x) > 0$

8. **Extremes:**

$$\textbf{relative maximum:} \quad \begin{cases} f^{(n)}(x_0) < 0, \ n \text{ even and} \\ f^{(k)}(x_0) = 0, \ \text{if } 1 \le k < n, \end{cases}$$

$$\textbf{relative minimum:} \quad \begin{cases} f^{(n)}(x_0) > 0, \ n \text{ even and} \\ f^{(k)}(x_0) = 0, \ \text{if } 1 \le k < n, \end{cases}$$

9. **Points of inflection:**

$$f''(x_W) = 0 \text{ and } f^{(n)}(x_W) \neq 0, n \text{ odd and } f^{(k)}(x_W) = 0, \ (2 < k < n).$$

12.7.3 Extreme value problems

In extreme value problems the extreme value is required for a certain problem, e.g., smallest amount of waste, highest energy gain (energy conversion efficiency), lowest consumption of energy, etc.

1. First determine the function that describes the problem.

   ▷ In practice, model approximation function is often chosen.

2. The zeros of the derivative yield the possible extreme points.

$$\text{Positions of extremes:} \qquad f'(x_i) = 0$$

   ▷ The zeros of the derivative yield the positions of extremes x_i, not the extreme values $f(x_i)$!

3. By means of the higher derivatives, check whether minima, maxima, or saddle points occur at these positions.

$$\textbf{Relative maximum:} \quad \begin{cases} f^{(n)}(x_0) < 0, \ n \text{ even, and} \\ f^{(k)}(x_0) = 0, \ \text{if } 1 \le k < n, \end{cases}$$

$$\textbf{Relative minimum:} \quad \begin{cases} f^{(n)}(x_0) > 0, \ n \text{ even, and} \\ f^{(k)}(x_0) = 0, \ \text{if } 1 \le k < n, \end{cases}$$

   ▷ In case of a maximum, the first nonvanishing higher derivative must be **negative**.

4. Calculate the function values of the found maxima (minima) as well as the boundary values of the function (boundary extremes). The greatest (smallest) value is the extreme value required.

Reduction for the determination of extremes.

Function	Extremes at x_0 with	Maxima/minima for
$f(x)$	$f'(x_0) = 0$	$f''(x_0) < 0 \quad (> 0)$
$\dfrac{1}{f(x)}$	$f'(x_0) = 0$	$-f''(x_0) < 0 \quad (> 0)$
$\dfrac{f(x)}{g(x)}$	$gf' - fg' = 0$	$gf'' - fg'' < 0 \quad (> 0)$
$\sqrt{f(x)}$	$f'(x_0) = 0 \quad (f(x_0) \neq 0)$	$f''(x_0) < 0 \quad (> 0)$

▷ For $+\sqrt{f(x_0)} = 0$, x_0 is an absolute minimum.

▷ If $f''(x_0) = 0$, an extreme is also possible, and then the higher derivatives must be considered.

□ Cut a rectangle with the largest possible area out of a semicircle.
Area of the rectangle: $f(x) = 2x\sqrt{r^2 - x^2} = 2\sqrt{r^2x^2 - x^4}$

Extremes: $(r^2x^2 - x^4)' = 2xr^2 - 4x^3 = 0 \rightarrow x_0 = r/\sqrt{2}$ (see the table above)

Maximum: $(2xr^2 - 4x^3)' = 2r^2 - 12x_0^2 = -4r^2 < 0$

Boundary values: $f(0) = f(r) = 0$, $f(x_0) = r^2$ (no boundary extremes)

The maximal area of the rectangle is $A = r^2$.

12.7.4 *Calculus of errors*

Every measured real quantity y has an error Δy (see the section on statistics in Chapter 21).
Absolute error: Δy = measured value minus exact value.
Relative error: $\Delta y / y$.
Percentage error: $\Delta y / y \cdot 100\%$.
In case of minor errors, the absolute and relative errors are approximately:

$$\Delta y \approx f(x)'\Delta x\,, \qquad \frac{\Delta y}{y} \approx \left(f'(x)\frac{x}{y}\right)\frac{\Delta x}{x}\,.$$

The **highest possible absolute error of a sum** is equal to the sum of the absolute errors:

$$y = f + g \rightarrow |\Delta y| \leq |\Delta f| + |\Delta g|\,.$$

The **highest possible relative error of a product or quotient** is equal to the sum of the relative errors:

$$y = f \cdot g, \quad y = \frac{f}{g} \rightarrow \left|\frac{\Delta y}{y}\right| \leq \left|\frac{\Delta f}{f}\right| + \left|\frac{\Delta g}{g}\right|\,.$$

The **relative error of a power** equals n times the relative error of the basis function:

$$y = f^n \rightarrow \left|\frac{\Delta y}{y}\right| = n\left|\frac{\Delta f}{f}\right|\,.$$

▷ Propagation of errors in a process $f(x)$: input x, output y, input error Δx or $\dfrac{\Delta x}{x}$, output error Δy or $\dfrac{\Delta y}{y}$.

12.7.5 Determination of zeros according to Newton's method

● If the derivative is known, then the zeros of a function $f(x)$ can be determined iteratively according to Newton:

$$x_{i+1} = x_i - \frac{f(x_i)}{f'(x_i)}, \quad i = 0, 1, 2, \ldots$$

Geometric interpretation: The $(i+1)$-th improved approximation to the zero is the zero of the tangent to the i-th value.

1. Estimate value x_0 of the zero.
2. Put $x_i = x_0$.
3. Calculate x_{i+1} according to the formula and set $x_i = x_{i+1}$.
4. Repeat the procedure until the error is sufficiently small.
5. Stop condition: $|x_{i+1} - x_i| < 10^{-n}|x_{i+1}|$, where n is the number of required exact places after the decimal point.

▷ For very slow convergence, the stop condition can be satisfied although the zero is not yet reached. Always look at the graph of the function.

▷ Sufficient condition for the convergence of Newton's method:

$$\left| \frac{f(x)\, f''(x)}{[f'(x)]^2} \right| \leq q\,, \quad q < 1\,, \quad x \in U$$

for all x values in the neighborhood U of the zero, which includes all points x_i.

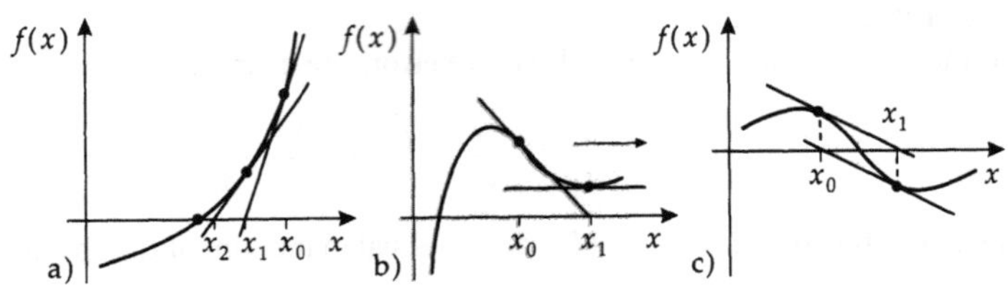

Newton's method Counterexample for Newton's method: (b) method diverges, (c) method oscillates

In some cases, the method converges very slowly or not at all. Then the starting value is too far from the zero x_0, or there is a multiple zero at x_0.

▷ If the zero x_e is a simple zero, i.e., $f(x_e) = 0$ and $f'(x_e) \neq 0$, then Newton's method is locally quadratically convergent (see **order of convergence**):

$$\varepsilon_{n+1} = \varepsilon_n^2 \frac{f''(x_e)}{2f'(x_e)}\,.$$

In the case of multiple zeros Newton's method converges only linearly:

$$\varepsilon_{n+1} = A\varepsilon_n \,, \quad |A| < 1 \,.$$

▷ The method may also oscillate or jump to regions outside the domain of definition of the function investigated (e.g. $y = \sqrt{x}$).

13

Differential geometry

The study of plane and space curves (and surfaces) using methods of differential calculus.

13.1 Plane curves

13.1.1 Representation of curves

A plane curve can be defined by means of the following methods:

Implicit representation	$F(x, y) = 0$
Explicit representation	$y = f(x)$
In parametric form	$x = x(t),\ y = y(t)$
In polar coordinates	$r = f(\phi)$

Positive direction of a curve
in parametric form: The direction in which the point $[(x(t), y(t))]$ of the curve [−] moves if the parameter value t increases.
In polar coordinates: The direction in which the angle ϕ increases.
Explicit representation: The direction of the x-axis (parameter x).

13.1.2 Differentiation by implicit representation

Partial derivatives:

$$F_x = \frac{\partial F(x, y)}{\partial x}, \qquad F_y = \frac{\partial F(x, y)}{\partial y}.$$

First derivative in explicit representation:

$$\frac{\mathrm{d}F(x,y)}{\mathrm{d}x} = F_x + F_y\frac{\mathrm{d}y}{\mathrm{d}x} = 0.$$

$$y' = \frac{\mathrm{d}y}{\mathrm{d}x} = -\frac{F_x}{F_y}.$$

13.1.3 Differentiation by parametric representation

Differentiation with respect to the parameter:

$$\frac{\mathrm{d}x}{\mathrm{d}t} = \dot{x} \qquad \frac{\mathrm{d}y}{\mathrm{d}t} = \dot{y}.$$

First derivative of a function by parametric representation:

$$y' = \frac{\mathrm{d}y}{\mathrm{d}x} = \frac{\mathrm{d}y/\mathrm{d}t}{\mathrm{d}x/\mathrm{d}t} = \frac{\dot{y}}{\dot{x}}.$$

Second derivative of a function by parametric representation:

$$y'' = \frac{\mathrm{d}^2y}{\mathrm{d}x^2} = \frac{\mathrm{d}}{\mathrm{d}x}\frac{\dot{y}}{\dot{x}} = \frac{\mathrm{d}(\dot{y}/\dot{x})/\mathrm{d}t}{\mathrm{d}x/\mathrm{d}t} = \frac{\dot{x}\ddot{y} - \ddot{x}\dot{y}}{\dot{x}^3}.$$

13.1.4 Differentiation by polar coordinates

Differentiation with respect to the angle ϕ:

$$\frac{\mathrm{d}r}{\mathrm{d}\phi} = \dot{r} \qquad \frac{\mathrm{d}^2r}{\mathrm{d}\phi^2} = \ddot{r}$$

First derivative of a curve by polar coordinates:

$$y' = \frac{\dot{y}}{\dot{x}} = \frac{\dot{r}\sin\phi + r\cos\phi}{\dot{r}\cos\phi - r\sin\phi}$$

Second derivative of a curve by polar coordinates:

$$y'' = \frac{\dot{x}\ddot{y} - \ddot{x}\dot{y}}{\dot{x}^3} = \frac{r^2 + 2\dot{r}^2 - r\ddot{r}}{(\dot{r}\cos\phi - r\sin\phi)^3}$$

13.1.5 Differential of arc of a curve

Differential of arc of a curve:

explicitly:	$\mathrm{d}s = \sqrt{1+y'^2}\mathrm{d}x$
in parametric form:	$\mathrm{d}s = \sqrt{\dot{x}^2+\dot{y}^2}\mathrm{d}t$
by polar coordinates:	$\mathrm{d}s = \sqrt{r^2+\dot{r}^2}\mathrm{d}\phi$

□ 1. $y = \mathrm{e}^x : \ \mathrm{d}s = \sqrt{1+\mathrm{e}^{2x}}\mathrm{d}x$

2. $x = t^2, \ y = t : \ \mathrm{d}s = \sqrt{4t^2+1}\mathrm{d}t$

3. $r = a\phi : \ \mathrm{d}s = a\sqrt{1+\phi^2}\mathrm{d}\phi$

13.1.6 Tangent, normal

Tangent: The secant PP' at a point P of the curve as $P' \to P$. The positive direction of the curve is also the positive direction of the tangent.
Normal: The straight line through P perpendicular to the tangent at P. The positive direction of the normal results from a 90° rotation of the positive direction of the tangent in mathematically positive rotational direction.
Equation of the tangent and the normal at the point (x_0, y_0):

Curve	Tangent equation	Normal equation
$F(x, y) = 0$	$F_x(x - x_0) + F_y(y - y_0) = 0$	$\dfrac{x - x_0}{F_x} = \dfrac{y - y_0}{F_y}$
$y = f(x)$	$y - y_0 = y'(x - x_0)$	$y - y_0 = -\dfrac{1}{y'}(x - x_0)$
$x(t), y(t)$	$\dfrac{y - y_0}{\dot{y}} = \dfrac{x - x_0}{\dot{x}}$	$\dot{x}(x - x_0) + \dot{y}(y - y_0) = 0$

Tangent intercept: Length of the tangent from the point (x_0, y_0) to where the tangent and the x-axis intersect, or the length of the perpendicular to the position vector r, traversing the pole:

$$T_l = \left|\frac{y}{y'}\sqrt{1 + y'^2}\right| \qquad T_l = \left|\frac{r}{\dot{r}}\sqrt{r^2 + \dot{r}^2}\right| .$$

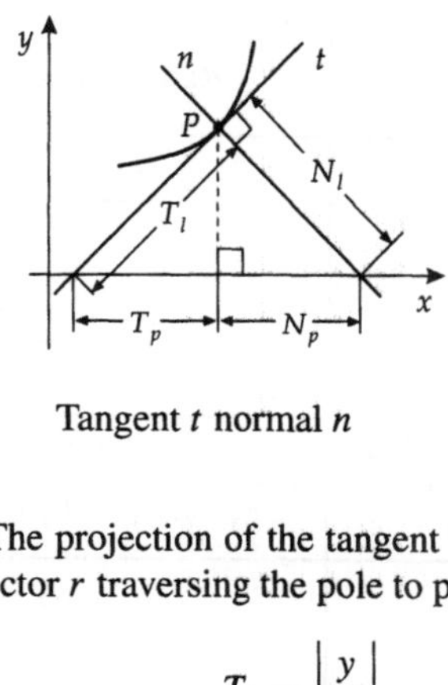

Tangent t normal n

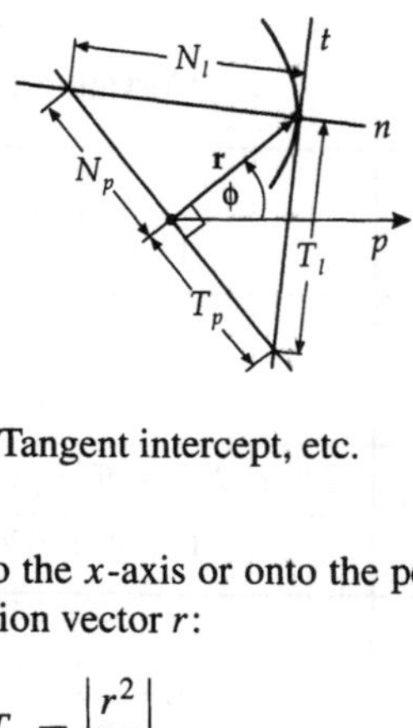

Tangent intercept, etc.

Subtangent: The projection of the tangent onto the x-axis or onto the perpendicular to the position vector r traversing the pole to position vector r:

$$T_p = \left|\frac{y}{y'}\right| , \qquad T_p = \left|\frac{r^2}{\dot{r}}\right| .$$

Normal intercept: Length of the normal from the point (x_0, y_0) to the intersection point with the x-axis, or the length of the perpendicular to the position vector r, traversing through the pole:

$$N_l = \left|y\sqrt{1 + y'^2}\right| , \qquad N_l = \left|\sqrt{r^2 + \dot{r}^2}\right| .$$

Subnormal: The projection of the normal onto the x-axis or onto the perpendicular to the position vector r, going through the pole:

$$N_p = \left|yy'\right| , \qquad N_p = |\dot{r}| .$$

□ $y = x^2 - 3x - 4$:
Equation of the tangent at the point $(0, -4)$:
$f'(x) = y' = 2x - 3,\ y'(0) = -3$
$y - (-4) = -3(x - 0) \rightarrow y_t = -3x - 4$
Equation of the normal at the point $(0, -4)$: $y_n = \frac{1}{3}x - 4$

Subtangent: $T_p = \left|\frac{y}{y'}\right| = \left|\frac{-4}{-3}\right| = \frac{4}{3}$

Tangent intercept: $T_l = \left|\frac{y}{y'}\sqrt{1 + y'^2}\right| = \left|\frac{-4}{-3}\sqrt{10}\right| \approx 4.22$

Normal intercept: $N_l = \left|y\sqrt{1 + y'^2}\right| = |-4|\sqrt{10} \approx 12.7$

Subnormal: $N_p = |yy'| = |(-4)(-3)| = 12.$

13.1.7 Curvature of a curve

Descriptive definition: The curvature of a curve is its local deviation from a straight line!
The curvature of a curve is the limit of the ratio of angle α between the positive directions of the tangents at the points P and P' to the arc length s as $P \rightarrow P'$:

$$K = \lim_{P \rightarrow P'} \frac{\Delta\alpha}{\Delta s} = \frac{d\alpha}{ds}\,.$$

Curvature of a curve in the various representations:

Curve	Curvature K	Radius of curvature R
$y = f(x)$	$\dfrac{y''}{(1 + y'^2)^{3/2}}$	$\dfrac{(1 + y'^2)^{3/2}}{\lvert y''\rvert}$
$x(t), y(t)$	$\dfrac{\dot{x}\ddot{y} - \ddot{x}\dot{y}}{(\dot{x}^2 + \dot{y}^2)^{3/2}}$	$\dfrac{(\dot{x}^2 + \dot{y}^2)^{3/2}}{\lvert\dot{x}\ddot{y} - \ddot{x}\dot{y}\rvert}$
$F(x, y) = 0$	$\dfrac{2F_xF_yF_{xy} - F_x^2F_{yy} - F_y^2F_{xx}}{(F_x^2 + F_y^2)^{3/2}}$	$\dfrac{(F_x^2 + F_y^2)^{3/2}}{\lvert 2F_xF_yF_{xy} - F_x^2F_{yy} - F_y^2F_{xx}\rvert}$
$r = r(\phi)$	$\dfrac{r^2 + 2\dot{r}^2 - r\ddot{r}}{(\dot{r}^2 + r^2)^{3/2}}$	$\dfrac{(\dot{r}^2 + r^2)^{3/2}}{\lvert r^2 + 2\dot{r}^2 - r\ddot{r}\rvert}$
$r(t), \phi(t)$	$\dfrac{\dot{\phi}(2\dot{r}^2 + r^2\dot{\phi}^2) + r(\dot{r}\ddot{\phi} - \dot{\phi}\ddot{r})}{(\dot{r}^2 + r^2\dot{\phi}^2)^{3/2}}$	$\dfrac{(\dot{r}^2 + r^2\dot{\phi}^2)^{3/2}}{\lvert\dot{\phi}(2\dot{r}^2 + r^2\dot{\phi}^2) + r(\dot{r}\ddot{\phi} - \dot{\phi}\ddot{r})\rvert}$

Possible curvatures:

1. zero at all points $\Longrightarrow$ straight line.
2. equal to a constant at all points $\Longrightarrow$ circle.
3. changes from point to point $\Longrightarrow$ general case of a curve.

▷ If $f'(x) \rightarrow \infty$, it is practical to choose another representation!

□ Curvature of the spiral of Archimedes $r = a\phi$: $\dot{r} = a, \ddot{r} = 0$,

$$K = \frac{a^2\phi^2 - 0 + 2a^2}{(a^2 + a^2\phi^2)^{3/2}} = \frac{\phi^2 + 2}{a(1 + \phi^2)^{3/2}}.$$

$K < 0$ Right-handed curvature: The convex side of the curve points in the direction of the positive normal of the curve, i.e., the center of curvature lies on the negative side of the normal of the curve.

$K > 0$ Left-handed curvature: The concave side of the curve points in the direction of the positive normal of the curve, i.e., the center of curvature lies on the positive side of the normal of the curve.

The spiral of Archimedes has a left-handed curvature ($K > 0$) when $a > 0$. The shaded branch shown in the figure of the spiral of Archimedes (Section 5.40) has a negative curvature, $K < 0$, right-handed curvature.

Circle of curvature: Touches the curve at one point such that the second derivatives (curvatures) coincide (contact of the second order).

Radius of curvature (radius of the circle of curvature): Equal to the absolute reciprocal value of curvature K:

$$R = \frac{1}{|K|}.$$

Coordinates of the center of curvature (x_K, y_K) for $y = f(x)$, $(x(t), y(t))$, $r = r(\phi)$ and $F(x, y) = 0$:

x_K	y_K
$x - y'\dfrac{1 + y'^2}{y''}$	$y + \dfrac{1 + y'^2}{y''}$
$x - \dot{y}\dfrac{\dot{x}^2 + \dot{y}^2}{\dot{x}\ddot{y} - \dot{y}\ddot{x}}$	$y + \dot{x}\dfrac{\dot{x}^2 + \dot{y}^2}{\dot{x}\ddot{y} - \dot{y}\ddot{x}}$
$r\cos\phi - \dfrac{(\dot{r}\sin\phi + r\cos\phi)(\dot{r}^2 + r^2)}{r^2 - r\ddot{r} + 2\dot{r}^2}$	$r\sin\phi + \dfrac{(\dot{r}\cos\phi - r\sin\phi)(\dot{r}^2 + r^2)}{r^2 - r\ddot{r} + 2\dot{r}^2}$
$x + \dfrac{F_x(F_x^2 + F_y^2)}{2F_xF_yF_{xy} - F_x^2F_{yy} - F_y^2F_{xx}}$	$y + \dfrac{F_y(F_x^2 + F_y^2)}{2F_xF_yF_{xy} - F_x^2F_{yy} - F_y^2F_{xx}}$

▷ The formulas can also be written as (R: radius of curvature)

$$x_K = x - R\frac{\mathrm{d}y}{\mathrm{d}s}, \qquad y_K = y + R\frac{\mathrm{d}x}{\mathrm{d}s}$$

where R is the radius of curvature.

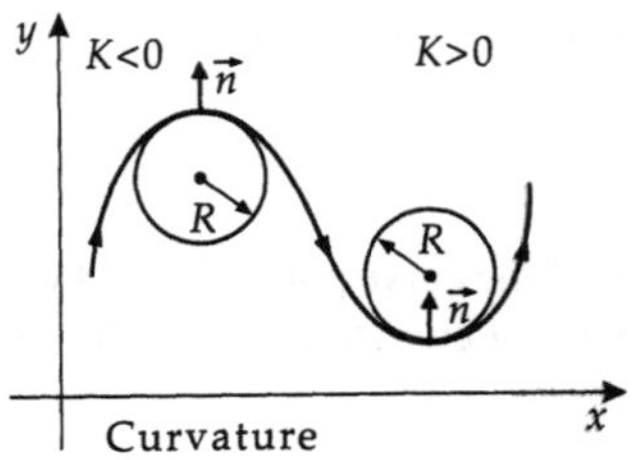

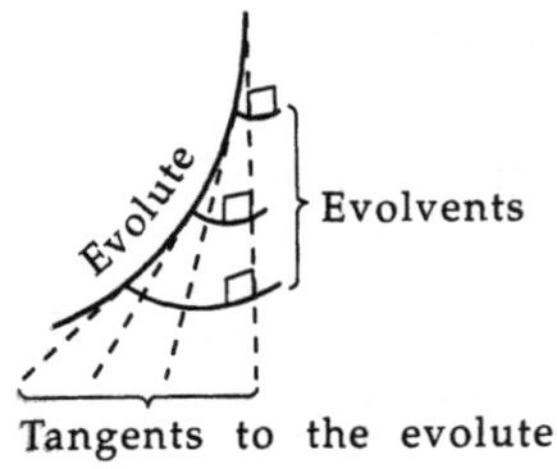

13.1.8 Evolutes and evolvents

Evolvent: The original curve.
Evolute: Curve consisting of all centers of curvature of a given curve (the evolvent).

- The normals to the evolvent are the tangents to the evolute.
- The evolute is the envelope of the normals to the evolvent.

□ Evolute of the parabola $y = \frac{1}{2}x^2$:
$x_k = x - (1 + x^2)x = -x^3$,
$y_k = y + (1 + x^2) = \frac{3}{2}x^2 + 1 = \frac{3}{2}x_k^{3/2} + 1.$

□ Circle $r = r_0$, $\dot{r} = 0$:
$x_k = r\cos\phi - r\cos\phi = 0$, $y_k = r\sin\phi - r\sin\phi = 0$
The evolute of a circle is the center of the circle.

13.1.9 Points of inflection, vertices

Points of inflection: Those points of a curve at which the sign of the curvature changes. The curve does not lie on one side of the tangent, but the tangent intersects the curve.

- At the point of inflection, it holds that $K = 0$ and $R \to \infty$ (necessary condition).

Curve	Possible points of inflection for
$y = f(x)$	$y'' = 0$
$x(t), y(t)$	$\dot{x}\ddot{y} - \dot{y}\ddot{x} = 0$
$r = f(\phi)$	$r^2 + 2\dot{r}^2 - r\ddot{r} = 0$
$F(x, y) = 0$	$2F_xF_yF_{xy} - F_x^2F_yy - F_y^2F_xx = 0$

Vertex: Those points of the curve at which the curvature has a maximum or minimum. Vertices are determined by calculating the extreme values of the curvature.
Necessary condition for a vertex of the curve $y = f(x)$:

$$3y'y''^2 = (1 + y'^2)y''' .$$

13.1.10 Singular points

Singular points:

(a) **Double points**: Points at which the curve intersects itself.

(b) **Isolated points**: Points that lie outside the curve.

(c) **Cusps**: Points at which the curve changes direction sign (equal tangents).

(d) **Points of tangency**: Points in which the curve touches itself.

(e) **Salient points**: Points at which the curve abruptly changes direction (different tangents).

(f) **Asymptotic points**: Those points about which the curve twists infinitely.

Determination of singular points of type (a) to (d):
Investigation of a curve of form $F(x, y) = 0$:
Necessary conditional equations:

$$F = 0 \text{ and } F_x = 0 \text{ and } F_y = 0,$$

and at least one of the three derivatives of the second order does not vanish:

$$F_{xx} \neq 0 \text{ or } F_{yy} \neq 0 \text{ or } F_{xy} \neq 0.$$

The properties of the multiple point depend on the sign of

$$\Delta = F_{xx} \cdot F_{yy} - F_{xy} \cdot F_{yx}\ .$$

1. $\Delta > 0$: Isolated point;

2. $\Delta = 0$: Cusp or point of tangency, slope of the tangent

$$m = -\frac{F_{xy}}{F_{yy}}.$$

3. $\Delta < 0$: Double point, slope of the tangent m from

$$F_{yy}m^2 + 2F_{xy}m + F_{xx} = 0.$$

▷ Other types of representations must be transformed to the implicit representation.

□ $F(x, y) = x^3 + y^3 - x^2 - y^2 = 0$:
$F_x = x(3x - 2)$, $F_y = y(3y - 2) \to (0, 0)$, $(0, 2/3)$, $(2/3, 0)$, $(2/3, 2/3)$,
but only the first point lies on the curve!
$F_{xx}(0, 0) = -2 \neq 0$, $\Delta = 4 > 0 \to$ isolated point.

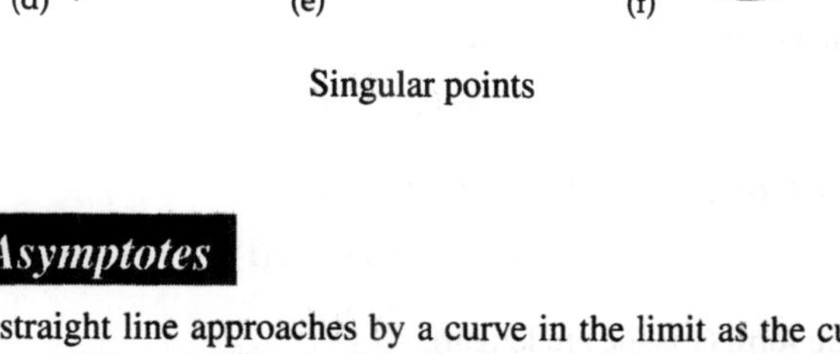

Singular points

13.1.11 Asymptotes

Asymptote: A straight line approaches by a curve in the limit as the curve approaches infinity. This is possible only if one of the branches of the curve moves to an infinite distance from the origin.

Determination in parametric representation: Determination of the values t_i for which it holds that $x(t) \to \infty$ or $y(t) \to \infty$.
Horizontal asymptote $y = a$: $x(t_i) \to \infty,\ y(t_i) = a < \infty$.
Vertical asymptote $x = a$: $x(t_i) = a < \infty,\ y(t_i) \to \infty$.
Straight line $y = mx + b$: $x(t_i) \to \infty,\ y(t_i) \to \infty$, if the limits

$$m = \lim_{t \to t_i} \frac{y(t)}{x(t)} \text{ and } b = \lim_{t \to t_i} (y(t) - m \cdot x(t))$$

exist and are finite.

□ $x = \dfrac{1}{\cos t},\ y = \tan t - t$: $t_1 = \pi/2,\ t_2 = -\pi/2$ etc.
$x(t_1) \to \infty,\ y(t_1) \to \infty$

$$m = \lim_{t \to \pi/2} (\sin t - t \cos t) = 1$$

$$b = \lim_{t \to \pi/2} (\tan t - t - \frac{1}{\cos t}) = \lim_{t \to \pi/2} \frac{\sin t - t \cos t - 1}{\cos t} = -\frac{\pi}{2}$$

Asymptote $y = mx + b = x - \dfrac{\pi}{2}$.

▷ Other types of representation must be transformed.

13.1.12 *Envelope of a family of curves*

The geometric loci of the boundary points of a one-parameter family of curves in the form of

$$F(x, y, a) = 0$$

are again a curve (or several curves), consist of the singular points, or are the envelope of the family of curves.
Equation of the envelope from:

$$F(x, y, a) = 0 \qquad \text{and} \qquad \frac{\partial}{\partial a} F(x, y, a) = 0.$$

by elimination of parameter a.

13.2 Space curves

13.2.1 *Representation of space curves*

Space curves can be defined by means of the following methods:

(a) Intersection of two surfaces in space:

$$F(x, y, z) = 0, \qquad G(x, y, z) = 0\,.$$

(b) Parametric form (t is an arbitrary parameter):

$$x(t), \quad y(t), \quad z(t)\,.$$

(c) Parametric form (s is the arc length):

$$x(s), \quad y(s), \quad z(s)\,.$$

(d) Vector equation (of the parametric form):

$$\vec{\mathbf{r}}(t) = x(t)\vec{\mathbf{e}}_x + y(t)\vec{\mathbf{e}}_y + z(t)\vec{\mathbf{e}}_z = (x(t), y(t), z(t)) .$$

□ Helical line: $x(t) = a\cos t$, $y(t) = a\sin t$, $z(t) = bt$.

Positive direction: Direction of increasing values of the parameter t along a space curve given in parametric form.

Derivative of a space curve in vector form:

$$\frac{\mathrm{d}}{\mathrm{d}t}\vec{\mathbf{r}} = \dot{\vec{\mathbf{r}}} = (\dot{x}(t), \dot{y}(t), \dot{z}(t)) .$$

Arc length of a space curve:

$$s = \int_{t_0}^{t} \sqrt{\dot{x}^2 + \dot{y}^2 + \dot{z}^2}\mathrm{d}t.$$

□ Helical line:

$$s = \int_0^t \sqrt{a^2\sin^2 t + a^2\cos^2 t + b^2}\mathrm{d}t = \int_0^t \sqrt{a^2 + b^2}\mathrm{d}t = t\sqrt{a^2 + b^2} = tc.$$

Representation in parametric form: $x(s) = a\cos(s/c)$, $y(s) = a\sin(s/c)$, $z(s) = bs/c$.

13.2.2 Moving trihedral

At every point of a space curve P (except the singular points), three lines and three planes that are mutually perpendicular can be defined.

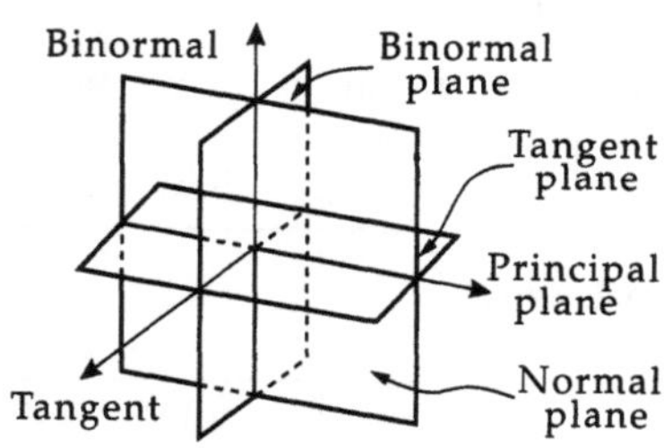

Moving trihedral

Tangent: Limiting position of the secant through P and P' for the limit $P' \to P$.
Normal plane: Perpendicular to the tangent.
Tangent plane: The limiting position of a plane passing through three neighboring points P, P' and P'' of the curve for limits $P' \to P$ and $P'' \to P$.
Principal normal: Intersection line of the normal plane and the tangent plane.
Binormal: Perpendicular to the tangent plane.
Binormal plane: Perpendicular to the principal normal.
Moving trihedral: Consists of tangent, normal, and binormal vectors.
Tangent vector $\vec{\mathbf{T}}$: Unit vector along the positive direction of the tangent;
Normal vector $\vec{\mathbf{N}}$: Unit vector in the principal normal in the direction of the concave side of the curve.

Binormal vector $\vec{\mathbf{B}}$: Unit vector perpendicular to the normal vector and to the tangent vector.

$$\vec{\mathbf{B}} = \vec{\mathbf{T}} \times \vec{\mathbf{N}}$$

Equations of the moving trihedral at the point $\vec{\mathbf{r}}_0 = (x_0, y_0, z_0)$ for space curves in parametric representation:

Vector equation	Equation in coordinates
Tangent	
$\vec{\mathbf{r}} = \vec{\mathbf{r}}_0 + \lambda\dot{\vec{\mathbf{r}}}_0$	$\dfrac{x - x_0}{\dot{x}_0} = \dfrac{y - y_0}{\dot{y}_0} = \dfrac{z - z_0}{\dot{z}_0}$
Normal plane	
$(\vec{\mathbf{r}} - \vec{\mathbf{r}}_0)\dot{\vec{\mathbf{r}}}_0 = 0$	$\dot{x}_0(x - x_0) + \dot{y}_0(y - y_0) + \dot{z}_0(z - z_0) = 0$
Tangent plane	
$(\vec{\mathbf{r}} - \vec{\mathbf{r}}_0)\dot{\vec{\mathbf{r}}}_0\ddot{\vec{\mathbf{r}}}_0 = 0$	$\begin{vmatrix} x - x_0 & y - y_0 & z - z_0 \\ \dot{x}_0 & \dot{y}_0 & \dot{z}_0 \\ \ddot{x}_0 & \ddot{y}_0 & \ddot{z}_0 \end{vmatrix} = 0$
Binormal	
$\vec{\mathbf{r}} = \vec{\mathbf{r}}_0 + \lambda(\dot{\vec{\mathbf{r}}}_0 \times \ddot{\vec{\mathbf{r}}}_0)$	$\dfrac{x - x_0}{\dot{y}_0\ddot{z}_0 - \ddot{y}_0\dot{z}_0} = \dfrac{y - y_0}{\dot{z}_0\ddot{x}_0 - \ddot{z}_0\dot{x}_0} = \dfrac{z - z_0}{\dot{x}_0\ddot{y}_0 - \ddot{x}_0\dot{y}_0} = 0$
Binormal plane	
$(\vec{\mathbf{r}} - \vec{\mathbf{r}}_0)\dot{\vec{\mathbf{r}}}_0(\dot{\vec{\mathbf{r}}}_0 \times \ddot{\vec{\mathbf{r}}}_0) = 0$	$\begin{vmatrix} x - x_0 & y - y_0 & z - z_0 \\ \dot{x}_0 & \dot{y}_0 & \dot{z}_0 \\ \dot{y}_0\ddot{z}_0 - \ddot{y}_0\dot{z}_0 & \dot{z}_0\ddot{x}_0 - \ddot{z}_0\dot{x}_0 & \dot{x}_0\ddot{y}_0 - \ddot{x}_0\dot{y}_0 \end{vmatrix} = 0$
Principal normal	
$\vec{\mathbf{r}} = \vec{\mathbf{r}}_0 + \lambda\dot{\vec{\mathbf{r}}}_0 \times (\dot{\vec{\mathbf{r}}}_0 \times \ddot{\vec{\mathbf{r}}}_0)$	$\dfrac{x - x_0}{\dot{y}_0(\dot{x}_0\ddot{y}_0 - \ddot{x}_0\dot{y}_0) - \dot{z}_0(\dot{z}_0\ddot{x}_0 - \ddot{z}_0\dot{x}_0)} = \dfrac{y - y_0}{\dot{z}_0(\dot{y}_0\ddot{z}_0 - \ddot{y}_0\dot{z}_0) - \dot{x}_0(\dot{x}_0\ddot{y}_0 - \ddot{x}_0\dot{y}_0)} = \dfrac{z - z_0}{\dot{x}_0(\dot{z}_0\ddot{x}_0 - \ddot{z}_0\dot{x}_0) - \dot{y}_0(\dot{y}_0\ddot{z}_0 - \ddot{y}_0\dot{z}_0)}$

The following elements are simplified in a parametric representation with the arc lengths as parameters:

Element	Vector equation	Equation in coordinates
Binormal plane	$(\vec{\mathbf{r}} - \vec{\mathbf{r}}_0)\ddot{\vec{\mathbf{r}}}_0 = 0$	$\ddot{x}_0(x - x_0) + \ddot{y}_0(y - y_0) + \ddot{z}_0(z - z_0) = 0$
Principal normal	$\vec{\mathbf{r}} = \vec{\mathbf{r}}_0 + \lambda\ddot{\vec{\mathbf{r}}}_0$	$\dfrac{x - x_0}{\ddot{x}_0} = \dfrac{y - y_0}{\ddot{y}_0} = \dfrac{z - z_0}{\ddot{z}_0}$

Equations of the moving trihedral at the point $P(x_0, y_0, z_0)$ for space curves given in terms of the intersection line of two surfaces $F(x, y, z) = 0$, $G(x, y, z) = 0$:
Tangent:

$$\frac{x - x_0}{F_y G_z - F_z G_y} = \frac{y - y_0}{F_z G_x - F_x G_z} = \frac{z - z_0}{F_x G_y - F_y G_x}$$

Normal plane:

$$\begin{vmatrix} x - x_0 & y - y_0 & z - z_0 \\ F_x & F_y & F_z \\ G_x & G_y & G_z \end{vmatrix} = 0$$

13.2.3 Curvature

Curvature of a curve: Measures the deviation from a straight line. Precise definition:

$$K = \frac{\mathrm{d}\vec{\mathbf{T}}}{\mathrm{d}s}\,.$$

Radius of curvature: Reciprocal value of the curvature:

$$R = \frac{1}{K}\,.$$

Calculation of the curvature for $\vec{\mathbf{r}} = \vec{\mathbf{r}}(s)$:

$$K = \left|\frac{\mathrm{d}^2\vec{\mathbf{r}}}{\mathrm{d}s^2}\right| = \sqrt{\ddot{x}^2 + \ddot{y}^2 + \ddot{z}^2}$$

(derivatives with respect to s).
Calculation of the curvature for $\vec{\mathbf{r}} = \vec{\mathbf{r}}(t)$:

$$\begin{aligned} K^2 &= \frac{(\dot{\vec{\mathbf{r}}})^2(\ddot{\vec{\mathbf{r}}})^2 - (\dot{\vec{\mathbf{r}}}\ddot{\vec{\mathbf{r}}})^2}{(\dot{\vec{\mathbf{r}}})^6} \\ &= \frac{(\dot{x}^2 + \dot{y}^2 + \dot{z}^2)(\ddot{x}^2 + \ddot{y}^2 + \ddot{z}^2) - (\dot{x}\ddot{x} + \dot{y}\ddot{y} + \dot{z}\ddot{z})^2}{(\dot{x}^2 + \dot{y}^2 + \dot{z}^2)^3} \end{aligned}$$

(derivatives with respect to t).

▷ The radius of curvature of a space curve, as defined here, is always positive.

□ Curvature of the helix: $\vec{\mathbf{r}}(s) = (a\cos(s/c), a\sin(s/c), bs/c)$
$K = \sqrt{(a^2/c^4) + 0} = a/c^2$.

13.2.4 Torsion of a curve

Torsion of a curve: Measures the deviation from a plane curve.
Definition of torsion:

$$T = \frac{\mathrm{d}\vec{\mathbf{B}}}{\mathrm{d}s}\,.$$

Radius of torsion: Reciprocal value of the torsion:

$$\tau = \frac{1}{T}\,.$$

Calculation of torsion for $\vec{\mathbf{r}} = \vec{\mathbf{r}}(s)$:

$$T = R^2 \left(\frac{\mathrm{d}\vec{\mathbf{r}}}{\mathrm{d}s} \frac{\mathrm{d}^2\vec{\mathbf{r}}}{\mathrm{d}s^2} \frac{\mathrm{d}^3\vec{\mathbf{r}}}{\mathrm{d}s^3} \right) = \frac{1}{\ddot{x}^2 + \ddot{y}^2 + \ddot{z}^2} \begin{vmatrix} \dot{x} & \dot{y} & \dot{z} \\ \ddot{x}_0 & \ddot{y}_0 & \ddot{z}_0 \\ \dddot{x}_0 & \dddot{y}_0 & \dddot{z}_0 \end{vmatrix} .$$

Calculation of torsion for $\vec{\mathbf{r}} = \vec{\mathbf{r}}(t)$:

$$T = \frac{R^2}{(\dot{\vec{\mathbf{r}}})^6} \left(\frac{\mathrm{d}\vec{\mathbf{r}}}{\mathrm{d}s} \frac{\mathrm{d}^2\vec{\mathbf{r}}}{\mathrm{d}s^2} \frac{\mathrm{d}^3\vec{\mathbf{r}}}{\mathrm{d}s^3} \right) = \frac{R^2}{(\dot{x}^2 + \dot{y}^2 + \dot{z}^2)^3} \begin{vmatrix} \dot{x} & \dot{y} & \dot{z} \\ \ddot{x}_0 & \ddot{y}_0 & \ddot{z}_0 \\ \dddot{x}_0 & \dddot{y}_0 & \dddot{z}_0 \end{vmatrix} .$$

□ Torsion of the helix:

$$T = \frac{(a^2+b^2)^2}{a} \frac{ab\cos^2 t + ab\sin^2 t}{[(-a\sin t)^2 + (a\cos t)^2 + b^2]^3} = \frac{b}{a^2+b^2} = \frac{b}{c} .$$

The helix has a constant torsion.

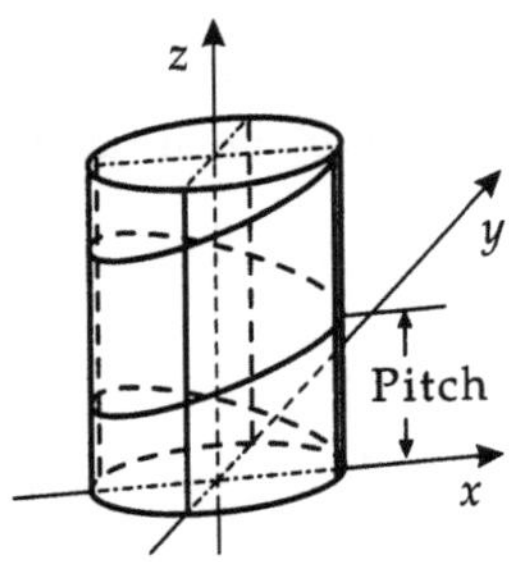

Helix

▷ Torsion can be positive or negative, depending on whether the sense of the torsion is clockwise or counterclockwise.

13.2.5 Frenet formulas

Frenet formulas: Relationships between the elements of the moving trihedral and their derivatives with respect to arc length s:

$$\frac{\mathrm{d}\vec{\mathbf{T}}}{\mathrm{d}s} = \frac{\vec{\mathbf{N}}}{R}, \quad \frac{\mathrm{d}\vec{\mathbf{N}}}{\mathrm{d}s} = \frac{\vec{\mathbf{T}}}{R} - \frac{\vec{\mathbf{B}}}{\tau}, \quad \frac{\mathrm{d}\vec{\mathbf{B}}}{\mathrm{d}s} = -\frac{\vec{\mathbf{N}}}{\tau},$$

where R denotes the radius of curvature and τ the radius of torsion.

13.3 Surfaces

13.3.1 Representation of a surface

A surface can be defined by means of the following methods:

Implicit representation	$F(x, y, z) = 0$
Explicit representation	$z = f(x, y)$
Parametric form	$x = x(u, v),\ y = y(u, v),\ z = z(u, v)$
Vector form	$\vec{\mathbf{r}} = \vec{\mathbf{r}}(u, v)$

□ Equation of the sphere: $x^2 + y^2 + z^2 - R^2 = 0$.
In parametric form: $x = R \cos u \sin v$, $y = R \sin u \sin v$, $z = R \cos v$.

Curvilinear coordinates: Parameter values u, v.

Coordinate lines: Space curves on the surface for fixed values of u (v-lines) or fixed values of v (u-lines); they cover the surface with a network of lines.

For $z = f(x, y)$, the coordinate lines are the intersection lines of the surface and the planes with constant x values and y values.

□ Parametric representation of a sphere (globe):
u is the geographical longitude,
v is the geographical latitude.

13.3.2 Tangent plane and normal to the surface

Tangent plane: Plane spanned by all possible tangents to the plane at the point P if the point is not a conical point.

$$\vec{\mathbf{t}}_u = \frac{\partial \vec{\mathbf{r}}}{\partial u}, \qquad \vec{\mathbf{t}}_v = \frac{\partial \vec{\mathbf{r}}}{\partial v}.$$

Normal to the surface: Straight line passing through point P of a plane perpendicular to the tangent plane.

Normal vector: Normalized vector product of the tangent vectors:

$$\vec{\mathbf{n}} = \frac{\vec{\mathbf{t}}_u \times \vec{\mathbf{t}}_v}{|\vec{\mathbf{t}}_u \times \vec{\mathbf{t}}_v|}.$$

Equation of the tangent plane at point $\vec{\mathbf{r}}_0 = (x_0, y_0, z_0)$:

Surface	Tangent plane
$F(x, y, z) = 0$	$F_x(x - x_0) + F_y(y - y_0) + F_z(z - z_0) = 0$
$z = f(x, y)$	$z - z_0 = \frac{\partial z}{\partial x}(x - x_0) + \frac{\partial z}{\partial y}(y - y_0)$
$x(u, v), y(u, v) z(u, v)$	$\begin{vmatrix} x - x_0 & y - y_0 & z - z_0 \\ x_u & y_u & z_u \\ x_v & y_v & z_v \end{vmatrix} = 0$
$\vec{\mathbf{r}} = \vec{\mathbf{r}}(u, v)$	$(\vec{\mathbf{r}} - \vec{\mathbf{r}}_0) \cdot \left(\frac{\partial \vec{\mathbf{r}}}{\partial u} \times \frac{\partial \vec{\mathbf{r}}}{\partial v} = 0 \right)$

Equation of the normal to the surface at point $\vec{r}_0 = (x_0, y_0, z_0)$:

Surface	Normal to the surface
$F(x, y, z) = 0$	$\dfrac{x - x_0}{F_x} = \dfrac{y - y_0}{F_y} = \dfrac{z - z_0}{F_z}$
$z = f(x, y)$	$\dfrac{x - x_0}{\partial z/\partial x} = \dfrac{y - y_0}{\partial z/\partial y} = \dfrac{z - z_0}{-1}$
$x(u, v), y(u, v)z(u, v)$	$\dfrac{x - x_0}{y_u z_v - y_v z_u} = \dfrac{y - y_0}{z_u x_v - z_v x_u} = \dfrac{z - z_0}{x_u y_v - x_v y_u}$
$\vec{r} = \vec{r}(u, v)$	$\vec{r} = \vec{r}_0 + \lambda \left(\dfrac{\partial \vec{r}_0}{\partial u} \times \dfrac{\partial \vec{r}_0}{\partial v} \right)$

The partial derivatives with respect to u and v are denoted by the indices u and v, respectively.

□ Sphere: $x^2 + y^2 + z^2 - R^2 = 0$.
Tangent plane: $2x_0(x - x_0) + 2y_0(y - y_0) + 2z_0(z - z_0) = 0$,
or $x_0 x + y_0 y + z_0 z - R^2 = 0$.
Normal to the surface: $\dfrac{x - x_0}{2x_0} = \dfrac{y - y_0}{2y_0} = \dfrac{z - z_0}{2z_0}$,
or $\dfrac{x}{x_0} = \dfrac{y}{y_0} = \dfrac{z}{z_0}$.

13.3.3 *Singular points of the surface*

Singular points of the surface or **conical points**: Point of a surface, given in the form of $F(x, y, z) = 0$, with the following properties:

$$F_x = F_y = F_z = F(x, y, z) = 0.$$

All tangents passing through the singular point form a cone of the second order.
If all six possible second-order partial derivatives vanish, then the result is a conical point of higher order.

14

Infinite series

Series expansions are used frequently in practice, for example, for the integration in terms of series expansions or for approximation formulas (see Chapter 5 on functions and series expansions as well as the tables in this chapter).

14.1 Series

Series are **sum sequences**, (s_n)

$$s_1 = a_1 , \quad s_2 = a_1 + a_2 , \quad s_3 = a_1 + a_2 + a_3 , \quad \ldots ,$$

for which the n-th term of the sequence (s_n) consists of the sum of the first n terms of a sequence (a_n)

$$s_n = a_1 + a_2 + a_3 + \cdots + a_n = \sum_{i=1}^{n} a_i = s_{n-1} + a_n .$$

The **infinite series** is obtained as $n \to \infty$:

$$a_1 + a_2 + \cdots + a_n + \cdots = \sum_{i=1}^{\infty} a_i \quad .$$

Partial sums: The finite sums s_n of the infinite series.
Remainder R_n of a convergent series: Difference between the sum s and the partial sum s_n.
Main problems for infinite series:

- Investigating the series for convergence.
- Determinating value s of the sum.

▷ Not only can the summands be numbers, but they may also contain variables and consist of functions or other mathematical objects (matrices)!

14.2 Criteria of convergence

Convergence of the series to the **sum** s, exists if the **sum** sequence (s_n) converges to s:

$$\lim_{n\to\infty} s_n = s = \sum_{i=1}^{\infty} a_i \ .$$

Divergence of the series: The partial sum s_n has no limit as $n \to \infty$.

□ **Geometric series**: The quotient q of two successive terms of the sequence is a constant value:

$$aq^0 + aq^1 + aq^2 + \cdots + aq^{n-1} + \cdots = \sum_{i=1}^{\infty} aq^{i-1} = a \cdot \sum_{i=1}^{\infty} q^{i-1} \ .$$

Partial sum s_n of a geometric series:

$$s_n = a \sum_{i=1}^{n} q^{i-1} = a \cdot \frac{1-q^n}{1-q} \quad , \quad q \neq 1 \ .$$

● The geometric series converges to the sum s if the absolute value of q is smaller than one:

$$s = \frac{a}{1-q} \quad , \quad |q| < 1 \ .$$

□ $a = 1,\ q = \dfrac{1}{2}$:

$$s_n = 1 + \frac{1}{2} + \frac{1}{4} + \frac{1}{8} + \cdots + \frac{1}{2^{n-1}} = \frac{1-(\frac{1}{2})^n}{1-\frac{1}{2}} = 2 \cdot \left(1 - \frac{1}{2^n}\right)$$

converges to

$$s = \frac{1}{1-\frac{1}{2}} = 2 \ .$$

● **Necessary condition** for the convergence of a sum: The terms a_i of the sequence form a **null sequence**:

$$\lim_{n\to\infty} a_n = 0 \ .$$

▷ This minimum condition is not sufficient in every case.

□ The harmonic series

$$s_n 1 + \frac{1}{2} + \frac{1}{3} + \cdots = \sum_{k=1}^{\infty} \frac{1}{k} \quad \text{is divergent.}$$

▷ If the condition for a null sequence is not met, then the series is always divergent.

● **Test for divergence:**

$$\lim_{n\to\infty} a_n \neq 0 \quad \Rightarrow \quad \sum_{i=1}^{\infty} a_i \quad \text{is divergent.}$$

□ The series $s = 1 + 1 + 1 + \cdots$ is divergent.

In order to consider the convergence of a series, other series can be taken into consideration whose convergence or divergence is already known (comparison criterion).

- **Majorant criterion** (comparison criterion): The series

$$\sum_{i=1}^{\infty} a_i$$

is absolutely convergent if there is a convergent series

$$\sum_{i=1}^{\infty} b_i\,, \qquad b_i \geq 0 \text{ for } i = 1, 2, 3, \ldots$$

whose terms b_i—from an index i onward—are greater than or equal to the absolute values of the terms a_i of the series investigated,

$$b_i \geq |a_i|\,.$$

- **Leibniz's criterion for series with alternating signs,**

$$\sum_{i=1}^{\infty} (-1)^{i+1} a_i = a_1 - a_2 + a_3 - a_4 + - \cdots\,,$$

is convergent if a_i forms a monotonically decreasing, non-negative null sequence.

Estimation of error for alternating series:

$$|s - s_n| \leq a_{n+1} \quad ,$$

where s is the limit of the series.

Absolute convergence of a series $\sum a_i$:
The series of absolute values $\sum |a_i|$ is convergent.
Absolute convergence guarantees that the terms of the series can be rearranged without changing the sum of the series.

□ The series

$$\sum_{n=1}^{\infty} \frac{(-1)^{n-1}}{n} = 1 - \frac{1}{2} + \frac{1}{3} - \frac{1}{4} \pm \cdots$$

is convergent, but not absolutely convergent. For example, there is a rearrangement of terms that diverges to $+\infty$:

$$\begin{aligned} &1 - \frac{1}{2} + \frac{1}{3} - \frac{1}{4} + \\ &+ \left(\frac{1}{5} + \frac{1}{7}\right) - \frac{1}{6} + \\ &+ \left(\frac{1}{9} + \frac{1}{11} + \frac{1}{13} + \frac{1}{15}\right) - \frac{1}{8} + \cdots \\ &+ \left(\frac{1}{2^n + 1} + \frac{1}{2^n + 3} + \cdots + \frac{1}{2^{n+1} - 1}\right) - \frac{1}{2n + 2} + \cdots\,. \end{aligned}$$

Proof of absolute convergence:

- **Ratio test** (d'Alembert)
  A sufficient, but not a necessary, convergence criterion.

  (a) A series $\sum_{i=1}^{\infty} a_i$ is absolutely convergent if

$$\lim_{n\to\infty} \left|\frac{a_{n+1}}{a_n}\right| < 1\,.$$

(b) The series a_n is divergent for

$$\lim_{n\to\infty}\left|\frac{a_{n+1}}{a_n}\right| > 1 .$$

(c) If the limit is equal to one, then no proposition is possible.

□ The series

$$s_n = \frac{1}{2} + \frac{2}{2^2} + \frac{3}{2^3} + \cdots + \frac{n}{2^n} + \cdots$$

is convergent since

$$\lim_{n\to\infty} \frac{n+1}{2^{n+1}} : \frac{n}{2^n} = \frac{1}{2} < 1 .$$

● **Root test**
A sufficient, but not a necessary, convergence criterion.

(a) A series $\sum_{i=1}^{\infty}$ is absolutely convergent if

$$\lim_{n\to\infty} \sqrt[n]{|a_n|} < 1 .$$

(b) The series a_n is divergent for

$$\lim_{n\to\infty} \sqrt[n]{|a_n|} > 1 .$$

(c) If the limit is equal to one, then no proposition is possible.

□ The series

$$s_n = \frac{1}{2} + \left(\frac{2}{3}\right)^4 + \left(\frac{3}{4}\right)^9 + \cdots \left(\frac{n}{n+1}\right)^{n^2}$$

is convergent since

$$\lim_{n\to\infty} \sqrt[n]{|a_n|} = \lim_{n\to\infty} \sqrt[n]{\left(\frac{n}{n+1}\right)^{n^2}} = \lim_{n\to\infty} \left(\frac{1}{1+\frac{1}{n}}\right)^n = \frac{1}{\mathrm{e}} < 1 .$$

▷ The ratio test as well as the root test are sufficient but not necessary conditions for convergence.

14.2.1 Special number series

$$\sum_{k=1}^{n} k = \frac{n(n+1)}{2} \qquad \sum_{k=1}^{n}(2k-1) = n^2$$

$$\sum_{k=1}^{n} k^2 = \frac{n(n+1)(2n+1)}{6} \qquad \sum_{k=1}^{n} k^3 = \frac{n^2(n+1)^2}{4}$$

$$\mathrm{e} = 1 + \frac{1}{1!} + \frac{1}{2!} + \frac{1}{3!} + \cdots \qquad \frac{1}{\mathrm{e}} = 1 - \frac{1}{1!} + \frac{1}{2!} - \frac{1}{3!} + - \cdots$$

$$\ln 2 = 1 - \frac{1}{2} + \frac{1}{3} - \frac{1}{4} + - \cdots \qquad 1 = \frac{1}{1\cdot 2} + \frac{1}{2\cdot 3} + \frac{1}{3\cdot 4} + \frac{1}{4\cdot 5} + \cdots$$

$$\frac{\pi}{4} = 1 - \frac{1}{3} + \frac{1}{5} - \frac{1}{7} + \frac{1}{9} - + \cdots$$ **Leibniz's series:**

$$\frac{\pi^2}{6} = \frac{1}{1^2} + \frac{1}{2^2} + \frac{1}{3^2} + \frac{1}{4^2} + \cdots \qquad \frac{\pi^2}{12} = \frac{1}{1^2} - \frac{1}{2^2} + \frac{1}{3^2} - \frac{1}{4^2} + - \cdots$$

$$\frac{\pi^2}{8} = \frac{1}{1^2} + \frac{1}{3^2} + \frac{1}{5^2} + \frac{1}{7^2} + \cdots$$

14.3 Taylor and MacLaurin series

Now we will discuss functions and variables, and not only with number sequences. **Function value** $f(x)$ can be approximated by a point on the tangent $f'(x)$ at x_0. The approximation is exact for polynomials of the first degree (straight lines); for other functions, the closer x lies to x_0, i.e., the smaller $|x - x_0|$ is, the better the approximation.

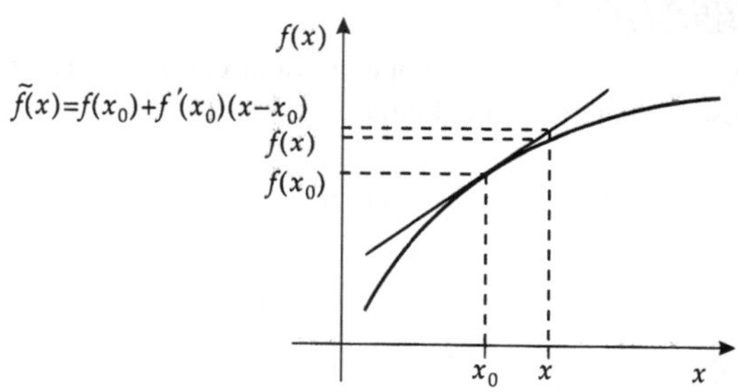

Explanation of the tangent formula

$$f(x) \approx f(x_0) + f'(x_0)\cdot(x - x_0)\,.$$

The generalization of this approximation formula follows from the mean value theorem in differential calculus.

14.3.1 Taylor's formula

- **Taylor's formula**: If a function $f(x)$ is $(n+1)$-times differentiable in the interval (a, b), and if its n-th derivative is also continuous at the end-points a and b, then the function can be represented in the following form:

$$f(x) = f(x_0) + \frac{f'(x_0)}{1!} \cdot (x - x_0) + \frac{f''(x_0)}{2!} \cdot (x - x_0)^2 + \cdots + \frac{f^{(n)}(x_0)}{n!} \cdot (x - x_0)^n + R_n(x) ,$$

where x and x_0 are in the interval (a, b).

Taylor's polynomial $P_n(x)$ of $f(x)$ at the point of expansion x_0 is a polynomial of degree n that coincides with $f(x)$ in the first n derivatives.

$$P_n(x) = f(x_0) + \frac{f'(x_0)}{1!} \cdot (x - x_0) + \frac{f''(x_0)}{2!} \cdot (x - x_0)^2 + \cdots + \frac{f^{(n)}(x_0)}{n!} \cdot (x - x_0)^n .$$

Lagrange's remainder R_n: The difference between the function $f(x)$ and Taylor's polynomial $P_n(x)$:

$$R_n(x) = f(x) - P_n(x) = \frac{f^{(n+1)}(x^*)}{(n+1)!} \cdot (x - x_0)^{n+1} , \quad |x^* - x| \leq |x_0 - x| .$$

x^* lies between x_0 and x!

▷ The smaller the remainder R_n, the better is the approximation of the function $f(x)$ by Taylor's polynomial.
Taylor's formula allows the calculation of function values with any degree of accuracy. The number of terms, and thus the degree of the polynomial, necessary for the required accuracy depends essentially on the distance $|x - x_0|$ from point x_0 to point x.
The greater $|x - x_0|$, the more terms must be taken into account. Usually, the remainder cannot be given exactly because point x^* is unknown. But it is often enough to estimate the upper limit of the error.

14.3.2 Taylor series

The **Taylor series** of $f(x)$ with the point of expansion x_0 arises if the degree n of Taylor's polynomial is allowed to increase indefinitely.

$$\begin{aligned} f(x) &= f(x_0) + \frac{1}{1!} f'(x_0) \cdot (x - x_0) + \\ &\quad + \frac{1}{2!} f''(x_0) \cdot (x - x_0)^2 + \cdots + \frac{1}{n!} f^{(n)}(x_0) \cdot (x - x_0)^n + \cdots \\ &= \sum_{n=0}^{\infty} \frac{1}{n!} f^{(n)}(x_0) \cdot (x - x_0)^n \quad . \end{aligned}$$

□ Expanding the exponential function about $x_0 = 0$ yields:

$$e^x = 1 + \frac{x}{1!} + \frac{x^2}{2!} + \frac{x^3}{3!} + \cdots + \frac{x^n}{n!} + \cdots .$$

The **MacLaurin series** of $f(x)$ is: The Taylor series, but for the special point of expansion $x_0 = 0$.

$$\begin{aligned} f(x) &= f(0) + \frac{1}{1!} f'(0) \cdot x + \frac{1}{2!} f''(0) \cdot x^2 + \cdots + \frac{1}{n!} f^{(n)}(0) \cdot x^n + \cdots \\ &= \sum_{n=0}^{\infty} \frac{1}{n!} f^{(n)}(0) \cdot x^n . \end{aligned}$$

Convergence behavior of the Taylor series: For $x = x_0$ the Taylor series converges trivially to $f(x_0)$. Otherwise, the series does not have to be convergent. Even in case of convergence, the limit does not have to be equal to the function value $f(x)$, as is shown by the following example:

□ The function

$$f(x) = \begin{cases} e^{-1/x^2} & \text{for} \quad x \neq 0 \\ 0 & \text{for} \quad x = 0 \end{cases}$$

is differentiable arbitrarily often with $f^{(n)}(0) = 0$ for all n. Hence, the Taylor series with the point of expansion $x_0 = 0$ converges to 0 for all x.

The **Taylor series converges** to the corresponding value $f(x)$ if for the remainder it holds that:

$$\lim_{n\to\infty} R_n = \lim_{n\to\infty} \frac{f^{(n+1)}(x^*)}{(n+1)!} \cdot (x - x_0)^{n+1} = 0\,.$$

In this case we say that $f(x)$ can be represented by its Taylor series.

Function in two variables:

Expansion of a function $f(x, y)$ in two variables in a Taylor series at the point $x = a$, $y = b$:

$$f(a+h, b+k) = f(a,b) + \left(h\frac{\partial}{\partial x} + k\frac{\partial}{\partial y}\right) f(x,y)\Big|_{\substack{x=a\\y=b}} + \cdots$$
$$+\frac{1}{n!}\left(h\frac{\partial}{\partial x} + k\frac{\partial}{\partial y}\right)^n f(x,y)\Big|_{\substack{x=a\\y=b}} + \cdots$$

with

$$\left(h\frac{\partial}{\partial x} + k\frac{\partial}{\partial y}\right)^2 f(x,y) = h^2\frac{\partial^2 f(x,y)}{\partial x^2} + 2hk\frac{\partial^2 f(x,y)}{\partial x \partial y} + k^2\frac{\partial^2 f(x,y)}{\partial y^2}\,.$$

14.4 Power series

Power series with the point of expansion x_0:

$$a_0 + a_1 \cdot (x - x_0) + a_2 \cdot (x - x_0)^2 + \cdots + a_n \cdot (x - x_0)^n + \cdots = \sum_{n=0}^{\infty} a_n \cdot (x - x_0)^n\,.$$

Taylor series are special cases of general power series.

If a power series converges for the x-values of an interval, a function $f(x)$ in the interval can be defined by the limits: To every x of the interval, the limit of the infinite series is assigned as the function value $f(x)$.

14.4.1 Test of convergence for power series

The power series $p(x) = \sum a_n \cdot (x - x_0)^n$ always converges for the point of expansion $x = x_0$; this is the trivial case, $p(x_0) = a_0$.

The **radius of convergence** r about the point of expansion can be given if the series is convergent also for values not equal to x_0.

The **radius of convergence** r of the power series is the least upper bound (supremum) of the numbers $|x - x_0|$ for which

$$\sum_{n=0}^{\infty} a_n \cdot (x - x_0)^n$$

is convergent. Here, $0 \leq r \leq \infty$.
The series is divergent for all x-values with $|x - x_0| > r$.
Range of convergence: The symmetric interval $(x_0 - r, x_0 + r)$ about the point of expansion x_0.

- **Propositions on convergence:**

$$|x - x_0| < r \;:\; \sum a_n \cdot (x - x_0)^n, \text{ the power series is absolutely convergent.}$$

$$|x - x_0| > r \;:\; \sum a_n \cdot (x - x_0)^n, \text{ the power series is divergent.}$$

$$|x - x_0| = r \;:\; \text{No generally applicable proposition is possible.}$$

Formulas for calculating the radius of convergence r:

- The **ratio test** for the radius of convergence r:

$$r = \lim_{n\to\infty} \left| \frac{a_n}{a_{n+1}} \right| , \quad \text{where} \quad r \to \infty \quad \text{for} \quad \left| \frac{a_{n+1}}{a_n} \right| \to 0 .$$

▷ This formula *cannot always* be applied although it meets the conditions of the ratio test.

□ Let every second a_n equal zero, then a_n/a_{n+1} is not defined!.

- The **root test** yields for the radius of convergence r:

$$r = \lim_{n\to\infty} \frac{1}{\sqrt[n]{|a_n|}} .$$

▷ These formulas can be used only if the corresponding limit exists. If both limits exist, then the results for r coincide.

14.4.2 Properties of convergent power series

- **Linear combinations** of power series may be performed **term-by-term** within the common range of convergence if the ranges of convergence of the power series overlap:

$$c \cdot \sum a_n \cdot (x - x_0)^n + d \cdot \sum b_n \cdot (x - x_0)^n = \sum (c \cdot a_n + d \cdot b_n)(x - x_0)^n .$$

- A series of functions $\sum f_n(x)$ can be integrated or differentiated term-by-term if the series converges uniformly. A simple (*pointwise*) convergence for each x is not sufficient!

For every given precision ϵ, uniformly convergent power series can be aborted after some number of N summands, independent of x. Then the error is less than or equal to ε.
Uniform convergence of power series: A power series converges uniformly in any closed and bounded subinterval $[x_0 - r_1, x_0 + r_1]$ of the range of convergence if r_1 is smaller than the radius of convergence r: $0 < r_1 < r$.
Absolute convergence: In the interior of the range of convergence, the power series converges absolutely.

- **Interchangeability of the differentiation and limiting process:**
Every power series can be differentiated term-by-term within its range of convergence:

$$f'(x) = \frac{\mathrm{d}}{\mathrm{d}x}\left[\sum_{n=0}^{\infty} a_n \cdot (x - x_0)^n\right] = \sum_{n=1}^{\infty} a_n \cdot \frac{\mathrm{d}}{\mathrm{d}x}(x - x_0)^n$$
$$= \sum_{n=1}^{\infty} n \cdot a_n \cdot (x - x_0)^{n-1} = \sum_{n=0}^{\infty} (n+1) \cdot a_{n+1} \cdot (x - x_0)^n .$$

The resulting power series has the same radius of convergence as the original power series.

- **Term-by-term integration:**
If the power series

$$f(x) = \sum_{n=0}^{\infty} a_n \cdot (x - x_0)^n .$$

is convergent with the radius of convergence r, then the series

$$g(x) = C + \sum_{n=0}^{\infty} \frac{a_n}{n+1} \cdot (x - x_0)^{n+1}$$
$$= C + a_0 \cdot (x - x_0) + \frac{a_1}{2} \cdot (x - x_0)^2 + \cdots , \qquad C = \text{ const},$$

is also convergent with the same radius of convergence r.

- Within the common range of convergence, two (or more) power series can be **added**, **subtracted**, and **multiplied** term-by-term. Sum, difference and product are at least convergent within the common range of convergence.
- **Continuity**: A power series is continuous (at least) within its range of convergence.
- **Identity theorem**:
If the limits of two power series are equal to each other for $|x - x_0| < r$,

$$\sum_{n=0}^{\infty} a_n \cdot (x - x_0)^n = \sum_{n=0}^{\infty} b_n \cdot (x - x_0)^n ,$$

where r is the minimum of the two radii of convergence, then all derivatives are also equal to each other:

$$\sum_{n=1}^{\infty} n \cdot a_n \cdot (x - x_0)^{n-1} = \sum_{n=1}^{\infty} n \cdot b_n \cdot (x - x_0)^{n-1} ,$$
$$\sum_{n=2}^{\infty} n(n-1) \cdot a_n \cdot (x - x_0)^{n-2} = \sum_{n=2}^{\infty} n(n-1) \cdot b_n \cdot (x - x_0)^{n-2} , \qquad \text{etc.}$$

Thus, it follows that the coefficients a_n and b_n are equal to each other for $x = x_0$.

$$a_0 = b_0 , \quad a_1 = b_1 , \quad a_2 = b_2 \; \ldots$$

- **Representation of a function:** If a function is represented by a power series, then the power series is the Taylor series for the expansion point.

14.4.3 Inversion of power series

The inversion of the power series

$$y = a_1x + a_2x^2 + a_3x^3 + a_4x^4 + a_5x^5 + a_6x^6 + \dots$$

yields the power series

$$x = b_1y + b_2y^2 + b_3y^3 + b_4y^4 + b_5y^5 + b_6y^6 + \dots$$

with the conditional equations:

$$a_1b_1 = 1$$
$$a_1^3b_2 = -a_2$$
$$a_1^5b_3 = 2a_2^2 - a_1a_3$$
$$a_1^7b_4 = 5a_1a_2a_3 - 5a_2^3 - a_1^2a_4$$
$$a_1^9b_5 = 6a_1^2a_2a_4 + 3a_1^2a_3^2 - a_1^3a_5 + 14a_2^4 - 21a_1a_2^2a_3$$
$$a_1^{11}b_6 = 7a_1^3a_2a_5 + 84a_1a_2^3a_3 + 7a_1^3a_3a_4 - 28a_1^2a_2a_3^2 - a_1^4a_6 - 28a_1^2a_2^2a_4 - 42a_2^5\,.$$

14.5 Special expansions of series and products

14.5.1 Binomial series

$$(a \pm x)^n = a^n \pm \binom{n}{1}a^{n-1}\cdot x + \binom{n}{2}a^{n-2}\cdot x^2 \pm \binom{n}{3}a^{n-3}\cdot x^3 + \cdots$$
$$(n > 0 \,:\, |x| \le |a|; \quad n < 0 \,:\, |x| < |a|)$$

$$(1 \pm x)^n = 1 \pm \binom{n}{1}x + \binom{n}{2}x^2 \pm \binom{n}{3}x^3 + \binom{n}{4}x^4 \pm \cdots$$
$$(n > 0 \,:\, |x| \le 1; \quad n < 0 \,:\, |x| < 1)$$

14.5.2 Special binomial series

$$(a \pm x)^2 = a^2 \pm 2ax + x^2$$
$$(a \pm x)^3 = a^3 \pm 3a^2x + 3ax^2 \pm x^3$$
$$(a \pm x)^4 = a^4 \pm 4a^3x + 6a^2x^2 \pm 4ax^3 + x^4$$
$$(1 \pm x)^{-1} = 1 \mp x + x^2 \mp x^3 + x^4 \mp \cdots \quad (|x| < 1)$$
$$(1 \pm x)^{-2} = 1 \mp 2x + 3x^2 \mp 4x^3 + 5x^4 \mp \cdots \quad (|x| < 1)$$
$$(1 \pm x)^{-3} = 1 \mp \frac{1}{1\cdot 2}(2\cdot 3x \mp 3\cdot 4x^2 + 4\cdot 5x^3 \mp 5\cdot 6x^4 + \cdots) \quad (|x| < 1)$$
$$(1 \pm x)^{-4} = 1 \mp \frac{1}{1\cdot 2\cdot 3}(2\cdot 3\cdot 4x \mp 3\cdot 4\cdot 5x^2 + 4\cdot 5\cdot 6x^3 \mp 5\cdot 6\cdot 7x^4 + \cdots) \quad (|x| < 1)$$
$$(1 \pm x)^{-5} = 1 \mp \frac{1}{1\cdot 2\cdot 3\cdot 4}(2\cdot 3\cdot 4\cdot 5x \mp 3\cdot 4\cdot 5\cdot 6x^2 + 4\cdot 5\cdot 6\cdot 7x^3 \mp \cdots) \quad (|x| < 1)$$
$$(1 \pm x)^{1/2} = 1 \pm \frac{1}{2}x - \frac{1\cdot 1}{2\cdot 4}x^2 \pm \frac{1\cdot 1\cdot 3}{2\cdot 4\cdot 6}x^3 - \frac{1\cdot 1\cdot 3\cdot 5}{2\cdot 4\cdot 6\cdot 8}x^4 \pm \cdots \quad (|x| \le 1)$$

$$(1 \pm x)^{1/3} = 1 \pm \frac{1}{3}x - \frac{1\cdot 2}{3\cdot 6}x^2 \pm \frac{1\cdot 2\cdot 5}{3\cdot 6\cdot 9}x^3 - \frac{1\cdot 2\cdot 5\cdot 8}{3\cdot 6\cdot 9\cdot 12}x^4 \pm \cdots \quad (|x| \le 1)$$

$$(1 \pm x)^{1/4} = 1 \pm \frac{1}{4}x - \frac{1\cdot 3}{4\cdot 8}x^2 \pm \frac{1\cdot 3\cdot 7}{4\cdot 8\cdot 12}x^3 - \frac{1\cdot 3\cdot 7\cdot 11}{4\cdot 8\cdot 12\cdot 16}x^4 \pm \cdots \quad (|x| \le 1)$$

$$(1 \pm x)^{3/2} = 1 \pm \frac{3}{2}x + \frac{3\cdot 1}{2\cdot 4}x^2 \mp \frac{3\cdot 1\cdot 1}{2\cdot 4\cdot 6}x^3 + \frac{3\cdot 1\cdot 1\cdot 3}{2\cdot 4\cdot 6\cdot 8}x^4 \mp \cdots \quad (|x| \le 1)$$

$$(1 \pm x)^{5/2} = 1 \pm \frac{5}{2}x + \frac{5\cdot 3}{2\cdot 4}x^2 \pm \frac{5\cdot 3\cdot 1}{2\cdot 4\cdot 6}x^3 - \frac{5\cdot 3\cdot 1\cdot 1}{2\cdot 4\cdot 6\cdot 8}x^4 \mp \cdots \quad (|x| \le 1)$$

$$(1 \pm x)^{-1/2} = 1 \mp \frac{1}{2}x + \frac{1\cdot 3}{2\cdot 4}x^2 \mp \frac{1\cdot 3\cdot 5}{2\cdot 4\cdot 6}x^3 + \frac{1\cdot 3\cdot 5\cdot 7}{2\cdot 4\cdot 6\cdot 8}x^4 \mp \cdots \quad (|x| < 1)$$

$$(1 \pm x)^{-1/3} = 1 \mp \frac{1}{3}x + \frac{1\cdot 4}{3\cdot 6}x^2 \mp \frac{1\cdot 4\cdot 7}{3\cdot 6\cdot 9}x^3 + \frac{1\cdot 4\cdot 7\cdot 10}{3\cdot 6\cdot 9\cdot 12}x^4 \mp \cdots \quad (|x| < 1)$$

$$(1 \pm x)^{-1/4} = 1 \mp \frac{1}{4}x + \frac{1\cdot 5}{4\cdot 8}x^2 \mp \frac{1\cdot 5\cdot 9}{4\cdot 8\cdot 12}x^3 + \frac{1\cdot 5\cdot 9\cdot 13}{4\cdot 8\cdot 12\cdot 16}x^4 \mp \cdots \quad (|x| < 1)$$

$$(1 \pm x)^{-3/2} = 1 \mp \frac{3}{2}x + \frac{3\cdot 5}{2\cdot 4}x^2 \mp \frac{3\cdot 5\cdot 7}{2\cdot 4\cdot 6}x^3 + \frac{3\cdot 5\cdot 7\cdot 9}{2\cdot 4\cdot 6\cdot 8}x^4 \mp \cdots \quad (|x| < 1)$$

14.5.3 Series of exponential functions

$$e^x = 1 + \frac{x}{1!} + \frac{x^2}{2!} + \frac{x^3}{3!} + \frac{x^4}{4!} + \cdots + \frac{x^n}{n!} + \cdots \quad (|x| < \infty)$$

$$a^x = 1 + \frac{\ln a}{1!}x + \frac{(\ln a)^2}{2!}x^2 + \frac{(\ln a)^3}{3!}x^3 + \frac{(\ln a)^4}{4!}x^4 + \cdots \quad (|x| < \infty)$$

$$e^x = e^a\left[1 + \frac{x-a}{1!} + \frac{(x-a)^2}{2!} + \frac{(x-a)^3}{3!} + \cdots\right] \quad (|x| < \infty)$$

$$e^x = \cfrac{1}{1 - \cfrac{x}{1 + \cfrac{x}{2 - \cfrac{x}{3 + \cfrac{x}{2 - \cfrac{x}{5 + \cdots}}}}}}, \qquad e^x = 1 + \cfrac{x}{1 - \cfrac{x}{2 + \cfrac{x}{3 - \cfrac{x}{2 + \cfrac{x}{5 - \cdots}}}}}$$

$$e^{\sin x} = 1 + \frac{x}{1!} + \frac{x^2}{2!} - \frac{3x^4}{4!} - \frac{8x^5}{5!} - \frac{3x^6}{6!} + \frac{56x^7}{7!} + \cdots \quad (|x| < \infty)$$

$$e^{\cos x} = e\left[1 - \frac{x^2}{2!} + \frac{4x^4}{4!} - \frac{31x^6}{6!} + \cdots\right] \quad (|x| < \infty)$$

$$e^{\tan x} = 1 + \frac{x}{1!} + \frac{x^2}{2!} + \frac{3x^3}{3!} + \frac{9x^4}{4!} + \frac{37x^5}{5!} + \cdots \quad \left(|x| < \frac{\pi}{2}\right)$$

$$e^x \sin x = x + x^2 + \frac{x^3}{3} - \frac{x^5}{30} - \frac{x^6}{90} + \cdots + \frac{2^{n/2}\sin(n\pi/4)x^n}{n!} + \cdots \quad (|x| < \infty)$$

$$e^x \cos x = 1 + x - \frac{x^3}{3} - \frac{x^4}{6} + \cdots + \frac{2^{n/2}\cos(n\pi/4)x^n}{n!} + \cdots \quad (|x| < \infty)$$

14.5.4 Series of logarithmic functions

$$\ln x = \frac{x-1}{x} + \frac{1}{2}\left(\frac{x-1}{x}\right)^2 + \frac{1}{3}\left(\frac{x-1}{x}\right)^3 + \cdots \quad \left(x > \frac{1}{2}\right)$$

$$\ln x = (x-1) - \frac{1}{2}(x-1)^2 + \frac{1}{3}(x-1)^3 - \frac{1}{4}(x-1)^4 + - \cdots \quad (0 < x \le 2)$$

$$\ln x = 2\left[\left(\frac{x-1}{x+1}\right) + \frac{1}{3}\left(\frac{x-1}{x+1}\right)^3 + \frac{1}{5}\left(\frac{x-1}{x+1}\right)^5 + \frac{1}{7}\left(\frac{x-1}{x+1}\right)^7 + \cdots\right] (x > 0)$$

$$\ln(ax) = \ln a + \frac{x-a}{a} - \frac{(x-a)^2}{2a^2} + \frac{(x-a)^3}{3a^3} - + \cdots \quad (0 < x \le 2a)$$

$$\ln(1+x) = x - \frac{x^2}{2} + \frac{x^3}{3} - \frac{x^4}{4} + - \ldots \quad (-1 < x \le 1)$$

$$\ln(1-x) = -\left[x + \frac{x^2}{2} + \frac{x^3}{3} + \frac{x^4}{4} + \ldots\right] \quad (-1 \le x < 1)$$

$$\ln(a+x) = \ln a + 2\left[\frac{x}{2a+x} + \frac{1}{3}\left(\frac{x}{2a+x}\right)^3 + \frac{1}{5}\left(\frac{x}{2a+x}\right)^5 + \cdots\right]$$
$$(a > 0, -a < x < \infty)$$

$$\ln\left(\frac{1+x}{1-x}\right) = 2\left[x + \frac{x^3}{3} + \frac{x^5}{5} + \frac{x^7}{7} + \cdots\right] \quad (|x| < 1)$$

$$\ln\left(\frac{x+1}{x-1}\right) = 2\left[\frac{1}{x} + \frac{1}{3x^3} + \frac{1}{5x^5} + \frac{1}{7x^7} + \cdots\right] \quad (|x| > 1)$$

$$\ln|\sin x| = \ln|x| - \frac{x^2}{6} - \frac{x^4}{180} - \frac{x^6}{2835} - \cdots - \frac{2^{2n-1}B_n x^{2n}}{n(2n)!} - \cdots \quad (0 < |x| < \pi)$$

$$\ln\cos x = -\frac{x^2}{2} - \frac{x^4}{12} - \frac{x^6}{45} - \frac{17x^8}{2520} \cdots - \frac{2^{2n-1}(2^{2n}-1)B_n x^{2n}}{n(2n)!} - \cdots \quad \left(|x| < \frac{\pi}{2}\right)$$

$$\ln|\tan x| = \ln|x| + \frac{x^2}{3} + \frac{7x^4}{90} + \frac{62x^6}{2835} + \cdots + \frac{2^{2n}(2^{2n-1}-1)B_n x^{2n}}{n(2n)!} + \cdots$$
$$\left(0 < |x| < \frac{\pi}{2}\right)$$

$$\frac{\ln(1+x)}{1+x} = x - (1 + \frac{1}{2})x^2 + (1 + \frac{1}{2} + \frac{1}{3})x^3 - \cdots \quad (|x| < 1)$$

$B_n, n = 1, 2, \ldots$ are the Bernoulli's numbers obtained from the power series expansion of the following function:

$$\frac{x}{e^x - 1} = 1 \cdots$$

14.5.5 Series of trigonometric functions

$$\sin x = x - \frac{x^3}{3!} + \frac{x^5}{5!} - \frac{x^7}{7!} + - \cdots \quad (|x| < \infty)$$

$$\sin x = x\left(1 - \frac{x^2}{\pi^2}\right)\left(1 - \frac{x^2}{2^2\pi^2}\right)\left(1 - \frac{x^2}{3^2\pi^2}\right)\cdots \quad (|x| < \infty)$$

$$\sin(ax) = \sin a + (x-a)\cos a - \frac{(x-a)^2}{2!}\sin a - \frac{(x-a)^3}{3!}\cos a + \frac{(x-a)^4}{4!}\sin a + \cdots \quad (|x| < \infty)$$

$$\sin(x+a) = \sin a + x\cos a - \frac{x^2 \sin a}{2!} - \frac{x^3 \cos a}{3!} + \frac{x^4 \sin a}{4!} + \ldots \quad (|x| < \infty)$$

$$\cos x = 1 - \frac{x^2}{2!} + \frac{x^4}{4!} - \frac{x^6}{6!} + \cdots \quad (|x| < \infty)$$

$$\cos x = \left(1 - \frac{4x^2}{\pi^2}\right)\left(1 - \frac{4x^2}{3^2\pi^2}\right)\left(1 - \frac{4x^2}{5^2\pi^2}\right)\cdots \quad (|x| < \infty)$$

$$\cos(x+a) = \cos a - x\sin a - \frac{x^2 \cos a}{2!} + \frac{x^3 \sin a}{3!} + \frac{x^4 \cos a}{4!} - \ldots \quad (|x| < \infty)$$

$$\tan x = x + \frac{1}{3}x^3 + \frac{2}{15}x^5 + \frac{17}{315}x^7 + \frac{62}{2835}x^9 + \cdots + \frac{2^{2n}(2^{2n}-1)}{(2n)!}B_n x^{2n-1} + \cdots \quad \left(|x| < \frac{\pi}{2}\right)$$

$$\cot x = \frac{1}{x} - \frac{x}{3} - \frac{x^3}{45} - \frac{2x^5}{945} - \cdots - \frac{2^{2n}}{(2n)!}B_n x^{2n-1} - \cdots \quad (0 < |x| < \pi)$$

$$\sec x = 1 + \frac{x^2}{2} + \frac{5x^4}{24} + \frac{61x^6}{720} + \cdots + \frac{E_n x^{2n}}{(2n)!} + \cdots \quad \left(|x| < \frac{\pi}{2}\right)$$

$$\csc x = \frac{1}{x} + \frac{x}{6} + \frac{7x^3}{360} + \frac{31x^5}{15120} + \cdots + \frac{2(2^{2n-1}-1)B_n x^{2n-1}}{(2n)!} + \cdots \quad (0 < |x| < \pi)$$

$B_n, n = 1, 2, \ldots$ are the Bernoulli's numbers (see previous section). $E_n = 1, 2, \ldots$ are the so-called Euler's numbers.

14.5.6 Series of inverse trigonometric functions

$$\arcsin x = x + \frac{1}{2\cdot 3}x^3 + \frac{1\cdot 3}{2\cdot 4\cdot 5}x^5 + \frac{1\cdot 3\cdot 5}{2\cdot 4\cdot 6\cdot 7}x^7 + \cdots \quad (|x| < 1)$$

$$\arccos x = \frac{\pi}{2} - \left[x + \frac{1}{2\cdot 3}x^3 + \frac{1\cdot 3}{2\cdot 4\cdot 5}x^5 + \frac{1\cdot 3\cdot 5}{2\cdot 4\cdot 6\cdot 7}x^7 + \cdots\right] \quad (|x| < 1)$$

$$\arctan x = x - \frac{x^3}{3} + \frac{x^5}{5} - \frac{x^7}{7} + - \cdots \quad (|x| < 1)$$

$$\arctan x = +\frac{\pi}{2} - \left[\frac{1}{x} - \frac{1}{3x^3} + \frac{1}{5x^5} - \frac{1}{7x^7} + - \cdots\right] \quad (x > 1)$$

$$\arctan x = -\frac{\pi}{2} - \left[\frac{1}{x} - \frac{1}{3x^3} + \frac{1}{5x^5} - \frac{1}{7x^7} + - \cdots\right] \quad (x < -1)$$

$$\operatorname{arccot} x = \frac{\pi}{2} - \left[x - \frac{x^3}{3} + \frac{x^5}{5} - \frac{x^7}{7} + - \cdots\right] \quad (|x| < 1)$$

$$\operatorname{arcsec} x = \arccos\frac{1}{x} = \frac{\pi}{2} - \left[\frac{1}{x} + \frac{1}{2\cdot 3x^3} + \frac{1\cdot 3}{2\cdot 4\cdot 5x^5} + \cdots\right] \quad (|x| > 1)$$

$$\operatorname{arccsc} x = \arcsin\frac{1}{x} = \frac{1}{x} + \frac{1}{2\cdot 3x^3} + \frac{1\cdot 3}{2\cdot 4\cdot 5x^5} + \cdots \quad (|x| > 1)$$

14.5.7 Series of hyperbolic functions

$$\sinh x = x + \frac{x^3}{3!} + \frac{x^5}{5!} + \frac{x^7}{7!} + \cdots \quad (|x| < \infty)$$

$$\cosh x = 1 + \frac{x^2}{2!} + \frac{x^4}{4!} + \frac{x^6}{6!} + \cdots \quad (|x| < \infty)$$

$$\tanh x = x - \frac{1}{3}x^3 + \frac{2}{15}x^5 - \frac{17}{315}x^7 + \cdots + \frac{(-1)^{n+1}2^{2n}(2^{2n}-1)}{(2n)!}B_n x^{2n-1} + - \cdots$$

$$\left(|x| < \frac{\pi}{2}\right)$$

$$\coth x = \frac{1}{x} + \frac{1}{3}x - \frac{1}{45}x^3 + \frac{2}{945}x^5 - + \cdots + \frac{(-1)^{n+1}2^{2n}}{(2n)!}B_n x^{2n-1} + - \cdots \quad (0 < |x| < \pi)$$

$$\operatorname{sech} x = 1 - \frac{x^2}{2} + \frac{5x^4}{24} - \frac{61x^6}{720} + \cdots \frac{(-1)^n E_n x^{2n}}{(2n)!} + \cdots \quad (|x| < \frac{\pi}{2})$$

$$\operatorname{csch} x = \frac{1}{x} - \frac{x}{6} + \frac{7x^3}{360} - \frac{31x^5}{15120} + \cdots \frac{(-1)^n 2(2^{2n-1}-1)B_n x^{2n-1}}{(2n)!} + \cdots \quad (0 < |x| < \pi)$$

14.5.8 Series of area hyperbolic functions

$$\operatorname{Arsinh} x = x - \frac{1}{2\cdot 3}x^3 + \frac{1\cdot 3}{2\cdot 4\cdot 5}x^5 - \frac{1\cdot 3\cdot 5}{2\cdot 4\cdot 6\cdot 7}x^7 + - \cdots \quad (|x| < 1)$$

$$\operatorname{Arcosh} x = \ln(2x) - \frac{1}{2\cdot 2x^2} - \frac{1\cdot 3}{2\cdot 4\cdot 4x^4} - \frac{1\cdot 3\cdot 5}{2\cdot 4\cdot 6\cdot 6x^6} - \cdots \quad (x > 1)$$

$$\operatorname{Artanh} x = x + \frac{x^3}{3} + \frac{x^5}{5} + \frac{x^7}{7} + \cdots \quad (|x| < 1)$$

$$\operatorname{Arcoth} x = \frac{1}{x} + \frac{1}{3x^3} + \frac{1}{5x^5} + \frac{1}{7x^7} + \cdots \quad (|x| > 1)$$

14.5.9 Partial fraction expansions

$$\cot x = \frac{1}{x} + 2x\left[\frac{1}{x^2-\pi^2} + \frac{1}{x^2-4\pi^2} + \frac{1}{x^2-9\pi^2} + \cdots\right]$$

$$\csc x = \frac{1}{x} - 2x\left[\frac{1}{x^2-\pi^2} - \frac{1}{x^2-4\pi^2} + \frac{1}{x^2-9\pi^2} - \cdots\right]$$

$$\sec x = 4\pi\left[\frac{1}{\pi^2-4x^2} - \frac{3}{9\pi^2-4x^2} + \frac{5}{25\pi^2-4x^2} - \cdots\right]$$

$$\tan x = 8x\left[\frac{1}{\pi^2-4x^2} + \frac{1}{9\pi^2-4x^2} + \frac{1}{25\pi^2-4x^2} + \cdots\right]$$

$$\sec^2 x = 4\left[\frac{1}{(\pi-2x)^2} + \frac{1}{(\pi+2x)^2} + \frac{1}{(3\pi-2x)^2} + \frac{1}{(3\pi+2x)^2} + \cdots\right]$$

$$\csc^2 x = \frac{1}{x^2} + \frac{1}{(x-\pi)^2} + \frac{1}{(x+\pi)^2} + \frac{1}{(x-2\pi)^2} + \frac{1}{(x+2\pi)^2} + \cdots$$

$$\coth x = \frac{1}{x} + 2x\left[\frac{1}{x^2+\pi^2} + \frac{1}{x^2+4\pi^2} + \frac{1}{x^2+9\pi^2} + \cdots\right]$$

$$\operatorname{csch} x = \frac{1}{x} - 2x\left[\frac{1}{x^2+\pi^2} - \frac{1}{x^2+4\pi^2} + \frac{1}{x^2+9\pi^2} - \cdots\right]$$

$$\operatorname{sech} x = 4\pi\left[\frac{1}{\pi^2+4x^2} - \frac{3}{9\pi^2+4x^2} + \frac{5}{25\pi^2+4x^2} - \cdots\right]$$

$$\tanh x = 8x\left[\frac{1}{\pi^2+4x^2} + \frac{3}{9\pi^2+4x^2} + \frac{5}{25\pi^2+4x^2} + \cdots\right]$$

14.5.10 Infinite products

$$\sin x = x\left(1-\frac{x^2}{\pi^2}\right)\left(1-\frac{x^2}{4\pi^2}\right)\left(1-\frac{x^2}{9\pi^2}\right)\cdots$$

$$\cos x = \left(1-\frac{4x^2}{\pi^2}\right)\left(1-\frac{4x^2}{9\pi^2}\right)\left(1-\frac{4x^2}{25\pi^2}\right)\cdots$$

$$\sinh x = x\left(1+\frac{x^2}{\pi^2}\right)\left(1+\frac{x^2}{4\pi^2}\right)\left(1+\frac{x^2}{9\pi^2}\right)\cdots$$

$$\cosh x = \left(1+\frac{4x^2}{\pi^2}\right)\left(1+\frac{4x^2}{9\pi^2}\right)\left(1+\frac{4x^2}{25\pi^2}\right)\cdots$$

$$\frac{1}{\Gamma(x)} = xe^{\gamma y}\left\{\left(1+\frac{x}{1}\right)e^{-x}\right\}\left\{\left(1+\frac{x}{2}\right)e^{-x/2}\right\}\left\{\left(1+\frac{x}{3}\right)e^{-x/3}\right\}\cdots$$

where $\gamma = 0.5772156649\ldots =$ Euler's constant. $\Gamma(x)$ is the gamma function which was defined in Section 5.24).

$$J_0(x) = \left(1-\frac{x^2}{\lambda_1^2}\right)\left(1-\frac{x^2}{\lambda_2^2}\right)\left(1-\frac{x^2}{\lambda_3^2}\right)\cdots$$

where the values λ_i are the positive roots of $J_0(x) = 0$. $J_0(x)$ and $J_1(x)$ are Bessel functions of the first kind, see Section 16.8.

$$J_1(x) = x\left(1-\frac{x^2}{\lambda_1^2}\right)\left(1-\frac{x^2}{\lambda_2^2}\right)\left(1-\frac{x^2}{\lambda_3^2}\right)\cdots$$

where the values λ_i are the positive roots of $J_1(x) = 0$.

$$\frac{\sin x}{x} = \cos\frac{x}{2}\,\cos\frac{x}{4}\,\cos\frac{x}{8}\,\cos\frac{x}{16}\cdots$$

Wallis's product: $\dfrac{\pi}{2} = \dfrac{2}{1}\cdot\dfrac{2}{3}\cdot\dfrac{4}{3}\cdot\dfrac{4}{5}\cdot\dfrac{6}{5}\cdot\dfrac{6}{7}\cdots.$

($J_0(x)$, $J_1(x)$ are Bessel functions of the first kind, $\Gamma(x)$ is the gamma function.)

15

Integral calculus

15.1 Definition and integrability

15.1.1 Primitive

- **Integration** is the inverse of **differential calculus**.

Differential calculus: $y = f(x) \rightarrow y' = f'(x)$.

Integral calculus: $y' = f'(x) \rightarrow y = f(x) = \int f'(x)\mathrm{d}x + c$.

Primitive $F(x)$ of a given function $f(x)$: Function defined in the same domain as $f(x)$ and whose derivative $F'(x)$ is equal to $f(x)$.
Integration of a function $f(x)$: The **primitive** $F(x)$ of $f(x)$ is required such that the differentiation of $F(x)$ yields the original function $f(x)$. Notation:

$$F(x) = \int f(x)\mathrm{d}x + c \quad \text{with} \quad F'(x) = f(x) .$$

Integrand: Notation for $f(x)$,
Integration variable: Notation for the independent variable of the function.

▷ The stylized S as the integration sign originates from the definition of the integral as a sum.

- For every integrable function $f(x)$ there are an **infinite number** of primitives $F(x)+c$ differing merely in an additive constant $c \in \mathbb{R}$,

$$\int f(x)\mathrm{d}x = \int F'(x)\mathrm{d}x = \int dF(x) = F(x) + c, \quad c \in \mathbb{R}$$

The **integration constant** c is an arbitrary number in primitives, since the derivative of every constant c vanishes.

Geometric interpretation: All primitives have the same slope at all points x_0, since they have equal derivatives.

The primitives can be transformed into each other by means of parallel displacement in y direction.

▷ Omitting the integration constant can cause errors!

15.1.2 Definite and indefinite integrals

Indefinite integral:

$$I = \int f(x)\mathrm{d}x + c\,.$$

Integrals whose integration constant c is indefinite (nor fixed); they are represented by a family of functions.

The **integration constant** c can be fixed uniquely by giving the **limits of integration.**

Particular integral

$$P(x) = \int_a^x f(\tilde{x})\mathrm{d}\tilde{x}\,, \quad \text{or} \quad P(x) = \int_x^b f(\tilde{x})\mathrm{d}\tilde{x}\,.$$

A definite integral with variable upper or lower limit is a function of the variable (upper or lower) integration limit.

▷ Because one of the limits of integration is x, the integration variable must not also be called x. It must be formally renamed.

Definite integral:

$$A = \int_a^b f(x)\mathrm{d}x\,,$$

both limits of integration are fixed; the definite integral is a number.

The expression appearing below (above) the integration sign is called the lower (upper) limit of integration.

● **Fundamental theorem of differential and integral calculus:** If $F(x)$ is a primitive of $f(x)$, i.e. $F'(x) = f(x)$, then it holds that:

Particular integral: $\int\limits_a^x f(\tilde{x})\mathrm{d}\tilde{x} = F(t)|_a^x = F(x) - F(a)$

Definite integral: $\int\limits_a^b f(x)\mathrm{d}x = F(x)|_a^b = F(b) - F(a)$

Calculation of a definite integral: Integrate a function $f(x)$ (the integrand) by

1. determining a primitive of function $F(x)$ whose differentiation yields the original function again, $F'(x) = f(x)$.

2. calculating the function value of the primitive $F(b)$ at the upper limit b, and then

3. subtracting the value of the primitive $F(a)$ at the lower limit a.

4. The integration constant cancels out in integrating:

$$F(b) + c - (F(a) + c) = F(b) - F(a)\,.$$

□ $\int_1^2 (4x - 3x^2)\mathrm{d}x = (2x^2 - x^3)|_1^2 = (8 - 8) - (2 - 1) = -1$

▷ The **particular integral** is a function of x, while the **definite integral** yields a **definite** fixed number.

15.1.3 Geometrical interpretation

● **Calculation of areas**: The definite integral corresponds to the **area** A between the function $f(x)$ and the x-axis in the range $x = a$ and $x = b > a$ provided the continuous function does not assume negative values between the two limits of integration:

$$A = \int_a^b f(x)\mathrm{d}x = F(b) - F(a) \qquad (a < b \text{ and } f(x) \geq 0 \text{ for } a \leq x \leq b).$$

● **Function of area**: The variable **area** $A(x)$ under a curve $f(x) > 0$ corresponds to the integral the function with lower limit $\tilde{x} = a$, that is, to the **particular** integral of $f(x)$:

$$A(x) = \int_a^x f(\tilde{x})\mathrm{d}\tilde{x} = F(x) - F(a) \qquad (a < x \text{ and } f(x) \geq 0 \text{ for } x \geq a).$$

▷ This applies only in this simple form if the function $f(x)$ is positive in the entire interval $[a, b]$, or for all $x \geq a$.

● If the function $f(x)$ is negative in the entire interval $[a, b]$, then the integral is also negative.

$$f(x) < 0, \quad x \in [a, b] \rightarrow \int_a^b f(x)\mathrm{d}x < 0 \quad \text{if } b > a.$$

▷ In case of negative functions, the area between $f(x)$ and the x-axis can be determined by the absolute value of the integral.

▷ If the function $f(x)$ is positive as well as negative in the entire interval, then the areas below the x-axis are subtracted in integrating.

□ The integral of $\sin x$ over the interval $[0, 2\pi]$ vanishes.

Usually, ($f(x)$ is also negative). The definite integral is the difference of the areas above and below the x-axis.

The **area** under a curve can be approximated by several rectangles:

Upper sum: The area patches go above the curve,

Lower sum: The area patches remain below the curve.

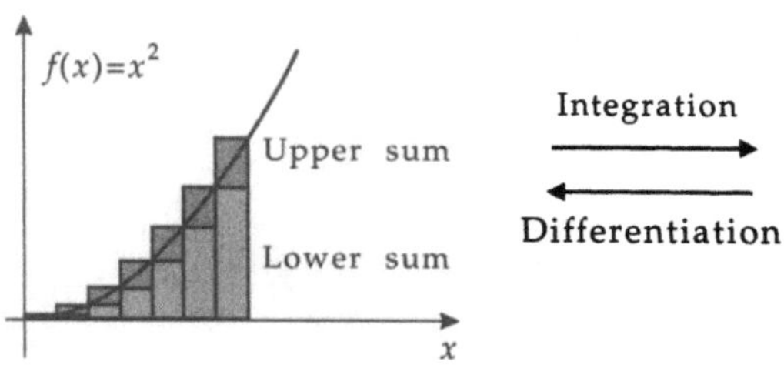

Upper/lower sum

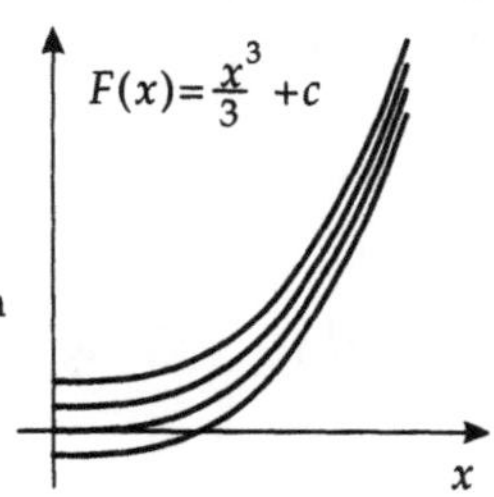

Areas below/above the x-axis

● **Calculation of an area**: Subdivide the interval $[a, b]$ into arbitrary subintervals Δx_i. If s_i and S_i are the greatest and the least upper bound of the function values in the subinterval i, then:

$$\textbf{Upper sum:} \quad U_n = \sum_{i=1}^{n} S_i \Delta x_i$$

$$\textbf{Lower sum:} \quad L_n = \sum_{i=1}^{n} s_i \Delta x_i$$

are the areas of the rectangles that rise above or lie below the curve, respectively.

▷ The upper and lower sums can also be negative if the function values are negative.

Lower and upper integrals: Denote the limits of the upper and lower sums for an infinite number of subintervals (if the limits exist).

▷ The upper and lower sums are the foundation of numerical integration according to the rectangular rule.

15.1.4 Rules for integrability

Riemann's integrability of $f(x)$: The **upper and lower integrals** of a function $f(x)$ over the interval $[a, b]$ are equal to each other.

□ The function $y = x^2$ is integrable over $[0, 1]$:
Subdivision into n equal intervals of length, $h = 1/n$.
Since the function is monotonically increasing, $s_i = f(x_{i-1})$ and $S_i = f(x_i)$ and

$$\lim_{n\to\infty} L_n = \lim_{n\to\infty} h \sum_{i=0}^{n-1} x_i^2 = \lim_{n\to\infty} \frac{1}{n-1} \sum_{i=1}^{n} \left(\frac{i}{n}\right)^2 = \lim_{n\to\infty} \frac{n(n-1)(2n-1)}{6n^3} = \frac{1}{3},$$

$$\lim_{n\to\infty} U_n = \lim_{n\to\infty} h \sum_{i=1}^{n} x_i^2 = \lim_{n\to\infty} \frac{1}{n} \sum_{i=1}^{n} \left(\frac{i}{n}\right)^2 = \lim_{n\to\infty} \frac{n(n+1)(2n+1)}{6n^3} = \frac{1}{3}.$$

The lower integral is equal to the upper integral; i.e., $y = x^2$ is integrable.

● **Functions continuous** in a closed interval are integrable over this interval.

□ x^n, $\sin x$, $\cos x$, e^x, $\ln x$, $|x|$, and algebraic compositions thereof are integrable (see the detailed tables in Chapter 25).

▷ Integrable functions do not necessarily have to be continuous!

● **Bounded functions** with a finite number of jumps are **integrable**.

□ The signum function $y = \text{sgn}x$ is integrable everywhere.

□ The absolute value function $y = |x|$ is integrable everywhere.

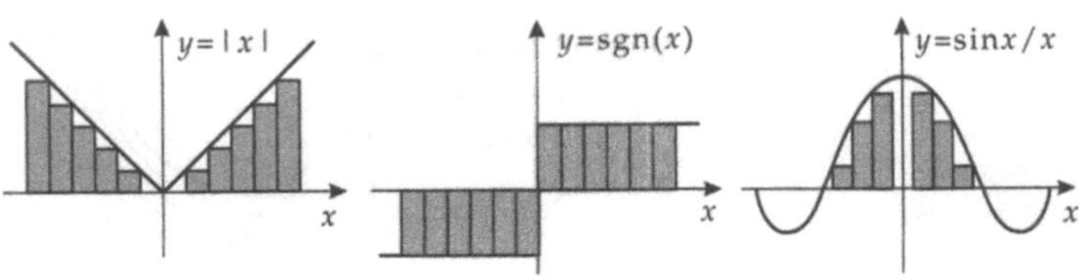

Integrable functions with jumps/salient points

- Unbounded functions, e.g., functions with poles, are not integrable at the pole. With such functions the integration must not include a pole (see also section on improper integrals below).

□ $\int_{-1}^{1} \frac{1}{x} \mathrm{d}x$ cannot be solved in this way because $f(x)$ has a pole at $x = 0$.

15.1.5 Improper integrals

There are two types of **improper integrals**:

(a) Integrals with infinite integration intervals: The limits of integration are improper numbers, $+\infty$, or $-\infty$.

- If one **limit of integration** is infinite, then

$$\int_a^{\infty} f(x)\mathrm{d}x = \lim_{b\to\infty} \int_a^b f(x)\mathrm{d}x$$

$$\int_{-\infty}^{b} f(x)\mathrm{d}x = \lim_{a\to\infty} \int_{-a}^b f(x)\mathrm{d}x$$

$$\int_{-\infty}^{\infty} f(x)\mathrm{d}x = \lim_{a\to\infty} \int_{-a}^a f(x)\mathrm{d}x$$

First calculate the integral with finite limits, then form the limit.

□ $\int_0^{\infty} \mathrm{e}^{-E/T}\mathrm{d}E = \lim_{b\to\infty} \int_0^b \mathrm{e}^{-E/T}\mathrm{d}E = -\frac{1}{T} \lim_{b\to\infty} \left(\mathrm{e}^{-b/T} - \mathrm{e}^0\right) = \frac{1}{T}, \quad$ if $T > 0$.

□ $\int_1^{\infty} \frac{\mathrm{d}x}{x^2} = \lim_{b\to\infty} \int_1^b \frac{\mathrm{d}x}{x^2} = \lim_{b\to\infty} \left[-\frac{1}{x}\right]_1^b = \lim_{b\to\infty} \left(-\frac{1}{b} + 1\right) = 1.$

□ $\int_a^{\infty} \frac{\mathrm{d}x}{x^n} = \lim_{b\to\infty} \int_a^b \frac{\mathrm{d}x}{x^n} = \lim_{b\to\infty} \left[\frac{-1}{(n-1)x^{n-1}}\right]_a^b$

$= \lim_{b\to\infty} \left(\frac{-1}{(n-1)b^{n-1}} - \frac{-1}{(n-1)a^{n-1}}\right) = \frac{a^{1-n}}{n-1} \quad$ for $n > 1, a > 0$.

▷ Note the sign in forming the limit!

(b) Integrals with points of discontinuity in the integrand, at which the integrand assumes improper values, $+\infty$, or $-\infty$.

- If the integrand has a **point of discontinuity** at u, then calculate the integral to the left and to the right of it:

$$\int_a^b f(x)\mathrm{d}x = \lim_{\varepsilon\to 0} \int_a^{u-\varepsilon} f(x)\mathrm{d}x + \lim_{\varepsilon\to 0} \int_{u+\varepsilon}^b f(x)\mathrm{d}x$$

□ $\int_0^1 \frac{\mathrm{d}x}{\sqrt{1-x^2}} = \lim_{\varepsilon\to 0} \int_0^{1-\varepsilon} \frac{\mathrm{d}x}{\sqrt{1-x^2}} = \lim_{\varepsilon\to 0} [\arcsin x]_0^{1-\varepsilon}$

$= \lim_{\varepsilon\to 0} (\arcsin(1-\varepsilon) - \arcsin 0) = \arcsin 1 = \frac{\pi}{2}.$

□ $$\int_0^b \frac{\mathrm{d}x}{x^n} = \lim_{a\to 0}\int_a^b \frac{\mathrm{d}x}{x^n} = \lim_{a\to 0}\left[\frac{-1}{(n-1)x^{n-1}}\right]_a^b$$

$$= \lim_{a\to 0}\left(\frac{-1}{(n-1)b^{n-1}} - \frac{-1}{(n-1)a^{n-1}}\right) = \frac{b^{1-n}}{1-1} \quad \text{for } n < 1, b > 0.$$

Cauchy principal value:

$$\int_a^b f(x)\mathrm{d}x = \lim_{\varepsilon\to 0}\left(\int_a^{u-\varepsilon} f(x)\mathrm{d}x + \int_{u+\varepsilon}^b f(x)\mathrm{d}x\right)$$

can exist evenly though the improper integral is divergent.

□ $$\int_{-1}^2 \frac{1}{x}\mathrm{d}x = \lim_{\varepsilon\to 0}\left(\int_\varepsilon^2 \frac{1}{x}\mathrm{d}x + \int_{-1}^{-\varepsilon} \frac{1}{x}\mathrm{d}x\right)$$

$$= \lim_{\varepsilon\to 0}(\ln 2 - \ln\varepsilon + \ln|-\varepsilon| - \ln|-1|) = \ln 2.$$

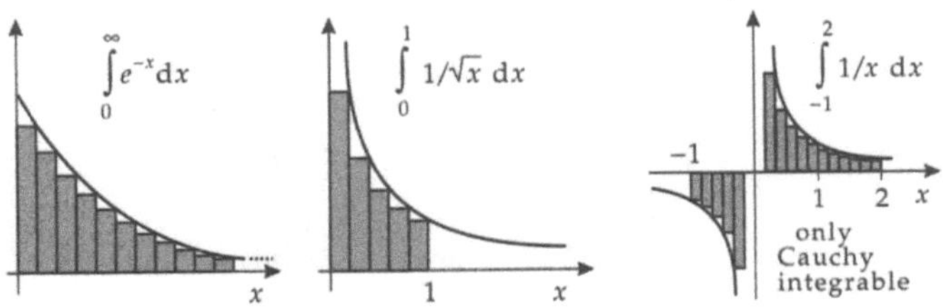

Improper integrals

Convergent improper integral: The limit does exist.
Divergent improper integral: The limit does not exist.

15.2 Integration rules

15.2.1 Rules for indefinite integrals

- With all **indefinite integrals**, an **integration constant** results:

$$\int f(x)\mathrm{d}x = F(x) + c\,.$$

- **Integration** and **differentiation** cancel each other:

$$\frac{d}{dx}\int_a^x f(x')\mathrm{d}x' = f(x)\,.$$

- **Constant rule**:

$$\int c\cdot f(x)\mathrm{d}x = c\cdot\int f(x)\mathrm{d}x\,.$$

A constant factor can be factored out of the integral: [−]

- **Sum rule**:

$$\int (f(x) + g(x))\mathrm{d}x = \int f(x)\mathrm{d}x + \int g(x)\mathrm{d}x \,.$$

The integral of a sum is equal to the sum of the integrals: [−]

- Generalization for a finite number of summands:

$$\int \sum_{i=1}^{n} f_i(x)\mathrm{d}x = \sum_{i=1}^{n} \int f_i(x)\mathrm{d}x \,.$$

15.2.2 Rules for definite integrals

Definite integrals: Calculation by substituting the limits into the primitive of function and subtracting the value of the primitive at the lower limit of the value at the upper limit:

$$\int_a^b f(x)\mathrm{d}x = F(x)|_a^b = F(b) - F(a) \,.$$

- **Change of sign** of the integral by interchanging the limits of integration:

$$\int_a^b f(x)\mathrm{d}x = -\int_b^a f(x)\mathrm{d}x \,.$$

- If the **upper and lower limits are equal**, the integral is zero:

$$\int_a^a f(x)\mathrm{d}x = 0 \,.$$

- Definite integrals can be subdivided into a finite number of **subintervals**, e.g., two:

$$\int_a^b f(x)\mathrm{d}x = \int_a^c f(x)\mathrm{d}x + \int_c^b f(x)\mathrm{d}x \,.$$

- Generalization to n subintervals:

$$\int_{a_0}^{a_n} f(x)\mathrm{d}x = \sum_{i=1}^{n} \int_{a_{i-1}}^{a_i} f(x)\mathrm{d}x \,.$$

- Changing the **name** of the **integration variables** does not change the value of the integral.

$$\int_a^b f(x)\mathrm{d}x = \int_a^b f(z)\mathrm{d}z \,.$$

The value of a definite integral is determined only by the limits and by the function $f(x)$.

15.2.3 Table of integration rules

Integration constant	$\int f(x)\mathrm{d}x = F(x) + c$
Integration ↔ differentiation	$\frac{\mathrm{d}}{\mathrm{d}x}\int_a^x f(\tilde{x})\mathrm{d}\tilde{x} = f(x)$
Factor rule	$\int c \cdot f(x)\mathrm{d}x = c \cdot \int f(x)\mathrm{d}x$
Sum rule	$\int (f(x) \pm g(x))\mathrm{d}x = \int f(x)\mathrm{d}x \pm \int g(x)\mathrm{d}x$
Power rule	$\int x^r \mathrm{d}x = \frac{x^{r+1}}{r+1}, \quad r \in \mathbb{R},\ r \neq -1$
Principal theorem	$\int_a^b f(x)\mathrm{d}x = F(x)\rvert_a^b = F(b) - F(a)$
Interchange rule	$\int_a^b f(x)\mathrm{d}x = -\int_b^a f(x)\mathrm{d}x$
Equal limits	$\int_a^a f(x)\mathrm{d}x = 0$
Interval rule	$\int_a^b f(x)\mathrm{d}x = \int_a^c f(x)\mathrm{d}x + \int_c^b f(x)\mathrm{d}x$
Integration by parts	$\int f(x) \cdot g'(x)\,\mathrm{d}x = f(x) \cdot g(x) - \int f'(x) \cdot g(x)\,\mathrm{d}x$
Substitution rule	$\int f(g(x))g'(x)\mathrm{d}x = \int f(z)\mathrm{d}z;\ z = g(x)$
Logarithmic integration	$\int \frac{f'(x)}{f(x)}\mathrm{d}x = \ln\lvert f(x)\rvert + c$
Integration of inverse function	$\int f^{-1}(x)\mathrm{d}x = xf^{-1}(x) - F\left(f^{-1}(x)\right) + c$
Particular form of integrand	$\int f(x)f'(x)\mathrm{d}x = \frac{1}{2}(f(x))^2 + c$

15.2.4 Integrals of some elementary functions

The table of elementary indefinite integrals can be obtained by inverting the differential calculus.

▷ To avoid confusion, the inverse trigonometric functions are denoted by lowercase letters, and the area hyperbolic functions by capital letters.

▷ $\int 1/(1-x^2)\mathrm{d}x$ cannot be integrated over $x = \pm 1$ (integrand diverges).

▷ Integrals of elementary functions do not always become elementary functions themselves. They often cannot be given in closed form, but can only be calculated numerically.

□ Sine integral function:

$$\mathrm{Si}(x) = \int_0^x \frac{\sin t}{t}\mathrm{d}t$$

cannot be given in closed form. See Section 15.3 on integration in terms of series expansion.

Integral	Result	Integral	Result
$\int 0\,dx$	c	$\int \sin x\,dx$	$-\cos x + c$
$\int 1\,dx$	$x + c$	$\int \cos x\,dx$	$\sin x + c$
$\int x\,dx$	$\frac{1}{2}x^2 + c$	$\int \tan x\,dx$	$-\ln\lvert\cos x\rvert + c$ $(x \neq (2k+1)\frac{\pi}{2})$
$\int x^2\,dx$	$\frac{1}{3}x^3 + c$	$\int \cot x\,dx$	$\ln\lvert\sin x\rvert + c$ $(x \neq k\pi)$
$\int x^3\,dx$	$\frac{1}{4}x^4 + c$	$\int \frac{1}{\sin^2 x}\,dx$	$-\cot x + c$ for $\sin x \neq 0$
$\int x^n\,dx$	$\frac{1}{n+1}x^{n+1} + c$ $(n \in \mathbb{N}, n \neq 1)$	$\int \frac{1}{\cos^2 x}\,dx$	$\tan x + c$ for $\cos x \neq 0$
$\int \frac{1}{x}\,dx$	$\ln\lvert x\rvert + c$ $(x \neq 0)$	$\int \sinh x\,dx$	$\cosh x + c$
$\int \frac{1}{x^2}\,dx$	$-\frac{1}{x} + c$ $(x \neq 0)$	$\int \cosh x\,dx$	$\sinh x + c$
$\int \frac{1}{x^n}\,dx$	$-\frac{1}{(n-1)x^{n-1}} + c$ $(n \in \mathbb{N}, n \neq 1)$	$\int \tanh x\,dx$	$\ln(\cosh x) + c$
$\int \sqrt{x}\,dx$	$\frac{2}{3}\sqrt{x^3} + c$ $(x \geq 0)$	$\int \coth x\,dx$	$\ln\lvert\sinh x\rvert + c$ $(x \neq 0)$
$\int \sqrt[3]{x}\,dx$	$\frac{3}{4}\sqrt[3]{x^4} + c$	$\int \frac{1}{\sinh^2 x}\,dx$	$-\coth x + c$ $(x \neq 0)$
$\int \frac{1}{\sqrt{x}}\,dx$	$2\sqrt{x} + c$ $(x > 0)$	$\int \frac{1}{\cosh^2 x}\,dx$	$\tanh x + c$
$\int x^r\,dx$	$\frac{1}{r+1}x^{r+1} + c$ $(x > 0, r \in \mathbb{R}, r \neq -1)$	$\int \frac{1}{1+x^2}\,dx$	$\arctan x + c$ $-\text{arccot}\, x + c_2$
$\int e^x\,dx$ $x \neq 0$	$e^x + c$	$\int \frac{1}{a^2+x^2}\,dx$	$\frac{1}{a}\arctan\frac{x}{a}$ $(a \neq 0)$
$\int a^x\,dx$	$\frac{1}{\ln a}a^x + c$ $(a > 0)$	$\int \frac{1}{1-x^2}\,dx$	$\begin{cases} \text{Artanh}\, x + c = \frac{1}{2}\ln\left(\frac{1+x}{1-x}\right) + c \; (\lvert x\rvert < 1) \\ \text{Arcoth}\, x + c = \frac{1}{2}\ln\left(\frac{x+1}{x-1}\right) + c \; (\lvert x\rvert > 1) \end{cases}$
$\int \ln\lvert x\rvert\,dx$	$x(\ln\lvert x\rvert - 1) + c$ $(x \neq 0)$	$\int \frac{1}{\sqrt{1+x^2}}\,dx$	$\text{Arsinh}\, x + c$ $= \ln(x + \sqrt{x^2+1} + c$
$\int \log_a x\,dx$	$\frac{x}{\ln a}(\ln x - 1) + c$ $(a, x > 0)$	$\int \frac{1}{\sqrt{1-x^2}}\,dx$	$\arcsin x + c_1$ $= \arccos x + c_2$ $(\lvert x\rvert < 1)$
$\int e^{ax}\,dx$	$\frac{1}{a}e^{ax} + c$ $(a \neq 0)$	$\int \frac{1}{\sqrt{x^2-1}}\,dx$	$\text{Arcosh} x + c$ $= \ln(x + \sqrt{x^2-1}) + c$ $(\lvert x\rvert > 1)$

15.3 Integration methods

▷ In contrast to differentiation, no universally valid rules can be given for the integration of arbitrary functions!

Frequently used methods:

(1) Integral rational functions (polynomials) are integrated term-by-term:

$$\int \left(a_n x^n + \cdots + a_1 x + a_0\right) \mathrm{d}x = \frac{a_n}{n+1} x^{n+1} + \cdots + \frac{a_1}{2} x^2 + a_0 x + c\,.$$

(2) Transformation of the integrand into a sum of several functions, i.e., separation of the integral into a sum of integrals.

□ $\int (x+1)(4x^2+7)\mathrm{d}x = \int 4x^3\mathrm{d}x + \int 4x^2\mathrm{d}x + \int 7x\,\mathrm{d}x + \int 7\mathrm{d}x$

(3) Constant in the argument:
if $\int f(x)\mathrm{d}x$ is known, but $f(ax)$, $f(x+b)$, or $f(ax+b)$ has to be integrated, then

$$\int f(ax)\mathrm{d}x = \frac{1}{a} F(ax) + c\,,$$

$$\int f(ax+b)\mathrm{d}x = \frac{1}{a} F(ax+b) + c\,.$$

□ $\int \mathrm{e}^{x+b}\mathrm{d}x = \mathrm{e}^{x+b} + c$

$\int \cos(ax)\mathrm{d}x = \frac{1}{a}\sin(ax) + c$

(4) Inversion of logarithmic differentiation.
If the integrand has the form $\frac{f'(x)}{f(x)}$, then the integral is equal to the logarithm of the **denominator**:

$$\int \frac{f'(x)}{f(x)}\mathrm{d}x = \ln|f(x)| + c\,.$$

▷ Take the logarithm of $f(x)$, **not** of $f'(x)$!

(5) Particular form of the integrand:

$$\int f(x)f'(x)\mathrm{d}x = \frac{1}{2}(f(x))^2 + c$$

15.3.1 Integration by substitution

● **Substitution rule**: If $f(x)$ is continuous, and $g(x)$ is continuously differentiable and invertible, then

$$\int_a^b f(g(x))\mathrm{d}x = \int_{g(a)}^{g(b)} f(z)\frac{\mathrm{d}x}{\mathrm{d}z}\mathrm{d}z = \int_{g(a)}^{g(b)} f(z)\frac{1}{g'(x)}\mathrm{d}z\,.$$

▷ At the end, do not forget back substitution (the variable x has to be replaced by the substitution function solved for x: $x = g^{-1}(z)$), or the limits must be changed.

□ $$\int_1^3 6x\ln(x^2)\mathrm{d}x = \int_1^9 6x\ln z\frac{\mathrm{d}z}{2x} = 3\int_1^9 \ln z\,\mathrm{d}z =$$

$3(9\ln 9 - 9 - \ln 1 + 1) = 27\ln 9 - 24.$

$\left(\text{Substitution } z = g(x) = x^2,\ z' = \frac{\mathrm{d}z}{\mathrm{d}x} = 2x,\ \text{inverse function } x = \sqrt{z}\right)$

□ $$\int \frac{1}{5x-7}\mathrm{d}x = \frac{1}{5}\int \frac{1}{z}\mathrm{d}z = \frac{1}{5}\ln|z| + c = \frac{1}{5}\ln|5x-7| + c$$

(Substitution: $z = 5x - 7$)

□ $$\int \sin(3-7x)\mathrm{d}x = -\frac{1}{7}\int \sin z\mathrm{d}z = \frac{1}{7}\cos z + c = \frac{1}{7}\cos(3-7x) + c$$

(Substitution: $z = 3 - 7x$)

□ $$\int \frac{1}{\sqrt{x}+x}\mathrm{d}x = 2\int \frac{1}{z+z^2}z\mathrm{d}z = 2\int \frac{1}{1+z}\mathrm{d}z = 2\ln|z+1| + c =$$

$2\ln|\sqrt{x}+1| + c$

(Substitution: $z = \sqrt{x}$)

□ $$\int xe^{x^2}\mathrm{d}x = \frac{1}{2}\int e^z\mathrm{d}z = \frac{1}{2}e^z + c = \frac{1}{2}e^{x^2} + c$$

(Substitution: $z = x^2$)

□ $$\int \frac{1}{e^x + e^{-x}}\mathrm{d}x = \int \frac{1}{z+1/z}\frac{1}{z}\mathrm{d}z = \int \frac{1}{z^2+1}\mathrm{d}z = \arctan z + c = \arctan(e^x) + c$$

(Substitution: $z = e^x$)

□ $$\int \sqrt{1+x^2}\mathrm{d}x = \int \cosh^2 z\mathrm{d}z = \frac{1}{2}(z + \sinh z\cdot\cosh z) + c =$$

$\frac{1}{2}\left(\mathrm{Arsinh}x + x\cdot\sqrt{1+x^2}\right) + c$

(Substitution: $z = \mathrm{Arsinh}x$)

□ $$\int \frac{\mathrm{d}x}{\sin x} = \int \frac{1+z^2}{2z}\cdot\frac{2\mathrm{d}z}{1+z^2} = \int \frac{\mathrm{d}z}{z} = \ln|z| + c = \ln|\tan(x/2)| + c$$

(Substitution: $z = \tan(x/2)$

Integral	Substitution	Result
$\int f(ax+b)\mathrm{d}x$	$z=ax+b$	$\frac{1}{a}\int f(z)\mathrm{d}z$
$\int f(ax^2+bx+c)\mathrm{d}x$	$z=x+\frac{b}{2a}$	$\int f\left(az^2+c-b^2/4a\right)\mathrm{d}z$
$\int (f(x))^n f'(x)\mathrm{d}x$, $n\neq 1$	$z=f(x)$	$\frac{1}{n+1}[f(x)]^{n+1}+c$
$\int \frac{f'(x)}{f(x)}\mathrm{d}x$	$z=f(x)$	$\ln\lvert f(x)\rvert+c$
$\int f(g(x))g'(x)\mathrm{d}x$	$z=g(x)$	$\int f(z)\mathrm{d}z$
$\int f\left(\frac{1}{x}\right)\mathrm{d}x$	$z=\frac{1}{x}$	$-\int \frac{f(z)}{z^2}\mathrm{d}z$
$\int f\left(\sqrt{x}\right)\mathrm{d}x$	$z=\sqrt{x}$	$2\int z\cdot f(z)\mathrm{d}z$
$\int f\left(\sqrt[n]{ax+b}\right)\mathrm{d}x$	$z=\sqrt[n]{ax+b}$	$\int f(z)\frac{n}{a}z^{n-1}\mathrm{d}z$
$\int f\left(\sqrt{x^2+a^2}\right)\mathrm{d}x$	$z=\mathrm{Arsinh}\frac{x}{a}$ $\sqrt{x^2+a^2}=a\cosh z$	$\int f(a\cosh z)a\cosh z\mathrm{d}z$
$\int f\left(\sqrt{x^2-a^2}\right)\mathrm{d}x$	$z=\mathrm{Arcosh}\frac{x}{a}$ $\sqrt{x^2-a^2}=a\sinh z$	$\int f(a\sinh z)a\sinh z\mathrm{d}z$
$\int f\left(\sqrt{a^2-x^2}\right)\mathrm{d}x$	$z=\arcsin\frac{x}{a}$ $\sqrt{a^2-x^2}=a\cos z$	$\int f(a\cos z)a\cos z\mathrm{d}z$
$\int f(\sin x;\cos x)\mathrm{d}x$	$z=\tan\frac{x}{2}$	$\int f\left(\frac{2z}{1+z^2};\frac{1-z^2}{1+z^2}\right)\frac{2}{1+z^2}\mathrm{d}z$
$\int f\left(a^x\right)\mathrm{d}x$	$z=a^x$	$\frac{1}{\ln a}\int f(z)\frac{1}{z}\mathrm{d}z$
$\int f(\sinh x;\cosh x)\mathrm{d}x$	$z=\mathrm{e}^x$	$\int f\left(\frac{z^2-1}{2z};\frac{z^2+1}{2z}\right)\frac{1}{z}\mathrm{d}z$

▷ The limits of integration must be in the domain of definition of the substitution.

▷ The new limits of integration are calculated via the substitution equation.

▷ After substitution, the integral obtained may have to be treated further by substitution, integration by parts, or with the help of partial fraction decomposition.

□ $\int \frac{x^3}{\sqrt{x^2-1}}\mathrm{d}x=\int \frac{\cosh^3 z}{\sinh z}\sinh z\,\mathrm{d}z=\int \cosh^3 z\,\mathrm{d}z=\int (1+\sinh^2 z)\cosh z\,\mathrm{d}z=$

$\int (1+v^2)\mathrm{d}v=\sinh z+\frac{1}{3}\sinh^3 z+c=\sqrt{x^2-1}+\frac{1}{3}(x^2-1)^{3/2}+c$

(First substitution: $z=\mathrm{Arcosh}x$; second substitution: $v=\sinh z$)

□ $I=\int_2^8 (3x+4)\mathrm{d}x=\int u\frac{\mathrm{d}u}{3}=\frac{1}{6}u^2=\frac{1}{6}(3x+4)^2\Big|_2^8=114\,,\qquad u=3x+4$

or

$$I = \int_2^8 (3x+4)\mathrm{d}x = \int_{10}^{28} u\frac{\mathrm{d}u}{3} = \frac{1}{6}u^2\Big|_{10}^{28} = \frac{1}{6}(28^2 - 10^2) = 114$$

▷ Occasionally, the goal is reached faster by simple substitutions than by the above named trigonometric (hyperbolic) substitutions.

□ $\int x\sqrt{x^2-4}\,\mathrm{d}x = \frac{1}{2}\int \sqrt{z}\mathrm{d}z = \frac{1}{2}\frac{2}{3}z^{3/2} = \frac{1}{3}\left(x^2-4\right)^{3/2}$

(Substitution: $z = x^2 - 4$, $\mathrm{d}z = 2x\mathrm{d}x$)

□ $\int \sin^3 x \cos x \mathrm{d}x = \int z^3 \mathrm{d}z = \frac{1}{4}z^4 + c = \frac{1}{4}\sin^4 x + c$

(Substitution: $z = \sin x$)

□ $\int \frac{x^3}{\sqrt{1+x^4}}\mathrm{d}x = \frac{1}{2}\int \frac{z\,\mathrm{d}z}{z} = \frac{1}{2}\int \mathrm{d}z = \frac{1}{2}z + c = \frac{1}{2}\sqrt{1+x^4} + c$

(Substitution: $z = \sqrt{1+x^4}$)

□ $\int \tan x \mathrm{d}x = \int \frac{\sin x}{\cos x}\mathrm{d}x = -\int \frac{\mathrm{d}z}{z} = -\ln|z| + c = -\ln|\cos x| + c$

(Substitution: $z = \cos x$)

□ $\int \frac{\cos x}{\sqrt{1+\sin^2 x}}\mathrm{d}x = \int \frac{\mathrm{d}z}{\sqrt{1+z^2}} = \mathrm{Arsinh}(z) + c = \mathrm{Arsinh}(\sin x) + c$

(Substitution: $z = \sin x$)

▷ Other integrals often occurring in practice are shown in the table of integrals in Chapter 25.

Euler substitution for the particular integral

$$I = \int f(\sqrt{ax^2+bx+c})\mathrm{d}x$$

Case	Substitution	Differential
$a > 0$	$\sqrt{ax^2+bx+c} = x\sqrt{a} + z$ $x = \dfrac{z^2 - c}{b - 2z\sqrt{a}}$	$\mathrm{d}x = 2\dfrac{-z^2\sqrt{a} + bz - c\sqrt{a}}{(b - 2z\sqrt{a})^2}\mathrm{d}z$
$c > 0$	$\sqrt{ax^2+bx+c} = xz + \sqrt{c}$ $x = \dfrac{2z\sqrt{c} - b}{a - z^2}$	$\mathrm{d}x = 2\dfrac{a\sqrt{c} - bz + z^2\sqrt{c}}{(a - z^2)^2}\mathrm{d}z$
Real roots x_1, x_2	$\sqrt{ax^2+bx+c} = z(x - x_1)$ $x = \dfrac{z^2 x_1 - a x_2}{z^2 - a}$	$\mathrm{d}x = 2\dfrac{az(x_2 - x_1)}{(z^2 - a)^2}\mathrm{d}z$

15.3.2 Integration by parts

Integration by parts: The inverse of the product rule for differentiation, with symbolic notation:

$$(uv)' = u'v + v'u \quad \rightarrow \quad \int uv' = uv - \int u'v\,.$$

- **Integration by parts**: Integration by two partial integrations:

$$\int f(x)g'(x)\mathrm{d}x = g(x)f(x) - \int g(x)f'(x)\mathrm{d}x\,.$$

Application of the rule in particular if the integrand is a **product of functions.**

▷ The derivative $f'(x)$ should be a function simpler than $f(x)$.

▷ It should be easy to integrate the function $g'(x)$, and the result should not be more complicated.

□ $\int x\mathrm{e}^x \mathrm{d}x = x\mathrm{e}^x - \int \mathrm{e}^x \mathrm{d}x = x\mathrm{e}^x - \mathrm{e}^x + c$

▷ If necessary, integration by parts can be applied several times in succession.

□ $\int x^2\mathrm{e}^x \mathrm{d}x = x^2\mathrm{e}^x - 2\int x\mathrm{e}^x \mathrm{d}x = x^2\mathrm{e}^x - 2x\mathrm{e}^x + 2\mathrm{e}^x + c$

Sometimes, the introduction of a product leads to the goal.

□ $\int \ln x\mathrm{d}x = \int 1 \ln x\mathrm{d}x = x\ln x - \int x\frac{1}{x}\mathrm{d}x = x\ln x - x + c$

▷ Rules:

1. Integrate the first factor,
2. write the resulting product,
3. differentiate the second factor,
4. write the resulting product with a minus sign under the integral.

▷ The **minus sign** before the integral is often forgotten, in particular, when integration by parts is applied several times.

Sometimes the original integral is obtained again in integration by parts. Then, the equation is solved according to this integral.

□ $\int \sin x \cos x\mathrm{d}x = \sin^2 x - \int \cos x \sin x\mathrm{d}x$

$\rightarrow \quad \int \sin x \cos x\mathrm{d}x = \frac{1}{2}\sin^2 x + c$

Special cases of integration by parts:

(a) The integrand is a product of a polynomial $p(x)$ of degree n and one of the functions e^x, $\sin x$, $\cos x$, $\sinh x$, or $\cosh x$:
Multiple (n-times) integration by parts for $f(x) = p(x)$.

□ $$\int x^2 \sin x\mathrm{d}x = -x^2\cos x + \int 2x\cos x\mathrm{d}x$$
$$= -x^2\cos x + 2x\sin x - \int 2\sin x\mathrm{d}x$$
$$= -x^2\cos x + 2x\sin x + 2\cos x + c.$$

(b) The integrand is a product of two of the functions e^x, $\sin x$, $\cos x$, $\sinh x$, or $\cosh x$:
Twofold product integration by parts leads back to the original integral; then the equation can be solved for this integral. The following relations may be used if necessary:

$$\sin^2 x + \cos^2 x = 1, \quad \cosh^2 x - \sinh^2 x = 1.$$

□ $\int e^x \sin x dx = e^x \sin x - \int e^x \cos dx$

$= e^x \sin x - e^x \cos x - \int e^x \sin x dx$

$\rightarrow \int e^x \sin x dx = \frac{1}{2} e^x (\sin x - \cos x) + c.$

▷ For a product of e^x and a hyperbolic function, replace the latter by

$$\cosh x = \frac{1}{2}(e^x + e^{-x}), \quad \text{or} \quad \sinh x = \frac{1}{2}(e^x - e^{-x}) .$$

and then substitute $z = e^x$.

(c) The integrand is a product of a rational function and a logarithmic inverse trigonometric, or area hyperbolic function:
Set the rational function equal to $g'(x)$.

□ $\int \frac{1}{x^2} \text{Arsinh} x dx = -\frac{1}{x} \text{Arsinh} x + \int \frac{dx}{x\sqrt{1+x^2}}$

$= -\frac{1}{x} \text{Arsinh} x + \int \frac{dz}{z^2 - 1} = -\frac{1}{x} \text{Arsinh} x - \text{Arcoth} \sqrt{x^2+1} + c, x > 1$

(Substitution $z = \sqrt{x^2+1}$).

15.3.3 *Integration by partial fraction decomposition*

Fractional rational functions that occur, for example, in the application of Laplace transformations can often be integrated via partial fraction decomposition.

1. The integrand is split [–] into an integral rational function $P(x)$ and a proper fractional rational function $R(x)$ [e.g., by polynomial division]:

$$\frac{f(x)}{g(x)} = P(x) + R(x) = P(x) + \frac{Z(x)}{N(x)} ,$$

where the degree of the polynomial of the numerator $Z(x)$ is smaller than that of the denominator $N(x)$.

▷ The polynomials in the numerator and in the denominator must be relatively prime.

▷ Important for transient processes in electronics and automatic control technology, Laplace transformation.

2. Determine all zeros x_i of the denominator $N(x)$ (also the complex zeros):

$$N(x) = a \prod (x - x_i)^{m_i} ,$$

where individual zeros x_i may occur as multiple (m_i-fold) zeros.

3. Proper fractional rational functions $R(x)$ are split into a sum of partial fractions where for every zero it holds that:
Simple real zero at x_0:

$$R(x) = \frac{Z(x)}{N(x)} = \frac{A}{x - x_0} + \cdots$$

Two real zeros at x_0, x_1:

$$R(x) = \frac{Z(x)}{N(x)} = \frac{A}{x_0 - x_1}\left(\frac{1}{x - x_0} - \frac{1}{x - x_1}\right) + \cdots$$

Double real zero at x_0:

$$R(x) = \frac{Z(x)}{N(x)} = \frac{A_2}{(x - x_0)^2} + \frac{A_1}{x - x_0} + \cdots$$

n**-fold real zero** at x_0:

$$R(x) = \frac{Z(x)}{N(x)} = \sum_{i=1}^{n} \frac{A_i}{(x - x_0)^i} + \cdots$$

Simple complex zero at $x_0 = s_0 \pm \mathrm{j}t_0$:

$$R(x) = \frac{Z(x)}{N(x)} = \frac{Ax + B}{x^2 - 2s_0x + s_0^2 + t_0^2} + \cdots$$

Double complex zero at $x_0 = s_0 \pm \mathrm{j}t_0$:

$$R(x) = \frac{A_2x + B_2}{\left(x^2 - 2s_0x + s_0^2 + t_0^2\right)^2} + \frac{A_1x + B_1}{x^2 - 2s_0x + s_0^2 + t_0^2} + \cdots$$

n**-fold complex zero** at $x_0 = s_0 \pm \mathrm{j}t_0$:

$$R(x) = \sum_{i=1}^{n} \frac{A_ix + B_i}{\left(x^2 - 2s_0x + s_0^2 + t_0^2\right)^i} + \cdots$$

4. Determination of the constants A_i and B_i of the decomposition by multiplication by the least common denominator, then equating the coefficients (equating the factors before the terms x^i on both sides of the equation) or application of the substitution method (substitution of different x values, e.g., the predetermined zeros of $N(x)$). For many equations, the Gaussian elimination method can be used (see Chapter **??** on systems of equations).

□ $R(x) = \dfrac{x + 1}{x^2 - 3x + 2} = \dfrac{A}{x - 1} + \dfrac{B}{x - 2} \to x + 1 = A(x - 2) + B(x - 1)$

$\to A + B = 1, \quad -2A - B = 1 \to A = -2, \quad B = 3$

$R(x) = -\dfrac{2}{x - 1} + \dfrac{3}{x - 2}$

□ $R(x) = \dfrac{3x^2 - 20x + 20}{(x - 2)^3(x - 4)} = \dfrac{A_1}{x - 2} + \dfrac{A_2}{(x - 2)^2} + \dfrac{A_3}{(x - 2)^3} + \dfrac{A_4}{x - 4}$

$\to 3x^2 - 20x + 20 = A_1(x - 2)^2(x - 4) + A_2(x - 2)(x - 4) + A_3(x - 4) + A_4(x - 2)^3$

Substitution method:
$x = 2$: $-8 = -2A_3 \to A_3 = 4$
$x = 4$: $-12 = 8A_4 \to A_4 = -3/2$
$x = 0$: $20 = -16A_1 + 8A_2 - 4A_3 - 8A_4$
$x = 1$: $3 = -3A_1 + 3A_2 - 3A_3 - A_4$
$\to A_1 = 3/2$, $A_2 = 6$

$R(x) = \dfrac{3}{2(x - 2)} + \dfrac{6}{(x - 2)^2} + \dfrac{4}{(x - 2)^3} - \dfrac{3}{2(x - 4)}$

5. Integration of the individual terms:

Integral rational function $P(x)$:

$$\int P(x)\mathrm{d}x = \int \left(a_n x^n + \cdots + a_1 x^1 + a_0\right)\mathrm{d}x = \frac{a_n}{n+1}x^{n+1} + \cdots + \frac{a_1}{2}x^2 + a_0 x$$

Simple real zero at x_0:

$$\int \frac{A}{x-x_0}\mathrm{d}x = A\ln|x-x_0|$$

Two real zeros x_0, x_1:

$$\int \frac{A}{x_0-x_1}\left(\frac{1}{x-x_0} - \frac{1}{x-x_1}\right)\mathrm{d}x = \frac{A}{x_0-x_1}\ln\left|\frac{x-x_0}{x-x_1}\right|$$

Double real zero at x_0:

$$\int\left(\frac{A_2}{(x-x_0)^2}\mathrm{d}x + \frac{A_1}{x-x_0}\mathrm{d}x\right) = -\frac{A_2}{x-x_0} + A_1\ln|x-x_0|$$

n**-fold real zero** at x_0:

$$\int\sum_{i=1}^{n}\frac{A_i}{(x-x_0)^i}\mathrm{d}x = -\sum_{i=1}^{n-1}\frac{A_{i+1}}{i(x-x_0)^i} + A_1\ln|x-x_0|$$

One pair of complex zeros at $x_0 = s_0 \pm \mathrm{j}t_0$:

$$\int\frac{Ax+B}{\left(x^2-2s_0x+s_0^2+t_0^2\right)}\mathrm{d}x =$$

$$\frac{A}{2}\ln|x^2-2s_0x+s^2+t^2| + \frac{As_0+B}{t_0}\cdot\arctan\left(\frac{x-s_0}{t_0}\right)$$

Double pair of complex zeros at $x_0 = s_0 \pm \mathrm{j}t_0$: Partial integration (the other partial integration can be performed as in the case of a simple complex zero):

$$\int\frac{A_2x+B_2}{(x^2-2s_0x+s_0^2+t_0^2)^2}\mathrm{d}x = \int\frac{A_2x+B_2}{X^2}\mathrm{d}x =$$

$$\frac{(A_2s_0+B_2)x - A_2s_0^2 - A_2t_0^2 - B_2s_0}{2t_0^2X} + \frac{A_2s_0+B_2}{2t_0^3}\cdot\arctan\left(\frac{x-s_0}{t_0}\right)$$

n**-fold pair of complex zeros**: The integration is performed with the help of the following recursion formulas:

$$\int\frac{Ax+B}{(x^2-2sx+s^2+t^2)^n}\mathrm{d}x = \int\frac{Ax+B}{X^n}\mathrm{d}x =$$

$$\frac{(As+B)x - As^2 - At^2 - Bs}{2t^2(n-1)X^{n-1}} + \frac{(As+B)(2n-3)}{2t^2(n-1)}\int\frac{\mathrm{d}x}{X^{n-1}}$$

$$\int\frac{\mathrm{d}x}{(x^2-2sx+s^2+t^2)^n} = \int\frac{\mathrm{d}x}{X^n} = \frac{x-s}{2t^2(n-1)X^{n-1}} + \frac{2n-3}{2t^2(n-1)}\int\frac{\mathrm{d}x}{X^{n-1}}$$

with $X = x^2 - 2sx + s^2 + t^2$.

▷ Many integrals are calculated explicitly in the table of integrals in Chapter 25.

□ $R(x) = \frac{N(x)}{Z(x)} = \frac{2x^2 - 2x + 4}{x^3 - x^2 + x - 1}, \quad Z(x) = (x-1)(x^2+1)$

Setup: $R(x) = \frac{A}{x-1} + \frac{Bx + C}{x^2+1}$

Equating the coefficients:

$2x^2 - 2x + 4 = A(x^2+1) + (Bx + C)(x-1)$

$= x^2(A + B) + x(C - B) + A - C$

$\to A = 2, \; B = 0, \; C = -2$

Integration: $\int \left(\frac{2}{x-1} - \frac{2}{x^2+1} \right) \mathrm{d}x = 2 \ln|x-1| - 2 \cdot \arctan x$

Summary of the partial fraction decomposition and of the integration of partial fractions for the different types of zeros x_0 of the denominator $N(x)$ function:

Zero at x_0 of $N(x)$	Setup for the partial fraction	Integration
Simple, real	$\frac{A}{x - x_0}$	$A \ln \lvert x - x_0 \rvert$
Two simple, real	$\frac{A}{x_0 - x_1}\left(\frac{1}{x - x_0} - \frac{1}{x - x_1}\right)$	$\frac{A}{x_0 - x_1} \ln \left\lvert \frac{x - x_0}{x - x_1} \right\rvert$
Double, real	$\frac{A}{(x - x_0)^2} + \frac{B}{x - x_0}$	$-\frac{A}{x - x_0} + B \ln \lvert x - x_0 \rvert$
n-fold, real	$\sum_{i=1}^{n} \frac{A_i}{(x - x_0)^i}$	$-\sum_{i=2}^{n} \frac{A_i}{(i-1)(x - x_0)^{i-1}}$ $+ A_1 \ln \lvert x - x_0 \rvert$
Simple, complex ($x_0 = s_0 \pm \mathrm{j}t_0$)	$\frac{Ax + B}{\left(x^2 - 2s_0 x + s_0^2 + t_0^2\right)}$	$\frac{A}{2} \ln \lvert x^2 - 2s_0 x + s^2 + t^2 \rvert$ $+ \frac{As_0 + B}{t_0} \cdot \arctan\left(\frac{x - s_0}{t_0}\right)$
n-fold, complex ($x_0 = s_0 \pm \mathrm{j}t_0$)	$\sum_{i=1}^{n} \frac{A_i x + B_i}{\left(x^2 - 2s_0 x + s_0^2 + t_0^2\right)^i}$	recursively, see simple, complex

15.3.4 Integration by series expansion

Power series expansion of the integrand with the radius of convergence r:

$$f(x) = \sum_{k=0}^{\infty} a_k \cdot x^k = a_0 + a_1 x + a_2 x^2 + a_3 x^3 + \cdots \quad \left(a_k = \frac{1}{k!} f^{(k)}(0), \; |x| < r \right)$$

Subsequent integration of the individual terms of the power series:

$$\int f(x)\mathrm{d}x = \sum_{k=0}^{\infty} a_k \frac{x^{k+1}}{k+1} = a_0 x + \frac{a_1}{2} x^2 + \frac{a_2}{3} x^3 + \cdots$$

This is usually possible for indefinite as well as for definite integrals!

▷ The limits of integration must lie within the radius of convergence r!

□ $\int \sin\sqrt{x}\,\mathrm{d}x$:

Power series: $\sin x = x - \frac{x^3}{3!} + \frac{x^5}{5!} - \cdots$, $\sin\sqrt{x} = \sqrt{x} - \frac{x^{3/2}}{3!} + \frac{x^{5/2}}{5!} - \cdots$,

Integration: $\int \sin\sqrt{x}\,\mathrm{d}x = \frac{2x^{3/2}}{3} - \frac{2x^{5/2}}{5\cdot 3!} + \frac{2x^{7/2}}{7\cdot 5!} - \cdots$

□ $\int_0^{0.5} \sqrt{1+x^2}\mathrm{d}x$ accurate to two places:

Power series for $z = x^2$: $\sqrt{1+z} = (1+z)^{1/2} = 1 + \frac{1}{2}z - \frac{1}{8}z^2 + \frac{3}{48}z^3 - \cdots$

$\rightarrow \sqrt{1+x^2} = 1 + \frac{1}{2}x^2 - \frac{1}{8}x^4 + \frac{3}{48}x^6 - \cdots$

Integration: $\int \sqrt{1+x^2}\mathrm{d}x = x + \frac{1}{6}x^3 - \frac{1}{40}x^5 + \frac{3}{336}x^7 - \cdots$

$\int_0^{0.5} \sqrt{1+x^2}\mathrm{d}x = 0.5 + 0.021 - -0.001 + \cdots \approx 0.520.$

Frequently occurring nonelementary integrals of elementary functions:

Sine integral:

$$\mathrm{Si}(x) = \int_0^x \frac{\sin t}{t}\mathrm{d}t = x - \frac{x^3}{18} + \frac{x^5}{600} - \frac{x^7}{35280} + \cdots + \frac{(-1)^i x^{2i+1}}{(2i+1)\cdot(2i+1)!} + \cdots$$

Cosine integral (Euler's constant $C = 0.57721566\ldots$):

$$\mathrm{Ci}(x) = \int_x^\infty \frac{\cos t}{t}\mathrm{d}t = C + \ln|x| - \frac{x^2}{4} + \frac{x^4}{96} - \frac{x^6}{4320} + \cdots + \frac{(-1)^i x^{2i}}{2i\cdot(2i)!} + \cdots$$

Exponential integral:

$$\mathrm{Ei}(x) = \int_{-\infty}^x \frac{\mathrm{e}^t}{t}\mathrm{d}t = C + \ln|x| + x + \frac{x^2}{4} + \frac{x^3}{18} + \frac{x^4}{96} + \cdots + \frac{x^i}{i\cdot i!} + \cdots$$

Logarithmic integral:

$$\mathrm{Li}(x) = \int_0^x \frac{\mathrm{d}t}{\ln t} =$$

$$C + \ln|\ln x| + \ln x + \frac{(\ln x)^2}{4} + \frac{(\ln x)^3}{18} + \frac{(\ln x)^4}{96} + \cdots + \frac{(\ln x)^i}{i\cdot i!} + \cdots$$

Gauss's error integral:

$$F(x) = \frac{1}{\sqrt{2\pi}}\int_{-\infty}^x \mathrm{e}^{-t^2/2}\mathrm{d}t =$$

$$\frac{1}{2} + \frac{1}{\sqrt{2\pi}}\left(x - \frac{x^3}{6} + \frac{x^5}{40} - \frac{x^7}{336} + \cdots + \frac{(-1)^i x^{2i+1}}{2^i\cdot i!\cdot(2i+1)} + \cdots\right)$$

Elliptic integral of the first kind: for $|k| < 1$, it holds that

$$F(k, 2\pi) = \int_0^{2\pi} \frac{\mathrm{d}t}{\sqrt{1-k^2\sin^2 t}} =$$

$$\frac{\pi}{2}\left(1 + \frac{1}{4}k^2 + \frac{9}{64}k^4 + \frac{25}{256}k^6 + \frac{1225}{16384}k^8 + \cdots + \left(\frac{(2i)!}{2^{2i}(i!)^2}\right)^2 k^{2i} + \ldots\right)$$

Elliptic integral of the second kind (circumference of an ellipse with eccentricity k):

$$U = 4aE(k, 2\pi) = a\int_0^{2\pi} \sqrt{1 - k^2 \sin^2 t}\, dt =$$

$$2\pi a\left(1 - \frac{1}{4}k^2 - \frac{3}{64}k^4 - \frac{5}{256}k^6 - \frac{175}{16384}k^8 - \cdots - \left(\frac{(2i)!}{2^{2i}(i!)^2}\right)^2 \frac{k^{2i}}{2i-1} - \cdots\right)$$

□ Circumference of an ellipse with semi-axes $a = 25$ cm and $b = 20$ cm:
Eccentricity: $k = \sqrt{1 - b^2/a^2} = \sqrt{1 - 20^2/25^2} = 0.6$
Circumference of the ellipse:
$U \approx 2\pi \cdot 25\,(1 - (-0.09) - (-0.00608) - (-0.00091) - \cdots)$ cm ≈ 141.8 cm.

15.4 Numerical integration

Integrals whose analytical solution is difficult if not impossible can be calculated numerically by splitting the integral into a finite sum:

$$\int_a^b f(x)\mathrm{d}x = h\sum_{i=0}^{N} c_i \cdot f(a + ih) + F(a, b, h) \approx h\sum_{i=0}^{N} c_i \cdot f(a + ih)\,,$$

with constants c_i, which depend on the method, and

$$h = \frac{b-a}{N}.$$

$F(a, b, N)$ is the error of approximation. N is the number of subdivisions of the interval and h is the width of the intervals. The greater N, the better the approximation, and the longer the calculation time.

▷ If the subdivision is too narrow (N is too large), rounding errors may falsify the result.

Increase the number of subdivisions N until the value of the integral no longer changes as far as the significant digits are concerned:

$$\left|\frac{I(2N) - I(N)}{I(2N)}\right| < 10^{-n} \quad (n = \text{ number of significant digits})$$

▷ n should not exceed the number of digits of the data type used (single accuracy: $n = 8$; double accuracy: $n = 16$).

▷ The quality of the approximation for a definite integral depends on:

1. the order of the error $O(h^n)$,
2. the step-size of the decomposition h,
3. the smoothness of the integrand.

15.4.1 *Rectangular rule*

Approximation by the upper sum or the lower sum (rectangles):

$$\int_a^b f(x)\mathrm{d}x = \frac{b-a}{N}\sum_{i=1}^{N} f(a + ih) + O(h)$$

$$= \frac{b-a}{N} \sum_{i=1}^{N} f(a+(i-1)h) + O(h) .$$

This is exact for constant functions.

15.4.2 Trapezoidal rule

Approximation of the area to be calculated by a trapezoid:

$$\int_a^b f(x)\mathrm{d}x \approx \frac{b-a}{2}(f(b)+f(a)) .$$

Subdivision of the integral into N subintervals of width h and N-fold use of the trapezoidal formula (summed trapezoidal formula):

$$\int_a^b f(x)\mathrm{d}x = \frac{b-a}{2N}\left(f(a)+f(b)+2\sum_{i=1}^{N-1} f(a+ih)\right) + O(h^k)$$

Exact for linear functions.

15.4.3 Simpson's rule

Simpson's approximation of the integrand by a polynomial of the second degree (1/3 rule):

$$\int_a^b f(x)\mathrm{d}x = \frac{b-a}{6}\left[f(a)+f(b)+4f\left(\frac{a+b}{2}\right)\right] + O(h^5) .$$

This is exact for polynomials up to and not including the third degree.
Use of N subintervals: In each subinterval the function is approximated by a polynomial of the second degree:

$$\int_a^b f(x)\mathrm{d}x = \frac{b-a}{3N}\left(f(a)+f(b)+4\sum_{i=1}^{N/2} f(a+(2i-1)h) \right.$$
$$\left. +2\sum_{i=1}^{N/2-1} f(a+2ih)\right) + O(h^4)$$

▷ The interval must be subdivided into an even number N of segments.
Simpson's 3/8 rule:

$$\int_a^b f(x)\mathrm{d}x = \frac{b-a}{8}\left(f(a)+3\cdot f((2a+b)/3)+3\cdot f((a+2b)/3)+f(b)+O(h^5)\right)$$

▷ Program sequence for Simpson's integration of the function f with $n+1$ abscissas.

```
BEGIN Simpson
INPUT a,b (limits of integration)
INPUT n (number of abscissas)
h := (b - a)/n
m := n
FOR i=0 TO n DO
    x[i] := a + i*h
ENDDO
```

```
INPUT f(x[i]), i=0...n
IF (n is odd AND n > 1) THEN
    dummy := f(x[n-3]) + 3*(f(x[n-2]) + f(x[n-1])) + f(x[n])
    int := int + 3*h*dummy/8
    m := n - 3
ENDIF
IF (m > 1) THEN
    sum1 := 0
    sum2 := 0
    FOR i=1 TO m-1 STEP 2 DO
       sum1 := sum1 + f(x[i])
    ENDDO
    FOR i=2 TO m-2 STEP 2 DO
       sum2 := sum2 + f(x[i])
    ENDDO
    int := int + h*(f(x[0]) + 4*sum1 + 2*sum2 + f(x[m]))/3
ENDIF
OUTPUT int
END Simpson
```

15.4.4 Romberg integration

The idea of the **Romberg integration** is to estimate the integration error in addition to a more refined subdivision of the intervals in the trapezoidal rule and to take this into account in the calculation of the integral (extrapolation). Thus, the order of the error term is increased, and fewer iteration steps are usually necessary to achieve the required accuracy.
Romberg's calculation procedure is:

$$\int_a^b f(x)\mathrm{d}x \approx I_{j,k} = \frac{4^{k-1} I_{j+1,k-1} - I_{j,k-1}}{4^{k-1} - 1} \;.$$

The index j denotes the number of subdivisions of the interval in using the trapezoidal rule, and k measures the order of the error of approximation.
Program sequence:

1. Calculate the integral for one interval ($I_{1,1}$), using the trapezoidal formula in a subroutine.

2. Start a loop over i and calculate the integral $I_{i,1}$ for 2^i intervals, using the trapezoidal formula in a subroutine.

3. Calculate the approximated integral $I_{j,k}$ in a loop over $k = 2, i+1$ with $j = 2+i-k$ according to the Romberg relation given above.

4. Close the i-loop if the maximum number of iteration steps is reached or if the error is sufficiently small:

$$\left| \frac{I_{1,i+1} - I_{1,i}}{I_{1,i}} \right| < 10^{-n} \quad (n = \text{ number of significant digits})$$

▷ Tabulated data usually cannot be calculated by the Romberg integration, since the step size must be continuously bisected. The main application is for functions given analytically.

▷ Program sequence for the numerical integration of f according to the Romberg scheme. In the subroutine `Trapezoidal rule` the integral is calculated with n points of the abscissa, and then it is transferred again to `Romberg`.

```
BEGIN Romberg
INPUT a,b (limits of integration)
INPUT eps
INPUT maxit (maximum number of iterations)
n := 1
CALL trapezoidal rule(n,a,b,integral)
int[1,1] := integral
epsa := 1.1*eps
i := 0
WHILE (epsa > eps AND i < maxit) DO
    i := i + 1
    n := power(2,i)
    CALL trapezoidal rule(n,a,b,integral)
    int[i+1,1] := integral
    FOR k = 2 TO i+1 DO
       j := 2 + i - k
       int[j,k] := power(4,(k-1))*int[j+1,k-1] - int[j,k-1]
       int[j,k] := int[j,k]/(power(4,(k-1)) - 1)
    ENDDO
    epsa := abs((int[1,i+1] - int[1,1])/(int[1,i+1]))*100
ENDDO
END Romberg
```

```
BEGIN trapezoidal rule
INPUT n, a, b
sum := 0
step := (b-a)/n
FOR i=1 TO n-1 DO
    x := a + i*step
    sum := sum + f(x)
ENDDO
integral := (b-a)*(f(a) + 2*sum + f(b))/2/n
OUTPUT integral
END trapezoidal rule
```

15.4.5 Gaussian quadrature

Use of the mean value theorem of integral calculus: The definite integral corresponds to the product of the length of the interval $(b - a)$ and the function value $f(c)$ at the best possible intermediate point c that has to be chosen optimally:

$$\int_a^b f(x)\mathrm{d}x = (b-a)\cdot f(c), \quad c \in [a,b] .$$

Gaussian quadrature: Choice of n appropriate non-equidistant points of the abscissa x_i and weights g_i such that the quadrature formula for a polynomial of the highest degree $2n-1$ is exact:

$$\int_a^b f(x)\mathrm{d}x = \sum_{i=1}^{n} g_i \cdot f(x_i) + R_n(f) .$$

Here, $R_n(f) = 0$ if f is a polynomial of degree $\leq 2n-1$.
Advantage: Gaussian quadratures are very accurate, even for only a few points of the abscissa.

▷ Gaussian quadratures require the calculation of function values at precisely predetermined points of the abscissa. In most cases, they are therefore not convenient for tabulated data.

Gauss-Legendre rule: Exact for Legendre polynomials of order n. For $n=2$:

$$\int_{-1}^1 f(x)\mathrm{d}x = f(-1/\sqrt{3}) + f(1/\sqrt{3}) .$$

▷ For Gauss-Legendre rules, the integration is in the interval $[-1, 1]$.

Transformation into an arbitrary interval $[a, b]$ is possible with the substitution

$$x = \frac{b-a}{2}z + \frac{b+a}{2} .$$

For $n=2$:

$$\begin{aligned}\int_a^b f(x)\mathrm{d}x &= \frac{b-a}{2}\int_{-1}^1 f\left(\frac{b-a}{2}z + \frac{b+a}{2}\right)\mathrm{d}z \\ &\approx \frac{b-a}{2}\left(f\left(-\frac{b-a}{2\sqrt{3}} + \frac{b+a}{2}\right) + f\left(\frac{b-a}{2\sqrt{3}} + \frac{b+a}{2}\right)\right) .\end{aligned}$$

n	i	x_i	g_i
1	1	0	2
2	1	−0.5773503	1
	2	0.5773503	1
3	1	−0.7745967	0.5555556
	2	0	0.8888889
	3	0.7745967	0.5555556
4	1	−0.8611363	0.3478548
	2	−0.3399810	0.6521455
	3	0.3399810	0.6521455
	4	0.8611363	0.3478548

Table points of the abscissa x_i and weight factors g_i for the Gauss-Legendre rule to the fourth order

▷ For long integration intervals, the Gauss-Legendre rule can be applied effectively to subintervals with subsequent summation of subintervals, without enlarging the number of points on the abscissa!

□ Gauss-Legendre with $n = 2$ points of the abscissa (!) for the normal distribution:

$$\frac{1}{\sqrt{2\pi}} \int_0^2 e^{-x^2/2} dx = \frac{1}{\sqrt{2\pi}} \int_{-1}^1 e^{-(1+x)^2/2} dx$$
$$\approx \frac{1}{\sqrt{2\pi}} \left(e^{-(1-1/\sqrt{3})^2/2} + e^{-(1+1/\sqrt{3})^2/2} \right) \approx 0.4798 \, .$$

Deviation from the exact value (0.4772): Only 0.5%!

Gauss-Legendre integration: For improper integrals with exponentially decreasing integrands (e.g., for Boltzmann distributions).

15.4.6 Table of numerical integration methods

In the estimation of error, ξ is a point in the interval $[x, x+h]$.

Method	Points of the abscissa for one application	Error	Applicability	Remarks
Trapezoid	2	$\sim h^3 f''(\xi)$	frequently	
Simpson (1/3)	3	$\sim h^5 f^{(4)}(\xi)$	frequently	
Simpson (3/8)	4	$\sim h^5 f^{(4)}(\xi)$	frequently	
Romberg	3		$f(x)$ known	not for tabulated data
Gaussian	≥ 2		$f(x)$ known	not for tabulated data

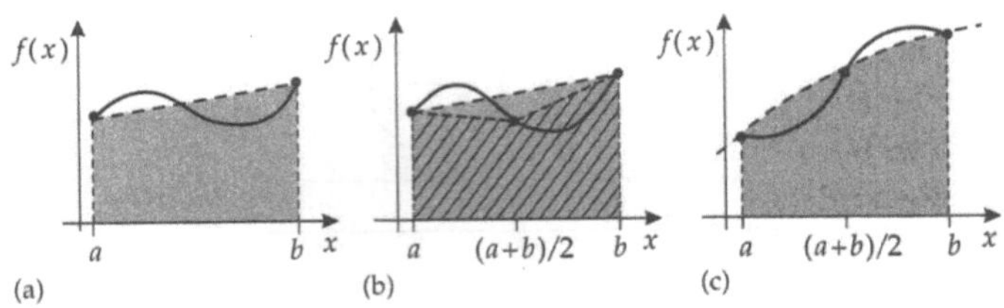

Numerical integration methods:
(a) Trapezoidal rule, (b) Romberg method, (c) Simpson rule

▷ The Romberg approximation I_{11} is identical to Simpson's $\frac{1}{3}$ formula

▷ Program sequence for the integration of a function f given at discrete points. The points of the abscissa for which the function values are known need not be equidistant.

```
BEGIN Integration
INPUT n (number of segments)
INPUT x[i], f(x[i]), i = 0...n
h := x[1] - x[0]
k := 1
int := 0
FOR j = 1 TO n DO
```

```
    hfuture := x[j+1] - x[j]
    IF (h = hfuture) THEN
       IF (k = 3) THEN
          int := int + 2*h*(f(x[j-1]) + 4*f(x[j-2])
             + f(x[j-3]))/6
          k := k - 1
       ELSEIF
          k := k + 1
       ENDIF
    ELSEIF
       IF (k = 1) THEN
          int := int + h*(f(x[j]) + f(x[j-1]))/2
       ELSEIF
          IF (k = 2) THEN
             int := int + 2*h*(f(x[j]) + 4*f(x[j-1])
             + f(x[j-2]))/6
          ELSEIF
             dummy := f(x[j]) + 3*(f(x[j-1]) + f(x[j-2]))
             + f(x[j-3])
             int := int + 3*h*dummy/8
          ENDIF
          k := 1
       ENDIF
    ENDIF
h := hfuture
ENDDO
END integration
```

▷ The integration of a function f given by a table of values (t_i, y_i), $i = 0, \ldots, n$, can also be done as follows: Calculate the interpolating cubic spline function s with the boundary condition s three times continuously differentiable in t_1 and t_{n-1}. The integration of s is done by segments using the Simpson's rule. This choice of the boundary condition guarantees that polynomials of highest degree 3 are integrated exactly by the method. (The program sequences "spline function," "boundary condition," and "spline" are given in the section on spline function.)
BEGIN integration of tabulated data

```
CALL spline function
CALL boundary condition (c,u,v,w,t,4,n);
                {compare spline interpolation}

integral=0
FOR i=1 TO n DO
BEGIN
 h:=t[i]-t[i-1]
integral:=integral+h/6*(y[i-1]+
          4*spline(t[i-1]+h/2,y,c,n)+y[i]);
  ENDDO

OUTPUT integral

END integration of tabulated data
```

15.5 Mean value theorem of integral calculus

- **Mean value theorem of integral calculus:** If $f(x)$ is continuous in $[a, b]$, then there is one point c in the interval $[a, b]$ for which it holds that

$$\int_a^b f(x)\mathrm{d}x = (b-a)f(c), \quad \text{for one } c \in [a, b].$$

▷ The definite integral $\int_a^b f(x)\mathrm{d}x$ with $f(x) \geq 0$ can also be represented by **one** rectangle with the length of the side $(b-a)$ long and $f(c)$ high; the problem is to find the correct point $x = c$.

Linear mean value (integral mean value) $f(c)$ of the function values in the interval $[a, b]$:

$$f(c) = \frac{1}{b-a}\int_a^b f(x)\mathrm{d}x, \quad c \in [a, b].$$

□ For a linear function we have $c = (a+b)/2$.

Root-mean-square value: Defined by the integral of the square of the function $f(x)$:

$$M_{quad} = \sqrt{\frac{1}{b-a}\int_a^b (f(x))^2\,\mathrm{d}x}.$$

Application of the mean value theorem:

Numerical integration using the Gaussian method of order n:

The function values are calculated not at the end-points of the interval $[a, b]$ but at appropriate x values in the interior of $[a, b]$, such that the integral is exact up to a polynomial of degree n.

- **Generalized first mean value theorem of integral calculus**: If $f(x)$ and $g(x)$ are continuous in the interval $[a, b]$, and if $g(x)$ does not change sign in the interval, then it holds that

$$\int_a^b f(x)g(x)\mathrm{d}x = f(c)\int_a^b g(x)\mathrm{d}x, \quad c \in (a, b)\,.$$

- **Generalized second mean value theorem of integral calculus**: If $f(x)$ is monotonic and bounded, and if $g(x)$ is integrable in the interval $[a, b]$, then it holds that

$$\int_a^b f(x)g(x)\mathrm{d}x = f(a)\int_a^c g(x)\mathrm{d}x + f(b)\int_c^b g(x)\mathrm{d}x, \quad c \in [a, b]\,.$$

15.6 Line, surface, and volume integrals

15.6.1 Arc length (rectification)

- **Line element** of a curve, according to Pythagoras:

$$\mathrm{d}s = \sqrt{\mathrm{d}x^2 + \mathrm{d}y^2}\,.$$

Thus, the **arc length** of a curve is

$$s=\int \mathrm{d}s=\int \sqrt{1+\frac{\mathrm{d}y^2}{\mathrm{d}x^2}}\,\mathrm{d}x=\int \sqrt{1+(f'(x))^2}\,\mathrm{d}x\ .$$

□ Circumference of a unit circle: $y=\sqrt{1-x^2}$, $y'=-\frac{x}{\sqrt{1-x^2}}$,

$$s=2\int_{-1}^{1}\sqrt{1+\frac{x^2}{1-x^2}}\,\mathrm{d}x=2\int_{-1}^{1}\frac{\mathrm{d}x}{\sqrt{1-x^2}}=2\ \arcsin x|_{-1}^{1}=2\pi .$$

15.6.2 Area

The **area** between a curve $f(x)$ and the x-axis is calculated by means of the integral

$$I=\int_a^b f(x)\mathrm{d}x=F(b)-F(a)\ .$$

▷ If f is negative in the integral, the integral becomes negative as well.
The sign of the integral also depends on which integration limit is the greater one.

Function value	Limits of integration	Value of the integral
$f(x)>0$	$a<b$	$I>0$
$f(x)<0$	$a<b$	$I<0$
$f(x)>0$	$a>b$	$I<0$
$f(x)<0$	$a>b$	$I>0$

Magnitude of the area:

1. Calculate the zeros $x_1, x_2, \ldots, x_n$ and set $x_0=a$, $x_{n+1}=b$.
2. Subdivide the integral in surface patches ranging from x_i to x_{i+1}.
3. Integrate the partial integrals.
4. Summate the absolute values of the various integrations:

$$A=\left|\int_a^b |f(x)|\mathrm{d}x\right|=\sum_{i=1}^{n}\left|\int_{x_i}^{x_{i+1}} f(x)\mathrm{d}x\right|=\sum_{i=1}^{n}|F(x_{i+1})-F(x_i)|\ .$$

□ Area between the cosine curve and the x-axis in the interval $[0,\pi]$:

$$A=\int_0^{\pi/2}\cos x\mathrm{d}x+\left|\int_{\pi/2}^{\pi}\cos x\mathrm{d}x\right|=\sin\pi/2+|-\sin\pi/2+\sin\pi|=1+|-1|=2.$$

(However, the integral of $\cos x$ without the absolute value is:
$I=\int_0^{\pi}\cos x\mathrm{d}x=0$.)

● In the case of even functions with the limits of integration/symmetric about the y-axis, only one side must be integrated:

$$I=\int_{-a}^{a} f(x)\mathrm{d}x=2\int_0^a f(x)\mathrm{d}x\quad (f\ \text{even})\ .$$

□ Parabola:

$$\int_{-2}^{2} x^2 \mathrm{d}x = 2\int_{0}^{2} x^2 \mathrm{d}x = 2\left.\frac{x^3}{3}\right|_0^2 = \frac{16}{3}.$$

● In the case of odd functions with the limits of integration symmetric about the y-axis, the integral is equal to zero,

$$I = \int_{-a}^{a} f(x)\mathrm{d}x = 0 \qquad A = 2\left|\int_{0}^{a} |f(x)|\mathrm{d}x\right| \quad (f \text{ odd}) .$$

□ $y = x^3$: $\int_{-\pi}^{\pi} x^3 \mathrm{d}x = 0$

● Magnitude of an area in the case of even or odd functions with the limits of integration symmetric about the y-axis:

$$A = \left|\int_{-a}^{a} |f(x)|\mathrm{d}x\right| = 2\left|\int_{0}^{a} |f(x)|\mathrm{d}x\right| \quad (f(x) \text{ even or odd function}) .$$

● Area between two functions:

$$A = \left|\int_{a}^{b} |f(x) - g(x)|\mathrm{d}x\right| = \sum_{i=0}^{n}\left|\int_{x_i}^{x_{i+1}} (f(x) - g(x))\mathrm{d}x\right| ,$$

where $x_0 = a$, $x_{n+1} = b$ and x_i $(i = 1, \ldots, n)$ are the points of intersection of the two functions.

▷ It is recommended to sketch the graph of the integrand.

● Area between a curve and the y-axis: Corresponds to the integration of the inverse function $U(y)$:

$$A = \int_{f(a)}^{f(b)} U(y)\mathrm{d}y .$$

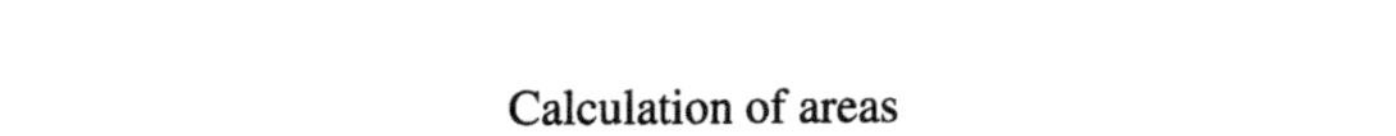

Calculation of areas

15.6.3 Solid of rotation (solid of revolution)

Solid of rotation (solid of revolution): Arises if a function $y = f(x)$ or an inverse function $U(y)$ is rotated about an axis.

▷ The axis is not necessarily a coordinate axis.

□ Hyperboloids of revolution oriented obliquely in space: Symmetry axis 45° $(x = y)$.

Volume of the solid of rotation: Integrating over all circular disks.

$$\text{Rotation about the } x\text{-axis:} \qquad V_x = \pi \int f(x)^2 \mathrm{d}x .$$

Rotation about the y-axis: $V_y = \pi \int U(y)^2 \mathrm{d}y$.

□ $y = x^2$: $V_x = \pi \int x^4 \mathrm{d}x = (\pi/5)x^5$,

Volume of a paraboloid of rotation with height h:

$$V_y = \pi \int_0^h \left(\sqrt{y}\right)^2 \mathrm{d}y = \frac{\pi h^2}{2}.$$

- **Surface of a solid of rotation**, or **lateral surface**: Corresponds to an integration of all circumferences of circles along the curve.

Rotation about the x-axis: $A_{Lx} = 2\pi \int f(x)\mathrm{d}s = 2\pi \int f(x)\sqrt{1 + f'(x)^2}\mathrm{d}x$.

Rotation about the y-axis: $A_{Ly} = 2\pi \int U(y)\mathrm{d}s = 2\pi \int U(y)\sqrt{1 + U'(y)^2}\mathrm{d}y$.

□ Surface of a spherical zone: $y = \sqrt{r^2 - x^2}$,

$$A_{Lx} = 2\pi \int_0^h \sqrt{r^2 - x^2}\sqrt{1 + \frac{x^2}{r^2 - x^2}}\mathrm{d}x = 2\pi r \int_0^h \mathrm{d}x = 2\pi r h.$$

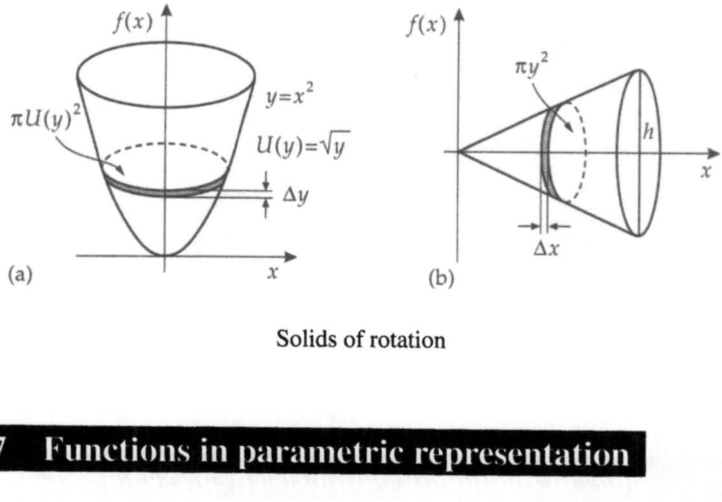

Solids of rotation

15.7 Functions in parametric representation

Parametric representation of a curve with parameter t:

$$x = x(t) , \qquad y = y(t) .$$

Representation in polar coordinates with angle ϕ:

$$x = r(\phi)\cos\phi , \qquad y = r(\phi)\sin\phi .$$

15.7.1 Arc length in parametric representation

Arc length in parametric representation and in polar coordinates:

$$s = \int \sqrt{\dot{x}^2 + \dot{y}^2}\mathrm{d}t , \qquad s = \int \sqrt{r^2 + \dot{r}^2}\mathrm{d}\phi , \qquad \dot{r} = \frac{\mathrm{d}r}{\mathrm{d}\phi} .$$

□ Arc length of a circular segment:

$$r = r_0, \quad \dot{r} = 0, \quad s = \int_0^{\alpha} \sqrt{r_0^2 + 0}\, \mathrm{d}\phi = \alpha r_0.$$

15.7.2 Sector formula

The surface differential and the area between the curve and the origin are given in polar coordinates:

$$\mathrm{d}A = \frac{1}{2} r \cdot r \mathrm{d}\phi\,, \qquad A = \frac{1}{2} \int r^2 \mathrm{d}\phi\,.$$

- **Leibniz's sector formula**: For functions in parametric representation, the area between the curve and the origin is calculated via

$$A = \frac{1}{2} \int (x\dot{y} - \dot{x}y)\mathrm{d}t.$$

□ The area of lemniscate $r = a\sqrt{\cos 2\phi}$:

$$A = 4\frac{1}{2}a^2 \int_0^{\pi/4} \cos 2\phi \, \mathrm{d}\phi = 2a^2 \frac{1}{2} \sin(2\phi)\Big|_0^{\pi/4} = a^2.$$

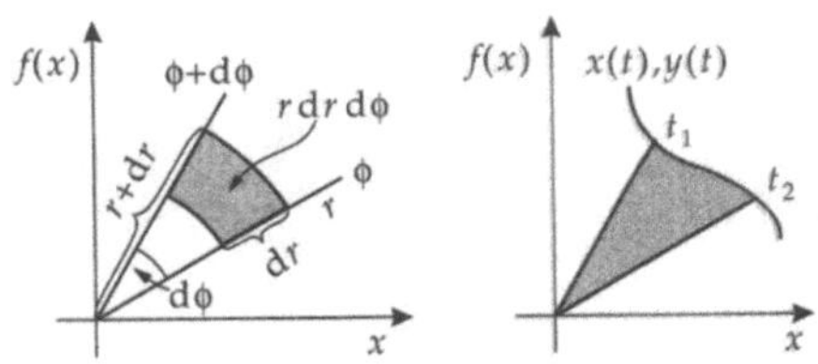

Sector formula

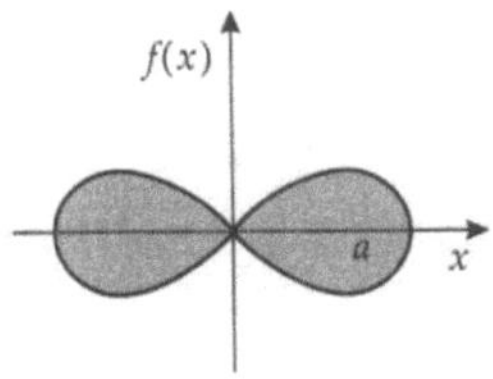

Lemniscate

15.7.3 Solids of rotation in parametric representation

Calculation of the lateral surface (**complanation**) and calculation of the volume (**cubature**) for a solid of revolution in parametric representation:

Axis of rotation	Lateral surface	Volume
x-axis	$A_{Lx} = 2\pi \int y\sqrt{\dot{x}^2 + \dot{y}^2}\mathrm{d}t$	$V_x = \pi \int y^2 \dot{x} \mathrm{d}t$
y-axis	$A_{Ly} = 2\pi \int x\sqrt{\dot{x}^2 + \dot{y}^2}\mathrm{d}t$	$V_y = \pi \int x^2 \dot{y} \mathrm{d}t$

Lateral surface and volume of a solid of revolution in polar coordinates:

Axis of rotation	Lateral surface	Volume
x-axis	$2\pi \int r \sin\phi \sqrt{r^2 + \dot{r}^2} \mathrm{d}\phi$	$\pi \int r^2 \sin^2\phi (\dot{r}\cos\phi - r\sin\phi) \mathrm{d}\phi$
y-axis	$2\pi \int r \cos\phi \sqrt{r^2 + \dot{r}^2} \mathrm{d}\phi$	$\pi \int r^2 \cos^2\phi (\dot{r}\sin\phi + r\cos\phi) \mathrm{d}\phi$

15.8 Multiple integrals and their applications

15.8.1 *Definition of multiple integrals*

Double integral: Limit of a double sum of surface elements over a function of two independent variables $f(x, y)$ (integral in two dimensions), defined in analogy to the single integral:

$$\iint f(x, y) \mathrm{d}y \, \mathrm{d}x = \lim_{n,m \to \infty} \sum_{i=1}^{n} \sum_{j=1}^{m} f(x_i, y_j) \Delta x_i \Delta y_j \ .$$

Surface differential: $\mathrm{d}A = \mathrm{d}x \, \mathrm{d}y$.
The double integral consists of an outer and inner integral. It is calculated by means of two successive ordinary integrations.

$$\underbrace{\int_{x=a}^{b} \underbrace{\int_{y=u(x)}^{o(x)} f(x, y) \, \mathrm{d}y}_{\text{inner integral}} \, \mathrm{d}x}_{\text{outer integral}}$$

In polar coordinates, the surface element is $\mathrm{d}A = r \, \mathrm{d}r \, \mathrm{d}\phi$:

$$\int_{\phi=\phi_1}^{\phi_2} \int_{r=0}^{r(\phi)} f(r, \phi) \, r \, \mathrm{d}r \, \mathrm{d}\phi \ .$$

Triple integral: Calculated by three successive ordinary integrations. Depending on the form of the volume to be integrated, choose the appropriately adapted coordinates or suitable volume elements.

Cartesian coordinates	Cylindrical coordinates	Spherical coordinates
$\iiint \underbrace{\mathrm{d}x \, \mathrm{d}y \, \mathrm{d}z}_{\mathrm{d}V}$	$\iiint \underbrace{r \, \mathrm{d}r \, \mathrm{d}\phi \, \mathrm{d}z}_{\mathrm{d}V}$	$\iiint \underbrace{r^2 \sin\theta \, \mathrm{d}r \, \mathrm{d}\phi \, \mathrm{d}\theta}_{\mathrm{d}V}$

□ Volume of a sphere:

$$\int_{r=0}^{R} \int_{\phi=0}^{2\pi} \int_{\theta=0}^{\pi} r^2 \sin\theta \, \mathrm{d}r \, \mathrm{d}\phi \, \mathrm{d}\theta = \frac{R^3}{3} 2\pi \int_{\theta=0}^{\pi} \sin\theta \mathrm{d}\theta = \frac{4\pi}{3} R^3.$$

Substitution rule for multiple integrals: Calculation of an integral in arbitrary coordinates, u, v and w, defined by

$$x = x(u, v, w), \quad y = y(u, v, w), \quad z = z(u, v, w) \ .$$

Separation of the domain of integration into volume elements in terms of the coordinate surfaces $u = \text{const}$, $v = \text{const}$, and $w = \text{const}$:

$$dV = |D| \, du \, dv \, dw \; ,$$

with the functional determinant (**Jacobian determinant**)

$$D = \begin{vmatrix} \frac{\partial x}{\partial u} & \frac{\partial x}{\partial v} & \frac{\partial x}{\partial w} \\ \frac{\partial y}{\partial u} & \frac{\partial y}{\partial v} & \frac{\partial y}{\partial w} \\ \frac{\partial z}{\partial u} & \frac{\partial z}{\partial v} & \frac{\partial z}{\partial w} \end{vmatrix} \; .$$

The substitution of the integral is possible for $D \neq 0$:

$$\iiint f(x, y, z) \, dV = \iiint f(u, v, w) |D| \, dw \, dv \, du \; .$$

▷ Numerical calculation of multiple integrals: Monte-Carlo methods, particularly efficient in the case of large n (n = number of integrations).

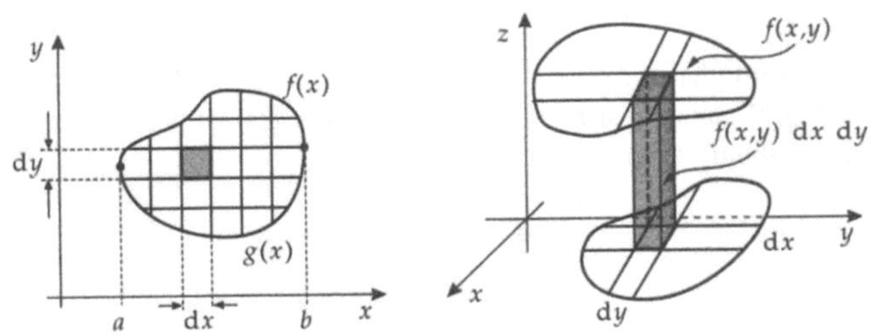

Multiple integrals

15.8.2 Calculation of areas

Area between two curves ($f(x)$ and $g(x)$) in Cartesian coordinates:

$$A = \int_{x=a}^{b} \int_{g(x)}^{f(x)} dy \, dx = \int_{x=a}^{b} (f(x) - g(x)) dx \; .$$

□ Circular segment:

$$A = \int_0^h \int_0^{\sqrt{r^2 - x^2}} dy \, dx = \int_0^h \sqrt{r^2 - x^2} dx = \frac{1}{2} \left(h\sqrt{r^2 - h^2} + r^2 \arcsin \frac{h}{r} \right) \; .$$

Area between two curves in polar coordinates ($r = r(\phi)$):

$$A = \int_{\phi = \phi_1}^{\phi_2} \int_{r = g(\phi)}^{f(\phi)} r \, dr \, d\phi \; .$$

□ Circular segment ($r(\phi) = r$):

$$A = \int_0^\alpha \int_0^r r \, dr \, d\phi = \int_0^\alpha \frac{r^2}{2} d\phi = \alpha \frac{r^2}{2} .$$

15.8.3 Center of mass of arcs

Center of mass of arcs: In three different representations of the arc in Cartesian coordinates, in parametric representation, and in polar coordinates.
x_S: x coordinate of the center of mass; y_S: y coordinate of the center of mass; s: length of the curve.

$y = f(x)$	$x = x(t),\ y = y(t)$	$r = r(\phi)$
$x_S = \frac{1}{s}\int x\sqrt{1+y'^2}\mathrm{d}x$	$x_S = \frac{1}{s}\int x\sqrt{\dot{x}^2+\dot{y}^2}\mathrm{d}t$	$x_S = \frac{1}{s}\int r\sqrt{r^2+\dot{r}^2}\cos\phi\mathrm{d}\phi$
$y_S = \frac{1}{s}\int y\sqrt{1+y'^2}\mathrm{d}x$	$y_S = \frac{1}{s}\int y\sqrt{\dot{x}^2+\dot{y}^2}\mathrm{d}t$	$y_S = \frac{1}{s}\int r\sqrt{r^2+\dot{r}^2}\sin\phi\mathrm{d}\phi$
$s = \int\sqrt{1+y'^2}\mathrm{d}x$	$s = \int\sqrt{\dot{x}^2+\dot{y}^2}\mathrm{d}t$	$s = \int\sqrt{r^2+\dot{r}^2}\mathrm{d}\phi$

15.8.4 Moment of inertia of an area

Moment of inertia of an area:
I_x: Moment of inertia about the x-axis
I_y: Moment of inertia about the y-axis

	Equatorial moment of inertia	
$y = f(x)$	$I_x = \int y^2\sqrt{1+y'^2}\mathrm{d}x$	$I_y = \int x^2\sqrt{1+y'^2}\mathrm{d}x$
$x(t),\ y(t)$	$I_x = \int y^2\sqrt{\dot{x}^2+\dot{y}^2}\mathrm{d}t$	$I_y = \int x^2\sqrt{\dot{x}^2+\dot{y}^2}\mathrm{d}t$
$r = r(\phi)$	$I_x = \int r^2\sin^2\phi\sqrt{r^2+\dot{r}^2}\mathrm{d}\phi$	$I_y = \int r^2\cos^2\phi\sqrt{r^2+\dot{r}^2}\mathrm{d}\phi$

I_p: Moment of inerta with respect to a point

	Polar moment of inertia
$y = f(x)$	$I_p = \int (x^2+y^2)\sqrt{1+y'^2}\mathrm{d}x$
$x = x(t),\ y = y(t)$	$I_p = \int (x^2+y^2)\sqrt{\dot{x}^2+\dot{y}^2}\mathrm{d}t$
$r = r(\phi)$	$I_p = \int r^2\sqrt{r^2+\dot{r}^2}\mathrm{d}\phi$

▷ The polar moment of inertia is always equal to the sum of the equatorial moments of inertia, $I_p = I_x + I_y$.

15.8.5 Center of mass of areas

Center of mass of an area: x_S: x coordinate of the center of mass; y_S: y coordinate of the center of mass; A: area.

$y = f(x)$	$x_S = \frac{1}{A}\iint x\mathrm{d}y\,\mathrm{d}x$	$y_S = \frac{1}{A}\iint y\mathrm{d}y\,\mathrm{d}x$	$A = \iint \mathrm{d}x\,\mathrm{d}y$
$r = r(\phi)$	$x_S = \frac{1}{A}\iint r^2\cos\phi\mathrm{d}r\,\mathrm{d}\phi$	$y_S = \frac{1}{A}\iint r^2\sin\phi\mathrm{d}r\,\mathrm{d}\phi$	$A = \iint r\,\mathrm{d}r\,\mathrm{d}\phi$

15.8.6 Moment of inertia of planes

Moment of inertia of planes:
I_x: Moment of inertia about the x-axis
I_y: Moment of inertia about the y-axis
I_p: Moment of inertia with respect to a point
I_{xy}: Centrifugal moment of inertia

	Equatorial moment of inertia	
$y = f(x)$	$I_x = \iint y^2\mathrm{d}x\,\mathrm{d}y$	$I_y = \iint x^2\mathrm{d}x\,\mathrm{d}y$
$r = r(\phi)$	$I_x = \int r^3\sin^2\phi\mathrm{d}r\,\mathrm{d}\phi$	$I_y = \int r^3\cos^2\phi\mathrm{d}r\,\mathrm{d}\phi$

	Polar moment of inertia
$y = f(x)$	$I_p = \iint (x^2 + y^2)\mathrm{d}x\,\mathrm{d}y$
$r = r(\phi)$	$I_p = \int r^3\mathrm{d}r\,\mathrm{d}\phi$

▷ In general, $I_p = I_x + I_y$.
Centrifugal moment of inertia:

$$I_{xy} = \int xy\mathrm{d}x\mathrm{d}y \,.$$

15.8.7 Center of mass of a body

Center of mass of a body in Cartesian coordinates:

$$x_s = \frac{1}{V}\iiint x\mathrm{d}x\mathrm{d}y\mathrm{d}z \,,$$

$$y_s = \frac{1}{V}\iiint y\mathrm{d}x\mathrm{d}y\mathrm{d}z \,,$$

$$z_s = \frac{1}{V}\iiint z\mathrm{d}x\mathrm{d}y\mathrm{d}z\,.$$

$$V = \iiint \mathrm{d}x\mathrm{d}y\mathrm{d}z \qquad \text{(Volume)}$$

□ Center of mass of a hemisphere: $V = \frac{2\pi}{3}R^3$,

$$z_S = \frac{1}{V}\int_{\phi=0}^{2\pi}\int_{r=0}^{R}\int_{z=0}^{\sqrt{R^2-r^2}} zr\,\mathrm{d}z\,\mathrm{d}r\,\mathrm{d}\phi = \frac{3}{2\pi R^3}\frac{2\pi}{2}\int_{r=0}^{R} r(R^2-r^2)\mathrm{d}r$$
$$= \frac{3}{2R^3}\frac{R^4}{4} = \frac{3}{8}R\,.$$

15.8.8 Moment of inertia of a body

Moment of inertia of a body in Cartesian coordinates:

$$I_z = \iiint (x^2+y^2)\mathrm{d}x\mathrm{d}y\mathrm{d}z$$

Reference axis is the z-axis.

15.8.9 Center of mass of rotational solids

Center of mass of solid of revolution:

$$z_S = \frac{1}{V}\iiint zr\,\mathrm{d}r\,\mathrm{d}z\,\mathrm{d}\phi, \quad V = \iiint r\,\mathrm{d}r\,\mathrm{d}z\,\mathrm{d}\phi\,.$$

Axis of rotation is the z-axis.
For reasons of symmetry, the center of mass always lies on the z-axis.

15.8.10 Moment of inertia of rotational solids

Moment of inertia of solid of revolution:

$$I = \iiint r^3\mathrm{d}r\,\mathrm{d}z\,\mathrm{d}\phi\,.$$

The z-axis is the axis of rotation.

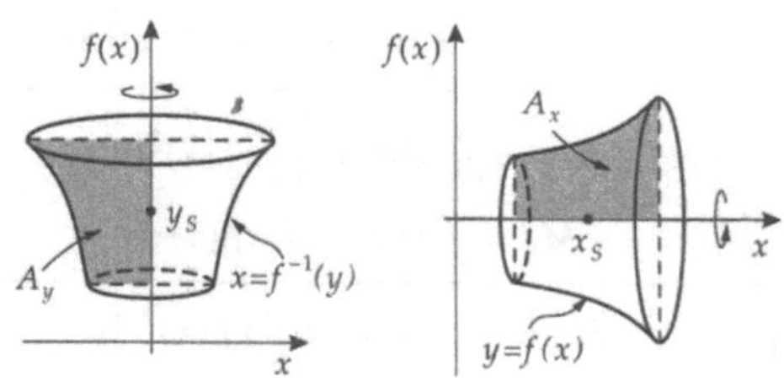

Center of mass of a rotational body.

15.9 Technical applications of integral calculus

15.9.1 Static moment, center of mass

Static moment M of a mass point: The product of mass m and distance r from the rotational axis:

$$M = rm.$$

For extended bodies, it holds that

$$M = \int r \, \mathrm{d}m.$$

Center of mass of an extended body: The point at which all static moments are at equilibrium:

$$r_S = \frac{1}{m} \int r \, \mathrm{d}m,$$

where r_S is the distance from the center of mass to the rotational axis.

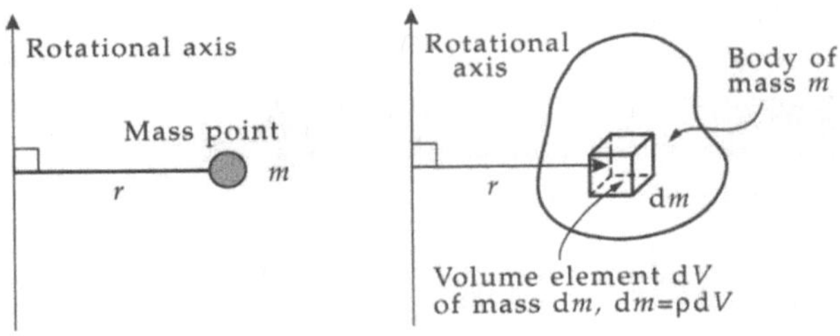

Calculation of moment

Objects which have homogeneous mass: For moments M_x and M_y and the center of mass with respect to the x- and y-axes, the following table applies:

Object	Moments	Center of mass position
Curve of length s	$M_x = \int y\mathrm{d}s = \int f(x)\sqrt{1+f'(x)^2}\mathrm{d}x$ $M_y = \int x\mathrm{d}s = \int x\sqrt{1+f'(x)^2}\mathrm{d}x$	$x_S = M_y/s$ $y_S = M_x/s$
Area A	$M_x = \int y\mathrm{d}A = \frac{1}{2}\int f(x)^2\mathrm{d}x$ $M_y = \int x\mathrm{d}A = \int xf(x)\,\mathrm{d}x$	$x_S = M_y/A$ $y_S = M_x/A$
Rotational solid with the volume V	$M_y = \int x\mathrm{d}V = \pi\int xf(x)^2\,\mathrm{d}x$ $M_x = \int y\mathrm{d}V = 0$	$x_S = M_y/V$ $y_S = 0$

⊳ The center of mass usually does not lie on the curve.

Guldin's first rule: The area of the **lateral surface** of a solid of rotation is equal to the product of length s of the generating curve on one side of the axis of rotation and the circumference of the circle described by the **center of mass of the generating curve** under a full revolution about the axis of rotation.

$$A_{Lx} = 2\pi y_S s \ , \quad A_{Ly} = 2\pi x_S s \ .$$

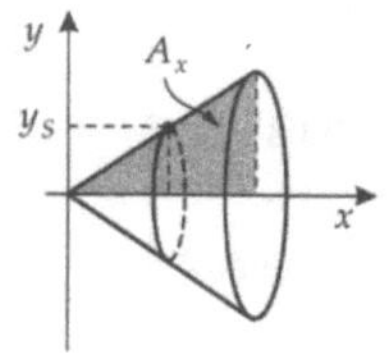

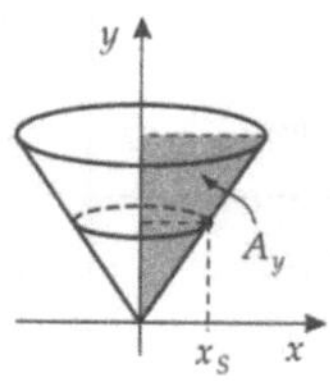

Guldin's second rule: The **volume** of a solid of rotation is equal to the product of the area of the generating plane on one side of the axis of rotation and the circumference of the circle described by the **center of mass of the generating plane** under a full revolution about the axis of rotation.

$$V_x = 2\pi y_S A_x \ , \quad V_y = 2\pi x_S A_y \ .$$

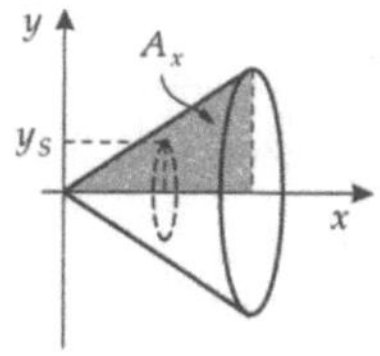

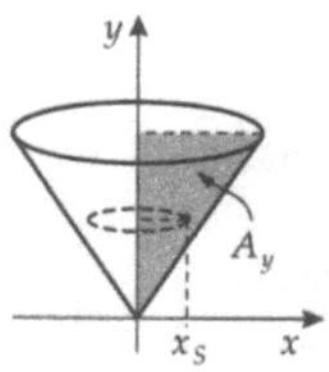

▷ The first rule involves the center of mass of the curve; the second rule involves the center of mass of the plane.

▷ In practice, Guldin's rules are used to determine the centers of mass if the arc length s and V_x, A_x, or A_{Lx} are known for the solid of rotation.

15.9.2 Mass moment of inertia

Mass moment of inertia M of rigid bodies with homogeneous mass density of total mass m and center of mass S:

Type	Rotational axis	Moment
Rod	perpendicular through S	$\frac{1}{12}ml^2$
(length l)	perpendicular through an end-point	$\frac{1}{3}ml^2$
Plate	perpendicular through S	$\frac{1}{12}m(a^2+b^2)$
(length of the sides a, b, c)	parallel to b through S	$\frac{1}{12}ma^2$
Circle	perpendicular through S	$\frac{1}{2}mr^2$
(radius r)	on the circle through S	$\frac{1}{4}mr^2$
Circular ring	perpendicular through S	mr^2
(radius r)	parallel through S	$\frac{1}{2}mr^2$
Cuboid	parallel to c through S	$\frac{1}{2}m(a^2+b^2)$
(lengths of sides a, b, c)	parallel to c, passing midpoint of b	$\frac{1}{2}m(4a^2+b^2)$
Circular cylinder	axis of the cylinder	$\frac{1}{2}mr^2$
(radius r, height h)	perpendicular to axis of the cylinder through S	$\frac{1}{12}m(h^2+3r^2)$
Circular cone	axis of the cone	$\frac{3}{10}mr^2$
(radius r, height h)	perpendicular to axis of the cone through S	$\frac{3}{80}m(h^2+4r^2)$
Sphere	through S	$\frac{2}{5}mr^2$
(radius r)	tangential	$\frac{7}{5}mr^2$
Hollow sphere	through S	$\frac{2}{5}m\frac{r_a^5-r_i^5}{r_a^3-r_i^3}$
(outer radius r_a, inner radius r_i)	tangential	$m\frac{7r_a^5+5r_a^2r_i^3-2r_i^5}{5(r_a^3-r_i^3)}$
Ellipsoid (semi-axes a, b, c)	parallel to c through S	$\frac{1}{5}m(a^2+b^2)$

Mass moment of inertia of a mass point: The product of mass m and the square of distance a^2 from the reference axis:

$$J = a^2m\ .$$

Equatorial moment of area: The reference axes x, y lying in the plane of the area.
Polar moment of area: Referred to a point in the plane of the area.

Reference to	Moment of inertia
x-axis (equatorial)	$J_x = y^2 m$
y-axis (equatorial)	$J_y = x^2 m$
point (polar)	$J_p = r^2 m = (x^2 + y^2)m = J_x + J_y$

Mass moment of inertia of a solid of rotation: With respect to the rotational axis, homogeneously covered by mass of density ρ:

$$J = \int r^2 \mathrm{d}m = \rho \int r^2 \mathrm{d}V .$$

▷ r is the distance from the rotational axis, not the position vector.
Area moments of inertia I of plane areas and mass inertia moments of solids of rotation:

Reference axis	Mass moment of inertia
x-axis	$J_x = \rho \int y^2 \mathrm{d}V = \frac{1}{2}\pi\rho \int f(x)^4 \mathrm{d}x$
y-axis	$J_y = \rho \int x^2 \mathrm{d}V = \frac{1}{2}\pi\rho \int U(y)^4 \mathrm{d}y$

- **Steiner's theorem**: The moment of inertia of a solid body about an arbitrary axis I_A corresponds to the moment of inertia about a parallel axis through the center of mass J_S plus the product of mass m and the square of the distance between the two axes a:

$$J_A = J_S + ma^2 .$$

▷ The calculation of the moment of inertia about an arbitrary point proceeds via the simpler calculation of the moment of inertia for rotational axes through the center of mass.

□ Moment of inertia of a sphere about a rotational axis at a distance $a = R$ from the center of the sphere:

$$I_S = 2\int_{\phi=0}^{2\pi}\int_{r=0}^{R}\int_{z=0}^{\sqrt{R^2-r^2}} r^3 \,\mathrm{d}z\,\mathrm{d}r\,\mathrm{d}\phi = 4\pi \int_{r=0}^{R} r^3\sqrt{R^2-r^2}\mathrm{d}r$$
$$= 4\pi\left(-\frac{1}{5}R^5 + \frac{R^2}{3}R^3\right) = \frac{8\pi}{15}R^5$$
$$I_A = I_S + Va^2 = \frac{8\pi}{15}R^5 + \frac{4\pi}{3}R^3 \cdot R^2 = \frac{28\pi}{15}R^5$$

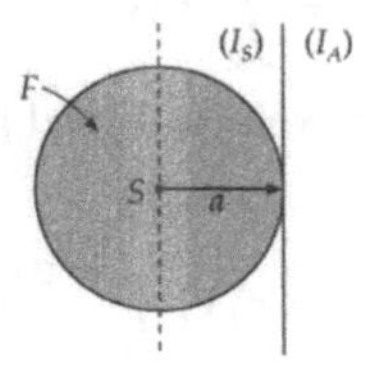

15.9.3 Statics

Supporting forces of a beam of length l applied to two supports:

$$F_A = \frac{F_G(l-a)}{l}, \qquad F_B = \frac{F_G a}{l},$$

with the total weight force F_G and the distance a between the center of mass and bearing A:

$$F_G = \int_0^l f(x)\mathrm{d}x, \qquad a = \frac{1}{F_G}\int_0^l x f(x)\mathrm{d}x.$$

Cutting forces: Crossforce F_Q and moment M at position x, taking into account the supporting forces and the supporting moments, are:

$$F_Q(x) = A - \int_0^x f(\tilde{x})\mathrm{d}\tilde{x}, \qquad M(x) = xF_Q(x) + \int_0^x \tilde{x} f(\tilde{x})\mathrm{d}\tilde{x}.$$

The increase of the moment corresponds to the cross force; the increase of the cross force corresponds to the weight force at position x:

$$\frac{\mathrm{d}M}{\mathrm{d}x} = F_Q, \qquad \frac{\mathrm{d}F_Q}{\mathrm{d}x} = -f(x).$$

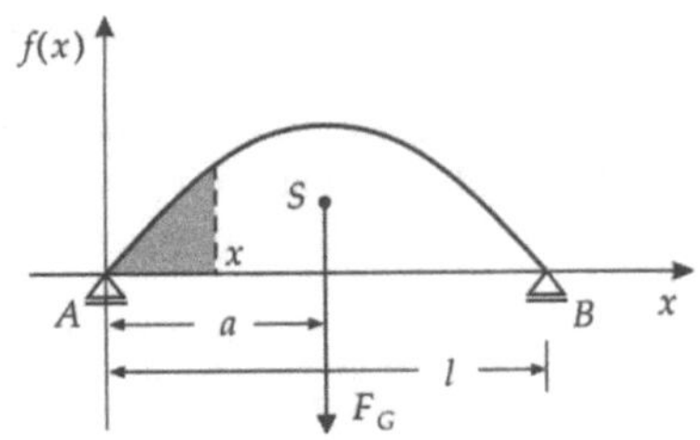

Beam on two supports

15.9.4 Calculation of work

Electrical work with alternating voltage $u(t)$ and alternating current $i(t)$ with circular frequency ω, period $T = 2\pi/\omega$, and phase shift ϕ:

$$W_{\text{current}} = \int_0^T u(t)i(t)\mathrm{d}t = u_0 i_0 \int_0^T \sin(\omega t)\sin(\omega t + \phi)\mathrm{d}t = \frac{u_0 i_0}{2} T\cos\phi.$$

Work of a spring, with path-dependent force $F(s) = ks$ dependent on path (Hooke's law, where k is the spring constant):

$$W_{\text{spring}} = \int_0^l F(s)\mathrm{d}s = \frac{kl^2}{2}.$$

Expansion work of an ideal gas of volume V and pressure $p = p_1 V_1/V$ (Boyle and Mariotte law):

$$W_{\text{gas}} = \int_{V_1}^{V_2} p(V)\mathrm{d}V = p_1 V_1 \int_{V_1}^{V_2} \frac{\mathrm{d}V}{V} = p_1 V_1 \ln\left(\frac{V_2}{V_1}\right).$$

15.9.5 Mean values

Mean values are important for time averaging (e.g., of periodic functions).
Linear mean value: Follows from the mean value theorem of integral calculus:

$$M_{\text{lin}} = \frac{1}{b-a} \int_a^b f(x)\mathrm{d}x \; .$$

▷ M_{lin} yields zero for periodic functions oscillating about the abscissa (e.g., $f(x) = \sin x$ or $\cos x$) in case of integration over a multiple of the period.

Root-mean-square value:

$$M_{\text{quadr.}} = \sqrt{\frac{1}{b-a} \int_a^b (f(x))^2 \mathrm{d}x} \; .$$

□ Mean values for $y = \sin x$ between 0 and π:

$$M_{\text{lin}} = \frac{1}{\pi - 0} \int_0^\pi \sin x \mathrm{d}x = \frac{1}{\pi}(-\cos x)|_0^\pi$$
$$= \frac{1}{\pi}(-\cos\pi - (-\cos 0)) = \frac{1}{\pi}(1+1) = \frac{2}{\pi},$$
$$M_{\text{quadr.}} = \sqrt{\frac{1}{\pi - 0} \int_0^\pi \sin^2 x \mathrm{d}x} = \sqrt{\frac{1}{2\pi}(-\sin x \cos x + x)\Big|_0^\pi}$$
$$= \sqrt{\frac{1}{2\pi}(-\sin\pi\cos\pi + \pi - (-\sin 0 \cos 0 + 0))} = \sqrt{\frac{1}{2\pi}\pi} = \frac{1}{\sqrt{2}}.$$

Application in electrical engineering:
Mean value of a periodic quantity of the time-dependent current $i(t) = i_0 \sin(2\pi t/T)$ of period T:

$$|\bar{i}| = \frac{1}{T} \int_0^T |i(t)| \mathrm{d}t = \frac{2}{\pi} i_0 \approx 0.637 \cdot i_0.$$

Mean effective value of the time-dependent current $i(t) = i_0 \sin(2\pi t/T)$ of period T:

$$i_{\text{eff}} = \sqrt{\frac{1}{T} \int_0^T (i(t))^2 \mathrm{d}t} = \frac{i_0}{\sqrt{2}} \approx 0.707 \cdot i_0.$$

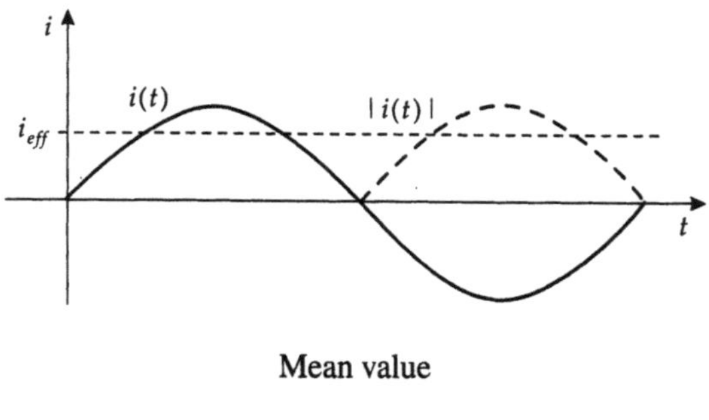

Mean value

16

Vector analysis

16.1 Fields

Scalar field: A scalar quantity (number) $U(x, y, z) = U(\vec{\mathbf{r}})$ is assigned to each point $P(x, y, z)$ in space or in a plane.

□ Temperature, density, electrical potential, atmospheric pressure, elevation above mean sea level.

Equipotential surface: Surface on which $U(x, y, z)$ has a constant value c.
Level line: $U(x, y) = c$ in the two-dimensional case.

□ The potential of an electric point charge has concentric spherical shells as equipotential surfaces.
The potential of a charged wire has concentric enveloping cylinders as equipotential surfaces.
The potential between the parallel plates of a capacitor has parallel planes as equipotential surfaces.

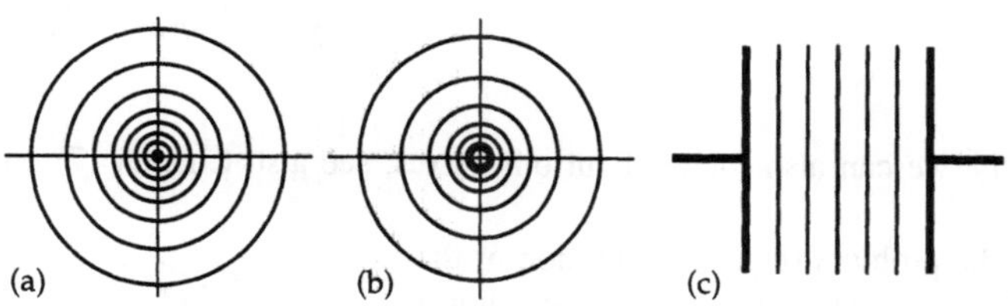

Section through the equipotential surfaces at $z = 0$: (a) point charge, (b) charged wire, (c) parallel plate capacitor

Vector field: A vector $\vec{\mathbf{A}}(x, y, z) = \vec{\mathbf{A}}(\vec{\mathbf{r}})$ is assigned to every point in space $P(x, y, z)$.

□ Gravitational field, magnetic field strength, and flow rate.

▷ Every coordinate A_x, A_y, A_z of the vector field
$\vec{\mathbf{A}}(x, y, z) = A_x(x, y, z)\,\vec{\mathbf{e}}_x + A_y(x, y, z)\,\vec{\mathbf{e}}_y + A_z(x, y, z)\,\vec{\mathbf{e}}_z$
is a scalar field.

Field line: Curve of all points (x, y, z) whose vector $\vec{\mathbf{A}}(x, y, z)$ is the tangential vector to the curve. Thus, for every point, the direction in which the field line runs to the next point is determined by its vector. Up to points with $\vec{\mathbf{A}} = \vec{\mathbf{0}}$, every point belongs to one and only one field line. Field lines do not intersect.

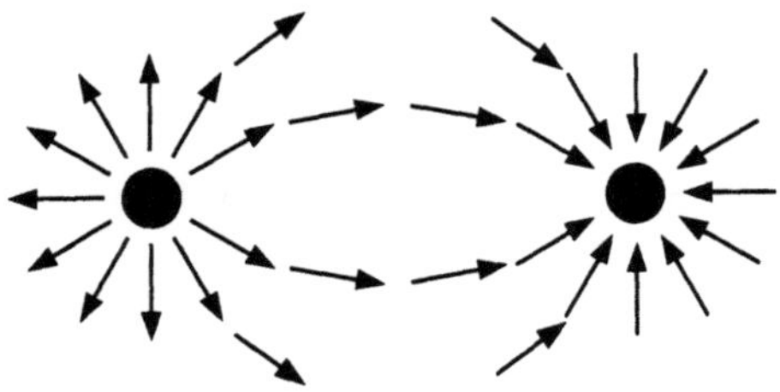

Field lines of an electric dipole

Stationary fields $\vec{\mathbf{A}}(x, y, z)$, $U(x, y, z)$ depend only on the point in space.
Variable fields $\vec{\mathbf{A}}(x, y, z, t)$, $U(x, y, z, t)$ also depend on further quantities, such as time t.

16.1.1 Symmetries of fields

Symmetries of fields make it possible to limit a study to certain domains in space or—with an appropriate choice of the coordinate system—to certain coordinates.
It is often better to transform the coordinate system to appropriate axes by means of **rotation**. The rotation through an angle α about the z-axis transforms the coordinates (x, y, z) to the rotated coordinates (x', y', z'):

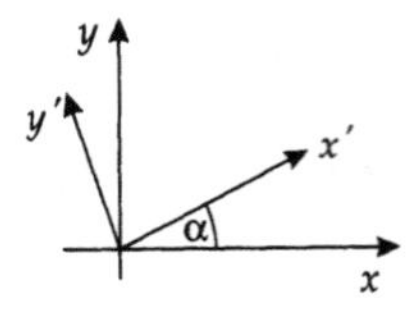

$$x' = x\cos\alpha + y\sin\alpha$$
$$y' = -x\sin\alpha + y\cos\alpha$$
$$z' = z$$
$$x = x'\cos\alpha - y'\sin\alpha$$
$$y = x'\sin\alpha + y'\cos\alpha$$

Analogously, we can also rotate about other axes; see also Chapter (7) on coordinate systems.

As a rule, the problem (e.g., the direction of the axis of a dipole or a tube) dictates a certain choice of axes.
Fields depending on sums or differences of coordinates, such as
$U = c(ax + by)^n$
can be rotated in such a way that they depend only on one coordinate (see figure).

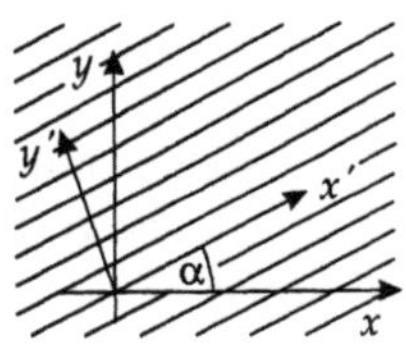

□ The potential $U(x, y, z) = \frac{D}{4}(x^2 + y^2 + 2z^2) + \frac{D}{2}xy$ can be transformed by a rotation through $\alpha = \frac{\pi}{4} = 45°$,

$$x' = \frac{1}{\sqrt{2}}x + \frac{1}{\sqrt{2}}y\,, \qquad y' = -\frac{1}{\sqrt{2}}x + \frac{1}{\sqrt{2}}y\,, \qquad z' = z\,,$$

$$x = \frac{1}{\sqrt{2}}x' - \frac{1}{\sqrt{2}}y'\,, \qquad y = \frac{1}{\sqrt{2}}x' + \frac{1}{\sqrt{2}}y'\,, \qquad z = z'\,,$$

into the following form:

$$U = \frac{D}{4}(x+y)^2 + \frac{D}{2}z^2 = \frac{D}{4}(\sqrt{2}x')^2 + \frac{D}{2}z'^2 = \frac{D}{2}(x'^2 + z'^2).$$

Rotating again through $-\frac{\pi}{2} = -90°$ about the x-axis, the result is $X = x'$, $Y = -z'$, $Z = y'$, and we can use the formula below for cylindrical coordinates (ρ, φ, z):

$$U = \frac{D}{2}(X^2 + Y^2) = \frac{D}{2}\rho^2 = U(\rho)$$

This is the potential of a two-dimensional harmonic oscillator.

Plane fields: The field depends on only two coordinates (e.g., $\vec{\mathbf{A}}(x, y)$) or even on one coordinate (e.g., $U(z)$).

▷ If such a representation can be achieved, the calculation is greatly simplified because derivations with respect to the missing coordinates vanish.

□ The gravitational field near the earth surface is plane.
The potential field between the plates of a capacitor is plane.

Spherical symmetry or **central symmetry**: The field depends only on the distance

$$r = \sqrt{x^2 + y^2 + z^2}$$

from point (x, y, z) to the origin. It is best to choose **spherical coordinates** (r, ϑ, φ) (see also Chapter (7) on coordinate systems). The coordinates are converted as follows:

$$\begin{aligned} x &= r\cos\varphi\sin\vartheta, & r &= \sqrt{x^2+y^2+z^2}, \\ y &= r\sin\varphi\sin\vartheta, & \vartheta &= \arccos\frac{z}{\sqrt{x^2+y^2+z^2}}, \\ z &= r\cos\vartheta, & \varphi &= \begin{cases} \operatorname{arccot}(x/y) & \text{for } y \geq 0 \\ \pi + \operatorname{arccot}(x/y) & \text{for } y < 0. \end{cases} \end{aligned}$$

□ The electric field of a point charge is spherically symmetric.
The gravitational force from a point-shaped or spherical body (sun, earth) is spherically symmetric.

Cylindrical symmetry or **axial symmetry**: The field depends only on distance r between point (x, y, z) and z-axis. It is best to use **cylindrical coordinates** (ρ, φ, z) (see also Chapter 7 on coordinate systems). The coordinates are converted as follows:

$$\begin{aligned} x &= \rho\cos\varphi, & \rho &= \sqrt{x^2+y^2}, \\ y &= \rho\sin\varphi, & \varphi &= \begin{cases} \operatorname{arccot}(x/y) & \text{for } y \geq 0 \\ \pi + \operatorname{arccot}(x/y) & \text{for } y < 0. \end{cases} \\ z &= z \end{aligned}$$

□ The magnetic field around a rectilinear long conductor wire is cylindrically symmetric.
The field of flow in a rectilinear tube is cylindrically symmetric.

16.2 Differentiation and integration of vectors

For a detailed description of differentiation and integration rules, see Chapters 12 and 15 on differential and integral calculus.
In the following we assume that the unit (base) vectors along the three axes (x, y, z), $\vec{\mathbf{e}}_x$, $\vec{\mathbf{e}}_y$, $\vec{\mathbf{e}}_z$, are fixed; they are not affected by differentiation and integration.

● The unit vectors in spherical or cylindrical coordinates are locally defined. In calculations, they often depend on time.

Differentiation of a vector: The differentiation of individual components with respect to one parameter:

$$\frac{\mathrm{d}\vec{\mathbf{A}}(t)}{\mathrm{d}t} = \frac{\mathrm{d}A_x(t)}{\mathrm{d}t}\vec{\mathbf{e}}_x + \frac{\mathrm{d}A_y(t)}{\mathrm{d}t}\vec{\mathbf{e}}_y + \frac{\mathrm{d}A_z(t)}{\mathrm{d}t}\vec{\mathbf{e}}_z .$$

The resulting field is a vector field.

□ Newton's second law of motion:
A force causes a change in velocity

$$\vec{\mathbf{F}} = m\vec{\mathbf{a}} = \frac{\mathrm{d}\vec{\mathbf{p}}}{\mathrm{d}t} = m\frac{\mathrm{d}\vec{\mathbf{v}}}{\mathrm{d}t} ,$$

$$F_x\,\vec{\mathbf{e}}_x + F_y\,\vec{\mathbf{e}}_y + F_z\,\vec{\mathbf{e}}_z = m\frac{\mathrm{d}v_x}{\mathrm{d}t}\vec{\mathbf{e}}_x + m\frac{\mathrm{d}v_y}{\mathrm{d}t}\vec{\mathbf{e}}_y + m\frac{\mathrm{d}v_z}{\mathrm{d}t}\vec{\mathbf{e}}_z .$$

We can calculate Newton's second law of motion separately for each component and then add all acceleration components.

Differentiation of a scalar product involves application of the product rule:

$$\frac{\mathrm{d}(\vec{\mathbf{A}}\cdot\vec{\mathbf{B}})}{\mathrm{d}t} = \frac{\mathrm{d}\vec{\mathbf{A}}}{\mathrm{d}t}\cdot\vec{\mathbf{B}} + \vec{\mathbf{A}}\cdot\frac{\mathrm{d}\vec{\mathbf{B}}}{\mathrm{d}t} .$$

The resulting field is a scalar field.

Differentiation of a cross product involves application of the product rule:

$$\frac{\mathrm{d}(\vec{\mathbf{A}}\times\vec{\mathbf{B}})}{\mathrm{d}t} = \frac{\mathrm{d}\vec{\mathbf{A}}}{\mathrm{d}t}\times\vec{\mathbf{B}} + \vec{\mathbf{A}}\times\frac{\mathrm{d}\vec{\mathbf{B}}}{\mathrm{d}t} .$$

The resulting field is a vector field.

□ Differentiation of angular momentum $\vec{\mathbf{L}}$ with respect to time:

$$\begin{aligned}\frac{\mathrm{d}}{\mathrm{d}t}\vec{\mathbf{L}} &= \frac{\mathrm{d}}{\mathrm{d}t}(\vec{\mathbf{r}}\times\vec{\mathbf{p}})\\ &= \frac{\mathrm{d}\vec{\mathbf{r}}}{\mathrm{d}t}\times\vec{\mathbf{p}} + \vec{\mathbf{r}}\times\frac{\mathrm{d}\vec{\mathbf{p}}}{\mathrm{d}t} .\end{aligned}$$

Because of $\frac{\mathrm{d}\vec{\mathbf{r}}}{\mathrm{d}t} = \vec{\mathbf{v}} = \frac{\vec{\mathbf{p}}}{m}$. the vectors are parallel in the first cross product. Thus, the first product vanishes, and it holds that:

$$\frac{\mathrm{d}}{\mathrm{d}t}\vec{\mathbf{L}} = \vec{\mathbf{r}}\times\vec{\mathbf{F}}$$

In central force fields, it holds that $\vec{\mathbf{F}}(r) = F(r) \cdot \frac{\vec{\mathbf{r}}}{r}$, such that the second term also vanishes.

Here, the vanishing of the derivative means that the rotational momentum is maintained.

Differentiation of a triple scalar product involves multiple application of the product rule:

$$\frac{\mathrm{d}[\vec{\mathbf{A}} \cdot (\vec{\mathbf{B}} \times \vec{\mathbf{C}})]}{\mathrm{d}t} = \frac{\mathrm{d}\vec{\mathbf{A}}}{\mathrm{d}t} \cdot \left(\vec{\mathbf{B}} \times \vec{\mathbf{C}}\right) + \vec{\mathbf{A}} \cdot \left(\frac{\mathrm{d}\vec{\mathbf{B}}}{\mathrm{d}t} \times \vec{\mathbf{C}}\right) + \vec{\mathbf{A}} \cdot \left(\vec{\mathbf{B}} \times \frac{\mathrm{d}\vec{\mathbf{C}}}{\mathrm{d}t}\right) .$$

The resulting field is a scalar field.

Integration of a vector involves integration by components:

$$\int \vec{\mathbf{A}}(t)\mathrm{d}t = \int A_x(t)\mathrm{d}t\,\vec{\mathbf{e}}_x + \int A_y(t)\mathrm{d}t\,\vec{\mathbf{e}}_y + \int A_z(t)\mathrm{d}t\,\vec{\mathbf{e}}_z .$$

Curvilinear integral of a vector field: Let $\vec{\mathbf{s}}(t) = x(t)\,\vec{\mathbf{e}}_x + y(t)\,\vec{\mathbf{e}}_y + z(t)\,\vec{\mathbf{e}}_z$ be the parametrization of curve C with the parameter t, $t_0 \le t \le t_1$. It holds that:

$$\begin{aligned}\int_C \vec{\mathbf{A}}(x, y, z)\mathrm{d}\vec{\mathbf{s}} &= \int_{t_0}^{t_1} \vec{\mathbf{A}}(x(t), y(t), z(t)) \cdot \frac{\mathrm{d}\vec{\mathbf{s}}(t)}{\mathrm{d}t}\mathrm{d}t \\ &= \int_{t_0}^{t_1} \left(A_x\frac{\mathrm{d}x}{\mathrm{d}t} + A_y\frac{\mathrm{d}y}{\mathrm{d}t} + A_z\frac{\mathrm{d}z}{\mathrm{d}t}\right)\mathrm{d}t .\end{aligned}$$

Multiple integrals: Integration over various independent parameters, with successive integration. The integration can be in any order if the limits of integration do not depend on the parameters:

$$\begin{aligned}\int_F f(x, y)\,\mathrm{d}F &= \int_{x_0}^{x_1}\int_{y_0}^{y_1} f(x, y)\,\mathrm{d}x\,\mathrm{d}y \\ &= \int_{x_0}^{x_1}\left(\int_{y_0}^{y_1} f(x, y)\mathrm{d}y\right)\mathrm{d}x = \int_{y_0}^{y_1}\left(\int_{x_0}^{x_1} f(x, y)\mathrm{d}x\right)\mathrm{d}y .\end{aligned}$$

Volume integals and surface integrals are multiple integrals.

- In many cases, such as the integral over the area of a triangle, the limits of integration depend on the parameters.

Vector surface element $\mathrm{d}\vec{\mathbf{F}}$: The vector which has the magnitude of area element $\mathrm{d}F$ and points in the direction of normal vector $\vec{\mathbf{n}}$. The normal vector $\vec{\mathbf{n}}$ is perpendicular to the surface F of an enclosed domain of space at the point of the surface element; it points "outward" (away from the domain of space enclosed by the surface). **The flux integral** of a vector $\vec{\mathbf{A}}$ over a closed surface F of a domain of space is

$$\oint_F \vec{\mathbf{A}} \cdot \mathrm{d}\vec{\mathbf{F}} = \oint_F \vec{\mathbf{A}} \cdot \vec{\mathbf{n}}\,\mathrm{d}F , \quad \text{where } \vec{\mathbf{n}} \text{ is the normal vector.}$$

Volume element in case of coordinate transformation: Let q_1, q_2, q_3 be the coordinates of an orthogonal coordinate system and $\vec{\mathbf{e}}_1, \vec{\mathbf{e}}_2, \vec{\mathbf{e}}_3$ the corresponding unit vectors. The position vector is

$$\begin{aligned}\vec{\mathbf{r}} &= x(q_1, q_2, q_3)\,\vec{\mathbf{e}}_x + y(q_1, q_2, q_3)\,\vec{\mathbf{e}}_y + z(q_1, q_2, q_3)\,\vec{\mathbf{e}}_z \\ &= r_1(x, y, z)\vec{\mathbf{e}}_1 + r_2(x, y, z)\vec{\mathbf{e}}_2 + r_3(x, y, z)\vec{\mathbf{e}}_3.\end{aligned}$$

With scale factors $h_i = \left|\frac{d\vec{r}}{dq_i}\right|$, $i = 1, 2, 3$, the volume element becomes

$$dV = dx\, dy\, dz = h_1 dq_1\, h_2 dq_2\, h_3 dq_3 = h_1 h_2 h_3\, dq_1 dq_2 dq_3\,.$$

□ **Volume element** in

Cartesian coordinates: $dV = dx\, dy\, dz.$

Spherical coordinates: $dV = dr\, r d\vartheta\, r \sin\vartheta d\varphi = r^2 dr\, d\cos\vartheta\, d\varphi.$

Cylindrical coordinates: $dV = d\rho\, \rho d\varphi\, dz = \rho d\rho\, d\varphi dz.$

16.2.1 Scale factors in general orthogonal coordinates

Let q_1, q_2, q_3 be orthogonal coordinates with unit vectors $\vec{\mathbf{e}}_1, \vec{\mathbf{e}}_2, \vec{\mathbf{e}}_3$.

Scale factor h_i: The absolute value of the partial derivative of position vector $\vec{\mathbf{r}}$ with respect to the coordinate q_i:

$$h_i = \left|\frac{\partial \vec{\mathbf{r}}}{\partial q_i}\right|, \qquad i = 1, 2, 3\,.$$

The direction vector of the partial derivative is unit vector $\vec{\mathbf{e}}_i$

$$\frac{\partial \vec{\mathbf{r}}}{\partial q_i} = h_i \vec{\mathbf{e}}_i, \qquad \vec{\mathbf{e}}_i = \frac{\partial \vec{\mathbf{r}}}{\partial q_i} \cdot \left|\frac{\partial \vec{\mathbf{r}}}{\partial q_i}\right|^{-1} = e_i^x\, \vec{\mathbf{e}}_x + e_i^y\, \vec{\mathbf{e}}_y + e_i^z\, \vec{\mathbf{e}}_z\,, \qquad i = 1, 2, 3,$$

where e_i^x, e_i^y, e_i^z are the x, y, and z coordinates (Cartesian coordinates) of $\vec{\mathbf{e}}_i$. The coordinates A_i of vector $\vec{\mathbf{A}}$ are obtained by projecting the vector (in Cartesian coordinates) onto the unit vectors $\vec{\mathbf{e}}_i$:

$$A_i = \vec{\mathbf{A}} \cdot \vec{\mathbf{e}}_i = A_x \cdot e_i^x + A_y \cdot e_i^y + A_z \cdot e_i^z.$$

Scale factors in various coordinate systems:

Cartesian coordinates, $q_1 = x,\ q_2 = y,\ q_3 = z$:

$$x = x, \quad y = y, \quad z = z,$$
$$h_1 = 1, \quad h_2 = 1, \quad h_3 = 1.$$

Spherical coordinates, $q_1 = r,\ q_2 = \vartheta,\ q_3 = \varphi$:

$$x = r\cos\varphi\sin\vartheta, \quad y = r\sin\varphi\sin\vartheta, \quad z = r\cos\vartheta,$$
$$h_1 = 1, \quad h_2 = r, \quad h_3 = r\sin\vartheta.$$

Cylindrical coordinates, $q_1 = \rho,\ q_2 = \varphi,\ q_3 = z$:

$$x = \rho\cos\varphi, \quad y = \rho\sin\varphi, \quad z = z,$$
$$h_1 = 1, \quad h_2 = \rho, \quad h_3 = 1.$$

Elliptic cylindrical coordinates, $q_1 = u,\ q_2 = v,\ q_3 = z$:

$$x = a\cosh u\cos v, \quad y = a\sinh u\sin v, \quad z = z,$$
$$h_1 = a\sqrt{\sinh^2 u + \sin^2 v}, \quad h_2 = a\sqrt{\sinh^2 u + \sin^2 v}, \quad h_3 = 1.$$

Parabolic cylindrical coordinates, $q_1 = u,\ q_2 = v,\ q_3 = z$:

$$x = \frac{1}{2}(u^2 - v^2), \quad y = uv, \quad z = z,$$
$$h_1 = \sqrt{u^2 + v^2}, \quad h_2 = \sqrt{u^2 + v^2}, \quad h_3 = 1.$$

Bipolar coordinates, $q_1 = u,\ q_2 = v,\ q_3 = z$:

$$x = \frac{a \sinh v}{\cosh v - \cos u}, \qquad y = \frac{a \sin u}{\cosh v - \cos u}, \qquad z = z,$$
$$h_1 = \frac{a}{|\cosh v - \cos u|}, \quad h_2 = \frac{a}{|\cosh v - \cos u|}, \quad h_3 = 1.$$

16.2.2 Differential operators

Operator: A rule of assignment that assigns every point of the domain of definition of the inverse image set to one and only one point of the image set.

▷ The image and inverse image sets need not be identical; for example, an operator can map a scalar field into a vector field and vice versa.

The differential operator $\frac{\mathrm{d}}{\mathrm{d}t}$ assigns [−] to any function $f(t)$ $\left[\text{its derivative } \frac{\mathrm{d}f}{\mathrm{d}t}\right]$.

The partial differential operator $\frac{\partial}{\partial y}$ differentiates a function $f(x, y, z, \ldots)$ in several variables with respect to a variable y. The other variables are treated as constants.

Nabla operator: Formally a vector whose corresponding components are the partial differential operators:

$$\vec{\nabla} = \vec{\mathbf{e}}_x \cdot \frac{\partial}{\partial x} + \vec{\mathbf{e}}_y \cdot \frac{\partial}{\partial y} + \vec{\mathbf{e}}_z \cdot \frac{\partial}{\partial z}\,.$$

Multiple partial derivative: The multiple application of the partial differential operator:

$$\frac{\partial^n}{\partial z^n} U(x, y, z) = \underbrace{\frac{\partial}{\partial z} \cdots \frac{\partial}{\partial z}}_{n \text{ times}} U(x, y, z)\,; \quad \text{for example, for } n = 2\text{:} \ \frac{\partial^2}{\partial y^2} U = \frac{\partial}{\partial y}\left(\frac{\partial U}{\partial y}\right).$$

results in the n-th derivative of U.

Laplace operator: The scalar product of the nabla operator by itself; summed over all partial second derivatives (see also Section 16.7):

$$\Delta = \vec{\nabla} \cdot \vec{\nabla} = \frac{\partial^2}{\partial x^2} + \frac{\partial^2}{\partial y^2} + \frac{\partial^2}{\partial z^2}\,.$$

▷ The Laplace operator is a scalar and not a vector. (See also Section 16.7 on the Laplace operator).

Mixed partial derivative: Successive partial differentiation with respect to distinct coordinates.

● If the function $f(x, y, z)$ to be differentiated is twice differentiable continuously, then the order of differentiation can be interchanged, e.g.:

$$\frac{\partial}{\partial x}\left(\frac{\partial f}{\partial y}\right) = \frac{\partial}{\partial y}\left(\frac{\partial f}{\partial x}\right).$$

Biharmonic operator: Two-fold application of the Laplace operator:

$$\nabla^4 U = \vec{\nabla}^2(\vec{\nabla}^2 U) = \Delta(\Delta U) = \Delta^2 U$$
$$\nabla^4 = \frac{\partial^4}{\partial x^4} + \frac{\partial^4}{\partial y^4} + \frac{\partial^4}{\partial z^4} + 2\frac{\partial^2}{\partial x^2}\frac{\partial^2}{\partial y^2} + 2\frac{\partial^2}{\partial y^2}\frac{\partial^2}{\partial z^2} + 2\frac{\partial^2}{\partial x^2}\frac{\partial^2}{\partial z^2}$$

16.3 Gradient and potential

Gradient $\vec{\nabla}$ of a scalar field U assigns to each point of field $U(\vec{\mathbf{r}})$ a vector

$$\vec{\mathbf{A}} = \vec{\nabla} U = \operatorname{grad} U$$

that points in the direction of the greatest increase of U. Its magnitude is the derivative of the field in that direction.

▷ The gradient is perpendicular to the equipotential surfaces.

$\vec{\mathbf{A}} = \vec{\nabla} U$ is a vector field.

Potential: If $\vec{\mathbf{A}} = -\operatorname{grad} U$, then U is called the potential of $\vec{\mathbf{A}}$.

A vector field $\vec{\mathbf{A}}$ is called **conservative** if the line integral $\int_C \vec{\mathbf{A}} d\vec{\mathbf{r}}$ depends only on the end-point of the line and not the integration path.

▷ Every conservative field $\vec{\mathbf{A}}$ has a potential U with $\vec{\mathbf{A}} = -\operatorname{grad} U$, and every gradient field is conservative.

Integrability conditions for a conservative vector field:

$$\frac{\partial A_x}{\partial y} = \frac{\partial A_y}{\partial x}, \qquad \frac{\partial A_y}{\partial z} = \frac{\partial A_z}{\partial y}, \qquad \frac{\partial A_x}{\partial z} = \frac{\partial A_z}{\partial x}.$$

▷ These integrability conditions correspond to $\operatorname{rot}\vec{\mathbf{A}} = 0$ (see Section 16.6 on rotation).

Since for every gradient field the path integral depends only on the end-points, we can write:

$$\int \operatorname{grad} U \, d\vec{\mathbf{s}} = \int dU .$$

▷ This means that $\operatorname{grad} U \cdot d\vec{\mathbf{r}}$ is a **total differential** of U.

□ The electric field $\vec{\mathbf{E}} = \dfrac{Q}{\varepsilon F} \vec{\mathbf{e}}_z$ in a plate capacitor is conservative.

The electric field of a charged wire

$$\vec{\mathbf{E}} = \frac{Q}{2\pi\varepsilon r} \frac{x\,\vec{\mathbf{e}}_x + y\,\vec{\mathbf{e}}_y}{x^2 + y^2}$$

is conservative.

The magnetic field of a live wire

$$\vec{\mathbf{B}} = \frac{\mu I}{2\pi\varepsilon r} \frac{-y\,\vec{\mathbf{e}}_x + x\,\vec{\mathbf{e}}_y}{x^2 + y^2}$$

is not a conservative field. The circular integral $\oint \vec{\mathbf{B}} \, d\vec{\mathbf{s}}$ on a circle around the wire yields the value $\mu_0 I$.

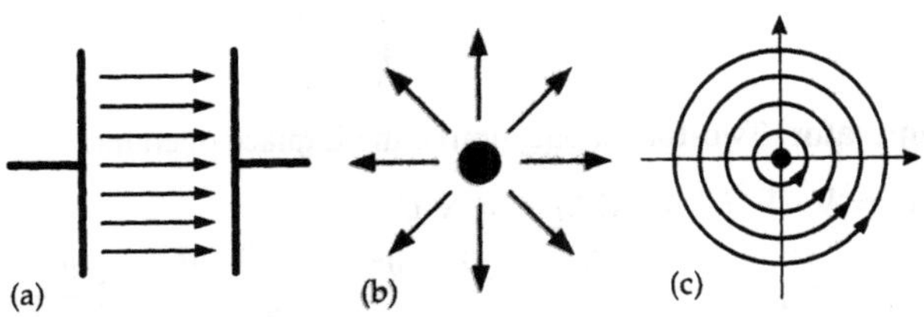

Field lines. (a) Electric field in a plate capacitor, (b) Electric field of a charged wire, (c) Magnetic field of a live wire

Representation of the gradient in various coordinate systems:
Cartesian coordinates (x, y, z):

$$\operatorname{grad} U = \vec{\nabla} U = \frac{\partial U}{\partial x}\vec{\mathbf{e}}_x + \frac{\partial U}{\partial y}\vec{\mathbf{e}}_y + \frac{\partial U}{\partial z}\vec{\mathbf{e}}_z .$$

Spherical coordinates (r, ϑ, φ):

$$\operatorname{grad} U = \frac{\partial U}{\partial r}\vec{\mathbf{e}}_r + \frac{1}{r}\frac{\partial U}{\partial \vartheta}\vec{\mathbf{e}}_\vartheta + \frac{1}{r\sin\vartheta}\frac{\partial U}{\partial \varphi}\vec{\mathbf{e}}_\varphi .$$

Cylindrical coordinates (ρ, φ, z):

$$\operatorname{grad} U = \frac{\partial U}{\partial \rho}\vec{\mathbf{e}}_\rho + \frac{1}{\rho}\frac{\partial U}{\partial \varphi}\vec{\mathbf{e}}_\varphi + \frac{\partial U}{\partial z}\vec{\mathbf{e}}_z .$$

General orthogonal coordinates (q_1, q_2, q_3):

$$\operatorname{grad} U = \frac{1}{h_1}\frac{\partial U}{\partial q_1}\vec{\mathbf{e}}_1 + \frac{1}{h_2}\frac{\partial U}{\partial q_2}\vec{\mathbf{e}}_2 + \frac{1}{h_3}\frac{\partial U}{\partial q_3}\vec{\mathbf{e}}_3 , \qquad h_i = \left|\frac{\mathrm{d}\vec{\mathbf{r}}}{\mathrm{d}q_i}\right| , \quad i = 1, 2, 3 .$$

□ Gradient of the electric field of a point charge with charge number Z, elementary charge e, and permittivity ε:

$$U = \frac{Ze}{4\pi\varepsilon r} = \frac{a}{r} , \qquad a = \frac{Ze}{4\pi\varepsilon}, \quad r = \sqrt{x^2 + y^2 + z^2} .$$

Gradient in Cartesian coordinates:

$$\begin{aligned}\operatorname{grad} U &= -\frac{2ax}{2\sqrt{x^2+y^2+z^2}^3}\vec{\mathbf{e}}_x - \frac{2ay}{2\sqrt{x^2+y^2+z^2}^3}\vec{\mathbf{e}}_y - \frac{2az}{2\sqrt{x^2+y^2+z^2}^3}\vec{\mathbf{e}}_z \\ &= -\frac{a}{\sqrt{x^2+y^2+z^2}^3}\left(x\,\vec{\mathbf{e}}_x + y\,\vec{\mathbf{e}}_y + z\,\vec{\mathbf{e}}_z\right) = -\frac{a}{r^3}\vec{\mathbf{r}} .\end{aligned}$$

Gradient in spherical coordinates:

$$\operatorname{grad} U = -\frac{a}{r^2}\vec{\mathbf{e}}_r + 0\cdot\vec{\mathbf{e}}_\vartheta + 0\cdot\vec{\mathbf{e}}_\varphi = -\frac{a}{r^3}\vec{\mathbf{r}} .$$

Two-dimensional case, Cartesian coordinates:

$$\operatorname{grad} U = \frac{\partial U}{\partial x}\vec{\mathbf{e}}_x + \frac{\partial U}{\partial y}\vec{\mathbf{e}}_y .$$

Two-dimensional case, polar coordinates:

$$\operatorname{grad} U = \frac{\partial U}{\partial \rho}\vec{\mathbf{e}}_\rho + \frac{1}{\rho}\frac{\partial U}{\partial \varphi}\vec{\mathbf{e}}_\varphi .$$

n-dimensional case, Cartesian coordinates:

$$\operatorname{grad} U = \frac{\partial U}{\partial x_1}\vec{\mathbf{e}}_1 + \frac{\partial U}{\partial x_2}\vec{\mathbf{e}}_2 + \cdots + \frac{\partial U}{\partial x_n}\vec{\mathbf{e}}_n .$$

Computational rules for gradients:
U, V are scalar fields, c is a constant, $\vec{\mathbf{a}}$ is a constant vector, and $\vec{\mathbf{r}}$ is the position vector:

$$\operatorname{grad} c = \vec{\mathbf{0}}, \qquad \operatorname{grad}(\vec{\mathbf{a}}\cdot\vec{\mathbf{r}}) = \vec{\mathbf{a}},$$

$$\operatorname{grad}(c\,U) = c\operatorname{grad} U, \quad \operatorname{grad}(U + c) = \operatorname{grad} U,$$

$$\operatorname{grad}(U+V)=\operatorname{grad}U+\operatorname{grad}V,\quad \operatorname{grad}(UV)=V\operatorname{grad}U+U\operatorname{grad}V,$$

$$\operatorname{grad}f(U)=\frac{\partial f}{\partial U}\operatorname{grad}U,\qquad \operatorname{grad}U^n=nU^{n-1}\operatorname{grad}U.$$

16.4 Directional derivative and vector gradient

Directional derivative: The derivative of a scalar field at point $\vec{\mathbf{r}}$ in the direction of the direction vector $\vec{\mathbf{n}}$ normalized to unity

$$\frac{\partial U}{\partial \vec{\mathbf{n}}}=\lim_{\Delta t\to 0}\frac{U(\vec{\mathbf{r}}+\Delta t\,\vec{\mathbf{n}})-U(\vec{\mathbf{r}})}{\Delta t}.$$

In Cartesian coordinates, $\vec{\mathbf{r}}=x\,\vec{\mathbf{e}}_x+y\,\vec{\mathbf{e}}_y+z\,\vec{\mathbf{e}}_z$, the directional derivative can be written as the scalar product of the direction vector and the gradient of the field.

$$\frac{\partial U}{\partial \vec{\mathbf{n}}}=\vec{\mathbf{n}}\cdot\operatorname{grad}U=n_x\frac{\partial U}{\partial x}+n_y\frac{\partial U}{\partial y}+n_z\frac{\partial U}{\partial z}$$

□ The electric field of a wire is:

$$U(x,y,z)=\frac{\sigma_e}{4\pi\varepsilon}\ln(x^2+y^2)=a\ln(x^2+y^2).$$

The gradient is:

$$\operatorname{grad}U=\frac{2a\,x}{x^2+y^2}\,\vec{\mathbf{e}}_x+\frac{2a\,y}{x^2+y^2}\,\vec{\mathbf{e}}_y+0\,\vec{\mathbf{e}}_z\ .$$

The directional derivative in the direction of the $\vec{\mathbf{r}}$-vector is:

$$\begin{aligned}\vec{\mathbf{n}}\operatorname{grad}U&=\frac{x\,\vec{\mathbf{e}}_x+y\,\vec{\mathbf{e}}_y+z\,\vec{\mathbf{e}}_z}{\sqrt{x^2+y^2+z^2}}\frac{2a}{x^2+y^2}\cdot\left(x\,\vec{\mathbf{e}}_x+y\,\vec{\mathbf{e}}_y\right)\\&=\frac{x^2+y^2}{x^2+y^2}\frac{2a}{\sqrt{x^2+y^2+z^2}}=\frac{2a}{r}\end{aligned}$$

with $r=\sqrt{x^2+y^2+z^2}$. The same relation is obtained if the system is considered in spherical coordinates and the $\vec{\mathbf{e}}_r$-component is calculated.
In spherical coordinates, the potential is $U=2a\ln(r\sin\vartheta)$.

$$(\operatorname{grad}U)_r=\frac{\partial U}{\partial r}=\frac{2a}{r\sin\vartheta}\cdot\sin\vartheta=\frac{2a}{r}$$

Directional derivative of a vector field and vector gradient.
The directional derivative of a vector field in the direction of the direction vector $\vec{\mathbf{n}}$ normalized to unity is

$$(\vec{\mathbf{n}}\operatorname{grad})\vec{\mathbf{A}}=\frac{\partial\vec{\mathbf{A}}}{\partial\vec{\mathbf{n}}}=\lim_{\Delta t\to 0}\frac{\vec{\mathbf{A}}(\vec{\mathbf{r}}+\Delta t\,\vec{\mathbf{n}})-\vec{\mathbf{A}}(\vec{\mathbf{r}})}{\Delta t}.$$

Generalized to non-normalized vectors $\vec{\mathbf{a}}=a\vec{\mathbf{n}}$, $(\vec{\mathbf{a}}\operatorname{grad})\vec{\mathbf{A}}$ can be represented as the multiplication of the directional derivative by the magnitude a of $\vec{\mathbf{a}}$:

$$(\vec{\mathbf{a}}\operatorname{grad})\vec{\mathbf{A}}=a(\vec{\mathbf{n}}\operatorname{grad})\vec{\mathbf{A}}\ .$$

In Cartesian coordinates, $\vec{\mathbf{r}} = x\,\vec{\mathbf{e}}_x + y\,\vec{\mathbf{e}}_y + z\,\vec{\mathbf{e}}_z$, one can write:

$$(\vec{\mathbf{a}}\,\mathrm{grad}\,)\vec{\mathbf{A}} = (\vec{\mathbf{a}}\cdot\mathrm{grad}\,A_x)\,\vec{\mathbf{e}}_x + (\vec{\mathbf{a}}\cdot\mathrm{grad}\,A_y)\,\vec{\mathbf{e}}_y + (\vec{\mathbf{a}}\cdot\mathrm{grad}\,A_z)\,\vec{\mathbf{e}}_z\,.$$

In general, one can write (see Section 16.5 for divergence and Section 16.6 for rotation):

$$(\vec{\mathbf{a}}\,\mathrm{grad}\,)\vec{\mathbf{A}} = \frac{1}{2}\Big(\mathrm{rot}\,(\vec{\mathbf{A}}\times\vec{\mathbf{a}}) + \mathrm{grad}\,(\vec{\mathbf{a}}\cdot\vec{\mathbf{A}}) + \vec{\mathbf{a}}\,\mathrm{div}\,\vec{\mathbf{A}} - \vec{\mathbf{A}}\,\mathrm{div}\,\vec{\mathbf{a}} \\ -\vec{\mathbf{a}}\times\mathrm{rot}\,\vec{\mathbf{A}} - \vec{\mathbf{A}}\times\mathrm{rot}\,\vec{\mathbf{a}}\Big)\,.$$

Vector gradient: The operator $\mathrm{grad}\,\vec{\mathbf{A}}$ that assigns the vector $(\vec{\mathbf{a}}\,\mathrm{grad}\,)\vec{\mathbf{A}}$ to every vector $\vec{\mathbf{a}}$. In Cartesian coordinates, this assignment can be written as multiplication of a vector by a matrix:

$$(\vec{\mathbf{a}}\,\mathrm{grad}\,)\vec{\mathbf{A}} = \begin{pmatrix} \dfrac{\partial A_x}{\partial x} & \dfrac{\partial A_x}{\partial y} & \dfrac{\partial A_x}{\partial z} \\ \dfrac{\partial A_y}{\partial x} & \dfrac{\partial A_y}{\partial y} & \dfrac{\partial A_y}{\partial z} \\ \dfrac{\partial A_z}{\partial x} & \dfrac{\partial A_z}{\partial y} & \dfrac{\partial A_z}{\partial z} \end{pmatrix} \begin{pmatrix} a_x \\ a_y \\ a_z \end{pmatrix}\,.$$

Thus, the vector gradient can be represented as a matrix (tensor):

$$\mathrm{grad}\,\vec{\mathbf{A}} = \begin{pmatrix} \dfrac{\partial A_x}{\partial x} & \dfrac{\partial A_x}{\partial y} & \dfrac{\partial A_x}{\partial z} \\ \dfrac{\partial A_y}{\partial x} & \dfrac{\partial A_y}{\partial y} & \dfrac{\partial A_y}{\partial z} \\ \dfrac{\partial A_z}{\partial x} & \dfrac{\partial A_z}{\partial y} & \dfrac{\partial A_z}{\partial z} \end{pmatrix}\,.$$

It is particularly important in the study of electricity in mechanical engineering (stress tensor, strain tensor).

16.5 Divergence and Gaussian integral theorem

Divergence or the source of a field $\vec{\mathbf{A}}$ at point (x, y, z) is the balance of vector flux $\vec{\mathbf{A}}$ through surface F of an infinitesimally small volume V about point (x, y, z); i.e., the difference between inflow and outflow:

$$\mathrm{div}\vec{\mathbf{A}} = \lim_{V\to 0}\frac{1}{V}\oint_F \vec{\mathbf{A}}\cdot\vec{\mathbf{n}}\,\mathrm{d}F$$

($\vec{\mathbf{n}}$ is the normal vector perpendicular to the surface element).

$\mathrm{div}\vec{\mathbf{A}} = 0$: inflow equals outflow solenoidal field.
$\mathrm{div}\vec{\mathbf{A}} > 0$: excess of outflow source.
$\mathrm{div}\vec{\mathbf{A}} < 0$: excess of inflow sink.

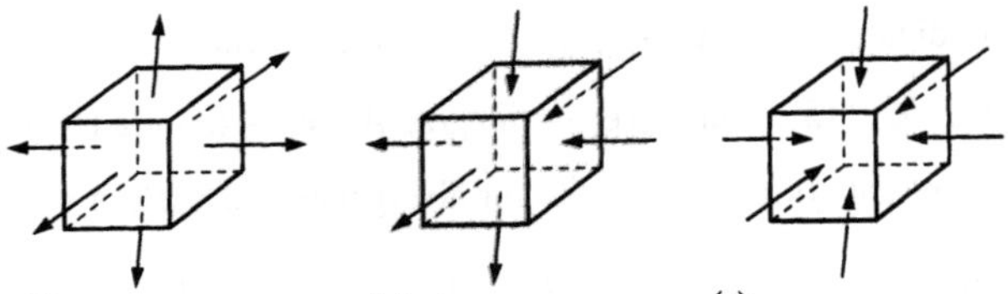

Schematic representation of the flux for (a) div$\vec{\mathbf{A}} > 0$, (b) div$\vec{\mathbf{A}} = 0$, and (c) div$\vec{\mathbf{A}} < 0$

□ Divergence of $\vec{\mathbf{r}}$ at the origin in a small spherical volume with radius R:

$$\begin{aligned}\operatorname{div}\vec{\mathbf{r}} &= \lim_{V\to 0}\frac{1}{V}\oint_F \vec{\mathbf{r}}\cdot\vec{\mathbf{n}}\,\mathrm{d}F = \lim_{R\to 0}\frac{1}{\frac{4}{3}\pi R^3}\oint_F \vec{\mathbf{r}}\cdot\vec{\mathbf{e}}_r\,\mathrm{d}F \\ &= \lim_{R\to 0}\frac{3}{4\pi R^3}\oint_F r\,\mathrm{d}F = \lim_{R\to 0}\frac{3}{4\pi R^3} R\,4\pi R^2 = 3\,.\end{aligned}$$

The **potential field** div$\vec{\mathbf{A}}$ can be regarded as the scalar product of the $\vec{\nabla}$ differential operator with a vector field $\vec{\mathbf{A}}$. It is therefore a **scalar field**.

□ In electrostatics, the divergence of the electric field $\vec{\mathbf{E}}$ (vector field) is equal to the (scalar) density field ϱ of charges:

$$\operatorname{div}\vec{\mathbf{E}} = \frac{\varrho}{\varepsilon} \quad \text{(third Maxwell's equation).}$$

The divergence of a volume V is $\int_V \operatorname{div}\vec{\mathbf{A}}\,\mathrm{d}V$.

● **Gaussian theorem**: The divergence of a volume V is equal to the integral of the flow through the surface F:

$$\int_V \operatorname{div}\vec{\mathbf{A}}\,\mathrm{d}V = \oint_F \vec{\mathbf{A}}\cdot\mathrm{d}\vec{\mathbf{F}} = \oint_F \vec{\mathbf{A}}\cdot\vec{\mathbf{n}}\,\mathrm{d}F\,,$$

where $\vec{\mathbf{n}}$ is the normal vector perpendicular to the surface element $\mathrm{d}F$ and pointing "outward," in other words, everything flowing out of a volume must cross the surface!

▷ This theorem applies to all dimension. In two dimensions one must read "area" and "circumference" instead of "volume" and "surface."

Green's theorem in the plane: A special case of the Gaussian theorem for two dimensions. If U and V are scalar fields, then it holds that

$$\int_F \left(\frac{\partial U}{\partial x} + \frac{\partial V}{\partial y}\right)\mathrm{d}x\,\mathrm{d}y = \oint_C (U\,\mathrm{d}y - V\,\mathrm{d}x)\,,$$

where C is the boundary of surface F.

This theorem can be derived easily from the Gaussian theorem if we consider a two-dimensional field $\vec{\mathbf{A}}$ with $A_x = U$ and $A_y = V$.

● **Gaussian theorem**: If $\vec{\mathbf{A}} = \frac{a}{r^3}\vec{\mathbf{r}}$, then the divergence of a volume V is given by:

$$\oint_F \vec{\mathbf{A}}\cdot\vec{\mathbf{n}}\,\mathrm{d}F = \begin{cases} 4\pi a, & \text{if the origin } (0,0,0) \text{ lies inside } V \\ 0, & \text{if the origin lies outside } V \end{cases}$$

▷ $\vec{\mathbf{A}}$ can be written as the gradient of function $U = \frac{a}{r}$.

As stated above, the Gaussian theorem applies only to the three-dimensional case. In the two-dimensional case, the Gaussian theorem holds for $\vec{\mathbf{A}} = \dfrac{a}{2r^2}\vec{\mathbf{r}}$ or $U = \dfrac{a}{2}\ln|r|$.

▷ The source strength of $\vec{\mathbf{A}}$ is divergent at origin (0, 0, 0) and is zero at all other points. The corresponding scalar functions U with ($\vec{\nabla}U = \vec{\mathbf{A}}$) are **Green's function** of the Poisson equation. See also Section 16.8 on the determination of a vector field from its sources and sinks.

Representation of $\vec{\nabla}\vec{\mathbf{A}}$ in the three-dimensional case:

Cartesian coordinates (x, y, z):

$$\operatorname{div}\vec{\mathbf{A}} = \vec{\nabla}\cdot\vec{\mathbf{A}} = \frac{\partial A_x}{\partial x} + \frac{\partial A_y}{\partial y} + \frac{\partial A_z}{\partial z}\,.$$

Spherical coordinates (r, ϑ, φ):

$$\operatorname{div}\vec{\mathbf{A}} = \frac{1}{r^2}\frac{\partial}{\partial r}(r^2 A_r) + \frac{1}{r\sin\vartheta}\frac{\partial}{\partial\vartheta}(\sin\vartheta\, A_\vartheta) + \frac{1}{r\sin\vartheta}\frac{\partial}{\partial\varphi}A_\varphi\,.$$

Cylindrical coordinates (ρ, φ, z):

$$\operatorname{div}\vec{\mathbf{A}} = \frac{1}{\rho}\left(\frac{\partial}{\partial\rho}(\rho A_\rho) + \frac{\partial}{\partial\varphi}A_\varphi\right) + \frac{\partial}{\partial z}(A_z)\,.$$

General orthogonal coordinates (q_1, q_2, q_3) (with $\left(h_i = \left|\dfrac{d\vec{\mathbf{r}}}{dq_i}\right|\right)$:

$$\operatorname{div}\vec{\mathbf{A}} = \frac{1}{h_1h_2h_3}\left(\frac{\partial}{\partial q_1}(A_1h_2h_3) + \frac{\partial}{\partial q_2}(A_2h_1h_3) + \frac{\partial}{\partial q_3}(A_3h_1h_2)\right)\,.$$

□ Divergence of the gravitational field on the earth surface for the approximation $\vec{\mathbf{A}} = -g\dfrac{\vec{\mathbf{r}}}{r} = -g\vec{\mathbf{e}}_r = \dfrac{-g}{\sqrt{x^2+y^2+z^2}}\left(x\,\vec{\mathbf{e}}_x + y\,\vec{\mathbf{e}}_y + z\,\vec{\mathbf{e}}_z\right)$.

In spherical coordinates: $\operatorname{div}\dfrac{-g\vec{\mathbf{r}}}{r} = -g\dfrac{1}{r^2}\dfrac{\partial r^2}{\partial r} = -g\dfrac{2r}{r^2} = \dfrac{-2g}{r}$.

In Cartesian coordinates, use the quotient rule with $r = \sqrt{x^2+y^2+z^2}$:

$$\operatorname{div}\frac{-g\vec{\mathbf{r}}}{r} = -g\left(\frac{r - x^2/r}{r^2} + \frac{r - y^2/r}{r^2} + \frac{r - z^2/r}{r^2}\right)$$

$$= -g\left(\frac{3r}{r^2} + \frac{x^2+y^2+z^2}{r^3}\right) = -g\left(\frac{3}{r} - \frac{r^2}{r^3}\right) = \frac{-2g}{r}\,.$$

Two-dimensional case, Cartesian coordinates:

$$\operatorname{div}\vec{\mathbf{A}} = \frac{\partial A_x}{\partial x} + \frac{\partial A_y}{\partial y}\,.$$

Two-dimensional case, polar coordinates:

$$\operatorname{div}\vec{\mathbf{A}} = \frac{1}{\rho}\left(\frac{\partial}{\partial\rho}(\rho A_\rho) + \frac{\partial}{\partial\varphi}A_\varphi\right)\,.$$

n-dimensional case, Cartesian coordinates:

$$\operatorname{div}\vec{\mathbf{A}} = \frac{\partial A_1}{\partial x_1} + \frac{\partial A_2}{\partial x_2} + \cdots + \frac{\partial A_n}{\partial x_n} .$$

Calculating rules for divergences
($\vec{\mathbf{A}}$, $\vec{\mathbf{B}}$ are vector fields, U is a scalar field, $\vec{\mathbf{a}}$ is a constant vector, c is a constant):

$$\operatorname{div}\vec{\mathbf{a}} = 0, \qquad \operatorname{div}(\vec{\mathbf{a}}\,U) = \vec{\mathbf{a}} \cdot \operatorname{grad} U,$$

$$\operatorname{div}(\vec{\mathbf{A}} + \vec{\mathbf{B}}) = \operatorname{div}\vec{\mathbf{A}} + \operatorname{div}\vec{\mathbf{B}}, \qquad \operatorname{div}(\vec{\mathbf{A}} + \vec{\mathbf{a}}) = \operatorname{div}\vec{\mathbf{A}},$$

$$\operatorname{div}(U\vec{\mathbf{A}}) = U \operatorname{div}\vec{\mathbf{A}} + \vec{\mathbf{A}} \operatorname{grad} U, \qquad \operatorname{div}(c\vec{\mathbf{A}}) = c \operatorname{div}\vec{\mathbf{A}},$$

$$\operatorname{div}(\vec{\mathbf{A}} \times \vec{\mathbf{B}}) = \vec{\mathbf{B}} \operatorname{rot}\vec{\mathbf{A}} - \vec{\mathbf{A}} \operatorname{rot}\vec{\mathbf{B}}, \qquad \operatorname{div}(\vec{\mathbf{A}} \times \vec{\mathbf{a}}) = \vec{\mathbf{a}} \operatorname{rot}\vec{\mathbf{A}}.$$

Successive application of div grad and div rot:

$$\operatorname{div} \operatorname{grad} U = \Delta U \quad \text{(see also Section 16.7 for the Laplace operator)},$$

$$\operatorname{div} \operatorname{rot}\vec{\mathbf{A}} = 0.$$

16.6 Rotation and Stokes's theorem

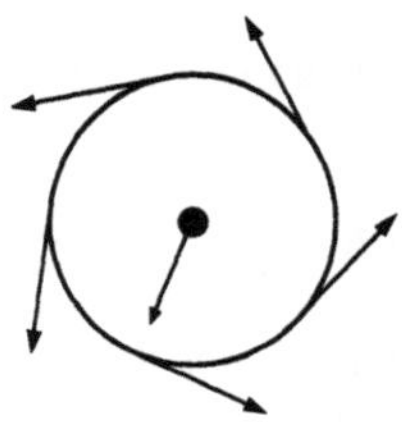

The rotation of a vector field, rot $\vec{\mathbf{A}}$, defines "vortices," i.e., closed field lines of a vector field. The rotation is defined as the line integral along the boundary C of an infinitesimal surface F:

$$\vec{\mathbf{n}} \cdot \operatorname{rot}\vec{\mathbf{A}} = \lim_{F\to 0} \frac{1}{F} \oint_C \vec{\mathbf{A}} \cdot \mathrm{d}\vec{\mathbf{s}} = \lim_{F\to 0} \frac{\Gamma}{F} ,$$

where $\vec{\mathbf{n}}$ is the normal vector on the surface F. Γ is called the **circulation** of the vector field.

□ Line integral $\oint \vec{\mathbf{A}} \cdot \mathrm{d}\vec{\mathbf{s}}$ in a circle around the origin for the velocity field of a disk rotating at angular velocity ω
(velocity field $\vec{\mathbf{A}} = -y\,\omega\,\vec{\mathbf{e}}_x + x\,\omega\,\vec{\mathbf{e}}_y$, surface $F = \pi R^2$,
circular line $\vec{\mathbf{s}}(t) = R \cos t\,\vec{\mathbf{e}}_x + R \sin t\,\vec{\mathbf{e}}_y$, $0 \le t \le 2\pi$):

$$\oint \vec{\mathbf{A}} \cdot \mathrm{d}\vec{\mathbf{s}} = \int_{t_0}^{t_1} \left[A_x \frac{\mathrm{d}x}{\mathrm{d}t} + A_y \frac{\mathrm{d}y}{\mathrm{d}t} \right] \mathrm{d}t$$

$$= \int_0^{2\pi} [(-\omega R \sin t)(-R \sin t) + (\omega R \cos t)(R \cos t)]\,\mathrm{d}t$$

$$= \omega R^2 \int_0^{2\pi} \left[\sin^2 t + \cos^2 t\right] \mathrm{d}t = 2\pi\omega R^2 .$$

$$\vec{\mathbf{n}} \cdot \operatorname{rot}\vec{\mathbf{A}} = \lim_{F\to 0} \frac{1}{F} \oint \vec{\mathbf{A}} \cdot \mathrm{d}\vec{\mathbf{s}} = \lim_{R\to 0} \frac{2\pi\omega R^2}{\pi R^2} = 2\omega .$$

rot$\vec{\mathbf{A}}$ is perpendicular to the plane of rotation; its magnitude is 2ω.

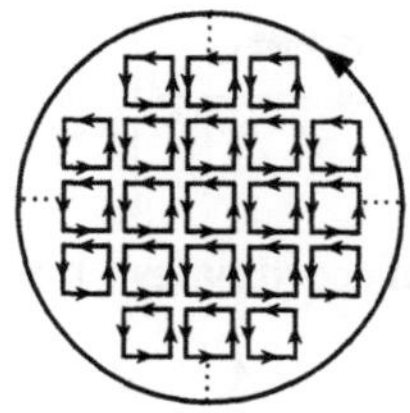

- **Stokes's integral theorem**: The integral over the vertices $\operatorname{rot}\vec{\mathbf{A}}$ in a surface F is equal to the circulation of the vector field along boundary C of the surface:

$$\int_F \operatorname{rot}\vec{\mathbf{A}}\cdot\vec{\mathbf{n}}\,\mathrm{d}F = \oint_C \vec{\mathbf{A}}\cdot\mathrm{d}\vec{\mathbf{s}},$$

where $\vec{\mathbf{n}}$ is again the normal vector on the surface element.
In other words:
The rotation is effective only at the boundary. The rotations in the interior are at equilibrium.

Stokes's integral theorem holds in all dimensions, but the formulation (via the so-called Pfaffian forms, which are not discussed here) is inconvenient.
A consequence of Stokes's integral theorem is the following theorem, in which F is the surface of the spatial volume V:

$$\int_V \operatorname{rot}\vec{\mathbf{A}}\,\mathrm{d}V = \oint_F \vec{\mathbf{A}}\times\vec{\mathbf{n}}\,\mathrm{d}F\ .$$

- A conservative field has no vortices and conversely, an irrotational field is conservative:

$$\operatorname{rot}\operatorname{grad} U = 0\ .$$

□ Electric fields in a plate capacitor and around a charged wire are conservative and thus irrotational.
The magnetic field around a live wire is not conservative; it contains vortices.
See also the representations regarding conservative fields in Section 16.3 on the gradient.

Symbolic notation of the rotation as a cross product of the nabla operator:

$$\operatorname{rot}\vec{\mathbf{A}} = \vec{\nabla}\times\vec{\mathbf{A}}$$

Cartesian coordinates $(x, y, z) = (x_1, x_2, x_3)$:

$$\operatorname{rot}\vec{\mathbf{A}} = \left(\frac{\partial A_z}{\partial y}-\frac{\partial A_y}{\partial z}\right)\vec{\mathbf{e}}_x + \left(\frac{\partial A_x}{\partial z}-\frac{\partial A_z}{\partial x}\right)\vec{\mathbf{e}}_y + \left(\frac{\partial A_y}{\partial x}-\frac{\partial A_x}{\partial y}\right)\vec{\mathbf{e}}_z$$

$$= \begin{vmatrix} \vec{\mathbf{e}}_x & \vec{\mathbf{e}}_y & \vec{\mathbf{e}}_z \\ \dfrac{\partial}{\partial x} & \dfrac{\partial}{\partial y} & \dfrac{\partial}{\partial z} \\ A_x & A_y & A_z \end{vmatrix}\ .$$

Spherical coordinates (r, ϑ, φ):

$$\operatorname{rot}\vec{\mathbf{A}} = \frac{1}{r\sin\vartheta}\left(\frac{\partial \sin\vartheta\, A_\varphi}{\partial\vartheta}-\frac{\partial A_\vartheta}{\partial\varphi}\right)\vec{\mathbf{e}}_r + \frac{1}{r}\left(\frac{1}{\sin\vartheta}\frac{\partial A_r}{\partial\varphi}-\frac{\partial r A_\varphi}{\partial r}\right)\vec{\mathbf{e}}_\vartheta$$
$$+\frac{1}{r}\left(\frac{\partial r A_\vartheta}{\partial r}-\frac{\partial A_r}{\partial\vartheta}\right)\vec{\mathbf{e}}_\varphi$$

$$= \frac{1}{r^2 \sin\vartheta} \begin{vmatrix} \vec{\mathbf{e}}_r & r\vec{\mathbf{e}}_\vartheta & r\sin\vartheta\,\vec{\mathbf{e}}_\varphi \\ \dfrac{\partial}{\partial r} & \dfrac{\partial}{\partial\vartheta} & \dfrac{\partial}{\partial\varphi} \\ A_r & rA_\vartheta & r\sin\vartheta\,A_\varphi \end{vmatrix}.$$

Cylindrical coordinates (ρ, φ, z):

$$\operatorname{rot}\vec{\mathbf{A}} = \left(\frac{1}{\rho}\frac{\partial A_z}{\partial\varphi} - \frac{\partial A_\varphi}{\partial z}\right)\vec{\mathbf{e}}_\rho + \left(\frac{\partial A_\rho}{\partial z} - \frac{\partial A_z}{\partial\rho}\right)\vec{\mathbf{e}}_\varphi + \left(\frac{\partial \rho A_\varphi}{\partial\rho} - \frac{1}{\rho}\frac{\partial A_\rho}{\partial\varphi}\right)\vec{\mathbf{e}}_z$$

$$= \frac{1}{\rho}\begin{vmatrix} \vec{\mathbf{e}}_\rho & \rho\vec{\mathbf{e}}_\varphi & \vec{\mathbf{e}}_z \\ \dfrac{\partial}{\partial\rho} & \dfrac{\partial}{\partial\varphi} & \dfrac{\partial}{\partial z} \\ A_\rho & \rho A_\varphi & A_z \end{vmatrix}.$$

General orthogonal coordinates (q_1, q_2, q_3):

$$\operatorname{rot}\vec{\mathbf{A}} = \frac{1}{h_2h_3}\left(\frac{\partial h_3A_3}{\partial q_2} - \frac{\partial h_2A_2}{\partial q_3}\right)\vec{\mathbf{e}}_1 + \frac{1}{h_1h_3}\left(\frac{\partial h_1A_1}{\partial q_3} - \frac{\partial h_3A_3}{\partial q_1}\right)\vec{\mathbf{e}}_2$$

$$+\frac{1}{h_1h_2}\left(\frac{\partial h_2A_2}{\partial q_1} - \frac{\partial h_1A_1}{\partial q_2}\right)\vec{\mathbf{e}}_3$$

$$= \frac{1}{h_1h_2h_3}\begin{vmatrix} h_1\vec{\mathbf{e}}_1 & h_2\vec{\mathbf{e}}_2 & h_3\vec{\mathbf{e}}_3 \\ \dfrac{\partial}{\partial q_1} & \dfrac{\partial}{\partial q_2} & \dfrac{\partial}{\partial q_3} \\ h_1A_1 & h_2A_2 & h_3A_3 \end{vmatrix}, \qquad h_i = \left|\frac{\partial\vec{\mathbf{r}}}{\partial q_i}\right|, \quad i = 1, 2, 3\,.$$

□ Rotation of the magnetic field in a homogeneous electron beam of current density $\vec{\mathbf{i}} = i\,\vec{\mathbf{e}}_z$.

In cylindrical coordinates:

$\vec{\mathbf{B}} = \dfrac{\mu_0}{2\pi}\dfrac{1}{\rho}\cdot\pi\rho^2 i\vec{\mathbf{e}}_\varphi = a\rho\vec{\mathbf{e}}_\varphi$ with $a = \dfrac{\mu_0 i}{2}$,

$\operatorname{rot}\vec{\mathbf{B}} = 0\vec{\mathbf{e}}_\rho + 0\vec{\mathbf{e}}_\varphi + \dfrac{1}{\rho}\left(\dfrac{\partial\rho\cdot a\rho}{\partial\rho} - 0\right)\vec{\mathbf{e}}_z = \dfrac{1}{\rho}a\,2\rho\,\vec{\mathbf{e}}_z = 2a\,\vec{\mathbf{e}}_z = \mu_0\vec{\mathbf{i}}.$

In Cartesian coordinates:

$\vec{\mathbf{B}} = \dfrac{\mu_0}{2\pi}\dfrac{1}{x^2+y^2}\cdot\pi(x^2+y^2)i\cdot\left(-y\,\vec{\mathbf{e}}_x + x\,\vec{\mathbf{e}}_y\right) = a\left(-y\,\vec{\mathbf{e}}_x + x\,\vec{\mathbf{e}}_y\right)$

with $a = \dfrac{\mu_0 i}{2}$,

$$\operatorname{rot}\vec{\mathbf{B}} = 0\,\vec{\mathbf{e}}_x + 0\,\vec{\mathbf{e}}_y + \left(\frac{\partial}{\partial x}ax - \frac{\partial}{\partial y}(-ay)\right)\vec{\mathbf{e}}_z = 2a\,\vec{\mathbf{e}}_z = \mu_0\vec{\mathbf{i}}$$

This is a Maxwell's equation for the stationary case.

Calculating rules for rotations

($\vec{\mathbf{A}}$, $\vec{\mathbf{B}}$ are vector fields, U is a scalar field, $\vec{\mathbf{a}}$ is a constant vector, c is a constant):

$$\operatorname{rot}(c\vec{\mathbf{A}}) = c\operatorname{rot}\vec{\mathbf{A}}, \qquad \operatorname{rot}(\vec{\mathbf{a}}U) = -\vec{\mathbf{a}}\times\operatorname{grad}U,$$

$$\operatorname{rot}(\vec{\mathbf{A}}+\vec{\mathbf{B}}) = \operatorname{rot}\vec{\mathbf{A}} + \operatorname{rot}\vec{\mathbf{B}}, \qquad \operatorname{rot}(\vec{\mathbf{A}} + \operatorname{grad}U) = \operatorname{rot}\vec{\mathbf{A}},$$

$$\operatorname{rot}(U\vec{\mathbf{A}}) = U\operatorname{rot}\vec{\mathbf{A}} - \vec{\mathbf{A}} \times \operatorname{grad} U, \quad \operatorname{rot}(U \operatorname{grad} U) = \vec{\mathbf{0}},$$

$$\operatorname{rot}(\vec{\mathbf{A}} \times \vec{\mathbf{B}}) = (\vec{\mathbf{B}}\operatorname{grad})\vec{\mathbf{A}} - (\vec{\mathbf{A}}\operatorname{grad})\vec{\mathbf{B}} + \vec{\mathbf{A}}\operatorname{div}\vec{\mathbf{B}} - \vec{\mathbf{B}}\operatorname{div}\vec{\mathbf{A}}.$$

It holds that Δ is the Laplace operator:

$$\operatorname{rot}\operatorname{rot}\vec{\mathbf{A}} = \operatorname{grad}\operatorname{div}\vec{\mathbf{A}} - \Delta\vec{\mathbf{A}}.$$

16.7 Laplace operator and Green's formulas

The Laplace operator $\Delta = \vec{\nabla} \cdot \vec{\nabla}$ is the scalar product of the nabla operator multiplied by itself.

Representation of the Laplace operator:

Cartesian coordinates (x, y, z):

$$\Delta = \vec{\nabla} \cdot \vec{\nabla} = \frac{\partial^2}{\partial x^2} + \frac{\partial^2}{\partial y^2} + \frac{\partial^2}{\partial z^2}.$$

Spherical coordinates (r, ϑ, φ):

$$\Delta = \frac{1}{r^2}\frac{\partial}{\partial r}\left(r^2\frac{\partial}{\partial r}\right) + \frac{1}{r^2 \sin\vartheta}\frac{\partial}{\partial\vartheta}\left(\sin\vartheta\frac{\partial}{\partial\vartheta}\right) + \frac{1}{r^2\sin^2\vartheta}\frac{\partial^2}{\partial\varphi^2}.$$

Cylindrical coordinates (ρ, φ, z):

$$\Delta = \frac{1}{\rho}\frac{\partial}{\partial\rho}\left(\rho\frac{\partial}{\partial\rho}\right) + \frac{1}{\rho^2}\frac{\partial^2}{\partial\varphi^2} + \frac{\partial^2}{\partial z^2}.$$

General orthogonal coordinates (q_1, q_2, q_3):

$$\Delta = \frac{1}{h_1h_2h_3}\left[\frac{\partial}{\partial q_1}\left(\frac{h_2h_3}{h_1}\frac{\partial}{\partial q_1}\right) + \frac{\partial}{\partial q_2}\left(\frac{h_3h_1}{h_2}\frac{\partial}{\partial q_2}\right) + \frac{\partial}{\partial q_3}\left(\frac{h_1h_2}{h_3}\frac{\partial}{\partial q_3}\right)\right].$$

□ Calculation of the source strength for the gravitational potential inside the earth's sphere:

$$U(r) = -\gamma 2\pi\sigma\left(R^2 - \frac{1}{3}r^2\right) = ar^2 + b = a\left(x^2 + y^2 + z^2\right) + b$$

with $a = \frac{2}{3}\gamma\pi\sigma$ and $b = -2\gamma\pi\sigma R^2$.

In Cartesian coordinates (x, y, z):

$$\Delta U = a\left(\frac{\partial^2 x^2}{\partial x^2} + \frac{\partial^2 y^2}{\partial y^2} + \frac{\partial^2 z^2}{\partial z^2}\right) = a(2 + 2 + 2) = 6a.$$

In spherical coordinates (r, ϑ, φ):

$$\Delta U = a\frac{1}{r^2}\frac{\partial}{\partial r}\left(r^2\frac{\partial r^2}{\partial r}\right) = a\frac{1}{r^2}\frac{\partial}{\partial r}2r^3 = a\frac{6r^2}{r^2} = 6a.$$

Two-dimensional case, Cartesian coordinates:

$$\Delta = \frac{\partial^2}{\partial x^2} + \frac{\partial^2}{\partial y^2}.$$

Two-dimensional case, polar coordinates:

$$\Delta = \frac{1}{\rho}\frac{\partial}{\partial\rho}\left(\rho\frac{\partial}{\partial\rho}\right) + \frac{1}{\rho^2}\frac{\partial^2}{\partial\varphi^2} .$$

n-dimensional case, Cartesian coordinates:

$$\Delta = \frac{\partial^2}{\partial x_1^2} + \cdots + \frac{\partial^2}{\partial x_n^2} = \sum_{i=1}^{n}\frac{\partial^2}{\partial x_i^2} .$$

Δ can be applied to scalar fields by component also to vector fields.
Representation in the vector field:

$$\Delta\vec{\mathbf{A}} = \text{grad div}\vec{\mathbf{A}} - \text{rot rot}\vec{\mathbf{A}}.$$

Representation in the scalar field:

$$\Delta U = \text{div grad}\, U .$$

ΔU describes the source strength of the potential field U.
Green's formulas: Green's formulas are based essentially on the principle of integration by parts. U, V are scalar functions, V is the integration volume, F is the surface of the volume.
Green's first formula: According to the product rule,

$$(uv)' = u'v + uv' \text{ with } u = U \text{ and } v = \text{grad}\, V ,$$

we integrate by parts and apply the Gaussian integral theorem:

$$\int_V \left[U\,\Delta V + (\text{grad}\, U)\cdot(\text{grad}\, V)\right] \mathrm{d}V = \oint_F \left[U\,\text{grad}\, V\right]\cdot\vec{\mathbf{n}}\,\mathrm{d}F .$$

where $\vec{\mathbf{n}}$ is the normal vector on the outer surface.
Green's second formula: Apply Green's first formula twice, with the arguments interchanged:

$$\int_V [U\,\Delta V - V\,\Delta U]\,\mathrm{d}V = \oint_F \left[U\,\text{grad}\, V - V\,\text{grad}\, U\right]\cdot\vec{\mathbf{n}}\,\mathrm{d}F .$$

The importance of Green's formulas lies in their use in the analytic solution of potential equations subject to special boundary conditions.
Sources of potential fields: The Laplace operator connects the source strength of the gradient field with the potential field.
The Laplace equation applies to the source-free space. $\Delta U = 0$.
The Poisson equation applies when sources $\Delta U = -4\pi\varrho$ are present, where ϱ is the source density.
Both equations are used as field equations to determine the potential field U (subject to certain boundary conditions).
They are of particular importance in electrostatics and electrodynamics.

□ The distant field of the dipole $U = \dfrac{a\cos\vartheta}{r^2}$, $a = \dfrac{w}{4\pi\varepsilon}$, satisfies the Laplace equation:

$$\Delta U = \frac{a}{r^2}\frac{\partial}{\partial r}\left(r^2\frac{\partial(\cos\vartheta\, r^{-2})}{\partial r}\right) + \frac{a}{r^2\sin\vartheta}\frac{\partial}{\partial\vartheta}\left(\sin\vartheta\frac{\partial(\cos\vartheta\, r^{-2})}{\partial\vartheta}\right)$$

$$= \cos\vartheta \frac{a}{r^2}\frac{\partial}{\partial r}\left(-2\frac{1}{r}\right) + \frac{a}{r^4 \sin\vartheta}\frac{\partial}{\partial\vartheta}\left(-\sin^2\vartheta\right)$$

$$= \cos\vartheta \frac{2a}{r^4} + \frac{-2a\sin\vartheta\cos\vartheta}{r^4\sin\vartheta} = \frac{2a\cos\vartheta}{r^4} - \frac{2a\cos\vartheta}{r^4} = 0\,.$$

16.7.1 Combinations of div, rot, and grad; calculation of fields

Combinations of div, rot, and grad with a scalar field U and a vector field $\vec{\mathbf{A}}$. U as well as $\vec{\mathbf{A}}$ are twice partial differentiable continuously.
The divergence applied to the gradient yields the Laplace operator:

$$\text{div grad}\, U = \Delta U\,.$$

Rotational fields are source-free:

$$\text{div rot}\vec{\mathbf{A}} = 0\,.$$

Conservative fields are irrotational:

$$\text{rot grad}\, U = \vec{\mathbf{0}}\,.$$

The gradient of a divergence yields the Laplace operator and mixed differential terms:

$$\text{grad div}\vec{\mathbf{A}} = \text{rot rot}\vec{\mathbf{A}} + \Delta\vec{\mathbf{A}}\,.$$

Calculation of vector fields from sources and rotations:
The combinations of vector operators mentioned above can be used to calculate the vector fields from the sources and rotations. Find a vector field $\vec{\mathbf{B}}$ with

$$\text{div}\,\vec{\mathbf{B}} = \varrho, \qquad \text{rot}\,\vec{\mathbf{B}} = \vec{\mathbf{j}}\,.$$

Usually, we first find a specific vector field that satisfies the required conditions. Then it is superposed by a source-free and irrotational field to satisfy specific boundary conditions.
Source-free and irrotational fields can be represented as gradient fields (irrotational) of a potential field U. The potential field must satisfy the **Laplace equation**:

$$\Delta U = \text{div grad}\, U = 0\,.$$

[for the boundary condition]. There are two possibilities [–]:

1. **Dirichlet problem**: The volume of U on the boundary of the considered volume is fixed.

□ A grounded sphere lies in the electrostatic field of a charge distribution.

2. **Neumann's problem**: The directional derivative of the field perpendicular to the boundary is fixed.

□ A conductor, not grounded, is located in the electrostatic field of a charge distribution.
The Laplace equation for certain boundary conditions can often be solved only numerically. In this context we refer to the section on partial differential equations. Field $\vec{\mathbf{B}}$ is obtained by $\vec{\mathbf{B}} = -\text{grad}\, U$.
Irrotational but not source-free field: Investigate a potential field P for which the **Poisson equation** is solved.

$$-\Delta P = -\text{div grad}\, P = \varrho\,.$$

A solution is obtained by

$$P(\vec{\mathbf{r}}) = \frac{1}{4\pi} \int_V \frac{\varrho(\vec{\mathbf{r}}\,')}{|\vec{\mathbf{r}} - \vec{\mathbf{r}}\,'|} \mathrm{d}V' \,.$$

▷ Function $U = \frac{a}{r}$ folded with density ϱ (with $a = 1$) (Gaussian theorem, see Section 16.5 on the divergence) is a Green's function for the Poisson equation due to the solution properties of the differential equation $-\Delta P = \varrho$ given above.

The solution for the boundary conditions is obtained by superposing a corresponding irrotational and source-free field U. Field $\vec{\mathbf{B}}$ is obtained by

$$\vec{\mathbf{B}} = -\mathrm{grad}\,(P + U).$$

Source-free but not irrotational field: It can be concluded from $\mathrm{div}\,\vec{\mathbf{B}} = 0$ that vector field $\vec{\mathbf{B}}$ can be written as the rotation of a vector field $\vec{\mathbf{A}}$:

$$\mathrm{div}\,\vec{\mathbf{B}} = \mathrm{div}\ \mathrm{rot}\,\vec{\mathbf{A}} = 0\,.$$

For $\vec{\mathbf{A}}$ it is required that $\mathrm{div}\,\vec{\mathbf{A}} = 0$ (pure rotational field). Then, it holds that

$$\mathrm{rot}\,\vec{\mathbf{B}} = \mathrm{rot}\ \mathrm{rot}\,\vec{\mathbf{A}} = -\Delta\vec{\mathbf{A}} = \vec{\mathbf{j}}\,,$$

leading again to a Poisson equation for the vector components $\vec{\mathbf{A}}$. A solution is

$$\vec{\mathbf{A}}(\vec{\mathbf{r}}) = \frac{1}{4\pi} \int_V \frac{\vec{\mathbf{j}}(\vec{\mathbf{r}}\,')}{|\vec{\mathbf{r}} - \vec{\mathbf{r}}\,'|} \mathrm{d}V' \,.$$

The solution for the boundary conditions is obtained by superposing a corresponding irrotational and source-free field U. Field $\vec{\mathbf{B}}$ is obtained by

$$\vec{\mathbf{B}} = \mathrm{rot}\,\vec{\mathbf{A}} - \mathrm{grad}\,U.$$

Field with sources and vortices: The problem of an irrotational but not source-free field yields a potential field P as the solution. Then the problem of a source-free but not irrotational field is solved, and we obtain vector field $\vec{\mathbf{A}}$. The solution for the boundary conditions is determined by the superposition with a corresponding irrotational and source-free field U. Then, field $\vec{\mathbf{B}}$ is:

$$\vec{\mathbf{B}} = \mathrm{rot}\,\vec{\mathbf{A}} - \mathrm{grad}\,(U + P).$$

16.8 Summary

Vector operators:

Nabla operator: $\vec{\nabla} = \vec{\mathbf{e}}_x \cdot \frac{\partial}{\partial x} + \vec{\mathbf{e}}_y \cdot \frac{\partial}{\partial y} + \vec{\mathbf{e}}_z \cdot \frac{\partial}{\partial z}\,.$

Operator	Abbreviation	Symbol	Argument	Result	Interpretation
Gradient	grad	$\vec{\nabla}U$	scalar	vector	greatest increase
Divergence	div	$\vec{\nabla}\cdot\vec{\mathbf{A}}$	vector	scalar	sources
Rotation	rot	$\vec{\nabla}\times\vec{\mathbf{A}}$	vector	vector	vortices
Laplace	Δ	$(\vec{\nabla}\cdot\vec{\nabla})U$	scalar	scalar	potential field
operator		$(\vec{\nabla}\cdot\vec{\nabla})\vec{\mathbf{A}}$	vector	vector	sources

Integral theorems:

$$\int \vec{\nabla} U \, d\vec{s} = \int dU$$

$$\int_V \vec{\nabla} \cdot \vec{\mathbf{A}} \, dV = \oint_F \vec{\mathbf{A}} \cdot d\vec{\mathbf{F}} = \oint_F \vec{\mathbf{A}} \cdot \vec{\mathbf{n}} \, dF$$

$$\int_F \left(\frac{\partial U}{\partial x} + \frac{\partial V}{\partial y} \right) dx \, dy = \oint_C (U \, dy - V \, dx)$$

$$\int_F \vec{\nabla} \times \vec{\mathbf{A}} \cdot \vec{\mathbf{n}} \, dF = \oint_C \vec{\mathbf{A}} \cdot d\vec{s}$$

$$\int_V \vec{\nabla} \times \vec{\mathbf{A}} \, dV = \oint_F \vec{\mathbf{A}} \times \vec{\mathbf{n}} \, dF$$

$$\int_V \left[U \, \Delta V + (\vec{\nabla} U) \cdot (\vec{\nabla} V) \right] dV = \oint_F \left[U \, \vec{\nabla} V \right] \cdot \vec{\mathbf{n}} \, dF$$

$$\int_V [U \, \Delta V - V \, \Delta U] \, dV = \oint_F \left[U \, \vec{\nabla} V - V \, \vec{\nabla} U \right] \cdot \vec{\mathbf{n}} \, dF$$

17

Complex variables and functions

17.1 Complex numbers

Imaginary unit: $j^2 = -1$.

▷ In mathematics, the imaginary unit is written as i. To avoid confusion with the notation for the current (i), we write the imaginary unit as j.

□ With the help of the imaginary unit, the solutions of the equation $x^2 + 1 = 0$ read $x_{1/2} = \pm j$. They can be represented as products of a real number and j:

$$x_1 = 1 \cdot j \quad \text{and} \quad x_2 = -1 \cdot j\,.$$

Powers of j,

$$\begin{aligned} &j^1 = j\,, \ j^2 = -1\,, \\ &j^3 = j^2 \cdot j = -j\,, \ j^4 = j^2 \cdot j^2 = 1\,, \\ &j^5 = j, \ldots \end{aligned}$$

⇒ For arbitrary $n \in \mathbb{N}$:

$$\begin{aligned} j^{4n} &= 1\,, \ j^{1+4n} = j\,, \\ j^{2+4n} &= -1\,, \ j^{3+4n} = -j\,. \end{aligned}$$

17.1.1 Imaginary numbers

Imaginary number: The product of a real number b and the imaginary unit j: $c = bj$. Notations: $bj, jb, b \cdot j, j \cdot b$.

- The square of an imaginary number is **always** a **negative** real number:

$$(bj)^2 = b^2 \cdot j^2 = b^2 \cdot (-1) = -b^2 < 0 \ \ (b \neq 0)\,.$$

Set of imaginary numbers $\mathbb{I} = \{c \mid c = jb, b \in \mathbb{R}\}$.

17.1.2 Algebraic representation of complex numbers

Complex number z: The sum of a real number x and an imaginary number $\mathrm{j}y$: $z = x + \mathrm{j}y$.

□ The solutions of the quadratic equation

$$x^2 - 6x + 13 = 0$$

are complex numbers. Application of the quadratic formula $p - q$ yields:

$$x_{1/2} = 3 \pm \sqrt{9 - 13} = 3 \pm \sqrt{-4} = 3 \pm 2\mathrm{j} .$$

Real part of a complex number $\mathrm{Re}(z) = x$;
Imaginary part of a complex number $\mathrm{Im}(z) = y$.

● Two complex numbers, z_1, z_2, are **equal**, $z_1 = z_2$, **if and only if** $x_1 = x_2$ **and** $y_1 = y_2$; i.e., the real part **and** the imaginary part must coincide.

Set of complex numbers $\mathbb{C} = \{z \mid z = x + \mathrm{j}y;\, x, y \in \mathbb{R}\}$: can be regarded as Cartesian product $\mathbb{R}^2 = \mathbb{R} \times \mathbb{R}$ with ordered pairs of real numbers (x, y), $x, y \in \mathbb{R}$ as elements z, $z \in \mathbb{C}$.

● $\mathbb{R}$ and $\mathbb{I}$ are proper **subsets** of $\mathbb{C}$, $\mathbb{R} \subset \mathbb{C}$, $\mathbb{I} \subset \mathbb{C}$.

● The inclusions $\mathbb{I} \subset \mathbb{C} \supset \mathbb{R} \supset \mathbb{Q} \supset \mathbb{Z} \supset \mathbb{N}$ apply.

Real number: A complex number with a **vanishing imaginary part**:
$x = x + \mathrm{j} \cdot 0 = z$, $\mathrm{Re}(z) = x$, $\mathrm{Im}(z) = 0$.

Imaginary number: A complex number with a **vanishing real part**:
$\mathrm{j} \cdot y = 0 + \mathrm{j} \cdot y = z$, $\mathrm{Re}(z) = 0$, $\mathrm{Im}(z) = y$.

17.1.3 Cartesian representation of complex numbers

● Regard real and imaginary parts of a complex number as the **Cartesian coordinates** of a point P in the (x, y) plane:

$$z = x + \mathrm{j}y \Leftrightarrow P(z) = (x, y) .$$

● **To every** complex number z **one and only one** image point $P(z)$ can be assigned.

Phasor $\underline{z}$, arrow between origin to $P(z)$.

▷ The arrow is **not** a vector. It is subject to different calculation rules (see Section 17.2 on the multiplication of complex numbers).

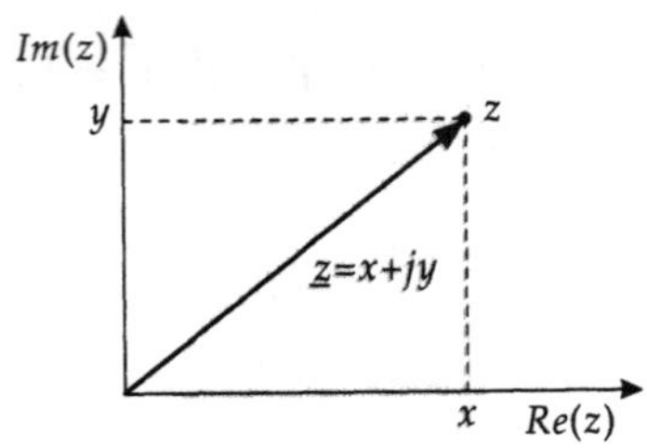

Representation of a complex number in the Gaussian number plane

Complex plane or **Gaussian number plane**: Notation for the (x, y)-plane.
Real axis, x axis;
Imaginary axis, y axis.

● Real numbers lie on the real axis; imaginary numbers lie on the imaginary axis.

17.1.4 Conjugate complex numbers

Conjugate complex number to $z = x + jy$: $z^* = x - jy$.
Other notation: $z^* = \bar{z}$.

- It holds that $\mathrm{Re}(z^*) = \mathrm{Re}(z)$ and $\mathrm{Im}(z^*) = -\mathrm{Im}(z)$.

▷ Change of sign in the imaginary part!
$\Rightarrow$ z^* is obtained from z by the substitution $j \to -j$.

□ The conjugate complex number for $z = 3 + 2j$ is $z^* = 3 - 2j$.

- For two complex numbers z_1, z_2 **conjugate to each other** it holds that:

$$z_1 = z_2^* \quad \text{and} \quad z_2 = z_1^* .$$

- The image points of two complex numbers conjugate to each other are **mirror-symmetric** about the real axis.

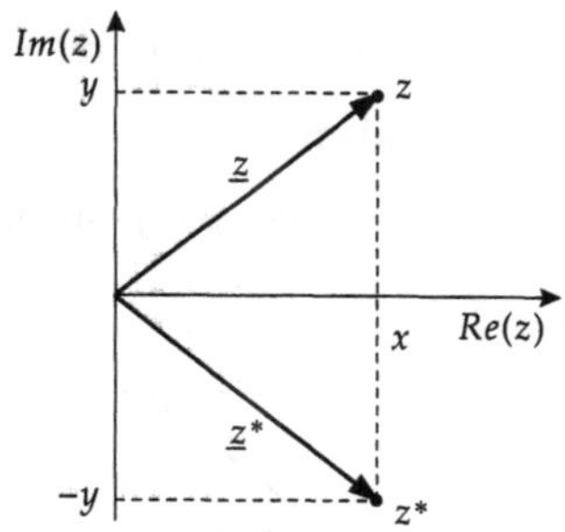

Conjugate complex numbers
in the Gaussian number plane

- Double reflection on the real axis reproduces the original point,

$$(z^*)^* = z .$$

- A complex number with $z = z^*$ is real.

17.1.5 Absolute value of a complex number

Absolute value or **modulus of a complex number**: $|z| = \sqrt{x^2 + y^2}$.

□ The absolute value of the complex number $z = 6 + 3j$ is $|z| = \sqrt{6^2 + 3^2} = \sqrt{36 + 9} = \sqrt{45}$.

- In the complex plane $|z|$ is the **Euclidean distance** between the image point and the origin.

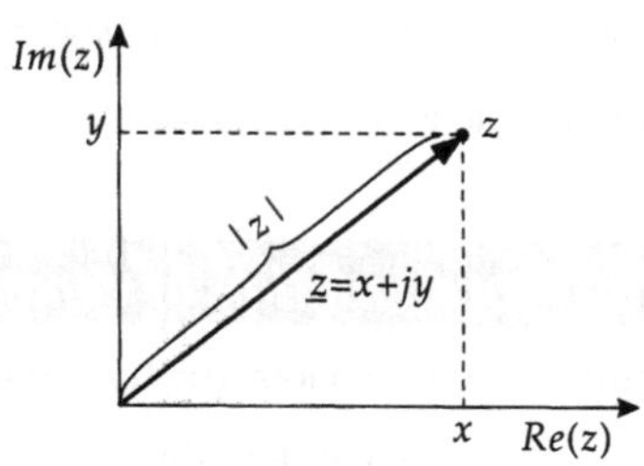

Absolute value of a complex
number

- $|z|$ is the **length** of the phasor $\underline{z}$.
- $|z| \geq 0$, always.
- The absolute value may be calculated also according to $|z| = \sqrt{z \cdot z^*}$ (see Section 17.2 on the multiplication of complex numbers).

▷ The absolute value of a complex number (the length of the phasor $\underline{z} = x + \mathrm{j}y$) corresponds to the length of vector $\vec{\mathbf{a}} = (x, y)$ in $\mathbb{R}^2$: $|z| = |\vec{\mathbf{a}}| = \sqrt{\vec{\mathbf{a}} \cdot \vec{\mathbf{a}}} = \sqrt{x^2 + y^2}$.

17.1.6 Trigonometric representation of complex numbers

Polar coordinates r, ϕ in the complex plane:
Cartesian components:

$$x = r \cdot \cos\phi\,, \quad y = r \cdot \sin\phi$$

Trigonometric form of complex numbers:

$$z = x + \mathrm{j}y = r(\cos\phi + \mathrm{j} \cdot \sin\phi)\,.$$

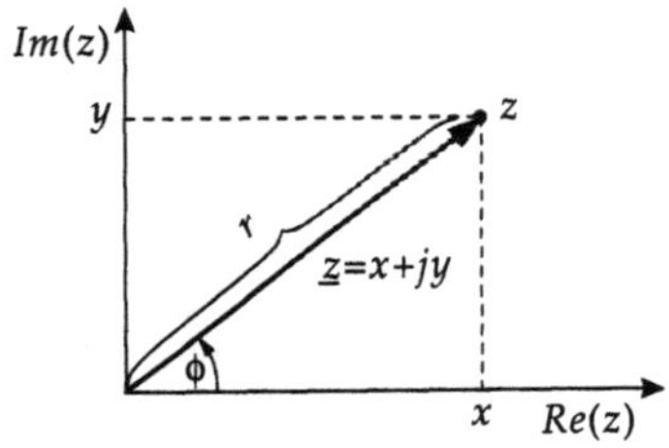

Trigonometric representation of a complex number

Radial coordinate r is equal to the **absolute value** of the complex number, $r = |z|$.
▷ $r \in \mathbb{R}$, also $r^* = r$.
Argument: **Angle** or **phase** ϕ of z.

- ϕ is **infinitely** multiple-valued: Every rotation about 2π leads to the **same** image point.

Principal value of ϕ is $\phi \in (0, 2\pi)$ or $\phi \in (-\pi, \pi)$.
[in trigonometric representation] The conjugate complex number z^* [−] is obtained by inverting the sign of ϕ:

$$z^* = r[\cos(-\phi) + \mathrm{j} \cdot \sin(-\phi)] = r(\cos\phi - \mathrm{j} \cdot \sin\phi)\,.$$

The absolute value r remains unchanged.

17.1.7 Exponential representation of complex numbers

Euler's formula (see Section 17.3 on trigonometric functions in the complex domain):

$$\mathrm{e}^{\mathrm{j}\phi} = \cos\phi + \mathrm{j} \cdot \sin\phi\,.$$

⇒ **Exponential form of complex numbers**,

$$z = r \cdot \mathrm{e}^{\mathrm{j}\phi}\,.$$

Conjugate complex number z^* in exponential representation:

$$z^* = [r\,\mathrm{e}^{\mathrm{j}\phi}]^* = r^* \cdot [\mathrm{e}^{\mathrm{j}\phi}]^* = r \cdot \mathrm{e}^{-\mathrm{j}\phi} .$$

▷ The **complex exponential function** $\mathrm{e}^{\mathrm{j}\phi}$ has a period $2\pi\mathrm{j}$:
$\mathrm{e}^{\mathrm{j}\phi+k\cdot 2\pi\mathrm{j}} = \mathrm{e}^{\mathrm{j}[\phi+k\cdot 2\pi]} = \mathrm{e}^{\mathrm{j}\phi}, k \in \mathbb{R}$.

● Special values of $\mathrm{e}^{\mathrm{j}\phi}$:

$$\mathrm{e}^{\mathrm{j}\pi/2} = \cos\left(\frac{\pi}{2}\right) + \mathrm{j}\sin\left(\frac{\pi}{2}\right) = \mathrm{j},$$

$$\mathrm{e}^{\mathrm{j}2\pi/3} = \cos\left(\frac{2\pi}{3}\right) + \mathrm{j}\sin\left(\frac{2\pi}{3}\right) = -\frac{1}{2} + \frac{\mathrm{j}}{2}\sqrt{3},$$

$$\mathrm{e}^{\mathrm{j}\pi} = \cos\pi + \mathrm{j}\sin\pi = -1,$$

$$\mathrm{e}^{\mathrm{j}4\pi/3} = \cos\left(\frac{4\pi}{3}\right) + \mathrm{j}\sin\left(\frac{4\pi}{3}\right) = -\frac{1}{2} - \frac{\mathrm{j}}{2}\sqrt{3},$$

$$\mathrm{e}^{\mathrm{j}3\pi/2} = \cos\left(\frac{3\pi}{2}\right) + \mathrm{j}\sin\left(\frac{3\pi}{2}\right) = -\mathrm{j}.$$

17.1.8 Transformation from Cartesian to trigonometric representation

● Cartesian representation → trigonometric representation (exponential representation):
$z = x + \mathrm{j}y \quad\rightarrow\quad z = r\cos\phi + \mathrm{j}r\sin\phi = r\mathrm{e}^{\mathrm{j}\phi}$,
where $r = \sqrt{x^2+y^2}$, $\tan\phi = \dfrac{y}{x}$. Determination for the angle of the principal value of ϕ:
1st quadrant, $x > 0, y > 0$: $\phi = \arctan(y/x)$;
2nd and 3rd quadrant, $x < 0$: $\phi = \arctan(y/x) + \pi$;
4th quadrant, $x > 0, y < 0$: $\phi = \arctan(y/x) + 2\pi$.

□ For $z \in \mathbb{R}$:
$\phi = 0$, if $x > 0$,
$\phi = \pi$, if $x < 0$, and
ϕ is indeterminate if $x = 0$ (origin).

□ For $z \in \mathbb{I}$:
$\phi = \pi/2$, if $y > 0$,
$\phi = 3\pi/2$, if $y < 0$, and
ϕ is indeterminate, if $y = 0$.

□ Rewriting the complex number $z = \sqrt{3} - \mathrm{j}$ (given in Cartesian representation) as a trigonometric representation (exponential representation):
$r = \sqrt{3+1} = 2$, $\tan\phi = -1/\sqrt{3}$, $\phi = -\pi/6 = -30°$.
Since z lies in the fourth quadrant ($x = \sqrt{3} > 0$, $y = -1 < 0$):
the principal value of ϕ is $\phi = -\pi/6 + 2\pi = 11\pi/6 = 330°$,
$\Rightarrow\ z = 2[\cos(11\pi/6) + \mathrm{j}\sin(11\pi/6)] = 2 \cdot \mathrm{e}^{\mathrm{j}5\pi/6}$.

● Trigonometric representation (exponential representation) ⇒ the Cartesian representation,
$z = r\cos\phi + \mathrm{j}r\sin\phi = r\mathrm{e}^{\mathrm{j}\phi} \quad\rightarrow\quad z = x + \mathrm{j}y$,
where $x = r\cdot\cos\phi$, $y = r\cdot\sin\phi$.

□ Rewriting the complex number $z = 1 \cdot e^{\pi/4}$ (given in exponential representation) as a Cartesian representation:
$x = 1 \cdot \cos(\pi/4) = \cos 45^o = 1/\sqrt{2}$, $y = 1 \cdot \sin(\pi/4) = \sin 45^o = 1/\sqrt{2}$,
$z = (1/\sqrt{2}) + (j/\sqrt{2}) = (1+j)/\sqrt{2}$.

17.1.9 Riemann sphere

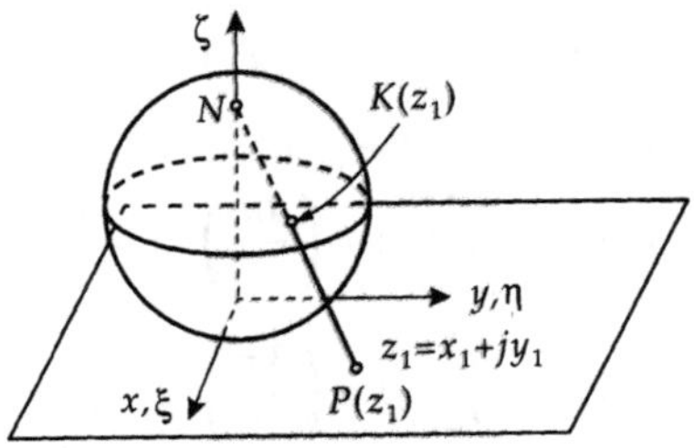

Riemann sphere

Riemann sphere: The sphere with radius 1/2 about point (0, 0, 1/2) in the Euclidean coordinate system (ξ, η, ζ) (with ξ axis = x-axis of the complex plane, η axis = y-axis of the complex plane).
The spherical surface satisfies the equation

$$\xi^2 + \eta^2 + \left(\zeta - \frac{1}{2}\right)^2 = \frac{1}{4}.$$

- The complex plane can be mapped onto the surface of the **Riemann sphere**:
Mapping of a point $P(z_1) = (x_1, y_1, 0)$, $z_1 = x_1 + jy_1$, in the complex plane onto a point $K(z_1) = (\xi_1, \eta_1, \zeta_1)$ on the surface of the Riemann sphere:
Intersection point between the line joining $P(z_1)$ and the north pole of the Riemann sphere, with the surface of the sphere of complex numbers.
If $P(z_1) = (x_1, y_1, 0)$ is given, then it holds that

$$K(z_1) = \left(\frac{x_1}{1+x_1^2+y_1^2}, \frac{y_1}{1+x_1^2+y_1^2}, \frac{x_1^2+y_1^2}{1+x_1^2+y_1^2}\right).$$

If $K(z_1) = (\xi_1, \eta_1, \zeta_1) \neq N$ is given, then it holds that

$$P(z_1) = \left(\frac{\xi_1}{1-\zeta_1}, \frac{\eta_1}{1-\zeta_1}, 0\right),$$

where $z_1 = (\xi_1 + j\eta_1)/(1-\zeta_1)$.
Point $z = \infty$ is formally assigned to the north pole N.

▷ ∞ is not a complex number. But by definition, the following calculating rules hold:

$$z + \infty = \infty, \quad z \cdot \infty = \infty, \quad \frac{z}{\infty} = 0, \quad \frac{z}{0} = \infty.$$

17.2 Elementary arithmetical operations with complex numbers

▷ In $\mathbb{C}$ there is no **axiom of order**. The inequalities $z_1 > z_2$ or $z_1 < z_2$ are not practical (but $|z_1| > |z_2|$, $|z_1| < |z_2|$).

● Since $\mathbb{R} \subset \mathbb{C}$, the calculating rules for complex numbers must coincide with the calculating rules for real numbers in the real domain.

17.2.1 *Addition and subtraction of complex numbers*

● The complex numbers $z_1 = x_1 + \mathrm{j}y_1$, $z_2 = x_2 + \mathrm{j}y_2$ are added or subtracted by adding and subtracting the real and imaginary parts **separately**:

$$z_1 + z_2 = (x_1 + x_2) + \mathrm{j} \cdot (y_1 + y_2),$$
$$z_1 - z_2 = (x_1 - x_2) + \mathrm{j} \cdot (y_1 - y_2).$$

▷ If complex numbers are given in trigonometric representation, then they must first be transformed to Cartesian representation.

▷ Geometric interpretation: Complex numbers are added and subtracted **like** two-dimensional vectors according to the parallelogram rule.

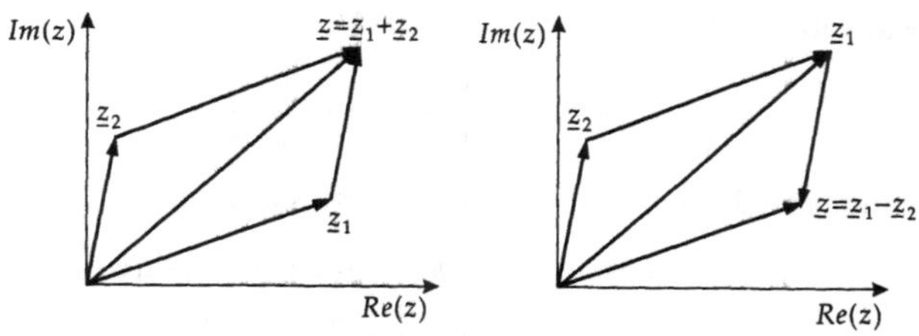

Geometrical addition and subtraction of two complex numbers

□ Addition of numbers $z_1 = 1 + 3\mathrm{j}$, $z_2 = 2 + \mathrm{j}/2$,
$z_1 + z_2 = (1 + 2) + \mathrm{j}(3 + 1/2) = 3 + 7\mathrm{j}/2$.

□ Subtraction of number $z_2 = 4 - 2\mathrm{j}$ from the number $z_1 = -3 + 6\mathrm{j}$,
$z_1 - z_2 = (-3 - 4) + \mathrm{j}(6 - (-2)) = -7 + 8\mathrm{j}$.

17.2.2 *Multiplication and division of complex numbers*

● In the Cartesian representation, the complex numbers $z_1 = x_1 + \mathrm{j}y_1$, $z_2 = x_2 + \mathrm{j}y_2$ are multiplied by terms like two bracketed expressions in the real domain, using $\mathrm{j}^2 = -1$:

$$z_1 \cdot z_2 = (x_1 + \mathrm{j}y_1) \cdot (x_2 + \mathrm{j}y_2) = (x_1x_2 - y_1y_2) + \mathrm{j} \cdot (x_1y_2 + x_2y_1).$$

▷ The multiplication of two complex numbers z_1, z_2 (phasors $\underline{z_1}$, $\underline{z_2}$) is not identical to the scalar product of the corresponding two-dimensional vectors, $\vec{\mathbf{a}}_1 = (x_1, y_1)$, $\vec{\mathbf{a}}_2 = (x_2, y_2)$: $\vec{\mathbf{a}}_1\vec{\mathbf{a}}_2 = x_1x_2 + y_1y_2$.

□ Absolute value of a complex number z as the positive root of the product $z \cdot z^*$:
$z \cdot z^* = x^2 + y^2 = |z|^2 \Rightarrow |z| = \sqrt{z \cdot z^*}$.

□ Multiplication of complex numbers $z_1 = 3 + \mathrm{j}$, $z_2 = 2 - 2\mathrm{j}$:
$z_1 \cdot z_2 = (3 + \mathrm{j}) \cdot (2 - 2\mathrm{j}) = (3 \cdot 2 - 1 \cdot (-2)) + \mathrm{j} \cdot (3 \cdot (-2) + 1 \cdot 2) = 8 - 4\mathrm{j}$.

● In Cartesian representation, the complex numbers $z_1 = x_1 + \mathrm{j}y_1, z_2 = x_2 + \mathrm{j}y_2 \neq 0$ are divided as follows:

$$\frac{z_1}{z_2} = \frac{x_1x_2 + y_1y_2}{x_2^2 + y_2^2} + \mathrm{j} \cdot \frac{x_2y_1 - x_1y_2}{x_2^2 + y_2^2} .$$

▷ Division by zero is not permissible!

□ Division of $z_1 = 6 + 2\mathrm{j}$ by $z_2 = 1 + \mathrm{j}$,
$z_1/z_2 = (6 + 2)/2 + \mathrm{j}(2 - 6)/2 = 4 - 2\mathrm{j}$.

● Practical calculation of the quotient of two complex numbers by means of
$z_1/z_2 = (z_1 \cdot z_2^*)/(z_2 \cdot z_2^*) = (z_1 \cdot z_2^*)/|z_2|^2$
and application of the multiplication rule.

● **Reciprocal value of j:**

$$\frac{1}{\mathrm{j}} = \frac{1 \cdot (-\mathrm{j})}{\mathrm{j} \cdot (-\mathrm{j})} = \frac{-\mathrm{j}}{1} = -\mathrm{j} .$$

● Negative powers of j:

$$\mathrm{j}^{-1} = -\mathrm{j} , \quad \mathrm{j}^{-2} = (-\mathrm{j})^2 = -1 ,$$
$$\mathrm{j}^{-3} = \mathrm{j}^{-1}\mathrm{j}^{-2} = \mathrm{j} , \quad \mathrm{j}^{-4} = \mathrm{j}^{-2}\mathrm{j}^{-2} = 1 , \ldots$$

● **Reciprocal value of z:**

$$\frac{1}{z} = \frac{z^*}{|z|^2} = \frac{x}{x^2 + y^2} - \mathrm{j}\frac{y}{x^2 + y^2} .$$

● Multiplication and division in trigonometric representation:

$$\begin{aligned} z_1 \cdot z_2 &= r_1(\cos\phi_1 + \mathrm{j}\sin\phi_1) \cdot r_2(\cos\phi_2 + \mathrm{j}\sin\phi_2) \\ &= r_1r_2\,(\cos(\phi_1 + \phi_2) + \mathrm{j}\sin(\phi_1 + \phi_2)) , \\ \frac{z_1}{z_2} &= r_1(\cos\phi_1 + \mathrm{j}\sin\phi_1)/r_2(\cos\phi_2 + \mathrm{j}\sin\phi_2) \\ &= \frac{r_1}{r_2}\,(\cos(\phi_1 - \phi_2) + \mathrm{j}\sin(\phi_1 - \phi_2)) . \end{aligned}$$

● Multiplication in exponential representation:
Multiplication of the absolute values, addition of the arguments:

$$z_1 \cdot z_2 = r_1\mathrm{e}^{\mathrm{j}\phi_1} \cdot r_2\mathrm{e}^{\mathrm{j}\phi_2} = r_1\,r_2 \cdot \mathrm{e}^{\mathrm{j}(\phi_1 + \phi_2)} .$$

Division in exponential representation:
Division of the absolute values, subtraction of the arguments:

$$\frac{z_1}{z_2} = \frac{r_1}{r_2} \cdot \mathrm{e}^{\mathrm{j}(\phi_1 - \phi_2)} .$$

● The multiplication of a complex number z by a real number $z_1 = \lambda + \mathrm{j} \cdot 0, \lambda > 0$, corresponds to a stretching ($\lambda > 1$) or a compression ($\lambda < 1$) of the phasor by a factor λ:

$$\lambda \cdot z = \lambda r \cdot \mathrm{e}^{\mathrm{j}\phi} .$$

▷ Only the absolute value of the complex number changes!

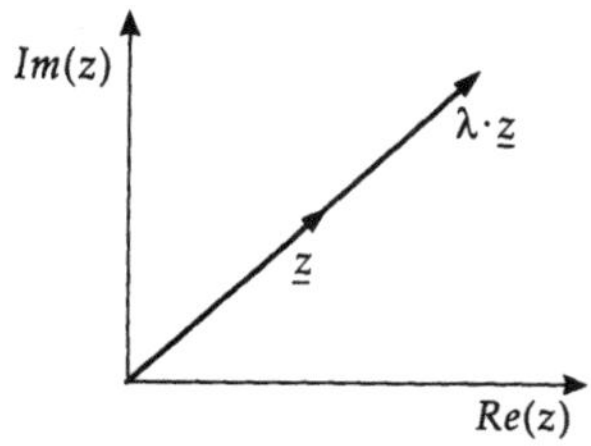

Multiplication of a complex number by a real number λ ($\lambda \approx 2$)

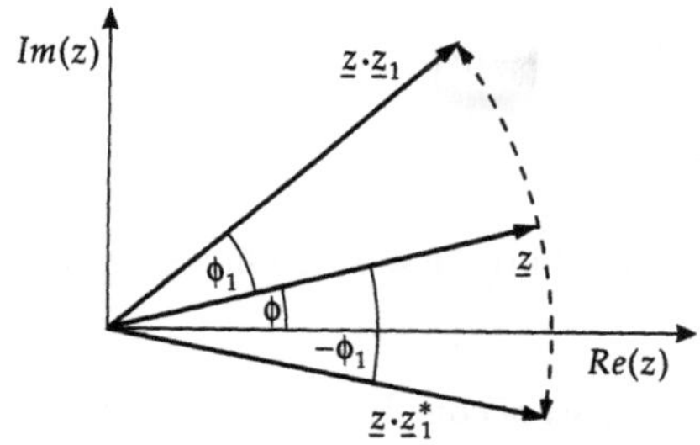

Multiplication of a complex number by a complex number of absolute value one

- Multiplication of a complex number $z = r \cdot \mathrm{e}^{\mathrm{j}\phi}$ by a complex number $z_1 = 1 \cdot \mathrm{e}^{\mathrm{j}\phi_1}$ of absolute value one corresponds to a rotation of the phasor $\underline{z}$ through the angle ϕ_1:

$$z_1 \cdot z = r\ \mathrm{e}^{\mathrm{j}(\phi+\phi_1)} .$$

$\phi_1 > 0$: Rotation in the mathematically **positive** sense (counterclockwise).
$\phi_1 < 0$: Rotation in the mathematically **negative** sense (clockwise).

▷ Only the argument of the complex number changes!

□ Multiplication of z by the imaginary unit $\mathrm{j} = \mathrm{e}^{\mathrm{j}\pi/2}$ corresponds to a rotation of $\underline{z}$ through 90°, $z \cdot \mathrm{j} = r\mathrm{e}^{\mathrm{j}(\phi+\pi/2)}$.

- Multiplication of a complex number $z = r\ \mathrm{e}^{\mathrm{j}\phi}$ by a general complex number $z_1 = r_1\ \mathrm{e}^{\mathrm{j}\phi_1}$:
Rotation of the phasor $\underline{z}$ through ϕ_1 and **stretching** by r_1,

$$z_1 \cdot z = rr_1\ \mathrm{e}^{\mathrm{j}(\phi+\phi_1)} .$$

These operations are interchangeable; it is possible to stretch first, and then to rotate. Since multiplication is commutative, $z_1 \cdot z = z \cdot z_1$, one can also start with z_1.

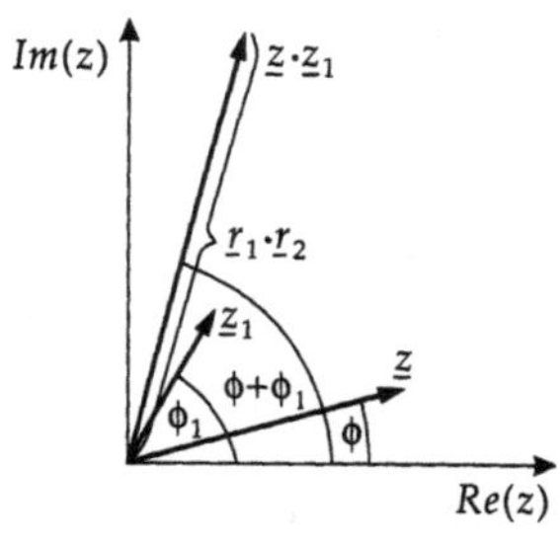

Multiplication of two complex numbers

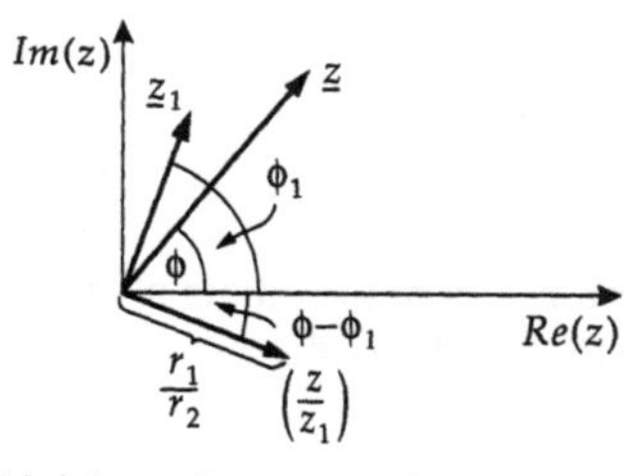

Division of two complex numbers

- Division of two complex numbers z, z_1, reducible to the multiplication:

$$z/z_1 = (r/r_1) \cdot \mathrm{e}^{\mathrm{j}(\phi-\phi_1)} .$$

Stretching the phasor $\underline{z}$ by $1/r_1$, **rotation** through $-\phi_1$.

□ Division of z by the imaginary unit $\mathrm{j} = \mathrm{e}^{\mathrm{j}\pi/2}$ corresponds to a rotation of $\underline{z}$ through -90°, $z/\mathrm{j} = r\ \mathrm{e}^{\mathrm{j}(\phi-\pi/2)}$.

- **Associative law** of addition and multiplication:

$$z_1 + (z_2 + z_3) = (z_1 + z_2) + z_3 ,$$
$$z_1(z_2 z_3) = (z_1 z_2) z_3 .$$

- **Commutative law** of addition and multiplication:

$$z_1 + z_2 = z_2 + z_1 \,,$$
$$z_1 z_2 = z_2 z_1 \,.$$

- **Distributive law** of addition and multiplication:

$$z_1(z_2 + z_3) = z_1 z_2 + z_1 z_3 \,.$$

- The set of complex numbers $\mathbb{C}$ forms a **field.**

17.2.3 Exponentiation in the complex domain

Exponentiation means the repeated multiplication of a complex number by itself, i.e., a repeated **rotational stretching** of the phasor.

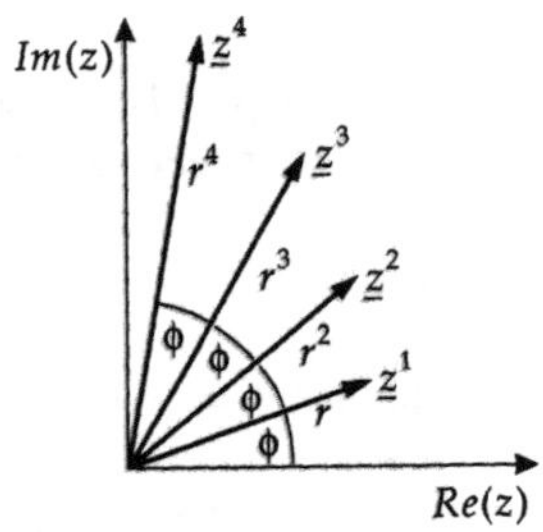

Geometrical exponentiation

***n*-th power of a complex number**: $n \in \mathbb{N}$, in exponential representation

$$z^n = (r \cdot \mathrm{e}^{\mathrm{j}\phi})^n = r^n \cdot \mathrm{e}^{\mathrm{j}n\phi} \,.$$

The absolute value is raised to the n-th power, and the argument is multiplied by n.

□ The fourth power of the complex number $z = 3 \cdot \mathrm{e}^{\mathrm{j}\pi/8}$ is $z^4 = 3^4 \cdot \mathrm{e}^{\mathrm{j}4\pi/8} = 81 \cdot \mathrm{e}^{\mathrm{j}\pi/2}$.

▷ It is useful to change complex numbers to **exponential form** before exponentiation. The trigonometric form is derived from Moivre's formula. The Cartesian form follows from the binomial theorem.

- **Moivre's formula**:

$$(\cos\phi + \mathrm{j}\cdot\sin\phi)^n = \cos(n\phi) + \mathrm{j}\cdot\sin(n\phi) \,.$$

- Powers of z in Cartesian representation:

$$(x + \mathrm{j}y)^2 = x^2 + 2\mathrm{j}xy + \mathrm{j}^2y^2 = x^2 - y^2 + \mathrm{j}\cdot 2xy$$
$$(x + \mathrm{j}y)^3 = x^3 - 3xy^2 + \mathrm{j}(3x^2y - y^3)$$
$$(x + \mathrm{j}y)^4 = x^4 - 6x^2y^2 + y^4 + \mathrm{j}(4x^3y - 4xy^3) \,.$$

- **Fundamental theorem of algebra**:
An algebraic equation of degree n

$$a_n z^n + a_{n-1} z^{n-1} + \cdots + a_1 z + a_0 = 0$$

has **exactly** n solutions (roots) in $\mathbb{C}$, where multiple zeroes must be counted according to their multiplicity.

▷ This theorem becomes clear with the help of **product representation**:

$$a_n z^n + a_{n-1} z^{n-1} + \cdots + a_1 z + a_0 = a_n (z - z_1)(z - z_2) \cdots (z - z_n) ,$$

where $z_1, z_2, \ldots, z_n$ represent the n **zeros of the polynomial**, i.e., the n solutions of the algebraic equation.

▷ For **real** coefficients a_i $(i = 0, 1, \ldots, n)$, complex solutions always occur **pairwise** as conjugate complex numbers; i.e., if z_1 is a solution, then z_1^* is also a solution.

17.2.4 Taking the root in the complex domain

n**-th root** z **of a complex number** $a \in \mathbb{C}$, satisfies the equation $z^n = a$.

● Solutions of the equation $z^n = a$, with $z = r\mathrm{e}^{\mathrm{j}\phi}$ and $a = a_0 \mathrm{e}^{\mathrm{j}\alpha} = a_0 \mathrm{e}^{\mathrm{j}(\alpha + 2\pi l)}$, $l \in \mathbb{Z}$ (periodicity of the complex exponential function):
There are exactly n different solutions of the equation $z^n = a$ (fundamental theorem of algebra):

$$z_l = r\mathrm{e}^{\mathrm{j}\phi_l} , \quad r = a_0^{1/n} , \quad \phi_l = \frac{\alpha + 2\pi l}{n} , \quad l = 0, 1, \ldots, n-1 .$$

▷ In the complex plane, the image points of z_l lie on the circle $\left[\text{with radius } R = a_0^{1/n}\right]$ about the origin [−] form the vertices of a regular n-gon (see Section 3.7 on polygons).

● For $a \in \mathbb{R}$ it holds that: If z is a solution of the equation $z^n = a$, then z^* is also a solution.

n**-th roots of unity**. Solution of $z^n = 1$.

□ **Third roots of unity**: Solution of $z^3 = 1$:
$\Rightarrow a = 1 , \ a_0 = 1 , \ \alpha = 0,$
$\Rightarrow \phi_l = 2\pi l/3 , \ l = 0, 1, 2 ,$
$\Rightarrow z_0 = 1 , \ z_1 = \mathrm{e}^{\mathrm{j}2\pi/3} , \ z_2 = \mathrm{e}^{\mathrm{j}4\pi/3} .$

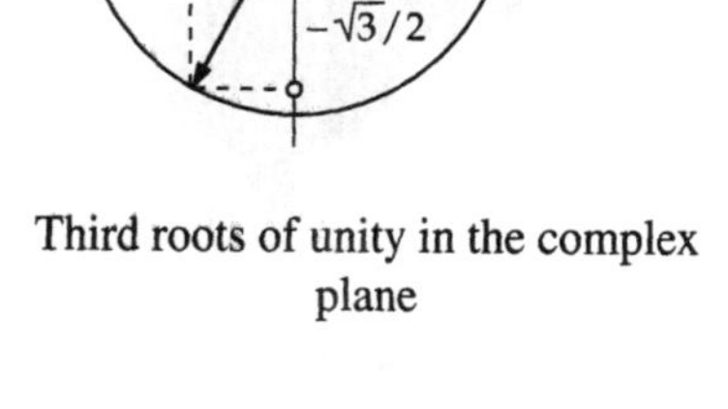

Third roots of unity in the complex plane

17.3 Elementary functions of a complex variable

Complex-valued function of a complex variable:
Assignment $f : \mathbb{C} \to \mathbb{C} , \ z \mapsto f(z)$.

17.3.1 Sequences in the complex domain

Complex sequence $z_1, z_2, \ldots, z_n, \ldots$: Infinite sequence of complex numbers.
Notation: $\{z_i\}$.

Limit z of a complex sequence: A sequence converges toward limit $z = \lim_{i\to\infty} z_i$, if for every arbitrarily chosen $\varepsilon > 0$ there is an index $N(\varepsilon) > 0$ such that $|z_m - z| < \varepsilon$ for all indices $m \geq N(\varepsilon)$; that is, from m onward, in the complex plane all z_i with $i > m$ lie in a circle about z with radius ε.

□ For the sequence $\left\{z^{1/i}\right\}$ the limit is $\lim_{i\to\infty}\left(z^{1/i}\right) = 1$.

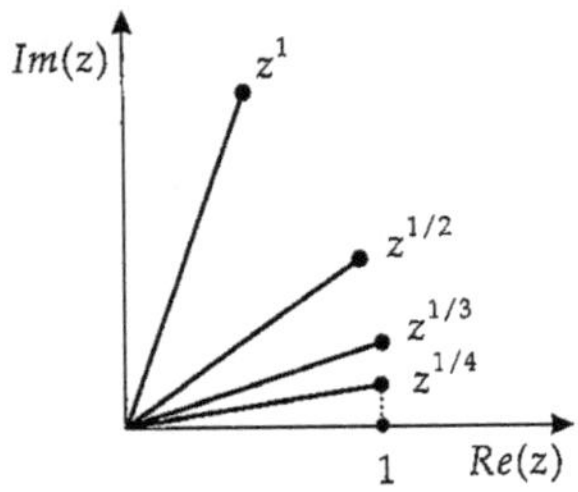

Convergence of the complex sequence $\left\{z^{1/i}\right\}$

□ **Mandelbrot set**:
Consider sequence $\{z_i\}$ iteratively constructed by the operation

$$z_{i+1} = z_i^2 + c\,, \quad z_0 = 0\,,$$

for a fixed $c \in \mathbb{C}$. The **Mandelbrot set** consists of all numbers c for which all iterates also belong to the set. For numbers c not lying in the Mandelbrot set, the iterates tend to infinity, $|z_i| \to \infty$ $(i \to \infty)$. The representation of the Mandelbrot set (i.e. all numbers c yielding finite iterates) under the above iteration rule in the complex plane yields a structure known as an "apple doll."

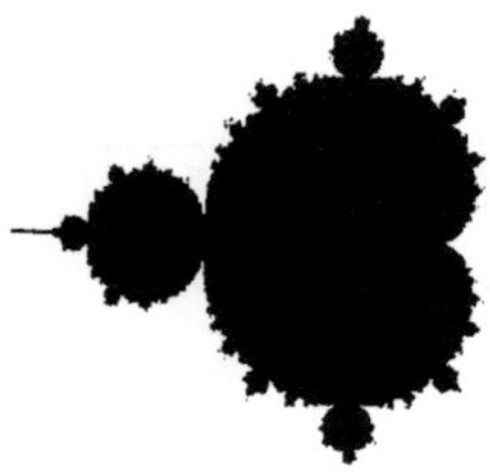

Mandelbrot set

□ **Julia sets**:
Consider the same iteration rule as for the Mandelbrot set for the construction of the sequence $\{z_i\}$. But now, $c \in \mathbb{C}$ is chosen to be fixed, and the starting value z_0 of the iteration is varied. A **Julia set** consists of all numbers z_0 whose iterates do not tend to infinity. Each c yields another Julia set; thus, there exists an infinite number of

such sets. The representation in the complex plane is as for the Mandelbrot set, but now, all z_0 must be plotted for fixed c, and the iterates of z_0 remain finite.

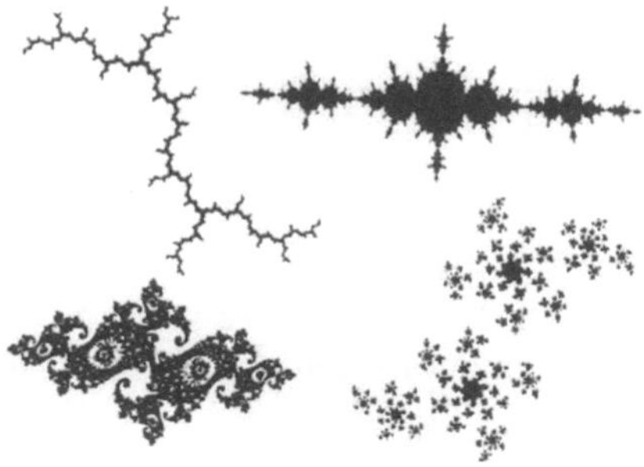

Julia sets

17.3.2 Series in the complex domain

Complex series $z_1 + z_2 + \cdots + z_n + \cdots$: The sum of all terms of the sequence $\{z_i\}$.
Notation: $z_1 + z_2 + \cdots + z_n + \cdots = \sum_{i=1}^{\infty} z_i$.
Partial sum of a complex series: $s_n = z_1 + z_2 + \cdots z_n = \sum_{i=1}^{n} z_i$.
Convergence of a series:
A complex series converges toward the number s if the **sequence of the partial sums** converges toward $s = \lim_{i\to\infty} s_i$.
Absolute convergence of a series: A complex series $\sum_i z_i$ converges **absolutely** if the series $\sum_i |z_i|$ converges.
Complex series with variable terms $a_1(z) + a_2(z) + \cdots + a_n(z) + \cdots$: The terms of the sequence $\{a_i\}$ are **functions** of a complex variable.
Notation: $a_1(z) + a_2(z) + \cdots + a_n(z) + \cdots = \sum_{i=1}^{\infty} a_i(z)$.
Uniform convergence of a complex series $\sum_i a_i(z)$ in a domain $M \subset \mathbb{C}$: If the series $\sum c_i$ with $c_i \geq |a_i(z)|$ converges for all $z \in M$ (Weierstrass's criterion).
Compact convergence of a complex series $\sum_i a_i(z)$ in a domain $G \subset \mathbb{C}$: If the series converges uniformly on every compact subset $M \subset G$.
Complex power series $a_0 + a_1 z + a_2 z^2 + \cdots + a_n z^n + \cdots$, $a_i \in \mathbb{C}$.
Notation: $a_0 + a_1 z + a_2 z^2 + \cdots + a_n z^n + \cdots = \sum_{i=0}^{\infty} a_i z^i$.

- A power series converges either absolutely for all $z \in \mathbb{C}$ (in the entire complex plane) or absolutely inside a certain **circle of convergence**, and diverges outside the circle of convergence.

Circle of convergence: Circle [−] about the origin [in the complex plane] with the **radius of convergence** R.

□ The power series $1 + z + z^2 + \ldots$ has the radius of convergence $R = 1$.

- A power series converges compactly at its circle of convergence.

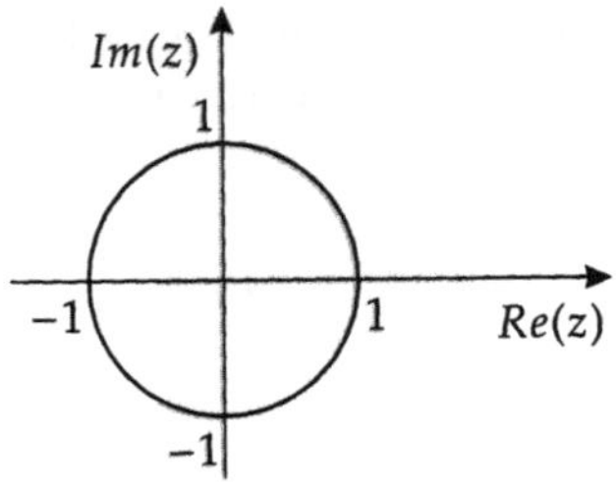

Circle of convergence of the series $1 + z + z^2 + \cdots$

17.3.3 Exponential function in the complex domain

Exponential function in the complex domain: As in the real domain, it is defined via the power series expansion:

$$e^z = \sum_{n=0}^{\infty} \frac{z^n}{n!} = 1 + \frac{z}{1!} + \frac{z^2}{2!} + \frac{z^3}{3!} \cdots .$$

- The power series of the exponential function converges absolutely and compactly in the entire complex plane.
- **Period** of the complex exponential function: $2\pi j$.

17.3.4 Natural logarithm in the complex domain

Natural logarithm of a complex number $z = r\,e^{j(\phi+2\pi l)}$, $0 \le \phi < 2\pi$, $l \in \mathbb{Z}$:

$$\ln z = \ln r + j(\phi + 2\pi l) .$$

▷ Since $l \in \mathbb{Z}$, $\ln z$ is **infinitely multiple-valued**.
Principal value $\operatorname{Ln} z$ of $\ln z$: the value of $\ln z$ for $l = 0$:

$$\operatorname{Ln} z = \ln r + j\phi .$$

Secondary values: The values of $\ln z$ for $l = \pm 1, \pm 2, \ldots$.

▷ $\ln z$ is defined **for every** complex number $z \neq 0$, thus **also** for **negative real** numbers, contrary to the real domain, where $\ln x$ is defined only for positive x!

□ $\ln(-5) = \ln(5e^{j\pi}) = \ln 5 + j(\pi + 2\pi l)$, $l \in \mathbb{Z}$.
Principal value: $\operatorname{Ln}(-5) = \ln 5 + j\pi$.

Special cases:

1. z is a positive real number, $z = x$: $\Rightarrow \phi = 0$

$$\Rightarrow \ln z = \ln x + j2\pi l .$$

□ $\ln 1 = j2\pi l$.

2. z is a negative real number, $z = -x$ $(x > 0)$: $\Rightarrow \phi = \pi$

$$\Rightarrow \ln z = \ln(-x) = \ln x + j(2l + 1)\pi .$$

□ $\ln(-1) = j(2l + 1)\pi$.

3. z is a positive imaginary number, $z = \mathrm{j}y$: $\Rightarrow \phi = \pi/2$

$$\Rightarrow \ln z = \ln(\mathrm{j}y) = \ln y + \mathrm{j}\left(2l + \frac{1}{2}\right)\pi \,.$$

□ $\ln \mathrm{j} = \mathrm{j}(2l + 1/2)\pi$.

4. z is a negative imaginary number $z = -\mathrm{j}y$ $(y > 0)$: $\Rightarrow \phi = 3\pi/2$

$$\Rightarrow \ln z = \ln(-\mathrm{j}y) = \ln y + \mathrm{j}\left(2l + \frac{3}{2}\right)\pi \,.$$

□ $\ln(-\mathrm{j}) = \mathrm{j}(2l + 3/2)\pi$.

17.3.5 General power in the complex domain

General power in the complex domain, $a^z = \mathrm{e}^{z \ln a}$, $a \in \mathbb{C}$, $z \in \mathbb{C}$: Infinitely multiple-valued function, exactly like the logarithm in the complex domain.

Principal value of the general power, $\mathrm{e}^{z \,\mathrm{Ln}\, a}$.

□ If $a = 1 + \mathrm{j} = \sqrt{2}\mathrm{e}^{\mathrm{j}\pi/4}$, $z = 2 - 2\mathrm{j}$, then the principal value of the general power a^z is given by

$$a^z = \mathrm{e}^{(2-2\mathrm{j})(\ln\sqrt{2}+\mathrm{j}\pi/4)} = \mathrm{e}^{2\ln\sqrt{2}+\pi/2+\mathrm{j}(\pi/2-2\ln\sqrt{2})} = 2\,\mathrm{e}^{\pi/2}\cdot\mathrm{e}^{\mathrm{j}(\pi/2-\ln 2)} \,.$$

17.3.6 Trigonometric functions in the complex domain

Sine function in the complex domain: Defined by the power series expansion (analogous to the real sine function):

$$\sin z = \sum_{n=0}^{\infty}(-1)^n \frac{z^{2n+1}}{(2n+1)!} = z - \frac{z^3}{3!} + \frac{z^5}{5!} - \cdots \,.$$

Cosine function in the complex domain: Defined by the power series expansion (analogous to the real cosine function):

$$\cos z = \sum_{n=0}^{\infty}(-1)^n \frac{z^{2n}}{(2n)!} = 1 - \frac{z^2}{2!} + \frac{z^4}{4!} - \cdots \,.$$

- **Euler's formula**:

$$\begin{aligned}
\mathrm{e}^{\mathrm{j}z} &= \sum_{n=0}^{\infty}\frac{(\mathrm{j}z)^n}{n!} = \sum_{n=0}^{\infty}\left(\mathrm{j}^{2n}\frac{z^{2n}}{(2n)!} + \mathrm{j}^{2n+1}\frac{z^{2n+1}}{(2n+1)!}\right) \\
&= \sum_{n=0}^{\infty}(-1)^n\frac{z^{2n}}{(2n)!} + \mathrm{j}\sum_{n=0}^{\infty}(-1)^n\frac{z^{2n+1}}{(2n+1)!} \\
&= \cos z + \mathrm{j}\sin z \,, \\
\mathrm{e}^{-\mathrm{j}z} &= \cos z - \mathrm{j}\sin z \,.
\end{aligned}$$

▷ For $z = \phi \in \mathbb{R}$ we obtain the form of Euler's formula used in the exponential representation of the complex numbers.

- Trigonometric functions, expressed by the complex exponential function:

$$\sin z = \frac{\mathrm{e}^{\mathrm{j}z} - \mathrm{e}^{-\mathrm{j}z}}{2\mathrm{j}} \,, \qquad \cos z = \frac{\mathrm{e}^{\mathrm{j}z} + \mathrm{e}^{-\mathrm{j}z}}{2} \,.$$

- Real part and imaginary part of the trigonometric functions:

$$\begin{aligned}\sin z = \sin(x + \mathrm{j}y) &= \sin x \cos(\mathrm{j}y) + \cos x \sin(\mathrm{j}y)\\ &= \sin x \cosh y + \mathrm{j} \cos x \sinh y \,,\\ \cos z = \cos(x + \mathrm{j}y) &= \cos x \cos(\mathrm{j}y) - \sin x \sin(\mathrm{j}y)\\ &= \cos x \cosh y - \mathrm{j} \sin x \sinh y \,.\end{aligned}$$

- **Addition theorems**:

$$\begin{aligned}\sin(z + w) &= \sin z \cos w + \cos z \sin w \,,\\ \cos(z + w) &= \cos z \cos w - \sin z \sin w \,,\\ \sin^2 z + \cos^2 z &= 1 \,.\end{aligned}$$

- **Period** of the sine and cosine: 2π.
- **Zeros**:
  $\sin z = 0 \;\Leftrightarrow\; z = 0, \pm\pi, \pm 2\pi, \pm 3\pi, \ldots,$
  $\cos z = 0 \;\Leftrightarrow\; z = \pm\pi/2, \pm 3\pi/2, \pm 5\pi/2, \ldots,$
  as in the real domain, no further zeros occur due to a continuation into the complex domain.

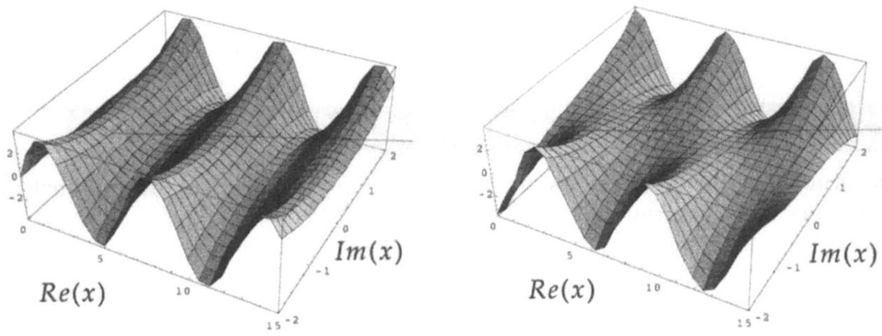

Real and imaginary parts of the complex sine

Tangent in the complex domain:

$$\tan z = \frac{\sin z}{\cos z} \,.$$

Cotangent in the complex domain:

$$\cot z = \frac{1}{\tan z} = \frac{\cos z}{\sin z} \,.$$

- Tangent and cotangent, expressed by the complex exponential function:

$$\tan z = -\mathrm{j}\frac{\mathrm{e}^{\mathrm{j}z} - \mathrm{e}^{-\mathrm{j}z}}{\mathrm{e}^{\mathrm{j}z} + \mathrm{e}^{-\mathrm{j}z}} \,, \qquad \cot z = \mathrm{j}\frac{\mathrm{e}^{\mathrm{j}z} + \mathrm{e}^{-\mathrm{j}z}}{\mathrm{e}^{\mathrm{j}z} - \mathrm{e}^{-\mathrm{j}z}} \,.$$

- Real and imaginary parts of the tangent:

$$\tan z = \tan(x + \mathrm{j}y) = \frac{\sin(2x) + \mathrm{j} \sinh(2y)}{\cos(2x) + \cosh(2y)} \,.$$

- **Period** of the tangent and cotangent: π.

17.3.7 Hyperbolic functions in the complex domain

Hyperbolic sine in the complex domain: Defined by the power series expansion (analogous to the real hyperbolic sine):

$$\sinh z = \sum_{n=0}^{\infty} \frac{z^{2n+1}}{(2n+1)!} = z + \frac{z^3}{3!} + \frac{z^5}{5!} + \cdots .$$

Hyperbolic cosine in the complex domain: Defined by the power series expansion (analogous to the real hyperbolic cosine):

$$\cosh z = \sum_{n=0}^{\infty} \frac{z^{2n}}{(2n)!} = 1 + \frac{z^2}{2!} + \frac{z^4}{4!} + \cdots .$$

- Hyperbolic functions, expressed by the complex exponential function:

$$\sinh z = \frac{e^z - e^{-z}}{2}, \qquad \cosh z = \frac{e^z + e^{-z}}{2} .$$

- Connection between complex trigonometric and hyperbolic functions:
$\sin(jz) = j\sinh z$, $\sinh(jz) = j\sin z$, $\cos(jz) = \cosh z$, $\cosh(jz) = \cos z$.
- Real part and imaginary part of the hyperbolic functions:

$$\begin{aligned}\sinh z = \sinh(x + jy) &= \sinh x \cosh(jy) + \cosh x \sinh(jy)\\ &= \sinh x \cos y + j\cosh x \sin y ,\\ \cosh z = \cosh(x + jy) &= \cosh x \cosh(jy) + \sinh x \sinh(jy)\\ &= \cosh x \cos y + j\sinh x \sin y .\end{aligned}$$

- **Addition theorems**:

$$\begin{aligned}\sinh(z + w) &= \sinh z \cosh w + \cosh z \sinh w ,\\ \cosh(z + w) &= \cosh z \cosh w + \sinh z \sinh w ,\\ \cosh^2 z - \sinh^2 z &= 1 .\end{aligned}$$

- **Period** of the hyperbolic sine and cosine: 2π.
- **Zeros**:
$\sinh z = 0 \Leftrightarrow z = 0, \pm j\pi, \pm j2\pi, \pm j3\pi, \ldots,$
$\cosh z = 0 \Leftrightarrow z = \pm j\pi/2, \pm j3\pi/2, \pm j5\pi/2, \ldots .$

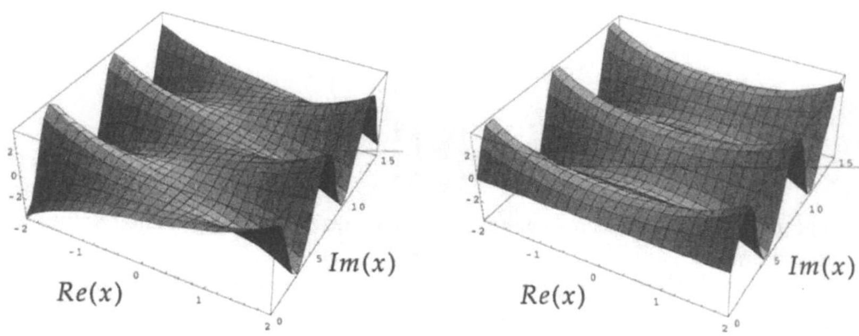

Real and imaginary parts of the complex hyperbolic sine

Hyperbolic tangent in the complex domain:

$$\tanh z = \frac{\sinh z}{\cosh z} .$$

Hyperbolic cotangent in the complex domain:

$$\coth z = \frac{1}{\tanh z} = \frac{\cosh z}{\sinh z} .$$

- Hyperbolic tangent and cotangent, expressed by the complex exponential function:

$$\tanh z = \frac{e^z - e^{-z}}{e^z + e^{-z}} , \qquad \coth z = \frac{e^z + e^{-z}}{e^z - e^{-z}} .$$

- Relation between the complex tangent and the complex hyperbolic tangent:

$$\tan(jz) = j \tanh z , \quad \tanh(jz) = j \tan z .$$

- Real part and imaginary part of the hyperbolic tangent:

$$\tanh z = \tanh(x + jy) = \frac{\sinh(2x) + j \sin(2y)}{\cosh(2x) + \cos(2y)} .$$

- **Period** of the hyperbolic tangent and cotangent: π.

17.3.8 *Inverse trigonometric, inverse hyperbolic functions in the complex domain*

Inverse trigonometric functions and the **inverse hyperbolic functions** in the complex domain are defined analogously to those in the real domain,

$$\begin{aligned}
&\arcsin z = w, && \text{if } z = \sin w; && \operatorname{Arsinh} z = w, && \text{if } z = \sinh w;\\
&\arccos z = w, && \text{if } z = \cos w; && \operatorname{Arcosh} z = w, && \text{if } z = \cosh w;\\
&\arctan z = w, && \text{if } z = \tan w; && \operatorname{Artanh} z = w, && \text{if } z = \tanh w;\\
&\operatorname{arccot} z = w, && \text{if } z = \cot w; && \operatorname{Arcoth} z = w, && \text{if } z = \coth w.
\end{aligned}$$

- Similar to the logarithm, the inverse hyperbolic functions are infinitely multiple-valued, since

$$\operatorname{Arsinh} z = \ln\left(z + \sqrt{z^2 + 1}\right) ,$$

$$\operatorname{Arcosh} z = \ln\left(z + \sqrt{z^2 - 1}\right) ,$$

$$\operatorname{Artanh} z = \frac{1}{2} \ln\left(\frac{1 + z}{1 - z}\right) ,$$

$$\operatorname{Arcoth} z = \frac{1}{2} \ln\left(\frac{z + 1}{z - 1}\right) .$$

  The **principal values** are obtained from the corresponding principal values of the logarithms.
- The inverse trigonometric functions are infinitely multiple-valued, since the connection between the trigonometric functions and the hyperbolic functions yields the following in the complex domain:

$$\arcsin z = -j \ln\left(jz + \sqrt{1 - z^2}\right) ,$$

$$\arccos z = -j \ln\left(z + \sqrt{z^2 - 1}\right) ,$$

$$\operatorname{arctan} z = -\frac{j}{2} \ln\left(\frac{1+jz}{1-jz}\right),$$

$$\operatorname{arccot} z = \frac{j}{2} \ln\left(\frac{jz+1}{jz-1}\right).$$

The **principal values** are obtained from the corresponding principal values of the logarithms.

17.4 Applications of complex functions

17.4.1 Representation of oscillations in the complex plane

Harmonic oscillation: $y(t) = A \cdot \sin(\omega t + \phi)$.

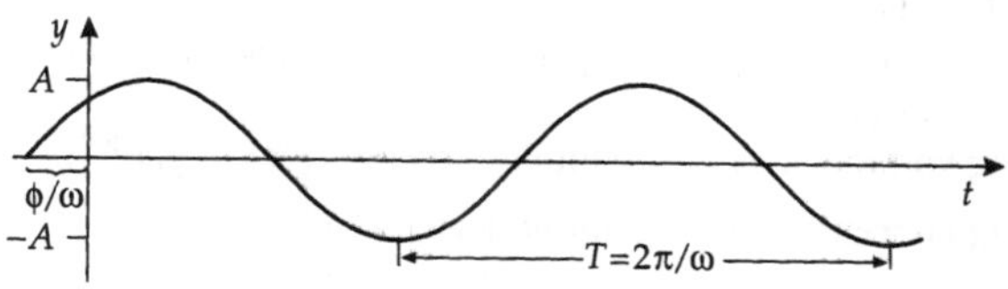

Harmonic oscillation

Notations:

- **time** t;
- **oscillation amplitude or mode** (in alternating current technology) $A > 0$;
- **angular velocity** or **angular frequency** $\omega > 0$;
- **phase** or **phase angle** $\omega t + \phi$;
- **zero phase** ϕ;
- **period or oscillation period** T, oscillation frequency $T = \dfrac{1}{f} = \dfrac{2\pi}{\omega}$.

▷ The frequency indicates how often the oscillation is repeated per unit of time.

□ Oscillations $y(t)$: Mechanical oscillation of a spring-mass system, alternating voltage, alternating current, oscillating circuit.

▷ **Cosine oscillations** can be reduced to phase-shifted sine oscillations (and vice versa): $y(t) = A\cos(\omega t + \phi) = A\sin(\omega t + \phi + \pi/2)$
(sine oscillation with the zero phase angle increased by $\pi/2$).

- Representation of the oscillation in terms of a **phasor diagram** in the complex plane:
  Time $t = 0$: phasor $\underline{z}(0) = A\,\mathrm{e}^{\mathrm{j}\phi}$.
  Time $t > 0$: $\underline{z}$ is rotated through the angle ωt about the origin (rotation of $\underline{z}$ with angular velocity ω); new phasor: $\underline{z}(t) = A\,\mathrm{e}^{\mathrm{j}(\omega t+\phi)}$.
  Oscillation: y-component of the phasor (imaginary part of the complex number z corresponding to the phasor $\underline{z}$):
  $y = A\sin(\omega t + \phi) = \mathrm{Im}(z)$.

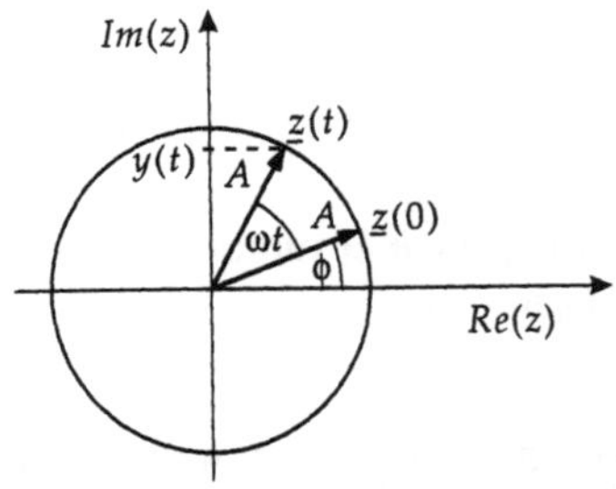

Oscillation in the phasor diagram

▷ Rotation of the phasor $\underline{z}$: Rotation through all function values of the sine oscillation.

Complex amplitude, $\underline{A} = A\,\mathrm{e}^{\mathrm{j}\phi}$ (time-independent part of z): Determines the initial position of $\underline{z}$ in the phasor diagram.

Time function, $\mathrm{e}^{\mathrm{j}\omega t}$ (time-dependent part of z): Describes the rotation of $\underline{z}$ about the origin of the complex plane.

□ **Alternating voltage**: $u(t) = u_0\sin(\omega t + \phi_u) \to \underline{u}(t) = \underline{u}_0\mathrm{e}^{\mathrm{j}\omega t}$.
Alternating current: $i(t) = i_0\sin(\omega t + \phi_i) \to \underline{i}(t) = \underline{i}_0\mathrm{e}^{\mathrm{j}\omega t}$.

17.4.2 Superposition of oscillations of equal frequency

Superposition of oscillations $y_1 = A_1 \cdot \sin(\omega t + \phi_1)$ and $y_2 = A_2 \cdot \sin(\omega t + \phi_2)$:

(a) Transition to the phasor representation:

$$y_1 \to \underline{z}_1 = \underline{A}_1 \cdot \mathrm{e}^{\mathrm{j}\omega t}$$
$$y_2 \to \underline{z}_2 = \underline{A}_2 \cdot \mathrm{e}^{\mathrm{j}\omega t} .$$

(b) Addition of the phasors:

$$\underline{z} = \underline{z}_1 + \underline{z}_2 = (\underline{A}_1 + \underline{A}_2) \cdot \mathrm{e}^{\mathrm{j}\omega t} .$$

(c) Resulting oscillation:

$$y = \mathrm{Im}(\underline{z}) = A \cdot \sin(\omega t + \phi) ,$$

$A = |\underline{A}_1 + \underline{A}_2|$, $\tan\phi = \mathrm{Im}(\underline{A}_1 + \underline{A}_2)/\mathrm{Re}(\underline{A}_1 + \underline{A}_2)$.

□ Superposition of the alternating voltages $u_1(t) = 10\mathrm{V} \cdot \sin(15t/\mathrm{s})$,
$u_2(t) = 20\mathrm{V} \cdot \cos(15t/\mathrm{s}) = 20\mathrm{V} \cdot \sin(15t/\mathrm{s} + \pi/2)$,

(a) Phasor representation:

$$u_1 \to \underline{u}_1 = \underline{u}_{01} \cdot \mathrm{e}^{\mathrm{j}\omega t}$$
$$u_2 \to \underline{u}_2 = \underline{u}_{02} \cdot \mathrm{e}^{\mathrm{j}\omega t} ,$$

with $\underline{u}_{01} = 10\text{V}$, $\underline{u}_{02} = 20\text{V} \cdot e^{j\pi/2} = j20\text{V}$, $\omega = 15/\text{s}$, also

$$u_1 \to \underline{u}_1 = 10\text{V} \cdot e^{j15t/s}$$
$$u_2 \to \underline{u}_2 = 20j\text{V} \cdot e^{j15t/s} ,$$

(b) Addition of the phasors:

$$\underline{u} = \underline{u}_1 + \underline{u}_2 = (10 + 20j)\text{V} \cdot e^{j15t/s} .$$

(c) Resulting oscillation:

$$u = \text{Im}(\underline{u}) = u_0 \cdot \sin(\omega t + \phi) ,$$

with $u_0 = |10 + 20j| = \sqrt{500}$ V, $\omega = 15/\text{s}$, $\tan\phi = \text{Im}(10 + 20j)/\text{Re}(10 + 20j) = 20/10 = 2$; thus $\phi = 63.435°$; thus

$$u = \sqrt{500} \cdot \sin(15t/\text{s} + 63.435°) .$$

17.4.3 Loci

Complex-valued function of a real variable,
Assignment $z : \mathbb{R} \to \mathbb{C}$, $t \mapsto z(t)$, $z(t) = x(t) + j \cdot y(t)$.

▷ Real and imaginary parts of z are functions of the **same** variable t.

● A **complex-valued function** z **of a real variable** t is a special case of a **complex-valued function** f **of a complex variable** z with a purely real argument.

□ Phasor $\underline{z}(t) = \underline{A}e^{j\omega t}$ which describes an oscillation, is a complex-valued function of the real variable t.

Locus: Trajectory described by the head of the phasor $\underline{z}(t)$ in the complex plane, while parameter t runs through the interval $[a, b]$.

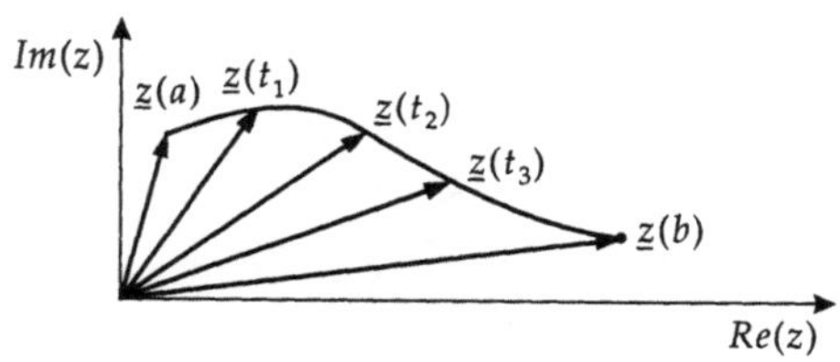

Point curve

● To every t belongs **exactly one** phasor, i.e., one point of the curve.

▷ The locus is the **geometric picture** of a complex valued function of a real variable.

▷ The locus can also be described by **parametric equations**

$$x = x(t) \ , \quad y = y(t) \ , \quad a \le t \le b \ .$$

□ **Network function**, the dependence of a complex electrical quantity on a real parameter (electrical engineering), such as:

(a) **Series connection** of an ohmic resistance and a coil:
Dependence of **complex impedance** on angular frequency:

$$\underline{Z} = \underline{Z}(\omega) = R + j\omega L \quad (\omega \ge 0) .$$

(b) **Parallel connection** of an ohmic resistance and capacitance:
Dependence of the **complex admittance** on the angular frequency:

$$\underline{Y} = \underline{Y}(\omega) = \frac{1}{R} + j\omega C \quad (\omega \geq 0) .$$

□ **Locus of a straight line** $g(t) = z_0 + z \cdot t$:
For $z_0 = 1 - 2j$, $z = 3 + 2j$ the locus of the straight line $g(t)$ is as follows.

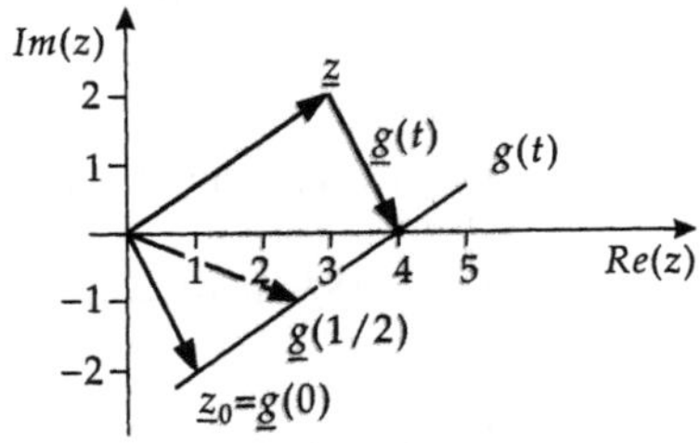

Point curve of the straight line g(t)

17.4.4 Inversion of loci

Inversion: The transition from the complex number z to its **reciprocal value** $w = 1/z$:

$$z \longrightarrow w = \frac{1}{z} ,$$
$$z = r\, e^{j\phi} \longrightarrow w = 1/z = 1/r\, e^{-j\phi} ,$$

Inversion: Forming the reciprocal value of the absolute value, changing the sign of the angle; i.e., reflection of the phasor $\underline{z}$ in the real axis, stretching of the phasor by a factor of $1/r^2$.

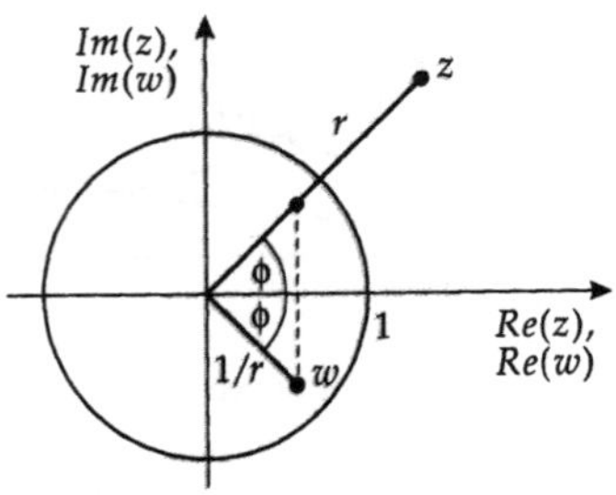

Inversion

□ **Complex impedance**: $\underline{Z}$.
Complex electrical admittance: $\underline{Y} = 1/\underline{Z}$.
Impedance: $|\underline{Z}|$.

Inverted locus: Results from the pointwise inversion of a locus.

● **Rules for inversion** of lines and circles:

z-plane		w-plane
1. **Straight line** through the origin	→	**Straight line** through the origin
2. **Straight line not** through the origin	→	**Circle** through the origin

3. **Central circle** → **Central circle**
4. **Circle** through the origin → **Line not** through the origin
5. **Circle not** through the origin → **Circle not** through the origin

▷ To the origin $z = 0$ (south pole of the Riemann number sphere in the z plane) is assigned the "infinitely far point" $w = \infty$ (north pole of the Riemann number sphere in the w-plane).

● 1. To the point $P(z)$ with the smallest absolute value $|z|$ is assigned to point $P(w)$ with the largest absolute value $|w|$.

● 2. Points $P(z)$ above the real axis become points $P(w)$ below the real axis.

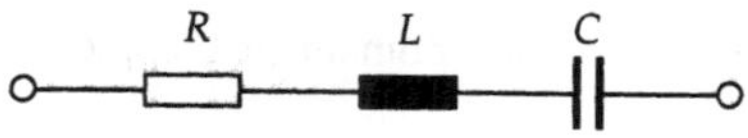

Series oscillator

□ Application in electrical engineering: Inversion of the impedance locus.
Series oscillator circuit:
Complex impedance: $\underline{Z}(\omega) = R + \mathrm{j}(\omega L - 1/\omega C)\ , \quad \omega \geq 0.$
Inversion: Admittance
$\underline{Y}(\omega) = 1/\underline{Z}(\omega) = 1/(R + \mathrm{j}(\omega L - 1/\omega C))\ , \quad \omega \geq 0.$

Second inversion rule ⇒ the admittance locus must be a circle through the origin.
According to item 1: Point $P(z)$ with the shortest distance from the origin in the z plane: $\underline{Z}(\omega_0) = R$ ($\omega_0 = 1/\sqrt{LC}$) $\Rightarrow \underline{Y}(\omega_0) = 1/R$ is point $P(w)$ with the farthest distance from the origin in the w-plane.
According to item 2: Point $\underline{Z}(\omega)$ above the real axis becomes point $\underline{Y}(\omega)$ below the real axis.

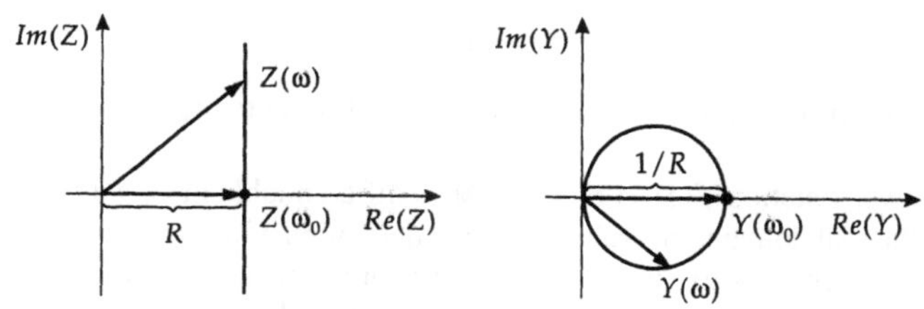

Impedance locus Admittance locus

17.5 Differentiation of functions of a complex variable

17.5.1 Definition of the derivative in the complex domain

The definition of the derivative in the complex domain corresponds to that in the real domain.

Differentiability of a function $f : \mathbb{C} \to \mathbb{C}\ , \ z \mapsto f(z)$ at $z_0 \in \mathbb{C}$: The limit

$$\lim_{z \to z_0} \frac{f(z) - f(z_0)}{z - z_0} = \frac{\mathrm{d}f}{\mathrm{d}z}(z_0)$$

exists.

Derivative of f, function $f' : \mathbb{C} \to \mathbb{C}$, $z \mapsto f'(z)$ with

$$f'(z_0) = \frac{\mathrm{d}f}{\mathrm{d}z}(z_0) .$$

Higher derivatives in the complex domain are defined recursively,
$f'' = (f')' , \ldots, f^{(n)} = (f^{(n-1)})'$.

17.5.2 Differentiation rules in the complex domain

The differentiation rules in the complex domain are completely analogous to those in the real domain.
Let the functions f and g be differentiable at z_0. Then the following applies

- **Sum rule in the complex domain**: $f + g$ is differentiable at z_0,

$$(f + g)'(z_0) = f'(z_0) + g'(z_0) .$$

- **Product rule in the complex domain**: $f \cdot g$ is differentiable at z_0,

$$(f \cdot g)'(z_0) = f'(z_0)g(z_0) + f(z_0)g'(z_0) .$$

- **Quotient rule in the complex domain**: If $g(z_0) \neq 0$, then f/g is differentiable at z_0,

$$\left(\frac{f}{g}\right)'(z_0) = \frac{f'(z_0)g(z_0) - f(z_0)g'(z_0)}{g^2(z_0)} .$$

- **Chain rule in the complex domain**: Let $h = g \circ f : \mathbb{C} \to \mathbb{C}$. If f is differentiable at z_0 and g is differentiable at $f(z_0)$, then also $h = g \circ f$ is also differentiable at z_0,

$$h'(z_0) = (g \circ f)'(z_0) = g'(f(z_0)) \cdot f'(z_0) .$$

- The constant function $f(z) = a$, $a \in \mathbb{C}$, is differentiable in $\mathbb{C}$, $f'(z) = 0$ for all $z \in \mathbb{C}$.
- The power function $f(z) = z^n$, $n \in \mathbb{N}$, is differentiable in $\mathbb{C}$, $f'(z) = n \cdot z^{n-1}$.

▷ As in the real domain, this rule allows the derivation of arbitrary functions by means of their power series expansion: Power series are compactly convergent on their circle of convergence. For compactly convergent series, differentiation and summation can be interchanged.

□ **Differentiation of the exponential function and the sine function in the complex domain**:

$$(\mathrm{e}^z)' = \sum_{n=0}^{\infty} \frac{nz^{n-1}}{n!} = \sum_{n=1}^{\infty} \frac{z^{n-1}}{(n-1)!} = \sum_{n=0}^{\infty} \frac{z^n}{n!} = \mathrm{e}^z ,$$

$$(\sin z)' = \left(\sum_{n=0}^{\infty} (-1)^n \frac{z^{2n+1}}{(2n+1)!}\right)' = \sum_{n=0}^{\infty} (-1)^n \frac{(2n+1)z^{2n}}{(2n+1)!}$$

$$= \sum_{n=0}^{\infty} (-1)^n \frac{z^{2n}}{(2n)!} = \cos z ,$$

etc.

- A function f is called **analytic, holomorphic** or **regular** at point z_0 if it is differentiable in a neighborhood G of z_0.

17.5.3 Cauchy-Riemann differentiability conditions

Interpret a function $f : \mathbb{C} \to \mathbb{C}$, $z \mapsto f(z) = u(z) + \mathrm{j}v(z)$ as a function $f : \mathbb{R}^2 \to \mathbb{R}^2$, $z = (x, y) \mapsto f(z) = (u(x, y), v(x, y))$,
$u(x, y) = \mathrm{Re}(f(z))$, $v(x, y) = \mathrm{Im}(f(z))$.

- **Cauchy-Riemann differentiability conditions**:
  The function $f = u + \mathrm{j}v : \mathbb{C} \to \mathbb{C}$, $z \mapsto f(z) = u(x, y) + \mathrm{j}v(x, y)$ is given. If f is differentiable at $z_0 = (x_0, y_0)$, then the partial derivatives exist:

$$\frac{\partial u}{\partial x}(x_0, y_0)\,,\ \frac{\partial u}{\partial y}(x_0, y_0)\,,\ \frac{\partial v}{\partial x}(x_0, y_0)\,,\ \frac{\partial v}{\partial y}(x_0, y_0)\,,$$

  and the **Cauchy-Riemann differential equations** apply.

$$\frac{\partial u}{\partial x}(x_0, y_0) = \frac{\partial v}{\partial y}(x_0, y_0)\,,$$
$$\frac{\partial u}{\partial y}(x_0, y_0) = -\frac{\partial v}{\partial x}(x_0, y_0)\,.$$

  Thus,

$$f'(z_0) = \frac{\partial u}{\partial x}(x_0, y_0) + \mathrm{j}\frac{\partial v}{\partial x}(x_0, y_0) = \frac{\partial v}{\partial y}(x_0, y_0) - \mathrm{j}\frac{\partial u}{\partial y}(x_0, y_0)\,.$$

17.5.4 Conformal mapping

Preservation of angle: A function f, continuous at z_0, **preserves the angle** at z_0 if for all loci

$$z_1(t)\,,\ z_2(t)\,,\ t \in \mathbb{R},$$

with

$$z_1(0) = z_2(0) = z_0,$$

which have tangents at z_0; also have the image curves

$$w_1(t) = f(z_1(t))\,,\ w_2(t) = f(z_2(t))$$

in $f(z_0)$ tangents, and if the angles between the pairs of tangents coincide.

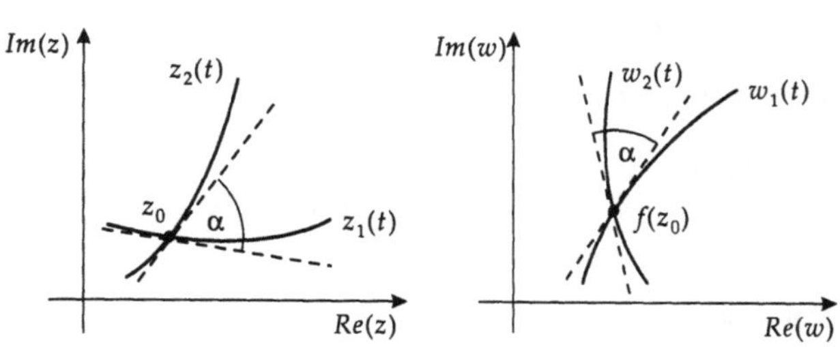

Explaining the term “angle-preserving”

Scale-preserving: A function f, continuous at z_0, is said to be **scale-preserving** if $|f'(z_0)| = \gamma > 0$.
Local conformity: A function f is said to be **locally conformal** at z_0 if it is angle-preserving and scale-preserving at z_0.
Local conformity of the first kind: In the case of angle-preserving not only the magnitude but also the **sense** of the angle is preserved.
Local conformity of the second kind: The magnitude of the angle is preserved, but the sense of the angle is reversed.

□ The function $f(z) = z + (1 + \mathrm{j})$ is locally conformal of the first kind at $z_0 = 1 + \mathrm{j}$. Considering the curves $z_1(t) = t + z_0$, $z_2(t) = \mathrm{j}t + z_0$, then $w_1(t) = f(z_1(t)) = (t + 2) + 2\mathrm{j}$, $w_2(t) = f(z_2(t)) = 2 + \mathrm{j}(t + 2)$. The angle between z_1, and z_2 and between w_1 and w_2 is $\pi/2$; thus, the sense is preserved.

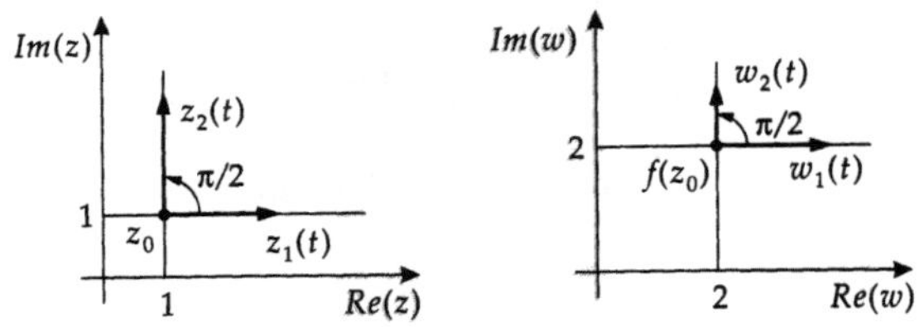

Local conformity of the first kind

□ The function $f(z) = z^*$ is locally conformal of the second kind at $z_0 = 2 + \mathrm{j}$. For the curves $z_1(t) = t + z_0$, $z_2(t) = \mathrm{j}t + z_0$, one has $w_1(t) = (t + 2) - \mathrm{j}$, $w_2(t) = 2 - \mathrm{j}(t + 1)$. Thus, the sense of the angle has been reversed from $\pi/2$ to $-\pi/2$.

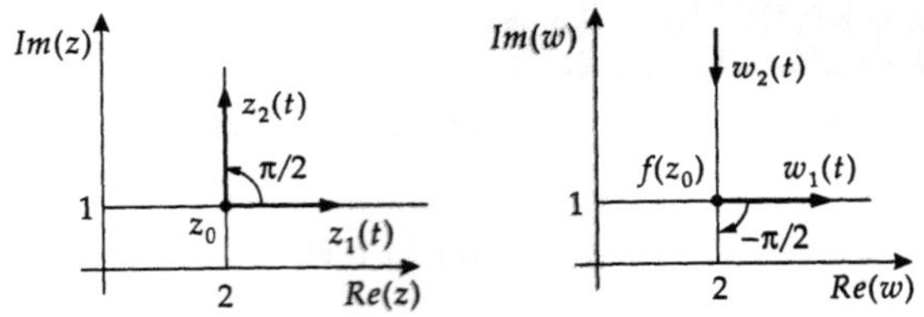

Local conformity of the second kind

Conformal mapping: A function $f : \mathbb{C} \to \mathbb{C}$ maps a domain $G \subset \mathbb{C}$ (global) **conformally** onto the domain $H = f(G) \subset \mathbb{C}$ if

(1) f is **locally** conformal at every point $z_0 \in G$, and

(2) $f : G \to H$ is a **one-to-one mapping**.

Simple conformal mapping:

(1) **Linear functions**: $f(z) = az + b$,
$a = r \cdot \mathrm{e}^{\mathrm{j}\phi} \in \mathbb{C}$, $b \in \mathbb{C}$. These functions can be divided into the following operations:

a. **Rotation** through ϕ, $w = \mathrm{e}^{\mathrm{j}\phi} z$,

b. **Stretching** by r, $v = rw$, and

c. **Parallel displacement** by b, $f(z) = v + b$.

(2) **Inversion** $f(z) = 1/z$.

Conformal mapping was discussed above under **loci**.

(3) **Fractional linear functions**:
$f(z) = (az + b)/(cz + d)$, $(ad - bc \neq 0, c \neq 0)$.
These functions cause:

a.) **parallel displacement** by d/c, $w = z + d/c$,
b.) **inversion,** $v = 1/w = c/(cz + d)$,
c.) **rotation and stretching,** $u = (bc - ad)v/c^2$, and
d.) **parallel displacement** by a/c, $f(z) = u + a/c$.

17.6 Integration in the complex plane

17.6.1 Complex curvilinear integrals

Complex curvilinear integral of the function $f : \mathbb{C} \to \mathbb{C}$, $z \mapsto f(z)$ along the curve $C = \{z | z(t) \in \mathbb{C}, t \in [a, b] \subset \mathbb{R}\}$:
For $N + 1$ points $z_0 = z(a), z_1, z_2, \ldots, z_N = z(b)$ on the curve C and N intermediate points c_i each lying on the curve between z_{i-1} and z_i, the complex curvilinear integral is defined as the limit

$$\int_C f(z)\,\mathrm{d}z = \lim_{\max|z_i - z_{i-1}| \to 0} \sum_{i=1}^{N} f(c_i)(z_i - z_{i-1}) \,.$$

The limit must not depend on the choice of the $N + 1$ decomposition points of the curve C or on the N intermediate points c_i.

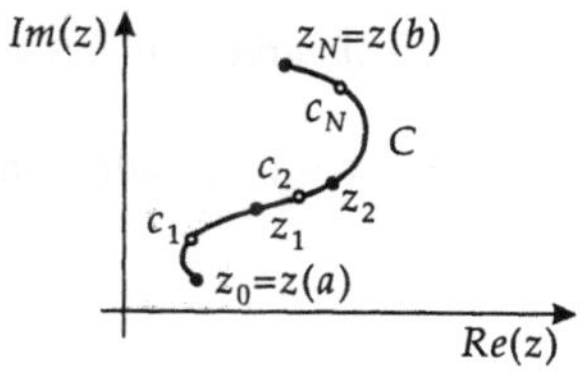

Definition of the curvilinear integral

- The curvilinear integral exists always for smooth curves C and functions f continuous on C.
- The complex curvilinear integral is related to the real curvilinear integral by means of the decomposition of terms:

$$\begin{aligned}\int_C f(z)\,\mathrm{d}z &= \int_C \{u(x, y) + \mathrm{j}v(x, y)\}\,\mathrm{d}(x + \mathrm{j}y) \\ &= \int_C u(x, y)\,\mathrm{d}x - \int_C v(x, y)\,\mathrm{d}y \\ &\quad + \mathrm{j}\left(\int_C u(x, y)\,\mathrm{d}y + \int_C v(x, y)\,\mathrm{d}x\right) .\end{aligned}$$

The known properties of real curvilinear integrals are transformed:

- Linearity:

$$\int_C (f(z) + g(z))\ \mathrm{d}z = \int_C f(z)\ \mathrm{d}z + \int_C g(z)\ \mathrm{d}z\,,$$

$$\int_C \lambda f(z)\ \mathrm{d}z = \lambda \int_C f(z)\ \mathrm{d}z\,.$$

- Inversion of the direction of integration:

$$\int_C f(z)\ \mathrm{d}z = -\int_{-C} f(z)\ \mathrm{d}z\,,$$

  where $-C$ means that the curve C is traversed in the opposite sense.

- Subdivision into partial integrals:
  If $a < c < b$, $C = \{z|z(t) \in \mathbb{C}, t \in [a, b] \subset \mathbb{R}\}$,
  $C_1 = \{z|z(t) \in \mathbb{C}, t \in [a, c] \subset \mathbb{R}\}$, $C_2 = \{z|z(t) \in \mathbb{C}, t \in [c, b] \subset \mathbb{R}\}$, then

$$\int_C f(z)\mathrm{d}z = \int_{C_1} f(z)\mathrm{d}z + \int_{C_2} f(z)\mathrm{d}z\,.$$

17.6.2 Cauchy's integral theorem

- **Independence of the complex integration on the path**:
  Let C be a smooth curve in a simply connected domain $G \subset \mathbb{C}$. If
  $f : \mathbb{C} \to \mathbb{C}\,,\ z \mapsto f(z)$ is a function continuous on G, then the curvilinear integral

$$\int_C f(z)\ \mathrm{d}z$$

  is independent on the special path of curve C if and only if f is an analytic function in G.

▷ To calculate the integral one can choose the path of integration which simplifies the calculation.

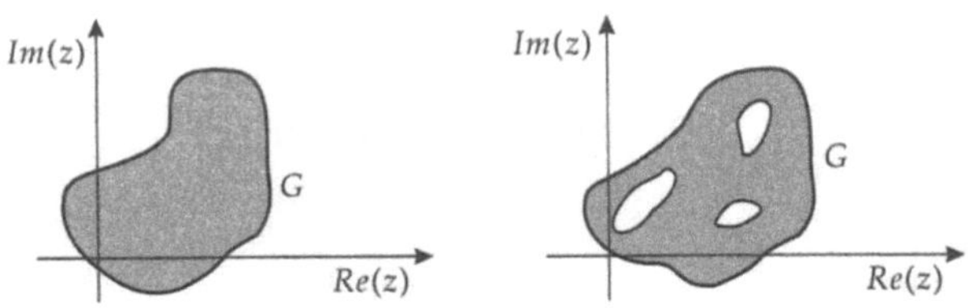

Domains with simple and multiple connections

This theorem forms the basis for **Cauchy's integral theorem**.

- **Cauchy's integral theorem**:
  If $f : \mathbb{C} \to \mathbb{C}\,,\ z \mapsto f(z)$ is a function analytic in a simply connected domain $G \subset \mathbb{C}$, then for every closed curve $C \subset G$,

$$\oint_C f(z)\ \mathrm{d}z = 0\,.$$

▷ $\oint_C f(z)\,dz$ denotes the integral along a **closed** curve; it is also called a **contour integral**.

Under certain conditions, the converse is also valid:

- **Morera's theorem**:
  if $f : \mathbb{C} \to \mathbb{C}\,,\ z \mapsto f(z)$ is unique and continuous in a simply connected domain $G \subset \mathbb{C}$, and if the integral

$$\oint_C f(z)\,dz = 0$$

vanishes for every closed curve $C \subset G$, then f is analytic.

17.6.3 Primitive functions in the complex domain

Primitive $F(z)$ function $f(z)$:
Given a simply connected domain G, a function $f : \mathbb{C} \to \mathbb{C}\,,\ w \mapsto f(w)$ analytic in G, and a curve $C \subset G$ with an arbitrary, but fixed initial point $a \in G$ and an end-point $z \in G$, the **primitive of the function** f is the function

$$F : \mathbb{C} \to \mathbb{C}$$
$$z \mapsto F(z) = \int_C f(w)\,dw\,.$$

The integral is independent of the path, since f is analytic, but the value of the integral still depends on the initial point a.

Indefinite integral of the function f: The totality of all primitives (for all points a).

▷ The integration rules for the indefinite integration of the elementary complex functions correspond to those in the real domain.

17.6.4 Cauchy's integral formulas

- Let f be an analytic function in a simply connected domain G, and let $C \subset G$ be a closed smooth curve. Then,

$$f(z) = \frac{1}{2\pi j}\oint_C \frac{f(w)}{w-z}\,dw\,,$$
$$f'(z) = \frac{1}{2\pi j}\oint_C \frac{f(w)}{(w-z)^2}\,dw\,,$$
$$f''(z) = \frac{2}{2\pi j}\oint_C \frac{f(w)}{(w-z)^3}\,dw\,,$$
$$f^{(n)}(z) = \frac{n!}{2\pi j}\oint_C \frac{f(w)}{(w-z)^{n+1}}\,dw\,,$$

if $z \in G$ lies inside C, and

$$f(z) = \frac{1}{2\pi j}\oint_C \frac{f(w)}{w-z}\,dw = 0\,,$$

if $z \in G$ lies outside C.

▷ If z lies inside C, then the function $f(w)/(w-z)$ is analytic inside C except for point z.

Cauchy's integral formulas allow for the calculation of the function f and the derivative of every order at **every** point z inside C, if f is known on C. Thus, from knowledge of the function on the boundary of the domain we can determine the function inside the interior of the domain.

If z lies outside C, then the function $f(w)/(w-z)$ is analytic everywhere inside C. According to Cauchy's integral theorem, $f(z)$ must vanish in that case.

17.6.5 Taylor series of an analytic function

Cauchy's integral formulas allow for the expansion of an analytic function into a Taylor series.

Let G be a simply connected domain, b the center of the circle $C \subset G$, and a a point inside the circle. The following series expansion (geometric series) applies to $w \in C$:

$$\begin{aligned}\frac{1}{w-a} &= \frac{1}{(w-b)-(a-b)} = \frac{1}{(w-b)}\,\frac{1}{(1-(a-b)/(w-b))} \\ &= \frac{1}{w-b}\left(1+\frac{a-b}{w-b}+\frac{(a-b)^2}{(w-b)^2}+\cdots+\frac{(a-b)^n}{(w-b)^n}+\cdots\right),\end{aligned}$$

since $|a-b| < |w-b|$.

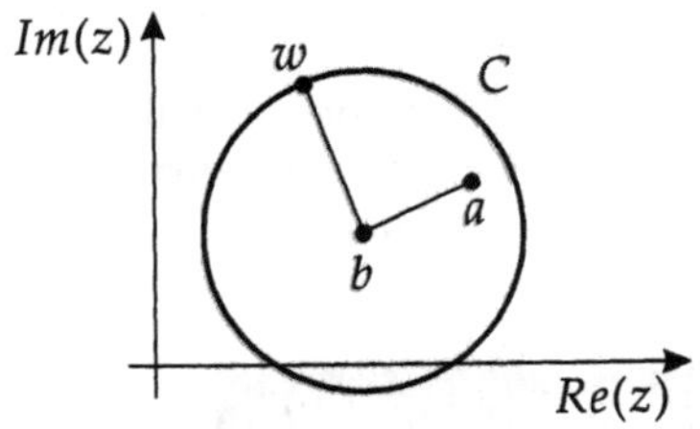

Series expansion of $1/(w-a)$

Let f be a function analytic in G. Substitution of the series expansion into the first Cauchy's integral formula for f:

$$\begin{aligned}f(a) &= \frac{1}{2\pi j}\oint_C \frac{f(w)}{w-a}\,dw \\ &= \frac{1}{2\pi j}\oint_C \frac{f(w)}{w-b}\,dw + \frac{a-b}{2\pi j}\oint_C \frac{f(w)}{(w-b)^2}\,dw \\ &\quad + \frac{(a-b)^2}{2\pi j}\oint_C \frac{f(w)}{(w-b)^3}\,dw + \cdots + \frac{(a-b)^n}{2\pi j}\oint_C \frac{f(w)}{(w-b)^{n+1}}\,dw \;\ldots\,.\end{aligned}$$

Complex Taylor series: The expansion of function f about point b; follows from the above equation given above with the help of Cauchy's integral formulas:

$$f(a) = f(b) + f'(b)(a-b) + \frac{f''(b)}{2!}(a-b)^2 + \cdots + \frac{f^{(n)}}{n!}(a-b)^n + \cdots\,.$$

▷ The expansion holds for all points a inside circle C.

17.6.6 Laurent series

Let the function $f : \mathbb{C} \to \mathbb{C}$, $z \mapsto f(z)$ be analytic on the **circular ring** $0 < r_1 \leq |z - a| \leq r_2$ about point a.
Laurent series: Unique expansion of function f inside the circular ring,

$$f(z) = \sum_{i=-\infty}^{\infty} a_i (z - a)^i ,$$

with the coefficients

$$a_i = \frac{1}{2\pi \mathrm{j}} \oint_C \frac{f(w)}{(w - a)^{i+1}} \, \mathrm{d}w .$$

C is an arbitrary closed, smooth curve located entirely inside the circular ring.
Regular part of the Laurent series: The part of the series with $i \geq 0$ (ordinary power series).
Principal part of the Laurent series: The part of the series with $i < 0$.

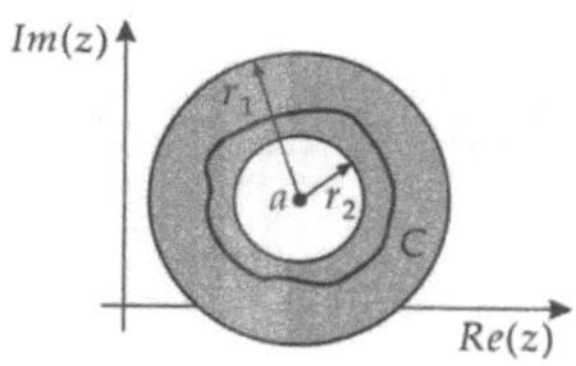

Circular ring about a

17.6.7 Classification of singular points

Regular point: A point at which a function f is analytic.
Singular point: A point at which a function f is not defined.
Isolated singularity: A singular point about which a circle can be constructed in such a way that it contains no further singular points.
Classification of isolated singularities:

1. **Removable singularity**: An isolated singularity a of function f in whose vicinity the Laurent series of f contains no negative powers of $(z - a)$ $(a_i = 0, i < 0)$.

   In this case the limit $f(a) = \lim_{z \to a} f(z)$ exists. From the Laurent expansion it follows that $f(a) = a_0$. This equation is considered as the **definition** of the function value of f at the point a. Thus, the function becomes analytic at a, a becomes a regular point, and the singularity is removed.

   □ The function $f(z) = \sin z/z$ is not defined for $z = 0$. But the Laurent series is

   $$\frac{\sin z}{z} = \frac{1}{z}\left(z - \frac{z^3}{3!} + \frac{z^5}{5!} - \cdots\right) = 1 - \frac{z^2}{3!} + \frac{z^4}{5!} - \cdots ,$$

   and thus $\lim_{z \to 0} \sin z/z = 1$. With the definition $f(0) = 1$, the singularity at $z = 0$ is removed.

2. **Pole of order** n: An isolated singularity a of the function f in whose vicinity the Laurent series of f contains no coefficients a_i with $i < -n$. One has $f(a) = \infty$.

□ The function $f(z) = 1/(z-a)^3$ has the Laurent expansion

$$f(z) = \sum_{i=-\infty}^{\infty} a_i (z-a)^i = 1(z-a)^{-3} ,$$

i.e., $a_i = 0$ for all $i \neq -3$, $a_{-3} = 1$. The function has a pole of order three at a.

3. **Essential singularity**: An isolated singularity a of function f such that in its vicinity the Laurent series of f contains an infinite number of coefficients a_i with $i < 0$.

□ Function $f(z) = \mathrm{e}^{1/z}$ has an essential singularity at $z = 0$, since its Laurent series is

$$\mathrm{e}^{1/z} = \sum_{i=0}^{\infty} \frac{1}{i!z^i} = \sum_{i=-\infty}^{0} a_i z^i \text{ , with } a_i = 1/(-i)! \,.$$

17.6.8 Residue theorem

Let f be an analytic function in G except for an isolated singularity a.

Residue of the analytic function f at point a: The coefficient a_{-1} of the Laurent series:

$$\mathrm{Res}(f, a) = a_{-1} = \frac{1}{2\pi \mathrm{j}} \oint_C f(w)\, \mathrm{d}w \,.$$

- **Residue theorem**:
  Let C be a closed curve and $f : \mathbb{C} \to \mathbb{C}$, $z \mapsto f(z)$ a function analytic inside C except for the isolated singularities $a_1, a_2, \ldots, a_n$. Then:

$$\oint_C f(z)\, \mathrm{d}z = 2\pi \mathrm{j} \sum_{i=1}^{n} \mathrm{Res}(f, a_i) \,.$$

Calculation of residues:

- Let the function f have a pole of order n at a. Then:

$$\mathrm{Res}(f, a) = \frac{1}{(n-1)!} \left(\frac{\mathrm{d}^{n-1}}{\mathrm{d}z^{n-1}} \left[(z-a)^n f(z) \right] \right)_{z=a} \,.$$

- Let the functions f and g be analytic in the vicinity of a with $f(a) \neq 0$, $g(a) = 0$, $g'(a) \neq 0$. Then:

$$\mathrm{Res}\left(\frac{f}{g}, a \right) = \frac{f(a)}{g'(a)} \,.$$

17.6.9 Inverse Laplace transformation

If $F(s)$ is analytic for all $s \in \mathbb{C}$ with $\operatorname{Re}(s) > c$, $c \in \mathbb{R}$, and if

$$\int_{c-j\infty}^{c+j\infty} |F(s)||ds| < \infty$$

and

$$\lim_{|s| \to \infty} F(s) = 0$$

uniformly with respect to the argument of s, then
$F(s)$ is the **Laplace transform** of the function

$$f(t) = \frac{1}{2\pi j} \int_{c-j\infty}^{c+j\infty} e^{st} F(s)\, ds\ .$$

$f(t)$ is the **inverse Laplace transform.**

18

Differential equations

18.1 General definitions

Differential equation: An equation containing an unknown function y, its n-th derivatives y', y'', ... $y(n)$, and independent variables.

□ Oscillation of a spring with spring constant k and mass point m (dots denote the derivative with respect to time, $\dot{x} = \frac{dx}{dt}$):

$$\ddot{x}(t) + \frac{k}{m} x(t) = 0 .$$

□ Electromagnetic oscillating circuit with inductance L, capacitance C, ohmic resistance R, and external voltage U_e:

$$L \cdot \ddot{I}(t) + R \cdot \dot{I}(t) + \frac{1}{C} \cdot I(t) = \dot{U}_e(t)$$

Ordinary differential equation: The unknown function y depends on only *one* independent variable: $y = f(x)$.

□ Newton's second law of motion for a free point particle of mass m:

$$m \cdot \ddot{x}(t) = 0 .$$

□ Harmonic oscillator with angular frequency ω:

$$\ddot{x}(t) + \omega^2 \cdot x(t) = 0 .$$

□ Loaded beam ($f(x)$: External force, $p(x)$, $q(x)$, $r(x)$: Material constants):

$$\left(p(x) \cdot u''(x)\right)'' - \left(q(x) \cdot u'(x)\right)' + r(x) \cdot u(x) = f(x) .$$

Partial differential equation: The unknown function depends on several independent variables,

$$y = f(x_1, \ldots, x_n) .$$

□ Poisson equation for electrostatic potentials for a given external charge distribution $\rho(x, y, z)$:

$$\Delta\varphi = \frac{\partial^2\varphi}{\partial x^2} + \frac{\partial^2\varphi}{\partial y^2} + \frac{\partial^2\varphi}{\partial z^2} = \rho(x, y, z) .$$

□ Laplace equation (the Poisson equation for a neutral space):

$$\Delta\varphi = 0 .$$

□ One-dimensional wave equation (propagation of electromagnetic waves in the vacuum):

$$\frac{\partial^2}{\partial x^2} y(x, t) - \frac{1}{c^2}\frac{\partial^2}{\partial t^2} y(x, t) = 0 .$$

Forms of ordinary differential equations:

Implicit form: $F(x, y, y', \ldots, y^{(n)}) = 0$,
Explicit form: $y^{(n)} = \tilde{F}(x, y, y', \ldots, y^{(n-1)})$

▷ The implicit form can often lead to an explicit form by solving for $y^{(n)}$.

□ $0 = F(x, y, y', y'', y''') = a(x) + b(x) \cdot y^2 + c(x) \cdot y' + d(x) \cdot y'' + e(x) \cdot y'''$
$y''' = -a(x)/e(x) - y^2 \cdot b(x)/e(x) - y' \cdot c(x)/e(x) - y'' \cdot d(x)/e(x)$

Every function $y = f(x)$ satisfying the differential equation $F(x, y, y', \ldots, y^{(n)}) = 0$ is a solution of the differential equation.

Order of the differential equation: The highest occurring order of the derivatives of y.

□ $y'' + x^3 + xy = 0$ is of the second order.

Degree of the differential equation: The degree of the highest power of y or its derivatives.

□ $y'' + y^3 + xy = 0$ is of the third degree.

Solution of the differential equation is every function $y = f(x)$ satisfied by the differential equation $F(x, y, y', \ldots, y^{(n)}) = 0$.

Integration of the differential equation: The determination of the solution of the differential equation.

General solution: The general solution of an n-th order differential equation has n free parameters.

Particular or special solution: A solution of an n-th order differential equation that fulfills n given conditions.

Initial-value problem: Required is a solution of an n-th order differential equation whose n free parameters are determined by the function and its derivatives at a particular point $x_0(i)$ $y(x_0), y'(x_0), \ldots, y^{(n-1)}(x_0)$ (**initial condition**).

Boundary-value problem: Required is a solution of an n-th order differential equation whose n free parameters are fixed by n **boundary conditions** at the end-points a, b of an interval $a \leq x \leq b$.

Linear differential equation: Differential equation of the first degree:

$$y^{(n)} + a_1 y^{(n-1)} + \cdots + a_{n-1} y'$$

Perturbation term of a differential equation: All terms containing neither y nor its derivatives $y', \ldots, y^{(n)}$.

Inhomogeneous differential equation: The perturbation term is different from nonzero.

□ $y' + x \cdot y = 4x + 3$

Homogeneous differential equation: The perturbation term is equal to zero.

□ $y' + x \cdot y = 0$

□ General solution of the oscillator equation:

$$y(t) = A \cdot \cos(\omega t) + B \cdot \sin(\omega t)\ .$$

□ The special solution with $y(0) = 1$ and $y'(0) = 0$ is:

$$y(t) = \cos(\omega t)\ .$$

18.2 Geometric interpretation

Family of curves: The curves obtained by varying the free parameters of the general solution of a differential equation. It is usually given by an implicit equation

$$F(x, y, c) = 0\ .$$

□ Circle with radius R:

$$F(x, y, R) = x^2 + y^2 - R^2 = 0\ .$$

Setup of differential equations: Multiple differentiation of the equation for the family of curves and elimination of the free parameters.

□ $F(x, y, R) = x^2 + y^2 - R^2 = 0$:

$$\Rightarrow \quad 2x + 2yy' = 0.$$

Integral curve: A particular solution of the differential equation, i.e., a special curve from the family of curves.

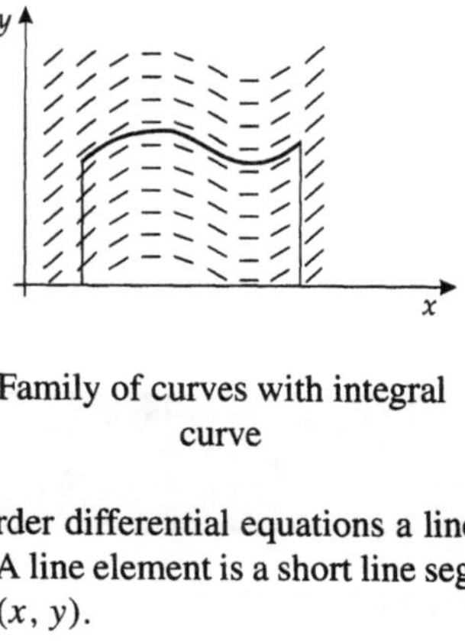

Family of curves with integral curve

Directional field: For first-order differential equations a line element is determined for every point in the x-y-plane. A line element is a short line segment through a point (x, y) with a given gradient $y' = f(x, y)$.

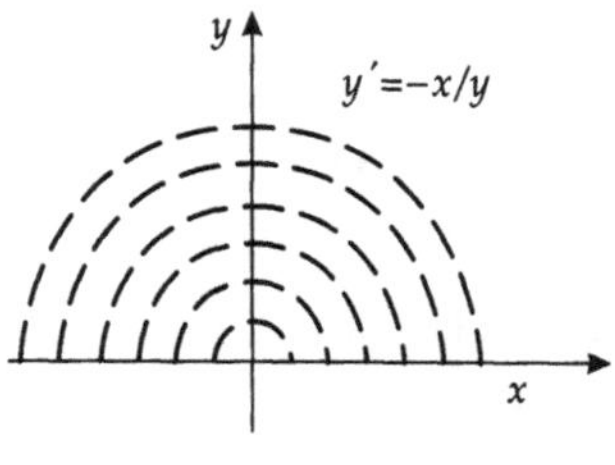

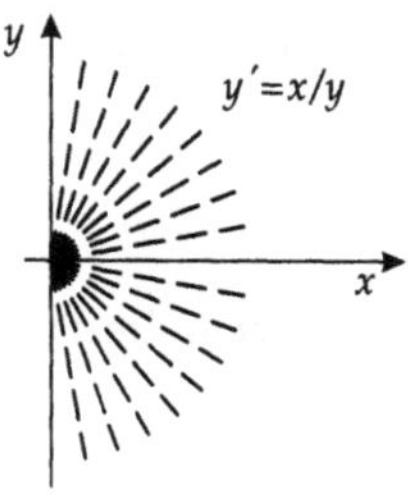

Directional fields of the differential equations $y' = -x/y$ and $y' = y/x$

Isocline: Curve connecting all points having the same direction as the line elements ($y' =$ const.).

▷ Isoclines can be used in approximations for the solutions of differential equations: They must be crossed by the solutions with the slope y'.

Graphical solution: Solving the differential equation by determining the isoclines.

Isocline equation: The equation $y' = C$, i.e., $C = f(x, y)$ solved for y.

□ $y' = 5y^2 + 3x^5$:

$$y' = C = 5y^2 + 3x^5 \quad \Rightarrow \quad y = \sqrt{\frac{C - 3x^5}{5}} \, .$$

Isogonal trajectory: Curve with a constant angle α at every point with respect to the corresponding line element. Given by the equation

$$y' = \frac{f(x, y) + \tan\alpha}{1 - f(x, y)\tan\alpha} \, .$$

Orthogonal trajectory: The special case of $\alpha = 90°$, given by the equation

$$y' = -\frac{1}{f(x, y)} \, .$$

□ If the field lines of a charge distribution are given, then the potential surfaces are the orthogonal trajectories.

□ For the family of curves $F(x, y, R) = x^2 + y^2 - R^2, \quad y' = -\frac{x}{y}$:

Differential equation of the family: $2x + 2yy' = 0$.

Isoclines: $y' = R, \quad y = -\frac{x}{R}$.

Orthogonal trajectory: $y' = \frac{y}{x}$.

Isogonal trajectory: $y' = \dfrac{-\frac{x}{y} + \tan\alpha}{1 + \frac{x}{y}\tan\alpha} = \dfrac{y\tan\alpha - x}{y + x\tan\alpha}$.

18.3 Solution methods for first-order differential equations

18.3.1 Separation of variables

Applicable if the representation

$$\frac{dy}{dx} = y' = f(x) \cdot g(y)$$

is possible.

● The solution is determined by solving the equation

$$\int \frac{dy}{g(y)} = \int f(x)dx$$

for y. Take all terms containing x or y to opposite sides of the equation (separation), then integrate and solve for y.

▷ This is possible only if $g(y) \neq 0$.

□ $y' = 3x^2y;\quad \int \frac{dy}{y} = \int 3x^2 dx \quad \Rightarrow \quad \ln|y| = x^3 + C \quad \Rightarrow \quad y = K \cdot e^{x^3}$, with $K = \pm e^C$.

18.3.2 Substitution

Reduce the differential equation by transferring the quantities x, y to a solvable differential equation. Solve the transformed differential equation by means of back substitution.

□ $y' = f(ax + by + c)$

Let $u = ax + by + c \Rightarrow u' = a + bf(u) \Rightarrow \displaystyle\int \frac{du}{a + bf(u)} = x + C$

▷ We first obtain $x(u)$. This is solved for u, then we obtain y according to $y = (u - ax - c)/b$.

□ Homogeneous differential equation

$$y' = f\left(\frac{y}{x}\right) \quad ; \quad u = \frac{y}{x} \quad \Rightarrow \quad u' = \frac{f(u) - u}{x}$$

$$x(u) = C \cdot \exp\left\{\int \frac{du}{f(u) - u}\right\}$$

▷ Solve for u and back-substitute.

18.3.3 Exact differential equations

Can be represented in the form

$$f(x, y)dx + g(x, y)dy = 0 \quad \text{with} \quad \frac{\partial f}{\partial y} = \frac{\partial g}{\partial x}.$$

▷ $\dfrac{\partial f}{\partial y} = \dfrac{\partial g}{\partial x}$ is called the integrability condition.

Solution:

$$\int f(x, y)dx + \int \left[g(x, y) - \int \frac{\partial f}{\partial y} dx\right] dy = C = \text{const.}$$

▷ Then the linear differential equation $f dx + g dy$ is the total differential du of a function $u(x, y)$.

□ $3y + yy' + 3xy' + x = 0$

$\Leftrightarrow (x + 3y)dx + (3x + y)dy = 0$

Solution:

$$\int (x + 3y)dx + \int \left(3x + y - \int 3dx\right) dy = C$$

$$y = -3x \pm \sqrt{8x^2 - C}$$

18.3.4 Integrating factor

Integrating factor: Function by which the differential equation must be multiplied to obtain an exact differential equation.

□ $e^y \cdot (\tan(x) - y') = 0$

Integrating factor $\cos(x)$

$$e^y \cdot (\sin(x) - \cos(x)y') = 0$$
$$\Leftrightarrow e^y \sin(x)dx + (-e^y \cos(x))dy = 0$$

Solution: $y = \ln\left(-\dfrac{C}{\cos(x)}\right)$

18.4 Linear differential equations of the first order

Linear differential equation: Function y and its derivative ly occur linearly only.

- Can be represented in the form:

$$y' + a(x) \cdot y = b(x).$$

Perturbation term: The term $b(x)$.
Homogeneous equation: The case of $b(x) = 0$.
Inhomogeneous equation: The case of $b(x) \neq 0$.

▷ Homogeneous linear differential equations of the first order can be solved by separating the variables.

$$y = C \cdot e^{-\int a(x)dx}$$

□ $y' + 2\sin(x)\cos(x)y = 0$

$$y = C \cdot e^{-2\int \sin(x)\cos(x)dx} = C \cdot e^{-\sin^2(x)}$$

18.4.1 *Variation of the constants*

- A solution of an inhomogeneous differential equation for y is obtained from the solution of the corresponding homogeneous differential equation y_H by "**variation of the constants**":

$$y' + a(x) \cdot y = b(x); \quad y_H' + a(x) \cdot y_H = 0;$$

$$y = \psi(x) \cdot y_H(x) = \psi(x) \cdot Ce^{-\int a(x)dx},$$

with $\psi(x) = \int \frac{b(x)}{y_H(x)}dx$

□ $y' + 2\sin(x)\cos(x) \cdot y = e^{\cos^2(x)}$.
It follows:

$$y_H(x) = C \cdot e^{-\sin^2(x)} \quad \text{(see above example)}$$
$$\psi(x) = \int \frac{b(x)}{y_H(x)}dx = \frac{1}{C} \cdot \int \frac{e^{\cos^2(x)}}{e^{-\sin^2(x)}}dx$$
$$= \frac{1}{C} \cdot \int e^{\cos^2(x)+\sin^2(x)}dx = \frac{1}{C} \cdot \int e^1 dx = \frac{x}{C} \cdot e$$
$$y(x) = \psi(x) \cdot y_H(x) = \frac{1}{C}xe^1 Ce^{-\sin^2(x)}$$
$$= xe^{1-\sin^2(x)} = xe^{\cos^2(x)}$$

□ $y' - 2xy = x^3$.

Homogeneous differential equation: $y'_H - 2xy_H = 0$

$$y_H(x) = Ce^{x^2}$$

$$\psi(x) = \frac{1}{C}\int \frac{x^3}{e^{x^2}}dx$$

$$= \frac{1}{C}\left(-\frac{1}{2}e^{-x^2}\left(1+x^2\right)\right) = -\frac{1}{2C}\left(1+x^2\right)\cdot e^{-x^2}$$

$$y(x) = \psi(x)\cdot y_H(x) = -\frac{1}{2}\left(1+x^2\right)$$

18.4.2 General solution

● The general solution y of an inhomogeneous linear differential equation is obtained from a particular solution y_P of the differential equation and the solution y_H of the corresponding homogeneous differential equation according to:

$$y = y_P + y_H \ .$$

□ The general solution of the first equation reads:

$$y = xe^{\cos^2(x)} + Ce^{-\sin^2(x)} \ .$$

□ The general solution of the second equation reads:

$$y = Ce^{x^2} - \frac{1}{2}\left(1+x^2\right) \ .$$

18.4.3 Determination of a particular solution

▷ It is often practical to guess a particular solution by setting up a solution function with several parameters in the differential equation.

▷ The solution should be set up in accordance with the perturbation term (i.e., a similar or cognate type of function).

□ $y' + \frac{1}{x}y = 1 + x$

Setup: $y_P(x) = ax + bx^2$

Solution: $a = \frac{1}{2}$, $b = \frac{1}{3}$, $y_P = \frac{1}{2}x + \frac{1}{3}x^2$.

18.4.4 Linear differential equations of the first order with constant coefficients

● This is the special case of $a(x) = a_0 = \text{const.}$

$$y' + a_0 \cdot y = b(x)$$

▷ Solution as described above by variation of the constants or by guessing the particular solution.

● Guessing the particular solution often leads to the goal more quickly.

□ $y' + a \cdot y = \cos(x)$
First setup: $y_P(x) = A\cos(x) + B\sin(x)$
Solution: $A = \dfrac{a}{1+a^2}$, $B = \dfrac{1}{1+a^2}$

$$y_P(x) = \frac{1}{1+a^2}\,(a\cos(x) + \sin(x))$$

Perturbation term $b(x)$	Recommended setup
$e^{kx}\sin(mx)$ or $e^{kx}\cos(mx)$	$e^{kx}\,(A\cdot\sin(mx) + B\cdot\cos(mx))$
$e^{kx}(a_0 + a_1x + \cdots + a_nx^n)$	$e^{kx}(A_0 + A_1x + \ldots + A_nx^n)$
$(a_0 + a_1x + \cdots + a_sx^s)\cdot\sin(mx)$ $+(b_0 + b_1x + \cdots + b_tx^t)\cdot\cos(mx)$	$(A_0 + A_1x + \cdots + A_sx^s)\cdot\sin(mx)$ $+(B_0 + B_1x + \cdots + B_tx^t)\cdot\cos(mx)$

18.5 Some specific equations

18.5.1 Bernoulli differential equation

$$y' + a(x)y = b(x)y^n \qquad (n \neq 1)$$

Substitution:

$$u = y^{1-n}, \quad u' = (1-n)y^{-n}y'\,.$$

Results in:

$$u' + (1-n)a(x)\cdot u = (1-n)b(x)\,.$$

Solution as described in Section 18.4 on linear differential equations of the first order.

□ $y' + 2x^2y + x^3y^2 = 0 \quad (n = 2)$

$$u = \frac{1}{y} \quad , \quad u' = -\frac{1}{y^2}\cdot y'$$

Thus: $u' - 2xu - x^3 = 0$

$$u(x) = -\frac{1+x^2}{2}$$
$$y(x) = -\frac{2}{1+x^2}$$

18.5.2 Riccati differential equation

$$y' = a(x) + b(x)y + c(x)y^2$$

is used to optimize feedback control circuits.

▷ For $a(x) = 0$ the Riccati equation becomes a Bernoulli equation with $n = 2$.

- The substitution $y(x) = -w'(x)/(c(x)\cdot w(x))$ changes the Riccati differential equation to a linear differential equation of the second order for $w(x)$:

$$w''(x) - w'(x)\cdot\left(\frac{c'(x)}{c(x)} + b(x)\right) + a(x)\cdot c(x)\cdot w(x) = 0\,.$$

▷ It is often better to guess the specific solution by means of a suitable first setup.

□ $y' = 4x^5 + \left(5x^4 + \frac{2}{x}\right)\cdot y + \frac{2}{x^2}y$

Substitution leads to:

$$w'' - 5x^4 w' + 8x^3 w = 0\,.$$

See also Section 18.7 on linear differential equations of the second order.

18.6 Differential equations of the second order

18.6.1 Simple special cases

The following special cases can be reduced by substitution to a differential equation of the first order.

- $y'' = f(x)$ is solvable by means of a twofold integration,

$$y(x) = \int\left(\int f(x)\mathrm{d}x\right)\mathrm{d}x + C_1x + C_2\,.$$

□
$$y'' = 5x^6 - 8x^5 + 7x^3 - 4x^2 + 3x - 20$$
$$y'(x) = \frac{5}{7}x^7 - \frac{4}{3}x^6 + \frac{7}{4}x^4 - \frac{4}{3}x^3 + \frac{3}{2}x^2 - 20x + C_1$$
$$y(x) = \frac{5}{56}x^8 - \frac{4}{21}x^7 + \frac{7}{20}x^5 - \frac{1}{3}x^4 + \frac{1}{2}x^3 - 10x^2 + C_1x + C_2$$

□ $y'' = \mathrm{e}^{-\lambda x}$

$$y'(x) = -\frac{1}{\lambda}\mathrm{e}^{-\lambda x} + C_1 \quad ; \quad y(x) = \frac{1}{\lambda^2}\mathrm{e}^{-\lambda x} + C_1x + C_2$$

□ $y'' = \dfrac{1}{1+x^2}$

$$y'(x) = \arctan(x) + C_1$$
$$y(x) = x\cdot\arctan(x) - \frac{1}{2}\cdot\ln(1+x^2) + C_1x + C_2$$

- $y'' = f(y')$ is solvable by means of the substitution $z = y'$

$$z' = f(z) \to x = \int\frac{\mathrm{d}z}{f(z)} + C_1 \to \text{solving for } z\,,$$
$$y(x) = \int z(x)\mathrm{d}x + C_2\,.$$

□ $y'' = 1 + y'^2$

$$z' = 1 + z^2 \to x = \int \frac{\mathrm{d}z}{1+z^2} + C = \arctan(z) + C$$

$$z(x) = \tan(x - C_1)$$

$$y(x) = \int z(x)\mathrm{d}x = \int \tan(x - C_1)\mathrm{d}x = -\ln|\cos(x - C_1)| + C_2$$

□ $$y''(x) = \frac{1}{\cos y}, \qquad z' = \frac{1}{\cos z} \to z = \arcsin(x + C_1)$$

$$y(x) = (x + C_1)\arcsin(x + C_1) + \sqrt{1 - (x + C_1)^2}\,.$$

- $y'' = f(y)$ is solvable by means of the substitution $z = y' \Rightarrow y'' = \frac{\mathrm{d}z}{\mathrm{d}x} = \frac{\mathrm{d}z}{\mathrm{d}y}\frac{\mathrm{d}y}{\mathrm{d}x} = z\frac{\mathrm{d}z}{\mathrm{d}y}$

$$z\frac{\mathrm{d}z}{\mathrm{d}y} = f(y) \Rightarrow z = \sqrt{2\left(\int f(y)\mathrm{d}y\right) - C_1}$$

$$x = \int \frac{\mathrm{d}y}{\sqrt{2\left(\int f(y)\mathrm{d}y\right) - C_1}} + C_2 \to \text{solving for } y.$$

□ $y'' + \omega^2 y = 0$ (oscillator equation)

$$z = \sqrt{2\left(-\omega^2\int y\mathrm{d}y - C_1\right)} = \sqrt{-\omega^2 y^2 - 2C_1}$$

$$x = \int \frac{\mathrm{d}y}{\sqrt{-\omega^2y^2 - 2C_1}} + C_2 = \frac{1}{\omega}\arcsin\left(\frac{\omega y}{\sqrt{-2C_1}}\right) + C_2$$

$$y = \sqrt{-2C_1}\sin(\omega x - C_2)\text{(real solution only for } C_1 \le 0 \text{ only)}$$

- $y'' = f(x, y')$ is solvable by means of the substitution $z = y'$

$$z' = f(x, z); \qquad y = \int z(x)\mathrm{d}x + C\,.$$

differential equation of the first order, solvable by means of the methods described above.

18.7 Linear differential equations of the second order

$$y'' + p(x)y' + q(x)y = f(x)$$

- If f, p, and q are continuous, then for arbitrary a_0, a_1 this differential equation has exactly one solution that satisfies the two initial conditions $y_0(x_0) = a_0$, $y_0'(x_0) = a_1$.
- The general solution of the differential equation of the second order therefore contains two and only two integration constants. Thus, the differential equation itself (without boundary conditions) has exactly two linearly independent solutions.

Wronskian determinant: The determinant

$$W(y_1, \ldots, y_n; x) = \begin{vmatrix} y_1(x) & \ldots & y_n(x) \\ y_1'(x) & \ldots & y_n'(x) \\ \vdots & & \vdots \\ y_1^{(n-1)}(x) & \ldots & y_n^{(n-1)}(x) \end{vmatrix}$$

is called the Wronskian determinant of the functions $y_1(x), \ldots, y_n(x)$.

- The solutions $y_1, \ldots, y_n$ of a certain differential equation of order n are linearly dependent if and only if there is a point x_0 with $W(y_1, \ldots, y_n; x_0) = 0$.

▷ In this case, $W(y_1, \ldots, y_n; x_0) \equiv 0$ holds even in the entire domain of definition of the Wronskian determinant $W(y_1, \ldots, y_n; x)$.

This theorem allows to determine immediately whether two solutions of a linear differential equation of the second order are linearly independent or not.

18.7.1 Homogeneous linear differential equation of the second order

In this case, the second linearly independent solution can be generated from a known specific solution.

- If y_1 is a solution of the above linear differential equation of the second order, given above, with $f(x) \equiv 0$, then the second linearly independent solution is given by

$$y_2(x) = y_1(x) \cdot \int \frac{\exp\{-\int p(x)\mathrm{d}x\}}{[y_1(x)]^2}\mathrm{d}x .$$

□ $y'' - \frac{2x}{1-x^2}y' + \frac{2}{1-x^2}y = 0$

Special solution: $y_1 = x$, where $y' = 1$, $y'' = 0$. Thus:

$$\begin{aligned} y_2(x) &= x \cdot \int \frac{\exp\{\int \frac{2x}{1-x^2}\mathrm{d}x\}}{x^2}\mathrm{d}x = x \cdot \int \frac{\exp\{-\ln(1-x^2)\}}{x^2}\mathrm{d}x \\ &= x \cdot \int \frac{1}{(1-x^2)\cdot x^2}\mathrm{d}x = x \cdot \int \left(\frac{1}{x^2} + \frac{1}{1-x^2}\right)\mathrm{d}x \\ &= -1 + \frac{x}{2}\ln\left(\frac{1+x}{1-x}\right) . \end{aligned}$$

General solution:

$$y(x) = C_1 x + C_2 \cdot \left[\frac{x}{2}\ln\left(\frac{1+x}{1-x}\right) - 1\right] .$$

18.7.2 Inhomogeneous linear differential equations of the second order

If the two linearly independent solutions y_1, y_2 of the corresponding homogeneous linear differential equation are known, then a special solution of the inhomogeneous differential equation can be calculated.

● If $y_1(x)$, $y_2(x)$ are linearly independent solutions of the differential equation

$$y'' + p(x)y' + q(x)y = 0,$$

then a solution y_0 of the inhomogeneous differential equation

$$y'' + p(x)y' + q(x)y = f(x)$$

is given by

$$y_0(x) = -y_1(x) \cdot \int \frac{y_2(x) \cdot f(x)}{W(y_1, y_2; x)} \mathrm{d}x + y_2(x) \cdot \int \frac{y_1(x) \cdot f(x)}{W(y_1, y_2; x)} \mathrm{d}x,$$

where

$$W(y_1, y_2; x) = y_1(x) \cdot y_2'(x) - y_1'(x) \cdot y_2(x)$$

denotes the Wronskian determinant of the functions y_1, y_2 (see above).

▷ Since it is assumed that y_1 and y_2 are linearly independent, it holds that $W(y_1, y_2; x) \neq 0$ in the entire interval.

□ $y'' - \frac{2x}{1-x^2} y' + \frac{2}{1-x^2} y = \frac{1}{x-x^3}$

From the above, it follows that: $y_1 = x$, $y_2 = \frac{x}{2} \ln\left(\frac{1+x}{1-x}\right) - 1$. Thus,

$$W(y_1, y_2; x) = \begin{vmatrix} x & \frac{x}{2} \ln\left(\frac{1+x}{1-x}\right) - 1 \\ 1 & \frac{1}{2} \ln\left(\frac{1+x}{1-x}\right) + \frac{x}{1-x^2} \end{vmatrix} = \frac{1}{1-x^2}.$$

$$\begin{aligned} y_0(x) &= -x \cdot \int \left(\frac{1}{2} \ln\left(\frac{1+x}{1-x}\right) - \frac{1}{x}\right) \mathrm{d}x + \left[\frac{x}{2} \ln\left(\frac{1+x}{1-x}\right) - 1\right] \cdot \int 1 \mathrm{d}x \\ &= -x \cdot \int \ln\left(\frac{1+x}{1-x}\right) \mathrm{d}x + x \cdot \int \frac{1}{x} \mathrm{d}x + \frac{x^2}{2} \ln\left(\frac{1+x}{1-x}\right) - x \\ &= -\frac{x}{2} \ln\left(\frac{1+x}{1-x}\right) - \frac{x}{2} \ln(1-x^2) + x \ln x + \frac{x}{2} \ln\left(\frac{1+x}{1-x}\right) - x \\ &= -\frac{x}{2} \ln(1-x^2) + x \ln x - x = x \cdot \ln\left(\frac{x}{\sqrt{1-x^2}}\right) - x. \end{aligned}$$

18.7.3 Reduction of special differential equations of the second order to differential equations of the first order

Differential equation	Substitution	Setup
$y'' = f(y')$	$y' = u$ $y'' = u'$	$u' = f(u)$ • separation of variables $\int \frac{\mathrm{d}u}{f(u)} = x + c$ (solving for u: $u = u(x)$) • back substitution $y' = u(x)$ • integration $\int u(x)\,\mathrm{d}x$
$y'' = f(y)$	$y' = u$ $y'' = \frac{\mathrm{d}u}{\mathrm{d}y}\frac{\mathrm{d}y}{\mathrm{d}x}$ $= \frac{\mathrm{d}u}{\mathrm{d}y}u$	$u\frac{\mathrm{d}u}{\mathrm{d}y} = f(y)$ • separation of variables $u = \pm\sqrt{2\int f(y)\,\mathrm{d}y}$ • back substitution $y' = \pm\sqrt{2\int f(y)\,\mathrm{d}y}$ • integration by separation of variables
$y'' = f(y, y')$	$y' = u$ $y'' = \frac{\mathrm{d}u}{\mathrm{d}y}u$	$u\frac{\mathrm{d}u}{\mathrm{d}y} = f(y, u)$ further approach depends on the type of the function $f(y, u)$
$y'' = f(x, y')$	$y' = u$ $y'' = u'$	$u' = f(x, u)$ further approach depends on the type of the function $f(x, u)$

18.7.4 Linear differential equations of the second order with constant coefficients

This is the important special case $p(x) = p = \text{const.}$, $q(x) = q = \text{const.}$

$$y'' + py' + qy = f(x)$$

- The homogeneous linear differential equation of the second order with constant coefficients can always be solved by setting $y = \mathrm{e}^{\lambda x}$. There are three possible cases:
(1) $p^2 - 4q > 0$

$$y_1(x) = \exp\left\{\left(-\frac{p}{2} + \frac{1}{2}\sqrt{p^2 - 4q}\right)\cdot x\right\}$$

$$y_2(x) = \exp\left\{\left(-\frac{p}{2} - \frac{1}{2}\sqrt{p^2 - 4q}\right)\cdot x\right\}$$

(2) $p^2 - 4q = 0$

$$y_1(x) = e^{-\frac{p}{2}\cdot x} \qquad y_2(x) = x \cdot e^{-\frac{p}{2}\cdot x}$$

(3) $p^2 - 4q < 0$

$$y_1(x) = e^{-\frac{p}{2}x} \cdot \sin\left(x \cdot \frac{1}{2}\sqrt{4q - p^2}\right)$$

$$y_2(x) = e^{-\frac{p}{2}x} \cdot \cos\left(x \cdot \frac{1}{2}\sqrt{4q - p^2}\right)$$

- The inhomogeneous equation is solved by means of the method described above.

$$\lambda_1 = -\frac{p}{2} + \frac{1}{2}\sqrt{p^2 - 4q}\,; \quad \lambda_2 = -\frac{p}{2} - \frac{1}{2}\sqrt{p^2 - 4q}\,; \quad \lambda = -\frac{p}{2}\,;$$
$$\omega = \frac{1}{2}\sqrt{4q - p^2}$$

(1) $p^2 - 4q > 0$

$$y_0(x) = \frac{e^{\lambda_1 \cdot x}}{\lambda_1 - \lambda_2} \cdot \int e^{-\lambda_1 \cdot x} \cdot f(x)\mathrm{d}x + \frac{e^{\lambda_2 \cdot x}}{\lambda_2 - \lambda_1} \cdot \int e^{-\lambda_2 \cdot x} \cdot f(x)\mathrm{d}x$$

(2) $p^2 - 4q = 0$

$$y_0(x) = -e^{\lambda \cdot x} \cdot \int x e^{-\lambda \cdot x} \cdot f(x)\mathrm{d}x + x e^{\lambda \cdot x} \cdot \int e^{-\lambda \cdot x} \cdot f(x)\mathrm{d}x$$

(3) $p^2 - 4q < 0$

$$y_0(x) = \frac{1}{\omega} \cdot e^{-\lambda x} \cdot \sin(\omega \cdot x) \int e^{\lambda \cdot x} \cos(\omega x) f(x)\mathrm{d}x$$
$$- \frac{1}{\omega} \cdot e^{-\lambda x} \cdot \cos(\omega \cdot x) \int e^{\lambda \cdot x} \sin(\omega x) f(x)\mathrm{d}x$$

▷ Since the occurring integrals are always of type $\int e^{kx} \cdot F(x)\mathrm{d}x$ or $\int \sin(kx) \cdot F(x)\mathrm{d}x$, $\int \cos(kx) \cdot F(x)\mathrm{d}x$ the solutions can be found for many perturbation terms $f(x)$in Chapters 19 and 20 on Fourier transformations or Laplace transformations.

□ The damped harmonic oscillator:
Equation of motion for a spring with spring constant k, mass point m and damping β:

$$m\ddot{x} = -\beta\dot{x} - kx\,; \qquad p = \frac{\beta}{m}\,; \qquad q = \frac{k}{m}$$

$$p^2 - 4q = \frac{\beta^2}{m^2} - 4\frac{k}{m} = \frac{1}{m^2} \cdot \left(\beta^2 - 4km\right)$$

For weak damping, $\beta^2 < 4km$, the oscillation is damped,

$$\lambda = -\frac{p}{2} = -\frac{\beta}{2m}; \, \omega = \frac{1}{2}\sqrt{4q - p^2} = \sqrt{\frac{k}{m} - \left(\frac{\beta}{2m}\right)^2}$$

$$x(t) = e^{-\lambda \cdot t} \cdot (C_1 \cos\omega t + C_2 \sin\omega t)$$

For strong damping, $\beta^2 > 4km$, the "creeping case" exists,

$$k_1 = \frac{\beta}{2m} + \sqrt{\left(\frac{\beta}{2m}\right)^2 - \frac{k}{m}};\ k_2 = \frac{\beta}{2m} - \sqrt{\left(\frac{\beta}{2m}\right)^2 - \frac{k}{m}},\ (k_1, k_2 > 0)$$

$$x(t) = C_1 \cdot e^{-k_1 t} + C_2 \cdot e^{-k_2 t}$$

In the "aperiodic borderline case," $\beta^2 = 4km$, it holds that $\left(\lambda = \sqrt{\frac{k}{m}}\right)$

$$x(t) = C_1 \cdot e^{-\lambda t} + C_2 \cdot t \cdot e^{-\lambda t}$$

□ Electromagnetic oscillating circuit with capacitance C, inductance L, and ohmic resistance R:

$$L \cdot \ddot{Q} + R \cdot \dot{Q} + \frac{1}{C} \cdot Q = 0\,.$$

This is the same equation as in the previous example; it now holds that $\lambda = \frac{R}{2L}$,

$$\omega = \frac{1}{2} \cdot \sqrt{\frac{1}{LC} - \left(\frac{R}{2L}\right)^2},\ k_1 = \frac{R}{2L} + \sqrt{\left(\frac{R}{2L}\right)^2 - \frac{1}{LC}},$$

$$k_2 = \frac{R}{2L} - \sqrt{\left(\frac{R}{2L}\right)^2 - \frac{1}{LC}}\,.$$

□ Electromagnetic oscillating circuit with external voltage

$$U(t) = L \cdot \ddot{Q}(t) + R \cdot \dot{Q}(t) + \frac{1}{C} Q(t)$$

$$U(t) = U_0 \cdot \cos(2\pi f t), \quad R = 1\ \text{k}\Omega, \quad L = 500\ \text{mH},$$

$$C = 1\ \mu\text{F}, \quad U_0 = 10\ \text{V}, \quad f = 50\ \text{Hz}$$

Thus,

$$p = \frac{R}{L} = 2000\ \text{Hz}; \quad q = \frac{1}{LC} = 2 \cdot 10^6\ (\text{Hz})^2$$

$$p^2 - 4q = 4 \cdot 10^6\ \text{s}^{-2} - 8 \cdot 10^6\ \text{s}^{-2} = -4 \cdot 10^6\ \text{s}^{-2} < 0$$

$$\lambda = \frac{R}{2L} = 1000\ \text{Hz}$$

$$\omega = \sqrt{\frac{1}{LC} - \left(\frac{R}{2L}\right)^2} = \sqrt{2 \cdot 10^6\ \text{s}^{-2} - 10^6\ \text{s}^{-2}} = 10^3\ \text{s}^{-1}$$

$$= 1\ \text{kHz}$$

We know the solutions of the homogeneous system from the above calculations:

$$Q_1(t) = e^{-\lambda t} \cos\omega t \quad ; \quad Q_2(t) = e^{-\lambda t} \sin\omega t$$

$$Q_0(t) = \frac{e^{-\lambda t}}{\omega} \cdot \left\{ \sin\omega t \cdot \int \frac{U_0}{L} e^{-\lambda t} \cos\omega t \cos 2\pi f t\, dt \right.$$

$$\left. - \cos\omega t \cdot \int \frac{U_0}{L} e^{-\lambda t} \sin\omega t \cos 2\pi f t\, dt \right\}$$

$$= \frac{U_0 e^{-\lambda t}}{\omega L} \cdot \left\{ \sin \omega t \cdot e^{-\lambda t} \left[\frac{(\omega - 2\pi f)\sin(\omega - 2\pi f)t + \lambda \cos(\omega - 2\pi f)t}{2(\lambda^2 + (\omega - 2\pi f)^2)} \right. \right.$$

$$\left. + \frac{(\omega + 2\pi f)\sin(\omega + 2\pi f)t + \lambda \cos(\omega + 2\pi f)t}{2(\lambda^2 + (\omega + 2\pi f)^2)} \right]$$

$$- \cos \omega t \cdot e^{-\lambda t} \left[\frac{\lambda \sin(\omega - 2\pi f)t - (\omega - 2\pi f)\cos(\omega - 2\pi f)t}{2(\lambda^2 + (\omega - 2\pi f)^2)} \right.$$

$$\left. \left. + \frac{\lambda \sin(\omega + 2\pi f)t - (\omega - 2\pi f)\cos(\omega + 2\pi f)t}{2(\lambda^2 + (\omega + 2\pi f)^2)} \right] \right\}$$

$$= \frac{U_0}{2\omega L} \left[\left(\frac{\omega - 2\pi f}{\lambda^2 + (\omega - 2\pi f)^2} + \frac{\omega + 2\pi f}{\lambda^2 + (\omega - 2\pi f)^2} \right) \cdot \cos(2\pi f t) \right.$$

$$\left. + \left(\frac{1}{\lambda^2 + (\omega - 2\pi f)^2} + \frac{1}{\lambda^2 + (\omega + 2\pi f)^2} \right) \cdot \sin(2\pi f t) \right]$$

$$= 0.3554\mu\text{As} \cdot \cos(100\pi t/\text{s}) + 0.3042\mu\text{As} \cdot \sin(100\pi t/\text{s}) .$$

This corresponds to a current

$$I(t) = \dot{Q}(t) = -0.1117\text{mA} \cdot \sin(100\pi t/\text{s}) + 0.0956\text{mA} \cdot \cos(100\pi t/\text{s}) .$$

Since $\lambda = 1000\ \text{s}^{-1}$, the factor $e^{-\lambda t}$ occurring in Q_1 and Q_2 causes these contributions to decay by a factor of $\approx 5 \cdot 10^{-5}$ within 0.01 s; the circuit oscillates with only frequency f.

18.8 Differential equations of the n-th order

Reduction of the order: In certain cases a differential equation of the n-th order can be reduced to a differential equation of the $(n-1)$-th order by means of the substitution $y \to z = \frac{dy}{dx}$ (see section on differential equations of the second order).

Power series approach:

$$y(x) = a_0 + a_1 x + a_2 x^2 + \cdots + a_n x^n = \sum_{\nu=0}^{n} a_\nu x^\nu$$

Then, it holds that:

$$y'(x) = a_1 + 2a_2 x + \cdots + n a_n x^{n-1} = \sum_{\nu=0}^{n-1} (\nu + 1) a_{\nu+1} x^\nu$$

$$y''(x) = 2a_2 + 6a_3 x + \cdots + n(n-1) a_n x^{n-2} = \sum_{\nu=0}^{n-2} (\nu + 1)(\nu + 2) a_{\nu+2} x^\nu$$

$$\vdots$$

Substitution into the differential equation. A comparison of coefficients yields the values $a_0, a_1, \ldots$ Suitable approximations for small x and large n.

For **linear** differential equations, the limit $n \to \infty$ can often be executed. This results in an exact solution.

▷ Many important functions are defined in this manner.

(a) Hermitian differential equation

$$y'' - 2xy' + \lambda y = 0$$

Setup:

$$y(x) = a_0 + a_1 x + a_2 x^2 + a_3 x^3 + \cdots = \sum_{\nu=0}^{\infty} a_\nu x^\nu$$

$$y'(x) = a_1 + 2a_2 x + 3a_3 x^2 + \cdots = \sum_{\nu=0}^{\infty} (\nu+1) a_{\nu+1} x^\nu$$

$$y''(x) = 2a_2 + 3 \cdot 2a_3 x + 4 \cdot 3a_4 x^2 + \cdots$$
$$= \sum_{\nu=0}^{\infty} (\nu+1)(\nu+2) a_{\nu+2} x^\nu$$
$$\vdots$$

Substitution:

$$y'' - 2xy' + \lambda y$$
$$= \sum_{\nu=0}^{\infty} \left[(\nu+1)(\nu+2)a_{\nu+2} - 2x(\nu+1)a_{\nu+1} + \lambda a_\nu\right] x^\nu$$
$$= \sum_{\nu=0}^{\infty} \left[(\nu+1)(\nu+2)a_{\nu+2} - 2\nu a_\nu + \lambda a_\nu\right] x^\nu$$
$$= \sum_{\nu=0}^{\infty} \left[(\nu+1)(\nu+2)a_{\nu+2} + (\lambda - 2\nu) a_\nu\right] x^\nu$$

Comparison of coefficients:

$$a_{\nu+2} = \frac{2\nu - \lambda}{(\nu+1)(\nu+2)} a_\nu \; .$$

a_0 and a_1 can be chosen arbitrarily (integration constants!). The two fundamental solutions follow for $a_0 = 1$, $a_1 = 0$ as well as for $a_0 = 0$, $a_1 = 1$:

$$y_1 = 1 - \frac{\lambda}{2!}x^2 - \frac{(4-\lambda)\cdot\lambda}{4!}x^4 - \frac{(8-\lambda)(4-\lambda)\cdot\lambda}{6!}x^6 - \cdots$$

$$y_2 = x + \frac{2-\lambda}{3!}x^3 + \frac{(6-\lambda)(2-\lambda)}{5!}x^5$$
$$+ \frac{(10-\lambda)(6-\lambda)(2-\lambda)}{7!}x^7 + \cdots$$

Hermitian polynomials:

▷ If λ is an even integer, $\lambda = 0, 2, 4, 6, \ldots$, then the series terminates, and one obtains, alternating for y_1 and y_2, polynomials of degree n, the so-called "**Hermitian polynomials**" (following the accepted convention, we choose the polynomials in such a way that the coefficient of x^n is equal to 2^n):

$$H_0(x) = 1$$
$$H_1(x) = 2x$$
$$H_2(x) = 4x^2 - 2$$
$$H_3(x) = 8x^3 - 12x$$
$$H_4(x) = 16x^4 - 48x^2 + 12$$
$$\vdots$$

▷ For each $\lambda = 0, 2, 4, 6, \ldots$ there exists a second solution of the Hermitian differential equation, but these are not polynomials.

(b) Legendre differential equation

$$y'' - \frac{2x}{1-x^2}y' + \frac{\lambda(\lambda+1)}{1-x^2}y = 0$$

□ To solve by means of a power series setup, we multiply by $(1 - x^2)$:

$$(1 - x^2)y'' - 2xy' + \lambda(\lambda+1)y = 0$$

Setup: $y(x) = \sum\limits_{\nu=0}^{\infty} a_\nu x^\nu$

Substitution:

$$\sum_{\nu=0}^{\infty} \left[(\nu+2)(\nu+1)a_{\nu+2} - \nu(\nu-1)a_\nu - 2\nu a_\nu + \lambda(\lambda+1)a_\nu\right] x^\nu = 0$$

Coefficients (a_0, a_1 arbitrary):

$$a_{\nu+2} = \frac{\nu(\nu+1) - \lambda(\lambda+1)}{(\nu+2)(\nu+1)} a_\nu$$

Solutions (y_1 for $a_0 = 1, a_1 = 0$; y_2 for $a_0 = 0, a_1 = 1$):

$$y_1^{(\lambda)} = 1 - \frac{\lambda(\lambda+1)}{2!}x^2 + \frac{\lambda(\lambda-2)(\lambda+1)(\lambda+3)}{4!}x^4 - \frac{\lambda(\lambda-2)(\lambda-4)(\lambda+1)(\lambda+3)(\lambda+5)}{6!}x^6 \pm \cdots$$

$$y_2^{(\lambda)}(x) = x - \frac{(\lambda-1)(\lambda+2)}{3!}x^3 + \frac{(\lambda-1)(\lambda-3)(\lambda+2)(\lambda+4)}{5!}x^5 - \frac{(\lambda-1)(\lambda-3)(\lambda-5)(\lambda+2)(\lambda+4)(\lambda+6)}{7!}x^7 \pm \cdots$$

Legendre polynomials:

For $\lambda = 0, 2, 4, 6, \ldots$ **one** of the two solutions is a polynomial of degree $n = \frac{\lambda}{2}$. Then, the other solution is **not** a polynomial. Normalize the polynomials so that $L_n(x) = 1$ for $x = 1$. They are written:

$$P_0(x) = 1$$
$$P_1(x) = x$$
$$P_2(x) = \frac{1}{2} \cdot (3x^2 - 1)$$

$$P_3(x) = \frac{1}{2} \cdot (5x^3 - 3x)$$
$$P_4(x) = \frac{1}{8} \cdot (35x^4 - 30x^2 + 3)$$
$$\vdots$$

▷ The example of the Legendre differential equation is given for the case of $\lambda = 1$. Substitution into the equation yields:

$$\begin{aligned} y_1(x) &= 1 - x^2 + \frac{1 \cdot (-1) \cdot 2 \cdot 4}{4!} x^4 - \frac{1 \cdot (-1) \cdot (-3) \cdot 2 \cdot 4 \cdot 6}{6!} x^6 \\ &= 1 - x^2 - \frac{1}{3} x^4 - \frac{1}{5} x^6 \dots \\ &= x \cdot \left(-x - \frac{x^3}{3} + \frac{x^5}{5} - \frac{x^7}{7} + \cdots \right) + 1 \\ &= (-1) \cdot \left[\frac{x}{2} \cdot \left\{ 2 \cdot \left(x + \frac{x^3}{3} + \frac{x^5}{5} + \frac{x^7}{7} + \cdots \right) \right\} - 1 \right] \end{aligned}$$

The expression in braces is just the series expansion of $\ln[(1+x)(1-x)]$ (see Chapter 5 on functions); thus, the expression in the brackets is the second solution mentioned in the example. The first solution mentioned in the example is

$$y_2 = x - \frac{0 \cdot 2}{3!} x^3 + \frac{0 \cdot (-2) \cdot 2 \cdot 4}{5!} x^5 + \cdots = x\,.$$

(c) Bessel function of the first and second kind
Bessel differential equation:

$$\frac{d^2}{dx^2} J_n(x) + \frac{1}{x} \frac{d}{dx} J_n(x) + \left(1 - \frac{n^2}{x^2} \right) J_n(x) = 0\,.$$

Boundary condition: $\lim\limits_{x \to 0} J_n(x)$ should be finite.
Power series:

$$J_n(x) = x^k \cdot \left(a_0 + a_1 x + a_2 x^2 + \cdots \right) = x^k \sum_{\nu=0}^{n} a_\nu x^\nu\,.$$

As $x \to 0$, the following holds:

$$J_n(x) \to x^k \cdot a_0\,.$$

Substitution:

$$\begin{aligned} 0 &= \frac{d^2}{dx^2} J_n(x) + \frac{1}{x} \frac{d}{dx} J_n(x) + \left(1 - \frac{n^2}{x^2} \right) J_n(x) \\ &= k \cdot (k-1) \cdot a_0 x^{k-2} + \frac{1}{x} \cdot k \cdot a_0 x^{k-1} + a_0 x^k - a_0 n^2 x^{k-2} \\ &= a_0 \left(k^2 - k + k - n^2 \right) x^{k-2} + a_0 x^k \to a_0 \cdot \left(k^2 - n^2 \right) x^{k-2} \end{aligned}$$

Thus: $n = k$ ($n = -k$ excluded since otherwise $J_n(x)$ is divergent as $x \to 0$).
Thus:

$$\begin{aligned} J_n(x) &= x^n \cdot \left(a_0 + a_1 x + a_2 x^2 + \cdots \right) \\ J_n'(x) &= n x^{n-1} a_0 + (n+1) x^n a_1 + (n+2) x^{n+1} + a_2 \dots \\ J_n''(x) &= n(n-1) x^{n-2} a_0 + (n+1) n x^{n-1} a_1 + (n+2)(n+1) x^n a_2 + \cdots \end{aligned}$$

Substitution:

$$\begin{aligned}0 &= a_0 \cdot \left(n(n-1)x^{n-2} + \frac{1}{x} \cdot n \cdot x^{n-1} + x^n - n^2x^{n-2}\right)\\ &\quad + a_1 \cdot \left((n+1)nx^{n-1} + \frac{1}{x}(n+1)x^n + x^{n+1} - n^2x^{n-1}\right) + \ldots\\ &= x^{n-2} \cdot \left(a_0\left(n(n-1) + n - n^2\right)\right)\\ &\quad + x^{n-1} \cdot \left(a_1 \cdot \left(n(n+1) + (n+1) - n^2\right)\right)\\ &\quad + x^n \cdot \left(a_0 + a_2\left((n+2)(n+1) + (n+2) - n^2\right)\right)\\ &\quad + x^{n+1}\left(a_3\left((n+3)(n+2) + (n+3) - n^2\right) + a_1\right) + \cdots\end{aligned}$$

Every coefficient of x^ν must vanish, thus:

$$\begin{aligned}0 &= a_0 \cdot \left(n^2 - n + n - n^2\right) = a_0 \cdot 0\\ &\Rightarrow a_0 \quad \text{is arbitrary}\\ 0 &= a_1 \cdot \left(n(n+1) + n + 1 - n^2\right) = a_1 \cdot (2n+1)\\ &\Rightarrow a_1 = 0\\ 0 &= a_2 \cdot \left((n+2)(n+1) + (n+2) - n^2\right) + a_0\\ &\Rightarrow a_2 = -\frac{a_0}{4n+4}.\end{aligned}$$

Usually:

$$\begin{aligned}a_{m+2} \cdot \left((n+m+2)(n+m+1) + (n+m+2) - n^2\right) + a_m &= 0\\ a_{m+2} &= -\frac{a_m}{(m+2)(2n+m+2)}.\end{aligned}$$

Finally, continued substitution leads to:

$$a_{2m} = \frac{(-1)^m \cdot a_0}{2^{2m} \cdot m! \cdot \frac{(m+n)!}{n!}}.$$

If we choose $a_0 = (2^n \cdot n!)^{-1}$, then the **Bessel function of the first kind** follows:

$$\begin{aligned}J_n(x) &= \left(\frac{x}{2}\right)^n \cdot \sum_{m=0}^{\infty} \frac{(-1)^m}{m!(m+n)!} \cdot \left(\frac{x}{2}\right)^{2m}\\ &= \sum_{m=0}^{\infty} \frac{(-1)^m}{m!(m+n)!} \cdot \left(\frac{x}{2}\right)^{2m+n}\end{aligned}$$

For integer values of n:

$$J_{-n}(x) = (-1)^n J_n(x).$$

In this case a second, linearly independent solution of the Bessel differential equation is the **Bessel function of the second kind** $Y_n(x)$ defined by

$$Y_n(x) = \lim_{m \to n} \frac{J_m(x)\cos(m\pi) - J_{-m}(x)}{\sin(m\pi)}.$$

For positive integers n the general solution of the Bessel differential equation is

$$y(x) = c_1 J_n(x) + c_2 Y_n(x).$$

Bessel function of the first kind with half-integral order: Can be expressed by sine and cosine functions:

$$J_{1/2}(x) = \sqrt{\frac{2}{\pi x}} \sin x$$

$$J_{-1/2}(x) = \sqrt{\frac{2}{\pi x}} \cos x$$

$$J_{3/2}(x) = \sqrt{\frac{2}{\pi x}} \left(\frac{\sin x}{x} - \cos x\right)$$

$$J_{-3/2}(x) = \sqrt{\frac{2}{\pi x}} \left(\frac{\cos x}{x} + \sin x\right)$$

Spherical Bessel function $j_u(x)$:

$$j_u(x) = \sqrt{\frac{\pi}{2x}} J_{u+1/2} , \qquad n = 0, \pm 1, \pm 2, \ldots$$

(d) Gaussian differential equation also called (**hypergeometric differential equation**)

$$x \cdot (1 - x) \cdot \frac{\mathrm{d}^2 y}{\mathrm{d}x^2} + [c - (a + b + 1)x] \frac{\mathrm{d}y}{\mathrm{d}x} - aby = 0$$

Setup:

$$y(x) = x^\alpha \cdot \sum_{\nu=0}^{\infty} d_\nu x^\nu$$

Substitution:

$$0 = x(1 - x) \cdot x^\alpha \cdot \sum_{\nu=0}^{\infty} d_\nu (\nu + \alpha)(\nu + \alpha - 1) x^{\nu-2}$$
$$+ (c - (a + b + 1)x)\, x^\alpha \sum_{\nu=0}^{\infty} d_\nu (\nu + \alpha) x^{\nu-1} - abx^\alpha \sum_{\nu=0}^{\infty} d_\nu x^\nu .$$

Leads to the following:

$$d_0 \alpha (c + \alpha - 1) x^{\alpha-1}$$
$$+ \sum_{\nu=0}^{\infty} \Big(d_{\nu+1} \cdot (\nu + \alpha + 1)(\nu + \alpha + c)$$
$$- d_\nu \cdot (\nu + \alpha + a)(\nu + \alpha + b) \Big) x^{\nu+\alpha} = 0 .$$

Comparison of coefficients yields the **index equation**:

$$\alpha \cdot (c + \alpha - 1) = 0$$
$$\text{and} \quad d_{\nu+1} = \frac{(\nu + \alpha + a)(\nu + \alpha + b)}{(\nu + \alpha + c)(\nu + \alpha + 1)} \cdot d_\nu .$$

This yields:

$$\alpha = 0 \quad \text{or} \quad \alpha = 1 - c;$$
$$d_\nu = d_0 \cdot \frac{(a+\alpha)_\nu \cdot (b+\alpha)_\nu}{\nu! \cdot (c+\alpha)_\nu}$$
$$= d_0 \frac{a(a+1)\cdot\dots\cdot(a+\nu-1)\cdot b(b+1)\cdot\dots\cdot(b+\nu-1)}{\nu!\cdot c(c+1)\cdot\dots\cdot(c+\nu-1)}$$

with $(a)_\nu = a \cdot (a+1) \cdot \dots \cdot (a+\nu)$, $(a)_0 = 1$.
This results in:

$$y(x) = d_0 \cdot x^\alpha \sum_{\nu=0}^{\infty} \frac{(a+\alpha)_\nu \cdot (b+\alpha)_\nu}{\nu!(c+\alpha)_\nu} \cdot x^\nu$$
$$= d_0 \cdot x^\alpha \cdot {}_2F_1(a+\alpha, b+\alpha; c+\alpha; x),$$

where ${}_2F_1(p_1, p_2; q_1; x)$ denotes the hypergeometric function. Hence, the general solution of the Gaussian differential equation reads

$$y(x) = A \cdot {}_2F_1(a, b; c; x) + B \cdot x^{1-c} \cdot {}_2F_1(a+1-c, b+1-c; 1; x)$$

▷ The second solution applies to the case of $c \neq 2, 3, 4, \dots$. If $c = 2, 3, 4, \dots$, then a solution exists, also, but it has a much more complicated form. It must be determined from the first solution according to the method described in Section 18.7 on linear differential equations of the second order.

18.9 Systems of coupled differential equations of the first order

● Every ordinary differential equation of the n-th order can be changed into a system of differential equations of the first order. Therefore, the **systems of differential equations of the first order** represent the general case.

□ $y'' + 5y'^2 + 7y^5 = 6x^3$
Putting $z = y' \Rightarrow y'' = z'$,
we obtain the coupled system:

$$z' + 5z^2 + 7y^5 = 6x^3$$
$$y' - z = 0$$

□ $y^{(6)} + 8x^2 y^{(4)} + \ln y = 0$
Putting $a(x) = y'(x)$, $b(x) = a'(x) = y''(x)$, $c(x) = b'(x) = y'''(x)$, $d(x) = c'(x) = y^{(4)}(x)$, $e(x) = d'(x) = y^{(5)}(x)$, $f(x) = e'(x) = y^{(6)}(x)$.
It follows that:

$$f(x) + 8x^2 d(x) + \ln y = 0$$
$$a(x) - y'(x) = 0$$
$$b(x) - a'(x) = 0$$
$$\vdots$$
$$f(x) - e'(x) = 0$$

□ **Coupled oscillating circuit** (R_i = ohmic resistance, L_i = inductance, C_i = capacitance, $L_{1,2}$ = mutual inductance. Let the applied voltage be $U = U_0 \sin \omega t$.

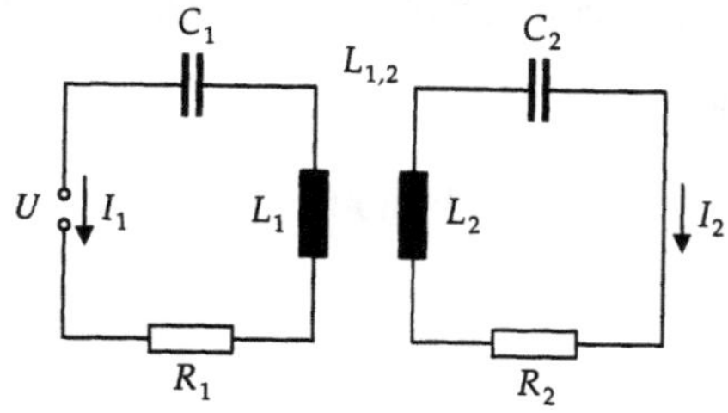

Coupled oscillating circuit

$$L_1\ddot{I}_1(t) + L_{1,2}\ddot{I}_2(t) + R_1\dot{I}_1(t) + \frac{1}{C_1}I_1(t) = U_0\omega\cos\omega t$$

$$L_2\ddot{I}_2(t) + L_{1,2}\ddot{I}_1(t) + R_2\dot{I}_2(t) + \frac{1}{C_2}I_2(t) = 0\,.$$

By substituting $f_1(t) = I_1(t)$; $f_2(t) = \dot{I}_1(t)$; $f_3(t) = I_2(t)$; $f_4(t) = \dot{I}_2(t)$, we obtain the system of coupled first-order differential equations:

$$\dot{f}_1(t) = f_2(t)$$

$$\dot{f}_2(t) = \frac{L_2}{L_1L_2 - L_{1,2}^2} \cdot \left(-\frac{1}{C_1}f_1(t) - R_1f_2(t) + \frac{L_{1,2}}{L_2C_2}f_3(t) + \frac{L_{1,2}R_2}{L_2}f_4(t) + U_0\omega\cos\omega t\right)$$

$$\dot{f}_3(t) = f_4(t)$$

$$\dot{f}_4(t) = \frac{L_1}{L_1L_2 - L_{1,2}^2} \cdot \left(\frac{L_{1,2}}{L_1C_1}f_1(t) + \frac{L_{1,2}R_1}{L_1}f_2(t) - \frac{1}{C_2}f_3(t) - R_2f_4(t) - \frac{L_{1,2}}{L_1}U_0\omega\cos\omega t\right)\,.$$

▷ We find in Section 18.12 on numerical methods that this procedure allows many practical applications. The most common algorithms for the numerical integration of differential equations are based on the method of reducing the order of differential equations.

▷ For analytic calculations the inverse of this method is more practical, i.e., a system of coupled differential equations can be transformed into a differential equation of higher order that can be solved by means of the known method. **However, this is usually not possible.**

18.10 Systems of linear homogeneous differential equations with constant coefficients

These systems are of the form:

$$y_1' = a_{11}y_1 + a_{12}y_2 + \cdots + a_{1n}y_n$$
$$\vdots$$
$$y_n' = a_{n1}y_1 + a_{n2}y_2 + \cdots + a_{nn}y_n$$

Setup: $y_n(x) = C_n \cdot e^{\lambda x} \Rightarrow y_n'(x) = \lambda C_n \cdot e^{\lambda x}$
Substitution:

$$\begin{aligned}(a_{11}-\lambda)y_1 + a_{12}y_2 + \cdots + a_{1n}y_n &= 0\\ a_{21}y_1 + (a_{22}-\lambda)y_2 + \cdots + a_{2n}y_n &= 0\\ &\vdots\\ a_{n1}y_1 + a_{n2}y_2 + \cdots + (a_{nn}-\lambda)y_n &= 0\,.\end{aligned}$$

Thus,

$$(\mathbf{A} - \lambda\mathbf{1})\,\vec{\mathbf{y}} = 0, \qquad \text{with} \quad \vec{\mathbf{y}} = \begin{pmatrix} y_1 \\ \vdots \\ y_n \end{pmatrix}.$$

- Nontrivial solutions exist only if $\det(\mathbf{A} - \lambda\mathbf{1}) \neq 0$.

Characteristic polynomial: Polynomial that results from multiplying the equation

$$\det(\mathbf{A} - \lambda\mathbf{1}) = 0\,.$$

□ The system of differential equations

$$\begin{aligned} y_1' &= y_1 \qquad\; + y_3\\ y_2' &= \qquad\; y_2 - y_3\\ y_3' &= 5y_1 + y_2 + y_3\end{aligned}$$

has the characteristic polynomial

$$P(\lambda) = (1-\lambda)(\lambda^2 - 2\lambda - 3) = -\lambda^3 + 3\lambda^2 + \lambda - 3\,.$$

This results in a polynomial of degree n for λ which possesses (usually complex) zeros $\lambda_1, \ldots, \lambda_n$. If these zeros are pairwise different, then the vectors

$$\vec{\mathbf{y}}_i = \vec{\mathbf{C}}_i \cdot e^{\lambda_i x}\,, \quad i = 1, \ldots, n$$

form a basis of the solution space. The various coefficient vectors $\vec{\mathbf{C}}_i$ determine themselves by substitution into the corresponding equation.

$$(\mathbf{A} - \lambda\mathbf{1})\,\vec{\mathbf{C}}_i = 0\,;$$

i.e.,

$$\begin{aligned}
(a_{11}-\lambda)C_i^{(1)} + a_{12}C_i^{(2)} + \cdots + a_{1n}C_i^{(n)} &= 0\\
a_{21}C_i^{(1)} + (a_{22}-\lambda)C_i^{(2)} + \cdots + a_{2n}C_i^{(n)} &= 0\\
&\vdots\\
a_{n1}C_i^{(1)} + a_{n2}C_i^{(2)} + \cdots + (a_{nn}-\lambda)C_i^{(n)} &= 0
\end{aligned}$$

Thus, the general solution reads:

$$\vec{\mathbf{y}}(x) = \sum_{i=1}^{n} B_i \cdot \vec{\mathbf{C}}_i \cdot e^{\lambda_i \cdot x},$$

where the B_i must be determined from the initial conditions.

□

$$\begin{aligned}
y_1' &= y_1 + y_3\\
y_2' &= y_2 - y_3\\
y_3' &= 5y_1 + y_2 + y_3
\end{aligned}$$

$$\det(\mathbf{A}-\lambda\mathbf{1}) = \begin{vmatrix} 1-\lambda & 0 & 1\\ 0 & 1-\lambda & -1\\ 5 & 1 & 1-\lambda \end{vmatrix} = (1-\lambda)(\lambda^2-2\lambda-3)\,.$$

Thus: $\lambda_1 = +1, \lambda_2 = -1, \lambda_3 = +3$.
The solution vectors follow from:

$$(\mathbf{A}-\lambda_i\mathbf{1})\,\vec{\mathbf{C}}_i = 0$$

$\lambda_1 = +1$:

$$\begin{pmatrix} 0 & 0 & 1\\ 0 & 0 & -1\\ 5 & 1 & 0 \end{pmatrix} \cdot \begin{pmatrix} C_1^{(1)}\\ C_1^{(2)}\\ C_1^{(3)} \end{pmatrix} = 0 \quad \Rightarrow \vec{\mathbf{C}}_1 = B_1 \cdot \begin{pmatrix} 1\\ -5\\ 0 \end{pmatrix}$$

$\lambda_1 = -1$:

$$\begin{pmatrix} 2 & 0 & 1\\ 0 & 2 & -1\\ 5 & 1 & 2 \end{pmatrix} \cdot \begin{pmatrix} C_2^{(1)}\\ C_2^{(2)}\\ C_2^{(3)} \end{pmatrix} = 0 \quad \Rightarrow \vec{\mathbf{C}}_2 = B_2 \cdot \begin{pmatrix} 1\\ -1\\ -2 \end{pmatrix}$$

$\lambda_1 = +3$:

$$\begin{pmatrix} -2 & 0 & 1\\ 0 & -2 & -1\\ 5 & 1 & -2 \end{pmatrix} \cdot \begin{pmatrix} C_3^{(1)}\\ C_3^{(2)}\\ C_3^{(3)} \end{pmatrix} = 0 \quad \Rightarrow \vec{\mathbf{C}}_3 = B_3 \cdot \begin{pmatrix} 1\\ -1\\ 2 \end{pmatrix}.$$

Thus, the general solution reads:

$$\begin{aligned}
y_1(x) &= B_1 e^x + B_2 e^{-x} + B_3 e^{3x}\\
y_2(x) &= -5B_1 e^x - B_2 e^{-x} - B_3 e^{3x}\\
y_3(x) &= -2B_2 e^{-x} + 2B_3 e^{3x}\,.
\end{aligned}$$

If the values of the three functions are given at $x = 0$, e.g., $y_1(0) = 1$, $y_2(0) = 2$, $y_3(0) = 3$, then the special solution follows from

$$\begin{pmatrix} 1 & 1 & 1 \\ -5 & -1 & -1 \\ 0 & -2 & 2 \end{pmatrix} \cdot \begin{pmatrix} B_1 \\ B_2 \\ B_3 \end{pmatrix} = \begin{pmatrix} 1 \\ 2 \\ 3 \end{pmatrix} .$$

Thus: $B_1 = -\frac{3}{4}$, $B_2 = \frac{1}{8}$, $B_3 = \frac{13}{8}$, and

$$\begin{aligned} y_1(x) &= -\frac{3}{4}e^x + \frac{1}{8}e^{-x} + \frac{13}{8}e^{3x} \\ y_2(x) &= \frac{15}{4}e^x - \frac{1}{8}e^{-x} - \frac{13}{8}e^{3x} \\ y_3(x) &= \qquad\quad -\frac{1}{4}e^{-x} + \frac{13}{4}e^{3x} \end{aligned}$$

18.11 Partial differential equations

Partial differential equation: Equation containing a function of several variables as well as derivatives of the function with respect to the variables,

$$F\left(y, x_1, x_2, \ldots, \frac{\partial y}{\partial x_1}, \frac{\partial y}{\partial x_2}, \ldots, \frac{\partial^2 y}{\partial x_1^2}, \frac{\partial^2 y}{\partial x_2^2}, \ldots\right) = 0$$

Order: The order of the highest occurring derivative of y.

□ $\frac{\partial^3 y}{\partial x_1 \partial x_2 \partial x_3} + x_1^5 \cdot \frac{\partial^3 y}{\partial x_1^2 \partial x_2} \cdot y - y^5 \cdot \frac{\partial y}{\partial x_3} + 3x_3^5 y = 0$ is of the third order.

Degree: The highest occurring power of y or of one of its derivatives.

□ $\frac{\partial^3 y}{\partial x_1 \partial x_3 \partial x_5} + 5x_1x_2x_3x_4x_5x_6y^2 + \left(\frac{\partial^2 y}{\partial x_4^2}\right)^6 = 0$ is of degree six.

Linear partial differential equation: Partial differential equation of the first degree.

Homogeneous partial differential equation: If no free term (independent of y or its derivatives) occurs.

□ The Laplace equation for the uncharged space

$$\Delta\varphi(x, y, z) = \frac{\partial^2 \varphi}{\partial x^2} + \frac{\partial^2 \varphi}{\partial y^2} + \frac{\partial^2 \varphi}{\partial z^2} = 0$$

is homogeneous (and linear).

Inhomogeneous partial differential equation: Terms independent of y and its derivatives occur.

□ The Poisson equation for a given charge distribution is inhomogeneous:

$$\Delta\varphi = \rho(\vec{\mathbf{x}}) = \frac{Q}{\frac{4}{3}\pi R^3} \cdot H(R - |\vec{\mathbf{x}}|) = \begin{cases} 3Q/4\pi R^3, \text{ if } |\vec{\mathbf{x}}| < R \\ 0 \text{ otherwise} \end{cases}$$

(homogeneously charged sphere of radius R, where H is the step function).

- **Integration constant:** In the case of partial differential equations, the integration constants are replaced by free functions of the independent variables.

□ Wave equation for oscillation systems in mechanical engineering and for electromagnetic waves

$$\frac{\partial^2 f(x,t)}{\partial x^2} - \frac{1}{c^2}\frac{\partial^2 f(x,t)}{\partial t^2} = 0 .$$

General solution:

$$f(x,t) = f_1(x+ct) + f_2(x-ct) .$$

▷ The case $c^2 < 0$, i.e., $c = \mathrm{j} \cdot a$ with $a \in \mathbb{R}$ is permissible!

□ $a \cdot \dfrac{\partial f(x,t)}{\partial x} + b \cdot \dfrac{\partial f(x,t)}{\partial t} = 0.$

Solution:

$$f(x,t) = g(b \cdot x - a \cdot t) .$$

▷ $a = 0$ and $b = 0$ are permissible!

18.11.1 Solution by separation

Separation setup: In many cases the setup

$$y(x_1, x_2, \ldots, x_n) = X_1(x_1) \cdot X_2(x_2) \cdot \cdots \cdot X_n(x_n)$$

is helpful (separation setup).

□ **Wave equation**

$$\frac{\partial^2 f(x,t)}{\partial x^2} - \frac{1}{c^2}\frac{\partial^2 f(x,t)}{\partial t^2} = 0 .$$

Boundary conditions: $y(0,t) = 0$, $y(L,t) = 0$.

Initial values: $y(x,0) = f(x)$; $\dfrac{\partial y}{\partial t}(x,t)|_{t=0} = 0$.

Setup: $y(x,t) = X(x) \cdot T(t)$.

Substitution:

$$T(t) \cdot \frac{\mathrm{d}^2 X}{\mathrm{d}x^2} - \frac{1}{c^2} X(x) \cdot \frac{\mathrm{d}^2 T}{\mathrm{d}t^2} = 0$$

$$\frac{\mathrm{d}^2 X}{\mathrm{d}x^2}/X = \frac{\mathrm{d}^2 T}{\mathrm{d}t^2}/(c^2 T) .$$

This can be satisfied only if both sides are equal to a constant value. The constants are $-k^2$.

Thus:

$$X''(x) + k^2 X(x) = 0 \quad ; \quad \ddot{T}(t) + c^2 k^2 T(t) = 0 .$$

Solution:

$$X(x) = A\cos kx + B\sin kx$$

$$T(t) = C\cos kct + D\sin kct .$$

The values of A, B, C, and D follow from the boundary and initial values:

$$y(x,t) = (A\cos kx + B\sin kx) \cdot (C\cos kct + D\sin kct) .$$

Boundary condition at $x = 0$:

$$0 = y(0, t) = A \cdot (C \cos kct + D \sin kct) ;$$

thus, $A = 0$.
Boundary condition at $x = L$:

$$0 = y(L, t) = B \sin kL \cdot (C \cos kct + D \sin kct) ;$$

thus, $B \neq 0$, since otherwise $y \equiv 0$. Thus, it follows that:

$$\sin kL = 0 \Rightarrow kL = \pi \cdot n \Rightarrow k = \frac{\pi \cdot n}{L}, n \in \mathbb{N} .$$

And, the solutions for y read:

$$y_n(x, t) = b_n \cdot \sin \frac{\pi n}{L} x \cdot \left(C \cos \frac{\pi n c}{L} t + D \sin \frac{\pi n c}{L} t \right) .$$

Initial condition:

$$\begin{aligned} 0 &= \frac{\partial y}{\partial t}(x, t)|_{t=0} \\ &= b_n \cdot \sin \frac{\pi n}{L} x \cdot \left(-\frac{\pi n c}{L} \cdot C \sin \frac{\pi n c}{L} t + \frac{\pi n c}{L} \cdot D \cos \frac{\pi n c}{L} t \right) |_{t=0} \\ &= b_n \cdot \frac{2\pi n c}{L} \cdot D \quad \Rightarrow D = 0 . \end{aligned}$$

Thus,

$$y_n(x, t) = b_n \cdot \sin \frac{\pi n}{L} x \cdot \cos \frac{\pi n c}{L} t .$$

General solution:

$$y(x, t) = \sum_{n=1}^{\infty} y_n(x, t) = \sum_{n=1}^{\infty} b_n \cdot \sin \frac{\pi n}{L} x \cdot \cos \frac{\pi n c}{L} t .$$

Initial condition:

$$f(x) = y(x, 0) = \sum_{n=1}^{\infty} b_n \cdot \sin \frac{\pi n}{L} x .$$

In Section 19.1 on Fourier series it will be shown that

$$b_n = \frac{1}{L} \int_0^{2L} f(x) \cdot \sin \left(\frac{\pi n}{L} x \right) \mathrm{d}x .$$

Thus, the solution reads:

$$y(x, t) = \sum_{n=1}^{\infty} \left[\frac{1}{L} \int_0^{2L} f(x) \cdot \sin \left(\frac{\pi n}{L} x \right) \mathrm{d}x \right] \cdot \sin \frac{\pi n}{L} x \cdot \cos \frac{\pi n c}{L} t .$$

□ **Laplace equation in spherical coordinates**
In Chapter 16, the Laplace operator Δ is represented in various coordinate systems. In spherical coordinates (r, ϑ, φ), the following holds:

$$\begin{aligned} \Delta &= \frac{\partial^2}{\partial r^2} + \frac{2}{r}\frac{\partial}{\partial r} + \frac{1}{r^2}\left(\frac{\partial^2}{\partial \vartheta^2} + \cot \vartheta \frac{\partial}{\partial \vartheta} + \frac{1}{\sin^2 \vartheta} \frac{\partial^2}{\partial \varphi^2} \right) \\ &= \frac{1}{r}\frac{\partial^2}{\partial r^2} \cdot r + \frac{1}{r^2 \sin^2 \vartheta} \frac{\partial}{\partial \vartheta} \left(\sin \vartheta \frac{\partial}{\partial \vartheta} \right) + \frac{1}{r^2 \sin^2 \vartheta} \frac{\partial^2}{\partial \varphi^2} . \end{aligned}$$

The Laplace equation $\Delta f(r, \vartheta, \varphi) = 0$ therefore reads:

$$0 = \frac{1}{r}\frac{\partial^2}{\partial r^2} \cdot (rf(r, \vartheta, \varphi)) + \frac{1}{r^2 \sin^2 \vartheta}\frac{\partial}{\partial \vartheta}\left(\sin \vartheta \frac{\partial f(r, \vartheta, \varphi)}{\partial \vartheta}\right) + \frac{1}{r^2 \sin^2 \vartheta}\frac{\partial^2 f(r, \vartheta, \varphi)}{\partial \varphi^2} .$$

Separation setup: $f(r, \vartheta, \varphi) = R(r) \cdot \Omega(\vartheta, \varphi)$

$$0 = \Omega(\vartheta, \varphi) \cdot \left\{ r \frac{\partial^2}{\partial r^2}(rR(r)) \right\} + R(r) \cdot \left\{ \frac{1}{\sin \vartheta}\left[\frac{\partial}{\partial \vartheta}\left(\sin \vartheta \frac{\partial \Omega(\vartheta, \varphi)}{\partial \vartheta}\right)\right] + \frac{1}{\sin^2 \vartheta}\frac{\partial^2 \Omega(\vartheta, \varphi)}{\partial \varphi^2} \right\} .$$

Thus,

$$\frac{r \frac{\partial^2}{\partial r^2}(rR(r))}{R(r)} = -\frac{\frac{\partial}{\partial \vartheta}\left(\sin \vartheta \frac{\partial \Omega(\vartheta, \varphi)}{\partial \vartheta}\right) + \frac{1}{\sin \vartheta}\frac{\partial^2 \Omega(\vartheta, \varphi)}{\partial \varphi^2}}{\sin \vartheta \, \Omega(\vartheta, \varphi)} = a .$$

Hence,

$$r \frac{\mathrm{d}^2}{\mathrm{d}r^2}(rR(r)) = aR$$

$$\frac{1}{\sin \vartheta}\frac{\partial}{\partial \vartheta}\left(\sin \vartheta \frac{\partial \Omega}{\partial \vartheta}\right) + \frac{1}{\sin^2 \vartheta}\frac{\partial^2 \Omega}{\partial \varphi^2} = -a\Omega(\vartheta, \varphi) .$$

Further separation: $\Omega(\vartheta, \varphi) = \Theta(\vartheta) \cdot \Phi(\varphi)$

$$\Phi(\varphi)\frac{1}{\sin \vartheta}\frac{\mathrm{d}}{\mathrm{d}\vartheta}\left(\sin \vartheta \frac{\mathrm{d}\Theta(\vartheta)}{\mathrm{d}\vartheta}\right) + a\Theta(\vartheta)\Phi(\varphi) = -\frac{\Theta(\vartheta)}{\sin^2 \vartheta} \cdot \frac{\mathrm{d}^2\Phi(\varphi)}{\mathrm{d}\varphi^2} .$$

Thus,

$$-\frac{\frac{\mathrm{d}^2\Phi(\varphi)}{\mathrm{d}\varphi^2}}{\Phi(\varphi)} = \frac{\sin \vartheta \frac{\mathrm{d}}{\mathrm{d}\vartheta}\left(\sin \vartheta \frac{\mathrm{d}\Theta(\vartheta)}{\mathrm{d}\vartheta}\right)}{\Theta(\vartheta)} + a \cdot \sin^2 \vartheta = b .$$

Thus,

$$\frac{\mathrm{d}^2\Phi(\varphi)}{\mathrm{d}\varphi^2} = -b\Phi(\varphi)$$

$$\frac{1}{\sin \vartheta}\frac{\mathrm{d}}{\mathrm{d}\vartheta}\left(\sin \vartheta \frac{\mathrm{d}\Theta(\vartheta)}{\mathrm{d}\vartheta}\right) - \frac{b}{\sin^2 \vartheta}\Theta(\vartheta) = -a\Theta(\vartheta) .$$

Since we must require $f(r, \vartheta, \varphi + 2\pi) = f(r, \vartheta, \varphi)$ (the solution must be unique), it follows that $b \geq 0$; that is, $b = m^2, m \in \mathbb{Z}$,

$$\Phi(\varphi) = C \cdot \mathrm{e}^{im\varphi}$$

Putting $a = l(l+1)$ (this is always possible), then

$$\frac{1}{\sin \vartheta}\frac{\mathrm{d}}{\mathrm{d}\vartheta}\left(\sin \vartheta \frac{\mathrm{d}\Theta(\vartheta)}{\mathrm{d}\vartheta}\right) - \frac{m^2}{\sin^2 \vartheta}\Theta(\vartheta) = -l(l+1)\Theta(\vartheta) .$$

With

$$\cos\vartheta = z \Rightarrow \sin\vartheta = \sqrt{1-z^2}, \quad \frac{\mathrm{d}}{\mathrm{d}\vartheta} = -\sqrt{1-z^2}\frac{\mathrm{d}}{\mathrm{d}z},$$

it follows that:

$$0 = \frac{1}{1-z^2}(-1)\sqrt{1-z^2}\frac{\mathrm{d}}{\mathrm{d}z}\left(\sqrt{1-z^2}\cdot(-1)\sqrt{1-z^2}\frac{\mathrm{d}}{\mathrm{d}z}\Theta(z)\right)$$
$$-\frac{m^2\Theta(z)}{1-z^2} = -l(l+1)\Theta(z),$$

i.e.,

$$(1-z^2)\frac{\mathrm{d}^2\Theta(z)}{\mathrm{d}z^2} - 2z\frac{\mathrm{d}\Theta(z)}{\mathrm{d}z} + \left\{l(l+1) - \frac{m^2}{1-z^2}\right\}\Theta(z) = 0.$$

For $m = 0$ this is exactly the Legendre differential equation. For $m \neq 0$, the differential equation is called the *associated* Legendre differential equation and its solutions are the *associated* Legendre **polynomials**. These can be calculated from the Legendre polynomials according to

$$P_l^m(x) = \left(1-x^2\right)^{m/2}\cdot\frac{\mathrm{d}^m}{\mathrm{d}x^m}P_l(x).$$

Thus:

$$\Omega(\vartheta,\varphi) = P_l^m(\cos\vartheta)\cdot \mathrm{e}^{im\varphi}.$$

Normalization:

$$\int \Omega^*_{l'm'}\Omega_{lm}\sin\vartheta\, d\vartheta\, d\varphi = \delta_{ll'}\delta_{mm'}.$$

Thus, putting:

$$\Omega(\vartheta,\varphi) = \sqrt{\frac{(2l+1)(l-m)!}{4\pi(l+m)!}}P_l^m(\cos\vartheta)\cdot \mathrm{e}^{im\varphi} = Y_{lm}(\vartheta,\varphi)$$

we obtain Y_{lm} as the surface spherical harmonics.
Radial equation:

$$l(l+1)R(r) = r^2\frac{\mathrm{d}^2R(r)}{\mathrm{d}r^2} + 2r\frac{\mathrm{d}R(r)}{\mathrm{d}r}.$$

Setup: $R(r) = r^\alpha$; this yields $l(l+1) = \alpha(\alpha+1)$; i.e., $\alpha = l$ or $\alpha = -l-1$.
Thus, the following holds:

$$f(r,\vartheta,\varphi) = \left(C_1 r^l + C_2 r^{-l-1}\right)\cdot Y_{lm}(\vartheta,\varphi).$$

18.12 Numerical integration of differential equations

18.12.1 Euler method

Approximation of the derivative by the difference quotient for explicit differential equations of the first order.

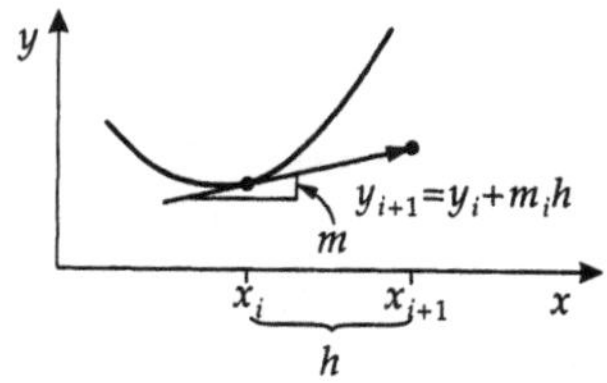

Euler method, h is step size,
m_i is slope

Geometric interpretation: The new function value is approximated by the tangent at the old function value.

$$y' = f(x, y) \quad \rightarrow \quad \frac{y(x_i + h) - y(x_i)}{h} \approx f(x_i, y(x_i)) + O(h)$$
$$\rightarrow \quad y(x_i + h) \approx y(x_i) + f(x_i, y(x_i)) \cdot h \,.$$

Error of approximation, $O(h)$: An infinitesimal quantity of order h, i.e., proportional to step size h.

Start with the initial condition $(x_0, y(x_0))$ and calculate the following function values $y(x_i) = y(x_0 + ih)$ according to the above equation.

Step size h can be bisected continuously until

$$\left| \frac{y(x_i + h/2) - y(x_i + h)}{y(x_i + h/2)} \right| < 10^{-n} \qquad (n = \text{ the number of significant places})$$

Advantage: The method is fast and simple to program.
Disadvantage: The method is not very reliable.

▷ The Euler method can easily lead to wrong results!

▷ Program sequence for the Euler method.

```
BEGIN Euler
INPUT x0, y0, xf, h
x := x0
y := y0
WHILE (x < xf ) DO
   y := y + h*f(x,y)
   x := x + h
   OUTPUT x, y
ENDDO
END Euler
```

18.12.2 Heun method

Predictor-corrector method: Use the Euler method to determine an estimated value (predictor step). In the corrector step the average values of the slope at the old position and those of the predictor step.

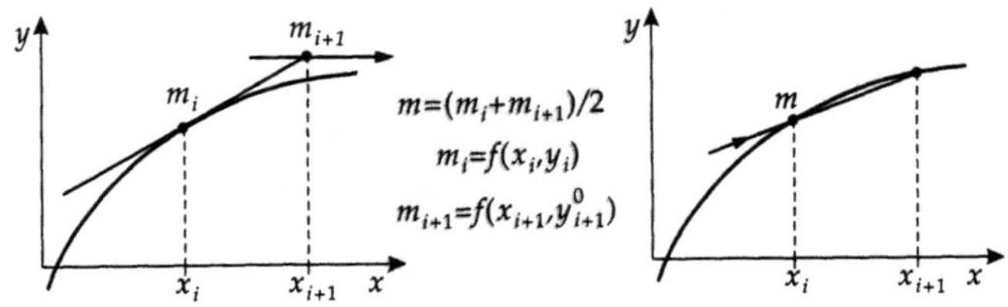

Heun method, $m_1 m_{ii}$ $m_i + 1$: Slope

Geometric interpretation: The new function value is approximated by the mean value of the tangents at the old and (preliminary) new function values.

$$\text{Predictor}: \quad y^0_{i+1} = y_i + f(x_i, y_i)h$$

$$\text{Corrector}: \quad y_{i+1} = y_i + \frac{h}{2}\left(f(x_i, y_i) + f(x_{i+1}, y^0_{i+1}\right)$$

Iteration: In the corrector step, y_{i+1} appears on both sides of the equation. Now, the old value y^0_{i+1} can be replaced iteratively by the new value on the right-hand side until the procedure converges; i.e.,

$$\left|\frac{y^j_{i+1} - y^{j-1}_{i+1}}{y^j_{i+1}}\right| < \varepsilon ,$$

or up to a required number of iterations j.

▷ This additional iteration does not always lead to better results, particularly in the case of large step sizes h.

▷ Program sequence for the Heun method.

```
BEGIN Heun
INPUT x0, y0, h, eps, imax, xf
x := x0
y := y0
WHILE ( x < xf ) DO
   s := y + h*f(x,y) (predictor)
   x1 := x + h
   y1 := y + s*h
   i := 0
   del := 2*eps
   WHILE ( del > eps AND i < imax ) DO (corrector)
      i := i + 1
      s1 := f(x1,y1)
      yneu := y + h*(s + s1)/2
      del := abs((yneu - y1/yneu)*100
      y1 := yneu
   ENDDO
   IF ( del > eps ) THEN
      OUTPUT ''no convergence''
   ENDIF
   x := x1
   y := yneu
   OUTPUT x, y
```

```
ENDDO
END Heun
```

18.12.3 Modified Euler method

Geometric interpretation: To calculate the next function value, use the slope at intermediate point $x_{i+1/2}$ instead of the slope at point x_i.

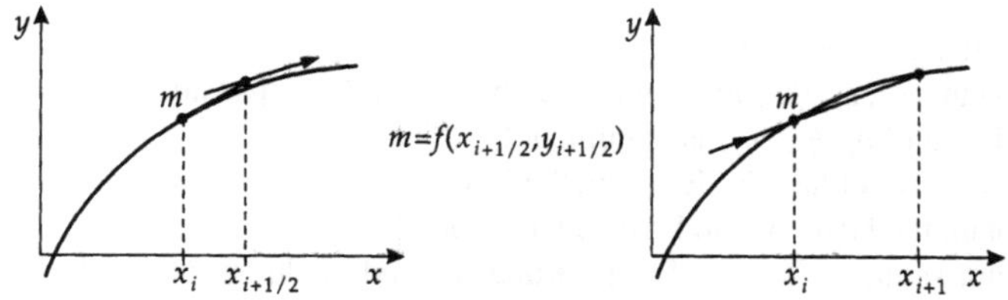

Modified Euler method; m = slope

$$y_{i+1/2} = y_i + h \cdot f(x_i, y_i)$$
$$y'_{i+1/2} = f(x_{i+1/2}, y_{i+1/2})$$
$$y_{i+1} = y_i + h \cdot f(x_{i+1/2}, y_{i+1/2})$$

- This method is more precise than the classical Euler method because it uses the derivative in the middle of the interval, which often represents a better approximation of the mean value of the derivatives in the interval.

▷ Since y_{i+1} no longer appears on both sides of the equation, an iteration is no longer possible.

▷ Program sequence for the modified Euler method.

```
BEGIN Euler (mod.)
INPUT x0, y0, xf, h
x := x0
y := y0
WHILE (x < xf ) DO
   x1 := x + h/2
   y1 := y + f(x,y)*h/2
   y := y + h*f(x1,y1)
   x := x + h
   OUTPUT x, y
ENDDO
END Euler (mod.)
```

18.12.4 Runge-Kutta methods

The best compromise between programming effort, calculating time, and numerical accuracy is the **Runge-Kutta method** of the fourth order.

Principle: Calculation of a "representative slope" of the function in the interval (averaging the slopes at various extrapolated points),

$$s(x_i) = a_1 \cdot k_1 + a_2 \cdot k_2 + \cdots + a_n \cdot k_n ,$$

with the function values

$$\begin{aligned}
k_1 &= f(x_i, y_i)\\
k_2 &= f(x_i + h\xi_1, y_i + h\psi_{1,1}k_1)\\
k_3 &= f(x_i + h\xi_2, y_i + h\psi_{2,1}k_1 + h\psi_{2,2}k_2)\\
&\vdots\\
k_n &= f(x_i + h\xi_n, y_i + h\psi_{n-1,1}k_1 + \cdots + h\psi_{n-1,n-1}k_n)\,.
\end{aligned}$$

New y value: $y : +1 = h_s(x_i)$.

The constants $a_i, \xi_i, \psi_{i,j}$ are determined by the Taylor expansion of f.

▷ The choice of $a_i, \xi_i, \psi_{i,j}$ is not unique!

● For $n = 1$ we obtain the Euler method; for $n = 2, a_1 = a_2 = 1/2, \xi_1 = \psi_{1,1} = 1$ we obtain the Heun method with a single corrector step.

▷ Program sequence for the Runge-Kutta method of the second order according to Ralston.

```
BEGIN Ralston
INPUT x0, y0, xf, h
x := x0
y := y0
WHILE (x < xf ) DO
   k1 := f(x,y)
   x1 := x + h*3/4
   y1 := y + k1*h*3/4
   k2 := f(x1,y1)
   y := y + k1/3 + 2*k2/3
   x := x + h
   OUTPUT x, y
ENDDO
END Ralston
```

Classical Runge-Kutta method of the fourth order: The next value is calculated according to

$$\begin{aligned}
y_{i+1} &= y_i + \frac{h}{6}(k_1 + 2k_2 + 2k_3 + k_4) + \hat{}(h^4)\\
k_1 &= f(x_i, y_i)\\
k_2 &= f\left(x_i + \frac{1}{2}h, y_i + \frac{1}{2}hk_1\right)\\
k_3 &= f\left(x_i + \frac{1}{2}h, y_i + \frac{1}{2}hk_2\right)\\
k_4 &= f\left(x_i + h, y_i + hk_3\right)
\end{aligned}$$

Adaptation of step size: The step size can be bisected continuously until again

$$\left|\frac{y(x_i + h/2) - y(x_i + h)}{y(x_i + h/2)}\right| < 10^{-n} \qquad (n = \text{ number of significant places})$$

or a fixed value is chosen before.

▷ Program sequence for the classical Runge-Kutta method of the fourth order.

```
BEGIN Runge-Kutta
INPUT x0, y0, xf, h
x := x0
y := y0
WHILE (x < xf ) DO
   k1 := f(x,y)
   x1 := x + h/2
   y1 := y + k1*h/2
   k2 := f(x1,y1)
   x2 := x1
   y2 := y + k2*h/2
   k3 := f(x2,y2)
   x3 := x + h
   y3 := y + k3*h
   k4 := f(x3,y3)
   y := y + (1/6)*(k1 + 2*k2 + 2*k3 + k4)*h
   x := x + h
   OUTPUT x, y
ENDDO
END Runge-Kutta
```

Adaptation of step size for the Runge-Kutta method of the fourth order: In the case of the classical Runge-Kutta method of the fourth order a rule to determine the optimal step size can be given. The following Q factor must be calculated:

$$Q = \left| \frac{k_3 - k_2}{k_2 - k_1} \right|$$

If the step size is too large, then the chosen approximation is too far from the "true" solution, and the more we proceed with the calculation, the farther away we get from the "true" solution. Therefore, Q should be <0.1. On the other hand, the step size must not be too small because then the rounding errors accumulate too much, i.e., Q should be > 0.025. Algorithm to determine the optimal step size:

1. Start the first calculation step with the chosen step size h.
2. Calculate the Q factor.
3. If Q >0.1, put $h \leftarrow h/2$ and repeat the last step.
4. If Q <0.025, accept the calculated approximation, but in the next step we put $h \leftarrow h \cdot 2$.
5. If $0.025 \leq Q \leq 0.1$, continue with the same step size.

▷ Program sequence for the classical Runge-Kutta method of the fourth order with adapted step size.

```
BEGIN Runge-Kutta method with adaption of step size
INPUT x0, y0, xf, h
x := x0
y := y0
WHILE (x < xf ) DO
   k1 := f(x,y)
   x1 := x + h/2
   y1 := y + k1*h/2
   k2 := f(x1,y1)
   x2 := x1
   y2 := y + k2*h/2
   k3 := f(x2,y2)
   q := ABS((k3-k2)/(k2-k1))
   IF ( q < 0.1 ) THEN
      x3 := x + h
      y3 := y + k3*h
      k4 := f(x3,y3)
      y := y + (1/6)*(k1 + 2*k2 + 2*k3 + k4)*h
      x := x + h
      OUTPUT x, y
      IF ( q < 0.025 ) THEN
         h := h*2
      ENDIF
   ELSE
      h := h/2
   ENDIF
ENDDO
END Runge-Kutta method with adaption of step size.
```

▷ A frequently used method for the numerical integration of ordinary differential equations is the **Stoer-Bulirsch-Gragg method**. If accuracy requirements are not too high, the **Dormand-Prince-method** can also be used.

▷ Program sequence for the Dormand-Prince method of the fifth order with step-size adaptation; may also be transferred to systems of differential equations. For error limits of 10^{-4} and 10^{-7} this is the best method of this type presently known

```
BEGIN DORMAND-PRINCE
INPUT x0,y0,x,f,h
INPUT eps {error limit for local truncation error;
                                   10^-7 ≤ eps ≤ 10^-4 }
{determination of machine precision }
uround:=1.0E-6;
WHILE((1+uround)>1) DO uround:=uround*0.1;
{initialization of variables}
```

$c[1]:=0;$

$c[2]:=\frac{1}{5};\quad a[2,1]:=\frac{1}{5};$

$c[3]:=\frac{3}{10};\quad a[3,1]:=\frac{3}{40};\quad a[3,2]:=\frac{9}{40};$

$c[4]:=\frac{4}{5};\quad a[4,1]:=\frac{44}{45};\quad a[4,2]:=\frac{-56}{15};\quad a[4,3]:=\frac{32}{9};$

$c[5]:=\frac{8}{9};\quad a[5,1]:=\frac{19372}{6561};\quad a[5,2]:=\frac{-25360}{2187};\quad a[5,3]:=\frac{64448}{6561};\quad a[5,4]:=\frac{-212}{729};$

$c[6]:=1;\quad a[6,1]:=\frac{9017}{3168};\quad a[6,2]:=\frac{-355}{33};\quad a[6,3]:=\frac{46732}{5247};\quad a[6,4]:=\frac{49}{176};$

$a[6,5]:=\frac{-5103}{18656};$

$c[7]:=1;\quad a[7,1]:=\frac{35}{384};\quad a[7,2]:=0;\quad a[7,3]:=\frac{500}{1113};\quad a[7,4]:=\frac{125}{192};$

$a[7,5]:=\frac{-2187}{6784};\quad a[7,6]:=\frac{11}{84};$

$b[1]:=\frac{71}{57600};\quad b[2]:=0;\quad b[3]:=\frac{-71}{16695};\quad b[4]:=\frac{71}{1920};\quad b[5]:=\frac{-17253}{339200};$

$b[6]:=\frac{22}{525};\quad b[7]:=\frac{-1}{40};$

```
x:=x0;
FAIL:=FALSE;
accepted:=TRUE;
maxit:=0;
REPEAT
{embedded Runge-Kutta method according to Dormand-Prince}
dy:=0;
   FOR j:=1 TO 7 DO
      dy:=0;
FOR i:=1 TO j-1 DO
         dy0:=dy0+a[j,i]*k[i]
      ENDDO
k[j]:=f(x+c[j]*h,y0+h*dy0);
dy:=dy+b[j]*k[j];
   ENDDO
y1:=y0+h*dy0;
{estimate local error; fix new step size}
denominator:=max(1.0E-5,abs(y1),abs(y0),2*uround/eps);
error:=ABS(h*dy/denominator);
```

$\texttt{factor:=max(0.1,min(5,1.1}\sqrt[5]{\frac{\texttt{error}}{\texttt{eps}}}\ \texttt{)}$

```
hnew:=h/factor;
IF(error<eps) THEN
      y0:=y1;
x:=x+h;
OUTPUT x,y0;
IF (NOT accepted) THEN hnew:=min(h,hnew) ENDIF
accepted:=TRUE;
```

```
maxit:=0
   ELSE    accepted:=FALSE;
maxit:=maxit+1;
FAIL:=(maxit=20) or (x+0.1*h=x);
   ENDIF;
h:=hnew;
UNTIL(x>xf) OR FAIL;
END DORMAND-PRINCE 5(4)
```

▷ As a check the equations

$$c[j] = \sum_{i=1}^{j-1} a[j,i], \quad j = 1, \ldots, 7 \quad \text{and} \quad \sum_{i=1}^{7} b[j] = 0\,.$$

can be used.

General method: program flow:

1. Define the right side of the differential equation $f(x, y)$ in a function routine for arbitrary (x, y) values.
2. Read in the initial condition (x_0, y_0) and the step size h.
3. Calculate the next values y_{i+1} in a loop over i, using the above formulas.
4. Present the result graphically or in a table.
5. Test the program with an analytically solvable differential equation. If necessary, bisect the step size h.

TABLE 18.1. Table of numerical methods for the integration of ordinary differential equations.

Method	Starting values	Iterative	Error	Programming effort	Remarks
Euler	1	no	$O(h)$	small	for fast estimate
Heun	1	yes	$O(h^2)$	moderate	
Euler, modified	1	no	$O(h^2)$	small	
Runge-Kutta, second order	1	no	$O(h^2)$	moderate	minimizes the rounding error
Runge-Kutta, fourth order	1	no	$O(h^4)$	moderate	often used
Dormand-Prince fifth order	1	no	$O(h^5)$	moderate	best known explicit Runge-Kutta method; accuracy requirements are not too high

18.12.5 Runge-Kutta method for systems of differential equations

The following system of first-order differential equations is given:

$$\frac{\mathrm{d}}{\mathrm{d}t}x = f(x, y, t)\,, \quad \frac{\mathrm{d}}{\mathrm{d}t}y = g(x, y, t)\,, \quad x(t_0) = x_0,\ y(t_0) = y_0\,.$$

Apply the fourth-order Runge-Kutta method for both differential equations:

$$\begin{aligned}
x_{i+1} &\approx x_i + \frac{1}{6}(k_1 + 2k_2 + 2k_3 + k_4)\\
k_1 &= hf(x_i, y_i, t_i)\\
k_2 &= hf\left(x_i + \frac{k_1}{2}, y_i + \frac{l_1}{2}, t_i + \frac{h}{2}\right)\\
k_3 &= hf\left(x_i + \frac{k_2}{2}, y_i + \frac{l_2}{2}, t_i + \frac{h}{2}\right)\\
k_4 &= hf(x_i + k_3, y_i + l_3, t_i + h)\\
y_{i+1} &\approx y_i + \frac{1}{6}(l_1 + 2l_2 + 2l_3 + l_4)\\
l_1 &= hg(x_i, y_i, t_i)\\
l_2 &= hg\left(x_i + \frac{k_1}{2}, y_i + \frac{l_1}{2}, t_i + \frac{h}{2}\right)\\
l_3 &= hg\left(x_i + \frac{k_2}{2}, y_i + \frac{l_2}{2}, t_i + \frac{h}{2}\right)\\
l_4 &= hg(x_i + k_3, y_i + l_3, t_i + h)
\end{aligned}$$

This method can also be applied for differential equations of the second order, since these can be transformed to systems of first-order differential equations:

$$\begin{aligned}
&y'' + f(x, y)y' + g(x, y)y = h(x, y)\\
&\rightarrow \quad y' = u, \quad u' = -f(x, y)u - g(x, y)y + h(x, y)
\end{aligned}$$

18.12.6 Difference method for the solution of partial differential equations

Difference method: Short explanation, using the example of the Laplace equation.

□ Temperature of a homogeneous plate whose boundaries are maintained at temperatures T_l, T_r, T_o, T_u:

$$\frac{\partial^2 T(x, y)}{\partial x^2} + \frac{\partial^2 T(x, y)}{\partial y^2} = 0\,.$$

● This equation also describes the potential curve in a square plane whose boundaries are maintained at potentials $\varphi_l, \varphi_r, \varphi_o, \varphi_u$.

Grid: Function $T(x, y)$ is represented by its values at discrete grid points x_i, y_j:

$$T(x, y) \rightarrow T(x_i, y_j) \equiv T_{ij}\,.$$

Difference quotient: The differential quotient is approximated by the difference quotient:

$$\frac{\partial T(x, y)}{\partial x} \approx \frac{T(x + \Delta x, y) - T(x, y)}{\Delta x}\,,$$

on the grid:

$$\frac{\partial T(x,y)}{\partial x} \approx \frac{T_{i+1,j} - T_{i,j}}{\Delta x},$$
$$\frac{\partial T(x,y)}{\partial y} \approx \frac{T_{i,j+1} - T_{i,j}}{\Delta y},$$
$$\frac{\partial^2 T(x,y)}{\partial x^2} \approx \frac{1}{\Delta x}\cdot\left(\frac{T_{i+1,j} - T_{i,j}}{\Delta x} - \frac{T_{i,j} - T_{i-1,j}}{\Delta x}\right) = \frac{T_{i+1,j} - 2T_{i,j} + T_{i-1,j}}{\Delta x^2},$$
$$\frac{\partial^2 T(x,y)}{\partial y^2} \approx \frac{1}{\Delta y}\cdot\left(\frac{T_{i,j+1} - T_{i,j}}{\Delta y} - \frac{T_{i,j} - T_{i,j-1}}{\Delta y}\right) = \frac{T_{i,j+1} - 2T_{i,j} + T_{i,j-1}}{\Delta y^2}.$$

Laplace equation on the grid ($\Delta x = \Delta y$):

$$\frac{T_{i+1,j} - 2T_{i,j} + T_{i-1,j}}{\Delta x^2} + \frac{T_{i,j+1} - 2T_{i,j} + T_{i,j-1}}{\Delta y^2} = 0.$$

Difference equation for the Laplace equation on the grid:

$$T_{i+1,j} + T_{i-1,j} + T_{i,j+1} + T_{i,j-1} - 4T_{i,j} = 0 \qquad i,j = 1,\ldots,N.$$

Boundary values: The values $T_{0,j}, T_{i,0}, T_{i,N+1}, T_{N+1,j}$ are the given boundary values (e.g., the temperature at the boundaries of the plate or the potential at the boundaries of the region).

● The last equation is a system of linear equations of N^2 equations for the N^2 unknowns $T_{i,j}$, which can be solved according to known methods.

Liebmann method: Iterative solution of the system of equations for T by solving for $T_{i,j}$:

$$T_{i,j} = \frac{T_{i+1,j} + T_{i-1,j} + T_{i,j+1} + T_{i,j-1}}{4},$$

and substituting the values obtained for the calculation of the following $T_{i,j}$ (all values not yet calculated are put equal to zero for the time being). Multiple iteration **can** lead to convergence of the method.

□

$$T_{1,1} = \frac{0 + T_{0,1} + 0 + T_{1,0}}{4}$$
$$T_{2,1} = \frac{0 + T_{1,1} + 0 + T_{2,0}}{4}$$
$$T_{1,2} = \frac{0 + T_{0,2} + 0 + T_{1,1}}{4}$$
$$\vdots$$

Confine this procedure until all values of $T_{i,j}$ are calculated. Then, start with the values thus calculated again until the required accuracy is reached, i.e., until the change of values in the next step is smaller than the required toleration limit ε.

▷ This method must not always converge!

Relaxation accelerate convergence. Calculate a modified value $\tilde{T}_{i,j}^{\text{new}}$ from every determined value $T_{i,j}^{\text{new}}$ by means of a relaxation factor λ, $1 \le \lambda \le 2$ according to

$$T_{i,j}^{\text{new}} \to \tilde{T}_{i,j}^{\text{new}} = \lambda T_{i,j}^{\text{new}} + (1-\lambda) T_{i,j}^{\text{old}}.$$

□ Square plate with a fixed temperature at the boundary ($N = 3$):

			200°C			
	□	$T_{1,4}$	$T_{2,4}$	$T_{3,4}$	□	
	$T_{0,3}$	$T_{1,3}$	$T_{2,3}$	$T_{3,3}$	$T_{4,3}$	
150°C	$T_{0,2}$	$T_{1,2}$	$T_{2,2}$	$T_{3,2}$	$T_{4,2}$	100°C
	$T_{0,1}$	$T_{1,1}$	$T_{2,1}$	$T_{3,1}$	$T_{4,1}$	
	□	$T_{1,0}$	$T_{2,0}$	$T_{3,0}$	□	
			0°C			

Heated plate

Boundary values: $T_{i,4} = 200°\text{C}$, $T_{4,j} = 100°\text{C}$, $T_{0,j} = 150°\text{C}$, $T_{i,0} = 0°\text{C}$.

$$T_{1,1} = \frac{0 + 150 + 0 + 0}{4} = 37.5$$

Relaxation ($\lambda = 1.5$)

$$\tilde{T}_{1,1} \leftarrow 1.5 \cdot 37.5 + (1 - 1.5) \cdot 0 = 56.25$$

$$T_{2,1} = \frac{0 + 56.25 + 0 + 0}{4} = 14.0625$$

Relaxation ($\lambda = 1.5$)

$$\tilde{T}_{2,1} \leftarrow 1.5 \cdot 14.0625 + (1 - 1.5) \cdot 0 = 21.09375$$

etc.

After the first iteration:

□	200	200	200	□
150	160.3	148.9	194.0	100
150	77.34	36.91	68.37	100
150	56.25	21.09	45.41	100
□	0	0	0	□

After the ninth iteration the error is smaller than 1%:

□	200	200	200	□
150	157.2	152.1	139.4	100
150	126.4	112.2	104.7	100
150	68.0	66.6	67.8	100
□	0	0	0	□

18.12.7 Method of finite elements

This is a short explanation of the basic idea of this method which is of extreme importance in the mathematics of engineering.

Finite elements: The domain of definition of the required function is divided into small subdomains on which the required function is approximated (e.g., by polynomials).

- In the last section, the method of finite differences was described for a square grid. It is difficult to use it for an irregular grid. But especially in technology, it is important to choose a grid as dense as possible where the function changes greatly (where it changes only slightly, the grid should be less dense to save memory and to keep the rounding errors small). It must also be possible to adapt the grid during calculation. This is much easier with the method of finite elements.

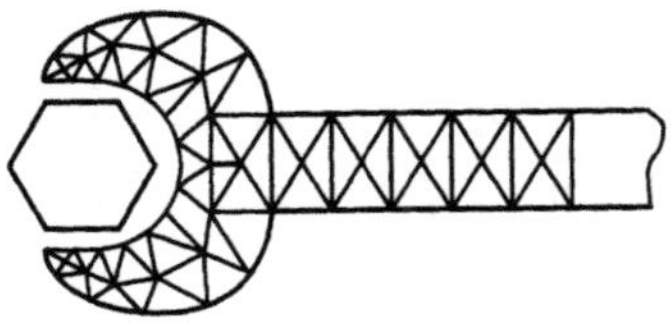

Example of decomposing of an object into finite elements

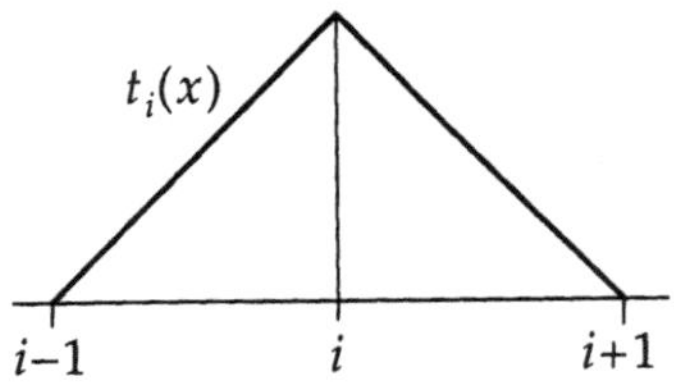

Example of a test function in element i

Partition: In case of one dimension a division of the interval is a partition (for simplicity, we chose the interval [0, 1]); often it holds that:

$$0 = x_0 < x_1 < \cdots < x_N < x_{N+1} = 1 .$$

Test function: Function by which the required function is approximated in each of the elements (= subintervals of the partition). The shape of the test function should be as simple as possible, and should be nonzero for as few elements as possible.

Representation of the function $u(x)$ as a sum over all elements:

$$u(x) = \sum_{\text{Element } i} u_i \cdot t_i(x)$$

Continuity: The test functions first contain indefinite parameters, which are fixed by the requirement of continuity of $u(x)$. (If we choose more complicated test functions than those sketched, we can also require the differentiability of u, etc.)

▷ What remains to be done is the determination of the coefficients. When these are calculated, the solution of the investigated differential equation is given by the above representation of u.

Ritz variational method: The determination of an approximate solution in terms of an optimal approximation of the "true" solution by functions of an appropriate function space (e.g., polynomials or angle functions, etc.).

Associated functional of the differential equation: Determined by multiplying by a general test function and integration over the domain (= the domain of definition). Then, the substitution of the representation of u into the equation for the functional of the differential equation leads to a system of linear equations for the determination of the coefficients u_i.

- Thus, the solution of the differential equation can be determined by solving a system of linear equations.

▷ Because the test function is chosen in such a way that it vanishes for as many elements as possible, the corresponding matrix is usually a narrow-band matrix, which, for example, can be solved by using the **Gauss-Seidel method**.

□ Let us look at the problem of an oscillating string (with an external load),

$$-pu''(x) + qu(x) = f(x) .$$

Rewrite into a functional ($v(x)$ test function):

$$\int_0^1 \left(-pu''(x) + qu(x)\right) \cdot v(x)\mathrm{d}x = \int_0^1 f(x) \cdot v(x)\mathrm{d}x$$

$$-p \cdot \int_0^1 u''(x)v(x)\mathrm{d}x + q \cdot \int_0^1 u(x)v(x)\mathrm{d}x = \int_0^1 f(x) \cdot v(x)\mathrm{d}x$$

$$\underbrace{-pu'(x)v(x)\Big|_0^1}_{=0} +$$

$$p \cdot \int_0^1 u'(x)v'(x)\mathrm{d}x + q \cdot \int_0^1 u(x)v(x)\mathrm{d}x = \int_0^1 f(x) \cdot v(x)\mathrm{d}x .$$

Put

$$u(x) = \sum_{i=1}^{N} u_i \cdot t_i(x) , \quad v(x) = t_k(x) , \quad k = 1, \ldots, N .$$

Thus:

$$p \cdot \int_0^1 \sum_{i=1}^{N} u_i t_i'(x) t_k'(x)\mathrm{d}x + q \cdot \int_0^1 \sum_{i=1}^{N} u_i t_i(x) t_k(x)\mathrm{d}x = \int_0^1 f(x) \cdot t_k(x)\mathrm{d}x$$

$$\sum_{i=1}^{N} u_i \cdot \underbrace{\left(p \cdot \int_0^1 t_i'(x)t_k'(x)\mathrm{d}x + q \cdot \int_0^1 t_i(x)t_k(x)\mathrm{d}x\right)}_{a_{ki}} = \underbrace{\int_0^1 f(x) \cdot t_k(x)\mathrm{d}x}_{b_k}$$

$$\Leftrightarrow \quad \sum_{i=1}^{N} a_{ki} u_i = b_k$$

Element matrix: Matrix $\mathbf{A} = (a_{ki})$.
Element vector: Vector $\vec{\mathbf{b}} = (b_k)$.

▷ With an appropriate choice of the test function, **A** is a band matrix, i.e., $a_{ki} = 0$, if $|k - i| > n$, with certain n depending on the choice of the trial function and the arrangement of the elements.

Solution of the differential equation: Obtained by solving the system of linear equations for u_i.

19

Fourier transformation

19.1 Fourier series

19.1.1 Introduction

Fourier series play an important role in the description of many technical problems, in mechanical engineering, signalling, telecommunication and control engineering. There, general periodic solutions occur due to the linearity of movement equations, as linear combinations of these eigenfunctions. However, in the mathematical sense, the latter represent the Fourier series.

19.1.2 Definition and coefficients

Periodic functions: If a periodic function $f(t)$ is shifted by an integral multiple of its period T, then it assumes the same value as at the unshifted point:

$$f(t+nT)=f(t)\,, \qquad n=0,\pm1,\pm2,\pm3,\ldots.$$

▷ If a function has period T, it is also called a T-periodic function.

□ Periodic function. The sine function is periodic with period 2π; it is 2π-periodic. Thus,

$$\sin(t+n2\pi)=\sin(t)\,, \qquad n=0,\pm1,\pm2,\pm3,\ldots.$$

Fourier series: The expansion of 2π-periodic, piecewise monotonic and continuous functions into a series of trigonometric functions:

$$f(t) = \frac{a_0}{2} + \sum_{n=1}^{\infty} (a_n \cos(nt) + b_n \sin(nt))$$

with the **Fourier coefficients**

$$a_n = \frac{1}{\pi} \int_{-\pi}^{\pi} f(t) \cos(nt)\, dt \qquad n = 0, 1, 2, \ldots$$

and

$$b_n = \frac{1}{\pi} \int_{-\pi}^{\pi} f(t) \sin(nt)\, dt \qquad n = 1, 2, 3, \ldots .$$

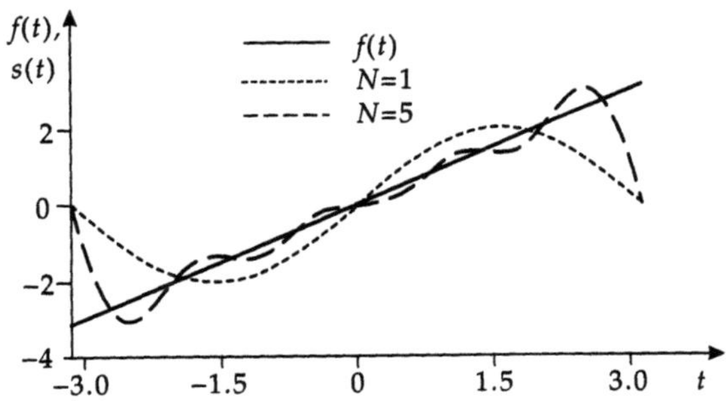

Sawtooth function in the interval $[-\pi, \pi]$ and its expansion into a Fourier series $s(t)$ with $N = 1$ and $N = 5$

□ This is the Fourier series of the 2π-periodic function $f(t) = t$ in $[-\pi, \pi]$, where $[-\pi, \pi]$ is the fundamental interval of periodicity and an abbreviation $-\pi < t < \pi$. Thus, the function can be determined for all t by repeating the function 2π-periodically on this fundamental interval. An example of a Fourier series is the sawtooth function,

$$f(t) = 2\left(\frac{\sin(t)}{1} - \frac{\sin(2t)}{2} + \frac{\sin(3t)}{3} - \cdots\right),$$

i.e., $a_0 = a_n = 0$ and $b_n = (-1)^{n+1} 2/n$ for $n = 1, 2, \ldots$.

▷ The Fourier coefficients of 2π-periodic functions follow from the orthogonality of the trigonometric functions in the interval $[-\pi, \pi]$:

$$\int_{-\pi}^{\pi} \left\{ \begin{array}{c} \sin(mt)\cdot\sin(nt) \\ \cos(mt)\cdot\cos(nt) \end{array} \right\} dt = \pi\,\delta_{mn} \qquad \text{and}$$

$$\int_{-\pi}^{\pi} \{\sin(mt) \cdot \cos(nt)\}\; dt = 0 \qquad \text{for all } m, n .$$

Here, δ_{mn} is the Kronecker symbol with $\delta_{mn} = 0$ for $m \neq n$ and $\delta_{mn} = 1$ for $m = n$.

□ Calculation of the orthogonality of the sine function in the interval $[-\pi, \pi]$: For the product of the two sine functions the following trigonometric formula holds:

$$\sin(mt) \cdot \sin(nt) = \frac{1}{2}(\cos((m-n)t) - \cos((m+n)t)) .$$

This decomposition can be substituted into the integral:

$$\int_{-\pi}^{\pi} \{\sin(mt) \cdot \sin(nt)\} \, dt$$
$$= \frac{1}{2} \int_{-\pi}^{\pi} \{\cos((m-n)t) - \cos((m+n)t)\} \, dt \, .$$

Now, a case distinction must be made: Case 1: $m \neq n$:

$$\frac{1}{2}\int_{-\pi}^{\pi} \cos((m-n)t) \, dt - \frac{1}{2}\int_{-\pi}^{\pi} \cos((m+n)t) \, dt$$
$$= \frac{1}{2}\frac{\sin((m-n)t)}{m-n}\Big|_{-\pi}^{\pi} - \frac{1}{2}\frac{\sin((m+n)t)}{m+n}\Big|_{-\pi}^{\pi} = 0 \, .$$

Case 2: $m = n$:

$$\frac{1}{2}\int_{-\pi}^{\pi} 1 \, dt - \frac{1}{2}\int_{-\pi}^{\pi} \cos(2mt) \, dt = \frac{1}{2}t\Big|_{-\pi}^{\pi} - \frac{1}{2}\frac{\sin(2mt)}{2m}\Big|_{-\pi}^{\pi} = \pi \, .$$

The orthogonality of the sine function follows immediately from both cases.

19.1.3 Condition of convergence

Dirichlet conditions: Criteria for the convergence of the Fourier series:

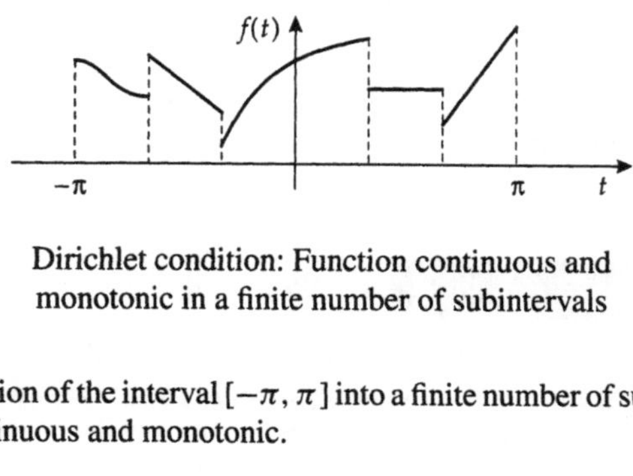

Dirichlet condition: Function continuous and monotonic in a finite number of subintervals

(a) Decomposition of the interval $[-\pi, \pi]$ into a finite number of subintervals in which $f(t)$ is continuous and monotonic.

(b) At the point of discontinuity t_0.

$$f_-(t_0) = \lim_{t \to t_0-} f(t)$$

(left limit) and

$$f_+(t_0) = \lim_{t \to t_0+} f(t)$$

(right limit) exist.

The Fourier series is the arithmetic mean at the jump discontinuity

$$\bar{f}(t_0) = \frac{f_+(t_0) + f_-(t_0)}{2} \, .$$

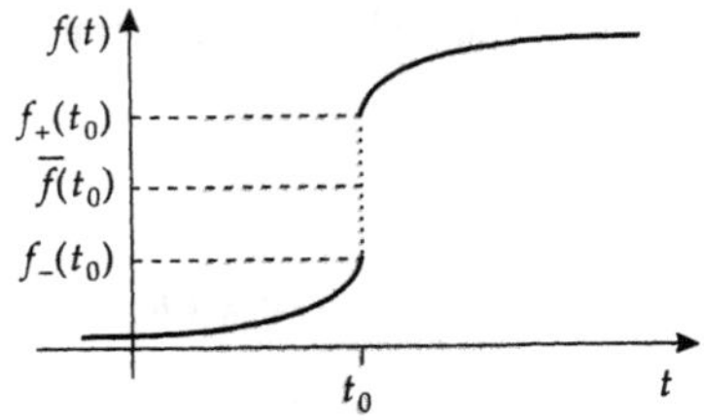

At the jump discontinuity, the Fourier series is the mean of the left and right limits

- **Bessel inequality**: The sum of the squares of the first N coefficients of the Fourier series is bounded upwards:

$$\frac{a_0^2}{2} + \sum_{k=1}^{N} \left(a_k^2 + b_k^2\right) \leq \frac{1}{\pi} \int_{-\pi}^{\pi} f^2(t) \, \mathrm{d}t \, .$$

The difference of the two terms represents the mean square error between the function $f(x)$ and its approximation by trigonometric functions. From this inequality we can conclude that the mean square error is positive or zero.

- **Parseval's equation**: Limiting value ($N \to \infty$) of the Bessel inequality:

$$\frac{a_0^2}{2} + \sum_{k=1}^{\infty} \left(a_k^2 + b_k^2\right) = \frac{1}{\pi} \int_{-\pi}^{\pi} f^2(t) \, \mathrm{d}t \, .$$

If the function $f(t)$ can be represented by a Fourier series, the mean square error vanishes in the limiting process $N \to \infty$.

19.1.4 Extended interval

Fourier series for T-periodic functions: The restriction on the interval $[-\pi, \pi]$ is thus abolished. Introducing the angular frequency $\omega = 2\pi/T$ (T = period) and substituting $t \to \omega t$ in the coefficients and the Fourier series, one obtains the form of the Fourier series that holds for a function of period T:

$$f(t) = \frac{a_0}{2} + \sum_{n=1}^{\infty} \left(a_n \cos(\omega n t) + b_n \sin(\omega n t)\right)$$

and the coefficients:

$$a_n = \frac{2}{T} \int_{-T/2}^{T/2} f(t) \cos(\omega n t) \, \mathrm{d}t \qquad n = 0, 1, 2, \ldots$$

and

$$b_n = \frac{2}{T} \int_{-T/2}^{T/2} f(t) \sin(\omega n t) \, \mathrm{d}t \qquad n = 1, 2, 3, \ldots .$$

The partial oscillation in the series expansion that has the smallest angular frequency ω is called the fundamental oscillation; all other oscillations with multiples of ω are called harmonics.

▷ 1. Since the function is periodic (with a period of T), we can also take interval $[-T/2, T/2]$. Then the limits of integration range from $-T/2$ to $T/2$.

2. In electronics, the Fourier series can be regarded as an expansion of periodically changeable currents (voltages) in terms of fundamental oscillations and harmonics. Then the coefficient a_0 is added to the interpretation of the double *direct-current component* (*direct-voltage component*) in the expansion.

3. Variable t can be interpreted not only as time, but also as position ($t \to x$). Then the angular frequency assumes the role of the wave number ($\omega \to k$), and the period T becomes the wavelength λ. Thus, the following holds for the wave number $k = 2\pi/\lambda$:

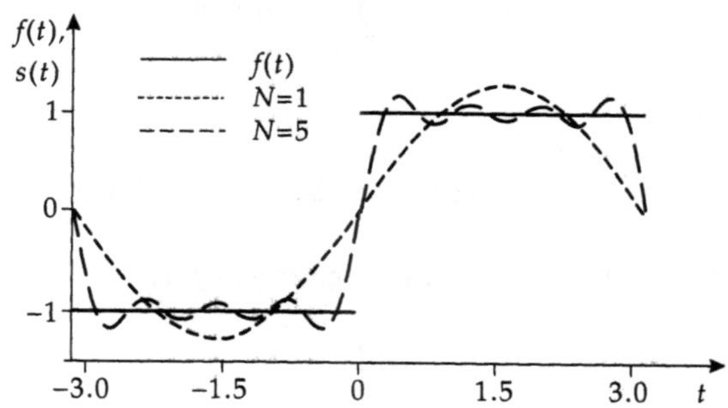

Step function $f(t)$ in the interval $[-\pi, \pi]$ and its expansion into a Fourier series $s(t)$ with $N = 1$ and $N = 7$

☐ Fourier series of the step function

$$f(t) = \begin{cases} C & \text{for } 0 < t \leq T/2 \\ -C & \text{for } -T/2 \leq t < 0 \end{cases}$$

with period T:

$$f(t) = \frac{4C}{\pi}\left(\frac{\sin(1\cdot\omega t)}{1} + \frac{\sin(3\cdot\omega t)}{3} + \frac{\sin(5\cdot\omega t)}{5} + \cdots\right)$$

with coefficients $a_0 = a_m = b_{2m} = 0$ and

$$b_{2m-1} = \frac{4C}{\pi(2m-1)}$$

for $m = 1, 2, \ldots$.

□ Value of the above Fourier series at the point of discontinuity $t_0 = 0$: The Fourier series vanishes, as the sine function vanishes for $t = 0$. For the arithmetic mean we obtain: $(f_-(t=0) + f_+(t=0))/2 = (-C+C)/2 = 0$.

▷ Numerical calculation of the Fourier series in the above example: `sum = 0`

```
j = 1
FOR i =1 TO Nmax/2
BEGIN
sum = sum + sin(j * w * t)/j
j = j+2
END
s(t) = sum * 4 * c/ π
```

Nmax is the upper summation index; it is odd.

19.1.5 Symmetries

Even functions $(f(-t) = f(t))$ are expanded only in terms of cosine,

$$f(t) = \frac{a_0}{2} + \sum_{n=1}^{\infty} a_n \cos(\omega n t)$$

with the coefficients

$$a_n = \frac{4}{T}\int_0^{T/2} f(t)\cos(\omega n t)\,\mathrm{d}t\,, \qquad n = 0, 1, 2, \ldots .$$

The coefficients b_n vanish due to the even symmetry of $f(t)$.

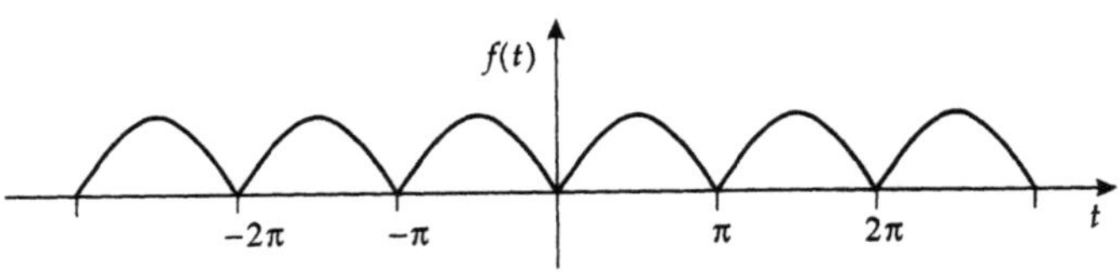

Function $f(t) = |\sin(t)|$

□ Expansion of the even periodic function $f(t) = |\sin(t)|$ with period $T = 2\pi$ into a Fourier series. For the coefficients it holds that $(\omega = 1)$:

$$\begin{aligned} a_n &= \frac{2}{\pi}\int_0^{\pi} \sin(t)\cos(nt)\,\mathrm{d}t \\ &= \frac{1}{\pi}\int_0^{\pi} \{\sin((n+1)t) - \sin((n-1)t)\}\,\mathrm{d}t \\ &= \begin{cases} 0 & \text{for odd } n \\ \dfrac{-4}{\pi(n^2-1)} & \text{for even } n\,, \end{cases} \end{aligned}$$

where $n > 0$. For $n = 0$ we obtain $a_0 = 4/\pi$. Thus, the Fourier series takes the form

$$f(t) = \frac{2}{\pi} - \frac{4}{\pi}\sum_{m=1}^{\infty} \frac{\cos(2mt)}{4m^2-1}\,.$$

Odd functions ($f(-t) = -f(t)$) are expanded in terms of sine only:

$$f(t) = \sum_{n=0}^{\infty} b_n \sin(\omega n t)$$

with the coefficients

$$b_n = \frac{4}{T} \int_0^{T/2} f(t) \sin(\omega n t)\, dt \; .$$

▷ For the Fourier series of even functions as well as for odd functions, the integrand of the integrals referring to the corresponding Fourier coefficients is an even function. Thus, it is sufficient to integrate over the half interval $[0, T/2]$ instead of $[-T/2, T/2]$:

$$\frac{2}{T} \int_{-T/2}^{T/2} \cdots \, dt = \frac{4}{T} \int_0^{T/2} \cdots \, dt \; .$$

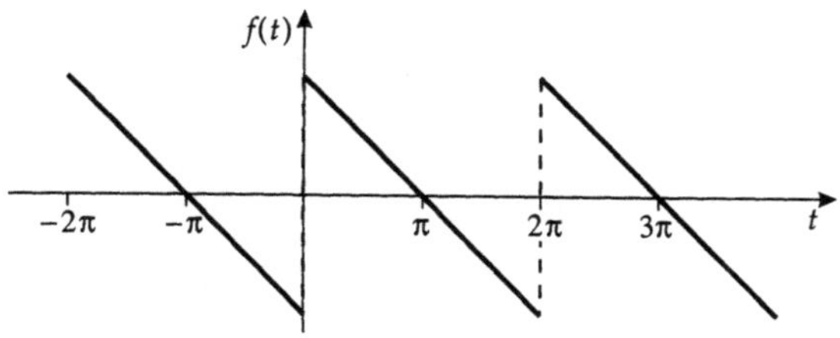

Sawtooth function with negative slope

□ Expansion of the odd function

$$f(t) = \begin{cases} \dfrac{\pi - t}{2} & \text{if } 0 < t < \pi \\[2ex] \dfrac{-\pi - t}{2} & \text{if } -\pi < t < 0 \end{cases}$$

with the period $T = 2\pi$ into a Fourier series. For the coefficients we obtain ($\omega = 1$) with the help of the product integration:

$$\begin{aligned} b_n &= \frac{2}{\pi} \int_0^{\pi} \frac{\pi - t}{2} \sin(nt)\, dt \\ &= \frac{1}{n} - \frac{1}{\pi n^2} \sin(nt) \Big|_0^{\pi} = \frac{1}{n} \, , \end{aligned}$$

where $n > 0$, and $n = 0$ yields $b_0 = 0$. The Fourier series of this function therefore reads:

$$f(t) = \sum_{n=1}^{\infty} \frac{\sin(nt)}{n} \; .$$

19.1.6 Fourier series in complex and spectral representation

The Fourier series

$$s(t) = \frac{a_0}{2} + \sum_{n=1}^{\infty} (a_n \cos(\omega n t) + b_n \sin(\omega n t))$$

can also be written in two other forms:

Spectral representation:

$$s(t) = \frac{a_0}{2} + \sum_{n=1}^{\infty} A_n \sin(\omega n t + \varphi_n)$$

with coefficients:

$$A_n = \sqrt{a_n^2 + b_n^2}, \quad \tan \varphi_n = \frac{a_n}{b_n}.$$

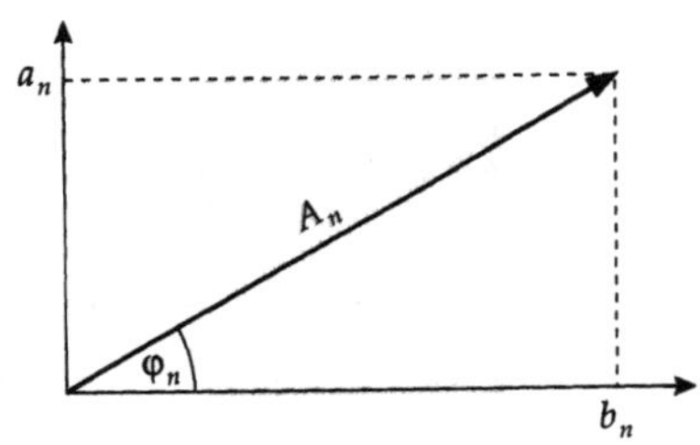

Geometric representation of the relation between a_n, b_n and A_n, φ_n

Amplitude spectrum: Graphical representation of the discrete amplitudes A_n as a function of n.

Phase spectrum: Graphical representation of the phase shifts φ_n as a function of n.

Complex form for T-periodic functions:

$$s(t) = \sum_{n=-\infty}^{\infty} c_n e^{j\omega n t}$$

with the coefficients

$$c_n = \frac{1}{T} \int_{-T/2}^{T/2} f(t) e^{-j\omega n t} \, dt.$$

Relation between the coefficients a_n, b_n, and c_n:

$$c_n = \begin{cases} a_0/2 & n = 0 \\ (a_n - jb_n)/2 & n > 0 \\ (a_{-n} + jb_{-n})/2 & n < 0 \end{cases} \quad \text{or} \quad \left. \begin{aligned} a_n &= c_n + c_{-n} \\ b_n &= j(c_n - c_{-n}) \end{aligned} \right\} n > 0.$$

The values $\omega_n = \omega n$ are called the spectrum of $f(t)$.

19.1.7 Formulas for the calculation of Fourier series

$$1 = \frac{4}{\pi}\left[\sin\frac{\pi t}{k} + \frac{1}{3}\sin\frac{3\pi t}{k} + \frac{1}{5}\sin\frac{5\pi t}{k} + \cdots\right] \qquad (0 < t < k)$$

$$t = \frac{2k}{\pi}\left[\sin\frac{\pi t}{k} - \frac{1}{2}\sin\frac{2\pi t}{k} + \frac{1}{3}\sin\frac{3\pi t}{k} - \cdots\right] \qquad (-k < t < k)$$

$$t = \frac{k}{2} - \frac{4k}{\pi^2}\left[\cos\frac{\pi t}{k} + \frac{1}{3^2}\cos\frac{3\pi t}{k} + \frac{1}{5^2}\cos\frac{5\pi t}{k} + \cdots\right] \qquad (0 < t < k)$$

$$t^2 = \frac{2k^2}{\pi^3}\left[\left(\frac{\pi^2}{1} - \frac{4}{1}\right)\sin\frac{\pi t}{k} - \frac{\pi^2}{2}\sin\frac{2\pi t}{k} + \left(\frac{\pi^2}{3} - \frac{4}{3^3}\right)\sin\frac{3\pi t}{k}\right.$$
$$\left. - \frac{\pi^2}{4}\sin\frac{4\pi t}{k} + \left(\frac{\pi^2}{5} - \frac{4}{5^3}\right)\sin\frac{5\pi t}{k} + \cdots\right] \qquad (0 < t < k)$$

$$t^2 = \frac{k^2}{3} - \frac{4k^2}{\pi^2}\left[\cos\frac{\pi t}{k} - \frac{1}{2^2}\cos\frac{2\pi t}{k} + \frac{1}{3^2}\cos\frac{3\pi t}{k}\right.$$
$$\left. - \frac{1}{4^2}\cos\frac{4\pi t}{k} + \cdots\right] \qquad (-k < t < k)$$

$$\frac{\pi}{4} = 1 - \frac{1}{3} + \frac{1}{5} - \frac{1}{7} + \cdots$$
$$\frac{\pi^2}{6} = 1 + \frac{1}{2^2} + \frac{1}{3^2} + \frac{1}{4^2} + \cdots$$
$$\frac{\pi^2}{12} = 1 - \frac{1}{2^2} + \frac{1}{3^2} - \frac{1}{4^2} + \cdots$$
$$\frac{\pi^2}{8} = 1 + \frac{1}{3^2} + \frac{1}{5^2} + \frac{1}{7^2} + \cdots$$
$$\frac{\pi^2}{24} = \frac{1}{2^2} + \frac{1}{4^2} + \frac{1}{6^2} + \frac{1}{8^2} + \cdots$$

19.1.8 Fourier expansion of simple periodic functions

$$f(t) = \frac{2c}{T} + \frac{2}{\pi}\sum_{n=1}^{\infty}\frac{(-1)^n}{n}\sin\frac{2n\pi c}{T}\cos\frac{2n\pi t}{T}$$

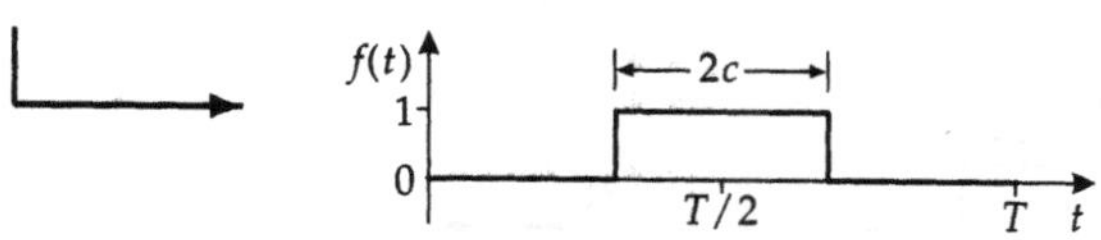

$$f(t)=\frac{4}{\pi}\sum_{n=1,3,5,..}\frac{1}{n}\sin\frac{2n\pi t}{T}$$

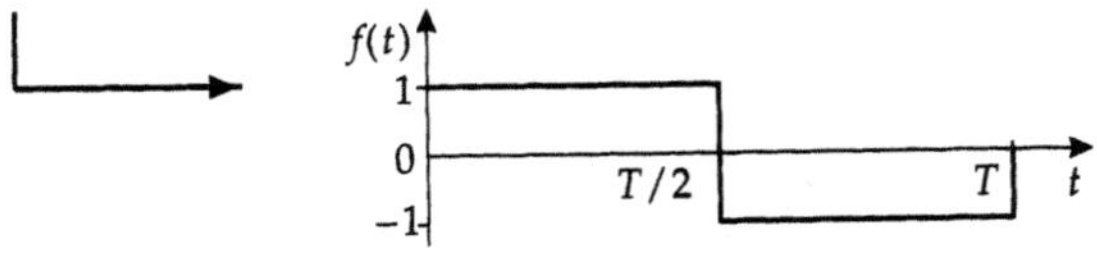

$$f(t)=\frac{4}{\pi}\sum_{n=1,3,5,..}\frac{(-1)^{(n-1)/2}}{n}\cos\frac{2n\pi t}{T}$$

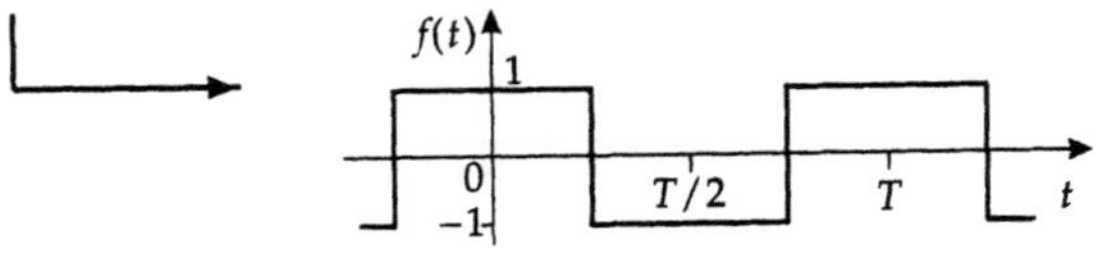

$$f(t)=\frac{c}{T}+\frac{2}{\pi}\sum_{n=1,2,3,..}\frac{1}{n}\sin\left(\frac{n\pi c}{T}\right)\cos\left(\frac{2n\pi t}{T}\right)$$

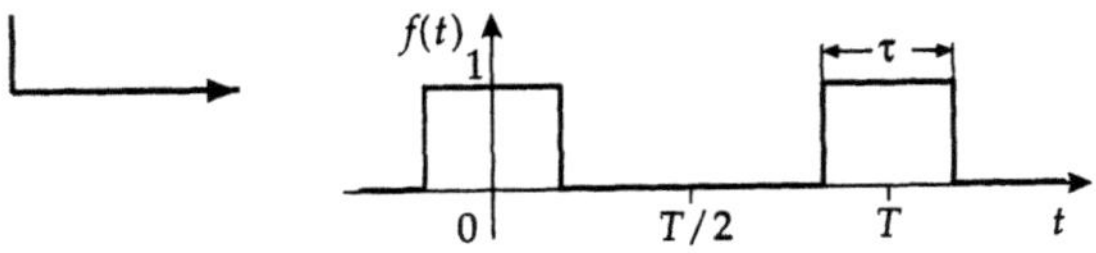

$$f(t)=\frac{2}{\pi}\sum_{n=1}^{\infty}\frac{(-1)^n}{n}\left(\cos\frac{2n\pi c}{T}-1\right)\sin\frac{2n\pi t}{T}$$

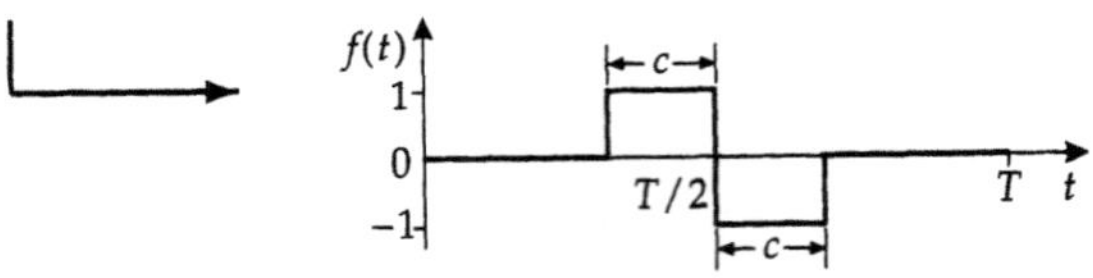

$$f(t)=\frac{4}{T}\sum_{n=1}^{\infty}\sin\frac{n\pi}{2}\,\frac{\sin(n\pi c/T)}{n\pi c/T}\sin\frac{2n\pi t}{T}$$

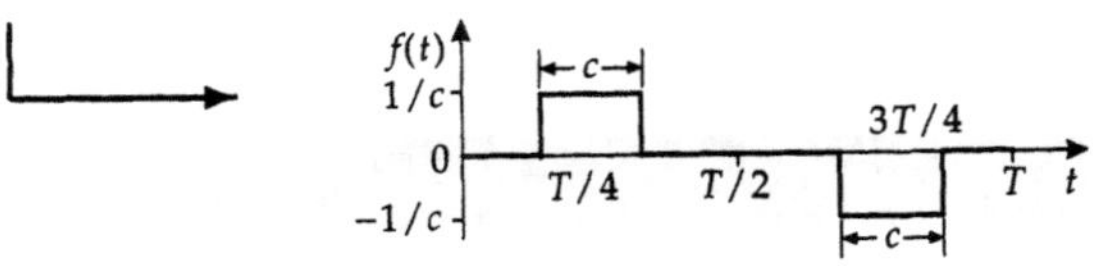

$$f(t)=\frac{4}{\pi}\sum_{n=1}^{\infty}\frac{1}{n}\sin\frac{n\pi}{4}\sin(n\pi a)\sin\frac{2n\pi t}{T};\quad\left(a=\frac{c}{T}\right)$$

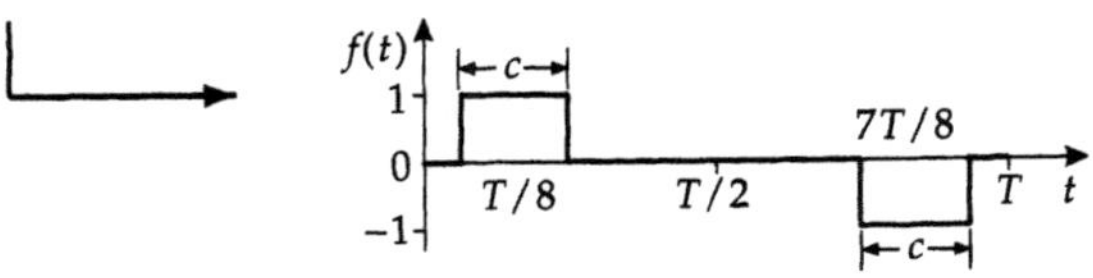

$$f(t) = \frac{8}{\pi^2} \sum_{n=1,3,5,..} \frac{(-1)^{(n-1)/2}}{n^2} \sin \frac{2n\pi t}{T}$$

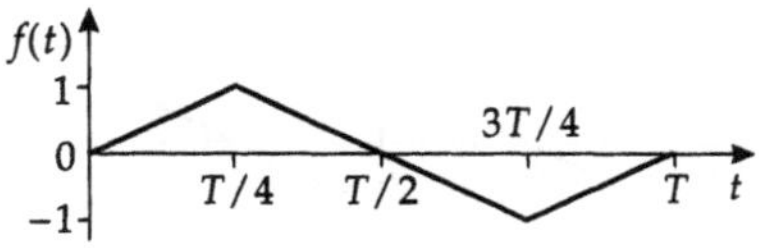

$$f(t) = \frac{8}{\pi^2} \sum_{n=1,3,5..} \frac{1}{n^2} \cos \frac{2n\pi t}{T}$$

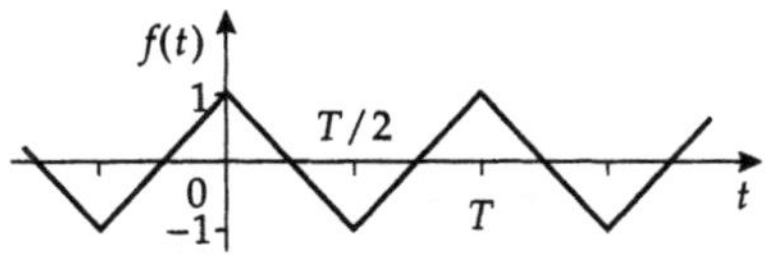

$$f(t) = \frac{32}{3\pi^2} \sum_{n=1}^{\infty} \frac{1}{n^2} \sin \frac{n\pi}{4} \sin \frac{2n\pi t}{T} \quad ; \quad \left(a = \frac{c}{T}\right)$$

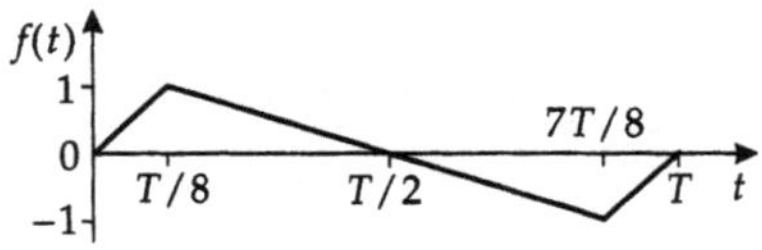

$$f(t) = \frac{9}{\pi^2} \sum_{n=1}^{\infty} \frac{1}{n^2} \sin \frac{n\pi}{3} \sin \frac{2n\pi t}{T}; \quad \left(a = \frac{c}{T}\right)$$

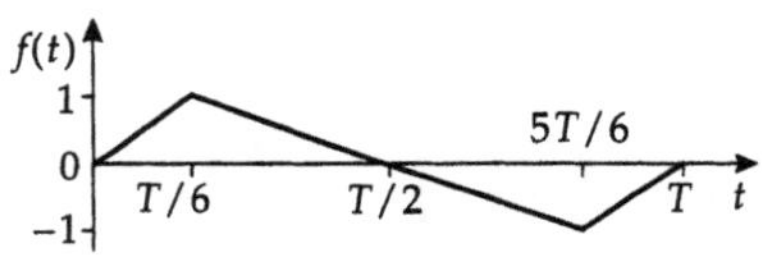

$$f(t) = \frac{1}{2} - \frac{1}{\pi} \sum_{n=1}^{\infty} \frac{1}{n} \sin \frac{2n\pi t}{T}$$

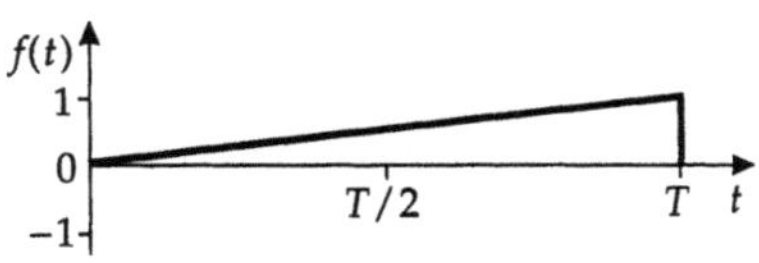

$$f(t) = \frac{2}{\pi} \sum_{n=1}^{\infty} \frac{(-1)^{n+1}}{n} \sin \frac{2n\pi t}{T}$$

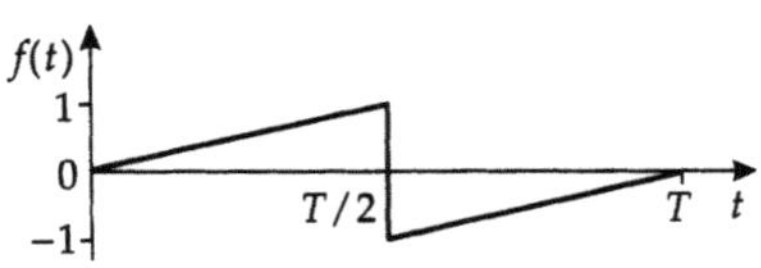

$$f(t) = \frac{2}{\pi} \sum_{n=1,3,5,..} \frac{1}{n} \sin \frac{2n\pi t}{T}$$

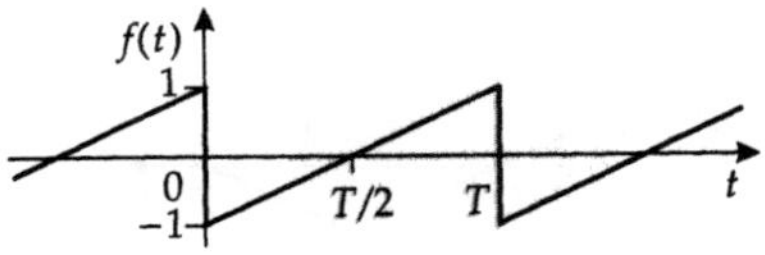

$$f(t) = \frac{1}{2} - \frac{4}{\pi^2} \sum_{n=1,3,5\cdots} \frac{1}{n^2} \cos \frac{2n\pi t}{T}$$

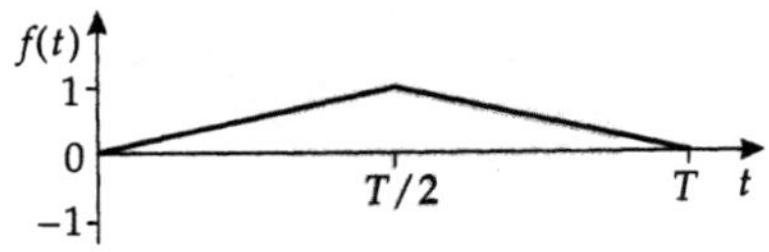

$$f(t) = \frac{1}{2}(1+a) + \frac{2}{\pi^2(1-a)} \sum_{n=1}^{\infty} \frac{1}{n^2} \left[(-1)^n \cos(n\pi a) - 1\right] \cos \frac{2n\pi t}{T}; \quad (a = c/T)$$

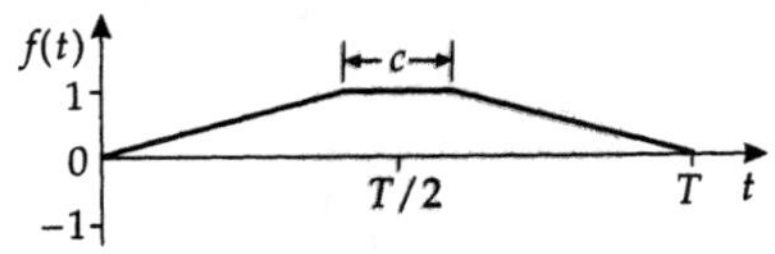

$$f(t) = \frac{4}{a\pi} \sum_{n=1,3,5,..} \frac{1}{n^2} \sin(na) \sin\left(\frac{2n\pi t}{T}\right), \qquad \left(a = \frac{2\pi c}{T}\right)$$

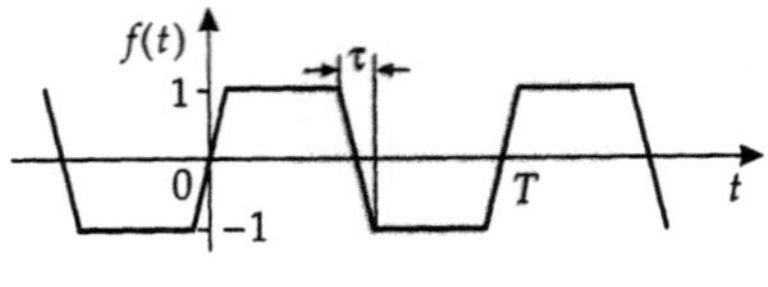

$$f(t) = \frac{1}{2} - \frac{4}{\pi^2(1-2a)} \sum_{n=1,3,5,..} \frac{1}{n^2} \cos(n\pi a) \cos \frac{2n\pi t}{T}; \quad \left(a = \frac{c}{T}\right)$$

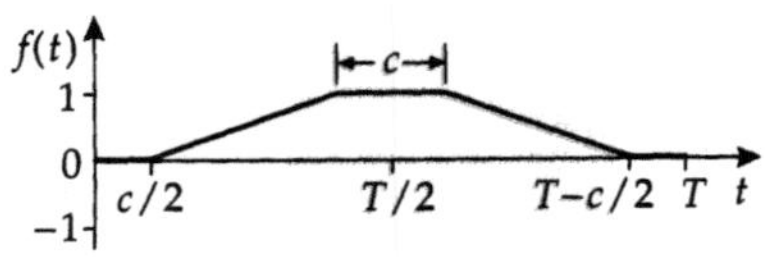

$$f(t) = \frac{2}{\pi}\sum_{n=1}^{\infty}\frac{(-1)^{n-1}}{n}\left[1+\frac{\sin(n\pi a)}{n\pi(1-a)}\right]\sin\frac{2n\pi t}{T}; \quad \left(a=\frac{c}{T}\right)$$

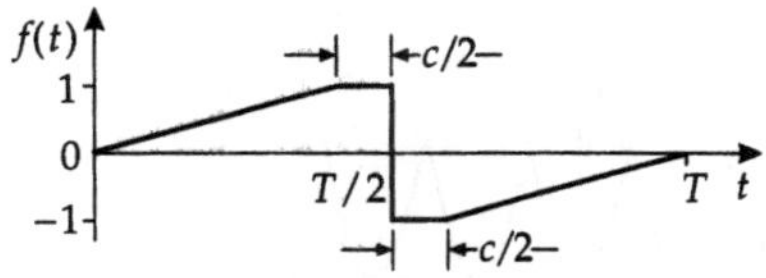

$$f(t) = -\frac{2}{\pi}\sum_{n=1}^{\infty}\frac{(-1)^{n-1}}{n}\left[1+\frac{1+(-1)^n}{n\pi(1-2a)}\sin(n\pi a)\right]\sin\frac{2n\pi t}{T}; \ \left(a=\frac{c}{T}\right)$$

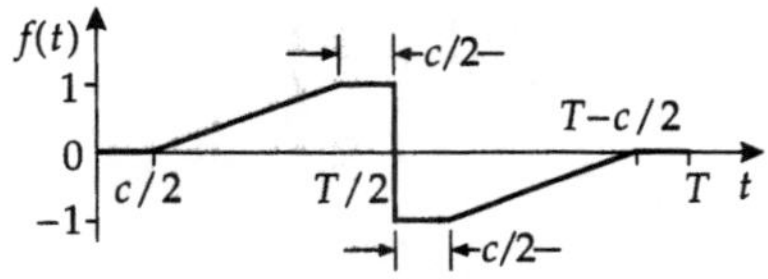

$$\begin{aligned} f(t) &= t(\pi-t)(\pi+t), \quad (-\pi<t<\pi) \\ &= 12\left(\frac{\sin t}{1^3}-\frac{\sin 2t}{2^3}+\frac{\sin 3t}{3^3}-\cdots\right) \end{aligned}$$

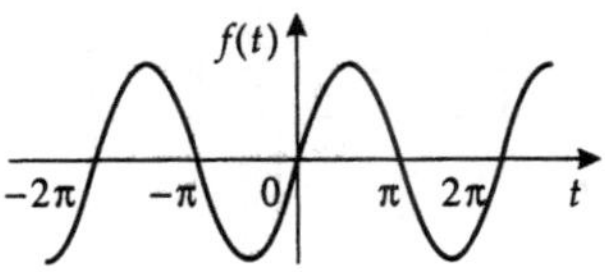

$$\begin{aligned} f(t) &= \begin{cases} t(\pi-t) & (0<t<\pi) \\ -t(\pi-t) & (-\pi<t<0) \end{cases} \\ &= \frac{8}{\pi}\left(\frac{\sin t}{1^3}+\frac{\sin 3t}{3^3}+\frac{\sin 5t}{5^3}+\cdots\right) \end{aligned}$$

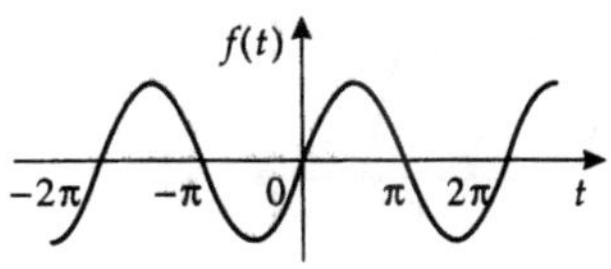

$$\begin{aligned} f(t) &= t(\pi-t), \quad (0<t<\pi) \\ &= \frac{\pi^2}{6}-\left(\frac{\cos 2t}{1^2}+\frac{\cos 4t}{2^2}+\frac{\cos 6t}{3^2}+\cdots\right) \end{aligned}$$

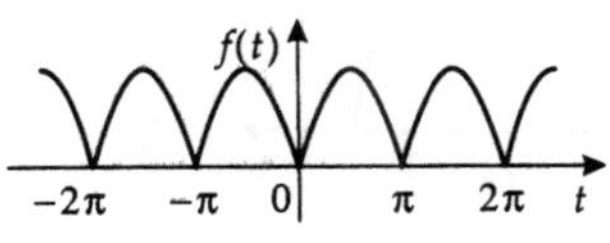

$$f(t) = t^2, \quad (-\pi < t < \pi)$$
$$= \frac{\pi^2}{3} - 4\left(\frac{\cos t}{1^2} - \frac{\cos 2t}{2^2} + \frac{\cos 3t}{3^2} - \cdots\right)$$

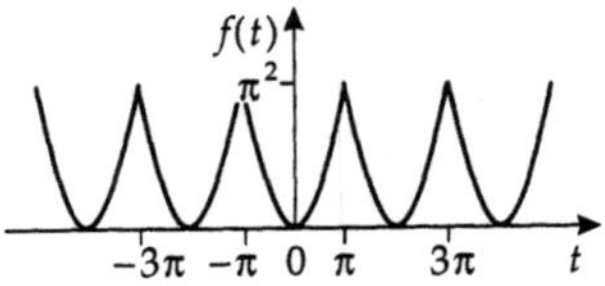

$$f(t) = |\sin t|, \quad (-\pi < t < \pi)$$
$$= \frac{2}{\pi} - \frac{4}{\pi}\left(\frac{\cos 2t}{1\cdot 3} + \frac{\cos 4t}{3\cdot 5} + \frac{\cos 6t}{5\cdot 7} + \cdots\right)$$

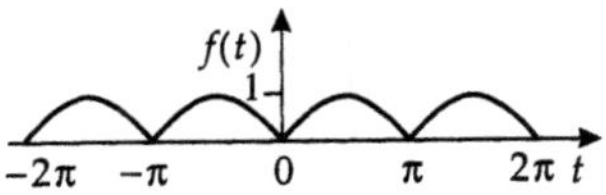

$$f(t) = \begin{cases} \sin t & (0 < t < \pi) \\ 0 & (\pi < t < 2\pi) \end{cases}$$
$$= \frac{1}{\pi} + \frac{1}{2}\sin t - \frac{2}{\pi}\left(\frac{\cos 2t}{1\cdot 3} + \frac{\cos 4t}{3\cdot 5} + \frac{\cos 6t}{5\cdot 7} + \cdots\right)$$

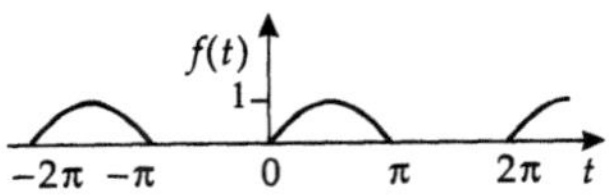

$$f(t) = \frac{2}{\pi} - \frac{4}{\pi}\left(\frac{1}{1\cdot 3}\cos(2\omega t) + \frac{1}{3\cdot 5}\cos(4\omega t) + \frac{1}{5\cdot 7}\cos(6\omega t) + \cdots\right)$$

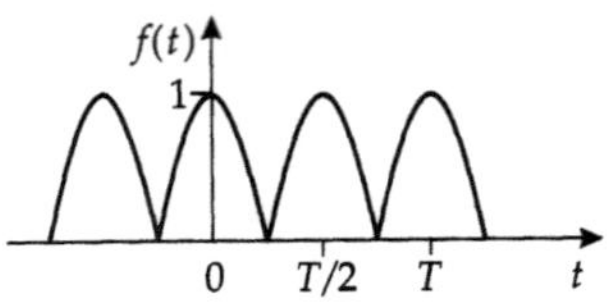

$$f(t) = \frac{1}{\pi} + \frac{2}{\pi}\left(\pi\cos(\omega t) + \frac{1}{1\cdot 3}\cos(2\omega t) + \frac{1}{3\cdot 5}\cos(4\omega t) + \cdots\right)$$

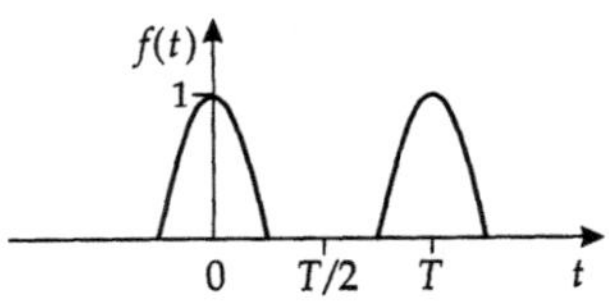

$$f(t) = \begin{cases} \cos t & (0 < t < \pi) \\ -\cos t & (-\pi < t < 0) \end{cases}$$

$$= \frac{8}{\pi}\left(\frac{\sin 2t}{1\cdot 3} + \frac{2\sin 4t}{3\cdot 5} + \frac{3\sin 6t}{5\cdot 7} + \cdots\right)$$

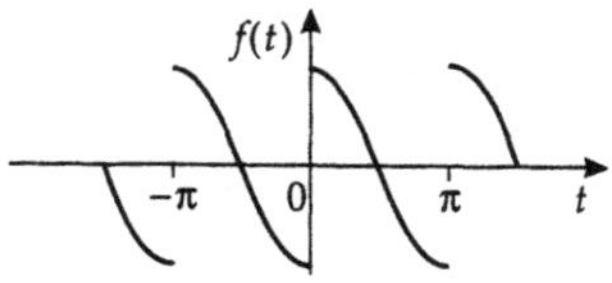

$$f(t) = \frac{3\sqrt{3}}{\pi}\left(\frac{1}{2} - \frac{1}{2\cdot 4}\cos(3\omega t) - \frac{1}{5\cdot 7}\cos(6\omega t) - \frac{1}{8\cdot 10}\cos(9\omega t) - \cdots\right)$$

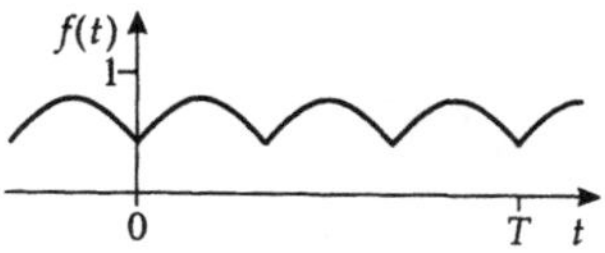

$$f(t) = \frac{2}{\pi}\sum_{n=0}^{\infty}\frac{\gamma\cos((2n+1)\omega t) + (2n+1)\sin((2n+1)\omega t)}{\gamma^2 + (2n+1)^2}; \quad (\gamma = T/2\pi\tau)$$

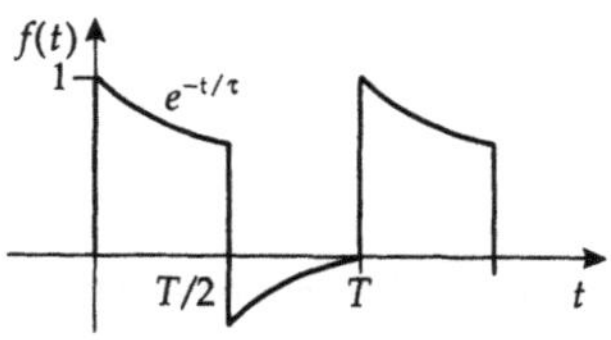

19.1.9 Fourier series (table)

$$f(t) = \frac{1}{12}t(t-\pi)(t-2\pi), \quad (0 \le t \le 2\pi)$$

$$= \frac{\sin t}{1^3} + \frac{\sin 2t}{2^3} + \frac{\sin 3t}{3^3} + \cdots$$

$$f(t) = \frac{1}{6}\pi^2 - \frac{1}{2}\pi t + \frac{1}{4}t^2, \quad (0 \le t \le 2\pi)$$

$$= \frac{\cos t}{1^2} + \frac{\cos 2t}{2^2} + \frac{\cos 3t}{3^2} + \cdots$$

$$f(t) = \frac{1}{90}\pi^4 - \frac{1}{12}\pi^2 t^2 + \frac{1}{12}\pi t^3 - \frac{1}{48}t^4, \quad (0 \le t \le 2\pi)$$

$$= \frac{\cos t}{1^4} + \frac{\cos 2t}{2^4} + \frac{\cos 3t}{3^4} + \cdots$$

$$f(t) = \sin \mu t, \quad (-\pi < t < \pi, \quad \mu \ne \text{integer number})$$

$$= \frac{2\sin\mu\pi}{\pi}\left(\frac{\sin t}{1^2-\mu^2} - \frac{2\sin 2t}{2^2-\mu^2} + \frac{3\sin 3t}{3^2-\mu^2} - \cdots\right)$$

$$f(t) = \cos \mu t, \quad (-\pi < t < \pi, \quad \mu \neq \text{integer number})$$
$$= \frac{2\mu \sin \mu\pi}{\pi}\left(\frac{1}{2\mu^2} + \frac{\cos t}{1^2 - \mu^2} - \frac{\cos 2t}{2^2 - \mu^2} + \frac{\cos 3t}{3^2 - \mu^2} - \cdots\right)$$

$$f(t) = \arctan[(a \sin t)/(1 - a \cos t)], \quad (-\pi < t < \pi, \quad |a| < 1)$$
$$= a \sin t + \frac{a^2}{2} \sin 2t + \frac{a^3}{3} \sin 3t + \cdots$$

$$f(t) = \frac{1}{2} \arctan[(2a \sin t)/(1 - a^2)], \quad (-\pi < t < \pi, \quad |a| < 1)$$
$$= a \sin t + \frac{a^3}{3} \sin 3t + \frac{a^5}{5} \sin 5t + \cdots$$

$$f(t) = \frac{1}{2} \arctan[(2a \cos t)/(1 - a^2)], \quad (-\pi < t < \pi, \quad |a| < 1)$$
$$= a \cos t - \frac{a^3}{3} \cos 3t + \frac{a^5}{5} \cos 5t - \cdots$$

$$f(t) = \ln|\sin(t/2)|, \quad (0 < t < \pi)$$
$$= -\left(\ln 2 + \frac{\cos t}{1} + \frac{\cos 2t}{2} + \frac{\cos 3t}{3} + \cdots\right)$$

$$f(t) = \ln|\cos(t/2)|, \quad (-\pi < t < \pi)$$
$$= -\left(\ln 2 - \frac{\cos t}{1} + \frac{\cos 2t}{2} - \frac{\cos 3t}{3} + \cdots\right)$$

$$f(t) = e^{\mu t}, \quad (-\pi < t < \pi)$$
$$= \frac{2 \sinh \mu\pi}{\pi}\left(\frac{1}{2\mu} + \sum_{n=1}^{\infty} \frac{(-1)^n(\mu \cos nt - n \sin nt)}{\mu^2 + n^2}\right)$$

$$f(t) = \ln(1 - 2a \cos t + a^2), \quad (-\pi < t < \pi, \quad |a| < 1)$$
$$= -2\left(a \cos t + \frac{a^2}{2} \cos 2t + \frac{a^3}{3} \cos 3t + \cdots\right)$$

$$f(t) = \sinh \mu t, \quad (-\pi < t < \pi)$$
$$= \frac{2 \sinh \mu\pi}{\pi}\left(\frac{\sin t}{1^2 + \mu^2} - \frac{2 \sin 2t}{2^2 + \mu^2} + \frac{3 \sin 3t}{3^2 + \mu^2} - \cdots\right)$$

$$f(t) = \cosh \mu t, \quad (-\pi < t < \pi)$$
$$= \frac{2\mu \sinh \mu\pi}{\pi}\left(\frac{1}{2\mu^2} - \frac{\cos t}{1^2 + \mu^2} + \frac{\cos 2t}{2^2 + \mu^2} - \frac{\cos 3t}{3^2 + \mu^2} + \cdots\right)$$

19.2 Fourier integrals

19.2.1 Introduction

Fourier integrals are the generalization of Fourier series. With them, we can also describe nonperiodic functions. For example, differential equations can often be changed to a simpler form by Fourier transformations so that they can be solved by algebraic transformations.

19.2.2 Definition and coefficients

Fourier integral: Integral expansion of nonperiodic functions in terms of trigonometric functions (continuous spectrum): Can be obtained from the Fourier series by shifting the end-points of the interval boundaries toward infinity ($T \to \pm\infty$).

$$f(t) = \int_0^\infty (a(\omega)\cos(\omega t) + b(\omega)\sin(\omega t))\, d\omega.$$

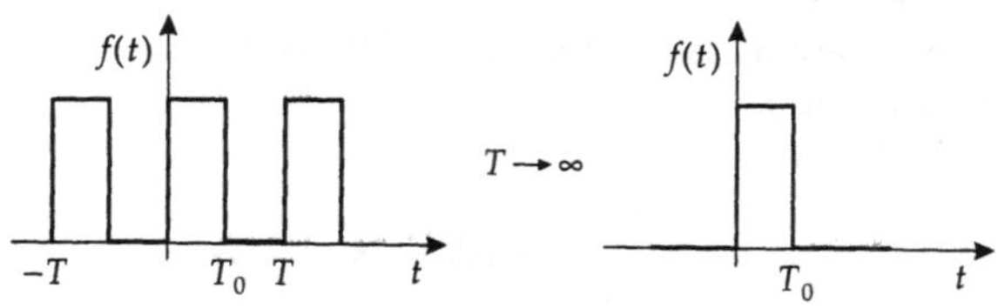

The step function loses its periodicity through the limit $T \to \infty$

Coefficients of the Fourier integral:

$$a(\omega) = \frac{1}{\pi}\int_{-\infty}^{\infty} f(\tau)\cos(\omega\tau)\, d\tau$$

and

$$b(\omega) = \frac{1}{\pi}\int_{-\infty}^{\infty} f(\tau)\sin(\omega\tau)\, d\tau\ .$$

▷ Notation: Sometimes a factor $1/\sqrt{\pi}$ instead of $1/\pi$ is in front of the integral that determines the coefficients. In this case, the factor $1/\sqrt{\pi}$ must also be in front of the Fourier integral!

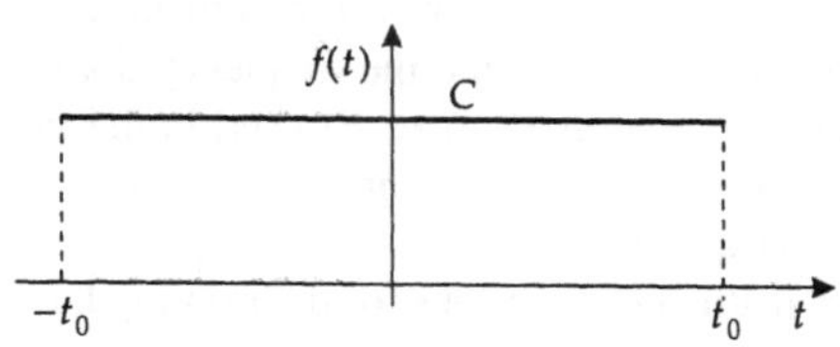

Nonperiodic step function

□ Expansion of the function

$$f(t) = \begin{cases} c & |t| \le t_0 \\ 0 & |t| > t_0 \end{cases}$$

into a Fourier integral:

$$f(t) = \int_0^{\infty} a(\omega) \cos(\omega t)\, \mathrm{d}\omega \,,$$

Because $b(\omega) = 0$ applies (for reasons of symmetry). For $a(\omega)$ we obtain:

$$a(\omega) = \frac{2}{\pi} \int_0^{t_0} c \cos(\omega \tau)\, \mathrm{d}\tau \,.$$

The integration yields:

$$a(\omega) = \frac{2}{\pi}\, c\, \frac{\sin(\omega t_0)}{\omega} \,.$$

19.2.3 Conditions for convergence

Dirichlet conditions must be satisfied in each subinterval for the Fourier integral (see Section 19.1, Fourier series).
Absolute integrability, i.e.,

$$\int_{-\infty}^{\infty} |f(t)|\, \mathrm{d}t < \infty \,.$$

- **Criterion of Dirichlet-Jordan**, sufficient condition.
  If $f(t)$ is of limited variation in each finite subinterval and has at most a finite number of jumps, and if $f(t)$ is absolutely integrable, then $f(t)$ can be expanded into a Fourier integral in continuous pieces; at the jump discontinuities t_0, the Fourier integral is $(f_+(t_0) + f_-(t_0))/2$ (arithmetic mean).

19.2.4 Complex representation, Fourier sine and cosine transformation

Complex representation of the Fourier integrals:

$$f(t) = \frac{1}{2\pi} \int_{-\infty}^{\infty} F(\omega) \mathrm{e}^{\mathrm{j}\omega t}\, \mathrm{d}\omega \quad \text{and} \quad F(\omega) = \int_{-\infty}^{\infty} f(t) \mathrm{e}^{-\mathrm{j}\omega t}\, \mathrm{d}t \,.$$

Notation: Instead of $F(\omega)$ and $f(t)$ we can also write $F[f(t)]$ and $F^{-1}[F(\omega)]$. The function $F(\omega)$ is called the Fourier transform (spectral function) of $f(t)$, and the transition from $f(t)$ to $F(\omega)$ is called the Fourier transformation (operator F). The Fourier transformation maps the time domain onto the frequency domain by assigning function $f(t)$ in the time domain to function $F(\omega) = F[f(t)]$ in the frequency domain. Due to the symmetry of $f(t)$ and $F(\omega)$, the transition from $F(\omega)$ to $f(t)$ is called the **inverse Fourier transformation** (operator F^{-1}).

▷ Here, too there are distinct conventions for distributing the factor $1/(2\pi)$:

a. Factor $1/\sqrt{2\pi}$ is in front of the integral of $f(t)$ as well as in front of that of the integral of $F(\omega)$.

b. There is only one factor $1/(2\pi)$ in front of the integral of $F(\omega)$.

c. There is only one factor $1/(2\pi)$ in front of the integral of $f(t)$.

Here, convention c.) is used.

Continuous amplitude spectrum: Corresponds to the amplitude spectrum of the Fourier series:

$$A(\omega) = \sqrt{(\mathrm{Re}F(\omega))^2 + (\mathrm{Im}F(\omega))^2} .$$

Continuous phase spectrum: Corresponds to the phase spectrum of the Fourier series:

$$\varphi(\omega) = \arctan \frac{\mathrm{Im}F(\omega)}{\mathrm{Re}F(\omega)} .$$

Fourier cosine transformation (cosine transformation): For even functions $f(t) = f(-t)$ the following relation holds:

$$f(t) = \frac{2}{\pi} \int_0^\infty F_C(\omega) \cos(\omega t)\, \mathrm{d}\omega ,$$

$$F_C(\omega) = \int_0^\infty f(t) \cos(\omega t)\, \mathrm{d}t .$$

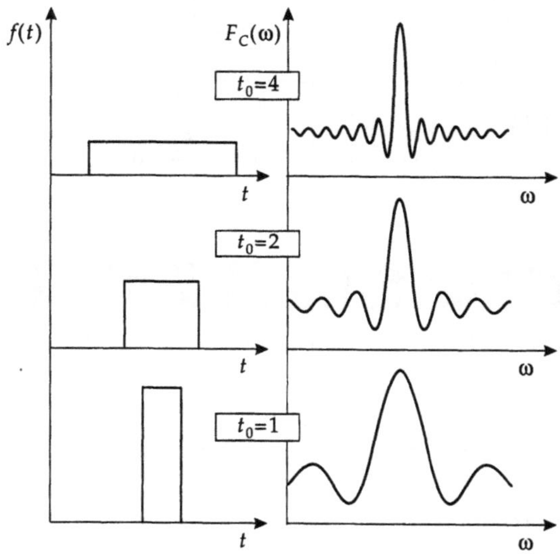

Step function $f(t)$ for different time periods t_0 (left) and the associated cosine transforms $F_C(\omega)$ (right)

□ Cosine transformation of

$$f(t) = \begin{cases} \frac{1}{t_0} & 0 < |t| < t_0 \\ 0 & |t| > t_0 \end{cases} .$$

Thus, for $F_C(\omega)$ the following holds:

$$F_C(\omega) = \int_0^{t_0} \frac{1}{t_0} \cos(\omega t)\, d\omega = \frac{\sin(t_0\omega)}{t_0\omega} .$$

Fourier sine transformation (sine transformation): An expansion in terms of sine functions used for odd functions $f(-t) = -f(t)$:

$$f(t) = \frac{2}{\pi} \int_0^\infty F_S(\omega) \sin(\omega t)\, \mathrm{d}\omega,$$

$$F_S(\omega) = \int_0^\infty f(t) \sin(\omega t)\, \mathrm{d}t .$$

19.2.5 Symmetries

With $F[f(t)] = F(\omega)$ (Fourier transform) the following symmetry relations hold ($*$ = complex conjugation):

If $f(t)$	Then
real	$F(-\omega) = (F(\omega))^*$
imaginary	$F(-\omega) = -(F(\omega))^*$
even	$F(\omega)$ even
odd	$F(\omega)$ odd
real and even	$F(\omega)$ real and even
real and odd	$F(\omega)$ imaginary and odd
imaginary and even	$F(\omega)$ imaginary and even
imaginary and odd	$F(\omega)$ real and odd

19.2.6 Convolution and some calculating rules

Convolution of two functions: The time integral over the product of one function and the other shifted function.

$$(f_1 * f_2)(t) = \int_{-\infty}^{\infty} f_1(\tau) f_2(t-\tau)\, \mathrm{d}\tau$$

□ $f_1(t) = \begin{cases} \sin(t) & \text{if } t \geq 0 \\ 0 & \text{if } t < 0 \end{cases}$

and $f_2(t) = \begin{cases} \cos(t) & \text{if } t \geq 0 \\ 0 & \text{if } t < 0 . \end{cases}$

According to the definition we obtain for the convolutions of f_1 and f_2:

$$(f_1 * f_2)(t) = \int_{-\infty}^{\infty} f_1(\tau) \cdot f_2(t-\tau)\, \mathrm{d}\tau$$

$$= \int_0^t \sin(\tau) \cdot \cos(t-\tau)\, \mathrm{d}\tau$$

$$= \frac{t}{2}\sin(t) .$$

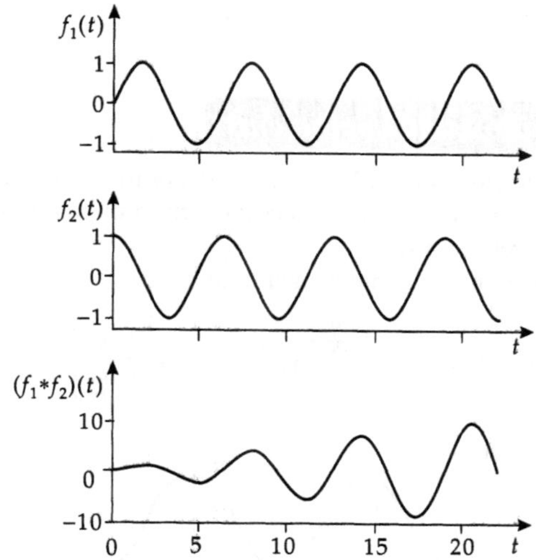

Functions $f_1(t)$ and $f_2(t)$ their convolution $(f_1 * f_2)(t)$

- **Convolution theorem**: The Fourier transform of the convolution of functions f_1 and f_2 is equal to the product of the Fourier transform of f_1 and f_2:

$$F[(f_1 * f_2)(t)] = F[f_1(t)] \cdot F[f_2(t)] .$$

- **Translation theorem**: The Fourier transform of a function shifted by time t_0 is equal to the product of the Fourier transform of the Fourier transforms f_1 and f_2 by the factor $\mathrm{e}^{-\mathrm{j}\omega t_0}$:

$$F[f(t - t_0)] = F[f(t)]\mathrm{e}^{-\mathrm{j}\omega t_0} .$$

- **Linearity theorem**: The Fourier transform of the sum of functions is equal to the sum of the Fourier transforms of these functions:

$$F[f(t) + g(t)] = F[f(t)] + F[g(t)] .$$

- **Similarity theorem**: The Fourier transform of a function with which a similarity transformation was carried out ($t \to at$) is equal to the Fourier transform of the original function with the similarity transformation $\omega \to \omega/a$ divided by the absolute value of factor a:

$$F[f(at)] = \frac{1}{a}G(\omega/a) \qquad a > 0 .$$

19.3 Discrete Fourier transformation (DFT)

The discrete Fourier transformation has the same applicability range as the Fourier integral. It is used for those functions whose values are given only for discrete grid points.

19.3.1 Definition and coefficients

Discrete Fourier transformation: Discrete version of the continuous Fourier transformation. It is used for functions whose values are given or can be scanned only in the interval $[0, T]$ at discrete points.
Expansion of the function into a trigonometric sum:

$$f(t_k) = T_N(t_k) = \frac{1}{N}\sum_{i=0}^{N-1} c_i e^{2\pi j i t_k/T} = \frac{1}{N}\sum_{i=0}^{N-1} c_i e^{2\pi j i k/N}$$

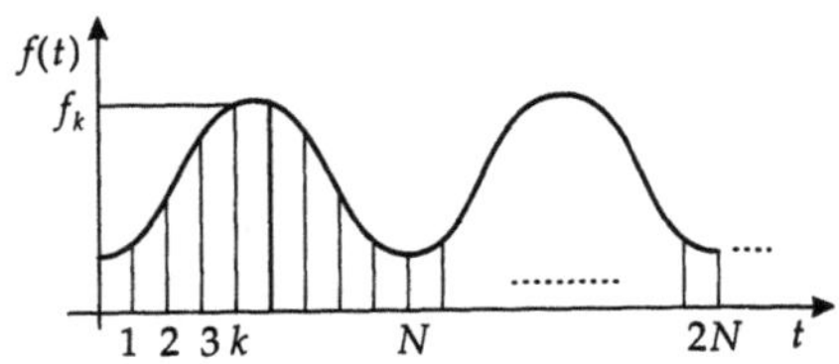

Schematic representation of a discrete periodic function $f(t_k)$

with $t_k = k\Delta t$ and $T = N\Delta t$, where Δt is the grid interval. In measurement engineering, Δt is also called the scanning period. Correspondingly, the scanning frequency F_0 is called the reciprocal of the scanning period: $F_0 = 1/\Delta t$.
Coefficient of the discrete Fourier transformation: Compare the coefficient of the Fourier series with the Fourier integrals:

$$c_i = \sum_{m=0}^{N-1} f(t_m) e^{-2\pi j i m/N}.$$

Notation: Often, we simply write f_k instead of $f(t_k)$.

▷ 1. The pre-factor $1/N$ can be divided into three types, just as in the case of the Fourier integral (see note on the complex representation of the Fourier integrals). Here, the pre-factor is to come before the sum of f_k.

2. The coefficients c_i follow from the orthogonality of the functions $e^{2\pi j i k/N}$ $(i, k = 0, 1, 2, \ldots, N-1)$:

$$\frac{1}{N}\sum_{i=0}^{N-1} e^{2\pi j i m/N} e^{-2\pi j i n/N} = \delta_{m,n}\,.$$

3. The discrete Fourier transformation can also be regarded as the trigonometric interpolation of the function $f(t_k)$, since at the grid points it holds that $f(t_k) = T_N(t_k)$, and for arbitrary points t from $[0, T]$: $f(t) \approx T_N(t)$.

4. The vector $\{c_i\}$ is called the discrete Fourier transform of $f(t_k)$, and the transition $c_i \to T_N(t_k)$ the discrete inverse Fourier transformation.

5. The same symmetries hold for the discrete Fourier transformation as for the continuous Fourier transformation.

▷ Numerical calculation of the discrete Fourier transformation: Since PASCAL does not support complex arithmetic, the exponential function is divided into a real and an imaginary part:

$$T_N(t_k) = \frac{1}{N}\sum_{i=0}^{n-1} c_i \mathrm{e}^{2\pi \mathrm{j} ik/n}$$

$$= \frac{1}{N}\sum_{i=0}^{n-1} (c_i \cos(2\pi ik/N) + \mathrm{j}c_i \sin(2\pi ik/N)) .$$

Pseudocode for the calculation of $T_N(t_k)$:

```
BEGIN discrete Fourier transformation
INPUT n
INPUT f[i], i=0...n-1
omega := 2*pi/n
FOR k = 0 TO n-1 DO
    FOR m = 0 TO n-1 DO
       angle := k*omega*m
       real[k] := real[k] + f[m]*cos(angle)/n
       imag[k] := imag[k] - f[m]*sin(angle)/n
    ENDDO
ENDDO
OUTPUT real[i], i=0...n
OUTPUT imag[i], i=0...n
END discrete Fourier transformation
```

If the coefficients c_i are complex, these must also be divided into a real and an imaginary part.

19.3.2 Shannon scanning theorem

- **Shannon scanning theorem**: If an analog measuring signal $f(t)$ is scanned with the scanning period Δt, is scanned at times $t_k = k\,\Delta t$, $k = 0, 1, 2, \ldots, N-1$, $(f(t) \to f(t_k) = f_k)$, then the shape of the signal $f(t)$ can be reconstructed from the scanning values f_k if the highest frequency F_m occurring in the measuring signal $f(t)$ is smaller than half the scanning frequency $F_0/2 = 1/(2\Delta t)$:

$$F_m < \frac{F_0}{2} .$$

- **Nyquist frequency**: Half the scanning frequency $1/(2\Delta t)$, which is the highest frequency that can be detected with the scanning period Δt.
- **Aliasing**: If the measuring signal contains frequencies higher than the Nyquist frequency, these are mapped onto frequency range $F < F_0/2$, so that the original measuring signal $f(t)$ can no longer be determined completely from scanning values f_k.

▷ Before scanning a measuring signal, all frequencies higher than the Nyquist frequency can be filtered out by an antialiasing filter so that the Shannon scanning theorem applies again.

19.3.3 Discrete sine and cosine transformation

Discrete sine transformation: Used for functions that vanish at the interval borders. Important for the solution of differential equations with the corresponding boundary condition:

$$f_k = \frac{2}{N} \sum_{i=1}^{N-1} s_i \sin(\pi i k/N) .$$

Coefficients of the discrete sine transformation:

$$s_i = \sum_{l=1}^{N-1} f_l \sin(\pi l i/N) ,$$

i.e., the sine transformation is its own inverse transformation.

Discrete cosine transformation: Used for functions whose derivation vanishes at the interval borders:

$$f_k = \frac{2}{N} \sum_{i=0}^{N-1} c_i \cos(\pi i k/N) .$$

Coefficients of the discrete cosine transformation: The calculation proceeds in several partial steps (N even):

a. Calculation of auxiliary coefficients $\tilde{c}_i$ as if the cosine transformation were its own inverse:

$$\tilde{c}_i = \sum_{l=0}^{N-1} f_l \cos\left(\frac{\pi l i}{N}\right) .$$

b. Relation between $\tilde{c}_i$ and the required coefficients c_i is:

$$\tilde{c}_0 = 2c_0 + \frac{2}{N} \sum_{m\ (e)} c_m ,$$

$$\tilde{c}_i = c_i + \frac{2}{N} \sum_{m\ (o)} c_m , \qquad \text{if } i \text{ odd,}$$

$$\tilde{c}_i = c_i + \frac{2}{N} \sum_{m\ (e)} c_m , \qquad \text{if } i \text{ even, } i \neq 0,$$

with the abbreviations (e) for even and (o) for odd.

c. Calculation of the unknown sums on the right side:

$$\sum_{m\ (\mathrm{e})} c_m = \frac{N}{2}\tilde{c}_0 - N(A_1 - A_2)\,,$$

$$\sum_{m\ (\mathrm{o})} c_m = A_1 - \sum_{m\ (\mathrm{e})} c_m\,,$$

where

$$A_1 = \sum_{i\ (\mathrm{e})} \tilde{c}_i \quad \text{and} \quad A_2 = \sum_{i\ (\mathrm{o}),\, i\neq 0} \tilde{c}_i\,.$$

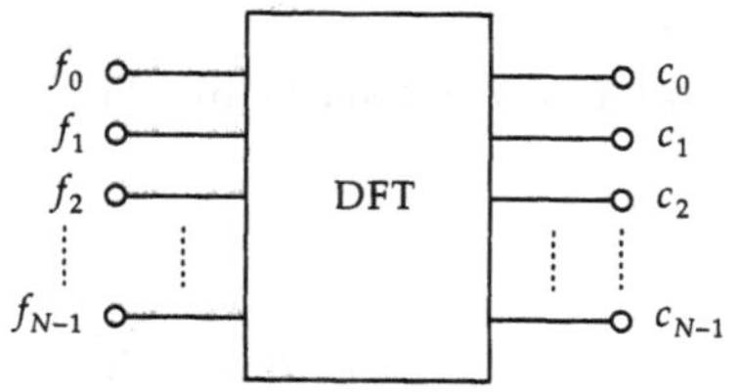

Symbolic representation of the discrete Fourier transformation

▷ For the sine and cosine DFT, the argument of the cosine and sine reads $(\pi ik/N)$ instead of $(2\pi ik/N)$. This is related to the fact that even or odd functions are considered for the above two transformations so that not just the whole interval $[-\pi, \pi)$ must be considered, but only the subinterval $[0, \pi)$.

19.3.4 Fast Fourier transformation (FFT)

Fast Fourier transformation (FFT): Fast algorithm for the calculation of the discrete Fourier transformation.

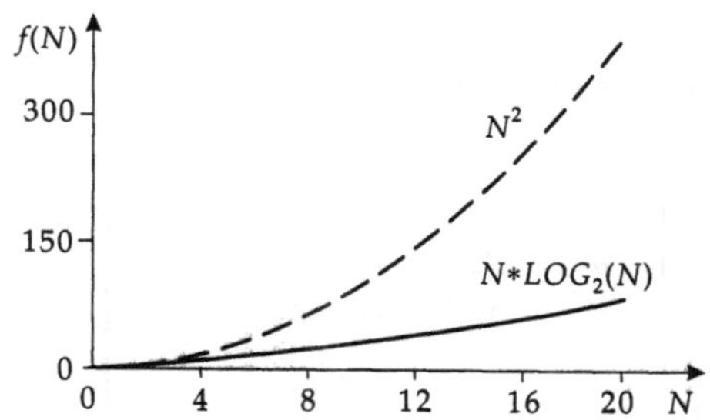

The functions N^2 and $N \log_2 N$

The number of calculating operations is proportional only to $N \log_2 N$, compared with $\propto N^2$ in the direct calculation of the sum.

FFT algorithm

a. Splitting the coefficients c_k of the discrete Fourier transformation into even and odd Fourier transforms:

$$\begin{aligned} c_k &= \sum_{i=0}^{N-1} f_i e^{-2\pi j i k/N} \\ &= \sum_{i=0}^{N/2-1} f_i e^{-2\pi j i k/N} + \sum_{i=0}^{N/2-1} f_{i+N/2} e^{-2\pi j k(i+N/2)/N} \\ &= \sum_{i=0}^{N/2-1} \left(f_i + e^{-j\pi k} f_{i+N/2}\right) e^{-2\pi j i k/N} . \end{aligned}$$

Consider c_k separately for even (e) and odd (o) values of k:

even:

$$\begin{aligned} c_{2k} &= \\ &= \sum_{i=0}^{N/2-1} \left(f_i + f_{i+N/2}\right) e^{-2\pi j i k/(N/2)} = \sum_{i=0}^{N/2-1} \left(f_i + f_{i+N/2}\right) W^{2ki} \end{aligned}$$

and odd:

$$\begin{aligned} c_{2k+1} &= \\ &= \sum_{i=0}^{N/2-1} \left(f_i - f_{i+N/2}\right) W^i e^{-2\pi j i k/(N/2)} = \sum_{i=0}^{N/2-1} \left(f_i - f_{i+N/2}\right) W^i W^{2ki} . \end{aligned}$$

Here, W is the complex number $e^{-2\pi j/N}$. Both equations also can be represented in matrix notation. For **even** c_k values, we then obtain:

$$\begin{pmatrix} c_0 \\ c_2 \\ c_4 \\ \vdots \\ c_{N-2} \end{pmatrix} = \begin{pmatrix} 1 & 1 & 1 & \cdots & 1 \\ 1 & W^2 & W^4 & \cdots & W^{N-2} \\ 1 & W^4 & W^8 & \cdots & W^{2(N-2)} \\ \vdots & \vdots & \vdots & \ddots & \vdots \\ 1 & W^{(N-2)} & W^{2(N-2)} & \cdots & W^{(N-2)^2} \end{pmatrix} \begin{pmatrix} f_0 + f_{N/2} \\ f_1 + f_{N/2+1} \\ f_2 + f_{N/2+2} \\ \vdots \\ f_{N/2-1} + f_{N-1} \end{pmatrix}$$

and, analogously, for **odd** c_k values:

$$\begin{pmatrix} c_1 \\ c_3 \\ c_5 \\ \vdots \\ c_{N-1} \end{pmatrix} = \begin{pmatrix} 1 & W & W^2 & \cdots & W^{N/2-1} \\ 1 & W^3 & W^6 & \cdots & W^{3(N/2-1)} \\ 1 & W^5 & W^{10} & \cdots & W^{5(N/2-1)} \\ \vdots & \vdots & \vdots & \ddots & \vdots \\ 1 & W^{(N-1)} & W^{2(N-1)} & \cdots & W^{(N-1)(N/2-1)} \end{pmatrix} \begin{pmatrix} f_0 - f_{N/2} \\ f_1 - f_{N/2+1} \\ f_2 - f_{N/2+2} \\ \vdots \\ f_{N/2-1} - f_{N-1} \end{pmatrix} .$$

$c_k^{(e)} = c_{2k}$ (**even** values of k) is thus the Fourier transform of function $f_i^{(e)} = f_i + f_{i+N/2}$,

$c_k^{(o)} = c_{2k+1}$ (**odd** values of k) is the Fourier transform of function $f_i^{(o)} = (f_i - f_{i+N/2})W^i$, with $i = 0, 1, 2, \ldots, N/2$ in both cases.

▷ Due to this splitting, the number of calculating operations is reduced from $\propto N^2$ to $\propto 2(N/2)^2 = N^2/2$.

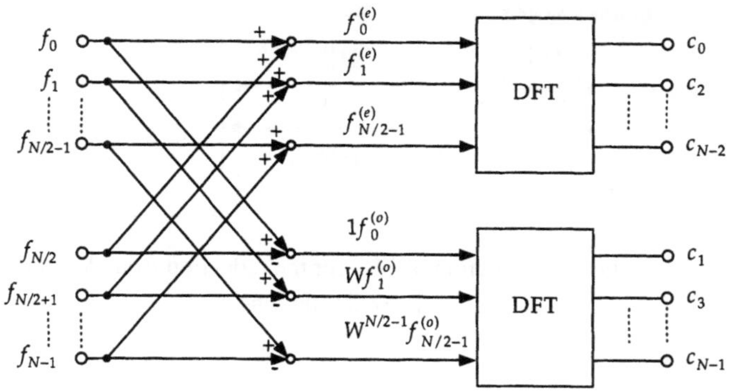

Symbolic representation of FFT: the step from N to $N/2$

b. **Recursion**: The splitting into even and odd functions is repeated for functions $f_i^{(e)}$ and $f_i^{(o)}$ with the upper index $[(N/2)/2] - 1 = [N/4] - 1$. Recursion which is defined by further repetition, is terminated when the upper index becomes $(N/N) - 1 = 0$, while the Fourier transform $c^{(o)(e)(e)\dots(o)}$ is simply the value of the function $f^{(o)(e)(e)\dots(o)}$. If the relation between the sequence of the even (e) and odd (o) splittings of the function f_k and the initial sequence of the discrete function values f_k is known, this recursion scheme yields the Fourier transform of f_k. The required relation results from

c. **Inversion of the sequence of bits**:

(a) First, invert the sequence of the even and odd splittings of f_k into $f^{(o)(e)(e)\dots(o)}$:

$$(o)(e)(e)\dots(o) \rightarrow (o)\dots(e)(e)(o)\ .$$

(b) Then, assign the values zero and one to (e) and (o), respectively.

(c) For each sequence of even and odd splittings of f_k the result of this operation yields the value of k in the dual system.

Thus, the Fourier transform is determined.

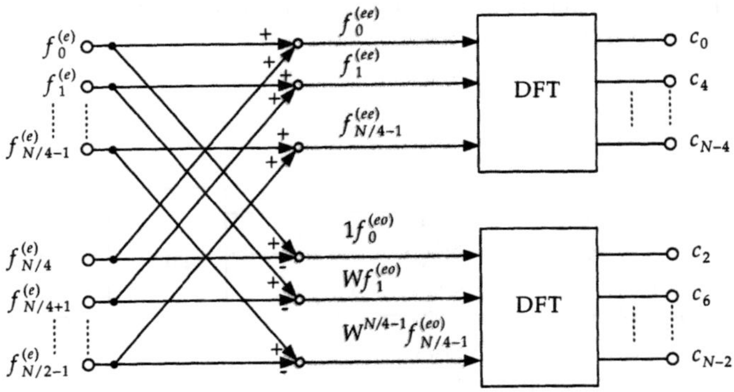

Symbolic representation of FFT for the reduction from $N/2$ to $N/4$ for the even functions $f_i^{(e)}$

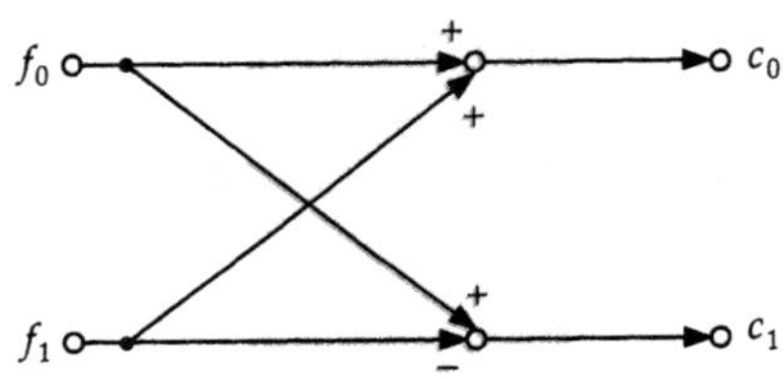

Symbolic representation of FFT for $N = 2$

□ Changing a sequence of even and odd splittings of f_k into a decimal value of k. Let the dimension of the problem be $N = 32$. Thus, the discrete function f_k can be split five times into even and odd values. Let $f^{(o)(e)(o)(o)(e)}$ be one of the 32 resulting functions. To determine the corresponding value of k ($k = 0, 1, 2, \ldots, 31$), the sequence of the even and odd splittings must be inverted following the above rule:

$$\text{oeooe} \rightarrow \text{eooeo} .$$

Then, the zero must be assigned to the even splitting, and the one to the odd:

$$\text{eooeo} \rightarrow 01101$$

This is the required k value in the dual system. The corresponding decimal value is $k = 1 \times 1 + 0 \times 2 + 1 \times 4 + 1 \times 8 + 0 \times 16 = 13$.

▷ The above recursion scheme is called the Sande-Tukey FFT algorithm. It is characterized by the fact that recursion starts with the Fourier transform.
If recursion takes place in inverted order, i.e., the algorithm starts with the function values in the inverted order of bits, this approach is called the Cooley-Tukey FFT algorithm.

▷ A condition for both algorithms is that N must be a multiple of 2 ($N = 2^m$, m integer). If this is not the case, f_k can be filled with zeros so that this criterion is met. But there are also modified FFT algorithms to which the limitation that N must be an integral multiple of 2 does not apply.

▷ Pseudocode for the Sande-Tukey FFT algorithm with $N = 2^m$:

```
BEGIN Fast Fourier transform
INPUT n
INPUT x[i], y[i], i=0...n-1
n2 := n
FOR k = 1 TO m DO
   n1 := n2
   n2 := n2/2
   angle := 0
   argument := 2*pi/n1
   FOR j = 0 TO n2-1 DO
      c := cos(angle)
      s := -sin(angle)
      FOR i = j TO n-1 STEP n1 DO
         l := i+n2
         xdum := x[i] - x[l]
         x[i] := x[i] + x[l]
         ydum := y[i] - y[l]
         y[i] := y[i] + y[l]
         x[l] := xdum*c - ydum*s
         y[l] := ydum*c + xdum*s
      ENDDO
      angle := (j + 1)*argument
   ENDDO
ENDDO
j := 0
FOR i = 0 TO n-2 DO
   IF (i < j) THEN
      xdum := x[j]
      x[j] := x[i]
      x[i] := xdum
      ydum := y[j]
      y[j] := y[i]
      y[i] := ydum
   ENDIF
   k := n/2
   WHILE (k < j+1) DO
      j := j - k
      k := k/2
   ENDDO
   j := j + k
ENDDO
FOR i = 0 TO n-1 DO
   x[i] := x[i]/n
   y[i] := y[i]/n
ENDDO
END Fast Fourier transform
```

19.3.5 Particular pairs of Fourier transforms

$$f(t) \quad \circ\!\!-\!\!\bullet \quad F(\omega)$$

$$\begin{cases} 1 & |t| < b \\ 0 & |t| > b \end{cases} \quad \circ\!\!-\!\!\bullet \quad \frac{2\sin b\omega}{\omega}$$

$$\frac{1}{t^2+b^2} \quad \circ\!\!-\!\!\bullet \quad \frac{\pi \mathrm{e}^{-b\omega}}{b}$$

$$\frac{t}{t^2+b^2} \quad \circ\!\!-\!\!\bullet \quad -\frac{\pi \mathrm{j}\omega}{b}\mathrm{e}^{-b\omega}$$

$$t^n f(t) \quad \circ\!\!-\!\!\bullet \quad \mathrm{j}^n \frac{\mathrm{d}^n F}{\mathrm{d}\omega^n}$$

$$f(bt)\mathrm{e}^{\mathrm{j}pt} \quad \circ\!\!-\!\!\bullet \quad \frac{1}{b}F\left(\frac{\omega-p}{b}\right)$$

$$f^{(n)}(t) \quad \circ\!\!-\!\!\bullet \quad \mathrm{j}^n\omega^n F(\omega)$$

19.3.6 Fourier transforms (table)

H(x)	Heaviside function
Γ(x)	Gamma function
$J_n(x)$	Bessel function of the first kind of order n (n = integer)
$J_{n+1/2}(x)$	Bessel function of the first kind of half integral order

$$f(t) \quad \circ\!\!-\!\!\bullet \quad F(\omega)$$

$$1 \quad \circ\!\!-\!\!\bullet \quad 2\pi\delta(\omega)$$

$$H(t) \quad \circ\!\!-\!\!\bullet \quad \pi\delta(\omega) + \frac{1}{\mathrm{j}\omega}$$

$$\operatorname{sgn}(t) \quad \circ\!\!-\!\!\bullet \quad \frac{2}{\mathrm{j}\omega}$$

$$\delta(t) \quad \circ\!\!-\!\!\bullet \quad 1$$

$$\sin\omega_0 t \quad \circ\!\!-\!\!\bullet \quad \mathrm{j}\pi\delta(\omega+\omega_0) - \mathrm{j}\pi\delta(\omega-\omega_0)$$

$$H(t)\sin(\omega_0 t) \quad \circ\!\!-\!\!\bullet \quad \frac{\pi}{2\mathrm{j}}\delta(\omega+\omega_0) - \frac{\pi}{2\mathrm{j}}\delta(\omega-\omega_0) + \frac{\omega}{\omega_0^2-\omega^2}$$

$$H(t)\mathrm{e}^{-at}\sin\omega_0 t \quad \circ\!\!-\!\!\bullet \quad \frac{\omega_0}{(\mathrm{j}\omega+a)^2+\omega_0^2} \quad \text{for } \mathrm{Re}\{a\} > 0$$

$$\cos\omega_0 t \quad \circ\!\!-\!\!\bullet \quad \pi\delta(\omega-\omega_0) + \pi\delta(\omega-\omega_0)$$

$$H(t)\cos\omega_0 t \quad \circ\!\!-\!\!\bullet \quad \frac{\pi}{2}\delta(\omega-\omega_0) + \frac{\pi}{2}\delta(\omega+\omega_0) + \mathrm{j}\frac{\omega}{\omega_0^2-\omega^2}$$

$$H(t)\mathrm{e}^{-at}\cos\omega_0 t \quad \circ\!\!-\!\!\bullet \quad \frac{\mathrm{j}\omega+a}{(\mathrm{j}\omega+a)^2+\omega_0^2} \quad \text{for } \mathrm{Re}\{a\} > 0$$

$$\frac{1}{\sqrt{2\pi}}\frac{\sin at}{t} \quad \circ\!\!-\!\!\bullet \quad \begin{cases} \sqrt{\dfrac{\pi}{2}} & (|\omega| < a) \\ 0 & (|\omega| > a) \end{cases}$$

$$f(t) = \begin{cases} \dfrac{1}{\sqrt{2\pi}}e^{jvt} & (p < t < q) \\ 0 & (t < p,\ t > q) \end{cases} \quad \circ\!\!-\!\!\bullet \quad \frac{j}{\sqrt{2\pi}}\frac{e^{jp(v+\omega)} - e^{jq(v+\omega)}}{(v+\omega)}$$

$$f(t) = \begin{cases} \dfrac{1}{\sqrt{2\pi}}e^{-ct+jvt} & (t > 0,\ c > 0) \\ 0 & (t < 0) \end{cases} \quad \circ\!\!-\!\!\bullet \quad \frac{j}{\sqrt{2\pi}}\frac{1}{(v+\omega+jc)}$$

$$\frac{1}{\sqrt{2\pi}}e^{-pt^2} \quad \mathrm{Re}(p) > 0 \quad \circ\!\!-\!\!\bullet \quad \frac{1}{\sqrt{2p}}e^{-\omega^2/4p}$$

$$\frac{1}{\sqrt{2\pi}}\cos pt^2 \quad \circ\!\!-\!\!\bullet \quad \frac{1}{\sqrt{2p}}\cos\left[\frac{\omega^2}{4p} - \frac{\pi}{4}\right]$$

$$\frac{1}{\sqrt{2\pi}}\sin pt^2 \quad \circ\!\!-\!\!\bullet \quad \frac{1}{\sqrt{2p}}\cos\left[\frac{\omega^2}{4p} + \frac{\pi}{4}\right]$$

$$\frac{1}{\sqrt{2\pi}}|t|^{-p} \quad (0 < p < 1) \quad \circ\!\!-\!\!\bullet \quad \sqrt{\frac{2}{\pi}}\frac{\Gamma(1-p)\sin(p\pi/2)}{|\omega|^{(1-p)}}$$

$$\frac{1}{\sqrt{2\pi}}\frac{e^{-a|t|}}{\sqrt{|t|}} \quad \circ\!\!-\!\!\bullet \quad \frac{(\sqrt{(a^2+\omega^2)}+a)^{1/2}}{\sqrt{a^2+\omega^2}}$$

$$\frac{1}{\sqrt{2\pi}}\frac{\cosh at}{\cosh \pi t} \quad (-\pi < a < \pi) \quad \circ\!\!-\!\!\bullet \quad \sqrt{\frac{2}{\pi}}\frac{\cos(a/2)\cosh(\omega/2)}{\cosh\omega + \cos a}$$

$$\frac{1}{\sqrt{2\pi}}\frac{\sinh at}{\sinh \pi t} \quad (-\pi < a < \pi) \quad \circ\!\!-\!\!\bullet \quad \frac{1}{\sqrt{2\pi}}\frac{\sin a}{\cosh\omega + \cos a}$$

$$\begin{cases} \dfrac{1}{\sqrt{2\pi}}\dfrac{1}{\sqrt{a^2-t^2}} & (|t| < a) \\ 0 & (|t| > a) \end{cases} \quad \circ\!\!-\!\!\bullet \quad \sqrt{\frac{\pi}{2}}J_0(a\omega)$$

$$\frac{1}{\sqrt{2\pi}}\frac{\sin(b\sqrt{a^2+t^2})}{\sqrt{a^2+t^2}} \quad \circ\!\!-\!\!\bullet \quad \begin{cases} 0 & (|\omega| > b) \\ \sqrt{\dfrac{\pi}{2}}J_0(a\sqrt{b^2-\omega^2}) & (|\omega| < b) \end{cases}$$

$$\begin{cases} \dfrac{1}{\sqrt{2\pi}}P_n(t) & (|t| < 1) \\ 0 & (|t| > 1) \end{cases} \quad \circ\!\!-\!\!\bullet \quad \frac{j^n}{\sqrt{\omega}}J_{n+1/2}(\omega)$$

$$\begin{cases} \dfrac{1}{\sqrt{2\pi}}\dfrac{\cos(b\sqrt{a^2-t^2})}{\sqrt{a^2-t^2}} & (|t| < a) \\ 0 & (|t| > a) \end{cases} \quad \circ\!\!-\!\!\bullet \quad \sqrt{\frac{\pi}{2}}J_0(a\sqrt{a^2+b^2})$$

$$\begin{cases} \dfrac{1}{\sqrt{2\pi}} \dfrac{\cosh(b\sqrt{a^2 - t^2})}{\sqrt{a^2 - t^2}} & (|t| < a) \\ 0 & (|t| > a) \end{cases} \quad \circ\!\!-\!\!\bullet \quad \sqrt{\frac{\pi}{2}} J_0(a\sqrt{a^2 - b^2})$$

19.3.7 Particular Fourier sine transforms

$H(x)$ Heaviside function
$\Gamma(x)$ Gamma function

$$f(t) \quad \circ\!\!-\!\!\bullet \quad F_s(\omega)$$

$$f(t) = \begin{cases} 1 & (0 < t < b) \\ 0 & (t > b) \end{cases} \quad \circ\!\!-\!\!\bullet \quad \frac{1 - \cos b\omega}{\omega}$$

$$t^{-1} \quad \circ\!\!-\!\!\bullet \quad \frac{\pi}{2}$$

$$\frac{t}{t^2 + b^2} \quad \circ\!\!-\!\!\bullet \quad \frac{\pi}{2} \mathrm{e}^{-b\omega}$$

$$\mathrm{e}^{-bt} \quad \circ\!\!-\!\!\bullet \quad \frac{\omega}{\omega^2 + b^2}$$

$$H(t)\mathrm{e}^{-at} \quad \circ\!\!-\!\!\bullet \quad \frac{1}{a + \mathrm{j}\omega} \quad \mathrm{Re}\{a\} > 0$$

$$H(t)\mathrm{e}^{-a|t|} \quad \circ\!\!-\!\!\bullet \quad \frac{2a}{a^2 + \omega^2} \quad \mathrm{Re}\{a\} > 0$$

$$t^{n-1}\mathrm{e}^{-bt} \quad \circ\!\!-\!\!\bullet \quad \frac{\Gamma(n) \sin(n \arctan^{-1} \omega/b)}{(\omega^2 + b^2)^{n/2}}$$

$$t\mathrm{e}^{-bt^2} \quad \circ\!\!-\!\!\bullet \quad \frac{\sqrt{\pi}}{4b^{3/2}} \omega \mathrm{e}^{-\omega^2/4b}$$

$$t^{-1/2} \quad \circ\!\!-\!\!\bullet \quad \sqrt{\frac{\pi}{2\omega}}$$

$$t^{-n} \quad \circ\!\!-\!\!\bullet \quad \frac{\pi \omega^{n-1} \csc(n\pi/2)}{2\Gamma(n)}, \quad (0 < n < 2)$$

$$\frac{\sin bt}{t} \quad \circ\!\!-\!\!\bullet \quad \frac{1}{2} \ln\left(\frac{\omega + b}{\omega - b}\right)$$

$$\frac{\sin bt}{t^2} \quad \circ\!\!-\!\!\bullet \quad \begin{cases} \pi\omega/2 & (\omega < b) \\ \pi b/2 & (\omega > b) \end{cases}$$

$$\frac{\cos bt}{t} \quad \circ\!\!-\!\!\bullet \quad \begin{cases} 0 & (\omega < b) \\ \pi/4 & (\omega = b) \\ \pi/2 & (\omega > b) \end{cases}$$

$$\arctan(t/b) \quad \circ\!\!-\!\!\bullet \quad \frac{\pi}{2\omega}\mathrm{e}^{-b\omega}$$

$$\csc bt \quad \circ\!\!-\!\!\bullet \quad \frac{\pi}{2b}\tanh\frac{\pi\omega}{2b}$$

$$\frac{1}{\mathrm{e}^{2t}-1} \quad \circ\!\!-\!\!\bullet \quad \frac{\pi}{4}\coth\left(\frac{\pi\omega}{2}\right) - \frac{1}{2\omega}$$

19.3.8 Particular Fourier cosine transforms

$\Gamma(x)$ Gamma function

$$f(t) \quad \circ\!\!-\!\!\bullet \quad F_c(\omega)$$

$$f(t) = \begin{cases} 1 & (0 < t < b) \\ 0 & (t > b) \end{cases} \quad \circ\!\!-\!\!\bullet \quad \frac{\sin b\omega}{\omega}$$

$$\frac{1}{t^2+b^2} \quad \circ\!\!-\!\!\bullet \quad \frac{\pi \mathrm{e}^{-b\omega}}{2b}$$

$$\mathrm{e}^{-bt} \quad \circ\!\!-\!\!\bullet \quad \frac{b}{\omega^2+b^2}$$

$$t^{n-1}\mathrm{e}^{-bt} \quad \circ\!\!-\!\!\bullet \quad \frac{\Gamma(n)\cos(n\tan\omega/b)}{(\omega^2+b^2)^{n/2}}$$

$$\mathrm{e}^{-bt^2} \quad \circ\!\!-\!\!\bullet \quad \frac{1}{2}\sqrt{\frac{\pi}{b}}\mathrm{e}^{-\omega^2/4b}$$

$$t^{-1/2} \quad \circ\!\!-\!\!\bullet \quad \sqrt{\frac{\pi}{2\omega}}$$

$$t^{-n} \quad \circ\!\!-\!\!\bullet \quad \frac{\pi\omega^{n-1}\sec(n\pi/2)}{2\Gamma(n)}, \quad (0 < n < 1)$$

$$\ln\left(\frac{t^2+b^2}{t^2+c^2}\right) \quad \circ\!\!-\!\!\bullet \quad \frac{\mathrm{e}^{-c\omega}-\mathrm{e}^{-b\omega}}{\pi\omega}$$

$$\frac{\sin bt}{t} \quad \circ\!\!-\!\!\bullet \quad \begin{cases} \pi/2 & (\omega < b) \\ \pi/4 & (\omega = b) \\ 0 & (\omega > b) \end{cases}$$

$$\sin bt^2 \quad \circ\!\!-\!\!\bullet \quad \sqrt{\frac{\pi}{8b}}\left(\cos\frac{\omega^2}{4b} - \sin\frac{\omega^2}{4b}\right)$$

$$\cos bt^2 \quad \circ\!\!-\!\!\bullet \quad \sqrt{\frac{\pi}{8b}}\left(\cos\frac{\omega^2}{4b} + \sin\frac{\omega^2}{4b}\right)$$

$$\mathrm{sech} bt \quad \circ\!\!-\!\!\bullet \quad \frac{\pi}{2b}\mathrm{sech}\frac{\pi\omega}{2b}$$

$$\frac{\cosh(\sqrt{\pi}t/2)}{\cosh(\sqrt{\pi}t)} \quad \circ\!\!-\!\!\bullet \quad \sqrt{\frac{\pi}{2}}\frac{\cosh(\sqrt{\pi}\omega/2)}{\cosh(\sqrt{\pi}\omega)}$$

$$\frac{e^{-b\sqrt{t}}}{\sqrt{t}} \quad \circ\!\!-\!\!\bullet \quad \sqrt{\frac{\pi}{2\omega}}[\cos(2b\sqrt{\omega}) - \sin(2b\sqrt{\omega})]$$

19.4 Wavelet transformation

19.4.1 Signals

Signal: Effect that propagates from a (physical) object and can be described mathematically. The mathematical description of a signal may be in terms of a number sequence or prearranged symbols.

□ Noise is converted by using a microphone and analog-to-digital converter into a function or a numerical sequence which describes pressure fluctuations of the air.

□ A picture can be converted into two-dimensional color and brightness distribution by using an electric camera.

□ An earthquake can be converted into number sequences that describe the motion of the surface of the earth surface at various points by using a network of measuring stations.

Signal analysis: The characterization of a signal by means of quantities appropriate to the (physical) problem. Mathematically: Mapping of the number sequence or function describing the signal onto another number sequence or function that describes characteristics of the signal. Some information may be lost during the mapping process (**reduction**).

Synthesis: The inversion of signal analysis, i.e., the reconstruction of the signal.

Typical applications of signal analysis:

- Determination of physical quantities from measurements
- **Recognition of images and speech**
- **Data compression** of a signal for the efficient transmission of communication.

□ A noise may be decomposed with respect to the tones which occur (piece of music) or with respect to its phonemes (speech).
For the compression of speech it is being tried to discriminate properties of a speech signal important to the human ear and to filter out the unimportant.

□ Systems for image recognition decompose an image signal into the objects represented.
In the compression of images, it is easier for example, to leave out the exact details of a surface than the contours of the body.

□ A three-dimensional picture of an earthquake can be reconstructed from the sum of all seismographic data. Thus, we obtain information about the epicenter and the energy of the earthquake.

Linear algebra as a mathematical tool of signal analysis:

- Signals are regarded as **vectors** in a multidimensional vector space.
- Analysis through multiplication by appropriate matrices (**operators**).
- The **scalar product** gives the distance of two signals (similar signals are closely adjacent).

□ **Fourier analysis** decomposes a signal according to frequency bands. Appropriate operators allow the filtering out of unimportant frequencies and to amplify others. Advantage: Linear operations can be achieved in the form of electronic circuits.

19.4.2 Linear signal analysis

Signal function: Representation of a signal as a continuous function $f(x)$ of one (or more) variables x.

Signal vector: Representation of a signal as a vector $f_i, i = 1, \ldots, n$, with n components. A signal vector is formed by a signal function through **discretization**:

$$f_i = f(x_i)$$

at fixed points x_i.

Linear signal analysis: Decomposition of a signal vector $f_i, i = 1, \ldots, n$, into a linear combination of given **basis vectors** (transformation matrix) $g_i^{(k)}$, $k = 1, \ldots, m$ with **coefficients** $\hat{f}_k$ (**linear superposition**):

$$f_i = \sum_{k=1}^{m} \hat{f}_k g_i^{(k)}.$$

The different basis vectors represent different properties of the signal. The signal is determined by the coefficients $\hat{f}_k$; the aim of signal analysis is to determine these coefficients.

□ Fourier analysis where the basis functions are sine and cosine functions for certain frequencies ω. Fourier coefficients $\hat{f}(\omega)$ indicate the presence of an oscillation of frequency ω.

Dual basis: Set of vectors $\hat{f}_i^{(k)}, i = 1, \ldots, m$, having the property that the scalar products of the signal vector $\mathbf{f}$ with vectors $\hat{\mathbf{f}}^{(k)}$

$$a_k = \sum_{i=1}^{n} f_i \hat{f}_i^{(k)}$$

yield the coefficients of the signal; i.e., that the signal analysis is performed in precisely this manner.

Self-dual basis: Basis in which the basis vectors and the dual basis vectors are identical.

□ Fourier analysis and Fourier synthesis are both performed by sine and cosine functions; thus both bases are identical.

□ Rectangle functions as basis: A rectangle function $g^{(t,d)}$ of width d at the point t is a function that has the value one between $t - d/2$ and $t + d/2$ and the value zero outside this interval. If we choose these functions as the basis, the coefficients $g^{(t,d)}$ indicate that the signal was present at the time t for the duration of d.
This basis is "overcomplete" because, for example, every rectangle function may be represented by halves, which are rectangle functions again.

Linear independence of the basis: The property that no basis vector can be represented as a linear combination of the other basis vectors.

● If the basis is linearly independent, then the coefficients of a signal are determined uniquely.

Orthonormal basis: A set of basis vectors $f_i^{(k)}$ in which every vector is normalized and orthogonal to any other vector, that is,

$$\sum_{i=1}^{n} f_i^{(k)} f_i^{(k')} = \begin{cases} 0 & \text{for } k \neq k' \\ 1 & \text{otherwise.} \end{cases}$$

- An orthonormal basis is linearly independent and self-dual. The scalar product of two vectors does not change in an orthogonal transformation of the basis.

Overcomplete basis: Several different coefficient vectors can be assigned to one signal.

□ **Fourier analysis**: Characterization of a signal $f(t)$ by the (angular) frequencies ω occurring in it:

$$\tilde{f}(\omega) = \frac{1}{\sqrt{2\pi}} \int \mathrm{d}t\, e^{-\mathrm{j}\omega t}\, f(t)$$

$$f(t) = \frac{1}{\sqrt{2\pi}} \int \mathrm{d}\omega\, e^{\mathrm{j}\omega t}\, \tilde{f}(\omega)$$

The Fourier basis is **self-dual.**

19.4.3 Symmetry transformations

The behavior of the coefficients of a signal analysis under a symmetry transformation characterizes the basis used. Basis functions can often be transformed into each other by symmetry transformations.

Elementary symmetry transformations:

Translation: The shift of a signal by the amount d, represented by the transformation

$$(T_d f)(x) = f(x + d)$$

with the **translation operator** T_d.

□ Rectangle functions of equal width as basis functions can be transformed into each other by translations.

Dilatation: The stretching (compression) of a signal by the factor α, represented by the transformation

$$(D_\alpha f)(x) = f(\alpha x)$$

with the **dilatation operator** D_α.

□ Every element $f^{(\omega)}(t) = e^{\mathrm{j}\omega t}$ of the Fourier basis can be transformed into any other one by dilatation. In particular, the entire basis may be generated from a single element, e.g., $f^{(0)}$:

$$f^{(\omega)}(t) = (D_\omega f^{(0)})(t).$$

□ Behavior of the Fourier transform $\tilde{f}(k)$ of a function under translation:

$$g(x) = (T_d f)(x) = f(x + d) \quad \Longleftrightarrow \quad \tilde{g}(k) = e^{-\mathrm{j}kd}\, \tilde{f}(k)\ .$$

Because this **phase factor** $\mathrm{e}^{-\mathrm{j}kd}$ is difficult to determine in a measurement, information about the time when a certain frequency occurs in the signal is difficult to obtain from the Fourier transform.

□ A piece of music consists of consecutive tones of various frequencies. The Fourier transform provides good information as to the times when these frequencies occur, but it is more difficult to determine their sequence from the Fourier transform.

19.4.4 Time-frequency analysis and Gabor transformation

Time-frequency analysis: Characterization of a signal with respect to the frequencies occurring which it contains and with respect to the points in time at which these frequencies occur, by dividing the signal into time intervals (windows) and Fourier transformations.
Gabor transformation: Linear transformation of a signal multiplied by a **window function** $g_\alpha^b(t)$

$$(\mathcal{G}_b^\alpha f)(\omega) = \int \mathrm{d}t\, g_\alpha^b(t)\, e^{-\mathrm{j}\omega t} f(t)$$

$$g_\alpha^b(t) = \frac{1}{2\sqrt{\pi\alpha}} e^{-(t-b)^2/4\alpha} .$$

The window function is chosen so that it extracts the signal outside the window of mean square width $2\sqrt{\alpha}$ by the point b. Every window function $g_\alpha^b(t)$ can be generated from any other by dilatation **and** translation:

$$g_\alpha^b(t) = T_d\, D_{1/\sqrt{\alpha}}\, g_1^1(t).$$

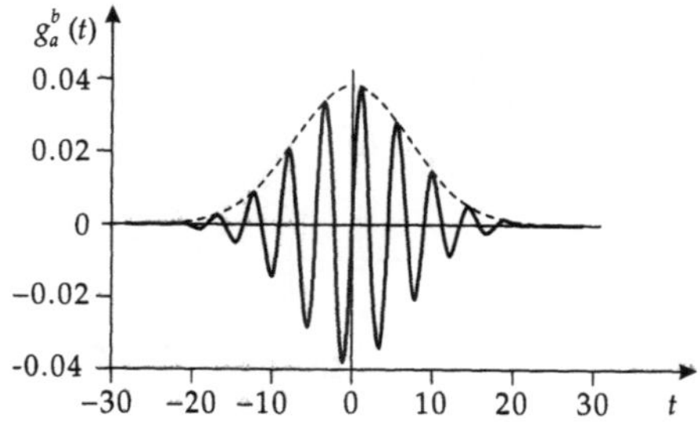

Basis function of the Gabor transformation.
Dotted line: envelope $g_\alpha^b(t)$

In contrast to the case of pure position or frequency representations, both symmetry operations are therefore permissible in the Gabor transformation.
Parameters of the Gabor transformation:

- Width $\sqrt{\alpha}$ of the window,
- Position b of the window,
- Angular frequency ω of the signal component.

The Gabor transformation is overdetermined, i.e., it is not necessary to know $(\mathcal{G}_b^\alpha f)(\omega)$ for all values of α, b, and ω.
The Gabor transform constitutes a midway between time and frequency representation. It is therefore interesting how it can also be generated from the pure frequency representation $\tilde{f}(\omega)$:

$$(\mathcal{G}_b^\alpha f)(\omega) = \frac{1}{2\sqrt{\pi\alpha}}\, e^{-\mathrm{j}b\omega}\, (\mathcal{G}_\omega^{1/4\alpha}\, \tilde{f})(b).$$

The Gabor transformation can therefore be applied to both time and frequency representation.

● In Fourier representation, the Gabor transform is obtained by setting a window of the width $\sqrt{1/\alpha}$ around frequency ω and transforming back to the position space with respect to b. Thus, the roles of b and ω are interchanged; the width in Fourier space is inversely proportional to the width in position space.

- The Gabor transform $(\mathcal{G}_b^\alpha f)(\omega)$ of a function $f(t)$ yields information about the function within a time window of width $2\sqrt{\alpha}$ about b and within a frequency window of width $\sqrt{1/\alpha}$ about frequency ω.

The inverse proportionality of the width in position space and the width in Fourier space is a general property of signal analysis that does not depend on the form of the window function:

- **Uncertainty relation**: The more precisely a property of a signal is to be localized in position space the broader a corresponding window must be chosen in the Fourier space.
  The two widths are inversely proportional to each other.

Time-frequency representation: Two-dimensional diagram with time t plotted on one axis and frequency ω on the other. In this diagram the Gabor transformation is represented by a rectangular time-frequency window of constant area 2. In particular, the side lengths of the rectangle depend only on the parameter α and not on frequency ω.

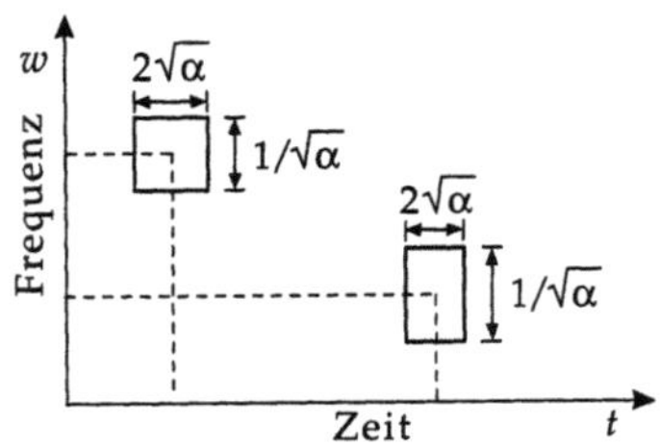

Gabor transformation in
time-frequency representation

The two limits $\alpha \to 0$ and $\alpha \to \infty$ result in the time and frequency representation of function f. We therefore call the Gabor transformation a **mixed time-frequency representation.**

19.4.5 Wavelet transformation

The term "wavelet" generalizes the Gabor transformation. This utilizes the fact that the Gabor functions g_α^b can be generated from a single function by dilatation and translation.

Integral wavelet transformation: The transformation

$$(W_\psi f)(b,a) = \frac{1}{\sqrt{|a|}} \int \mathrm{d}t\ f(t)\, \psi_{b,a}(t)$$

$$\psi_{b,a}(t) = \overline{\psi\left(\frac{t-b}{a}\right)}$$

for $a, b \in \mathbb{R}$, $a \neq 0$. Parameter b gives the position and a gives the **scale** (extension) of the basis functions $\psi_{b,a}$. The function ψ is called a **wavelet** (small localized wave) if it is possible to re-calculate the original function $f(x)$ from $(W_\psi f)(b,a)$:

$$f(t) = \frac{1}{C_\psi} \int_{-\infty}^{\infty} (W_\psi f)(b,a)\, \psi_{b,a}(t) \frac{\mathrm{d}a\,\mathrm{d}b}{a^2}$$

with a constant C_ψ as a function of the wavelet.

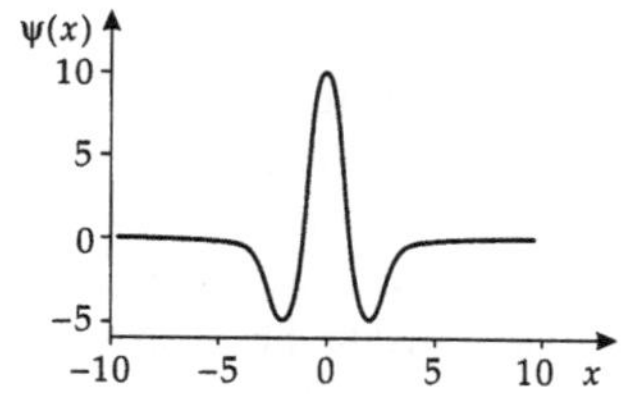

Second derivative of a Gauss curve is a simple wavelet.

- The wavelet transformation is a linear signal analysis whose basis may be generated by translation and dilatation from a single function $\psi(x)$, the wavelet.

The general wavelet basis is "overcomplete." It is therefore only necessary to know the value of $(W_\psi f)(b,a)$ if a and b have certain values. This is achieved by dividing the time-frequency representation into rectangles of equal size, according to the uncertainty relation.

Dyadic wavelet transformation: Integral wavelet transformation whose basis functions are generated from a wavelet ψ by

- dyadic dilatations, i.e., dilatations by a factor of 2^i with an integer i that gives the scale,
- binary translations, i.e., for a given dilatation scale i translation by $2^i j$ with integer j.

$$\psi_j^{(i)}(t) = \frac{1}{\sqrt{2^i}} \psi\left(\frac{t - 2^i j}{2^i}\right)$$

$$(W_\psi f)_{i,j} = \int \mathrm{d}t\; f(t)\, \overline{\psi_{i,j}(t)}$$

Every basis function is characterized by its width (scale) and its position. The different basis functions are obtained by repeated doubling and halving of the width and displacements by integral multiples of the width.

In the time-frequency representation every basis function represents approximately a rectangle of width 2^i in space and therefore, according to the uncertainty relation, to width 2^{-i} in frequency. The windows are chosen so that they do not overlap much.

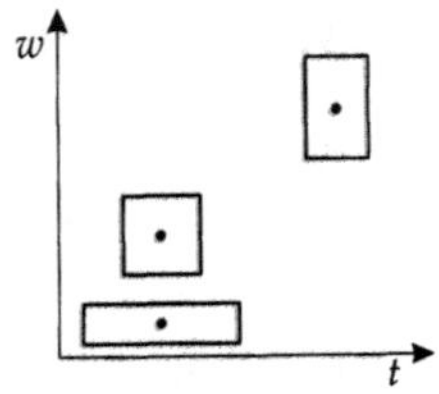

Wavelet transformation in time-frequency representation. The ratios of the sides of the rectangles are adjusted to the chosen frequency

Multi-resolution analysis: Decomposition of a signal according to the scales occurring in it. In wavelet transformation, this is accomplished by means of the scaling parameter a or i.

□ Decomposition of a photo (one of the authors of this book) with scales by Daubechies 6 Wavelets: Original: scales 1-2, 3-4, and 5-8, respectively

● The dyadic wavelet transformation allows us to obtain simultaneous information on the scale (frequency) of a property of a signal and on its position (point in time).

The uncertainty relation is avoided by the fact that information is available only for dyadic scales and positions.

Reconstruction of the signal function $f(t)$ by means of the dual basis $\tilde{\psi}_{i,j}$:

$$f(t) = \sum_i \sum_j (W_\psi f)_{i,j}\, \tilde{\psi}_{i,j}(t).$$

$\mathcal{R}$-wavelet: A wavelet whose dual basis can be generated again by means of dyadic translations and binary dilatations from only one function $\tilde{\psi}$. Then, $\tilde{\psi}$ is called the **dual wavelet**.

Haar function $\psi_H(t)$: The function

$$\psi_H(t) = \begin{cases} 1 & \text{for} \quad 0 \le t < \frac{1}{2} \\ -1 & \text{for} \quad \frac{1}{2} \le t < 1 \\ 0 & \text{otherwise.} \end{cases}$$

Simplest example of a wavelet function. The wavelets generated in this way

$$\psi_j^{(i)}(t) = \begin{cases} 1/\sqrt{2^i} & \text{for} \quad 2^i j \le t < j + 2^i/2 \\ -1/\sqrt{2^i} & \text{for} \quad j + 2^i/2 \le t < j + 2^i \\ 0 & \text{otherwise} \end{cases}$$

indicate that the signal has changed at point in time $2^i j$ for a duration of 2^i.

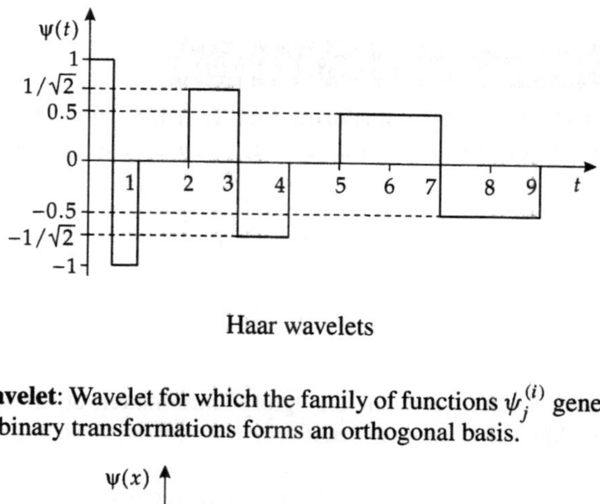

Haar wavelets

Orthogonal wavelet: Wavelet for which the family of functions $\psi_j^{(i)}$ generated by dyadic dilatations and binary transformations forms an orthogonal basis.

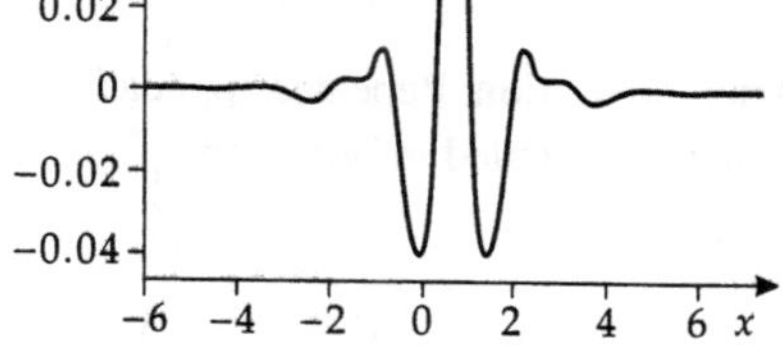

Battle-Lemarié wavelet, an orthogonal wavelet

Compact carrier: The property of a wavelet to be different from zero only on part of the time axis.

- The Haar function is the simplest example of an orthogonal wavelet with compact carrier. Thus, it is also self-dual, so that reconstruction can proceed with the same basis functions.

Other orthogonal wavelets: Construction according to I. Daubechies (1988).

Daubechies wavelets: Orthogonal wavelets with compact carrier. Irregular functions without closed representation but with particularly good numerical properties.

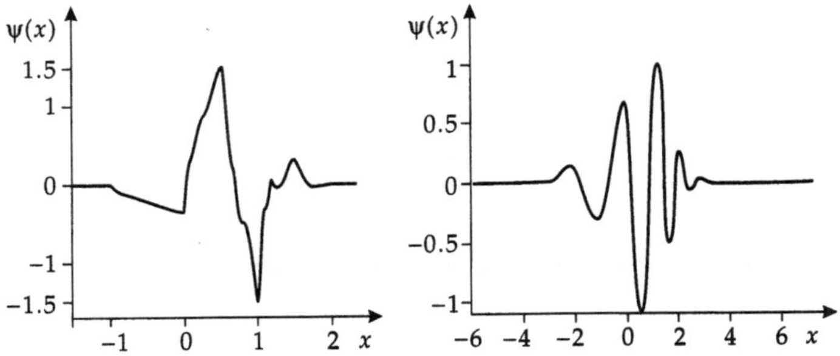

Daubechies wavelets of type $N = 2$ and $N = 7$

19.4.6 Discrete wavelet transformation

Discrete partial Haar wavelet transformation: Transformation of the signal vector f_i, $i = 1, \ldots, n$ into the detail vector d_i, $i = 1, \ldots, n/2$, and the smoothed vector s_i, $i = 1, \ldots, n/2$:

$$s_i = \frac{1}{\sqrt{2}}\,(f_{2i-1} + f_{2i}) \,,$$
$$d_i = \frac{1}{\sqrt{2}}\,(f_{2i-1} - f_{2i}).$$

- The smoothed vector s_i contains the "gross" information of the signal (averaged over two neighboring times); the detail vector d_i contains the complementary "fine" information.

The original signal may be reconstructed from both of vectors:

$$f_{2i-1} = s_i + d_i \,,$$
$$f_{2i} = s_i - d_i \,.$$

Discrete Haar-wavelet transformation: Repeated application of the partial Haar wavelet transformation: In every step the detailed information $d_i^{(n)}$ is split off and the smoothed function is transformed again:

$$s_i^{(n+1)} = \frac{1}{\sqrt{2}}\,(s_{2i-1}^{(n)} + s_{2i}^{(n)}),$$
$$d_i^{(n+1)} = \frac{1}{\sqrt{2}}\,(s_{2i-1}^{(n)} - s_{2i}^{(n)}).$$

The original signal is transformed into a sequence $\{d_i^{(n)}\}$ of detail vectors of **scale** n. Every detail vector contains information about the properties of the signal at scale 2^n.
The discrete Haar wavelet transformation can be generalized to arbitrary wavelets: Every wavelet is characterized by two sets of coefficients a_j and b_j. Then, the discrete wavelet transformation reads:

$$s_i^{(n+1)} = \sum_j a_j s_{2i+j}^{(n)} ,$$

$$d_i^{(n+1)} = \sum_j b_j s_{2i+j}^{(n)} .$$

- For large n the discrete wavelet transformation converges to the integral wavelet transformation.
  The discrete wavelet transformation of a signal recorded on a fine scale is a good approximation for the integral wavelet transformation of the signal.

Reconstruction formula: Inversion of the discrete wavelet transformation:

$$s_i^{(n)} = \sum_j \left(p_{i-2j} s_j^{(n+1)} + q_{i-2j} d_j^{(n+1)} \right) .$$

The coefficients p_i and q_i are connected with the coefficients a_j and b_j. They are characteristic for the wavelet.
The discrete wavelet transformation and its reconstruction formula lead to very fast numerical algorithms which together are called **fast wavelet transformation**. Every step of the wavelet transformation consists numerically of a **linear filter** followed by a **decimation** of the signal vector.

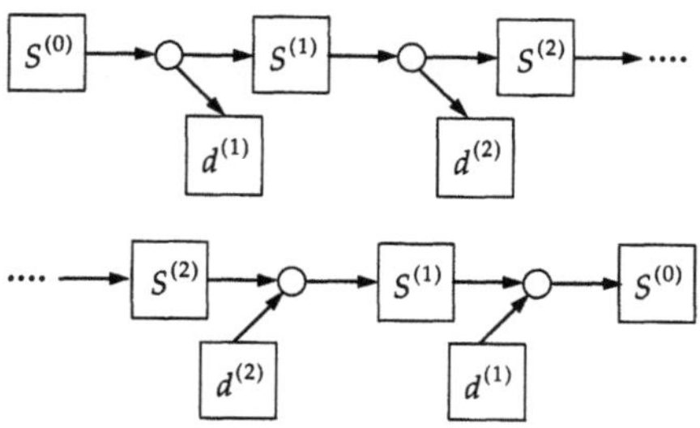

Discrete wavelet transformation and its inversion in a calculating scheme

The **calculation of a wavelet function** proceeds with the help of the reconstruction formula by setting a single detail coefficient $d_j^{(n)}$ to one. The reconstructed signal function referring to this coefficient is precisely the wavelet.

□ Signal compression by wavelets: Instead of transmitting the entire signal we transmit only the largest wavelet coefficient of the signal (according to the absolute value). For many signals even 10% of the coefficients are enough to obtain a reasonable representation of the signal. Practical applications are the compression of video and audio information.

□ Numerics with wavelets: For the numerical solution of integral and differential equations, wavelets are often better suited than position or frequency representations because they may be a better fit to corners and spikes appearing in a non-linear system. A related method is the **multigrid** for which the grid quantity is adapted to the behavior of the function.

Data compression by wavelets. The same photo with 5%, 2%, and 1% of the highest coefficients respectively.

20

Laplace and z transformations

20.1 Introduction

Laplace transformation, $\mathcal{L}\{f(t)\}$: Of practical importance for simplifying the process of solving linear differential equations with constant coefficients by solving an *algebraic* equation in the image space instead of solving the differential equation directly. The procedure consists of three steps:

1. **Laplace transformation:**
   Transformation of the **differential equation** into an **algebraic** equation.

2. **Algebraic solution in the image domain:**
   In the image domain the algebraic equation is solved for the unknown $F(s)$, the so-called image function of the required solution.

3. **Inverse transformation:**
   The inverse Laplace transformation $\mathcal{L}^{-1}$ of the obtained image function (i.e., the result of step 2) is performed to obtain the **original function**, i.e., the final solution of the differential equation in the original domain. Partial fraction separation is often necessary in this context.

Simplification for practical use: In Section 20.5 extensive transformation tables are given for the Laplace transformation from the original domain to the image domain as well as for the inverse transformation from the image function to the original solution function for almost all cases with practical application.

▷ The three-step procedure of the Laplace transformation is analogous to the former use of logarithm tables, e.g., for the multiplication (or taking the root, exponentiation) of two numbers:

The process of multiplying two numbers x and y can be converted to the addition of logarithms by $\log(xy) = \log(x) + \log(y)$. The procedure consists of the following three steps, analogous to the Laplace transformation:

1. Transformation of the factors into their logarithms;

2. Addition of the logarithms;

3. Forming the antilogarithm for the sum value, i.e., the inverse transformation of the logarithm obtained from step 2.

20.2 Definition of the Laplace transformation

The **Laplace transformation** assigns the image function (original function) $f(t)$:

$$\mathcal{L}\{f(t)\} = F(s) = \int_0^\infty f(t)\mathrm{e}^{-st}\,\mathrm{d}t$$

The new variable $s = \delta + \mathrm{j}\omega$ is usually *complex*, $s \in C$.
Laplace transform: Image function $F(s)$ of $f(t)$.
Correspondence: Symbolic notation for the pair of functions, the original function $f(t)$ and the image function $F(s)$:

$$f(t) \quad \circ\!\!-\!\!\bullet \quad F(s)\ .$$

Conditions for convergence for the Laplace transformation:
The **original function** (time function) $f(t)$ must vanish for $t < 0$, $f(t < 0) = 0$, it must be known completely for $t \geq 0$, and must be integrable on the interval $(0, \infty)$.
Damping factor: e^{-st} causes the integral to be convergent for as many original functions as possible, i.e., an exponential growth limit of the original function $f(t)$:

$$|f(t)| \leq K\mathrm{e}^{ct}\ .$$

Under these conditions, the integral converges for $\mathrm{Re}\, s > c$.

▷ The Laplace transformation transforms differential equations in the variable t (physical interpretation: time) to algebraic equations with the variable s. Since s is time-independent, it is a constant in case of integration over t.

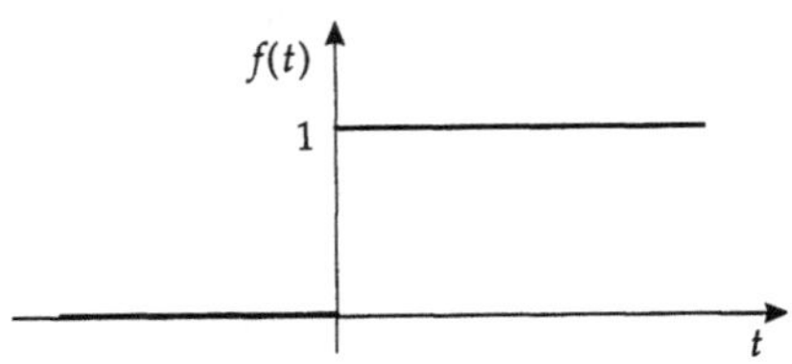

The step function

Step function: Defined as follows:

$$H(t) = \begin{cases} 0 & \text{for } t \leq 0 \\ 1 & \text{for } t > 0 \end{cases}$$

▷ Other common notations for the step function $H(t)$ are $\sigma(t)$, $\Theta(t)$, and $E(t)$. The function $\Theta(t)$ should not be confused here with the theta function.

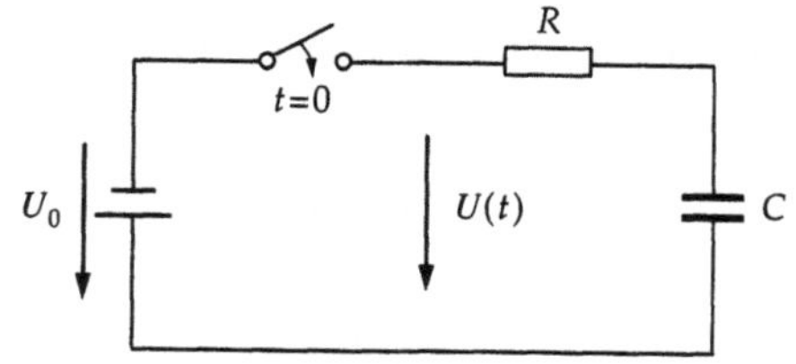

RC-element

□ The voltage at the RC-element is

$$u(t) = U_0 \cdot H(t) \,.$$

□ Laplace transform of the step function $H(t)$:

$$\mathcal{L}\{H(t)\} = \int_0^\infty U_0 H(t)\, \mathrm{e}^{-st}\, \mathrm{d}t = U_0 \int_0^\infty \mathrm{e}^{-st}\, \mathrm{d}t.$$

If Re $s > 0$, then the integral is defined with the result:

$$\mathcal{L}\{U_0 H(t)\} = \left[-\frac{U_0}{s}\mathrm{e}^{-st}\right]_0^\infty = -\frac{U_0}{s}(0-1) = \frac{U_0}{s}\,,$$

or written as a correspondence:

$$U_0 H(t) \quad \circ\!\!-\!\!\bullet \quad \frac{U_0}{s}\,.$$

The result implies that the Laplace transform of a step function with a constant is equal to the constant divided by s.

Inverse Laplace transformation: Represents the inversion of the Laplace transformation; i.e., the inverse Laplace transformation maps the image function $F(s)$ onto an original function $f(t)$ in the time domain:

$$\mathcal{L}^{-1}\{F(s)\} = f(t)\,,$$

or as a correspondence:

$$F(s) \quad \bullet\!\!-\!\!\circ \quad f(t)\,.$$

□ The following correspondence holds:

$$U_0 H(t) \quad \circ\!\!-\!\!\bullet \quad \frac{U_0}{s}\,,$$

or in other notation: $\mathcal{L}\{U_0 H(t)\} = U_0/s$. Thus, we obtain the following for the inverse Laplace transformation:

$$\frac{U_0}{s} \quad \bullet\!\!-\!\!\circ \quad U_0 H(t)$$

or also $\mathcal{L}^{-1}\{U_0/s\} = U_0 H(t)$.

20.3 Transformation theorems

- **Differentiation theorem for the first derivative**: The differentiation operation in the original domain becomes a multiplication by a time-independent variable s and a subtraction of a constant f_0 (initial value of f at time $t = 0$) in the image domain:

$$\mathcal{L}\{\mathrm{d}f/\mathrm{d}t\} = sF(s) - f_0 .$$

□ Relation between the voltage $u(t)$ and the current $i(t)$ of a coil with inductance L is given by:

$$u(t) = L\frac{\mathrm{d}i(t)}{\mathrm{d}t} .$$

For time $t = 0$, let the initial value of the current be i_0. By means of the differentiation theorem for the first derivative, the voltage in the image domain becomes:

$$U(s) = LsI(s) - Li_0 .$$

- **Differentiation theorem for the second derivative**: The operation of a twofold differentiation in the original domain is converted to a multiplication by the square of the variable s, a subtraction of the initial condition multiplied by s, and a subtraction of the initial condition $\mathrm{d}f/\mathrm{d}t|_{t=0} = \mathrm{d}f/\mathrm{d}t|_0$:

$$\mathcal{L}\left\{\mathrm{d}^2 f/\mathrm{d}t^2\right\} = s^2F(s) - sf_0 - \left.\frac{\mathrm{d}f}{\mathrm{d}t}\right|_0 .$$

□ The image function of the term in the original domain

$$m\frac{\mathrm{d}^2x(t)}{\mathrm{d}t^2}$$

with the initial conditions: $x(t = 0) = x_0$ and $\mathrm{d}x/\mathrm{d}t|_{t=0} = \mathrm{d}x/\mathrm{d}t|_0$ must be determined. According to the differentiation theorem for the second derivative we obtain:

$$\mathcal{L}\left\{m\frac{\mathrm{d}^2x}{\mathrm{d}t^2}\right\} = m\left[s^2X(s) - sx_0 - \left.\frac{\mathrm{d}x}{\mathrm{d}t}\right|_0\right] .$$

- **Differentiation theorem for the n-th derivative**: The image function of the n-th derivative of the original function is equal to the image function of the original function $f(t)$, multiplied by s^n, minus a polynomial of degree $(n-1)$ in the variable s. The coefficients of the polynomial are the initial values $(t = 0)$ of the original function and, successively of its derivatives, so that the last coefficient contains the $(n-1)$-th derivative:

$$\mathcal{L}\left\{f^{(n)}(t)\right\} = s^n F(s) - s^{n-1} f_0 - \cdots - sf_0^{(n-2)} - f_0^{(n-1)}$$

with $f_0^{(n)} = \lim_{t\to+0} f^{(n)}(t)$.
The upper index in parentheses (n) of the function f denotes its n-th derivative.

▷ The differentiation theorem is the central theorem that allows transformation of a differential equation (differentiation in the original domain) into an algebraic equation (multiplication in the image domain).

- **Linearity theorem**: The Laplace transformation of a sum is equal to the sum of the Laplace transforms; constant factors can be moved ahead of the Laplace transformation.

$$\mathcal{L}\{af(t) + bg(t)\} = a\mathcal{L}\{f(t)\} + b\mathcal{L}\{g(t)\} .$$

□ Laplace transformation of the function $f(t) = 3t - 5t^2 + 3\cos(t)$. With the help of the table of transforms, we find:

$$\mathcal{L}\left\{3t - 5t^2 + 3\cos(t)\right\} = 3\mathcal{L}\{t\} - 5\mathcal{L}\left\{t^2\right\} + 3\mathcal{L}\{\cos(t)\}$$
$$= 3\frac{1}{s^2} - 5\frac{2}{s^3} + 3\frac{s}{s^2+1}$$
$$= \frac{3s^4 + 3s^3 - 10s^2 + 3s - 10}{s^3(s^2+1)} .$$

- **Differentiation theorem for the image function**: The n-th derivative of the image function is equal to the Laplace transform of the original function $f(t)$ multiplied by $(-t)^n$:

$$F^{(n)}(s) = \mathcal{L}\left\{(-t)^n f(t)\right\} .$$

□ For the original function, $f(t) = \sinh(t)$, the following correspondence holds (see the table of transforms in Section 20.5):

$$f(t) = \sinh(t) \quad \circ\!\!-\!\!\bullet \quad F(s) = \frac{1}{s^2-1} .$$

If the differentiation theorem is applied to the function $g(t) = tf(t)$ $(n = 1)$, then we obtain the Laplace transform of g:

$$\mathcal{L}\{t\sinh(t)\} = (-1)^1 F'(s) = \frac{2s}{(s^2-1)^2} .$$

- **Integration theorem**: The image function of the integral over the original function, $\int_0^t f(u)\,du$, is equal to the image function, $F(s)$ of the original function multiplied by $1/s$:

$$\mathcal{L}\left\{\int_0^t f(u)\,du\right\} = \frac{1}{s}F(s) .$$

□ For a constant, the following correspondence holds (see above):

$$f_1(t) = U_0 E(t) \quad \circ\!\!-\!\!\bullet \quad F_1(s) = \frac{U_0}{s} .$$

With the help of the integration theorem the Laplace transform of the function $f_2(t) = U_0 t$ can thus be determined, since this function is given as the integral of $f_1(t)$:

$$f_2(t) = \int_0^t f_1(t)\,dt = \int_0^t U_0\,dt = U_0 t .$$

Thus, the integration theorem yields:

$$F_2(s) = \mathcal{L}\left\{\int_0^t f_1(t)\,dt\right\} = \frac{1}{s}F_1(s) = \frac{U_0}{s^2} .$$

▷ In the original domain differentiation and integration are at equilibrium:

$$\frac{d}{dt}\int_0^t f(u)\,du = f(t) .$$

This is also the case for the image domain:

$$\underset{\substack{\uparrow\\ \text{differentiation}}}{s} \cdot \underset{\substack{\uparrow\\ \text{integration}}}{\frac{1}{s}} \cdot F(s) = F(s) .$$

- **Convolution**: The integral over the product of two original functions, $f_1(t)$ and $f_2(t)$, where the second original function $f_2(t)$ is shifted in time:

$$(f_1 * f_2)(t) = \int_0^t f_1(u) f_2(t-u)\, du \,.$$

This **convolution integral** or **convolution product** is denoted symbolically as $(f_1 * f_2)(t)$.

▷ This definition of the convolution differs from that for the Fourier integrals by the choice of the integration limits, which in the latter case have been $-\infty$ and ∞.

▷ Like a product, convolution is commutative, associative, and distributive; that is why the term convolution product is sometimes used.

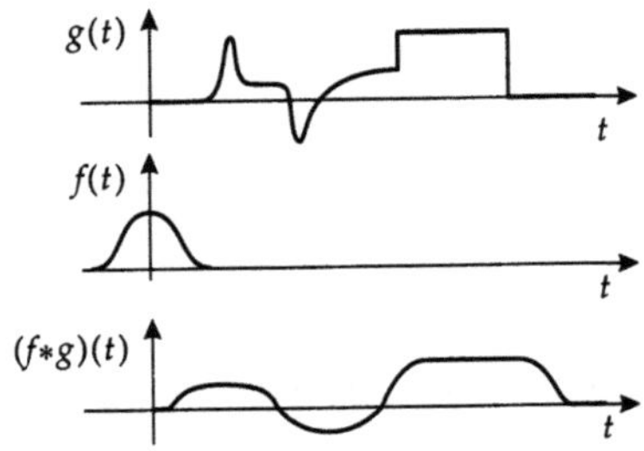

Two examples of convolution

- **Convolution theorem**: The Laplace transform of the convolution of two original functions, $f_1(t)$ and $f_2(t)$, is equal to the product of the Laplace transform of these two original functions:

$$\mathcal{L}\left\{\int_0^t f_1(u) f_2(t-u)\, du\right\} = \mathcal{L}\{(f_1 * f_2)(t)\} = \mathcal{L}\{f_1(t)\} \cdot \mathcal{L}\{f_2(t)\} \,.$$

▷ In practice, the convolution theorem is used to determine the original function from an image function $F(s)$ which can be factorized into two functions in the image domain, $F(s) = F_1(s)F_2(s)$. This leads to the following procedure:

1. Factorization of the image function: $F(s) = F_1(s)F_2(s)$.
2. Determination of the original functions $f_1(t)$ and $f_2(t)$ the image functions $F_1(s)$ and $F_2(s)$, with the help of the table of transforms.
3. The convolution of $f_1(t)$ and $f_2(t)$ in the original domain yields the required original function $f(t) = (f_1 * f_2)(t)$ belonging to the image function $F(s)$.

□ To determine the original function $f(t)$ belonging to the image function

$$F(s) = 1/((s^2+1)s) :$$

1. The separation of $F(s)$ into $F_1(s)$ and $F_2(s)$ yields:

$$F_1(s) = \frac{1}{s^2+1} \qquad \text{and} \qquad F_2(s) = \frac{1}{s} \,.$$

2. By means of the table of transforms we obtain the original functions $f_1(t)$ and $f_2(t)$:

$$f_1(t) = \mathcal{L}^{-1}\{F_1(s)\} = \sin(t)$$

and

$$f_2(t) = \mathcal{L}^{-1}\{F_2(s)\} = 1\ .$$

3. The required solution is the convolution of $f_1(t)$ and $f_2(t)$:

$$\begin{aligned} f(t) = (f_1 * f_2)(t) &= \int_0^t f_1(u) f_2(t-u)\, \mathrm{d}u \\ &= \int_0^t \sin(u)\cdot 1\, \mathrm{d}u \\ &= [-\cos(u)]_0^t = 1 - \cos(t)\ , \end{aligned}$$

i.e., the required original function belonging to the image function $F(s)$ is $f(t) = 1 - \cos(t)$.

- **Shift theorem for a shift to the right**: The Laplace transform of an original function shifted in time by a to the right is equal to the Laplace transform of the unshifted original function multiplied by the factor e^{-as}:

$$\mathcal{L}\{f(t-a)\} = \mathrm{e}^{-as}\mathcal{L}\{f(t)\}\ .$$

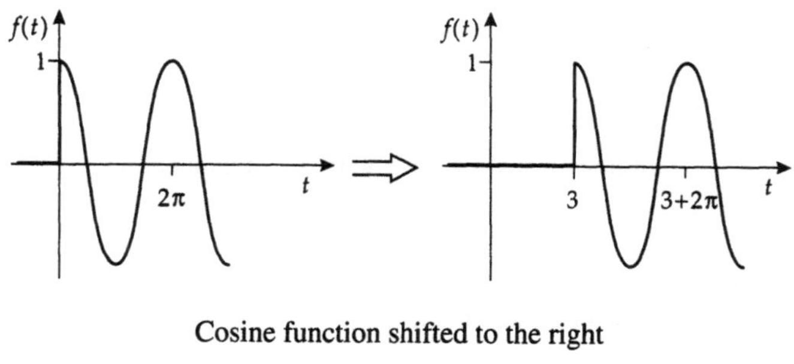

Cosine function shifted to the right

□ Laplace transformation of the cosine function shifted to the right by the time interval 3. With $f_1(t) = \cos(t)$ and the shifted cosine function $f_2(t) = f_1(t-3) = \cos(t-3)$, it follows that:

$$\mathcal{L}\{f_2(t)\} = \mathrm{e}^{-3s}\mathcal{L}\{f_1(t)\} = \mathrm{e}^{-3s}\mathcal{L}\{\cos(t)\} = \mathrm{e}^{-3s}\frac{s}{s^2+1}\ ,$$

where the Laplace transformation of the cosine was taken from the table of transforms.

- **Shift theorem for a shift to the left**: The Laplace transform of an original function shifted in time by a to the left is equal to the difference between the Laplace transform of the unshifted function and the integral $\int_0^a f(t)\mathrm{e}^{-st}\,\mathrm{d}t$, where the difference must be multiplied by the factor e^{as}:

$$\mathcal{L}\{f(t+a)\} = \mathrm{e}^{as}\left(F(s) - \int_0^a f(t)\mathrm{e}^{-st}\,\mathrm{d}t\right)\ , \qquad (a>0)\ .$$

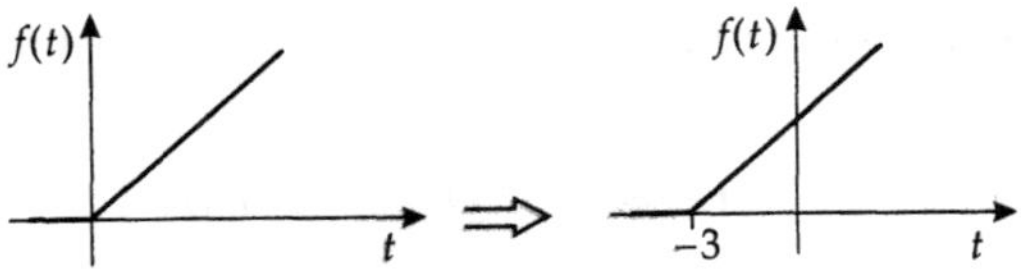

Straight line shifted to the left

☐ Laplace transform of the function $f(t) = t$ shifted to the left by three units, which has the Laplace transform $F(s) = 1/s^2$ (see the table of transforms). With the help of the shift theorem for shift to the left the following holds:

$$\mathcal{L}\{f(t+3)\} = \mathcal{L}\{t+3\} = e^{3s}\left(F(s) - \int_0^3 t e^{-st}\, dt\right)$$
$$= e^{3s}\left(\frac{1}{s^2} - \left[\left(\frac{-st-1}{s^2}\right)e^{-st}\right]_0^3\right) = \frac{3s+1}{s^2} .$$

● **Similarity theorem**: The Laplace transform of an original function subjected to a similarity transformation ($t \to at$) is equal to the Laplace transform of the original function with the argument (s/a), divided by a:

$$\mathcal{L}\{f(at)\} = \frac{1}{a}F(s/a) , \qquad a > 0 .$$

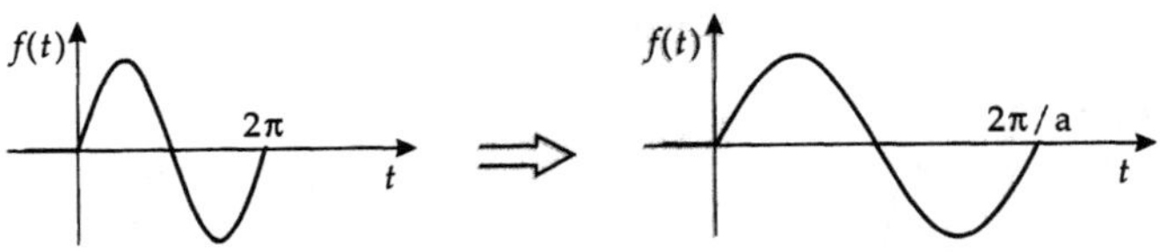

Similarity transformation with the sine function

☐ Calculation of the Laplace transform of $f(t) = \sin(\omega t)$ where the correspondence for the sine is taken from the table of transforms: $\mathcal{L}\{\sin(t)\} = F(s) = 1/(s^2+1)$. With the similarity theorem we obtain:

$$\mathcal{L}\{\sin(\omega t)\} = \frac{1}{\omega}F(s/\omega) = \frac{1}{\omega}\,\frac{1}{(s/\omega)^2+1} = \frac{\omega}{s^2+\omega^2} .$$

● **Damping theorem**: The Laplace transformation of an original function damped by the factor e^{-bt} is equal to the Laplace transform of the original function with the argument $s+b$ ($s \to s+b$):

$$\mathcal{L}\left\{e^{-bt} f(t)\right\} = F(s+b) .$$

☐ Required is the Laplace transform of the original function $f(t) = \sin(t)$ damped by the factor e^{-2t}, with the correspondence

$$\sin(t) \;\circ\!\!-\!\!\bullet\; \frac{1}{s^2+1} .$$

The damping theorem yields:

$$\mathcal{L}\left\{e^{-2t}\sin(t)\right\} = F(s+2) = \frac{1}{(s+2)^2+1} = \frac{1}{s^2+4s+5} .$$

- **Division theorem**: The Laplace transform of the quotient of an original function $f(t)$ and time t is equal to the integral of the image function of the original function $f(t)$, with limits s, the variable in the image domain, and infinity:

$$\mathcal{L}\left\{\frac{f(t)}{t}\right\} = \int_s^\infty F(u)\, du\,.$$

A condition for the convergence of this integral one is that the limit $\lim_{t\to 0} f(t)/t$ exists.

□ The Laplace transform of the original function

$$g(t) = f(t)/t = (1 - e^{-t})/t$$

must be calculated.
According to L'Hôpital's rule, the limit is:

$$\lim_{t\to 0}\frac{1-e^{-t}}{t} = \lim_{t\to 0}\frac{(1-e^{-t})'}{t'} = \lim_{t\to 0}\frac{e^{-t}}{1} = 1\,.$$

Thus, the condition for the application of the division theorem is valid, and the image function $F(s) = 1/s - 1/(s+1)$ is found in the table of transforms for the original function $f(t) = 1 - e^{-t}$:

$$\begin{aligned}
\mathcal{L}\left\{\frac{1-e^{-t}}{t}\right\} &= \int_s^\infty F(u)\, du \\
&= \lim_{M\to\infty}\int_s^M \left(\frac{1}{u} - \frac{1}{u+1}\right) du \\
&= \lim_{M\to\infty} [\ln(u) - \ln(u+1)]_s^M \\
&= \lim_{M\to\infty}\left[\ln\left(1+\frac{1}{s}\right) - \ln\left(1+\frac{1}{M}\right)\right] \\
&= \ln\left(1+\frac{1}{s}\right)\,.
\end{aligned}$$

- **Laplace transform of a periodic function**: The Laplace transform of a T-periodic function $f(t)$ is given by the following formula:

$$\mathcal{L}\{f(t)\} = \frac{1}{1-e^{-sT}}\int_0^T f(t)e^{-st}\, dt\,.$$

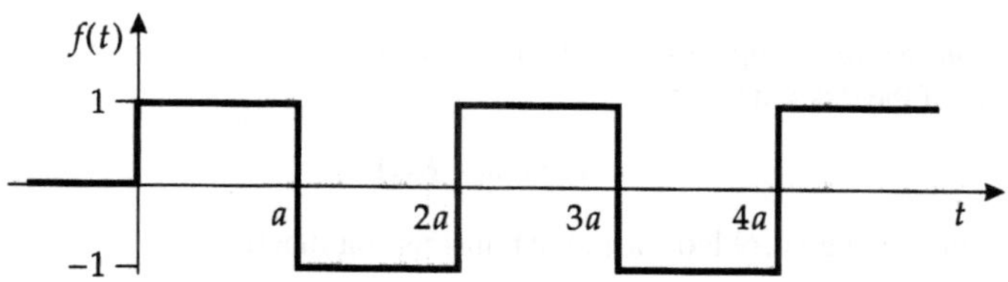

Rectangle function

□ The Laplace transform of the rectangle function

$$f(t) = \begin{cases} 1 & \text{for the case } 0 < t < a \\ -1 & \text{for the case } a < t < 2a. \end{cases}$$

with period $T = 2a$ is to be calculated:
With the theorem for periodic functions we obtain:

$$F(s) = \frac{1}{1 - \mathrm{e}^{-2as}} \left(\int_0^a 1\mathrm{e}^{-st}\,\mathrm{d}t + \int_a^{2a} (-1)\mathrm{e}^{-st}\,\mathrm{d}t \right) .$$

The first integral yields:

$$\int_0^a 1\mathrm{e}^{-st}\,\mathrm{d}t = \left[\frac{\mathrm{e}^{-st}}{-s} \right]_0^a = \frac{1 - \mathrm{e}^{-as}}{s} .$$

Analogously, the second integral yields:

$$\int_a^{2a} (-1)\mathrm{e}^{-st}\,\mathrm{d}t = \frac{\mathrm{e}^{-2as} - \mathrm{e}^{-as}}{s} .$$

Thus, we obtain for $F(s)$:

$$F(s) = \frac{1 - 2\mathrm{e}^{-as} + \mathrm{e}^{-2as}}{s(1 - \mathrm{e}^{-2as})} .$$

The numerator and the denominator can be represented as binomials, namely, the numerator as $(1 - \mathrm{e}^{-as})^2$ and the denominator as $(1 + \mathrm{e}^{-as})(1 - \mathrm{e}^{-as})$. Thus, the Laplace transform of the rectangle function assumes the following form:

$$F(s) = \frac{1 - \mathrm{e}^{-as}}{s(1 + \mathrm{e}^{-as})} = \frac{1}{s} \tanh\left(\frac{as}{2}\right) .$$

- **Limit theorem for the initial value**: The initial value of the original function at time $t = 0$ results if the limit $s \to \infty$ is performed for the product of the corresponding image function and the variable s:

$$f_0 = \lim_{t \to 0} f(t) = \lim_{s \to \infty} [sF(s)] .$$

- **Limit theorem for the final value**: The final value of the original function for time $t \to \infty$ is equal to the limit of the product of the corresponding image function and the variable s as $s \to 0$:

$$f_\infty = \lim_{t \to \infty} f(t) = \lim_{s \to 0} [sF(s)] .$$

▷ The limit theorems apply only if the limits exist.

□ A jump of the desired value

$$w(t) = w_0 \cdot E(t)$$

is switched to a control loop with the transmission function:

$$G(s) = \frac{X(s)}{W(s)} = \frac{k_R k_S}{(1 + sT_S) + k_R k_S} .$$

The final value of the controlled quantity $x(t \to \infty)$ is required. For $x(t)$ in the image domain the following holds:

$$X(s) = G(s)W(s) = \frac{k_R k_S}{(1 + sT_S) + k_R k_S} \cdot \frac{w_0}{s} .$$

By means of the final-value theorem we obtain for $x(t \to \infty)$:

$$x(t \to \infty) = \lim_{s \to 0} sX(s) = \lim_{s \to 0} \frac{k_R k_S w_0}{(1 + sT_S) + k_R k_S} = \frac{k_R k_S}{1 + k_R k_S} w_0 < w_0 .$$

The controlled quantity x does not reach the desired value w_0. The constants k_R and k_S are the amplifier factors of the control loop, and T_S is the time constant.

20.4 Partial fraction separation

Partial fraction separation is needed to calculate the original function of fractional rational image functions

$$F(s) = \frac{Z(s)}{N(s)} = \frac{b_0 + b_1 s + \cdots + b_{n-1} s^{n-1}}{a_0 + a_1 s + \cdots + a_n s^n}$$

where $Z(s)$ is the numerator polynomial and $N(s)$ is the denominator polynomial. For this, the image function is first separated into a sum of partial fractions that are transformed term by term into the original domain using the linearity theorem (see also functions and integral calculus).

20.4.1 *Partial fraction separation with simple real zeros*

Partial fraction separation with simple real zeros: Required is the partial fraction decomposition of fractional rational image functions whose denominator consists of polynomials with only simple real zeros:

$$N(s) = a_0 + a_1 s + \cdots + a_n s^n = (s - \alpha_1)(s - \alpha_2) \cdots (s - \alpha_n) .$$

Then, the partial fraction separation has the following form:

$$\begin{aligned} F(s) &= \frac{b_0 + b_1 s + \cdots + b_{n-1} s^{n-1}}{(s - \alpha_1)(s - \alpha_2) \cdots (s - \alpha_n)} \\ &= \frac{r_1}{s - \alpha_1} + \frac{r_2}{s - \alpha_2} + \cdots + \frac{r_n}{s - \alpha_n} \end{aligned}$$

in which the coefficients r_i, $i = 1, 2, \ldots, n$ must still be calculated. This is done in two steps:

1. Multiply the left and the right sides by $(s - \alpha_i)$:

$$\begin{aligned} (s - \alpha_i)F(s) &= \frac{(s - \alpha_i)(b_0 + b_1 s + \cdots + b_{n-1} s^{n-1})}{(s - \alpha_1) \cdots (s - \alpha_{i-1})(s - \alpha_i)(s - \alpha_{i+1}) \cdots (s - \alpha_n)} \\ &= \frac{(s - \alpha_i) r_1}{s - \alpha_1} + \cdots + r_i + \cdots + \frac{(s - \alpha_i) r_n}{s - \alpha_n} . \end{aligned}$$

2. Put $s = \alpha_i$ to eliminate all coefficients r except r_i:

$$r_i = \left. \frac{b_0 + b_1 s + \cdots + b_{n-1} s^{n-1}}{(s-\alpha_1)\cdots(s-\alpha_{i-1})(s-\alpha_{i+1})\cdots(s-\alpha_n)} \right|_{s=\alpha_i}$$

With this method all coefficients r_i can be calculated:

□ The current

$$I(s) = \frac{5(s+2)}{(s+10)} \cdot \frac{U_0}{sR}$$

is given in the image domain. Its partial fraction separation is required

$$\frac{5(s+2)}{(s+10)} \cdot \frac{U_0}{sR} = \frac{r_1}{s} + \frac{r_2}{s+10} .$$

According to the above procedure, first multiply by s; then put $s = 0$. This yields r_1:

$$r_1 = \left. \frac{U_0 5(s+2)}{R(s+10)} \right|_{s=0} = \frac{10U_0}{10R} = \frac{U_0}{R} .$$

Accordingly, multiplication by the factor $(s+10)$ and putting $s = -10$ yields the coefficient r_2:

$$r_2 = \left. \frac{U_0 5(s+2)}{Rs} \right|_{s=-10} = \frac{-40U_0}{-10R} = \frac{4U_0}{R} .$$

Thus, the required partial fraction separation reads:

$$I(s) = \frac{U_0}{Rs} + \frac{4U_0}{R(s+10)} .$$

20.4.2 Partial fraction decomposition with multiple real zeros

Partial fraction decomposition with multiple real zeros: The denominator $N(s)$ of the function in the image domain has a multiple real zero:

$$N(s) = a_0 + a_1 s + \cdots + a_n s^n = (s-\alpha_1)^k (s-\alpha_{k+1}) \cdots (s-\alpha_n) .$$

Thus, α_1 is a k-fold zero of $N(s)$, and the remaining zeros $\alpha_{k+1} \cdots \alpha_n$ are simple. Then, the partial fraction decomposition of $F(s)$ has the following form:

$$\begin{aligned} F(s) &= \frac{b_0 + b_1 s + \cdots + b_{n-1} s^{n-1}}{(s-\alpha_1)^k (s-\alpha_{k+1}) \cdots (s-\alpha_n)} \\ &= \frac{r_1}{s-\alpha_1} + \frac{r_2}{(s-\alpha_1)^2} + \cdots + \frac{r_k}{(s-\alpha_1)^k} + \frac{r_{k+1}}{s-\alpha_{k+1}} + \cdots + \frac{r_n}{s-\alpha_n} \end{aligned}$$

The calculation is similar to that for image functions with simple real zeros. We shall explain the procedure by means of the following example.

□ Required is the partial fraction decomposition of the function in the image domain

$$F(s) = \frac{2(s^3+5)}{s^3(s+1)}$$

with the *3-fold zero* $s_0 = 0$. The partial fraction decomposition of this expression reads:

$$F(s) = \frac{r_4}{s+1} + \frac{r_3}{s^3} + \frac{r_2}{s^2} + \frac{r_1}{s} .$$

To determine r_4 and r_3, proceed as in the case of simple real zeros.
To calculate r_4, multiply by $s+1$ and put $s=-1$:

$$r_4 = \left.\frac{2(s^3+5)}{s^3}\right|_{s=-1} = \frac{8}{-1} = -8$$

and correspondingly for r_3:

$$r_3 = \left.\frac{2(s^3+5)}{s+1}\right|_{s=0} = \frac{10}{1} = 10 .$$

Continuing in the same manner to determine r_2, we obtain:

$$\frac{2(s^3+5)}{s(s+1)} = \frac{r_4 s^2}{s+1} + \frac{r_3}{s} + r_2 + r_1 s .$$

If $s=0$, the term on the left-hand side and the second term on the right-hand side become infinite. However, if we add the two divergent terms and substitute the result for r_3, an expression is found in which $s=0$ is permitted:

$$r_2 = \left.\left\{\frac{2(s^3+5)}{s(s+1)} - \frac{10}{s}\right\}\right|_{s=0} = \left.\frac{2(s^2-5)}{s+1}\right|_{s=0} = \frac{-10}{1} = -10 .$$

Analogous to the calculation of r_1, we must combine the terms divergent for $s \to 0$ and substitute the known values of r_3 and r_2 to calculate r_2. This yields $r_1 = 10$. Altogether, we thus obtain the following partial fraction decomposition:

$$F(s) = \frac{-8}{s+1} + \frac{10}{s^3} + \frac{-10}{s^2} + \frac{10}{s} .$$

▷ Calculating partial fraction decompositions with multiple real zeros, the partial fractions must be calculated in the correct order, as in the above example. Those partial fractions that contain separations with the highest order of the multiple zero are taken first.

20.4.3 Partial fraction decomposition with complex zeros

Partial fraction decomposition with complex zeros: Calculation of the partial fraction decomposition of fractional rational functions with complex zeros in the denominator $N(s)$:

$$N(s) = (s^2 + \gamma s + \delta)(s - \alpha_3) \cdots (s - \alpha_n) ,$$

where the equation $s^2 + \gamma s + \delta = 0$ has no real solution ($\delta > \gamma^2/4$). The two conjugate complex solutions are:

$$s_{1/2} = -\frac{\gamma}{2} \pm \mathrm{j}\sqrt{\delta - \frac{\gamma^2}{4}} = d \pm \mathrm{j}c ,$$

where $d = -\gamma/2$ and $c = \sqrt{\delta - \gamma^2/4}$. Thus, the factorized form of $N(s)$ is:

$$N(s) = (s - d - \mathrm{j}c)(s - d + \mathrm{j}c)(s - \alpha_3)\cdots(s - \alpha_n)\,.$$

Then, the partial fraction decomposition of the image function $F(s)$ becomes:

$$\begin{aligned} F(s) &= \frac{b_0 + b_1 s + \cdots + b_{n-1}s^{n-1}}{(s - d - \mathrm{j}c)(s - d + \mathrm{j}c)(s - \alpha_3)\cdots(s - \alpha_n)} \\ &= \frac{r_1}{s - d - \mathrm{j}c} + \frac{r_2}{s - d + \mathrm{j}c} + \frac{r_3}{s - \alpha_3} + \cdots + \frac{r_n}{s - \alpha_n}\,. \end{aligned}$$

The calculation of the coefficients r_1 and r_2 is done with the methods explained under simple real zeros.

□ In the image domain, the current

$$I(s) = \frac{4}{s(s^2 + 2s + 5)}$$

is given. Required is its partial fraction decomposition. The term $s^2 + 2s + 5$ has the complex zeros $s_{1/2} = -1 \pm \mathrm{j}2$ (p-q formula). Thus, $I(s)$ can be separated into partial fractions:

$$I(s) = \frac{4}{s(s+1-\mathrm{j}2)(s+1+\mathrm{j}2)} = \frac{r_1}{s+1-\mathrm{j}2} + \frac{r_2}{s+1+\mathrm{j}2} + \frac{r_3}{s}\,.$$

Now, the coefficients r_1, r_2 and r_3 must be determined. r_3 belongs to the simple real zero. To calculate it, multiply the equation by s on both sides, and then put $s = 0$:

$$r_3 = \frac{4}{(1-\mathrm{j}2)(1+\mathrm{j}2)} = \frac{4}{5}\,.$$

Proceed analogously to calculate r_1. Multiply both sides by $s + 1 - \mathrm{j}2$, then put $s = -1 + \mathrm{j}2$. This yields:

$$r_1 = \frac{4}{s(s+1+\mathrm{j}2)} = \frac{4}{(-1+\mathrm{j}2)(4\mathrm{j})} = \frac{1}{(-2-\mathrm{j})} = -\frac{2-\mathrm{j}}{(2+\mathrm{j})(2-\mathrm{j})} = -\frac{2}{5} + \mathrm{j}\frac{1}{5}\,.$$

r_2 is obtained by multiplying both sides of $I(s)$ by $s+1+\mathrm{j}2$ and choosing $s = -1-\mathrm{j}2$:

$$r_2 = -\frac{2}{5} - \mathrm{j}\frac{1}{5}\,.$$ r_2 is the complex conjugate of r_1.

20.5 Linear differential equations with constant coefficients

As mentioned in the introduction to this chapter, the Laplace transformation can be used to find special solutions of linear differential equations with constant coefficients. The procedure, which consists of three steps, is represented schematically below:

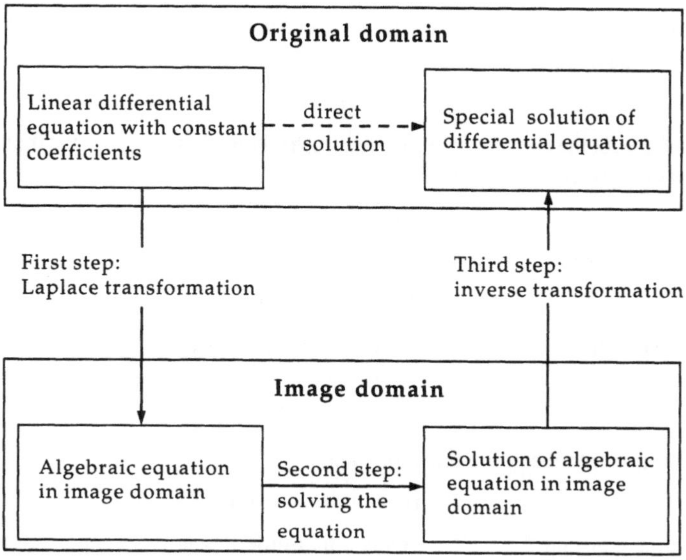

20.5.1 Laplace transformation: linear differential equation of the first order with constant coefficients

The differential equation of the first order, whose solution $f(t)$ is required:

$$f'(t) + af(t) = h(t)$$

with the constant coefficient a, is required.
To find the solution $f(t)$, carry out the following procedure consisting of three steps.
Step 1: Transformation of the first-order differential equation in the image domain with the help of the Laplace transformation:

$$[sF(s) - f(0)] + aF(s) = H(s) \ .$$

Here $f(0)$ is the initial condition for the original function $f(t)$ at time $t = 0$.
Step 2: Algebraic solution in the image domain:

$$F(s) = \frac{H(s) + f(0)}{s + a} \ .$$

Step 3: Inverse transformation of the solution $F(s)$ in the image domain to the original domain $f(t)$, using the table of transforms.

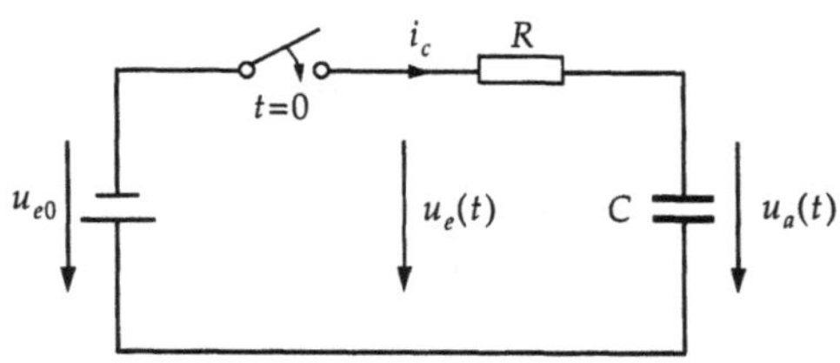

Circuit with resistance and capacitor

□ The time-dependence of voltage $u_a(t)$ at capacitor C is to be calculated with the initial condition

$$u_a(t=0) = u_{a0} \ .$$

At time $t = 0$, the constant input voltage u_{e0} is applied; in this case the input voltage $u_e(t)$ is described by step function $E(t)$:

$$u_e(t) = u_{e0} E(t) \ .$$

For the voltages we obtain the equation:

$$i_C(t) R + u_a(t) = u_e(t) \ .$$

The relation between current $i_C(t)$ and output voltage $u_a(t)$ is:

$$i_C(t) = C \frac{\mathrm{d}u_a(t)}{\mathrm{d}t} = C \dot{u}_a(t) \ .$$

Substitution into the above equation yields:

$$T \dot{u}_a(t) + u_a(t) = u_e(t)$$

with time constant $T = RC$. This is a first-order linear differential equation with constant coefficients for the output voltage $u_a(t)$. The solution is formed by means of the Laplace transformation in three steps:

1. Transformation of the differential equation to the image domain:

$$T\,[sU_a(s) - u_{a0}] + U_a(s) = u_{e0}\,\frac{1}{s} \ .$$

Here, $U_a(s)$ is the output voltage in the image domain, namely $\mathcal{L}\{u_a(t)\}$.

2. Solution of the differential equation in the image domain:

$$U_a(s) = \frac{T u_{a0}}{1+sT} + \frac{u_{e0}}{(1+sT)s} \ .$$

3. Inverse transformation of $U_a(s)$ from the image domain to the original domain with the help of the table of transforms yields the required voltage $u_a(t)$ in the original domain:

$$\begin{aligned} u_a(t) &= u_{a0}\,\mathcal{L}^{-1}\left\{\frac{T}{1+sT}\right\} + u_{e0}\,\mathcal{L}^{-1}\left\{\frac{1}{(1+sT)s}\right\} \\ &= u_{a0}\,\mathrm{e}^{-t/T} + u_{e0}\left(1-\mathrm{e}^{-t/T}\right) \ . \end{aligned}$$

This result agrees with the solution that is obtained in the original domain by decomposition of variables.

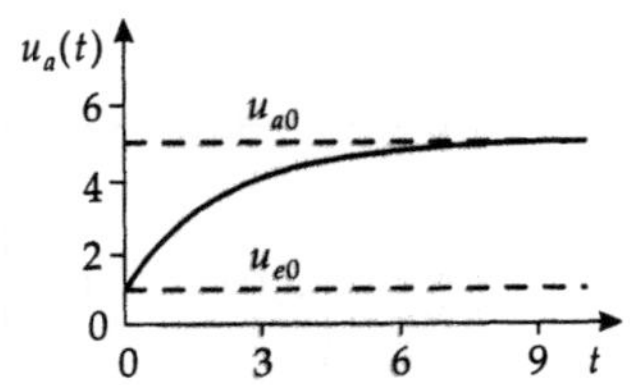

Graphical representation of the output voltage $u_a(t)$

20.5.2 Laplace transformation: linear differential equation of the second order with constant coefficients

The differential equation of the second order, for which the solution is required, reads:

$$f''(t) + af'(t) + bf(t) = h(t)$$

It is formed by means of the following three steps:
Step 1: Transformation of the differential equation to the image domain:

$$\left[s^2 F(s) - sf(0) - f'(0)\right] + a\left[sF(s) - f(0)\right] + bF(s) = H(s)$$

with the initial conditions for the original function

$$f(t = 0) = f(0) \qquad \text{and} \qquad f'(t = 0) = f'(0) \, .$$

Step 2: Solution of the algebraic equation in the image domain:

$$F(s) = \frac{H(s) + f(0)(s + a) + f'(0)}{s^2 + as + b} \, .$$

Step 3: Inverse transformation of the solution function in the image domain yields the required solution function in the original domain, using the table of transforms.

□ A mass m is attached to the end of a perpendicularly suspended spring. The spring is suspended from the ceiling of an elevator which moves upward at a constant acceleration a.

The position-time behavior of the mass in the reference frame fixed to the elevator will be delivered.

Newton's second law of motion provides a linear differential equation of the second order with constant coefficients:

$$m\ddot{x}(t) = -kx(t) + maH(t)$$

with unit jump function $H(t)$. Rewriting the equation yields:

$$\ddot{x}(t) + \omega^2 x(t) = aH(t) \, .$$

The spring satisfies Hooke's law with spring constant k. The angular frequency ω is given by $\omega^2 = k/m$.

Initial condition: At time $t = 0$, the mass should not be displaced, and should not move: $x(0) = \dot{x}(0) = 0$.

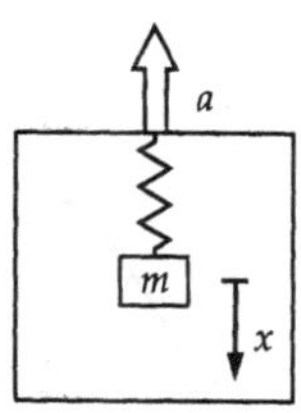

A mass attached to a spring accelerated upward by an elevator

The solution is determined in three steps with the help of the Laplace transformation:

1. Transformation from the original to the image domain:

$$\left[s^2X(s) - sx(0) - \dot{x}(0)\right] + \omega^2 X(s) = \mathcal{L}\{aH(t)\} = \frac{a}{s}\,.$$

2. Algebraic solution of the differential equation in the image domain:

$$X(s) = \underbrace{\frac{1}{s^2+\omega^2}}_{F_1(s)} \cdot \underbrace{\frac{a}{s}}_{F_2(s)} = F_1(s)\ F_2(s)\,.$$

3. Inverse transformation from the image domain to the original domain:
   $X(s)$ is the product of the two functions $F_1(s)$ and $F_2(s)$ in the image domain; therefore, the convolution theorem can be used for the solution. The original functions $f_1(t)$ and $f_2(t)$ belonging to $F_1(s)$ and $F_2(s)$, respectively, must be determined by means of the tables of transforms:

$$f_1(t) = \mathcal{L}^{-1}\{F_1(s)\} = \frac{\sin(\omega t)}{\omega}\,,$$
$$f_2(t) = \mathcal{L}^{-1}\{F_2(s)\} = aH(t)\,.$$

   According to the convolution theorem, the required solution in the original domain is the convolution product of $f_1(t)$ and $f_2(t)$:

$$\begin{aligned} x(t) = (f_1 * f_2)(t) &= \int_0^t f_1(u) f_2(t-u)\,\mathrm{d}u \\ &= \int_0^t \frac{\sin(\omega u)}{\omega} aE(t-u)\,\mathrm{d}u = \frac{a}{\omega}\int_0^t \sin(\omega u)\,\mathrm{d}u \\ &= \frac{a}{\omega}\left[-\frac{\cos(\omega u)}{\omega}\right]_0^t = \frac{a}{\omega^2}[1-\cos(\omega t)] \\ &= \frac{ma}{k}[1-\cos(\omega t)]\,. \end{aligned}$$

   Solution $x(t)$ describes a harmonic oscillation of the spring with the maximum displacement $2ma/k$.

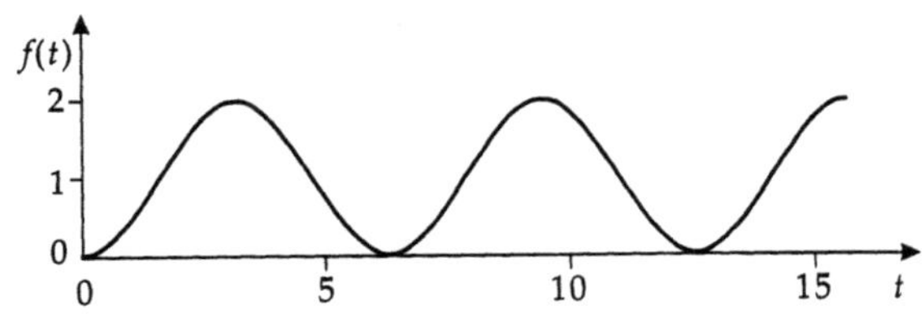

Displacement $x(t)$ of mass m as a function of time

▷ The methods described for differential equations of the first and second order with constant coefficients can be generalized to become differential equations of higher order with the help of the differentiation theorem.

20.5.3 Example: linear differential equations

□ The object is to calculate the water level $h(t)$ in a container with base A and with an inflow and an outflow. The inflow and outflow are given by $q(t) = we^{-t}$ and $kh(t)$, respectively, where w and k are proportionality constants. The conservation of mass leads to a linear differential equation of the first order for the water level $h(t)$.

$$A\frac{dh(t)}{dt} + kh(t) = q(t) = we^{-t} .$$

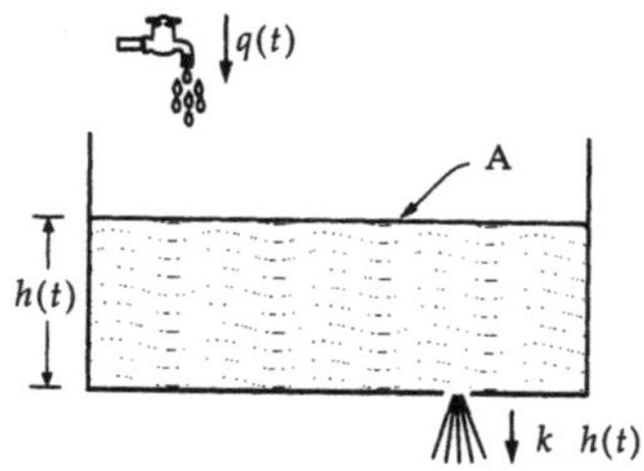

A water container with inflow and outflow

The initial condition is: $h(t = 0) = h_0 = 4$. The numerical values of the constants are:

$$k = \frac{1}{2}\,\mathrm{m}^2/\mathrm{s}\,, \quad w = 4\,\mathrm{m}^3/\mathrm{s}\,, \quad \text{and} \quad A = \frac{1}{4}\mathrm{m}^2 .$$

The solution proceeds in three steps, using the Laplace transformation:

1. Transformation of the differential equation to the image domain:

$$A\,[sH(s) - h_0] + kH(s) = Q(s) = \frac{w}{s+1} ,$$

with numerical values:

$$\frac{1}{4}\,[sH(s) - 4] + \frac{1}{2}H(s) = \frac{4}{s+1} .$$

2. Algebraic solution in the image domain:

$$\frac{1}{4}(s+2)H(s) = \frac{4}{s+1} + 1 = \frac{s+5}{s+1} ,$$

from which it follows that:

$$H(s) = \frac{4(s+5)}{(s+1)(s+2)} .$$

A partial fraction decomposition must be performed for $H(s)$:

$$H(s) = \frac{4(s+5)}{(s+1)(s+2)} = \frac{A}{s+1} + \frac{B}{s+2} .$$

To calculate A, multiply this equation by $(s+1)$, and put $s=-1$:

$$A = \left.\frac{4(s+5)}{s+2}\right|_{s=-1} = \frac{4(-1+5)}{-1+2} = 16\,.$$

Correspondingly, we obtain for B:

$$B = \left.\frac{4(s+5)}{s+1}\right|_{s=-2} = \frac{4(-2+5)}{-2+1} = -12\,.$$

Thus, $H(s)$ also has the form:

$$H(s) = \frac{16}{s+1} - \frac{12}{s+2}\,.$$

3. Inverse transformation to the original domain with the help of the tables of transforms:

$$h(t) = \mathcal{L}^{-1}\left\{\frac{16}{s+1}\right\} - \mathcal{L}^{-1}\left\{\frac{12}{s+2}\right\} = 16\mathrm{e}^{-t} - 12\mathrm{e}^{-2t}\,.$$

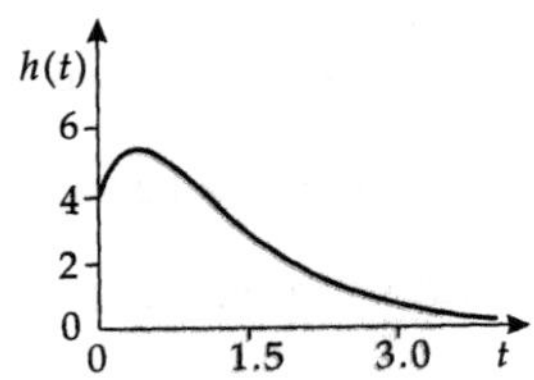

Water level $h(t)$ in the container as a function of time

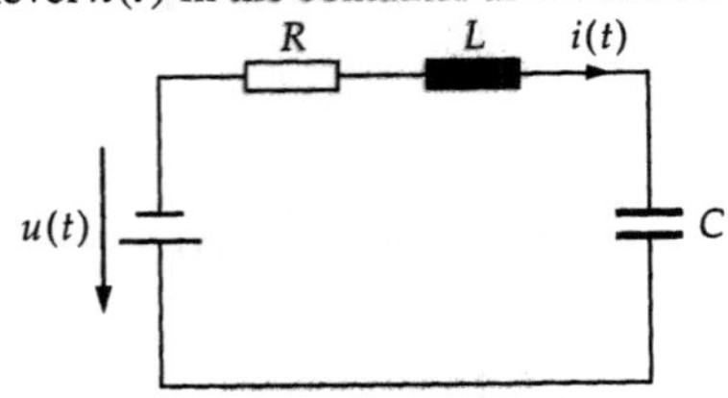

Circuit with resistance, coil, and capacitor

□ For the circuit represented in the above figure (right) the current $i(t)$ is to be calculated. The initial conditions are $i(t=0)=0$ and $u_C(t=0)=0$. The circuit is described by the differential equation

$$L\frac{\mathrm{d}i(t)}{\mathrm{d}t} + Ri(t) + \frac{1}{C}\int_0^t i(t)\,\mathrm{d}t + u_C(0) = u(t)$$

The constants are the inductance $L = 2$ H, the resistance $R = 4\ \Omega$, and the capacitance $C = 1/4$ F. The applied voltage should increase linearly in time:

$$u(t) = \frac{U_0}{T}t = 8\frac{\mathrm{V}}{\mathrm{s}}t\,.$$

Thus, the differential equation becomes:

$$2\frac{\mathrm{d}i(t)}{\mathrm{d}t} + 4i(t) + 4\int_0^t i(t)\,\mathrm{d}t = 8t\ .$$

Again, the solution is done in three steps using the Laplace transformation:

1. Transformation to the image domain:

$$2sI(s) + 4I(s) + 4\frac{I(s)}{s} = \frac{8}{s^2}\ .$$

2. Solving the algebraic equation in the image domain:

$$I(s) = \frac{4}{s(s^2+2s+2)}\ .$$

To determine the original function $i(t)$ a partial fraction decomposition of $I(s)$ must be performed. For this, the zeros of the quadratic form must be calculated first:

$$s^2 + 2s + 2 = 0\ .$$

With the help of the p-q formula we can find the two complex conjugate zeros

$$s_{1/2} = -1 \pm \mathrm{j}\ .$$

Thus, the partial fraction decomposition is

$$I(s) = \frac{4}{s(s+1-\mathrm{j})(s+1+\mathrm{j})} = \frac{A}{s} + \frac{B}{s+1-\mathrm{j}} + \frac{C}{s+1+\mathrm{j}}\ .$$

The coefficients A, B, and C must be calculated. A is obtained by multiplying the above equation by s and choosing $s = 0$:

$$A = \frac{4}{(1-\mathrm{j})(1+\mathrm{j})} = 2\ .$$

To calculate B, we multiply by $s+1-\mathrm{j}$ and put $s = -1+\mathrm{j}$:

$$B = \left.\frac{4}{s(s+1+\mathrm{j})}\right|_{s=-1+\mathrm{j}} = \frac{4}{(-1+\mathrm{j})2\mathrm{j}} = \frac{2}{-1-\mathrm{j}}$$
$$= \frac{2(-1+\mathrm{j})}{(-1-\mathrm{j})(-1+\mathrm{j})} = -1+\mathrm{j}\ .$$

Every complex number $z = r + \mathrm{j}i$ can be written in terms of polar coordinates: $z = |z|\mathrm{e}^{\vartheta}$ with the absolute value of the complex number $|z| = \sqrt{r^2+i^2}$ and the polar angle ϑ in the complex plane. In polar coordinates B becomes

$$B = -1+\mathrm{j} = \sqrt{2}\,\mathrm{e}^{3/4\pi\mathrm{j}}\ .$$

Correspondingly, C is obtained by multiplying by $s+1+\mathrm{j}$ and choosing $s = -1-\mathrm{j}$:

$$C = -1-\mathrm{j} = \sqrt{2}\,\mathrm{e}^{-3/4\pi\mathrm{j}}\ ,$$

which is the complex conjugate of B. Thus, for $I(s)$ we obtain:

$$I(s) = \frac{2}{s} + \frac{\sqrt{2}\,\mathrm{e}^{3/4\pi\mathrm{j}}}{s+1-\mathrm{j}} + \frac{\sqrt{2}\,\mathrm{e}^{-3/4\pi\mathrm{j}}}{s+1+\mathrm{j}}\ .$$

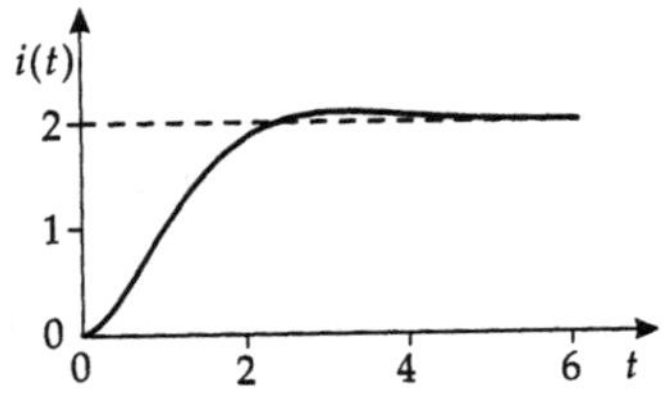

Current $i(t)$ as a function of time

3. The original function $i(t)$ is determined from the image function $I(s)$ with the help of the table of transforms:

$$\begin{aligned} i(t) &= \mathcal{L}^{-1}\left\{\frac{2}{s}\right\} + \mathcal{L}^{-1}\left\{\frac{\sqrt{2}\,e^{3/4\pi j}}{s+1-j}\right\} + \mathcal{L}^{-1}\left\{\frac{\sqrt{2}\,e^{-3/4\pi j}}{s+1+j}\right\} \\ &= 2 + \sqrt{2}\,e^{3/4\pi j}e^{(-1+j)t} + \sqrt{2}\,e^{-3/4\pi j}e^{(-1-j)t} \\ &= 2 + \sqrt{2}\,e^{-t}\left[e^{\,j(t+3/4\pi)} + e^{-j(t+3/4\pi)}\right] \\ &= 2 + 2\sqrt{2}\,e^{-t}\cos(t+3/4\pi)\,. \end{aligned}$$

In the last step the Euler formula was used.

20.5.4 Laplace transforms (table)

$\Gamma(x)$	Gamma function
$\Psi(x)$	Psi function, $\psi(x) = \frac{d}{dx}\ln\Gamma(x) = \Gamma'(x)/\Gamma(x)$
$H(x)$	Step function, Heaviside function
$L_n(x)$	Laguerre polynomial of order n
$H_n(x)$	Hermitian polynomial of order n
$\mathrm{erf}(x)$	Error function
$\mathrm{erf}_c(x)$	Conjugated error function
$E_n(x)$	Exponential integral function, n-th Schlömilch function
$Si(x)$	Sine integral function, integral sine
$Ci(x)$	Cosine integral function, integral cosine
$J_n(x)$	Bessel function of the first kind of order n
$I_n(x)$	Modified Bessel function of the first kind of order n, Bessel function of the first kind with imaginary argument, hyperbolic Bessel function
$K_n(x)$	modified Bessel function of the second kind of order n, Bessel function of the second kind with imaginary argument, MacDonald function

$F(s)$		$f(t)$
1	●—○	$\delta(t)$ (Dirac δ function)
$\frac{1}{s}$	●—○	1 (Step function)
$\frac{1}{s^2}$	●—○	t
$\frac{1}{s^3}$	●—○	$\frac{t^2}{2}$
$\frac{1}{s^n}$ $(n = 1, 2, 3, \ldots)$	●—○	$\frac{t^{n-1}}{(n-1)!}$

$\frac{1}{\sqrt{s}}$		$\bullet\!-\!\circ$	$\frac{1}{\sqrt{\pi t}}$
$s^{-3/2}$		$\bullet\!-\!\circ$	$2\sqrt{t/\pi}$
$s^{-(n+1/2)}$	$(n = 1, 2, 3, \ldots)$	$\bullet\!-\!\circ$	$\frac{2^n t^{n-1/2}}{1 \cdot 3 \cdot 5 \cdots (2n-1)\sqrt{\pi}}$
$\frac{\Gamma(k)}{s^k}$	$(k > 0)$	$\bullet\!-\!\circ$	t^{k-1}
$\frac{1}{s+a}$		$\bullet\!-\!\circ$	e^{-at}
$\frac{1}{(s+a)^2}$		$\bullet\!-\!\circ$	te^{-at}
$\frac{1}{(s+a)^n}$	$(n = 1, 2, 3, \cdots)$	$\bullet\!-\!\circ$	$\frac{t^{n-1}e^{-at}}{(n-1)!}$
$\frac{\Gamma(k)}{(s+a)^k}$	$(k > 0)$	$\bullet\!-\!\circ$	$t^{k-1}e^{-at}$
$\frac{1}{s(s+a)}$	$(a \neq 0)$	$\bullet\!-\!\circ$	$\frac{1-e^{-at}}{a}$
$\frac{1}{s(s+a)^2}$	$(a \neq 0)$	$\bullet\!-\!\circ$	$-\frac{1}{a^2}(1+at)(1+e^{-at})$
$\frac{1}{(s+a)(s+b)}$	$(a \neq b)$	$\bullet\!-\!\circ$	$\frac{e^{-at}-e^{-bt}}{b-a}$
$\frac{1}{(s+a)(s+b)(s+c)}$	$(a \neq b \neq c)$	$\bullet\!-\!\circ$	$-\frac{(b-c)e^{-at}+(c-a)e^{-bt}+(a-b)e^{-ct}}{(a-b)(b-c)(c-a)}$
$\frac{1}{s^2(s+a)^2}$		$\bullet\!-\!\circ$	$\frac{1}{a^2}(e^{-at}+at-1)$
$\frac{s}{(s+a)^2}$		$\bullet\!-\!\circ$	$(1-at)e^{-at}$
$\frac{s}{(s+a)(s+b)}$	$(a \neq b)$	$\bullet\!-\!\circ$	$\frac{ae^{-at}-be^{-bt}}{a-b}$
$\frac{s}{(s+a)^3}$		$\bullet\!-\!\circ$	$\left(t-\frac{at^2}{2}\right)e^{-at}$
$\frac{s}{(s+a)(s+b)^2}$	$(a \neq 0)$	$\bullet\!-\!\circ$	$-\frac{1}{(a-b)^2}(ae^{-at}+[b(a-b)t-a]e^{-bt})$
$\frac{s}{(s+a)(s+b)(s+c)}$	$(a \neq b \neq c)$	$\bullet\!-\!\circ$	$-\frac{a(b-c)e^{-at}+b(c-a)e^{-bt}+c(a-b)e^{-ct}}{(a-b)(b-c)(c-a)}$
$\frac{s^2}{(s+a)^3}$		$\bullet\!-\!\circ$	$\left(\frac{at^2}{2}-2at+1\right)e^{-at}$
$\frac{s^2}{(s+a)(s+b)^2}$	$(a \neq b)$	$\bullet\!-\!\circ$	$\frac{(a^2e^{-at}+[b^2(a-b)t-2ab+b^2]e^{-bt}}{(a-b)^2}$
$\frac{s^2}{(s+a)(s+b)(s+c)}$	$(a \neq b \neq c)$	$\bullet\!-\!\circ$	$-\frac{a^2(b-c)e^{-at}+b^2(c-a)e^{-bt}+c^2(a-b)e^{-ct}}{(a-b)(b-c)(c-a)}$
$\frac{1}{s^2+a^2}$	$(a \neq 0)$	$\bullet\!-\!\circ$	$\frac{1}{a}\sin at$
$\frac{s\sin b+a\cos b}{s^2+a^2}$		$\bullet\!-\!\circ$	$\sin(at+b)$
$\frac{s}{s^2+a^2}$		$\bullet\!-\!\circ$	$\cos at$
$\frac{s\cos b+a\sin b}{s^2+a^2}$		$\bullet\!-\!\circ$	$\cos(at+b)$
$\frac{1}{s^2-a^2}$	$(a \neq 0)$	$\bullet\!-\!\circ$	$\frac{1}{a}\sinh at$

$$\frac{s}{s^2-a^2} \quad \bullet\!\!-\!\!\circ \quad \cosh at$$

$$\frac{1}{s(s^2+a^2)} \quad (a \neq 0) \quad \bullet\!\!-\!\!\circ \quad \frac{1}{a^2}(1-\cos at)$$

$$\frac{1}{s^2(s^2+a^2)} \quad (a \neq 0) \quad \bullet\!\!-\!\!\circ \quad \frac{1}{a^3}(at-\sin at)$$

$$\frac{1}{(s^2+a^2)^2} \quad (a \neq 0) \quad \bullet\!\!-\!\!\circ \quad \frac{1}{2a^3}(\sin at - at\cos at)$$

$$\frac{s}{(s^2+a^2)^2} \quad (a \neq 0) \quad \bullet\!\!-\!\!\circ \quad \frac{t}{2a}\sin at$$

$$\frac{1}{s(s^2+4a^2)} \quad (a \neq 0) \quad \bullet\!\!-\!\!\circ \quad \frac{1}{2a}\sin^2(at)$$

$$\frac{s^2}{(s^2+a^2)^2} \quad (a \neq 0) \quad \bullet\!\!-\!\!\circ \quad \frac{1}{2a}(\sin at + at\cos at)$$

$$\frac{s^2-a^2}{(s^2+a^2)^2} \quad \bullet\!\!-\!\!\circ \quad t\cos at$$

$$\frac{s^2+2a^2}{s(s^2+4a^2)} \quad \bullet\!\!-\!\!\circ \quad \cos^2(at)$$

$$\frac{s}{(s^2+a^2)(s^2+b^2)} \quad (a^2 \neq b^2) \quad \bullet\!\!-\!\!\circ \quad \frac{\cos at - \cos bt}{b^2-a^2}$$

$$\frac{1}{(s+a)^2+b^2} \quad (b \neq 0) \quad \bullet\!\!-\!\!\circ \quad \frac{1}{b}e^{-at}\sin bt$$

$$\frac{1}{(s+a)^2-b^2} \quad (b \neq 0) \quad \bullet\!\!-\!\!\circ \quad \frac{1}{b}e^{-at}\sinh(bt)$$

$$\frac{s+a}{(s+a)^2+b^2} \quad \bullet\!\!-\!\!\circ \quad e^{-at}\cos bt$$

$$\frac{s+a}{(s+a)^2-b^2} \quad \bullet\!\!-\!\!\circ \quad e^{-at}\cosh(bt)$$

$$\frac{3a^2}{s^3+a^3} \quad \bullet\!\!-\!\!\circ \quad e^{-at} - e^{at/2}\left(\cos\frac{at\sqrt{3}}{2} - \sqrt{3}\sin\frac{at\sqrt{3}}{2}\right)$$

$$\frac{4a^3}{s^4+4a^4} \quad \bullet\!\!-\!\!\circ \quad \sin at\cosh at - \cos at\sinh at$$

$$\frac{s}{s^4+4a^4} \quad (a \neq 0) \quad \bullet\!\!-\!\!\circ \quad \frac{1}{2a^2}\sin at\sinh at$$

$$\frac{1}{s^4-a^4} \quad (a \neq 0) \quad \bullet\!\!-\!\!\circ \quad \frac{1}{2a^3}(\sinh at - \sin at)$$

$$\frac{s}{s^4-a^4} \quad (a \neq 0) \quad \bullet\!\!-\!\!\circ \quad \frac{1}{2a^2}(\cosh at - \cos at)$$

$$\frac{8a^3s^2}{(s^2+a^2)^3} \quad \bullet\!\!-\!\!\circ \quad (1+a^2t^2)\sin at - at\cos at$$

$$\frac{1}{s}\left(\frac{s-1}{s}\right)^n \quad \bullet\!\!-\!\!\circ \quad L_n(t)$$

$$\frac{s}{(s+a)^{3/2}} \quad \bullet\!\!-\!\!\circ \quad \frac{1}{\sqrt{\pi t}}e^{-at}(1-2at)$$

$$\sqrt{s+a}-\sqrt{s+b} \quad \bullet\!\!-\!\!\circ \quad \frac{1}{2\sqrt{\pi t^3}}(e^{-bt}-e^{-at})$$

$$\frac{1}{\sqrt{s}+a} \quad \bullet\!\!-\!\!\circ \quad \frac{1}{\sqrt{\pi t}} - ae^{a^2t}\operatorname{erfc} a\sqrt{t}$$

$$\frac{\sqrt{s}}{s-a^2} \quad \bullet\!\!-\!\!\circ \quad \frac{1}{\sqrt{\pi t}} + ae^{a^2t}\operatorname{erf} a\sqrt{t}$$

$$\frac{\sqrt{s}}{s+a^2} \quad \bullet\!\!-\!\!\circ \quad \frac{1}{\sqrt{\pi t}} - \frac{2a}{\sqrt{\pi}}e^{-a^2t}\int_0^{a\sqrt{t}} e^{\lambda^2}\,d\lambda$$

$$\frac{1}{\sqrt{s}(s-a^2)} \quad (a \neq 0) \quad \bullet\!\!-\!\!\circ \quad \frac{1}{a}\mathrm{e}^{a^2 t}\operatorname{erf} a\sqrt{t}$$

$$\frac{1}{\sqrt{s}(s+a^2)} \quad (a \neq 0) \quad \bullet\!\!-\!\!\circ \quad \frac{2}{a\sqrt{\pi}}\mathrm{e}^{-a^2 t}\int_0^{a\sqrt{t}} \mathrm{e}^{\lambda^2}\,d\lambda$$

$$\frac{b^2-a^2}{(s-a^2)(b+\sqrt{s})} \quad \bullet\!\!-\!\!\circ \quad \mathrm{e}^{a^2 t}(b - a\operatorname{erf} a\sqrt{t}) - b\mathrm{e}^{b^2 t}\operatorname{erfc} b\sqrt{t}$$

$$\frac{1}{\sqrt{s}(\sqrt{s}+a)} \quad \bullet\!\!-\!\!\circ \quad \mathrm{e}^{a^2 t}\operatorname{erfc}(a\sqrt{t})$$

$$\frac{1}{(s+a)\sqrt{s+b}} \quad (a \neq b) \quad \bullet\!\!-\!\!\circ \quad \frac{1}{\sqrt{b-a}}\mathrm{e}^{-at}\operatorname{erf}(\sqrt{b-a}\sqrt{t})$$

$$\frac{b^2-a^2}{\sqrt{s}(s-a^2)(\sqrt{s}+b)} \quad (a \neq 0) \quad \bullet\!\!-\!\!\circ \quad \mathrm{e}^{a^2 t}\left[\frac{b}{a}\operatorname{erf}(a\sqrt{t}) - 1\right] + \mathrm{e}^{b^2 t}\operatorname{erfc} b\sqrt{t}$$

$$\frac{(1-s)^n}{s^{n+1/2}} \quad \bullet\!\!-\!\!\circ \quad \frac{n!}{(2n)!\sqrt{\pi t}}H_{2n}(\sqrt{t})$$

$$\frac{(1-s)^n}{s^{n+3/2}} \quad \bullet\!\!-\!\!\circ \quad \frac{n!}{(2n+1)!\sqrt{\pi}}H_{2n+1}(\sqrt{t})$$

$$\frac{\sqrt{s+2a}}{\sqrt{s}} - 1 \quad \bullet\!\!-\!\!\circ \quad a\mathrm{e}^{-at}[I_1(at) + I_0(at)]$$

$$\frac{1}{\sqrt{s+a}\sqrt{s+b}} \quad \bullet\!\!-\!\!\circ \quad \mathrm{e}^{-(a+b)t/2} I_0\left(\frac{a-b}{2}t\right)$$

$$\frac{\Gamma(k)}{(s+a)^k(s+b)^k} \quad (a \neq b)\ (k > 0) \quad \bullet\!\!-\!\!\circ \quad \sqrt{\pi}\left(\frac{t}{a-b}\right)^{k-1/2}\mathrm{e}^{-(a+b)t/2} I_{k-1/2}\left(\frac{a-b}{2}t\right)$$

$$\frac{1}{(s+a)^{1/2}(s+b)^{3/2}} \quad \bullet\!\!-\!\!\circ \quad t\mathrm{e}^{-(a+b)t/2}\left[I_0\left(\frac{a-b}{2}t\right) + I_1\left(\frac{a-b}{2}t\right)\right]$$

$$\frac{\sqrt{s+2a}-\sqrt{s}}{\sqrt{s+2a}+\sqrt{s}} \quad \bullet\!\!-\!\!\circ \quad \frac{1}{t}\mathrm{e}^{-at} I_1(at)$$

$$\frac{(a-b)^k}{(\sqrt{s+a}+\sqrt{s+b})^{2k}} \quad (k > 0) \quad \bullet\!\!-\!\!\circ \quad \frac{k}{t}\mathrm{e}^{-(a+b)t/2} I_k\left(\frac{a-b}{2}t\right)$$

$$\frac{(\sqrt{s+a}+\sqrt{s})^{-2\nu}}{\sqrt{s}\sqrt{s+a}} \quad (a \neq 0)(\nu > -1) \quad \bullet\!\!-\!\!\circ \quad \frac{1}{a^\nu}\mathrm{e}^{-at/2} I_\nu(at/2)$$

$$\frac{1}{\sqrt{s^2+a^2}} \quad \bullet\!\!-\!\!\circ \quad J_0(at)$$

$$\frac{(\sqrt{s^2+a^2}-s)^\nu}{\sqrt{s^2+a^2}} \quad (\nu > -1) \quad \bullet\!\!-\!\!\circ \quad a^\nu J_\nu(at)$$

$$\frac{1}{(s^2+a^2)^k} \quad (k > 0) \quad \bullet\!\!-\!\!\circ \quad \frac{\sqrt{\pi}}{\Gamma(k)}\left(\frac{t}{2a}\right)^{k-1/2} J_{k-1/2}(at)$$

$$(\sqrt{s^2+a^2}-s)^k \quad (k > 0) \quad \bullet\!\!-\!\!\circ \quad \frac{ka^k}{t}J_k(at)$$

$$\frac{(s-\sqrt{s^2-a^2})^\nu}{\sqrt{s^2-a^2}} \quad (\nu > -1) \quad \bullet\!\!-\!\!\circ \quad a^\nu I_\nu(at)$$

$$\frac{1}{(s^2-a^2)^k} \quad (a \neq 0) \quad (k > 0) \quad \bullet\!\!-\!\!\circ \quad \frac{\sqrt{\pi}}{\Gamma(k)}\left(\frac{t}{2a}\right)^{k-1/2} I_{k-1/2}(at)$$

Jump function

$$\frac{1}{s}\mathrm{e}^{-ks} \quad \bullet\!\!-\!\!\circ \quad \sigma(t-k)$$

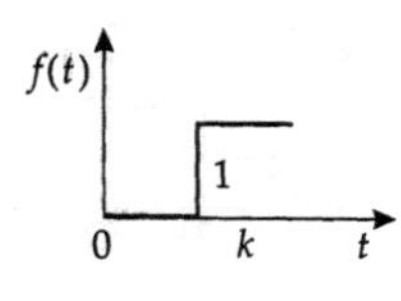

$$\frac{1}{s^2}\mathrm{e}^{-ks} \quad \bullet\!\!-\!\!\circ \quad (t-k)\sigma(t-k)$$

$$\frac{1}{s^{\mu}}\mathrm{e}^{-ks}(\mu > 0) \quad \bullet\!\!-\!\!\circ \quad \frac{(t-k)^{\mu-1}}{\Gamma(\mu)}\sigma(t-k)$$

Rectangular curves

$$\frac{1-\mathrm{e}^{-ks}}{s} \quad \bullet\!\!-\!\!\circ \quad \sigma(t)-\sigma(t-k)$$

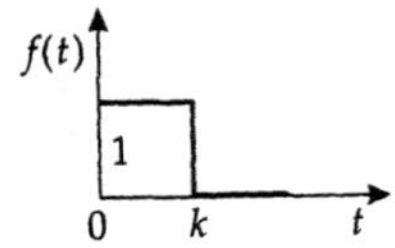

$$\frac{1}{s(1-\mathrm{e}^{-ks})} = \frac{1+\coth(ks/2)}{2s} \quad \bullet\!\!-\!\!\circ \quad \sum_{n=0}^{\infty}\sigma(t-nk)$$

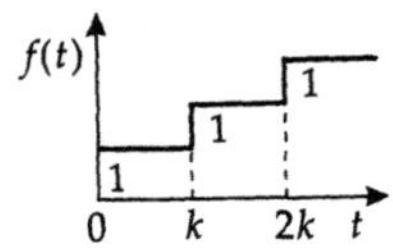

$$\frac{1}{s(\mathrm{e}^{ks}-a)} \quad \bullet\!\!-\!\!\circ \quad \sum_{n=1}^{\infty}a^{n-1}\sigma(t-nk)$$

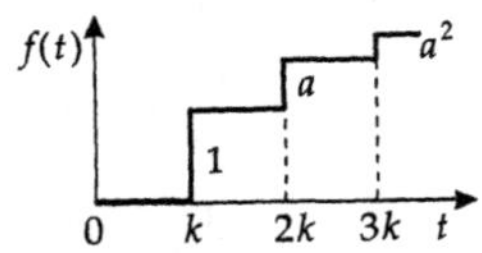

$$\frac{1}{s}\tanh(ks) \quad \bullet\!\!-\!\!\circ \quad \sigma(t)+2\sum_{n=1}^{\infty}(-1)^n\sigma(t-2nk)$$

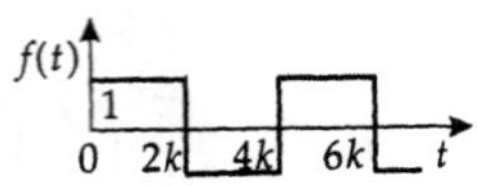

$$\frac{1}{s(1+\mathrm{e}^{-ks})} \quad \bullet\!\!-\!\!\circ \quad \sum_{n=0}^{\infty}(-1)^n\sigma(t-nk)$$

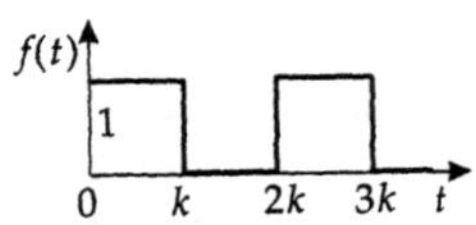

$$\frac{1}{s\sinh(ks)} \quad \bullet\!\!-\!\!\circ \quad 2\sum_{n=0}^{\infty}\sigma[(t-(2n+1)k]$$

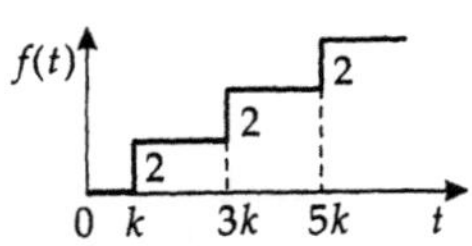

$$\frac{1}{s\cosh(ks)} \quad \bullet\!\!-\!\!\circ \quad 2\sum_{n=0}^{\infty}(-1)^n\sigma[(t-(2n+1)k]$$

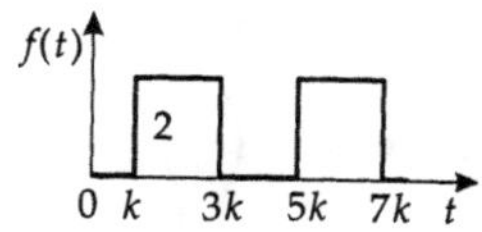

$$\frac{1}{s}\coth(ks) \quad \bullet\!\!-\!\!\circ \quad \sigma(t)+2\sum_{n=1}^{\infty}\sigma(t-2nk)$$

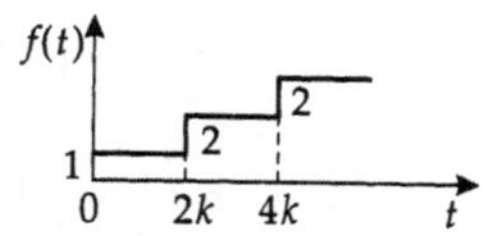

Triangular curves

$$\frac{1}{s^2}\tanh(ks) \quad \bullet\!\!-\!\!\circ \quad t\sigma(t)+2\sum_{n=1}^{\infty}(-1)^n(t-2nk)\sigma(t-2nk)$$

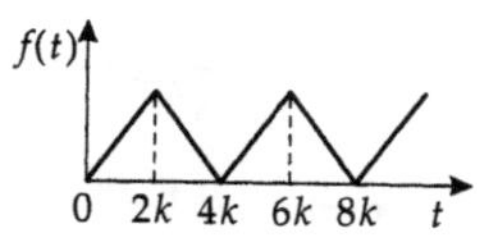

Sinus pulse (two-way rectification)

$$\frac{k}{s^2+k^2}\coth\frac{\pi s}{2k} \quad \bullet\!\!-\!\!\circ \quad |\sin kt|$$

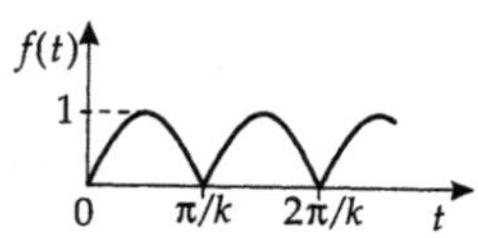

Sinus pulse (one-way rectification)

$$\frac{1}{(s^2+1)(1-\mathrm{e}^{-\pi s})} \quad \bullet\!\!-\!\!\circ \quad \sum_{n=0}^{\infty}(-1)^n\sigma(t-n\pi)\sin t$$

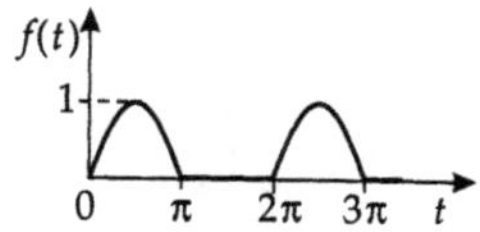

$$\frac{1}{s}\mathrm{e}^{-k/s} \quad \bullet\!\!-\!\!\circ \quad J_0(2\sqrt{kt})$$

$$\frac{1}{\sqrt{s}}\mathrm{e}^{-k/s} \quad \bullet\!\!-\!\!\circ \quad \frac{1}{\sqrt{\pi t}}\cos(2\sqrt{kt})$$

$$\frac{1}{\sqrt{s}}\mathrm{e}^{k/s} \quad \bullet\!\!-\!\!\circ \quad \frac{1}{\sqrt{\pi t}}\cosh(2\sqrt{kt})$$

$$\frac{1}{s^{3/2}}\mathrm{e}^{-k/s} \quad \bullet\!\!-\!\!\circ \quad \frac{1}{\sqrt{\pi k}}\sin(2\sqrt{kt})$$

$\frac{1}{s^{3/2}}e^{k/s}$		$\bullet\!\!-\!\!\circ$	$\frac{1}{\sqrt{\pi k}}\sinh(2\sqrt{kt})$
$\frac{1}{s^\mu}e^{-k/s}$	$(\mu > 0)$	$\bullet\!\!-\!\!\circ$	$\left(\frac{t}{k}\right)^{(\mu-1)/2} J_{\mu-1}(2\sqrt{kt})$
$\frac{1}{s^\mu}e^{k/s}$	$(\mu > 0)$	$\bullet\!\!-\!\!\circ$	$\left(\frac{t}{k}\right)^{(\mu-1)/2} I_{\mu-1}(2\sqrt{kt})$
$e^{-k\sqrt{s}}$	$(k > 0)$	$\bullet\!\!-\!\!\circ$	$\frac{k}{2\sqrt{\pi t^3}}\exp\left(-\frac{k^2}{4t}\right)$
$\frac{1}{s}e^{-k\sqrt{s}}$	$(k \geq 0)$	$\bullet\!\!-\!\!\circ$	$\operatorname{erfc}\frac{k}{2\sqrt{t}}$
$\frac{1}{\sqrt{s}}e^{-k\sqrt{s}}$	$(k \geq 0)$	$\bullet\!\!-\!\!\circ$	$\frac{1}{\sqrt{\pi t}}\exp\left(-\frac{k^2}{4t}\right)$
$\frac{1}{s^{3/2}}e^{-k\sqrt{s}}$	$(k \geq 0)$	$\bullet\!\!-\!\!\circ$	$2\sqrt{\frac{t}{\pi}}\exp\left(-\frac{k^2}{4t}\right) - k\operatorname{erfc}\frac{k}{2\sqrt{t}} = 2\sqrt{t}\,\mathrm{j}\operatorname{erfc}\frac{k}{2\sqrt{t}}$
$\frac{1}{s^{1+(3/2)n}}e^{-k\sqrt{s}}$	$(n = 0, 1, 2, \cdots,\ k \geq 0)$	$\bullet\!\!-\!\!\circ$	$(4t)^{n/2}\mathrm{j}^n\operatorname{erfc}\frac{k}{2\sqrt{t}}$
$s^{(n-1)/2}e^{-k\sqrt{s}}$	$(n = 0, 1, 2, \cdots,\ k > 0)$	$\bullet\!\!-\!\!\circ$	$\frac{\exp(-k^2/4t)}{2^n\sqrt{\pi t^{n+1}}} H_n\left(\frac{k}{2\sqrt{t}}\right)$
$\frac{e^{-k\sqrt{s}}}{a+\sqrt{s}}$	$(k \geq 0)$	$\bullet\!\!-\!\!\circ$	$\frac{1}{\sqrt{\pi t}}\exp\left(-\frac{k^2}{4t}\right) - ae^{ak}e^{a^2t}\operatorname{erfc}\left(a\sqrt{t}+\frac{k}{2\sqrt{t}}\right)$
$\frac{ae^{-k\sqrt{s}}}{s(a+\sqrt{s})}$	$(k \geq 0)$	$\bullet\!\!-\!\!\circ$	$-e^{ak}e^{a^2t}\operatorname{erfc}\left(a\sqrt{t}+\frac{k}{2\sqrt{t}}\right) + \operatorname{erfc}\frac{k}{2\sqrt{t}}$
$\frac{e^{-k\sqrt{s}}}{\sqrt{s}(a+\sqrt{s})}$	$(k \geq 0)$	$\bullet\!\!-\!\!\circ$	$e^{ak}e^{a^2t}\operatorname{erfc}\left(a\sqrt{t}+\frac{k}{2\sqrt{t}}\right)$
$\frac{e^{-k\sqrt{s(s+a)}}}{\sqrt{s(s+a)}}$	$(k \geq 0)$	$\bullet\!\!-\!\!\circ$	$e^{-at/2}I_0\left(\frac{a}{2}\sqrt{t^2-k^2}\right)\sigma(t-k)$
$\frac{e^{-k\sqrt{s^2+a^2}}}{\sqrt{s^2+a^2}}$	$(k \geq 0)$	$\bullet\!\!-\!\!\circ$	$J_0(a\sqrt{t^2-k^2})\sigma(t-k)$
$\frac{e^{-k\sqrt{s^2-a^2}}}{\sqrt{s^2-a^2}}$	$(k \geq 0)$	$\bullet\!\!-\!\!\circ$	$I_0(a\sqrt{t^2-k^2})\sigma(t-k)$
$\frac{e^{-k(\sqrt{s^2+a^2}-s)}}{\sqrt{s^2+a^2}}$	$(k \geq 0)$	$\bullet\!\!-\!\!\circ$	$J_0(a\sqrt{t^2+2kt})$
$e^{-ks} - e^{-k\sqrt{s^2+a^2}}$	$(k > 0)$	$\bullet\!\!-\!\!\circ$	$\frac{ak}{\sqrt{t^2-k^2}}J_1(a\sqrt{t^2-k^2})\sigma(t-k)$
$e^{-k\sqrt{s^2-a^2}} - e^{-ks}$	$(k > 0)$	$\bullet\!\!-\!\!\circ$	$\frac{ak}{\sqrt{t^2-k^2}}I_1(a\sqrt{t^2-k^2})\sigma(t-k)$
$\frac{a^\nu e^{-k\sqrt{s^2+a^2}}}{\sqrt{s^2+a^2}(\sqrt{s^2+a^2}+s)^\nu}$	$(\nu > -1,\ k \geq 0)$	$\bullet\!\!-\!\!\circ$	$\left(\frac{t-k}{t+k}\right)^{\nu/2} J_\nu(a\sqrt{t^2-k^2})\sigma(t-k)$
$\frac{1}{s}\ln s$		$\bullet\!\!-\!\!\circ$	$-E - \ln t$ $(E \approx 0.5772\ldots$ Euler constant)
$\frac{1}{s^k}\ln s$	$(k > 0)$	$\bullet\!\!-\!\!\circ$	$\frac{t^{k-1}}{\Gamma(k)}[\Psi(k) - \ln t]$
$\frac{\ln s}{s-a}$	$(a > 0)$	$\bullet\!\!-\!\!\circ$	$e^{at}[\ln a + E_1(at)$

$\dfrac{\ln s}{s^2+1}$	$\bullet\!\!-\!\!\circ$	$\cos t\,\mathrm{Si}(t) - \sin t\,\mathrm{Ci}(t)$
$\dfrac{s\ln s}{s^2+1}$	$\bullet\!\!-\!\!\circ$	$-\sin t\,\mathrm{Si}(t) - \cos t\,\mathrm{Ci}(t)$
$\dfrac{1}{s}\ln(1+ks)\quad (k>0)$	$\bullet\!\!-\!\!\circ$	$E_1\left(\dfrac{t}{k}\right)$
$\ln\dfrac{s+a}{s+b}$	$\bullet\!\!-\!\!\circ$	$\dfrac{1}{t}(\mathrm{e}^{-bt}-\mathrm{e}^{-at})$
$\dfrac{1}{s}\ln(1+k^2s^2)\quad (k>0)$	$\bullet\!\!-\!\!\circ$	$-2\mathrm{Ci}\left(\dfrac{t}{k}\right)$
$\dfrac{1}{s}\ln(s^2+a^2)\quad (a>0)$	$\bullet\!\!-\!\!\circ$	$2\ln a - 2\mathrm{Ci}(at)$
$\dfrac{1}{s^2}\ln(s^2+a^2)\quad (a>0)$	$\bullet\!\!-\!\!\circ$	$\dfrac{2}{a}[at\ln a + \sin at - at\,\mathrm{Ci}(at)]$
$\ln\dfrac{s^2+a^2}{s^2}$	$\bullet\!\!-\!\!\circ$	$\dfrac{2}{t}(1-\cos at)$
$\ln\dfrac{s^2-a^2}{s^2}$	$\bullet\!\!-\!\!\circ$	$\dfrac{2}{t}(1-\cosh at)$
$\arctan\dfrac{k}{s}$	$\bullet\!\!-\!\!\circ$	$\dfrac{1}{t}\sin kt$
$\dfrac{1}{s}\arctan\dfrac{k}{s}$	$\bullet\!\!-\!\!\circ$	$\mathrm{Si}(kt)$
$\mathrm{e}^{k^2s^2}\operatorname{erfc} ks\quad (k>0)$	$\bullet\!\!-\!\!\circ$	$\dfrac{1}{k\sqrt{\pi}}\exp\left(-\dfrac{t^2}{4k^2}\right)$
$\dfrac{1}{s}\mathrm{e}^{k^2s^2}\operatorname{erfc} ks\quad (k>0)$	$\bullet\!\!-\!\!\circ$	$\operatorname{erf}\dfrac{t}{2k}$
$\mathrm{e}^{ks}\operatorname{erfc}\sqrt{ks}\quad (k>0)$	$\bullet\!\!-\!\!\circ$	$\dfrac{\sqrt{k}}{\pi\sqrt{t}(t+k)}$
$\dfrac{1}{\sqrt{s}}\operatorname{erfc}\sqrt{ks}\quad (k\ge 0)$	$\bullet\!\!-\!\!\circ$	$\dfrac{1}{\sqrt{\pi t}}\sigma(t-k)$
$\dfrac{1}{\sqrt{s}}\mathrm{e}^{ks}\operatorname{erfc}\sqrt{ks}\quad (k\ge 0)$	$\bullet\!\!-\!\!\circ$	$\dfrac{1}{\sqrt{\pi(t+k)}}$
$\operatorname{erf}\dfrac{k}{\sqrt{s}}$	$\bullet\!\!-\!\!\circ$	$\dfrac{1}{\pi t}\sin 2k\sqrt{t}$
$\dfrac{1}{\sqrt{s}}\mathrm{e}^{k^2/s}\operatorname{erfc}\dfrac{k}{\sqrt{s}}$	$\bullet\!\!-\!\!\circ$	$\dfrac{1}{\sqrt{\pi t}}\mathrm{e}^{-2k\sqrt{t}}$
$K_0(ks)\quad (k>0)$	$\bullet\!\!-\!\!\circ$	$\dfrac{1}{\sqrt{t^2-k^2}}\sigma(t-k)$
$K_0(k\sqrt{s})\quad (k>0)$	$\bullet\!\!-\!\!\circ$	$\dfrac{1}{2t}\exp\left(-\dfrac{k^2}{4t}\right)$
$\dfrac{1}{s}\mathrm{e}^{ks}K_1(ks)\quad (k>0)$	$\bullet\!\!-\!\!\circ$	$\dfrac{1}{k}\sqrt{t(t+2k)}$
$\dfrac{1}{\sqrt{s}}K_1(k\sqrt{s})\quad (k>0)$	$\bullet\!\!-\!\!\circ$	$\dfrac{1}{k}\exp\left(-\dfrac{k^2}{4t}\right)$
$\dfrac{1}{\sqrt{s}}\mathrm{e}^{k/s}K_0\left(\dfrac{k}{s}\right)\quad (k>0)$	$\bullet\!\!-\!\!\circ$	$\dfrac{2}{\sqrt{\pi t}}K_0(2\sqrt{2kt})$
$\pi\mathrm{e}^{-ks}I_0(ks)\quad (k>0)$	$\bullet\!\!-\!\!\circ$	$\dfrac{1}{\sqrt{t(2k-t)}}[\sigma(t)-\sigma(t-2k)]$
$\mathrm{e}^{-ks}I_1(ks)\quad (k>0)$	$\bullet\!\!-\!\!\circ$	$\dfrac{k-t}{\pi k\sqrt{t(2k-t)}}[\sigma(t)-\sigma(t-2k)]$
$\mathrm{e}^{as}E_1(as)\quad (a>0)$	$\bullet\!\!-\!\!\circ$	$\dfrac{1}{t+a}$

$\frac{1}{a} - s e^{as} E_1(as) \quad (a > 0)$ ●—○ $\frac{1}{(t+a)^2}$

$a^{1-n} e^{as} E_n(as)$
$(a > 0\,;\; n = 0, 1, 2, \ldots)$ ●—○ $\frac{1}{(t+a)^n}$

$\left[\frac{\pi}{2} - \mathrm{Si}(s)\right] \cos s + \mathrm{Ci}(s) \sin s$ ●—○ $\frac{1}{t^2+1}$

$\frac{1}{as^2} - \frac{e^{-as}}{s(1-e^{-as})}$ ●—○ **Sawtooth oscillation**

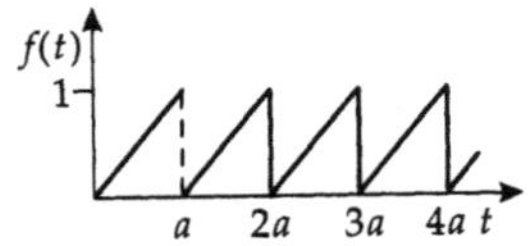

$\frac{e^{-as}(1-e^{-\epsilon s})}{s}$ ●—○ **Pulse function**

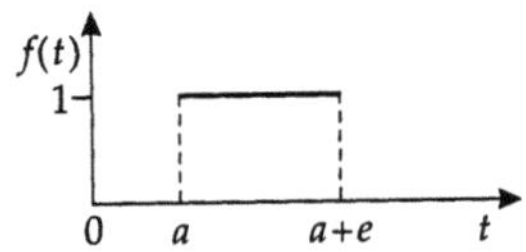

$\frac{e^{-s} + e^{-2s}}{s(1-e^{-s})^2}$ ●—○ $F(t) = n^2 \quad (n \le t < n+1,\; n = 0, 1, 2, \cdots)$

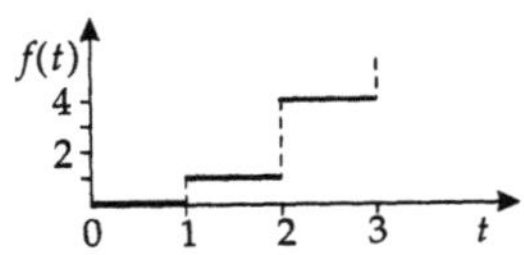

Sinus pulse

$\frac{\pi a(1+e^{-as})}{a^2 s^2 + \pi^2}$ ●—○ $F(t) = \begin{cases} \sin(\pi t/a) & (0 \le t \le a) \\ 0 & (t > a) \end{cases}$

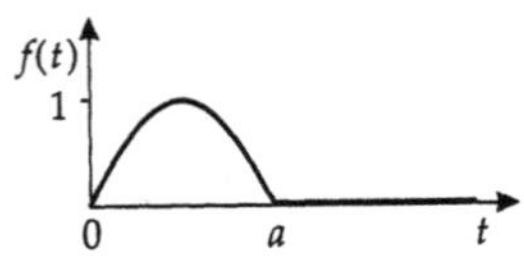

20.6 z transformation

The **z transformation** is used for continuous time-dependent functions whose values are scanned at periodically recurring intervals. Thus, the z transformation is the discrete analog of the Laplace transformation.

20.6.1 *Definition of the z transformation*

Pulse recurrence function: Definition by means of the sequence of values $f(kT) = f_k$, $k = 0, 1, 2, \ldots$ obtained by scanning the measuring signal $f(t)$ with scanning period T

(scanning frequency: $F = 1/T$):

$$f^*(t) = \sum_{k=0}^{\infty} f_k \delta(t - kT) ,$$

where $\delta(t)$ is the Dirac δ function. Contrary to the sequence of values f_k, the pulse recurrence function is defined for all values of the variable t.

▷ The pulse recurrence function $f^*(t)$ follows uniquely from the sequence of values f_k and the scanning period T, while the sequence of values f_k does not follow uniquely from the pulse recurrence function $f^*(t)$.

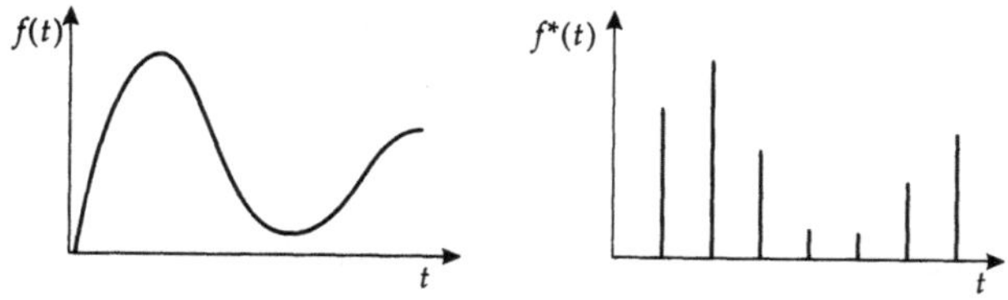

The measuring signal $f(t)$ and its pulse recurrence function $f^*(t)$

The Laplace transformation of the pulse recurrence function $f^*(t)$ is:

$$\mathcal{L}\left\{f^*(t)\right\} = F^*(s) = \int_0^{\infty} f^*(t) e^{-st}\, dt$$

$$= \int_0^{\infty} \sum_{k=0}^{\infty} f_k \delta(t - kT) e^{-st}\, dt = \sum_{k=0}^{\infty} f_k e^{-kTs} ,$$

where the property of the δ function

$$\int_{-\infty}^{\infty} \delta(t - t_0) f(t)\, dt = f(t_0)$$

has been used. Thus, in the Laplace transformation of pulse recurrence functions the integration becomes a summation.

z transformation: Follows with the substitution

$$z = e^{Ts}$$

from the Laplace transform $F^*(s)$ of the pulse recurrence function:

$$\mathcal{Z}\{f_k\} = F(z) = \sum_{k=0}^{\infty} f_k z^{-k} .$$

Here, the operator notation $z f_k$ has been used for the z transformation of the sequence of values f_k (original sequence).

Correspondence notation: Analogous to the Laplace transformation:

$$f_k \quad ○\!\!-\!\!\bullet \quad F(z) .$$

The image function $F(z)$ is assigned to the original sequence f_k.

Relation between the Laplace transformation and the z-transformation:

$$\mathcal{Z}\{f_k\} = \mathcal{L}\left\{f^*(t)\right\} ,$$

The z transformation of the sequence of values f_k is therefore equal to the Laplace transform of the pulse recurrence function $f^*(t)$:

20.6.2 Convergence conditions for the z transformation

- **Convergence conditions for the z transformation**: If the following holds for the sequence of values:

$$|f_k| < Mm^k$$

with the two positive constants M and m, then the series of the z transforms

$$\sum_{k=0}^{\infty} f_k z^{-k}$$

converges outside a circle with radius R in the complex plane:

$$|z| > R .$$

then, all singular points of the z transform lie inside this circle ($|z| \leq R$).

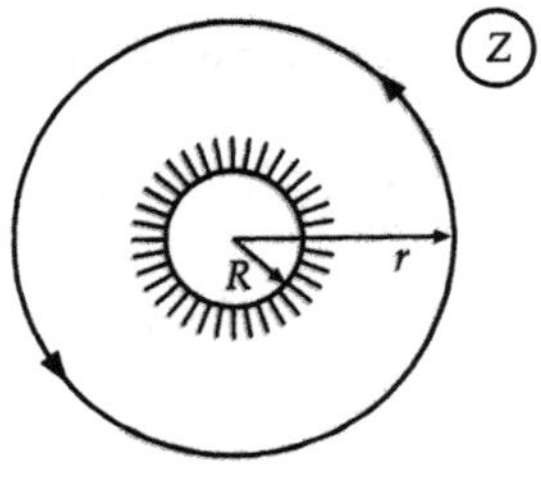

Domain of convergence of the z transform in the complex plane

□ Let the sequence of values be constant:

$$f_k = 1 .$$

Then the z transform is the geometric series:

$$\mathcal{Z}\{f_k\} = F(z) = \sum_{k=0}^{\infty} 1z^{-k} = \frac{1}{1-z^{-1}} = \frac{z}{z-1} ,$$

or in correspondence notation:

$$f_k = 1 \quad \circ\!\!-\!\!\bullet \quad F(z) = \frac{1}{1-z^{-1}} = \frac{z}{z-1} .$$

In this case, the radius of convergence is $R = 1$, and the z transform converges for $|z| > 1$.

▷ The z transform is a special Laurent series for which the coefficients f_k for exponents with a positive sign are vanishing. Thus, the convergence conditions follow directly from the convergence conditions of the Laurent series.

20.6.3 Inversion of the z transformation

Inverse z transformation $\mathcal{Z}^{-1}$:

$$\mathcal{Z}^{-1}\{F(z)\} = f_k\,.$$

Under most practical conditions, the inverse z transformation can be determined either directly or analogously as in the case of the Laplace transformation with the help of partial fraction decomposition.
In the following, possibilities for calculating the inverse z transformation are given in cases for which the inverse transformation cannot be found in the tables:

1. If we regard the z transform as a Laurent series, the inverse z transformation follows from the coefficients of that series:

$$\mathcal{Z}^{-1}\{F(z)\} = f_k = \frac{1}{2\pi \mathrm{j}} \oint F(z) z^{k-1}\, dz\,.$$

The curvilinear integral must be taken over an arbitrarily closed curve that is larger than the radius of convergence R of $F(z)$. The evaluation of the curvilinear integral can be performed with the help of the residue theorem of function theory.

2. If the complex number z is represented by polar coordinates in the complex plane $z = |z|e^{\mathrm{j}\varphi}$ ($|z|$: absolute value of z; φ: polar angle of z), then the formula for f_k yields:

$$\mathcal{Z}^{-1}\{F(z)\} = f_k = \frac{|z|^n}{2\pi} \int_{-\pi}^{\pi} F\left(|z|e^{\mathrm{j}\varphi}\right) e^{\mathrm{j}n\varphi}\, d\varphi\,.$$

3. The function

$$F(z^{-1}) = \sum_{k=0}^{\infty} f_k z^k$$

represents a power series with ascending powers of z. We therefore obtain the following for its coefficients f_k, according to the Taylor expansion:

$$f_k = \frac{1}{k!}\left[\frac{d^k F(z^{-1})}{dz^k}\right]_{z=0}\,.$$

20.6.4 Calculating rules

- **Shift to the right**: The z transform of the original sequence f_k, shifted by n points to the right, is equal to the sum of the z transform of the unshifted original sequence and the sum $\sum_{l=1}^{n} f_{-l} z^l$, multiplied by z^{-n}.

$$\mathcal{Z}\{f_{k-n}\} = z^{-n}\left[F(z) + \sum_{l=1}^{n} f_{-l} z^l\right]\,.$$

- **Shift to the left**: The z transform of the original sequence shifted by n points to the left is equal to the difference of image function of the unshifted original sequence and the sum $\sum_{l=0}^{n-1} f_l z^{-l}$, multiplied by z^n.

$$\mathcal{Z}\{f_{k+n}\} = z^n \left[F(z) - \sum_{l=0}^{n-1} f_l z^{-l} \right] .$$

- **Forward difference**: The z transform of the forward difference $f_{k+1} - f_k$ of the original sequence is equal to the z transform of the original sequence at point k, multiplied by the factor $(z-1)$, minus the value of the original sequence at point $k = 0$, multiplied by z:

$$\mathcal{Z}\{f_{k+1} - f_k\} = (z-1)F(z) - zf_0 .$$

- **Backward difference**: The z transform of the backward difference $f_k - f_{k-1}$ of the original sequence is equal to the z transform of the original sequence at point k, multiplied by the factor $(z-1)/z$, minus the value of the original sequence at point $k = -1$:

$$\mathcal{Z}\{f_k - f_{k-1}\} = \frac{z-1}{z} F(z) - f_{-1} .$$

▷ If we divide the forward or backward difference by the scanning period T, then we obtain the numerical approximation of the first derivative to two-point accuracy. The above difference formulas are, therefore, the analog of the differentiation theorems for the Laplace transformation.
Further difference formulas can be easily calculated with the help of the two shift theorems.

- **Summation**: The z transform of the sum of the terms of the original sequence

(a) up to the upper summation index n is equal to the z transform of the original sequence at point n, multiplied by factor $z/(z-1)$.

$$\mathcal{Z}\left(\sum_{l=0}^{n} f_l \right) = \frac{z}{z-1} F(z) ;$$

(b) up to the upper summation index $n-1$ is equal to the z transform of the original sequence at point n multiplied by factor $1/(z-1)$:

$$\mathcal{Z}\left(\sum_{l=0}^{n-1} f_l \right) = \frac{1}{z-1} F(z) .$$

- **Damping**: The z transform of the original sequence f_k damped by β^k is equal to the z transform of the original sequence with the argument $z\beta^{-1}$:

$$\mathcal{Z}\left\{ f_k \beta^k \right\} = F(z\beta^{-1}) ,$$

where β is an arbitrary complex number.

- **Differentiation of the image function**: The negative value of the first derivative of the image function, multiplied by z, is equal to the z transform of the original sequence f_k, multiplied by k:

$$\mathcal{Z}\{kf_k\} = -z \frac{dF(z)}{dz} .$$

- **Integration of the image function**:

$$z\left\{k^{-1}f_k\right\} = -\int_0^z \frac{F(z')}{z'}\mathrm{d}z'$$

Convolution of sequences: The convolution of two sequences f_k and g_k means the sum:

$$\sum_{l=0}^{k} f_l g_{k-l} \; .$$

- **Convolution**: The z transform of the convolution of two sequences f_k and g_k is equal to the product of the z transform of these two sequences f_k and g_k:

$$\mathcal{Z}\left\{\sum_{l=0}^{k} f_l g_{k-l}\right\} = F(z)\cdot G(z) \; .$$

- **Initial value theorem**: The value of the original sequence at point $k = 0$ is obtained by letting the variable z in the image function go to infinity along an arbitrary path in the complex plane:

$$f_0 = \lim_{z\to\infty} F(z) \; .$$

- **Final value theorem**: Limit of the original sequence $\lim_{k\to\infty} f_k$ is equal to the limit $\lim_{z\to 1+0}$ of the image function of f_k multiplied by the factor $(z-1)$:

$$\lim_{k\to\infty} f_k = \lim_{z\to 1+0}(z-1)F(z) \; .$$

20.6.5 Calculating rules for the z transformation

Name of the operation	Operation with the number sequences	Operation with the z transforms
Multiplication by a constant	$\alpha f_k \ (\alpha \in \mathbb{R})$	$\alpha F(z)$
Sum of sequences	$f_k \pm g_k$	$F(z) \pm G(z)$
Shift to the right	f_{k-n}	$z^{-n}\left[F(z) + \sum_{l=1}^{n} f_{-l} z^{l}\right]$
Shift to the left	f_{k+n}	$z^{n}\left[F(z) - \sum_{l=0}^{n-1} f_l z^{-l}\right]$
Difference formation	$f_k - f_{k-1}$ $f_{k+1} - f_k$	$\frac{z-1}{z} F(z) - f_{-1}$ $(z-1)F(z) - f_0 z$
Summation $(f = 0,\ t < 0)$	$\sum_{l=0}^{k} f_l$ $\sum_{l=0}^{k-1} f_l$	$\frac{z}{z-1} F_z(z)$ $\frac{1}{z-1} F_z(z)$
Damping	$f_k \beta^k$	$F(z\beta^{-1})$
Differentiation of the image function	$k f_k$	$-zF'(z)$
Integration of the image function	$k^{-1} f_k$	$-\int_0^z \frac{F(z')}{z'}\,\mathrm{d}z'$
Convolution $f * g \ (f, g = 0,\ t < 0)$	$\sum_{l=0}^{k} f_l g_{k-l}$	$F(z) \cdot G(z)$
	$\mathrm{e}^{\alpha kT} f_k \ (\alpha \in \mathbb{R})$	$F(\mathrm{e}^{-\alpha T} z)$
	$\mathrm{e}^{-\alpha kT} f_k \ (\alpha \in \mathbb{R})$	$F(\mathrm{e}^{\alpha T} z)$

20.6.6 Table of z transforms

$H(x)$ Step function, Heaviside function

$$f_k \quad \bullet\!\!-\!\!\circ \quad F(z)$$

$$(kT = t, k = t/T)$$

$$\delta_{k,0} \quad \bullet\!\!-\!\!\circ \quad 1$$

$$\delta_{k,m} \quad \bullet\!\!-\!\!\circ \quad \frac{1}{z^m}$$

$$H(kT) \quad \bullet\!\!-\!\!\circ \quad \frac{z}{z-1}$$

$$H(kT-T) \quad \bullet\!\!-\!\!\circ \quad \frac{1}{z-1}$$

$$H(kT-mT) \quad \bullet\!\!-\!\!\circ \quad \frac{z}{z^m(z-1)}$$

$$H(kT) - H(kT-T) \quad \bullet\!\!-\!\!\circ \quad 1$$

$$H(kT) - H(kT-2T) \quad \bullet\!\!-\!\!\circ \quad 1 + \frac{1}{z}$$

$$H(kT-mT) - H(kT-\overline{m+1}T) \quad \bullet\!\!-\!\!\circ \quad \frac{1}{z}{}^{m}$$

$\frac{T}{kT}H(kT-T)$	$\bullet\!\!-\!\!\circ$	$\ln\left(\frac{z}{z-1}\right)$
kT	$\bullet\!\!-\!\!\circ$	$\frac{Tz}{(z-1)^2}$
kT^2	$\bullet\!\!-\!\!\circ$	$\frac{T^2 z(z+1)}{(z-1)^3}$
kT^3	$\bullet\!\!-\!\!\circ$	$\frac{T^3 z(z^2+4z+1)}{(z-1)^4}$
kT^n	$\bullet\!\!-\!\!\circ$	$(-1)^n \lim_{\chi\to 0}\frac{\partial^n}{\partial\chi^n}\left(\frac{z}{z-e^{-\chi T}}\right)$
$1-a^{\omega kT}$	$\bullet\!\!-\!\!\circ$	$\frac{z(1-a^{\omega T})}{(z-1)(z-a^{\omega T})}$
$a^{\omega kT}$	$\bullet\!\!-\!\!\circ$	$\frac{z}{(z-a^{\omega T})}$
$kTa^{\omega kT}$	$\bullet\!\!-\!\!\circ$	$\frac{Tza^{\omega T}}{(z-a^{\omega T})^2}$
$kT^2a^{\omega kT}$	$\bullet\!\!-\!\!\circ$	$\frac{T^2a^{\omega T}z(z+a^{\omega T})}{(z-a^{\omega T})^3}$
$kT^3a^{\omega kT}$	$\bullet\!\!-\!\!\circ$	$T^3a^{\omega T}z\frac{z^2+4za^{\omega T}+a^{2\omega T}}{(z-a^{\omega T})^4}$
$kT^n a^{\omega kT}\quad (n=1,2,\ldots)$	$\bullet\!\!-\!\!\circ$	$\frac{\partial^n}{\partial\omega^n}\left(\frac{z}{z-a^{\omega T}}\right)=T^n z a^{\omega T}\frac{z^{n-1}+\cdots}{(z-a^{\omega T})^{n+1}}$
$\sin\omega kT$	$\bullet\!\!-\!\!\circ$	$\frac{z\sin\omega T}{z^2-2z\cos\omega T+1}$
$\cos\omega kT$	$\bullet\!\!-\!\!\circ$	$\frac{z(z-\cos\omega T)}{z^2-2z\cos\omega T+1}$
$\sin(\omega kT+\phi)$	$\bullet\!\!-\!\!\circ$	$z\frac{z\sin\phi+\sin(\omega T-\phi)}{z^2-2z\cos\omega T+1}$
$\sinh\omega kT$	$\bullet\!\!-\!\!\circ$	$\frac{z\sinh\omega T}{z^2-2z\cosh\omega T+1}$
$\cosh\omega kT$	$\bullet\!\!-\!\!\circ$	$\frac{z(z-\cosh\omega T)}{z^2-2z\cosh\omega T+1}$
$kT\sin\omega kT$	$\bullet\!\!-\!\!\circ$	$Tz\frac{(z^2-1)\sin\omega T}{(z^2-2z\cos\omega T+1)^2}$
$kT\cos\omega kT$	$\bullet\!\!-\!\!\circ$	$Tz\frac{(z^2+1)\cos\omega T-2z}{(z^2-2z\cos\omega T+1)^2}$
$kT\sin(\omega kT+\phi)$	$\bullet\!\!-\!\!\circ$	$Tz\frac{z^2\sin(\omega T+\phi)-2z\sin\phi-\sin(\omega T-\phi)}{(z^2-2z\cos\omega T+1)^2}$
$e^{-\alpha kT}\sin\omega kT$	$\bullet\!\!-\!\!\circ$	$\frac{ze^{-\alpha T}\sin\omega T}{z^2-2ze^{-T}\cos\omega T+e^{-2\alpha T}}$
$e^{-\alpha kT}\cos\omega kT$	$\bullet\!\!-\!\!\circ$	$\frac{z(z-e^{-\alpha T}\cos\omega T)}{z^2-2ze^{-T}\cos\omega T+e^{-2\alpha T}}$
$e^{-\alpha kT}\sin(\omega kT+\phi)$	$\bullet\!\!-\!\!\circ$	$z\frac{z\sin\phi+e^{-\alpha T}\sin(\omega T-\phi)}{z^2-2ze^{-\alpha T}\cos\omega T+e^{2-\alpha T}}$
$e^{-\alpha kT}\sinh\omega kT$	$\bullet\!\!-\!\!\circ$	$\frac{ze^{-\alpha T}\sinh\omega T}{z^2-2ze^{-\alpha T}\cosh\omega T+e^{-2\alpha T}}$
$e^{-\alpha kT}\cosh\omega kT$	$\bullet\!\!-\!\!\circ$	$\frac{z(z-e^{-\alpha T}\cosh\omega T)}{z^2-2ze^{-\alpha T}\cosh\omega T+e^{-2\alpha T}}$
$-\frac{1}{a}[\delta_{k,0}-a^k]$	$\bullet\!\!-\!\!\circ$	$\frac{1}{z-a}$
$\frac{1}{(a-b)}\left[a^{k-1}-b^{k-1}\right]$	$\bullet\!\!-\!\!\circ$	$\frac{1}{(z-a)(z-b)}$

$$\frac{1}{(a-b)}\left[a^k - b^k\right] \quad \bullet\!\!-\!\!\circ \quad \frac{z}{(z-a)(z-b)}$$

$$\frac{1}{(a-b)}\left[(a-c)a^{k-1} - (b-c)b^{k-1}\right] \quad \bullet\!\!-\!\!\circ \quad \frac{z-c}{(z-a)(z-b)}$$

$$\frac{1}{(a-b)}\left[a^{k+1} - b^{k+1}\right] \quad \bullet\!\!-\!\!\circ \quad \frac{z^2}{(z-a)(z-b)}$$

$$\frac{1}{k!} \quad \bullet\!\!-\!\!\circ \quad e^{1/z}$$

$$\frac{1}{(2k)!} \quad \bullet\!\!-\!\!\circ \quad \cosh(z^{-1/2})$$

$$e^{k\ln a} = a^k \quad \bullet\!\!-\!\!\circ \quad \frac{z}{z-a}$$

$$(-1)^k \quad \bullet\!\!-\!\!\circ \quad \frac{z}{z+1}$$

$$k\beta^{k-1} \quad \bullet\!\!-\!\!\circ \quad \frac{z}{(z-\beta)^2}$$

$$\binom{k}{n-1} \quad (n = 1, 2, \ldots) \quad \bullet\!\!-\!\!\circ \quad \frac{z}{(z-1)^n}$$

$$\frac{e^{k\ln\beta}}{\beta^{n-1}}\binom{k}{n-1} \quad \bullet\!\!-\!\!\circ \quad \frac{z}{(z-\beta)^n}, \ \beta \neq 0, \ n = 1, 2, \cdots$$

21

Probability theory and mathematical statistics

Probability theory: Mathematical discipline that applies and describes the regularities concerning events.
While an individual random phenomenon may defy exact mathematical description, large numbers of random phenomena can be covered by mathematical laws. Probability theory provides the foundation for mathematical statistics.
Mathematical statistics: Theory of numerical description and examination of large numbers of events in nature and society.
Descriptive statistics: Elementary part of mathematical statistics. Deals with the description, processing, subdivision and representation of empirical data. Essential task of mathematical statistics: To draw conclusions as to the totality of material from a set of results based on observations (**random sampling**).

21.1 Combinatorics

Complexions: The object of combinatorics; complexions investigate the arrangement or composition of a finite number of elements (e.g., persons, objects, numbers) identified by symbols.
Basic notions of combinatorics: Permutations, combinations, variations.

- Complexions in which n given elements are adjacent to each other in random order and called permutations of the n elements.

Number of permutations of n different elements:

$$P_n = n!$$

Number of permutations of n given elements with n_1 equal elements of a first kind, n_2 equal elements of a second kind, ..., n_k equal elements of a k-th kind:

$$P_n^{(n_1,n_2,\ldots,n_k)} = \frac{n!}{n_1!n_2!\ldots n_k!}, \quad \sum_{i=1}^{k} n_i = n.$$

□ 3 students may choose $3! = 6$ different seating orders on 3 chairs.

● Complexions consisting of k elements that can be formed from n different elements **without** taking into account their arrangement are called **combination** of n elements of order k.

Number of combinations without the repetition of n elements:

$$C_n^k = \binom{n}{k}, \quad 1 \leq k \leq n.$$

Number of combinations of order k with repetition of n elements:

$${}^W C_n^k = \binom{n+k-1}{k}.$$

□ 6 out of 49 is a combination of order 6.

$$C_{49}^6 = \binom{49}{6} = \frac{49 \cdot 48 \cdot 47 \cdot 46 \cdot 45 \cdot 44}{1 \cdot 2 \cdot 3 \cdot 4 \cdot 5 \cdot 6} = 13983816$$

● **Variations** of order k from n elements: Complexions consisting of k elements that can be formed from n different elements **taking into account** their arrangement are called variations of order k of n elements.

Number of variations without repetition of n elements:

$$V_n^k = \frac{n!}{(n-k)!}, \quad 1 \leq k \leq n.$$

Number of variations of order k with repetition of n elements:

$${}^W V_n^k = n^k.$$

□ Maximum number of flare signals when 4 different colors are available, and each signal consists of 2 multicolored signals fired one after the other:

$$V_4^2 = \frac{4!}{2!} = 12.$$

21.2 Random events

21.2.1 Basic notions

Random test, random experiment: Test whose result is influenced by random factors and which is therefore uncertain to a degree.

□ Dice, lottery, roulette, random number generator, statistical measuring error, estimating function of a random sample.

Elementary event w_i: Directly possible result of a random trial.

Set of events Ω: Set of elementary events.

Discrete set of events: Set of events whose elements can be mapped onto the set of natural numbers.

Continuous set of events: Set of events whose elements are non-denumerable and can be mapped only onto the set of real numbers.

□ The numbers on a die $E = (1, 2, 3, 4, 5, 6)$ are a **discrete** set of events; the set of real numbers in the interval $X = [0, 1]$ are a **continuous** set of events.

Random event A: An event that may occur as the result of a random trial (subset of Ω); element of the field of events.

□ All numbers of a die greater than 3 or less than 3;
the intervals $x_1 = [0, \frac{1}{3})$, $x_2 = [\frac{1}{3}, \frac{2}{3})$, $x_3 = [\frac{2}{3}, 1]$.

Certain event Ω: An event that must occur (Ω as a subset of Ω).
Impossible event ϕ: An event that cannot occur (empty set ϕ as a subset of Ω).
Event $\overline{A}$ complementary (opposite) to A: Event that occurs if and only if A does not occur.

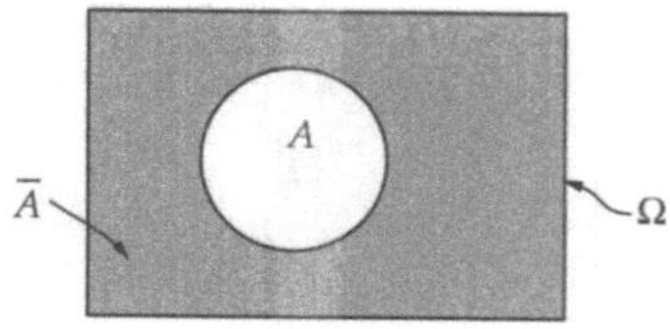

Event $\overline{A}$ complementary to A

□ A: All even numbers of a die (2, 4, 6); $\overline{A}$: all odd numbers of a die (1, 3, 5).

Field of events $\mathcal{E}_l$: Set of all interesting events of a trial for the following holds:
1. $\Omega \in \mathcal{E}$
2. $A \in \mathcal{E} \rightarrow \overline{A} \in \mathcal{E}$
3. $A, B \in \mathcal{E} \rightarrow ((A \cup B \in \mathcal{E}) \wedge (A \cap B \in \mathcal{E}))$

□ The field of events of 6 incompatible events w_i, $i = 1, 2, \ldots, 6$, consists of 2^6 random events (example: dice):
$w_1, w_2, \cdots, w_6$, i.e., $\binom{6}{1} = 6$ random events, the elementary events, $w_i \cup w_j, i \neq j$, i.e., $\binom{6}{2} = 15$ random events, another $\binom{6}{3}$, $\binom{6}{4}$, $\binom{6}{5}$ events plus the certain event and the impossible event, a total of 64 random events.

21.2.2 Event relations and event operations

A is **contained in** B, $A \subset B$, if every elementary event of A is also an elementary event of B (or A implies B).

▷ For any random event A: $A \subset A$; $A \subset \Omega$.

Equality of A and B: $A = B$, if $A \subset B$ as well as $B \subset A$.

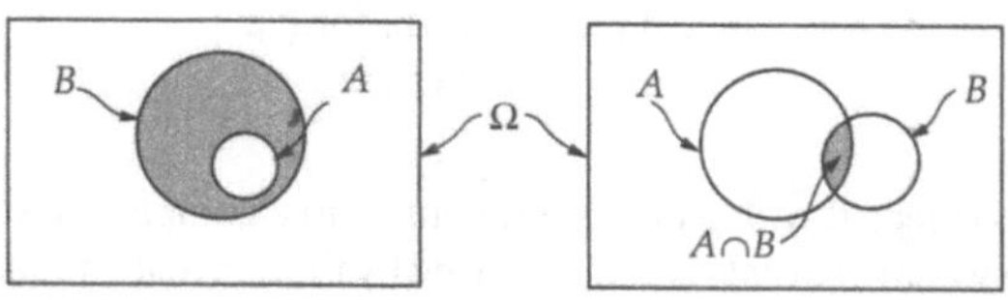

A implies B
Product of A and B

- **Product of the events A and B, $A \cap B$**: An event that occurs if and only if A as well as B occur.

▷ Product of the events A_i $(i = 1, 2, \ldots, n)$: $A_1 \cap A_2 \cap \cdots \cap A_n = \cap_{i=1}^n A_i$: Event that occurs if and only if all A_i occur. Read: A_1 and A_2 and ... and A_n.

□ $A = \{3, 4, 5, 6\},\ B = \{1, 2, 3, 4\},\ A \cap B = \{3, 4\}$

Incompatible events (mutually exclusive events) A, B: Events for which $A \cap B = \phi$; A and B do not possess common elementary events. Otherwise, A and B are called compatible events.

▷ Pairwise incompatible events A_i $(i = 1, 2, \ldots, n)$ are events for which: $A_i \cap A_j = \phi,\ i \neq j$.

● **Sum of events A and B**, $A \cup B$: Event that occurs if and only if A or B occur, i.e., if at least one of the events A or B occur.

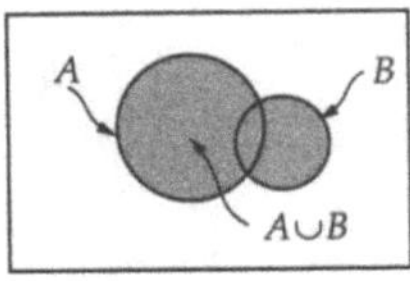

Sum of A and B

▷ Here, the "or" is not an exclusive or; it allows A and B to occur.

▷ Sum of the events A_i $(i = 1, 2, \ldots, n)$: $A_1 \cup A_2 \cup \cdots \cup A_n = \cup_{i=1}^n A_i$, an event that occurs if and only if at least one A_i occurs.

□ $A = \{2, 4, 6\},\ B = \{1, 3, 5\},\ A \cup B = \{1, 2, 3, 4, 5, 6\}$.

▷ Commutative, associative and distributive laws apply to the product and sum of random events.

$$A \cap B = B \cap A \qquad A \cup B = B \cup A$$
$$A \cap (B \cap C) = (A \cap B) \cap C \qquad A \cup (B \cup C) = (A \cup B) \cup C$$
$$A \cap (B \cup C) = (A \cap B) \cup (A \cap C) \qquad A \cup (B \cap C) = (A \cup B) \cap (A \cup C)$$

Complete system of events A_i $(i = 1, 2, \ldots, n)$: Events A_i such that

$$A_i \cap A_j = \phi,\ i \neq j,\ i, j = 1, 2, \ldots, n$$

and

$$\cup_{i=1}^n A_i = \Omega\ .$$

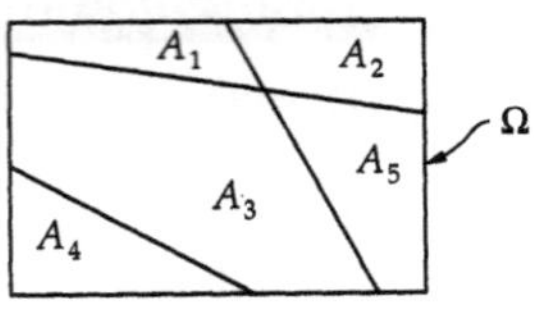

A_1, A_2, A_3, A_4, A_5 form a complete system

□ The events connected with a die experiment form a complete system of events.

▷ A complete system of events is characterized by the fact that one and only one of these events occurs as the result of the unique execution of the corresponding experiment. Thus, for example, all elementary events always form a complete system with respect to an experiment.

Properties of events $A \subset A \qquad A \subset \Omega \qquad \overline{\overline{A}} = A$

Product	Sum
$A \cap A = A$	$A \cup A = A$
$A \cap B = B \cap A$	$A \cup B = B \cup A$
$A \cap (B \cap C) = (A \cap B) \cap C$	$A \cup (B \cup C) = (A \cup B) \cup C$
$A \cap (B \cup C) = (A \cap B) \cup (A \cap C)$	$A \cup (B \cap C) = (A \cup B) \cap (A \cup C)$
$A \cap \overline{A} = \phi$	$A \cup \overline{A} = \Omega$
$A \cap \Omega = A$	$A \cup \phi = A$
$A \cap \phi = \phi$	$A \cup \Omega$
$\overline{A \cap B} = \overline{A} \cup \overline{B}$	$\overline{A \cup B} = \overline{A} \cap \overline{B}$

21.2.3 Structural representation of events

Structural representation: Representation of a random event as a sum or product of subevents; representation of results of a multistage random experiment.

□ A directly observable test result (elementary event) is a random event composed of subevents.

Subevents: Results of partial steps of a random experiment.

▷ There are random events for which several structural representations can be given.

▷ A tree of events may be used to determine a structural representation and to calculate probabilities.

Tree of events: Set of nodes connected by edges. Subevents (possible intermediate results) constitute nodes; edges constitute relations between these events.

Branch: Connection of "starting nodes" and final nodes (possible final results).

Conditions to be met when setting up a tree of events:

(a) Random events following a node form a complete system of events.

(b) Starting from a rooted node, only one edge may end in every following node.

(c) Along the branches, the events must be connected with "and."

(d) Every branch represents an elementary event of the experiment, which is obtained as the product of the subevents on this branch.

□ Tree of events for a three-stage random experiment

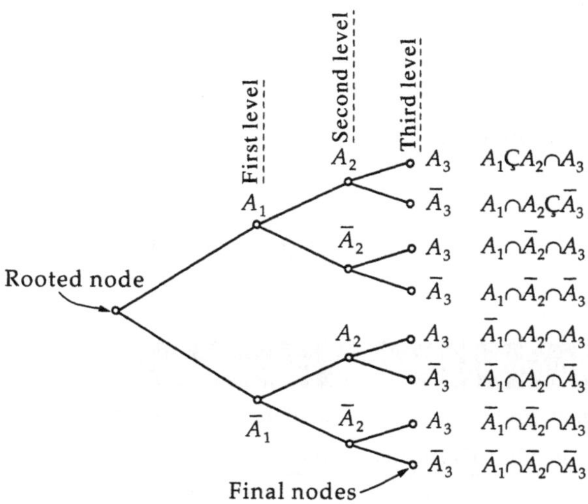

▷ The sequence of nodes must not necessarily constitute a sequence in time. For processes proceeding parallel in time, the sequence of nodes is an arbitrary sequence of observations of the results of the partial experiments.

21.3 Probability of events

Relative frequency $k_n(A)$: Value for the occurrence of the event A during n independent repetitions of a random experiment.

▷ During long series experiments under constant conditions, the relative frequencies fluctuate about a fixed unknown value p.

Stability of relative frequency: The property that deviations from p decrease as the number of experimental repetitions increases.

Probability (axiomatic definition according to Kolmogorov): Unique assignment of a real number $p = P(A)$ to a random event $A \in \mathcal{E}_l$.

Axiom I: $P(A)$ is a nonnegative real number ≤ 1: $0 \leq P(A) \leq 1$.

Axiom II: The probability of the certain event is 1: $P(\Omega) = 1$.

Axiom III: The probabilty of a sum of pairwise incompatible (disjoint) events A_1, A_2, $A_3, \ldots$ is equal to the sum of the probabilities of the events: $P(A_1 \cup A_2 \cup A_3 \cup \cdots) = P(A_1) + P(A_2) + P(A_3) + \cdots$.

21.3.1 *Properties of probabilities*

- For the probability of a complementary event $\overline{A}$ belonging to A

$$P(\overline{A}) = 1 - P(A).$$

- For the probability of the impossible event ϕ

$$P(\phi) = 0.$$

- If A implies B $(A \subset B)$ then

$$P(A) \leq P(B).$$

- If A_i, $i = 1, 2, \ldots, n$, form a complete system of events, then

$$P(A_i \cap A_j) = 0, \quad i \neq j, \quad i, j = 1, 2, \ldots, n$$

and

$$P(\cup_{i=1}^{n} A_i) = \sum_{i=1}^{n} P(A_i) = 1.$$

21.3.2 *Methods to calculate probabilities*

Classical method: Probability as quotient of numbers of elementary events.

$$P(A) = \frac{\text{number of elementary events favoring the occurrence of } A}{\text{number of all equally probable elementary events with equal probability}} = \frac{m}{n}$$

Geometric method: Probability as quotient of the size of a set of geometrical elements.

$$P(A) = \frac{\text{size of the partial set of (geometrical) elements favoring the occurrence of } A}{\text{size of the possible (geometrical) element}} = \frac{m_g}{m_G}$$

Statistical method: Probability as quotient of numbers of repetitions of experiments.

$$P(A) \approx k_n(A) = \frac{\text{number of repetitions of the experiment with result } A}{\text{number of independent repetitions of the trial}} = \frac{m}{n}$$

21.3.3 Conditional probabilities

Conditional probabilities $P(B|A)$: Probability for the occurrence of event B under the condition (assumption) that event A has occurred.

$$P(B|A) = \frac{P(A \cap B)}{P(A)}, \quad P(A) > 0$$

▷ If event B implies event A ($A \subset B$) then $P(B|A) = 1$.

▷ $P(B|A)$ and $P(B)$ are usually different from each other.
If $P(B|A) = P(B)$, then the probability of B is not affected by the occurrence of event A.

● A and B are **(stochastically) independent** if $P(B|A) = P(B)$ and $P(A|B) = P(A)$; otherwise A and B are **stochastically dependent**.

● If A and B are independent events, then the events $\overline{A}$ and B, A and $\overline{B}$, and $\overline{A}$ and $\overline{B}$ are also independent events.

Total probability for the occurrence of event B, $P(B)$: Probability, dependent on $P(A_i)$, where A_i form a complete system of events, and on the conditional probabilities $P(B|A_i)$, according to the **formula of total probability**

$$P(B) = \sum_{i=1}^{n} P(A_i)P(B|A_i).$$

Conditional probability for the occurrence of event A_i under the condition that event B has already occurred: According to the **formula of Bayes**

$$P(A_i|B) = \frac{P(A_i)P(B|A_i)}{P(B)}$$

where $P(B)$ is calculated from the formula of total probability.

21.3.4 Calculating with probabilities

Multiplication theorem: Theorem to calculate the probability of a product of random events.

Addition theorem: Theorem to calculate the probability of a sum of random events.

Events		A, B	$A_1, A_2, \ldots, A_n$
Product	independent events	$P(A \cap B) = P(A) \cdot P(B)$	$P(\cap_{i=1}^n A_i) = \prod_{i=1}^n P(A_i)$
Product	dependent events	$P(A \cap B) = P(A) \times P(B/A)$ $= P(B) \cdot P(A/B)$	$P(\cap_{i=1}^n A_i) = P(A_1) \cdot P(A_2/A_1)$ $\times P(A_3/A_1 \cap A_2) \cdot \ldots$ $\times P(A_n/A_1 \cap A_2 \cap \ldots \cap A_{n-1})$
Sum	pairwise incompatible events	$P(A \cup B) = P(A) + P(B)$	$P(\cup_{i=1}^n A_i) = \sum_{i=1}^n P(A_i)$
Sum	compatible events	$P(A \cup B) = P(A) + P(B)$ $-P(A \cap B) = 1 - P(\overline{A} \cap \overline{B})$	$P(\cap_{i=1}^n A_i) = 1 - P(\cap_{i=1}^n \overline{A_i})$ $= 1 - (1-p)^n$ if A_i is independent and $P(A_i)$ $= p$ for all i

□ Experiment with a die: Let the classes be defined as:

$$A_1 = (1, 2), \quad A_2 = (3, 4), \quad A_3 = (5, 6)\ .$$

Two throws of a die yields a pair of numbers (E_1, E_2) from the set of all possible pairs of numbers.

$$\begin{array}{llllll}
(1,1) & /\,(1,2) & /\,(1,3) & /\,(1,4) & /\,(1,5) & /\,(1,6) \\
(2,1) & /\,(2,2) & /\,(2,3) & /\,(2,4) & /\,(2,5) & /\,(2,6) \\
(3,1) & /\,(3,2) & /\,(3,3) & /\,(3,4) & /\,(3,5) & /\,(3,6) \\
(4,1) & /\,(4,2) & /\,(4,3) & /\,(4,4) & /\,(4,5) & /\,(4,6) \\
(5,1) & /\,(5,2) & /\,(5,3) & /\,(5,4) & /\,(5,5) & /\,(5,6) \\
(6,1) & /\,(6,2) & /\,(6,3) & /\,(6,4) & /\,(6,5) & /\,(6,6)
\end{array}$$

The event A_i occurs if and only if one of the two numbers thrown is contained in the set A_i.

$$A_1 = \begin{array}{llllll}
(1,1) & /\,(1,2) & /\,(1,3) & /\,(1,4) & /\,(1,5) & /\,(1,6) \\
(2,1) & /\,(2,2) & /\,(2,3) & /\,(2,4) & /\,(2,5) & /\,(2,6) \\
(3,1) & /\,(3,2) & & & & \\
(4,1) & /\,(4,2) & & & & \\
(5,1) & /\,(5,2) & & & & \\
(6,1) & /\,(6,2) & & & &
\end{array}$$

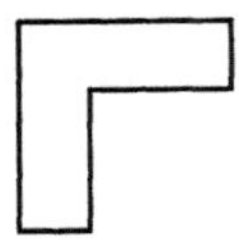

$$A_3 = \begin{array}{llllll}
 & & & & /\,(1,5) & /\,(1,6) \\
 & & & & /\,(2,5) & /\,(2,6) \\
 & & & & /\,(3,5) & /\,(3,6) \\
 & & & & /\,(4,5) & /\,(4,6) \\
(5,1) & /\,(5,2) & /\,(5,3) & /\,(5,4) & /\,(5,5) & /\,(5,6) \\
(6,1) & /\,(6,2) & /\,(6,3) & /\,(6,4) & /\,(6,5) & /\,(6,6)
\end{array}$$

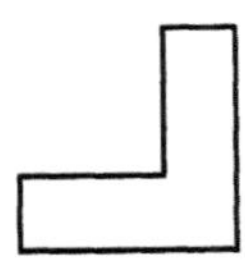

Thus it is possible that two events, e.g., A_1, A_3, may occur simultaneously. Thus, the total number of possible events is exactly $6 \times 6 = 36$. Therefore, the probability of throwing an element of the subset of events A_1 is

$$\begin{aligned}
P(A_1) = {} & P(1,1) + P(1,2) + P(1,3) + P(1,4) + P(1,5) + P(1,6) \\
& + P(2,1) + P(2,2) + P(2,3) + P(2,4) + P(2,5) + P(2,6) \\
& + P(3,1) + P(4,1) + P(5,1) + P(6,1) \\
& + P(3,2) + P(4,2) + P(5,2) + P(6,2) \\
= {} & 20 \cdot \frac{1}{36}\ ,
\end{aligned}$$

since the probability of the elementary event (E_i, E_j) is $P(i, j) = 1/36$. Analogously, we have $P(A_2) = 20/36$ and $P(A_3) = 20/36$.
The probability $P(A_1 \cap A_3)$ that the result is A_1 **and** A_3 can be calculated according to the multiplication theorem

$$P(A_1 \cap A_3) = P(A_1) \cdot P(A_3|A_1),$$

provided A_1 has already been met, i.e., (E_1, E_2) is already an element of the A_1. Then, the probability $P(A_3|A_1)$ that among the total of 20 events from A_1 exactly those have been thrown that are also contained in set A_3 is

$$\begin{aligned} P(A_3|A_1) &= \tilde{P}(1,5) + \tilde{P}(1,6) + \tilde{P}(2,5) + \tilde{P}(2,6) \\ &\quad + \tilde{P}(5,1) + \tilde{P}(6,1) + \tilde{P}(5,2) + \tilde{P}(6,2) \\ &= 8 \cdot 1/20, \end{aligned}$$

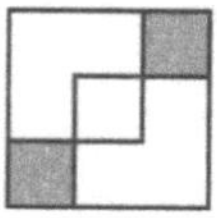

because the probability for the elementary event (E_i, E_j) with i, j from A_1 is $\tilde{P}(i, j) = 1/20$ in this case.
It follows that

$$P(A_1 \cap A_3) = (20/36) \cdot (8/20) = 8/36\,.$$

If we calculate this explicitly, we obtain the correct result:

$$\begin{aligned} P(A_1 \cap A_3) &= P(5,1) + P(6,1) + P(5,2) + P(6,2) \\ &\quad + P(1,5) + P(2,5) + P(1,6) + P(2,6) \\ &= 8/36\,. \end{aligned}$$

The probability $P(A_1 \cup A_3)$ that the result of the throw lies in A_1 **or** A_3 is calculated according to the addition theorem:

$$\begin{aligned} P(A_1 \cup A_3) &= P(A_1) + P(A_3) - P(A_1 \cap A_3) \\ &= 20/36 + 20/36 - 8/36 = 32/36\,. \end{aligned}$$

This is correct, since only the four events $(3, 3)$, $(3, 4)$, $(4, 3)$ and $(4, 4)$ do not belong to sets A_1 and A_3.

21.4 Random variables and their distributions

Random variable, $X, Y, \ldots; X_1, X_2, \ldots$: Functions whose domain of definition is a set Ω of elementary events and whose range of values is a subset of real numbers. (Quantity whose values or realizations, denoted by $x, y, \ldots; x_1, x_2, \ldots$, are influenced by chance.)
Discrete random variable: Random variable that can assume a finite number or a denumerably infinite number of events.
Continuous random variable: Random variable that can assume any real numerical value of an interval.
Probability distribution of a random variable, **distribution**: Statement that facilitates the complete characterization of the random variable.

▷ Distribution may be given by a distribution function or by the individual probabilities (in case of a discrete random variable) or the density function (in case of a continuous random variable).

▷ There are three forms of specifying a distribution: analytical, graphical and by means of a distribution table.

□ Example: One throw of a die:
X ... number of points
x_i ... realizations, $x_i = i$, $i = 1, 2, 3, 4, 5, 6$
p_i ... individual probabilities, $p_i = P(X = i) = \frac{1}{6}$
Distribution table:

x_i	1	2	3	4	5	6
p_i	$\frac{1}{6}$	$\frac{1}{6}$	$\frac{1}{6}$	$\frac{1}{6}$	$\frac{1}{6}$	$\frac{1}{6}$

Graphical representation:

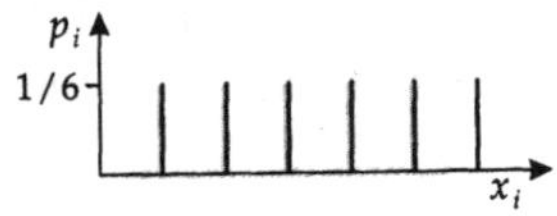

21.4.1 *Individual probability, density function and distribution function*

Individual probability p_i: The probability that a discrete random variable X takes just the value x_i.

$$P(X = x_i) = P(x_i) = p_i$$

Density function $f(x)$, **probability density, distribution density** of a continuous random variable X: Non-negative function $f(x)$ for which the following holds:

$$P(a \leq X < b) = \int_a^b f(x)\mathrm{d}x = F(b) - F(a), \; a < b, \; a, b \,\text{real}.$$

▷ **Interpretation** of the probability that X takes a value of $[a, b)$.
Analytical: The difference of $F(x)$ (distribution function of the random variable X) at the points $x = a$ and $x = b$.
Geometrical: Measuring the area enclosed by the graph of the density function, the x axis, and the straight lines $x = a$ und $x = b$.

Distribution function $F(x)$: Probability that the (discrete or continuous) random variable X takes a value less than x.

$$F(x) = P(X < x)$$

▷ The distribution function of a discrete random variable is a step function.

□ Example: One throw of a die.

$$P(X = i) = p_i = \frac{1}{6}.$$

X is a uniformly distributed discrete random variable.

$$F(x)=\begin{cases} 0 & \text{for} \quad x \le 1 \\ \frac{1}{6} & \text{for} \quad 1 < x \le 2 \\ \frac{2}{6} & \text{for} \quad 2 < x \le 3 \\ \frac{3}{6} & \text{for} \quad 3 < x \le 4 \\ \frac{4}{6} & \text{for} \quad 4 < x \le 5 \\ \frac{5}{6} & \text{for} \quad 5 < x \le 6 \\ 1 & \text{for} \quad 6 < x \end{cases}$$

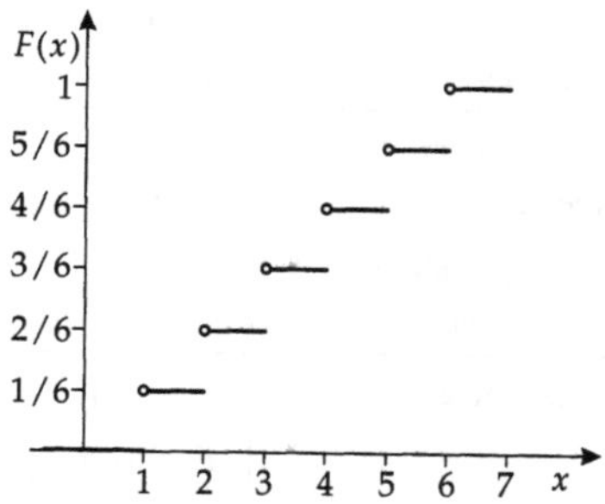

Special case: $F(3) = P(X < 3) = P(X = 2) + P(X = 1) = p_2 + p_1 = \frac{1}{3}$.

Summary:

Discrete random variable	Continuous random variable
Distribution function $F(x) = P(X < x)$	
$F(x) = \sum_{x_i < x} P(X = x_i)$	$F(x) = \int_{-\infty}^{x} f(t)\mathrm{d}t$ $P(X = x) = 0$
$P(a \le X < b) = F(b) - F(a)$	
Individual probabilities p_i $p_i = P(X = x_i)$ $\sum_i p_i = 1$	Density function $f(x)$ $f(x) \ge 0$ for all $x \in \mathbb{R}$ $\int_{-\infty}^{\infty} f(x)\mathrm{d}x = 1$ $\dfrac{\mathrm{d}F(x)}{\mathrm{d}x} = f(x)$

21.4.2 *Parameters of distributions*

Parameters of a probability distribution: Parameters that characterize the random variable.

⊳ The values of a random variable scatter about a fixed point, the **dispersion center**.

Position parameters: Parameters that specify something pertaining to the dispersion about the dispersion center.

Parameter	Discrete random variable	Continuous random variable
mathematical expectation (expected value) $M(X) = m$	$M(X) = \sum_i x_i p_i$ if $\sum_i \|x_i\| p_i < \infty$	$M(X) = \int_{-\infty}^{\infty} x f(x) \mathrm{d}x$ if $\int_{-\infty}^{\infty} \|x\| f(x) \mathrm{d}x < \infty$
modal value M_0	$M_0 = x_i$, if $P(X = x_{i-1} \leq P(X = x_i)$ $> P(X = x_{i+1})$ or $P(X = x_{i-1} < P(X = x_i)$ $\geq P(X = x_{i+1})$	real number M_0, for which $f(x)$ has a relative or absolute maximum
median M_e	$\sum_{x_i \leq M_e} P(X = x_i) \geq \frac{1}{2}$ and $\sum_{x_i \leq M_e} P(X = x_i) \leq \frac{1}{2}$	$\int_{-\infty}^{M_e} f(x)\,\mathrm{d}x = F(M_e) = \frac{1}{2}$
dispersion (variance) $D(X) = \sigma^2$	$D(X) = M\left((X - M(X))^2\right) = M(X^2) - (M(X))^2$ $D(X) = \sum_i (x_i - m)^2 p_i$	$D(X) = \int_{-\infty}^{\infty} (x - m)^2 f(x)\,\mathrm{d}x$
standard deviation	$\sigma = \sqrt{D(X)}$	
probable deviation E		$\int_{m-E}^{m+E} f(x)\,\mathrm{d}x = F(m + E)$ $-F(m - E) = \frac{1}{2}$
range S	$S = x_{max} - x_{min}$ if $x_{max} - x_{min} < \infty$ x_{max} greatest, x_{min} least value of X with $P(X = x_{max}) < 0$, $P(X = x_{min} > 0$ interval of the values of X with $f(x) = 0$ for all $x \not\subset I$	$I = [c_{min}, x_{max}]$ least
skewness γ (according to Pearson)	$\gamma = \dfrac{m - M_0}{\sigma}$ if the distribution is unimodal	

▷ The expected value is often denoted by $< x >$.

● Calculating rules for expected values:

$M(c) = c$ $\quad M(cX) = cM(X)$ $\quad M((cX^k)) = c^k M(X^k)$

$M(X \pm Y) = M(X) \pm M(Y)$ $\quad M(XY) = M(X)M(Y)$ if X, Y are independent.

● Calculating rules for dispersions:

$D(c) = 0$ $\quad D(cX) = c^2 D(X)$

$D(X + c) = D(X)$ $\quad D(X \pm Y) = D(X) + D(Y)$ if X, Y are independent.

Standardization: Transformation that changes a random variable X with expected value m and variance σ^2 into a random variable Z with expected value 0 and variance 1.

$$Z = \frac{X - m}{\sigma}, \quad M(Z) = 0, \quad D(Z) = 1$$

21.4.3 Special discrete distribution

Special distribution: Types of distributions with which unknown distributions occurring in practice can be approximated.

▷ The decision about a justified approximation of a distribution for a given unknown distribution (with respect to parameters) is made by means of statistical tests.

Zero-one distribution: Distribution of a discrete random variable X with values $x_1 = 1$ and $x_2 = 0$; it is used in random experiments when it is of interest only whether a certain event A occurs or not $(\bar{A})$.

Individual probabilities:

$$P(1) = p \quad \text{and} \quad P(0) = 1 - p, \; 0 < p < 1.$$

Distribution function:

$$F(x) = \begin{cases} 0 & \text{for} \quad x \leq 0 \\ 1 - p & \text{for} \quad 0 < x \leq 1 \\ 1 & \text{for} \quad 1 < x. \end{cases}$$

The parameter of distribution is p.

Expected value: $m = p$.

Variance: $\sigma^2 = p(1 - p)$.

□ A component does or does not function during a certain time interval.

Uniform discrete distribution: Distribution of a discrete random variable X with the values $x_i, i = 1, 2, \ldots, n$ and uniformly distributed probabilities; it is used if the conditions of the corresponding random experiment allow the application of the classical method to calculate probabilities.

□ Game of dice

Individual probabilities: $P(x_i) = \dfrac{1}{n}$.

Distribution function:

$$F(x) = \begin{cases} 0 & \text{for} \quad x \leq x_i \\ \dfrac{1}{n} & \text{for} \quad x_i < x \leq x_{i+1}, \; i = 1, 2, \ldots, n \\ 1 & \text{for} \quad x_n < x. \end{cases}$$

The parameter of distribution is n.

Expected value: $m = \dfrac{1}{n} \displaystyle\sum_{i=1}^{n} x_i$.

Variance: $\sigma^2 = \dfrac{1}{n} \displaystyle\sum_{i=1}^{n} x_i^2 - \left(\dfrac{1}{n} \sum_{i=1}^{n} x_i \right)^2$.

Hypergeometric distribution: Used if the universe is divided into two classes of elements.

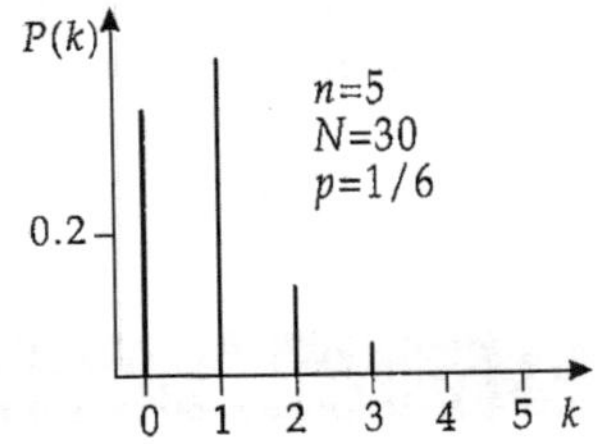

Individual probabilities:

$$P(k) = \binom{pN}{k}\binom{N(1-p)}{n-k}\Big/\binom{N}{n}; \qquad p \cdot N = \text{integer} .$$

Distribution function:

$$F(k) = \sum_{i=0}^{k} P(i) = \sum_{i=0}^{k} \binom{pN}{i}\binom{N(1-p)}{n-i}\Big/\binom{N}{n} .$$

The parameters of distribution are p, N and sample size n.
Expected value: $m = n \cdot p$.
Variance: $\sigma^2 = n \cdot p(1-p)[(N-n)/(N-1)]$.

□ Given a total set of $N = N_1 + N_2$ units, N_1 of which are specially marked (e.g., different colors, efficiency, etc.), so that $p = N_1/N_2$ indicates what portion of the total set consists of the specially marked units. The probability that among $n \leq N$ randomly chosen units (sample) there are exactly k specially marked units is given by the hypergeometric distribution.

▷ After each withdrawal of a unit, the portion of marked elements in the universe may change. Therefore, for each of the n withdrawals different probabilities to obtain a marked element apply. Thus, the hypergeometric distribution simulates the draw of lottery tickets **without** replacement of the drawn ticket (see **urn model**), in contrast to the **binomial distribution**.

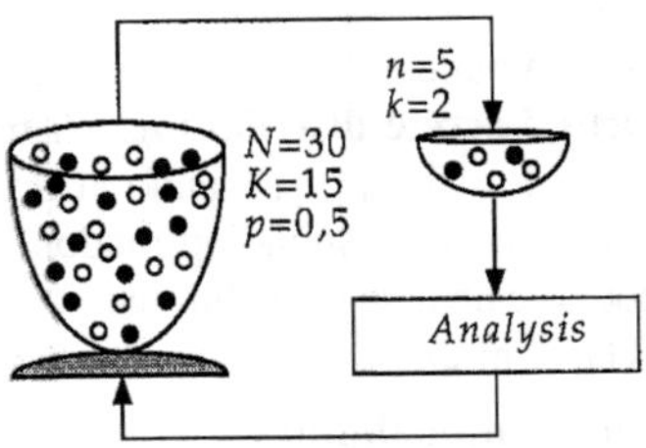

□ The probability for k defective units in a sample of size n in a total set of N units, where p is the fraction of the defective pieces in the total set, i.e., $p \cdot N$ is the number of defective units.

Cumulative hypergeometric distribution: The sum of probabilities for a maximum of k defective units:

$$\sum_{i=0}^{k} P(i) \qquad p \cdot N = \text{integer}.$$

● Limits of the hypergeometric distribution:

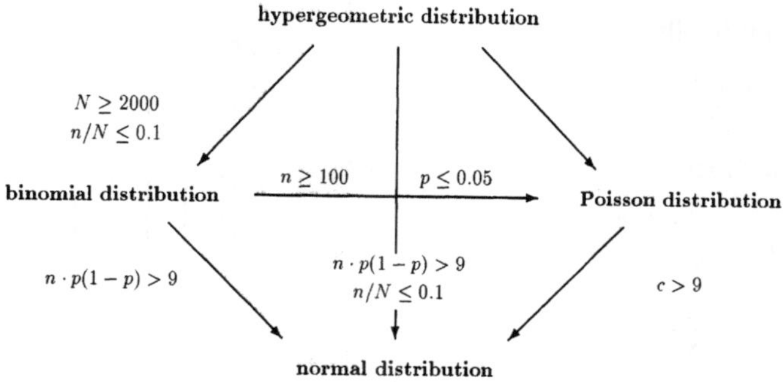

1

Binomial distribution: Results from the hypergeometric distribution, if the number N of the elements of the universe becomes very large ($N \to \infty$) and the sample size n remains small.

It is used if n repetitions of a random experiment are performed independently from each other, and if in every repetition the event A occurs with $P(A) = p$ and event $\overline{A}$ occurs with $P(\overline{A}) = 1 - p$, (Bernoulli trial). Therefore, the binomial distribution describes exactly the drawing of lots from an urn **with** replacement of the drawn lots. The binomial distribution describes the probability of finding exactly k marked units among n chosen units.

Individual probabilities:

$$P(k) = \binom{n}{k} p^k (1-p)^{n-k} .$$

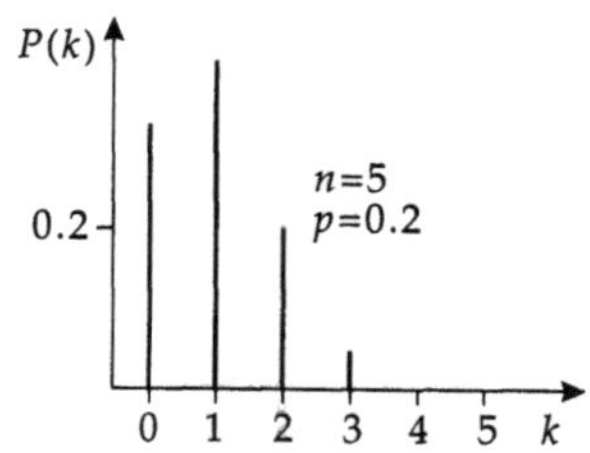

The parameters of distribution are the marked fraction p and the sample size n.

Distribution function:

$$F(k) = \sum_{i=0}^{k} \binom{n}{i} p^i (1-p)^{n-i} .$$

Expected value: $m = n \cdot p$.

Variance: $\sigma^2 = n \cdot p(1-p)$.

▷ It is important that the withdrawal of a unit changes the probability of drawing a marked unit only slightly or not at all.

□ In quality control with a large number of items N and a small number of random samples n, the binomial distribution yields the probability of finding k defective units in n samples.

Poisson distribution: Results from the binomial distribution if the random sample size n is very large ($N \to \infty$) and the marked fraction p is very small ($p \to 0$), but finite.

Individual probabilities:

$$P(k) = \frac{(np)^k}{k!} \cdot e^{-np} = \frac{c^k}{k!} \cdot e^{-c}$$

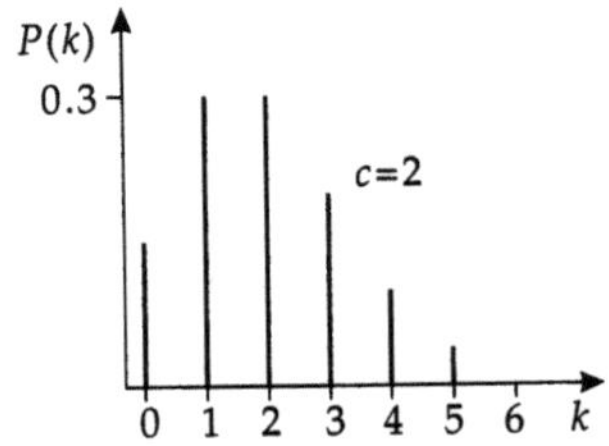

Distribution function:

$$F(k) = \sum_{i=0}^{k} \frac{(np)^i}{i!} \cdot e^{-np}$$

$$= \sum_{i=0}^{k} \frac{c^i}{i!} \cdot e^{-c} .$$

The parameter of the distribution is $c = n \cdot p$.
Expected value: $m = n \cdot p$.
Variance: $\sigma^2 = n \cdot p$.
Thus, quantity $c = p \cdot n$ is approximately constant, and parameter n is eliminated. Therefore, the Poisson distribution is suitable for problems in which the extent of the test n is large ($n \geq 1500p$, $c = np \leq 10$), while the exact numbers are not known and insignificant.

□ The daily production of electronic components is usually very high ($n \geq 1500$). With a small number of errors ($p \rightarrow c = pn \leq 10$), the daily production of defective items is described by a Poisson distribution.

Table of the binomial distribution

$$P(X = k) = \binom{n}{k} p^k (1-p)^{n-k} = b(k; n, p)$$

For $p > 0.5$ the relation $b(k; n, p) = b(n - k; n, 1 - p)$ must be used.

n	k	p									
		0.010	0.020	0.050	0.100	0.150	0.200	0.250	0.300	0.400	0.500
1	0	0.990	0.980	0.950	0.900	0.850	0.800	0.750	0.700	0.600	0.500
1	1	0.010	0.020	0.050	0.100	0.150	0.200	0.250	0.300	0.400	0.500
2	0	0.980	0.960	0.902	0.810	0.722	0.640	0.562	0.490	0.360	0.250
2	1	0.020	0.039	0.095	0.180	0.255	0.320	0.375	0.420	0.480	0.500
2	2			0.003	0.010	0.023	0.040	0.062	0.090	0.160	0.250
3	0	0.970	0.941	0.857	0.729	0.614	0.512	0.422	0.343	0.216	0.125
3	1	0.029	0.058	0.135	0.243	0.325	0.384	0.422	0.441	0.432	0.375
3	2		0.001	0.007	0.027	0.057	0.096	0.141	0.189	0.288	0.375
3	3				0.001	0.003	0.008	0.016	0.027	0.064	0.125
4	0	0.961	0.922	0.815	0.656	0.522	0.410	0.316	0.240	0.130	0.062
4	1	0.039	0.075	0.171	0.292	0.368	0.410	0.422	0.412	0.346	0.250
4	2	0.001	0.002	0.014	0.049	0.098	0.154	0.211	0.265	0.346	0.375
4	3				0.004	0.011	0.026	0.047	0.076	0.154	0.250
4	4					0.001	0.002	0.004	0.008	0.026	0.062
5	0	0.951	0.904	0.774	0.590	0.444	0.328	0.237	0.168	0.078	0.031
5	1	0.048	0.092	0.204	0.328	0.392	0.410	0.396	0.360	0.259	0.156
5	2	0.001	0.004	0.021	0.073	0.138	0.205	0.264	0.309	0.346	0.312
5	3			0.001	0.008	0.024	0.051	0.088	0.132	0.230	0.312
5	4					0.002	0.006	0.015	0.028	0.077	0.156
5	5							0.001	0.002	0.010	0.031
6	0	0.941	0.886	0.735	0.531	0.377	0.262	0.178	0.118	0.047	0.016
6	1	0.057	0.108	0.232	0.354	0.399	0.393	0.356	0.303	0.187	0.094
6	2	0.001	0.006	0.031	0.098	0.176	0.246	0.297	0.324	0.311	0.234
6	3			0.002	0.015	0.041	0.082	0.132	0.185	0.276	0.312
6	4				0.001	0.005	0.015	0.033	0.060	0.138	0.234
6	5						0.002	0.004	0.010	0.037	0.094
6	6								0.001	0.004	0.016
7	0	0.932	0.868	0.698	0.478	0.321	0.210	0.133	0.082	0.028	0.008
7	1	0.066	0.124	0.257	0.372	0.396	0.367	0.311	0.247	0.131	0.055
7	2	0.002	0.008	0.041	0.124	0.210	0.275	0.311	0.318	0.261	0.164
7	3			0.004	0.023	0.062	0.115	0.173	0.227	0.290	0.273
7	4				0.003	0.011	0.029	0.058	0.097	0.194	0.273
7	5					0.001	0.004	0.012	0.025	0.077	0.164
7	6							0.001	0.004	0.017	0.055
7	7									0.002	0.008

n	k	p									
		0.010	0.020	0.050	0.100	0.150	0.200	0.250	0.300	0.400	0.500
8	0	0.923	0.851	0.663	0.430	0.272	0.168	0.100	0.058	0.017	0.004
8	1	0.075	0.139	0.279	0.383	0.385	0.336	0.267	0.198	0.090	0.031
8	2	0.003	0.010	0.051	0.149	0.238	0.294	0.311	0.296	0.209	0.109
8	3			0.005	0.033	0.084	0.147	0.208	0.254	0.279	0.219
8	4				0.005	0.018	0.046	0.087	0.136	0.232	0.273
8	5					0.003	0.009	0.023	0.047	0.124	0.219
8	6						0.001	0.004	0.010	0.041	0.109
8	7								0.001	0.008	0.031
8	8									0.001	0.004
9	0	0.914	0.834	0.630	0.387	0.232	0.134	0.075	0.040	0.010	0.002
9	1	0.083	0.153	0.299	0.387	0.368	0.302	0.225	0.156	0.060	0.018
9	2	0.003	0.013	0.063	0.172	0.260	0.302	0.300	0.267	0.161	0.070
9	3		0.001	0.008	0.045	0.107	0.176	0.234	0.267	0.251	0.164
9	4			0.001	0.007	0.028	0.066	0.117	0.172	0.251	0.246
9	5				0.001	0.005	0.017	0.039	0.074	0.167	0.246
9	6					0.001	0.003	0.009	0.021	0.074	0.164
9	7							0.001	0.004	0.021	0.070
9	8									0.004	0.018
9	9										0.002
10	0	0.904	0.817	0.599	0.349	0.197	0.107	0.056	0.028	0.006	0.001
10	1	0.091	0.167	0.315	0.387	0.347	0.268	0.188	0.121	0.040	0.010
10	2	0.004	0.015	0.075	0.194	0.276	0.302	0.282	0.233	0.121	0.044
10	3		0.001	0.010	0.057	0.130	0.201	0.250	0.267	0.215	0.117
10	4			0.001	0.011	0.040	0.088	0.146	0.200	0.251	0.205
10	5				0.001	0.008	0.026	0.058	0.103	0.201	0.246
10	6					0.001	0.006	0.016	0.037	0.111	0.205
10	7						0.001	0.003	0.009	0.042	0.117
10	8								0.001	0.011	0.044
10	9									0.002	0.010
10	10										0.001

n	k	p									
		0.010	0.020	0.050	0.100	0.150	0.200	0.250	0.300	0.400	0.500
15	0	0.860	0.739	0.463	0.206	0.087	0.035	0.013	0.005		
15	1	0.130	0.226	0.366	0.343	0.231	0.132	0.067	0.031	0.005	
15	2	0.009	0.032	0.135	0.267	0.286	0.231	0.156	0.092	0.022	0.003
15	3		0.003	0.031	0.129	0.218	0.250	0.225	0.170	0.063	0.014
15	4			0.005	0.043	0.116	0.188	0.225	0.219	0.127	0.042
15	5			0.001	0.010	0.045	0.103	0.165	0.206	0.186	0.092
15	6				0.002	0.013	0.043	0.092	0.147	0.207	0.153
15	7					0.003	0.014	0.039	0.081	0.177	0.196
15	8					0.001	0.003	0.013	0.035	0.118	0.196
15	9						0.001	0.003	0.012	0.061	0.153
15	10							0.001	0.003	0.024	0.092
15	11								0.001	0.007	0.042
15	12									0.002	0.014
15	13										0.003
15	14										
15	15										
20	0	0.818	0.668	0.358	0.122	0.039	0.012	0.003	0.001		
20	1	0.165	0.272	0.377	0.270	0.137	0.058	0.021	0.007		
20	2	0.016	0.053	0.189	0.285	0.229	0.137	0.067	0.028	0.003	
20	3	0.001	0.006	0.060	0.190	0.243	0.205	0.134	0.072	0.012	0.001
20	4		0.001	0.013	0.090	0.182	0.218	0.190	0.130	0.035	0.005
20	5			0.002	0.032	0.103	0.175	0.202	0.179	0.075	0.015
20	6				0.009	0.045	0.109	0.169	0.192	0.124	0.037
20	7				0.002	0.016	0.055	0.112	0.164	0.166	0.074
20	8					0.005	0.022	0.061	0.114	0.180	0.120
20	9					0.001	0.007	0.027	0.065	0.160	0.160
20	10						0.002	0.010	0.031	0.117	0.176
20	11							0.003	0.012	0.071	0.160
20	12							0.001	0.004	0.035	0.120
20	13								0.001	0.015	0.074
20	14									0.005	0.037
20	15									0.001	0.015
20	16										0.005
20	17										0.001

Table of the Poisson distribution $P(X = k) = \frac{\lambda^k}{k!} e^{-\lambda}$

k/λ	0.1	0.2	0.3	0.4	0.5	0.6
0	0.904837	0.818731	0.740818	0.670320	0.606531	0.548812
1	0.090484	0.163746	0.222245	0.268128	0.303265	0.329287
2	0.004524	0.016375	0.033337	0.053626	0.075816	0.098786
3	0.000151	0.001092	0.003334	0.007150	0.012636	0.019757
4	0.000004	0.000055	0.000250	0.000715	0.001580	0.002964
5		0.000002	0.000015	0.000057	0.000158	0.000356
6			0.000001	0.000004	0.000013	0.000036
7					0.000001	0.000003

k/λ	0.7	0.8	0.9	1.0	2.0	3.0
0	0.496585	0.449329	0.406570	0.367879	0.135335	0.049787
1	0.347610	0.359463	0.365913	0.367879	0.270671	0.149361
2	0.121663	0.143785	0.164661	0.183940	0.270671	0.224042
3	0.028388	0.038343	0.049398	0.061313	0.180447	0.224042
4	0.004968	0.007669	0.011115	0.015328	0.090224	0.168031
5	0.000696	0.001227	0.002001	0.003066	0.036089	0.100819
6	0.000081	0.000164	0.000300	0.000511	0.012030	0.050409
7	0.000008	0.000019	0.000039	0.000073	0.003437	0.021604
8	0.000001	0.000002	0.000004	0.000009	0.000859	0.008102
9				0.000001	0.000191	0.002701
10					0.000038	0.000810
11					0.000007	0.000221
12					0.000001	0.000055
13						0.000013
14						0.000003
15						0.000001

k/λ	4.0	5.0	6.0	7.0	8.0	9.0
0	0.018316	0.006738	0.002479	0.000912	0.000335	0.000123
1	0.073263	0.033690	0.014873	0.006383	0.002684	0.001111
2	0.146525	0.084224	0.044618	0.022341	0.010735	0.004998
3	0.195367	0.140374	0.089235	0.052129	0.028626	0.014994
4	0.195367	0.175467	0.133853	0.091226	0.057252	0.033737
5	0.156293	0.175467	0.160623	0.127717	0.091604	0.060727
6	0.104196	0.146223	0.160623	0.149003	0.122138	0.091090
7	0.059540	0.104445	0.137677	0.149003	0.139587	0.117116
8	0.029770	0.065278	0.103258	0.130377	0.139587	0.131756
9	0.013231	0.036266	0.068838	0.101405	0.124077	0.131756
10	0.005292	0.018133	0.041303	0.070983	0.099262	0.118580
11	0.001925	0.008242	0.022529	0.045171	0.072190	0.097020
12	0.000642	0.003434	0.011264	0.026350	0.048127	0.072765
13	0.000197	0.001321	0.005199	0.014188	0.029616	0.050376
14	0.000056	0.000472	0.002228	0.007094	0.016924	0.032384
15	0.000015	0.000157	0.000891	0.003311	0.009026	0.019431
16	0.000004	0.000049	0.000334	0.001448	0.004513	0.010930
17	0.000001	0.000014	0.000118	0.000596	0.002124	0.005786
18		0.000004	0.000039	0.000232	0.000944	0.002893
19		0.000001	0.000012	0.000085	0.000397	0.001370
20			0.000004	0.000030	0.000159	0.000617
21			0.000001	0.000010	0.000061	0.000264
22				0.000003	0.000022	0.000108
23				0.000001	0.000008	0.000042

21.4.4 Special continuous distributions

Special continuous distributions: Probability densities and the corresponding distribution functions of **continuous** random variables (e.g., measured values of the capacitance of a capacitor) or **idealized** analytic functions for the probability distribution of **discrete** random variables; can be used exactly as **sample function** in random sampling analyses.

Uniform continuous distribution, rectangular distribution: Distribution of a random variable which may assume any value of an interval $[a, b]$ and for which the probability of assuming a value of a subinterval of $[a, b]$ is the same for all subintervals of equal length.

▷ A uniformly distributed continuous random variable can be given for cases in which the geometrical method of calculating the probability may be applied.

Probability density:

$$f(x) = \begin{cases} \dfrac{1}{b-a} & \text{for} \quad a \le x \le b \\ 0 & \text{for} \quad x < a, x > b;\ a < b;\ a, b\, \text{real.} \end{cases}$$

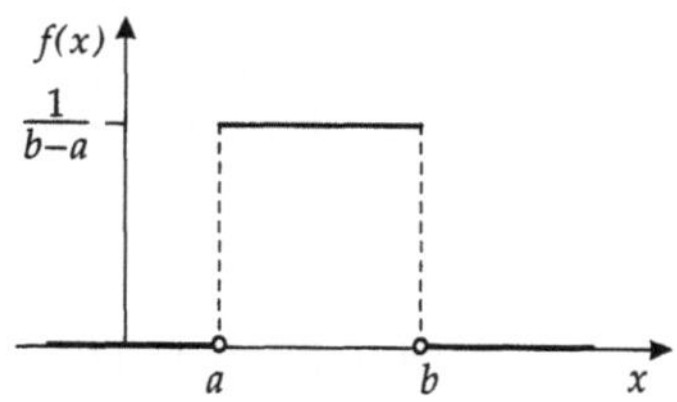

Distribution function:

$$F(x) = \begin{cases} 0 & \text{for} \quad x \le a \\ \dfrac{x-a}{b-a} & \text{for} \quad a < x \le b \\ 1 & \text{for} \quad b < x. \end{cases}$$

The parameters of distribution are a and b.

Expected value: $m = \dfrac{a+b}{2}$

Variance: $\sigma^2 = \dfrac{(b-a)^2}{12}$

Normal distribution, Gaussian distribution: Distribution of random variables (measured values), which can be regarded as the superpositions of defective quantities of approximately equal strength.

Probability density:

$$f_n(x) = \frac{1}{\sqrt{2\pi}\sigma} e^{\{-(x-m)^2/2\sigma^2\}} .$$

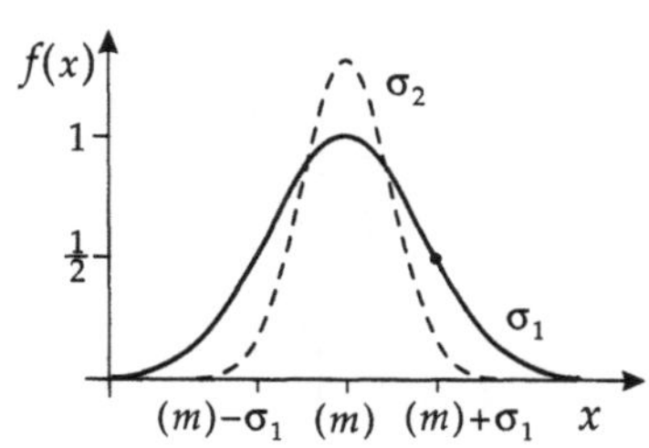

Distribution function:

$$F_n(x) = \frac{1}{\sqrt{2\pi}\sigma} \int_{-\infty}^{x} e^{\{-(y-m)^2/2\sigma^2\}} dy .$$

The parameters of the distribution are the expected value m and the variance σ^2.

Expected value: m.

Variance: σ^2.

▷ σ is the distance between the abscissas of the inflection points of the bell curve and the most probable value.

● **Central-limit theorem**: The **sum** of n independent random variables subject to the same distribution will always approach a normal distribution curve as n increases.

□ With good approximation, measuring errors are usually distributed normally due to the multiple superposition of various errors. The same applies to the production of components with constant properties such as resistors, capacitors, screw lengths, wavelengths, etc.

▷ For the expected zero ($m = 0$) and $\sigma = 1$, we obtain the **standard normal distribution**.

$$f_n(x; c, \sigma) = \frac{1}{\sigma} f_{\text{sn}}\left(\frac{x-m}{\sigma}\right)$$
$$F_n(x; c, \sigma) = F_{\text{sn}}\left(\frac{x-m}{\sigma}\right)$$

α-quantile x_α: Value of the random variables x_α for which it follows that

$$P(y \le x_\alpha) = \alpha.$$

The probability of obtaining a value $y \le x_\alpha$ amounts to just α.
Relation between the parameters of normal distribution and special quantiles:
0.16-quantile: $T_{0.16} = m - \sigma$,
0.84-quantile: $T_{0.84} = m + \sigma$,
0.5-quantile: $T_{05} = m$.

□ About 68% of all events of normal distribution lie in the range of standard deviation around the expected value $m \pm \sigma$.
Generalization:

$$P(|X-m| < \sigma) \approx 0.683$$
$$P(|X-m| < 2\sigma) \approx 0.954$$
$$P(|X-m| < 3\sigma) \approx 0.997.$$

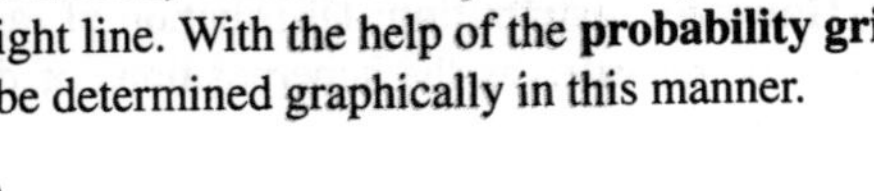

Probability grid: Coordinate system in which the α-quantile is plotted against α. The quantile axis is recalled with respect to a given special distribution function F (e.g., Gaussian normal distribution) in such a way that a representation of this distribution function yields a straight line. With the help of the **probability grid**, the parameters of a distribution can also be determined graphically in this manner.

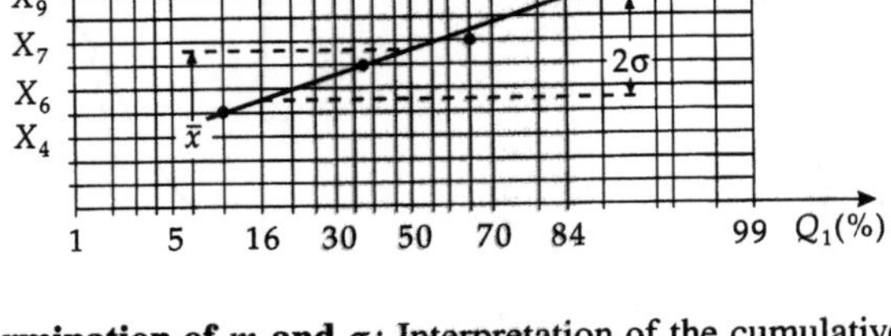

Graphical determination of m and σ: Interpretation of the cumulative frequencies of measured values $Q_l = \sum_i^l h(X_i)$ as a quantile of measured values X_l and plotting of these cumulative frequencies Q_l against measured values X_l on the **probability grid**.

(A few values are usually sufficient.) Draw a straight line through the entered points. On this straight line, read the characteristics of the special 0.16, 0.84 and 0.5 quantities given by the normal distribution, and calculate m and σ from the above relations (see normal distribution).

Standard normal distribution, Gaussian normal distribution: Special case of the normal distribution for $m = 0$ and $\sigma = 1$.

Probability density:

$$f_{sn}(x) = \frac{1}{\sqrt{2\pi}} \exp^{-x^2/2} .$$

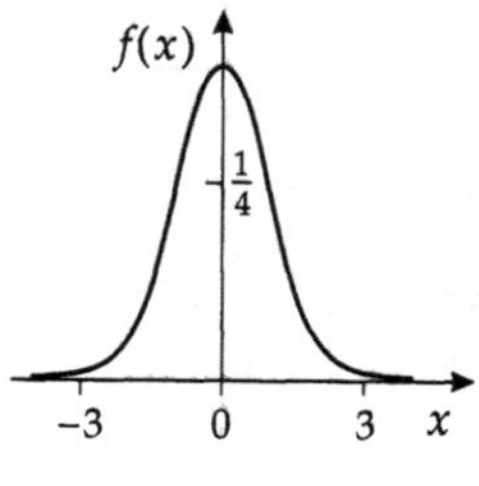

Distribution function:

$$F_{sn}(x) = \frac{1}{\sqrt{2\pi}} \int_{-\infty}^{x} \exp^{-y^2/2} dy .$$

Expected value: $m = 0$.
Variance: $\sigma^2 = 1$.

▷ By replacing

$$f_{sn}(x) \to f_n(x) = \frac{1}{\sigma} f_{sn}\left(\left(\frac{x-m}{\sigma}\right)^2\right) ,$$

We obtain the normal distribution.

- The α-quantiles of standard normal distribution t_α **and** of normal distribution T_α are both given by distribution function F_{sn}:

$$\alpha = F_{sn}(t_\alpha) = F_{sn}\left(\frac{T_\alpha - m}{\sigma}\right) .$$

- The relation between the quantiles is

$$T_\alpha = m + t_\alpha \sigma .$$

▷ The normal distribution is the borderline case of binomial distribution for a large size n and with a parameter value of $p = 0.5$ ($n \to \infty$, $p \equiv 0.5$). $p = 0.5$ means an error of 50% in the sense of a random sample examination of attributes.

Standard normal distribution ($f_{sn}(-x) = f_{sn}(x)$)					
x	$f_{sn}(x)$	x	$f_{sn}(x)$	x	$f_{sn}(x)$
0.0	0.3989	1.5	0.1295	3.0	0.0044
0.5	0.3520	2.0	0.0539	3.5	0.0008
1.0	0.2419	2.5	0.0175	4.0	0.0001

Percentile $t = t_{F_{sn}}$ of the standard normal distribution f_{sn}							
$F_{sn}(t)$	t	$F_{sn}(t)$	t	$F_{sn}(t)$	t	$F_{sn}(t)$	t
0.05	−1.644	0.30	−0.524	0.55	0.125	0.80	0.841
0.10	−1.281	0.35	−0.385	0.60	0.253	0.85	1.036
0.15	−1.036	0.40	−0.253	0.65	0.385	0.90	1.281
0.20	−0.841	0.45	−0.125	0.70	0.524	0.95	1.644
0.25	−0.674	0.50	0.000	0.75	0.674		

Error function, also called **Gaussian probability integral**, erf(x): The area under the bell curve between the variable limits $-x\sqrt{2}$ and $+x\sqrt{2}$:

$$\operatorname{erf}(x) = \frac{1}{\sqrt{2\pi}} \int_{-x\sqrt{2}}^{+x\sqrt{2}} e^{-y^2/2} dy = \frac{2}{\sqrt{2\pi}} \int_0^{x\sqrt{2}} e^{-y^2/2} dy .$$

▷ Approximation for $x \geq 2$:

$$1 - \frac{a}{x e^{x^2}}$$

$a = 0.515 \quad 2 \le x \le 3$
$a = 0.535 \quad 3 \le x \le 4$
$a = 0.545 \quad 4 \le x \le 7$
$a = 0.56 \quad 7 \le x \le \infty$

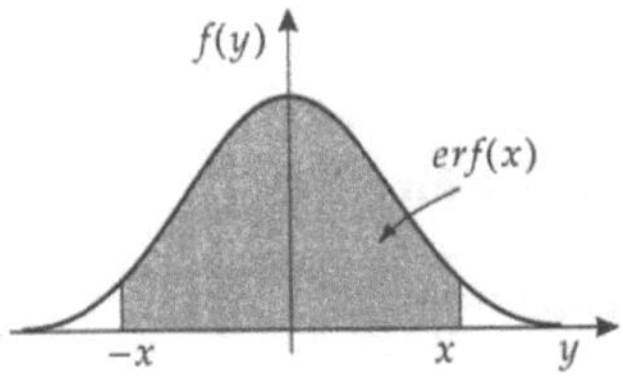

Table of the probability density of the standard normal distribution

$$f_{\text{sn}}(x) = \frac{1}{\sqrt{2\pi}} e^{\frac{x^2}{2}}$$
$$f_{\text{sn}}(-x) = f_{\text{sn}}(x)$$
$$f(x) = \frac{1}{\sigma} f_{\text{sn}}\left(\frac{x-m}{\sigma}\right)$$

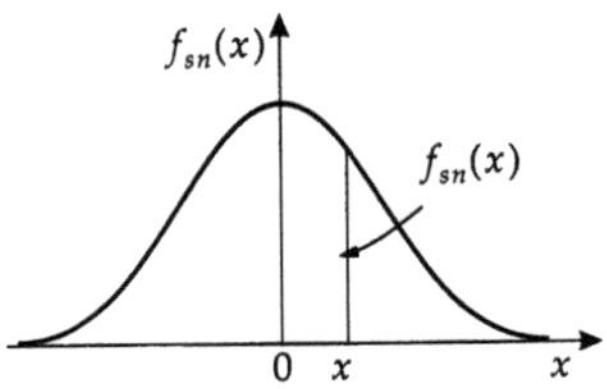

x	0	1	2	3	4	5	6	7	8	9	x
0.0	0.39894	0.39892	0.39886	0.39876	0.39862	0.39844	0.39822	0.39797	0.39767	0.39733	0.0
0.1	0.39695	0.39654	0.39608	0.39559	0.39505	0.39448	0.39387	0.39322	0.39253	0.39181	0.1
0.2	0.39104	0.39024	0.38940	0.38853	0.38762	0.38667	0.38568	0.38466	0.38361	0.38251	0.2
0.3	0.38139	0.38023	0.37903	0.37780	0.37654	0.37524	0.37391	0.37255	0.37115	0.36973	0.3
0.4	0.36827	0.36678	0.36526	0.36371	0.36213	0.36053	0.35889	0.35723	0.35553	0.35381	0.4
0.5	0.35207	0.35029	0.34849	0.34667	0.34482	0.34294	0.34105	0.33912	0.33718	0.33521	0.5
0.6	0.33322	0.33121	0.32918	0.32713	0.32506	0.32297	0.32086	0.31874	0.31659	0.31443	0.6
0.7	0.31225	0.31006	0.30785	0.30563	0.30339	0.30114	0.29887	0.29659	0.29431	0.29200	0.7
0.8	0.28969	0.28737	0.28504	0.28269	0.28034	0.27798	0.27562	0.27324	0.27086	0.26848	0.8
0.9	0.26609	0.26369	0.26129	0.25888	0.25647	0.25406	0.25164	0.24923	0.24681	0.24439	0.9
1.0	0.24197	0.23955	0.23713	0.23471	0.23230	0.22988	0.22747	0.22506	0.22265	0.22025	1.0
1.1	0.21785	0.21546	0.21307	0.21069	0.20831	0.20594	0.20357	0.20121	0.19886	0.19652	1.1
1.2	0.19419	0.19186	0.18954	0.18724	0.18494	0.18265	0.18037	0.17810	0.17585	0.17360	1.2
1.3	0.17137	0.16915	0.16694	0.16474	0.16256	0.16038	0.15822	0.15608	0.15395	0.15183	1.3
1.4	0.14973	0.14764	0.14556	0.14350	0.14146	0.13943	0.13742	0.13542	0.13344	0.13147	1.4
1.5	0.12952	0.12758	0.12566	0.12376	0.12188	0.12001	0.11816	0.11632	0.11450	0.11270	1.5
1.6	0.11092	0.10915	0.10741	0.10567	0.10396	0.10226	0.10059	0.09893	0.09728	0.09566	1.6
1.7	0.09405	0.09246	0.09089	0.08933	0.08780	0.08628	0.08478	0.08329	0.08183	0.08038	1.7
1.8	0.07895	0.07754	0.07614	0.07477	0.07341	0.07206	0.07074	0.06943	0.06814	0.06687	1.8
1.9	0.06562	0.06438	0.06316	0.06195	0.06077	0.05959	0.05844	0.05730	0.05618	0.05508	1.9
2.0	0.05399	0.05292	0.05186	0.05082	0.04980	0.04879	0.04780	0.04682	0.04586	0.04491	2.0
2.1	0.04398	0.04307	0.04217	0.04128	0.04041	0.03955	0.03871	0.03788	0.03706	0.03626	2.1
2.2	0.03547	0.03470	0.03394	0.03319	0.03246	0.03174	0.03103	0.03034	0.02965	0.02898	2.2
2.3	0.02833	0.02768	0.02705	0.02643	0.02582	0.02522	0.02463	0.02406	0.02349	0.02294	2.3
2.4	0.02239	0.02186	0.02134	0.02083	0.02033	0.01984	0.01936	0.01888	0.01842	0.01797	2.4
2.5	0.01753	0.01709	0.01667	0.01625	0.01585	0.01545	0.01506	0.01468	0.01431	0.01394	2.5
2.6	0.01358	0.01323	0.01289	0.01256	0.01223	0.01191	0.01160	0.01130	0.01100	0.01071	2.6
2.7	0.01042	0.01014	0.00987	0.00961	0.00935	0.00909	0.00885	0.00861	0.00837	0.00814	2.7
2.8	0.00792	0.00770	0.00748	0.00727	0.00707	0.00687	0.00668	0.00649	0.00631	0.00613	2.8
2.9	0.00595	0.00578	0.00562	0.00545	0.00530	0.00514	0.00499	0.00485	0.00470	0.00457	2.9

Table of the distribution function of the standard normal distribution

$F_{sn} = \frac{1}{\sqrt{2\pi}} \int_{-\infty}^{x} e^{-\frac{t^2}{2}}\, dt$

$F_{sn}(-x) = 1 - F_{sn}(x)$

$P(X < b) = F_{sn} = \left(\frac{b-m}{\sigma}\right)$

$P(a \le X < b) = F_{sn}\left(\frac{b-m}{\sigma}\right) - F_{sn}\left(\frac{a-m}{\sigma}\right)$

$P(|X - m| < b) = 2F_{sn}\left(\frac{b}{\sigma}\right) - 1$

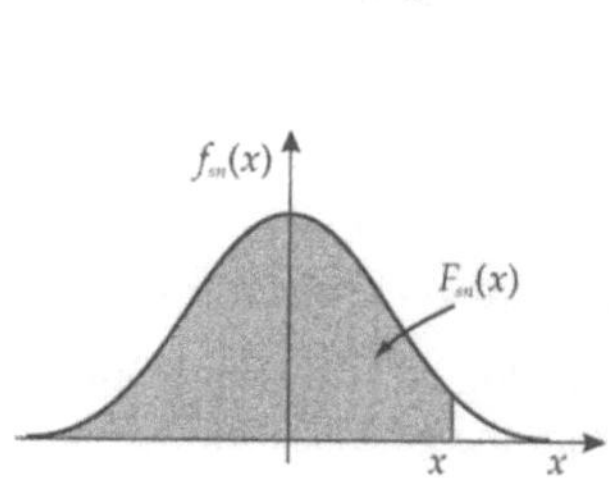

x	0	1	2	3	4	5	6	7	8	9	x
0.0	0.500000	0.503989	0.507978	0.511966	0.515953	0.519938	0.523922	0. 527903	0.531883	0.535856	0.0
0.1	0.539828	0.543795	0.547758	0.551717	0.555670	0.559618	0.563560	0. 567495	0.571424	0.575345	0.1
0.2	0.579260	0.583166	0.587064	0.590954	0.594835	0.598706	0.602568	0. 606420	0.610261	0.614092	0.2
0.3	0.617911	0.621720	0.625616	0.629300	0.633072	0.636831	0.640576	0. 644309	0.648027	0.651733	0.3
0.4	0.655422	0.659097	0.662757	0.666402	0.670031	0.673645	0.677242	0. 680822	0.684386	0.687933	0.4
1.0	0.841345	0.843752	0.846136	0.848495	0.850330	0.853141	0.855428	0. 857690	0.859929	0.862143	1.0
1.1	0.864334	0.866500	0.868643	0.870762	0.872857	0.874928	0.876976	0. 879000	0.881000	0.882977	1.1
1.2	0.884930	0.886861	0.888768	0.890651	0.892512	0.894350	0.896165	0. 897958	0.899727	0.901475	1.2
1.3	0.903200	0.904902	0.906582	0.908241	0.909877	0.911492	0.913085	0. 914656	0.916207	0.917736	1.3
1.4	0.919243	0.920730	0.922196	0.923642	0.925066	0.926471	0.927855	0. 929219	0.930563	0.931889	1.4
1.5	0.933193	0.934478	0.935744	0.936922	0.938220	0.939429	0.940620	0. 941792	0.942907	0.944083	1.5
1.6	0.945201	0.946301	0.947384	0.948449	0.949497	0.950528	0.951543	0. 952540	0.953521	0.954486	1.6
1.7	0.955434	0.956367	0.957284	0.958185	0.959070	0.959941	0.960796	0. 961636	0.962462	0.963273	1.7
1.8	0.964070	0.964852	0.965620	0.966375	0.967116	0.967843	0.968557	0. 969258	0.969946	0.970621	1.8
1.9	0.971283	0.971933	0.972571	0.973197	0.973810	0.974412	0.975002	0. 975581	0.976138	0.976704	1.9
2.0	0.977250	0.977784	0.978308	0.978822	0.979325	0.979818	0.980301	0. 980774	0.981237	0.981691	2.0
2.1	0.982136	0.982571	0.982997	0.983414	0.983823	0.984222	0.984614	0. 984997	0.985371	0.985738	2.1
2.2	0.986097	0.986447	0.986791	0.987126	0.987454	0.987776	0.988089	0. 988396	0.988696	0.988989	2.2
2.3	0.989276	0.989556	0.989830	0.990097	0.990358	0.990613	0.990862	0. 991106	0.991344	0.991576	2.3
2.4	0.991802	0.992024	0.992240	0.992451	0.992656	0.992857	0.993053	0. 993244	0.993431	0.993613	2.4
2.5	0.993790	0.993963	0.994132	0.994297	0.994457	0.994614	0.994766	0. 994915	0.995060	0.995201	2.5
2.6	0.995339	0.995473	0.995604	0.995731	0.995855	0.995975	0.996093	0. 996207	0.996319	0.996427	2.6
2.7	0.996533	0.996636	0.996736	0.996833	0.996928	0.997020	0.997110	0. 997197	0.997282	0.997365	2.7
2.8	0.997445	0.997523	0.997599	0.997673	0.997744	0.997814	0.997882	0. 997948	0.998012	0.998074	2.8
2.9	0.998134	0.998193	0.998250	0.998305	0.998359	0.998411	0.998462	0. 998511	0.998559	0.998605	2.9
	0	1	2	3	4	5	6	7	8	9	
3.0	0.998650	0.999032	0.999313	0.999517	0.999663	0.999767	0.999841	0. 999892	0.999928	0.999952	3.0

Log-normal distribution: Normal distribution for the logarithms of the random variable x.

Probability density:

$$f_{\log}(x) = \frac{1}{\sqrt{2\pi}\sigma x} \exp^{-(\ln x - \mu)^2/2\sigma^2}, \quad x > 0\,.$$

Distribution function:

$$F_{\log}(x) = \int_0^x \frac{1}{\sqrt{2\pi}\sigma y} \exp^{-(\ln y - \mu)^2/2\sigma^2} \mathrm{d}y, \quad x > 0\,.$$

Expected value: $m = e^{\mu+\sigma^2/2}$.
Variance: $e^{2\mu+\sigma^2}\left(e^{\sigma^2} - 1\right)$.

▷ The log-normal distribution results from the normal distribution through the replacement:

$$f_{\mathrm{n}}(x) \to f_{\log}(x) = \frac{1}{x} f_{\mathrm{n}}(\ln x)\,.$$

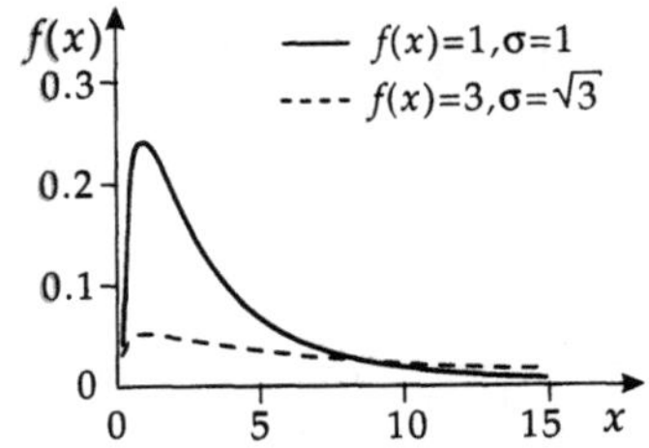

Exponential distribution: Distribution of the distance between two measured values following a Poisson distribution with the parameter $\lambda = c$.
Probability density:

$$f_{\exp}(x) = \lambda e^{-\lambda x}, \quad \lambda > 0, x \geq 0\,.$$

Distribution function:

$$F_{\exp}(x) = 1 - e^{-\lambda x}\,.$$

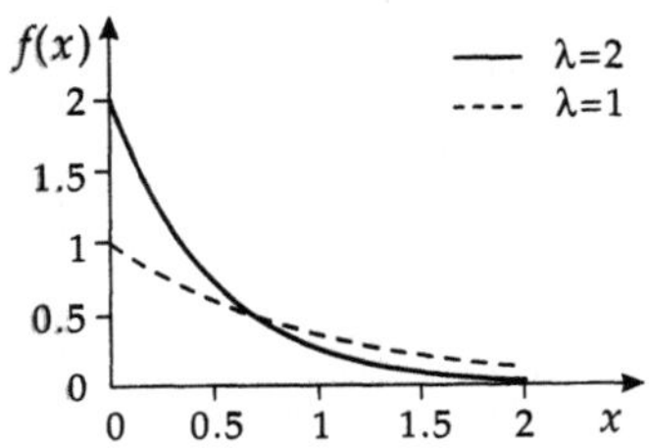

Expected value: $m = 1/\lambda$.
Variance: $\sigma^2 = 1/\lambda^2$.

▷ In the sense of time processes (failure of technical components per time unit are subjected to, infection with a rare disease per time unit, etc.) only the **lifetimes of non-aging units** obeys the exponential distribution.

▷ A condition for an exponential distribution is that the events must be truly **random** and follow the Poisson distribution; see **Weibull distribution**.

□ Let the number of events within a time interval be Poisson distributed (e.g., radioactive decay per second, failure of electronic components per year, infections with a certain disease per year, etc.) The time intervals between two such events are exponentially distributed.

□ Given an urn with $N \to \infty$ spheres, a small fraction $p = N_1/N$, $N_1 \ll N$ from one experiment to another is black; while all others are of different color. If we take a very large number of spheres $n \geq 10/p$ from this urn (performing a random experiment), then the number of black spheres drawn is approximately Poisson distributed.
If this random experiment is performed several times, the difference in the number of black spheres is approximately according to an exponential distribution.

Weibull distribution: Extension of the exponential distribution to events that are not purely random and do not exactly follow the Poisson distribution.

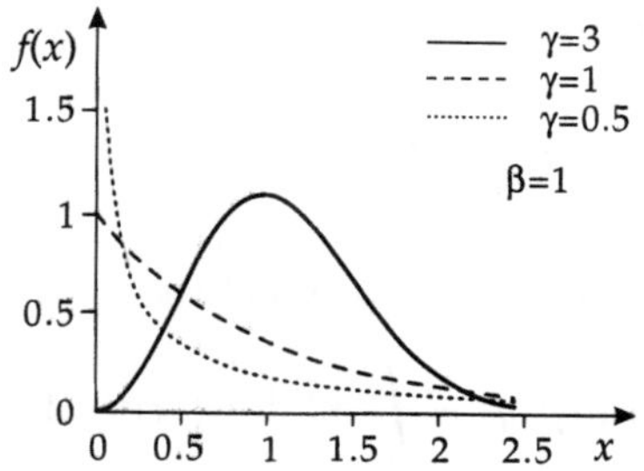

Probability density:

$$f(x) = \frac{\gamma}{\beta}\left(\frac{x-\alpha}{\beta}\right)^{\gamma-1} e^{-((x-\alpha)/\beta)^{\gamma}} \quad , \quad x \geq \alpha \, .$$

Distribution function:

$$F(x) = 1 - e^{-((x-\alpha)/\beta)^{\gamma}} \, .$$

Expected value: $m = \beta\Gamma(1+1/\gamma)+\alpha$.
Variance: $\sigma^2 = \beta^2\{\Gamma(1+2/\gamma) - [\Gamma(1+1/\gamma)]^2\}$.

▷ The **gamma function** $\Gamma(x)$ is defined as follows:

$$\Gamma(x) := \int_0^{\infty} e^{-t} t^{x-1} dt \, .$$

▷ In the sense of processes involving time, such as the failure of a clutch, we also speak of the **lifetime** of **aging objects**. This is the same as a dependence on external influences, such as wear and tear.

□ Lifetime with aging: Wear of working parts.

▷ The Weibull distribution and the exponential distribution are closely connected with the concept of **reliability**.

Beta distribution: Used to describe continuous random variables whose values have a lower bound a and an upper bound b.

▷ The beta distribution is used in mathematical statistics.

Probability density:

$$f(x) = \begin{cases} \dfrac{(x-a)^{p-1}(b-x)^{q-1}}{(b-a)^{p+q-1}B(p,q)} & \text{for} \quad a \leq x \leq b \\ 0 & \text{for} \quad x < a, x > b, \; p > 0, q > 0, \; a < b. \end{cases}$$

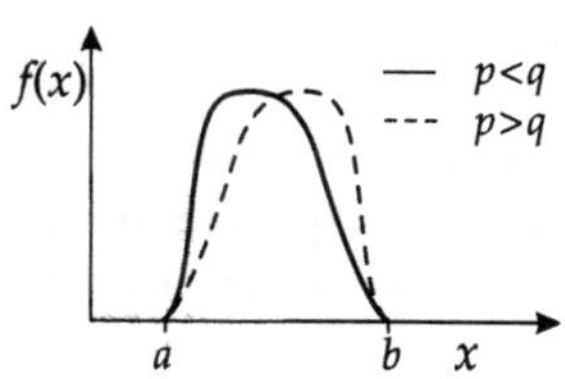

Distribution function:

Can usually not be determined because the resulting integral cannot be solved in closed form.

▷ Special tables are used for the beta distribution and for the determination of values of the beta function.

The parameters of distribution are p, q, a, b.

Expected value: $m = \dfrac{aq + bp}{p + q}$

Variance: $\sigma^2 = \dfrac{(b-a)^2 pq}{(p+q)^2(p+q+1)}$

▷ The **beta function** $B(p,q)$ is defined as follows:

$$B(p,q) = \int_0^1 x^{p-1}(1-x)^{q-1}\,\mathrm{d}x.$$

▷ The connection between beta and gamma function is:

$$B(p,q) = \frac{\Gamma(p)\Gamma(q)}{\Gamma(p+q)}.$$

▷ For $p = q = 1$ the beta distribution becomes a uniformly continuous distribution.

21.5 Limit theorems

21.5.1 Laws of large numbers

Laws of large numbers: Mathematical expression for the fact that the relative frequency of an event occurring in a large number of independent events stabilizes with the random result, i.e., it fluctuates about a fixed value, the probability of the event; rules result from the action of a large number of independent random factors.
Stochastic convergence: Convergence behavior of sequences of random variables.
Convergence in probability: Convergence of the probability of a sequence of random variables to a value.
Convergence with probability 1: Probability of convergence of a sequence of random variables to a value.

▷ Convergence with probability 1 expresses a stronger convergence behavior than convergence in probability.

▷ Random variables are subject to the strong or weak law of large numbers.

Chebyshev's inequality: For unknown distribution but known expected value m and known variance σ^2, this inequality estimates the probability that the random variable X assumes values from certain intervals:

$$P(|X - m| > \varepsilon) \le \frac{\sigma^2}{\varepsilon^2}, \quad (\varepsilon > 0).$$

● **Bernoulli's law of large numbers**:
In a series of n independent trials (performed according to Bernoulli's trials) the relative frequency of an event A converges stochastically to the probability p of the occurrence of A.

Bernoulli trials:
n independent trials,
the results are A or $\overline{A}$,

$$P(A) = p, \quad P(\overline{A}) = 1 - p.$$

▷ The probability of a random event is not the limit of the relative frequency, but the sequence of random variables is subject to the law of large numbers.

- **Poisson's law of large numbers**:
  In a series of n independent trials (following Poisson's trials) the relative frequency of an event A converges stochastically to the arithmetic mean of the probabilities of the occurrence of A.

Poisson trials:
n independent trials; the results are either A or $\overline{A}$,
$P(A) = p_i$ in the i-th trial; p_n becomes very small if the trial series is enlarged,
$n - p_n$ remains constant for increasing n.

- **Chebyshev's law of large numbers**:
  The arithmetic mean of a sequence of pairwise independent random variables with bounded variance converges stochastically to the arithmetic mean of their mathematical expectations.

▷ Bernoulli's law of large numbers is a special case of the Poisson's law of large numbers which in turn is a special case of Chebyshev's law. Further generalizations are the laws of Borel, Chintchin, and Kolmogorov.

21.5.2 Limit theorems

Limit theorems: Theorems concerning the stochastic convergence of a sequence of distribution functions (global limit theorems) and of individual probabilities or density functions (local limit theorems).

Limiting distribution function, limiting distribution: The limit of a sequence of distribution functions.

□ The distribution function of a normal random variable is the limiting distribution function of a binomial random variable.

- **Poisson's limit theorem**: Connection between binomial distribution and Poisson distribution; statement about the limiting behavior of the individual probabilities of binomial random variables for large n and small p in terms of a Poisson distribution:

$$\lim_{n\to\infty,\, p\to 0} \binom{n}{k} p^k (1-p)^{n-k} = \frac{\lambda^k}{k!} e^{-\lambda}, \quad k = 0, 1, 2, \ldots; \ \lambda = np = \text{const.}$$

▷ The approximation is of practical importance in reducing the work of calculating binomial random variables; it can also be used for $np < 10$ and $n > 1500p$.

- **Central-limit theorem, Lindeberg–Levy limit theorem**:
  A random variable is approximately normal with the expected value nm and variance $n\sigma^2$ if it can be regarded as a sum of a large number of stochastically independent random variables all of which satisfy the same distribution function with expected value m and variance σ^2.

- **Lyapunov's limit theorem**:
  A random variable Y is approximately normal with the parameters $m = \sum_{k=1}^{n} m_k$ and $\sigma^2 = \sum_{k=1}^{n} \sigma_k^2$ if it can be represented as a sum of a large number n of independent summands (random variables X_k with expected values m_k and variances σ_k^2) each of which contributes insignificantly to the sum.

▷ In practice, test of an approximate normal random variable by: recognizing a straight line on normal probability paper after plotting the random sample values, or application of a test of goodness-of-fit test used in mathematical statistics.

- **Limit theorem of Moivre–Laplace**:
  In a large number of (Bernoulli) trials the distribution function of the standardized binomial random variable converges stochastically to the distribution function of the

standardized normal distribution (global statement):

$$\lim_{n\to\infty} P\left(a \le \frac{X - np}{\sqrt{np(1-p)}} < b\right) = \frac{1}{\sqrt{2\pi}} \int_a^b e^{-\frac{x^2}{2}}\, dx.$$

Local statement (simplified): Individual probabilities of a binomial distribution with parameters n and p converge to the corresponding values of the density function f_n of a normal distribution with parameters $m = np$ and $\sigma^2 = np(1-p)$:

$$P(X = k) = \binom{n}{k} p^k (1-p)^{n-k} \approx \frac{1}{\sqrt{np(1-p)}} f_n\left(\frac{k - np}{\sqrt{np(1-p)}}\right).$$

▷ While Poisson's limit theorem provides a good approximation to the individual probabilities of binomial random variables only for small values of p, according to the limit theorem of Moivre–Laplace their approximation by normal distribution can be done for every p, $0 < p < 1$ and sufficiently large n.
Rule of thumb for applicability:

$$np > 4, \quad n(1-p) > 4.$$

21.6 Multidimensional random variables

***n*-dimensional random variable, *n*-dimensional random vector**: n-tuple of random variables $(X_1, X_2, \ldots, X_n)$ that results if n real numbers $x_1, x_2, \ldots, x_n$ are assigned uniquely to every elementary event of a random trial.

□ By means of a two-dimensional random variable (X_1, X_2), a pair of values (x_1, x_2) is assigned uniquely to every elementary event.

▷ Further considerations are restricted to two-dimensional random variables; in general, the results can usually be transferred to n-dimensional random variables.

21.6.1 Distribution functions of two-dimensional random variables

Distribution function $F(x, y)$ of (X, Y): Expression defined by

$$F(x, y) = P(X < x, Y < y), \ (x, y) \in \mathbb{R}^2.$$

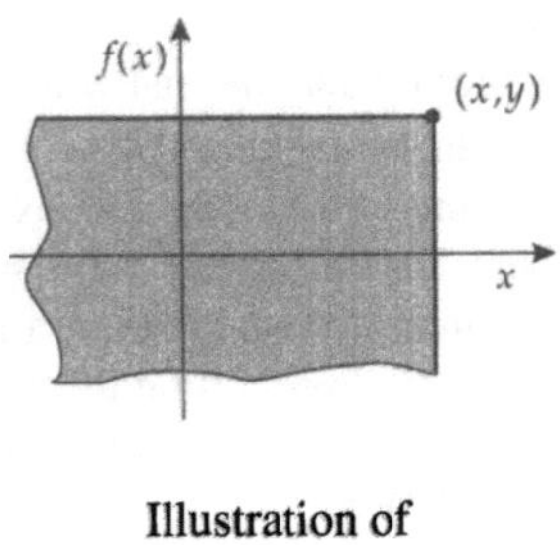

Illustration of
$P(X < x, Y < y)$

▷ Two-dimensional random variables are discrete (continuous) if X and Y are discrete (continuous), otherwise they are mixed.

Marginal distribution of X or Y: Distribution from which the values of X or Y and the corresponding probabilities can be determined.

▷ Marginal distributions can be given in terms of the distribution functions of X and Y or in terms of the individual probabilities and the density functions of discrete and continuous random variables, respectively.

Marginal distribution functions $F_1(x)$ and $F_2(y)$ of X and Y:

$$F_1(x) = P(X < x) = P(X < x, Y < \infty)$$
$$F_2(x) = P(Y < y) = P(X < \infty, Y < y).$$

● For a two-dimensional random variable (X, Y) with the distribution function $F(x, y)$ and $x_1 < x_2$, $y_1 < y_2$ the following holds:

$$P(x_1 \leq X < x_2, y_1 \leq Y < y_2) = F(x_2, y_2) - F(x_2, y_1) - F(x_1, y_2) + F(x_1, y_1).$$

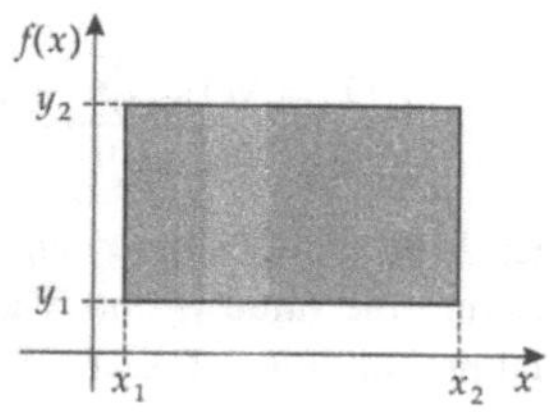

Illustration of $P(x_1 \leq X < x_2, y_1 \leq Y < y_2)$

● $F(x, y)$ is distribution function of a two-dimensional random variable if:
$F(x, y)$ is monotonically increasing in x and y,

$$\lim_{x\to-\infty} F(x, y) = \lim_{y\to-\infty} F(x, y) = 0, \qquad \lim_{(x,y)\to(\infty,\infty)} F(x, y) = 1,$$

$F(x, y)$ is at least left-hand continuous in x and y,

$$F(x_2, y_2) - F(x_2, y_1) - F(x_1, y_2) + F(x_1, y_1) \geq 0 \text{ for arbitrary } x_1 < x_2,\ y_1 < y_2 .$$

21.6.2 Two-dimensional discrete random variables

Two-dimensional discrete random variable (X, Y): A random variable that can take only a finite number or denumerable infinite number of pairs of values (x_i, y_k), $i, k = 1, 2, \ldots$.

▷ Complete description of (X, Y) in terms of pairs of values (x_i, y_k) and individual probabilities $P(X = x_i, Y = y_k)$.

Distribution table of a two-dimensional discrete random variable:

X/Y	y_1	$y_2 \cdots$	$y_n \cdots$	$p_{i.}$
x_1	p_{11}	$p_{12} \cdots$	$p_{1n} \cdots$	$p_{1.}$
x_2	p_{21}	$p_{22} \cdots$	$p_{2n} \cdots$	$p_{2.}$
.	.	.	.	.
.	.	.	.	.
.	.	.	.	.
x_n	p_{n1}	$p_{n2} \cdots$	$p_{nn} \cdots$	$p_{n.}$
.	.	.	.	.
.	.	.	.	.
.	.	.	.	.
$p_{.k}$	$p_{.1}$	$p_{.2} \cdots$	$p_{.n} \cdots$	$p_{..} = 1$

Marginal-distribution individual probabilities of X and Y:

$$p_{i.} = P(X = x_i) = \sum_{(k)} p_{ik}$$

$$p_{.k} = P(Y = y_k) = \sum_{(i)} p_{ik}$$

with $p_{i.}$ the probability assumes X takes the value x_i and Y assumes an arbitrary value. $p_{.k}$ is the probability that Y assumes the value y_k and X assumes an arbitrary value.

▷ For the marginal individual probabilities of X and Y the following holds:

$$\sum_{(i)} p_{i.} = \sum_{(k)} p_{.k} = 1, \quad \sum_{(i)} \sum_{(k)} p_{ik} = 1\,.$$

Distribution function of (X, Y):

$$F(x, y) = P(X < x, Y < y) = \sum_{x_i < x} \sum_{y_k < y} P(X = x_i, Y = y_k).$$

Conditional individual probabilities, conditional distribution
of X for fixed $Y = y_k$:

$$P(X = x_i | Y = y_k) = \frac{P(X = x_i, Y = y_k)}{P(Y = y_k)} = \frac{p_{ik}}{p_{.k}},$$

of Y for fixed $X = x_i$:

$$P(Y = y_k | X = x_i) = \frac{P(X = x_i, Y = y_k)}{P(X = x_i)} = \frac{p_{ik}}{p_{i.}}\,.$$

21.6.3 Two-dimensional continuous random variables

Two-dimensional continuous random variable (X, Y): A random variable for which a function $f(x, y) \geq 0$ exists and for $a < b$ and $c < d$:

$$P(a \leq X < b, c \leq Y < d) = \int_a^b \int_c^d f(x, y)\, \mathrm{d}y\, \mathrm{d}x\,.$$

$f(x, y)$ is called **density function** or **distribution density** of (X, Y).

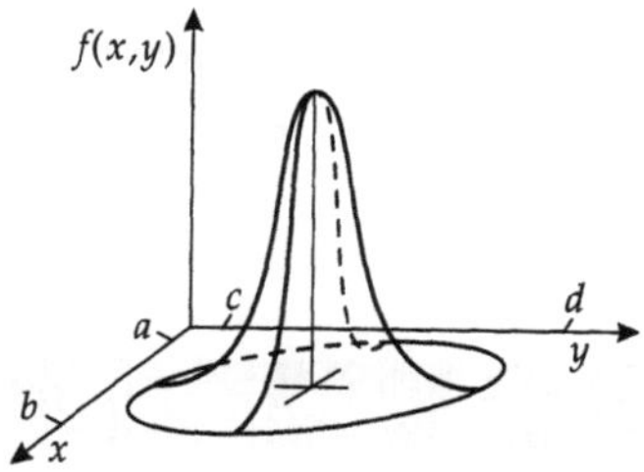

Graph of the density function of a two-dimensional random variable

Marginal distribution density of X and Y:

$$f_1(x) = \int_{-\infty}^{\infty} f(x, y)\,\mathrm{d}y$$

$$f_2(y) = \int_{-\infty}^{\infty} f(x, y)\,\mathrm{d}x.$$

Distribution function of (X, Y):

$$F(x, y) = P(X < x, Y < y) = \int_{-\infty}^{x} \int_{-\infty}^{y} f(u, v)\,\mathrm{d}v\,\mathrm{d}u.$$

▷ For $F(x, y)$ and $f(x, y)$ the following relation holds:

$$\frac{\partial^2 F(x, y)}{\partial x \partial y} = f(x, y).$$

Conditional density function f, **conditional distribution densities**, and **conditional distribution functions** F
of X for a fixed value $Y = y$:

$$f(x|y) = \frac{f(x, y)}{f_2(y)}, \quad F(x|y) = \frac{F(x, y)}{F_2(y)},$$

of Y for a fixed value $X = x$:

$$f(y|x) = \frac{f(x, y)}{f_1(x)}, \quad F(y|x) = \frac{F(x, y)}{F_1(x)}$$

with the marginal distribution functions

$$F_1(x) = F(x, \infty) = P(X < x, Y < \infty) = \int_{-\infty}^{x} \int_{-\infty}^{\infty} f(u, y)\mathrm{d}y\,\mathrm{d}u$$

$$F_2(x) = F(\infty, y) = P(X < \infty, Y < y) = \int_{-\infty}^{\infty} \int_{-\infty}^{y} f(x, v)\mathrm{d}v\mathrm{d}x\ .$$

21.6.4 Independence of random variables

Independence of X and Y: If for the distribution function it holds that

$$F(x, y) = F_1(x) \cdot F_2(y),$$

if for the individual probabilities and density functions it holds that

$$P(X = x_i, Y = y_k) = P(X = x_i) \cdot P(Y = y_k) \quad \text{or} \quad p_{ik} = p_{i.} \cdot p_{.k},$$

and $f(x, y) = f_1(x) \cdot f_2(y)$, respectively.

21.6.5 Parameters of two-dimensional random variables

Expected value (m_x, m_y) for discrete and continuous random variables (X, Y) with the individual probabilities $P(X = x_i, Y = y_k) = p_{ik}$, $i, k = 1, 2, \ldots$ and the density function $f(x, y)$ with

$$m_x = \sum_{(i)}\sum_{(k)} x_i p_{ik} = \sum_{(i)} x_i p_{i.}$$

$$m_y = \sum_{(i)}\sum_{(k)} y_k p_{ik} = \sum_{(k)} y_k p_{.k}$$

for discrete random variables (X, Y), and

$$m_x = \int_{-\infty}^{\infty}\int_{-\infty}^{\infty} x f(x, y)\,\mathrm{d}y\,\mathrm{d}x = \int_{-\infty}^{\infty} x f_1(x)\,\mathrm{d}x$$

$$m_y = \int_{-\infty}^{\infty}\int_{-\infty}^{\infty} y f(x, y)\,\mathrm{d}x\,\mathrm{d}y = \int_{-\infty}^{\infty} y f_2(y)\,\mathrm{d}y$$

for continuous random variables (X, Y).

Variance (σ_x^2, σ_y^2) for discrete and continuous random variables (X, Y) with

$$\sigma_x^2 = \sum_{(i)}\sum_{(k)} (x_i - m_x)^2 p_{ik} = \sum_{(i)} (x_i - m_x)^2 p_{i.}$$

$$\sigma_y^2 = \sum_{(i)}\sum_{(k)} (y_k - m_y)^2 p_{ik} = \sum_{(k)} (y_k - m_y)^2 p_{.k}$$

for discrete random variables (X, Y), and

$$\sigma_x^2 = \int_{-\infty}^{\infty}\int_{-\infty}^{\infty} (x - m_x)^2 f(x, y)\,\mathrm{d}y\,\mathrm{d}x = \int_{-\infty}^{\infty} (x - m_x)^2 f_1(x)\,\mathrm{d}x$$

$$\sigma_y^2 = \int_{-\infty}^{\infty}\int_{-\infty}^{\infty} (y - m_y)^2 f(x, y)\,\mathrm{d}x\,\mathrm{d}y = \int_{-\infty}^{\infty} (y - m_y)^2 f_2(y)\,\mathrm{d}y$$

for continuous random variables (X, Y).

Covariance of (X, Y), $\mathrm{cov}(X, Y)$: Expected value M of the product of the deviations of the corresponding expected value from the random variable:

$$\mathrm{cov}(X, Y) = M[(X - m_x)(Y - m_y)].$$

▷ Calculating formulae:

$$\mathrm{cov}(X, Y) = \sum_{i}\sum_{k} (x_i - m_x)(y_k - m_y) p_{ik}$$

in the discrete case and

$$\mathrm{cov}(X, Y) = \int_{-\infty}^{\infty}\int_{-\infty}^{\infty} (x - m_x)(y - m_y) f(x, y)\,\mathrm{d}x\,\mathrm{d}y$$

in the continuous case.

Correlation coefficients of (X, Y), ρ_{xy}: A measure of the strength and direction of the linear dependence of X and Y,

$$\rho_{xy} = \frac{\operatorname{cov}(X, Y)}{\sigma_x \sigma_y}, \quad -1 \leq \rho_{xy} \leq 1.$$

▷ X and Y are **uncorrelated** if $\rho_{xy} = 0$.

▷ The sign of the correlation coefficients shows whether the correlation is positive or negative.
Positive correlation: If an increase (decrease) of the values of a random variable leads to an increase (a decrease) of the values of the other random variable.
Negative correlation: If an increase (decrease) of the values of a random variable leads to a decrease (increase) of the values of the other random variable.

- If X and Y are independent random variables, then $\rho_{xy} = 0$.
- X and Y are linearly dependent if and only if $\rho_{xy} = \pm 1$; $Y = aX + b, \quad a, b \in \mathbb{R}$; geometric interpretation: the values of (X, Y) lie on a straight line.

21.6.6 Two-dimensional normal distribution

Most important (continuous) distribution of a special type of distribution.
Density function of the two-dimensional normal distribution of (X, Y) with the parameters $m_x, m_y, \sigma_x^2, \sigma_y^2, \rho_{xy}$:

$$f(x, y) = \frac{1}{2\pi\sigma_x\sigma_y\sqrt{1-\rho_{xy}^2}} \times \exp\left\{ -\frac{1}{2(1-\rho_{xy}^2)} \left[\frac{(x-m_x)^2}{\sigma_x^2} - 2\rho_{xy}\frac{(x-x_m)(y-y_m)}{\sigma_x\sigma_y} + \frac{(y-m_y)^2}{\sigma_y^2} \right] \right\},$$

$$-\infty < x < \infty \quad , \quad -\infty < y < \infty.$$

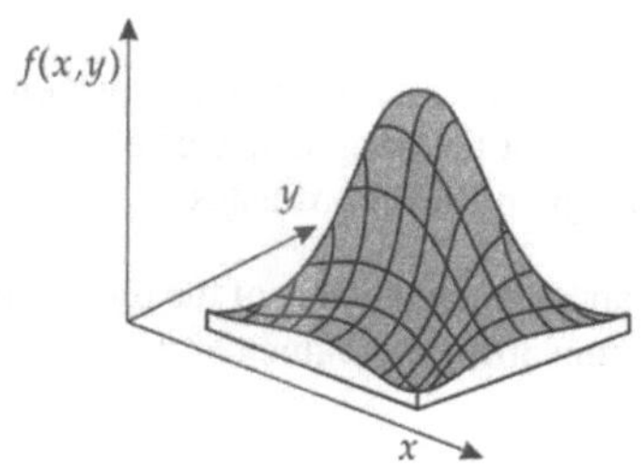

Density function of a
two-dimensional normal
distributed random value (X, Y)

▷ If the coordinate axes are placed parallel to the principal axes of the dispersion then $\rho_{xy} = 0$ and it holds that

$$f(x, y) = \frac{1}{2\pi\sigma_x\sigma_y} e^{-\frac{(x-x_m)^2}{2\sigma_x^2} - \frac{(y-y_m)^2}{2\sigma_y^2}}.$$

Calculation of probabilities by means of a standardized two-dimensional normal distribution

$$P(a \leq X < b, c \leq Y < d) = \left[F_{\text{sn}}\frac{b - m_x}{\sigma_x} - F_{\text{sn}}\frac{a - m_x}{\sigma_x}\right] + \left[F_{\text{sn}}\frac{d - m_y}{\sigma_y} - F_{\text{sn}}\frac{c - m_y}{\sigma_y}\right].$$

Dispersion grid: Graphical representation of the probability distribution in the $k\sigma$ limits.

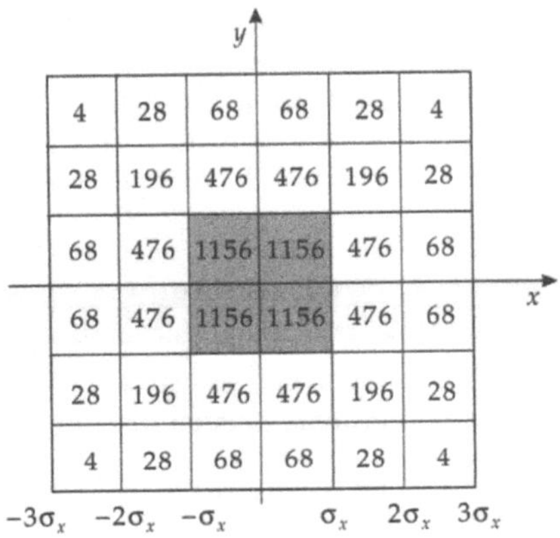

Dispersion grid

▷ The values in the squares of the dispersion grid divided by 100 show the probability (in %) that the random variable will assume a value from the corresponding square.

Approximate determination of probabilities by means of a dispersion grid by

- true-to-scale transfer of the interesting region for which the probability is to be considered, taking into account the congruence of expected values for the occurrence of the event and the origin of the coordinates,
- determination of the approximated value of the probability for the occurrence of the event by adding and converting the values of the covered squares in the dispersion grid or parts thereof.

21.7 Basics of mathematical statistics

As an example, quality control with **random sampling** provides reject rates that correspond to the exact number of rejects only with a **certain probability**. Statistics offer various methods which, for certain boundary conditions, allow statements about the **expected value** (mean value) and the **dispersion** (deviation from the mean value) of the random value to be considered (e.g., a sample or measurement/series of measurements), thus facilitating an **estimation** of the **error** relative to the actual value.

21.7.1 Description of measurements

Measured quantity, measured variable, characteristic: Terms which mean the property to be determined by a measurement, statistical survey, sampling, or random experiment, such as:

Discrete variables:

□ Numbers 1 to 6 on a die, sides of a coin (head or tail).

Continuous variables:

□ Measured values for the capacitance of a capacitor or the resistance of a resistor.

Nominal characteristic: Characteristic for which properties can be distinguished only by name.

□ Colors

Ordinal characteristic: Characteristic distinguished by a quantitative hierarchy.

□ Numbers on dice, frequency values of a high-frequency circuit.

Measuring result, measured value, actual value: The value of one or several measurable variables following a measurement; usually not exactly reproducible, but fluctuating about a **mean value** or **true value**.

▷ In many cases, this fluctuation (**measuring error**) can be described by the so-called **normal distribution.**

□ This could be, for example, the length of a screw in a manufacturing process, the result of a numerical random number generator, the energy of a particle in an ideal gas, or the rainfall within a 24-hour period.

Series of measurements: Compilation of several measured values, automatically producing a **primary list**.

Absolute frequency $\tilde{F}(K_i) = \tilde{F}_i$: Number of measured values in a series of measurements with a certain measured value $K_i = x_i$, or a certain **class** of measured values.

Frequency table, frequency distribution: Combination of measured values in various classes (e.g., intervals) K_i $(i = 1, \ldots, k)$, into which the measured values can be divided; tables for the map of such classes showing the number of measured values $\tilde{F}(K_i)$ they contain.

Class K_i: The set of several measuring results with certain properties that may be grouped under index i.

□ Measured values in various intervals, defective and nondefective units in a random sampling.

Histogram, bar graph: Graphical representation of a frequency table.

□ Sample, frequency table, and histogram of a die experiment:

x_1	x_2	x_3	x_4	x_5	x_6	x_7	x_8	x_9	x_{10}	x_{11}	x_{12}	x_{13}	x_{14}	x_{15}	x_{16}
4	6	1	4	3	5	6	3	1	5	2	6	3	1	4	6

Measured value x	Frequency $\tilde{F}(x)$
1	3
2	1
3	3
4	3
5	2
6	4

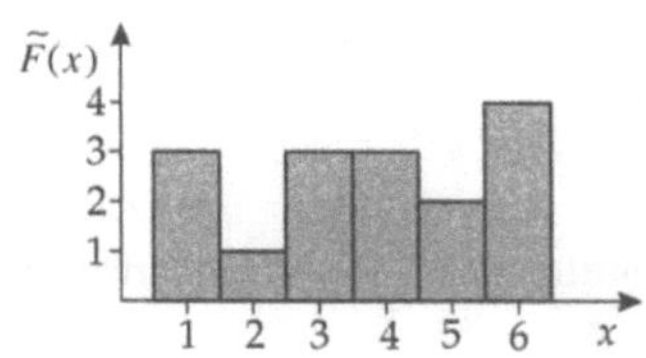

Relative frequency: Absloute frequency divided by the total number of measured values n:

$$\tilde{f}_i = \frac{\tilde{F}_i}{n} \Rightarrow \sum_{i=1}^{k} \tilde{f}_i = 1, \quad n = \sum_{i=1}^{k} \tilde{F}_i \,.$$

True value, desired value: Value about which all measuring results fluctuate. It is the result of measurement **without** error (ideal case).
Measuring error: Deviation of a measured value from the true value. We distinguish between **systematic** and **statistical** errors, depending on the cause.
Arithmetic mean value (arithmetic mean), empirical expected value: Approximate value for the true value of a series of measurements. Frequently, the equally weighted mean of n measured values is often given as:

$$\bar{x} = \frac{1}{n} \sum_{i=1}^{n} x_i \,.$$

▷ Often, the true value is equated to the mean value. But that is usually **not** correct; the mean value is only an **approximation** of the true value.
▷ To distinguish the empirical expected value of a measurement from the expected value of a **distribution**, a different notation is used. m is the expected value of a known particular distribution, $\bar{x}$ or $\bar{x}_n$ is the empirical mean resulting from a measurement, or also the estimated expected value.
▷ The parameters of the mean value can be defined in different ways; see **position parameter**.

Empirical standard deviation: Measure of the dispersion due to measuring errors. Fluctuation of the measured values about the true value,

$$\sqrt{\overline{(\Delta x)^2}} \equiv \sqrt{\frac{1}{n-1} \sum_{i=1}^{n} (x_i - \bar{x})^2} \,.$$

▷ In some cases, the standard deviation is also called variance.
▷ Dispersion measures can be defined in different ways; see **dispersion parameters**.

21.7.2 Types of error

Systematic errors: Deviations due, for example, to experimental uncertainties (e.g., a wrong calibration of the measuring instrument); can only be avoided to a certain extent.
Statistical or random errors: Deviations due to uncontrollable perturbations (the influence of temperature, change in atmospheric pressure, etc.) or by the randomness of the event (e.g., radioactive decay; in principle, statements about decay can only be made in terms of the probability at which a decay may occur within a given time).
True error Δx_{iw}: Deviation of i-th measurement from the "true value." Usually unknown, as in x_w

$$\Delta x_{iw} = x_i - x_w$$

Absolute error: Measuring error referring to an individual (i-th) measurement,

$$v_i = x_i - \bar{x} \,.$$

Apparent error:

$$v_i = x_1 - \overline{x}$$

Mean error, linear dispersion:

$$d_x = \bar{v_i} = \frac{1}{n}\sum_{i=1}^{n} |x_i - \bar{x}| .$$

Relative error, v_{rel}: Absolute error divided by the mean value; a nondimensional quantity:

$$v_{\mathrm{rel}} = \frac{v_i}{\bar{x}} = \frac{x_i - \bar{x}}{\bar{x}} .$$

Percentage error, $v_{\%}$: Relative error in percent, $v_{\%} = v_{\mathrm{rel}} \cdot 100\%$.
Absolute maximum error, $\Delta z_{\max}$: Upper error threshold of a quantity $z = f(x, y)$, depending on the parameters x and y containing errors,

$$\Delta z_{\max} = \left|\frac{\partial}{\partial x} f(\bar{x}, \bar{y}) \Delta x\right| + \left|\frac{\partial}{\partial y} f(\bar{x}, \bar{y}) \Delta y\right| .$$

Relative maximum error, $\Delta z_{\max}/\bar{z}$: Absolute maximum error divided by the mean value.
Mean error of the individual measurement, $\overline{\Delta x}$:

$$\sigma_n = \overline{\Delta x} = \sqrt{\frac{1}{(n-1)}\sum_{i=1}^{n}(x_i - \bar{x})^2}, \quad \bar{x} \text{ is the arithmetic mean value.}$$

Mean error of the mean value, $\overline{\Delta \bar{x}}$:

$$\bar{\sigma}_n = \overline{\Delta \bar{x}} = \sqrt{\frac{1}{n(n-1)}\sum_{i=1}^{n}(x_i - \bar{x})^2}, \quad \bar{x} \text{ is the arithmetic mean value.}$$

- The mean error $\overline{\Delta \bar{x}}$ of the mean value $\bar{x}$ is equal to the mean error $\overline{\Delta x}$ of the individual measurement x_i, divided by the root of the number of measurements:

$$\overline{\Delta \bar{x}} = \frac{\overline{\Delta x}}{\sqrt{n}} .$$

Error propagation in the individual measurement:

$$\overline{\Delta f(x_0, y_0)} = \left.\frac{\partial f(x, y)}{\partial x}\right|_{x_0, y_0} \overline{\Delta x} + \left.\frac{\partial f(x, y)}{\partial y}\right|_{x_0, y_0} \overline{\Delta y} .$$

Error propagation of the mean value error:

$$\overline{\overline{\Delta f(x_0, y_0)}} = \sqrt{\left(\left.\frac{\partial f(x, y)}{\partial x}\right|_{x_0, y_0} \overline{\Delta x}\right)^2 + \left(\left.\frac{\partial f(x, y)}{\partial y}\right|_{x_0, y_0} \overline{\Delta y}\right)^2} .$$

21.8 Parameters for describing distributions of measured values

21.8.1 Position parameter, means of series of measurements

Arithmetic mean, often simply called mean value: Equally weighted mean of n measured values

$$\bar{x} = \frac{1}{n}\sum_{i=1}^{n} x_i = \frac{1}{n}\sum_{j=1}^{k} \tilde{F}_j \cdot x_j = \sum_{j=1}^{k} \tilde{f} \cdot x_j \ ;$$

i.e., the n measured values are distributed over $k \le n$ different x_j-values at frequency $\tilde{F}_j$.

- **Centroid property**: The sum of deviations of the measured values of the primary list from the arithmetic mean is identically zero by definition.

$$\sum_{i}^{n}(x_i - \bar{x}) \equiv 0$$

- **Linearity of the arithmetic mean**:

$$\overline{(ax + b)} = a\bar{x} + b \ ,$$

where a, b are constants, x is the measured variable.

- **Quadratic minimum property**: The sum of the **square** of the distances of all measured values x_i from the mean value $\bar{x}$ is minimal:

$$\sum_{i}^{n}(x_i - \bar{x})^2 = \text{ minimum.}$$

▷ This property is a basic element of the **calculus of observation.**

- **Union of measurements**: The mean of the total measurement with N measured values is the sum of the mean values of the partial measurements, weighted by the relative portion of measured points $N_i / \sum N_i = N_i/N$:

$$\bar{x} = \sum \bar{x}_i \cdot \frac{N_i}{N} = \sum \bar{x}_i N_i / \sum N_i \ .$$

- If the series of measurements is in the form of a frequency distribution, then

$$\bar{x} = \frac{1}{\sum_i^k \tilde{F}_i} \sum_{i=1}^{k} x_i \tilde{F}_i \ .$$

In this case, the x_i are the mid-values of classes K_i $(i = 1, \ldots, k)$.

Quantile, percentile of order p: Measured value that is **not** exceeded by a portion p of all measured values from the primary list and that does **not** fall below a portion $1 - p$; parameter for describing the position of the individual measured values with respect to each other.

Empirical median, $\tilde{x}$: Special case of a **percentile**; the value that **bisects** the number of N measured values of the primary list arranged by magnitude.

Empirical median for an even number of measured values:

$$\tilde{x} = \frac{x_{N/2} + x_{N/2+1}}{2}$$

Median for an odd number of measured values:

$$\tilde{x} = x_{(N+1)/2}$$

▷ The empirical median is used mainly in the following cases:

(a) No classes at the boundaries of the ordered primal list.

(b) Extreme measured values ("mavericks") occur that would falsify the result.

(c) Changes in the measured values above and below the mean value must not influence its value.

- The sum of the absolute values of the deviations of all measured values x_i from the median $\tilde{x}$ is smaller than the sum of the deviations from every other value a:

$$\sum_{i=1}^{N} |x_i - \tilde{x}| < \sum_{i=1}^{N} |x_i - a| \quad \begin{cases} \text{for all} \quad a \neq \tilde{x}\,, & \text{if } N \text{ odd} \\ \text{for all} \quad x_{N/2} \leq a \leq x_{N/2+1}\,, & \text{if } N \text{ even}\,. \end{cases}$$

Frequency, absolute frequency $\tilde{F}_i \equiv \tilde{F}(X_i)$: The number of occurrences of $x = X_i$ in the series of measurements, where a measured variable x can take the same value X_i several times during a series of measurements. The set of the possible measured values X_j forms a classification of the measuring results, where $1 \leq j \leq k$ with $k \leq n$ and

$$\sum_{j}^{k} \tilde{F}(X_i) = n\,.$$

Empirical mode, most probable value x_m: The most common in a sequence of measured values.

Limiting points: Measured values with a frequency greater than that of their neighboring values.

▷ For series of measurements with several limiting points, there are also several most probable values. Every limiting region must be considered separately.

Quadratic mean:

$$x_{\text{quad}} = \sqrt{\frac{1}{n}\sum_{i=1}^{n} x_i^2}\,.$$

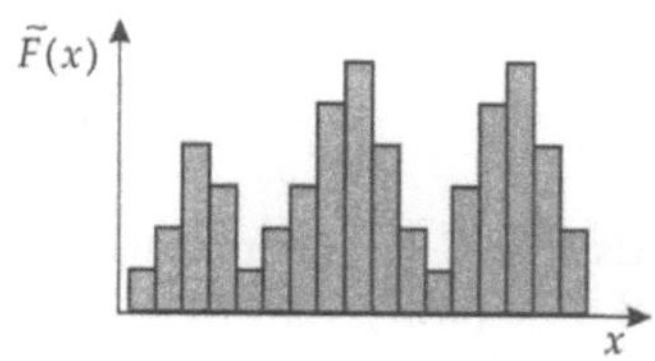

Geometric mean:

$$\hat{x} = \sqrt[n]{\prod_{i=1}^{n} x_i} = (x_1 \cdot x_2 \cdot \cdots \cdot x_n)^{(1/n)}\,.$$

▷ The geometric mean is used mainly for quantities in which geometric sequences occur as regularities.

□ **Mean average growth** rate or **rate of increase** of process in time, radioactive decay, life expectancy of components:

$$\hat{x} = (x_1 \cdot x_2 \cdot \cdots \cdot x_N)^{(1/N)}, \quad x_i > 0\,.$$

- The logarithm of the geometric mean is equal to the arithmetic mean of the logarithm of all measured values,

$$\ln \hat{x} = \frac{1}{n}(\ln x_1 + \cdots + \ln x_n)\,.$$

Growth rate: The average percentage expansion from x_n to x_{n+1} (in percent of total amount A, where $x_n = a_n/A$):

$$\overline{W} \equiv \sqrt[n-1]{\frac{x_n}{x_1}} \cdot 100\% .$$

Rate of increase: The average percentage development by $\bar{R}$ percent,

$$\bar{R} \equiv \left(\sqrt[n-1]{\frac{x_n}{x_1}} - 1\right) \cdot 100\% .$$

▷ If there is no percentage expansion exists, then the absolute values x_1, x_n can be substituted for $a_1 = x_1 \cdot A$, $a_n = x_n \cdot A$.

Harmonic mean:

$$x_h = N \Big/ \sum_{i=1}^{N} \frac{1}{x_i} .$$

● **Theorem of Cauchy**: There exists the following hierarchy of mean values x_{quad}, x_h, $\hat{x}$, and $\bar{x}$:

$$x_{\min} \leq x_h \leq \hat{x} \leq \bar{x} \leq x_{\text{quad}} \leq x_{\max} .$$

21.8.2 Dispersion parameter

Empirical range, range of variation, spread: Difference between the largest and the smallest measured values:

$$\Delta x_{\max} \equiv x_{\max} - x_{\min} .$$

▷ Mostly used when there is only a small number of measured values. Used in statistical quality control with control cards.

Empirical mean absolute deviation about value C:

$$\overline{|\Delta x|_C} = \frac{1}{N} \sum_{i=1}^{N} |x_i - C| .$$

▷ Usually, $C = \tilde{x}$ (empirical median) or $C = \bar{x}$ (arithmetic mean) are used.

▷ If a frequency table ordered according to classes exists, the class midpoints are inserted as measurable quantities x_i.

Empirical mean square deviation, standard deviation, empirical dispersion:

$$\sigma_n = \sqrt{\overline{(\Delta x)^2}} := \sqrt{\frac{1}{n-1} \sum_{i=1}^{n} (x_i - \bar{x})^2} .$$

● If the series of measurements is presented in the form of a frequency distribution, then:

$$\sigma_n = \sqrt{\overline{(\Delta x)^2}} := \sqrt{\frac{1}{n-1} \sum_{i=1}^{k} (x_i - \bar{x})^2 \tilde{F}(x_i), \quad n = \sum \tilde{F}(x_i)} .$$

▷ In the case of a division into classes, the class midpoints are often used instead of unknown measured values.

Empirical variance, σ_n^2: The square of the standard deviation, usually referred to as **variance**.

▷ The empirical dispersion σ_n is an **unbiased** estimate of the dispersion of an underlying probability function over the universe; see **sampling methods**.

Relative measure of dispersion, variation coefficient, variability coefficient: Percentage of the dispersion measure, related to the arithmetic mean:

$$\overline{(\Delta x)^2}_{\text{rel}} \equiv \frac{\overline{(\Delta x)^2}}{\bar{x}} \cdot 100\%$$

21.9 Special distributions

21.9.1 Frequency distributions

Prime notation: List of all measured values in a series of measurements; the same measuring results can occur repeatedly.

□ In the production of N capacitors with a capacitance $C = 100\ \mu\text{F}$, the value of each component is usually not exactly $100\ \mu\text{F}$, but fluctuates about this value. We say: the value satisfies a characteristic distribution about the desired value $C = 100\ \mu\text{F}$. To understand this kind of distribution and the nature of the underlying probability process in greater detail, we determine the so-called **relative frequency distribution** and compare it with the special probability functions derived from known probability structures. (For example, hypergeometric distribution can be traced back to the simple **urn model**.)
In our example, the individual measurable quantity is the capacitance of every capacitor. These measured values form the so-called **primary list**:

Capacitor number	1	2	3	4	5	6	...	N
Capacitance in μF	101.1	99.6	101.4	103.3	98.0	99.5	...	C_N

Class K_i: Set of several elements (measuring results) of a primary list with certain properties grouped under index i.

□ In the daily production of n capacitors of a given capacitance C a classification can be made by dividing the capacitance into $N = 8$ interval ranges ($N = 8$ classes).

Class	Interval Limits	Class	End-point of interval
K_1	$C < 92.5$	K_5	$100.0 \le C < 102.5$
K_2	$92.5 \le C < 95.0$	K_6	$102.5 \le C < 105.0$
K_3	$95.0 \le C < 97.5$	K_7	$105.0 \le C < 107.5$
K_4	$97.5 \le C < 100.0$	K_8	$107.5 \le C$

▷ It is not always necessary to define classes. In the case of discrete measured values $x = X_i$ repeated in the primary list, these can of course, be regarded as a class of their own $K_i = X_i$.

Class midpoint, interval midpoint: Arithmetic mean of interval limits of a class.

▷ It is more exact to form the arithmetic mean of all measured values within a class. But the individual measured values are sometimes not known, or their determination could take too much time and effort. Therefore, the midpoint of the interval is usually a good approximation.

Frequency $\tilde{F}_i := \tilde{F}(K_i)$: The number of measuring results from the primary list in the class K_i.

▷ For measured values repeatedly occurring in the primary list, a discrete measured value can also be regarded as a class!

Frequency table: Representation of each class by showing the corresponding number (**frequency**) of measured values in the form of a table.

□ Frequency table of daily production, related to the capacitance of capacitors, could look as follows:

K_i	K_1	K_2	K_3	K_4	K_5	K_6	K_7	K_8	Sum
$\tilde{F}(K_i)$	133	43789	189345	281321	255128	206989	26923	155	1003783

Frequency distribution, frequency histogram: Graphical representation of a frequency table.

Relative frequency: Relative frequency of class K_i for n measured values:

$$\tilde{f}_i = \frac{\tilde{F}_i}{n} .$$

Relative or normalized frequency distribution $\tilde{f}_i$:

$$\sum_{i=1}^{n} \tilde{f}_i = 1 .$$

The relative frequency can be represented graphically in terms of a histogram.

▷ Other diagrams such as pie graphs are often used to provide a clear representation.

□ Histogram and pie graph for the above example:

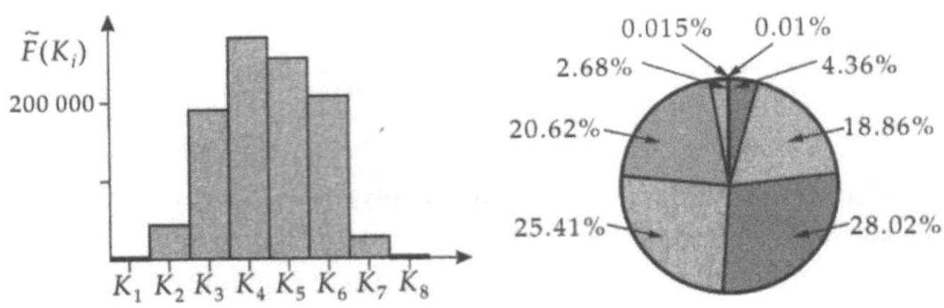

● When we divide the (relative) frequency by a constant factor c, the arithmetic mean remains:

$$\frac{\sum_i^n x_i \cdot \tilde{F}(x_i)/c}{\sum_i^n \tilde{F}(x_i)/c} \equiv \bar{x} .$$

21.9.2 Distribution of random sample functions

Sample function: Mapping of n measured values $x_1, \ldots, x_n$ of a random sample onto a value $W_n(x_1, \ldots, x_n)$ used to estimate or check a property of the universe. For example, the value of a sample function W_n may be just the approximation for a **distribution parameter**:

$$W = \lim_{n\to\infty} W_n \,.$$

Distribution of a sample function $f(W_n)$: Results from the multiple repetition of the selection process, such as the repeated drawing of n lots; probability density for the value of random sample function W_n.

χ^2**-function**: The summed square of n measured values of a random sample,

$$\chi^2(x_1, \ldots, x_n) = \sum_i^n x_i^2 \,.$$

χ^2**-distribution (Helmert-Pearson)**: A distribution $f_\chi(Y_n; n)$, which for the measured quantity $Y_n(x_1, \ldots, x_n)$ is

$$W_n = Y_n(x_1, \ldots, x_n) := \chi^2 = \sum_{i=1}^n x_i^2$$

if the individual measured values x_i $(i = 1, \ldots, n)$ are distributed according to a **standard normal distribution**,

$$f(x_i) = f_{sn}(x_i) \,.$$

Probability density:

$$f_\chi(Y_n; n) = \frac{1}{2^{n/2}\Gamma(n/2)} Y_n^{n/2-1} \mathrm{e}^{-Y_n/2} \,.$$

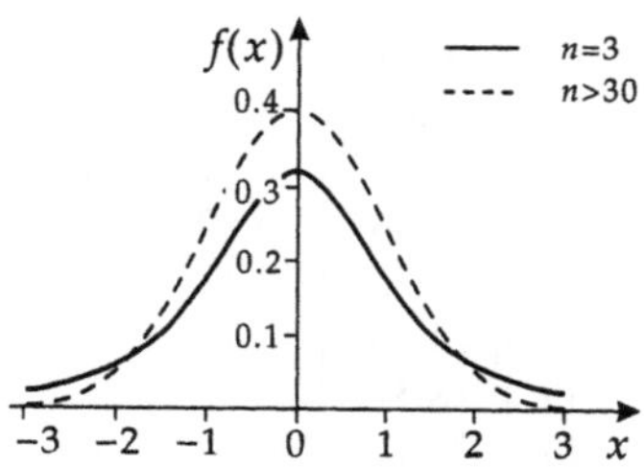

Distribution function:

$$F_\chi(Z_n; n) = \int_{-\infty}^{Z_n} f(Y_n)\mathrm{d}Y_n \,.$$

The parameter of distribution is the random sample size n.
Expected value: $m = n$.
Variance: $\sigma^2 = 2n$.

▷ If the n measured values x_i satisfy a more general normal distribution about the known expected value m with variance σ^2, then they may also be substituted by

$$\frac{x_i - m}{\sigma} \,.$$

The distribution of Y_n is also $f_\chi(Y_n; n)$.

▷ If the expected value m is **not** known, then it can be substituted by the arithmetic mean of sample $\bar{x}$. But then the distribution of Y_n is $f_\chi(Y_n; n-1)$.

▷ The value

$$Y_n/n = \frac{1}{n}\sum_i^n (x_i - m)^2$$

is an estimated function for the variance (root-mean-square deviation) if x is normally distributed about m.

▷ The percentiles of the χ^2 distribution $t_{\chi;\alpha;n}$ with

$$F_\chi(t_{\chi;\alpha;n}) = \alpha$$

are used to define the prediction and confidence intervals in the interval of estimation of variance parameters σ in normal distributions by means of the **estimation theory**. Here, Y_n/n serves as the estimator.
Analogously, in the **theory of testing** such percentiles are also used to determine the validity of an estimated σ^2-value.

Percentile $t_{\chi;F_\chi;n}$ of the χ^2 distribution f_χ

$n\backslash F_\chi$	0.01	0.025	0.05	0.1	0.9	0.95	0.975	0.99
1	0.000	0.000	0.004	0.016	2.71	3.84	5.02	6.63
2	0.020	0.051	0.103	0.211	4.61	5.99	7.38	9.21
3	0.115	0.216	0.352	0.584	6.25	7.81	9.35	11.35
4	0.297	0.484	0.711	1.064	7.78	9.49	11.14	13.28
5	0.554	0.831	1.15	1.61	9.24	11.07	12.83	15.08
6	0.872	1.24	1.64	2.20	10.64	12.59	14.45	16.81
7	1.24	1.69	2.17	2.83	12.01	14.06	16.01	18.47
8	1.65	2.18	2.73	3.49	13.36	15.51	17.53	20.09
9	2.09	2.70	3.33	4.17	14.68	16.92	19.02	21.67
10	2.56	3.25	3.94	4.87	15.99	18.31	20.48	23.21
11	3.05	3.82	4.57	5.58	17.27	19.67	21.92	24.72
12	3.57	4.40	5.23	6.30	18.55	21.03	23.34	26.22
13	4.11	5.01	5.89	7.04	19.81	22.36	24.74	27.69
14	4.66	5.63	6.57	7.79	21.06	23.68	26.12	29.14
15	5.23	6.26	7.26	8.55	22.31	25.00	27.49	30.58
16	5.81	6.91	7.96	9.31	23.54	26.30	28.85	32.00
17	6.41	7.56	8.67	10.09	24.77	27.59	30.19	33.41
18	7.01	8.23	9.39	10.86	25.99	28.87	31.53	34.81
19	7.63	8.91	10.12	11.65	27.20	30.14	32.85	36.19
20	8.26	9.59	10.85	12.44	28.41	31.41	34.17	37.57
25	11.52	13.12	14.61	16.47	34.38	37.65	40.65	44.31
30	14.95	16.79	18.49	20.60	40.26	43.77	46.98	50.89
35	18.51	20.57	22.46	24.80	46.06	49.80	53.20	57.34
40	22.17	24.43	26.51	29.05	51.81	55.76	59.34	63.69
50	29.71	32.36	34.76	37.69	63.17	67.51	71.42	76.15
60	37.49	40.48	43.19	46.46	74.40	79.08	83.30	88.38
70	45.44	48.76	51.74	55.33	85.53	90.53	95.02	100.4
80	53.54	57.15	60.39	64.28	96.58	101.9	106.6	112.3
90	61.75	65.65	69.13	73.29	107.6	113.2	118.1	124.1
100	70.07	74.22	77.93	82.36	118.5	124.3	129.6	135.8

***t* distribution, Student's distribution**: Distribution $f_T(T_n; n)$ resulting for the measurable quantity

$$T_n \equiv \frac{Z}{\sqrt{Y_n/n}}$$

if Z satisfies the standard normal distribution and Y_n satisfies the $f_\chi(Y_n; n)$-distribution.
Probability density:

$$f_T(T_n; n) = \frac{\Gamma((n+1)/2)}{\sqrt{n\pi}\,\Gamma(n/2)}\left(1 + \frac{T_n^2}{n}\right)^{-(n+1)/2} .$$

Distribution function:

$$F_T(U_n; n) = \int_{-\infty}^{U_n} f_T(T_n) dT_n \, .$$

The parameter of distribution is the random sample size n.
Expected value: $m = 0$.
Variance: $\sigma^2 = n/(n-2)$.

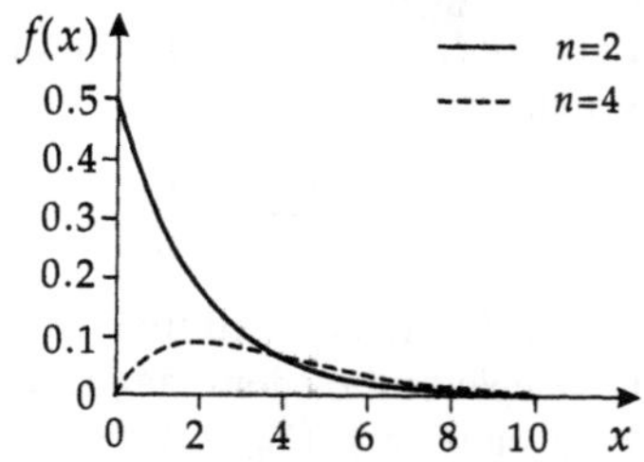

Percentile $t_{T;F_T;n}$ of the t distribution f_T					
$n \backslash F_T$	0.9	0.95	0.975	0.99	0.995
1	3.078	6.314	12.71	31.82	63.66
2	1.886	2.920	4.303	6.965	9.925
3	1.638	2.353	3.182	4.541	5.841
4	1.533	2.132	2.776	3.747	4.604
5	1.476	2.015	2.571	3.365	4.032
6	1.440	1.943	2.447	3.143	3.707
7	1.415	1.895	2.365	2.998	3.499
8	1.397	1.860	2.306	2.896	3.355
9	1.383	1.833	2.262	2.821	3.250
10	1.372	1.812	2.228	2.764	3.169
11	1.363	1.796	2.201	2.718	3.106
12	1.356	1.782	2.179	2.681	3.055
13	1.350	1.771	2.160	2.650	3.012
14	1.345	1.761	2.145	2.624	2.977
15	1.341	1.753	2.131	2.602	2.947
16	1.337	1.746	2.120	2.583	2.921
17	1.333	1.740	2.110	2.567	2.898
18	1.330	1.734	2.101	2.552	2.878
19	1.328	1.729	2.093	2.539	2.861
20	1.325	1.725	2.086	2.528	2.845
25	1.316	1.708	2.060	2.485	2.787
35	1.306	1.690	2.030	2.438	2.724
50	1.299	1.676	2.009	2.403	2.678
100	1.290	1.660	1.984	2.364	2.626
200	1.286	1.653	1.972	2.345	2.601
500	1.283	1.648	1.965	2.334	2.586

▷ The **standardized arithmetic mean of a sample** as a variable

$$T_s = \frac{\bar{x} - m}{\sigma_n}\sqrt{n} = \frac{\bar{x} - m}{\sqrt{[1/(n-1)]\sum_i^n (x_i - \bar{x})^2}}\sqrt{n}$$

satisfies the $f_T(T_s; n-1)$ distribution.

▷ For as $n \to \infty$, the t distribution can be substituted by the standard normal distribution.

Fisher's distribution, F-distribution: Distribution $f_F(F; n_1, n_2)$ resulting for a measurable quantity

$$F_{n_1,n_2} := \frac{Y_{n_1}(1)/n_1}{Y_{n_2}(2)/n_2}$$

if Y_{n_1} and Y_{n_2} are obtained from two mutually independent random sample draws, and both satisfy a $f_\chi(Y_{n_i}; n_i)$-distribution $(i = 1, 2)$.

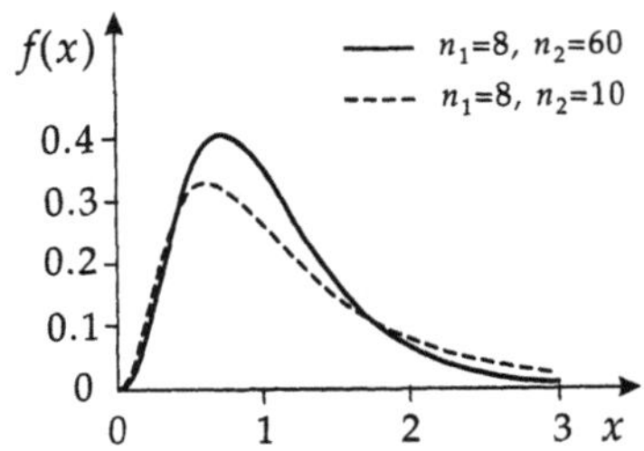

Probability density:

$$f_F(F_{n_1,n_2}; n_1, n_2) = \left\{\Gamma\left(\frac{n_1+n_2}{2}\right) \Big/ \left[\Gamma\left(\frac{n_1}{2}\right)\cdot\Gamma\left(\frac{n_2}{2}\right)\right]\right\}\left(\frac{n_1}{n_2}\right)^{n_1/2}$$
$$\times \frac{F^{n_1/2-1}}{[1+(n_1F/n_2)]^{(n_1+n_2)/2}} \quad , \quad F \ge 0 .$$

Distribution function:

$$F_F(\tilde{F}_{n_1,n_2}; n_1, n_2) = \int_0^{\tilde{F}} f_F(F; n_1, n_2)\mathrm{d}F .$$

The sizes n_1 and n_2 of the two random samples n_1 and n_2 are the **numerator parameter** and the **denominator parameter** (**degrees of freedom**) of the F distribution.
Expected value: $m = n_2/(n_2 - 2)$ $(n_2 \ge 3)$.
Variance: $\sigma^2 = [2n_2^2(n_1 + n_2 - 2)]/[n_1(n_2 - 2)^2(n_2 - 4)]$ $(n_2 \ge 5)$.

▷ For the special configuration $n_1 = 1$ and $n_2 = N$ the F-distribution satisfies the $\sqrt{F}$ $f_T(\sqrt{F}; N)$ distribution.

▷ For the special configuration $n_1 = N$ and $n_2 \ge 200\ n \cdot F$ the F distribution asymptotically satisfies (i.e., as $n_2 \to \infty$) the $f_\chi(n \cdot F; N)$-distribution.

● The F-percentiles $t_{F;\alpha}$ are defined by

$$F_F(t_{F;\alpha;n_1,n_2}) = \alpha .$$

They have the symmetry

$$t_{F;\alpha;n_1,n_2} = [t_{F;1-\alpha;n_2,n_1}]^{-1} .$$

21.10 Analysis by means of random sampling (theory of testing and estimating)

Methods for determining certain attributes by taking random samples from a distribution of characteristic.
Random sampling, statistical survey: A small set n is selected from a totality or universe N. Example: From a total population or universe $N = 600{,}000$ inhabitants, a smaller set $n = 3000$ inhabitants is selected whose characteristics are examined. Thus, the results from n are representative for the universe N.

▷ Another example of random sampling is the recording of a series of measurements pertaining to a measurable quantity that can be physically observed, or pertaining to a technical quantity. In such cases, the universe has an extent of $N = \infty$.

▷ In the case of an opinion poll, the "characteristics" consist of the opinions of the persons interviewed. They are usually **nominal characteristics.**

Random sample size: n units selected from N, the total set (universe) of all existing units.

Random sample attribute: The characteristics for which the units of the random sample are surveyed.

□ Such attributes may include: defective/non-defective pieces; the colors red, green or blue; the position of a sample function (**critical range, acceptance range**); opinions in an opinion poll.

Random sample variable: The generic term for quantities whose value is determined by means of information derived from the sample.

□ For example, sample variables may include attributes and estimation functions or test functions.

Parameter of a special distribution W: Value that defines the shape and/or position of a special distribution. The notation is

$$f(x) = f(x; W) ,$$

i.e., f is a special probability function (probability density) for the random variable x with parameter W.

□ Mean value m and variance σ^2 are parameters of normal distribution;
sample size n is a parameter of the χ^2 and the t distribution;
the fraction of marked units p is a parameter of the hypergeometric, the binomial, and the Poisson distributions.

21.10.1 Estimation methods

Estimation methods: Methods for estimating the parameters of a distribution (e.g., coefficients of measure) of random variables by sampling.

Point estimation: Estimation of a value for a sample function by sampling.

Sample function: Mapping of n samples $x_1, \ldots, x_n$ onto a value $S_n(x_1, \ldots, x_n)$; random variable of a sample.

□ Simple examples are the number of defective pieces in the production of goods or the mean value of a series of measurements of a physical quantity.

▷ Common examples of sample functions are estimators and test functions.

Estimators: Mapping of n samples onto an **estimate** $W_n(x_1, \ldots, x_n)$ for a parameter W of an assumed distribution; a special case of a sample function.

□ Examples of estimators are the arithmetic mean of a sample of size n

$$W_n = \bar{x}_n = \frac{1}{n}\sum_i^n x_i$$

and the empirical variance

$$W_n = \sigma_n^2 = \frac{1}{n-1}\sum (x_i - \bar{x})^2 .$$

Realization of an estimator: Value of an estimator after sampling.

□ Drawing of lots, throwing dice, taking a series of measurements, etc.

Unbiased estimation function: Estimator whose expected value for several samples is given by the real value of the required parameters:

$$\lim_{n\to\infty} W_n = W, \quad \text{or} \quad m(W_n) = W \quad \text{for all } n\,.$$

Median estimating function: Estimator with equal probabilities for underestimating and overestimating:

$$P(W_n \le W) = 0.5\,.$$

Simply consistent estimator: Estimation function for which

$$\lim_{n\to\infty} P(|W_n - W| \le \varepsilon) = 1$$

$\varepsilon > 0$ ($\varepsilon \to 0$); i.e., $n \to \infty$ value W_n derived from the sample tends to the actual parameter value W, and the variance (dispersion) tends to zero, $\sigma^2 \to 0$.

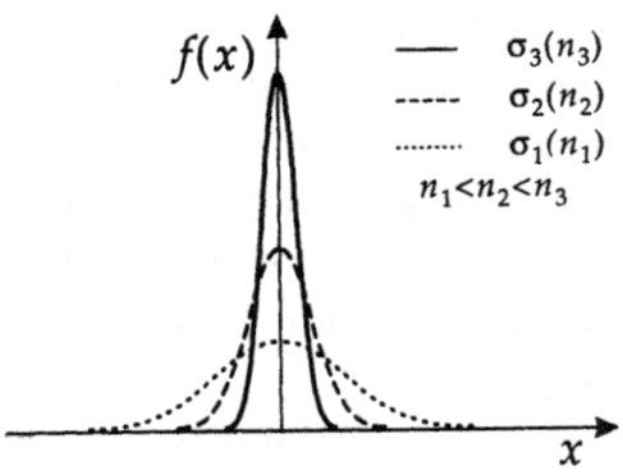

□ For example, the arithmetic mean $\bar{x}$ of the sample drawn from a normally distributed universe tends to the expected value m of a normal distribution.

Consistency in the mean square error:

$$\lim_{n\to\infty} m((W_n - W)^2) = 0\,.$$

▷ The consistency in the mean square error implies simple consistency, but the reverse is not necessarily true!

Mean squared error of W_n: $\overline{W} = \overline{(W_n - W)^2}$.

Efficiency: Of two unbiased estimators $W_n(1)$ and $W_n(2)$ the more **efficient** is the function whose variance is smaller for a given sample size n:

$$\sigma^2(W_n, 1) < \sigma^2(W_n, 2) \Rightarrow \quad W_n(1) \text{ is more efficient,}$$

$$\sigma^2(W_n, 2) < \sigma^2(W_n, 1) \Rightarrow \quad W_n(2) \text{ is more efficient;}$$

i.e., the expected dispersion about the true value W is smaller.

Absolutely efficient estimator: Estimator with the smallest variance.

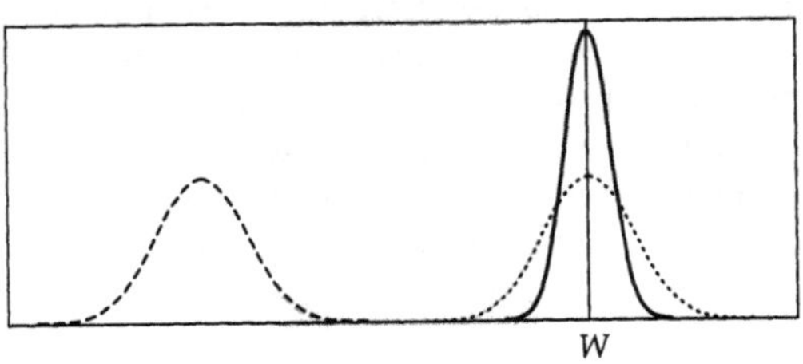

Estimators: absolutely efficient: —
unbiased:······ biased: - - - -

● **Inequality of Rao and Cramer**: If W_n is any estimator for the distribution parameter W, and if the universe is $f(x; W)$ or $P(x; W)$ distributed, then it holds that:

$$\sigma^2(W_n) \ge 1/[n\,\langle(\partial \ln f(x; W)/\partial W)\rangle]\,,$$

where x is continuous; otherwise, it is

$$\sigma^2(W_n) \geq 1/\left[n \left\langle (\partial \ln P(x; W)/\partial W) \right\rangle\right],$$

where x is discrete.

Sufficiency: An estimator is **sufficient** if it takes into account all of the information of the n samples x_i, $(i = 1, 2, \ldots, n)$.

▷ Only very few estimators are sufficient. For example, the median is **not** sufficient because it only takes into account the value **in the middle** of an ordered sample.

▷ Estimators calculated according to the **maximum likelihood method** are sufficient.

Factorization criterion: For sufficiency, the following criterion is necessary and sufficient

$$\begin{matrix} f(x_1, \ldots, x_n; W) \\ P(x_1, \ldots, x_n; W) \end{matrix} = g(W_n; W) \cdot \tilde{g}(x_1, \ldots, x_n);$$

i.e., the probability for the special sample configuration $x_1, \ldots, x_n$ is the product of the probability distribution of the estimating function g, depending on parameter W, and a parameter-independent distribution $\tilde{g}$, the sample configuration.

BAN (best asymptotically normal distributed estimator): Estimator W_n for parameter W with the characteristic that, for large n $(n \to \infty)$, W_n converges to a normal distribution with $\langle W_n \rangle = W$ and with the lower bound of the variance according to the Rao-Cramer inequality.

□ Maximum likelihood estimators are BAN estimators.

BLU (best linear unbiased estimator): Estimator W_n for parameter W that is unbiased, absolutely efficient (lowest variance), and is a linear function of the x_i.

21.10.2 Construction principles for estimators

Construction principle: Procedure to determine suitable estimators; depending on procedure and application, we distinguish moment, maximum likelihood, χ^2, and percentile estimators.

21.10.3 Method of moments

Moment of a distribution of a measured value or a sample: Parameters of a frequency distribution.

k-th moment: $m_k := (1/n) \sum_i^n x_i^k$.

k-th central moment: $mz_k := (1/n) \sum_i^n (x_i - \bar{x})^k$.

Moment estimator: The moments of special probability functions can usually be related to their parameters. Parameters can be expressed in terms of moments, leading to an estimator for the unknown parameters.

□ Exponential distribution: $W = m = \dfrac{1}{\lambda} \Rightarrow W_n(\lambda) = 1/\left[(1/n) \sum_i^n x_i\right]$.

Gaussian distribution: $W = \sigma^2$, $W = m \Rightarrow W_n = \sigma_n^2$, $W_n = \bar{x}_n$.

▷ General characteristics of moment estimators:

- always consistent,

- at least asymptotically unbiased,

- always asymptotically normally distributed,

- often not absolutely efficient,
- often not sufficient.

21.10.4 Maximum likelihood method

Maximum likelihood method: Method for determining estimators and their parameters.
Likelihood function $L(a)$: The probability or probability density for the occurrence of a sample configuration $x_1, \ldots, x_n$, provided that the probability density $f(x; a)$ with parameter a is known,

$$L(a) := f(x_1; a) \cdot \cdots \cdot f(x_n; a) .$$

Maximum likelihood estimator $\tilde{a}(x_1, \ldots, x_n)$: The value of the parameter a for which $L(a)$ is maximal. It is defined by the condition

$$\frac{\mathrm{d}\ln L(a)}{\mathrm{d}a} \equiv 0 \quad \text{or} \quad \frac{\mathrm{d}L(a)}{\mathrm{d}a} \equiv 0 ,$$

and by inversion to a function of the sample variables $x_1, \ldots, x_n$.

▷ Taking the logarithm is often chosen because this simplifies the defining equation for $\tilde{a}$.

▷ For several parameters, the defining equations result from the system of equations

$$\frac{\partial \ln L(a_1, \ldots, a_n)}{\partial a_i} \equiv 0 \quad i = 1, \ldots, n .$$

□ Thus, for the estimation of the expected value of a Poisson distribution, we obtain the estimating function

$$W_n = \frac{1}{n} \sum_{i=1}^{n} x_i = \bar{x}_n .$$

▷ Properties of the maximum likelihood estimator:

–consistent,

–at least asymptotically unbiased,

–at least asymptotically absolutely efficient,

–sufficient,

–best asymptotically normally distributed (BAN) estimator.

21.10.5 Method of least squares

Method of least squares: The parameters a of given function approach $y(x; a)$ that optimally describe selection of two-dimensional measuring points (measuring curves) (x_i, y_i). See **calculus of adjustment**.

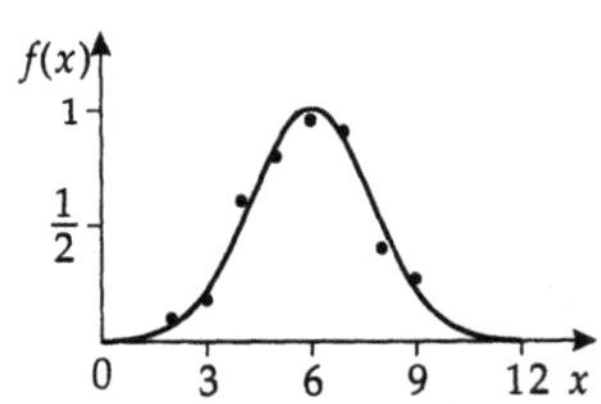

Minimal principal of least squares according to Gauss: For a given first approach of the function $y(x)$ and n given pairs of values (x_i, y_i), the optimal parameter configuration of the approximation function is defined by

$$\left\{\sum_{i}^{n}(y_i - y(x_i; a))^2\right\} = \text{Min}_a ,$$

where a is the parameter of the first approach.

Least squares estimator: If the **normal equations** of the minimum are solved, we obtain estimators for the parameter of the first approach.

Characteristics for a linear first approach of the estimator/estimating line:

–linear and unbiased.

21.10.6 χ^2-minimum method

χ^2**-function**: A function for the determination of the best parameter value for a given type of distribution; for a sample of size n the sample variables x_i, $i = 1, \ldots, n$ are combined in a table of relative frequencies $\tilde{f} = \tilde{f}(K_j)$ with a given classification K_j, $j = 1, \ldots, k$; then, the χ^2 function reads

$$\chi^2 \equiv \left\{\sum_{j}^{k} \frac{\tilde{f} - f(X_j; a)}{f(X_j; a)}\right\} ,$$

where X_j denotes the **midpoint** of the interval of the j-th class.

▷ The χ^2-function measures the deviation of the relative frequency distribution $\tilde{f}_i$ from an ideal distribution $f(x; a)$ for a given parameter value a.

▷ Division into classes is usually unnecessary for discrete measured values; these are used instead of the midpoints of the intervals.

χ^2**-minimum principle**: The method of least squares for determining the statistical distribution of measured values; as with the maximum likelihood estimator, the minimum of the χ^2-function with respect to the variable a defines the parameter estimating value $\tilde{a}$ and thus the estimator $f(x; \tilde{a})$ for a given type of distribution $f(x; a)$.

$$\chi^2 := \text{Min}_a .$$

▷ The χ^2-method can also be used for maximum likelihood distributions which depend on several parameters $a_1, \ldots, a_k$.

▷ The χ^2-method is also used for **testing procedures** in which even the **type** of the distribution is **unknown**.

● Asymptotically (for a large sample size n) the χ^2-estimator and the maximum likelihood estimator coincide.

21.10.7 Method of quantiles, percentiles

The quantile (percentile) x_α of the order α (distribution function $F(x_\alpha := \alpha)$) of the total universe is estimated by the percentile of sample X_α.

▷ Properties of the percentile estimator:

– unbiased,

- asymptotically normally distributed,
- seldom absolutely efficient.

● The variance of the percentile estimators is

$$\sigma^2(X_\alpha) = \frac{\alpha(1-\alpha)}{n(f(X_\alpha))^2} .$$

21.10.8 Interval estimation

To give the goodness of a point estimation, two types of intervals are defined: **Fluctuation** or **predicting, interval** and the **confidence interval**.

□ A typical example of an interval estimation is quality control: a consignment of goods with $N = 10{,}000$ electronic components must be inspected to determine what portion $p = N_{\text{error}}/N$ is defective; the fraction of the defective devices must be determined. Of course, not all components are tested, but only a small sample e.g., $n = 20$. The rate of rejects p is estimated by means of this sampling as a point estimation. Object of the interval estimation is to determine the probability at which p falls into a certain interval, i.e., the goodness of the estimation.

Realization of an estimating function, w_n: The value of the estimating function W_n after a single sampling of size n.

□ In the above example of quality control, the probability for the number of defective components is given by the **hypergeometric distribution**. The estimation of the proportion of rejects p among all N pieces from the rejected portion in the random sample w_n is the value w_n of the realization of the estimating function $W_n = p_n = n_{\text{error}}/n$.

Critical region K_W: Interval into which the realization w_n of an estimating function W_n falls with probability α. The characteristic "**critical**" is connected with the theory of testing in which a hypothesis is **rejected** as soon as the realization of an estimating/testing function falls into this region. Usually, $\alpha \ll 1$.

□ The probability of obtaining a maximum of k rejects is precisely the cumulative probability or the so-called **cumulative hypergeometric distribution**. If p_n is greater than a given limit of $\tilde{p}$, then the consignment of goods should be rejected. Thus, the interval $p > \tilde{p}$ defines the critical range.

Prediction interval, fluctuation interval, acceptance region: The range into which the realization w_n of an estimating function W_n falls with probability $(1-\alpha)$. The prediction interval is the range in which the realization of an estimating function is found with high probability.

□ Analogously to the above example, the range $p \leq \tilde{p}$ is precisely the prediction interval or **acceptance region**, since in this case the consignment of goods is accepted; see testing methods.

● The prediction interval is complementary to the critical range.

Proposition probability: The probability $(1-\alpha)$ with which the realization w_n falls into the prediction interval.

Boundary point of the prediction interval $v_{n;\alpha}$: The **percentile** of order α of a distribution of the estimation value w_n.

We distinguish three types:

(a) **Two-sided**:

$$v_{n;\alpha/2} \leq W_n \leq v_{n;1-\alpha/2}, \quad P(v_{n;\alpha/2} \leq W_n \leq v_{n;1-\alpha/2}) = 1-\alpha .$$

(b) **One-sided upper**:

$$W_n \leq v_{n;1-\alpha}\,, \quad P(W_n \leq v_{n;1-\alpha}) = 1-\alpha\,.$$

(c) **One-sided lower**:

$$W_n \geq v_{n;\alpha}\,, \quad P(W_n \geq v_{n;1-\alpha}) = 1-\alpha\,, \quad \text{bounded prediction interval.}$$

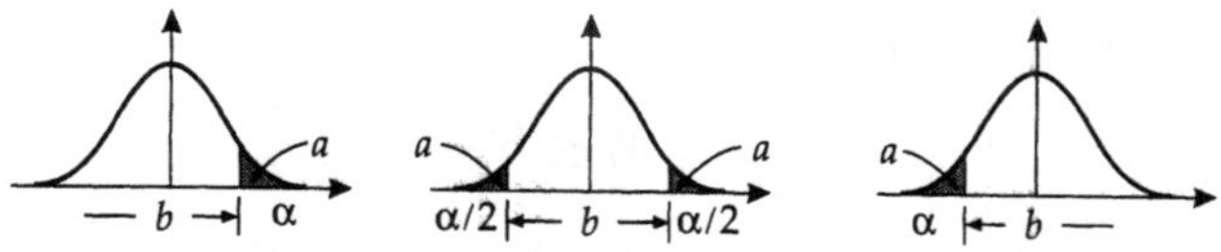

a: critical range, b: prediction interval

Confidence interval: The interval **about the realization** w_n of W_n, which includes the parameter W with the probability $(1-\alpha)$ (**confidence level**).
Boundary point of the confidence interval $V_{n;\alpha}$: The bounds of the interval about a realization w_n of W_n within which the parameter W is contained with probability $(1-\alpha)$.

▷ In contrast to the prediction interval, whose position and width is fixed, the confidence interval changes its location and width depending on the value of the realization w_n.

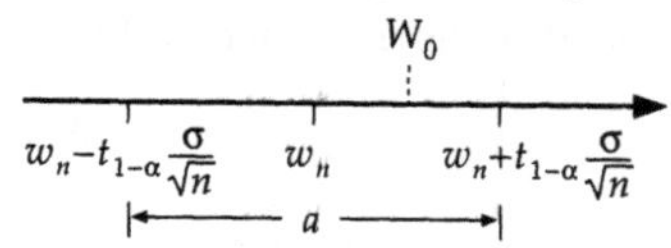

Confidence interval for W_0, α

Confidence level: Probability $(1-\alpha)$ with which the confidence interval about a realization of W_n, i.e., about a value w_n, includes parameter W.
Rule for constructing the confidence bounds $V_{n;\alpha}$: The bounds of the prediction interval $v_{n;\alpha}$ are usually reversibly unique one-to-one functions of the parameter value W of an assumed type of distribution of the universe $f(x, W)$.
After the inversion:

$$v_{n;\alpha} = y = f(x) \to x = f^{-1}(y) = v_{n;\alpha}^{-1} = V_{n;1-\alpha}$$

the **upper** bound of prediction becomes a **lower** bound of confidence interval due to the inversion of the inequality

$$W_n \leq v_{n;1-\alpha/2}(W) \quad \to \quad V_{n;\alpha/2} := v_{n;1-\alpha/2}^{-1}(W_n) \leq W\,.$$

Analogously, we obtain an **upper** bound of confidence $V_{n;1-\alpha/2}$ by the **lower** bound of prediction,

$$v_{n;\alpha/2}(W) \leq W_n \quad \to \quad W \leq V_{n;1-\alpha/2} := v_{n;\alpha/2}^{-1}(W_n)\,,$$

and thus:

$$V_{n;\alpha/2}(W_n) := v_{n;1-\alpha/2}^{-1}(W_n) \leq W \leq V_{n;1-\alpha/2}(W_n) := v_{n;\alpha/2}^{-1}(W_n)\,.$$

One-sided upper/lower bounds of confidence:

$$W \leq V_{n;1-\alpha} := v_{n;\alpha}^{-1}(W_n)\,,$$
$$V_{n;\alpha} := v_{n;1-\alpha}^{-1}(W_n) \leq W\,.$$

21.10.9 Interval bounds for normal distribution

Estimation of $W = m$ for known σ^2:
The estimating function is

$$W_n = \bar{x}_n = \frac{1}{n}\sum_i^n x_i \,.$$

Two-sided prediction interval:

$$W - t_{1-\alpha/2}\frac{\sigma}{\sqrt{n}} \le W_n \le W + t_{1-\alpha/2}\frac{\sigma}{\sqrt{n}} \,,$$

where t_α is the percentile of order α of the standard normal distribution.
Two-sided confidence interval:

$$W_n - t_{1-\alpha/2}\frac{\sigma}{\sqrt{n}} \le W \le W_n + t_{1-\alpha/2}\frac{\sigma}{\sqrt{n}} \,.$$

One-sided prediction interval (upper):

$$W_n \le W + t_{1-\alpha}\frac{\sigma}{\sqrt{n}} \,.$$

One-sided confidence interval (lower):

$$W \ge W_n - t_{1-\alpha}\frac{\sigma}{\sqrt{n}} \,.$$

One-sided prediction interval (lower):

$$W_n \ge W - t_{1-\alpha}\frac{\sigma}{\sqrt{n}} \,.$$

One-sided confidence interval (upper):

$$W \le W_n + t_{1-\alpha}\frac{\sigma}{\sqrt{n}} \,.$$

Estimation of $\bar{x}$ for unknown σ^2:
The unknown parameter σ is replaced by the **empirical** value

$$\sigma = \sqrt{\frac{1}{n-1}\sum_i^n (x_i - \bar{x})^2}$$

and the **percentile** t_α of the normal distribution is replaced by the percentile $t_{T;\alpha;k}$ of the **t-distribution** with $k = n - 1$ degrees of freedom.
Estimation of σ^2 :
The estimating function is

$$W_n = \sigma_n^2 = \frac{1}{n-1}\sum_i^n (x_i - \bar{x})^2 \,.$$

Two-sided prediction interval:

$$\frac{\sigma^2}{n-1} t_{\chi;\alpha/2;n-1} \le W_n \le \frac{\sigma^2}{n-1} t_{\chi;1-\alpha/2;n-1} \,.$$

Two-sided confidence interval:

$$\frac{(n-1)W_n}{t_{\chi;1-\alpha/2;n-1}} \le \sigma^2 \le \frac{(n-1)W_n}{t_{\chi;\alpha/2;n-1}} \,.$$

One-sided prediction interval (upper):

$$W_n \leq \frac{\sigma^2}{n-1} t_{\chi;1-\alpha;n-1} \,.$$

One-sided confidence interval (lower):

$$\frac{(n-1)W_n}{t_{\chi;1-\alpha;n-1}} \leq \sigma^2 \,.$$

One-sided prediction interval (lower):

$$\frac{\sigma^2}{n-1} t_{\chi;\alpha/2;n-1} \leq W_n \,.$$

One-sided confidence interval (upper):

$$\sigma^2 \leq \frac{(n-1)W_n}{t_{\chi;\alpha/2;n-1}} \,.$$

21.10.10 *Prediction and confidence interval bounds for binomial and hypergeometric distributions*

Estimate of the parameter p of the binomial or hypergeometric distribution: In the sense of attribute sampling (see application under **quality test**), in which k marked components in a sample of size n are drawn, the estimator is given by the empirical portion:

$$W_n = k/n$$

in the sample. For the hypergeometric distribution, already under the conditions of $p = K/N$; i.e., the number of **all** marked components over the total number of elements.
Two-sided prediction interval (binomial distribution):

$$p < t_{1-\alpha/2}\sqrt{\frac{p(1-p)}{n}} \leq W_n \leq p + t_{1-\alpha/2}\sqrt{\frac{p(1-p)}{n}} \,.$$

A condition is that $n \geq 9/(p(1-p))$, since in this case W_n is approximately normally distributed about p with $\sigma^2 = p(1-p)/n$.
Two-sided confidence interval (binomial distribution):

$$\frac{W_n + t^2_{1-\alpha/2}/2n - t_{1-\alpha/2}\sqrt{W_n(1-W_n)/n + t^2_{1-\alpha/2}/4n^2}}{1 + t^2_{1-\alpha/2}/n}$$

$$\leq p \leq \frac{W_n + t^2_{1-\alpha/2}/2n + t_{1-\alpha/2}\sqrt{W_n(1-W_n)/n + t^2_{1-\alpha/2}/4n^2}}{1 + t^2_{1-\alpha/2}/n} \,.$$

▷ For the hypergeometric distribution with the total number N and the marked portion K, where $p = K/N$, we obtain the same expression, but with

$$t_{1-\alpha/2} \rightarrow t_{1-\alpha/2}\sqrt{\frac{N-n}{N-1}} \,.$$

21.10.11 Interval bounds for a Poisson distribution

Estimation of $c = n \cdot p$: For the following interval bounds it is assumed that $c \geq 9$. Only then, the Poisson distribution corresponds to a normal distribution with $m = c$ and $\sigma^2 = c$, and the percentiles of the normal distribution can be inserted.
The estimator is the variable x obeying precisely the Poisson and normal distribution (i.e., the sample size is $n = 1$).
Two-sided prediction interval:

$$c - t_{1-\alpha/2}\sqrt{c} \leq x \leq c - t_{1+\alpha/2}\sqrt{c}\,.$$

Two-sided confidence interval:

$$x + \frac{t^2_{1-\alpha/2}}{2} - t\sqrt{x + \frac{t^2_{1-\alpha/2}}{4}} \leq c \leq x + \frac{t^2_{1-\alpha/2}}{2} + t\sqrt{x + \frac{t^2_{1-\alpha/2}}{4}}\,.$$

Approximate two-sided confidence interval with $t^2_{1-\alpha/2} \ll x$:

$$x - t_{1-\alpha/2}\sqrt{x} \leq c \leq x + t_{1-\alpha/2}\sqrt{x}\,.$$

21.10.12 Determination of sample sizes n

In practice, the probability α (or $1 - \alpha/2$) and the confidence interval are often given. The task is to determine the necessary sample size n which guarantees that the parameter W is actually situated within the confidence bounds with a probability of at least $1 - \alpha$ (**proposition certainty**).
Given accuracy: Given the confidence limits, the deviation $|W - W_n|$, which must **not** be exceeded with the probability $1 - \alpha$. This is given by the absolute and relative error of estimation and is independent of the given proposition certainty $1 - \alpha$.
▷ The given accuracy is an additional requirement, apart from the proposition certainty.
Absolute/relative error of estimation v, $v_{\text{rel}} = v/W$: The value/relative value of the deviation $|W - W_n|$.
Requirements for the expected value $W = m$ (**heterograde investigation**): If the proposition certainty $1 - \alpha$ as well as the limits of accuracy are given by v and v_{rel}, respectively, then we obtain the following conditions under which we must choose the sample size n:

$$P(|m - W_n| \leq v) = P(|m - W_n|/m \leq v_{\text{rel}}) \equiv 1 - \alpha\,.$$

For a normal distribution this is the equivalent of:

$$|m - W_n| \leq v \leq t_{1-\alpha/2}\frac{\sigma}{\sqrt{n}},$$

and, thus, for n the following holds:

$$n \geq \frac{t^2_{1-\alpha/2}\sigma^2}{v^2}.$$

▷ Given v_{rel}, the corresponding conditions are the same if v is replaced by v_{rel}/W ($W = m$ or $W = p$).
Requirements for the fraction parameter p (**homograde investigation**): If the proposition certainty $1 - \alpha$ and the limits of accuracy are given by v or v_{rel}, respectively, then

we obtain the following conditions according to which we must choose the sample size n:

$$P(|p - W_n| \leq v) = P(|p - W_n|/p \leq v_{\text{rel}}) \equiv 1 - \alpha \,.$$

For a binomial distribution this is the equivalent of:

$$|p - W_n| \leq v \leq t_{1-\alpha/2}\sqrt{\frac{p(1-p)}{n}}\,,$$

and, thus, for n the following holds:

$$n \geq \frac{t_{1-\alpha/2}^2\, p(1-p)}{v^2}\,.$$

▷ Of course, these inequalities are just as valid for the hypergeometric distribution. But the quantities σ and t must be replaced as stated above!

▷ Given v_{rel}, the corresponding conditions are the same if v is replaced by v_{rel}/W ($W = m$ or $W = p$).

21.10.13 Test methods

Testing of hypotheses about the value of coefficients of measure of a given distribution and/or the type of distribution of characteristics.

□ A typical example: Given a daily production of $N = 10{,}000$ high-frequency circuits with a frequency of $\nu = 22{,}000$ Hz (according to the manufacturer's specifications). Usually, the frequencies of these circuits will not meet the exact value of 22,000 Hz, but will fluctuate about this or another value, satisfying a normal distribution to a good approximation.

Principle of a sampling: Establish a null hypothesis H_0 whose acceptance or rejection is decided by the sampling. For comparison, an alternative hypothesis H_1 can but must not necessarily be used.

Parameter hypothesis: Hypothesis on parameter W of a distribution $f(x; W)$ whose type is already known.

▷ Instead of the fixed parameter values W and $\tilde{W}$, whole sets of parameters such as intervals may be used as well. This leads to the so-called left or right parameter tests.

Distribution hypothesis: Hypothesis concerning distributions whose type (and often whose parameters) are not known.

□ Still following the example of industrial production, the assumption of a normal distribution for the frequency values already determines the type of distribution, namely, a **parameter test**. If the expected value m is now tested, then the **parameter hypothesis** reads H_0: $W(H_0) = m = 22{,}000$ Hz. H_1 should be the alternative $W(H_1) = m = 23{,}000$ Hz. For the sake of simplicity, the variance parameter σ^2 of the normal distributions is known for both hypotheses.

Test function, test variable, $W_n(x_1, \ldots, x_n)$: Random sample function with a known probability distribution.

▷ In the case of a parameter hypothesis the test functions should be unbiased, consistent, efficient, and sufficient.

□ The arithmetic mean can be chosen as a test function for the expected value of frequency.

Sampling plan: The prescription for performing a sampling inspection.

Simple algorithm of a sample:

Random sampling: (unitary/multistage, stratified/unstratified, with/without replacement)
↓
Analysis: determination of the value of one/more test variables
↓
application of the **decision rule**: acceptance/rejection of H_0

Power: The characteristic of the decision rule for separating measured values.
Error of the first kind: Rejection of H_0 although H_0 is true.
Probability for the error of the first kind: α.
Error of the second kind: Acceptance of H_0 although H_0 is false.
Probability for the error of the second kind: β.

▷ Always assume that one of the hypotheses is true (which, of course, is not necessarily the case). Depending on the case (H_0 true or false) and the decision, different probabilities are thus calculated, which are summarized in the table below (for simple parameter tests):

Value of the hypothesis:	$H_0(W)$ true	
location of W_n	W_n not in K_W	W_n in K_W
decision	accepted H_0	rejected H_0
value	true	false
probability $P =$	$OC(W)$ $= 1 - GF(W)$ $= 1 - \alpha$	$GF(W)$ $= 1 - OC(W)$ $= \alpha$

value of the hypothesis:	$H_0(W)$ false or $H_1(\tilde{W})$ true	
location of W_n	W_n not in K_W	W_n in K_W
decision	accepted H_0	rejected H_0
value	false	true
probability $P =$	$OC(\tilde{W})$ $= 1 - GF(\tilde{W})$ $= \beta$	$GF(\tilde{W})$ $= 1 - OC(\tilde{W})$ $= 1 - \beta$

□ In the above example, if m is actually 22,000 Hz; then the rejection of the hypothesis H_0 would be an error of the first kind. But if $m \neq 22{,}000$ Hz and it is decided H_0 is true, this would be an error of the second kind.

Critical region K_W: The region where, according to the hypothesis H_0 for the probability distribution with parameter W, the test function is only with probability α; hypothesis H_0 (the parameter $W(H_0)$) is rejected if the measured value for the test function is in the critical region. Therefore, α is the probability with which the null hypothesis is falsely rejected; it is therefore also called the **fiducial probability**.

Significance level, fiducial probability α: The probability with which a **measured** value W_n for the H_0 test function falls into the **critical region** K_W; the probability for the error of the first kind.

$$\alpha = P(W_n \in K_W)$$

□ In the example the critical region consists of precisely the boundary regions of the normal distribution of H_0. The bounds of the critical regions are defined by the percentiles of the normal distribution $T_{\alpha/2}$, $T_{1-\alpha/2}$. If the H_0 hypothesis is true, then the probability that the sample value W_n falls into the critical region is exactly α. If this happens, then the hypothesis H_0 is rejected, and the error (if any) is of the first kind. But if the test value W_n does **not** fall into the critical region then it is assumed that hypothesis H_0 is true and hypothesis H_1 is false. If H_0 is still false, the error (if any) is an error of the second kind.

▷ The figure clearly illustrates that the probability of an error of the second kind depends very much on the choice of the alternative hypothesis H_1, and that it is not influenced only by the significance level α of the H_0 hypothesis.

Acceptance probability

$1-\alpha$: Probability that the value of the H_0 test-function W_n does **not** fall into the critical region K_W, and it is therefore assumed that:

$$1-\alpha = P(W_n \notin K_W).$$

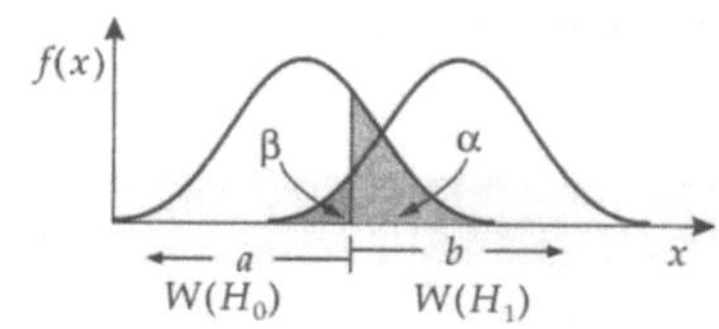

α: error of the first kind, β: error of the second kind, a: acceptance region, b: rejection region W_0

▷ $1-\alpha$ defines the so-called **prediction interval** from the estimation theory. The prediction interval is complementary to the critical region.

▷ If α is chosen too small, then the probability of a false assumption (error of the second kind) is high.

□ In the above figure, the acceptance region is given by the middle region about the H_0 expectation value.

Operating characteristic: Acceptance probability for value W_n of the H_0 test-function as a function of the H_0 parameter values $W(H_0)$ for given fixed bounds of the interval in the critical region.

$$OC(W) = P(W_n \notin K_W)$$

▷ The bounds of the interval of the critical regions are defined uniquely by the H_0 hypothesis $W = W(H_0)$ and the value of α (for example, if we vary the parameter W for fixed bounds of the critical region, then we obtain the operating characteristic).

▷ For different sample sizes n we obtain different characteristics.

Power function: The rejection probability for value W_n of the H_0 test-function as a function of the H_0 parameter value W:

$$PF(W) = P(W_n \in K_W)\ .$$

▷ The power function is the probability of an error of the first kind.

● The operation characteristic and power function are mutually 1-complementary.

$$OC(W) + PF(W) = 1\ .$$

▷ The alternative hypothesis H_1 with alternative parameter $\tilde{W}$, $OC(\tilde{W}) = 1 - PF(\tilde{W}) = \beta$ is just the probability for the error of the second kind, namely, that the alternative hypothesis is accepted and the null hypothesis is falsely rejected.

▷ For hypotheses with several permitted values for parameter W (W from a range of values), α, and β are only the upper and lower bounds for the given probabilities. Depending on the actual situation, such tests are also called two-sided, left-hand, or right-hand parameter tests.

21.10.14 Parameter tests

Simple parameter hypothesis: Hypothesis with a single value for distribution parameter $W = W_1$.

Composite parameter hypothesis: Hypothesis with several possibilities for distribution parameter $W = \{W_1, W_2, \ldots\}$.

□ A typical composite parameter hypothesis is $W > W_0$.

Two-sided test: Test of simple hypothesis against a composite alternative

□ H_0: $W = \tilde{W}$; H_1: $W > \tilde{W}$.

Left-hand test: H_0: $W \geq \tilde{W}$; H_1: $W < \tilde{W}$.

Right-hand test: H_0: $W \leq \tilde{W}$; H_1: $W > \tilde{W}$.

21.10.15 Parameter tests for a normal distribution

For some borderline cases and applications, many special distributions can be reduced to a normal distribution. Therefore, the following expressions for the bounds of acceptance regions or the shape of the power functions or the operating characteristics can be applied universally.

Hypotheses about m, σ^2 known:

Size n and fiducial probability α are given. t_α is the percentile of the standard normal distribution.

(a) H_0: $m = x_0$; H_1: $m = x_1$; $x_0 < x_1$.

Acceptance region:

$$W_n \leq \bar{x}_{1-\alpha} = x_0 + t_{1-\alpha}\frac{\sigma}{\sqrt{n}}\,.$$

Error of the second kind:

$$\beta = P\left(\frac{W_n - x_1}{\sigma}\sqrt{n} \leq \frac{\bar{x}_{1-\alpha} - x_1}{\sigma}\sqrt{n}\right) = PF\left(\frac{\bar{x}_{1-\alpha} - x_1}{\sigma}\sqrt{n}\right)\,.$$

(b) H_0: $m = x_0$; H_1: $m = x_1$; $x_0 > x_1$.

Acceptance region:

$$W_n \geq \bar{x}_\alpha = x_0 + t_\alpha\frac{\sigma}{\sqrt{n}}\,.$$

Error of the second kind:

$$\beta = P\left(\frac{W_n - x_1}{\sigma}\sqrt{n} \geq \frac{\bar{x}_\alpha - x_1}{\sigma}\sqrt{n}\right) = 1 - PF\left(\frac{\bar{x}_\alpha - x_1}{\sigma}\sqrt{n}\right)\,.$$

(c) Left-hand test: H_0: $m \geq x_0$; H_1: $m < x_0$.

Rejection region:

$$W_n \leq \bar{x}_\alpha = x_0 + t_\alpha\frac{\sigma}{\sqrt{n}}\,.$$

For all m, that is, for H_0 as well as for H_1, the power function is given by

$$PF(m) = P(W_n < \bar{x}_\alpha) = PF\left(\frac{\bar{x}_\alpha - m}{\sigma}\sqrt{n}\right) .$$

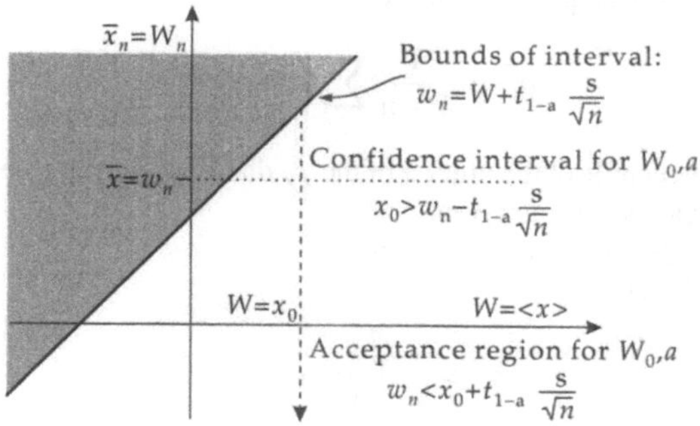

(d) Right-hand test: H_0: $m \le x_0$; H_1: $m > x_0$.
Rejection region:

$$W_n \ge \bar{x}_{1-\alpha} = x_0 + t_{1-\alpha}\frac{\sigma}{\sqrt{n}} .$$

For all m, that is, for H_0 as well as for H_1, the power function is given by

$$PF(m) = P(W_n > \bar{x}_{1-\alpha}) = 1 - PF\left(\frac{\bar{x}_{1-\alpha} - m}{\sigma}\sqrt{n}\right) .$$

(e) Two-sided test: H_0: $m = x_0$, H_1: $m \ne x_0$.
Rejection region:

$$|W_n - x_0| > t_{1-\alpha/2}\frac{\sigma}{\sqrt{n}} .$$

Acceptance region:

$$x_0 - t_{1-\alpha/2}\frac{\sigma}{\sqrt{n}} \le W_n \le x_0 + t_{1-\alpha/2}\frac{\sigma}{\sqrt{n}} .$$

Power function:

$$\begin{aligned} PF(m) &= P\left(W_n > x_0 + t_{1-\alpha/2}\frac{\sigma}{\sqrt{n}}\right) + P\left(W_n < x_0 - t_{1-\alpha/2}\frac{\sigma}{\sqrt{n}}\right) \\ &= 1 - PF\left(\frac{x_0 - m}{\sigma}\sqrt{n} + t_{1-\alpha/2}\right) + PF\left(\frac{x_0 - m}{\sigma}\sqrt{n} - t_{1-\alpha/2}\right) . \end{aligned}$$

Hypotheses about m, σ^2 unknown: The tested quantity is the mean value of the random sample of size n:

$$W_n = \bar{x}_n .$$

Instead of σ^2 the empirical variance is usually employed:

$$\sigma_n^2 = \frac{1}{1-n}\sum_i^n (x_i - \bar{x}_n)^2 ,$$

and instead of the percentiles t_α those of the t-distribution, according to Helmert-Pearson, $t_{T;n-1;\alpha}$ are used.

▷ For $n-1>30$, the percentiles of the standard normal distribution are sufficient.

Hypotheses about σ^2 of the normal distribution:

The test function is the empirical variance:

$$\sigma_n^2 = \frac{1}{1-n}\sum_i^n (x_i - \bar{x}_n)^2 .$$

▷ $t_{\chi;\alpha;n}$ is the percentile of the order α of the f_χ distribution with n degrees of freedom. The significance level α is given.

(a) H_0: $\sigma = \sigma_0$; H_1: $\sigma = \sigma_1$; $\sigma_0 < \sigma_1$.

Rejection region for H_0:

$$\sigma_n^2 > \frac{\sigma_0^2}{n-1} t_{\chi;1-\alpha;n-1} .$$

(b) H_0: $\sigma = \sigma_0$; H_1: $\sigma = \sigma_1$; $\sigma_0 > \sigma_1$. Rejection region for H_0:

$$\sigma_n^2 < \frac{\sigma_0^2}{n-1} t_{\chi;\alpha;n-1} .$$

(c) Left-hand test: H_0: $\sigma \geq \sigma_0$ H_1: $\sigma < \sigma_0$.

Rejection region:

$$\sigma_n^2 < \frac{\sigma_0^2}{n-1} t_{\chi;\alpha;n-1} .$$

(d) Right-hand test: H_0: $\sigma \leq \sigma_0$; H_1: $\sigma > \sigma_0$. Rejection region:

$$\sigma_n^2 > \frac{\sigma_0^2}{n-1} t_{\chi;1-\alpha;n-1} .$$

(e) Two-sided test: H_0: $\sigma = \sigma_0$; H_1: $\sigma \neq \sigma_0$. Rejection region:

$$\sigma_n^2 < \frac{\sigma_0^2}{n-1} t_{\chi;\alpha/2;n-1}$$

and

$$\sigma_n^2 > \frac{\sigma_0^2}{n-1} t_{\chi;1-\alpha/2;n-1} .$$

▷ The power functions are calculated with the noncentral f_χ-distribution.

21.10.16 Hypotheses about the mean value of arbitrary distributions

If the variance of the distribution is known, the arithmetic mean value $\bar{x}_n$ of the sample for a large sample size n will approximately follow the normal distribution about an expected value m with $\sigma \to \sigma/2$. Then, the expected value m is tested by multiple sampling and with the interval bounds as given for the usual normal distribution, but with the replacement $\sigma \to \sigma/n$.

21.10.17 Hypotheses about p of binomial and hypergeometric distributions

Tests of this kind can be reduced to tests of normal distribution for $n > 9/[p(1-p)]$ with

$$\sigma \to \frac{p(1-p)}{n}$$

(drawing with replacement: binomial distribution)
or for $n > 9/[p(1-p)]$ **and** $n/N < 0.1$

$$\sigma \to \frac{p(1-p)}{n}\frac{N-n}{N-1}$$

(drawing without replacement: hypergeometric distribution).

21.10.18 Tests of goodness of fit

Test of goodness of fit: Method for comparing different types of distributions; used if the type of the distributions and not the parameter must be determined.
χ^2-test of fit: Method to test hypotheses about distribution by means of the χ^2-function.
▷ The χ^2-test is suitable only for a large sample size n.
Procedure:

1. Propose the H_0 hypothesis: Assume a special type of distribution (e.g., normal distribution) $f(x; W)$. Give the significance level α.

2. Draw up a frequency table $H_i = H(K_i)$ from the sample. The classes K_i must be defined in such a way that the frequency of each class is at least 10.

3. Calculate the probabilities $P(K_i; W)$ of classes K_i according to the given special distribution $f(x; W)$:

$$P(K_i; W) = \int_{x_i}^{x_i+\Delta x_i} f(x; W)\mathrm{d}x \ ,$$

   where x_i and $x + \Delta x_i$ are the bounds of the interval of class K_i with the interval width Δx_i.

4. Estimate the parameters W of the given probability distribution $P(K_i; W)$ with the help of the frequency table. The estimation value of the parameters $\tilde{W}$ is defined by the minimum of the χ^2-function:

$$\chi^2 = \sum_i^k \frac{[\tilde{F}_i - nP(K_i; \tilde{W})]^2}{nP(K_i; \tilde{W})} = \mathrm{Min}_W \ .$$

5. Calculate the tested quantity: the tested quantity is just the χ^2-function whose value was already calculated under step 4.

6. Apply the decision rule:
   Reject H_0 if the value for χ^2 exceeds the percentile of the order $1-\alpha$ of the χ^2-distribution:

$$\chi^2 > t_{\chi;k-1-l} \, .$$

▷ The number of the χ^2 degrees of freedom to be inserted is $k-1-l$, where k is the number of classes, and l is the number of parameters estimated under step 4.

Kolmogorov-Smirnov-test of fit: A procedure to measure goodness of fit in which the theoretical **distribution function** must be known completely for all parameters.

▷ Also very good for small sample size n.

▷ Applicable only for continuous distributions.

Procedure:

1. Propose the hypothetical distribution function H_0: $F = F(x; W)$. Give the significance level α.

2. Arrange the sample variable according to its magnitude

$$x_1 \leq x_2 \leq \cdots \leq x_n$$

   and determine the empirical distribution function $F_n(x)$

$$F_n(x) = \begin{cases} 0 & \text{for} \quad x < x_1 \\ i/n & \text{for} \quad x_i \leq x < x_{i+1} \\ 1 & \text{for} \quad x \geq x_n \end{cases}$$

3. Calculate the tested quantity, maximum absolute difference:

$$D_n = \sup_x |F(x) - F_n(x)| \, .$$

4. Apply the decision procedure: Reject H_0, if $D_n > D_{\alpha;n}$.

Bounds $D_{\alpha;n}$ are plotted on the following table:

$\alpha\backslash n$	1	2	3	4	5	6	7	8	9	10
0.20	0.900	0.684	0.565	0.494	0.446	0.410	0.381	0.358	0.339	0.322
0.15	0.925	0.726	0.597	0.525	0.474	0.436	0.405	0.381	0.360	0.342
0.10	0.950	0.776	0.642	0.564	0.510	0.470	0.438	0.411	0.388	0.368
0.05	0.975	0.842	0.708	0.624	0.565	0.521	0.486	0.457	0.432	0.410
0.01	0.995	0.929	0.828	0.733	0.669	0.618	0.577	0.543	0.514	0.490

$\alpha\backslash n$	11	12	13	14	15	16	17	18	19	20
0.20	0.307	0.295	0.284	0.274	0.266	0.258	0.250	0.244	0.237	0.231
0.15	0.326	0.313	0.302	0.292	0.283	0.274	0.266	0.259	0.252	0.246
0.10	0.352	0.338	0.325	0.314	0.304	0.295	0.286	0.278	0.272	0.264
0.05	0.391	0.375	0.361	0.349	0.338	0.328	0.318	0.309	0.301	0.294
0.01	0.468	0.450	0.433	0.418	0.404	0.392	0.381	0.371	0.363	0.356

21.10.19 Application: acceptance/rejection test

Estimating the acceptance/rejection rate of a tested lot by means of arbitrary (representative) sampling instead of testing the entire batch.

Hypergeometric distribution: Gives the probability of finding k defective items among n samples of the population N.

$$P_{\text{error}}(k; n, N, p) = \binom{pN}{k}\binom{N(1-p)}{n-k}\Big/\binom{N}{n}.$$

Number of rejects: Number of defective items (known) $= p \cdot N$.
Rejection rate p: The proportion of defective items in the population.

▷ In terms of a test method, the condition "number of rejects $\leq p{\cdot}N$" can be regarded as the hypothesis, and p can be regarded as the estimation parameter of the probability distribution.

Proposition certainty, acceptance probability P_k: The probability of finding less than or exactly k defective items (probability sum):

$$P_k = \sum_{i=1,k} P_{\text{error}}(i; n, N, p) .$$

▷ With the probability P_k it is correct to reject a consignment if the number of the defective items in the sample is $n_{\text{error}} > k$. For very large numbers of items ($N \to \infty$), the Poisson distribution can also be used.

AQL value (Acceptable Quality Level): Value for P_k fixed by agreement between customer and manufacturer (usually $P_k(p) = 90\%$) in which no more than k defects are allowed in a sample size n.

▷ Of course, as the reject rate p of the total set decreases, the acceptance probability increases. The manufacturer will try to remain far below the number of rejects $p \cdot N$ used to calculate the AQL value.

● The AQL value is the significance level of the quality control test, see **method of testing**.

Acceptance curve: Mapping of the reject proportion in the total set $p = N_{\text{error}}/N$ onto acceptance probability P_k for a given sample size n and a maximum number of rejects k in the sample.

▷ If the AQL value for P_k is given, the required sample size n and the maximum permissible number of rejects n_{error} can be read from the corresponding curve.

□ As an example, the following figure shows two characteristic curves for **one** sample size, but different values for the maximum permissible number of rejects k_1 and k_2.

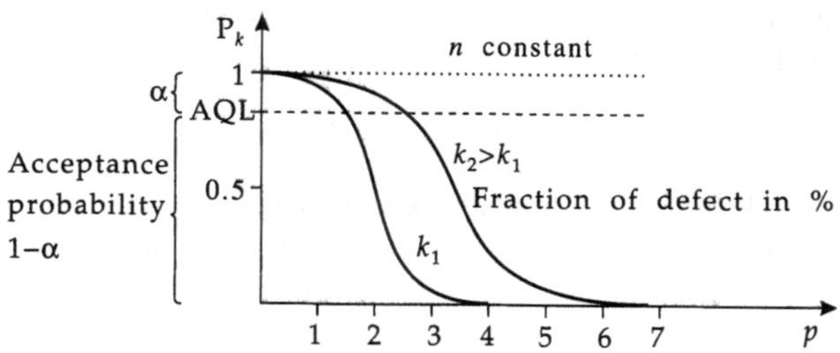

21.11 Reliability

Time-dependent events (e.g., radioactive decay, failure of an electrical component) can be described in practical terms by means of some special quantities.

Lifetime: Distance in time between the failures of objects. The distribution of failures in time can be purely random (nonaging objects), or it can be subject to external influences (aging objects).

Nonaging objects: Objects with infinite **lifetimes** whose failure is purely random and satisfies a distribution based on a purely combinatorial random principle (**urn model, Poisson distribution, exponential distribution**). They are not subject to an aging process such as external signs of wear and tear.

▷ The failures of nonaging objects satisfy a Poisson distribution in time. The time intervals between failures satisfy the exponential distribution.

□ In good approximation, electronic components such as resistors, capacitors and integrated circuits are nonaging objects (if used properly without undue loads from excessive current or voltage).

□ Objects with infinite "lifetimes" are also in areas other than technology. For example, an infection with a rare disease satisfies a Poisson distribution to a good approximation; the time interval between several infections satisfies the exponential distribution.

Aging objects: Objects with finite **lifetimes** subject to an aging process. Thus, aging may influence the random decaying process and thus also the distribution of failures (see **Weibull distribution**).

▷ The failure of aging objects does not satisfy a Poisson distribution. To describe the time interval between failures, a special type of distribution must be considered.

□ Typical examples of aging objects are motors, tires and tools.

▷ The time intervals between failures can often be described by a superposition of several exponential distributions.

▷ The lifetime of aging objects can sometimes be described by the **Weibull distribution.**

▷ Exponential and Weibull distributions are special cases of **reliability**.

Reliability $Z(t)$: Mean number of components $N(t)$ still functioning after time t, relative to the initial number N_0; general setup

$$Z(t) = \frac{N(t)}{N_0} = \mathrm{e}^{\left(-\int_0^t \lambda(t')\mathrm{d}t'\right)}$$

for the description of aging processes as a function of time.

$Z(t)$ is the probability that a part has **not** yet failed after time t.

Failure probability $F(t)$: Mean number of parts that has failed after time t $N_0 - N(t)$, relative to the initial number N_0:

$$F(t) = 1 - Z(t) .$$

$F(t)$ is the probability that a part has failed after time t.

Failure density ρ: Mean number of failures per time at time unit t, relative to the initial number N_0:

$$\rho(t) = \frac{\mathrm{d}F(t)}{\mathrm{d}t} = -\frac{\mathrm{d}Z(t)}{\mathrm{d}t} = \lambda(t)Z(t) .$$

▷ The integral of the failure density is just the number of failures, relative to the initial number N_0:

$$\int_0^t \rho(t')\mathrm{d}t' = -\int_0^t \frac{\mathrm{d}Z(t')}{\mathrm{d}t'}\mathrm{d}t' = -(Z(t) - Z(0)) = 1 - Z(t) = F(t) .$$

Failure rate: The mean number of failures per time unit, relative to the number of functioning parts pieces $N(t)$:

$$\lambda(t) = -\frac{1}{N(t)}\frac{\mathrm{d}N(t)}{\mathrm{d}t} = -\frac{1}{Z(t)}\frac{\mathrm{d}Z}{\mathrm{d}t} = \frac{\rho(t)}{Z(t)} .$$

Mean Time To Failure, MTTF:

$$\mathrm{MTTF} = \int_0^\infty Z(t)\mathrm{d}t .$$

- The probability that the total system still functions after time t is equal to the product of the reliabilities of the individual system:

$$Z_{\text{total}} = Z_1 Z_2 \cdots Z_n .$$

Analysis by means of random sampling
Nonaging objects:

$$\lambda_{\text{total}} = \lambda_1 + \lambda_2 + \cdots + \lambda_n .$$

- Provided that rate λ and time t are small, an approximation for the failure rate is the number of failures per initial number and operating time.

$$\lambda \approx \frac{(1 - N(t))}{N_0 \cdot t} = \frac{\text{failures}}{\text{initial number} \cdot \text{operating time}}$$

- For nonaging objects, $Z(t)$ is the exponential distribution ($\lambda = \text{const.}$), and the failure time corresponding to $1/\lambda$.
  Typical failure rates (λ in fit = failure/10^9 h):

wrap connection	0.0025
glimmer capacitor	1
high-frequency coil	1
metal film resistor	1
paper capacitor	2
transistor	200
light-emitting diode (50% loss of luminosity)	500

21.12 Computation of adjustment, regression

Regression line, **trend curve**, $y = f(x)$: Represents approximately the shape of the pairs of measured values (x_i, y_i) (two-dimensional random events).

Proposed regression function
$f(x; a, b, \ldots)$: Parameter-dependent set-up for the regression line or trend curve.

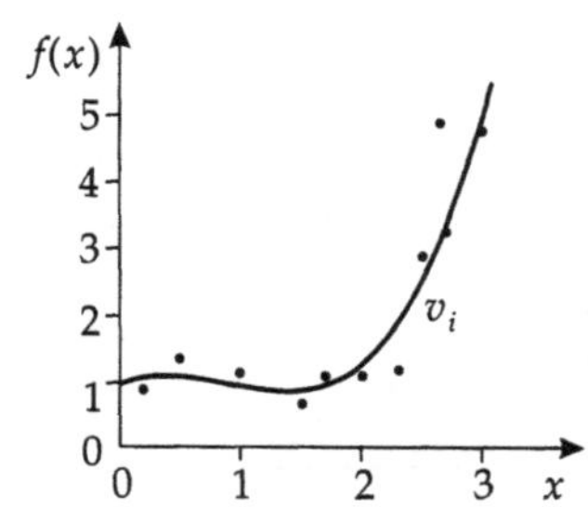

Parameters of the curve: Parameter of the function chosen for the regression line.

Trend of expansion, **general tendency**, **tendency of expansion**, **trend**: Rough categorization of the functional expansion of a series of a measurements $y_i(x_i)$ by **slope** (first derivative) and **curvature** (second derivation):

(a) progressively increasing (concave)	$y' > 0$,	$y'' > 0$
(b) linearly increasing (linear)	$y' > 0$,	$y'' = 0$
(c) degressively increasing (convex)	$y' > 0$,	$y'' < 0$
(d) progressively decreasing (concave)	$y' < 0$,	$y'' > 0$
(e) linearly decreasing (linear)	$y' < 0$,	$y'' = 0$
(f) degressively decreasing (convex)	$y' < 0$,	$y'' < 0$

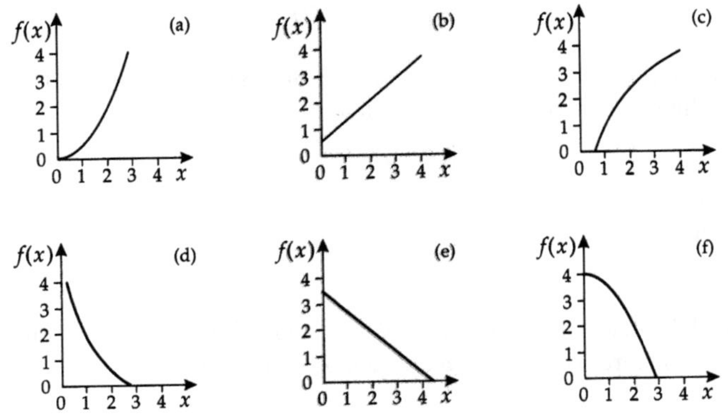

▷ Depending on setup of the trend function, we distinguish between **polynomial regression, power regression** and **exponential regression.**

▷ The most common functions are:

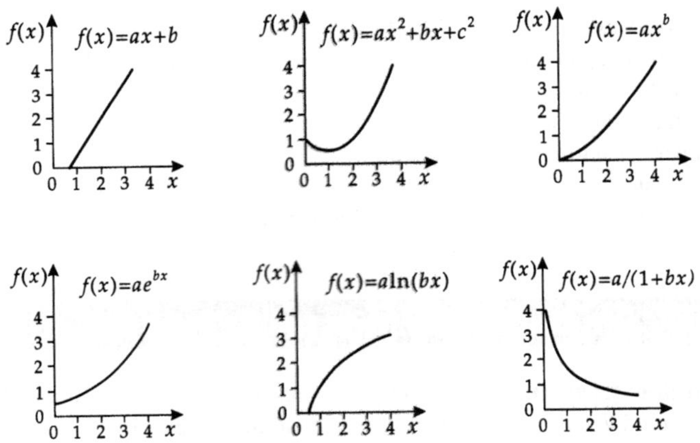

Connection		Parameter
$y = ax + b$	linear function, regression line	a, b
$y = ax^2 + bx + c$	quadratic function, regression parabola	a, b, c
$y = ax^b$	power function	a, b
$y = ae^{bx}$	exponential function	a, b
$y = a \ln bx$	logarithmic function	a, b
$y = a/(1 + bx)$	fractional rational function	a, b

▷ Many setups must be changed to linear form. For example, the exponential function

$$y = ae^{bx}$$

can be converted to the linear form

$$\tilde{y} = \ln y = \ln a + bx = \tilde{a} + bx \; .$$

Performing a linear regression with the pairs of values $(\ln y_i, x_i)$, we obtain the quantities $\tilde{a}$, b and from these the parameters $a = e^{\tilde{a}}$, b.

▷ Caution is advisable when determining the mean errors of the parameters. Their determination is given by the rules of error propagation.

21.12.1 Linear regression, least squares method

Linear regression: Calculus of adjustment with a regression function as an integral rational function of the first order (**regression line**)

$$y := a_0 + a_1 x \; .$$

Sum of the squares of the distances: Summed squares of the vertical distances

$$\Delta := \sum v_i^2 = \sum_i (y_i - f(x_i))^2 \; .$$

Least squares principle, Gaussian minimal principle: Allows the unique calculation of the best parameter set for an approximation function for the expansion of measured values of a given regression curve; the sum of the squares of the distance Δ is minimal for the optimal approximation

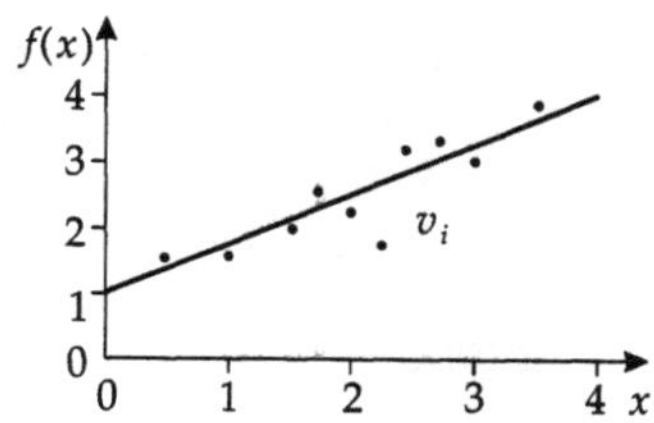

$$\frac{\partial \Delta}{\partial a_i} = 0 \; .$$

Parameter of the regression function a_i.

Normal equations: Equations to determine the parameters a_i of the regression function following the Gaussian minimal principle:

$$\frac{\partial \Delta}{a_i} \equiv 0, \quad 1 \le i \le n, \quad a_1, \ldots, a_n = \text{parameters of the regression function.}$$

Linear regression coefficients: Parameters of the regression line, solution of the normal equations for regression lines at a sample size N:

$$a_1 = \frac{N \sum_i x_i y_i - \sum_i x_i \sum_i y_i}{N \sum_i x_i^2 - (\sum_i x_i)^2} = \frac{\sum_i x_i y_i - \bar{y} \sum_i x_i}{\sum_i x_i^2 - \bar{x} \sum_i x_i} \, ,$$

$$a_0 = \frac{\sum_i x_i^2 \sum_i y_i - \sum_i x_i y_i \sum_i x_i}{N \sum_i x_i^2 - (\sum_i x_i)^2} = \bar{y} - a_1 \bar{x} \; .$$

Mean error of the linear regression coefficients:

$$\langle \Delta a_1 \rangle = N \cdot \sqrt{\frac{N}{N \sum_i x_i^2 - (\sum_i x_i)^2}} \, , \qquad \langle \Delta a_0 \rangle = N \cdot \sqrt{\frac{\sum_i x_i^2}{N \sum_i x_i^2 - (\sum_i x_i)^2}} \; .$$

Covariance: Measure for the goodness of the chosen regression line:

$$\sigma^2_{f(x)} := \frac{1}{N-1}\sum_{i=1}^{N}(y_i - f(x_i))^2 = \frac{N-1}{N-2}\left(\sigma_y^2 - \frac{\sigma_{xy}^2}{\sigma_x^2}\right) .$$

Vertical distance: Difference between the function value of the regression curve $y(x_i) = f(x_i)$ and the measured value $y_i(x_i)$:

$$v_i = y_i - f(x_i) .$$

▷ Program sequence for multidimensional linear regression of n data points $(\vec{\mathbf{x}}_i, y_i)$ in *dimen* dimensions. The output is matrix $\mathbf{A}$ of $dimen + 1 \times dimen + 1$ dimensions and a vector $\vec{\mathbf{c}}$ of $dimen + 1$ dimensions. The regression coefficients r_i are obtained as a solution of the system of equations $\mathbf{A}\vec{\mathbf{r}} = \vec{\mathbf{c}}$.

```
BEGIN multidimensional linear regression
INPUT n, dimen
INPUT x[i,k], i=1...dimen, k=1...n
INPUT y[i], i=1...n
x[0,i] := 1, i=1,...,n
FOR i = 1 TO dimen+1 DO
    FOR j = 1 TO i DO
       sum := 0
       FOR l = 1 TO n DO
          sum := sum + x[i-1,l]*x[j-1,l]
       ENDDO
       a[i,j] := sum
       a[j,i] := sum
    ENDDO
    sum := 0
    FOR l = 1 TO n DO
       sum := sum + y[l]*x[i-1,l]
    ENDDO
    c[i] := sum
ENDDO
OUTPUT a[i,k], i=1...dimen+1, k=1...dimen+1
OUTPUT c[i], i=1...dimen+1
END multidimensional linear regression
```

21.12.2 Regression of the n-th order

Regression parabola of the n-th order: Integral rational function of the n-th order with the coefficients $a_0, \ldots, a_n$ as parameters:

$$f(x) = \sum_{i=0}^{n} a_i x^i .$$

System of normal equations for the approximating function of the n-th order:

$$\sum_i y_i x_i^k = a_0 \sum_i x_i^k + a_1 \sum_i x_i^{k+1} + a_2 \sum_i x_i^{k+2} + \cdots + a_n \sum_i x_i^{k+n} ,$$

$k = 0, \ldots, n$; x_i, y_i known; a_i unknown.

▷ **Uniqueness of the minimization**: The system of normal equations for the regression has a solution if and only if the determinant of the coefficients does not vanish.

▷ Program sequence for the polynomial regression of n data points (x_i, y_i). The output is a matrix $\mathbf{A}$ of $dimen + 1 \times dimen + 1$ dimensions and a vector $\vec{\mathbf{c}}$ of $dimen + 1$ dimensions. The regression coefficients r_i are obtained as a solution of the system of equations $\mathbf{A}\vec{\mathbf{r}} = \vec{\mathbf{c}}$.

```
BEGIN Polynomial regression
INPUT n
INPUT order (order of the polynomial)
INPUT x[i], i=1...n
INPUT y[i], i=1...n
FOR i = 1 TO order+1 DO
   FOR j = 1 TO i DO
      k := i + j - 2
      sum := 0
      FOR l = 1 TO n DO
         sum := sum + power(x[l],k)
      ENDDO
      a[i,j] := sum
      a[j,i] := sum
   ENDDO
   sum := 0
   FOR l = 1 TO n DO
      sum := sum + y[l]*potenz(x[l],(i-1))
   ENDDO
   c[i] := sum
ENDDO
OUTPUT a[i,k], i=1...order+1, k=1...order+1
OUTPUT c[i], i=1...order+1
END Polynomial regression
```

22

Fuzzy logic

Classical logic: Two-valued logic. There are only two truth values for the truth content of a proposition (true = 1, false = 0).
Switching algebra (two-valued): Physical realization of logical conclusions by means of electronic components in two-valued logic. The truth value "true" is interpreted as 1 or, if a threshold value is not exceeded, as 0.
Fuzzy logic: Multivalued logic whose algebraic operations are defined by mathematical operators. Here, too, logical conclusions are based on rules, but what is important is the linguistic (colloquial) content of the proposition.
Fuzzy-logical inference: Conclusions based on substitution rules (*modus ponens*).
Fuzzy switching algebra: For fuzzy propositions in the sense of fuzzy set theory, computer processing is required so that fuzzy-logic conclusions can be drawn for electronic components which can determine and process truth values between 0 and 1.

22.1 Fuzzy sets

Membership function $y(x)$ in classical set theory assumes only the values 0 and 1:

$$\mu(x) = \begin{cases} 1 & \text{element } x \in G(\text{universe}) \\ 0 & \text{element } x \notin G. \end{cases}$$

Modeling of fuzzy sets: Membership categories are established between 0 and 1.

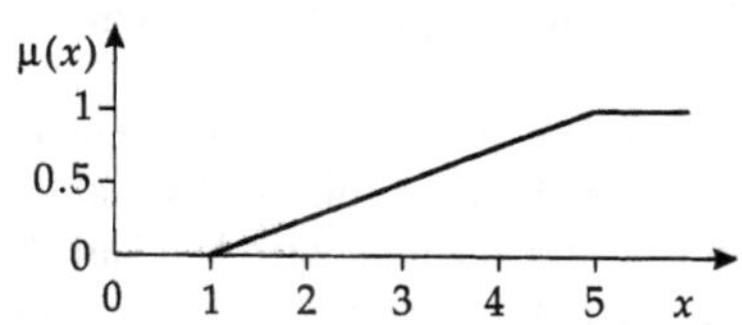

The **membership degree** of $x = 1, 2, 3, 4, 5$ in the **set** is $\mu(x) = 0.0, 0.25, 0.75, 1.0$.

- **Fuzzy subset** A of a set X is characterized by its **membership function**:

$$\mu_A\colon\ X \to [0, 1].$$

To every element x of X a number $\mu(x)$ in the interval $[0, 1]$ is assigned which represents the degree of membership of x in A.

X represents the universe, which must be chosen appropriately. Ordinary sets are interpreted as special fuzzy sets only the values 0 and 1 occur as membership values.
Equality of two fuzzy sets A and B: The values of their membership functions are equal, $A = B$, if $\mu_A(x) = \mu_B(x)$ for all $x \in X$.
Another form of writing this is:

$$A := \{\mu_A(x)/x | x \in X\}.$$

Singletons: Pair of values $\mu_A(x_i)/x_i$ with an ordinate value $\mu_A(x_i)$ and the abscissa value x_i. In terms of content, this corresponds to representation $(x_i, \mu_A(x_i)$ of points in $\mathbb{R}^2$.
For discrete and finite supporting sets $S_A = \{x_1, x_2, \ldots, x_n\}$ of A, the **representation in terms of a sum** holds:

$$A = \mu_A(x_1)/x_1 + \cdots + \mu_A(x_n)/x_n) := \sum_{i=1}^{n} \mu_A(x_i)/x_i.$$

The degree of membership $\mu_A(x)$ in a fuzzy set itself may be a fuzzy set there, as well.
Ultrafuzzy set A fuzzy set whose membership function itself is a fuzzy set.

▷ The plus sign characterizes the union and not the arithmetic sum.

If the universe X is infinite, the summation sign is replaced symbolically by the integral sign:

$$A = \int_{x \in X} \mu_A(x)/x \,\mathrm{d}x.$$

▷ Summation sign and integral sign must be regarded here as abbreviating symbols. These signs as well as the division sign are not used as operation signs.

22.2 Fuzzy concept

Fuzzy methods: Problems that can be written in linguistic form may be transformed to algorithmic calculation methods.
Linguistic elements can be represented in fuzzy sets by means of characteristics or graphs (see functional graphs, Section 22.3). Then linguistic propositions such as IF-THEN rules of algorithms become computation methods (see Sections 22.4, 22.5, and 22.6).
Fuzzy linguistics: linguistic variable and linguistic term: Let us use the example of "low temperature" as an explanation: The temperature is denoted as a parameter. It specifies the **linguistic variable** and represents the universe in fuzzy modeling. "Low" is the colloquial (linguistic) term of the parameter and is described by a fuzzy set (Section 22.3). Fuzzy sets such as "low," "medium," "high," etc., are **linguistic terms**, and their number depends on the problem at hand.

▷ We distinguish between linguistic terms based on equal linguistic variables or parameters and linguistic terms based on different variables.
In Section 22.3 a linguistic variable is denoted generally by x, where x stands, based, for such variables as temperature, pressure, volume, frequency, velocity, brightness,

age, degree of wear but also for such variables as medical, electrical, chemical, economic, etc.

22.3 Functional graphs for the modeling of fuzzy sets

Membership functions are modeled by means of graphs with values between 0 and 1; they represent a graded membership.
Kinds of representation:

- **Discrete representation**: A number of discrete pairs $\mu(x_i)/x_i$ (singletons) are given.
- **Parametric representation**: Analytic functions or piecewise continuous functions are given.

Modeling of a membership function by means of a triangular function:

$$\mu_A(x) = \begin{cases} 0 & \text{for } x < a_1 \\ \dfrac{x-a_1}{a_2-a_1} & \text{for } a_1 \le x \le a_2 \\ \dfrac{a_3-x}{a_3-a_2} & \text{for } a_2 \le x \le a_3 \\ 1 & \text{for } x_n < x. \end{cases}$$

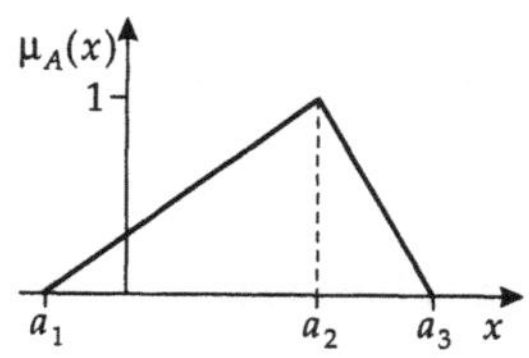

□ Often, symmetric triangular functions are used. If x represents the temperature the above asymmetric triangular function may be interpreted as "better for the temperature to be somewhat too low than too high (with respect to a_2)."

Γ-function as a membership function:

$$\mu_A(x) = \begin{cases} 0 & \text{for } x < a_1 \\ \dfrac{x-a_1}{a_2-a_1} & \text{for } a_1 \le x \le a_2 \\ 1 & \text{for } x_n < x\,. \end{cases}$$

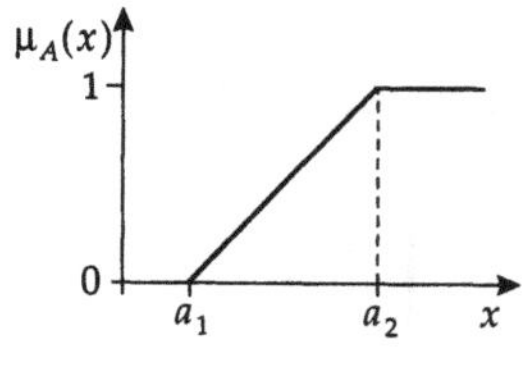

Smoothed gamma function:

$$\mu_A(x) = \begin{cases} 0 & \text{for } 0 \le x \le a \\ \dfrac{k(x-a)^2}{1+k(x-a)^2} & \text{for } a \le x \le \infty. \end{cases}$$

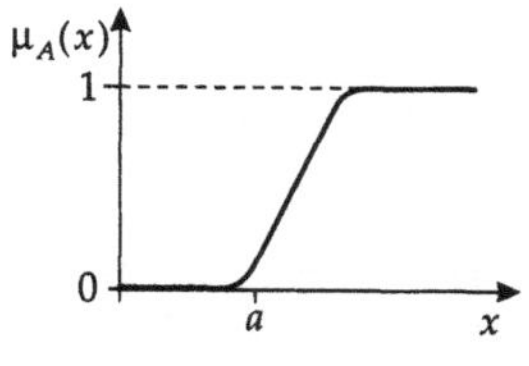

Alternatively:

$$\mu_A(x) = \begin{cases} 0 & \text{for } 0 \le x \le a \\ 1 - \mathrm{e}^{-k(x-a)^2} & \text{for } a > x, k > 0. \end{cases}$$

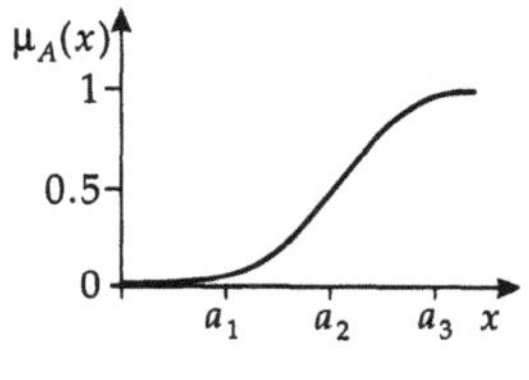

Zadeh's S-function:

$$\mu_A(x) = \begin{cases} 0 & \text{for} \quad x < a_1 \\ 2\left(\dfrac{x-a_1}{a_3-a_1}\right)^2 & \text{for} \quad a_1 \le x \le a_2 \\ 1-2\left(\dfrac{x-a_3}{a_3-a_1}\right)^2 & \text{for} \quad a_2 \le x \le a_3 \\ 1, \text{ with } a_2 = \dfrac{a_1+a_3}{2} & \text{for} \quad x > a_4. \end{cases}$$

▷ For a closed representation the arctan function is also suitable.

L-function:

$$\mu_A(x) = \begin{cases} 1 & \text{for} \quad x < a_1 \\ \dfrac{a_2-x}{a_2-a_1} & \text{for} \quad a_1 \le x \le a_2 \\ 0 & \text{for} \quad x > a_2\,. \end{cases}$$

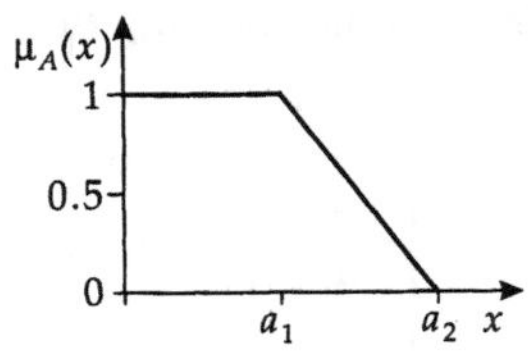

Smoothed L-function:

$$\mu_A(x) = \mathrm{e}^{-kx^2}, \quad k > 0$$

Alternatively:

$$\mu_A(x) = \frac{1}{1+kx^2}, \quad k > 1.$$

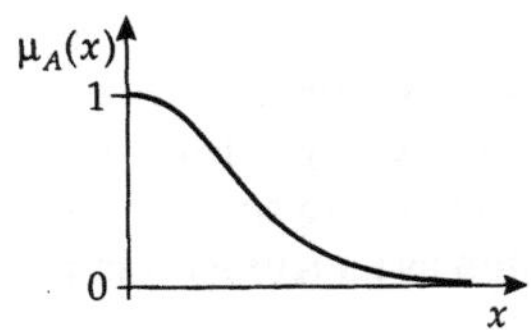

Modeling by means of a trapezoidal function:

$$\mu_A(x) = \begin{cases} 0 & \text{for} \quad x < a_1 \\ \dfrac{x-a_1}{a_2-a_1} & \text{for} \quad a_1 \le x \le a_2 \\ 1 & \text{for} \quad a_2 \le x \le a_3 \\ \dfrac{a_4-x}{a_4-a_3} & \text{for} \quad a_4 \le x \le a_4 \\ 0 & \text{for} \quad x > a_4\,. \end{cases}$$

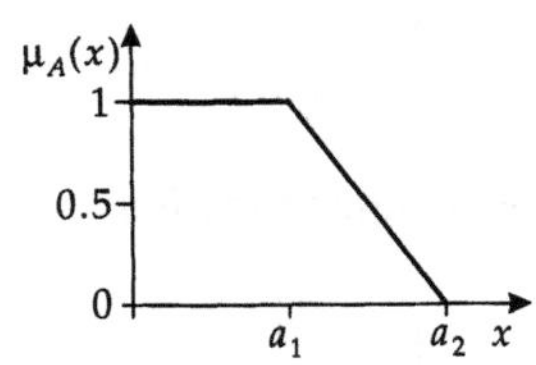

▷ For $a_2 = a_3 = a$, this graph becomes a graph of a symmetric triangular function.

Smoothed membership function:

$$\mu_A(x) = \mathrm{e}^{-k(x-a)^2}, \quad k > 0.$$

Alternatively:

$$\mu_A(x) = \frac{1}{1+k(x-a)^2}, \quad k > 1.$$

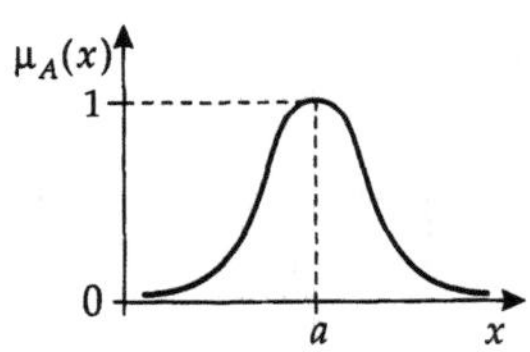

▷ The cosine half-wave offers another possibility for modeling a fuzzy set.

Function with sink:

$$\mu_A(x) = \begin{cases} 1 & \text{for} \quad -\infty \le x \le a_1 \\ \dfrac{a_2 - x}{a_2 - a_1} & \text{for} \quad a_1 \le x \le a_2 \\ 0 & \text{for} \quad a_2 \le x \le a_3 \\ \dfrac{x - a_3}{a_4 - a_3} & \text{for} \quad a_3 \le x \le a_4 \\ 1 & \text{for} \quad a_4 \le x \le \infty\,. \end{cases}$$

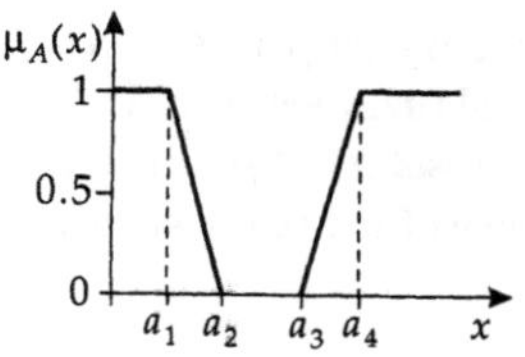

Smoothed sink function:

$$\mu_A(x) = 1 - \mathrm{e}^{-k(x-a)^2}, \quad k > 1,$$

Alternatively:

$$\mu_A(x) = \frac{k(x-a)^2}{1 + k(x-a)^2}, \quad k > 0.$$

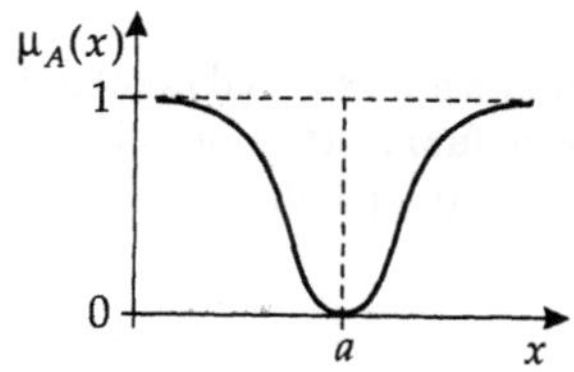

Modeling with a generalized trapezoidal function:

$$\mu_A(x) = \begin{cases} 0 & \text{for} \quad x < a_1 \\ \dfrac{b_2(x - a_1)}{a_2 - a_1} & \text{for} \quad a_1 \le x \le a_2 \\ \dfrac{(b_3 - b_2)(x - a_2)}{a_3 - a_2} + b_2 & \text{for} \quad a_2 \le x \le a_3 \\ 1 \text{ or } b_3 - b_4 & \text{for} \quad a_3 \le x \le a_4 \\ \dfrac{(b_4 - b_5)(a_5 - x)}{a_5 - a_4} + b_5 & \text{for} \quad a_4 \le x \le a_5 \\ \dfrac{b_5(a_6 - x)}{a_6 - a_5} & \text{for} \quad a_5 \le x \le a_6 \\ 0 & \text{for} \quad a_6 < x\,. \end{cases}$$

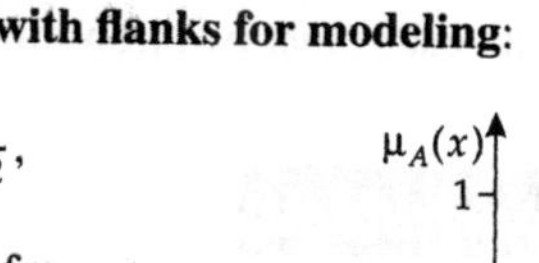

Membership function $F(x)$ with flanks for modeling:

$$F_1 = \mathrm{e}^{-x}, \quad F_2 = \frac{1}{1 + x^2},$$

$$\mu_A(x) = \begin{cases} F_1\left(\dfrac{a_1 - x}{\alpha}\right) & \text{for} \quad x < a_1 \\ 1 & \text{for} \quad a_1 \le x \le a_2 \\ F_2\left(\dfrac{x - a_2}{\beta}\right) & \text{for} \quad x > a_2\,. \end{cases}$$

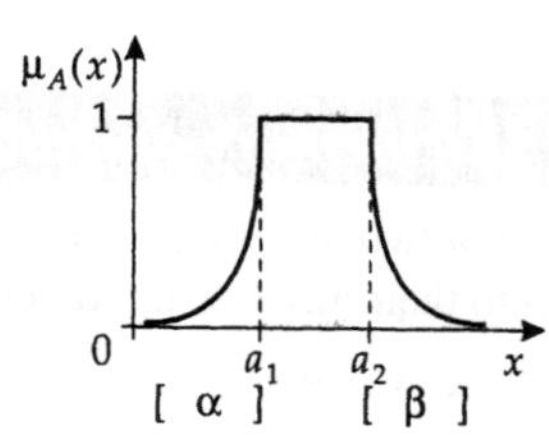

Summary:
Fuzzy and impressive information can be described by fuzzy sets and visualized by characteristic curves.

Support (or carrier) of the fuzzy set μ in A defined by

$$\text{supp}(\mu) = T(\mu): \ \{x | x \in A, \mu > 0\}.$$

Empty fuzzy set μ in A: if $\mu(x) = 0$ for all $x \in A$.
Universal fuzzy set μ in A: if $\mu(x) = 1$ for all $x \in A$.
Fuzzy subset μ_1 of μ_2: If $\mu_1(x) \leq \mu_2(x)$ for all $x \in A$; Notation $\mu_1 \subseteq \mu_2$.
Tolerance of the fuzzy set μ in A: Defined by

$$[a, b] = \{x | \mu(x) = 1\}.$$

Height $H(\mu)$ of the fuzzy set μ in A defined by

$$H(\mu) := \max\{\mu(x) | x \in A\}.$$

The fuzzy set μ is called normal if $H(\mu) = 1$, otherwise it is called subnormal.
Cut of a fuzzy set μ at height α (α–cut): If $\mu : A \to [0, 1]$ and $\alpha \in (0, 1]$, then $\mu_\alpha : A \to [0, 1]$ with

$$\mu_\alpha(x) = \begin{cases} 1 & \text{for} \quad \mu(x) \geq \alpha \\ 1 & \qquad \text{otherwise} \end{cases}$$

is called a cut of the fuzzy set μ in A.
Similarity of fuzzy sets μ_1 and μ_2:

1. $\mu_1, \mu_2 : A \to [0, 1]$ are called fuzzy-like if for every $\alpha \in (0, 1]$ there are numbers α_i with $\alpha < \alpha_i \leq 1, i = 1, 2$, such that

$$\text{supp}(\alpha_1\mu_1)_\alpha \subseteq \text{supp}(\mu_2)_\alpha, \quad \text{supp}(\alpha_2\mu_2)_\alpha \subseteq \text{supp}(\mu_1)_\alpha.$$

2. $\mu_1, \mu_2 : A \to [0, 1]$ are called strictly fuzzy-like if it holds that

$$\mu_1 \approx \mu_2 \quad \text{and} \quad 1 - \mu_1 \approx 1 - \mu_2.$$

▷ • μ_1 und μ_2 are fuzzy-like if they have the same tolerance,
supp$(\mu_1)_1$ =supp$(\mu_2)_2$, because the tolerance is just equal to the influence width of the α–cut of a fuzzy set at height 1, $\text{supp}(\mu)_1 = \{x \in A | \mu = 1\}$.

• Consequence: If the tolerance and the influence width of two fuzzy sets are equal to each other then the sets are strictly fuzzy-like.

22.4 Combination of fuzzy sets

Fuzzy sets can be combined with each other by means of operators.

22.4.1 *Elementary operations*

Union $A \cup B$ of two fuzzy sets A und B: Defined by the maximum operation MAX(. ; .) with respect to their membership functions $\mu_A(x)$ and $\mu_B(x)$:

$$C := A \cup B \text{ and } \mu_C := \text{MAX}(\mu_A(x), \mu_B(x)), \quad \text{for all} \quad x \in X,$$

where

$$\text{MAX}(a, b) = \begin{cases} a & \text{if} \quad a \geq b \\ b & \text{if} \quad a < b\,. \end{cases}$$

The union as the logical OR-connection can be represented in different forms:
OR-operation of two membership functions: Operation MAX.
In this representation $\mu_c(x)$ defines the maximum value of the corresponding membership functions $\mu_A(x)$ or $\mu_B(x)$.

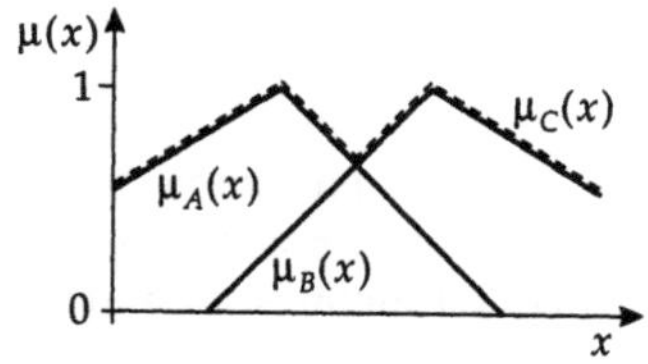

With the help of the **bounded**, **algebraic**, and **drastic sum**, other combinations can be defined to complement the formation of the union.
OR operation of two membership functions, bounded sum. Defined by $C := A \oplus B$ and $\mu_C := \text{MAX}\{1, \mu_A(x) + \mu_B(x)\}$.

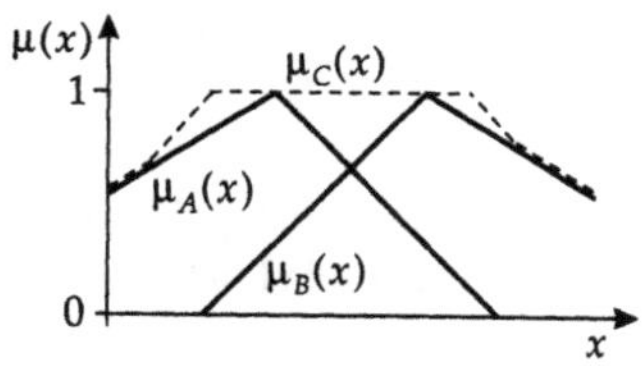

OR operation of two membership functions: Algebraic sum. Defined by

$$C := A + B \text{ and } \mu_C := \mu_A(x) + \mu_B(x) - \mu_A(x)\mu_B(x).$$

For every x, the membership function $\mu_C(x)$ describes the algebraic sum.
OR-operation of two membership functions, drastic sum. The membership function of the drastic sum is defined by:

$$C := A \diamond B \text{ and } \mu_C(x) := \begin{cases} \text{MAX}\{\mu_A(x), \mu_B(x)\} & \text{if} \quad \mu_A(x) = 0 \\ & \text{or} \quad \mu_B(x) = 0 \\ 1 & \text{otherwise.} \end{cases}$$

Intersection $A \cap B$ of two fuzzy sets A and B, defined by the minimum operation MIN(.,.) with respect to the membership functions $\mu_A(x)$ and $\mu_B(x)$:

$$C := A \cap B \text{ and } \mu_C(x) := \text{MIN}(\mu_A(x), \mu_B(x)), \quad \text{for all} \quad x \in X,$$

where

$$\text{MIN}(a, b) = \begin{cases} a & \text{for} \quad a \leq b \\ b & \text{for} \quad a > b. \end{cases}$$

Alternative notation:

$$A \cap B = \int_x \text{MIN}(\mu_A(x), \mu_B(x))/x.$$

The formation of intersection corresponds to the logical AND connection and can be interpreted by means of various representations:
AND operation of two membership functions, operation MIN.
The membership function $\mu_C(x)$ defines the minimum value of $\mu_A(x)$ or $\mu_B(x)$.

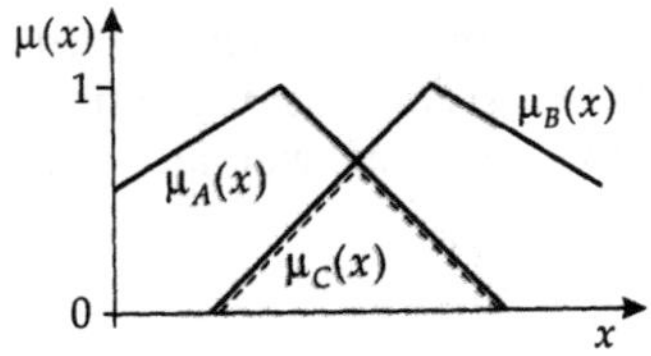

For all $x \in X$, analogous to the extended summation term for the union formation, there are corresponding extensions also for corresponding extensions of the intersection.
AND operation of two membership functions: Bounded product, defined by

$$C := A \cdot B \quad \text{and} \quad \mu_C(x) := \text{MAX}\{0, \mu_A(x) + \mu_B(x) - 1\}.$$

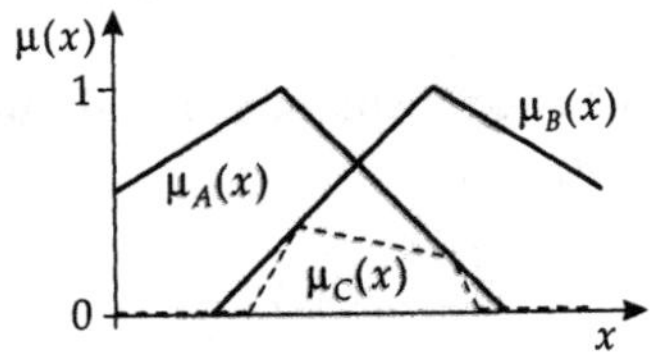

$\mu_C(X)$: Membership function of the bounded product

AND operation of two membership functions: Algebraic product, defined by:

$$C := A \bullet B \text{ and } \mu_C(x) := \mu_A(x)\mu_B(x).$$

The membership function $\mu_C(x)$ represents the algebraic product.
AND-operation of two membership functions: Drastic product.

$$C := A * B \text{ and } \mu_C(x) := \begin{cases} \text{MIN}\{\mu_A(x), \mu_B(x)\} & \text{if} \quad \mu_A = 1 \\ & \text{or} \quad \mu_B = 1 \\ 1 & \text{otherwise.} \end{cases}$$

Complement A^C of a fuzzy set A: Defined by the negation of its membership function,

$$\mu_{A^C} = 1 - \mu_A(x) \quad \text{or} \quad A^C = \int_X (1 - \mu_A(x))/x.$$

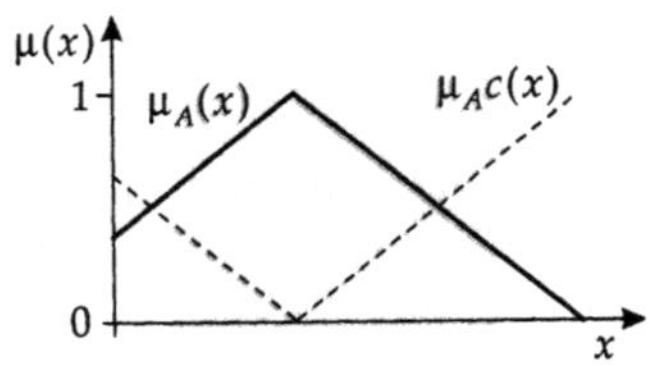

Graphical representation of the negation of a membership function

The negation of membership functions is described by the corresponding complements. $\mu_{A^C}(x)$ is the complement of membership function $\mu_A(x)$.

Difference of membership functions: The difference of two membership functions is given by

$$\mu_C(x) = \mu_A(x) - \mu_B(x) = \mathrm{MAX}(0, \mu_A(x)\mu_B(x)).$$

The range of values is defined by $\mu_C(x) \in [0, 1]$, if $\mu_A(x)$ and $\mu_B(x) \in [0, 1]$.

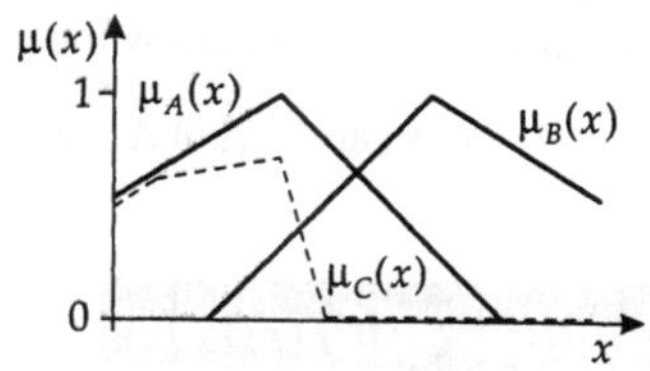

By forming the difference of $\mu_A(x)$ and $\mu_B(x)$, a new membership function $\mu_C(x)$ is defined with values in [0, 1].
Cut: For fuzzy α-cuts we obtain:

$(A \cup B)^{>\alpha} = A^{>\alpha} \cup B^{>\alpha}$	for the union,
$(A \cap B)^{>\alpha} = A^{>\alpha} \cap B^{>\alpha}$	for the intersection,
$(A^C)^{>\alpha} = A^{\leq 1-\alpha} = \{x \in X \mid m_A(x) \leq 1 - \alpha\}$	for the complement.

▷

Operator	Boolean logic	Fuzzy logic
AND	$C = A \wedge B$	$\mu_C = \mathrm{MIN}(\mu_A, \mu_B)$ with $C = A \cap B$
OR	$C = A \vee B$	$\mu_C = \mathrm{MAX}(\mu_A, \mu_B)$ with $C = \cup B$
NOT	$C = \neg A$	$\mu_C^C = 1 - \mu_C$, μ_C^C: complement of μ_C, and $\mu_A, \mu_B, \mu_C \in [0, 1]$.

22.4.2 Calculating rules for fuzzy sets

Simple calculating rules can be derived from the definitions of union, intersection, and complement. From the **union**, we obtain the following for abitrary fuzzy sets A, B, C, etc.:

Commutativity	:	$A \cup B = B \cup A$
Associativity	:	$A \cup (B \cup C) = (A \cup B) \cup C$
Idempotency	:	$A \cup A = A$
Monotonicity	:	$A \subseteq B \Rightarrow A \cup C \subseteq B \cup C$
Special rule	:	$A \cup \emptyset = A$ and $A \cup X = X$.

Correspondingly, for the **intersection**:

Commutativity	:	$A \cap B = B \cap A$
Associativity	:	$A \cap (B \cap C) = (A \cap B) \cap C$
Idempotency	:	$A \cap A = A$
Monotonicity	:	$A \subseteq B \Rightarrow A \cap C \subseteq B \cap C$
Special rule	:	$A \cap \emptyset = \emptyset$ and $A \cap X = A$.

The same distribution laws that apply to ordinary sets also apply to fuzzy sets.

$$A \cup (B \cap C) = (A \cup B) \cap (A \cup C),$$

$$A \cap (B \cup C) = (A \cap B) \cup (A \cap C).$$

Calculating rules for **complements**:

Idempotency	:	$A = A^{CC}$
Relation for inclusion	:	$A \subseteq B \Leftrightarrow B^C \subseteq A^C$
Union	:	$(A \cap B)^C = A^C \cup B^C$
Intersection	:	$(A \cup B)^C = A^C \cap B^C$.

Not all calculating rules that apply to ordinary sets are automatically applied to fuzzy sets. Due to the characteristics of fuzzy sets, the following possibilities exist:

$$A \cup A^C \neq X, \text{ and } \quad A \cap A^C \neq \emptyset.$$

22.4.3 *Rules for families of fuzzy sets*

For **families** $(A_j | j \in J)$ of fuzzy sets (J as index set), the following definitions hold: for all $x \in X$

Union:

$$C := \cup_{j \in J} A_j \quad \text{with} \quad \mu_C := \sup_{j \in J} \mu_{A_j}(x).$$

Intersection:

$$D := \cap_{j \in J} A_j \quad \text{with} \quad \mu_D(x) := \inf_{j \in J} \mu_{A_j}(x).$$

The above forms are direct generalizations of union and intersection.
Distributive law:

$$B \cap \cup_{j \in J} A_j = \cup_{j \in J}(B \cap A_j),$$

$$B \cup \cap_{j \in J} A_j = \cap_{j \in J}(B \cup A_j).$$

Monotonicity characteristics:

$$\cap_{j \in J} A_j \subseteq A_k \subseteq \cup_{j \in J} A_j \quad \text{for all} \quad k \in J\ ,$$

$$B \subseteq A_k \quad \text{for all} \quad k \in J \Rightarrow B \subseteq \cap_{j \in J} A_j\ ,$$

$$A_j \subseteq B \quad \text{for all} \quad k \in J \Rightarrow \cup_{j \in J} A_j \subseteq B.$$

22.4.4 *t norm and t conorm*

Apart from minimum and maximum operations there are other operations that can be used methodically for an aggregation of fuzzy sets. They have been suggested as operators for the intersection (**triangular (t) norms**) and for the union of fuzzy sets (**triangular (t) conorms**).

t norm: Binary operation t in [0, 1], map t: $[0, 1] \times [0, 1] \rightarrow [0, 1]$. The t norm is a two-valued function t in [0, 1]. This function is symmetric, associative, and monotonically increasing, it has 0 as the null element and 1 as the neutral element. For $x, y, z, v, w \in [0, 1]$ the following characteristics apply:

- **Symmetry**: $t(x, y) = t(y, x)$,
- **Associativity**: $t(x, t(y, z)) = t(t(x, y), z)$,
- Operations with the **neutral element** 1 and the **null element** 0:

$$t(x, 1) = x \quad \text{and for the sake of symmetry} \quad t(1, x) = x,$$
$$t(x, 0) = 0.$$

- **Monotonicity**: if $x \le v$ and $y \le w$, then $t(x, y) \le t(v, w)$.

By definition, an intersection $A \cap_t B$ can be introduced to a t norm for fuzzy sets:

$$C := A \cap_t B \text{ with } \mu_C(x) := t(\mu_A(x), \mu_B(x)) \quad \text{for all} \quad x \in X.$$

Due to symmetry and associativity of the intersection the following holds:

$$A \cap_t B = B \cap_t A,$$
$$A \cap_t (B \cap_t C) = (A \cap_t B) \cap_t C,$$
$$A \cap_t \emptyset, \quad A \cap_t X = A \text{ for } A, B, C \in F(X).$$

The above intersections are generated from t norms. Consequently, the following relations hold:

$$A \cap_t B \subseteq A \text{ and } A \cap_t B \subseteq B,$$

as well as

$$A \cap_t B \subseteq A \cap B.$$

For every intersection $\cap_t$ a **dual union** $\cup_t$ can be defined:

$$A \cap_t B := (A^C \cap_t B^C)^C.$$

From this, the **Morgan's rules** follow directly with respect to every t norm for A and B.

$$(A \cap_t B)^C = A^C \cup_t B^C, \qquad (A \cup_t B)^C = A^C \cap_t B^C.$$

The other class of general operators is called t conorms, also designated as s norms. These classes find application in forming the union of fuzzy sets. The maximum operator represents the class of t conorms.
t conorm (or **s conorm**): A map s: $[0, 1] \times [0, 1] \to [0, 1]$.
Characteristics of the t conorm:

- **Symmetry**: $s(x, y) = s(y, x)$,
- **Associativity**: $s(x, s(y, z)) = s(s(x, y), z)$,
- Special operations for the null element 0 and the neutral element 1:

  $s(x, 0) = s(0, x) = x, \quad s(x, 1) = 1,$
- **Monotonicity**: if $x \le v$ and $y \le w$, then: $s(u, v) \le s(v, w)$.

From the union $A \cup_t B$ we can define:

$$C := A \cap_t B \quad \text{with} \quad \mu_C(x) = s_t(\mu_A(x), \mu_B(x)) .$$

Furthermore, for every union $\cap_t$ between fuzzy sets:

$$A \cup_t B = B \cup_t A,$$
$$A \cup_t (B \cup_t C) = (A \cup_t B) \cup_t C,$$
$$A \cup_t \emptyset = A, \quad A \cup_t X = X.$$

A t conorm s is uniquely assigned to every t norm, and vice versa, i.e., there is a functional relationship between the t norm and the s norm,

$$s_t(x, y) = 1 - t(1 - x, 1 - y), \quad t_s(x, y) = 1 - s(1 - x, 1 - y).$$

22.4.5 Non-parametrized operators: t norms and s norms (t conorms)

Intersection t_i and union s_u:

$$t_i(x, y) = \min\{x, y\}, \quad s_u(x, y) = \max\{x, y\}.$$

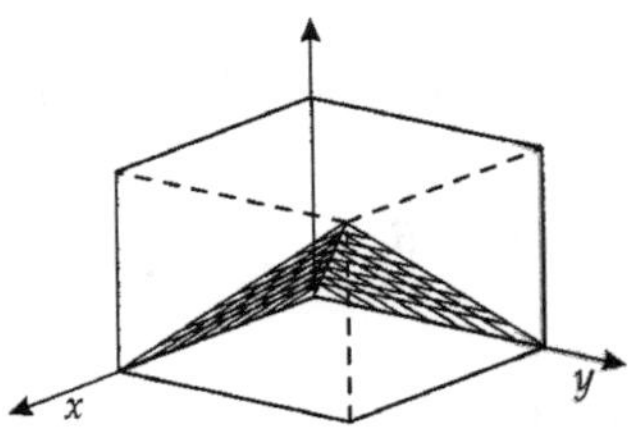

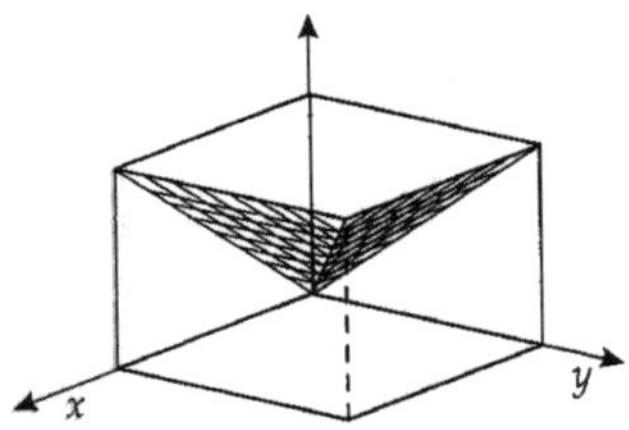

Hamacher product t_h and **Hamacher sum** s_h:

$$t_h(x, y) = \frac{xy}{x + y - xy}, \quad s_h(x, y) = \frac{x + y - 2xy}{1 - xy}.$$

Algebraic product t_a and **algebraic sum** s_a:

$$t_a(x, y) = xy, \quad s_a(x, y) = x + y - xy.$$

Einstein product t_e and **Einstein sum** s_e:

$$t_e(x, y) = \frac{xy}{1 + (1 - x)(1 - y)}, \quad s_e(x, y) = \frac{x + y}{1 + xy}.$$

Bounded product t_b and **bounded sum** s_b:

$$t_b(x, y) = \max\{0, x + y - 1\}, \quad s_b(x, y) = \min\{1, x + y\}.$$

Drastic product t_{dp} and **drastic sum** s_{ds}:

$$t_{dp}(x, y) = \begin{cases} \min\{x, y\} & \text{if} \quad x = 1 \text{ or } y = 1 \\ 0 & \text{if} \quad x, y < 1, \end{cases}$$

$$s_{ds}(x, y) = \begin{cases} \max\{x, y\} & \text{if} \quad x = 0 \text{ or } y = 0 \\ 1 & \text{if} \quad x, y > 0. \end{cases}$$

Order relations for t and s norms listed with respect to their return values:

$$t_{dp} \le t_b \le t_e \le t_a \le t_n \le t_i \le s_u \le s_h \le s_a \le s_e \le s_b \le s_{ds}.$$

22.4.6 Parametrized t and s norms

Yager ($p > 0$):

$$t_{\text{ya}}(x, y) = 1 - \min(1, ((1-x)^p + (1-y)^p)^{\frac{1}{p}}),$$

$$s_{\text{ya}}(x, y) = \min(1, (x^p + y^p)^{\frac{1}{p}}), \quad p \in \mathbb{R}^2.$$

Schweizer (1) ($p > 0$):

$$t_{\text{s1}}(x, y) = (\max(0, x^p + y^p - 1))^{\frac{1}{p}},$$

$$s_{\text{s1}}(x, y) = 1 - \max(0, ((1-x)^p + (1-y)^p - 1)^{\frac{1}{p}}).$$

Schweizer (2) ($p > 0$):

$$t_{\text{s2}}(x, y) = \left(\frac{1}{x^p} + \frac{1}{y^p} - 1\right)^{-\frac{1}{p}},$$

$$s_{\text{s2}}(x, y) = 1 - \left(\frac{1}{(1-x)^p} + \frac{1}{(1-y)^p} - 1\right)^{-\frac{1}{p}}.$$

Schweizer (3) ($p > 0$):

$$t_{\text{s3}}(x, y) = 1 - ((1-x)^p + (1-y)^p - (1-x)^p(1-y)^p)^{\frac{1}{p}},$$

$$s_{\text{s4}}(x, y) = (x^p + y^p - x^p y^p)^{\frac{1}{p}}.$$

Hamacher ($p \geq 0$):

$$t_{\text{h}}(x, y) = \frac{xy}{p + (1-p)(x + y - xy)},$$

$$s_{\text{h}}(x, y) = \frac{x + y - xy - xy(1-p)}{1 - xy(1-p)}.$$

Frank ($p > 0$, $p \neq 1$):

$$t_{\text{f}}(x, y) = \log_p \left[1 + \frac{(p^x - 1)(p^y - 1)}{p - 1}\right],$$

$$s_{\text{f}}(x, y) = \log_p \left[1 + \frac{(p^{1-x} - 1)(p^{1-y} - 1)}{p - 1}\right].$$

Dombi ($p > 0$):

$$t_{\text{do}}(x, y) = \frac{1}{1 + \left[\left(\frac{1-x}{p}\right)^p + \left(\frac{1-y}{y}\right)^p\right]^{\frac{1}{p}}},$$

$$s_{\text{do}}(x, y) = 1 - \frac{1}{1 + \left[\left(\frac{x}{1-x}\right)^p + \left(\frac{y}{1-y}\right)^p\right]^{\frac{1}{p}}}.$$

Weber ($p \geq -1$):

$$t_{\mathrm{w}}(x, y) = \max(0, (1+p)(x+y-1) - pxy),$$
$$s_{\mathrm{w}}(x, y) = \min(1, x+y+pxy).$$

Dubois ($0 \leq p \leq 1$):

$$t_{\mathrm{du}}(x, y) = \frac{xy}{\max(xy, p)},$$
$$s_{\mathrm{du}}(x, y) = 1 - \frac{(1-x)(1-y)}{\max((1-x), (1-y), p)}.$$

22.4.7 Compensatory operators

The requirements of some problems are not satisfactorily met by AND connections and OR connections because they lie between a pure AND and a pure OR.
Compensatory operators:
Lambda operator:

$$\mu_{A\lambda B}(x) = \lambda[\mu_a(x)\mu_B(x)] + (1-\lambda)[\mu_A(x) + \mu_B(x) - \mu_A(x)\mu_B(x)],$$

with $\lambda \in [0, 1]$. The value of λ determines where the operator lies between pure AND and pure OR.
Case $\lambda = 0$: We obtain the OR operator

$$\mu_{A\lambda B}(x)|_{\lambda=0} = \mu_A(x) + \mu_B(x) - \mu_A(x)\mu_B(x).$$

Case $\lambda = 1$: Pure AND operator

$$\mu_{A\lambda B}|_{\lambda=1} = \mu_A(x)\mu_B(x).$$

Gamma operator:

$$\mu_{A\gamma B}(x) = [\mu_A(x)\mu_B(x)]^{1-\gamma}[1 - (1-\mu_A(x))(1-\mu_B(x))]^{\gamma}, \quad \gamma \in [0, 1].$$

Case $\gamma = 1$: Pure OR operator

$$\begin{aligned}\mu_{A\gamma B}(x)|_{\gamma=1} &= 1 - (1-\mu_A(x))(1-\mu_B(x)) \\ & 1 - [1 - \mu_A(x) + \mu_B(x) + \mu_A(x)\mu_B(x)].\end{aligned}$$

Case $\gamma = 0$: Pure AND operator

$$\mu_{A\gamma B}(x)|_{\gamma=0} = \mu_A(x)\mu_B(x).$$

Application of the gamma operator to an arbitary number of fuzzy sets

$$\mu(x) = \left[\prod_{i=1}^{n} \mu_i(x)\right]^{1-\gamma} \times \left[1 - \prod_{i=1}^{n}(1-\mu_i(x))\right]^{\gamma}$$

and provided with weighting δ_i

$$\mu(x) = \left[\prod_{i=1}^{n} \mu_i(x)^{\delta_i}\right]^{1-\gamma} \times \left[1 - \prod_{i=1}^{n}(1-\mu_i(x))^{\delta_i}\right]^{\gamma}, \; x \in G, \sum_{i=1}^{n} \delta_i = 1, \gamma \in [0, 1].$$

22.5 Fuzzy relations

Cartesian product: Let X and Y be fuzzy universes; then the Cartesian product $X \times Y$ in the universe u is a fuzzy set:

$$u = X \times Y = \{(x, y) | x \in X \wedge y \in Y\}.$$

Fuzzy relation: A fuzzy relation R in u is a fuzzy subset R: $R \in F(u)$.
R can be described by a membership function $\mu_R(x, y)$ that assigns the membership degree μ from $|0, 1|$ to every element $(x, y) \in u$.

▷ A **cross product set** of n universes is called an n array fuzzy relation.

Consequence: The fuzzy sets considered so far are one-valued fuzzy relations, i.e., they are characteristics over a universe. A two-valued fuzzy relation can be interpreted as a surface over u.

Discrete finite universes: Representation of the two-valued fuzzy relation as a **fuzzy relation matrix**.

□ Color and ripeness relation: Modeling of the known relation in the form of a relation matrix between color x and ripeness y of a fruit with possible colors X ={green, yellow, red } and ripeness Y = {unripe, half-ripe, ripe}. Binary relation matrix with elements {0, 1}:

R:	unripe	half-ripe	ripe
green	1	0	0
yellow	0	1	0
red	0	0	1

thus

$$R = \begin{pmatrix} 1 & 0 & 0 \\ 0 & 1 & 0 \\ 0 & 0 & 1 \end{pmatrix}.$$

Interpretation:

IF a fruit is green ,	THEN it is unripe.
IF a fruit is yellow,	THEN it is half-ripe.
IF a fruit is red,	THEN it is ripe.

A fuzzy relation matrix with $\mu_R \in [0, 1]$: μ_R(green, unripe) = 1.0; μ_R(green, half-ripe) = 0.5; μ_R(green, ripe) = 0.0; μ_R(yellow, unripe) = 0.25 μ_R(yellow, half-ripe) = 1.0; μ_R(yellow, unripe) = 0.25; μ_R(red, unripe) = 0.0; μ_R(red, half-ripe) = 0.5; μ_R(red, ripe) = 1.0

$$R = \begin{pmatrix} 1 & 0.5 & 0 \\ 0.25 & 1 & 0.25 \\ 0 & 0.5 & 1 \end{pmatrix}.$$

▷ Calculating rule for the connection of fuzzy sets, such as μ_1 and μ_2, on different universes with the AND-connection, i.e., the MIN-operator:

$$\mu_R(x, y) = \mathrm{MIN}(\mu_1(x), \mu_2(y)) \text{ or } (\mu_1 \times \mu_2)(x, y) = \mathrm{MIN}(\mu_1(x), \mu_2(x)).$$

Interpretation: The result of the connection is a fuzzy relation R on the cross product set u with $(x, y) \in u$.

Cross product or **Cartesian product** of fuzzy sets: If X and Y are discrete sets, and hence $\mu_1(x)$, $\mu_2(y)$ can be represented as vectors, then

$$\mu_1 \times \mu_2 = \mu_1^T \circ \mu_2, \quad \mu_{R^{-1}}(x, y) := \mu_R(x, y) \quad \text{for all} \quad (x, y) \in G.$$

▷ The connection operator $\circ$ does not represent the common matrix product; the product and addition are replaced by the MIN-operator and MAX-operator, respectively.

Interpretation: The degree to which an inverse relation R^{-1} applies to the objects $\{x, y\}$ is always the same as the degree to which R applies to the objects $\{x, y\}$.

Calculating rule for the connection of fuzzy relations on the same product set: Binary fuzzy relations $R_1, R_2 : X \times Y \to [0, 1]$ and $(x, y) \in G$:

Rule for the **AND connection** via the **MIN operator**:

$$\mu_{R_1 \cap R_2}(x, y) = \text{MIN}(\mu_{R_1}(x, y), \mu_{R_2}(x, y)).$$

Rule for the **OR connection** via the **MAX operator**:

$$\mu_{R_1 \cup R_2}(x, y) = \text{MAX}(\mu_{R_1}(x, y), \mu_{R_2}(x, y)).$$

Composite or **relational product** $R \circ S$: Let be $R \in F(X \times Y)$ and $S \in F(X \times Z)$, but also, specifically, $R, S \in F(G)$ with $G \subseteq X \times Y$, then the composite or fuzzy relational product is $R \circ S$:

$$\mu_{R \circ S}(x, y) := \text{SUP}_{y \in Y}\{\text{MIN}(\mu_R(x, y), \mu_S(y, z))\} \quad \text{for all} \quad (x, z) \in X \times Z.$$

● Here, SUP stands for supremum, which is often replaced by MAX. Interpretation: Let R be a relation from X to Y, and S a relation from Y to Z.
Consequence: The composition $R \circ S$ of R and S is defined as the Max-Min product.

▷ 1. The above fuzzy composition product is called **MAX-MIN composition.**

2. **MAX-PRO composition**: The product is formed with an ordinary matrix multiplication.

3. **MAX average composition**: Multiplication is replaced by forming the average value.

Calculating rules for the concatenation of fuzzy relations R, S, T:

Associative law:

$$(R \circ S) \circ T = R \circ (S \circ T).$$

Distributive law for the union:

$$R \circ (S \cup T) = (R \circ S) \cup (R \circ T).$$

Distributive law (weakened) for the intersection:

$$R \circ (S \cap T) \subseteq (R \circ S) \cap (R \circ T).$$

Inverse:

$$(R \circ S)^{-1} = S^{-1} \circ R^{-1}.$$

Complement and inverse:

$$(R^{-1})^{-1} = R \text{ and } (R^C)^{-1} = (R^{-1})^C.$$

□ Let $X = \{\text{Tom, Harry}\}$ and $Y = \{\text{John, Jim}\}$. A fuzzy relation "similarity" between the members of X and Y could be:
similarity $= 0.8/(\text{Tom, John}) + 0.6/(\text{Tom, Jim}) + 0.2/(\text{Harry, John}) + 0.9/(\text{Harry, Jim})$.

Written as a relation matrix:

	John	Jim
Tom	0.8	0.6
Harry	0.2	0.9

Let be $Z = \{\text{Arthur, George}\}$ and let the similarity relation matrix μ_S between Y and Z let be given by:

	Arthur	George
John	0.5	0.9
Jim	0.4	1.0

Then, the relation matrix between X and Z follows from the Max-Min product by components, according to the following definition:

$$\begin{aligned}
\text{Matrix element}(1,1)&: \text{SUP}[\text{MIN}(0.8, 0.5), \text{MIN}(0.6, 0.4)] = \text{SUP}[0.5, 0.4] = 0.5,\\
\text{Matrix element}(1,2)&: \text{SUP}[\text{MIN}(0.8, 0.9), \text{MIN}(0.6, 1.0)] = \text{SUP}[0.8, 0.6] = 0.8,\\
\text{Matrix element}(2,1)&: \text{SUP}[\text{MIN}(0.2, 0.5), \text{MIN}(0.9, 0.4)] = \text{SUP}[0.2, 0.4] = 0.4,\\
\text{Matrix element}(2,2)&: \text{SUP}[\text{MIN}(0.2, 0.9), \text{MIN}(0.9, 1.0)] = \text{SUP}[0.2, 0.9] = 0.9.
\end{aligned}$$

Interpretation: The matrix $R \circ S$ gives the membership values among individual persons.

Inference compositions: IF-THEN rules in the form of relation matrices can be connected by means of the composition of inference.

▷ The known matrix multiplication must be replaced by: Product → MIN operator and addition → MAX operator.

▷ Fuzzy-logic inference with IF-THEN rule via of the composition $\mu_2 = \mu_1 \circ R$, the fuzzy set μ_2 represents the required conclusion. Expressed in terms of a formula:

$$\mu_2(y) = \text{MAX}(\text{MIN}(\mu_1(x), \mu_R(x, y)))$$

with $y \in Y$, $\mu_1 : X \to [0, 1]$, $\mu_2 : Y \to [0, 1]$ and $R : G \to [0, 1]$.

22.6 Fuzzy inference

Application of fuzzy relations for modeling fuzzy inference, i.e., fuzzy-logic inference to fuzzy information.

Fuzzy inference: Consists of one or several rules, called **implications**, a **fact**, and a **conclusion.**

▷ Fuzzy inference (approximate reasoning) usually cannot be described with classical logic.

Fuzzy implication: Consists of an IF-THEN rule:

1. The IF part of the rule, usually called the premise, is the condition.
2. The THEN part of the rule is the conclusion.
3. $\mu_2(y) = \text{MAX}_{x \in X}(\text{MIN}(\mu_1(x), \mu_R(x, y)))$ and $\mu_2 = \mu_1 \circ R_1$.

   Interpretation: μ_2 is the fuzzy inference image of μ_1 with respect to the fuzzy relation R.

▷ Fuzzy relations are suited for modeling fuzzy implications, such as IF-THEN rules.

Fuzzy inference: Inference based on rules with fuzzy propositions; in principle, this results in a "calculating prescription" for IF-THEN rules or groups of rules.

General fuzzy-inference scheme: For rules with several conditions

IF A_1 AND A_2 AND A_3 ... AND A_n THEN B

where $A_i : \mu_i : X_i \to [0, 1]$, $i = 1, \ldots, n$, and the membership function of the conclusion $B : \mu : Y \to [0, 1]$ with $R : X_1 \times X_2 \times \cdots \times X_n \times Y \to [0, 1]$, it holds for the actual event with the sharp values $x'_1, x'_2, \ldots, x'_n$ of the quantifiers X_i, $i = 1, \ldots, n$ and $y \in Y$:

$$\mu_{B'} = \mu_R(x'_1, x'_2, \ldots, x'_n, y) = \text{MIN}(\mu_1(x'_1), \ldots, \mu_n(x'_n), \mu_B(y)).$$

□ **Formation of a fuzzy relation:**

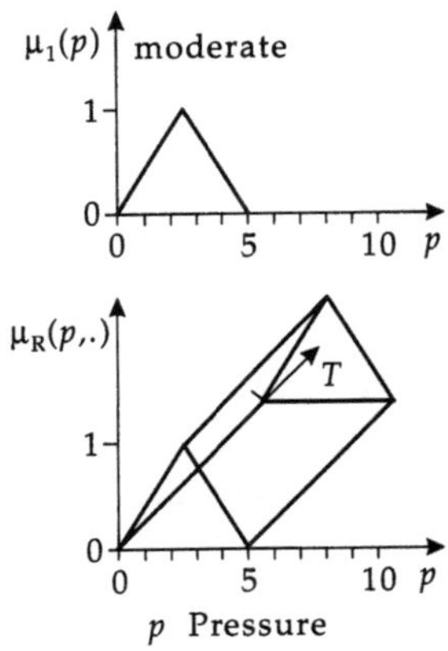

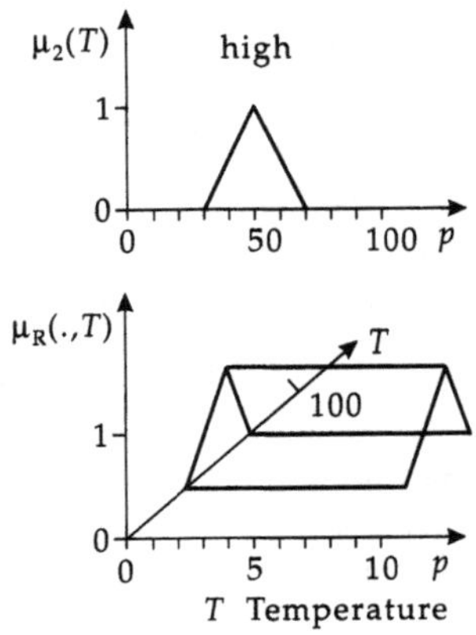

Cylindrical extension of the fuzzy set moderate pressure: $\tilde{\mu}_1(p, T) = \mu_1(p)$ for all $T \in X_1$ and $\mu_1 : X_1 \to [0, 1]$

Cylindrical extension of the fuzzy set high temperature: $\tilde{\mu}_2(p, T) = \mu_2(T)$ for all $p \in X_2$ and $\mu_2 : X_2 \to [0, 1]$; $\tilde{\mu}_1, \tilde{\mu}_2 : G \to [0, 1]$

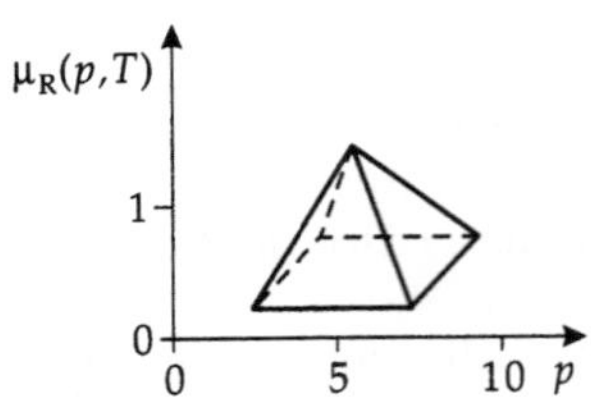

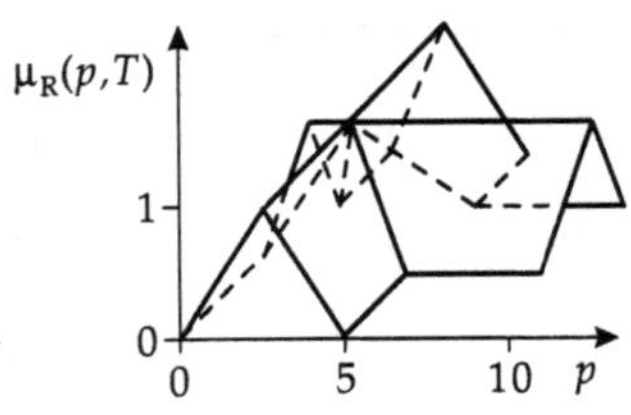

Fuzzy relation "moderate pressure **AND** high temperature" are combined with the **MIN operator**: $\mu_R(p, T) = \text{MIN}(\mu_1(p), \mu_2(T))$

Fuzzy relation "moderate pressure **OR** high temperature" are combined with the **MAX operator**: $\mu_R(p, T) = \text{MAX}(\mu_1(p), \mu_2(T))$

22.7 Defuzzification methods

Maximum criterion method:
From the interval in which the fuzzy set $\mu^{\text{Output}}_{x_1,\ldots,x_n}$ has a maximum degree of membership, an arbitrary value $\eta \in Y$ is selected.

Mean-of-maximum (MOM) method:
As a control value we take the mean of the maximum membership values, i.e., set Y is an interval; it should not be empty and is characterized by:

$$\mathrm{SUP}(\mu^{\mathrm{Output}}_{x_1,\dots,x_n}) := \{y \in Y \,|\, \mu_{x_1,\dots,x_n}(y) \geq \mu_{x_1,\dots,x_n}(y^*) \quad \text{for all} \quad y^* \in Y\}.$$

Consequence:

$$\eta_{\mathrm{MOM}} = \frac{1}{\int_{y \in \mathrm{SUP}(\mu^{Output}_{x_1,\dots,x_n})} \mathrm{d}y} \times \int_{y \in \mathrm{SUP}(\mu^{Output}_{x_1,\dots,x_n})} y \, \mathrm{d}y.$$

Center of gravity (COG):
The center of gravity of the area (density equal to 1) is calculated:

$$\eta_{\mathrm{COG}} = \frac{1}{\int_{y_{\mathrm{inf}}}^{y_{\mathrm{sup}}} \mu(y)\, \mathrm{d}y} \times \int_{y_{\mathrm{inf}}}^{y_{\mathrm{sup}}} \mu(y) y \, \mathrm{d}y.$$

Center of area (COA):
The position of an axis parallel to the ordinate is calculated such that the left and right sides of the area below the membership function become equal to each other,

$$\int_{y_{\mathrm{inf}}}^{\eta} \mu(y)\, \mathrm{d}y = \int_{\eta}^{y_{\mathrm{sup}}} \mu(y)\, \mathrm{d}y.$$

Basic Defuzzification Distribution (BADD):
A parametric method with $\gamma \in \mathbb{R}$ as parameter:

$$\eta_{\mathrm{BADD}} = \frac{1}{\int_{y_{\mathrm{inf}}}^{y_{\mathrm{sup}}} \mu(y)^{\gamma}\, \mathrm{d}y} \times \int_{y_{\mathrm{inf}}}^{y_{\mathrm{sup}}} \mu(y)^{\gamma} y \, \mathrm{d}y.$$

Also: $\eta_{\mathrm{COG}} = \eta_{\mathrm{BADD}},\ \gamma = 1,\ \eta_{\mathrm{MOM}} = \eta_{\mathrm{BADD}},\ \gamma \to 0$.
Extended Center of Area (XCOA):

$$\int_{y_{\mathrm{inf}}}^{\eta_{\mathrm{XCOA}}} \mu(y)^{\gamma}\, \mathrm{d}y = \int_{\eta_{\mathrm{XCOA}}}^{y_{\mathrm{sup}}} \mu(y)^{\gamma}\, \mathrm{d}y.$$

Center of Largest Area (CLA):
From the total set a significant subset is selected that is further evaluated with known methods, e.g., the Center of Gravity method or the Center of Area method.
Customizable Basic Defuzzification Distribution (CBADD):
If the exponent γ of the parametric defuzzification method is regarded as a function of y, then we directly obtain.

$$\eta_{\mathrm{CBADD}} = \frac{\int_{y_{\mathrm{inf}}}^{y_{\mathrm{sup}}} \mu(y)^{\gamma(y)} y \, \mathrm{d}y}{\int_{y_{\mathrm{inf}}}^{y_{\mathrm{sup}}} \mu(y)^{\gamma(y)}\, \mathrm{d}y}.$$

The CBADD method is a generalization of the BADD method. It is of interest if $\mu(y)$ itself must be especially weighted.
Modified Semi-Linear Defuzzification (MSLIDE):
MSLIDE is a parametric defuzzification method proportional to η_{COG} and η_{MOM} as special cases:

$$\eta_{\mathrm{MLSLIDE}} := (1 - \beta)\eta_{\mathrm{COG}} + \beta \eta_{\mathrm{MOM}}, \quad 0 \leq \beta \leq 1.$$

Other methods: clustering methods, knowledge-based methods, decision under fuzzy constraints.

22.8 Example: erect pendulum

Method of fuzzy control using the example of an erect pendulum. Basic structure of a fuzzy controller:

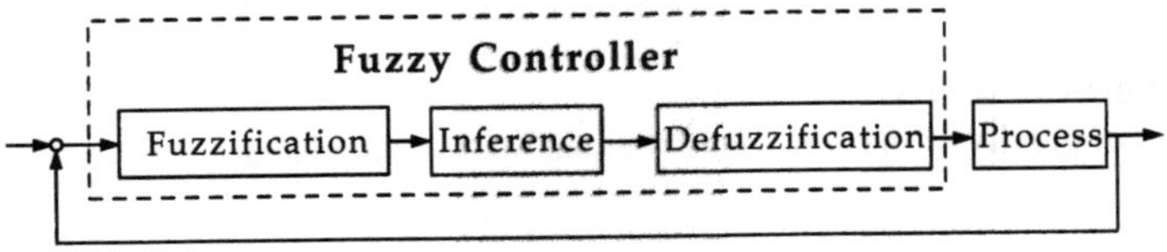

Fuzzy control of an erect pendulum (Mamdani control concept). The concept is based on the model of a human "control expert" (cognitive task). The expert formulates his knowledge in the form of linguistic rules.
⇒ Questioning of the expert is necessary,
⇒ Observation of the expert's behavior.
In general, **linguistic rules** consist of
– premise (specification of the values for the measuring quantities),
– conclusion (giving an appropriate control value).
Consequence: For all sets of values $X_1, X_2, \ldots, X_n$ (for the measuring quantities) and Y (for the control quantity) appropriate linguistic terms such as approximately zero, positive small, etc. must be established. But approximately zero, with regard to the measuring quantity ζ_1, may mean something entirely different than approximately zero for the measuring quantity ζ_2.
For set X_1 we can use the three linguistic terms "negative," "approximately zero," and "positive." Then, in mathematical modeling, a fuzzy set must be assigned to every one of these three linguistic terms (see below).
Crude partitioning:

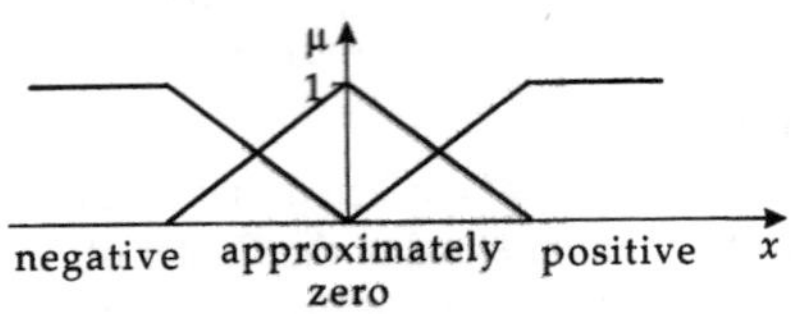

More refined partitioning:

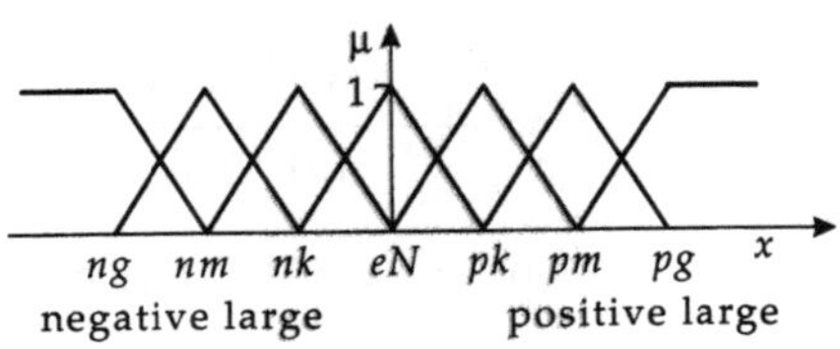

Range of values:
Values of angles $(-90° < 0 < 90°) : X_1 := [-90°, 90°]$,
Values of angular velocity $(-45°/\text{s} \leq \dot{\theta} \leq 45°/\text{s}) : X_2 := [-45°, 45°]$,
Values of force: $(-10\,\text{N} \leq F \leq 10\,\text{N}) : Y := [-10, 10]$.

Partitioning: **for the set X_1**

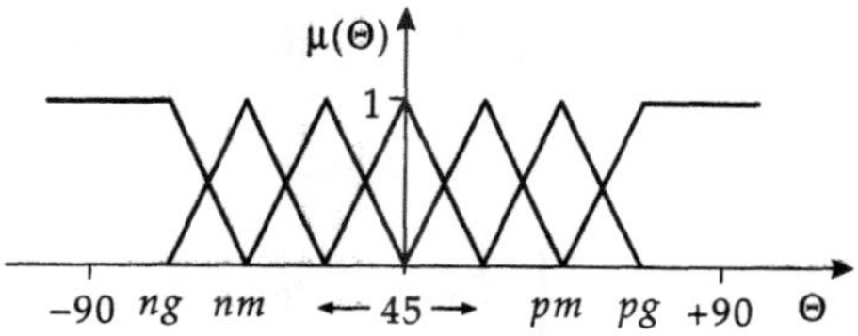

Partitioning: **for the set X_2**

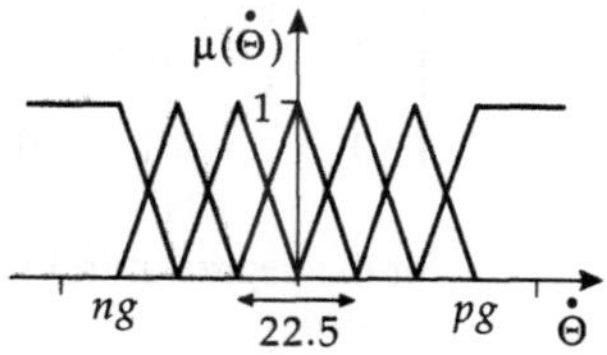

Partitioning: **for the set Y**

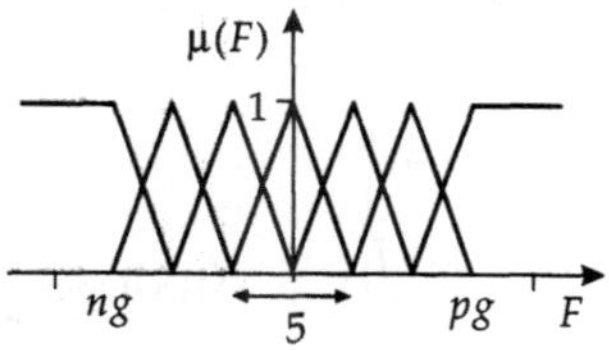

Basis of rules:
nl = negative large
nm = negative moderate
ns = negative small
az = approximately zero
ps = positive small
pm = positive medium
pl = positive large

	nl	nm	ns	az	ps	pm	pl
nl			ps	pl			
nm				pm			
ns	nm		ns	ps			
az	nl	nm	ns	az	ps	pm	pl
ps				ns	ps		pm
pm				nm			
pl				nl	ns		

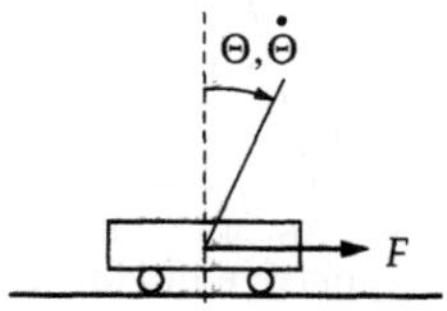

The basis of rules contains 19 rules.
Actual measuring values: Possible starting values.
Angle $\theta = 36°$.
Angular velocity $\dot{\theta} = -2.25°/s$.
Selection from 19 rules:

R_1: If θ is positive small and $\dot{\theta}$ is approximately zero then F is positive small.

R_2: If θ is positive moderate and $\dot{\theta}$ is approximately zero then F is positive moderate.

For this rule R_1 the premise is met at $\min\{0.4, 0.8\} = 0.4$, or more precisely:

$$\alpha = \min\{\mu^{(1)}(\theta); \mu^{(1)}(\dot{\theta})\} = \{0.4; 0.8\} = 0.4.$$

Consequence:

$$\mu^{\text{Output(R1)}}_{36;-2.25}(y) = \begin{cases} \frac{2}{5}y & \text{for} \quad 0 \le y \le 1 \\ \frac{2}{5} & \text{for} \quad 1 \le y \le 4 \\ 2 - \frac{2}{5}y & \text{for} \quad 4 \le y \le 5 \\ 0 & \quad\text{otherwise} \end{cases}$$

For the rule R_2: $\min\{0.6, 0.8\} = 0.6$ for the degree of truth to which the premise applies.
Or more precisely: $\alpha = \min\{\mu^{(2)}(\theta); \mu^{(2)}(\dot{\theta})\}$
Practical approach:

Evaluation of rule R_1

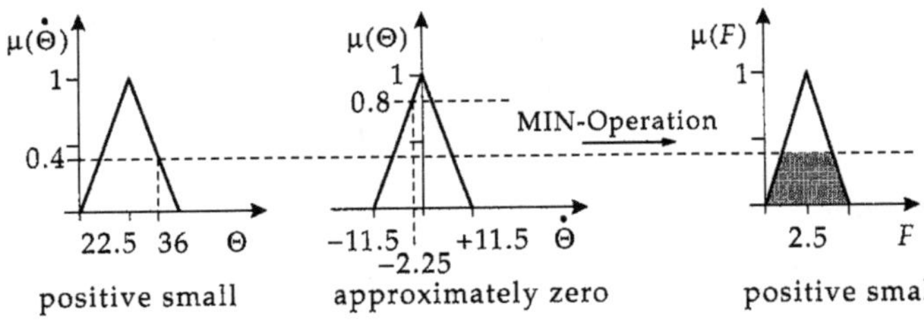

Evaluation of rule R_2

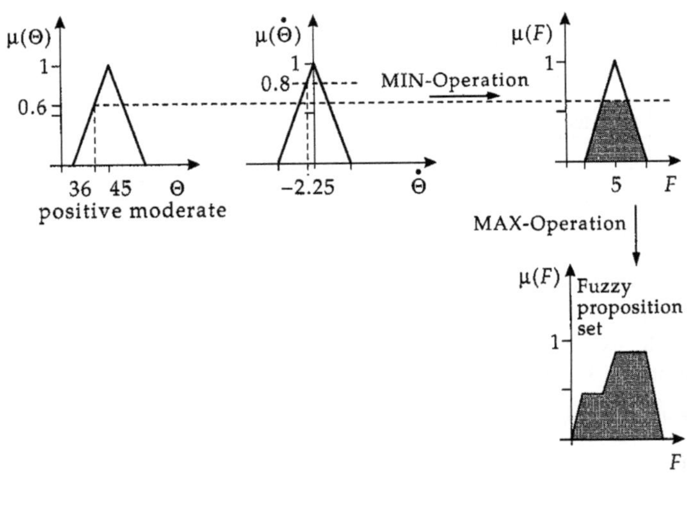

$$\mu^{\text{Output(R2)}}_{36;-2.25}(y) = \begin{cases} \frac{2}{5}y - 1 & \text{for} \quad 2.5 \le y \le 4 \\ \frac{3}{5} & \text{for} \quad 4 \le y \le 6 \\ 3 - \frac{2}{5}y & \text{for} \quad 6 \le y \le 7.5 \\ 0 & \text{otherwise} \end{cases}$$

For all other 17 rules the degree of truth zero for the premise is obtained; they therefore yield fuzzy sets that are constantly zero.

Evaluation of both rules (decision logic):

Approach:

Evaluation of the obtained fuzzy sets that are composed by means of forming the maximum (union) (see MAX-MIN composite in Section 22.5). The decision logic yields:

$$\mu^{\text{Output}}_{X_1,\ldots,X_n} : Y \to [0, 1];\ y \to \max_{r\in\{1,\ldots,k\}} \left\{\min\left[\mu^{(1)}_{i_{1,r}}(x_1), \ldots, \mu^{(n)}_{i_{1,r}}(x_n), \mu_{i_r}(y)\right]\right\}.$$

Fuzzy set after forming the maximum:

$$\mu^{\text{Output}}_{36;-2.25}(y) = \begin{cases} \frac{2}{5}y & \text{for} \quad 0 \le y \le 1 \\ \frac{2}{5} & \text{for} \quad 1 \le y \le 3.5 \\ \frac{2}{5}y - 1 & \text{for} \quad 3.5 \le y \le 4 \\ \frac{3}{5} & \text{for} \quad 4 \le y \le 6 \\ 3 - \frac{2}{5}y & \text{for} \quad 6 \le y \le 7.5 \\ 0 & \text{otherwise.} \end{cases}$$

The other rules do not yield any contribution since the fuzzy sets themselves are zero.

Defuzzification:

The decision logic yields no acute value for the control value but a fuzzy set. With this method we therefore obtain a transformation that assigns a fuzzy set $\mu^{\text{Output}}_{x_1,\ldots,x_n}$ of Y to any n-tuple $(x_1, \ldots, x_n) \in X_1 \times \ldots \times X_n$ of measured values. Defuzzification means that **one** control valuemust be measured.

The center of gravity method yields: $F = 3.95$.

The maximum criterion method yields : $F = 5.0$.

General approach: The trajectory should run such that the endpoint lies in the center.

Rules belonging to the defuzzification output:

1. The defuzzification output initiates an iteration process that ultimately leads to the middle of the rule area, i.e., it yields the control quantity = 0.
2. Stability checks must be performed.
3. Fuzzy Control

   → knowledge-based definition of a set of characteristics

   → design parameters:

- choice of fuzzy sets
- partitions
- type of conclusion
- defuzzification

→ structure set: Every (nonlinear) set of characteristics can be approximated with any required accuracy by choosingthese parameters.

▷ We distinguish the Mamdani controller and the Sugeno controller.

22.9 Fuzzy realizations

Fuzzy software development tools:
Professional software development tools for fuzzy systems are available that contain algorithms for fuzzification, inference, and defuzzification.

□ "Fuzzylib" offers DOS and WINDOWS applications for PCs, double precision algorithms for fuzzification, inference, and defuzzification. The number of fuzzy variables and fuzzy rules per fuzzy object is practically endless, and fuzzy objects can be arbitrarily combined. A fuzzy object can also be configured by means of a fuzzy description file in text format.

□ Other available software tools: SIEFUZZY, Fuzzy Control Manager (FCM).

▷ A ombination of neural network technology and fuzzy-logic technology offers new opportunities for generating intelligent software tools.

Fuzzy hardware:
For fuzzy sets to be realized, we must take into account the fact that the characteristics must be discretized. Straight flanks of fuzzy sets become staircase-like boundaries, and nonlinear graphs for fuzzy sets can be changed so that they are no longer fuzzy-like. The algebra of fuzzy sets is not boolean algebra.

□ The fuzzy modular system ZDIS-IC of T. Yamakawa on CMOS-basis is a modular system for a fuzzy controller with a module for membership functions, MIN-MAX module, and a defuzzification module for two input quantities and one output quantity.

▷ **Fuzzy SPS** (Fuzzy memory programmable control system) are available for use in industry.

23

Neural networks

23.1 Function and structure

Aritificial neural networks: The basic ideas for the construction of electronic neural networks originate in neurophysiology. The function of natural neurons is simulated in a greatly simplified manner. Thus, computer programs (or so-called neural hardware) may contain properties of the human brain such as **tolerance toward errors** or failure of some components, as well as an **ability to generalize**.

□ A typical example for generalization is the recognition of incomplete, distorted or noisy patterns.

● The ability to transfer acquired knowledge to new situations is one of the most important requirements of neural networks.

As in the biological model, (an electronic) neural network consists of a large number of "neurons." These simple processors are also called **nodes**.

23.1.1 Function

The state of every node is determined by the signals it receives from other nodes. To every connection between two nodes (called synapses, as in biology), a **weight** is assigned.
A signal transmitted by the synapse is multiplied by the approximate weight. As a rule, every node adds all incoming signals and assigns a simple function to the sum. Usually, this function is s-shaped and is therefore called a **sigmoid function**.

□ A common sigmoid function is the hyperbolic tangent.

The result can be transmitted directly as its own signal. Usually, every connection from a node j to a node i has its own weight w_{ij}.

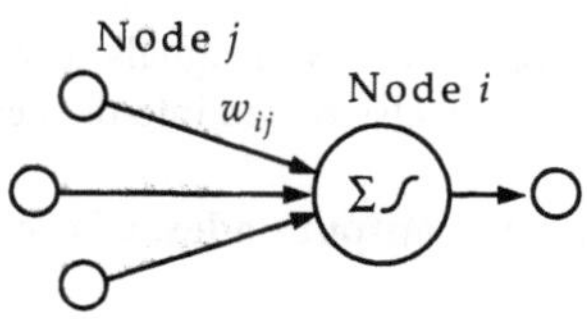

Technical model of a neuron

- The **weights** determine the response of a network to an input, thus acting as its **information memory**.
- The **nonlinearity of sigmoid functions** is crucial for the efficiency of neural networks. For mathematical reasons, differentiability, strictly monotonic behavior, and limited scope are necessary and practical.

▷ It has proven convenient to introduce **threshold parameters** that shift the argument of the sigmoid function. The threshold parameters are treated the same as weights of connections that originate in a node with a constant signal (usually 1).

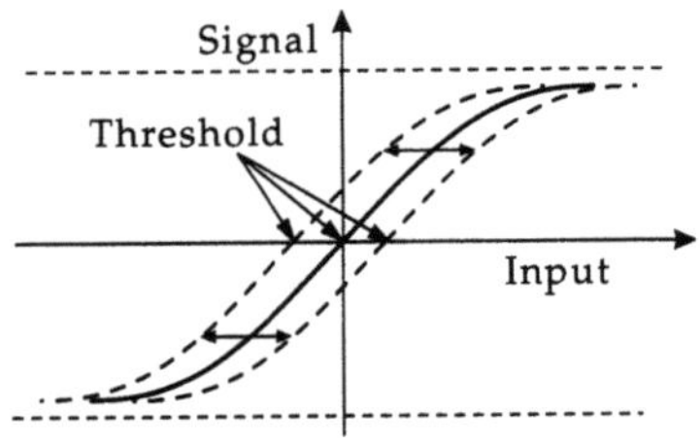

Effect of a sigmoid function with different threshold parameter values

Learning: Neural networks **learn** from example inputs with or without evaluating their outputs (**supervised learning** or **unsupervised learning**) by modifying their weights.

▷ With many learning techniques it is practical to assign (small) random values to the weights before learning begins.

23.1.2 Structure

Configuration of a network, also called **architecture**: The most important special case is the **feed-forward** structure in which the nodes are usually divided into **layers**. Every node transfers its signal only to nodes that lie in one of the subsequent layers (**multilayer perceptron**).
However, if at least one node is able to influence (directly or indirectly) its own input, then we speak of **recurrent networks**.

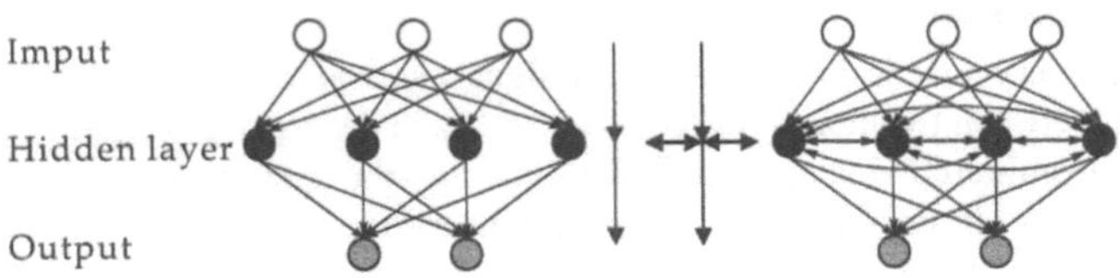

Completely connected feed-forward network with one hidden layer; Network with recurrence in the hidden layer

- **Input interfaces** for example, are given signals from **input nodes**, which have no other function. Information can also be fed into the network in the form of its **initial state**.
- **Output interfaces** are usually **output nodes**, whose signals are picked up directly by the user.

All nodes that are neither input nor output nodes are called **hidden nodes**, since the user does not deal with them directly.

23.2 Implementation of the neuron model

23.2.1 *Time-independent systems*

Most neural applications are based on the simple case of time-independent feed-forward networks. Here, for every input, every node assumes a fixed state that can be calculated directly.

- The various types of time-independent feed-forward networks are well tested, and easy to apply.

Time-independent signal processing: The **input signal** x_i of a node i is usually calculated, in the case of **time-independent systems**, from weights w_{ij} (always from node j to node i), and signals y_j of other nodes:

$$x_i = \sum_j w_{ij} y_j \ .$$

Then, the signal σ (in the case of a sigmoid function) is

$$y_i = \sigma_i(x_i) \ .$$

- For a feed-forward network, these equations can be solved directly by successively calculating all signals, beginning at the input.

□ For hidden nodes, the sigmoid function could, for example, be the hyperbolic tangent, and for output nodes the identity $\sigma_i(x_i) = x_i$.

□ **Typical connecting structure**: Complete connections from the input layer to the hidden layer and from the hidden layer to the output layer.

▷ The **precision** with which a feed-forward network with appropriate sigmoid functions (e.g., hyperbolic tangent) can reproduce a **continuous function** (i.e., a continuous assignment of input values to output values) can be improved arbitrarily by increasing the number of hidden nodes. As a rule, there is also no guarantee that a solution is found by means of the learning algorithm.

23.2.2 *Time-dependent systems*

Time-independent systems can often be used to treat **time-dependent data**, for example, by defining a **time window** (data of several time intervals are presented simultaneously) and treating the individual time steps as separate patterns. This approach is simple, but not very flexible.

Instead, we may employ the **dynamic behavior** of recurrent systems within a **time-dependent description.**

Time-dependent signal processing: In this case the **waveforms** are of interest. The operations performed by the nodes resembles those in time-independent systems.

- For time-dependent systems the states of the nodes must be determined by the (numerical) solution of the associated **differential equations.**

To stabilize the waveforms a **decay term** is usually introduced. For example, the signals may develop according to

$$\frac{\mathrm{d}y_i}{\mathrm{d}t} = -y_i(t) + \sigma_i(x_i(t)) \ ,$$

after being initialized at $t = 0$ with $y_i(0) = y_i^0$ (t formally regarded as non-dimensional).

▷ If the initial conditions are not given by the application $y_i^0 = 0$ is usually chosen.

● **Inputs** and **outputs** proceed in terms of waveforms.

The number of input and output nodes usually corresponds to the dimension of the input and output data. It is not necessary to present the data of several time intervals simultaneously.

▷ To integrate the differential equations, the simple **Euler method** of the first order is sufficient, where the differential quotients are replaced by the difference quotients.

▷ Since there is no regular dynamic for pure feed-forward architectures, a time-dependent formulation is not practical in this case.

□ **Typical connecting structure**: Complete connections from the input layer to the hidden layer **within the hidden layer** and from the hidden layer to the output layer.

23.2.3 *Application*

Computer simulation of neural networks is of growing importance in technology.

● An important area of application is the **classification of patterns**.

□ Examples are the recognition of hand-written signs or the assignment of workpieces on an assembly line by means of digitized video images.

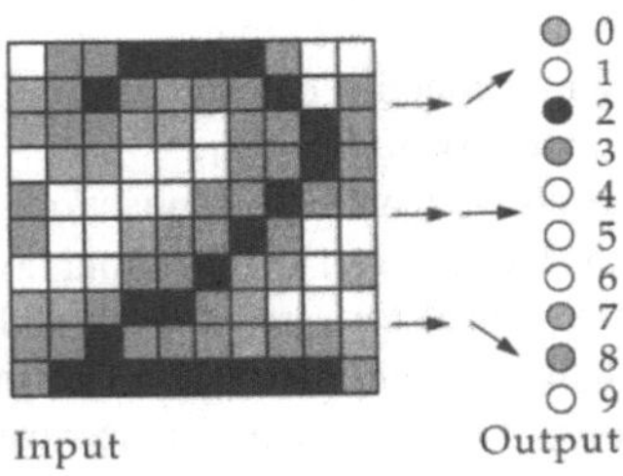

Example of a pattern classification

□ A common architecture for pattern classification assigns one output node to every possible pattern type.

● Neural networks can also be used in **control** systems.

□ Other examples include robotic control, or the determination of optimum roller pressure on sheet metal in a rolling mill.

23.3 Supervised learning

23.3.1 *Principle of supervised learning*

Supervised learning: The adjustment of the parameters (weights) of a network with respect to a **judgment criterion**, usually in the form of a so-called objective function.

● **Objective functions** or target functions are also called **deviation function** or **energy function** depending on the type: They evaluate the outputs of a network.

Lower values of the objective function represent higher quality so that the task is to find a set of parameters for which the objective function becomes minimal.

Gradient descent: Simplest method of supervised learning most often used for neural networks. The weights in learning step $n+1$ are calculated from the weights in learning step n. For this purpose all input patterns to be learned are presented successively to the network, and the corresponding outputs are evaluated according to the objective function. For a set of weights $\vec{\mathbf{W}}$ a definite value of the objective function $E\left(\vec{\mathbf{W}}(n)\right)$ is attained in learning step n.

- Since the gradient $\vec{\nabla}E\left(\vec{\mathbf{W}}(n)\right)$ indicates the direction of the **steepest ascent** of the objective function the weights are modified in small increments in the **opposite direction**.
- The magnitude of the increment is determined by the **learning rate** $\varepsilon > 0$.

$$\mathbf{W}(n+1) = \mathbf{W}(n) - \epsilon\vec{\nabla}E\left(\mathbf{W}(n)\right).$$

Hence, for the **modification of a weight** $\delta w_{ij}(n) = w_{ij}(n+1) - w_{ij}(n)$ one obtains:

$$\delta w_{ij} = -\epsilon\frac{\partial E}{\partial w_{ij}}.$$

($\partial E/\partial w_{ij}$ is the corresponding component of the gradient.)
Each run through the set of the patterns to be learned forms a **learning cycle**.
The learning process terminates when the value of the objective function is no longer clearly diminished, or when the problem is solved with the required precision.
Every one of the possibly numerous minima of the objective function has a basis of attraction in which the gradient descent tends to the corresponding values for the weights. But often there is only a **global minimum**, distinguished by the best output quality.

▷ The gradient descent can come to a halt in a **local minimum** that has a higher value of the objective function than the required global minimum or **optimum**. The simplest remedy is to restart learning (repeatedly if necessary) with altered initial weights.

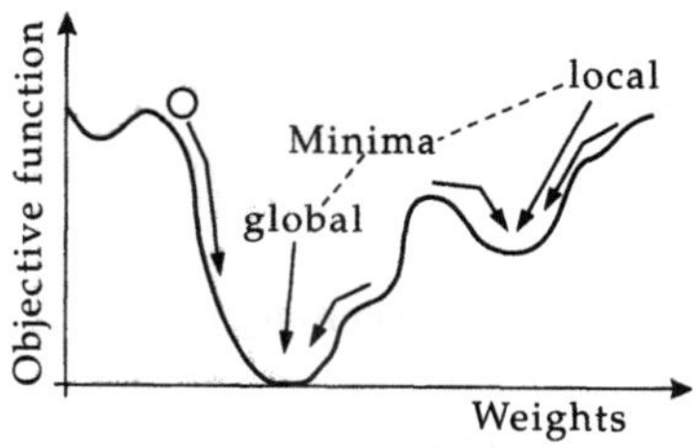

The gradient descent

▷ During the gradient descent, large **jumps**, in some cases to high values, can occur. **The objective function** may occur because the step size cannot be chosen arbitrarily small.

Learning rate: Determines the course of the gradient descent. A lower learning rate leads to smaller jumps in the objective function and sometimes also to a more exact adaptation. Higher learning rates can accelerate the learning process and facilitate the escape from local minima. An appropriate learning rate must be determined by trial and error.

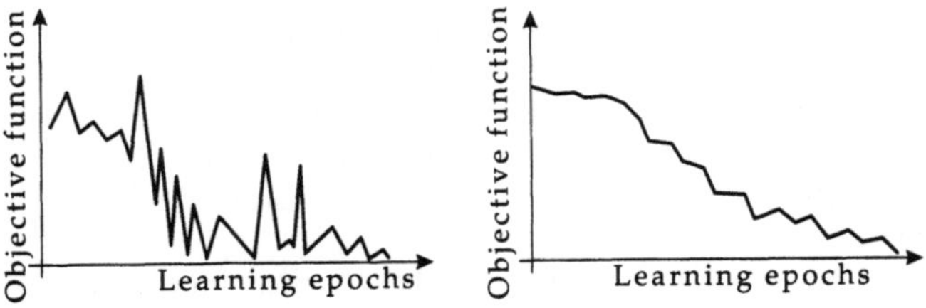

Objective function during learning (left: high learning rate; right: low learning rate)

Incremental learning: Adjustment of the weights after each presentation of a **single pattern**. For this, the gradient of a corresponding objective function is calculated once for each learning pattern. Thus, no "proper" gradient descent is performed in which all patterns to be learned are taken together (**batch learning**).

▷ For large data sets involving many repetitions, incremental learning may cause a considerable **reduction in learning effort**. In addition, the probability that the learning process comes to a halt in local minima is reduced, particularly if the order of the patterns is (randomly) mixed after every learning cycle.

23.3.2 *Standard backpropagation*

Standard backpropagation is the most important method of calculating the gradient of an **objective function** for **feed-forward time-independent networks**. It is based on passing **error signals** backward through the network.

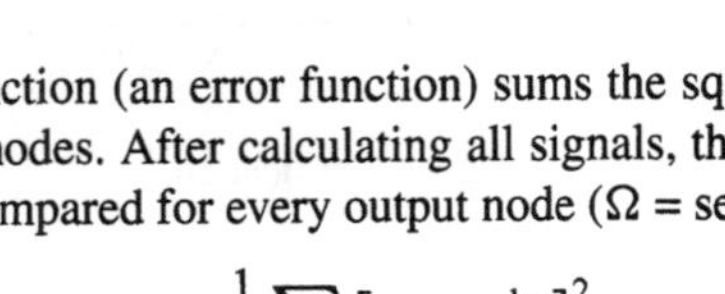

Error backpropagation

The typical objective function (an error function) sums the squares of the output errors at the individual output nodes. After calculating all signals, the actual signal y_i and the desired signal y^{des} are compared for every output node (Ω = set of output node):

$$E = \frac{1}{2} \sum_{i \in \Omega} \left[y_i - y_i^{\text{des}} \right]^2 .$$

The **weight changes** are calculated with the help of the error signals Δ_i according to

$$\delta w_{ij} = -\epsilon \, \frac{\partial E}{\partial w_{ij}} = -\epsilon \, \Delta_i \, y_j \; .$$

Error signals for output nodes ($i \in \Omega$) result directly from

$$\Delta_i = \sigma_i'(x_i) \left[y_i - y_i^{\text{des}} \right] .$$

For hidden nodes ($i \notin \Omega$):

$$\Delta_i = \sigma_i'(x_i) \sum_k w_{ki}\, \Delta_k \,.$$

(No error signals are needed for hidden nodes.)

- For **feed-forward** connections the error signals can be calculated **successively**, beginning at the output nodes.

▷ If several patterns must be taken into account at once (**batch learning**), then the corresponding error signals are calculated for each pattern. Their contributions to the individual weight changes are simply summed. This procedure corresponds to a **sum overall learning patterns** in the objective function.

▷ Supervised learning often works better with **linear output nodes** (sigmoid function = identity) because the error signals can be transmitted better.

▷ **Binary input data** should be chosen from $\{-1, +1\}$ (instead of $\{0, +1\}$) because null signals do not contribute to learning.

23.3.3 *Backpropagation through time*

This approach is the generalization of the standard backpropagation algorithm to time-dependent systems. The error function E is replaced by the error **functional**

$$F = \frac{1}{2\tau} \int_o^\tau \sum_{i\in\Omega} \left[y_i(t) - y_i^{\text{des}}(t)\right]^2 \, \mathrm{d}t \,.$$

The **target signals** are also shown as time-dependent, of course; τ is the length of the considered **time interval**.

Weight changes are obtained by means of **integration**:

$$\delta w_{ij} = -\epsilon \frac{\partial F}{\partial w_{ij}} = -\epsilon \frac{1}{\tau} \int_0^\tau \Delta_i\, \sigma_i'(x_i)\, y_j \, \mathrm{d}t \,.$$

Error signals of a time-dependent system are calculated (like the signals) by means of (numerical) solution of differential equations:

$$\frac{\mathrm{d}\Delta_i}{\mathrm{d}t} = -\frac{\partial E}{\partial y_i} + \Delta_i - \sum_k w_{ki}\, \sigma_k'(x_k)\, \Delta_k \,,$$

always with the boundary conditions $[\Delta_i]_{t=\tau} = 0$. For output nodes, $\partial E/\partial y_i = y_i - y_i^{\text{des}}$; for all other nodes $\partial E/\partial y_i = 0$.

- After forward propagation of the signals there is a **backward propagation** of the error signals through the network and **through time**.

□ For example, we obtain the following equation according to the Euler method for error signals from output nodes:

$$\Delta_i(t - \delta t) = \Delta_i(t) + \delta t \left[y_i(t) - y_i^{\text{des}}(t) - \Delta_i(t) + \sum_k w_{ki}\, \sigma_k'(x_k(t))\, \Delta_k(t) \right].$$

▷ Using the **Euler method** to calculate signals and error signals, we obtain a **discrete model** in which the gradient calculation is again exact.

Apart from the combination of many time intervals into single patterns, the learning process proceeds analogous to standard backpropagation.

For the numerical calculation of error signals in every time interval of forward propagation, all signals are temporarily stored.

▷ If there is sufficient memory it is practical to store the input signals x_i as well.

▷ Apart from the size of the network, the **time scale** should also be adjusted to the corresponding application. The simplest method is to normalize the data accordingly.

□ To draw a figure eight (through a network with one completely combined hidden layer and one output node each for horizontal and vertical components), a τ interval of 2π has proven practical.

▷ For the learning of complex connections it can be an advantage to consider first only a small portion of the **time interval** and then to increase it successively. Thus, difficulties with very high error signals after long propagations can be avoided. Alternatively, a "teacher forcing" method can be used.

Teacher forcing: Version of a learning method in which the output nodes are "forced" to their target values at all times.

- In teacher forcing, the **tendency** of the system to **deviate** from a forced output behavior is evaluated.

In the differential equations of the error signals for hidden nodes and in the equations for the components of the gradient, all **error signals for output nodes** (Δ_i with $i \in \Omega$) are replaced by

$$\Delta_i^{\mathrm{TF}} = \sigma_i(x_i) - y_i - \frac{\mathrm{d}y_i}{\mathrm{d}t} .$$

The term $\partial E/\partial y_i$ becomes redundant. The "original" error signals for the output nodes are no longer needed.

For some applications, teacher forcing may clearly improve the convergence of the learning process.

▷ Sometimes teacher forcing works better when the signals of the output nodes are not transmitted because the dynamics in the application (with necessarily free output signals) differ too much from the dynamics of the learning process.

▷ Even if the exact time derivatives $\mathrm{d}y_i/\mathrm{d}t$ of the output nodes are known analytically, it is better for the **Euler method** when appropriate difference quotients are used, too.

23.3.4 *Improved learning methods*

Momentum term: To avoid very slow convergence on the one hand and oscillations on the other, a **momentum term** can be added to the rule for the gradient descent. The momentum term has an **accelerating** and **stabilizing** effect. Using the preceding learning step, we thus obtain for the modification of weight:

$$\delta w_{ij}(n) = -\epsilon \frac{\partial E(n)}{\partial w_{ij}} + \alpha\, \delta w_{ij}(n-1) .$$

The momentum term is weighed by the coefficient α (with $0 \le \alpha < 1, \alpha = 0.9$). If the gradient does not change the **effective learning rate** converges to $\epsilon/(1-\alpha)$.

▷ The use of a momentum term is a common method of improving convergence. But with complicated problems, difficulties can arise because minima are missed or necessary noise components may be suppressed in the gradient descent.

Quickprop: Learning method that is essentially a **much simplified Newton method**. We use:

$$\delta w_{ij}(n) = \frac{\partial E(n)/\partial w_{ij}}{\partial E(n-1)/\partial w_{ij} - \partial E(n)/\partial w_{ij}} \, \delta w_{ij}(n-1) \, .$$

For situations in which $\delta w_{ij}(n-1) \approx 0$, a gradient term and—to restrict the weights—a decay term are added, such as:

$$\delta w_{ij}(n) = \left(\frac{\partial E(n)/\partial w_{ij}}{\partial E(n-1)/\partial w_{ij} - \partial E(n)/\partial w_{ij}} - \lambda \right) \delta w_{ij}(n-1) - \epsilon \, \frac{\partial E(n)}{\partial w_{ij}} \, .$$

The decay constant should be small, $0 < \lambda \ll 1$. The first step proceeds as a gradient descent. If $\delta w_{ij}(n) > \mu \, \delta w_{ij}(n-1)$ (with growth limit $\mu > 1$, $\mu = 1.75$), we use $\delta w_{ij}(n) = \mu \, \delta w_{ij}(n-1)$. If the signs of quickprop and gradient contributions differ, then it would be practical to use $\delta w_{ij}(n) = \delta w_{ij}(n-1)$. We set $\epsilon = 0$ if the component of the gradient changes sign; this avoids oscillations.

▷ For simple problems, quickprop can save much learning time (compared with the gradient descent). However, for more complicated problems it should be tested how useful this is. Experiment with different versions.

23.3.5 Hopfield network

Ideally, a Hopfield network **associates** a (**binary**) pattern of the learning set with a presented (incomplete or noisy) set. **All nodes are input as well as output nodes** (one for each pattern component).

● The output of the Hopfield network originates at the position of the input. Hopfield networks are **auto-associative memories**.

Symmetric weights ($w_{ij} = w_{ji}$) are assigned to the **connections** between the nodes. The nodes are usually **completely connected** to each other.

Because the system is recurrent, the recall of information proceeds **dynamically**. The symmetry of the weights guarantees that the system always runs to stable states. The areas of attraction of these stable states (**fixed points**) are called **attractors**.

● Stored information is represented by the location of these attractors.

In the recall of information the **data input** is in the form of the inital state. Every signal is set to the value of the corresponding pattern component.

▷ Because only **time-independent data** can be processed satisfactorily, the Hopfield network is regarded as a time-independent system.

Functioning of a Hopfield network: The influence of nodes with incorrect signals is often not great enough to distort the signals of other nodes. Then, the incorrect signals are corrected due to the influence of the other signals.

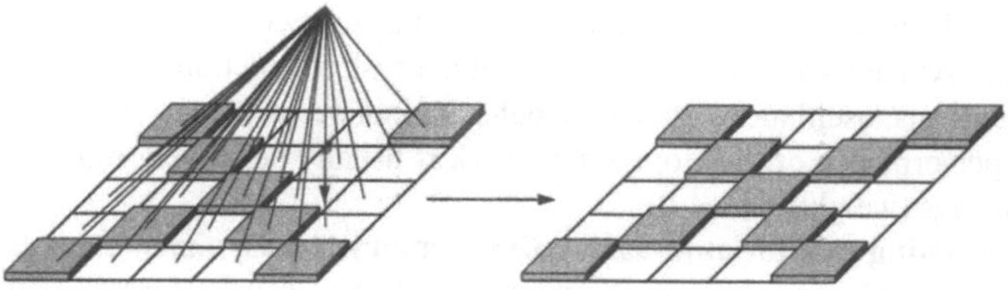

Functioning of a Hopfield network

Discrete Hopfield network: Simplest version. A signum function is used instead of a sigmoid function. The system works **in discrete time**.

To reach the fixed point

$$y_i = \operatorname{sgn}\left(\sum_j w_{ij} y_j\right) \qquad \text{for all } i ,$$

we iterate

$$y_i(t+1) = \operatorname{sgn}\left(\sum_j w_{ij} y_j(t)\right)$$

until no signal changes.

▷ Oscillations are avoided if the **recalculation of signals** proceeds sequentially, i.e., if after each calculation of the input signal of a node, its signal is adjusted directly.

The **weights** are **calculated directly** according to

$$w_{ij} \propto \sum_p \sum_{i,j} y_i^p y_j^p ,$$

where p runs through all patterns to be stored. The above signal products are usually called a **Hebb term**.

The factor of proportionality can be chosen freely. Note the symmetry $w_{ij} = w_{ji}$. The initialization of the weights and the implementation of several training cycles becomes unnecessary.

Continuous Hopfield network: To calculate a **fixed point**, solve the following differential equations for the continuous Hopfield:

$$\frac{\mathrm{d}y_i}{\mathrm{d}t} = -y_i(t) + \sigma_i(x_i(t))$$

(t is assumed to be nondimensional.)

When there is sufficient accuracy of the numerical integration, no oscillations occur.

▷ The size of the weights should be chosen so that the x_i lie predominantly in the nonlinear range of the sigmoid function.

● Every Hopfield network has a certain **storage capacity**. Large, completely connected networks with N nodes are able to store about $0.14\,N$ uncorrelated patterns.

▷ If more patterns are integrated into the training set than the network is able to store, the **recall quality** may break down completely. The **magnitude of the attractors** decreases with the number of stored patterns.

▷ Since the associated objective function (due to physical analogies, also called the **energy function**) considers only the **parities** of signal signs, Hopfield networks are able to store and reproduce only sign parities, i.e., **binary patterns**.

▷ Hopfield networks can be **diluted** by removing connections. The **memory capacity** is approximately proportional to the number of connections. Due to asymmetric connections, oscillations or chaotic behavior can occur.

▷ The performance of the Hopfield network is usually negatively influenced by **self-couplings** (weights $w_{ii} \neq 0$).

Iterative learning rule for auto-associative memory: An asymmetric iterative learning rule, in which well-represented patterns contribute less can be formed (similar to the backpropagation algorithm) from the **gradient** of a corresponding **deviation function**:

$$\delta w_{ij} = \epsilon \left[y_i^{\text{des}} - \sigma_i \left(\sum_k w_{ik} y_k^{\text{des}} \right) \right] \sigma_i' \left(\sum_k w_{ik} y_k^{\text{des}} \right) y_j^{\text{des}} .$$

As with the backpropagation algorithm the patterns are presented in **cycles**.

▷ In this manner (and with a corresponding discrete version), clearly **higher memory capacities** are achieved than with pure Hebb weights.

23.4 Unsupervised learning

23.4.1 *Principle of unsupervised learning*

In unsupervised learning, no data about a desired output behavior are available to the learning system.
There is no evaluation of outputs during the learning process.
Therefore, the desired behavior of the network must be fixed by means of the learning algorithm.
An important application of such systems is the **development** of various kinds of **data transformation**.

● **Redundancy** plays a decisive role in the unsupervised learning process because it represents the only indication contained in the data for the solution of the task.

▷ In some applications the use of the neural network is already terminated with the learning process because the resulting set of weights itself represents the solution of the problem.

Self-organizing systems usually have the task of finding an **assignment scheme** of two geometrically different spaces (input and output space), i.e., as far as possible, environment-preserving (**topology preserving**).

23.4.2 *Kohonen model*

The Kohonen model is probably the best known **self-organizing** neural system. Every component of an input vector (every input node) acts via connections with changeable weights on a number of active nodes to which a geometrical connection is assigned.

□ Example: One-dimensional (chain-like) arrangement of active nodes.

Typically, neighboring nodes in a trained network react to similar input values where every node is specialized to a certain region of the input space (topology preservation). The **resolution** (the number of "relevant" nodes) increases with the importance (number of "stimuli") of the corresponding area of the input space.

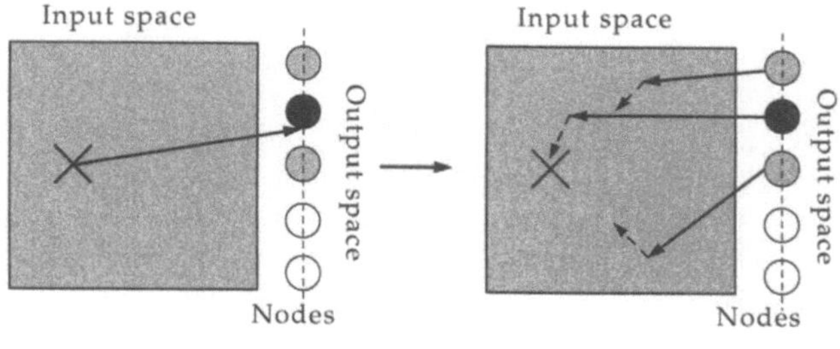

Learning by means of the Kohonen model

The **weights** are initialized with random values. For every **learning step** under an input vector $\vec{\mathbf{y}}$ the **node position** i^0 is determined for which the value

$$\sum_j (w_{i^0 j} - y_j)^2$$

is a **minimum**.

▷ This is a simplification of the method of choosing a node with the maximum signal after calculating "conventional" node signals.

If the **connections** of every node are viewed as a vector then

$$\left\| \vec{\mathbf{w}}_{i^0} - \vec{\mathbf{y}} \right\| \leq \left\| \vec{\mathbf{w}}_i - \vec{\mathbf{y}} \right\| \quad \text{for all } i.$$

Modification of weights: The connecting weights for all nodes from the "environment" of the node i^0 (here, the geometric connection appears) are displaced in the direction of the actual input:

$$\delta w_{ij} = \epsilon\, g(i, i^0) \left(y_j - w_{ij} \right).$$

The effect of the displacement depends on the distance of the positions assigned to the nodes i and i^0; the **intensity function** g must decrease as the distance increases.

□ An appropriate example of the intensity function would be:

$$g(i, i^0) = \exp\left(-\frac{(r_i - r_{i^0})^2}{2\alpha^2} \right),$$

where r_i denotes the position of the i-th output node, and α is the width of the affected area (here, Gauss bell curve).

Again, the **learning rate** ε gives the increment of learning.

During the learning process, very many different input vectors are presented to the network, which are usually chosen at random (but are usually not evenly distributed).

▷ The input and output spaces may have different dimensions.

□ The redundancy of multidimensional data must be reduced in a transformation to low-dimensional data.

▷ To obtain a useful coarse structure of the allocation, the learning process must be carried out with a high value of α at the beginning, which is reduced once the allocation becomes finer. Also the learning rate ε, chosen to be high at the beginning to create a structure quickly, is diminished in the course of learning. The choice of proper methods is always somewhat arbitrary.

□ For example, linear, exponential and hyperbolic methods are possible; the best choice must be determined by trial and error.

24

Computers

24.1 Operating systems

Operating system: Basic software of every computer system. It is called in any computer process and makes available fundamental functions such as the execution of programs (applications), the organization of the primary memory (working memory), the output on the display, the input with the keyboard, the organization of the file system on the storage media, or the selection of peripheral units (printer, modem, etc.).

Single-task operating system: An operating system that enables the simultaneous execution of only one program (application). To start a second application, it is only possible after termination of the first.

□ MS-DOS

● Single-task operating systems may be programmed very simply and efficiently and have only minor requirements for hardware.

Multitask operating system: An operating system that allows for the simultaneous execution of several distinct programs (processes).

□ UNIX, LINUX, OS/2, MVS (IBM), VMS (Vax), Windows NT, Windows 95.

▷ In multitasking a certain (short) **time slice** is assigned to each process during which this process is handled by the computer. After this time interval the next process is handled, then the following, until the first process again has its turn (**time sharing**).

● Multitasking places extreme requirements on the operating system. Thus, it must be ensured that an individual area of working memory is assigned to each process that cannot be overwritten by another process. Also the assignment of the time slice must be controlled.

● The process "display of a text page on a screen" should be handled with high priority compared to other running processes. Correspondingly, the time slices of all other processes should be interrupted until the display output is terminated.

Single-user operating systems: Operating systems designed for use by only one user (for the time being).

□ MS-DOS, OS/2

● For a single-user operating system there is no need to take measures for a partition of the resources of the system or for the simultaneous access to storage media for several users because the entire computer is available for one user.

Multiuser operating systems: Operating systems designed for the simultaneous use of one computer by several users.

□ UNIX, LINUX, MVS.

▷ The simultaneous access of several users is established by serial terminals or **networks**.

● Multiuser operating systems must require password protection of the user data, the correct distribution of system resources, and multitasking.

Survey of various categories of operating systems:

user	task	example
single	single	MS-DOS
single	multi	OS/2, Windows NT, Windows 95
multi	multi	UNIX, MVS, VMS

User interface (UI): Part of the operating system. Forms the interface between human and computer. Enables entering commands and calling programs via keyboard or mouse.

▷ The notion "user interface" is often used erroneously for operating system. In fact, the user interface represents only the small visible part of the operating system.

Graphical user interface (GUI): Intuitive user interface for operating with the mouse. Must enable a simple, uncomplicated contact with the computer, easy to learn. Does not require much previous knowledge. In most cases free to configure by the user.

□ Windows are a GUI for the operating system MS-DOS.

File system: Area on a storage medium allocated by the operating system to store files and programs. The format of the stored data is given. One distinguishes between:

- **File**: Set of data or program in the file system; specified by a unique **filename** for identification. The precise location of the file on the storage medium is stored on the medium in a designated area together with information concerning the file (file size, access rights, data of creation or modification, etc.), separate from the actual data.

- **Directory**: Partition of the file system into various groups to which files that must be stored are assigned. Like a file, a directory may contain an arbitary number of subdirectories partitioning the directory.

□ By setting up directions, data may be structured and ordered according to fields of application.

▷ The physical location of the files on the storage medium is independent of the logical directory structure. Only the references to the files are stored according to this structure.

□ Within a directory, a subdirectory or file is separated from the directory by "/" or "\" in UNIX or MS-DOS, respectively.
UNIX: `/usr/local/bin/ls`
DOS: `\dos\graphic\`

Directory tree: From a multiple subdivision of directories into subdirectories a **tree structure** results. The branches of the trees are the directories, which themselves branch again into directories.

Root directory: The starting point of the directory tree. The only directory that is not a subdirectory of another directory.

□ The root directory is specified by "/" in UNIX and by "\" in MS-DOS.

Parent directory: Directory entry occurring for every directory (apart from the root directory); it refers to the directory that contains the current directory as a subdirectory.

□ In most cases the parent directory of a directory is invoked by "..".

Path: Description of the route or trail of the logical location which a file or directory must take inside the directory structure. The path shows all the subdirectories that must be passed in succession.

Absolute path: The path to be followed from the root directory upward.

Relative path: The path to be followed from the *current* directory upward or downward.

● Every file is characterized uniquely by giving its name and the absolute path.

▷ In UNIX there is only one large directory tree that covers all file systems on all storage media (disks, CD-ROM, floppy disk drives). The subdirectory systems are mounted to defined places in the directory tree. In MS-DOS the directory tree is defined on each disk and each drive separately.

Current directory: The directory to which all relative paths are referred.

24.1.1 Introduction to MS-DOS

MS-DOS (Microsoft Disk Operating System), also called **DOS** for short, is the most widely used operating system for PCs.

Windows: Graphical user interface for the DOS operating system. Offers many extended posibilities far beyond DOS, but is always restricted by the limitations of DOS.

Windows 95:...

Most important characteristics of Windows...

FAT (File Allocation Table): Table created on every data medium. Identifies all sectors of the data medium either free or busy. The physical locations of all files on the data medium are stored together with the filename in the corresponding directory. A file is deleted by removing the corresponding entry from the directory (actually, only the first character of the filename is overwritten) and by releasing the sectors in the FAT.

▷ Physically, the file initially remains on the data medium. This allows for the recovery of files deleted accidentally.

□ Utility programs for the recovery of files deleted in error include PC-Tools, Norton Utilities, etc.

Current drive: Drive to whose file system all operations relate.

▷ Under DOS, every drive (floppy or hard disk) has its own file system and thus its own FAT.

Prompt: Symbol on the screen indicating that a DOS command must be entered.

□ Usually, the current directory of the current drive is displayed as a prompt.
`C:\DOS>`
In principle, the prompt can be configured freely.

Filename: In MS-DOS, restricted to eight characters (only letters, numbers, and "_") plus an extension of no more than 3 characters. Name and extension must be separated by a dot.

□ Valid filenames in MS-DOS: `TEST.BAT`, `programm.exe`, ...
Invalid filenames: `programme.exe`, `a+b=c.dat`, `test.dat.exe`

Wildcards ("*" and "?"): Represent parts of an incomplete or ambiguous filename. "*" represents all character combinations of the filename or extension beginning from where it occurs. "?" replaces *exactly one* character of the filename or extension.

▷ Characters following "*" are always ignored (in contrast to UNIX).

□ `*TEST.DAT` = `*.DAT`

▷ Name and extension are treated *separately.*

□ The identifier `XYZ*` does *not* include the file `XYZAB.DAT`. Here, `XYZ*.*` must be given explicitly.

Syntax for MS-DOS commands:

command arg1 arg2 ...

▷ Here, *command* is the name of the command, and *arg1, arg2, ...* are the arguments it contains.

▷ MS-DOS does not differentiate between capital and lowercase letters.

Most important commands in MS-DOS:

`DIR` [/*option*] [*drive*:] [*path*] [*filename*]

lists the files in the given drive/path.
Options:

w – multicolumn display of filenames.

`CD` *path*

Changes the directory specified by *path* and declares it to be the *current* directory. Relative and absolute paths may be entered.

`DEL` [*drive*:] [*path*] *filename*

Deletes the file given by drive, path, and name from the FAT. However, the file is not physically removed from the storage medium, it can in some cases be recovered with the help of appropriate utility programs. If a directory is given instead of a file, the content of this directory is deleted following a request for confirmation.

`COPY` [*drive1*:] [*path1*]*file1* [*drive2*:] [*path2*]*file2*

file1 is copies from given path1 to file2 in path2. Files which may already exist may be overwritten, but the source file is retained. If no destination file is given, the source file is written [with the same name] into the destination path [−]. If no destination path is given, the file is copied to the current directory.

`REN` [*drive1*:] [*path1*]*filename1* [*filename2*]

the file *filename1* in the given path to *filename2*. If a file *filename2* already exists in the given path, then the command ends with an error message.

24.1.2 Introduction to UNIX

In contrast to MS-DOS, there are numerous different UNIX versions all based on the same principle, but differing in details.

□ Unix-V (AT&T), BSD (Berkeley), Ultrix (VAX), AIX (IBM), Solarix (Sun), HP-UX (Hewlett-Packard), Linux for PC.

Shell: Text-oriented interface between user and computer. In the shell, UNIX commands can be called, directories can be changed, texts can be edited and programs can be executed.

▷ There are various types of shells that are basically identical but in some cases provide simplified operating options or an extended Shellscript language.

□ The following shells are available on nearly all UNIX systems:
Bourne-Shell (bsh), C-Shell (csh), Korn-shell (ksh), "Bourne-again" shell (bash).

▷ The question of which shell is the best is controversial. The most common and most widely used is the Korn shell.

Shellscript is designed for OS-related tasks (file system manipulation, network cooperation, etc.) and we therefore have some reservations in calling it a high-level programming language, although it certainly has most of the programming elements.

● On all UNIX systems, users help is available as a standard feature, stored on disk or CD-ROM, and always available in the form of the so-called **manual pages** (**"man pages"**). They are *not* a textbook about UNIX. They allow only queries about information (task, parameters, etc.) pertaining to most UNIX commands..

Access to the man pages:

`man` *command name*

□ `man ls`, `man cp`, ...

▷ UNIX systems often have a complete manual for UNIX and programming languages such as C or FORTRAN, but access and operation differ from system to system.

Syntax for UNIX commands:

command options file/directory G. reads:

command options Arg1 Arg2:

▷ In UNIX capital and lowercase letters *must* be differentiated. Thus, `Example` and `example` are two different files.

Options: Switches that can be set for most UNIX commands to activate or deactivate certain variations or modified versions of commands.

● Options are always located *between* the command and the file upon which the command acts.

● Options are specified by "-" followed by the desired option. Depending on the command, any number of options can be given in any order.

□ Example: `ls -l -a /usr/local/bin`

▷ As a rule, options consisting of only one letter can usually be combined.

□ `ls -la /usr/local/bin`

Wildcards: Analogous to MS-DOS. In addition, "*" may stand at the beginning of a filename.

□ `*fold` implies the filenames `twofold`, `threefold`, but also the filename `fold`.

Standard output: Output unit to which the UNIX commands direct their output.

□ Usually: Screen.

Standard input: Input unit from which UNIX programs receive data to be entered.

□ Usually: Keyboard.

Standard error output: Output unit to which UNIX programs write the *error messages* which may occur.

□ Usually: Screen.

▷ In UNIX, standard input, standard output, and standard error output can be *redirected*. Outputs or error messages can be written to a file, and input may be read from a file.

> Redirecting the standard **output**.

```
ls -l > output
```

Writes all files of the current directory to a file named `output`.

< Redirecting the standard **input**.

```
programm < input
```

The program reads all input data from the `input` file.

2 > Redirecting the standard error **input.**

`mv abc def 2> error`

Writes error messages (such as "file `abc` does not exist") to the `error` file.

Pipe "|": Buffer memory in UNIX. Allows the direct redirection of the standard output of one program to the standard input of another. This facilitates the concatenation of UNIX programs.

□ `ls -l | pg`

Directs the output of the `ls` command to the "pager" for a page display on the screen.

Most important commands in UNIX:

▷ A complete overview of all options can be accessed via the man page of the command in question.

`ls [-`*option*`] [`*path*`] [`*file*`]`

Lists the files in the given path. When no options are set, only the filename is displayed.

Options:

l – Complete file information (access rights, date of creation, etc.)

a – Also *hidden* files are listed (filenames beginning with a dot).

`cd` *path*

Changes to the path directory and declares it to be the *current* directory. Relative as well as absolute paths can be given.

`rm [-`*option*`] [`*path*`] [`*file*`]`

Deletes the given file or directory. It is impossible to recover the files because the data are physically removed from the disk.
Options:

i – Displays request for confirmation before deleting a file.

r – Recursive deletion of the specified directory. All subdirectories and files contained therein are removed.

Warning! rm–rf deletes all files in the file system.

f – Read-only files are also deleted without a request for confirmation.

`cp [-`*option*`]` *path1* `[`*file1*`]` *path2* `[`*file2*`]`

Copies the specified file(s) *file1* to *path2 without* erasing the source file. If *file2* is not specified, the file assumes the name of the source file.

Options:

R – The specified directory, including *all* subdirectories, is copied recursively to the directory *file2* that must exist.

`mv [-`*option*`]` *path1* `[`*file1*`]` *path2* `[`*file2*`]`

Moves the specified file(s) *file1* to *path2*. The source file is deleted! If *file2* is not specified, the file assumes the name of the source file.

Options:

R – The specified directory, including *all* subdirectories, is copied recursively to the directory *file2* that must exist.

▷ In UNIX, `mv` carries out the `REN` command of DOS.

`pg` [*path*] *file*

"Pager". Shows the content of readable ASCII files page by page.

24.2 High-level programming languages

High-level programming language: Language for the programming of computers, which orients itself on human thought patterns and is therefore relatively easy to learn. The translation of the commands to **machine code** is done either by a **compiler** or an **interpreter**.

Interpreter: Translates the commands from a high-level language to machine language *while the program is running*. All commands are translated in the order of their call.

- Programming in interpreter languages is very comfortable, since programming errors become visible immediately while the program is running, without first having to translate the entire code.
- Interpreter languages are very slow to execute commands because even with repeated calling of program modulus, all commands must be translated again.

▷ Today, interpreter languages are practically obsolete because **compilers** have become much faster and more efficient at debugging.

Compiler: Translates the program code of a high-level programming language (**source code**) into machine code (**object code**) *before* the program is executed. Detects errors in syntax and in the arrangement of contiguous commands. Data structures are checked for errors but not yet established in the memory. Various program modules (functions, subroutines) can be translated as independent units. The combination of the individual program modules is the task of the **linker** which connects individual previously compiled program sequences (**object modules**) with the data structures defined therein to form a contiguous program that can be executed.

- Apart from connecting different program modules, the linker also usually integrates so-called **library routines** into the program. They contain previously defined functions and subroutines (e.g., for input and output operations) which are often needed. Users can also create their own libraries.
- In many cases, compiler and linker are combined in one unit. After successful compilation, the linker is called automatically. If errors have occurred, the call for the linker is suppressed. Automatic linking can also be turned off upon request.

□ In UNIX, common compilers are f77 (FORTRAN), cc(C), g++(Gnu-C++). They already contain the linker, but it can be turned off via **compiler flags**:

`f77 -c prog.f`	*compiles prog.f without calling the linker*
`cc prog1.c prog2.c`	*compiles modules prog1.c and prog2.c and links them*

Debugger: Program for detecting persistent programming errors that do not cause errors in compiling and linking but which occur while the program is running. The debugger allows the program to execute command after command and to give out targeted contents of variables.

▷ To enable debugging, the program usually must be compiled with a special compiler flag.

□ `f77 -g prog.f`

□ Programming languages for the PC are often provided with an integrated debugger. In UNIX systems one must use stand-alone programs such as `dbx`.

24.2.1 Program structures

Modular programming language: Programming language that enables a program to be divided into many segments (**modules**) that are as independent as possible.
Data structures: Declarations of (combined) data types. They are stored in the working memory only through the definition of **variables**.

- The terms data structure and variables are often used as synonyms.

Scope: Range of validity of data structures. Most of the high-level programming languages distinguish at least between the following:

- **Global scope**: Data structures defined here are known in the entire program, i.e., also in all subprograms (modules).

- **Module scope**: All data structures are defined only inside a module. These data are outside of this area.

▷ Variables in different modules can be given the same name (**identifier**). The change of a local variable in one module does not affect the value of a variable of the same name in another module.

□ FORTRAN 77 does not recognize global variables. Variables to be used in several modules must be defined as **common** blocks.

Local variable: Variable declared in a subprogram. Valid only in this subprogram. In other subprograms, variables with an identical name can be declared.
Global variable: Variable defined outside of subprograms. Has the same value in *all* subprograms.

- It is possible to define a local variable to a global variable of the same name. In that case, only the local variable exists in the subprogram.

- The memory locations of global and local variables are completely separate. Global variables are **statically** reserved by the compiler **before** the program begins. Local variables, on the other hand, are placed **dynamically** in the so-called stack, while the program is running, and released again after the subprogram is exited, so that the other subprograms can utilize the same memory area for their data structures. Thus, the contents of local variables are as a rule **not reproducible** when a subprograms calls them again.

Subprogram: Program module as separate as possible. A program should be subdivided into subprograms in such a way that every subprogram performs a certain task. If this task consists of several subtasks, the subprogram itself should be divided into further subprograms.
Function: Subprogram that returns a value to the calling program module.

□ `y = sin(x)` assigns the result of the function `sin(x)` to variable `y`, where `x` is the **argument** of the function.

□ In C, all subprograms are functions.

Argument: Interface between the calling program module and the called subprogram or function. Data can be given to the subprogram or received from it.

▷ A user-defined function can usually contain any number of arguments, but this number must be retained after being defined once. In calling the subprogram or function, the number and mode of the passed arguments must be compatible with the definitions of the subprogram.

Call-by-value: Method of transferring an argument. During the call, a copy of the transferred value is placed in the local memory of the subprogram.

● A change of the variable within the subprogram has *no* effect on the value in the calling program module.

▷ A disadvantage of the call-by-value transfer of arguments is the fact that a copy of the passed object must always be made. This requires memory, and the memory content copied. For larger data structures or often called subprograms this may consume a considerable amount of computer time and memory.

□ Examples of languages with call-by-value transfer of arguments:
PASCAL, C, C++.

Call-by-reference: Most common mode of transferring arguments. The transferred value is not copied to a new memory, only the **address** of the memory area containing the data is transferred. Subprogram and calling program use the same memory area.

● A change of the argument in the subprogram modifies the value in the running program, unless the arguments are defined as constant in the subprogram. This is always a danger in the call-by-reference mode.

▷ Call-by-reference is the most efficient method of transferring data because no memory areas must be copied (besides that which contains the address of the value).

□ Call-by-reference can be used in all high-level programming languages. For example, it is a standard in FORTRAN. In PASCAL or C++, call-by-reference must be explicitly specified.

▷ There are *no* call-by-reference parameters in C. They must be simulated by **pointers**.

● The option to select between call-by-value and call-by-reference contributes essentially to the data integrity in the calling program module.

Stack: Area of memory for the temporary storage of local variables in program modules. Functions according to the principle of *last in-first out* (LIFO). Data are always defined **on top of** the stack and must be removed later on in reverse order.

▷ If a program contains many nested or recursive calls from modules, then the local variables are successively placed in the stack. If the stack is too small, the program terminates when the stack is full.

● For modular programming, local data structures are essential. Especially with large projects which usually involve more than one programmer, it must be guaranteed that variables in one module are not accidentally overwritten by variables of the same name in another module.

Encapsulation: Principle for achieving maximum data protection and integrity. A change of variable contents is possible only via interfaces well defined by the programmer of a program module. Every direct access to the variables from outside the program module is prohibited.

▷ Modular programming is a first step toward the encapsulation of data. However, it does not work with data structures that are needed in more than one module.

● The attempt to implement data encapsulation consistently leads to the principle of **object orientation**.

24.2.2 Object-oriented programming (OOP)

▷ In programming languages that are not object-oriented, such as C or PASCAL, data encapsulation can be achieved through disciplined programming. However, protection against unwanted data manipulation cannot be achieved. In OOP, this protection is forced.

Recursion: Self-calling subprogram, which, without suitable interruption criteria, inevitably causes the program to crash (stack overflow).

- Recursive programs can usually be replaced by iterative algorithms, but this involves much effort.

- Recursions are usually much more elegant than their iterative pendant.

Object-oriented programming language (OOP language): Programming language in which data structures are not assigned to a program module, but in which data (**members**) and program modules working with them (**methods** or **member functions**) are combined into **classes**.

Basic principles of OOP:

Data encapsulation
Inheritance
Polymorphism

□ Examples of object-oriented programming languages: **Smalltalk**, **C++**, and modifications and extensions of various high-level modular programming languages (e.g., PASCAL).

Class: Abstract construction that combines data structures and the program modules that process them. Data and methods always belong together and cannot be addressed separately. A class only becomes physically existent as a storage location through **instantiation**.

□ A class is the equivalent of a data type, while an instance is the equivalent of a variable of this data type.

Object or **Instance**: Physical implementation of a class by assigning a specifier and an associated allocation of the required memory. There may be any number of instances per class, all occupying **different** memory locations.

Methods: Subprograms and functions assigned to an object, which control access to the object's data. These modules are defined only in connection with the object.

● Outwardly, *methods are defined only by their task.* The concrete realization of this task is hidden.

▷ The nomenclature of various programming languages can vary, but the basic principles of OOP (encapsulation, inheritance, etc.) are always the same.

□ For example, in C++, methods are called *member functions.*

● Variables in objects are always local (private). Only methods have access to them. It is usually possible to make the member globally accessible. However, this contradicts the principle of encapsulation and should be avoided.

● With these methods it is ensured that only those data can be manipulated and defined which are designated by the programmer.

□ Define a MATRIX whose data are the dimensions and coefficients of a matrix. Access to these data is controlled by methods: Selecting/setting certain coefficients, matrix addition, matrix multiplication, etc.

Inheritance: Principle of OOP. It is possible to derive new classes from already existing classes which automatically assume (inherit) all properties (methods, data) of the **base**

class. However, new data structures and methods extending the application of the class beyond that of the base class are additionally assigned to the new class. Methods of the base class can be replaced (*downloaded*) by new methods in the derived class, so that the base methods can still be used, but their use must be *explicitly* arranged.

Related class: Class derived from the same base class. The base class is related to every one of its heirs and the heirs are related to each other.

● Inheritance does not contradict the encapsulation principle! Private data of a base class can be processed only via the methods of this class. Furthermore, in C++ it is possible to deliberately weaken encapsulation so that derived classes can change certain variables directly. Independent classes or other program modules still have no direct access.

Multiple inheritance: A class may be derived from more than one base class, and would thus inherit all methods of all base classes.

□ Let a VECTOR class be derived from the MATRIX class defined above. A vector is a matrix with column number 1. The addition of vectors is defined by components, as with matrices, so that this method does not have to be newly defined in VECTOR. However, multiplication of vectors is not defined (we are considering only column vectors). If the method of multiplication is defined in VECTOR as the formation of the scalar product, then the method of multiplication is rejected in MATRIX, and from now on, multiplication of vectors always means the scalar product.

Polymorphism: In case of related classes, the ability to call the right method for a special class automatically via the same function message.

□ This ability becomes important if, for example, certain methods of related classes inside loops need to be called to do outside but are implemented differently on the inside due to the differences of various classes.

Important techniques for implementing polymorphism:

Object conversion
Virtuality

Object conversion: Derived objects can always be converted to the type of their base objects and can therefore be invoked like a base object **on the same level**.

▷ In this case, the special properties of the derived objects that deviate from the base object (variable methods for the same task) are lost. To recover them, the following are necessary:

Virtuality of methods: In object conversion the ability to transfer certain properties (methods) of the derived object to the (converted) base object.

Abstract class: Class whose heirs define methods that do not exist in this class, but must be created *virtually* to ensure polymorphism.

● An abstract class *cannot* be instantiated. It can only serve as a common (virtual) base of derived objects.

Introduction to PASCAL

This section explains the most important linguistic elements of the PASCAL programming language.

To represent the syntax of individual commands or program fragments, `typescript font` is used for words and symbols that must be written exactly as intended. Uppercase letters are used for reserved words, and lowercase letters for predefined identifiers. PASCAL itself does not differentiate between lowercase and upper-case letters. Parts written in *italics* must be replaced by individual statements or identifiers, depending on the application. In our examples, concrete values are always used in such places.

24.3 Basic structure

The basic structure of a PASCALprogram consists of:

PROGRAM *program name* ;
Declaration section
BEGIN
  Statement section
END.

▷ Remember the period behind the last END !
The parts which still must be entered in italics will be discussed in the sections which follow.
A **Blank spaces** can be inserted anywhere in PASCAL (except within identifiers or reserved words). The end of a line is treated the same as a blank space. Thus, one statement may extend over several lines, and several statements may be written on a single line.
Comments can (and should) be inserted for better readability. The text written in braces or inside (* ... *) represents a comment and it is ignored by PASCAL.
□

```
{ This is a comment }
(* and here
a further comment *)
```

24.4 Variables and types

Before a variable or a subprogram can be used in PASCAL, the corresponding identifiers must be declared. The following sequence of declarations is prescribed:

Definition of constants
Definition of types
Declaration of variables
Declaration of functions and procedures

Some PASCAL dialects allow deviations from these rules, but every identifier must always be declared before it is used for the first time.
In declaring a variable, it must be stated which data type is later to be included in the variable.
Identifiers, as well as the *program name* consist of a letter followed by any number of letters or numbers. The reserved words of PASCAL are not allowed as identifiers.
□ Valid identifiers: i, x123, test2a
Invalid identifiers: 2ab, gmbh&co, program
Declaration of variables:

VAR

$name_{11}, name_{12}, \ldots name_{1i} : type_1;$

$$name_{21}, \ldots name_{2j} : type_2;$$

$$\vdots$$

$$name_{n1}, \ldots name_{nk} : type_n;$$

Here, several variables always receive the same *type* by listing the variables separately, then dividing by commas and entering the type after a colon.

▷ Remember the semicolons!

Type of a variable: Specifies the purpose of the variable. There are several predefined types, and there are instructions for defining other types.

24.4.1 *Integers*

`integer`: Predefined type, can accommodate integer values within a certain range (usually between −32768 and 32767 depending on the computers). Some `PASCAL` dialects also recognize other integer types with other ranges of values, such as `longint` (−2147483648 to 2147483647) or `shortint` (−128 to 127).

□

```
VAR
   n, number: integer;
```

Variables with the names `n` and `number` can accommodate integer values.

24.4.2 *Real numbers*

`real`: Predefined type, can accommodate values from a subset of real numbers. The values are calculated and stored only with a finite accuracy.

▷ Due to this finite accuracy, rounding errors often occur for "`real`" types. Particular caution is advisable for comparison operations with "`real`" numbers!

Some `PASCAL` dialects recognize other types with "real" numbers with other accuracies of representation and other ranges of numbers.

Real constants are given with a decimal point or in exponential notation.

□ Real constants:

program text	meaning
`3.14159`	3.14159
`6.022E23`	$6.022 \cdot 10^{23}$
`1.602E-19`	$1.602 \cdot 10^{-19}$
`-2E-4`	$-2 \cdot 10^{-4} = -0.0002$

24.4.3 *Boolean values*

`Boolean`: Predefined type, can accommodate only one of the values "`true`" or "`false`."

▷ Boolean values are used particularly in conditional statements (`IF`) or loops (`WHILE`, `REPEAT`).

24.4.4 ARRAYs

There is often a need to process data in linear order. The required data are usually selected via an index. In `PASCAL` it is possible to declare an `ARRAY`-type element of:

ARRAY [*lower index*. . *upper index*] OF *type*

[−] a defined array element can be selected by giving the index [In the statement section of the program,]

name [*index*]

□
```
PROGRAM example;
VAR
   i: integer;
   a: ARRAY[1..10] OF real; { array of 10 elements }
BEGIN
   FOR i:=1 TO 10 DO a[i]:=sqrt(i); { calculate the array
   elements }
   FOR i:=10 DOWNTO 1 DO writeln(a[i]) { output in inverse
   manner }
END.
```

Two-dimensional arrangements of the array elements (matrices):

ARRAY [u_1. . o_1, u_2. . o_2] OF *type*

▷ This is an abbreviation of

ARRAY [u_1. . o_1] OF ARRAY [u_2. . o_2] OF *type*

and

name [$index_1$, $index_2$]

is an abbreviation of

name [$index_1$] [$index_2$]

▷ Multidimensional ARRAYs are declared analogously.

□ There is access to an element of the ARRAY

`multimatrix: ARRAY[1..5,0..10,-3..3,5..8] OF integer`

by

`multimatrix[3,0,-1,7]` or by `multimatrix[3][0][-1][7]`

▷ It is permissible to copy a complete ARRAY by a single assignment, but no arithmetical operations are allowed!

□
```
PROGRAM example;
VAR a,b: ARRAY[1..10] OF integer;
BEGIN
   a[3]:=b[3]; (* Copy of single component *)
   a:=b (* Copy of the complete array *)
END.
```

Of course, the types of both arrays must be equal.

PACKED ARRAY: Also an array that can be accessed exactly the same as an ordinary ARRAY, but the computer uses space-saving memory. It is used especially in the form of PACKED ARRAY[1. . *length*] OF char to store character strings.

▷ In some PASCAL dialects ARRAYs and PACKED ARRAYs are identical.

24.4.5 Characters and character strings

char: Predefined type, can take a single character as its value. The character must be contained in the character set of the computer. In any case, the character set includes the digits, the lowercase and uppercase letters of the alphabet as well as some special characters.

Character constants are always enclosed by apostrophes.

□
```
PROGRAM example;
VAR c,d: char;
BEGIN
   c:='4'; d:='A'; c:=d;
   writeln(c)
END.
```

▷ In the assignment `c:='4'`, the **character** 4 is assigned and **not the number** 4 is assigned.

Character string: Several characters enclosed by apostrophes, may be assigned directly to a `PACKED ARRAY OF char`.

▷ The dimensioning of the `ARRAY`s must agree with the length of the character string!

□
```
PROGRAM example;
VAR string: PACKED ARRAY[1..10] OF char;
BEGIN
   string:='Frankfurt ';
   writeln(string)
END.
```

24.4.6 RECORD

`ARRAY`: Combines several components of the same type. The component is selected via an index.

`RECORD`: Combines several components of any type. The component is selected via its name.

`RECORD`
   $component_{11}, \ldots component_{1j}$: $type_1$;
   ⋮
   $component_{n1}, \ldots component_{nk}$: $type_n$
`END`

Thus, a `RECORD` is defined by the given components, each possessing the specified type. In the statement section a `RECORD` component is selected by stating its name:

Record name . Component name

□
```
PROGRAM example;
VAR
   circle1,circle2: RECORD
      center: RECORD
         x,y: integer
      END;
      radius: integer
   END;
BEGIN
   circle1.center.x:=100; (* assignment to component *)
   circle1.center.y:=150;
   circle1.radius:=20;
```

```
    circle2:=circle1 (* assignment to the complete record
    *)
END.
```

As with ARRAYs, complete RECORDs can also be assigned provided that the types are identical.

▷ The names of components must not necessarily be different from other names. Of course, the different components of a **single** RECORD must have different names.

▷ ARRAYs and RECORDs may be nested arbitrarily.

24.4.7 Pointers

Pointer: Reference to a variable. A pointer to a variable does not contain the variable itself but only information on **where** to find the variable in the memory. When declaring a pointer variable one has to give the type of the variable to which the pointer points must be stated.

↑*type*

is a pointer type that refers to the given *type*.

- For access to the variable to which a pointer points, the arrow of the pointer variable is up.
- With

  new(*pointer variable*)

  a new variable of the type to which the pointer variable points is created in the computer memory. Then, the pointer variable points to this variable.
- With

  dispose(*pointer variable*)

  a variable of the type to which the pointer variable points can be removed. Then, there is no more access to this variable.

▷ In most computers, ^ is used instead of ↑. However, in printed listings ↑ is more common.

▷ With the declaration of a pointer variable only the memory location of the pointer variable itself is allocated. The pointer does not yet point to a practical memory location.

▷ We must carefully distinguish between the pointer variable and the variable to which the pointer points.

□
```
PROGRAM example;
VAR
   p1int,p2int: ↑integer;
BEGIN
   new(p1int); new(p2int); (* allocate memory *)
   p1int↑:=42; p2int↑:=-3;
   writeln(p1int↑:3,p2int↑:3);
   p2int↑:=p1int↑; (* variable assignment *)
   p1int↑:=5;
   writeln(p1int↑:3,p2int↑:3);
   p2int:=p1int; (* pointer assignment *)
   p1int↑:=19;
   writeln(p1int↑:3,p2int↑:3);
END.
```

The program creates the output

```
42 -3
 5 42
19 19
```

After the statement called by "variable assignment," the two pointers point as before to **distinct variables** which now, however, have **the same content**. Therefore, in the subsequent assignment of the variable to which `p1int` points, a new value may be assigned that is independent of the other variable.

After the statement called "pointer assignment," the two pointers point to **the same variable**. Therefore, the assignment to the variable to which `p1int` points influences also the content of `p2int↑`. `p1int↑` and `p2int↑` are different notations for the same variable.

NIL: Special value that can be assigned to any pointer variable. It indicates that the pointer points to "nowhere."

24.4.8 Self-defined types

If types created with ARRAY and RECORD are used several times, it is advisable to declare a new type first and to use the name of the new type in the variable declaration.

Type declaration:

TYPE
    $name_1$=$type_1$;
    ⋮
    $name_n$=$type_n$;

▷ The type declaration must be made **before** the variable declaration.

□ Use of type declarations:

```
PROGRAM example;
TYPE
   whole number = integer;
   vector = ARRAY[1..3] OF real;
   matrix = ARRAY[1..3] OF vector;
   tabell = ARRAY[1..100] OF RECORD
      city: PACKED ARRAY[1..10] OF char;
      zip: integer
   END;
VAR
   a,b: matrix;
   x:   vector;
   r:   vector;
   symbols: tabell;
   i,j: integer;
BEGIN
   a:=b; (* assign the complete matrix *)
   FOR i:=1 TO 3 DO BEGIN
      x[i]:=0;
      FOR j:=1 TO 3 DO BEGIN
         x[i]:=x[i]+a[i,j]*r[j]
      END
```

```
    END;
    symbols[15].city:='Boston ';
    symbols[15].zip:=23456???
END.
```

24.5 Statements

The *statement part* of a PASCAL program consists of statements separated by semicolons.

▷ The semicolons separate consecutive statements. This does not mean that every statement must end in a semicolon! In some cases a final semicolon is even prohibited, for example after the statement between THEN and ELSE.

During the execution of the program, the statements are normally executed in sequence. It is possible to deviate from this natural sequence by using conditional and loop statements.

24.5.1 Assignments and expressions

Assignment:

variable : =*expression*

assigns the value of an expression to the variable. The value must be of the same type as the variable. Exception: An `integer` value can be assigned to a `real` variable. Conversely, however, the assignment of a `real` value to an `integer` variable is impossible. In such cases the type must be converted by means of functions [(e.g., `round` or `trunc`)] designated for this purpose [−].

Expression: Quantity of a defined type, formed from constants and variables by arithmetic, Boolean, or comparison operators.

Arithmetic expression: An expression of the `real` or `integer` type. An arithmetic expression can be formed by using the four basic arithmetic operations, functions, and appropriate brackets. The basic arithmetic operations are represented by the symbols +, -, *, and /. The usual "point before bar" calculating rule applies; a deviation is possible by using brackets.

The `integer` values are divided by the `DIV` operator, which yields the integer part of the division as a result. The remainder of the division is obtained by using the `MOD` operator.

□ `22 DIV 5` result: 4
`22 MOD 5` result: 2

□ Arithmetic expressions in assignments:

```
k:=(i+3) DIV 2;
x:=2*a*b;
x:=-p/2+sqrt(p*p/4-q);
x:=sin(phi)/cos(phi)
```

Standard functions in PASCAL			
Name	Type of argument	Type of result	Function
`sqrt`	`real`	`real`	$\sqrt{x}$
`sin`	`real`	`real`	$\sin(x)$
`cos`	`real`	`real`	$\cos(x)$
`arctan`	`real`	`real`	$\arctan(x)$
`exp`	`real`	`real`	e^x
`ln`	`real`	`real`	$\ln(x)$
`sqr`	`integer`	`integer`	x^2
	`real`	`real`	
`abs`	`integer`	`integer`	$\|x\|$
	`real`	`real`	
`round`	`real`	`integer`	rounding to nearest integer
`trunc`	`real`	`integer`	truncation of digits after decimal point
`int`	`real`	`real`	truncation of digits after decimal point without type conversion

▷ Some frequently required functions, such as exponentiation, the tangent function, or the hyperbolic functions cannot be found among the standard functions. However, they can be defined easily. See the section on procedures and functions.

Boolean expression: An expression of `Boolean` type. A Boolean expression is formed in comparison operations or through connecting simple Boolean expressions by means of the Boolean operators NOT, AND, and OR.

By means of **comparison operators**, expressions of type `real` or `integer` type can be compared to each other.

Comparison operators	
Operator	Meaning
=	$=$
<>	$\neq$
<	$<$
<=	$\leq$
>	$>$
>=	$\geq$

▷ The Boolean operators have higher priority than the comparison operators. In the case of Boolean connection of several comparisons, the comparisons must be written in brackets.

□ false: `IF x>5 AND x<10 THEN ...`
is interpreted as
`IF x>(5 AND x)<10 THEN ...`
true: `IF (x>5) AND (x<10) THEN ...`

24.5.2 Input and output

In every program, the data to be manipulated are the input, and the calculated results are the output. Here, we first discuss the input from the keyboard and the output on the screen.

The input value is accomplished with the `read` and `readln` procedures:

`read(`$variable_1$, $variable_2$, ..., $variable_n$`)`

This statement consecutively reads in the values for *variable*$_1$ to *variable*$_n$ with the keyboard and assigns them to the variables. Only variables of numerical type and character variables are allowed.

▷ `ARRAY`s cannot be read in by a single `read`. If a complete array must be read in, the array elements must be read in separately in a loop.

For several consecutive `read` statements, the values are taken successively from one or more input rows. If a `read` statement did not read all values of an input row, then the subsequent values are read by the next `read` statement.

If `readln` is used instead of `read` [after the input of the required values], the remaining input row is ignored for the subsequent `read` or `readln`. `readln` without an argument does not read in any value, but clears the input buffer [–] for subsequent inputs.

□ `read(a); read(b); readln`

is equivalent to

`readln(a,b)`

For the data output, the procedures `write` and `writeln` are available:

`write(`*expression*$_1$`,` *expression*$_2$`, ... ,` *expression*$_n$`)` .

This statement consecutively writes the values of *expression*$_1$ to *expression*$_n$ on the screen. By using `writeln`, an additional line is created after the last data unit, i.e., the next `write` or `writeln` statement will start the output in the next line.

Similar to `readln`, `writeln` can also be used without an argument, in which case it only creates a line feed.

▷ Only numerical, Boolean and character expressions are allowed as arguments of `write` and `writeln`. The output of complete `ARRAY`s must be output via a loop. Only `PACKED ARRAY OF char` can be output directly.

▷ Normally, the data are output consecutively without blank spaces between them. Thus, for numerical output values we obtain only a long sequence of digits, and we do not know where the individual data unit begins.

Formatted output: After the expression to be output a colon followed by a number can be introduced. This causes the expression to be output within a field of given size.

□ `writeln(125:6,-3:5,'abc':5)`

causes the following output:

`␣␣␣125␣␣␣-3␣␣abc`

(blank spaces are denoted by ␣)

24.5.3 Compound statements

In some places in the program only a single statement is allowed, for example, after `THEN` within a conditional statement. If in such places several statements must be executed, they can be combined with `BEGIN` and `END`, and they are treated by `PASCAL` as a single statement.

```
BEGIN
    statement1;
    statement2;
      :
    statementn
END
```

▷ No semicolons follow after the last statement before the `END`, since the semicolon is the separation sign **between** statements. On the other hand, a semicolon at that point does no harm either, and some programmers always insert it.

24.5.4 Conditional statements IF and CASE

Conditional statements can be used to execute statements or not, depending on a condition:

`IF` *condition* `THEN` *statement*

The *statement* is executed if the *condition* is `true`. The *condition* may be any Boolean expression.
It is also possible to execute one of two statements depending on a condition:

`IF` *condition* `THEN` $statement_1$ `ELSE` $statement_2$

If the *condition* is `true`, $statement_1$ is executed in the `THEN` part, otherwise $statement_2$ in the `ELSE` part is executed.
If several statements must be executed depending on a condition, they can be bracketed by `BEGIN` and `END`:

```
IF condition THEN BEGIN
    statement11 ;
    :
    statement1n
END
ELSE BEGIN
    statement21 ;
    :
    statement2k
END
```

▷ A semicolon **never** occurs before an `ELSE`!
In some cases we want to select one of many alternatives. The `CASE` statement evaluates an arithmetic expression and compares it with several constant values. If it is equal to one of these values a corresponding statement is executed:

```
CASE expression OF
constant1 : statement1 ;
constant2 : statement2 ;
:
constantn : statementn
END
```

Equivalent to this is a string of `IF`s:

```
auxiliary variable : =expression ;
IF auxiliary variable=constant1 THEN statement1
ELSE IF auxiliary variable=constant2 THEN statement2
:
ELSE IF auxiliary variable=constantn THEN statementn
```

▷ The *expression* and thus the *constants* must not be of the `real` type! Only types with an enumerable range of values are allowed.

24.5.5 *Loops* FOR, WHILE, *and* REPEAT

Loop instructions: Cause one or several statements to be executed consecutively several times.

In the case of the **count** loop the number of loop cycles must be known before entry of the loop. At the beginning, a count variable is placed to an initial value and is increased by 1 after every loop cycle. The last cycle of the loop has the given final value.

`FOR` *variable* `:=`*initial value* `TO` *final value* `DO` *statement*

The count variable is counted backward when `TO` is replaced by `DOWNTO`:

`FOR` *variable* `:=`*initial value* `DOWNTO` *final value* `DO` *statement*

▷ The control variable can only be incremented (counted upward) or decremented (counted downward) in increments of 1. Other increment sizes are not possible in `PASCAL`.

▷ The count variable must be of an enumerable type normally of the `integer` type. Real-valued types are **not** allowed!

▷ The body of the loop contains only one statement. If necessary, several statements can be combined with `BEGIN` and `END`.

□ Real-valued count variables of any increment size can be achieved as follows:

```
FOR i:=0 TO 9 DO BEGIN
   x:=start+increment size*i;
   writeln(x:5:2,sqrt(x):10:5)
END
```

If the number of loop cycles is not known in advance, then the `WHILE` or `REPEAT` loop must be used. Here, a condition decides whether an additional loop cycle is needed or not.

The `WHILE` loop tests **before** each loop cycle whether a condition is met. If this is the case, the body of the loop is executed; if not, the loop is terminated.

`WHILE` *condition* `DO` *statement*

▷ The body of the loop contains only one statement. If necessary, several statements can be combined with `BEGIN` and `END`.

▷ If the condition is already not met at the beginning, the body of the loop is not executed at all.

The `REPEAT` loop tests **after** each cycle of the loop whether a condition is met. If this is the case, the loop is terminated; if not, the body of the loop is executed once more.

`REPEAT`
    statement$_1$;
    ⋮
    statement$_n$
`UNTIL` *condition*

▷ Since the condition is tested after the body of the loop, the `REPEAT` loop is always executed at least once.

▷ Here, any number of statements may appear between `REPEAT` and `UNTIL`. Thus, it is not necessary to bracket several statements with `BEGIN` and `END`. No semicolon is necessary before `UNTIL` and before `END`, but it is not prohibited.

24.6 Procedures and functions

In large programs there are constantly recurring program modules which we do not want to program again every time. It is therefore possibile to program these modules only once as a **subprogram** and insert simple subprogram calls in the corresponding places in the main program.
Furthermore, it is practical to combine logically connected program modules within a subprogram even when it is called only once in the main program. This approach increases readability and makes a program less error-prone.

24.6.1 Procedures

Procedure declaration:

PROCEDURE *procedure name*;
declaration section
BEGIN
  statement section
END;

⊳ A procedure declaration is very similar to a complete program. Local types and variables and even local procedures and functions can be declared in the declaration section.

Very often, we want to transfer data to procedures. In such cases the procedure declaration is extended by a **parameter list**:

PROCEDURE *Name* ($parameter_{11}$, ...$parameter_{1i}$: $type_1$;
  $parameter_{21}$, ...$parameter_{2j}$: $type_2$;
  ⋮
  $parameter_{n1}$, ...$parameter_{nk}$: $type_n$);
declaration section
BEGIN
  statement section
END;

The names and types of the parameters are declared in the parameter list. This declaration is very similar to a variable declaration.

⊳ ARRAYs or RECORDs must not appear in a parameter list. If such a parameter is to be used, then a type must be defined whose name is used in the parameter list.

□
```
TYPE field=ARRAY[1..10] OF integer;
PROCEDURE hoo(i: integer; a: field);
```

When a procedure is called, the actual arguments are transferred in brackets and separated from each other by commas. Subsequently, we can access the actual arguments via the parameter list.

□
```
PROGRAM example;

PROCEDURE output(n: integer; x,y: integer);
VAR i: integer;
BEGIN
   FOR i:=1 TO n DO writeln(x:5,y:5)
END;
```

```
BEGIN (* of the main program *)
   output(2,1,2);
   output(1,2+3,1+1)
END.
```

24.6.2 Functions

Functions: Similar to procedures, but they return a value of the result. In the declaration of a function, the reserved word `PROCEDURE` is replaced by `FUNCTION`, and the type of the result of the function must be stated. This is done by a colon followed by the result type following the parameter list.

- In the statement section of the function the result of the function must be assigned. For that, the function name is used like a variable name, i.e., without an argument list.

□ Important arithmetic functions, not among the standard functions, can be defined as follows:

```
FUNCTION tan(x: real): real;
BEGIN
   tan:=sin(x)/cos(x)
END;

FUNCTION power(x,y: real): real;
BEGIN
   power:=exp(y*ln(x))
END;

FUNCTION sinh(x: real): real;
VAR dummy: real;
BEGIN
   dummy:=exp(x);
   sinh:=(dummy-1.0/dummy)/2.0
END;

FUNCTION cosh(x: real): real;
VAR dummy: real;
BEGIN
   dummy:=exp(x);
   cosh:=(dummy+1.0/dummy)/2.0
END;

FUNCTION tanh(x: real): real;
BEGIN
   tanh:=sinh(x)/cosh(x)
END;
```

24.6.3 Local and global variables, parameter passing

Local variable: A variable declared inside a function or procedure. A local variable is known only inside this function or procedure. There is no access to this variable from the main program or from other subprograms.

Global variable: A variable declared in declaration section of the main program. A global variable is known in the whole program, including all subprograms, and there is access to this variable from every subprogram.

▷ A local variable can be given the same name as a global variable. In this case, in the subprogram there is only access to the local variable under that name. There is no longer access to the global variable.

▷ With every call of a subprogram, the memory location of the local variable is allocated anew. The content of the local variable is lost after the return from the subprogram.

▷ The parameters of a subprogram also represent local variables whose content is defined with the call of the subprogram by the actual arguments transferred.

□
```
PROGRAM example;
VAR a,b,c: integer; (* global variable *)

PROCEDURE foo(a: integer); (* a is local *)
VAR b,d: integer; (* b and d are local *)
BEGIN
   b:=2*a; (* assignment to local variable *)
   c:=3*a; (* assignment to global variable *)
   d:=4*a;
   writeln('in procedure:   ',a:3,b:3,c:3,d:3)
END;

BEGIN (* of the main program *)
   a:=5; (* assignment to global variable *)
   b:=3;
   c:=-4;
   writeln('before procedure: ',a:3,b:3,c:3);
   foo(b);
   writeln('after procedure:',a:3,b:3,c:3)
END.
```

The program creates the following output:

```
before procedure:   5  3 -4
in procedure:     3  6  9 12
after procedure:  5  3  9
```

With the call of a subprogram, the value of an actual argument is normally copied into a local variable to which there is then access in the subprogram via the parameter name (**call by value**). This has the consequence that

1. for large ARRAYs or RECORDs as a parameter, the computer must copy a large amount of data, and

2. the assignments inside the subprogram to the parameters outside the subprogram have no effect.

In PASCAL there is a further access mechanism in which no copy of the variable is made, but where the variable transferred as an argument is called via the parameter name (**call by reference**).

In the parameter list the word VAR is placed in front of the parameters accessed via call by reference.

□
```
PROGRAM example;
VAR a,b: integer;

PROCEDURE hoo(VAR x: integer; y: integer);
BEGIN
   x:=1;
   y:=2;
END;

BEGIN (* of the main program *)
   a:=5;
   b:=6;
   hoo(a,b);
   writeln(a:3,b:3)
END.
```

The program creates the output

```
  1  6
```

In the procedure `x` was a `VAR` parameter, and this changes the variable `a`, which had been given as an argument, through the assignment `x:=1`. On the other hand, `y` was not a `VAR` parameter; and thus, `b` was not changed by the assignment `y:=2`.

▷ The arguments transferred to a `VAR` parameter can only be variables. Constants or arithmetic expressions are not allowed here.

In other programming languages both access possibilities do not always exist. `FORTRAN` always uses call by reference, while `C` always uses call by value.

24.7 Recursion

Recursion: Call of a procedure or a function by itself.
To avoid an infinite sequence of procedure calls, there must be a **start of recursion** for which no further call of the procedure is necessary.
Iteration: Calculation inside a **loop**, using the values calculated in the previous loop cycle.

▷ Every recursive program can be rewritten as an iterative program. Some problems can be solved much more simply and more naturally through recursion.

□ The **factorial** is defined by

$$n! = n \cdot (n-1)! \quad \text{if } n > 0, \qquad 0! = 1.$$

This can be rewritten immediately in a recursive function:

```
FUNCTION facto(n: integer): real;
BEGIN
   IF n>0 THEN facto:=facto(n-1)*n (* recursion *)
   ELSE IF n=0 THEN facto:=1 (* start of the recursion *)
   ELSE writeln('error');
END;
```

In this case an iterative solution is the simpler and preferred solution:

```
FUNCTION facto(n: integer): real;
VAR i: integer; f: real;
BEGIN
```

```
    f:=1;
    FOR i:=1 TO n DO f:=f*i;
    facto:=f;
END;
```

For the recursive solution, note that the local variable n is newly allocated with every call of facto, and that its value after every return is the same as before the call in question.

It is also a recursion when a procedure is not called directly by itself but indirectly via another procedure. The difficulty is that every procedure must be defined before its first use.

```
PROCEDURE a(n: integer);
BEGIN
    call of procedure b
END;

PROCEDURE b(k: integer);
BEGIN
    call of procedure a
END;
```

In this example, procedure b is not defined when b is called inside a. Reversing the order of a and b does not help, since in that case, a would be undefined inside b.

Remedy:

FORWARD declaration of the procedure before its first use.

```
PROCEDURE b(k: integer); FORWARD;

PROCEDURE a(n: integer);
BEGIN
    call of procedure b
END;

PROCEDURE b; (* now without parameter list *)
BEGIN
    call of procedure a
END;
```

▷ The parameter list and (for functions) the type of the function value are given **only** in the FORWARD declaration. If the procedure is actually defined, these declarations are not repeated again.

Recursions are not allowed in all programming languages. PASCAL and C allow recursions, but they are prohibited in many FORTRAN dialects.

24.8 Basic algorithms

24.8.1 Dynamic data structures

Static data structure: A data structure that cannot be changed while a program is running.

□ ARRAYs and RECORDs are static data structures. The dimensions of an ARRAY cannot be changed. If the number of data which an ARRAY accommodates is not known during programming, then the dimensions of the ARRAY must be large enough for large applications. The disadvantage is that a large amount of storage may be wasted.

Dynamic data structure: A data structure that can be changed while the program is running. In PASCAL all dynamic data structures work **with pointers**. By means of the procedures new and dispose it is possible to allocate and to clear memory location while the program is running.

The most important dynamic data structures are linear **lists** (singly or doubly linked lists) and **trees**.

Linear list: One-dimensional arrangement of list elements.

Singly linked list: Every list element contains a pointer which points to the next list element.

Doubly linked list: Every list element contains pointers which point to the next and the preceding list element.

▷ The last element of the list has no successor. Thus, the corresponding pointer is set to the NIL value.

Ring list: A linear list whose last element contains a pointer which points back to the first element.

□ Insertion and deletion in a singly linked linear list.

```
TYPE element=RECORD
   key:  integer;
   next: ↑element;
END;
VAR start of list,dummy,position: ↑element;
:
(* insertion in list *)
new(dummy);
dummy↑.next:=position↑.next;
position↑.next:=dummy;
:
(* deletion of list *)
dummy:=position↑.next;
position↑.next:=dummy↑.next;
dispose(dummy);
:
(* printing of list elements *)
position:=start of list;
WHILE position<>NIL DO BEGIN
   writeln(position↑.key);
   position:=position↑.next;
END;
:
```

Here, we insert **after** the element to which the pointer position points, or one deletes the successor of the element to which the pointer position points.

24.8.2 Search

In most assignment or translation problems, searching is required.

□ Search for an entry in a glossary or dictionary.

A PASCAL compiler searches in a table of reserved words for a word read from an input file to decide whether it is a reserved word.

Example: Search for the zip code of a certain city.

In all following search algorithms, it is assumed that the data to be searched are in an ARRAY and declared by

```
TYPE searchfield = ARRAY[1..max] OF integer
```

In the following procedures the parameter n indicates the number of valid entries in the ARRAY.

Linear search: Comparison with all data; no assumption regarding data. Execution time $\propto n$.

```
FUNCTION find(key: integer; n: integer; a:
searchfield): integer;
VAR i,k: integer;
BEGIN
   i:=0; k:=1;
   WHILE (k<=n) AND (i=0) DO BEGIN
      if key=a[k] THEN i:=k;
      k:=k+1;
   END;
   find:=i;
END;
```

In case of a successful search, the index of the found element is returned as the function value; otherwise the function value is zero.

Binary search: Requires sorted data. Execution time $\propto \log n$. By means of a continued bisection of the search interval, the search is limited to fewer and fewer elements until the element is found.

```
FUNCTION find(key: integer; n: integer; a: searchfield):
integer;
BEGIN
   start:=1; end:=n;
   i:=0;
   REPEAT
      middle:=(start+end) DIV 2;
      IF a[middle]<key THEN end:=middle-1
      ELSE IF a[middle]>key THEN start:=middle
      ELSE i:=middle;
   UNTIL (start=end) or (i>0);
   IF a[start]=key THEN i:=start;
   find:=i;
END;
```

24.8.3 Sorting

The sorting of data according to a certain characteristic is a problem that occurs often.

□ Sorting the participants in a competition by their results.

Alphabetic sorting of names.

Alphabetic sorting for the index of a book.

Sorting of data gain quick access in a binary search.

In all the following sort algorithms, it is assumed that the data to be sorted are arranged in an ARRAY declared by

```
TYPE sortfield = ARRAY[1..max] OF integer
```

Furthermore, a procedure swap is used that reverses the contents of the two integer variables:

```
PROCEDURE swap(VAR x,y: integer);
VAR dummy: integer;
BEGIN
   dummy:=x;
   x:=y;
   y:=dummy;
END;
```

Sorting by direct selection: Very simple sort algorithm; execution time $\propto n^2$. If the first k elements are already sorted, then the smallest of the remaining elements is searched for and exchanged for the element at position $(k+1)$. Now, if k is incremented in a loop, then the whole array is sorted.

```
PROCEDURE sort(n: integer; VAR a: sortfield);
VAR k,i,j: integer;
BEGIN
   FOR k:=1 TO n-1 DO BEGIN
      FOR j:=k+1 TO n DO BEGIN
         IF a[j]<a[k] THEN swap(a[k],a[j]);
      END;
   END;
END;
```

Bubble sort: Also very easy to program; execution time $\propto n^2$. Inside a loop it is tested whether two adjacent elements are sorted in ascending order. If not, they are exchanged. As long as elements still must be exchanged, this is repeated for all elements.

```
PROCEDURE bubblesort(n: integer; VAR a: sortfield);
VAR
   i: integer;
   interchanged: Boolean;
BEGIN
   REPEAT
      interchanged:=false;
      FOR i:=1 TO n-1 DO BEGIN
         IF a[i]>a[i+1] THEN BEGIN
            swap(a[i],a[i+1]);
            interchanged:=true;
         END;
      END;
   UNTIL NOT exchanged;
END;
```

Quick sort: A very fast algorithm; execution time $\propto n \log n$. The data to be sorted are subdivided into two groups as equal as possible, so that all elements of the first group are smaller than all elements of the second group. Now, each group is sorted separately by means of recursion. When the sorted groups are linked again, then all data are to be sorted.

```
PROCEDURE quicksort(start,end: integer; VAR a: sortfield);
VAR i,j,x: integer;
BEGIN
   i:=start; j:=end;
   x:=a[(start+end) DIV 2]; (* subdivision of the groups *)
   REPEAT
      WHILE a[i]<x DO i:=i+1; (* search for a larger
                                 element *)
      WHILE a[j]>x DO j:=j-1; (* search for a smaller
                                 element *)
      IF i<=j THEN BEGIN
         swap(a[i],a[j]);
         i:=i+1; j:=j-1;
      END;
   UNTIL i>j;
   IF start<j THEN quicksort(start,j,a);
   IF i<end THEN quicksort(i,end,a);
END;
```

24.9 Computer graphics

A graphics output is very useful for illustrating results obtained by numerical means. A certain number of basic functions are necessary to create graphics output, but these functions are not contained in the normal PASCAL language. Often, there are graphics libraries to make such functions available. Unfortunately, the functions they contain vary a great deal. This section deals with functions and procedures made available by TURBO-PASCAL.

Pixel: Single dot on the matrix that makes up the computer screen. Every pixel is represented by one or more bits of computer memory. For black and white graphics, one bit per pixel is enough. For color graphics or several gray tones, more bits per pixel are needed. With p bits per pixel, 2^p different colors or gray tones can be represented.

24.9.1 *Basic functions*

There must be functions to place a single dot on the screen and to delete it, and to clear the whole screen. All other functions, e.g., the representation of straight lines or circles or to place text inside graphics, can basically be reduced to those functions. Among the many graphics procedures in TURBO-PASCAL are:

`PROCEDURE circle(x,y,r: integer)`

Draws a circle of a radius r about the center (x, y).

`PROCEDURE cleardevice`

Clears graphics screen.

`PROCEDURE closegraph`

Terminates graphics.

`FUNCTION getgraphmode: integer`

Determines the graphics mode.

```
FUNCTION getmaxx: integer
FUNCTION getmaxy: integer
```

Determines if the x and y coordinates are representable maximally.

```
PROCEDURE line(x1,y1,x2,y2: integer)
```

Straight line from (x_1, y_1) to (x_2, y_2).

```
PROCEDURE putpixel(x,y,color: integer)
```

Drawing of a single pixel in a given color.

Before the first use of graphics the procedure `initgraph` must be called. In `TURBO-PASCAL`, to enable these procedures and functions, it must be declared immediately after the program heading that graphics will be used:

```
PROGRAM program name;
USES graph;
  ⋮
```

Introduction to C

- ● C is the most popular high-level programming language for system programming. C has not only all elements of a high-level language, it also allows very machine-oriented programming and thus fast and efficient programs.

▷ The use of C in the style of a high-level language such as PASCAL is possible, but seldom effective. The art of C programming is the consistent utilization of characteristic language elements of C (such as pointers).

24.9.2 *Basic structures*

Basic rules:

1. Every C command must end in a semicolon ";".
2. Braces and preprocessor commands are *never* followed by a semicolon (exception: `struct, class`).
3. Every variable, constant, or function used must be declared *before* its first call, i.e., it must be specified with respect to the data type. If a function is called in a file in which it is not defined, it must be given a **prototype** form.

Function: Only kind of subprogram in C. Any number of arguments can be stated. A return value is always expected.

— Functions can also be called as commands by *not* using them in an assignment. The function is executed, but the return value is rejected.

The actual function can be defined in another program module or in a library.

Function prototype: Declaration of a function. Introduction of the function name including all parameters *before* the actual definition of the function. Necessary in case of programs divided into different files or for the incorporation of library functions.

□ The file math.h contains prototypes for mathematical functions (sin, cos, ...)

▷ In the case of parameters being transferred, the old C standard (Kernighan & Ritchie) allows type declaration within the function head. Here, the types must be defined subsequently:

```
int function(v,x)
char v;
float x;
{
    :
}
```

Header file: Combination of function prototypes in one file. Makes available to all program modules the function bodies defined elsewhere and still to be linked.

☐ The header file `stdio.h` contains the function prototypes for the standard input/output routines in C. The header must be inserted in every program file that uses input/output operations.

Comments: Realized in C by encapsulating the text of the comment in /*...*/.

Preprocessor: Program run prior to the actual compilation to remove comments, to replace short characters, or to enter declaration files. Searches for special **preprocessor commands**:

`#define`	global text replacement in the source program text, definition of macros
`#undef`	removal of a definition
`#include`	incorporation of text from other source files
`#pragma`	special instructions to the compiler

#define: Preprocessor command for the definition of macros and compiler switches. Parameter transfer is allowed:

☐ `#define MIN(x,y) ( ((x) >= (y)) ? (y) : (x) )`
defines an expression that yields the minimum of two numbers, independent of the data type.

▷ Macros defined by `#define` are *not* functions but are inserted by the preprocessor in the source code at the positions of their call. The parameters are not variable names but merely space holders. There is no testing for types.

#include: Preprocessor command to insert other source files, such as function prototypes used in different program sequences or prototypes of standard libraries.

☐ The prototypes of functions should be read in via `#include` from other files and not be declared separately in every source file.

Program block or **block**: Sequence of C statements enclosed by braces "{...}". Variables or functions declared within a block are local and therefore valid only in that block.

▷ A block can stand wherever a single C statement is also allowed.

● Declarations of variables are allowed only at the beginning of a block prior to the first executable statement.

Main program: Function `int main()` that is called at the beginning of a program. It is allowed to be defined only *once* in a program, but this can be anywhere in the program. The return value can be requested from the calling operating system for error processing.

☐ Example of a program:

```
  #include <stdio.h> /* insert the IO library */
int main()
{
    int i = 5;
    printf(''Hello world: %d'',i); /* output command */
}
```

24.9.3 Operators

Operators: Symbols to connect or process expressions.
Expression: Construction created by a combination of valid operations which returns the value of a defined data type.
Left value (lvalue): Expression to which a value can be assigured by "=", i.e., it may appear on the *left side* of a statement.

□ Variables (`i = 7`), pointers without reference:

```
  int y[10];
*(y+2) = 5;
```

Incrementation operator ++: A variable is raised by 1. C distinguishes between **prefixed** and **postfixed incrementation operators**.
Prefixed incrementation: + + *lvalue*, increment the expression and returns this incremented value.
Postfixed incrementation operator: *lvalue*, ++, the return value corresponds to the old (not incremented) value. The incrementation is visible only in the *next* use of the variable.

□
```
int i=5, j=5;
printf("'i++ = printf("'i =
```

Decrementation operator −−: A variable is lowered by 1. Like the incrementation operator, it can be postfixed.
Priority: The order in which operators are executed. Operators with higher priority are executed before those with lower priority. A change in priority requires **bracketing**.

▷ In the table on the next page the operators are arranged in the order of decreasing priority. Operators within the individual segments have equal priority.

Operators in C

.	structure	*structure.element*
->	structure (pointer)	*pointer->element*
[]	subscript	*pointer*[int- *expression*]
()	call of function	*expression* (*expression list*)
++	incrementation	*lvalue*
--	decrementation	*lvalue*
sizeof	size of object	sizeof *expression*
sizeof	size of datatype	sizeof(*datatype*)
++	postfixed incrementation	*lvalue* ++
--	postfixed decrementation	*lvalue* --
~	complement	~*expression*
!	logical negation	! *expression*
−	minus	− *expression*
+	plus	+ *expression*
&	address	&*lvalue*
*	indirection	* *expression*
()	type conversion	(*data type*) *expression*
*	multiplication	*expression* * *expression*
/	division	*expression* / *expression*
%	remainder	*expression* % *expression*
+	addition	*expression* + *expression*
−	subtraction	*expression* − *expression*
<<	left shift, bitwise	*expression* << *expression*
>>	right shift, bitwise	*expression* >> *expression*
<	less than	*expression* < *expression*
<=	less than or equal	*expression* <= *expression*
>	greater than	*expression* > *expression*
>=	greater than or equal	*expression* >= *expression*
==	equality	*expression* == *expression*
!=	non-equality	*expression* != *expression*
&	bitwise AND	*expression* & *expression*
^	bitwise XOR	*expression*^*expression*
|	bitwise OR	*expression* | *expression*
&&	logical AND	*expression* && *expression*
||	logical OR	*expression* || *expression*
? :	case distinction	*expression* ? *expression* : *expression*
=	assignment	*lvalue* = *expression*
*=	assign product	*lvalue* *= *expression*
/=	assign quotient	*lvalue* /= *expression*
%=	assign remainder	*lvalue* %= *expression*
+=	assign sum	*lvalue* += *expression*
−=	assign difference	*lvalue* −= *expression*
<<=	assign left shift	*lvalue* <<= *expression*
>>=	assign right shift	*lvalue* >>= *expression*
&=	assign bitwise AND	*lvalue* &= *expression*
|=	assign bitwise OR	*lvalue* |= *expression*
^=	assign bitwise XOR	*lvalue*^= *expression*

24.9.4 Data structures

Automatic variables: Variables whose associated memory location in the stack is requested during their declaration—while the program is running—and released again when they leave their scope of validity.
Static variables: The opposite of automatic variables. Even during the compilation process, static variables are assigned a fixed memory location which they keep as long as the program is running.

`static`: Declares a variable to be *static*.

`auto`: Declares a variable to be *automatic*.

▷ By definition, all variables defined inside blocks are *automatic*; all variables defined outside all blocks (file scope) are *static*.

Data type: Property that must be assigned to *every* variable in its declaration. We distinguish **simple** and **composite** (user-defined) data types.

● Simple data types: `int, char, long, float, double, void`.

`void`: Special data type which in pointer programming serves to represent a pointer of undefined type. In the declaration of functions, `void` indicates to the compiler that *no* parameters are transferred or no return value is expected.

▷ The call of a function declared with `void` in the parameter list is done without `void`, simply with empty brackets:

```
int test function(void)   ...
int i = test function();  /* and not: test function(void) */
```

▷ `int, char` and `long` are **integer** data types, while `float` and `double` represent **floating point types**.

▷ The range of values of simple data types is not standardized because it depends on the memory location (length) available for the data type in question. Only the `char`-types have a standard length of 8 bits.

data type	description	length (32 bit computer)	range
`int`	integers	4 bytes	$-2^{31} \dots 2^{31}-1$
`unsigned int`		4 bytes	$0 \dots 2^{32}-1$
`char`	bytes	1 bytes	$-128 \dots 127$
`unsigned char`		1 bytes	$0 \dots 255$
`long`	long integers	4 bytes	$-2^{31} \dots 2^{31}-1$
`unsigned long`		4 bytes	$0 \dots 2^{32}-1$
`float`	simple-accurate	4 bytes	
`double`	double-accurate	8 bytes	

Logical expression: In C equivalent to any integer data type `int, char, long`. True if it has a value != 0 ; false if == 0.

□ Declaration of variables:

```
int i,j,k;
char c = 0;
unsigned long l = -2;  invalid because unsigned required!!
```

● Variables must always be declared *at the beginning* of a program element (before the first row with executable code).

Type conversion: Always possible between simple data types. Achieved by using the data type as prefix:

(*simple data type*) *expression*

□
```
int m = 32.7; /* invalid because types are incompatible!
*/
int n = (int)32.7; /* valid because valid */
/ type conversion. */
```

Composite data type (struct): Construct of several single or other composite data types. Used for the declaration of variables, analogous to simple data types.

□
```
construct date
{
   unsigned int day, month, year;
}
```

Declaration of a variable:

```
construct date today;
```

● The elements of a composite data type are addressed via the **structure selection operator** ".".

```
printf(''date: %ud.%ud.%ud'',,today.day, today.month,
today.year);
```

union type: Composite data type that allows different data types to address the same memory area.

□
```
union demo {
int number;
construct { char byte3,byte2,byte1,byte0; } bytes;
}
```

A variable `union demo test;` can be addressed by `test.number` and by `test.bytes.byte2`.

typedef: Used to rename single and composite data types.

□ `typedef float REAL;` or `typedef union demo UDemo;` data types thus defined can be used for declaration like simple data types (without the prefix `construct, union, ...`).

● Data types defined by `typedef` are processed by the compiler as if they were identical to their defined data types. Thus the compiler, `union demo x;` and `UDemo y;` are of identical type.

Pointer: Most important data structure for effective C programming. Does not contain the actual value to be stored, but the memory address where the value is stored.

● Pointers are type-specific! A pointer pointing to an `int` variable differs from a pointer pointing to a `char` variable.

▷ The length of the data-type pointer depends on the size of the memory location to be addressed and on the architecture of the computer but does not depend on the type of the pointer!

Indirection (dereferencing) operator `*x`: Denotes the content of the memory cell to which the pointer points. `*x` can stand on the left or right of an assignment:

```
int x;
int* y;
*y = 5;
x = *y;
```

▷ After the declaration according to `int* x;` the pointer `x` is undefined first, i.e., it contains a random value. A description according to `*x = 5`, leads to an uncontrolled change of the memory cells allocated to other variables or—in the worst case—to an executable program. This leads to errors or crashes.

Address operator "&": Yields a pointer of the corresponding type pointing to a variable.

```
int* p1;
char* p2;
int i = 1;
char c = 2;
p1 = &i;
p2 = &c;
printf(''i = %d, c = %d'',*p1,*p2); /* yields: "'i = 1, c = 2"'
*/
```

▷ In the case of composite data types with high memory requirements, it is, for example, clearly faster to copy variables with pointers than to perform operations with normal variables for which the entire area of memory always must be copied.

Array: Connected list of variables. Individual elements of the array are invoked via the **index**.

▷ Indexing of an array *always* start at 0.

```
int field[5]; /* defines field[0] ... field[4] */
```

● The programmer must ensure that the dimension is not exceeded. The row `int j = field[5];` is translated without errors but the content of this memory cell is not defined because it may be used by other variables.

▷ An array variable is *equivalent* to a pointer of the corresponding type.

```
int* alsofield = &field[0]; or simpler
int* alsofield = field;
```

defines a data array `alsofield` *identical* with `field` with the same elements `alsofield[0]...alsofield[3]`.

▷ The first element of the array field can be addressed with `*field` as well as with `field[0]`.

Differences between array and pointer declaration:

1. Contrary to pointers, arrays *cannot* be assigned to each other even if they are of the same type and the same size.

2. In the declaration of an array a memory location is automatically requested that corresponds to the size of the array and the data type (**static** memory management).

□ `float y[6] = {1,2,3,4,5,6};`

Vector of dimension n: One-dimensional array of length n.

Strings: In C, accomplished via `char` arrays. Functions for processing strings are made available via `#include <string.h>`

▷ In C, strings are terminated by the null character "\0", i.e., they occupy an additional byte in the memory.

□
```
char* string=''This is a string'';
/* string[0] = 'D', string[19] = '\0' */
```

Multidimensional arrays: Memory area addressed by several indices. In C, this can be achieved up to any dimension.

□ `float x[2][3] = {{1,2,3},{4,5,6}};`

▷ Like a one-dimensional array, a multidimensional array is just as equivalent to a pointer of the corresponding type as a one-dimensional array:

□ `float* y = x;`
The (now one-dimensional) array `y` contains: `1,2,3,4,5,6`.
This addressing `y[i][j]` is *not permissible* because the pointer `y` has no information on the length of the vectors (3 in our example).

□ Equivalent formulation: `x[i][j]`≡`y[i*3+j]`.

Dynamical storage management: Array sizes are defined not during the compilation, but only while the program is running. For this purpose, C provides standard libraries for storage management:

```
#include <stdlib.h>
```

Reservation of memory: `char* String = malloc(10*sizeof(char));`
Release of memory: `free(String);`

24.9.5 Loops and branches

Commands for the programming of loops: `for, while, do...while, break, continue`

While loop:

```
while (expression)
    statement;
```

expression: terminating criterion.

statement: statement that must be executed in the loop.

▷ If more that one statement must be executed, these must be combined in a **block** ({...}).

The statements within the `while` loop are executed **as long as** the *expression* is logically true.

▷ The terminating criterion is tested at the beginning of the loop. If the *expression* is already logically false before the first looping, the program jumps to the first statement outside the loop, and the loop is never executed.

□
```
float x = 64.0;
while (x > 1.0) {
    x /= 2.0;
    printf(''x = %f'',x);
}
```

do-while loop:

```
do
    statement;
while (expression);
```

The loop is executed **until** the terminating criterion is met, i.e., until the *expression* is logically false.

▷ The `do-while` loop is executed *at least once* because the terminating criterion is tested only at the *end* of the loop.

for loop:

```
for (expression1;expression2;expression3)
    statement;
```

expression 1: initializes the loop variable (counter).

expression 2: terminating criterion.

expression 3: changes the loop variable.

□
```
int i;
for (i=0; i<10; i++) printf("'Nr. %d"',i);
```
▷ `for` loops can be handled very flexibly. For example, `while` loops can be simulated:
```
for (x = 64.0 ;x>1.0; ) {...}
```
break statement: Leads to the immediate termination of a loop, regardless of the terminating criterion.

▷ In the case of nested loops, **only that loop plane is exited** in which the break command was given.

□
```
for (i=0; i<10; i++)
    for (j=0; j<10; j++)
    {
        if (j == 5 && i == 5) break;
        printf(''i,j = %d,%d\n'',i,j);
    }
```
The pairs 5.5 bis 5.9 are not displayed because the `i` loop for `i==5` is always exited at `i==5`.

▷ In a `switch` construction the `break` statement is used to avoid the unintentional execution of `case` branches.

continue statement: Within loops it causes the immediate jump to the end of the loop, but the termination criterion is tested.

□
```
for (i=0; i<10; i++)
    for (j=0; j<10; j++)
    {
        if (j == 5 && i == 5) continue;
        printf(''i,j = %d,%d\n'',i,j);
    }
```
Only the pair 5.5 is displayed!

□ Commands for the programming of conditional branches: `if, if...else, switch`

if structure: Conditional execution of program sequences.

```
if (expression)
    statement1;
```

statement1 is executed only if *expression* is logically **true**.

▷ *statement* can be a single C command or a program block.

if-else structure:

```
if (expression)
    statement1;
else
    statement2;
```

statement2 is executed if *expression* is logically **false**.

switch structure: Needed for conditional branches with two or more alternatives:

```
switch (expression)
{
    case alternative1 : statement1; break;
    case alternative2 : statement2; break;
              ⋮
```

```
    case alternativen : statementn; break;
    default : statement;
}
```

⊳ The `break` command causes the immediate exit from the `switch` block after execution of the corresponding command. Without `break`, all other alternatives would be executed additionally!

Introduction to C++

- C++ is an object-oriented programming language and should always be used as such. However, it also contains a complete C compiler so that it can be regarded as a language higher than C.

▷ Originally, C++ had only been designed as a **precompiler** for C, i.e., a program that translates the C++ code to an original C code, which is then translated by a conventional C compiler.

- The difference between C++ and C consists only in the bookkeeping with regard to valid and invalid access to certain data structures. Of course, this could also be done by a careful programmer. But experience has shown that this is better left to the computer.

▷ In principle, C++ has very little in common with C because, due to the object-oriented language elements, a completely different program structure is required (the programmer is forced to be more careful). But the *syntax* of C++ is based on that of C, so that knowledge of C is essential.

24.9.6 Variables and constants

With respect to the declaration of variables and constants, C++ has the same syntax as C. Below, the main differences are listed.

1. The declaration of variables and `typedef` must no longer be at the beginning of a block; they may occur anywhere in the code.

   In C++, the declaration of a variable should *always* be connected with a value assignment (`int k = 1; int j = k;` instead of `int k,j; ...; k = 1; j = k;`).

   In C: `int i; for (i=1; i<10; i++) {...}`
   In C++: `for (int i=1; i<10; i++) {...}`

2. **Constants** are defined via the keyword `const`
   (`const float x = 0.5;`).

24.9.7 Overloading of functions

- In C++, a function is not only specified via its name, but also via the number and types of its arguments. Thus, within the same scope, several functions may occur with the same name, as long as they can be clearly distinguished by their arguments.

▷ On the other hand, the return value of a function does *not* contribute to identification!

Overloaded functions: C++ functions of the same name differ in their arguments.

```
int function();
int function(double);            // ok.
int function(double,char);       // ok.
char function(double,char);      // error because only the return value
                                 // is distinct
```

24.9.8 Overloading of operators

- In C++, operators are defined as functions with a fixed number of arguments:

□ `a + b` ≡ `operator+(a,b)`

▷ Operators behave like functions and can be overloaded if their arguments are not elementary types, but *classes*.

Binary operators: Operators with exactly *two* arguments.

□ +, -, *, /, but also <<, >>

▷ The number of arguments of an operator *cannot* be changed. New operators can also not be defined; it is possible only to overload already existing ones.

Function-call operator `()`: The only operator for which the number of arguments is not given (it is determined during definition). Allows use of a class instance, analogous to a function.

```
class
{
    float operator()(...arguments...)  ... ;
}
main()
{
    class instance;
    float y = instance(1,2,3,...);
}
```

24.9.9 Classes

Class: Data object in C++. Consists of data structures (members, variables, arrays, `construct`'s) and methods (**member functions**) for the manipulation of data.

▷ Classes are the backbone of a C++ program.

● Members and member functions can be declared to be **private** (only local access by member functions), **protected** (access also by member functions of derived classes), or **public** (to be accessed globally in the whole program).

▷ The key terms `private`, `protected`, and `public` can be used at any time and in any order.

```
class example                // the class name is example
{
    private:
    int i,j;                 // private integer variables
    public:
    void set(int,int);       // function sets i and j with
                             // the value to be passed.
    int quotient();          // function divides i by j
}
```

▷ Since the variables `i` and `j` are defined as `private` they *cannot* be manipulated from outside the class. Setting and changing variables can be done *only* by a member function.

The *definition* of the member function is done (anywhere) *after* the class declaration:

```
void example::set(int i_new,j_new)
{
    i = 2*(i_new/2);
    j = 2*((j_new+1)/2)-1;
}
int example::quotient()
```

```
{
    return i/j;
}
```

▷ The member function `set` ensures that `i` and `j` can assume only even values and `j` only odd values. If no other member functions are defined, it is impossible to assign an even value to variable `j`. This guarantees, for example, that the quotient is *always* defined (since `j=0` cannot occur).

Scope operator "::": Defines the scope of validity of a function or variable.

□ In the example, `int example::quotient()` defines the function `int quotient()` as a member function of the `example` class. Within this scope the specifier `quotient` must be unique.

24.9.10 Instantiation of classes

Like the name of a data type (`int, double, ...`) the name of a defined class can be used for the declaration of a variable. The object thus declared is called an **instance**.

```
example x;                 // Yields the instantiation of the class example
                           // with the identifier x.
x.set(1,2);                // Sets the variables i and j of instance x
int k = x.quotient();      // Yields k = 1/2 = 1
```

Structure operator ".": Addresses members and member functions of the instance of a class (`x.i, x.j, x.quotient(),...`).

24.9.11 `friend` functions

`friend`: Key term that defines any function as friend of a class. Friend functions are allowed access to all private members and member functions of the class.

- `friend` functions weaken the strictness of data encapsulation. Their use should therefore be avoided although in many cases there is no alternative.

```
class
{
    friend float f(class& a,float x);
    private:
    float b;
}
float f(class& a,float x)  return a.b*x;
```

- `friend` declarations should stand at the beginning of the class definition.

24.9.12 Operators as member functions

Operators between classes or between a class and an elementary data type can be defined either as **operator member function** or as `friend` operator:

Operator member function:

```
class x
{
    x operator+(x& b)  ... ;
};
```

`friend` **operators:**

```
class y
{
    friend y& operator+(y& a,y& b)  ... ;
};
```

- As the first argument operator member functions always have the instance of the class from which the operator is called. Therefore, only the second argument (or all further arguments) must be given.

24.9.13 Constructors

- If `x.quotient()` is called *prior* to `x.set()`, then the variables `i` and `j` do not have a defined value (a random bit sequence in the memory allocated to variables `i` and `j`). Thus, the quotient is *undefined*.

Constructor: Member function that is called during the instantiation of a class and is intended for the definition of the member contents.

- Constructors *always* carry the name of the class. They have *no* return value (not even `void`). Constructors can be *overloaded*.

```
class example
{
    private:
    int i,j;
    public:
    example();              // standard constructor
    example(int,int);       // overloaded constructor
    void set(int,int);      // the rest as before...
    int quotient();
}
```

Member initialization: Definition of member contents during the call of constructors.

```
<class>::<class>(...) : member1(value1), ... , memberN(valueN)
{...}
```

- Here, *value* may also be any already declared functions or complex expressions.

□
```
example::example() : i(0), j(1)
{...}
```

▷ Member initialization is the *only* possibility to initialize *constant* members.

▷ Member contents can also be assigned to the function body of the constructor. In contrast to member initialization, a value assignment for constant members is not possible here.

- Basically, *all* members should be defined as far as possible via member initialization.

□ (Bad) example:

```
example::example(int a,int b)
{
    i = 2*(a/2);
    j = 2*((b+1)/2)-1;
}
```

Now, the class is instantiated in the following syntax:

```
example x();                                // places i=0 and j=1
example y(1,2);                             // places i=1 and j=2
example z(x.quotient(),y.quotient());       // places i=0 and j=1
```

Copy constructor: Constructor for generating a copy of an object.

□ The copy constructor of a class is called during definitions (`example a = b`) or "call by value" transfer of an object to a function.

```
class example
{
        :
    example(example&);
        :
}
example::example(example& a)
{
    i = a.i;
    j = a.j;
}
main()
{
    example a(1,2);
    example b = a;      a is copied and assigned to b
}
```

● If no copy constructor is defined, then, as a standard, the contents of all members are copied. This may create problems in dynamical storage management!

24.9.14 Derived classes (inheritance)

Derived class: Given a class `a`. Then

```
class b : public a {...} ;
```

defines a class derived from class `a`. All `public` and `protected` members of `a` are accessible from `b`.

Virtual member functions: Always ensure the execution of the right member function in object conversions.

● If a reference or a pointer pointing to a derived instance is assigned to a reference or to a pointer pointing to the basic class, then in a normal case (without virtuality), the member function is called via this reference or via this pointer when the member function is called. If the overloaded member function is declared as `virtual` the object "remembers" its origin and calls the member function of the derived class.

Without virtuality:

```
class a
{
  void f() { cout << ''a''; }
}
class b : public a
{
  void f() { cout << ''b''; }
}
main() {
    b x;
    a& y = x;
    y.f();
} Yields: "'a"'
```

With virtuality:

```
class a
{
  virtual void f() { cout ... }
}
class b : public a
{
  virtual void f() { cout ... }
}
main() {
    b x;
    a& y = x;
    y.f();
} Yields: "'b"'
```

24.9.15 Class libraries

In C++, class definitions and the associated data encapsulation allow the very simple creation of completely independent data and program structures that can be used by a programmer in a program without having to know exactly how the routines work. It is enough to know the interface: the `public` members and member functions of the class hierarchies defined in the **class libraries.**

Class library: Collection of predefined classes for the implementation of certain types of problems. Users have no access to the data and the code.

● The classes defined in the class libraries cannot (and must not) be altered by the user. For the modifications of certain properties, derived versions of the library classes can be created and these can be used instead by employing the principles of overloading and inheritance. The original code is left untouched.

□ Example: Data type "complex numbers" with all conceivable combinations. The programmer simply defines complex numbers by:

```
  #include <complex.h>
complex y(1,3),z(0,1);        // corresponds to 1+3*i and 0+1*i = i
```

which makes it possible to perform all conceivable operations: `y*z, sqrt(z), ....`

▷ Some class libraries are already contained in standard C++, so that only the corresponding **header file** must be implemented. The use of all class libraries requires the linking of additional libraries (files that include the already compiled program code referring to the classes declared in the header file):

□ `xlC -l task.l programm.C`

Introduction to FORTRAN

- FORTRAN was developed late in the 1960s as a language for the fast solution of problems in natural sciences (FORmula TRANslator). Therefore, many of its elements are custom-tailored for the solution of mathematical problems. System-related programming, as is possible in C, for example, was not the intention in FORTRAN.

FORTRAN77: Most common FORTRAN version. Since it is available in most computers, it is easily portable between different types.

▷ FORTRAN77 was standardized in 1978 (ANSI and ISO standards).

FORTRAN90: Most recent development of FORTRAN. Comes close to a "true" high-level programming language such as PASCAL, C, MODULA. But FORTRAN90 cannot compete with such powerful object-oriented programming languages as C++.

▷ Here, only the standard FORTRAN77 language standard will be described.

24.9.16 Program structure

- Every row of a FORTRAN program is allowed to contain only one command.
- FORTRAN rows are allowed to have a length of up to 72 columns. From column 73 down everything is treated as a comment.
- Columns 1 to 5 are reserved for the statement number.
- Any character (except blank) in column 6 specifies a continuation row.
- FORTRAN commands can only start at column 7.
- Every FORTRAN block must end in the `end` command.
- The declaration of variables and the definition of `common` blocks must precede the first executable program row of every block.

Block: Main program or subprogram. The variables defined in a block are always local.
Main program: Must be placed at the top of a program. Should begin with the **program name** `program`.
Statement number: A number with up to 5 digits which can precede every FORTRAN command. Serves as a reference for loops or jumps. Must be within the first 5 columns of a FORTRAN row and must be unique within a block.
Continuation row: Program row which would exceed 72 columns and must be continued in the next row. This is specified by a character other than a blank in column 6 of the continuation row.

24.9.17 Data structures

- In FORTRAN, *all* variables are always **static**. The memory location of all global *and* local variables is already allocated during compilation.

  There is neither **automatic** nor **dynamical** storage management.
- There are only **local** variable names. Variables with a greater scope of validity must be defined via `common` blocks.

common block: Combination of several variables to form a block contiguous in the memory, which (if necessary) can be moved to subprograms where it makes the variables available.

`common` */name/variable1, variable2, ...*

▷ **Caution!** The compiler does **not** recognize different data types in different subprograms. Such errors inevitably lead to wrong variable contents and thus to programming errors that are usually difficult to detect.

▷ There are no recursive function calls in FORTRAN.

● There are *no* composite data types in FORTRAN.

□ Data types: `integer`, `real`, `double precision`, `complex`, `character`, `character*`*length*, `logical`

▷ The storage requirement of the data types `integer`, `real`, and `double precision` depends on the computer. Usually, the data type `real` requires four times the memory allocation that the data type `integer` requires, and correspondingly, `double precision` requires eight times as much.

● Thus, we also write: `real*4` instead of `real`, and `real*8` instead of `double precision`.

Constants:

Data type	Example
`integer`	`206; 0; -25`
`real`	`2.06; -0.075; .3; 1.e-7`
`double precision`	`2.06d0; -7.5d-2; 1.d-7`
`complex`	`(0.,1.); (2,2.e-5)`
`character`	`'a'; 'B'; 'c'`
`character*`*length*	`'abc'; 'What's that'`
`logical`	`.true., .false`

▷ The data type `complex` consists of two *single-precision* value. A double-precision complex data type is not included in standard FORTRAN77.

▷ Numerical calculations should always be performed in `double-precision` if the memory capacity of the program allows it. On the other hand, modern computers require no more calculating time for `double precision` than for `real`.

24.9.18 Type conversion

If "number-valued" data types `integer, real, double precision`, `complex` are assigned to values of another number-valued data type the latter one is automatically converted to the data type of the variable (implicit data conversion).

▷ The conversion `real`→`integer` is accomplished by simply truncating the digits after the decimal point.

Explicit type conversion	
`int()`	conversion to `integer`
`float()`	conversion to `real`
`dble()`	conversion to `double precision`
`cmplx()`	conversion to `complex`

□
```
integer i
real x
x = float(i)
```

Declaration of variables:

data type variable name

□
```
integer i,j
double precision x,y
character*128 satz
```

parameter statement: Assigns a name to a constant.

▷ Constants declared with `parameter` are *local*, as are all parameters in FORTRAN.

`parameter (`*name* `=` *value*`)`

▷ First, a data type must be assigned to the *name*.

□
```
real pi
parameter (pi = 3.1415926)
```

Conventions for variable names: Variable names must begin with a letter, otherwise they can contain only letters, numbers, and — at least in the case of the more recent compilers — the character "_" (underline).

● Capital and lowercase letters are *not* differentiated.

implicit statement: Connects a data type with the initial letter of a variable name. Must be at the beginning of a block (main program or subprogram).

▷ The use of *implicit* allows the use of variables without the need to declare them at the beginning of the program or subprogram.

□ `implicit integer (i-n), real (a-h,o-z)` assigns the data type `integer` to all variables beginning with the letters j, k, l, m or n.

● Deviations from this rule can usually be accomplished by means of explicit type assignment: `real mass`.

● Without `implicit`, the basic setting given in the example always applies. This can be switched off by means of `implicit none`. Now, *all* variables used must be declared.

▷ If possible, the use of `implicit` should be avoided because it is a source of frequent errors:

□
```
implicit integer (i-n),real (a-h,o-z)
p = 0.150
m = 0.938
E = p**2/(2.0*m)
```

This program leads to a division by zero because, `m` is declared as *integer* by means of *implicit*, and thus the numerical value 0.938 is converted to 0.

Arrays:

A one-dimensional array with subscripts $i_1, i_1+1, \ldots, i_2$ is defined by: *datatype name*

`dimension` *name*$(i_1:i_2)$

or shortened to:

datatype name$(i_1:i_2)$

● i_1 and i_2 must be of type `integer`.

▷ If the enumeration of the subscripts starts with 1, then the following definition is sufficient:

data type name(i_2) .

data statement: Initialization of variables and arrays.

`data` *name list1*`/`*values1*`, ...,` `/`*name list*n`/`*values*n

□
```
real x,y(5),z
data y,z/ 1,2,3,4,5,0/, x/5/
```

24.9.19 Operators

▷ In the following table the operators are arranged in order of descending priority. Operators within the individual sections have equal priority.

Operators in FORTRAN		
**	exponentiation	*expression1**expression2*
+	positive sign	+*expression*
−	negative sign	−*expression*
*	multiplication	*expression1*expression2*
/	division	*expression1/expression2*
+	addition	*expression1+expression2*
−	subtraction	*expression1−expression2*
`.eq.`	equal	*expression1* `.eq.` *expression2*
`.ne.`	unequal	*expression1* `.ne.` *expression2*
`.gt.`	greater than	*expression1* `.gt.` *expression2*
`.ge.`	greater than or equal to	*expression1* `.ge.` *expression2*
`.lt.`	less than	*expression1* `.lt.` *expression2*
`.le.`	less than or equal to	*expression1* `.le.` *expression2*
`.and.`	logical AND	*log. expression1* `.and.` *log. expression2*
`.or.`	logical OR	*log. expression1* `.or.` *log. expression2*

24.9.20 Loops and branches

goto statement: Causes an unconditional jump to the given statement number.

`goto` *statement number*

Calculated `goto`:

`goto` (*statement No1, statement No2, ...*) *integer left value*

Jumps to the statement number of the list fixed by the integer expression.
continue statement: "Empty" command without effect on the program. Usually used as a dummy command to define statement numbers.

```
      goto 123
123   continue
```

if statement: Conditional execution of a command.

`if` (*logical expression*) *statement*

statement is executed only if the logical expression has the value `.true.`.
Arithmetic `if` **statement**:

`if` (*numerical expression*) *statement No1, statement No2, statement No3*

Jumps to the *statement No1* if *numerical expression* < 0, to *statement No2* if $= 0$ and to *statement No3* if > 0.

▷ Due to rounding errors the arithmetic `if` may cause problems in case of `real` or `double precision` expressions, because in this case the numerical expression is never exactly = 0.

Block-if structure: Comprises one-sided decisions, alternative, and case distinction.

One-sided decision:

if (*logical expression*) then
statements
endif

▷ This corresponds to a simple `if` statement.

Alternative:

if (*logical expression*) then
statements1
else
statements2
endif

Case distinction:

if (*logical expression 1*) then
statements1
else if (*logical expression 2*) then
statements2
else
statements3
endif

▷ Any number of case distinctions can be made.

Counter-controlled loops:

do *statement number subscript=start*, it end, *step size*

▷ Omitting the step size corresponds to a step size of 1.

subscript: numerical, noncomplex variable, runs through the range of values defined by *start*, *end*, and *step size*.
start: numerical, noncomplex expression, defines the initial value of the counter *subscript*.
end: numerical, noncomplex expression, terminating condition, terminates the loop if *subscript*≥*end*.
step size: numerical, noncomplex expression, defines the increment of *subscript* in every loop.

Event-controlled loops: Can only be achieved in FORTRAN via `if-goto` constructs. There are no separate commands.

24.9.21 Subprograms

□ FORTRAN distinguishes the subprogram types `subroutine`, `function`, and `blockdata`.

▷ In contrast to C, FORTRAN distinguishes between functions that return a value and so-called procedures (`subroutines`) that do not return a value.

● All subprograms *always* end with the end statement.

▷ The return from a procedure or function to the calling program is accomplished by the `return` command. `return` may appear anywhere in the subprogram.

▷ `blockdata` subprograms must *not* contain a `return` command.

Procedures:

Definition: `subroutine` *name* `(`*variable list*`)`
Call: `call` *name* `(`*variable list*`)`

name: Name of the procedure corresponding to the FORTRAN convention.

variable list: List of variables of defined type that must be transferred or that have been transferred.

- The type of the transferred variables must be specified within the procedure. According to the ANSI standard the compiler does not recognize inconsistencies in type and number of the parameters and these can lead to disastrous errors.

▷ Most compilers allow for translation with the option to register such inconsistencies.

□
```
subroutine Demo(var1,var2)
integer var1
double precision var2
write(6,*) "'anything"'
return
end
```

functions:

Definition:
type specification `function` *name* `(`*variable list*`)`
or
`function` *name* `(`*variable list*`)`
type specification *name*

- Within the function, the *name* is treated like a variable to which a value can be assigned. This value serves as the return value of the function.

▷ Names of functions and procedures can be transferred to another function as a parameter. Within the calling program, this name must be declared as `external` for self-defined functions or as `intrinsic` for intrinsic functions.

□
```
program demo
intrinsic sin,cos
write (6,*) deriv(sin,0.0),deriv(cos,0.0)
end
function deriv(func,x)
real x,deriv,h
h = 1.0e-6
deriv = (func(x+h)-func(x))/h
return
end
```

`entry` **statement**: Defines other entry addresses within subprograms. The parameters of the subprogram must be identical to those of the `entry` command. The call of `entry` proceeds analogous to the call of the subprogram itself.

□
```
subroutine demo(var)
real var
  :
entry jump(var)
  :
end
```

```
call demo(1.0)
call jump(2.5)
```

`save` **statement**: Causes the storing of all local variable contents in a subprogram. In the next call of the subprogram, all variables have the same value as in the exit from the routine after the preceding call.

▷ Due to *static* storage management, the variable contents are retained even without `save`. However, while the program is running, parts of the program must be moved out for reasons of memory capacity. `save` ensures the protection of the contents of the variables.

● The use of `save` is **bad programming style** and **inappropriate** for a high-level programming language. **All** local variables should always be redefined in every call of the subprogram (that is why they are called local).

`blockdata` **subprograms**: Initialization of `common` blocks. Is called automatically once and only once *at the beginning* of the program.

● An explicit call of the `blockdata` subprogram is not possible.

▷ In `blockdata` subprograms, *no* executable statements are allowed. Only `data` statements and declarations of variables and `common` blocks are allowed.

□
```
real x(10),y(5),z
common /daten/x,y,z
integer i
write(6,*) (x(i),i=1,10),(y(i),i=1,5),z
end
blockdata init real x(10),y(5),z
common /daten/x,y,z
data x/1,2,3,4,5,6,7,8,9,10/
data y/1,2,3,4,5/, z/42/
end
```

Computer algebra

Computer algebra systems: Allow symbolic solutions of mathematical problems by means of transformations according to mathematical rules, as well as numerical calculations.

Formula manipulation: The manipulation of mathematical expressions for the purpose of simplification, for the solution of equations, the differentiation of functions, the calculation of the indefinite integral, the solution of differential equations, forming infinite series, etc. Solutions should be given in closed form.

Possibilities and simplification through algebra programs: In particular for tasks that are analytically (exactly) solvable, but which require much time "on paper" while being prone to errors.

Advantage compared with numerical algorithms (e.g., written in one of the programming languages such as C, C++, Fortran, etc.): Rounding errors do not occur. The solution can be a symbolic formula.

Limits: Imposed by time-consuming algorithms and the finite memory capacity of computers.

▷ Like reference books and tables, computer algebra systems are not 100% without errors!

Mathematica: Computer algebra system developed by Wolfram Research, Inc.

Maple: Computer algebra system developed at Waterloo University in Waterloo, Ontario, Canada, distributed by Waterloo Maple Software.

▷ **Mathcad**: Program used widely by engineers; it contains a truncated version of Maple.

▷ Maple and Mathematica are written in C. Mathematica particularly supports graphics applications, but it places greater demands on hardware. In spite of its extensive capabilities, Maple can be run on small computers starting at 2 Mbyte RAM and without a co-processor.

24.9.22 Structural elements of Mathematica

● Depending on the user interface Mathematica represents input/output as follows:

`In[1]:=`*input*

`Out[1]:=`*output*

Expressions, $name_0$`[`$name_1$,$name_2$,..., $name_n$`]`: The main structural elements of the system consist of the head $name_0$ (an operator or function) and the elements $name_i$ $(i = 1, \dots n)$ (operands or variables). The head as well as the elements of an expression can be expressions themselves.

□ `times[a,Power[b,-1]]` (short form `a/b`) is an expression with the operator `times` as head and one variable as well as further expressions as elements.

▷ Square brackets are used only for the representation of expressions.

`Part[`*expression,i*`]`

extracts the element with subscript i from the given expression.

Symbol: The name of an object, any sequence of letters, numbers, and the special character `$`. It must not begin with a number, and it differentiates between capital and lowercase letters.

- Symbols predefined by the system begin with a capital letter. User-defined symbols must therefore be written in lowercase.

Types of numbers:
`Integer`: Exact integer of arbitrary length, input in the form *nnnnn*.
`Rational`: Rational number (relatively prime fractions of the form `Integer/Integer`), input in the form of *ppppp/qqqqq*.
`Real`: Real floating-point number of any precision to be specified, input in the form *nnnnn.mmmmm*.
`Complex`: Complex number of the form *number+number* `I`, where the real and imaginary components can belong to any number type.
Test operation: Provide information on numbers or expressions:

```
Head[x]
```

determines the type of `x` and displays it.

```
NumberQ[x]
```

yields the Boolean value true if `x` is a number and false if it is not. Similar tests (self-explanatory): `IntegerQ[x]`, `EvenQ[x]`, `OddQ[x]`, `PrimeQ[x]`.
Number conversion:

```
N[x,n]
```

converts any real number x into a floating-point number with n significant digits.

```
Rationalize[x,dx]
```

approximates the number x by a rational number x with precision dx.

▷ Predefined special symbol (arbitrary precision): `PI` (for the number π), `E` (for the Euler number e), `INFINITY` (for ∞), and `I` (for the imaginary unit).

Some other important operators with their short forms:

`x+y`	`Plus[a,b]`	`x==y`	`Equal[x,y]`
`x y` or `x*y`	`Times[x,y]`	`x!=y`	`Unequal[x,y]`
`x^y`	`Power[x,y]`	`x>y`	`Greater[x,y]`
`x/y`	`Times[x,Power[y,-1]]`	`x>=y`	`GreaterEqual[x,y]`
`x->y`	`Rule[x,y]`	`x<y`	`Less[x,y]`
`x=y`	`Set[x,y]`	`x<=y`	`LessEqual[x,y]`

▷ `Set` is a statement for which the expression on the left side is represented by that on the right side (for example, a number is assigned to a variable).
On the other hand, `Rule` is a transformation rule that frequently occurs with the replacement operator as `x/.y->z` or `Replace[x,y->z]`. In this case, in the expression `x` all elements `y` are represented by the expression `z`.

- For these assignment or transformation operators the right side is evaluated immediately.

Delayed operators, `x:=y` or `SetDelayed[x,y]`, and `x:>y` or `RuleDelayed[x,y]`: Transformations or statements for which the right side is evaluated only when the left side is evaluated.

Lists, vectors, matrices

List: Combination of several objects into one new object. The objects are differentiated only by their position in the list.

□ `In[1]:=`*name*`=List[`$x_1, x_2, x_3, x_4, x_5, x_6$`]`
`Out[1]={`$x_1, x_2, x_3, x_4, x_5, x_6$`}`

▷ **Nested lists** of any nesting depth can be created by setting a list again as an element of a list.

Commands which access the elements of a list:

`First[`*name*`]`	selects the first element
`Last[`*name*`]`	selects the last element
`Part[`*name*`,n]` or *name*`[[n]]`	selects the n-th element
`Part[`*name*`,{n1,n2,...}]`	creates a list from the given elements
`Take[`*name*`,m]`	yields the list of the first m elements of *name*
`Take[`*name*`,{m,n}]`	yields the list of elements from m to n
`Drop[`*name*`,n]`	yields the list without the first n elements
`Drop[`*name*`,{m,n}]`	yields the list without the elements from m to n

With which the lists can be queried or changed:

`Position[`*name*`,x]`	gives the list of positions at which x occurs
`MemberQ[`*name*`,x]`	tests whether x is an element of the list
`FreeQ[`*name*`,x]`	tests whether x occurs nowhere in the list
`Append[`*name*`,x`	adds x at the end of the list
`Prepend[`*name*`,x]`	adds x at the beginning of the list
`Insert[`*name*`,x,i]`	inserts x at position i of the list
`Delete[`*name*`,{ i,j,...} ]`	deletes the elements with numbers $i, j, \ldots$
`ReplacePart[`*name*`,x,i]`	replaces the element at position i by x

Table`[f,{`*imax*`}]`: Operation that creates a list with $imax$ values of $f = f(i)$.

□ `In[1]:=Table[Binomial[7,`*i*`],{`*i*`,0,7}]]`
`Out[1]={1,7,21,35,35,21,7,1}` generates a table of binomial coefficients for $n = 7$.

Array`[v,`*n*`]`: Special operation that creates the **vector** `{v[1],v[2],...,v[`*n*`]}` as a one-stage list.
`Array[[a,{`*n*`,`*m*`}]` correspondingly creates a **matrix** a as a two-stage list with n rows and m columns.
Algebraic operations for the manipulation of matrices and vectors:

`c a`	matrix a is multiplied by a scalar c
`a.b`	the product of matrices a and b
`Det[a]`	the determinant of matrix a
`Inverse[a]`	the inverse matrix of a
`Transpose[a]`	the matrix to be transposed to a
`MatrixPower[a,n]`	the n-th power of matrix a

▷ Mathematica does not distinguish between column vectors and row vectors.
Mathematical standard functions:
`Exp[x], Log[x], Log[b,x], Sin[x], Cos[x], Tan[x], Cot[x], Sec[x], Csc[x], ArcSin[x], ArcCos[x], ArcTan[x], ArcCot[x], ArcSec[x], ArcCsc[x], Sinh[x], Cosh[x], Tanh[x], Coth[x], Sech[x], Csch[x], ArcSinh[x], ArcCosh[x], ArcTanh[x], ArcCoth[x], ArcSech[x], ArcCsc[x]`
Special functions:
Bessel functions $J_n(z)$ (`BesselJ[n,z]`) and $Y_n(z)$ (`BesselY[n,z]`), modified Bessel functions $I_n(z)$ (`BesselI[n,z]`) and $K_n(z)$ (`BesselK[n,z]`),
Legendre polynomials $P_n(x)$ `LegendreP[n,x]`),

Spherical harmonics $Y_l^m(\theta, \phi)$ (`SphericalHarmonicY[l,m,theta,phi]`), and other functions.

sample: The definition of a user-defined function:

`In[1]:=f[x_]:=`*function(x)*

creates a special function f "anything with the name x."

▷ The sample fixes a *structure*, it may thus stand for a whole class of expressions.

□ `In:=A[x_,n_]:=(1+x)∧n` defines a function of two variables. It can be used in the following:

```
  In[2]:=A[2,3]
Out[2]:=27
```

Functional operators: In keeping with the symbolic character of Mathematica the manipulation of a function that is treated like an expression.

□ **Inverse function**: `InverseFunction[f][x]` or `InverseFunction[f]`
Differentiation: As a map in the space of functions: `Derivative[1][f]` or `f'`
Nesting of a function (n-times applied to x): `Nest[f,x,n]` creates `f[f[...f[x]]...]`.
FixedPoint[f,x]: Searches for the solution of the equation $x = f(x)$.
Apply[f,{*Liste*}]: Creates the function applied to the list.
Map[f,{*Liste*}]: Creates the list of expressions formed by applying the function to the elements of the initial list: {`f[`*Element*$_1$`]`, `f[`*Element*$_2$`]`...}.

24.9.23 Structural elements of Maple

● According to the user interface of input/output is represented by

>input; → *output*

▷ An input ending in a colon rather than a semicolon is executed, but not represented.

Basic structure of an input: `obj`$_0$`(obj`$_1$`,obj`$_2$`,...,obj`$_n$`);`, where `obj`$_0$, is usually an operator, a statement or a function acting on the expressions in brackets.

Symbols: May consist of letters, numbers, and the underlining (_). Capital and lowercase letters are distinguished; a symbol is not allowed to begin with a number.

● The underlining should be avoided in user-defined symbols, since it is used for internal symbols.

▷ Whether a value has already been assigned a symbol can be determined by using `?Name`.

Type classification: Determines the membership of objects in object classes, i.e., the type of data structure or class of operators. Basic types:

'+'	'*'	'∧'	'='	'<>'	'<'	'≤'	'.'
'and'	'or'	'not'	exprseq	float	fraction	function	indexed
integer	list	procedure	series	set	string	table	uneval

▷ One object can be assigned to several types.

● The type of an object must be queried by `> whattype(`*obj*`);`.

□ `>whattype(25);` → `integer`

Types of numbers: A whole number, entered by means of

integer: *nnnnn* (arbitrary length).

fraction: Fractional numbers, represented by a fraction of two integers: ppp/qqq.

float: Floating point numbers represented as $nnnn.mmmm$ or, in scientific notation, as $n.mmm * 10 \wedge (pp)$.

▷ A fraction that can be reduced to an integer is not recognized as a symbolically represented `fraction`.

Number conversion, `eval(`*number*`);`: Converts rational numbers or numbers and results into floats with predetermined precision (pre-setting: 20 digits).

`convert(`*expression, form, opt*`)`: Usually converts expressions from one form (a certain type) into another, as if this is practical.

□ `>convert('FFA2' ,decimal,hex);` → 65442

Operators: Important arithmetical operators

`+,-,*,/,^`

and relational operators

`=, <, <=, >, >=, <>.`

Algebraic expressions: Constructions of the `algebraic` type consisting of (linked) symbols.

`subs(x=`*Wert*`,q);`: In the expression q, the variable x is replaced by a value.

□ `>q:=x^4-2*x^2-1:`
`>subs(x=2,q);` →7

`op(i,q);`: Extracts the i-th term from the expression q.

□ `>q:=x^4-2*x^2-1:`
`>op(2,q);` →-2*x²

Sequences: Sequence of expressions (separated by comma) in which the order is of significance:

`>seq(`$f(i)$`,i=1..n)`

creates the sequence $f(1), f(2), \dots, f(n)$.

▷ Simplified notation: `>`$f(i)$`$i=1..n;`

List: Results when a sequence f is placed in square brackets.

□ `>l=[i$i=1..6)]` → `l:=[1,2,3,4,5,6]`
`op(n,`*list*`);`: Accesses the n-th element of a list.

Vectors and matrices: Created with the help of the `table` and `array` structures.

`table(`*subscript function, list*`)`: Creates a table-like structure that contains equations as elements of the list.

`>op(T);` returns the table.
`>indices(T);` returns the subscripts.
`>entries(T);` provides a sequence of terms.

□
```
>T:=table([a,b,c)];  →  T:=table([
                                 1=a
                                 2=b
                                 3=c
                               ])

  >indices(T);      → [1],[2],[3]
  >entries(T);      → [a],[b],[c]
```

`array(`*subscript function, list*`);`: Creates special, possibly multidimensional arrays with a range of integer values for each dimension.

□ The one-dimensional array
`v:=array(1..n,[a(1),a(2),...,a(n)]);`
is interpreted as a vector.
The two-dimensional array
`A:=array(1..m,1..n,[a(1,1),...,a(1,n)],...,[a(m,1), ..., a(m,n)]]);`
is interpreted as a matrix.

Functions: Predefined, contained in large numbers, belong to the type `mathfunc`. Some elementary functions are:
`exp, log, ln, sin, cos, tan, cot, sec, csc, sinh, cosh, tanh, coth, sech, csch, arcsin, arccos, arctan, arccot, arcsinh, arccosh, arctanh, arccoth`
Special nonelementary functions are, including the Bessel function `BesselI(v,z)`, `BesselK(v,z)` and `BesselJ(v,z)`, the Gamma function `gamma(x)`, the integral exponential function `Ei(x)` and the Fresnel functions. Other functions, such as orthogonal polynomials, are provided in special packages.

▷ `?inifcns` provides a list of all available pre-defined functions.

Operators: Names of functions (without argument).

□ `>type(cos,operator);` and `>type(cos(x),function);` yield the value `true`; while reversely `>type(cos(x),operator);` and `>type(cos,function);` yield the value `false`.

User-defined functions: Can be created (as operators) by

`>F:=x->`*expression*`:`

By adding a variable as argument, they become a function. Inserting a numerical value

`>F(nn.mmmm);`

yields the function value.

Operator product: Applies the first operator to the second; it is created with the multiplication symbol for operators `@` (composition operator).

□ Let be `F:=x->cos(2*x)` and `G:=x->x∧2` then

`>(G@F(x);` $\rightarrow$ `cos`^{2}`(2x)` and

`>(F@G)(x);` $\rightarrow$ `cos(2x`^{2}`)`.

Differential operator `D`: Acts on functions in operator form.

`D[i](F)`

provides the derivative with respect to the i-th variable. Higher derivatives can be written as `(D@@n)(F)` (the n-th power of the operator).

▷ The derivative as a function can also be written as `D(F)(x)=diff(F(x),x)`.

Mapping: Functional operator for the application of an operator or a function to an expression or its components.

□ `>map(f,[a,b,c,d]);` $\rightarrow$ `[f(a),f(b),f(c),f(d)]`

Library functions: May be made available by means of the command `readlib(`*library package*`)`.

Information and help: Provided by the input

`>?`*command*`;` or `help(`*command*`);`

24.9.24 Algebraic expressions

Multiplication expressions: Expanding the powers and products in a polynomial:

Mathematica:

`Expand[`*polynomial*`]`

Maple:

`>expand(`p`,` q_1`,` q_2`, ...);`

expands the powers and products in the algebraic expression p without the (optional) expansion of the subexpressions q_i.

Factor decomposition:
Mathematica:

`Factor[`p`]`

decomposes the polynomial p over integral or rational numbers.
Maple:

`>`p`1:=factor(`p`);`

decomposes a polynomial into irreducible factors with respect to the body of rational numbers.
Operations on polynomials:
Mathematica:

`PolynomialGCD[`p_1`,`p_2`]`

determines the greatest common divisor of both polynomials.

`PolynomialLCM[`p_1`,`p_2`]`

determines the least common multiple.
Maple:

`>gcd(`p`,`q`);` and `>lcm(`p`,`q`);`

determine the greatest common divisor and the least common multiple of the polynomials.
Partial fraction decomposition:
Mathematica:

`Apart[`q/p`]`

decomposes the quotient of two polynomials over the body of rational numbers.
Maple:

`>convert(`p/q`,parfrac,`x`);`

as above.
Other manipulations: Simplification of complicated expressions, also of those nonpolynomial in nature:
Mathematica:

`Simplify[`*expression*`]`

Maple:

`>simplify(`*expression*`);`

24.9.25 Equations and systems of equations

Equations:
Mathematica: An equation is regarded as a logical expression.
□ `In[1]:=`$g = \mathrm{x}^2 + 2\mathrm{x} - 9 == 0$
Solution of equations: Equations can usually be solved symbolically up to equations of degree 4. Computer algebra programs try to offer solutions by means of simplification, even in case of very complicated expressions.
Mathematica:

- `Solve[`*equation*`,x]`

  provides a list of solutions for variable x.

- `NSolve[`*equation*`,x]`

  provides the numerical solution.

- `FindRoot[`$g\{x,\ x_s\}$`]`

  finds solution of transcendental equations (that can have an infinite number of solutions). x_s serves as the starting value in the search.

Maple:

- `>solve(`*expression*`);`

  solves an equation *expression*$(x) = 0$ of degree ≤ 4 with one unknown. If the coefficients are floats, then Maple solves the equation numerically.

- `>fsolve(`*expression*`);`

  offers a numerical solution for equations of any kind (e.g., transcendental equations) for which (as a rule) a real root is required. As an option a range for searching for the root may be given:

□ `>fsolve(x+arccoth(x)-4=0,x,1..1.5);` → 1.005020099

Systems of equations:
Mathematica:

- `Solve[`$\{l_1 == r_1, l_2 == r_2, \ldots\}$`,`$\{x, y, \ldots\}$`]`: Solves the given system of equations for the unknowns.

- `Eliminate[`$\{l_1 == r_1, \ldots\}$`,`$x, \ldots\}$`]`: Eliminates the elements $x, \ldots$ from the system of equations.

- `Reduce[`$\{l_1 == r_1, \ldots\}$`,`$x, \ldots\}$`]`: Simplifies the system of equations and provides all possible solutions.

Maple:

`>solve(`*equations,unknowns*`):`

Solves systems of equations where the first argument is to give all equations in braces and the second is to give the unknowns.

□ The system of equations

$$x_1 + x_2 = 4$$
$$-x_1^2 + x_2^2 = 8$$

with $x_1 = 1, x_2 = 3$ is solved by
`>solve(x1+x2=4,-x1∧2+x2∧2=8,x1,x2);`

▷ Numerical solutions (specifically for systems of equations that cannot be solved exactly) are obtained with the command `fsolve`.

24.9.26 Linear algebra

Systems of linear equations:
Mathematica: The general system of linear homogeneous or inhomogeneous equations reads:

`p.x==0` and `p.x==b`,

respectively, with the matrix `p=Array[p,{m,n}]`
and the vectors `x=Array[x,{n}]` and `b=Array[b,{m}]`.

```
x=Inverse[p].b
```

solves the inhomogeneous system in the special case $n = m$, $\det p \neq 0$.

▷ Systems with up to about 50 unknowns can be solved in a reasonably short time.

General case: Here a set of basis vectors of the null space of the matrix p can be generated with `null space[p]`. `LinearSolve[p,b]` solves the system of equations if this is possible.

Maple:

```
>with(linalg);
```

incorporates a package with special applications.

▷ Instead of using `array`, matrices and vectors are generated by using *linalg* by means of `matrix[m,n,`*elements*`]` or `vector[n,`*components*`]`.

```
>linsolve(A,c);
```

solves the system of linear equations $A \cdot \vec{x} = \vec{c}$ or returns the sequence `Null`.

```
>nullspace(A);
```

finds a basis in the null space of the matrix A.

Eigenvalues and eigenvectors:

Mathematica:

```
Eigenvalues[a]
```

determines the eigenvalues of matrix a.

```
Eigenvectors[a]
```

determines the eigenvectors of matrix a.

▷ By using `N[a]` instead of a, the numerical eigenvalues are obtained.

Maple:

Correspondingly, `>eigenvals(A);` and `>eigenvects(A);`.

24.9.27 Differential and integral calculus

Differentiation:

Mathematica:

```
D[f[x],{x,n}]
```

provides the n-th derivative of the function $f(x)$.

```
D[f,{x1,n1},{x2,n2},...]
```

provides the mixed multiple partial derivative. `Dt[f]` creates the total differential of the function f. `Dt[f,x]` provides the total derivative $\frac{\mathrm{d}f}{\mathrm{d}x}$.

Maple:

```
>D[i](f);
```

the operator of differentiation creates the derivative of the (operator) function with respect to the i-th variable.

▷ `D[i,j](f)` is equivalent to `D[i](D[j](f))` and `D[](f)=f`.

```
>diff(expression,x1,x2,...,xn);
```

performs the partial derivative of the algebraic expression with respect to the variables $x1, \ldots, xn$.

▷ Multiple differentiation can be represented with the sequential operator $:

□ `>diff(sin(x),x$5);` (≡`diff(sin(x),x,x,x,x,x)`)→`cos(x)`

Indefinite integrals:
Mathematica:

```
Integrate[f,x]
```

provides the indefinite integral $\int f(x)\mathrm{d}x$ without integration constants.
Maple:

```
>int(f,x);
```

▷ Only *very* few integrals are found; unfortunately, not even all integrals given in the table.

Definite integrals, multiple integrals:
Mathematica:

```
Integrate[f,{x,xa,xe}]
```

computes the definite integral of function $f(x)$ with the lower limit x_a and the upper limit x_e.

▷ In versions older than 2.2, Mathematica calculates:
`In[1]:=Integrate[1/x^4,{x,-1,1}]`
`Out[1]=` $-\frac{2}{3}$.
This is an error because the integral is improper!

`Integrate[f[x,y],{x,xa,xe},{y,ya,ye}]`, determines the double integral first over y, then over x.

▷ The limits y_a and y_e can be functions of x.

Numerical integration: Performed with the statement `NIntegrate`. In contrast to the symbolic method, a data list is used here.

`MinRecursion` and `MaxRecursion`: The minimum or maximum number, of recursion steps expected in problematic regions such as poles (pre-setting to 0 and 6).

□ `In[1]:=NIntegrate[Exp[-x^2],{x,-1000,1000}]`
creates a warning, because of the very large integration range and the "peak" at $x = 0$ and leads to a wrong result.
```
In[1]:=NIntegrate[Exp[-x^2],{x,-1000,1000},
MinRecursion->3, MaxRecursion->10]
```
solves the problem.

Maple:

```
>int(f,x=a..b);
```

if possible, provides a symbolic solution of the integral.

▷ For multiple integrals it can often be nested.

Numerical intepretation: Achieved by predetermining a command `evalf` with a precision of n digits:

```
>evalf(int(f(x),x=a..b),n);
```

`>readlib('evalf/int')`: Library call for using an adaptive Newton method.

□ `>' evalfint'(exp(-x^2),x=-1000..1000,10,_NCrule);`
`1.772453851`
The third argument gives the precision, the last gives the internal name of approximation method.

Differential equations:
Mathematica: Symbolic treatment if a solution in closed form is possible. Statements:

`DSolve[`*diff.eq.*`,y[x],x]`	solves the differential equation for $y[x]$
`DSolve[`*diff.eq.*`,y,x]`	provides the solution of the differential equation in the form of a pure function
`DSolve[`*diff.eq.*₁, *diff.eq.*₂,... `y,x]`	solves a system of ordinary differential equations

`NDSolve[`*Dgl*`,y,{x,xa,xb}]`

provides a numerical solution of the differential equation in the range between x_a and x_b.

□ Solution of the equation for Foucault's pendulum:

```
In[1]:=dg=NDSolve[{x''[t]==-x[t]=-x[t]+0.05y'[t],
y''[t]=-y[t]-
0.05x'[t],x[0]==0,y[0]==10,x'[0]==y'[0]==0},
{x,y},{t,0,40}]
Out[1]={{x->InterpolatingFunction[{0.,40.},<>],
y->InterpolatingFunction[{0.,40.},<>]}}
```

With

```
In[2]:=ParametricPlot{x[t],y[t]}/.dg,t,0,40,
AspectRatio->1]
```

the solution is plotted.

Maple:

`>dsolve(`*diff.eq.*`, y(x));`

solves differential equations and systems symbolically. The last argument can be an option:

`explicit`	provides the solution in explicit form, if possible
`laplace`	employs the Laplace transformation for the solution
`series`	employs the decomposition into power series for the solution
`numeric`	provides a procedure for calculating numerical solution values

□ Solution with initial conditions:

```
>dsolve({diff(y(x),x)-exp(x)-y(x)∧2,y(0)=0},y(x), series);
```

$$y(x) = x + \frac{1}{2}x^2 + \frac{1}{2}x^3 + \frac{7}{24}x^4 + \frac{31}{120}x^5 + O(x^6)$$

▷ Also for the option `numeric`, the statement contains the initial conditions. The Runge-Kutta method is used to calculate the solution.

24.9.28 Programming

User-generated program blocks: Procedures created by the user, employing loop and control structures, can be added to the existing libraries of Mathematica and Maple.

Mathematica:

`Do[`*expression,* {*i,i1,i2,di*}`]`

causes the evaluation of the expression, where i passes through the range of values from $i1$ (default:1) to $i2$ in steps of di (default:1).

`While[`*test, expression*`]`: Evaluates the expression as long as *test* has the value `True`.

`Module[{t1,t2,...},`*proced*`]`: Causes the variables or constants included in the list to be local with respect to their use in the module, but the values assigned to them are unknown outside the module.

□
```
In[1]:= sumq[n_]:=
        Module[{sum=1.},
               Do[sum=sum+N[Sqrt[i]],{i,2,n}];
               sum ];
```
Then, the call `sumq[30]` yields, for example, 112.083.

▷ The last example can be written as a function:
`sumq[n_]:=N[Apply[Plus,Table[Sqrt[i],{i,1,n}]],10]`
for a precision of 10 digits. Then, `sumq[30]` yields 112.0828452 .

Maple:
Loops are created with `for` or `while` followed by a statement of the kind `do` *statements* `do`:

`for` *i* `from` *n* `to` *m* `by` *di*, or
`for` *ind* `while` *bed*

Case distinctions, performed via

`if` *cond* `then` *anw*$_1$ `elif` *cond*$_i$ `then` *anw*$_i$ `fi` ... `else` *anw*$_n$ `fi`

Procedure statement for designing closed programs:

```
>proc(args)
>    local ...
>    options ...
>    anw
>end;
```

□ A procedure that defines the sum of the square roots of the first n natural numbers:
```
>sumqw:=proc(n)
>    local s,i;
>    s[0]:=0;
>    for i to n
>    do s[i]:=s[i-1]+sqrt(i) od;
>    evalf(s[n]);
>end;
```
The procedure can now be called with an argument n:
```
>sumqw(30);
                         112.0828452
```

24.9.29 Fitting curves and interpolation with Mathematica

Fitting of curves: Approximation of selected functions to a data set by means of the least-squares method.

`Fit[{`$y_1, y_2, \ldots$`},`*funct*`,x]`

with data list y_i, the range of values x, and *funct* as the list of selected functions, provides an approximation.

□ `Table[x^i,{i,0,n}]` as the list of approximation functions performs a fit to a polynomial of degree n.

Interpolation: Determination of an approximating function for a sequence of data points. `InterpolatingFunction` objects are created and structured similar to pure functions.

`Interpolation[{`$y_1, y_2, \ldots$`}]`

generates an approximating function with the values y_i for $x_i = i$.

`Interpolation[{{x1,y1},{x2,y2},...}]`

generates an approximation function for a sequence of points (x_i, y_i).

24.9.30 Graphics

Graphic objects in Mathematica: Representation of graphics objects on the basis of **graphic primitives** such as points (`Point`), lines (`Line`), and polygons (`Polygon`) of certain thickness and color:

`Point[{x,y}]`	point at position x, y
`Line[{x1,y1},{x2,y2},...}]`	line through the given points
`Rectangle[{xlu,ylu},{xro,yro]`	solid rectangle between (x_{lb}, y_{lb}) (bottom left) and (x_{ru}, y_{ru}) (top right)
`Polygon[{{x1,y1},{x2,y2,...}]`	solid polygon between the given corners
`Circle[{x,y},r,{α1,α2}]`	circle (circular arc) with radius r, centered at (x, y) (and boundary angles α_1, α_2)
`Circle[{x,y},{a,b}]`	ellipse with the semi-axes a and b
`Text[`*Text*`,{x,y}]`	text centered at the point (x, y)

▷ With `Graphics[`*list*`]`, a graph of the listed objects is generated.

Representation of functions:

Mathematica:

`Plotf[x],{x,xmin,xmax}]`

plots the function $f(x)$ in the range from $x = x_{min}$ to $x = x_{max}$.

`Plot[{f1[x],f2[x],...},{x,xmin,xmax}]`

generates several functions simultaneously.

□ `In[1]:=Plot[{BesselJ[0,z],BesselJ[2,z],BesselJ[4,z]},{z,0,10}]`
creates a graph of the Bessel functions $J_n(z)$ for $n = 0, 2, 4$.

Maple:

`>plot(`*funct, hb, vb, options*`);`

creates a two-dimensional graph. *funct* can have the following meaning:

- A real function $f(x)$ (also several functions, enclosed in braces).
- The operator of a function.
- The parametric representation of a real function in the form of `[u(t),v(t),t=a..b]` with (a, b) as the range of values.
- A list of numbers to be interpreted continuously as the (x, y)-coordinates of points.
- *hb* is the range of values in the form of $x = a \ldots b$, *vb* is the range of values of the dependent variable (it is essential that this is given for poles).

▷ One of the limits on the x-axis can take the value $\pm\infty$. In this case, the x-axis is represented as arctan.

□ `>plot({BesselJ(0,z),BesselJ(2,z),BesselJ(4,z)},z=0..10);`
creates a graph of the Bessel function $J_n(z)$ for $n = 0, 2, 4$.

Parametric representation:

□ *Mathematica*: `ParametricPlot[{t Cos[t],t Sin[t]},{t,0,3Pi}]`
Maple: `>plot([t*cos(t),t*sin(t),t=0..3*Pi]);`
create plots of the spiral of Archimedes.

3D graphics:

Mathematica:

`Plot3D[f[x,y],{x,`x_a`,`x_e`},{y,`y_a`,`y_e`}]`

represents a curved surface in three-dimensional space; requires a function of two variables $f(x, y)$ and the ranges of values.

`ParametricPlot3D[{`f_x`[t,u],`f_y`[t,u],`f_z`[t,u]},{t,`t_a`,`t_e`},`
`{u,`u_a`,`u_e`}]`

plots a surface given parametrically.

`ParametricPlot3D[{`f_x`[t],`f_y`[t],`f_z`[t]},{t,`t_a`,`t_e`}]`

plots a parametrically given curve in space.

Maple:

`>plot3d(`*funct*`,x=a..b,y=c..d);`

with a function of two independent variables *funct*; plots a surface in space.

```
>plot3d([u(s,t),v(s,t),w(s,t)],s=a..b,t=c..d);
```

plots a surface given parametrically.

```
>spacecurve([u(t),v(t),w(t)],t=a..b);
```

is available in the library package; plots a curve in space.

25

Tables of integrals

- ▷ For the integrals listed in the tables the constant C of integration is omitted.
- ▷ If a logarithm occurs in the primitive on the right, then the **logarithm of the absolute value of the argument** is always intended, even if it is not mentioned explicitly.

25.1 Integrals of rational functions

$$\int c\mathrm{d}x = cx \tag{25.1}$$

$$\int x\mathrm{d}x = \frac{x^2}{2} \tag{25.2}$$

$$\int x^n\mathrm{d}x = \frac{x^{n+1}}{n+1} \quad (n \neq -1) \tag{25.3}$$

25.1.1 Integrals with $P = ax + b, \quad a \neq 0$

$$\int P\mathrm{d}x = bx + \frac{ax^2}{2} \tag{25.4}$$

$$\int P^2\mathrm{d}x = \frac{P^3}{3a} \tag{25.5}$$

$$\int P^3\mathrm{d}x = \frac{P^4}{4a} \tag{25.6}$$

$$\int P^4 \mathrm{d}x = \frac{P^5}{5a} \tag{25.7}$$

$$\int P^n \mathrm{d}x = \frac{P^{n+1}}{a\,(n+1)} \quad (n \neq -1) \tag{25.8}$$

$$\int x P^n \mathrm{d}x = \frac{P^{n+2}}{a^2\,(n+2)} - \frac{b P^{n+1}}{a^2\,(n+1)} \quad (n \neq -1, -2) \tag{25.9}$$

25.1.2 Integrals with $x^m/(ax+b)^n$, $P = ax + b$, $a \neq 0$, $P \neq 0$

$$\int \frac{\mathrm{d}x}{P} = \frac{\ln|P|}{a} \tag{25.10}$$

$$\int \frac{\mathrm{d}x}{P^2} = -\frac{1}{aP} \tag{25.11}$$

$$\int \frac{\mathrm{d}x}{P^3} = -\frac{1}{2aP^2} \tag{25.12}$$

$$\int \frac{\mathrm{d}x}{P^4} = -\frac{1}{3aP^3} \tag{25.13}$$

$$\int \frac{x\mathrm{d}x}{P} = \frac{ax - b\ln|P|}{a^2} \tag{25.14}$$

$$\int \frac{x\mathrm{d}x}{P^2} = \frac{b}{a^2 P} + \frac{\ln|P|}{a^2} \tag{25.15}$$

$$\int \frac{x\mathrm{d}x}{P^3} = -\frac{b + 2ax}{2a^2 P^2} \tag{25.16}$$

$$\int \frac{x\mathrm{d}x}{P^4} = -\frac{b + 3ax}{6a^2 P^3} \tag{25.17}$$

$$\int \frac{x\mathrm{d}x}{P^n} = \frac{1}{a^2}\left(\frac{-1}{(n-2)P^{n-2}} + \frac{b}{(n-1)P^{n-1}}\right) \quad (n \neq 1, 2) \tag{25.18}$$

$$\int \frac{x^2\mathrm{d}x}{P} = -\frac{bx}{a^2} + \frac{x^2}{2a} + \frac{b^2\ln|P|}{a^3} \tag{25.19}$$

$$\int \frac{x^2\mathrm{d}x}{P^2} = \frac{x}{a^2} - \frac{b^2}{a^3 P} - \frac{2b\ln|P|}{a^3} \tag{25.20}$$

$$\int \frac{x^2\mathrm{d}x}{P^3} = \frac{4abx + 3b^2}{2a^3 P^2} + \frac{1}{a^3}\ln|P| \tag{25.21}$$

$$\int \frac{x^2 \mathrm{d}x}{P^n} = \frac{1}{a^3}\left[\frac{-1}{(n-3)P^{n-3}} + \frac{2b}{(n-2)P^{n-2}} - \frac{b^2}{(n-1)P^{n-1}}\right] \quad (n \neq 1, 2, 3) \tag{25.22}$$

$$\int \frac{x^3 \mathrm{d}x}{P} = \frac{x^3}{3a} - \frac{bx^2}{2a^2} + \frac{b^2 x}{a^3} - \frac{b^3}{a^4}\ln|P| \tag{25.23}$$

$$\int \frac{x^3 \mathrm{d}x}{P^2} = \frac{P^2}{2a^4} - \frac{3bP}{a^4} + \frac{b^3}{a^4 P} + \frac{3b^2}{a^4}\ln|P| \tag{25.24}$$

$$\int \frac{x^3 \mathrm{d}x}{P^3} = \frac{P}{a^4} + \frac{b^3}{2a^4 P^2} - \frac{3b^2}{a^4 P} - \frac{3b\ln|P|}{a^4} \tag{25.25}$$

$$\int \frac{x^3 \mathrm{d}x}{P^4} = \frac{b^3}{3a^4 P^3} - \frac{3b^2}{2a^4 P^2} + \frac{3b}{a^4 P} + \frac{\ln|P|}{a^4} \tag{25.26}$$

$$\int \frac{x^3 \mathrm{d}x}{P^n} = \frac{1}{a^4}\left(\frac{-1}{(n-4)P^{n-4}} + \frac{3b}{(n-3)P^{n-3}} - \frac{3b^2}{(n-2)P^{n-2}} + \frac{b^3}{(n-1)P^{n-1}}\right) \quad (n \neq 1, 2, 3, 4) \tag{25.27}$$

25.1.3 Integrals with $1/(x^n(ax+b)^m)$, $P = ax + b$ $b \neq 0$

$$\int \frac{\mathrm{d}x}{xP} = \frac{\ln|x/P|}{b} \tag{25.28}$$

$$\int \frac{\mathrm{d}x}{xP^2} = -\frac{1}{b^2}\left(\frac{ax}{P} + \ln\left|\frac{P}{x}\right|\right) + \frac{1}{b^2} \tag{25.29}$$

$$\int \frac{\mathrm{d}x}{xP^3} = \frac{a^2x^2}{2b^3P^2} - \frac{2ax}{b^3P} - \frac{1}{b^3}\ln\left|\frac{P}{x}\right| \tag{25.30}$$

$$\int \frac{\mathrm{d}x}{xP^4} = -\frac{a^3x^3}{3b^4P^3} + \frac{3a^2x^2}{2b^4P^2} - \frac{3ax}{b^4P} - \frac{1}{b^4}\ln\left|\frac{P}{x}\right| \tag{25.31}$$

$$\int \frac{\mathrm{d}x}{xP^n} = -\frac{1}{b^n}\left[\ln\left|\frac{P}{x}\right| - \sum_{i=1}^{n-1}\binom{n-1}{i}\frac{(-a)^i x^i}{iP^i}\right] \quad (n \geq 1) \tag{25.32}$$

$$\int \frac{\mathrm{d}x}{x^2P} = -\frac{1}{bx} + \frac{a}{b^2}\ln\left|\frac{P}{x}\right| \tag{25.33}$$

$$\int \frac{\mathrm{d}x}{x^2P^2} = -\frac{1}{b^2x} - \frac{a}{b^2P} - \frac{2a\ln|x/P|}{b^3} \tag{25.34}$$

$$\int \frac{\mathrm{d}x}{x^2P^3} = -\frac{1}{b^3x} - \frac{a}{2b^2P^2} - \frac{2a}{b^3P} - \frac{3a\ln|x/P|}{b^4} \tag{25.35}$$

$$\int \frac{dx}{x^2 P^4} = -\frac{1}{b^4 x} - \frac{a}{3b^2 P^3} - \frac{a}{b^3 P^2} - \frac{3a}{b^4 P} - \frac{4a \ln|x/P|}{b^5} \tag{25.36}$$

$$\int \frac{dx}{x^2 P^n} = -\frac{1}{b^{n+1}} \left[-\sum_{i=2}^{n} \binom{n}{i} \frac{(-a)^i \, x^{i-1}}{(i-1)\,(P)^{i-1}} + \frac{P}{x} - na \ln\left|\frac{P}{x}\right| \right] \quad (n \geq 2) \tag{25.37}$$

$$\int \frac{dx}{x^3 P} = -\frac{P^2}{2b^3 x^2} + \frac{2aP}{b^3 x} - \frac{a^2}{b^3} \ln\left|\frac{P}{x}\right| \tag{25.38}$$

$$\int \frac{dx}{x^3 P^2} = -\frac{P^2}{2b^4 x^2} + \frac{3aP}{b^4 x} - \frac{a^3 x}{b^4 P} - \frac{3a^2}{b^4} \ln\left|\frac{P}{x}\right| \tag{25.39}$$

$$\int \frac{dx}{x^3 P^3} = -\frac{1}{2b^3 x^2} + \frac{3a}{b^4 x} + \frac{a^2}{2b^3 P^2} + \frac{3a^2}{b^4 P} + \frac{6a^2 \ln|x/P|}{b^5} \tag{25.40}$$

$$\int \frac{dx}{x^3 P^4} = -\frac{1}{2b^4 x^2} + \frac{4a}{b^5 x} + \frac{a^2}{3b^3 P^3} + \frac{3a^2}{2b^4 P^2} + \frac{6a^2}{b^5 P} + \frac{10a^2 \ln|x/P|}{b^6} \tag{25.41}$$

$$\int \frac{dx}{x^3 P^n} = -\frac{1}{b^{n+2}} \left[-\sum_{i=3}^{n+1} \binom{n+1}{i} \frac{(-a)^i \, x^{i-2}}{(i-2)\, P^{i-2}} + \frac{a^2 P^2}{2x^2} - \frac{(n+1)aP}{x} - \frac{n(n+1)a^2}{2} \ln\left|\frac{P}{x}\right| \right] \quad (n \geq 3) \tag{25.42}$$

$$\int \frac{dx}{x^4 P} = -\frac{1}{3bx^3} + \frac{a}{2b^2 x^2} - \frac{a^2}{b^3 x} - \frac{a^3 \ln|x/P|}{b^4} \tag{25.43}$$

$$\int \frac{dx}{x^4 P^2} = -\frac{1}{3b^2 x^3} + \frac{a}{b^3 x^2} - \frac{3a^2}{b^4 x} - \frac{a^3}{b^4 P} - \frac{4a^3 \ln|x/P|}{b^5} \tag{25.44}$$

$$\int \frac{dx}{x^4 P^3} = -\frac{1}{3b^3 x^3} + \frac{3a}{2b^4 x^2} - \frac{6a^2}{b^5 x} - \frac{a^3}{2b^4 P^2} - \frac{4a^3}{b^5 P} - \frac{10a^3 \ln|x/P|}{b^6} \tag{25.45}$$

$$\int \frac{dx}{x^4 P^4} = -\frac{1}{3b^4 x^3} + \frac{2a}{b^5 x^2} - \frac{10a^2}{b^6 x} - \frac{a^3}{3b^4 P^3} - \frac{2a^3}{b^5 P^2} - \frac{10a^3}{b^6 P} - \frac{20a^3 \ln|x/P|}{b^7} \tag{25.46}$$

$$\int \frac{dx}{x^m P^n} = -\frac{1}{b^{m+n-1}} \sum_{i=0}^{m+n-2} \binom{m+n-2}{i} \frac{(-a)^i \, (P)^{m-i-1}}{(m-i-1)\, x^{m-i-1}} \tag{25.47}$$

If the denominator of the term after the summation sign vanishes, it is replaced by

$$\binom{m+n-2}{m-1} (-a)^{m-1} \ln\left|\frac{P}{x}\right| .$$

25.1.4 Integrals with $ax+b$ and $cx+d$ $\quad c \neq 0$

▷ **Abbreviation**: $A = bc - ad$

$$\int \frac{ax+b}{cx+d}\mathrm{d}x = \frac{ax}{c} + \frac{A}{c^2}\ln|cx+d| \tag{25.48}$$

$$\int \frac{\mathrm{d}x}{(ax+b)(cx+d)} = \frac{1}{A}\ln\left|\frac{cx+d}{ax+b}\right| \tag{25.49}$$

$$\int \frac{x\mathrm{d}x}{(ax+b)(cx+d)} = \frac{1}{A}\left[\frac{b}{a}\ln|ax+b| - \frac{d}{c}\ln|cx+d|\right] \quad (a \neq 0) \tag{25.50}$$

$$\int \frac{\mathrm{d}x}{(ax+b)^2(cx+d)} = \frac{1}{A}\left(\frac{1}{ax+b} + \frac{c}{A}\ln\left|\frac{cx+d}{ax+b}\right|\right) \quad (a, c \neq 0) \tag{25.51}$$

$$\int \frac{\mathrm{d}x}{(ax+b)^m(cx+d)^n} = -\frac{1}{(n-1)A}\left[\frac{1}{(ax+b)^{m-1}(cx+d)^{n-1}}\right. \tag{25.52}$$

$$\left. +(m+n-2)a\int \frac{\mathrm{d}x}{(ax+b)^m(cx+d)^{n-1}}\right] \quad (n \neq 1,\ A \neq 0)$$

$$\int \frac{(ax+b)^m}{cx+d}\mathrm{d}x = \frac{(ax+b)^m}{mc} + \frac{A}{c}\int \frac{(ax+b)^{m-1}}{cx+d}\mathrm{d}x \quad (m \neq 0) \tag{25.53}$$

$$\int \frac{(ax+b)^m}{(cx+d)^n}\mathrm{d}x = -\frac{1}{(n-1)c}\left[\frac{(ax+b)m}{(cx+d)^{n-1}} - ma\int \frac{(ax+b)^{m-1}}{(cx+d)^{n-1}}\mathrm{d}x\right] \quad (n \neq 1) \tag{25.54}$$

$$\int \frac{x^2\,\mathrm{d}x}{(ax+b)^2(cx+d)} = \frac{b^2}{a^2A(ax+b)} + \frac{1}{A^2}\left[\frac{d^2}{c}\ln|cx+d|\right. \tag{25.55}$$

$$\left. +\frac{b(bc-2ad)}{a^2}\ln|ax+b|\right] \quad (A, a \neq 0)$$

25.1.5 Integrals with $a+x$ and $b+x$ $\quad a \neq b$

$$\int \frac{x\mathrm{d}x}{(a+x)(b+x)^2} = \frac{b}{(a-b)(b+x)} - \frac{a}{(a-b)^2}\ln\left|\frac{a+x}{b+x}\right| \tag{25.56}$$

$$\int \frac{x^2\mathrm{d}x}{(a+x)(b+x)^2} = \frac{b^2}{(b-a)(b+x)} + \frac{a^2}{(a-b)^2}\ln|a+x| + \frac{b^2-2ab}{(b-a)^2}\ln|b+x| \tag{25.57}$$

$$\int \frac{\mathrm{d}x}{(a+x)^2(b+x)^2} = \frac{-1}{(a-b)^2}\left(\frac{1}{a+x} + \frac{1}{b+x}\right) + \frac{2}{(a-b)^3}\ln\left|\frac{a+x}{b+x}\right| \tag{25.58}$$

$$\int \frac{x\mathrm{d}x}{(a+x)^2(b+x)^2} = \frac{1}{(a-b)^2}\left(\frac{a}{a+x} + \frac{b}{b+x}\right) + \frac{a+b}{(a-b)^3}\ln\left|\frac{a+x}{b+x}\right| \tag{25.59}$$

$$\int \frac{x^2 \mathrm{d}x}{(a+x)^2(b+x)^2} = \frac{-1}{(a-b)^2}\left(\frac{a^2}{a+x}+\frac{b^2}{b+x}\right)+\frac{2ab}{(a-b)^3}\ln\left|\frac{a+x}{b+x}\right| \quad (25.60)$$

25.1.6 Integrals with $P = ax^2 + bx + c \qquad (a \neq 0)$

▷ **Abbreviation**: $A = 4ac - b^2$, $A \neq 0$

$$\int P\mathrm{d}x = \frac{x\left(6c+3bx+2ax^2\right)}{6} \quad (25.61)$$

$$\int P^2\mathrm{d}x = c^2x + bcx^2 + \frac{\left(b^2+2ac\right)x^3}{3} + \frac{abx^4}{2} + \frac{a^2x^5}{5} \quad (25.62)$$

$$\int \frac{\mathrm{d}x}{P} = \begin{cases} \dfrac{2}{\sqrt{A}}\arctan\left(\dfrac{2ax+b}{\sqrt{A}}\right), & A > 0 \\ \dfrac{1}{\sqrt{|A|}}\ln\left|\dfrac{2ax+b-\sqrt{|A|}}{2ax+b+\sqrt{|A|}}\right|, & A < 0 \end{cases} \quad (25.63)$$

$$\int \frac{\mathrm{d}x}{P^2} = \frac{2ax+b}{AP} + \frac{2a}{A}\int\frac{\mathrm{d}x}{P} \quad (25.64)$$

$$\int \frac{\mathrm{d}x}{P^3} = \frac{2ax+b}{2AP^2} + \frac{3a(2ax+b)}{A^2P} + \frac{6a^2}{A^2}\int\frac{\mathrm{d}x}{P} \quad (25.65)$$

$$\int \frac{\mathrm{d}x}{P^n} = \frac{2ax+b}{(n-1)AP^{n-1}} + \frac{(2n-3)\,2a}{(n-1)\,A}\int\frac{\mathrm{d}x}{P^{n-1}} \quad (n \neq 0) \quad (25.66)$$

25.1.7 Integrals with $x^n/(ax^2+bx+c)^m$, $P = ax^2 + bx + c \quad a \neq 0$

▷ **Abbreviation**: $A = 4ac - b^2 \quad A \neq 0$

$$\int \frac{x\,\mathrm{d}x}{P} = \frac{1}{2a}\ln|P| - \frac{b}{2a}\int\frac{\mathrm{d}x}{P} \quad (25.67)$$

$$\int \frac{x\,\mathrm{d}x}{P^2} = -\frac{bx+2c}{AP} - \frac{b}{A}\int\frac{\mathrm{d}x}{P} \quad (25.68)$$

$$\int \frac{x\,\mathrm{d}x}{P^3} = -\frac{bx+2c}{2AP^2} - \frac{3b(2ax+b)}{2A^2P} - \frac{3ab}{A^2}\int\frac{\mathrm{d}x}{P} \quad (25.69)$$

$$\int \frac{x\mathrm{d}x}{P^n} = -\frac{bx+2c}{(n-1)\,AP^{n-1}} - \frac{b\,(2n-3)}{(n-1)\,A}\int\frac{x}{P^{n-1}}\mathrm{d}x \quad (n \neq 1) \quad (25.70)$$

$$\int \frac{px+q}{P}\,\mathrm{d}x = \frac{p}{2a}\ln|P| + \frac{2aq-bp}{2a}\int\frac{\mathrm{d}x}{P} \quad (25.71)$$

$$\int \frac{px+q}{P^n}\,dx = \frac{(2aq-bp)x+bp-2cp}{(n-1)(2aq-bp)}(n-1)A\int \frac{dx}{P^{n-1}} \quad (n \neq 1) \tag{25.72}$$

$$\int \frac{x^2\,dx}{P} = \frac{x}{a} - \frac{b}{2a^2}\ln|P| + \frac{b^2-2ac}{2a^2}\int \frac{dx}{P} \tag{25.73}$$

$$\int \frac{x^2\,dx}{P^2} = \frac{(b^2-2ac)x+bc}{aAP} + \frac{2c}{A}\int \frac{dx}{P} \tag{25.74}$$

$$\int \frac{x^2\,dx}{P^3} = \frac{(b^2-2ac)x+bc}{2aAP^2} + \frac{4(b+x)+3(b^2-2ac)}{2aAP} + \frac{b^2+2ac}{A^2}\int \frac{dx}{P} \tag{25.75}$$

$$\int \frac{x^3\,dx}{P} = \frac{x^2}{2a} - \frac{bx}{a^2} + \frac{b^2-2ac}{2a^3}\ln|P| + \frac{3abc-b^3}{2a^3}\int \frac{dx}{P} \tag{25.76}$$

$$\int \frac{x^2dx}{P^n} = -\frac{x}{(2n-3)\,aP^{n-1}} + \frac{c}{(2n-3)\,a}\int \frac{dx}{P^n} - \frac{b\,(n-2)}{(2n-3)\,a}\int \frac{x\,dx}{P^{n-1}} \tag{25.77}$$

$$\int \frac{x^m dx}{P^n} = -\frac{x^{m-1}}{(2n-m-1)\,aP^{n-1}} + \frac{(m-1)\,c}{(2n-m-1)\,a}\int \frac{x^{m-2}dx}{P^n} - \frac{(n-m)\,b}{(2n-m-1)\,a}\int \frac{x^{m-1}dx}{P^n} \quad (m \neq 2n-1) \tag{25.78}$$

$$\int \frac{x^{2n-1}dx}{P^n} = \frac{1}{a}\int \frac{x^{2n-3}dx}{P^{n-1}} - \frac{c}{a}\int \frac{x^{2n-3}dx}{P^n} - \frac{b}{a}\int \frac{x^{2n-2}dx}{P^n} \tag{25.79}$$

25.1.8 Integrals with $1/(x^n(ax^2+bx+c)^m)$, $P = ax^2+bx+c \quad c \neq 0$

$$\int \frac{dx}{xP} = \frac{1}{2c}\ln\left|\frac{x^2}{P}\right| - \frac{b}{2c}\int \frac{dx}{P} \tag{25.80}$$

$$\int \frac{dx}{xP^n} = \frac{1}{2c\,(n-1)\,P^{n-1}} - \frac{b}{2c}\int \frac{dx}{P^n} + \frac{1}{c}\int \frac{dx}{xP^{n-1}} \tag{25.81}$$

$$\int \frac{dx}{x^2P} = \frac{b}{2c^2}\ln\left|\frac{P}{x^2}\right| - \frac{1}{cx} + \left(\frac{b^2}{2c^2} - \frac{a}{c}\right)\int \frac{dx}{P} \tag{25.82}$$

$$\int \frac{dx}{x^m P^n} = \frac{1}{(m-1)\,cx^{m-1}P^{n-1}} - \frac{(2n+m-3)\,a}{(m-1)\,c}\int \frac{dx}{x^{m-2}P^n} - \frac{(n+m-2)\,b}{(m-1)\,c}\int \frac{dx}{x^{m-1}P^n} \quad (m > 1) \tag{25.83}$$

$$\int \frac{dx}{(fx+g)P} = \frac{f}{2\left(cf^2-gfb+g^2a\right)}\ln\left|\frac{(fx+g)^2}{P}\right| + \frac{2ga-bf}{2\left(cf^2-gfb+g^2a\right)}\int \frac{dx}{P} \tag{25.84}$$

25.1.9 Integrals with $P = a^2 \pm x^2$

$$\int P \mathrm{d}x = a^2 x \pm \frac{x^3}{3} \tag{25.85}$$

$$\int P^2 \mathrm{d}x = a^4 x \pm \frac{2a^2x^3}{3} + \frac{x^5}{5} \tag{25.86}$$

$$\int P^3 \mathrm{d}x = a^6 x \pm a^4 x^3 + \frac{3a^2x^5}{5} \pm \frac{x^7}{7} \tag{25.87}$$

25.1.10 Integrals with $1/(a^2 \pm x^2)^n$, $P = a^2 \pm x^2$ $a \neq 0$

▷ **Abbreviations**:

$$Y = \begin{cases} \arctan\left(\frac{x}{a}\right) & \text{for } a^2 + x^2 \\ \frac{1}{2}\ln\left|\frac{a+x}{a-x}\right| & \text{for } a^2 - x^2 \text{ and } |x| < a \\ \frac{1}{2}\ln\left|\frac{a+x}{x-a}\right| & \text{for } a^2 - x^2 \text{ and } |x| > a \end{cases}$$

$$\int \frac{\mathrm{d}x}{P} = \frac{Y}{a} \tag{25.88}$$

$$\int \frac{\mathrm{d}x}{P^2} = \frac{x}{2a^2P} + \frac{Y}{2a^3} \tag{25.89}$$

$$\int \frac{\mathrm{d}x}{P^3} = \frac{x}{4a^2P^2} + \frac{3x}{8a^4P} + \frac{3Y}{8a^5} \tag{25.90}$$

$$\int \frac{\mathrm{d}x}{P^4} = \frac{x}{6a^2P^3} + \frac{5x}{24a^4P^2} + \frac{5x}{16a^6P} + \frac{5Y}{16a^7} \tag{25.91}$$

$$\int \frac{\mathrm{d}x}{P^{n+1}} = \frac{x}{2na^2P^n} + \frac{2n-1}{2na^2}\int \frac{\mathrm{d}x}{P^n} \quad (n \neq 0) \tag{25.92}$$

25.1.11 Integrals with $x^n/(a^2 \pm x^2)^m$, $P = a^2 \pm x^2$ $a \neq 0$

▷ **Abbreviations**:

$$Y = \begin{cases} \arctan\left(\frac{x}{a}\right) & \text{for } a^2 + x^2 \\ \frac{1}{2}\ln\left|\frac{a+x}{a-x}\right| & \text{for } a^2 - x^2 \text{ and } |x| < a \\ \frac{1}{2}\ln\left|\frac{a+x}{x-a}\right| & \text{for } a^2 - x^2 \text{ and } |x| > a \end{cases}$$

$$\int \frac{x\mathrm{d}x}{P} = \pm\frac{\ln|P|}{2} \tag{25.93}$$

$$\int \frac{x\mathrm{d}x}{P^2} = \mp\frac{1}{2P} \tag{25.94}$$

$$\int \frac{x\mathrm{d}x}{P^3} = \mp\frac{1}{4P^2} \tag{25.95}$$

$$\int \frac{x\mathrm{d}x}{P^4} = \mp\frac{1}{6P^3} \tag{25.96}$$

$$\int \frac{x\mathrm{d}x}{P^{n+1}} = \mp\frac{1}{2nP^n} \quad (n \neq 0) \tag{25.97}$$

$$\int \frac{x^2\mathrm{d}x}{P} = \pm(x - aY) \tag{25.98}$$

$$\int \frac{x^2\mathrm{d}x}{P^2} = -\frac{x}{2P} \pm \frac{Y}{2a} \tag{25.99}$$

$$\int \frac{x^2\mathrm{d}x}{P^3} = \mp\frac{x}{4P^2} \pm \frac{x}{8a^2P} \pm \frac{Y}{8a^3} \tag{25.100}$$

$$\int \frac{x^2\mathrm{d}x}{P^4} = \mp\frac{x}{6P^3} \pm \frac{x}{24a^2P^2} + \frac{x}{16a^4P} \pm \frac{Y}{16a^5} \tag{25.101}$$

$$\int \frac{x^2\mathrm{d}x}{P^{n+1}} = \mp\frac{x}{2nP^n} \pm \frac{1}{2n}\int \frac{\mathrm{d}x}{P^n} \quad (n \neq 0) \tag{25.102}$$

$$\int \frac{x^3\mathrm{d}x}{P} = \frac{x^2 \mp a^2\ln|P|}{2} \tag{25.103}$$

$$\int \frac{x^3\mathrm{d}x}{P^2} = \frac{a^2}{2P} + \frac{\ln|P|}{2} \tag{25.104}$$

$$\int \frac{x^3\mathrm{d}x}{P^3} = -\frac{a^2 + 2x^2}{4P^2} \tag{25.105}$$

$$\int \frac{x^3\mathrm{d}x}{P^4} = -\frac{a^2 + 3x^2}{12P^3} \tag{25.106}$$

$$\int \frac{x^3\mathrm{d}x}{P^{n+1}} = -\frac{1}{2\,(n-1)\,P^{n-1}} + \frac{a^2}{2nP^n} \quad (n > 1) \tag{25.107}$$

25.1.12 Integrals with $1/(x^n(a^2 \pm x^2)^m)$ $P = a^2 \pm x^2$ $a \neq 0$

▷ **Abbreviations:**

$$Y = \begin{cases} \arctan\left(\frac{x}{a}\right) & \text{for } a^2 + x^2 \\ \frac{1}{2}\ln\left|\frac{a+x}{a-x}\right| & \text{for } a^2 - x^2 \text{ and } |x| < a \\ \frac{1}{2}\ln\left|\frac{a+x}{x-a}\right| & \text{for } a^2 - x^2 \text{ and } |x| > a \end{cases}$$

$$\int \frac{\mathrm{d}x}{xP} = \frac{\ln\left|x^2/P\right|}{2a^2} \qquad (25.108)$$

$$\int \frac{\mathrm{d}x}{xP^2} = \frac{1}{2a^2P} + \frac{\ln\left|x^2/P\right|}{2a^4} \qquad (25.109)$$

$$\int \frac{\mathrm{d}x}{xP^3} = \frac{1}{4a^2P^2} + \frac{1}{2a^4P} + \frac{\ln\left|x^2/P\right|}{2a^6} \qquad (25.110)$$

$$\int \frac{\mathrm{d}x}{xP^4} = \frac{1}{6a^2P^3} + \frac{1}{4a^4P^2} + \frac{1}{2a^6P} + \frac{\ln\left|x^2/P\right|}{2a^8} \qquad (25.111)$$

$$\int \frac{\mathrm{d}x}{x^2P} = -\frac{a \pm xY}{a^3x} \qquad (25.112)$$

$$\int \frac{\mathrm{d}x}{x^2P^2} = -\frac{1}{a^4x} \mp \frac{x}{2a^4P} \mp \frac{3Y}{2a^5} \qquad (25.113)$$

$$\int \frac{\mathrm{d}x}{x^2P^3} = -\frac{1}{a^6x} \mp \frac{x}{4a^4P^2} \mp \frac{7x}{8a^6P} \mp \frac{15Y}{8a^7} \qquad (25.114)$$

$$\int \frac{\mathrm{d}x}{x^2P^4} = -\frac{1}{a^8x} \mp \frac{x}{6a^4P^3} \mp \frac{11x}{24a^6P^2} \mp \frac{19x}{16a^8P} \mp \frac{35Y}{16a^9} \qquad (25.115)$$

$$\int \frac{\mathrm{d}x}{x^3P} = -\frac{1}{2a^2x^2} \mp \frac{\ln\left|x^2/P\right|}{2a^4} \qquad (25.116)$$

$$\int \frac{\mathrm{d}x}{x^3P^2} = -\frac{1}{2a^4x^2} \mp \frac{1}{2a^4P} \mp \frac{\ln\left|x^2/P\right|}{a^6} \qquad (25.117)$$

$$\int \frac{\mathrm{d}x}{x^3P^3} = -\frac{1}{2a^6x^2} \mp \frac{1}{4a^4P^2} \mp \frac{1}{a^6P} + \frac{3\ln\left|P/x^2\right|}{2a^8} \qquad (25.118)$$

$$\int \frac{\mathrm{d}x}{x^3P^4} = -\frac{1}{2a^8x^2} \mp \frac{1}{6a^4P^3} - \frac{1}{2a^6P^2} \mp \frac{3}{2a^8P} \mp \frac{2\ln\left|x^2/P\right|}{a^{10}} \qquad (25.119)$$

$$\int \frac{\mathrm{d}x}{x^4P} = -\frac{1}{3a^2x^3} \pm \frac{1}{a^4x} + \frac{Y}{a^5} \qquad (25.120)$$

$$\int \frac{dx}{x^4 P^2} = -\frac{1}{3a^4x^3} \pm \frac{2}{a^6 x} + \frac{x}{2a^6 P} + \frac{5Y}{2a^7} \tag{25.121}$$

$$\int \frac{dx}{x^4 P^3} = -\frac{1}{3a^6x^3} + \frac{3}{a^8 x} + \frac{x}{4a^6 P^2} + \frac{11x}{8a^8 P} + \frac{35Y}{8a^9} \quad \text{(only } a^2 + x^2\text{)} \tag{25.122}$$

$$\int \frac{dx}{x^4 P^4} = -\frac{1}{3a^8x^3} + \frac{4}{a^{10} x} + \frac{x}{6a^6 P^3} + \frac{17x}{24a^8 P^2} + \frac{41x}{16a^{10} P} + \frac{105Y}{16a^{11}} \quad \text{(only } a^2 + x^2\text{)} \tag{25.123}$$

$$\int \frac{dx}{(b+cx)\,P} = \frac{1}{a^2c^2 \pm b^2}\left[c \ln|b+cx| - \frac{c}{2}\ln|P| \pm \frac{b}{a}Y\right] \quad a^2c^2 \neq b^2 \tag{25.124}$$

25.1.13 Integrals with $P = a^3 \pm x^3 \quad a \neq 0$

$$\int \frac{dx}{P} = \pm\frac{1}{6a^2}\ln\left|\frac{(a\pm x)^2}{a^2 \mp ax + x^2}\right| + \frac{1}{a^2\sqrt{3}}\arctan\left(\frac{2x \mp a}{a\sqrt{3}}\right) \tag{25.125}$$

$$\int \frac{1}{P^2}dx = \frac{x}{3a^3 P} + \frac{2}{3a^3}\int \frac{dx}{P} \tag{25.126}$$

$$\int \frac{x\,dx}{P} = -\frac{1}{6a}\ln\left|\frac{(a\pm x)^2}{a^2 \mp ax + x^2}\right| \pm \frac{1}{a\sqrt{3}}\arctan\left(\frac{2x \mp a}{a\sqrt{3}}\right) \tag{25.127}$$

$$\int \frac{x^2 dx}{P} = \pm\frac{1}{3}\ln|P| \tag{25.128}$$

$$\int \frac{x\,dx}{P^2} = \frac{x}{3a^3 P} + \frac{1}{3a^3}\int \frac{x\,dx}{P} \tag{25.129}$$

$$\int \frac{x^2 dx}{P^2} = \mp\frac{1}{3P} \tag{25.130}$$

$$\int \frac{x^3 dx}{P} = \pm x \mp a^3 \int \frac{dx}{P} \tag{25.131}$$

$$\int \frac{x^3 dx}{P^2} = \mp\frac{x}{3P} \pm \frac{1}{3}\int \frac{dx}{P} \tag{25.132}$$

$$\int \frac{dx}{xP} = \frac{1}{3a^3}\ln\left|\frac{x^3}{P}\right| \tag{25.133}$$

$$\int \frac{dx}{xP^2} = \frac{1}{3a^3 P} + \frac{1}{3a^6}\ln\left|\frac{x^3}{P}\right| \tag{25.134}$$

$$\int \frac{dx}{x^2 P} = -\frac{1}{a^3 x} \mp \frac{1}{a^3}\int \frac{x\,dx}{P} \tag{25.135}$$

$$\int \frac{dx}{x^2P^2} = -\frac{1}{a^6x} \mp \frac{x^2}{3a^6P} \mp \frac{4}{3a^6}\int \frac{x\,dx}{P} \tag{25.136}$$

$$\int \frac{dx}{x^3P} = -\frac{1}{2a^3x^2} \mp \frac{1}{a^3}\int \frac{dx}{P} \tag{25.137}$$

$$\int \frac{dx}{x^3P^2} = -\frac{1}{2a^6x^2} \mp \frac{x}{3a^6P} \pm \frac{5}{3a^6}\int \frac{dx}{P} \tag{25.138}$$

25.1.14 Integrals with $a^4 + x^4$ $(a > 0)$

$$\int \frac{dx}{a^4+x^4} = \frac{1}{4a^3\sqrt{2}} \ln\left|\frac{x^2+ax\sqrt{2}+a^2}{x^2-ax\sqrt{2}+a^2}\right| + \frac{1}{2a^3\sqrt{2}}\arctan\left(\frac{ax\sqrt{2}}{a^2-x^2}\right) \tag{25.139}$$

$$\int \frac{x\,dx}{a^4+x^4} = \frac{1}{2a^2}\arctan\left(\frac{x^2}{a^2}\right) \tag{25.140}$$

$$\int \frac{x^2dx}{a^4+x^4} = -\frac{1}{4a\sqrt{2}} \ln\left|\frac{x^2+ax\sqrt{2}+a^2}{x^2-ax\sqrt{2}+a^2}\right| + \frac{1}{2a\sqrt{2}}\arctan\left(\frac{ax\sqrt{2}}{a^2-x^2}\right) \tag{25.141}$$

$$\int \frac{x^3dx}{a^4+x^4} = \frac{1}{4}\ln\left|a^4+x^4\right| \tag{25.142}$$

$$\int \frac{dx}{x(a^4+x^4)} = \frac{1}{4a^4}\ln\left(\frac{x^4}{a^4+x^4}\right) \tag{25.143}$$

25.1.15 Integrals with $a^4 - x^4$ $(a > 0)$

$$\int \frac{dx}{a^4-x^4} = \frac{1}{4a^3}\ln\left|\frac{a+x}{a-x}\right| + \frac{1}{2a^3}\arctan\left(\frac{x}{a}\right) \tag{25.144}$$

$$\int \frac{x\,dx}{a^4-x^4} = \frac{1}{4a^2}\ln\left|\frac{a^2+x^2}{a^2-x^2}\right| \tag{25.145}$$

$$\int \frac{x^2dx}{a^4-x^4} = \frac{1}{4a}\ln\left|\frac{a+x}{a-x}\right| - \frac{1}{2a}\arctan\left(\frac{x}{a}\right) \tag{25.146}$$

$$\int \frac{x^3dx}{a^4-x^4} = -\frac{1}{4}\ln\left|a^4-x^4\right| \tag{25.147}$$

$$\int \frac{dx}{x(a^4-x^4)} = -\frac{1}{4a^4}\ln\left|\frac{a^4-x^4}{x^4}\right| \tag{25.148}$$

25.2 Integrals of irrational functions

25.2.1 Integrals with $x^{1/2}$ and $P = ax + b \quad a, b \neq 0$

▷ **Abbreviations**:

$$Y = \begin{cases} \arctan\left(\sqrt{ax/b}\right) & \text{for } a > 0,\ b > 0 \\ \frac{1}{2}\ln\frac{\sqrt{b}+\sqrt{-ax}}{\sqrt{b}-\sqrt{-ax}} & \text{for } a < 0,\ b > 0 \end{cases}$$

$$\int \frac{\sqrt{x}\,\mathrm{d}x}{P} = \frac{2\sqrt{x}}{a} - \frac{2\sqrt{b}Y}{a^{\frac{3}{2}}} \tag{25.149}$$

$$\int \frac{\sqrt{x}\,\mathrm{d}x}{P^2} = -\frac{\sqrt{x}}{aP} + \frac{Y}{a^{\frac{3}{2}}\sqrt{b}} \tag{25.150}$$

$$\int \frac{\sqrt{x}\,\mathrm{d}x}{P^3} = \frac{-\sqrt{x}}{2aP^2} + \frac{\sqrt{x}}{4abP} + \frac{Y}{4a^{\frac{3}{2}}b^{\frac{3}{2}}} \tag{25.151}$$

$$\int \frac{\sqrt{x}\,\mathrm{d}x}{P^4} = \frac{-\sqrt{x}}{3aP^3} + \frac{\sqrt{x}}{12abP^2} + \frac{\sqrt{x}}{8ab^2P} + \frac{Y}{8a^{\frac{3}{2}}b^{\frac{5}{2}}} \tag{25.152}$$

$$\int \frac{x^{\frac{3}{2}}\,\mathrm{d}x}{P} = \frac{-2b\sqrt{x}}{a^2} + \frac{2x^{\frac{3}{2}}}{3a} + \frac{2b^{\frac{3}{2}}Y}{a^{\frac{5}{2}}} \tag{25.153}$$

$$\int \frac{x^{\frac{3}{2}}\,\mathrm{d}x}{P^2} = \frac{2\sqrt{x}}{a^2} + \frac{b\sqrt{x}}{a^2P} - \frac{3\sqrt{b}Y}{a^{\frac{5}{2}}} \tag{25.154}$$

$$\int \frac{x^{\frac{3}{2}}\,\mathrm{d}x}{P^3} = \frac{b\sqrt{x}}{2a^2P^2} - \frac{5\sqrt{x}}{4a^2P} + \frac{3Y}{4a^{\frac{5}{2}}\sqrt{b}} \tag{25.155}$$

$$\int \frac{x^{\frac{3}{2}}\,\mathrm{d}x}{P^4} = \frac{b\sqrt{x}}{3a^2P^3} - \frac{7\sqrt{x}}{12a^2P^2} + \frac{\sqrt{x}}{8a^2bP} + \frac{Y}{8a^{\frac{5}{2}}b^{\frac{3}{2}}} \tag{25.156}$$

$$\int \frac{\mathrm{d}x}{\sqrt{x}P} = \frac{2Y}{\sqrt{a}\sqrt{b}} \tag{25.157}$$

$$\int \frac{\mathrm{d}x}{\sqrt{x}P^2} = \frac{\sqrt{x}}{bP} + \frac{Y}{\sqrt{a}b^{\frac{3}{2}}} \tag{25.158}$$

$$\int \frac{\mathrm{d}x}{\sqrt{x}P^3} = \frac{\sqrt{x}}{2bP^2} + \frac{3\sqrt{x}}{4b^2P} + \frac{3Y}{4\sqrt{a}b^{\frac{5}{2}}} \tag{25.159}$$

$$\int \frac{\mathrm{d}x}{\sqrt{x}P^4} = \frac{\sqrt{x}}{3bP^3} + \frac{5\sqrt{x}}{12b^2P^2} + \frac{5\sqrt{x}}{8b^3P} + \frac{5Y}{8\sqrt{a}b^{\frac{7}{2}}} \tag{25.160}$$

$$\int \frac{dx}{x^{\frac{3}{2}}P} = -\frac{2}{b\sqrt{x}} - \frac{2\sqrt{a}Y}{b^{\frac{3}{2}}} \tag{25.161}$$

$$\int \frac{dx}{x^{\frac{3}{2}}P^2} = -\frac{2}{b^2\sqrt{x}} - \frac{a\sqrt{x}}{b^2P} - \frac{3\sqrt{a}Y}{b^{\frac{5}{2}}} \tag{25.162}$$

$$\int \frac{dx}{x^{\frac{3}{2}}P^3} = -\frac{2}{b^3\sqrt{x}} - \frac{a\sqrt{x}}{2b^2P^2} - \frac{7a\sqrt{x}}{4b^3P} - \frac{15\sqrt{a}Y}{4b^{\frac{7}{2}}} \tag{25.163}$$

$$\int \frac{dx}{x^{\frac{3}{2}}P^4} = -\frac{2}{b^4\sqrt{x}} - \frac{a\sqrt{x}}{3b^2P^3} - \frac{11a\sqrt{x}}{12b^3P^2} - \frac{19a\sqrt{x}}{8b^4P} - \frac{35\sqrt{a}Y}{8b^{\frac{9}{2}}} \tag{25.164}$$

25.2.2 Integrals with $(ax+b)^{1/2}$ $P = ax + b$ $a \neq 0$

$$\int \sqrt{P}dx = \frac{2}{3a}\sqrt{P^3} \tag{25.165}$$

$$\int x\sqrt{P}dx = \frac{2(3ax-2b)}{15a^2}\sqrt{P^3} \tag{25.166}$$

$$\int x^2\sqrt{P}dx = \frac{2(15a^2x^2-12abx+8b^2)}{105a^3}\sqrt{P^3} \tag{25.167}$$

$$\int \frac{dx}{\sqrt{P}} = \frac{2\sqrt{P}}{a} \tag{25.168}$$

$$\int \frac{xdx}{\sqrt{P}} = \frac{2(ax-2b)}{3a^2}\sqrt{P} \tag{25.169}$$

$$\int \frac{x^2dx}{\sqrt{P}} = \frac{2(3a^2x^2-4abx+8b^2)}{15a^3}\sqrt{P} \tag{25.170}$$

$$\begin{aligned}\int \frac{dx}{x\sqrt{P}} &= \frac{1}{\sqrt{b}}\ln\left(\frac{\sqrt{P}-\sqrt{b}}{\sqrt{P}+\sqrt{b}}\right) \quad \text{for } b>0\\ &= \frac{2}{\sqrt{-b}}\arctan\sqrt{\frac{P}{-b}} \quad \text{for } b<0\end{aligned} \tag{25.171}$$

$$\int \frac{dx}{x^2\sqrt{P}} = -\frac{\sqrt{P}}{bx} - \frac{a}{2b}\int\frac{dx}{x\sqrt{P}} \tag{25.172}$$

$$\int \frac{\sqrt{P}dx}{x} = 2\sqrt{P} + b\int\frac{dx}{x\sqrt{P}} \tag{25.173}$$

$$\int \frac{\sqrt{P}dx}{x^2} = -\frac{\sqrt{P}}{x} + \frac{a}{2}\int\frac{dx}{x\sqrt{P}} \tag{25.174}$$

$$\int \frac{\mathrm{d}x}{x^n\sqrt{P}} = -\frac{\sqrt{P}}{(n-1)\,bx^{n-1}} - \frac{(2n-3)\,a}{(2n-2)\,b}\int \frac{\mathrm{d}x}{x^{n-1}\sqrt{P}} \tag{25.175}$$

$$\int P^{\frac{3}{2}}\mathrm{d}x = \frac{2P^{\frac{5}{2}}}{5a} \tag{25.176}$$

$$\int xP^{\frac{3}{2}}\mathrm{d}x = \frac{-14bP^{\frac{5}{2}} + 10P^{\frac{7}{2}}}{35a^2} \tag{25.177}$$

$$\int x^2P^{\frac{3}{2}}\mathrm{d}x = \frac{126b^2P^{\frac{5}{2}} - 180bP^{\frac{7}{2}} + 70P^{\frac{9}{2}}}{315a^3} \tag{25.178}$$

$$\int \frac{P^{\frac{3}{2}}\mathrm{d}x}{x} = 2b\sqrt{P} + \frac{2P^{\frac{3}{2}}}{3} + b^2\int \frac{\mathrm{d}x}{x\sqrt{P}} \tag{25.179}$$

$$\int \frac{\mathrm{d}x}{xP^{\frac{3}{2}}} = \frac{2}{b\sqrt{P}} + \frac{1}{b}\int \frac{\mathrm{d}x}{x\sqrt{P}} \tag{25.180}$$

$$\int \frac{\mathrm{d}x}{x^2P^{\frac{3}{2}}} = -\frac{3a}{b^2\sqrt{P}} - \frac{1}{bx\sqrt{P}} - \frac{3a}{2b^2}\int \frac{\mathrm{d}x}{x\sqrt{P}} \tag{25.181}$$

$$\int \frac{x\mathrm{d}x}{P^{\frac{3}{2}}} = \frac{4b + 2ax}{a^2\sqrt{P}} \tag{25.182}$$

$$\int \frac{x^2\mathrm{d}x}{P^{\frac{3}{2}}} = \frac{-16b^2 - 8abx + 2a^2x^2}{3a^3\sqrt{P}} \tag{25.183}$$

$$\int xP^{\pm n/2}\mathrm{d}x = \frac{2}{a^2}\left(\frac{P^{(4\pm n)/2}}{4\pm n} - \frac{bP^{(2\pm n)/2}}{2\pm n}\right) \tag{25.184}$$

$$\int x^2P^{\pm n/2}\mathrm{d}x = \frac{2}{a^3}\left(\frac{P^{(6\pm n)/2}}{6\pm n} - \frac{2bP^{(4\pm n)/2}}{4\pm n} + \frac{b^2P^{(2\pm n)/2}}{2\pm n}\right) \tag{25.185}$$

$$\int \frac{P^{n/2}\mathrm{d}x}{x} = \frac{2P^{n/2}}{n} + b\int \frac{P^{(n-2)/2}\mathrm{d}x}{x} \tag{25.186}$$

$$\int \frac{\mathrm{d}x}{xP^{n/2}} = \frac{2}{(n-2)\,bP^{(n-2)/2}} + \frac{1}{b}\int \frac{\mathrm{d}x}{xP^{(n-2)/2}} \tag{25.187}$$

$$\int \frac{\mathrm{d}x}{x^2P^{n/2}} = -\frac{1}{bxP^{(n-2)/2}} - \frac{na}{2b}\int \frac{\mathrm{d}x}{xP^{n/2}} \tag{25.188}$$

25.2.3 Integrals with $(ax+b)^{1/2}$ and $(cx+d)^{1/2}$, $a, c \neq 0$

▷ **Abbreviation**: $P = ax + b, \quad Q = cx + d \quad ad \neq bc$

$$\int \frac{\mathrm{d}x}{\sqrt{PQ}} = \frac{2}{\sqrt{ac}} \ln\left(\sqrt{aQ} + \sqrt{cP}\right) \quad (ac > 0) \tag{25.189}$$

$$= -\frac{2}{\sqrt{-ac}} \arctan \sqrt{-\frac{cP}{aQ}} \quad (ac < 0)$$

$$\int \frac{x\mathrm{d}x}{\sqrt{PQ}} = \frac{\sqrt{PQ}}{ac} - \frac{ad+bc}{2ac} \int \frac{\mathrm{d}x}{\sqrt{PQ}} \tag{25.190}$$

$$\int \frac{\mathrm{d}x}{\sqrt{P}Q} = \frac{2}{\sqrt{acd - bc^2}} \arctan\left(\frac{c\sqrt{P}}{\sqrt{acd - bc^2}}\right) \quad (bc^2 - acd < 0) \tag{25.191}$$

$$= \frac{1}{\sqrt{bc^2 - acd}} \ln\left(\frac{c\sqrt{P} - \sqrt{bc^2 - acd}}{c\sqrt{P} - \sqrt{bc^2 - acd}}\right) \quad (bc^2 - acd > 0)$$

$$\int \frac{\mathrm{d}x}{\sqrt{P}Q^{\frac{3}{2}}} = \frac{2\sqrt{P}}{(ad - bc)\sqrt{Q}} \tag{25.192}$$

$$\int \frac{\sqrt{P}\mathrm{d}x}{\sqrt{Q}} = \frac{1}{c}\sqrt{PQ} - \frac{bc - ad}{2c} \int \frac{\mathrm{d}x}{\sqrt{PQ}} \tag{25.193}$$

$$\int \frac{\sqrt{P}\mathrm{d}x}{Q} = \frac{2\sqrt{P}}{c} + \frac{(2cb - 2ad)}{c\sqrt{-\left(bc^2\right) + acd}} \arctan\left(\frac{c\sqrt{P}}{\sqrt{-\left(bc^2\right) + acd}}\right) \tag{25.194}$$

$$\int \frac{Q^n \mathrm{d}x}{\sqrt{P}} = \frac{2}{(2n+1)\,a}\left(\sqrt{P}Q^n - n\,(bd - ac) \int \frac{Q^{n-1}\mathrm{d}x}{\sqrt{P}}\right) \tag{25.195}$$

$$\int \frac{\mathrm{d}x}{\sqrt{P}Q^n} = -\frac{1}{(n-1)\,(bd - ac)}\left(\sqrt{P}Q^{n-1} + (n - 3/2)\,a \int \frac{\mathrm{d}x}{Q^{n-1}\sqrt{P}}\right) \tag{25.196}$$

$$\int \sqrt{P}Q^n \mathrm{d}x = \frac{2}{(2n+3)\,c}\left(-\frac{\sqrt{P}}{Q^{n-1}} + \frac{a}{2} \int \frac{\mathrm{d}x}{Q^{n-1}\sqrt{P}}\right) \tag{25.197}$$

$$\int \frac{\sqrt{P}\mathrm{d}x}{Q^n} = \frac{1}{(n-1)\,c}\left(-\frac{\sqrt{P}}{Q^{n-1}} + \frac{a}{2} \int \frac{\mathrm{d}x}{\sqrt{P}Q^{n-1}}\right) \quad (n \neq 1) \tag{25.198}$$

25.2.4 Integrals with $R = (a^2 + x^2)^{1/2}$ $a \neq 0$

$$\int R\mathrm{d}x = \frac{xR + a^2 \ln(x + R)}{2} \tag{25.199}$$

$$\int xR\mathrm{d}x = \frac{R^3}{3} \tag{25.200}$$

$$\int x^2 R\mathrm{d}x = \frac{-a^2xR + 2xR^3 - a^4\ln(x+R)}{8} \tag{25.201}$$

$$\int x^3 R\mathrm{d}x = -\frac{a^2R^3}{3} + \frac{R^5}{5} \tag{25.202}$$

$$\int \frac{R\mathrm{d}x}{x} = R - a\ln\frac{a+R}{x} \tag{25.203}$$

$$\int \frac{R\mathrm{d}x}{x^2} = -\frac{R}{x} + \ln(x+R) \tag{25.204}$$

$$\int \frac{R\mathrm{d}x}{x^3} = -\frac{R}{2x^2} - \frac{\ln[(a+R)/x]}{2a} \tag{25.205}$$

$$\int \frac{\mathrm{d}x}{R} = \ln(x+R) \tag{25.206}$$

$$\int \frac{x\mathrm{d}x}{R} = R \tag{25.207}$$

$$\int \frac{x^2\mathrm{d}x}{R} = \frac{xR - a^2\ln(x+R)}{2} \tag{25.208}$$

$$\int \frac{x^3\mathrm{d}x}{R} = -a^2R + \frac{R^3}{3} \tag{25.209}$$

$$\int \frac{\mathrm{d}x}{xR} = -\frac{\ln[(a+R)/x]}{a} \tag{25.210}$$

$$\int \frac{\mathrm{d}x}{x^2R} = -\frac{R}{a^2x} \tag{25.211}$$

$$\int \frac{\mathrm{d}x}{x^3R} = -\frac{R}{2a^2x^2} + \frac{\ln[(a+R)/x]}{2a^3} \tag{25.212}$$

$$\int R^3\mathrm{d}x = \frac{3a^2xR + 2xR^3 + 3a^4\ln(x+R)}{8} \tag{25.213}$$

$$\int xR^3\mathrm{d}x = \frac{R^5}{5} \tag{25.214}$$

$$\int x^2R^3\mathrm{d}x = \frac{1}{48}\left(-3a^4xR - 2a^2xR^3 + 8xR^5 - 3a^6\ln(x+R)\right) \tag{25.215}$$

$$\int x^3R^3\mathrm{d}x = -\frac{a^2R^5}{5} + \frac{R^7}{7} \tag{25.216}$$

$$\int \frac{R^3 \mathrm{d}x}{x} = a^2 R + \frac{R^3}{3} - a^3 \ln \frac{a+R}{x} \tag{25.217}$$

$$\int \frac{R^3 \mathrm{d}x}{x^2} = \frac{3xR}{2} - \frac{R^3}{x} + \frac{3a^2 \ln(x+R)}{2} \tag{25.218}$$

$$\int \frac{R^3 \mathrm{d}x}{x^3} = \frac{3R}{2} - \frac{R^3}{2x^2} - \frac{3a \ln[(a+R)/x]}{2} \tag{25.219}$$

$$\int \frac{\mathrm{d}x}{R^3} = \frac{x}{a^2 R} \tag{25.220}$$

$$\int \frac{x\mathrm{d}x}{R^3} = -\frac{1}{R} \tag{25.221}$$

$$\int \frac{x^2 \mathrm{d}x}{R^3} = -\frac{x}{R} + \ln(x+R) \tag{25.222}$$

$$\int \frac{x^3 \mathrm{d}x}{R^3} = R + \frac{a^2}{R} \tag{25.223}$$

$$\int \frac{\mathrm{d}x}{xR^3} = \frac{1}{a^2 R} - \frac{\ln[(a+R)/x]}{a^3} \tag{25.224}$$

$$\int \frac{\mathrm{d}x}{x^2 R^3} = -\frac{a^2 + 2x^2}{a^4 R x} \tag{25.225}$$

$$\int \frac{1}{x^3 R^3} \mathrm{d}x = -\frac{3}{2a^4 R} - \frac{1}{2a^2 x^2 R} + \frac{3 \ln[(a+R)/x]}{2a^5} \tag{25.226}$$

25.2.5 Integrals with $S = (x^2 - a^2)^{1/2}$ $a \neq 0$

$$\int S\mathrm{d}x = \frac{xS - a^2 \ln(x+S)}{2} \tag{25.227}$$

$$\int xS\mathrm{d}x = \frac{S^3}{3} \tag{25.228}$$

$$\int x^2 S\mathrm{d}x = \frac{-\left(a^2 x S\right) + 2xS^3 - a^4 \ln(x+S)}{8} \tag{25.229}$$

$$\int x^3 S\mathrm{d}x = \frac{a^2 S^3}{3} + \frac{S^5}{5} \tag{25.230}$$

$$\int \frac{S\mathrm{d}x}{x} = S - a \arccos(a/x) \tag{25.231}$$

$$\int \frac{S\mathrm{d}x}{x^2} = -\frac{S}{x} + \ln(x+S) \tag{25.232}$$

$$\int \frac{S\mathrm{d}x}{x^3} = -\frac{S}{2x^2} + \frac{\arccos(a/x)}{2a} \tag{25.233}$$

$$\int \frac{\mathrm{d}x}{S} = \ln(x+S) \tag{25.234}$$

$$\int \frac{x\mathrm{d}x}{S} = S \tag{25.235}$$

$$\int \frac{x^2\mathrm{d}x}{S} = \frac{xS + a^2\ln(x+S)}{2} \tag{25.236}$$

$$\int \frac{x^3\mathrm{d}x}{S} = a^2S + \frac{S^3}{3} \tag{25.237}$$

$$\int \frac{\mathrm{d}x}{xS} = \frac{\arccos(a/x)}{a} \tag{25.238}$$

$$\int \frac{\mathrm{d}x}{x^2S} = \frac{S}{a^2x} \tag{25.239}$$

$$\int \frac{\mathrm{d}x}{x^3S} = \frac{aS + x^2\arccos(a/x)}{2a^3x^2} \tag{25.240}$$

$$\int S^3\mathrm{d}x = \frac{-3a^2xS + 2xS^3 + 3a^4\ln(x+S)}{8} \tag{25.241}$$

$$\int xS^3\mathrm{d}x = \frac{S^5}{5} \tag{25.242}$$

$$\int x^2S^3\mathrm{d}x = \frac{1}{48}\left(8xS^5 + 2a^2xS^3 - 3a^4xS + 3a^6\ln(x+S)\right) \tag{25.243}$$

$$\int x^3S^3\mathrm{d}x = \frac{a^2S^5}{5} + \frac{S^7}{7} \tag{25.244}$$

$$\int \frac{S^3\mathrm{d}x}{x} = -a^2S + \frac{S^3}{3} + a^3\arccos(a/x) \tag{25.245}$$

$$\int \frac{S^3\mathrm{d}x}{x^2} = \frac{3xS}{2} - \frac{S^3}{2} - \frac{3a^2\ln(x+S)}{2} \tag{25.246}$$

$$\int \frac{S^3\mathrm{d}x}{x^3} = \frac{3S}{2} - \frac{S^3}{2x^2} - \frac{3a\arccos(a/x)}{2} \tag{25.247}$$

$$\int \frac{dx}{S^3} = -\frac{x}{a^2 S} \tag{25.248}$$

$$\int \frac{x\,dx}{S^3} = -\frac{1}{S} \tag{25.249}$$

$$\int \frac{x^2 dx}{S^3} = -\frac{x}{S} + \ln(x + S) \tag{25.250}$$

$$\int \frac{x^3 dx}{S^3} = \frac{-2a^2 + x^2}{S} \tag{25.251}$$

$$\int \frac{dx}{xS^3} = -\frac{1}{a^2 S} - \frac{\arccos(a/x)}{a^3} \tag{25.252}$$

$$\int \frac{dx}{x^2 S^3} = \frac{a^2 - 2x^2}{a^4 x S} \tag{25.253}$$

$$\int \frac{dx}{x^3 S^3} = -\frac{1}{2a^2 x^2 S} - \frac{3}{2a^4 S} - \frac{3\arccos(a/x)}{2a^5} \tag{25.254}$$

25.2.6 Integrals with $T = (a^2 - x^2)^{1/2}$ $\quad a \neq 0$

$$\int T\,dx = \frac{xT + a^2 \arcsin(x/a)}{2} \tag{25.255}$$

$$\int xT\,dx = -\frac{T^3}{3} \tag{25.256}$$

$$\int x^2 T\,dx = \frac{a^2 xT - 2xT^3 + a^4 \arcsin(x/a)}{8} \tag{25.257}$$

$$\int x^3 T\,dx = -\frac{a^2 T^3}{3} + \frac{T^5}{5} \tag{25.258}$$

$$\int \frac{T\,dx}{x} = T + a \ln[(a - T)/x] \tag{25.259}$$

$$\int \frac{T\,dx}{x^2} = -\frac{T}{x} - \arcsin(x/a) \tag{25.260}$$

$$\int \frac{T\,dx}{x^3} = -\frac{T}{2x^2} + \frac{\ln[(a + T)/x]}{2a} \tag{25.261}$$

$$\int \frac{dx}{T} = \arcsin(x/a) \tag{25.262}$$

$$\int \frac{x\mathrm{d}x}{T} = -T \tag{25.263}$$

$$\int \frac{x^2\mathrm{d}x}{T} = \frac{-xT + a^2 \arcsin(x/a)}{2} \tag{25.264}$$

$$\int \frac{x^3\mathrm{d}x}{T} = -a^2T + \frac{T^3}{3} \tag{25.265}$$

$$\int \frac{\mathrm{d}x}{xT} = -\frac{\ln[(a+T)/x]}{a} \tag{25.266}$$

$$\int \frac{\mathrm{d}x}{x^2T} = -\frac{T}{a^2x} \tag{25.267}$$

$$\int \frac{\mathrm{d}x}{x^3T} = -\frac{T}{2a^2x^2} - \frac{\ln[(a+T)/x]}{2a^3} \tag{25.268}$$

$$\int T^3\mathrm{d}x = \frac{3a^2xT + 2xT^3 + 3a^4 \arcsin(x/a)}{8} \tag{25.269}$$

$$\int xT^3\mathrm{d}x = -\frac{T^5}{5} \tag{25.270}$$

$$\int x^2T^3\mathrm{d}x = \frac{1}{48}\left(3a^4xT + 2a^2xT^3 - 8xT^5 + 3a^6 \arcsin(x/a)\right) \tag{25.271}$$

$$\int x^3T^3\mathrm{d}x = -\frac{a^2T^5}{5} + \frac{T^7}{7} \tag{25.272}$$

$$\int \frac{T^3\mathrm{d}x}{x} = a^2T + \frac{T^3}{3} - a^3\ln[(a+T)/x] \tag{25.273}$$

$$\int \frac{T^3\mathrm{d}x}{x^2} = -\frac{3xT}{2} - \frac{T^3}{x} - \frac{3a^2 \arcsin(x/a)}{2} \tag{25.274}$$

$$\int \frac{T^3\mathrm{d}x}{x^3} = -\frac{3T}{2} - \frac{T^3}{2x^2} + \frac{3a\ln[(a+T)/x]}{2} \tag{25.275}$$

$$\int \frac{\mathrm{d}x}{T^3} = \frac{x}{a^2T} \tag{25.276}$$

$$\int \frac{x\mathrm{d}x}{T^3} = \frac{1}{T} \tag{25.277}$$

$$\int \frac{x^2\mathrm{d}x}{T^3} = \frac{x}{T} - \arcsin(x/a) \tag{25.278}$$

$$\int \frac{x^3\mathrm{d}x}{T^3} = T + \frac{a^2}{T} \tag{25.279}$$

$$\int \frac{\mathrm{d}x}{xT^3} = \frac{1}{a^2T} - \frac{\ln[(a+T)/x]}{a^3} \tag{25.280}$$

$$\int \frac{\mathrm{d}x}{x^2T^3} = \frac{-a^2+2x^2}{a^4xT} \tag{25.281}$$

$$\int \frac{\mathrm{d}x}{x^3T^3} = \frac{3}{2a^4T} - \frac{1}{2a^2x^2T} - \frac{3\ln[(a+T)/x]}{2a^5} \tag{25.282}$$

25.2.7 Integrals with $(ax^2+bx+c)^{1/2}$ $X = ax^2+bx+c \quad a \neq 0$

▷ **Abbreviations**:

$$A = 4ac - b^2 \quad (A \neq 0)$$

$$\int \frac{\mathrm{d}x}{\sqrt{X}} = \begin{cases} \frac{1}{\sqrt{a}} \ln|2\sqrt{a}\sqrt{X} + 2ax + b| & \text{for} \quad a > 0 \\ \frac{1}{\sqrt{a}} \operatorname{Arsinh}\left(\frac{2ax+b}{\sqrt{A}}\right) & \text{for} \quad a > 0, A > 0 \\ -\frac{1}{\sqrt{|a|}} \arcsin\left(\frac{2ax+b}{\sqrt{|X|}}\right) & \text{for} \quad a < 0, A < 0 \end{cases} \tag{25.283}$$

$$\int \sqrt{X}\,\mathrm{d}x = \frac{2ax+b}{4a}\sqrt{X} + \frac{A}{8a}\int \frac{\mathrm{d}x}{\sqrt{X}} \tag{25.284}$$

$$\int \frac{\mathrm{d}x}{x\sqrt{X}} = \begin{cases} -\frac{\sqrt{c}}{\ln}\left|\frac{2\sqrt{c}\sqrt{X} + bx + 2c}{x}\right| & \text{for} \quad c > 0 \\ -\frac{1}{\sqrt{c}} \operatorname{Arsinh}\left(\frac{bx+2c}{\sqrt{A}x}\right) & \text{for} \quad c > 0, A > 0 \\ \frac{1}{\sqrt{|c|}} \arcsin\left(\frac{bx+2c}{\sqrt{|A|}x}\right) & \text{for} \quad c < 0, A < 0 \end{cases} \tag{25.285}$$

$$\int \frac{x\,\mathrm{d}x}{\sqrt{X}} = \frac{\sqrt{X}}{a} - \frac{b}{2a}\int \frac{\mathrm{d}x}{\sqrt{X}} \tag{25.286}$$

$$\int \frac{x^2\,\mathrm{d}x}{\sqrt{X}} = \frac{2ax-3b}{4a^2}\sqrt{X} + \frac{3b^2-4ac}{8a^2}\int \frac{\mathrm{d}x}{\sqrt{X}} \tag{25.287}$$

$$\int x\sqrt{X}\,\mathrm{d}x = \frac{1}{3a}\sqrt{X^3} - \frac{b(2ax+b)}{8a^2}\sqrt{X} - \frac{bA}{16a^2}\int \frac{\mathrm{d}x}{\sqrt{X}} \tag{25.288}$$

$$\int x^2\sqrt{X}\,\mathrm{d}x = \frac{1}{24a^2}(6ax-5b)\sqrt{X^3} + \frac{5b^2-4ac}{16a^2}\int \sqrt{X}\,\mathrm{d}x \tag{25.289}$$

$$\int \frac{\sqrt{X}\,\mathrm{d}x}{x} = \sqrt{X} + \frac{b}{2}\int \frac{\mathrm{d}x}{\sqrt{X}} + c\int \frac{\mathrm{d}x}{x\sqrt{X}} \tag{25.290}$$

$$\int \frac{\sqrt{X}\,\mathrm{d}x}{x^2} = -\frac{\sqrt{X}}{x} + a\int \frac{\mathrm{d}x}{\sqrt{X}} + \frac{b}{2}\int \frac{\mathrm{d}x}{x\sqrt{X}} \tag{25.291}$$

$$\int \frac{\mathrm{d}x}{x^2\sqrt{X}} = -\frac{\sqrt{X}}{cx} - \frac{b}{2c}\int \frac{\mathrm{d}x}{x\sqrt{X}} \qquad c \neq 0 \tag{25.292}$$

$$\int \sqrt{X^3}\,\mathrm{d}x = \frac{2ax+b}{8a}\sqrt{X^3} + \frac{3A}{16a}\int \sqrt{X}\,\mathrm{d}x \tag{25.293}$$

$$\int x\sqrt{X^3}\,\mathrm{d}x = \frac{1}{5a}\sqrt{X^5} - \frac{b}{2a}\int \sqrt{X^3}\,\mathrm{d}x \tag{25.294}$$

$$\int \frac{\mathrm{d}x}{\sqrt{X^3}} = \frac{4ax+2b}{A\sqrt{X}} \tag{25.295}$$

$$\int \frac{x\,\mathrm{d}x}{\sqrt{X^3}} = -\frac{2bx+4c}{A\sqrt{X}} \tag{25.296}$$

$$\int \frac{\mathrm{d}x}{x\sqrt{X^3}} = \frac{1}{c\sqrt{X}} + \frac{1}{c}\int \frac{\mathrm{d}x}{x\sqrt{X}} - \frac{b}{2c}\int \frac{\mathrm{d}x}{\sqrt{X^3}} \qquad c \neq 0 \tag{25.297}$$

25.3 Integrals of transcendental functions

25.3.1 Integrals with exponential functions

$$\int \mathrm{e}^{ax}\,\mathrm{d}x = \frac{\mathrm{e}^{ax}}{a} \tag{25.298}$$

$$\int x\,\mathrm{e}^{ax}\,\mathrm{d}x = \frac{\mathrm{e}^{ax}}{a^2}(ax-1) \tag{25.299}$$

$$\int x^2\,\mathrm{e}^{ax}\,\mathrm{d}x = \mathrm{e}^{ax}\left(\frac{x^2}{a} - \frac{2x}{a^2} + \frac{2}{a^3}\right) \tag{25.300}$$

$$\int x^n\,\mathrm{e}^{ax}\,\mathrm{d}x = \frac{1}{a}x^n\mathrm{e}^{ax} - \frac{n}{a}\int x^{n-1}\mathrm{e}^{ax}\,\mathrm{d}x \tag{25.301}$$

$$\int \frac{\mathrm{e}^{ax}}{x}\,\mathrm{d}x = \ln(x) + \frac{ax}{1\cdot 1!} + \frac{(ax)^2}{2\cdot 2!} + \frac{(ax)^3}{3\cdot 3!} + \cdots \tag{25.302}$$

The definite integral $\int_{-\infty}^{x} \frac{\mathrm{e}^t}{t}\,\mathrm{d}t$ is called the **exponential integral function** (Ei[x]).

$$\int \frac{e^{ax}}{x^n} dx = \frac{1}{n-1}\left(-\frac{e^{ax}}{x^{n-1}} + a \int \frac{e^{ax}}{x^{n-1}} dx\right) \quad (n \neq 1) \tag{25.303}$$

$$\int \frac{dx}{1+e^{ax}} = \frac{1}{a} \ln\left(\frac{e^{ax}}{1+e^{ax}}\right) \tag{25.304}$$

$$\int \frac{dx}{b + c\, e^{ax}} = \frac{x}{b} - \frac{1}{ab} \ln(b + c\, e^{ax}) \tag{25.305}$$

$$\begin{aligned} \int \frac{dx}{b\, e^{ax} + c\, e^{-ax}} &= \frac{1}{a\sqrt{bc}} \arctan\left(e^{ax}\sqrt{b/c}\right) \quad (bc > 0) \\ &= \frac{1}{2a\sqrt{-bc}} \ln\left(\frac{c + e^{ax}\sqrt{-bc}}{c - e^{ax}\sqrt{-bc}}\right) \quad (bc < 0) \end{aligned} \tag{25.306}$$

$$\int \frac{e^{ax} dx}{b + c\, e^{ax}} = \frac{1}{ac} \ln(b + c\, e^{ax}) \tag{25.307}$$

$$\int \frac{x\, e^{ax} dx}{(1+ax)^2} = \frac{e^{ax}}{a^2(1+ax)} \tag{25.308}$$

$$\int e^{ax} \ln(x) dx = \frac{e^{ax} \ln(x)}{a} - \frac{1}{a} \int \frac{e^{ax}}{x} dx \tag{25.309}$$

$$\int e^{ax} \sin(bx) dx = \frac{e^{ax}}{a^2+b^2}[a \sin(bx) - b \cos(bx)] \tag{25.310}$$

$$\int e^{ax} \cos(bx) dx = \frac{e^{ax}}{a^2+b^2}[a \cos(bx) + b \sin(bx)] \tag{25.311}$$

$$\int e^{ax} \sin^n x dx = \frac{e^{ax} \sin^{n-1} x}{a^2+n^2}[a \sin(x) - n \cos(x)] + \frac{n(n-1)}{a^2+n^2} \int e^{ax} \sin^{n-2} x dx \tag{25.312}$$

$$\int e^{ax} \cos^n x dx = \frac{e^{ax} \cos^{n-1} x}{a^2+n^2}[a \cos(x) + n \sin(x)] + \frac{n(n-1)}{a^2+n^2} \int e^{ax} \cos^{n-2} x dx \tag{25.313}$$

$$\begin{aligned} \int x\, e^{ax} \sin(bx) dx &= \frac{x\, e^{ax}}{a^2+b^2}[a \sin(bx) - b \cos(bx)] \\ &- \frac{e^{ax}}{(a^2+b^2)^2}[(a^2-b^2)\sin(bx) - 2ab \cos(bx)] \end{aligned} \tag{25.314}$$

$$\begin{aligned} \int x\, e^{ax} \cos(bx) dx &= \frac{x\, e^{ax}}{a^2+b^2}[a \cos(bx) + b \sin(bx)] \\ &- \frac{e^{ax}}{(a^2+b^2)^2}[(a^2-b^2)\cos(bx) + 2ab \sin(bx)] \end{aligned} \tag{25.315}$$

25.3.2 Integrals with logarithmic functions $(x > 0)$

$$\int \ln(x)\mathrm{d}x = x\ \ln(x) - x \tag{25.316}$$

$$\int [\ln(x)]^2\mathrm{d}x = x[\ln(x)]^2 - 2x\ \ln(x) + 2x \tag{25.317}$$

$$\int [\ln(x)]^3\mathrm{d}x = x[\ln(x)]^3 - 3x\ \ln(x)^2 + 6x\ \ln(x) - 6x \tag{25.318}$$

$$\int [\ln(x)]^n\mathrm{d}x = x[\ln(x)]^n - n\int [\ln(x)]^{n-1}\mathrm{d}x \quad (n \neq -1 \quad \text{for } n = -1 \text{ see } 326) \tag{25.319}$$

$$\int \ln[\sin(x)]\mathrm{d}x = x\ \ln(x) - x - \frac{x^3}{18} - \frac{x^5}{900} - \cdots - \frac{2^{2n-1}B_n x^{2n+1}}{2(2n+1)!} \cdots - \tag{25.320}$$

B_n are the **Bernoulli numbers.**

$$\int \ln[\cos(x)]\mathrm{d}x = -\frac{x^3}{6} - \frac{x^5}{60} - \frac{x^7}{315} - \cdots - \frac{2^{2n-1}(2^{2n}-1)B_n}{n(2n+1)!}x^{2n+1} - \cdots \tag{25.321}$$

$$\int \ln[\tan(x)]\mathrm{d}x = x\ \ln(x) - x + \frac{x^3}{9} + \frac{7x^5}{450} + \cdots + \frac{2^{2n}(2^{2n-1}-1)B_n}{n(2n+1)!}x^{2n+1} + \cdots \tag{25.322}$$

$$\int \sin[\ln(x)]\mathrm{d}x = \frac{x}{2}\,[\sin[\ln(x)] - \cos[\ln(x)]] \tag{25.323}$$

$$\int \cos[\ln(x)]\mathrm{d}x = \frac{x}{2}\,[\sin[\ln(x)] + \cos[\ln(x)]] \tag{25.324}$$

$$\int \mathrm{e}^{ax}\ln(x)\mathrm{d}x = \frac{1}{a}\mathrm{e}^{ax}\ \ln(x) - \frac{1}{a}\int \frac{\mathrm{e}^{ax}}{x}\mathrm{d}x \tag{25.325}$$

$$\int \frac{\mathrm{d}x}{\ln(x)} = \ln\ |\ln(x)| + \ln(x) + \frac{[\ln(x)]^2}{2\cdot 2!} + \frac{[\ln(x)]^3}{3\cdot 3!} + \ldots \tag{25.326}$$

The definite integral $\int_0^x \frac{\mathrm{d}t}{\ln(t)}$ is called the **logarithmic integral** (Li[x])

$$\int \frac{\mathrm{d}x}{x\ \ln(x)} = \ln[\ln(x)] \tag{25.327}$$

$$\int \frac{\mathrm{d}x}{[\ln(x)]^n} = -\frac{x}{(n-1)[\ln(x)]^{n-1}} + \frac{1}{n-1}\int \frac{\mathrm{d}x}{[\ln(x)]^{n-1}} \quad (n \neq 1) \tag{25.328}$$

$$\int \frac{\mathrm{d}x}{x^n\ln(x)} = \ln[\ln(x)] - (n-1)\ln(x) + \frac{(n-1)^2[\ln(x)]^2}{2\cdot 2!} - \frac{(n-1)^3[\ln(x)]^3}{3\cdot 3!} + \ldots \tag{25.329}$$

$$\int \frac{\mathrm{d}x}{x\,[\ln(x)]^n} = -\frac{1}{(n-1)[\ln(x)]^{n-1}} \quad (n \neq 1) \tag{25.330}$$

$$\int \frac{\mathrm{d}x}{x^n[\ln(x)]^m} = -\frac{1}{x^{n-1}(m-1)[\ln(x)]^{m-1}} - \frac{n-1}{m-1}\int \frac{\mathrm{d}x}{x^n[\ln(x)]^{m-1}} \quad (m \neq 1) \tag{25.331}$$

$$\int x \ln x \,\mathrm{d}x = \frac{1}{2}x^2\left(\ln x - \frac{1}{2}\right) \tag{25.332}$$

$$\int x^2 \ln x \,\mathrm{d}x = \frac{1}{3}x^3\left(\ln x - \frac{1}{3}\right) \tag{25.333}$$

$$\int x^n \ln(x)\mathrm{d}x = x^{n+1}\left(\frac{\ln(x)}{n+1} - \frac{1}{(n+1)^2}\right) \quad (n \neq -1) \tag{25.334}$$

$$\int x^n([\ln(x)]^m)\mathrm{d}x = \frac{x^{n+1}[\ln(x)]^m}{n+1} - \frac{m}{n+1}\int x^n[\ln(x)]^{m-1}\mathrm{d}x \quad (n, m \neq 1) \tag{25.335}$$

$$\int \frac{\ln x}{x}\,\mathrm{d}x = \frac{1}{2}(\ln x)^2 \tag{25.336}$$

$$\int \frac{[\ln(x)]^n}{x}\mathrm{d}x = \frac{[\ln(x)]^{n+1}}{n+1} \quad (n \neq -1) \tag{25.337}$$

$$\int \frac{\ln(x)}{x^n}\mathrm{d}x = -\frac{\ln(x)}{(n-1)x^{n-1}} - \frac{1}{(n-1)^2x^{n-1}} \quad (n \neq 1) \tag{25.338}$$

$$\int \frac{\mathrm{d}x}{x\ln x} = \ln|\ln x| \qquad (x \neq 1) \tag{25.339}$$

$$\int \frac{x^m\,\mathrm{d}x}{\ln x} = \ln|\ln x| + (m+1)\ln x + \frac{(m+1)^2}{2\cdot 2!}[\ln x]^2 + \frac{(m+1)^3}{3\cdot 3!}[\ln x]^3 + ..\,(x \neq 1) \tag{25.340}$$

$$\int x^m[\ln x]^n\,\mathrm{d}x = \frac{x^{m+1}[\ln x]^n}{m+1} - \frac{n}{m+1}\int x^m[\ln x]^{n-1}\,\mathrm{d}x \quad (m \neq -1) \tag{25.341}$$

$$\int \ln(x^2+a^2)\,\mathrm{d}x = x\ln(x^2+a^2) - 2x + 2a\arctan\left(\frac{x}{a}\right) \tag{25.342}$$

$$\int \ln(x^2-a^2)\,\mathrm{d}x = x\ln(x^2-a^2) - 2x + a\ln\left(\frac{x+a}{x-a}\right) \quad (x^2 > a^2) \tag{25.343}$$

$$\int \frac{[\ln(x)]^n}{x^m}\mathrm{d}x = -\frac{[\ln(x)]^n}{(m-1)x^{m-1}} + \frac{n}{m-1}\int \frac{[\ln(x)]^{n-1}}{x^m}\mathrm{d}x \quad (m \neq 1) \tag{25.344}$$

$$\int \frac{x^n\mathrm{d}x}{\ln(x)} = \int \frac{\mathrm{e}^{-\theta}}{\theta}\mathrm{d}\theta \ \text{ with } \theta = -(n+1)\ln(x) \tag{25.345}$$

$$\int \frac{x^n \mathrm{d}x}{[\ln(x)]^m} = -\frac{x^{n+1}}{(m-1)[\ln(x)]^{m-1}} + \frac{n+1}{m-1}\int \frac{x^n \mathrm{d}x}{[\ln(x)]^{m-1}} \quad (m \neq 1) \tag{25.346}$$

25.3.3 Integrals with hyperbolic functions $(a \neq 0)$

$$\int \sinh(ax)\mathrm{d}x = \frac{\cosh(ax)}{a} \tag{25.347}$$

$$\int \sinh^2(ax)\mathrm{d}x = \frac{1}{2a}\sinh(ax)\cosh(ax) - \frac{1}{2}x \tag{25.348}$$

$$\int \sinh^n(ax)\mathrm{d}x = \frac{1}{an}\sinh^{n-1}(ax)\cosh(ax) - \frac{n-1}{n}\int \sinh^{n-2}(ax)\mathrm{d}x \tag{25.349}$$

$(n > 0)$,

$$= \frac{1}{(n+1)}\sinh^{n+1}(ax)\cosh(ax) - \frac{n+2}{n+1}\int \sinh^{n+2}(ax)\mathrm{d}x$$

$(n < 0, n \neq -1)$

$$\int \cosh(ax)\mathrm{d}x = \frac{\sinh(ax)}{a} \tag{25.350}$$

$$\int \cosh^2(ax)\mathrm{d}x = \frac{1}{2a}\sinh(ax)\cosh(ax) + \frac{1}{2}x \tag{25.351}$$

$$\int \cosh^n(ax)\mathrm{d}x = \frac{1}{an}\sinh(ax)\cosh^{n-1}(ax) + \frac{n-1}{n}\int \cosh^{n-2}(ax)\mathrm{d}x \tag{25.352}$$

$(n > 0)$,

$$= -\frac{1}{(n+1)}\sinh(ax)\cosh^{n+1}(ax) + \frac{n+2}{n+1}\int \cosh^{n+2}(ax)\mathrm{d}x$$

$(n < 0, n \neq -1)$

$$\int \tanh(ax)\mathrm{d}x = \frac{\ln[\cosh(ax)]}{a} \tag{25.353}$$

$$\int \tanh^2 ax\mathrm{d}x = x - \frac{\tanh(ax)}{a} \tag{25.354}$$

$$\int \coth(ax)\mathrm{d}x = \frac{\ln[\sinh(ax)]}{a} \tag{25.355}$$

$$\int \coth^2(ax)\mathrm{d}x = x - \frac{\coth(ax)}{a} \tag{25.356}$$

$$\int x\sinh(ax)\mathrm{d}x = \frac{1}{a}x\cosh(ax) - \frac{1}{a^2}\sinh(ax) \tag{25.357}$$

$$\int x\cosh(ax)\mathrm{d}x = \frac{1}{a}x\sinh(ax) - \frac{1}{a^2}\cosh(ax) \tag{25.358}$$

$$\int \frac{dx}{\sinh(ax)} = \frac{1}{a} \ln\left[\tanh\left(\frac{ax}{2}\right)\right] \tag{25.359}$$

$$\int \frac{dx}{\cosh(ax)} = \frac{2}{a} \arctan(e^{ax}) \tag{25.360}$$

$$\int \sinh(ax)\sin(ax)dx = \frac{1}{2a}[\cosh(ax)\sin(ax) - \sinh(ax)\cos(ax)] \tag{25.361}$$

$$\int \sinh(ax)\sin(bx)\,dx = \frac{a\cosh(ax)\sin(bx) - b\sinh(ax)\cos(bx)}{a^2+b^2} \tag{25.362}$$

$$\int \cosh(ax)\cos(ax)dx = \frac{1}{2a}[\sinh(ax)\cos(ax) + \cosh(ax)\sin(ax)] \tag{25.363}$$

$$\int \cosh(ax)\cos(bx)\,dx = \frac{a\sinh(ax)\cos(bx) + b\cosh(ax)\sin(bx)}{a^2+b^2} \tag{25.364}$$

$$\int \sinh(ax)\sinh(bx)dx = \frac{1}{a^2-b^2}[a\sinh(bx)\cosh(ax) - b\cosh(bx)\sinh(ax)] \quad (a^2 \neq b^2) \tag{25.365}$$

$$\int \cosh(ax)\cosh(bx)dx = \frac{1}{a^2-b^2}[a\sinh(ax)\cosh(bx) - b\sinh(bx)\cosh(ax)] \quad (a^2 \neq b^2) \tag{25.366}$$

$$\int \sinh(ax)\cos(ax)dx = \frac{1}{2a}[\cosh(ax)\cos(ax) + \sinh(ax)\sin(ax)] \tag{25.367}$$

$$\int \sinh(ax)\cos(bx)\,dx = \frac{a\cosh(ax)\cos(bx) - b\sinh(ax)\cos(bx)}{a^2+b^2} \tag{25.368}$$

$$\int \cosh(ax)\sin(ax)dx = \frac{1}{2a}[\sinh(ax)\sin(ax) - \cosh(ax)\cos(ax)] \tag{25.369}$$

$$\int \cosh(ax)\cos(bx)\,dx = \frac{a\sinh(ax)\cos(bx) + b\cosh(ax)\sin(bx)}{a^2+b^2} \tag{25.370}$$

$$\int \cosh(ax)\sinh(bx)dx = \frac{1}{a^2-b^2}[a\sinh(bx)\sinh(ax) - b\cosh(bx)\cosh(ax)] \quad (a^2 \neq b^2) \tag{25.371}$$

$$\int \sinh(ax)\cosh(ax)\,dx = \frac{\sinh^2(ax)}{2a} \tag{25.372}$$

25.3.4 Integrals with inverse hyperbolic functions

$$\int \operatorname{Arsinh}\left(\frac{x}{a}\right) \mathrm{d}x = x \operatorname{Arsinh}\left(\frac{x}{a}\right) - \sqrt{x^2 + a^2} \tag{25.373}$$

$$\int \operatorname{Arcosh}\left(\frac{x}{a}\right) \mathrm{d}x = x \operatorname{Arcosh}\left(\frac{x}{a}\right) - \sqrt{x^2 - a^2} \tag{25.374}$$

$$\int \operatorname{Artanh}\left(\frac{x}{a}\right) \mathrm{d}x = x \operatorname{Artanh}\left(\frac{x}{a}\right) + \frac{a}{2} \ln(a^2 - x^2) \tag{25.375}$$

$$\int \operatorname{Arcoth}\left(\frac{x}{a}\right) \mathrm{d}x = x \operatorname{Arcoth}\left(\frac{x}{a}\right) + \frac{a}{2} \ln(x^2 - a^2) \tag{25.376}$$

25.3.5 Integrals with sine and cosine functions $(a \neq 0)$

$$\int \sin(ax)\mathrm{d}x = -\frac{\cos(ax)}{a} \tag{25.377}$$

$$\int \cos(ax)\mathrm{d}x = \frac{\sin(ax)}{a} \tag{25.378}$$

$$\int \sin^2(ax)\mathrm{d}x = \frac{1}{2}x - \frac{1}{4a}\sin(2ax) = \frac{1}{2}x - \frac{\sin(ax)\cos(ax)}{2a} \tag{25.379}$$

$$\int \cos^2(ax)\mathrm{d}x = \frac{1}{2}x + \frac{1}{4a}\sin(2ax) \tag{25.380}$$

$$\int \sin^3(ax)\mathrm{d}x = \frac{\cos^3(ax)}{3a} - \frac{\cos(ax)}{a} \tag{25.381}$$

$$\int \cos^3(ax)\mathrm{d}x = -\frac{\sin^3(ax)}{3a} + \frac{\sin(ax)}{a} \tag{25.382}$$

$$\int \sin^4(ax)\mathrm{d}x = \frac{\sin(4ax)}{32a} - \frac{\sin(2ax)}{4a} + \frac{3x}{8} \tag{25.383}$$

$$\int \cos^4(ax)\mathrm{d}x = \frac{\sin(4ax)}{32a} + \frac{\sin(2ax)}{4a} + \frac{3x}{8} \tag{25.384}$$

$$\int \sin^n(ax)\mathrm{d}x = -\frac{\sin^{n-1}(ax)\ \cos(ax)}{na} + \frac{n-1}{n}\int \sin^{n-2}(ax)\mathrm{d}x \quad (n \in \mathbb{N},\ n \neq 0) \tag{25.385}$$

$$\int \cos^n(ax)\mathrm{d}x = \frac{\cos^{n-1}(ax)\ \sin(ax)}{na} + \frac{n-1}{n}\int \cos^{n-2}(ax)\mathrm{d}x \quad (n \in \mathbb{N},\ n \neq 0) \tag{25.386}$$

$$\int x \sin(ax)\mathrm{d}x = \frac{\sin(ax)}{a^2} - \frac{x \cos(ax)}{a} \tag{25.387}$$

$$\int x \cos(ax)\mathrm{d}x = \frac{\cos(ax)}{a^2} + \frac{x \sin(ax)}{a} \tag{25.388}$$

$$\int x^2 \sin(ax)\mathrm{d}x = \frac{2x}{a^2} \sin(ax) - \left(\frac{x^2}{a} - \frac{2}{a^3}\right) \cos(ax) \tag{25.389}$$

$$\int x^2 \cos(ax)\mathrm{d}x = \frac{2x}{a^2} \cos(ax) + \left(\frac{x^2}{a} - \frac{2}{a^3}\right) \sin(ax) \tag{25.390}$$

$$\int x^3 \sin(ax)\mathrm{d}x = \left(\frac{3x^2}{a^2} - \frac{6}{a^4}\right) \sin(ax) - \left(\frac{x^3}{a} - \frac{6x}{a^3}\right) \cos(ax) \tag{25.391}$$

$$\int x^3 \cos(ax)\mathrm{d}x = \left(\frac{3x^2}{a^2} - \frac{6}{a^4}\right) \cos(ax) + \left(\frac{x^3}{a} - \frac{6x}{a^3}\right) \sin(ax) \tag{25.392}$$

$$\int x^n \sin(ax)\mathrm{d}x = -\frac{x^n}{a} \cos(ax) + \frac{n}{a} \int x^{n-1} \cos(ax)\mathrm{d}x \quad (n > 0) \tag{25.393}$$

$$\int x^n \cos(ax)\mathrm{d}x = \frac{x^n}{a} \sin(ax) - \frac{n}{a} \int x^{n-1} \sin(ax)\mathrm{d}x \quad (n > 0) \tag{25.394}$$

$$\int \sin(ax)\sin(bx)\mathrm{d}x = \frac{\sin[(a-b)x]}{2(a-b)} - \frac{\sin[(a+b)x]}{2(a+b)}, \quad (|a| \neq |b|) \tag{25.395}$$

$$\int \cos(ax)\cos(bx)\mathrm{d}x = \frac{\sin[(a-b)x]}{2(a-b)} + \frac{\sin[(a+b)x]}{2(a+b)}, \quad (|a| \neq |b|) \tag{25.396}$$

$$\int \frac{\sin(ax)\mathrm{d}x}{x} = ax - \frac{(ax)^3}{3 \cdot 3!} + \frac{(ax)^5}{5 \cdot 5!} - \frac{(ax)^7}{7 \cdot 7!} + \cdots \tag{25.397}$$

$$\int \frac{\cos(ax)\mathrm{d}x}{x} = \ln(ax) - \frac{(ax)^2}{2 \cdot 2!} + \frac{(ax)^4}{4 \cdot 4!} - \frac{(ax)^6}{6 \cdot 6!} + \cdots \tag{25.398}$$

$$\int \frac{\sin(ax)\mathrm{d}x}{x^2} = -\frac{\sin(ax)}{x} + a \int \frac{\cos(ax)\mathrm{d}x}{x} \tag{25.399}$$

$$\int \frac{\cos(ax)\mathrm{d}x}{x^2} = -\frac{\cos(ax)}{x} - a \int \frac{\sin(ax)\mathrm{d}x}{x} \tag{25.400}$$

$$\int \frac{\sin(ax)\mathrm{d}x}{x^n} = -\frac{\sin(ax)}{(n-1)x^{n-1}} + \frac{a}{n-1} \int \frac{\cos(ax)\mathrm{d}x}{x^{n-1}} \quad (n \neq 1) \tag{25.401}$$

$$\int \frac{\cos(ax)\mathrm{d}x}{x^n} = -\frac{\cos(ax)}{(n-1)x^{n-1}} - \frac{a}{n-1} \int \frac{\sin(ax)\mathrm{d}x}{x^{n-1}} \quad (n \neq 1) \tag{25.402}$$

$$\int \frac{dx}{\sin(ax)} = \int \csc(ax)dx = \frac{1}{a} \ln\left[\tan\left(\frac{ax}{2}\right)\right] = \frac{1}{a} \ln[\csc(ax) - \cot(ax)] \quad (25.403)$$

$$\int \frac{dx}{\cos(ax)} = \int \sec(ax)dx = \frac{1}{a} \ln\left[\tan\left(\frac{ax}{2} + \frac{\pi}{4}\right)\right] = \frac{1}{a} \ln[\sec(ax) - \tan(ax)] \quad (25.404)$$

$$\int \frac{dx}{\sin^2(ax)} = -\frac{1}{a} \cot(ax) \quad (25.405)$$

$$\int \frac{dx}{\cos^2(ax)} = \frac{1}{a} \tan(ax) \quad (25.406)$$

$$\int \frac{dx}{\sin^3(ax)} = \frac{1}{2a} \ln\left[\tan\left(\frac{ax}{2}\right)\right] - \frac{\cos(ax)}{2a \sin^2(ax)} \quad (25.407)$$

$$\int \frac{dx}{\cos^3(ax)} = \frac{1}{2a} \ln\left[\tan\left(\frac{\pi}{4} + \frac{ax}{2}\right)\right] + \frac{\sin(ax)}{2a \cos^2(ax)} \quad (25.408)$$

$$\int \frac{dx}{\sin^n(ax)} = -\frac{1}{a(n-1)} \frac{\cos(ax)}{\sin^{n-1}(ax)} + \frac{n-2}{n-1} \int \frac{dx}{\sin^{n-2}(ax)} \quad (n > 1) \quad (25.409)$$

$$\int \frac{dx}{\cos^n(ax)} = \frac{1}{a(n-1)} \frac{\sin(ax)}{\cos^{n-1}(ax)} + \frac{n-2}{n-1} \int \frac{dx}{\cos^{n-2}(ax)} \quad (n > 1) \quad (25.410)$$

$$\int \frac{x dx}{\sin(ax)} = \frac{1}{a^2}\left(ax + \frac{(ax)^3}{3 \cdot 3!} + \frac{7(ax)^5}{3 \cdot 5 \cdot 5!} + \frac{31(ax)^7}{3 \cdot 7 \cdot 7!} + \frac{127(ax)^9}{3 \cdot 5 \cdot 9!} + \cdots \right.$$
$$\left. + \frac{2(2^{2n-1} - 1)}{(2n+1)!} B_n (ax)^{2n+1} + \cdots \right) \quad (25.411)$$

B_n are the **Bernoulli numbers** (see Section 5.24).

$$\int \frac{x dx}{\cos(ax)} = \frac{1}{a^2}\left(\frac{(ax)^2}{2} + \frac{(ax)^4}{4 \cdot 2!} + \frac{5(ax)^6}{6 \cdot 4!} + \frac{61(ax)^8}{8 \cdot 6!} + \right. \quad (25.412)$$
$$\left. + \frac{1385(ax)^{10}}{10 \cdot 8!} + \cdots + \frac{E_n (ax)^{2n+2}}{(2n+2)(2n)!} + \cdots \right)$$

E_n are the **Euler numbers** (see Section 5.34).

$$\int \frac{x dx}{\sin^2(ax)} = -\frac{x}{a} \cot(ax) + \frac{1}{a^2} \ln[\sin(ax)] \quad (25.413)$$

$$\int \frac{x dx}{\cos^2(ax)} = \frac{x}{a} \tan(ax) + \frac{1}{a^2} \ln[\cos(ax)] \quad (25.414)$$

$$\int \frac{x dx}{\sin^n(ax)} = -\frac{x \cos(ax)}{(n-1)a \sin^{n-1}(ax)} - \frac{1}{(n-1)(n-2)a^2 \sin^{n-2}(ax)} \quad (25.415)$$
$$+ \frac{n-2}{n-1} \int \frac{x dx}{\sin^{n-2}(ax)} \quad (n > 2)$$

$$\int \frac{x\mathrm{d}x}{\cos^n(ax)} = \frac{x\sin(ax)}{(n-1)a\cos^{n-1}(ax)} - \frac{1}{(n-1)(n-2)a^2\cos^{n-2}(ax)} + \frac{n-2}{n-1}\int \frac{x\mathrm{d}x}{\cos^{n-2}(ax)} \quad (n>2) \tag{25.416}$$

$$\int \frac{\mathrm{d}x}{1+\sin(ax)} = -\frac{1}{a}\tan\left(\frac{\pi}{4}-\frac{ax}{2}\right) \tag{25.417}$$

$$\int \frac{\mathrm{d}x}{1+\cos(ax)} = \frac{1}{a}\tan\left(\frac{ax}{2}\right) \tag{25.418}$$

$$\int \frac{\mathrm{d}x}{1-\sin(ax)} = \frac{1}{a}\tan\left(\frac{\pi}{4}+\frac{ax}{2}\right) \tag{25.419}$$

$$\int \frac{\mathrm{d}x}{1-\cos(ax)} = -\frac{1}{a}\cot\left(\frac{ax}{2}\right) \tag{25.420}$$

$$\int \frac{x\mathrm{d}x}{1+\sin(ax)} = -\frac{x}{a}\tan\left(\frac{\pi}{4}-\frac{ax}{2}\right) + \frac{2}{a^2}\ln\left[\cos\left(\frac{\pi}{4}-\frac{ax}{2}\right)\right] \tag{25.421}$$

$$\int \frac{x\mathrm{d}x}{1+\cos(ax)} = \frac{x}{a}\tan\left(\frac{ax}{2}\right) + \frac{2}{a^2}\ln\left[\cos\left(\frac{ax}{2}\right)\right] \tag{25.422}$$

$$\int \frac{x\mathrm{d}x}{1-\sin(ax)} = \frac{x}{a}\cot\left(\frac{\pi}{4}-\frac{ax}{2}\right) + \frac{2}{a^2}\ln\left[\sin\left(\frac{\pi}{4}-\frac{ax}{2}\right)\right] \tag{25.423}$$

$$\int \frac{x\mathrm{d}x}{1-\cos(ax)} = -\frac{x}{a}\cot\left(\frac{ax}{2}\right) + \frac{2}{a^2}\ln\left[\sin\left(\frac{ax}{2}\right)\right] \tag{25.424}$$

$$\int \frac{\mathrm{d}x}{b+c\,\sin(ax)} = \frac{2}{a\sqrt{b^2-c^2}}\arctan\left[\frac{b\,\tan(ax/2)+c}{\sqrt{b^2-c^2}}\right] \quad (b^2>c^2), \tag{25.425}$$

$$= \frac{1}{a\sqrt{c^2-b^2}}\ln\left[\frac{b\,\tan(ax/2)+c-\sqrt{c^2-b^2}}{b\,\tan(ax/2)+c+\sqrt{c^2-b^2}}\right] \quad (b^2<c^2)$$

$$\int \frac{\mathrm{d}x}{b+c\,\cos(ax)} = \frac{2}{a\sqrt{b^2-c^2}}\arctan\left[\frac{(b-c)\,\tan(ax/2)}{\sqrt{b^2-c^2}}\right] \quad (b^2>c^2), \tag{25.426}$$

$$= \frac{1}{a\sqrt{c^2-b^2}}\ln\left[\frac{(c-b)\,\tan(ax/2)-\sqrt{c^2-b^2}}{(c-b)\,\tan(ax/2)+\sqrt{c^2-b^2}}\right] \quad (b^2<c^2)$$

$$\int \frac{\mathrm{d}x}{[1+\sin(ax)]^2} = -\frac{1}{2a}\tan\left(\frac{\pi}{4}-\frac{ax}{2}\right) - \frac{1}{6a}\tan^3\left(\frac{\pi}{4}-\frac{ax}{2}\right) \tag{25.427}$$

$$\int \frac{\mathrm{d}x}{[1+\cos(ax)]^2} = \frac{1}{2a}\tan\left(\frac{ax}{2}\right) + \frac{1}{6a}\tan^3\left(\frac{ax}{2}\right) \tag{25.428}$$

$$\int \frac{\mathrm{d}x}{[1-\sin(ax)]^2} = \frac{1}{2a}\cot\left(\frac{\pi}{4}-\frac{ax}{2}\right) + \frac{1}{6a}\cot^3\left(\frac{\pi}{4}-\frac{ax}{2}\right) \tag{25.429}$$

$$\int \frac{\mathrm{d}x}{[1-\cos(ax)]^2} = -\frac{1}{2a}\cot\left(\frac{ax}{2}\right) - \frac{1}{6a}\cot^3\left(\frac{ax}{2}\right) \tag{25.430}$$

$$\int \frac{\mathrm{d}x}{\sin(ax)[1 \pm \sin(ax)]} = \frac{1}{a}\tan\left(\frac{\pi}{4} \mp \frac{ax}{2}\right) + \frac{1}{a}\ln\left[\tan\left(\frac{ax}{2}\right)\right] \tag{25.431}$$

$$\int \frac{\mathrm{d}x}{\cos(ax)[1+\cos(ax)]} = \frac{1}{a}\ln\left[\tan\left(\frac{\pi}{4} + \frac{ax}{2}\right)\right] - \frac{1}{a}\tan\left(\frac{ax}{2}\right) \tag{25.432}$$

$$\int \frac{\mathrm{d}x}{\cos(ax)[1-\cos(ax)]} = \frac{1}{a}\ln\left[\tan\left(\frac{\pi}{4} + \frac{ax}{2}\right)\right] - \frac{1}{a}\cot\left(\frac{ax}{2}\right) \tag{25.433}$$

$$\int \frac{\mathrm{d}x}{1+\sin^2(ax)} = \frac{1}{2\sqrt{2}a}\arcsin\left[\frac{3\ \sin^2(ax)-1}{\sin^2(ax)+1}\right] \tag{25.434}$$

$$\int \frac{\mathrm{d}x}{1+\cos^2(ax)} = \frac{1}{2\sqrt{2}a}\arcsin\left[\frac{1-3\ \cos^2(ax)}{1+\cos^2(ax)}\right] \tag{25.435}$$

$$\int \frac{\mathrm{d}x}{1-\sin^2(ax)} = \int \frac{\mathrm{d}x}{\cos^2(ax)} = \frac{1}{a}\tan(ax) \tag{25.436}$$

$$\int \frac{\mathrm{d}x}{1-\cos^2(ax)} = \int \frac{\mathrm{d}x}{\sin^2(ax)} = -\frac{1}{a}\cot(ax) \tag{25.437}$$

$$\int \frac{\sin(ax)\mathrm{d}x}{1 \pm \sin(ax)} = \pm x + \frac{1}{a}\tan\left(\frac{\pi}{4} \mp \frac{ax}{2}\right) \tag{25.438}$$

$$\int \frac{\cos(ax)\mathrm{d}x}{1+\cos(ax)} = x - \frac{1}{a}\tan\left(\frac{ax}{2}\right) \tag{25.439}$$

$$\int \frac{\cos(ax)\mathrm{d}x}{1-\cos(ax)} = -x - \frac{1}{a}\cot\left(\frac{ax}{2}\right) \tag{25.440}$$

$$\int \frac{\sin(ax)\mathrm{d}x}{[1+\sin(ax)]^2} = -\frac{1}{2a}\tan\left(\frac{\pi}{4} - \frac{ax}{2}\right) + \frac{1}{6a}\tan^3\left(\frac{\pi}{4} - \frac{ax}{2}\right) \tag{25.441}$$

$$\int \frac{\cos(ax)\mathrm{d}x}{[1+\cos(ax)]^2} = \frac{1}{2a}\tan\left(\frac{ax}{2}\right) + \frac{1}{6a}\tan^3\left(\frac{ax}{2}\right) \tag{25.442}$$

$$\int \frac{\sin(ax)\mathrm{d}x}{[1-\sin(ax)]^2} = -\frac{1}{2a}\cot\left(\frac{\pi}{4} - \frac{ax}{2}\right) + \frac{1}{6a}\cot^3\left(\frac{\pi}{4} - \frac{ax}{2}\right) \tag{25.443}$$

$$\int \frac{\cos(ax)\mathrm{d}x}{[1-\cos(ax)]^2} = \frac{1}{2a}\cot\left(\frac{ax}{2}\right) - \frac{1}{6a}\cot^3\left(\frac{ax}{2}\right) \tag{25.444}$$

$$\int \frac{\sin(ax)\mathrm{d}x}{b+c\ \sin(ax)} = \frac{x}{c} - \frac{b}{c}\int \frac{\mathrm{d}x}{b+c\ \sin(ax)} \tag{25.445}$$

$$\int \frac{\cos(ax)\mathrm{d}x}{b+c\cos(ax)} = \frac{x}{c} - \frac{b}{c}\int \frac{\mathrm{d}x}{b+c\ \cos(ax)} \tag{25.446}$$

$$\int \frac{dx}{\sin(ax)[b+c\ \sin(ax)]} = \frac{1}{ab}\ln\left[\tan\left(\frac{ax}{2}\right)\right] - \frac{c}{b}\int \frac{dx}{b+c\ \sin(ax)} \tag{25.447}$$

$$\int \frac{dx}{\cos(ax)[b+c\ \cos(ax)]} = \frac{1}{ab}\ln\tan\left(\frac{ax}{2}+\frac{\pi}{4}\right) - \frac{c}{b}\int \frac{dx}{b+c\ \cos(ax)} \tag{25.448}$$

$$\int \frac{dx}{[b+c\ \sin(ax)]^2} = \frac{c\ \cos(ax)}{a(b^2-c^2)[b+c\ \sin(ax)]} + \frac{b}{b^2-c^2}\int \frac{dx}{b+c\ \sin(ax)} \tag{25.449}$$

$$\int \frac{dx}{[b+c\ \cos(ax)]^2} = \frac{c\ \sin(ax)}{a(c^2-b^2)[b+c\ \cos(ax)]} + \frac{b}{b^2-c^2}\int \frac{dx}{b+c\ \cos(ax)} \tag{25.450}$$

$$\int \frac{\sin(ax)dx}{[b+c\ \sin(ax)]^2} = \frac{b\ \cos(ax)}{a(c^2-b^2)[b+c\ \sin(ax)]} + \frac{b}{c^2-b^2}\int \frac{dx}{b+c\ \sin(ax)} \tag{25.451}$$

$$\int \frac{\cos(ax)dx}{[b+c\ \cos(ax)]^2} = \frac{b\ \sin(ax)}{a(b^2-c^2)[b+c\ \cos(ax)]} - \frac{c}{b^2-c^2}\int \frac{dx}{b+c\ \cos(ax)} \tag{25.452}$$

$$\int \frac{dx}{b^2+c^2\ \sin^2(ax)} = \frac{1}{ab\sqrt{b^2+c^2}}\arctan\left[\frac{\sqrt{b^2+c^2}\ \tan(ax)}{b}\right] \quad (b>0) \tag{25.453}$$

$$\int \frac{dx}{b^2+c^2\ \cos^2(ax)} = \frac{1}{ab\sqrt{b^2+c^2}}\arctan\left[\frac{b\ \tan(ax)}{\sqrt{b^2+c^2}}\right] \quad (b>0) \tag{25.454}$$

$$\int \frac{dx}{b^2-c^2\ \sin^2(ax)} = \frac{1}{ab\sqrt{b^2-c^2}}\arctan\left[\frac{\sqrt{b^2-c^2}\ \tan(ax)}{b}\right] \tag{25.455}$$

$$(b^2>c^2, b>0)$$

$$= \frac{1}{2ab\sqrt{c^2-b^2}}\ln\left[\frac{\sqrt{c^2-b^2}\ \tan(ax)+b}{\sqrt{c^2-b^2}\ \tan(ax)-b}\right] \quad (c^2>b^2, b>0)$$

$$\int \frac{dx}{b^2-c^2\ \cos^2(ax)} = \frac{1}{ab\sqrt{b^2-c^2}}\arctan\left[\frac{b\ \tan(ax)}{\sqrt{b^2-c^2}}\right] \quad (b^2>c^2, b>0) \tag{25.456}$$

$$= \frac{1}{2ab\sqrt{c^2-b^2}}\ln\left[\frac{b\ \tan(ax)-\sqrt{c^2-b^2}}{b\ \tan(ax)+\sqrt{c^2-b^2}}\right] \quad (c^2>b^2, b>0)$$

25.3.6 Integrals with sine and cosine functions $(a \neq 0)$

$$\int \sin(ax)\cos(ax)dx = \frac{\sin^2(ax)}{2a} \tag{25.457}$$

$$\int \sin(ax)\cos(bx)dx = -\frac{\cos[(a+b)x]}{2(a+b)} - \frac{\cos[(a-b)x]}{2(a-b)} \quad (a^2 \neq b^2) \tag{25.458}$$

$$\int \sin^2(ax)\cos^2(ax)\mathrm{d}x = \frac{x}{8} - \frac{\sin(4ax)}{32a} \tag{25.459}$$

$$\int \sin^n(ax)\cos(ax)\mathrm{d}x = \frac{1}{a(n+1)}\sin^{n+1}(ax) \quad (n \neq -1) \tag{25.460}$$

$$\int \sin(ax)\cos^n(ax)\mathrm{d}x = -\frac{1}{a(n+1)}\cos^{n+1}(ax) \quad (n \neq -1) \tag{25.461}$$

$$\int \sin^n(ax)\cos^m(ax)\mathrm{d}x = -\frac{\sin^{n-1}(ax)\cos^{m+1}(ax)}{a(n+m)} + \frac{n-1}{n+m}\int \sin^{n-2}(ax)\cos^m(ax)\mathrm{d}x \tag{25.462}$$

Lowering of the power of the sine $(n, m > 0)$

$$= \frac{\sin^{n+1}(ax)\cos^{m-1}(ax)}{a(n+m)} + \frac{m-1}{n+m}\int \sin^n(ax)\cos^{m-2}(ax)\mathrm{d}x$$

Lowering of the power of the cosine $(n, m > 0)$

$$\int \frac{\mathrm{d}x}{\sin(ax)\cos(ax)} = \frac{1}{a}\ln|\tan(ax)| \tag{25.463}$$

$$\int \frac{\mathrm{d}x}{\sin^2(ax)\cos(ax)} = \frac{1}{a}\left[\ln\left[\tan\left(\frac{\pi}{4}+\frac{ax}{2}\right)\right] - \frac{1}{\sin(ax)}\right] \tag{25.464}$$

$$\int \frac{\mathrm{d}x}{\sin(ax)\cos^2(ax)} = \frac{1}{a}\left[\ln\left[\tan\left(\frac{ax}{2}\right)\right] + \frac{1}{\cos(ax)}\right] \tag{25.465}$$

$$\int \frac{\mathrm{d}x}{\sin^3(ax)\cos(ax)} = \frac{1}{a}\left[\ln\left[\tan(ax)\right] - \frac{1}{2\sin^2(ax)}\right] \tag{25.466}$$

$$\int \frac{\mathrm{d}x}{\sin(ax)\cos^3(ax)} = \frac{1}{a}\left[\ln\left[\tan(ax)\right] + \frac{1}{2\cos^2(ax)}\right] \tag{25.467}$$

$$\int \frac{\mathrm{d}x}{\sin^2(ax)\cos^2(ax)} = -\frac{2}{a}\cot(2ax) \tag{25.468}$$

$$\int \frac{\mathrm{d}x}{\sin^2(ax)\cos^3(ax)} = \frac{1}{a}\left[\frac{\sin(ax)}{2\cos^2(ax)} - \frac{1}{\sin(ax)} + \frac{3}{2}\ln\tan\left(\frac{\pi}{4}+\frac{ax}{2}\right)\right] \tag{25.469}$$

$$\int \frac{\mathrm{d}x}{\sin^3(ax)\cos^2(ax)} = \frac{1}{a}\left[\frac{1}{\cos(ax)} - \frac{\cos(ax)}{2\sin^2(ax)} + \frac{3}{2}\ln\tan\left(\frac{ax}{2}\right)\right] \tag{25.470}$$

$$\int \frac{\mathrm{d}x}{\sin^n(ax)\cos(ax)} = -\frac{1}{a(n-1)\sin^{n-1}(ax)} + \int \frac{\mathrm{d}x}{\sin^{n-2}(ax)\cos(ax)} \quad (n \neq 1) \tag{25.471}$$

$$\int \frac{dx}{\sin(ax)\cos^n(ax)} = \frac{1}{a(n-1)\cos^{n-1}(ax)} + \int \frac{dx}{\sin(ax)\cos^{n-2}(ax)} \quad (n \neq 1) \tag{25.472}$$

$$\int \frac{dx}{\sin^n(ax)\cos^m(ax)} = \frac{1}{a(n-1)}\frac{1}{\sin^{n-1}(ax)\cos^{m-1}(ax)} + \frac{n+m-2}{n-1}\int \frac{dx}{\sin^{n-2}(ax)\cos^m(ax)} \tag{25.473}$$

Lowering of the power of the sine $(n > 1, m > 0)$

$$= \frac{1}{a(m-1)}\frac{1}{\sin^{n-1}(ax)\cos^{m-1}(ax)} + \frac{n+m-2}{m-1}\int \frac{dx}{\sin^{n}(ax)\cos^{m-2}(ax)}$$

Lowering of the power of the cosine $(m > 1, n > 0)$

$$\int \frac{\sin(ax)dx}{\cos^2(ax)} = \frac{1}{a\,\cos(ax)} \tag{25.474}$$

$$\int \frac{\sin(ax)dx}{\cos^3(ax)} = \frac{1}{2a\,\cos^2(ax)} + C = \frac{1}{2a}\tan^2(ax) + C_1 \tag{25.475}$$

$$\int \frac{\sin(ax)dx}{\cos^n(ax)} = \frac{1}{a(n-1)\cos^{n-1}(ax)} \tag{25.476}$$

$$\int \frac{\cos(ax)dx}{\sin^2(ax)} = -\frac{1}{a\,\sin(ax)} \tag{25.477}$$

$$\int \frac{\cos(ax)dx}{\sin^3(ax)} = -\frac{1}{2a\,\sin^2(ax)} + C = -\frac{\cot^2(ax)}{2a} + C_1 \tag{25.478}$$

$$\int \frac{\cos(ax)dx}{\sin^n(ax)} = -\frac{1}{a(n-1)\sin^{n-1}(ax)} \tag{25.479}$$

$$\int \frac{\sin^2(ax)dx}{\cos(ax)} = -\frac{1}{a}\sin(ax) + \frac{1}{a}\ln\left[\tan\left(\frac{\pi}{4} + \frac{ax}{2}\right)\right] \tag{25.480}$$

$$\int \frac{\sin^3(ax)dx}{\cos(ax)} = -\frac{1}{a}\left[\frac{\sin^2(ax)}{2} + \ln[\cos(ax)]\right] \tag{25.481}$$

$$\int \frac{\sin^n(ax)dx}{\cos(ax)} = -\frac{\sin^{n-1}(ax)}{a(n-1)} + \int \frac{\sin^{n-2}(ax)dx}{\cos(ax)} \quad (n \neq 1) \tag{25.482}$$

$$\int \frac{\cos^2(ax)dx}{\sin(ax)} = \frac{1}{a}\left[\cos(ax) + \ln\left[\tan\left(\frac{ax}{2}\right)\right]\right] \tag{25.483}$$

$$\int \frac{\cos^3(ax)dx}{\sin(ax)} = \frac{1}{a}\left[\frac{\cos^2(ax)}{2} + \ln[\sin(ax)]\right] \tag{25.484}$$

$$\int \frac{\cos^n(ax)\mathrm{d}x}{\sin(ax)} = \frac{\cos^{n-1}(ax)}{a(n-1)} + \int \frac{\cos^{n-2}(ax)\mathrm{d}x}{\sin(ax)} \quad (n \neq 1) \tag{25.485}$$

$$\int \frac{\sin^2(ax)\mathrm{d}x}{\cos^3(ax)} = \frac{1}{a}\left[\frac{\sin(ax)}{2\cos^2(ax)} - \frac{1}{2}\ln\left[\tan\left(\frac{\pi}{4} + \frac{ax}{2}\right)\right]\right] \tag{25.486}$$

$$\int \frac{\sin^2(ax)\mathrm{d}x}{\cos^n(ax)} = \frac{\sin(ax)}{a(n-1)\cos^{n-1}(ax)} - \frac{1}{n-1}\int \frac{\mathrm{d}x}{\cos^{n-2}(ax)} \tag{25.487}$$

$$\int \frac{\sin^3(ax)\mathrm{d}x}{\cos^2(ax)} = \frac{1}{a}\left[\cos(ax) + \frac{1}{\cos(ax)}\right] \tag{25.488}$$

$$\int \frac{\sin^3(ax)\mathrm{d}x}{\cos^n(ax)} = \frac{1}{a}\left[\frac{1}{(n-1)\cos^{n-1}(ax)} - \frac{1}{(n-3)\cos^{n-3}(ax)}\right] \quad (n \neq 1, n \neq 3) \tag{25.489}$$

$$\begin{aligned}\int \frac{\sin^n(ax)\mathrm{d}x}{\cos^m(ax)} &= \frac{\sin^{n+1}(ax)}{a(m-1)\cos^{m-1}(ax)} - \frac{n-m+2}{m-1}\int \frac{\sin^n(ax)\mathrm{d}x}{\cos^{m-2}(ax)} \\ &\quad (m \neq 1) \\ &= -\frac{\sin^{n-1}(ax)}{a(n-m)\cos^{m-1}(ax)} + \frac{n-1}{n-m}\int \frac{\sin^{n-2}(ax)\mathrm{d}x}{\cos^m(ax)} \quad (m \neq n) \\ &= \frac{\sin^{n-1}(ax)}{a(m-1)\cos^{m-1}(ax)} - \frac{n-1}{m-1}\int \frac{\sin^{n-1}(ax)\mathrm{d}x}{\cos^{m-2}(ax)} \quad (m \neq 1)\end{aligned} \tag{25.490}$$

$$\int \frac{\cos^2(ax)\mathrm{d}x}{\sin^3(ax)} = -\frac{1}{2a}\left[\frac{\cos(ax)}{\sin^2(ax)} - \ln\left[\tan\left(\frac{ax}{2}\right)\right]\right] \tag{25.491}$$

$$\int \frac{\cos^2(ax)\mathrm{d}x}{\sin^n(ax)} = -\frac{1}{(n-1)}\left[\frac{\cos(ax)}{a\sin^{n-1}(ax)} + \int \frac{\mathrm{d}x}{\sin^{n-2}(ax)}\right] \tag{25.492}$$

$$\int \frac{\cos^3(ax)\mathrm{d}x}{\sin^2(ax)} = -\frac{1}{a}\left[\sin(ax) + \frac{1}{\sin(ax)}\right] \tag{25.493}$$

$$\int \frac{\cos^3(ax)\mathrm{d}x}{\sin^n(ax)} = \frac{1}{a}\left[\frac{1}{(n-3)\sin^{n-3}(ax)} - \frac{1}{(n-1)\sin^{n-1}(ax)}\right] \quad (n \neq 1, n \neq 3) \tag{25.494}$$

$$\begin{aligned}\int \frac{\cos^n(ax)\mathrm{d}x}{\sin^m(ax)} &= -\frac{\cos^{n+1}(ax)}{a(m-1)\sin^{m-1}(ax)} - \frac{n-m+2}{m-1}\int \frac{\cos^n(ax)\mathrm{d}x}{\sin^{m-2}(ax)} \\ &\quad (m \neq 1) \\ &= \frac{\cos^{n-1}(ax)}{a(n-m)\sin^{m-1}(ax)} + \frac{n-1}{n-m}\int \frac{\cos^{n-2}(ax)\mathrm{d}x}{\sin^m(ax)} \quad (m \neq n) \\ &= -\frac{\cos^{n-1}(ax)}{a(m-1)\sin^{m-1}(ax)} - \frac{n-1}{m-1}\int \frac{\cos^{n-2}(ax)\mathrm{d}x}{\sin^{m-2}(ax)} \quad (m \neq 1)\end{aligned} \tag{25.495}$$

$$\int \frac{\mathrm{d}x}{\sin(ax) \pm \cos(ax)} = \frac{1}{a\sqrt{2}}\ln\left[\tan\left(\frac{ax}{2} \pm \frac{\pi}{8}\right)\right] \tag{25.496}$$

$$\int \frac{\sin(ax)\mathrm{d}x}{\sin(ax) \pm \cos(ax)} = \frac{x}{2} \mp \frac{1}{2a} \ln[\sin(ax) \pm \cos(ax)] \tag{25.497}$$

$$\int \frac{\cos(ax)\mathrm{d}x}{\sin(ax) \pm \cos(ax)} = \pm\frac{x}{2} + \frac{1}{2a} \ln[\sin(ax) \pm \cos(ax)] \tag{25.498}$$

$$\int \frac{\mathrm{d}x}{\sin(ax)[1 \pm \cos(ax)]} = \pm\frac{1}{2a[1 \pm \cos(ax)]} + \frac{1}{2a} \ln\left[\tan\left(\frac{ax}{2}\right)\right] \tag{25.499}$$

$$\int \frac{\mathrm{d}x}{\cos(ax)[1 \pm \sin(ax)]} = \mp\frac{1}{2a[1 \pm \sin(ax)]} + \frac{1}{2a} \ln\left[\tan\left(\frac{\pi}{4} + \frac{ax}{2}\right)\right] \tag{25.500}$$

$$\int \frac{\sin(ax)\mathrm{d}x}{\cos(ax)[1 \pm \cos(ax)]} = \frac{1}{a} \ln\left[\frac{1 \pm \cos(ax)}{\cos(ax)}\right] \tag{25.501}$$

$$\int \frac{\cos(ax)\mathrm{d}x}{\sin(ax)[1 \pm \sin(ax)]} = -\frac{1}{a} \ln\left[\frac{1 \pm \sin(ax)}{\sin(ax)}\right] \tag{25.502}$$

$$\int \frac{\sin(ax)\mathrm{d}x}{\cos(ax)[1 \pm \sin(ax)]} = \frac{1}{2a[1 \pm \sin(ax)]} \pm \frac{1}{2a} \ln\left[\tan\left(\frac{\pi}{4} + \frac{ax}{2}\right)\right] \tag{25.503}$$

$$\int \frac{\cos(ax)\mathrm{d}x}{\sin(ax)[1 \pm \cos(ax)]} = -\frac{1}{2a[1 \pm \cos(ax)]} \pm \frac{1}{2a} \ln\left[\tan\left(\frac{ax}{2}\right)\right] \tag{25.504}$$

$$\int \frac{\mathrm{d}x}{1 + \cos(ax) \pm \sin(ax)} = \pm\frac{1}{a} \ln\left[1 \pm \tan\left(\frac{ax}{2}\right)\right] \tag{25.505}$$

$$\int \frac{\mathrm{d}x}{b\ \sin(ax) + c\ \cos(ax)} = \frac{1}{a\sqrt{b^2 + c^2}} \ln\ \tan\left(\frac{ax + \theta}{2}\right) \tag{25.506}$$

$$\text{with} \quad \sin\theta = \frac{c}{\sqrt{b^2 + c^2}} \quad \text{and} \quad \tan\theta = \frac{c}{b}$$

$$\int \frac{\sin(ax)\mathrm{d}x}{b + c\ \cos(ax)} = -\frac{1}{ac} \ln[b + c\ \cos(ax)] \tag{25.507}$$

$$\int \frac{\cos(ax)\mathrm{d}x}{b + c\ \sin(ax)} = \frac{1}{ac} \ln[b + c\ \sin(ax)] \tag{25.508}$$

$$\int \frac{\mathrm{d}x}{b + c\ \cos(ax) + f\ \sin(ax)} = \int \frac{\mathrm{d}(x + \theta/a)}{b + \sqrt{c^2 + f^2}\ \sin(ax + \theta)}, \tag{25.509}$$

$$\text{with} \quad \sin(\theta) = \frac{c}{\sqrt{c^2 + f^2}}, \quad \tan(\theta) = \frac{c}{f}$$

$$\int \frac{\mathrm{d}x}{b^2 \cos^2(ax) + c^2 \sin^2(ax)} = \frac{1}{abc} \arctan\left[\frac{c}{b} \tan(ax)\right] \tag{25.510}$$

$$\int \frac{\mathrm{d}x}{b^2 \cos^2(ax) - c^2 \sin^2(ax)} = \frac{1}{2abc} \ln\left[\frac{c\ \tan(ax) + b}{c\ \tan(ax) - b}\right] \tag{25.511}$$

25.3.7 Integrals with tangent or cotangent functions $(a \neq 0)$

$$\int \tan(ax)\mathrm{d}x = -\frac{\ln[\cos(ax)]}{a} \tag{25.512}$$

$$\int \cot(ax)\mathrm{d}x = \frac{\ln[\sin(ax)]}{a} \tag{25.513}$$

$$\int \tan^2(ax)\mathrm{d}x = \frac{\tan(ax)}{a} - x \tag{25.514}$$

$$\int \cot^2(ax)\mathrm{d}x = -\frac{\cot(ax)}{a} - x \tag{25.515}$$

$$\int \tan^3(ax)\mathrm{d}x = \frac{1}{2a}\tan^2(ax) + \frac{1}{a}\ln[\cos(ax)] \tag{25.516}$$

$$\int \cot^3(ax)\mathrm{d}x = -\frac{1}{2a}\cot^2(ax) - \frac{1}{a}\ln[\sin(ax)] \tag{25.517}$$

$$\int \tan^n(ax)\mathrm{d}x = \frac{1}{a(n-1)}\tan^{n-1}(ax) - \int \tan^{n-2}(ax)\mathrm{d}x \tag{25.518}$$

$$\int \cot^n(ax)\mathrm{d}x = -\frac{1}{a(n-1)}\cot^{n-1}(ax) - \int \cot^{n-2}(ax)\mathrm{d}x \quad (n \neq 1) \tag{25.519}$$

$$\int x\tan(ax)\mathrm{d}x = \frac{ax^3}{3} + \frac{a^3x^5}{15} + \frac{2a^5x^7}{105} + \frac{17a^7x^9}{2835} + \cdots + \frac{2^{2n}(2^{2n}-1)B_n a^{2n-1}x^{2n+1}}{(2n+1)!} + \cdots \tag{25.520}$$

B_n are the **Bernoulli numbers**.

$$\int x\cot(ax)\mathrm{d}x = \frac{x}{a} - \frac{ax^3}{9} - \frac{a^3x^5}{225} - \cdots - \frac{2^{2n}B_n a^{2n-1}x^{2n+1}}{(2n+1)!} - \cdots \tag{25.521}$$

$$\int \frac{\tan(ax)\mathrm{d}x}{x} = ax + \frac{(ax)^3}{9} + \frac{2(ax)^5}{75} + \frac{17(ax)^7}{2205} + \cdots + \frac{2^{2n}(2^{2n}-1)B_n(ax)^{2n-1}}{(2n-1)(2n)!} + \cdots \tag{25.522}$$

$$\int \frac{\cot(ax)\mathrm{d}x}{x} = -\frac{1}{ax} - \frac{ax}{3} - \frac{(ax)^3}{135} - \frac{2(ax)^5}{4725} - \cdots - \frac{2^{2n}B_n(ax)^{2n-1}}{(2n-1)(2n)!} - \cdots \tag{25.523}$$

$$\int \frac{\tan^n(ax)\mathrm{d}x}{\cos^2(ax)} = \frac{1}{a(n+1)}\tan^{n+1}(ax) \quad (n \neq -1) \tag{25.524}$$

$$\int \frac{\cot^n(ax)\mathrm{d}x}{\sin^2(ax)} = -\frac{1}{a(n+1)}\cot^{n+1}(ax) \quad (n \neq -1) \tag{25.525}$$

$$\int \frac{\mathrm{d}x}{\tan(ax) \pm 1} = \pm\frac{x}{2} + \frac{1}{2a}\ln[\sin(ax) \pm \cos(ax)] \tag{25.526}$$

$$\int \frac{\tan(ax)\mathrm{d}x}{\tan(ax) \pm 1} = \frac{x}{2} \mp \frac{1}{2a} \ln[\sin(ax) \pm \cos(ax)] \tag{25.527}$$

$$\int \frac{\mathrm{d}x}{1 \pm \cot(ax)} = \int \frac{\tan(ax)\mathrm{d}x}{\tan(ax) \pm 1} \tag{25.528}$$

25.3.8 Integrals with inverse trigonometric functions $(a \neq 0)$

$$\int \arcsin\left(\frac{x}{a}\right) \mathrm{d}x = x \arcsin\left(\frac{x}{a}\right) + \sqrt{a^2 - x^2} \tag{25.529}$$

$$\int \arccos\left(\frac{x}{a}\right) \mathrm{d}x = x \arccos\left(\frac{x}{a}\right) - \sqrt{a^2 - x^2} \tag{25.530}$$

$$\int \arctan\left(\frac{x}{a}\right) \mathrm{d}x = x \arctan\left(\frac{x}{a}\right) - \frac{a}{2} \ln(a^2 + x^2) \tag{25.531}$$

$$\int \operatorname{arccot}\left(\frac{x}{a}\right) \mathrm{d}x = x \operatorname{arccot}\left(\frac{x}{a}\right) + \frac{a}{2} \ln(a^2 + x^2) \tag{25.532}$$

$$\int x \arcsin\left(\frac{x}{a}\right) \mathrm{d}x = \left(\frac{x^2}{2} - \frac{a^2}{4}\right) \arcsin\left(\frac{x}{a}\right) + \frac{x}{4}\sqrt{a^2 - x^2} \tag{25.533}$$

$$\int x \arccos\left(\frac{x}{a}\right) \mathrm{d}x = \left(\frac{x^2}{2} - \frac{a^2}{4}\right) \arccos\left(\frac{x}{a}\right) - \frac{x}{4}\sqrt{a^2 - x^2} \tag{25.534}$$

$$\int x \arctan\left(\frac{x}{a}\right) \mathrm{d}x = \frac{1}{2}(x^2 + a^2) \arctan\left(\frac{x}{a}\right) - \frac{ax}{2} \tag{25.535}$$

$$\int x \operatorname{arccot}\left(\frac{x}{a}\right) \mathrm{d}x = \frac{1}{2}(x^2 + a^2) \operatorname{arccot}\left(\frac{x}{a}\right) + \frac{ax}{2} \tag{25.536}$$

$$\int x^2 \arcsin\left(\frac{x}{a}\right) \mathrm{d}x = \frac{x^3}{3} \arcsin\left(\frac{x}{a}\right) + \frac{1}{9}(x^2 + 2a^2)\sqrt{a^2 - x^2} \tag{25.537}$$

$$\int x^2 \arccos\left(\frac{x}{a}\right) \mathrm{d}x = \frac{x^3}{3} \arccos\left(\frac{x}{a}\right) - \frac{1}{9}(x^2 + 2a^2)\sqrt{a^2 - x^2} \tag{25.538}$$

$$\int x^2 \arctan\left(\frac{x}{a}\right) \mathrm{d}x = \frac{x^3}{3} \arctan\left(\frac{x}{a}\right) - \frac{ax^2}{6} + \frac{a^3}{6} \tan(a^2 + x^2) \tag{25.539}$$

$$\int x^2 \operatorname{arccot}\left(\frac{x}{a}\right) \mathrm{d}x = \frac{x^3}{3} \operatorname{arccot}\left(\frac{x}{a}\right) + \frac{ax^2}{6} - \frac{a^3}{6} \ln(a^2 + x^2) \tag{25.540}$$

$$\int x^n \arctan\left(\frac{x}{a}\right) \mathrm{d}x = \frac{x^{n+1}}{n+1} \arctan\left(\frac{x}{a}\right) - \frac{a}{n+1} \int \frac{x^{n+1}}{a^2 + x^2} \mathrm{d}x \quad (n \neq -1) \tag{25.541}$$

$$\int x^n \operatorname{arccot}\left(\frac{x}{a}\right) dx = \frac{x^{n+1}}{n+1} \operatorname{arccot}\left(\frac{x}{a}\right) + \frac{a}{n+1} \int \frac{x^{n+1}}{a^2+x^2} dx \quad (n \neq -1) \tag{25.542}$$

$$\int \frac{\arcsin\left(\frac{x}{a}\right) dx}{x} = \frac{x}{a} + \frac{1}{2 \cdot 3 \cdot 3} \frac{x^3}{a^3} + \frac{1 \cdot 3}{2 \cdot 4 \cdot 5 \cdot 5} \frac{x^5}{a^5} + \frac{1 \cdot 3 \cdot 5}{2 \cdot 4 \cdot 6 \cdot 7 \cdot 7} \frac{x^7}{a^7} + \cdots \tag{25.543}$$

$$\int \frac{\arccos\left(\frac{x}{a}\right) dx}{x} = \frac{\pi}{2} \ln(x) - \frac{x}{a} - \frac{1}{2 \cdot 3 \cdot 3} \frac{x^3}{a^3} - \frac{1 \cdot 3}{2 \cdot 4 \cdot 5 \cdot 5} \frac{x^5}{a^5} - \frac{1 \cdot 3 \cdot 5}{2 \cdot 4 \cdot 6 \cdot 7 \cdot 7} \frac{x^7}{a^7} - \cdots \tag{25.544}$$

$$\int \frac{\arctan\left(\frac{x}{a}\right) dx}{x} = \frac{x}{a} - \frac{x^3}{3^2 a^3} + \frac{x^5}{5^2 a^5} - \frac{x^7}{7^2 a^7} + \cdots \quad (|x| < |a|) \tag{25.545}$$

$$\int \frac{\operatorname{arccot}\left(\frac{x}{a}\right) dx}{x} = \frac{\pi}{2} \ln(x) - \frac{x}{a} + \frac{x^3}{3^2 a^3} - \frac{x^5}{5^2 a^5} + \frac{x^7}{7^2 a^7} - \ldots \tag{25.546}$$

$$\int \frac{\arcsin\left(\frac{x}{a}\right) dx}{x^2} = -\frac{1}{x} \arcsin\left(\frac{x}{a}\right) - \frac{1}{a} \ln\left(\frac{a + \sqrt{a^2 - x^2}}{x}\right) \tag{25.547}$$

$$\int \frac{\arccos\left(\frac{x}{a}\right) dx}{x^2} = -\frac{1}{x} \arccos\left(\frac{x}{a}\right) + \frac{1}{a} \ln\left(\frac{a + \sqrt{a^2 - x^2}}{x}\right) \tag{25.548}$$

$$\int \frac{\arctan\left(\frac{x}{a}\right) dx}{x^2} = -\frac{1}{x} \arctan\left(\frac{x}{a}\right) - \frac{2}{a} \ln\left(\frac{a^2 + x^2}{x^2}\right) \tag{25.549}$$

$$\int \frac{\operatorname{arccot}\left(\frac{x}{a}\right) dx}{x^2} = -\frac{1}{x} \operatorname{arccot}\left(\frac{x}{a}\right) + \frac{1}{2a} \ln\left(\frac{a^2 + x^2}{x^2}\right) \tag{25.550}$$

$$\int \frac{\arctan\left(\frac{x}{a}\right) dx}{x^n} = -\frac{1}{(n-1)x^{n-1}} \arctan\left(\frac{x}{a}\right) + \frac{a}{n-1} \int \frac{dx}{x^{n-1}(a^2+x^2)} \quad (n \neq 1) \tag{25.551}$$

$$\int \frac{\operatorname{arccot}\left(\frac{x}{a}\right) dx}{x^n} = -\frac{1}{(n-1)x^{n-1}} \operatorname{arccot}\left(\frac{x}{a}\right) - \frac{a}{n-1} \int \frac{dx}{x^{n-1}(a^2+x^2)} \quad (n \neq 1) \tag{25.552}$$

25.4 Definite integrals

25.4.1 Definite integrals with algebraic functions

$$\int_1^\infty \frac{dx}{x^a} = \frac{1}{a-1} \qquad (a > 1) \tag{25.553}$$

$$\int_0^1 x^a(1-x)^\beta dx = 2\int_0^1 x^{2\alpha+1}(1-x^2)^\beta dx = \frac{\Gamma(\alpha+1)\,\Gamma(\beta+1)}{\Gamma(\alpha+\beta+2)} \tag{25.554}$$

$\Gamma(\alpha)$ is the **Gamma function**.

$$\int_0^\infty \frac{dx}{(1+x)x^a} = \frac{\pi}{\sin(a\pi)} \qquad (a < 1) \tag{25.555}$$

$$\int_0^\infty \frac{dx}{(1-x)x^a} = -\pi\cot(a\pi) \qquad (a < 1) \tag{25.556}$$

$$\int_0^\infty \frac{x^{a-1}}{1+x^b}dx = \frac{\pi}{b\sin(a\pi/b)} \qquad (0 < a < b) \tag{25.557}$$

$$\int_0^1 \frac{dx}{\sqrt{1-x^a}} = \frac{\sqrt{\pi}\,\Gamma(1/a)}{a\,\Gamma(2+a/2a)} \tag{25.558}$$

$$\int_0^1 \frac{dx}{1+2x\cos(a)+x^2} = \frac{a}{2\sin(a)} \qquad \left(0 < a < \frac{\pi}{2}\right) \tag{25.559}$$

$$\int_0^\infty \frac{dx}{1+2x\cos(a)+x^2} = \frac{a}{\sin(a)} \qquad \left(0 < a < \frac{\pi}{2}\right) \tag{25.560}$$

25.4.2 Definite integrals with exponential functions

$$\int_0^\infty e^{-ax}dx = \frac{1}{a} \qquad (a > 0) \tag{25.561}$$

$$\int_0^\infty x^n e^{-ax}dx = \frac{\Gamma(n+1)}{a^{n+1}} \qquad (a > 0, n > -1) \tag{25.562}$$

For integers $n > 0$ this integral is equal to $\frac{n!}{a^{n+1}}$.
$\Gamma(n)$ denotes the **Gamma function** (see Section 5.24).

$$\int_0^\infty e^{-a^2x^2}dx = \frac{\sqrt{\pi}}{2a} \qquad (a > 0) \tag{25.563}$$

$$\int_0^\infty x\,e^{-x^2}dx = \frac{1}{2} \tag{25.564}$$

$$\int_0^\infty x^2 e^{-a^2x^2} dx = \frac{\sqrt{\pi}}{4a^3} \quad (a > 0) \tag{25.565}$$

$$\int_0^\infty x^{2n} e^{-ax^2} dx = \frac{1 \cdot 3 \cdot 5 \cdots (2n-1)}{2^{n+1} a^n} \sqrt{\frac{\pi}{a}} \tag{25.566}$$

$$\int_0^\infty x^{2n+1} e^{-ax^2} dx = \frac{n!}{2a^{n+1}} \quad (a > 0, n > -1) \tag{25.567}$$

$$\int_0^\infty x^n e^{-ax^2} dx = \frac{\Gamma[(n+1)/2]}{2a^{(n+1)/2}} \quad (a > 0, n > -1) \tag{25.568}$$

$$\int_0^\infty e^{\left(-x^2 - \frac{a^2}{x^2}\right)} dx = \frac{e^{-2a}\sqrt{\pi}}{2} \tag{25.569}$$

$$\int_0^\infty e^{-ax} \sqrt{x} dx = \frac{1}{2a}\sqrt{\frac{\pi}{a}} \tag{25.570}$$

$$\int_0^\infty \frac{e^{-ax}}{\sqrt{x}} dx = \sqrt{\frac{\pi}{a}} \tag{25.571}$$

$$\int_0^\infty e^{-a^2x^2} \cos(bx) dx = \frac{\sqrt{\pi}}{2a} \cdot e^{-b^2/4a^2} \qquad (a > 0) \tag{25.572}$$

$$\int_0^\infty \frac{x dx}{e^x - 1} = \frac{\pi^2}{6} \tag{25.573}$$

$$\int_0^\infty \frac{x dx}{e^x + 1} = \frac{\pi^2}{12} \tag{25.574}$$

$$\int_0^\infty e^{-ax} \cos(mx) dx = \frac{a}{a^2 + m^2} \quad (a > 0) \tag{25.575}$$

$$\int_0^\infty e^{-ax} \sin(mx) dx = \frac{m}{a^2 + m^2} \quad (a > 0) \tag{25.576}$$

$$\int_0^\infty x e^{-ax} \sin(bx) dx = \frac{2ab}{(a^2 + b^2)^2} \quad (a > 0) \tag{25.577}$$

$$\int_0^\infty x e^{-ax} \cos(bx) dx = \frac{a^2 - b^2}{(a^2 + b^2)^2} \quad (a > 0) \tag{25.578}$$

$$\int_0^\infty \frac{e^{-ax} \sin(ax)}{x} dx = \operatorname{arccot}(a) = \arctan\left(\frac{1}{a}\right) \quad (a > 0) \tag{25.579}$$

$$\int_0^\infty e^{-x} \ln(x) dx = -C \approx -0.5772 \tag{25.580}$$

C is **Euler's constant** (see Section 15.3).

25.4.3 Definite integrals with logarithmic functions

$$\int_0^1 [\ln(x)]^n \mathrm{d}x = (-1)^n n! \tag{25.581}$$

$$\int_0^1 \left[\ln\left(\frac{1}{x}\right)\right]^{\frac{1}{2}} \mathrm{d}x = \frac{\sqrt{\pi}}{2} \tag{25.582}$$

$$\int_0^1 \left[\ln\left(\frac{1}{x}\right)\right]^{-\frac{1}{2}} \mathrm{d}x = \sqrt{\pi} \tag{25.583}$$

$$\int_0^1 \left[\ln\left(\frac{1}{x}\right)\right]^a \mathrm{d}x = \Gamma(a+1) \quad (-1 < a < \infty) \qquad (= a!\,,\ \text{if}\, a \in \mathbb{N}) \tag{25.584}$$

$$\int_0^1 x\ln(1-x)\mathrm{d}x = -\frac{3}{4} \tag{25.585}$$

$$\int_0^1 x\ln(1+x)\mathrm{d}x = \frac{1}{4} \tag{25.586}$$

$$\int_0^1 \ln[\ln(x)]\mathrm{d}x = -C = -0.5772\ldots \tag{25.587}$$

$$\int_0^1 \frac{\ln(x)}{x-1}\mathrm{d}x = \frac{\pi^2}{6} \tag{25.588}$$

$$\int_0^1 \frac{\ln(x)}{x+1}\mathrm{d}x = -\frac{\pi^2}{12} \tag{25.589}$$

$$\int_0^1 \frac{\ln(x)}{x^2-1}\mathrm{d}x = \frac{\pi^2}{8} \tag{25.590}$$

$$\int_0^1 \ln\left(\frac{1+x}{1-x}\right)\frac{\mathrm{d}x}{x} = \frac{\pi^2}{4} \tag{25.591}$$

$$\int_0^1 \frac{\ln(x)\mathrm{d}x}{\sqrt{1-x^2}} = -\frac{\pi}{2}\ln(2) \tag{25.592}$$

$$\int_0^1 \frac{\ln(1+x)}{x^2+1}\mathrm{d}x = \frac{\pi}{8}\ln(2) \tag{25.593}$$

$$\int_0^{\pi/2} \ln[\sin(x)]\mathrm{d}x = \int_0^{\pi/2} \ln[\cos(x)]\mathrm{d}x = -\frac{\pi}{2}\ln(2) \tag{25.594}$$

$$\int_0^{\pi} x \ln[\sin(x)]\mathrm{d}x = -\frac{\pi^2 \ln(2)}{2} \tag{25.595}$$

$$\int_0^{\pi/2} \sin(x) \ln[\sin(x)]\mathrm{d}x = \ln(2) - 1 \tag{25.596}$$

$$\int_0^{\pi} \ln[a \pm b\cos(x)]\mathrm{d}x = \pi \ln\left(\frac{a + \sqrt{a^2 - b^2}}{2}\right) \qquad (a \geq b) \tag{25.597}$$

$$\begin{aligned}\int_0^{\pi} \ln[a^2 - 2ab\cos(x) + b^2]\mathrm{d}x &= 2\pi \ln(a) \quad (a \geq b > 0) \\ &= 2\pi \ln(b) \quad (b \geq a > 0)\end{aligned} \tag{25.598}$$

$$\int_0^{\pi/2} \ln[\tan(x)]\mathrm{d}x = 0 \tag{25.599}$$

$$\int_0^{\pi/4} \ln[1 + \tan(x)]\mathrm{d}x = \frac{\pi}{8} \ln(2) \tag{25.600}$$

25.4.4 Definite integrals with trigonometric functions

$$\begin{aligned}\int_0^{\pi/2} \sin^{2\alpha+1}(x)\cos^{2\beta+1}(x)\mathrm{d}x &= \frac{\Gamma(\alpha+1)\Gamma(\beta+1)}{2\Gamma(\alpha+\beta+2)} \\ &= \frac{1}{2}B(\alpha+1, \beta+1) \quad (\alpha, \beta > -1) \\ &= \frac{\alpha!\beta!}{2(\alpha+\beta+1)!} \quad \text{(for positive integers } \alpha, \beta)\end{aligned} \tag{25.601}$$

$B(x, y) = \dfrac{\Gamma(x) \cdot \Gamma(y)}{\Gamma(x+y)}$ is the beta function or Euler integral of the first kind. $\Gamma(x)$ is the Gamma function or Euler integral of the second kind.
This formula is valid for arbitrary values of α and β; it is used to determine

$$\int_0^{\pi/2} \sqrt{\sin(x)}\mathrm{d}x, \quad \int_0^{\pi/2} \sqrt[3]{\sin(x)}\mathrm{d}x, \quad \int_0^{\pi/2} \frac{\mathrm{d}x}{\sqrt[3]{\cos(x)}} \quad \text{etc.}$$

$$\begin{aligned}\int_0^{\infty} \frac{\sin(ax)}{x}\mathrm{d}x &= \frac{\pi}{2} \quad \text{if } a > 0 \\ &= 0 \quad \text{if } a = 0 \\ &= -\frac{\pi}{2} \quad \text{if } a < 0\end{aligned} \tag{25.602}$$

$$\int_0^{\alpha} \frac{\cos(ax)\mathrm{d}x}{x} = \infty \qquad (\alpha \text{ arbitrary}) \tag{25.603}$$

$$\begin{aligned}\int_0^{\infty} \frac{\tan(ax)\mathrm{d}x}{x} &= \frac{\pi}{2} \quad \text{if } a > 0 \\ &= -\frac{\pi}{2} \quad \text{if } a < 0\end{aligned} \tag{25.604}$$

$$\int_0^{\pi} \sin^2(ax)\mathrm{d}x = \int_0^{\pi} \cos^2(ax)\mathrm{d}x = \frac{\pi}{2} \tag{25.605}$$

$$\int_0^{\pi} \sin(ax)\sin(bx)\mathrm{d}x = \int_0^{\pi} \cos(ax)\cos(bx)\mathrm{d}x = 0 \quad (a \neq b) \tag{25.606}$$

$$\int_0^{\infty} \frac{\cos(ax) - \cos(bx)}{x}\mathrm{d}x = \ln\left(\frac{b}{a}\right) \tag{25.607}$$

$$\begin{aligned} \int_0^{\infty} \frac{\sin(x)\cos(ax)}{x}\mathrm{d}x &= \frac{\pi}{2} \quad (|a| < 1) \\ &= \frac{\pi}{4} \quad (|a| = 1) \\ &= 0 \quad (|a| > 1) \end{aligned} \tag{25.608}$$

$$\int_0^{\infty} \frac{\sin(x)}{\sqrt{x}}\mathrm{d}x = \int_0^{\infty} \frac{\cos(x)}{\sqrt{x}}\mathrm{d}x = \sqrt{\frac{\pi}{2}} \tag{25.609}$$

$$\int_0^{\infty} \frac{x\ \sin(bx)}{a^2 + x^2}\mathrm{d}x = \pm\frac{\pi}{2}\ \mathrm{e}^{-|ab|} \tag{25.610}$$

The sign coincides with the sign of b.

$$\int_0^{\infty} \frac{\cos(ax)}{1 + x^2}\mathrm{d}x = \frac{\pi}{2}\ \mathrm{e}^{-|a|} \tag{25.611}$$

$$\int_0^{\infty} \frac{\sin^2(ax)}{x^2}\mathrm{d}x = \frac{\pi}{2}|a| \tag{25.612}$$

$$\int_{-\infty}^{+\infty} \sin(x^2)\mathrm{d}x = \int_{-\infty}^{+\infty} \cos(x^2)\mathrm{d}x = \sqrt{\frac{\pi}{2}} \tag{25.613}$$

$$\int_0^{\pi/2} \frac{\mathrm{d}x}{1 + a\ \cos(x)} = \frac{\cos^{-1}(a)}{\sqrt{1 - a^2}} \qquad (a < 0) \tag{25.614}$$

$$\int_0^{\pi} \frac{\mathrm{d}x}{a + b\ \cos(x)} = \frac{\pi}{\sqrt{a^2 - b^2}} \qquad (a > b \geq 0) \tag{25.615}$$

$$\int_0^{2\pi} \frac{\mathrm{d}x}{1 + a\ \cos(x)} = \frac{2\pi}{\sqrt{1 - a^2}} \qquad (a^2 < 1) \tag{25.616}$$

$$\int_0^{\pi/2} \frac{\sin(x)\mathrm{d}x}{\sqrt{1 - k^2\sin^2(x)}} = \frac{1}{2k}\ln\left(\frac{1+k}{1-k}\right) \qquad (|k| < 1) \tag{25.617}$$

$$\int_0^{\pi/2} \frac{\cos(x)\mathrm{d}x}{\sqrt{1 - k^2\sin^2(x)}} = \frac{1}{k}\arcsin(k) \qquad (|k| < 1) \tag{25.618}$$

$$\int_0^{\pi/2} \frac{\sin^2(x)\mathrm{d}x}{\sqrt{1 - k^2\sin^2(x)}} = \frac{1}{k^2}(K - E) \qquad (|k| < 1) \tag{25.619}$$

E and K are complete elliptic integrals: $E = E(k, \frac{\pi}{2})$, $K = F(k, \frac{\pi}{2})$.

$$\int_0^{\pi/2} \frac{\cos^2(x)\mathrm{d}x}{\sqrt{1-k^2\sin^2(x)}} = \frac{1}{k^2}E - (1-k^2)E \qquad (|k| < 1) \tag{25.620}$$

$$\int_0^{\pi} \frac{\cos(ax)\mathrm{d}x}{1-2b\ \cos(x)+b^2} = \frac{\pi b^a}{1-b^2} \qquad (\text{ for integers } a \geq 0, |b| < 1) \tag{25.621}$$

Index

addition theorems for trigonometric functions

$$\sin(\alpha \pm \beta) = \sin(\alpha)\cos(\beta) \pm \cos(\alpha)\sin(\beta), \qquad \cos(\alpha \pm \beta) = \cos(\alpha)\cos(\beta) \mp \sin(\alpha)\sin(\beta)$$

$$\tan(\alpha \pm \beta) = \frac{\tan(\alpha) \pm \tan(\beta)}{1 \mp \tan(\alpha)\tan(\beta)}, \qquad \cot(\alpha \pm \beta) = \frac{\cot(\alpha)\cot(\beta) \mp 1}{\cot(\beta) \pm \cot(\alpha)}$$

integer multiples in argument

$$\sin(2\alpha) = 2\sin(\alpha)\cos(\alpha), \qquad \sin(3\alpha) = 3\sin(\alpha) - 4\sin^3(\alpha)$$

$$\cos(2\alpha) = 2\cos^2(\alpha) - 1, \qquad \cos(3\alpha) = 4\cos^3(\alpha) - 3\cos(\alpha)$$

$$\tan(2\alpha) = \frac{2\tan(\alpha)}{1 - \tan^2(\alpha)}, \qquad \tan(3\alpha) = \frac{3\tan(\alpha) - \tan^3(\alpha)}{1 - 3\tan^2(\alpha)}$$

$$\cot(2\alpha) = \frac{\cot^2(\alpha) - 1}{2\cot(\alpha)}, \qquad \cot(3\alpha) = \frac{\cot^3(\alpha) - 3\cot(\alpha)}{3\cot^2(\alpha) - 1}$$

half argument $\mathrm{Int}(x)$ = next smaller integer

$$\sin\left(\frac{\alpha}{2}\right) = (-1)^m \sqrt{\frac{1 - \cos(\alpha)}{2}} \qquad m = \mathrm{Int}\left[\frac{180^\circ + |\alpha|}{360^\circ}\right]$$

$$\cos\left(\frac{\alpha}{2}\right) = (-1)^m \sqrt{\frac{1 + \cos(\alpha)}{2}} \qquad m = \mathrm{Int}\left[\frac{|\alpha|}{360^\circ}\right]$$

$$\tan\left(\frac{\alpha}{2}\right) = (-1)^m \sqrt{\frac{1 - \cos(\alpha)}{1 + \cos(\alpha)}} \qquad m = \mathrm{Int}\left[\frac{2\alpha}{180^\circ}\right]$$

addition of functions with equal argument

$$\cos(\alpha) \pm \sin(\alpha) = \sqrt{2}\sin(45^\circ \pm \alpha) = \sqrt{2}\cos(45^\circ \mp \alpha)$$

$$\cot(\alpha) + \tan(\alpha) = \frac{2}{\sin(2\alpha)}, \qquad \cot(\alpha) - \tan(\alpha) = 2\cot(2\alpha)$$

addition of functions of the same kind

$$\sin(\alpha) \pm \sin(\beta) = 2\sin\left(\frac{\alpha \pm \beta}{2}\right)\cos\left(\frac{\alpha \mp \beta}{2}\right)$$

$$\cos(\alpha) + \cos(\beta) = 2\cos\left(\frac{\alpha + \beta}{2}\right)\cos\left(\frac{\alpha - \beta}{2}\right)$$

$$\cos(\alpha) - \cos(\beta) = -2\sin\left(\frac{\alpha + \beta}{2}\right)\sin\left(\frac{\alpha - \beta}{2}\right)$$

special case

$$\alpha = x + y, \ \beta = x - y$$

$$\frac{\alpha + \beta}{2} = x, \quad \frac{\alpha - \beta}{2} = y$$

$$\tan(\alpha) \pm \tan(\beta) = \frac{\sin(\alpha \pm \beta)}{\cos(\alpha)\cos(\beta)}, \qquad \cot(\alpha) \pm \cot(\beta) = \pm\frac{\sin(\alpha \pm \beta)}{\sin(\alpha)\sin(\beta)}$$

products of functions

$$\sin(\alpha)\sin(\beta) = \frac{1}{2}\cos(\alpha - \beta) - \frac{1}{2}\cos(\alpha + \beta) \qquad \sin^2(\alpha) = \frac{1 - \cos(2\alpha)}{2}$$

$$\cos(\alpha)\sin(\beta) = \frac{1}{2}\sin(\alpha + \beta) - \frac{1}{2}\sin(\alpha - \beta) \qquad \cos^2(\alpha) = \frac{1 + \cos(2\alpha)}{2}$$

$$\cos(\alpha)\cos(\beta) = \frac{1}{2}\cos(\alpha + \beta) + \frac{1}{2}\cos(\alpha - \beta)$$

$$\tan(\alpha)\tan(\beta) = \frac{\cos(\alpha - \beta) - \cos(\alpha + \beta)}{\cos(\alpha - \beta) + \cos(\alpha + \beta)} = \frac{\tan(\alpha) + \tan(\beta)}{\cot(\alpha) + \cot(\beta)}$$

$$\tan(\alpha)\cot(\beta) = \frac{\sin(\alpha + \beta) + \sin(\alpha - \beta)}{\sin(\alpha + \beta) - \sin(\alpha - \beta)} = \frac{\tan(\alpha) + \cot(\beta)}{\cot(\alpha) + \tan(\beta)}$$

$$\cot(\alpha)\cot(\beta) = \frac{\cos(\alpha - \beta) + \cos(\alpha + \beta)}{\cos(\alpha - \beta) - \cos(\alpha + \beta)} = \frac{\cot(\alpha) + \cot(\beta)}{\tan(\alpha) + \tan(\beta)}$$

higher powers

$$\sin^3(\alpha) = \frac{3\sin(\alpha) - \sin(3\alpha)}{4}, \qquad \sin^4(\alpha) = \frac{3 - 4\cos(2\alpha) + \cos(4\alpha)}{8}$$

$$\cos^3(\alpha) = \frac{3\cos(\alpha) + \cos(3\alpha)}{4}, \qquad \cos^4(\alpha) = \frac{3 + 4\cos(2\alpha) + \cos(4\alpha)}{8}$$

quadratic functions, powers, roots

$$x^2 + px + q = 0 \quad x_{1/2} = -\frac{p}{2} \pm \sqrt{\frac{p^2}{4} - q}, \quad ax^2 + bx + c = 0 \quad x_{1/2} = -\frac{b}{2a} \pm \sqrt{\frac{b^2}{4a^2} - \frac{c}{a}}$$

$$(x \pm y)^2 = x^2 \pm 2xy + y^2, \qquad x^2 - y^2 = (x+y)(x-y)$$

$$(x \pm y)^3 = x^3 \pm 3x^2y + 3xy^2 \pm y^3, \qquad x^n \cdot y^m = (x \cdot y)^n \cdot y^{m-n} = x^{n-m} \cdot (x \cdot y)^m$$

$$(x \pm y)^n = \sum_{k=0}^{n} \binom{n}{k} (\pm 1)^k x^{n-k} y^k \qquad \binom{n}{k} = \frac{n!}{(n-k)!\,k!} \qquad (x^n)^m = x^{mn}$$

$$\sqrt[n]{x} = x^{1/n}, \quad \sqrt[n]{\sqrt[m]{x}} = \sqrt[n\cdot m]{x}, \quad \sqrt[n]{x} \cdot \sqrt[m]{x} = \sqrt[n\cdot m]{x^{n+m}}, \quad \sqrt{x^2} = |x|, \quad \sqrt{-1} = \mathrm{j}$$

exponential functions – formulas for e^x (except of e^{jx}) are also valid for a^x

$$a^x = e^{x\ln(a)}, \quad (e^x)^y = e^{xy}, \quad e^{x+y} = e^x \cdot e^y, \quad e^{x-y} = \frac{e^x}{e^y}, \quad e^{jx} = \cos(x) + j\sin(y)$$

logarithms $\log(x)$ is short for $\ln(x)$, $\lg(x)$, $\log_a(x)$

$$\log_a(x) = \frac{\ln(x)}{\ln(a)} = \frac{\lg(x)}{\lg(a)}, \qquad \lg(x) \approx 0.4343\ln(x), \quad \ln(x) \approx 2.3026\lg(x)$$

$$\log(x^y) = y \cdot \log(x), \qquad \lg\left(x \cdot 10^m\right) = \lg(x) + m$$

$$\log|x \cdot y| = \log|x| + \log|y|, \qquad \log\left|\frac{x}{y}\right| = \log|x| - \log|y|$$

hyperbolic functions

$$\sinh(x) = \frac{e^x - e^{-x}}{2}, \quad \cosh(x) = \frac{e^x + e^{-x}}{2}, \qquad \tanh(x) = \frac{\sinh(x)}{\cosh(x)} = \frac{1}{\coth(x)} = \frac{e^{2x} - 1}{e^{2x} + 1}$$

$$\cosh^2(x) = 1 + \sinh^2(x) = \frac{1}{2}(\cosh(2x) + 1), \qquad (\cosh(x) \pm \sinh(x))^n = e^{\pm nx}$$

$$\sinh(x \pm y) = \sinh(x)\cosh(y) \pm \cosh(x)\sinh(y), \qquad \sinh(2x) = 2\sinh(x)\cosh(x)$$

$$\cosh(x \pm y) = \cosh(x)\cosh(y) \pm \sinh(x)\sinh(y), \qquad \cosh(2x) = 2\cosh^2(x) - 1$$

trigonometry: angle – a period is $360° \mathrel{\hat{=}} 2\pi = 6.2832\ldots$

$$\alpha(\text{grad}) \approx 57.2958x(\text{rad}) \qquad 90° \mathrel{\hat{=}} \frac{\pi}{2}, \qquad x(\text{rad}) \approx 0.01745\alpha(\text{grad}) \qquad \pi \mathrel{\hat{=}} 180°$$

value	0°	30°	45°	60°	90°
$\sin(\alpha)$	0	$\frac{1}{2}$	$\frac{\sqrt{2}}{2}$	$\frac{\sqrt{3}}{2}$	1
$\cos(\alpha)$	1	$\frac{\sqrt{3}}{2}$	$\frac{\sqrt{2}}{2}$	$\frac{1}{2}$	0

value	0°	30°	45°	60°	90°
$\tan(\alpha)$	0	$\frac{1}{\sqrt{3}}$	1	$\sqrt{3}$	∞
$\cot(\alpha)$	∞	$\sqrt{3}$	1	$\frac{1}{\sqrt{3}}$	0

angle and complementary angle $90° - \alpha$

$$\sin(\alpha) = \cos(90° - \alpha) = -\sin(\alpha + 180°), \qquad \cos(\alpha) = \sin(90° - \alpha) = -\cos(\alpha + 180°)$$

$$\tan(\alpha) = \cot(90° - \alpha) = \tan(\alpha + 180°), \qquad \cot(\alpha) = \tan(90° - \alpha) = \cot(\alpha + 180°)$$

$$f(\alpha) = -f(-\alpha), \quad f = \sin, \quad \tan, \quad \cot, \qquad \cos(\alpha) = \cos(-\alpha)$$

relation between trigon. functions, $\sin(x)$, $\cos(x)$: x in rad

$$\sin^2(\alpha) + \cos^2(\alpha) = 1, \qquad \sin(x) = \frac{1}{2j}\left(e^{jx} - e^{-jx}\right), \qquad \cos(x) = \frac{1}{2}\left(e^{jx} + e^{-jx}\right)$$

$$\tan(\alpha) = \frac{\sin(\alpha)}{\cos(\alpha)} = \frac{1}{\cot(\alpha)}, \qquad \cot(\alpha) = \frac{\cos(\alpha)}{\sin(\alpha)} = \frac{1}{\tan(\alpha)}$$

Mathematical symbols (according to DIN 1302)

$=$	equal
$\neq$	unequal
$<$	less than
$\leq$	less or equal
$>$	more than
$\geq$	more or equal
$\approx$	approximately equal
$\sim$	proportional, similar
$\cong$	congruent
$\triangleq$	corresponds to
$\equiv$	identically equal
$\parallel$	parallel
$\perp$	orthogonal to
$\wedge$	logical and
$\vee$	logical or
$\Longrightarrow$	from ... follows
$\Longleftrightarrow$	equivalent to
lim	limiting value
$\rightarrow$	goes to, puts in relation
$\ldots$	and so on until
∞	infinity
j or i	imaginary unit
%	percent
$\{a, b, c\}$	set with elements
$\in$	element of
$\notin$	not element of
$\subset$	partial set of
$\mathbb{N}$	set of natural numbers
$\mathbb{Z}$	set of integer numbers
$\mathbb{Q}$	set of rational numbers
$\mathbb{R}$	set of real numbers
$\mathbb{C}$	set of complex numbers
(a, b)	ordered pair
$\overline{AB}$	segment AB
$[a; b]$	closed interval
Δx	difference of two x values
$\mathrm{d}x$	differential $\lim_{\Delta x \to 0}$
$\frac{\mathrm{d}y}{\mathrm{d}x}$	differential quotient
$f'(x)$	first derivative
$f''(x)$	second derivative
$\frac{\partial f}{\partial x}$	partial derivative
∇	Nabla operator
$\int \ldots \mathrm{d}x$	indefinite integral
$\int_a^b \ldots \mathrm{d}x$	definite integral

functions and calculation rules

$n!$	n factorial
$\sum x_i$	sum of all x_i
$\prod x_i$	product of all x_i
$\lvert x \rvert$	absolute value of x
$\mathrm{arc}(z)$	arc of z
z^*	complex conjug. of z
$[x]$	Gauss bracket
$\mathrm{int}(x)$	Gauss bracket
$\mathrm{frac}(x)$	remainder
$\mathrm{sgn}(x)$	sign
$\max(a, b, c)$	largest argument
$\min(a, b, c)$	smallest argument
$\sqrt{x}$	square root
$\sqrt[n]{x}$	nth root
$\log_a(x)$	logarithm with base a
$\ln(x)$	natural logarithm
$\lg(x)$	decadic logarithm
$\exp(x)$	exponential
$\mathrm{erf}(x)$	error function
$\mathrm{erfc}(x)$	conjug. error func. $1 - \mathrm{erf}(x)$
$\sin(x)$	sine
$\cos(x)$	cosine
$\tan(x)$	tangents
$\cot(x)$	cotangents
$\arcsin(x)$	arcus sine, etc.
$\sinh(x)$	hyperbolic sine, etc.
$\mathrm{Arsinh}\,(x)$	area sine hyperbolicus, etc.
$\lvert A \rvert$, $\det A$	determinant of a matrix

Numbers and natural constants (SI units)

$\pi = 3,1415927\ldots$	circle constant
$C = 0,577216\ldots$	Euler constant
$\varepsilon_0 = 8,8542 \cdot 10^{-12} \frac{\mathrm{As}}{\mathrm{Vm}}$	dielectric constant
$\mathrm{e} = 2,7182818\ldots$	Euler number
$g = 9,80665 \frac{\mathrm{m}}{\mathrm{s}^2}$	acceleration of free fall
$\mu_0 = 1,2566 \cdot 10^{-6} \frac{\mathrm{Vs}}{\mathrm{Am}}$	permeability

greek letters

α	A	Alpha
β	B	Beta
γ	Γ	Gamma
δ	Δ	Delta
ε, ϵ	E	Epsilon
ζ	Z	Zeta
η	H	Eta
ϑ, θ	Θ	Theta
ι	I	Jota
κ	K	Kappa
λ	Λ	Lambda
μ	M	My
ν	N	Ny
ξ	Ξ	Xi
o	O	Omikron
π	Π	Pi
ρ, ϱ	P	Rho
σ	Σ	Sigma
τ	T	Tau
υ	Υ	Ypsilon
φ, ϕ	Φ	Phi
χ	X	Chi
ψ	Ψ	Psi
ω	Ω	Omega

MIX
Papier aus verantwortungsvollen Quellen
Paper from responsible sources
FSC® C105338

If you have any concerns about our products,
you can contact us on
ProductSafety@springernature.com

In case Publisher is established outside the EU,
the EU authorized representative is:
Springer Nature Customer Service Center GmbH
Europaplatz 3, 69115 Heidelberg, Germany

Printed by Libri Plureos GmbH
in Hamburg, Germany

function $f(x)$, derivative $f'(x)=\dfrac{\mathrm{d}f}{\mathrm{d}x}$ and primitive $\int f(x)\mathrm{d}x = F(x)+c$.

$f'(x)$	$f(x)$	$F(x)$	$f'(x)$	$f(x)$	$F(x)$
0	c	cx	e^x	e^x	e^x
1	x	$\frac{1}{2}x^2$	$a^x \ln(a)$	a^x	$\frac{a^x}{\ln(a)}$
ax^{a-1}	x^a $a\neq -1$	$\frac{x^{a+1}}{a+1}$	$\frac{1}{x}$	$\ln(x)$	$x\ln(x)-x$
$-\frac{1}{x^2}$	$\frac{1}{x}$	$\ln\lvert x\rvert$	$\frac{1}{x\ln(a)}$	$\log_a(x)$	$\frac{x\ln(x)-x}{\ln(a)}$
$\cos(x)$	$\sin(x)$	$-\cos(x)$	$\frac{1}{\sqrt{1-x^2}}$	$\arcsin(x)$	$x\arcsin(x)+\sqrt{1-x^2}$
$-\sin(x)$	$\cos(x)$	$\sin(x)$	$\frac{-1}{\sqrt{1-x^2}}$	$\arccos(x)$	$x\arccos(x)-\sqrt{1-x^2}$
$\frac{1}{\cos^2(x)}$	$\tan(x)$	$-\ln\lvert\cos(x)\rvert$	$\frac{1}{1+x^2}$	$\arctan(x)$	$x\arctan(x)-\frac{1}{2}\ln(1+x^2)$
$\frac{-1}{\sin^2(x)}$	$\cot(x)$	$\ln\lvert\sin(x)\rvert$	$\frac{-1}{1+x^2}$	$\operatorname{arccot}(x)$	$x\operatorname{arccot}(x)+\frac{1}{2}\ln(1+x^2)$
$\cosh(x)$	$\sinh(x)$	$\cosh(x)$	$\frac{1}{\sqrt{x^2+1}}$	$\operatorname{Arsinh}(x)$	$x\operatorname{Arsinh}(x)-\sqrt{x^2+1}$
$\sinh(x)$	$\cosh(x)$	$\sinh(x)$	$\frac{1}{\sqrt{x^2-1}}$	$\operatorname{Arcosh}(x)$	$x\operatorname{Arcosh}(x)-\sqrt{x^2-1}$
$\frac{1}{\cosh^2(x)}$	$\tanh(x)$	$\ln(\cosh(x))$	$\frac{1}{1-x^2}$	$\operatorname{Artanh}(x)$	$x\operatorname{Artanh}(x)+\frac{1}{2}\ln(1-x^2)$
$\frac{-1}{\sinh^2(x)}$	$\coth(x)$	$\ln\lvert\sinh(x)\rvert$	$\frac{1}{1-x^2}$	$\operatorname{Arcoth}(x)$	$x\operatorname{Arcoth}(x)+\frac{1}{2}\ln(x^2-1)$

differentiation and integration rules

short	$f'=\frac{\mathrm{d}}{\mathrm{d}x}f(x)$	short	$\int_a^b f\mathrm{d}t=\int_a^b f(t)\mathrm{d}t$
integral	$\left(\int_a^x f\mathrm{d}t\right)'=f(x)$	fundamental theorem	$\int_a^x f\mathrm{d}t = F\rvert_a^x = F(x)-F(a)$
constant	$(c\cdot f)'=c\cdot f'$	constant	$\int_a^b c\cdot f\mathrm{d}t = c\cdot\int_a^b f\mathrm{d}t$
sum	$(f\pm g)'=f'\pm g'$	sum	$\int_a^b (f\pm g)\mathrm{d}t=\int_a^b f\mathrm{d}t\pm\int_a^b g\mathrm{d}t$
product	$(f\cdot g)'=f'\cdot g+f\cdot g'$	integration by parts	$\int_a^b f'\cdot g\mathrm{d}t = f\cdot g\rvert_a^b-\int_a^b f\cdot g'\mathrm{d}t$
chain rule	$(f(g(x)))'=\frac{\mathrm{d}f(g)}{\mathrm{d}g}\cdot g'(x)$	substitution	$\int_a^b f(g)\cdot g'(t)\mathrm{d}t=\int_{g(a)}^{g(b)} f(g)\mathrm{d}g$
logarithm. diff. $f,g>0$	$(f^g)'=f^g\left(g'\ln(f)+g\frac{f'}{f}\right)$	logarithm. integrat.	$\int_a^b \frac{f'}{f}\mathrm{d}t=\ln\lvert f(t)\rvert\Big\rvert_a^b$

further rules

quotient	$\left(\frac{u}{v}\right)'=\frac{u'\cdot v-u\cdot v'}{v^2}$	intervals	$\int_a^c f\mathrm{d}t+\int_c^b f\mathrm{d}t=\int_a^b f\mathrm{d}t$
inverse funct.	$(f^{-1}(x))'=\frac{1}{\frac{\mathrm{d}}{\mathrm{d}x}f\left(f^{-1}(x)\right)}$	limits $b=a$	$\int_a^a f\mathrm{d}t=0$
3-product	$(fgh)'=f'gh+fg'h+fgh'$	commutation	$\int_a^b f\mathrm{d}t=-\int_b^a f\mathrm{d}t$
3-chain	$(f(g(h(x))))'=f'(g)g'(h)h'(x)$	f odd $a=-b$	$f(t)=-f(-t):\int_{-b}^b f\mathrm{d}t=0$